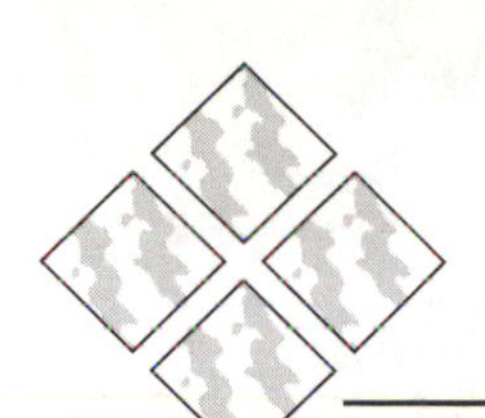

BASIC OP AMP MODULES

 Circuit	Block Diagram	 Gains
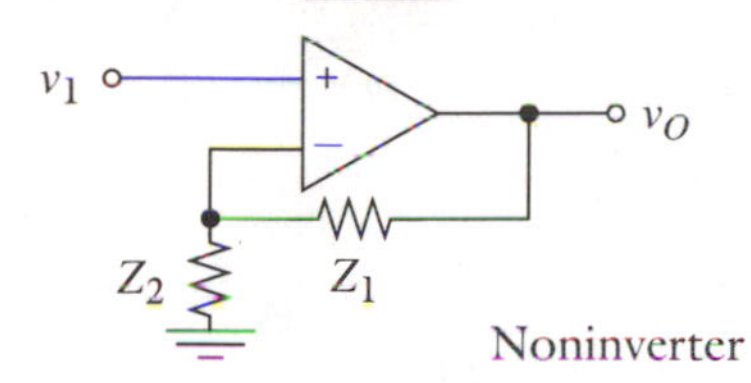 Noninverter	$v_1 \rightarrow [K] \rightarrow v_O$	$K = \dfrac{Z_1 + Z_2}{Z_2}$
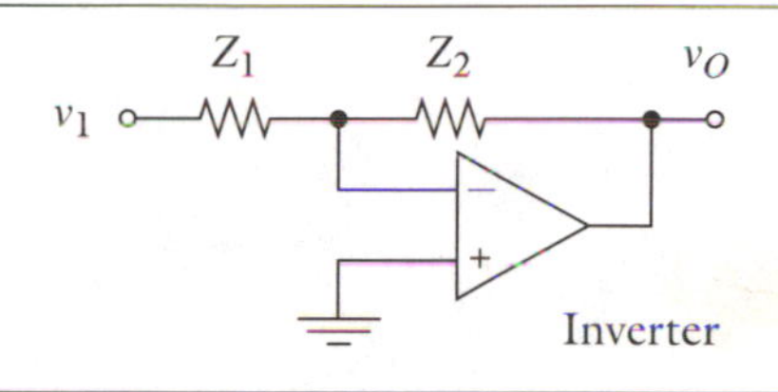 Inverter	$v_1 \rightarrow [K] \rightarrow v_O$	$K = -\dfrac{Z_2}{Z_1}$
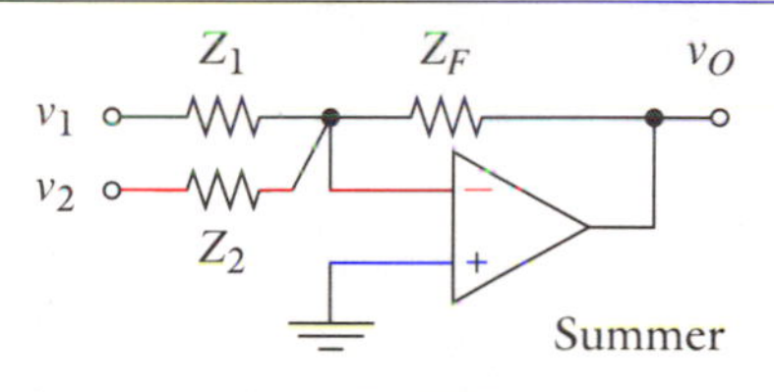 Summer	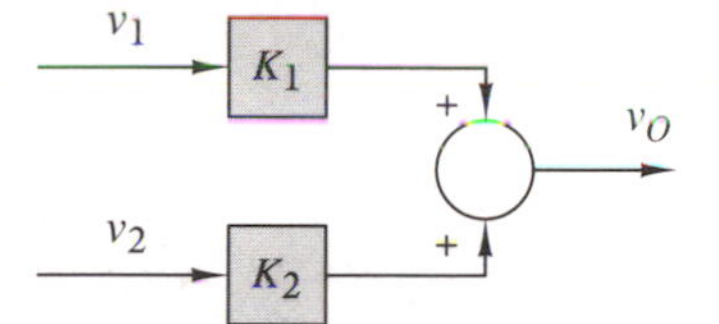	$K_1 = -\dfrac{Z_F}{Z_1}$ $K_2 = -\dfrac{Z_F}{Z_2}$
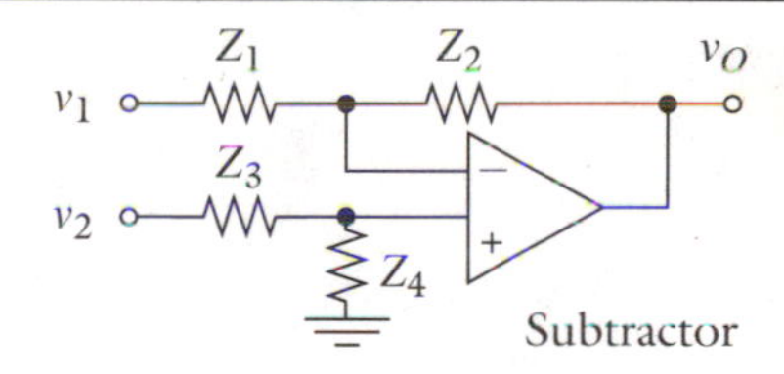 Subtractor	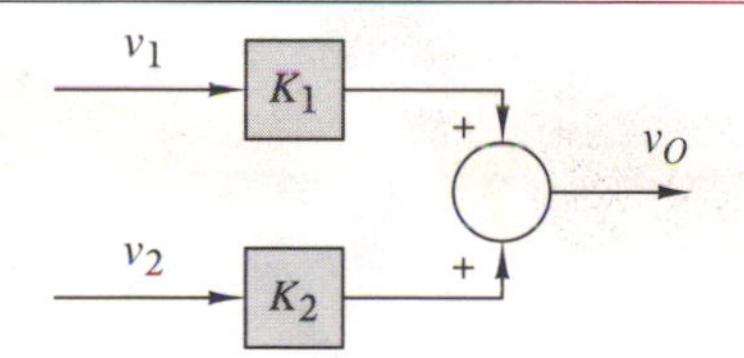	$K_1 = -\dfrac{Z_2}{Z_1}$ $K_2 = \left(\dfrac{Z_1 + Z_2}{Z_1}\right)\left(\dfrac{Z_4}{Z_3 + Z_4}\right)$
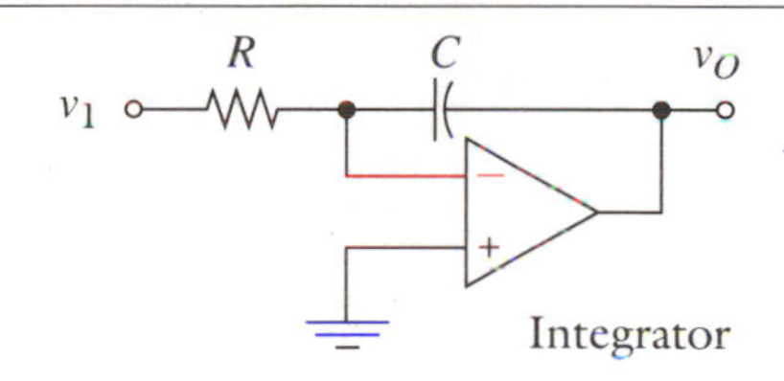 Integrator	$v_1 \rightarrow [K] \rightarrow [\int] \rightarrow v_O$	$K = -\dfrac{1}{RC}$
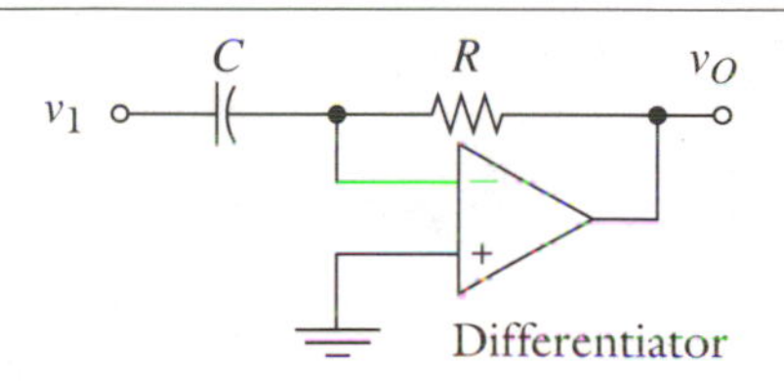 Differentiator	$v_1 \rightarrow [K] \rightarrow [\frac{d}{dt}] \rightarrow v_O$	$K = -RC$

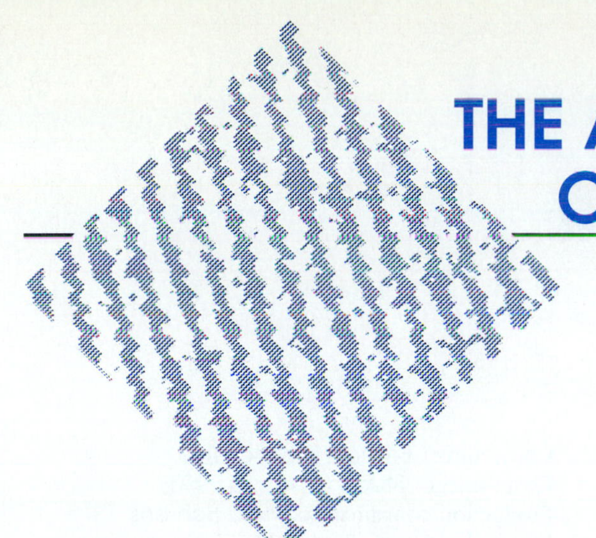

THE ANALYSIS AND DESIGN OF LINEAR CIRCUITS

Roland E. Thomas
Professor Emeritus
United States Air Force Academy

Albert J. Rosa
Professor and Chairman of Engineering
University of Denver

PRENTICE HALL, Englewood Cliffs, New Jersey 07632

Library of Congress Cataloging-in-Publication Data

Thomas, Roland E.
The analysis and design of linear circuits / Roland E. Thomas, Albert J. Rosa.
p. cm.
Includes bibliographical references and index.
ISBN 0-13-220005-8
1. Electric circuit analysis. 2. Electric circuits, Linear--Design and construction. I. Rosa, Albert J., II. Title.
TK454.T466 1994
621.319'2--dc20 93-23107
CIP

Publisher: Alan Apt
Development editor: Sondra Chavez
Interior and cover designer: Amy Rosen
Production editor: Jennifer Wenzel
Marketing manager: Tom McElwee
Copy editor: Brenda Melissaratos
Cover image: Stuart Simon © 1993
Production coordinator: Linda Behrens
Editorial assistant: Shirley McGuire

The author and publisher of this book have used their best efforts in preparing this book. These efforts include the development, research, and testing of the theories and programs to determine their effectiveness. The author and publisher make no warranty of any kind, expressed or implied, with regard to these programs or the documentation contained in this book. The author and publisher shall not be liable in any event for incidental or consequential damages in connection with, or arising out of, the furnishing, performance, or use of these programs.

Printed in the United States of America
10 9 8 7 6 5 4 3 2 1

ISBN 0-13-220005-8

TRADEMARK INFORMATION

MathCAD is a registered trademark of MathSoft, Inc.
MATLAB is a registered trademark of the MathWorks, Inc.
MICROCAP is a registered trademark of Spectrum Software, Inc.
PSpice is a registered trademark of MicroSim Corp.

Prentice-Hall International (UK) Limited, *London*
Prentice-Hall of Australia Pty. Limited, *Sydney*
Prentice-Hall Canada Inc., *Toronto*
Prentice-Hall Hispanoamerica, S. A., *Mexico*
Prentice-Hall of India. Private Limited, *New Delhi*
Prentice-Hall of Japan, Inc., *Tokyo*
Simon & Schuster Asia Pte. Ltd., *Singapore*
Editora Prentice-Hall do Brasil, Ltda., *Rio de Janeiro*

To our wives
Juanita and Kathleen

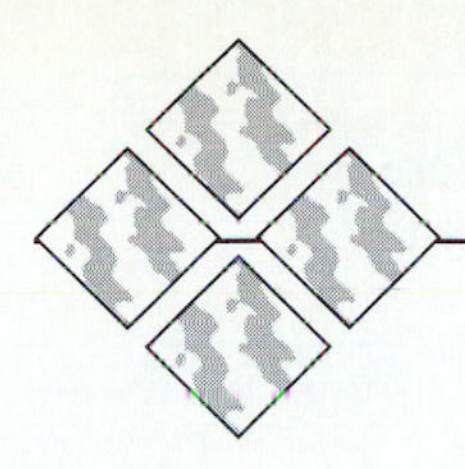

PREFACE

 WHY WE WROTE THIS BOOK

The Analysis and Design of Linear Circuits evolved during twenty-five years of teaching linear circuits. Our overarching objective is to provide an alternative to the many textbooks that focus on analysis and place primary emphasis on ac circuits. These classical books ignore the useful and exciting design applications made possible by the integrated circuit OP AMP and do not provide an adequate foundation for subsequent upper division courses in signal processing, control, and communication systems. While this book thoroughly covers the fundamentals of circuit analysis, it goes beyond analysis to introduce circuit design concepts using real world examples and constraints. This text introduces students to the signal processing point of view through the use of block diagrams and transform methods of describing signals and circuits. Our experience is that the design and signal processing flavor included in this book makes the subject of circuit analysis more interesting to students. The development of this book has been guided by the following three premises.

Design Approach: Design problems provide early motivation and another means to measure comprehension of basic concepts.

Pedagogy: Students learn best when objectives are explicitly stated.

Course Content: Basic circuit courses should emphasize techniques used in subsequent courses.

Class Testing

Class testing in three ABET accredited programs and fifteen years of using this approach has shown it eases the transition between the circuits class and signals and systems courses. More importantly, students take less time to come to a broad and meaningful understanding of circuit analysis and design when material is presented using this approach.

Steps taken in the development of this text to ensure accuracy and currency include:

- A total of ten instructors reviewed our manuscript three times.
- Two instructors, in addition to ourselves, verified the accuracy of examples, exercises, problems, and solutions.
- The entire manuscript was class tested at three schools for over two years.
- This text was modified, updated, and refined three times over the last two years.
- A developmental editor was assigned to edit and enhance the presentation.

We are anxious to offer *The Analysis and Design of Linear Circuits* for use by the electrical engineering community and welcome comments for improvement.

Key Features

The key features in this text include:

- significant introduction to circuit design and evaluation
- early introduction and frequent use of the OP AMP for problem solving
- distributed treatment of computer-aided circuit analysis using SPICE
- inclusion of structured pedagogical features, including chapter and course objectives, that enhance teachability
- use of realistic circuit parameters in examples, exercises, and problems
- a chapter devoted to signals prior to studying dynamic circuits
- emphasis on Laplace transform methods for handling dynamic circuits

Circuit Design

We believe teaching circuit analysis is enhanced by interweaving design examples and concepts in several chapters. Involving students early in design provides motivation as students find they can apply their knowledge to practical situations. Design is introduced for pedagogical reasons, however, it also permits faculty to include some material in a circuits course that meets the ABET Engineering Design criteria. To this end, we have included 35 worked-out Design Examples, nine Design Exercises, and 120 Design Problems which can be easily located by the Design Icon (❖). Students are introduced to circuit design in Chapter 3, where the idea of constructing a circuit to meet different constraints is first discussed. Introducing the OP AMP in Chapter 4 simplifies the design of analog signal processing functions. Some of the circuit design methods discussed in Chapters 4, 7, 11, 12, and 15 offer criteria for evaluating alternate designs, while others emphasize a modular approach to transfer function realization. Further understanding of network function analysis and design can be gained by adding the two-port concepts contained in Appendix E to the end of Chapter 11. The Design Table of Contents may be used to locate Design Examples and Design Exercises throughout the text.

The OP AMP

We believe the OP AMP is an important engineering tool that students should be exposed to early in their study of electrical engineering. Students are taught to use OP AMPs as a tool in solving design problems by breaking circuits down into more manageable modules. We also use OP AMPs to demonstrate dependent sources.

After being introduced in Chapter 4, the OP AMP is applied extensively throughout the remainder of the text as a premiere linear device for realizing various signal processing functions. The formal method of writing node equations for OP AMP circuits is presented in Chapter 5 for dc circuits and in Chapter 10 for dynamic circuits. OP AMPs occur frequently in sections throughout the text that emphasize applications and design, especially Chapter 15 on analog filters.

Computer-Aided Circuit Analysis

A distributed introduction to computer-aided circuit analysis and SPICE syntax begins in Chapter 5. Computer-aided circuit analysis material can be easily located by the blue strips on the outer edge of pages. Although SPICE syntax is used as a vehicle, the underlying principles apply to other circuit analysis programs such as MICRO-CAP®. Our purpose in introducing computer-aided analysis is not to simply teach students to solve circuits and produce numbers called answers. Rather, our goals are to:

1. identify the data and instructions that must be communicated to the program to define a circuit analysis problem
2. identify the circuit analysis conventions that must be understood to properly interpret the generated answers
3. show how to use computer-aided circuit analysis in conjunction with basic principles to calculate circuit characteristics or adjust element values to achieve specified characteristics

Pedagogy

We spent considerable time and energy to ensure that our book's pedagogy enhances teachability and assists students in understanding and retaining the course material. Each chapter begins with a brief introduction, including a few historical facts, and ends with a summary of the major chapter concepts. Other important pedagogical features of our book are discussed below.

Educational Objectives

A major pedagogical feature is the incorporation of explicitly stated objectives, based on a proven learning theory,[1] at several levels within our book. This framework of objectives allows students to identify the ultimate goals and clearly focus on the intermediate steps required to achieve those goals. The ability to correctly work the related problems in the text validates that students have mastered the stated objectives. **Course Objectives** are stated at the beginning of each of the three parts of the book. Course Objectives represent the type of learning a student should be

[1] Benjamin S. Bloom, *Taxonomy of Educational Objectives, Book 1 Cognitive Domain*, 1956 Longman Inc., New York.

able to demonstrate at major checkpoints in a course. To help students reach this level of achievement, we have defined **En Route Objectives (EROs)** aimed at each chapter's basic concepts and skills. EROs are listed at the end of each chapter and are immediately followed by problems that test the students' ability to meet the objective.

Examples, Exercises, and Problems

We have included over 275 worked-out examples to show how chapter concepts apply to specific problems. Forty-five of the examples are Application Notes and Design Examples that illustrate practical application and real world constraints. Immediate feedback is provided by over 260 in-text exercises that allow students to practice the analysis or design technique discussed in the preceding example. The examples and exercises are designed to help students develop the problem solving abilities needed to meet the En Route and Course Objectives.

The book includes over 1000 end-of-chapter problems. Most of these problems are directly related to En Route Objectives and can be used to validate the students' mastery of basic concepts. **Chapter Integrating Problems**, requiring the mastery of two or more EROs, are included to help students integrate their knowledge. The **Course Integrating Problems**, given at the end of Chapters 5, 11, and 16, require proficiency in EROs across several chapters and are designed to test the mastery of the Course Objectives. Integrating Problems are labeled as Analysis (A), Design (D), or Evaluation (E) depending on the principle purpose of the problem.

Text Design and Use of Second Color

This text is designed using a unique, open layout that permits the user to quickly locate important material such as examples, exercises, and valuable concepts. Key words and phrases are presented in blue italics. This feature serves as a quick reference of concepts and ideas contained in that portion of the text. It is also useful when reviewing material to prepare for an evaluation. A second color is used to enhance the students' learning process. For example, the second color is used in figures to: 1) identify the unknown response, 2) emphasize a reference direction, 3) show a circuit reduction operation, or 4) help differentiate between two plots made on the same axis. In addition, the second color is used to highlight important material in the book. Finally, we have used color to easily identify examples and exercises, and highlight information in chapter openers and summaries.

Circuit Parameters

We strike a balance in the examples, exercises, and homework problems between using numerical values and leaving circuit parameters in symbolic form. Using parameters in symbolic form complicates the algebra a bit, but points out that responses depend on both the input signal and the circuit parameters. It reminds students that in a design situation, circuit parameters are the unknowns. When numerical values are used or are unavoidable, as in second-order circuits, we use realistic values. Scientific calculators, and other computational aids such as MATLAB® or MathCAD®, eliminate the need to scale circuit parameters into numerically convenient, but patently impractical, ranges. Using realistic parameter values is also consistent with the circuit design examples.

Signals Chapter

The treatment of dynamic circuits begins with a separate chapter on the signal waveforms (Chapter 6) used in the rest of the book. Covering these signals separately, allows students to gain confidence and familiarity with basic waveforms prior to launching into the complexity of writing and solving circuit differential equations. The signals chapter treats the step function, exponential, and sinusoid waveforms as basic waveforms and discusses composite waveforms obtained by combining these basic signals. Partial signal descriptors such as peak, peak-to-peak, rms, and average values are also defined and illustrated. A concluding section on signal spectra provides students with a brief introduction to an important alternative way to view signals. The signals chapter precedes the introduction of energy storage devices such as capacitors, inductors, magnetically coupled coils, and transformers. Section 16-1 on Fourier Series could be added to the end of this chapter if further depth in signals is desired or it can be left in Chapter 16 to form a smooth transition from Laplace to Fourier transforms.

Laplace Transforms

We have found that an early introduction to Laplace transforms is more efficient because students need not master intermediate techniques which they must later abandon. It also better prepares students for subsequent courses such as signals and systems, electronics, communications, and controls than does the traditional approach emphasizing ac circuits. After an introduction to the Laplace transformation in Chapter 9, emphasis is placed on developing major circuit analysis tools such as the zero-input and zero-state responses, network functions, impulse and step responses, frequency response, convolution, Bode diagrams, and analog filter design. AC circuit analysis is not shortchanged by using this approach. We cover phasors, sinusoidal steady state analysis, impedance, power, balanced three-phase circuits, resonance, and other traditional ac topics. The difference is that ac circuit analysis is simply introduced after Laplace transforms as a special case of s-domain circuit analysis. We do not claim our approach is more fundamental than the classical approach of treating ac circuits prior to Laplace transforms. We only submit that in our experience, beginning students take less time to achieve a meaningful understanding of circuit analysis when Laplace transforms are covered first.

HOW TO USE THIS BOOK

This book contains sufficient material for a two semester sequence in basic circuit analysis, if most of the En Route Objectives (EROS) and Course Objectives are used. However, the book can also be used in courses of shorter duration by selecting a subset of EROs and proportionately lowering the level of achievement of the Course Objectives. Courses of differing lengths and emphasis can be designed while still retaining the integrity and advantages of the framework of objectives. A diagram of this course design framework can be found in the Instructor's Manual.

The EROs given at the end of each chapter provide a way to develop a course syllabus since they generally correlate to a text section. Instructors can usually cover one to three EROs in a 50 to 60 minute lecture. However, in a few cases, several sections must be covered before an objective can be mastered. In these cases, 90 or more minutes may be necessary to cover the material needed to master the objective.

Students should be encouraged to read the objectives before coming to class and study the related problems. The worked-out examples help students understand how to solve related problems. The exercises that follow examples allow students to verify whether they understand the concept being taught. Homework problems help achieve further mastery of Course Objectives. About one-third of the homework problems have answers in the back of the book.

The material in our book is divided into three logical parts: Part I - Resistance Circuits, Part II - Dynamic Circuits, and Part III - Applications. We have organized the book so many important circuit analysis concepts are reinforced by revisiting them in a new context. For example, circuit analysis is first introduced in Chapter 2 as the combination of device constraints (element i-v characteristics) and connection constraints (Kirchhoff's laws). We revisit this concept in Chapter 3 for nonlinear circuits, Chapter 4 for OP AMP circuit analysis, Chapter 5 for node and mesh analysis, Chapter 8 for circuit differential equations, Chapter 10 for s-domain circuit analysis, and Chapter 13 for ac circuit analysis. Important concepts like Thévenin and Norton theorems and source transformations are first introduced in Chapter 3 and are treated again in the s-domain in Chapter 10 and in Chapter 13 using phasors.

INSTRUCTIONAL AIDS

Three instructional aids are available to use in conjunction with this book. The **Instructor's Manual** contains:

1. a detailed explanation of the pedagogy found in the text
2. sample syllabi that are keyed to course objectives for:
   - electrical engineering classes of one or two semesters
   - electrical engineering classes of two or three quarters
   - service courses of one semester or two quarters
3. solutions to homework problems
4. solutions to laboratory assignments for the separately purchased Lab Manual
5. a sample final examination

The **Laboratory Manual**, which can be purchased separately by students, presents twenty well written lab assignments. Each lab assignment has been tested extensively over many years. The lab assignments are designed to correlate with various chapters in the book and use standard components and generic laboratory instruments such as oscilloscopes, multimeters, function generators, and power supplies. The lab assignments reinforce student skills as they master EROs and Course Objectives. Many of the lab assignments require students to run a computer-aided analysis using SPICE or MICROCAP® to verify a design or analytical solution.

A set of two hundred bound and perforated overhead **Transparency Masters** may be ordered separately. Key figures and tables, as well as a few select Design Examples, are included to enhance lectures.

ACKNOWLEDGMENTS

The authors acknowledge their indebtedness to students and colleagues who gave us help and encouragement during the preparation of this book. We particularly thank the following individuals who reviewed all or part of the manuscript and offered many valuable suggestions:

Doran G. Baker, *Utah State University*

William E. Bennett, *United States Naval Academy*

Maqsood A. Chaudhry, *California State University at Fullerton*

James G. Gottling, *Ohio State University*

K.S.P. Kumar, *University of Minnesota*

Jerry I. Lubell, *Mission Research Corporation*

Richard L. Moat, *Motorola*

William Rison, *New Mexico Institute of Mining and Technology*

Martin S. Roden, *California State University at Los Angeles*

Bruce F. Wollenberg, *University of Minnesota*

We express our gratitude to the many people at Prentice Hall who contributed to the production of this book. Special thanks go to our developmental editor, Sondra L. Chavez, for her detailed scrutiny of the original manuscript and to our publishing editor, Alan R. Apt, for his support and encouragement. The authors are indebted to the engineering administrative staff at the University of Denver, Louise Carlson and Kevin Kite, and to John Getty for his assistance in developing the associated laboratory program and manual.

Finally, we gratefully acknowledge the support of our wives, Juanita and Kathleen, who unceasingly encouraged us in this work.

Roland E. Thomas

Albert J. Rosa

❖ DESIGN CONTENTS

DESIGN EXAMPLES

DESIGN EXERCISES

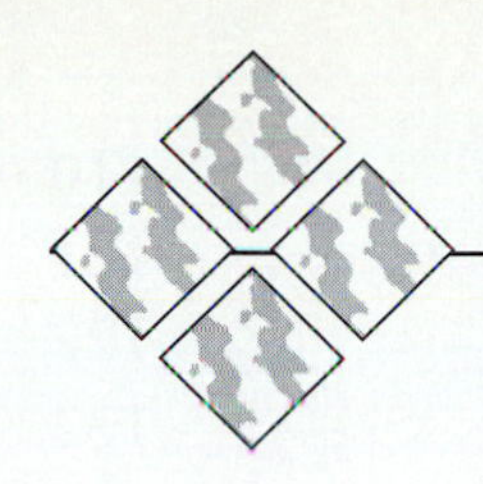

CONTENTS

❖ Design material included in this section.

PART TWO

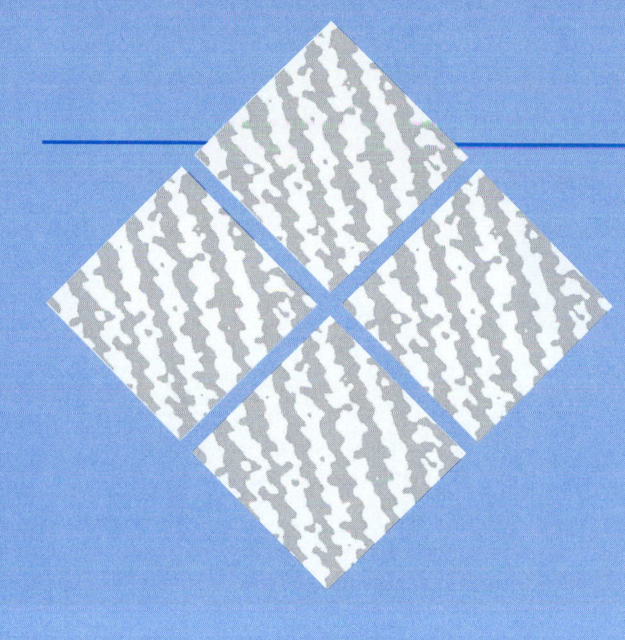

PART ONE

Resistance Circuits

Course Objectives

- **Analysis (A)**—Given a linear circuit with known input signals, select an analysis technique and determine output signals or input-output relationships.
- **Design (D)**—Devise a linear circuit or modify an existing circuit to obtain a specified output signal or a specified input-output relationship.
- **Evaluation (E)**—Given several circuits that perform the same signal-processing function, rank-order the circuits using criteria such as cost, parts count, or power dissipation.

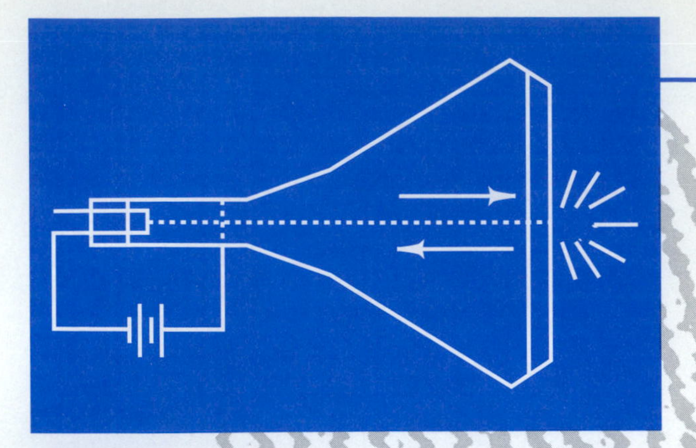

Chapter One

"The electromotive action manifests itself in the form of two effects which I believe must be distinguished from the beginning by a precise definition. I will call the first of these 'electric tension,' the second 'electric current.'"

André-Marie Ampère, 1820
French Mathematician/Physicist

Introduction

This book deals with the analysis and design of linear electric circuits. A circuit is an interconnection of electric devices that processes energy or information. Understanding circuits is important because energy and information are the underlying technological commodities in electrical engineering. The study of circuits provides a foundation for other areas of electrical engineering such as electronics, power systems, communication systems, and control systems.

This chapter describes the structure of the book, introduces basic notation, and defines the primary physical variables in electric circuits—voltage and current. André-Marie Ampère (1775–1836) was the first to recognize the importance of distinguishing between the electrical effects now called voltage and current. He also invented the galvanometer, the forerunner of today's voltmeter and ammeter. A natural genius who had mastered all of the then-known mathematics by age 12, he is best known for defining the mathematical relationship between electric current and magnetism. This relationship, now called Ampère's law, is one of the basic concepts of modern electromagnetics.

The first section of this chapter describes the pedagogical framework and terminology that must be understood to effectively use this book. This section describes how the learning objectives are structured to help the student develop the engineering problem-solving abilities needed to analyze and design circuits. The second section provides some of the standard scientific notation and conventions used throughout the book. The last section introduces electric voltage, current, and power—the physical variables used throughout the book to describe the signal-processing capabilities of linear circuits.

1-1 About This Book

The basic purpose of this book is to introduce the analysis and design of linear circuits. Circuits are important in electrical engineering because they process electrical signals that carry energy and information. For the present we can define a *circuit* as an interconnection of electrical devices, and a *signal* as a time-varying electrical quantity. For example, during a recording session a videocassette recorder (VCR) stores information in a magnetic medium on videotape. During playback electrical circuits within the VCR recover the information as electronic signals and process the signals to generate the images displayed on a television screen. In an electrical power system some form of stored energy is converted to electrical form and transferred to loads, where it is converted into the form required by the customer. Much of the work of our industrial society is accomplished through the generation, transfer, and conversion of electrical energy. The VCR and the electrical power system both involve circuits that process and transfer signals carrying energy and information.

In this text we are primarily interested in *linear circuits*. An important feature of a linear circuit is that the amplitude of the output signal is proportional to the input signal amplitude. The proportionality property of linear circuits greatly simplifies the process of circuit analysis and design. Most circuits are linear only within a restricted range of signal levels. When driven outside this range they become nonlinear, and proportionality no longer applies. Although we will treat a few examples of nonlinear circuits, our attention is focused on circuits operating within their linear range.

Our study also deals with interface circuits. For the purposes of this book we define an *interface* as a pair of accessible terminals at which signals may be observed or specified. The interface concept is especially important with integrated circuit (IC) technology. Integrated circuits involve many thousands of interconnections, but only a small number are accessible to the user. Creating systems using integrated circuits involves interconnecting large circuits at a few accessible terminals in such a way that the circuits are compatible. Ensuring compatibility often involves relatively small circuits whose purpose is to change signal levels or formats. Such interface circuits are intentionally introduced to ensure that the appropriate signal conditions exist at the connections between two larger circuits.

Course Objectives

This book has a well-defined set of learning objectives and related homework that can be used to verify your progress. The overarching *Course Objectives* are defined at three levels: analysis, design, and evaluation. In terms of signal processing, *analysis* involves determining the output signals of a given circuit with known input signals. Analysis has the compelling feature that a unique solution exists in linear circuits. Circuit analysis will occupy the bulk of our attention, since it provides the foundation for understanding the interaction of signals and circuits. *Design* involves devising circuits that perform a prescribed signal-processing function. In contrast to analysis, a design problem may have no solution or several solutions. The latter possibility leads to the *evaluation* function. Given several circuits that perform the same basic function, the alternative designs are rank-ordered using factors such as cost, power consumption, and part counts. In reality the engineer's role involves analysis, design, and evaluation, and the boundaries between these functions are often blurred. By explicitly stating our analysis, design, and evaluation objectives, it is our hope that this book will help to prepare you for that role.

The book is divided into three parts: (I) Resistance Circuits (Chapter 1–5), (II) Dynamic Circuits (Chapters 6–11), and (III) Applications (Chapters 12–16). The course objectives are listed at the beginning of the first chapter in each of the three parts of the book. The *course-integrating problems* given at the end of the last chapter in each part are designed to test your mastery of the course objectives.

En Route Objectives

The course objectives are built on a structured set of smaller goals called *En Route Objectives* (ERO). These smaller milestones are listed at the end of each chapter together with associated homework problems designed to test your understanding of basic concepts and techniques. Although they are listed at the end of the chapter, the enroute objectives are an integral part of the chapter and are keyed to one or more sections of the text. Collectively, the enroute objectives represent the basic knowledge and understanding needed to achieve the course objectives. At the very end of each chapter there are a number of *chapter-integrating problems*. These problems encompass several enroute objectives mostly from within that chapter and offer an opportunity to deal with more complex problems before attempting the even broader course-integrating problems. Course-integrating problems appear at the end of each part and require mastery of two or more EROs from different chapters usually from within that part. Both the course- and chapter-integrating problems are designated as analysis (A), design ❖ (D) or evaluation ❖❖ (E) depending on their primary purpose.

The book contains many worked *Examples* to help you understand how to apply the concepts needed to master the enroute objectives. These examples describe in detail the steps needed to obtain the final answer. They usually treat analysis problems, although design examples and application notes are included where appropriate. There are also in-text *Exercises*, which include only the problem statement and the final answer. These exercises provide an opportunity for immediate feedback on your ability to correctly apply the enroute concepts discussed in a particular section. You should work out the problem solution in detail and then check your results with the final answer given in the exercise.

1-2 SYMBOLS AND UNITS

Throughout this text we will use the international system (SI) of units. The SI units include six fundamental units: meter (m), kilogram (kg), second (s), ampere (A), kelvin (K) and candela (cd). All the other units can be derived from these six.

Like all disciplines, electrical engineering has its own terminology and symbology. The symbols used to represent some of the more important physical quantities and their units are listed in Table 1-1. It is not our purpose to define these quantities here, nor to offer this list as an item for memorization. Rather, the purpose of this table is simply to list in one place all of the quantities commonly used in this book.

Numerical values encountered in electrical engineering range over many orders of magnitude. Consequently, the system of standard decimal prefixes in Table 1-2 is used. These prefixes on the unit abbreviation of a quantity indicate the power of 10 that is applied to the numerical value of the quantity.

EXERCISE 1-1

Given the pattern in the statement 1 kΩ = 1 kilohm = 1×10^3 ohms, fill in the blanks in the folowing statements using the standard decimal prefixes.

(a) ____ = ____ = 5×10^{-3} watts.
(b) 10.0 dB = ____ = ____.
(c) 3.6 ps = ____ = ____.
(d) ____ = 0.03 microfarads = ____.
(e) ____ = ____ megahertz = ____ gigahertz = 6.6×10^9 hertz.

ANSWERS:
(a) 5.0 mW = 5 milliwatts.
(b) 10.0 decibels = 1.0 Bel.
(c) 3.6 picoseconds = 3.6×10^{-12} seconds.
(d) 30 nF or 0.03 μF = 30.0×10^{-9} farads.
(e) 6.6 GHz or 6600 MHz = 6600 megahertz = 6.6 gigahertz.

TABLE 1-1 SOME IMPORTANT QUANTITIES, THEIR SYMBOLS, AND THEIR UNIT ABBREVIATIONS

Quantity	Symbol	Unit	Unit Abbreviation
Time	t	second	s
Frequency	f	hertz	Hz
Radian frequency	ω	radian/sec	rad/s
Phase angle	θ,ϕ	degree or radian	° or rad
Energy	w	joule	J
Power	p	watt	W
Charge	q	coulomb	C
Current	i	ampere	A
Electric field	$\mathcal{E}$	volt/meter	V/m
Voltage	v	volt	V
Impedance	Z	ohm	Ω
Admittance	Y	siemen	S
Resistance	R	ohm	Ω
Conductance	G	siemen	S
Reactance	X	ohm	Ω
Suspectance	B	siemen	S
Inductance, self	L	henry	H
Inductance, mutual	M	henry	H
Capacitance	C	farad	F
Magnetic flux	φ	weber	wb
Flux linkages	λ	weber-turns	wb-t
Power ratio	$\log_{10}(p_2/p_1)$	Bel	B

1-3 CIRCUIT VARIABLES

The underlying physical quantities in the study of electronic systems are two basic variables, *charge* and *energy*. The concept of electrical charge explains the very strong electrical forces that occur in nature. To explain both attraction and repulsion, we say there are two kinds of charge—positive and negative. Like charges repel, while unlike charges attract. The symbol q is used to represent charge. If the amount of charge is varying with time we emphasize the fact by writing $q(t)$. In the international system (SI) charge is measured in *coulombs* (abbreviated C). The smallest quantity of charge in nature is an electron's charge ($q_E = 1.6 \times 10^{-19}$C). There are 6.24×10^{18} electrons in 1 coulomb.

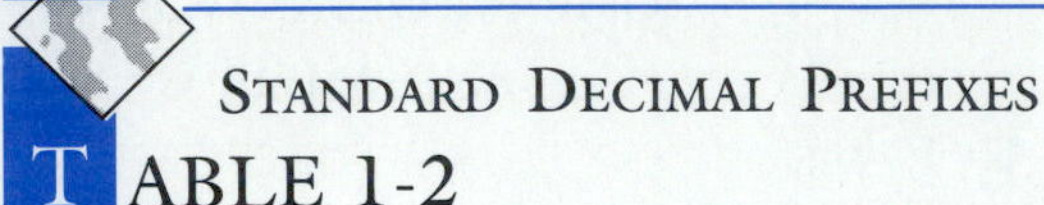

TABLE 1-2 STANDARD DECIMAL PREFIXES

Multiplier	Prefix	Abbreviation
10^{18}	exa	E
10^{15}	peta	P
10^{12}	tera	T
10^{9}	giga	G
10^{6}	mega	M
10^{3}	kilo	k
10^{-1}	deci	d
10^{-2}	centi	c
10^{-3}	milli	m
10^{-6}	micro	μ
10^{-9}	nano	n
10^{-12}	pico	p
10^{-15}	femto	f
10^{-18}	atto	a

Electrical charge is a rather cumbersome variable to work with in practice. Moreover, in many situations the charges are moving, and so we find it more convenient to measure the amount of charge passing a given point per unit time. To do this, we define, in differential form, a signal variable i called *current* as follows:

$$i = \frac{dq}{dt} \tag{1-1}$$

Current is a measure of the flow of electrical charge. It is the time rate of change of charge passing a given point. The physical dimensions of current are coulombs per second. In the SI system, the unit of current is the *ampere* (abbreviated A). That is,

$$1 \text{ coulomb/second} = 1 \text{ ampere}$$

Since there are two types of electrical charge, there is a bookkeeping problem associated with the direction assigned to the current. In electrical engineering it is customary to define the direction of current as the direction of the net flow of positive charges.

The concept of *voltage*, a second signal variable, is associated with the change in energy that would be experienced by a charge as it passes through a circuit. The symbol w is commonly used to represent energy. In the SI system of units, energy carries the units of *joules* (abbreviated J). If a small charge dq were to experience a change in energy dw in passing from point A to point B, then the voltage v between A and B is defined as the change in energy per unit charge. We can express this definition in differential form as

$$v = \frac{dw}{dq} \tag{1-2}$$

Voltage does not depend on the path followed by the charge dq in moving from point A to point B. Furthermore, there can be a voltage between two points even if there is no charge motion (i.e., no current), since voltage is a measure of how much energy dw would be involved if a charge dq were moved. The dimensions of voltage are joules per coulomb. The unit of voltage in the SI system is the *volt* (abbreviated V). That is,

$$1 \text{ joule/coulomb} = 1 \text{ volt}$$

A third signal variable *power* is defined as the time rate of change of energy:

$$p = \frac{dw}{dt} \tag{1-3}$$

The dimensions of power are joules per second, which in the SI system is called a *watt* (abbreviated W). In electrical situations, it is useful to have power expressed in terms of current and voltage. Using the chain rule Eq. (1-3) can be written as

$$p = \left(\frac{dw}{dq}\right)\left(\frac{dq}{dt}\right) \tag{1-4}$$

Now using Eqs. (1-1) and (1-2) we obtain

$$p = vi \tag{1-5}$$

The electrical power associated with a situation is determined by the product of voltage and current. The total energy transferred during the period from t_1 to t_2 is found by solving for dw in Eq. (1-3), and then integrating

$$w_T = \int_{w_1}^{w_2} dw = \int_{t_1}^{t_2} p\, dt \tag{1-6}$$

EXAMPLE 1-1

Determine the power in an electron beam in a TV picture tube if the electron beam is carrying 10^{14} electrons per second and is accelerated by a voltage of 50 kV (Fig. 1-1).

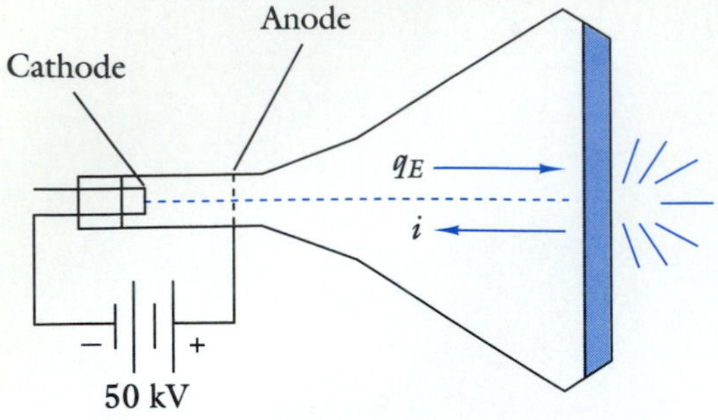

Figure 1-1

SOLUTION:

Since current is the rate of charge flow, we can find the net current by multiplying the charge of the electron, q_E, by the rate of electron flow, dn_E/dt.

$$i = q_E \frac{dn_E}{dt} = (1.6 \times 10^{-19})(10^{14}) = 1.6 \times 10^{-5} \text{ A}$$

Therefore, the beam power is found as

$$p = vi = (50 \times 10^3)(1.6 \times 10^{-5}) = 0.8 \text{ W}$$

EXAMPLE 1-2

The current through a circuit element is 50 mA. Find the total charge and the number of electrons transferred during a period of 100 ns.

SOLUTION:

The relationship between current and charge is given in Eq. (1-1) as

$$i = \frac{dq}{dt}$$

Since the current i is given, we obtain charge transferred by solving this equation for dq and then integrating

$$q_T = \int_{q_1}^{q_2} dq = \int_0^{10^{-7}} i\, dt$$
$$= \int_0^{10^{-7}} 50 \times 10^{-3} dt = 50 \times 10^{-10} C = 5 \text{ nC}$$

There are 6.24×10^{18} electrons/coulomb, so the number of electrons transferred is

$$n_E = (5 \times 10^{-9} \text{ C})(6.24 \times 10^{18} \text{electrons/C}) = 31.2 \times 10^{9} \text{electrons}$$

EXERCISE 1-2

A device dissipates 100 W of power. How much energy is delivered to the device in 10 seconds?

ANSWER:
1 kJ

EXERCISE 1-3

The graph in Fig. 1-2 (a) shows the charge $q(t)$ flowing past a point in a wire as a function of time.

(a) Find the current $i(t)$ at $t = 1, 2.5, 3.5, 4.5$, and 5.5 ms.

(b) Sketch the variation of $i(t)$ versus time.

ANSWERS:

(a) -10 nA, $+40$ nA, 0 nA, -20 nA, 0 nA.

(b) The variations in $i(t)$ are shown in Fig. 1-2 (b).

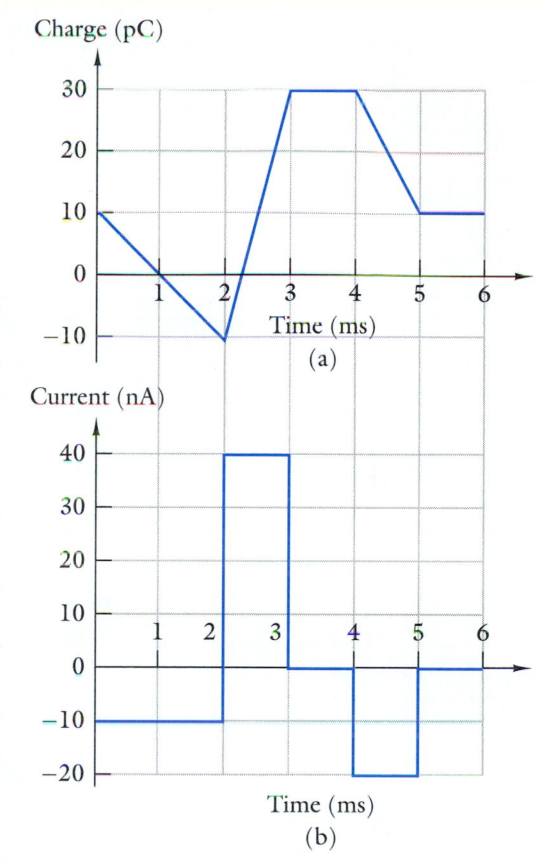

Figure 1-2

Signal References

So far we have defined three signal variables (current, voltage, and power) in terms of two basic variables (charge and energy). Charge and energy, like mass, length, and time, are basic concepts of physics that provide the scientific foundation for electrical engineering. However, engineering problems rarely involve charge and energy directly, but are usually stated in terms of the signal variables. The reason for this is quite simple. Current and voltage are much easier to measure and therefore are the most useful working variables in engineering practice.

Up to this point in the text, we have said a signal can be either a current or a voltage, and we have not made any great distinction between them. From this point forward, it is essential that the reader recognize that current and voltage are interrelated but quite different variables. Current is a measure of the time rate of charge passing a point. Since current indicates the direction of the flow of electrical charge, we think of current as a *through* variable. We think of voltage as an *across* variable because it inherently involves two points. Voltage is a measure of the net change in energy involved in moving a charge from one point to another. Voltage is not measured at a point, but rather between two points or across an element.

Figure 1-3 shows a notation used for assigning reference directions to current and voltage. The reference mark for current (the arrow below the wire) does not indicate the actual direction of the current. The actual direction may be reversing a million times per second. However, when the actual direction coincides with the reference direction, we say the current is positive. When the opposite occurs, we say that the current is negative. If the net flow of positive charge in Fig. 1-3 is to the right, we say the current $i(t)$ is positive. Conversely, if the current $i(t)$ is positive, then we know the net flow of positive charge is to the right.

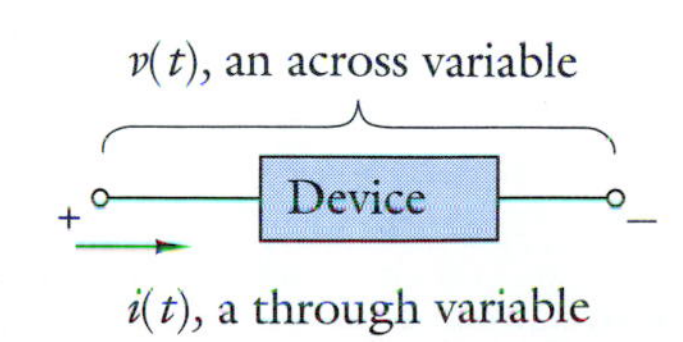

Figure 1-3 Two-terminal device voltage and current.

Similarly, the voltage reference marks (+ and − symbols) in Fig. 1-3 do not imply that the potential at the + terminal is always higher than the potential at the − terminal. However, when this is true, we will say the voltage across the device is positive. When the opposite is true, we say that the voltage is negative. The bracket in Fig. 1-3 is used here to clarify the situation and to remind us that voltage is an across variable. When we interconnect devices to form a circuit, it becomes impractical to draw all of the brackets that would be required.

The importance of relating the reference directions (the plus and minus voltage signs, and current arrows) to the actual direction of the current (in and out) and voltage (hi and lo) can be seen in determining the power associated with a device. Consider the device operating in the first quadrant of Fig. 1-4. The actual direction of the current is the same as the reference arrow drawn on the device. That is, the current goes "in" and comes "out" of the device in the same direction as the reference arrow. Also, the voltage is "hi" at the positive reference and "lo" at the negative reference. In the first quadrant the actual and reference directions agree and i and v are both positive. The power associated with this device is positive, since the product of the current and voltage is positive.

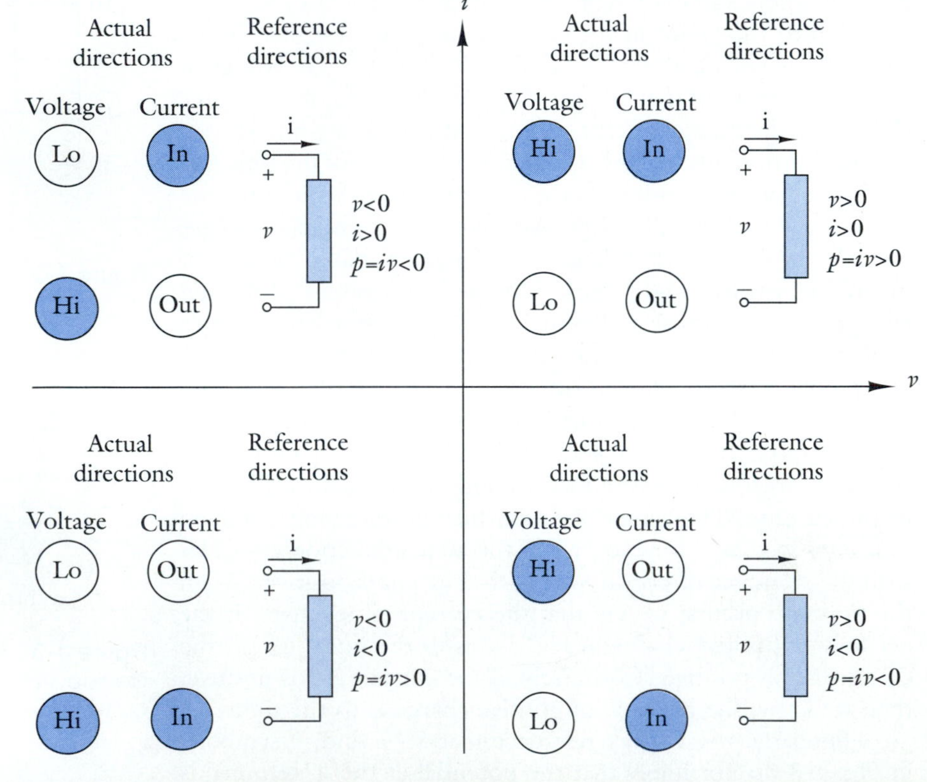

Figure 1-4 Actual and reference directions for voltage, current and power.

In the lower left, or third quadrant the actual and reference directions both disagree, so i and v are both negative and $p = iv$ is positive. A positive sign for the associated power indicates that the device absorbs or uses power. In the second and fourth quadrants the actual and reference direction disagrees for either voltage or current so i and v have opposite signs and $p = iv$ is negative. In this case the associated power is negative and the device is said to provide or deliver power. Note that the device reference directions are the same in all four quadrants; only the actual directions change between quadrants.

This is more than just an exercise in semantic bookkeeping. In Fig. 1-4 the current reference arrow enters the device at the terminal with the plus voltage reference mark. This is called the *passive convention* because $p = iv$ is positive when a device absorbs power, and is negative when it delivers power. Certain devices such as heaters (e.g., a toaster) can only absorb power. Their i-v characteristics must lie in the first and third quadrants in Fig. 1-4. On the other hand, the power associated with a battery is positive when it is charging (absorbing power) and negative when it is discharging (delivering power). In its charging mode a battery has its characteristics lying in the first and third quadrants, but in its discharging mode it has its characteristics lying in the second and fourth quadrants.

The passive convention is used throughout the book. It is also the convention used by computer circuit simulation programs such as SPICE and MICRO–CAP.[1] To properly interpret the results of circuit analysis it is important to remember that the arrows and plus/minus signs are reference directions for the signal variables, not indications of directions of actual signal flow. If current is in the same (opposite) direction as the reference arrow, then its numerical value is positive (negative). If the voltage polarity is the same (opposite) as the reference polarity, then its numerical value is positive (negative). Under the passive convention, if the current and voltage have the same (opposite) signs, then the associated power is positive (negative) and the device absorbs (delivers) power.

Ground

We have stated that voltage is an across variable defined and measured between two points. It is convenient to identify one of the points as a reference point commonly called *ground*. The volt-

[1]We will discuss circuit analysis programs such as SPICE and MICRO–CAP beginning with Chapter 5.

age at all other points in a circuit are defined with respect to this reference point. This is not an unusual idea. For example, the elevation of a mountain is the number of feet or meters between the top of the mountain and a reference level at mean sea level. The reference point for an electric circuit is not as obvious as the sea level reference for elevations. Therefore, we denote the circuit reference using one of the "ground" symbols shown in Fig. 1-5. The voltage at the ground point is defined to be zero volts just as an elevation at mean sea level is taken to be at zero feet elevation. We will return to the concept of ground in Chapter 2.

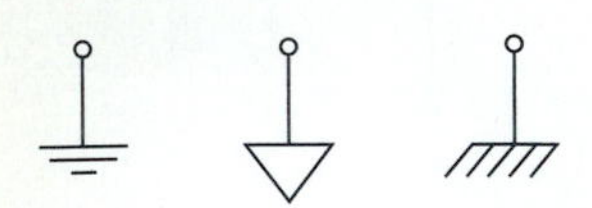

Figure 1-5 Ground symbols.

In the next chapter we will begin our study of electrical devices. The rectangular box in Fig. 1-3 represents any two-terminal device. Each device has its own *i-v* characteristics. The interconnection of these devices into a useful grouping is called a circuit. The analysis and design of circuits is the subject of this book.

Exercise 1-4

Using the passive sign convention the signal variables of a set of two-terminal electrical devices are observed to be the following:

	DEV 1	DEV 2	DEV 3	DEV 4	DEV 5
v	+10 V	?	−15 V	+5 V	?
i	−3 A	−3 A	+10 mA	?	−12 mA
p	?	+40 W	?	+10 mW	−120 mW

Find the magnitude and sign of the unknown signal variable and state whether the device is absorbing or delivering power.

ANSWERS:
Device 1: $p = -30$ W (delivering)
Device 2: $v = -13.3$ V (absorbing)
Device 3: $p = -150$ mW (delivering)
Device 4: $i = +2$ mA (absorbing)
Device 5: $v = +10$ V (absorbing) ■

EXAMPLE 1-3

Figure 1-6 shows a circuit formed by interconnecting five two-terminal devices. A voltage and current variable has been assigned to each device using the passive sign convention. The signal variables for each device are observed to be the following:

	DEV 1	DEV 2	DEV 3	DEV 4	DEV 5
v	+100 V	?	+25 V	+75 V	−75 V
i	?	+5 mA	+5 mA	?	+5 mA
p	−1 W	+0.5 W	?	0.75 W	?

(a) Find the missing signal variable for each device and state whether the device is absorbing or delivering power.

(b) Check your work by showing that the sum of the device powers is zero.

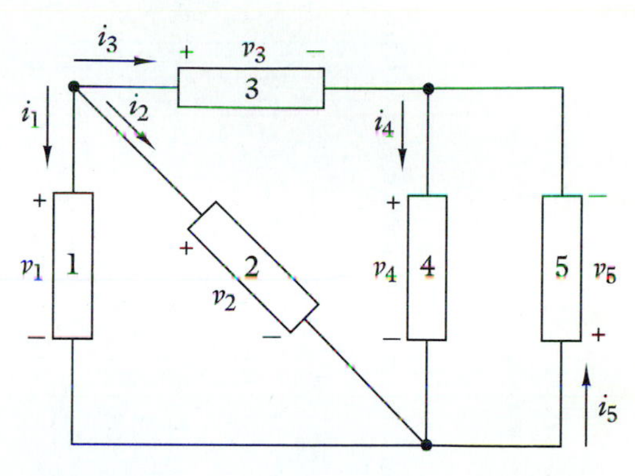

Figure 1-6

SOLUTION:

(a) We use $p = vi$ to solve for the missing variable, since two of the three signal variables are given for each device.
Device 1: $i_1 = p_1/v_1 = -1/100 = -10$ mA (delivering power).
Device 2: $v_2 = p_2/i_2 = 0.5/0.005 = 100$ V (absorbing power).
Device 3: $p_3 = v_3 i_3 = 25 \times 0.005 = 0.125$ W (absorbing power).
Device 4: $i_4 = p_4/v_4 = 0.75/75 = 10$ mA (absorbing power).
Device 5: $p_5 = v_5 i_5 = -75 \times 0.005 = -0.375$ W (delivering power).

(b) Summing the device powers yields

$$\begin{aligned} p_1 + p_2 + p_3 + p_4 + p_5 &= -1 + 0.5 + 0.125 + 0.75 - 0.375 \\ &= +1.375 - 1.375 = 0 \end{aligned}$$

This example shows the sum of the power absorbed by devices is equal in magnitude to the sum of the power supplied devices. This power balance must exist in the types of circuits treated in this book. The required power balance can be used as an overall check of calculations such as those in this example. In Chapter 3, we find that the power balance is a consequence of Tellegen's theorem.

EXERCISE 1-5

Figure 1-7 shows a circuit with six two-terminal devices and their assigned voltage and current variables. The signal variables for each element are observed to be the following:

	DEV 1	DEV 2	DEV 3	DEV 4	DEV 5	DEV 6
v	+7.5 V	+5 V	?	+7.5 V	?	+12.5 V
i	−3 A	?	+0.5 A	+1.5 A	+2.5 A	+1 A
p	?	+10 W	+1.25 W	?	−12.5 W	?

(a) Find the missing signal variable for each device and state whether the device is absorbing or delivering power.

(b) Use the power balance to check your work.

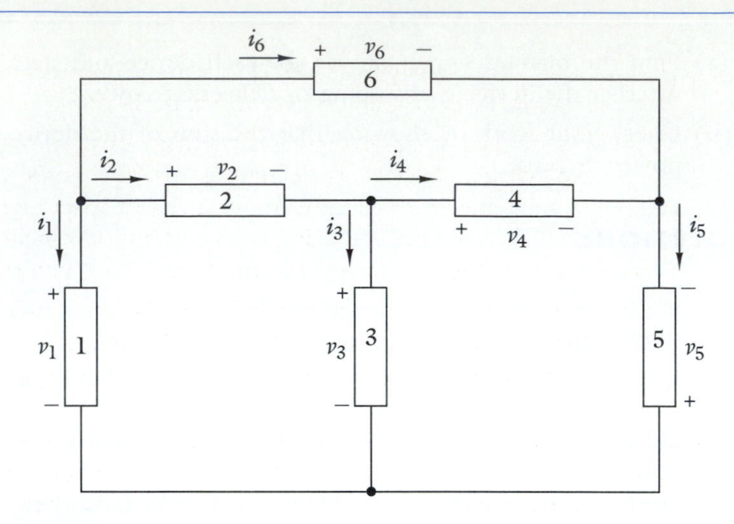

Figure 1-7

ANSWERS:

(a) $p_1 = -22.5$ W (delivering), $i_2 = +2$ A (absorbing), $v_3 = +2.5$ V (absorbing), $p_4 = +11.25W$ (absorbing), $v_5 = -5$ V (delivering), and $p_6 = +12.5$ W (absorbing).

(b) $35 - 35 = 0$. ■

SUMMARY

- Circuits are important in electrical engineering because they process signals that carry energy and information. A circuit is an interconnection of electrical devices. A signal is a time-varying electrical quantity. An interface is a pair of accessible terminals at which signals may be observed or specified.
- This book defines overall course objectives at the analysis, design, and evaluation levels. Analysis involves determining the output signals of a given circuit with known input signals. Design involves devising circuits that perform a prescribed signal-processing function. Evaluation involves appraising alternative circuit designs using criteria such as cost, power consumption, and parts count. The enroute objectives given at the end of each chapter define the basic knowledge and understanding needed to achieve the course objectives.
- Charge *(q)* and energy *(w)* are the basic physical quantities involved in electrical phenomena. Current *(i)*, voltage *(v)*, and power *(p)* are the derived signal variables used in circuit analysis and design. In the SI system, charge is measured

in coulombs (C), energy in joules (J), current in amperes (A), voltage in volts (V), and power in watts (W).

- Current is defined as dq/dt and is a measure of the flow of electric charge. Voltage is defined as dw/dq and is a measure of the energy required to move a charge from one point to another. Power is defined as dw/dt and is a measure of the rate at which energy is being transferred. Power is related to current and voltage as $p = vi$.
- Under the passive convention reference marks are assigned to two-terminal devices as shown in Fig. 1-3. The numerical value of a current is positive (negative) when the current through a device has the same (opposite) direction as the assigned reference arrow. The numerical value of a voltage is positive (negative) when the polarity of the voltage across a device agrees (disagrees) with the assigned reference marks. When current and voltage have the same (opposite) signs the device power is positive (negative) and the device is said to be absorbing (delivering) power.

EN ROUTE OBJECTIVES AND ASSOCIATED PROBLEMS

ERO 1-1 Electrical Quantities (Sect. 1-2)

Given an electrical quantity described in terms of words, scientific notation, or decimal prefix notation; convert the quantity to an alternate description.

1-1 Write the following quantities in symbolic form:
(a) four terahertz
(b) sixteen point six picoseconds
(c) zero point forty-seven microhenrys
(d) one hundred fifty kilohms
(e) five thousand six hundred milliamperes

1-2 Write out the meaning of the following symbols in words:
(a) 5 Grad/sec
(b) 2.02 J
(c) 640 nV
(d) 10 dB
(e) 1.1 MΩ

1-3 Express the following quantities using appropriate engineering prefixes, that is, state the numeric to the nearest standard prefix:
(a) 1,200,000 ohms
(b) 5555.56 hertz
(c) .000000001 farads

(d) 6200×10^{-9} amperes
(e) 2.45×10^{5} volts

1-4 A graphical display monitor can address 1024 x 1024 separate elements called pixels. If each pixel can be coded to display 24 bits of information, how many bits can be displayed on the entire screen? If a byte contains eight bits of information, how many bytes of data can be displayed on the entire screen?

1-5 Answer the following questions using scientific notation:
(a) How many microfarads are there in 1000 picofarads?
(b) How many kilohms are there in 10 megohms?
(c) How many femtoamps are there in 50.5 amperes?
(d) How many megahertz are there in 125 hertz?
(e) How many nanohenrys are there in 0.625 millihenrys?

1-6 A signal source provides 100 mW of power to a circuit. An element in the circuit called the load absorbs 10 mW of the power. Express the ratio of the source power delivered to the load power in decibels (dB), where dB = $10 \log_{10}$ (power ratio).

ERO 1-2 Circuit Variables (Sect. 1-3)

Given any two of the three signal variables (i, v, p) or the two basic variables (q, w), find the magnitude and direction (sign) of the unspecified variables.

1-7 Determine the number of electrons that flow through a discharge tube if all of the 2 joule/coulomb supply is dissipated in 10 nanoseconds at the cost of 3.2 watts.

1-8 Figure P1-8 shows a plot of the charge flowing in a wire as a function of time. Make a sketch of the corresponding current during the same time period.

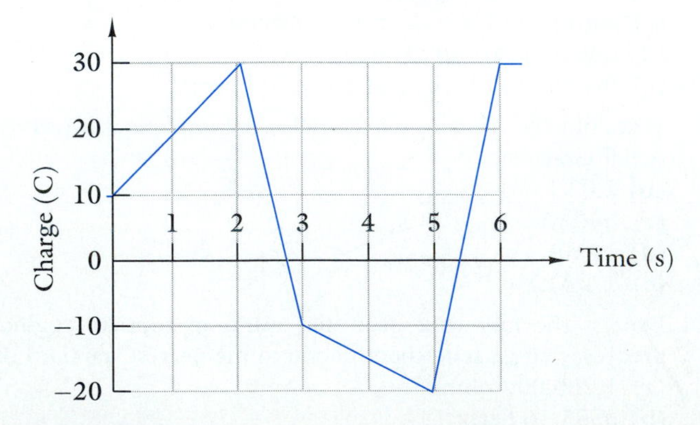

Figure P1-8

1-9 The charge flowing through a device varies as $q(t) = 3t^2$ C. Find the current through the device at $t = 0$ s, $t = 1$ s and $t = 3$ s.

1-10 The current through a device is given by $i(t) = 0.05t$ A. How many coulombs enter the device between time $t = 0$ s and $t = 5$ s?

1-11 The charge flowing in a wire varies as shown in the graph of Fig. P1-11. Make a sketch of the corresponding current.

1-12 An incandescent lamp absorbs 75 W when connected to a 120-V source. What is the current through the lamp? How much does it cost to operate the light for 8 hours if electricity costs 7.2 cents/kW-hr?

1-13 Find the current drawn from the 12-V automobile battery in Fig. P1-13 by a headlamp that absorbs 60 watts of power. How much energy can be stored in the battery if it is rated at 100 ampere-hour? For how long will the battery light the headlamp?

1-14 The current density in a copper wire is 100 A/cm^2. If copper has 5 x 10^{22} electrons per cm^3, what is the average velocity of the electrons flowing in the wire?

1-15 The current through a resistive heating element is 10 A. How much charge is transferred to the device per hour?

1-16 The voltage between two points is 100 V. How much energy is required to move a 1-C charge from one point to the other?

1-17 A two-terminal device has the current versus voltage characteristics shown in Fig. P1-17.

(a) What is the device power when operating at point A? Is the device absorbing or delivering power?

(b) Repeat (a) when the device is operating at the points B and C.

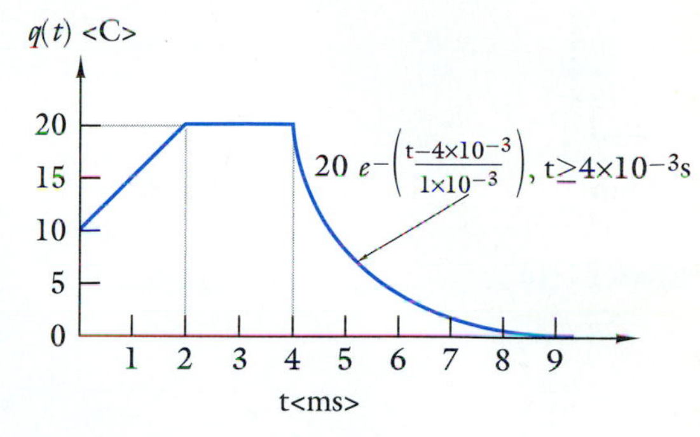

Figure P1-11

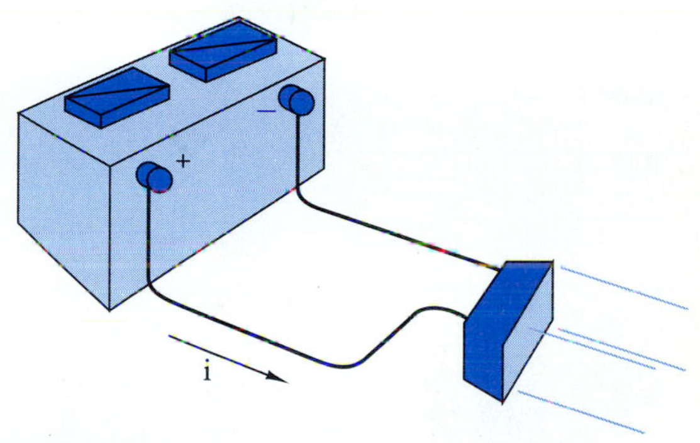

Figure P1-13

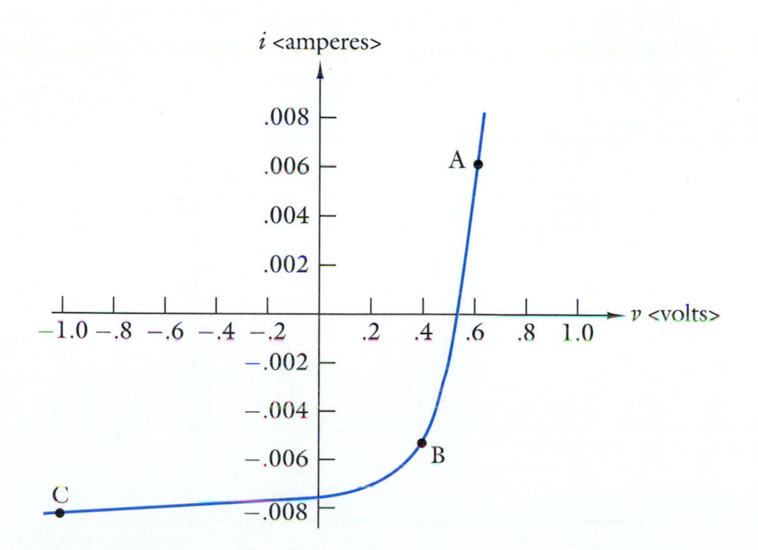

Figure P1-17

(c) Identify the operating point on the curve (it could be a point other than A, B or C) at which the device delivers the maximum power output.

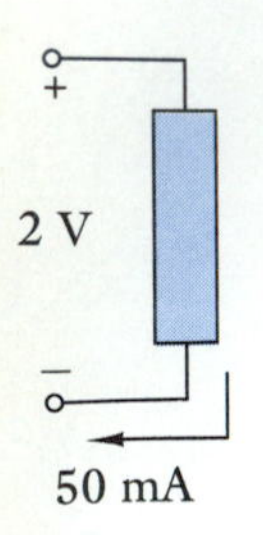

Figure P1-18

1-18 The values of the current through and voltage across a two- terminal device are shown in Fig. P1-18. What is the device power? Is the device absorbing or delivering power?

1-19 Suppose the direction of the current in Fig. P1-18 is reversed and the voltage is unchanged. Is the device absorbing or delivering power under these conditions?

1-20 For the current and voltage conditions shown in Fig. P1-18, how much energy is delivered by or absorbed by the device?

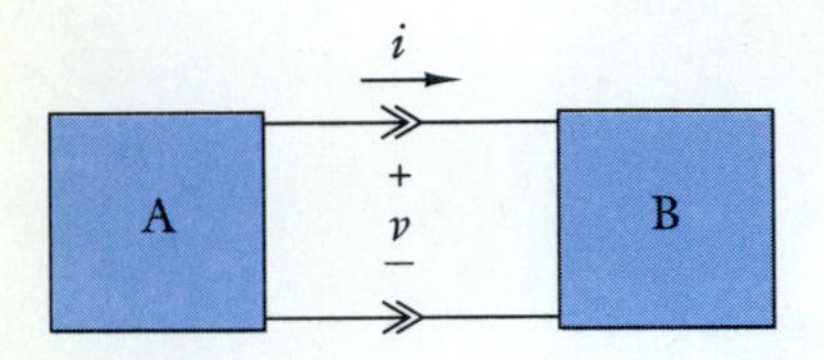

Figure P1-21

1-21 Two electrical devices are connected at an interface as shown in Fig. P1-21. Using the reference marks shown in the figure, find the power transferred and state whether the power is transferred from A to B or B to A when:

(a) $v = +12$ V and $i = -2.2$ A,
(b) $v = -33$ V and $i = -1.2$ mA,
(c) $v = +15$ V and $i = +40$ mA,
(d) $v = -37.5$ V and $i = -43$ mA.

1-22 Figure P1-22 shows an electric circuit with a voltage and current variable assigned to each of the six devices. The device signal variables are observed to be the following:

	DEV 1	DEV 2	DEV 3	DEV 4	DEV 5	DEV 6
v	+15 V	+10 V	?	+15 V	?	?
i	−3 A	?	+0.5 A	+2.5 A	+3.5 A	−1 A
p	?	+30 W	+2.5 W	?	−35 W	+10 W

Find the unknown signal variable associated with each device and state whether the device is absorbing or delivering power. Use the power balance to check your work.

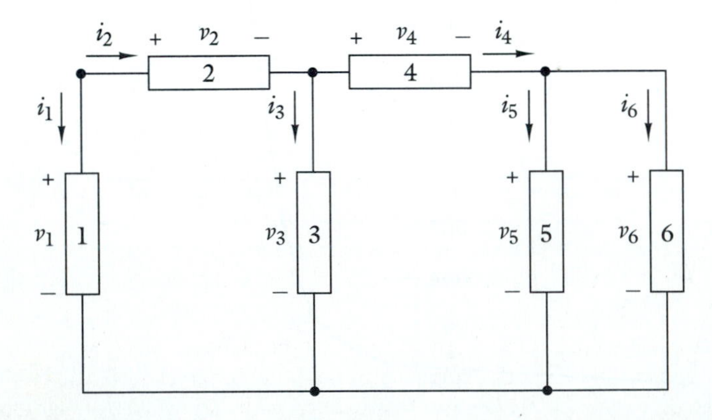

Figure P1-22

1-23 Figure P1-23 shows an electric circuit with a voltage and current variable assigned to each of the eight devices. The device signal variables are observed to be the following:

	DEV 1	DEV 2	DEV 3	DEV 4	DEV 5	DEV 6	DEV 7	DEV 8
v	+10 V	−20 V	+30 V	−10 V	+10 V	?	?	−7.5 V
i	+3 A	−2 A	?	−1 A	?	+2 A	−1 A	−1 A
p	?	?	−150 W	?	+20 W	+40 W	+2.5 W	?

Find the unknown signal variable associated with each device and state whether the device is absorbing or delivering power. Use the power balance to check your work.

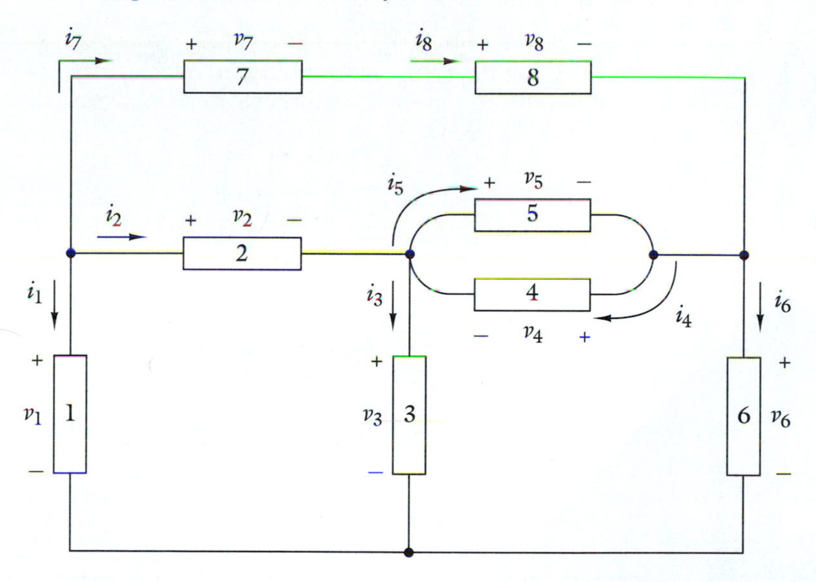

Figure P1-23

1-24 The voltage across and current through a two-terminal device are $v(t) = 5e^{-100t}$ V and $i(t) = -5e^{-100t}$ mA. How much power is delivered by the device at $t = 0$, $t = 5$ ms, and $t = 25$ ms? How much energy is delivered by the device during the interval $0 < t < \infty$?

1-25 The voltage across and current through a two-terminal device are $v(t) = 120 \cos 1000t$ V and $i(t) = 4 \sin 1000t$ A.

(a) Show that the device absorbs power during some time intervals and delivers power during others.

(b) Find the maximum power absorbed by the device and the maximum power delivered by the device. Hint: $\cos x \sin x = \frac{1}{2} \sin 2x$.

1-26 The voltage across a two-terminal light detector is +5 V. In complete darkness the current through the device is +10 nA. In full sunlight the current is +5 mA. Express the light/dark power ratio of the device in decibels (dB).

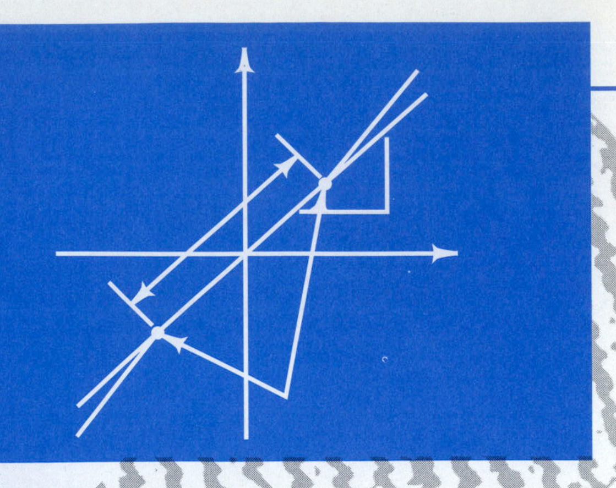

Chapter Two

"The equation S = A / L shows that the current of a voltaic circuit is subject to a change, by each variation originating either in the magnitude of a tension or in the reduced length of a part, which latter is itself again determined, both by the actual length of the part as well as its conductivity and its section."

Georg Simon Ohm, 1827
German Mathematician/Physicist

Basic Circuit Analysis

This chapter begins our study of the ideal models used to describe the physical devices in electric circuits. Foremost among these is the famous Ohm's law, which defines the model of a linear resistor. Georg Simon Ohm (1787–1854) discovered the law that now bears his name in 1827. His results drew heavy criticism and were not generally accepted for many years. Fortunately, the importance of his contribution was eventually recognized during his lifetime. He was honored by the Royal Society of London in 1841 and appointed a professor of physics at the University of Munich in 1849.

The first section deals with the element constraints derived from the ideal models of resistors, voltage sources, and current sources. The models for other linear devices such as amplifiers, inductors, capacitors, and transformers are introduced in subsequent chapters. When devices are interconnected to form a circuit they are subject to connection constraints that are based on fundamental conservation laws. These are known as Kirchhoff's laws and are discussed in the second section. The remainder of the chapter uses the combined element and connection constraints to develop a basic set of circuit analysis tools that include equivalent circuits, voltage and current division, and circuit reduction. The basic circuit analysis tools developed in this chapter are used frequently in the rest of the book and by practicing engineers. These analysis tools promote basic understanding because they involve working directly with the circuit model.

◇ 2-1 Element Constraints

A *circuit* is a collection of interconnected electrical devices that performs a useful function. An electrical *device* is a component that is treated as a distinct entity. The rectangular box in Fig. 2-1 is used to represent all of the two-terminal devices used to form circuits. A two-terminal device is described by its *i-v characteristics*, that is, the relationship between the voltage

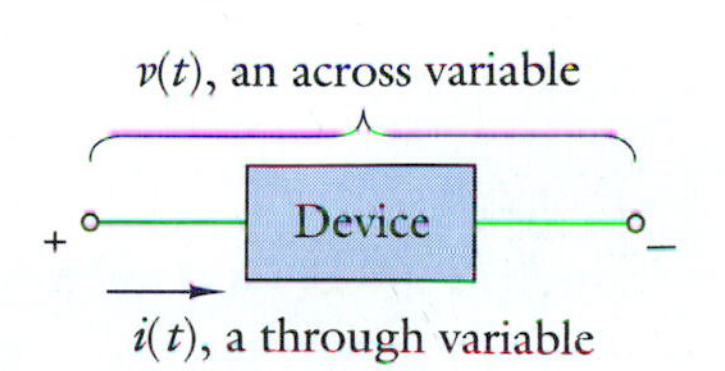

Figure 2-1 Two-terminal device voltage and current.

across and current through the device. In most cases the relationship is complicated and nonlinear so we use simpler linear models that adequately approximate the dominant features of a device.

To distinguish between a device (the real thing) and its model (an approximate stand-in), we call the model a circuit *element*. Thus, a device is an article of hardware described in manufacturers' catalogs and parts specifications. An element is a model described in textbooks on circuit analysis. This book is no exception, and a catalog of circuit elements will be introduced as we proceed.

The Linear Resistor

The first element in our catalog is a linear model of the device described in Fig. 2-2. The current-voltage characteristics of this device are nonlinear as shown by the thin blue line labeled "actual" in Fig. 2-2(b). A cubic equation would be required to write a reasonably accurate mathematical expression for this nonlinear curve. Using such a complex relationship would make circuit analysis rather difficult. If we look closely at the i-v characteristic, we can see a straight line is a good approximation when we limit the operating range of the device. The power rating of the device determines the range over which the i-v characteristics are a straight line through the origin. In sum, when the device is operated within its power rating we obtain a circuit element called the *linear resistor*

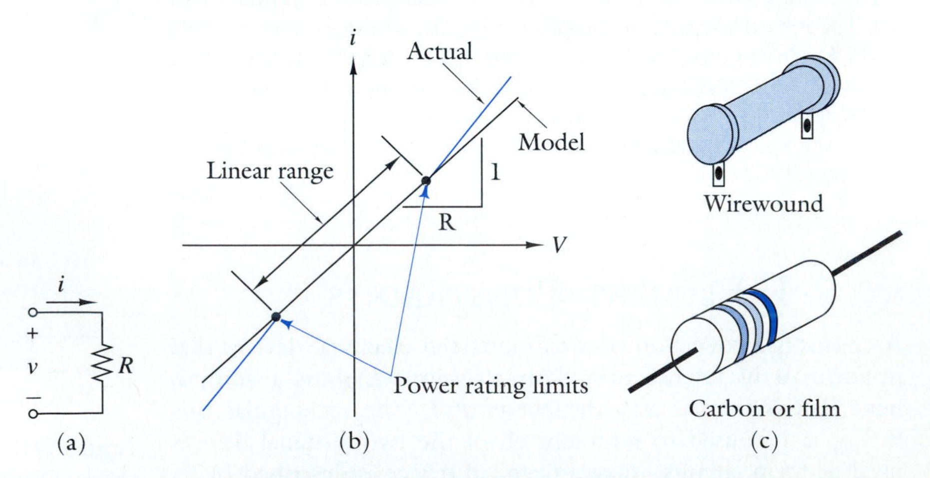

Figure 2-2 The resistor: (a) Circuit symbol. (b) i-v characteristics. (c) Actual discrete resistors.

For the passive sign convention used in Fig. 2-2(a), the equations describing this element are

$$v = Ri \qquad \text{or} \qquad i = Gv \tag{2-1}$$

Where R and G are positive constants related as

$$G = \frac{1}{R} \tag{2-2}$$

Equations (2-1) are collectively known as *Ohm's law*. The parameter R is called *resistance* and has the unit *ohms*, Ω. The parameter G is called *conductance*, with the unit *siemens*, S. In earlier literature the unit of conductance was cleverly called the mho, ℧ (ohm spelled backward and the ohm symbol upside down). Note that Ohm's law presumes the reference marks assigned to voltage and current follow the **passive sign convention**.

The Ohm's law model is represented graphically by the black straight-line in Fig. 2-2(b). The i-v characteristic for the Ohm's law model is linear and bilateral. *Linear* means the characteristic is a straight line through the origin. An important consequence of linearity is the voltage is always proportional to the current, and vice versa. *Bilateral* means the characteristic curve has odd symmetry about the origin.[1] Because of the bilateral property, reversing the polarity of the applied voltage reverses the direction but not the magnitude of the current, and vice versa. The net result is that we can connect a resistor into a circuit without regard to which terminal is which. This is important because devices such as diodes, transistors, operational amplifiers (OP AMPs), and sources are not bilateral, and we must keep very careful track of which terminal is which.

Figure 2-2(c) shows sketches of actual discrete resistor devices. Resistor fabrication techniques and physical characteristics are discussed in Appendix A.

The power associated with the resistor can be found from $p = vi$. Using Eq. (2-1) to eliminate v from this relationship yields

$$p = i^2 R \tag{2-3}$$

or using the same equations to eliminate i yields

$$p = v^2 G = \frac{v^2}{R} \tag{2-4}$$

Since the parameter R is positive, these equations tell us the power is always nonnegative. Under the passive sign convention this means the resistor **always absorbs power**.

[1]A curve $i = f(v)$ had odd symmetry if $f(-v) = -f(v)$.

EXAMPLE 2-1

A resistor functions as a linear element as long as the voltage and current are within the limits defined by its power rating. Suppose we have a resistor of 47 kΩ with a power rating of 0.25 W. Determine the maximum current and voltage that can be applied to the resistor and remain within its linear operating range.

SOLUTION:

Using Eq. (2-3) to relate power and current we obtain

$$I_{MAX} = \sqrt{\frac{P_{MAX}}{R}} = \sqrt{\frac{0.25}{47 \times 10^3}} = 2.31 \text{ mA}$$

Similarly, using Eq. (2-4) to relate power and voltage we obtain

$$V_{MAX} = \sqrt{RP_{MAX}} = \sqrt{47 \times 10^3 \times 0.25} = 108 \text{ V}$$

Open and Short Circuits

The next two circuit elements occur so often in electrical engineering they deserve a special introduction. Consider a resistor R with a voltage v applied across it. Let's calculate the current through the resistor for different values of resistance. If $v = 10$ V and $R = 1$ Ω, using Ohm's law we readily find that $i = 10$ A. If we increase the resistance to 100 Ω, we find i has decreased to 0.1 A, or 100 mA. If we continue to increase R to 1 MΩ, i becomes a very small 10 μA. Continuing this process, we arrive at a point when R is very nearly infinite and i just about zero. When i is zero, we call the special value of resistance; that is, $R = \infty$ Ω, an *open circuit*. Similarly, if we reduce R until it approaches zero, we find that the voltage is also very near zero. When v is zero, we call the special value of resistance; that is, $R = 0$ Ω, a *short circuit*. The circuit symbols for these two elements are shown in Fig. 2-3. The elements in a circuit are assumed to be interconnected using zero-resistance ideal wire, that is, using short circuits.

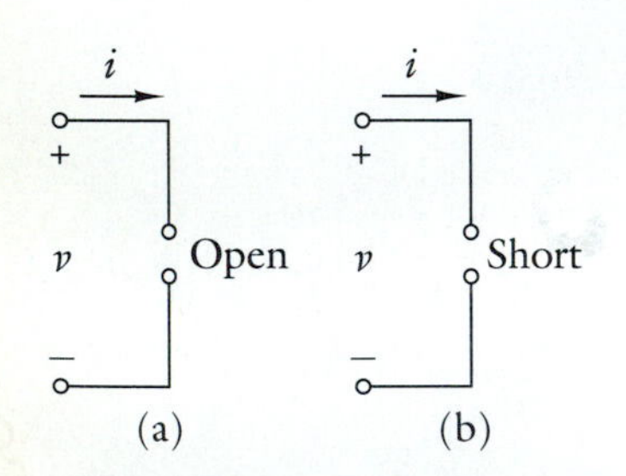

Figure 2-3
(a) Open-circuit symbol.
(b) Short-circuit symbol.

The Ideal Switch

A switch is a familiar device with many applications in electrical engineering. The *switch element* can be modeled as a combination of an open and a short circuit. Figure 2-4 shows the circuit

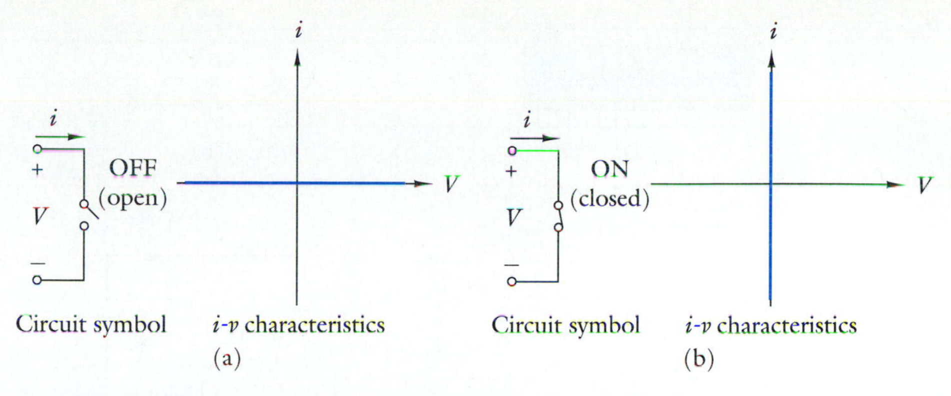

Figure 2-4 The ideal switch: (a) Switch OFF. (b) Switch ON.

symbol and i-v characteristics of a switch. When the switch is closed

$$v = 0 \qquad \text{and} \qquad i = \text{any value} \tag{2-5a}$$

and when it is open

$$i = 0 \qquad \text{and} \qquad v = \text{any value} \tag{2-5b}$$

When the switch is closed, the voltage across the device is zero and the element will pass any current that may result. When open, the current is zero and the element will withstand any voltage across its terminals. For the ideal switch element the power is zero, since the product $vi = 0$ when the switch is open ($i = 0$) or closed ($v = 0$). Actual switch devices have limitations such as the maximum current they can safely carry when closed, and the maximum voltage they can withstand when open. The switch is activated (opened or closed) by some external influence, such as a mechanical motion, temperature, or pressure.

Application Note

EXAMPLE 2-2

The *analog switch* is an important device found in analog-to-digital interfaces. Figure 2-5(a) shows an actual integrated circuit device in a dual-inline package (DIP). Figures 2-5(b), (c) and (d) show various circuit models of the device. In all models the switch is put in the closed state by applying a voltage to the terminal labeled "gate."

In the basic model in Fig. 2-5(a) the ideal switch closes when voltage is applied to the gate terminal and opens when no voltage is applied. The intermediate model in Fig. 2-5(c) includes two switches. When voltage is applied to the gate, the

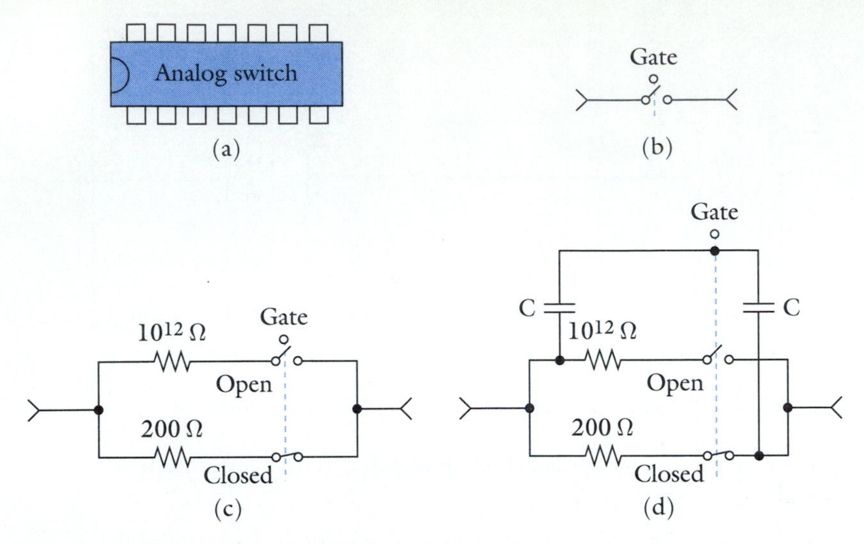

Figure 2-5 The analog switch: (a) Actual device. (b) Basic model. (c) Intermediate model. (d) Advanced model.

upper switch opens and the lower switch closes. In the closed state a resistance of 200 Ω is connected between the terminals of the switch. Any current carried by the closed switch must pass through this resistance causing an i^2R power loss. When voltage is removed from the gate the upper switch closes and the lower switch opens. In the open state a very high resistance (10^{12} Ω) is connected between the switch terminals. The voltage across the open switch causes a very small current through the high open resistance, although this current is usually insignificant. The advanced model in Fig. 2-5(d) includes the intermediate model plus two elements labeled "C" that represent capacitance. These dynamic circuit elements account for transient effects in the switch and will be studied in Chapter 7.

This example illustrates how a few basic circuit elements can be combined to model other electrical devices. It also suggests that no single model can serve in all applications. It is up to the engineer to select the simplest model that adequately represents the actual device in each application.

Ideal Sources

The signal and power sources required to operate electronic circuits are modeled using two elements: voltage sources and current sources. These sources can produce either constant or time varying signals. The circuit symbols and i-v characteristics of an ideal voltage source are shown in Fig. 2-6, while the circuit

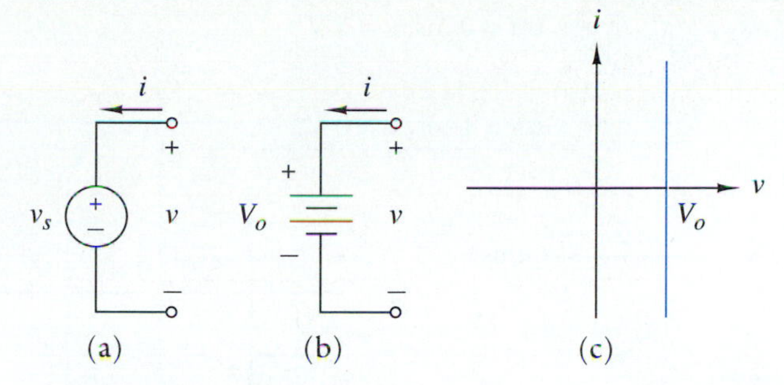

Figure 2-6 Ideal voltage source: (a) Time-varying source. (b) Constant source—Battery. (c) Constant source—*i-v* characteristics.

symbol and *i*-*v* characteristics of an ideal current source are shown in Fig. 2-7. The symbol in Fig. 2-6(a) represents either time-varying or a constant voltage source. The battery symbol in Fig. 2-6(b) is used exclusively for a constant voltage source. There is no separate symbol for a constant current source.

The *i*-*v* characteristics of an *ideal voltage source* in Fig. 2-6(c) are described by the following element equations:

$$v = v_S \qquad \text{and} \qquad i = \text{any value} \tag{2-6}$$

The element equations mean the ideal voltage source produces v_S volts across its terminals and will supply whatever current may be required by the circuit to which it is connected.

The *i*-*v* characteristics of an *ideal current source* in Fig. 2-7(b) are described by the following element equations:

$$i = i_S \qquad \text{and} \qquad v = \text{any value} \tag{2-7}$$

The ideal current source supplies i_S amperes in the direction of its arrow symbol and will furnish whatever voltage is required by the circuit to which it is connected. The voltage or current produced by these ideal sources is called a *forcing function* or a *driving function* because they represent inputs that cause circuit responses.

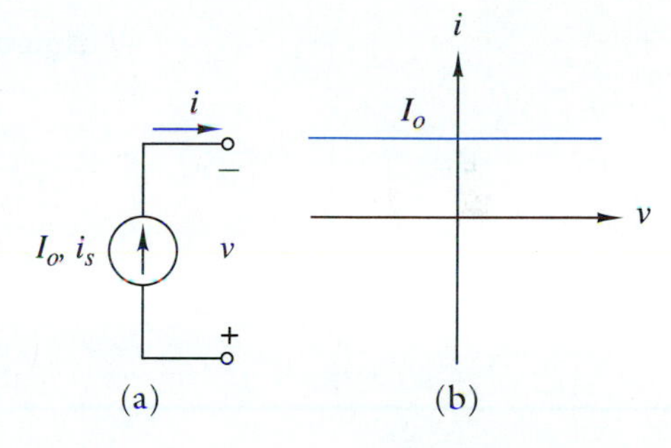

Figure 2-7 Ideal current source. (a) Time-varying or constant source. (b) Constant source—*i-v* characteristics.

EXAMPLE 2-3

Given an ideal voltage source with the time-varying voltage shown in Figure 2-8(a), sketch its *i*-*v* characteristics at time $t = 0$, 1, and 2 ms.

SOLUTION:

At any instant of time, the time-varying source voltage has only one value. We can treat the voltage and current at each instant of time as constants representing a snapshot of the source *i*-*v*

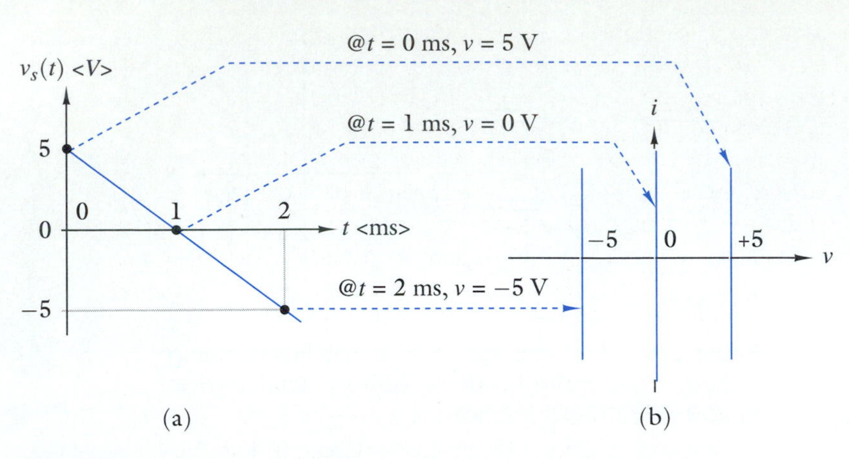

Figure 2-8

characteristics. For example, at $t = 0$, the i-v characteristics of the source are $v_S = 5$ V and i = any value. Figure 2-8(b) shows the i-v characteristics at the other instants of time. Curiously, the voltage source i-v characteristics at $t = 1$ ms ($v_S = 0$ and i = any value) are the same as those of a short circuit [see Eq. (2-5a) or Fig. 2-3(b)]. We will make use of this fact in the next chapter.

Practical Sources

In practice circuit analysis involves selecting an appropriate model for the actual device. A device may have several models as illustrated by the various models for the analog switch shown in Fig. 2-5. Similarly, voltage and current sources have several models that can be used depending on the nature of the application. Figure 2-9 shows the practical models for the voltage and current sources. These models are called practical because they more ac-

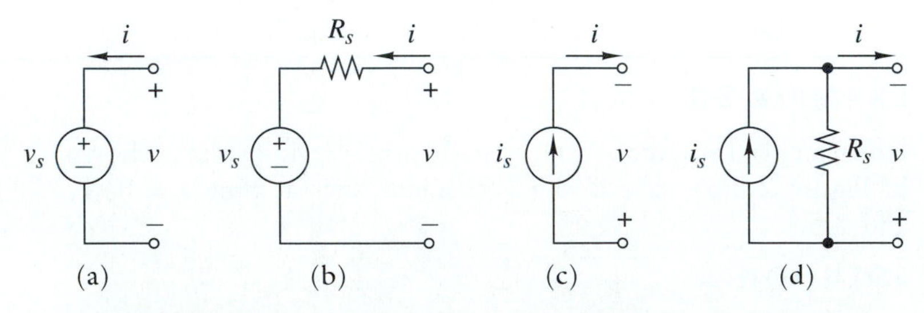

Figure 2-9 Ideal and practical sources: (a) Ideal voltage source. (b) Practical voltage source. (c) Ideal current source. (d) Practical current source.

curately represent the characteristics of real-world sources than do the ideal models. It is important to remember that models are interconnections of elements, not devices. For example, the resistance in a model does not always represent an actual resistor. As a case in point, the resistances R_s in the practical source models in Fig. 2-9 do not represent physical resistors but circuit elements used to account for resistive effects within the devices being modeled. Likewise, the intermediate model of the analog switch in Fig. 2.5(c) includes resistance, since the actual device behaves as if 200 Ω and 10^{12} Ω resistors were indeed there.

The linear resistor, open circuit, short circuit, ideal switch, ideal voltage source, and ideal current source are the initial entries in our catalog of circuit elements. In Chapter 4 we will develop models for active devices like the transistor and OP AMP. Models for dynamic elements like capacitors and inductors are introduced in Chapter 7. Additional device models will be treated in future courses as you proceed in your study of electronics and systems. The study of devices and their models is a continuing process.

 2-2 CONNECTION CONSTRAINTS

In the previous section, we dealt with individual devices and models. In this section, we turn our attention to the constraints introduced by interconnections of devices to form circuits. The laws governing circuit behavior are based on the meticulous work of the German scientist Gustav Kirchhoff (1824–1887). *Kirchhoff's laws* are derived from conservation laws as applied to circuits. They tell us that interconnecting devices to form a circuit forces the device currents and voltages to behave in certain ways. These conditions are called *connection constraints* because they are based only on the circuit connections and not on the specific devices in the circuit.

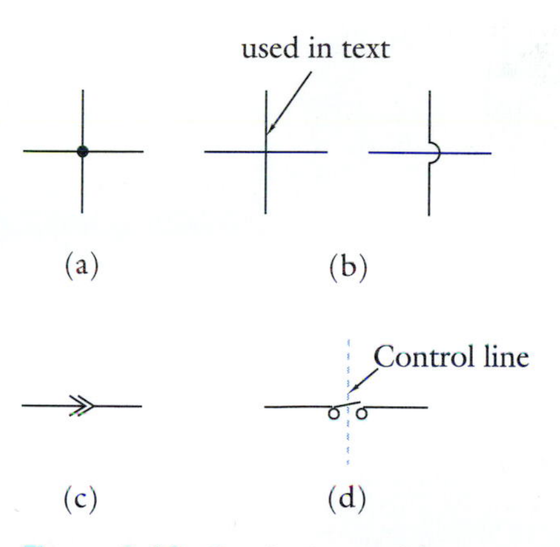

Figure 2-10 Symbols used in circuit diagrams. (a) Electrical connection. (b) Crossover with no connection. (c) Jack connection. (d) Control line.

In this book we will indicate that crossing wires are connected together (electrically tied together) using the dot symbol in Fig. 2-10(a). Sometimes crossing wires are not connected (electrically insulated) but pass over or under each other. Since we are restricted to drawing wires on a planar surface, we will indicate unconnected crossovers by not placing a dot at their intersection as indicated in the left of Fig. 2-10(b). Other books sometimes show unconnected crossovers using the semicircular "hopover" shown on the right of Fig. 2-10(b). In engineering systems two or more independent circuits are often connected to form a larger circuit, for example, the interconnection of two integrated circuit packages. The interconnecting of different circuits forms an *interface* between the circuits and may also be a contractual boundary between different manufacturers. The special jack or interface symbol in Fig. 2-10(c) is used in this book because interface connections represent divisions between circuits. On certain occasions a control line is required to show a mechanical or other nonelectrical dependency. Figure 2-10(d) indicates how this dependency is shown in this book.

The treatment of Kirchhoff's laws uses the following definitions:

A **circuit** is any collection of devices connected at their terminals.

A **node** is an electrical juncture of two or more devices.

A **loop** is a closed path formed by tracing through a sequence of devices without passing through any node more than once.

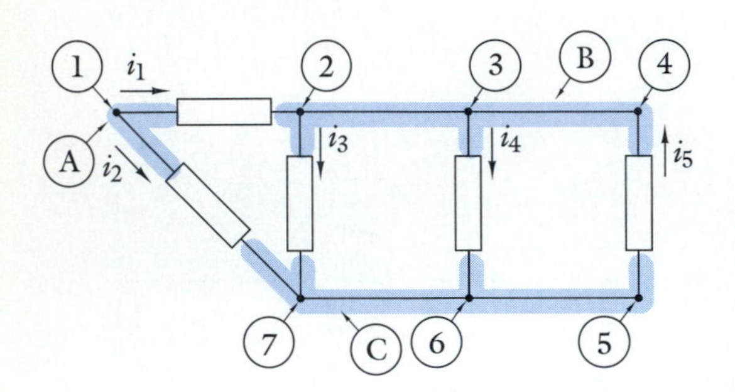

Figure 2-11 Circuit for demonstrating Kirchhoff's current law.

While it is customary to designate a juncture of two or more elements as a node, it is important to realize that a node is not confined to a point but includes all the wire from the point to each element. In the circuit of Fig. 2-11, there are only three different nodes: A, B, and C. The Points 2, 3, and 4, for example, are part of Node B, while the Points 5, 6, and 7 are all part of Node C.

Kirchhoff's Current Law

Kirchhoff's first law is based on the principle of conservation of charge. *Kirchhoff's current law* (abbreviated **KCL**) states:

The algebraic sum of the currents entering a node is zero at every instant.

In forming the algebraic sum of currents, we must take into account the current reference directions associated with the devices. If the current reference direction is into the node, we assign a positive sign in the algebraic sum to the corresponding current. If the reference direction is away from the node, we assign a negative sign. Applying this convention to the nodes in Fig. 2-11, we obtain the following KCL connection equations:

$$\begin{aligned} \text{Node A:} &\quad -i_1 - i_2 = 0 \\ \text{Node B:} &\quad i_1 - i_3 - i_4 + i_5 = 0 \\ \text{Node C:} &\quad i_2 + i_3 + i_4 - i_5 = 0 \end{aligned} \tag{2-8}$$

The KCL equation at Node A does not mean the currents are all negative. The minus signs in this equation simply means the reference direction for each current is directed away from Node A. Likewise, the equation at Node B could be written as

$$i_3 + i_4 = i_1 + i_5 \tag{2-9}$$

This form illustrates an alternative statement of KCL:

> The sum of the currents entering a node equals the sum of the currents leaving the node.

There are two signs associated with each current in the application of KCL. First is the sign given to a current in writing a KCL connection equation. This sign is determined by the orientation of the current reference direction relative to a node. The second sign is determined by the actual direction of the current relative to the reference direction. The next example illustrates these two signs.

EXAMPLE 2-4

Given $i_1 = +4$ A, $i_3 = +1$ A, $i_4 = +2$ A in the circuit shown in Fig. 2-11. Find i_2 and i_5.

SOLUTION:

Using the Node A constraint in Eq. (2-8) yields

$$-i_1 - i_2 = -(+4) - i_2 = 0$$

The sign outside the parentheses comes from the Node A KCL connection constraint in Eq. (2-8). The sign inside the parentheses comes from the actual direction of the current. Solving this equation for the unknown current, we find that $i_2 = -4$ A. The minus sign here indicates the actual direction of current i_2 is directed upward in Fig. 2-11, which is opposite to the reference direction assigned. Using this current in the second KCL equation in the Node B constraint in Eq. (2-8) we can write

$$i_1 - i_3 - i_4 + i_5 = (+4) - (+1) - (+2) + i_5$$

which yields the result $i_5 = -1$ A.

Again, the signs inside the parentheses are associated with the actual direction of the current and the signs outside come from the Node B KCL connection constraint in Eq. (2-8). The minus sign in the final answer means the current i_5 is directed in the opposite direction of its assigned reference direction. We can check our work by substituting the values found above into the Node C constraint in Eq. (2-8). These substitutions yield

$$+i_2 + i_3 + i_4 - i_5 = (-4) + (+1) + (+2) - (-1) = 0$$

as required by KCL. Given three currents, we determined all of the remaining currents in the circuit using only KCL without knowing what the elements are.

In Example 2-4 the unknown currents were found using only the KCL constraints at Nodes A and B. The Node C equation was shown to be valid, but it did not add any new information. In fact, if we look back at Eq. (2-8) we see that Node C equation is the negative of the sum of Node A and B equations. In other words, the KCL connection constraint at Node C is not independent of the previous two. This example illustrates the following general principle:

In a circuit containing N nodes there are only $N - 1$ independent KCL connection equations.

Current equations written at $N - 1$ nodes contain all of the independent connection constraints that can be derived from KCL. In general, to write these equations, we select one node as the reference or ground node, and then write KCL equations at the remaining $N - 1$ nonreference nodes.

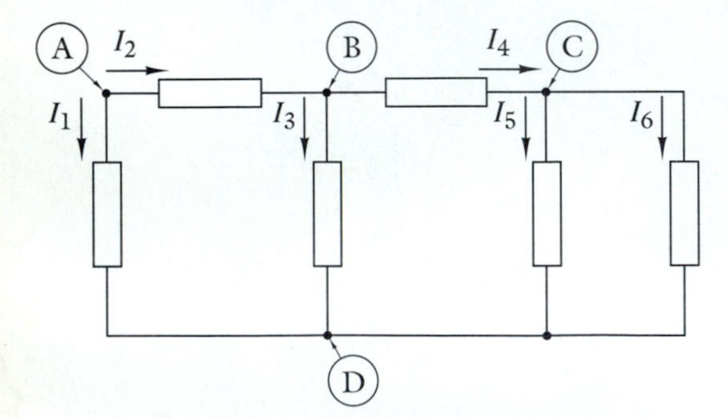

Figure 2-12

Exercise 2-1

Refer to Fig. 2-12.

(a) Write KCL equations at Nodes A, B, C, and D.

(b) Given: $I_1 = -1$ mA, $I_3 = 0.5$ mA, $I_6 = 0.2$ mA, find I_2, I_4, and I_5.

ANSWERS:

(a) Node A: $-I_1 - I_2 = 0$, Node B: $I_2 - I_3 - I_4 = 0$, Node C: $I_4 - I_5 - I_6 = 0$, Node D: $I_1 + I_3 + I_5 + I_6 = 0$.

(b) $I_2 = 1$ mA, $I_4 = 0.5$ mA, $I_5 = 0.3$ mA.

Kirchhoff's Voltage Law

The second of Kirchhoff's circuit laws is based on the principle of conservation of energy. *Kirchhoff's voltage law* (abbreviated **KVL**) states:

The algebraic sum of all of the voltages around a loop is zero at every instant.

For example, three loops are shown in the circuit of Fig. 2-13. In writing the algebraic sum of voltages, we must account for the assigned reference marks. As a loop is traversed a positive sign is assigned to a voltage when we go from a "+" to a "−" reference mark. When we go from "−" to "+" we use a minus sign. Traversing the three loops in Fig. 2-13 in the indicated clockwise direction yields the following KVL connection equations:

$$\begin{aligned} &\text{Loop 1:} \quad & -v_1 + v_2 + v_3 = 0 \\ &\text{Loop 2:} \quad & -v_3 + v_4 + v_5 = 0 \\ &\text{Loop 3:} \quad & -v_1 + v_2 + v_4 + v_5 = 0 \end{aligned} \tag{2-10}$$

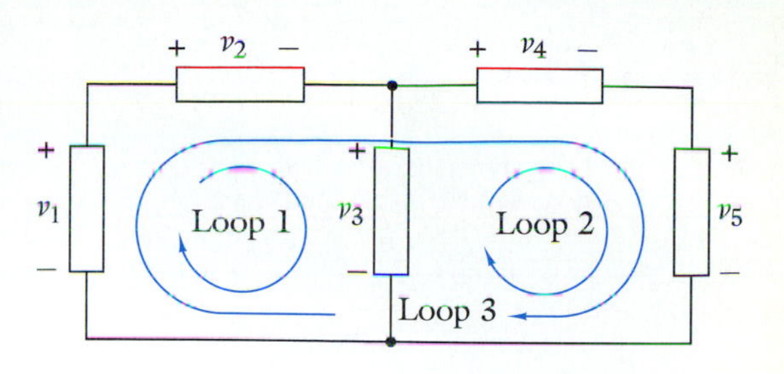

Figure 2-13 Circuit for demonstrating Kirchhoff's voltage law.

There are two signs associated with each voltage. The first is the sign given the voltage when writing the KVL connection equation. The second is the sign determined by the actual polarity of a voltage relative to its assigned reference polarity. The following example illustrates these concepts.

EXAMPLE 2-5

Given $v_1 = 5$ V, $v_2 = -3$ V and $v_4 = 10$ V in the circuit shown in Fig. 2-13, find v_3 and v_5.

SOLUTION:

With the given values inserted the Loop 1 KVL equation in Eq. (2-10) reads

$$-v_1 + v_2 + v_3 = -(+5) + (-3) + (v_3) = 0$$

The sign outside the parentheses comes from the Loop 1 KVL constraint in Eq. (2-10). The sign inside comes from the actual polarity of the voltage. This equation yields $v_3 = +8$ V. Using this value in the Loop 2 KVL constraint in Eq. (2-10) produces

$$-v_3 + v_4 + v_5 = -(+8) + (+10) + v_5 = 0$$

which yields $v_5 = -2$ V. The minus sign here means the actual polarity of v_5 is the opposite of its reference polarity assigned in Fig. 2-13. The results can be checked by substituting all of the above values into the Loop 3 KVL constraint in Eq. (2-10). These substitutions yield

$$-(+5) + (-3) + (+10) + (-2) = 0\,\text{V}$$

as required by KVL.

In Example 2-5 the unknown voltages were found using only the KVL constraints for Loops 1 and 2. The Loop 3 equation was shown to be valid, but it did not add any new information. In fact, if we look back at Eq. (2-10) we see that the Loop 3 equation is equal to the sum of the Loop 1 and 2 equations. In other words, the KVL connection constraint around Loop 3 is not independent of the previous two. This example illustrates the following general principle:

> In a circuit containing E two-terminal elements and N nodes there are only $E - N + 1$ independent KVL connection equations.

Voltage equations written around $E - N + 1$ **different loops** contain all of the independent connection constraints that can be derived from KVL. Loops are different if each contains at least one element that is not contained in any other loop. In simple circuits the open space between elements form $E - N + 1$ independent loops. In complicated circuits finding all of the loops can be a nontrivial problem.

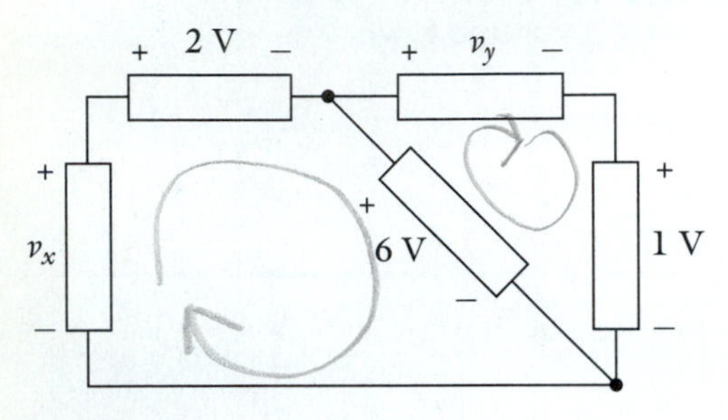

Figure 2-14

EXERCISE 2-2

Find the voltages v_x and v_y in Fig. 2-14.

ANSWERS:

$v_x = +8$ V; $v_y = +5$ V.

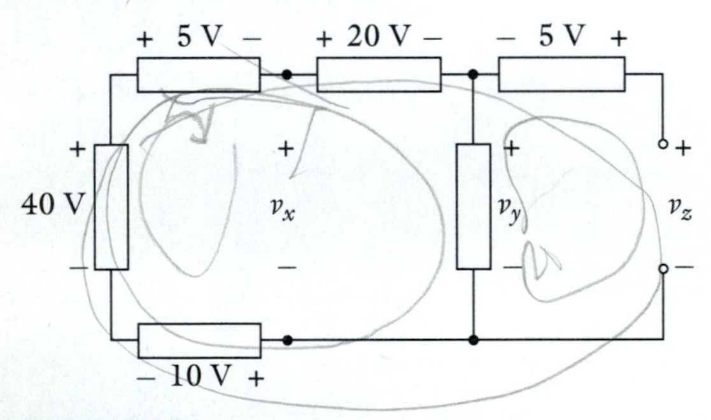

Figure 2-15

EXERCISE 2-3

Find the voltages v_x, v_y and v_z in Fig. 2-15.

ANSWERS:

$v_x = +25$ V; $v_y = +5$ V; $v_z = +10$ V. Note: KVL yields the voltage v_z even though it appears across an open circuit.

Parallel and Series Connections

Two types of connections occur so frequently in circuit analysis that they deserve special attention. Elements 1 and 2 in Fig. 2-16 are said to be connected in *parallel* because they are connected between two common nodes. As a result, applying KVL around the loop between the two elements shows that the voltages across the two elements are equal. That is, the KVL connection constraint around Loop A is

$$-v_1 + v_2 = 0 \tag{2-11}$$

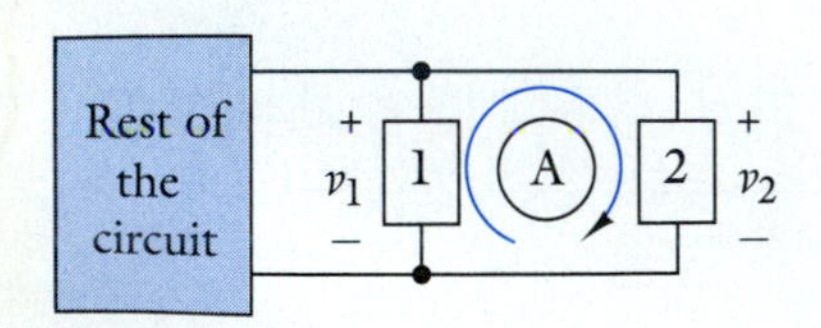

Figure 2-16 A parallel connection.

which yields $v_1 = v_2$. In other words, KVL demands equality of the voltage across elements connected in parallel. The parallel connection is not restricted to two elements. Any number of elements connected between two common nodes are in parallel and, as a result, the same voltage appears across all of them.

Two elements are said to be connected in *series* when they have one common node to which no other element is connected. Elements 1 and 2 in Fig. 2-17 are connected in series, since only these two elements are connected at Node A. Applying KCL at Node A yields

$$i_1 - i_2 = 0 \qquad \text{or} \qquad i_1 = i_2 \tag{2-12}$$

In a series connection KCL requires equal current through each element. Any number of elements can be connected in series. For example, Element 3 in Fig. 2-17 is connected in series with Element 2 at Node B and KCL requires $i_2 = i_3$. Therefore, in this circuit $i_1 = i_2 = i_3$ and we say Elements 1, 2, and 3 are connected in series and the same current exists in all three.

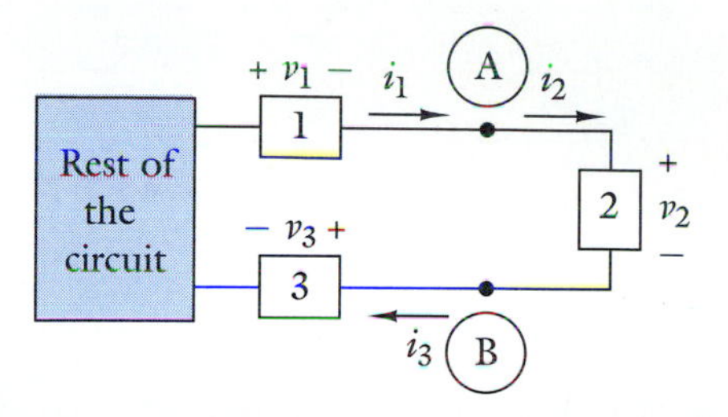

Figure 2-17 A series connection.

EXAMPLE 2-6

Identify the elements connected in parallel and in series in each of the circuits in Fig. 2-18.

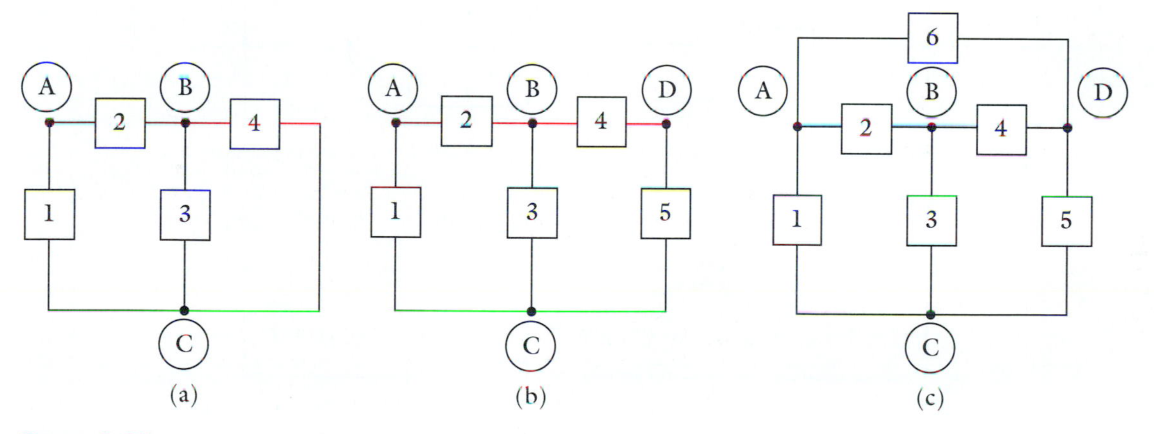

Figure 2-18

SOLUTION:

In Fig. 2-18(a) Elements 1 and 2 are connected in series at Node A and Elements 3 and 4 are connected in parallel between Nodes B and C. In Fig. 2-18(b) Elements 1 and 2 are connected in series at Node A as are Elements 4 and 5 at Node D. There are no elements connected in parallel in Fig. 2-18(b). In Fig. 2-18(c) there are no elements connected in either series or parallel. It is important to realize that elements need not be connected in either series or parallel.

EXERCISE 2-4

Identify the elements connected in series or parallel when a short circuit is connected between Nodes A and B in each of the circuits of Fig. 2-18.

ANSWERS:

Circuit in Fig. 2-18(a): Elements 1, 3, and 4 are all in parallel.
Circuit in Fig. 2-18(b): Elements 1 and 3 are in parallel; Elements 4 and 5 are in series.
Circuit in Fig. 2-18(c): Elements 1 and 3 are in parallel; Elements 4 and 6 are in parallel.

EXERCISE 2-5

Identify the elements in Fig. 2-19 that are connected in (a) parallel, (b) series, or (c) neither.

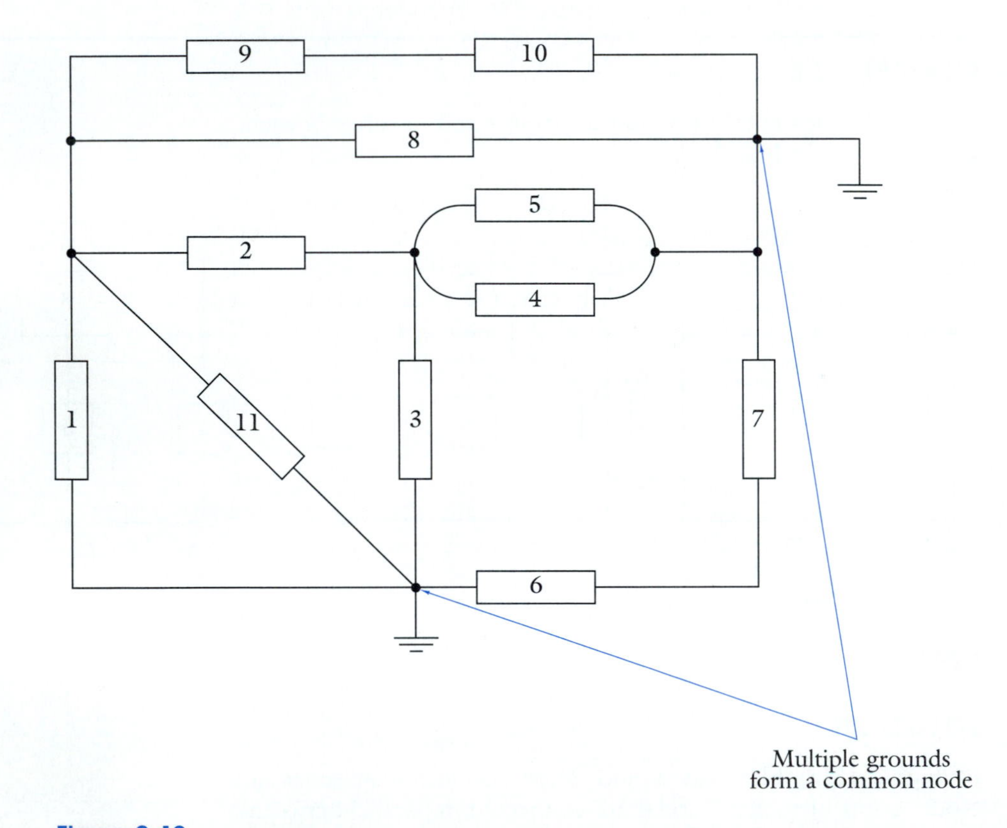

Figure 2-19

ANSWERS:

(a) The following elements are in parallel: 1, 8, and 11; 3, 4, and 5.
(b) The following elements are in series: 9 and 10; 6 and 7.
(c) Only element 2 is not in series or parallel with any other element.

COMMENTS:
The ground symbol indicates the reference node. When ground symbols are shown at several nodes, it means the nodes are connected by a short circuit and form a single node. ■

2-3 COMBINED CONSTRAINTS

The usual goal of circuit analysis is to determine the currents or voltages at various places in a circuit. This analysis is based on constraints of two distinctly different types. The element constraints are based on the models of the specific devices connected in the circuit. The connection constraints are based on Kirchhoff's laws and the circuit connections. The element equations are independent of the circuit in which the device is connected. Likewise, the connection equations are independent of the specific devices in the circuit. But taken together, the element and connection equations provide the data needed to analyze a circuit.

The study of *combined constraints* begins by considering the simple but important example in Fig. 2-20. This circuit is driven by the current source i_s and the resulting responses are current/voltage pairs (i_x, v_x) and (i_o, v_o). The reference marks for the response pairs have been assigned using the passive sign convention. To solve for all four responses we must write four equations. The first two are the element equations

$$\begin{aligned} i_x &= i_s \\ v_o &= Ri_o \end{aligned} \tag{2-13}$$

The first element equation states that the response current i_x and the input driving force i_s are equal in magnitude and direction. The second element equation is Ohm's law relating v_o and i_o under the passive sign convention.

The connection equations are obtained by applying Kirchhoff's laws. The circuit in Fig. 2-20 has two elements ($E = 2$) and two nodes ($N = 2$) so we need $E - N + 1 = 1$ KVL equation and $N - 1 = 1$ KCL equation. Selecting Node B as the reference node, we apply KCL at node A and apply KVL around the loop to write

$$\begin{aligned} \text{KCL:} \quad -i_x - i_o &= 0 \\ \text{KVL:} \quad -v_x + v_o &= 0 \end{aligned} \tag{2-14}$$

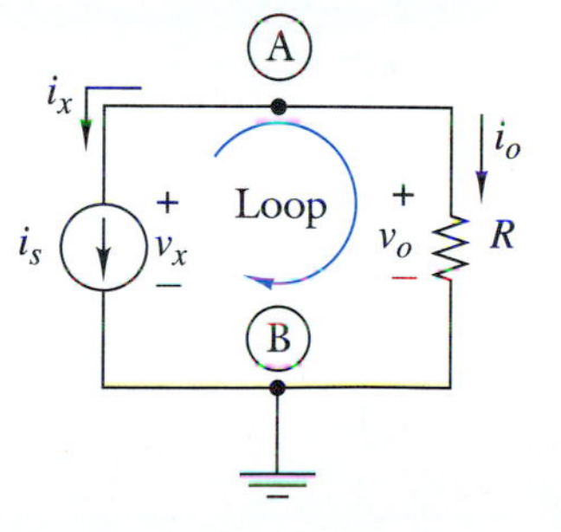

Figure 2-20 Circuit used to demonstrate combined constraints.

We now have two element constraints in Eq. (2-13) and two connection constraints in Eq. (2-14), so we can solve for all four responses in terms of the input driving force i_s. Combining the KCL connection equation and the first element equations yields $i_o = -i_x = -i_s$. Substituting this result into the second element equations (Ohm's law) produces

$$v_o = -Ri_s \tag{2-15}$$

The minus sign in this equation does not mean v_o is always negative. Nor does it mean the resistance is negative. It means that when the input driving force i_s is positive, then the response v_o is negative, and vice versa. This sign reversal is a result of the way we assigned reference marks at the beginning of our analysis. The reference marks defined the circuit input and outputs in such a way that i_s and v_o always have opposite signs. Put differently, Eq. (2-15) is an input-output relationship not a device i-v relationship.

EXAMPLE 2-7

(a) Find the responses i_x, v_x, i_o, and v_o in the circuit in Fig. 2-20 when $i_s = +2$ mA and R = 2 kΩ.

(b) Repeat for $i_s = -2$ mA.

SOLUTION:

(a) From Eq. (2-13) we have $i_x = i_s = +2$ mA and $v_o = 2000i_o$. From Eqs. (2-14) we have $i_o = -i_x = -2$ mA and $v_x = v_o$. Combining these results we get

$$v_x = v_o = 2000i_o = 2000(-0.002) = -4 \text{ V}$$

(b) In this case $i_x = i_s = -2$ mA, $i_o = -i_x = -(-0.002) = +2$ mA, and

$$v_x = v_o = 2000i_o = 2000(+0.002) = +4 \text{ V}$$

This example confirms that numerical values of the outputs v_x, v_o, and i_o always have the opposite sign of the input i_s.

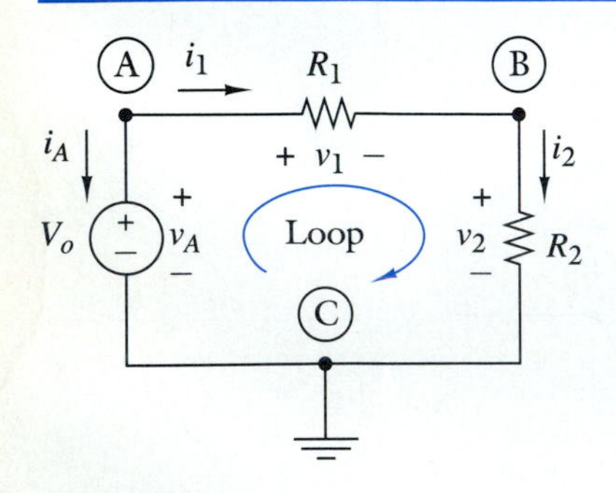

Figure 2-21 Circuit used to demonstrate combined constraints.

The circuit in Fig. 2-21 can be used to further illustrate the formulation of combined constraints. We first assign reference marks for all the voltages and currents using the passive sign convention. Then, using these definitions we can write the element constraints as

$$\begin{aligned} v_A &= V_o \\ v_1 &= R_1 i_1 \\ v_2 &= R_2 i_2 \end{aligned} \tag{2-16}$$

These equations describe the three devices and do not depend on how the devices are connected in the circuit.

The connection equations are obtained from Kirchhoff's laws. But before we can apply them we must label the different loops and nodes. The circuit contains $E = 3$ elements and $N = 3$ nodes so there are $E - N + 1 = 1$ independent KVL constraints and $N - 1 = 2$ independent KCL constraints. There is only one loop but there are three nodes in this circuit. We will select one node as the reference point and write KCL equations at the other two nodes. Any node can be chosen as the reference so we select Node C as the reference node and indicate this choice by drawing the ground symbol there. The connection constraints are

$$\begin{aligned}
&\text{KCL: Node A} \qquad && -i_A - i_1 = 0 \\
&\text{KCL: Node B} \qquad && i_1 - i_2 = 0 \qquad (2\text{-}17) \\
&\text{KVL: Loop} \qquad && -v_A + v_1 + v_2 = 0
\end{aligned}$$

These equations are independent of the specific devices in the circuit. They depend only on Kirchhoff's laws and the circuit connections.

This circuit has six unknowns: three element currents and three element voltages. Taken together, the element and connection equations give us six independent equations. In general, for a network with (N) nodes and (E) two-terminal elements, we can write $(N - 1)$ independent KCL connections equations, $(E - N + 1)$ independent KVL connection equations, and (E) element equations. The total number of equations generated is

Element	E
KCL	$N - 1$
KVL	$E - N + 1$
Total	$2E$

The grand total is then $(2E)$ combined connection and element equations, which is exactly the number of equations needed to solve for the voltage across and current through every element—a total of $(2E)$ unknowns.

EXAMPLE 2-8

Find all of the element currents and voltages in Fig. 2-21 for $V_o = 10$ V, $R_1 = 2000\ \Omega$, and $R_2 = 3000\ \Omega$.

SOLUTION:

Substituting the element constraints from Eq. (2-16) into the KVL connection constraint in Eq. (2-17) produces

$$-V_o + R_1 i_1 + R_2 i_2 = 0$$

This equation can be used to solve for i_1, since the second KCL connection equation requires that $i_2 = i_1$.

$$i_1 = \frac{V_o}{R_1 + R_2} = \frac{10}{2000 + 3000} = 2 \text{ mA}$$

By finding this current, we have determined every device current since the KCL connection equations collectively require that

$$-i_A = i_1 = i_2$$

since all three elements are connected in series. Substituting all of the known values into the element equations gives

$$v_A = 10 \text{ V} \qquad v_1 = R_1\, i_1 = 4 \text{ V} \qquad v_2 = R_2\, i_2 = 6 \text{ V}$$

Every element voltage and current has been found. Note the analysis strategy used. We first found all of the element currents and then used these values to find the voltages.

EXAMPLE 2-9

Consider the source-resistor-switch circuit of Fig. 2-22 with $V_o = 10$ V and $R_1 = 2000\ \Omega$. Find all of the element voltages and currents with the switch open and again with the switch closed.

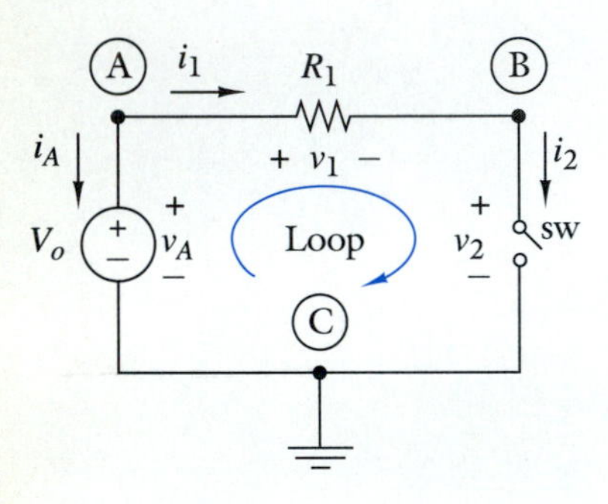

Figure 2-22

SOLUTION:

The connection equations for the circuit are

$$\begin{aligned} &\text{KCL: Node A} \qquad & -i_A - i_1 &= 0 \\ &\text{KCL: Node B} \qquad & i_1 - i_2 &= 0 \\ &\text{KVL: Loop} \qquad & -v_A + v_1 + v_2 &= 0 \end{aligned}$$

These connection equations are the same as those in Eq. (2-17) for the circuit in Fig. 2-21. The two circuits have the same connections but different devices. When the switch is open the current through the switch is zero, so $i_2 = 0$. The KCL connection equation at Node B requires $i_1 = i_2$; therefore, $i_1 = 0$. The element equation for R_1 yields $v_1 = R_1 i_1 = 0$, and the KVL connection equations yields $v_1 + v_2 = v_2 = v_A = 10$ V. When the switch is open, the current is zero and all of the source voltage appears across the switch. When the switch is closed in the switch element, equations require $v_2 = 0$. The KVL connection equation gives $v_1 + v_2 = v_1 = v_A = 10$ V. So that the current through the circuit is

$$i_1 = \frac{v_1}{R_1} = \frac{10}{2000} = 5 \text{ mA}$$

When the switch is closed all of the source voltage appears across the resistor R_1 rather than the switch.

EXERCISE 2-6

For the circuit of Fig. 2-23:

(a) Write a complete set of element equations.

(b) Write a complete set of connection equations.

(c) Solve the equations in (a) and (b) for all element currents and voltages.

ANSWERS:

(a) $v_A = 30$ V; $v_3 = 300\,i_3$; $v_1 = 100\,i_1$; $v_2 = 200\,i_2$.

(b) $-i_A - i_1 - i_3 = 0$; $+i_1 - i_2 = 0$; $-30 + v_3 = 0$; $v_1 + v_2 - v_3 = 0$.

(c) $v_A = 30$ V; $v_1 = 10$ V; $v_2 = 20$ V; $v_3 = 30$ V; $i_A = -200$ mA; $i_1 = i_2 = 100$ mA; $i_3 = 100$ mA. ■

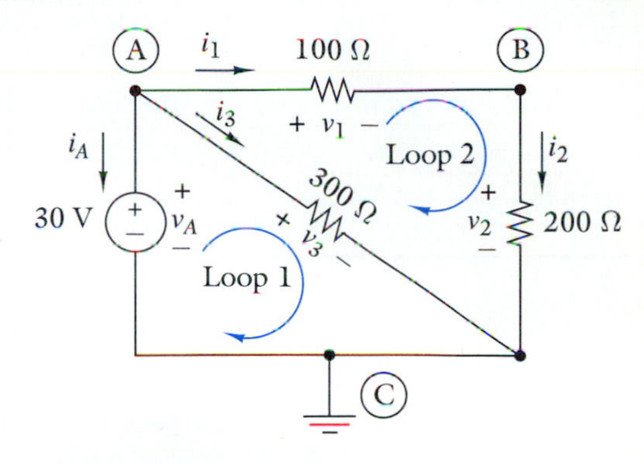

Figure 2-23

Assigning Reference Marks

In all previous examples and exercises the reference marks for the element currents (arrows) and voltages (+ and −) were given. When reference marks are not shown on a circuit diagram, they must be assigned by the person solving the problem. Beginners sometimes wonder how to assign reference marks when the actual voltage polarities and current directions are as yet unknown. It is important to remember the reference marks do not indicate the actual polarities and directions. They are benchmarks assigned in an arbitrary way at the beginning of the analysis. If it turns out the actual direction and reference direction agree, then the numerical value of the response will be positive. If they disagree, the numerical value will be negative. In other words, the sign of the answer together with arbitrarily assigned reference marks tell us the actual voltage polarity or current direction.

When assigning reference marks in this book we always follow the passive sign convention. This means that for any given two-terminal element we can arbitrarily assign either the + voltage reference mark or the current reference arrow, but not both. For example, we could assign the + voltage reference mark to either terminal of a two-terminal device. Once the voltage reference is assigned, however, the passive sign convention requires that the current reference arrow be directed into the device at the terminal with the + mark. On the other hand, we could start by assigning the current reference arrow into the device at either terminal. Once the current reference is assigned, however, the passive sign convention requires that the + voltage reference be assigned to the terminal selected.

By always following the passive sign convention we avoid possible confusion about the direction of power flow in a device. In addition, Ohm's law and other device i-v characteristics assume the voltage and current reference marks follow the passive sign convention. By sticking to this convention we also follow the practice used in computer circuit analysis programs such as SPICE and MICRO-CAP.

The next example illustrates the assignment of reference marks.

EXAMPLE 2-10

Find all of the element voltages and currents in Fig. 2-24(a).

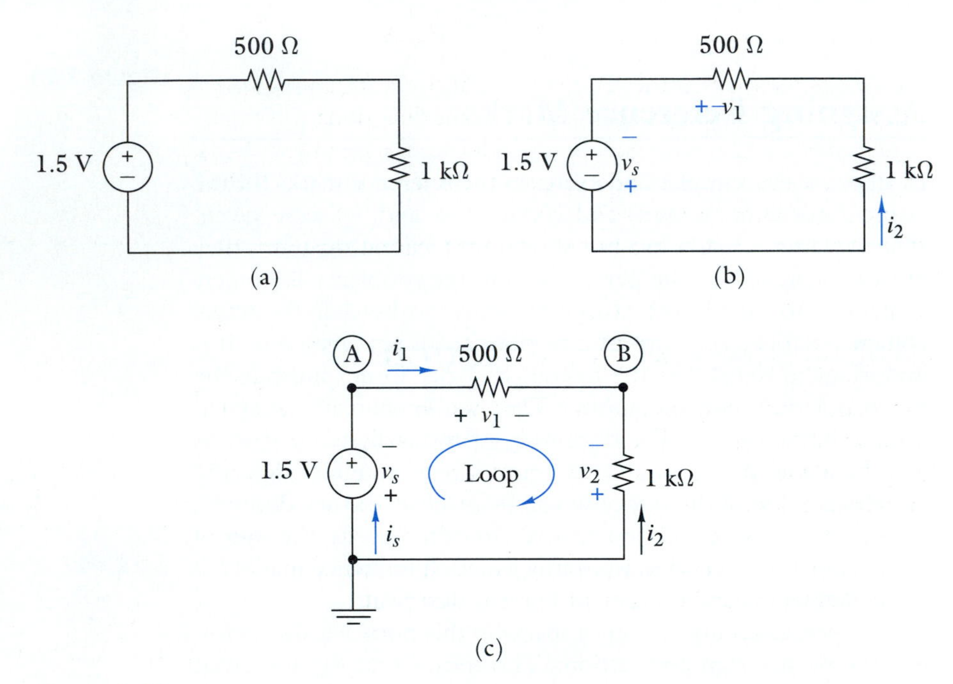

Figure 2-24

SOLUTION:

Since no reference marks are shown in Fig. 2-24(a), we assign references to two voltages and one current as shown in Fig. 2-22(b). Other choices are of course possible, but once these marks are selected, the passive sign convention dictates the reference marks for the remaining voltage and currents as shown in Fig. 2-22(c). Using all of these reference marks we write the element equations as

$$v_1 = 500 i_1$$

$$v_2 = 1000 i_2$$

$$v_S = -1.5 \text{ V}$$

Using the indicated reference node we write two KCL and one KVL equation.

$$\text{KCL: Node A} \qquad +i_S - i_1 = 0$$

$$\text{KCL: Node B} \qquad +i_1 + i_2 = 0$$

$$\text{KVL: Loop} \qquad +v_S + v_1 - v_2 = 0$$

Solving the combined element and connection equations yields

$$v_S = -1.5\text{ V}, \qquad i_S = +1.0\text{ mA}$$

$$v_1 = +0.5\text{ V}, \qquad i_1 = +1.0\text{ mA}$$

$$v_2 = -1.0\text{ V}, \qquad i_2 = -1.0\text{ mA}$$

These results show that the reference marks for v_1, i_s, and i_1 agree with the actual voltage polarities and current directions, while the minus signs on the other responses indicate disagreement. It is important to realize that this disagreement does not mean that assigned reference marks are wrong.

2-4 EQUIVALENT CIRCUITS

The analysis of a circuit can often be simplified by replacing part of the circuit with one that is equivalent but simpler. The underlying basis for two circuits to be equivalent is contained in their i-v relationships:

Two circuits are said to be equivalent if they have identical i-v characteristics at a specified pair of terminals.

Equivalent Resistance

In Fig. 2-25(a) two resistors are connected in series between a pair of terminals A and B. The objective is to simplify the circuit without altering the electrical behavior of the rest of the circuit.

The KVL equation around the loop from A to B is

$$v = v_1 + v_2 \tag{2-18}$$

Since the two resistors are connected in series the same current i exists in both. By applying Ohm's law we get, $v_1 = R_1 i$ and $v_2 = R_2 i$. Substituting these relationships into Eq. (2-18) and then simplifying yields

$$v = R_1\, i + R_2\, i = i(R_1 + R_2)$$

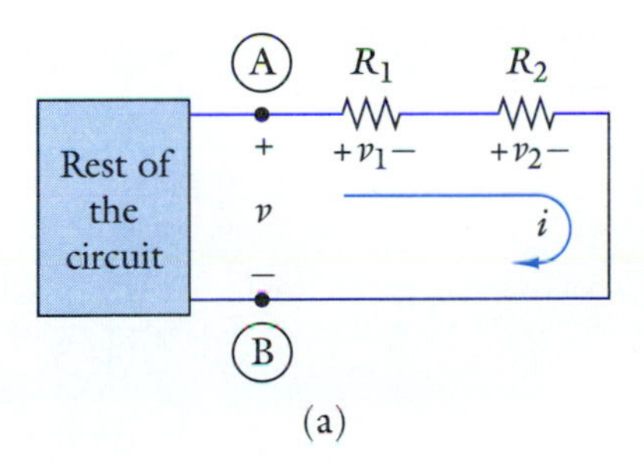

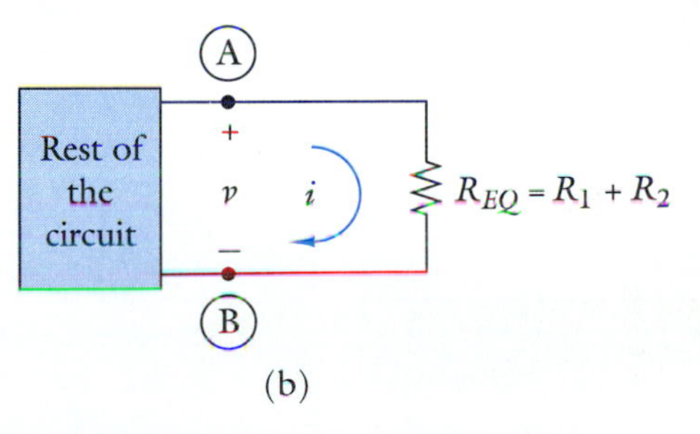

Figure 2-25 A series resistance circuit. (a) Original circuit. (b) Equivalent circuit.

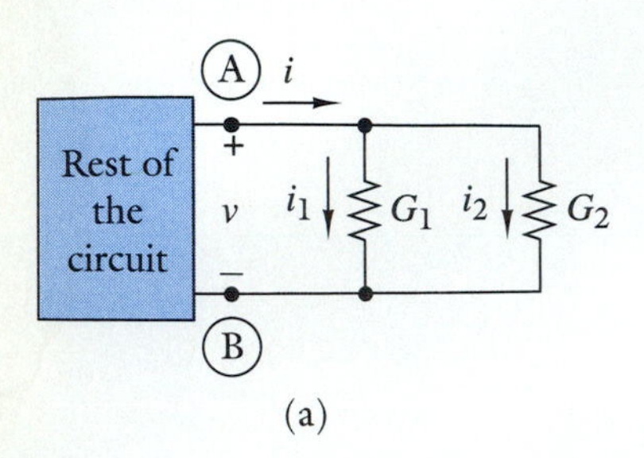

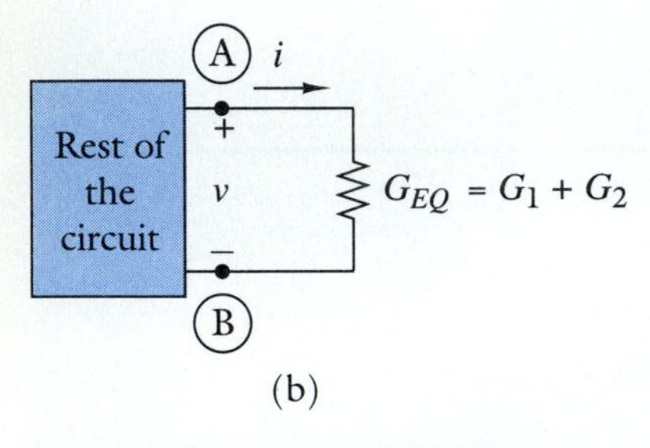

Figure 2-26 A parallel resistance circuit. (a) Original circuit. (b) Equivalent circuit.

We can write this equation as

$$v = iR_{EQ} \qquad \text{where} \qquad R_{EQ} = R_1 + R_2 \tag{2-19}$$

This result is significant because it means the response of the rest of the circuit would be unchanged if resistors R_1 and R_2 were replaced by a single equivalent resistance R_{EQ}, as shown in Fig. 2-25(b).

The dual[2] situation is shown in Fig. 2-26(a), where two conductances are connected in parallel between terminals A and B. Again the objective is to simplify the circuit without altering the behavior of the rest of the circuit.

A KCL equation at Node A produces

$$i = i_1 + i_2 \tag{2-20}$$

Since the conductances are connected in parallel, the voltage v appears across both. Applying Ohm's law we get $i_1 = G_1 v$, and $i_2 = G_2 v$. Substituting these relationships into Eq. (2-20) and then simplifying yields

$$i = vG_1 + vG_2 = v(G_1 + G_2)$$

So we can write

$$i = vG_{EQ}, \qquad \text{where} \qquad G_{EQ} = G_1 + G_2 \tag{2-21}$$

The response of the rest of the circuit would be unchanged if the two parallel conductances connected between terminals A and B were replaced by an equivalent conductance G_{EQ}, as shown in Fig. 2-26(b).

Since we usually do not think of resistors in terms of their conductances it is sometimes more useful to rewrite G_{EQ} in Eq. (2-21) in terms of resistance R_{EQ}. That is,

$$R_1 \parallel R_2 = R_{EQ} = \frac{1}{G_{EQ}} = \frac{1}{G_1 + G_2} = \frac{1}{\dfrac{1}{R_1} + \dfrac{1}{R_2}} = \frac{R_1 R_2}{R_1 + R_2} \tag{2-22}$$

where the symbol "$\parallel$" is shorthand for "in parallel." This product over the sum rule for two resistors in parallel is a convenient way to express their equivalent value.

Caution: The product over sum rule only applies to two resistors connected in parallel. When more than two resistors are

[2] In circuit analysis two seemingly different circuits can have identical solutions with the exception that the roles of voltage and current and resistance and conductance are interchanged. This interchangeability is called the principle of ***duality***. In later chapters we will see duality exhibited in other circuit parameters.

in parallel we must use the general result implied by Eq. (2-21) to obtain the equivalent resistance.

$$R_{EQ} = \frac{1}{G_{EQ}} = \frac{1}{\frac{1}{R_1} + \frac{1}{R_2} + \frac{1}{R_3} + \cdots} \tag{2-23}$$

EXAMPLE 2-11

Consider the circuit in Fig. 2-27(a). Find the equivalent resistance R_{EQ1} connected between terminals A-B and the equivalent resistance R_{EQ2} connected between terminals C-D.

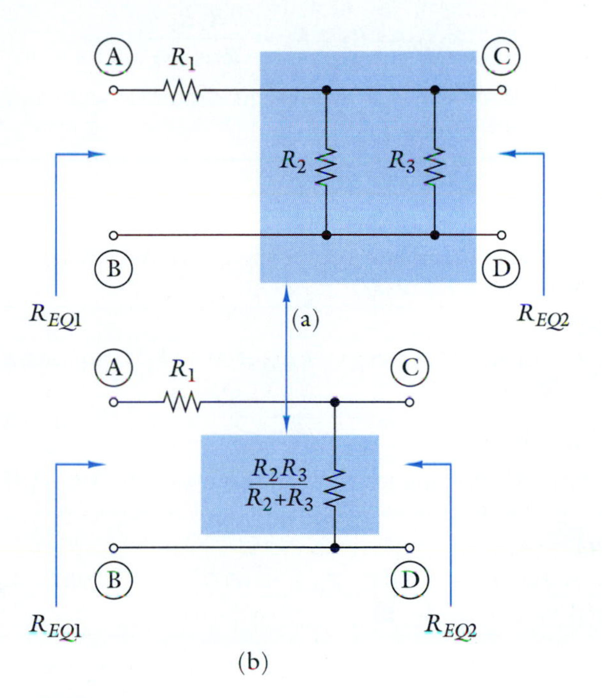

Figure 2-27

SOLUTION:

First note that resistors R_2 and R_3 are in parallel because they are connected between the same nodes. Applying Eq. (2-22), we obtain

$$R_2 \parallel R_3 = \frac{R_2 R_3}{R_2 + R_3}$$

As an interim step, we redraw the circuit as shown in Fig. 2-27(b). To find the equivalent resistance between terminals A and B, we note that R_1 and the equivalent resistance $R_2 \parallel R_3$ are connected in series. The total equivalent resistance R_{EQ1} between terminals A and B is

$$R_{EQ1} = R_1 + (R_2 \parallel R_3)$$

$$R_{EQ1} = R_1 + \frac{R_2 R_3}{R_2 + R_3}$$

$$R_{EQ1} = \frac{R_1 R_2 + R_1 R_3 + R_2 R_3}{R_2 + R_3}$$

Looking into terminals C-D yields a different result. In this case R_1 is not involved, since there is an open circuit (an infinite resistance) between terminals A and B. Therefore only $R_2 \parallel R_3$ is seen between terminals C-D, resulting in

$$R_{EQ2} = R_2 \parallel R_3 = \frac{R_2 R_3}{R_2 + R_3}$$

This example shows that an equivalent circuit depends upon the pair of terminals involved.

EXERCISE 2-7

Find the equivalent resistance between terminals A-C, B-D, A-D, and B-C in the circuit in Fig. 2-27.

ANSWERS:

$R_{A\text{-}C} = R_1$, $R_{B\text{-}D} = 0\,\Omega$ (a short), $R_{A\text{-}D} = R_1 + (R_2 \parallel R_3)$, and $R_{B\text{-}C} = R_2 \parallel R_3$.

EXERCISE 2-8

Find the equivalent resistance between terminals A-B, A-C, A-D, B-C, B-D, and C-D in the circuit of Fig. 2-28.

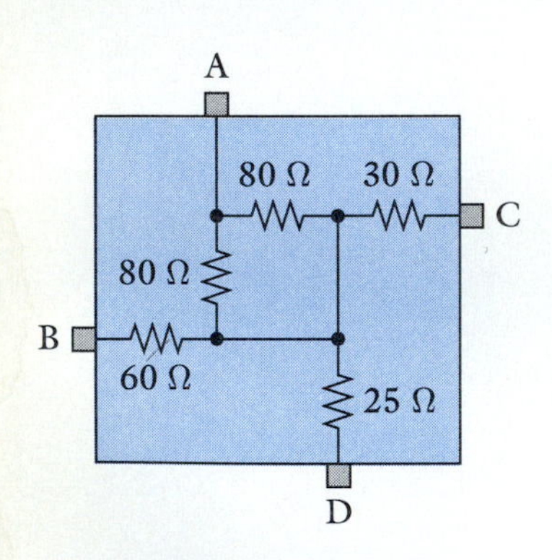

Figure 2-28

ANSWERS:

$R_{A\text{-}B} = 100\,\Omega$, $R_{A\text{-}C} = 70\,\Omega$, $R_{A\text{-}D} = 65\,\Omega$, $R_{B\text{-}C} = 90\,\Omega$, $R_{B\text{-}D} = 85\,\Omega$, and $R_{C\text{-}D} = 55\,\Omega$.

Equivalent Sources

Up to now we have considered sources as the ideal elements shown in Figs. 2-6 and 2-7. The practical source models shown in Fig. 2-9 consist of an ideal voltage source in series with a resistance and an ideal current source in parallel with a resistance. To continue the study of equivalent circuits, we now determine the conditions under which the practical voltage and the practical current sources are equivalent.

Figure 2-29 shows the two nonideal sources connected between terminals labeled A and B. A parallel analysis of these circuits will yield the conditions for equivalency at terminals A and

B. First applying Kirchhoff's laws as

Circuit A	Circuit B
KVL	KCL
$v_S = v_R + v$	$i_S = i_R + i$

Next using Ohm's law to write

Circuit A	Circuit B
$v_R = R_1 i$	$i_R = \dfrac{v}{R_2}$

Combining these results we find that the i-v characteristic of the sources equations at terminals A and B are

Circuit A	Circuit B
$i = -\dfrac{v}{R_1} + \dfrac{v_S}{R_1}$	$i = -\dfrac{v}{R_2} + i_S$

The graph of these two i-v characteristics are in Fig. 2-30. Using either the graph in Fig. 2-30 or the equations we conclude that the source will have the same i-v characteristics when

$$R_1 = R_2 = R \quad \text{and} \quad v_S = i_S R \qquad (2\text{-}24)$$

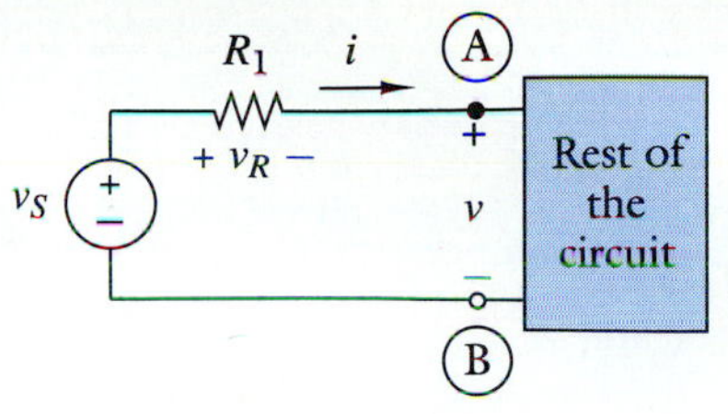

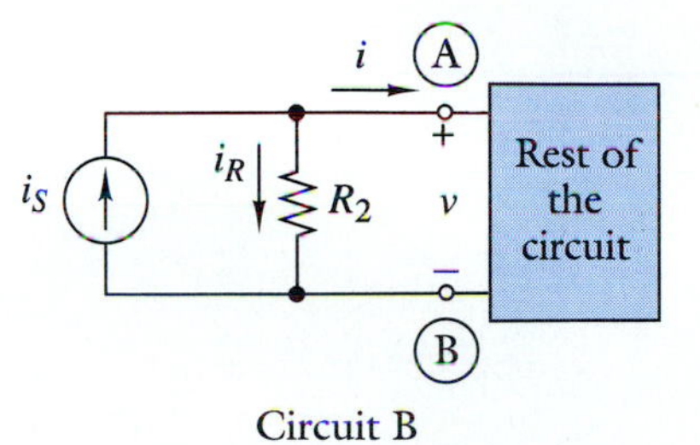

Figure 2-29 Equivalent circuits derived by source transformation.

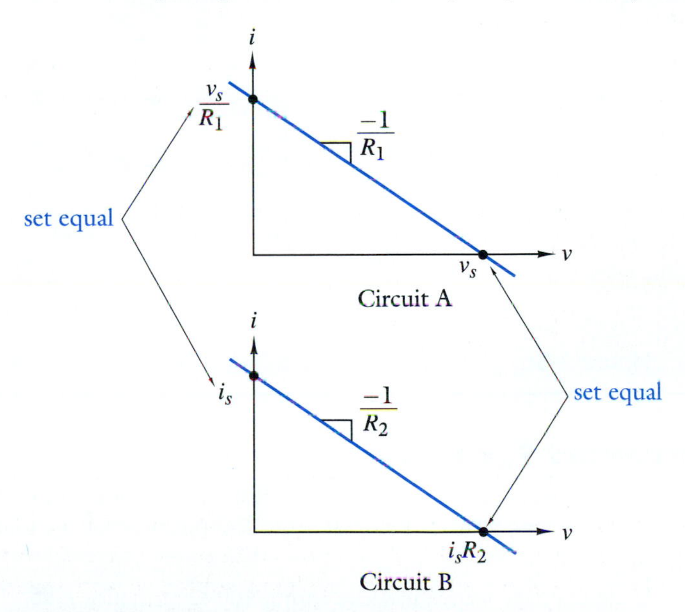

Figure 2-30 The *i*-*v* characteristics of practical sources.

When equivalency conditions in Eq. (2-24) are met, the rest of the circuit is unaffected when we replace a practical voltage source by a practical current source, or vice versa. Exchanging one practical source model for an equivalent model is called *source transformation*

The source transformation equivalency means that either model will deliver the same voltage and current to the rest of the circuit. It does not mean the two models are identical in every way. For example, when the rest of the circuit is an open circuit there is no current in the resistance of the practical voltage source, hence no i^2R power loss. But the practical current source model has a power loss because the open-circuit voltage is produced by the source current in the parallel resistance.

EXAMPLE 2-12

Convert the practical voltage source in Fig. 2-31(a) into an equivalent current source.

SOLUTION:

Using Eq. (2-24) we have

$$R_1 = R_2 = R = 10\ \Omega$$

$$i_S = v_S/R = 5\ \text{A}$$

The equivalent practical current source is shown in Fig. 2-31(b).

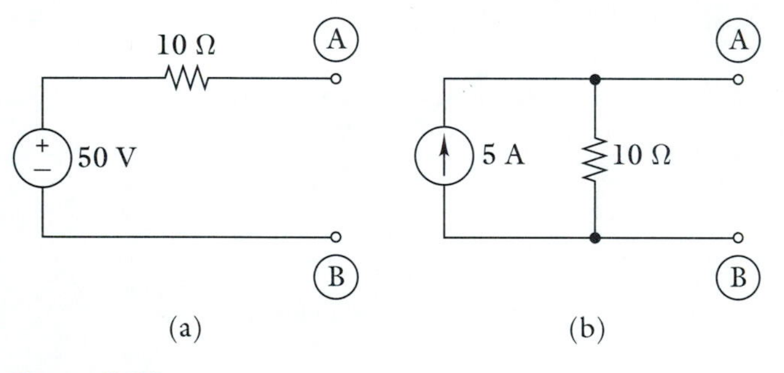

Figure 2-31

■ EXERCISE 2-9

A practical current source with a 2-mA ideal current source in parallel with a .002-S conductance. Find the equivalent practical voltage source.

ANSWER:

A 1-V ideal voltage source in series with a 500-Ω resistance. ■

$\Upsilon \leftrightarrows \Delta$ Transformations

In some circuits the methods of series and parallel equivalence do not apply because there are no resistors connected in series or parallel. Such nonseries/parallel circuits contain three-terminal subcircuits connected in the Δ-configuration or the Y-configuration shown in Fig. 2-32. Replacing a Δ-connected subcircuit by an equivalent Y-configuration, or vice versa, changes connections so the modified circuit has series or parallel resistors. As a result, series and parallel equivalence can again be used to further reduce the overall circuit.

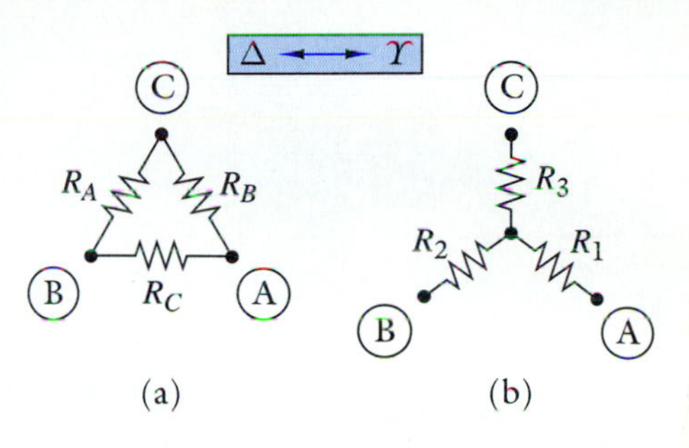

Figure 2-32 The Y-to-Δ transformation: (a) Δ configuration. (b) Y configuration.

Interchanging these equivalent subcircuits is called a Δ-Y transformation or a Y-Δ transformation, depending on the starting and ending configurations. Our immediate task is to find the conditions under which Δ and Y subcircuits are equivalent. By definition circuits are equivalent if they have the same i-v characteristics between specified terminals. The three-terminal Y and Δ subcircuits will have the identical i-v characteristics if the same equivalent resistance is seen between terminal pairs A-B, B-C, and C-A. Equating the equivalent resistance seen between these terminals yields the following requirements:

$$\begin{aligned} R_{A\text{-}B} &= \frac{R_C(R_A + R_B)}{R_A + R_B + R_C} = R_1 + R_2 \\ R_{B\text{-}C} &= \frac{R_A(R_B + R_C)}{R_A + R_B + R_C} = R_2 + R_3 \\ R_{C\text{-}A} &= \frac{R_B(R_C + R_A)}{R_A + R_B + R_C} = R_3 + R_1 \end{aligned} \tag{2-25}$$

Solving Eq. (2-25) for R_1, R_2, and R_3 yields the equations for a Δ to Y transformation:

$$\begin{aligned} R_1 &= \frac{R_B R_C}{R_A + R_B + R_C} \\ R_2 &= \frac{R_C R_A}{R_A + R_B + R_C} \\ R_3 &= \frac{R_A R_B}{R_A + R_B + R_C} \end{aligned} \tag{2-26}$$

Solving Eq. (2-25) for R_A, R_B, and R_C yields the equations for a Y to Δ transformation:

$$\begin{aligned} R_A &= \frac{R_1R_2 + R_2R_3 + R_1R_3}{R_1} \\ R_B &= \frac{R_1R_2 + R_2R_3 + R_1R_2}{R_2} \\ R_C &= \frac{R_1R_2 + R_2R_3 + R_1R_3}{R_3} \end{aligned} \tag{2-27}$$

The Y and Δ subcircuits are said to be *balanced* when $R_1 = R_2 = R_3 = R_Y$ and $R_A = R_B = R_C = R_\Delta$. Under balanced conditions the transformation equations reduce to $R_Y = R_\Delta/3$ and $R_\Delta = 3R_Y$.

It is important to realize the equivalency still applies when a circuit diagram does not show resistors in a geometric Y or Δ pattern. Sometimes the Y may appear as a T-connection or the Δ as a Π-connection. But regardless of how a circuit diagram is drawn, a Y-configuration involves three resistors connected at a single node. A Δ-configuration involves single resistors connected between each of three nodes. The next example illustrates using these transformations.

EXAMPLE 2-13

Find the equivalent resistance seen by the source in Fig. 2-33(a).

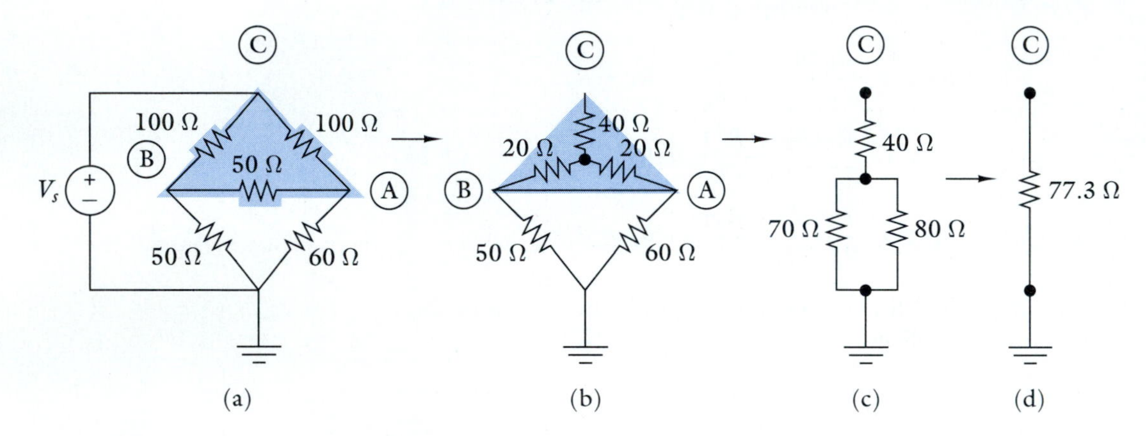

Figure 2-33

SOLUTION:

The circuit shown is called a *wheatstone bridge*. There are no series or parallel connected resistors, so a Δ-to-Y or a Y-to-Δ transformation can be used to find the resistance seen by the source. Using Eq. (2-26) we convert the Δ-connected subcircuit in the upper part of the bridge into the equivalent Y subcircuit shown in Fig. 2-33(b). The modified circuit has series and parallel connected resistors so it can easily be reduced to a single equivalent resistance as shown in Figs. 2-33(c) and (d).

The method given above is not the only way to reduce the circuit. We could have performed a Y-to-Δ transformation on the three leftmost resistors of the wheatstone bridge to obtain resistors connected in parallel. It is left to the reader to show that the alternative approach produces the same equivalent resistance shown in Fig. 2-33(d).

Summary of Equivalent Circuits

Figure 2-34 summarizes two-terminal equivalent circuits involving resistors and sources connected in series or parallel. We will make frequent use of the equivalencies in the first two rows of this figure. The first row shows that series or parallel equiva-

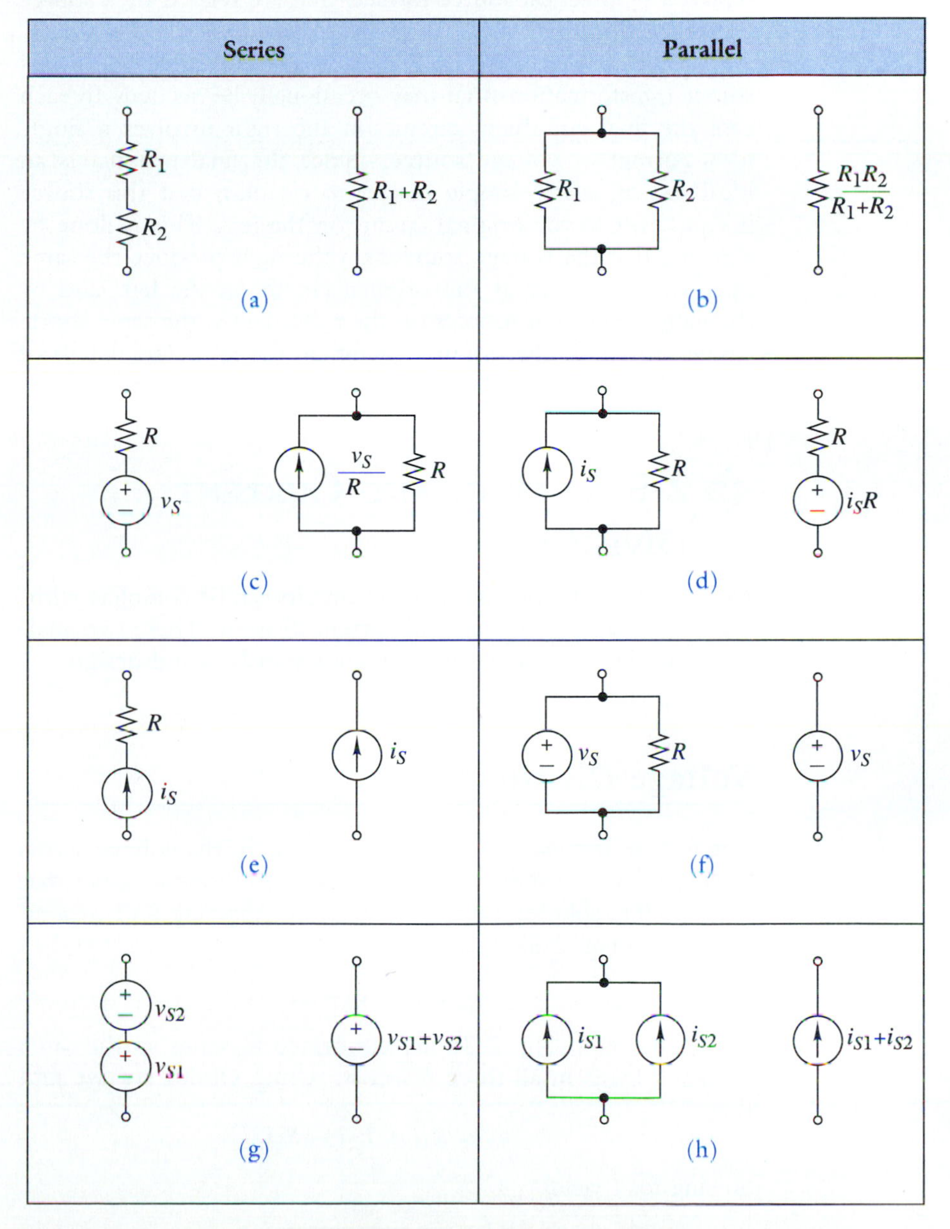

Figure 2-34 Summary of two-terminal equivalent circuits.

lence can be used to replace two resistors by a single equivalent resistance. The second row shows the source transformations used to replace one practical source model with its dual model. These transformations convert a practical voltage source into a practical current source, and vice versa. The concept of source conversion is important. In the next chapter we study Thévenin's and Norton's theorems, which show that entire circuits can be replaced by practical source models that are related by a source transformation.

The last two rows in Fig. 2-34 present some additional source transformations that may occasionally be needed. In each case the final equivalent circuit on the right involves a single ideal current or voltage source. Since the final product is an ideal source, it is a simple matter to establish that this source is equivalent to the original circuit on the left. This is done by showing that the voltage sources on the right produce the same open-circuit voltage as the original circuit on the left, and by showing the current sources on the right deliver the same short-circuit current as the original circuit on the left. The details of such a derivation are left as an exercise for the student.

2-5 VOLTAGE AND CURRENT DIVISION

No treatment of series and parallel circuits can be complete without a discussion of voltage and current division. These two analysis tools find wide application in circuit analysis and design.

Voltage Division

The *voltage division rule* allows us to solve for the voltage across each element in a series circuit. Figure 2-35 shows a circuit that lends itself to solution by voltage division. Applying KVL around the loop in Fig. 2-35 yields

$$v_S = v_1 + v_2 + v_3$$

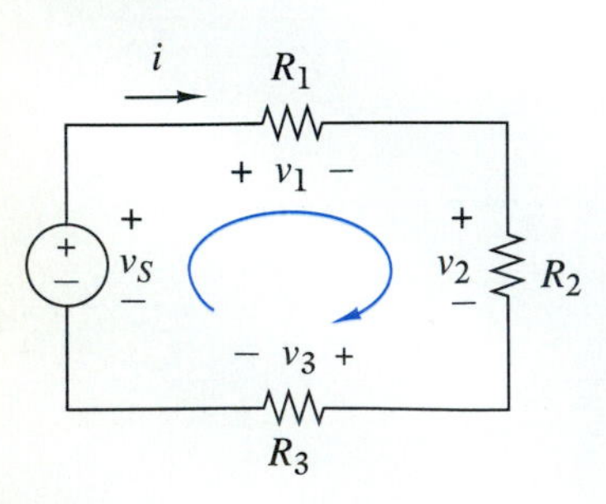

Figure 2-35 A voltage divider circuit.

The elements in Fig. 2-35 are connected in series so the same current i exists in all three resistors. Using Ohm's law we find that

$$v_S = R_1 i + R_2 i + R_3 i$$

Solving for i yields

$$i = \frac{v_S}{R_1 + R_2 + R_3}$$

Once the current in the series circuit is found, the voltage across each resistor is found using Ohm's law

$$v_1 = R_1 i = \left(\frac{R_1}{R_1 + R_2 + R_3}\right) v_S \quad (2\text{-}28)$$

$$v_2 = R_2 i = \left(\frac{R_2}{R_1 + R_2 + R_3}\right) v_S \quad (2\text{-}29)$$

$$v_3 = R_3 i = \left(\frac{R_3}{R_1 + R_2 + R_3}\right) v_S \quad (2\text{-}30)$$

Looking over these results we see an interesting pattern. In each case the element voltage is equal to its resistance divided by the equivalent series resistance in the circuit times the total voltage across the series circuit:

$$v_k = \left(\frac{R_k}{R_{EQ}}\right) v_{TOTAL} \quad (2\text{-}31)$$

In other words, the total voltage delivered is divided among the series resistors in proportion to their resistance over the equivalent resistance of the series connection. Once this pattern is recognized the analysis need not be repeated and we can immediately use the voltage division rule (Eq. 2-31) to find the voltage across any resistor in a series connection. Several examples will help to clarify this concept.

EXAMPLE 2-14

Find the voltage across the 330-Ω resistor in the circuit of Fig. 2-36.

SOLUTION:

Applying the voltage division rule we find that

$$v_o = \left(\frac{330}{100 + 560 + 330 + 220}\right) 24 = 6.55 \text{ V}$$

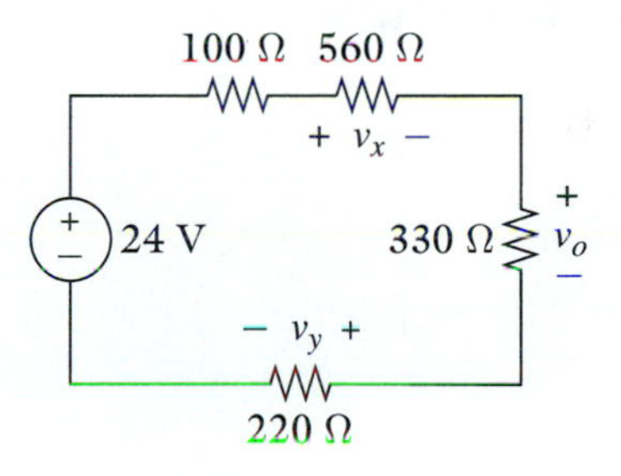

Figure 2-36

EXERCISE 2-10

Find the voltages v_x and v_y in Fig. 2-36.

ANSWERS:

$v_x = 11.1$ V; $v_y = 4.36$ V.

EXAMPLE 2-15

Select a value for the resistor R_x in Fig. 2-37 so $v_o = 8$ V.

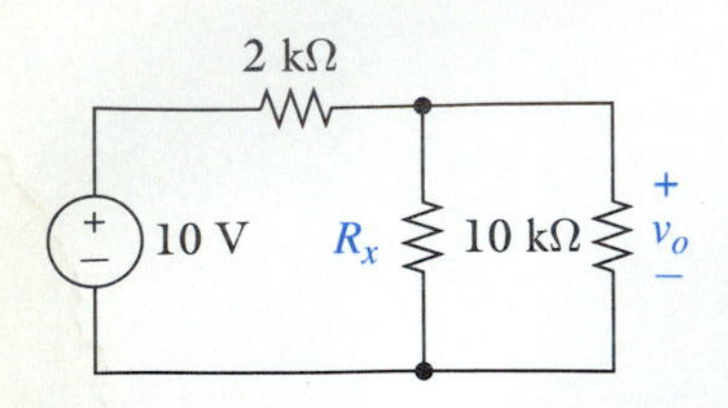

Figure 2-37

SOLUTION:

The unknown resistor is in parallel with the 10-kΩ resistor. Since voltages across parallel elements are equal, the voltage $v_o = 8$ V appears across both. We first define an equivalent resistance $R_{EQ} = R_x \parallel 10$ kΩ.

$$R_{EQ} = \frac{R_x \times 10000}{R_x + 10000}$$

We write the voltage division in terms of R_{EQ} as

$$v_o = 8 = \left(\frac{R_{EQ}}{R_{EQ} + 2000}\right) 10$$

which yields $R_{EQ} = 8$ kΩ. Finally, we substitute this value into the equation for R_{EQ} and solve for R_x. This operation yields $R_x = 40$ kΩ.

EXAMPLE 2-16

Use the voltage division rule to find the output voltage (v_o) of the circuit in Fig. 2-38.

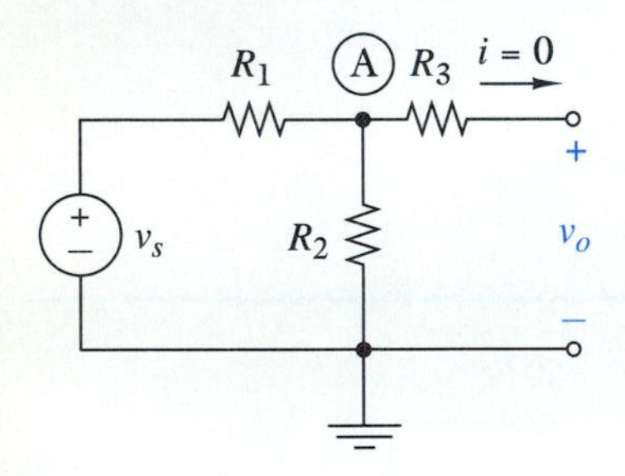

Figure 2-38

SOLUTION:

At first glance it appears the voltage division rule does not apply, since the resistors are not connected in series. However, the current through R_3 is zero, since the output of the circuit is an open circuit. Therefore, Ohm's law shows that $v_3 = R_3 i_3 = 0$. Applying KCL at Node A shows the same current exists in R_1 and R_2, since the current through R_3 is zero. Applying KVL around the output loop shows the voltage across R_2 must be equal to v_o since the voltage across R_3 is zero. In essence, it is as if R_1 and R_2 were connected in series. Therefore, voltage division yields the output voltage as

$$v_o = \left(\frac{R_2}{R_1 + R_2}\right) v_S$$

The reader should carefully review the logic leading to this result, since voltage division applications of this type occur frequently.

Application Note

EXAMPLE 2-17

The potentiometer operates on the principle of voltage division. This three-terminal device makes use of voltage (potential) division to meter out a fraction of the applied voltage. Figure 2-39

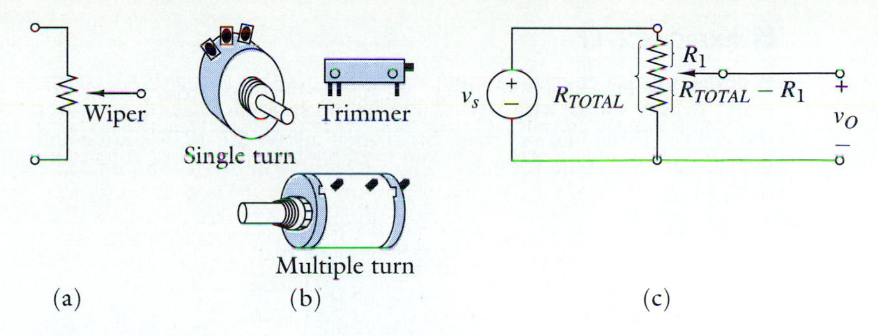

Figure 2-39 The potentiometer: (a) Circuit symbol. (b) Actual devices. (c) An application.

shows the circuit symbol of a potentiometer, sketches of three different types of actual potentiometers, and a typical application. Simply stated, a *potentiometer is an adjustable voltage divider*.

The voltage v_O in Fig. 2-39(c) can be adjusted by turning the shaft on the potentiometer to move the wiper arm contact. Using the voltage divider rule v_o is found as

$$v_O = \left(\frac{R_{TOTAL} - R_1}{R_{TOTAL}}\right) v_S \tag{2-32}$$

When we make R_1 zero by moving the wiper all the way to the top, we obtain

$$v_O = \left(\frac{R_{TOTAL} - 0}{R_{TOTAL}}\right) v_S = v_S \tag{2-33}$$

In other words, 100% of the applied voltage is delivered to the rest of the circuit. When the wiper is moved all the way to the bottom, we make R_1 equal to R_{TOTAL} and

$$v_O = \left(\frac{R_{TOTAL} - R_{TOTAL}}{R_{TOTAL}}\right) v_S = 0 \tag{2-34}$$

The other extreme delivers zero voltages. We can adjust the output voltage all the way from the applied voltage v_S to zero. Halfway between we intuitively expect to have half of the applied voltage. Setting $R_1 = 1/2R_{TOTAL}$ yields

$$v_O = \left(\frac{R_{TOTAL} - \frac{1}{2}R_{TOTAL}}{R_{TOTAL}}\right) v_S = \frac{v_S}{2} \tag{2-35}$$

as expected. The many applications of the potentiometer include volume controls, voltage balancing, and fine-tuning adjustment.

EXERCISE 2-11

A device with an equivalent input resistance of 50 Ω requires 9 V to operate. A 12-V source with a 10-Ω internal resistance is available to power the device. Design an interface circuit that allows the 12-V source and 9-V device to operate properly. Figure 2-40(a) illustrates the interface design task.

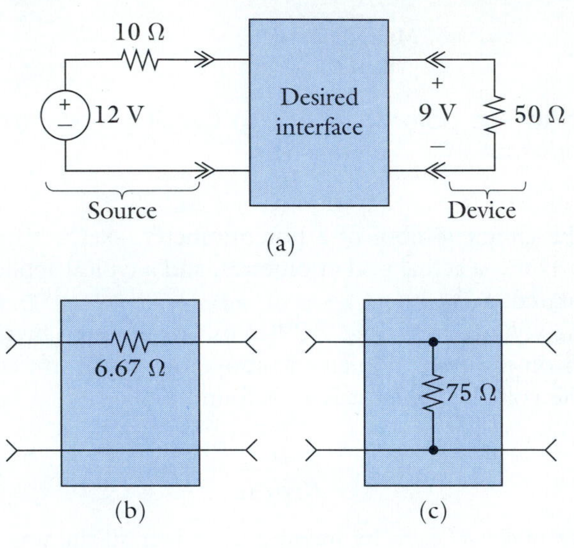

Figure 2-40

ANSWER:

Two possible solutions are shown in Fig. 2-40(b) and Fig. 2-40(c). Can you think of a reason for choosing one circuit over the other?

Current Division

Current division is the dual of voltage division. By duality we expect current division to allow us to solve for the current through each element in a parallel circuit. Figure 2-41 shows a parallel circuit that lends itself to solution by current division. Applying KCL at Node A yields

$$i_S = i_1 + i_2 + i_3$$

The voltage v appears across all three conductances since they are connected in parallel. So that using Ohm's law we can write

$$i_S = vG_1 + vG_2 + vG_3$$

and solve for v as

$$v = \frac{i_S}{G_1 + G_2 + G_3}$$

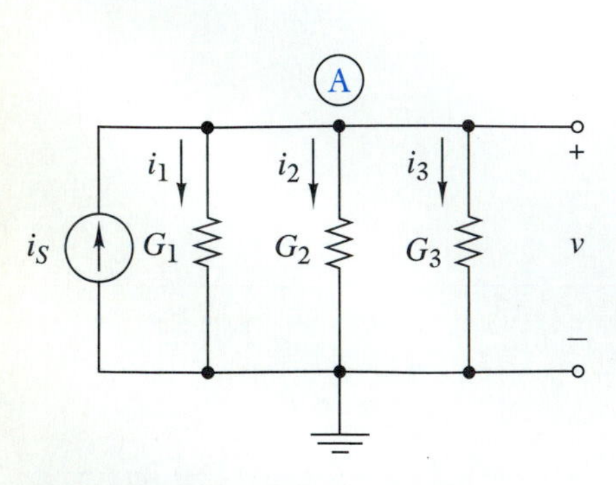

Figure 2-41 A current divider.

Given the voltage v, the current through any element is found using Ohm's law.

$$i_1 = vG_1 = \left(\frac{G_1}{G_1+G_2+G_3}\right) i_S \qquad (2\text{-}36)$$

$$i_2 = vG_2 = \left(\frac{G_2}{G_1+G_2+G_3}\right) i_S \qquad (2\text{-}37)$$

$$i_3 = vG_3 = \left(\frac{G_3}{G_1+G_2+G_3}\right) i_S \qquad (2\text{-}38)$$

The results show that the total current delivered is divided among the parallel resistors in proportion to their conductances over the equivalent conductances in the parallel connection. That is, the *current division rule* can be expressed as

$$i_k = \left(\frac{G_k}{G_{EQ}}\right) i_{TOTAL} \qquad (2\text{-}39)$$

Since we do not think of resistors in terms of conductance it is useful to express the current division rule in terms of two resistors in parallel. To find i_1 in Fig. 2-42 the current division yields

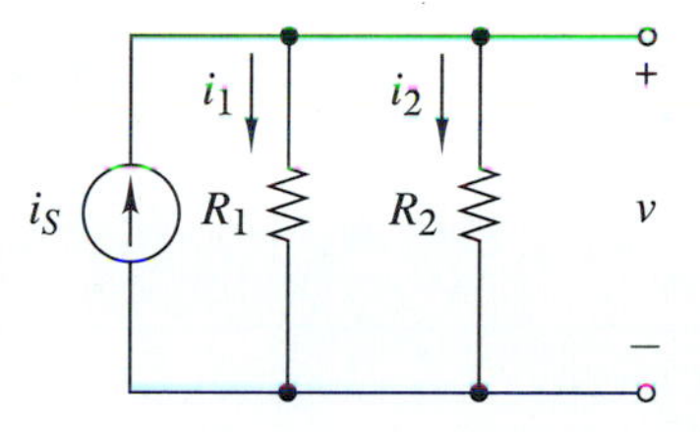

Figure 2-42 Two-resistor current divider.

$$i_1 = \left(\frac{G_1}{G_1+G_2}\right) i_S = \frac{\dfrac{1}{R_1}}{\dfrac{1}{R_1}+\dfrac{1}{R_2}} i_S = \left(\frac{R_2}{R_1+R_2}\right) i_S \qquad (2\text{-}40)$$

Similarly, solving for i_2 in Fig. 2-42 yields

$$i_2 = \left(\frac{G_2}{G_1+G_2}\right) i_S = \frac{\dfrac{1}{R_2}}{\dfrac{1}{R_1}+\dfrac{1}{R_2}} i_S = \left(\frac{R_1}{R_1+R_2}\right) i_S \qquad (2\text{-}41)$$

These results lead to the following rule. When a circuit can be reduced to two resistance paths in parallel, the current through a path equals the resistance in the **other** parallel path divided by the sum of the resistance in both paths times the total current entering the parallel combination.

Caution: Equations (2-40) and (2-41) only apply when the circuit is reduced to two parallel paths in which one path contains the unknown current and the other path is the equivalent resistance of all other paths.

EXAMPLE 2-18

Find the current i_x Fig. 2-43(a).

SOLUTION:

To find i_x, we reduce the circuit to two paths, a path containing i_x and a path equivalent to all other paths as shown in Fig. 2-43(b). Now we can use the two-path current divider rule as

$$i_x = \frac{6.67}{20 + 6.67} \times 5 = 1.25 \text{ A}$$

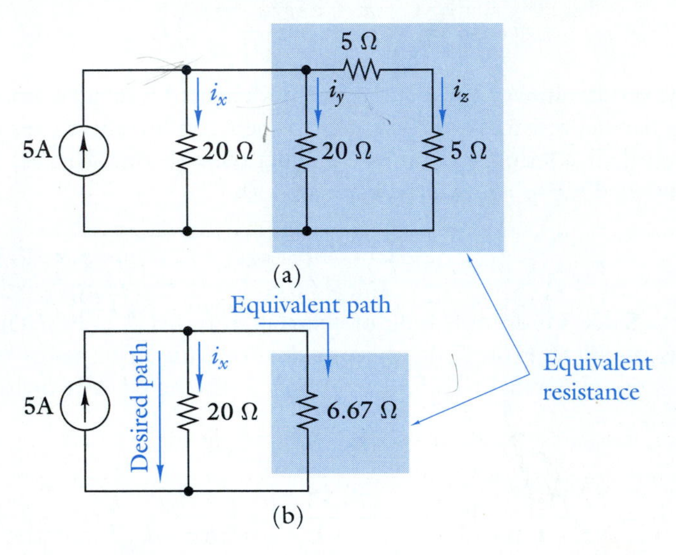

Figure 2-43

■ EXERCISE 2-12

(a) Find i_y and i_z in the circuit of Fig. 2-43(a).
(b) Show that the sum of i_x, i_y, and i_z equals the source current.

ANSWERS:

(a) $i_y = 1.25$ A, $i_z = 2.5$ A.
(b) $i_x + i_y + i_z = 5$ A. ■

■ EXERCISE 2-13

The circuit in Fig. 2-44 shows a delicate device that is modeled by a 90-Ω equivalent resistance. The device requires 1 mA of current to operate properly. A 1.5-mA fuse is inserted in series with the device to protect it from overheating. The resistance of the fuse is 10 Ω. Without the shunt resistance R_x the source would deliver 5 mA to the device, causing the fuse to blow. Inserting a shunt resistor R_x diverts a portion of the available source current around the fuse and device. Select a value for R_x so only 1 mA is delivered to the device.

ANSWER:

$R_x = 12.5$ Ω. ■

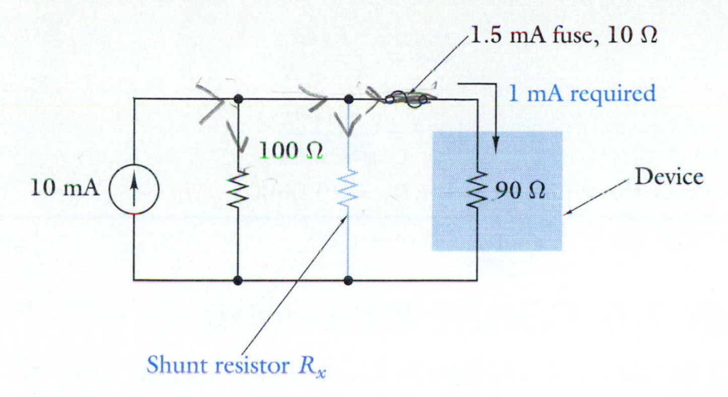

Figure 2-44

Application Note

EXAMPLE 2-19

The *D'Arsonval meter* shown in Fig. 2-45(a) is a device used to measure currents or voltages. In simple terms, a coil of wire is mounted between the poles of a permanent magnet so it is free to rotate. The coil current interacts with the magnet to produce a torque, which causes the coil to turn. The deflection of the pointer attached to the coil is linearly proportional to the current.

D'Arsonval movements are rated by the current necessary to produce full scale deflection of the pointer. Ratings range from 1 μA to 1 mA depending on the structure of the device. Instruments that can measure much larger currents are clearly required. To get around this limitation meter designers employ a very precise shunt resistance R_S, as shown in Fig. 2-45(b). This shunt resistance diverts a precisely known amount I_S of the measured current I_M around the coil, so the current I_{FS} remains within the meter rating. Different values of R_S allow the meter to measure different ranges of I_M using the same D'Arsonval movement.

For example, suppose currents up to 10 A are to be measured using a D'Arsonval movement with a full scale rating of 10 μA and a coil resistance R_M is 20 Ω. Using the circuit model in Fig. 2-45(b), the problem is to shunt current around the meter movement so I_{FS} = 10 μA when I_M = 10 A. In Fig. 2-45(b) R_M is the resistance of the meter movement, while R_S is the shunt resistance to be selected. From the model it is evident the problem can be solved using the two-path current division rule.

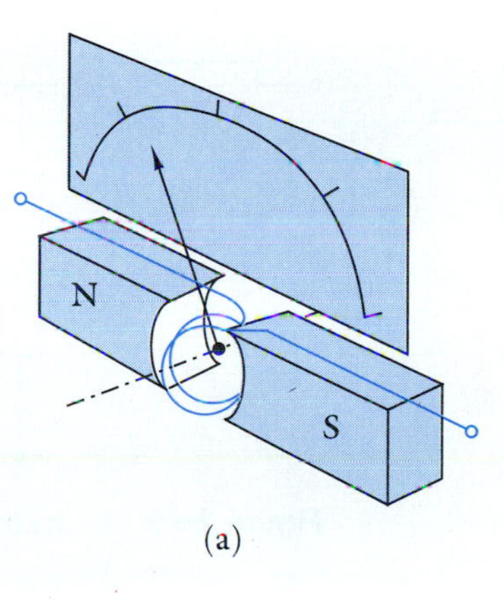

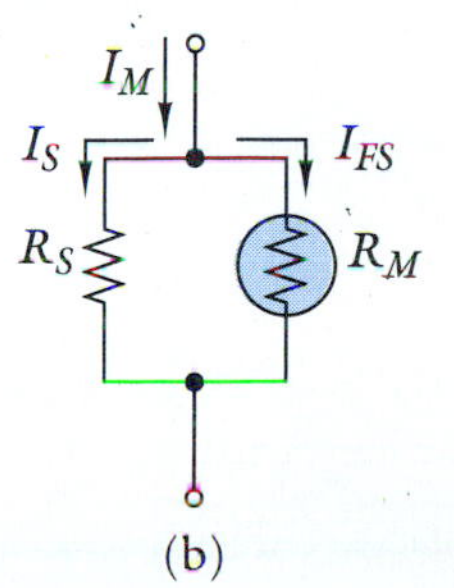

Figure 2-45 D'Arsonval meter: (a) Sketch of meter movement. (b) Equivalent circuit.

$$I_{FS} = \frac{R_S I_M}{R_M + R_S}$$

$$10^{-5} = \frac{R_S(10)}{20 + R_S}$$

Solving this equation yields $R_S = 20.00002\ \mu\Omega$.

2-6 CIRCUIT REDUCTION

The concepts of series/parallel equivalence, voltage/current division, and source transformations can be used to analyze ***ladder circuits*** of the type shown in Fig. 2-46. The basic analysis strategy is to reduce the circuit to a simpler equivalent in which the output is easily found by voltage or current division or Ohm's law. There is no fixed pattern to the reduction process, and much depends on the insight of the analyst. But in any case, with circuit reduction we work directly with the circuit model, and so the process gives us insight into circuit behavior.

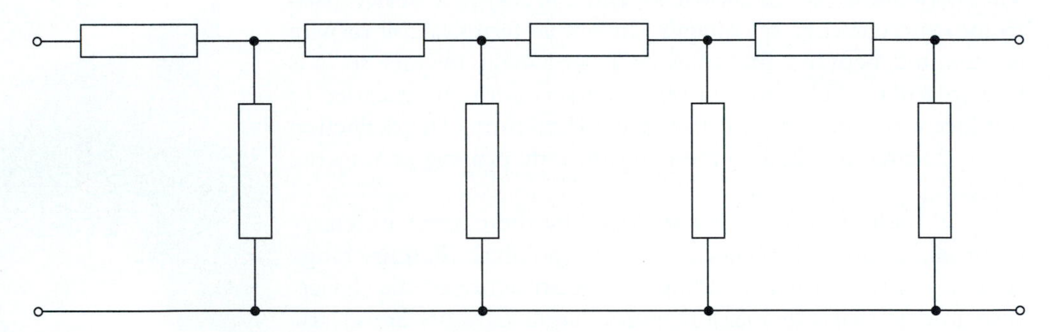

Figure 2-46 A ladder circuit.

When we apply circuit reduction it is important to remember that the unknown voltage exists between two nodes and the unknown current exists in a branch. The reduction process must not eliminate the required node pair or branch; otherwise the unknown voltage or current cannot be found. The next three examples illustrate circuit reduction. The final example shows that rearranging the circuit can simplify the analysis.

EXAMPLE 2-20

Use series and parallel equivalence to find the output voltage v_F and the input current i_S in the ladder circuit shown in Fig. 2-47(a).

SOLUTION:

Our approach is to combine parallel resistors and use voltage division to find v_F. Then combine all resistances into a single equivalent to find the input current i_S. Figure 2-47(b) shows the step required to determine the equivalent resistance between the terminals B and ground. The equivalent resistance of the parallel $2R$ and R resistors is

$$R_{EQ1} = \frac{R \times 2R}{R + 2R} = \frac{2}{3}R$$

The reduced circuit in Fig. 2-47(b) is a voltage divider. Notice the two nodes needed to find the voltage v_F have been retained. The unknown voltage is then found in terms of the source voltage as

$$v_F = \frac{\frac{2}{3}R}{\frac{2}{3}R + R} v_S = \frac{2}{5} v_S$$

The input current is found by combining the equivalent resistance found above with the remaining resistor R to obtain

$$R_{EQ2} = R + R_{EQ1}$$
$$= R + \frac{2}{3}R = \frac{5}{3}R$$

Application of series/parallel equivalence has reduced the ladder circuit to the single equivalent resistance shown in Fig. 2-47(c). Using Ohm's law the input current in

$$i_s = \frac{v_S}{R_{EQ2}} = \frac{3}{5}\frac{v_S}{R}$$

Notice that the reduction step between Figs. 2-47(b) and (c) eliminates Node B so the voltage v_F must be calculated before this reduction step is taken.

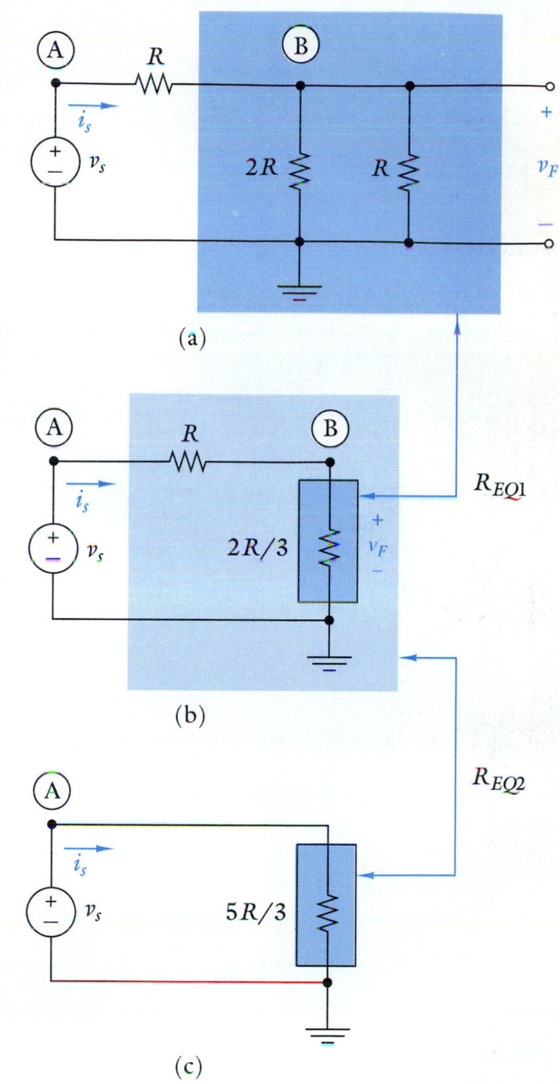

Figure 2-47

EXAMPLE 2-21

Use source transformations to find the output voltage v_F and the input current i_S in the ladder circuit shown in Fig. 2-48(a).

SOLUTION:

Figure 2-48 shows another way to reduce the circuit analyzed in Example 2-20. Breaking the circuit at points X and Y in Fig. 2-48(a) produces a voltage source v_S in series with a resistor R. Using source transformation this combination can be replaced

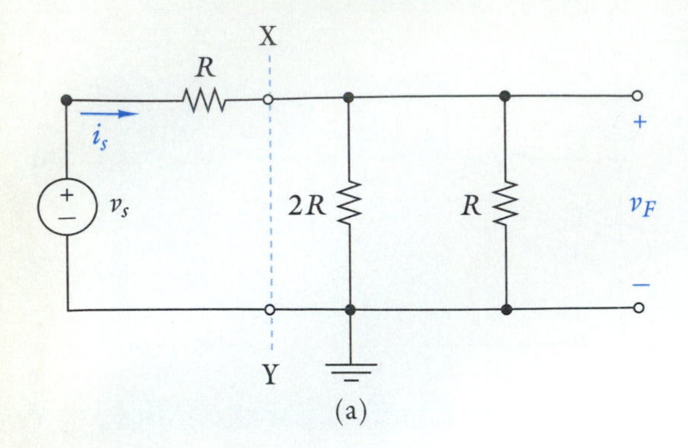

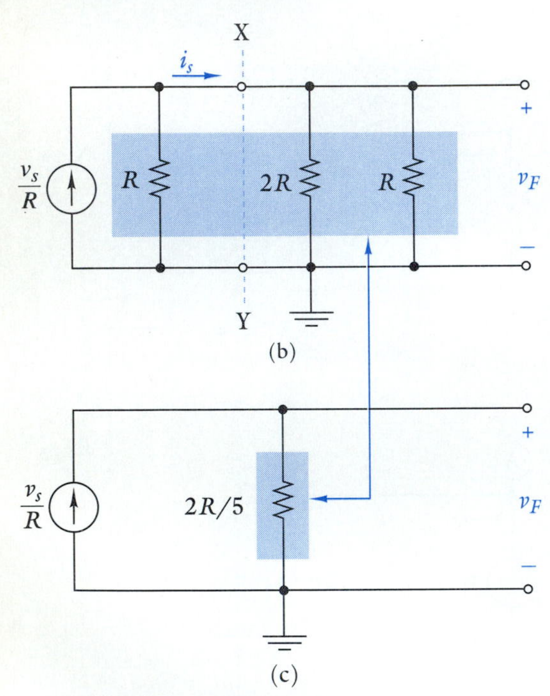

Figure 2-48

by an equivalent current source in parallel with the same resistor as shown in Fig. 2-48(b).

Caution: The current source v_S/R is *not* the input current i_S as is indicated in Fig. 2-48(b). Applying the two-path current division rule to the circuit in Fig. 2-48(b) yields the input current i_S as

$$i_S = \frac{R}{\frac{2}{3}R + R} \times \frac{v_S}{R} = \frac{v_S}{\frac{5}{3}R} = \frac{3}{5}\frac{v_S}{R}$$

The three parallel resistances in Fig. 2-48(b) can be combined into a single equivalent conductance without eliminating the node pair used to define the output voltage v_F. Using parallel equivalence we obtain

$$G_{EQ} = G_1 + G_2 + G_3$$
$$= \frac{1}{R} + \frac{1}{2R} + \frac{1}{R} = \frac{5}{2R}$$

which yields the equivalent circuit in Fig. 2-48(c). The current source v_S/R determines the current through the equivalent resistance in Fig. 2-48(c) so the output voltage is found using Ohm's law as

$$v_F = \frac{v_S}{R} \times \frac{2R}{5} = \frac{2}{5}v_s$$

Of course, these results are the same result obtained in Example 2-20, except here they were obtained using a different sequence of circuit reduction steps.

EXAMPLE 2-22

Find v_x in the circuit shown in Fig. 2-49(a).

SOLUTION:

In the two previous examples the unknown responses were circuit inputs and outputs. In this example the unknown voltage appears across a 10-Ω resistor in the center of the network. The approach is to reduce the circuit at both ends while retaining the 10-Ω resistor defining v_x. Applying a source transformation to the left of terminals X-Y and a series reduction to the two 10-ohm resistors on the far right yields the reduced circuit shown in Fig. 2-49(b). The two pairs of 20-Ω resistors connected in parallel can be combined to produce the circuit in Fig. 2-49(c). At this point there are several ways to proceed. For example, a source transformation at the points W-Z in Fig. 2-49(c) produces the circuit in Fig. 2-49(d). Using voltage division in Fig. 2-49(d) yields v_x.

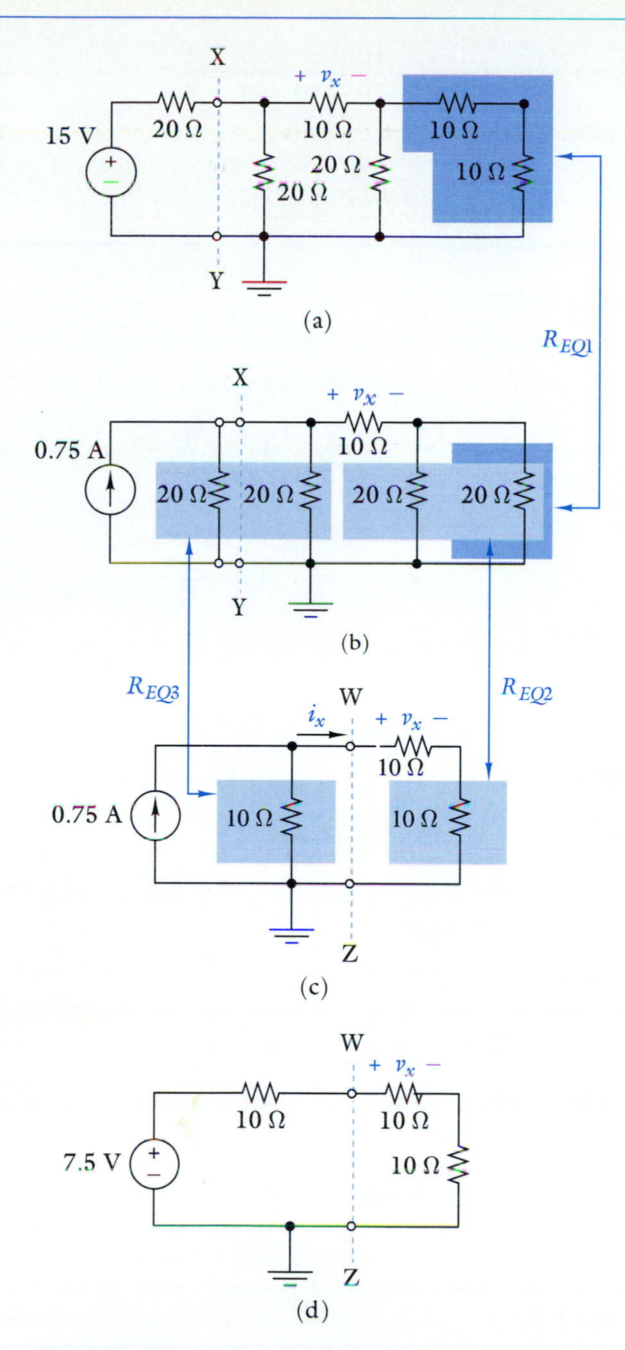

Figure 2-49

$$v_x = \frac{10}{10 + 10 + 10} \times 7.5 = 2.5 \text{ V}$$

A different approach is to use current division in Fig. 2-49(c) to find the current i_x.

$$i_x = \frac{10}{10+10+10} \times \frac{3}{4} = \frac{1}{4}\text{ A}$$

Then applying Ohm's law to obtain v_x

$$v_x = 10 \times \frac{1}{4} = 2.5\text{ V}$$

EXERCISE 2-14

Find v_x and i_x using circuit reduction on the circuit in Fig. 2-50.

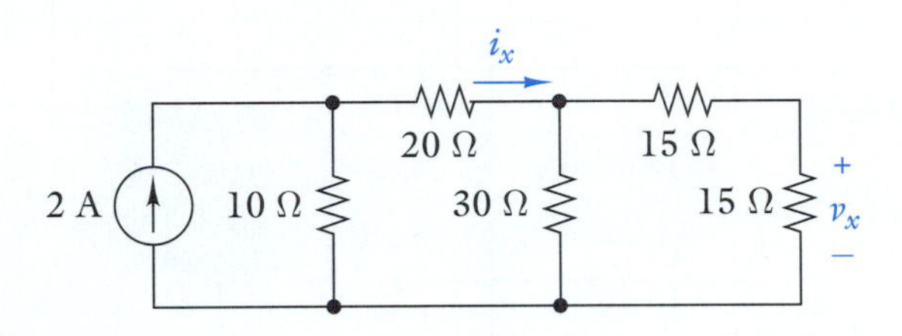

Figure 2-50

ANSWERS:

$v_x = 3.33$ V; $i_x = 0.444$ A.

EXERCISE 2-15

Find v_x and v_y using circuit reduction on the circuit in Fig. 2-51.

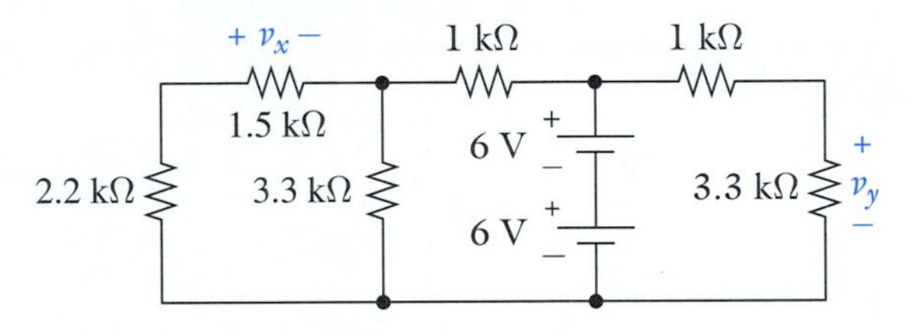

Figure 2-51

ANSWERS:

$v_x = -3.09$ V; $v_y = 9.21$ V.

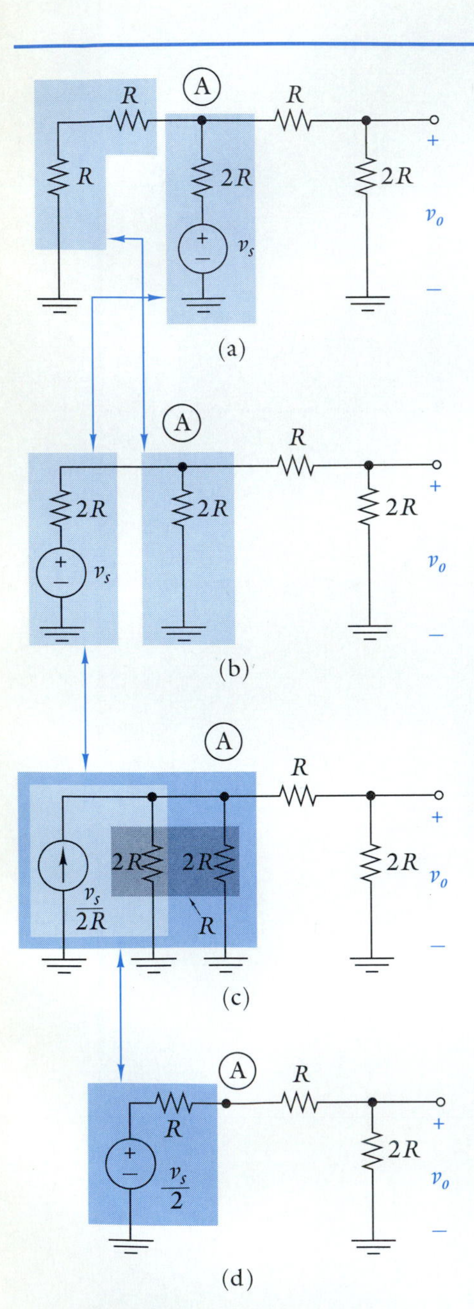

Figure 2-52

EXAMPLE 2-23

Using circuit reduction to find v_o in Fig. 2-52(a).

SOLUTION:

One way to solve this problem is to notice that the source branch and the left-most two-resistor branch are connected in parallel between Node A and ground. Switching the order of these branches and replacing the two resistors by their series equivalent yield the circuit of Fig. 2-52(b). A source transformation yields the circuit in Fig. 2-52(c). This circuit contains a current source $v_S/2R$ in parallel with two $2R$ resistances whose equivalent resistance is

$$R_{EQ} = 2R \parallel 2R = \frac{2R \times 2R}{2R + 2R} = R$$

Applying a source transformation to the current source $v_S/2R$ in parallel with R_{EQ} results in the circuit of Fig. 2-52(d) where

$$V_{EQ} = \left(\frac{v_s}{2R}\right) \times R_{EQ} = \left(\frac{v_s}{2R}\right) R = \frac{v_S}{2}$$

Finally, applying voltage division in Fig. 2-52(d) yields

$$v_O = \left(\frac{2R}{R + R + 2R}\right) \frac{v_S}{2} = \frac{v_S}{4}$$

EXERCISE 2-16

Find the voltage across the source in Fig. 2-53.

ANSWER:
$v_S = -0.225$ V.

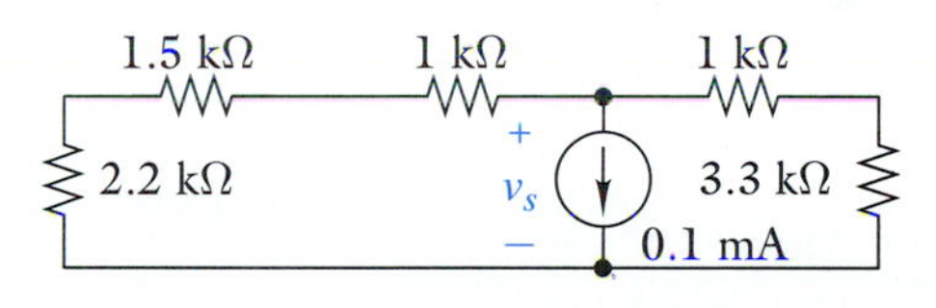

Figure 2-53

SUMMARY

- An electrical device is the real physical object, while a circuit element is a mathematical or graphical model that approximates major features of the device.
- Two-terminal circuit elements are represented by a circuit symbol and are characterized by a single constraint on the associated current and/or voltage.
- An electrical circuit is a collection of devices interconnected at their terminals. The interconnections form nodes and loops.
- A node is an electrical juncture of the terminals of two or more devices. A loop is a closed path formed by tracing through a sequence of devices without passing through any node more than once.
- Device interconnections in a circuit give rise to two connection constraints: Kirchhoff's current law (KCL)—*The algebraic sum of currents at any node is zero at every instant*—and Kirchhoff's voltage law (KVL)—*The algebraic sum of voltages around any loop is zero at every instant.*

- Two two-terminal elements are connected in parallel if they are connected between the same pair of nodes. The same voltage appears across any two elements connected in parallel.
- Two two-terminal elements are connected in series if they are connected at a node to which no other devices are connected. The same current exists in any two elements connected in series.
- Two circuits are said to be equivalent if they have the same i-v constraints at a specified pair of terminals.
- Series and parallel equivalence, and voltage and current division, are very important tools in circuit analysis and design.
- Source conversion changes a voltage source in series with a resistor into an equivalent current source in parallel with a resistor, or vice versa.
- Circuit reduction is a method of determining selected signal variables in series/parallel circuits. The method involves sequential application of the series/parallel equivalence rules, $\Delta - Y$ transformations, source conversion, and the voltage/current division rules. The reduction sequence used depends on the variables to be determined and the structure of the circuit and is not unique.

EN ROUTE OBJECTIVES AND ASSOCIATED PROBLEMS

ERO 2-1 Element Constraints (Sect. 2-1)

Given a two-terminal element with one or more signal variables specified, use the element i-v relationship to find the magnitude and direction of the unknown signal variables.

2-1 Determine the unspecified signal variables associated with each of the two-terminal elements shown in Fig. P2-1.

2-2 A variable resistor is connected to constant voltage source. Explain how the current through the resistor changes as its resistance varies from 0 Ω to ∞ Ω.

2-3 Find the current through, voltage across, and power dissipated in each of the devices in Fig. P2-3.

2-4 A fuse contains a metal link that is heated by the current through the device. The fusible link melts to produce an open circuit for

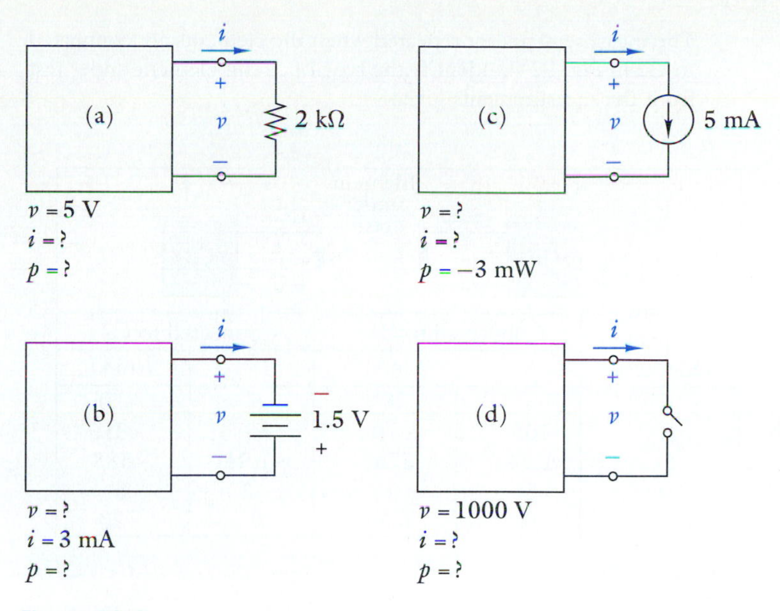

Figure P2-1

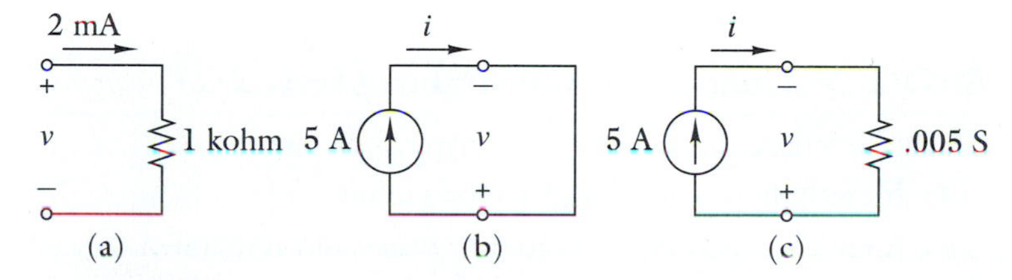

Figure P2-3

a predetermined current. When the fuse is intact the fusible link exhibits a fixed resistance R_{Fuse}. A basic model for a fuse is a resistor in series with a switch. Draw a sketch of the current through a 20-mA fuse versus time when $R_{Fuse} = 10\ \Omega$ and the voltage across the fuse increases linearly as a function of time as $v = t$ V.

2-5 A semiconductor diode is a two-terminal circuit element with the circuit symbol shown in Fig. P2-5. The diode i-v relationship is given below. Find i_D, and p_D, for $v_D = -0.8, -0.4, -0.2, -0.1, 0, 0.1, 0.2, 0.4$, and 0.8 V. Use these data to plot the i-v characteristics of the element.

$$i_D = 10^{-15}(e^{40v_D} - 1)$$

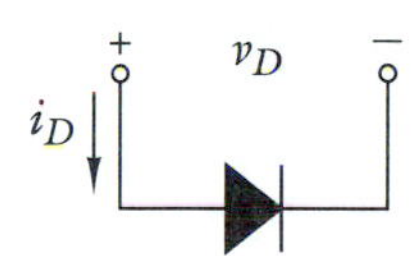

Figure P2-5

2-6 The i-v characteristics of a nonlinear resistor are $v = 100i + 0.2i^2$.

(a) Calculate v and p for $i = \pm 0.5, \pm 1, \pm 2, \pm 5$, and ± 10 A.

(b) If the operating range of the device is limited to $|\,i\,| < 0.5$ A, what is the maximum error in v when the device is approximated by a 100-Ω linear resistance?

2-7 The voltage across and current through two-terminal elements are measured when they are connected to circuit C1 in Fig. P2-7.

The measurements are repeated when the elements are connected to C2 in Fig. P2-7. Identify the type of circuit element under test from the measurements given.

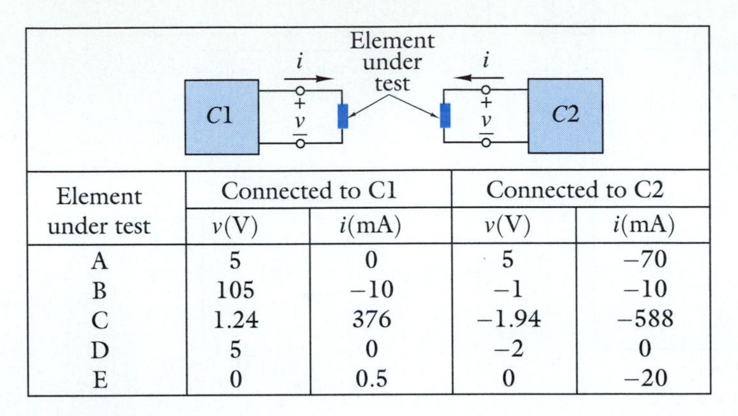

Element under test	Connected to C1		Connected to C2	
	v(V)	i(mA)	v(V)	i(mA)
A	5	0	5	−70
B	105	−10	−1	−10
C	1.24	376	−1.94	−588
D	5	0	−2	0
E	0	0.5	0	−20

Figure P2-7

ERO 2-2 Connection Constraints (Sect. 2-2)

Given a circuit consisting of two-terminal elements:

(a) Identify nodes and loops in the circuit.

(b) Identify elements connected in series and in parallel.

(c) Use Kirchhoff's laws (KCL and KVL) to find selected signal variables.

2-8 For each of the circuits in Fig. P2-8:
(a) Identify at least three loops.
(b) Identify elements connected in series or in parallel.

2-9 Repeat Problem 2-8 when Nodes A and B in each circuit are shorted together.

2-10 Repeat Problem 2-8 if Node A in each circuit is shorted to ground.

2-11 Repeat Problem 2-8 if Element 2 in each circuit is replaced by an open circuit and Nodes A and B are shorted together.

2-12 Repeat Problem 2-8 if Element 2 in each circuit is replaced by an open circuit and Node A is shorted to ground.

2-13 For the circuit in Fig. P2-13:
(a) How many nodes are in the circuit?
(b) Identify elements connected in series and in parallel.

2-14 Repeat Problem 2-13 when Line B is connected to ground.

2-15 Show that the circuit in Fig. P2-13 is the same as C3 in Fig. P2-8.

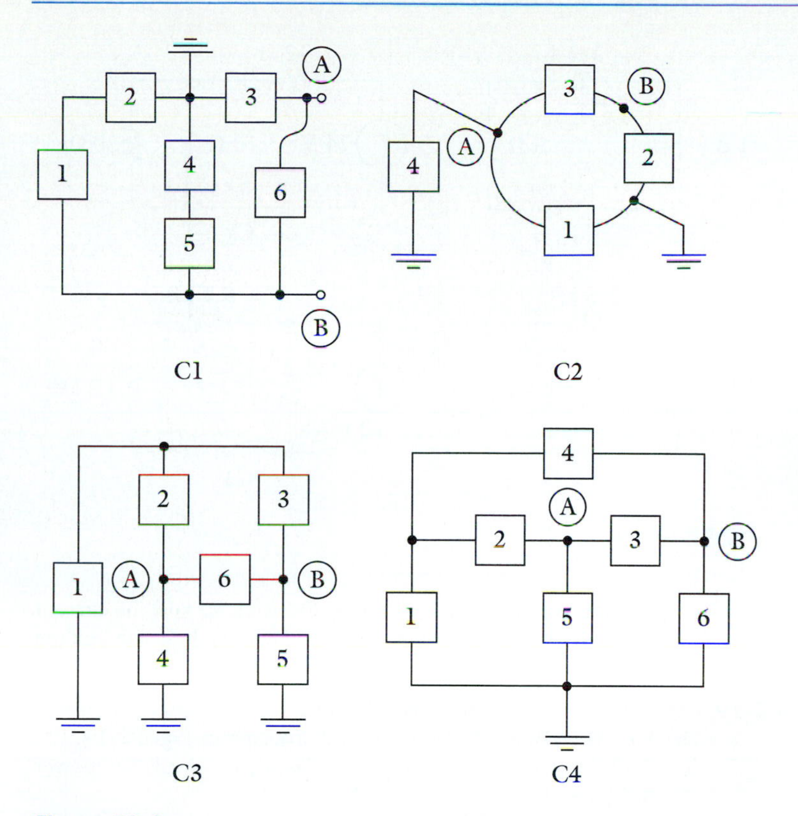

Figure P2-8

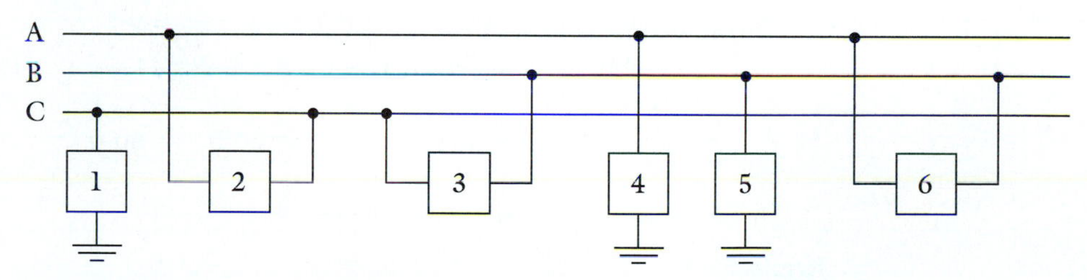

Figure P2-13

ERO 2-3 Combined Constraints (Sect. 2-3)

Given a circuit consisting of sources and linear resistors, use element constraints and connection constraints to find selected signal variables.

2-16 Find v_x in each of the circuits of Fig. P2-16.

2-17 Find i_x in each of the circuits of Fig. P2-16.

2-18 (a) Use KVL to find the voltage across each resistor in Fig. P2-18.

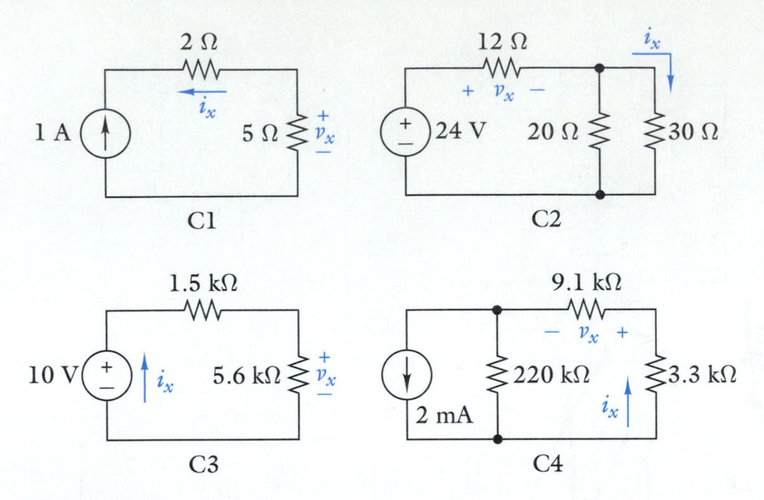

Figure P2-16

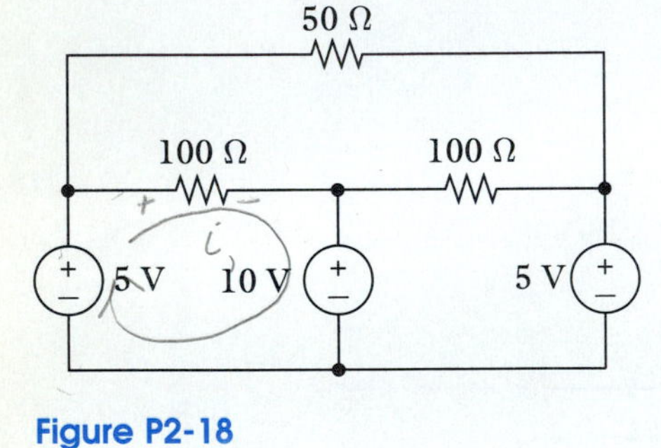

Figure P2-18

$5\text{ V} = 100 i_1 + 10\text{V}$

$i_1 = -50\text{mA}$

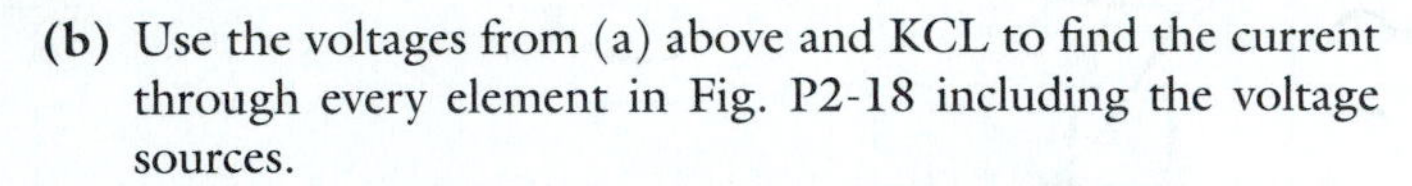

(b) Use the voltages from (a) above and KCL to find the current through every element in Fig. P2-18 including the voltage sources.

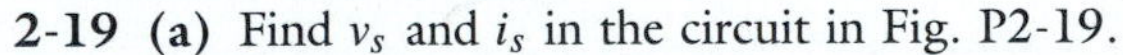

2-19 (a) Find v_s and i_s in the circuit in Fig. P2-19.
(b) Find the power absorbed by each resistor in Fig. P2-19.
(c) Verify that the sum of your answers in (b) equals the power produced by the source.

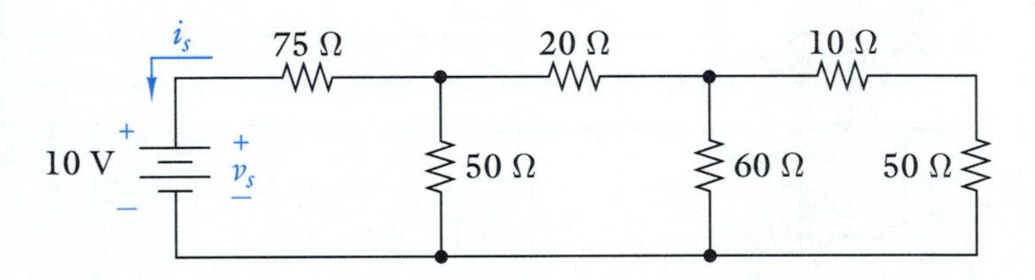

Figure P2-19

2-20 Find the voltage across the current sources and current through the voltage sources in the circuits shown in Fig. P2-20.

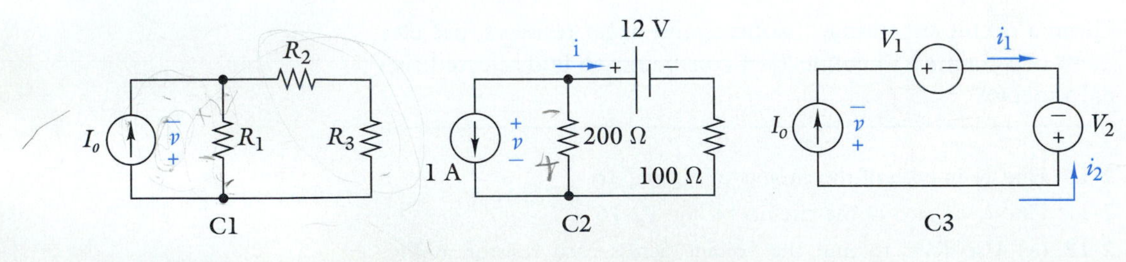

Figure P2-20

2-21 The device in the box in Fig. P2-21 has the i-v characteristics shown. Find the voltage across the 10-Ω resistor for $v_S = -2, 0.5,$ and 2 V.

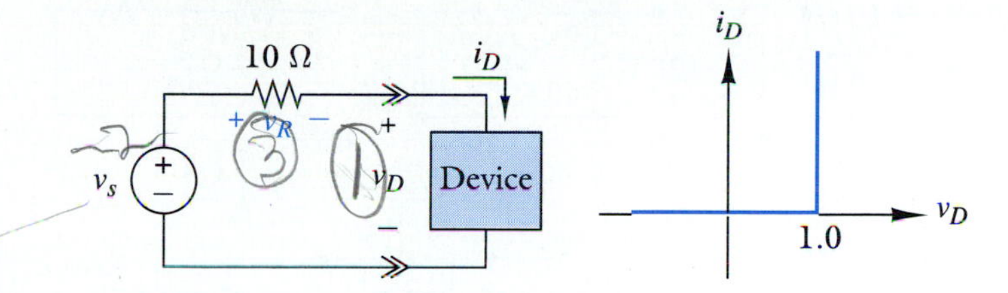

Figure P2-21

ERO 2-4 Equivalent Resistive Circuits and Source Transformations (Sect. 2-4)

(a) Given a circuit consisting of linear resistors, find the equivalent resistance circuit between a specified pair of terminals.

(b) Given a circuit consisting of a source-resistor combination find its equivalent circuit.

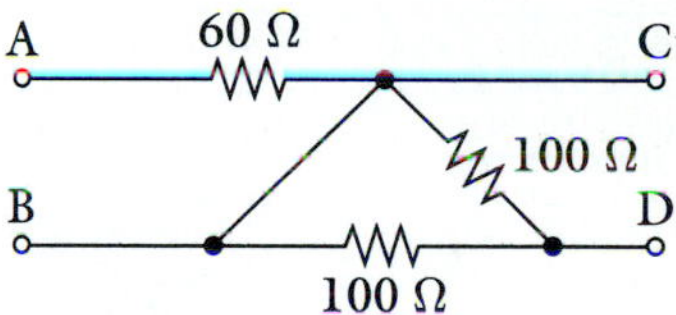

Figure P2-22

2-22 Find the equivalent resistance between terminals A-B, B-C, A-C, C-D, B-D, and A-D in the circuit of Fig. P2-22.

2-23 Find the equivalent resistance between terminals A-B, A-C, A-D, B-C, B-D, and C-D in the circuit of Fig. P2-23.

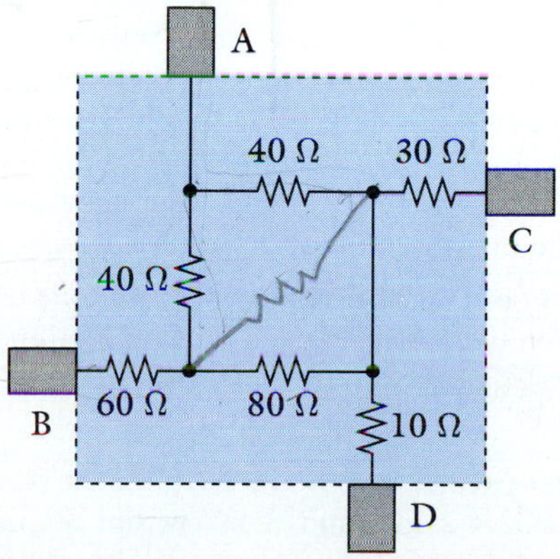

Figure P2-23

2-24 Find the equivalent resistance at terminals A-B in each circuit of Fig. P2-24.

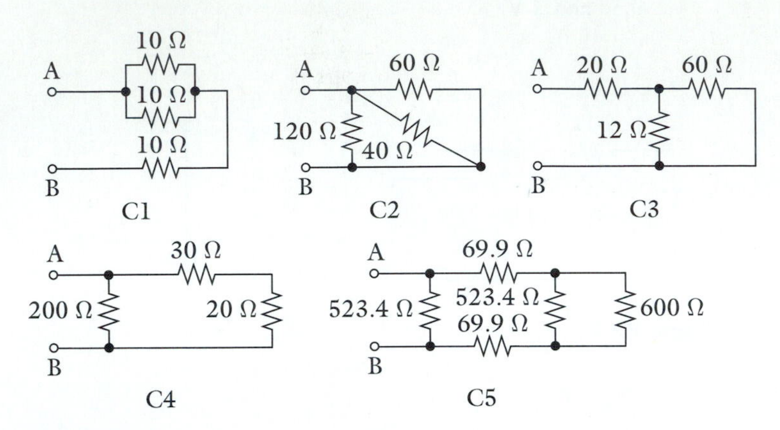

Figure P2-24

2-25 For each of the circuits in Fig. P2-25, find an equivalent circuit containing only one resistor and one source.

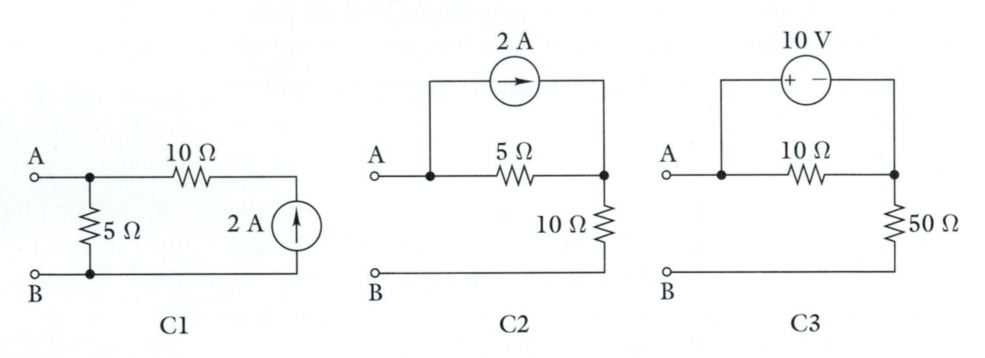

Figure P2-25

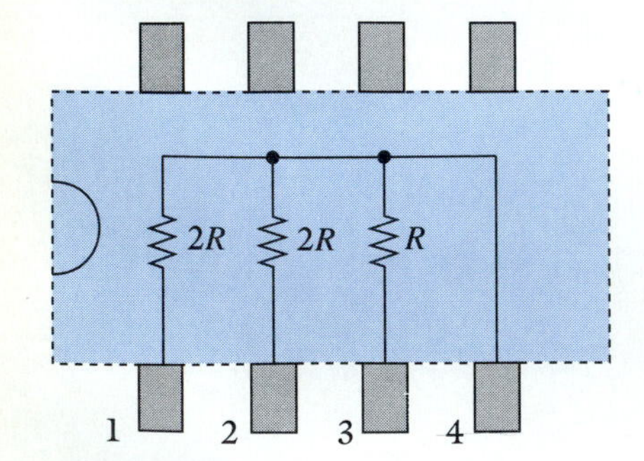

Figure P2-26

2-26 The circuit of Fig. P2-26 is an R-2R resistance array package. All of the following equivalent resistances can be obtained by making proper connections of the array **except for one:** $R/2$, $2R/3$, R, $8R/3$, $5R/3$, $2R$, $3R$, and $4R$. Show how to interconnect the terminals of the array to produce the equivalent resistances and identify the one that cannot be obtained using this array.

2-27 An engineer often makes approximations to save time and money. Simplify the circuits in Fig. P2-27 considering that the resistors have 5% tolerance (they are only accurate to ±5% of their stated value).

❖ **2-28** **(D)** Design circuits that produce equivalent resistances of 10 kΩ, 2 kΩ, 1 kΩ, 4.2 kΩ, and 3 kΩ within 5% using only 3.3-kΩ resistors. Minimize the number of resistors used and never use more than six 3.3-kΩ resistors.

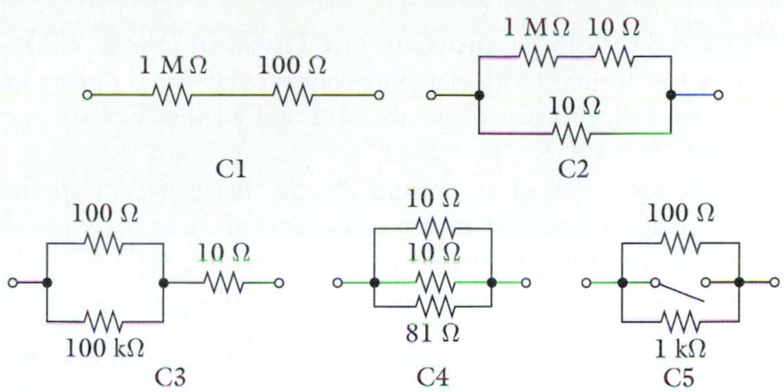

Figure P2-27

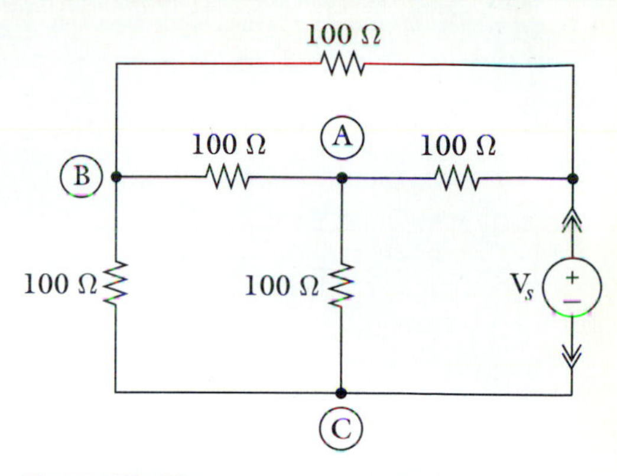

Figure P2-29

2-29 Use a Δ-to-Y or a Y-to-Δ transformation to find the equivalent resistance seen by the source in Fig. P2-29.

2-30 The circuit of Fig. P2-30 is the pin diagram of a resistance array. Find the equivalent resistance between the pins listed below.

Pins
10-11
11-12
2-3
1-2
1-13
1-6
5-7
4-5
4-12
1-7
8-13
12-13

2-31 Find the equivalent resistance seen between terminals A-B in Fig. P2-31. Each resistor is 45 Ω.

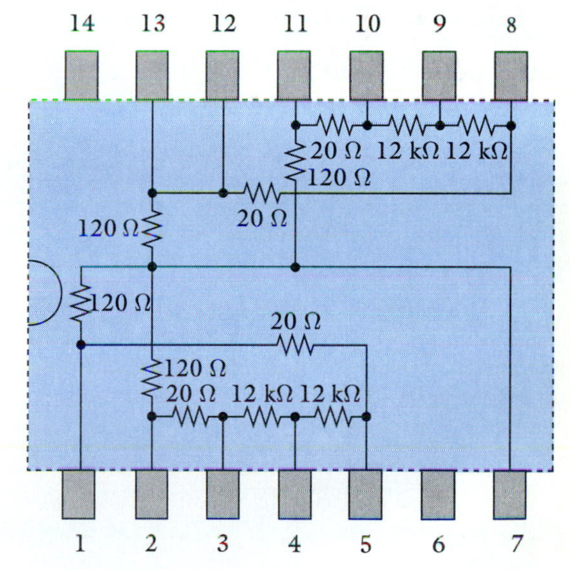

Figure P2-30

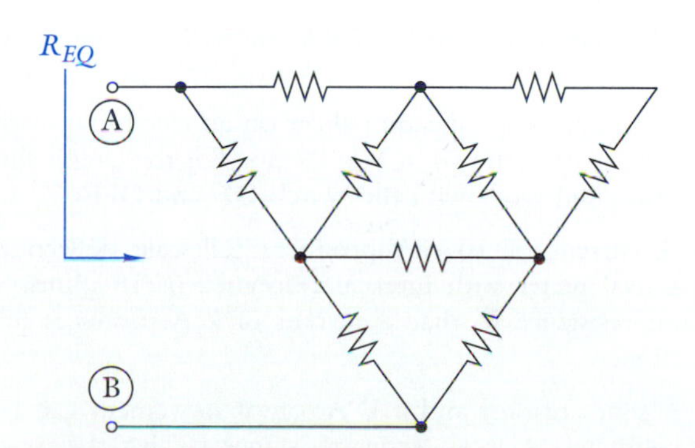

Figure P2-31

ERO 2-5 Voltage and Current Division (Sect. 2-5)

(a) Given a circuit with elements connected in series or parallel, use voltage and current division to find specified voltages or currents.

(b) Design a voltage or current divider that delivers a specific voltage or current under constraints such as parts count, specific values, or cost.

2-32 Find the indicated signal variable using voltage or current division in each of the circuits of Fig. P2-32.

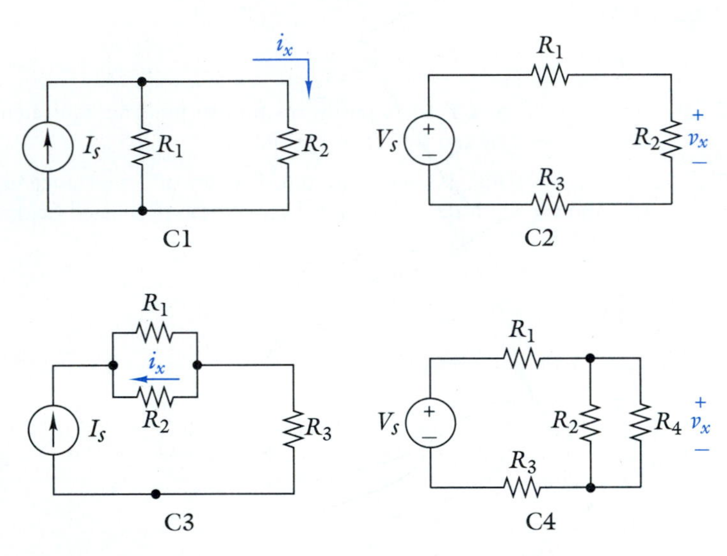

Figure P2-32

2-33 Find v_x in each of the circuits of Fig. P2-33.

2-34 Find the indicated signal variable using voltage and current division in each circuit of Fig. P2-34.

2-35 Find the indicated signal variable in each of the circuits of Fig. P2-35.

2-36 A manufacturer's specification sheet on an electronic switch lists the characteristics shown in Fig. P2-36. For the circuit shown in the figure, find v_{OUT} with the switch ON and OFF.

❖ **2-37** **(D)** A current of 10 mA produces full-scale deflection in a D'Arsonval meter with internal resistance of 10 ohms. Select a shunt resistance so that a current of 2 A produces full-scale deflection.

❖ **2-38** **(D)** A series resistor and a D'Arsonval movement can be used as a voltmeter. Select a series resistance so that the movement described in Problem 2-37 will measure 10 V at full scale.

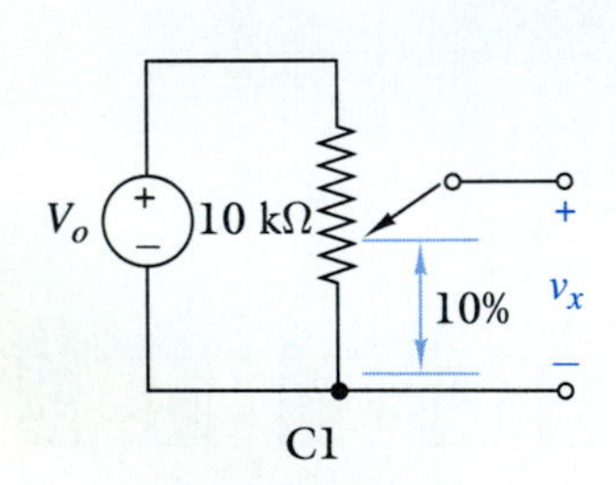

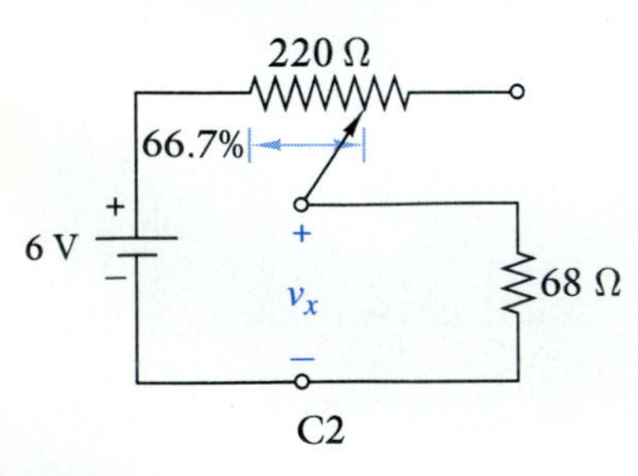

Figure P2-33

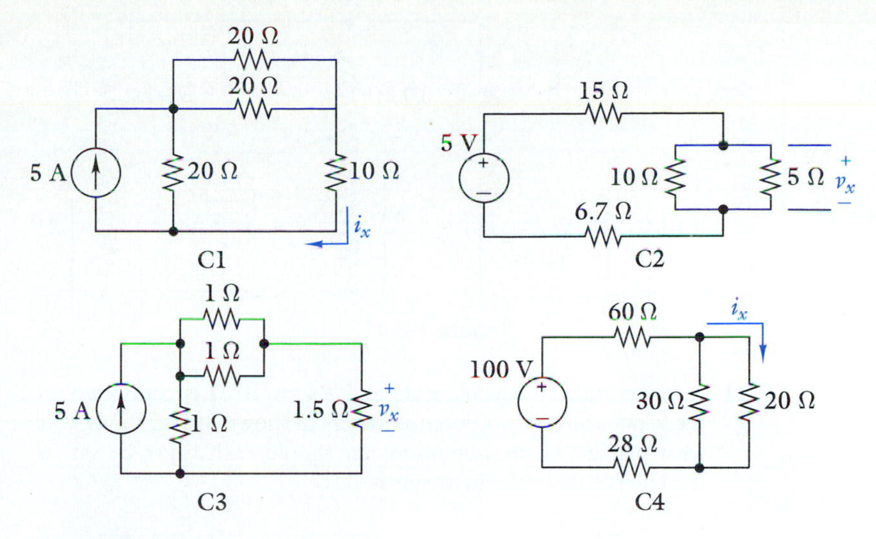

Figure P2-34

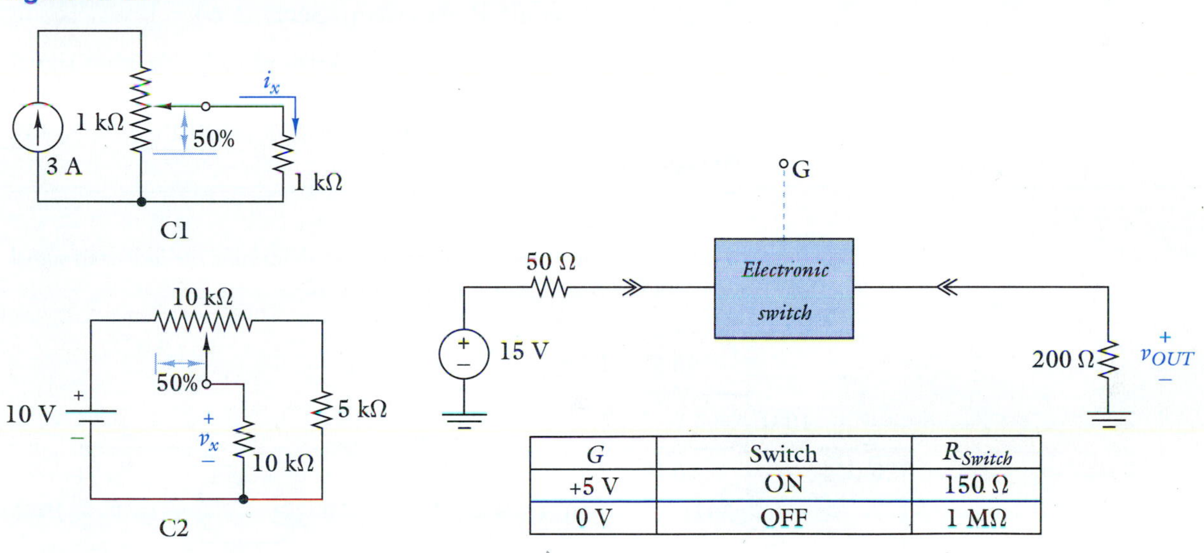

G	Switch	R_{Switch}
+5 V	ON	150 Ω
0 V	OFF	1 MΩ

Figure P2-35

Figure P2-36

2-39 (D)

(a) Design the interface circuit in Fig. P2-39 so the current through the 2-kΩ load is 1 mA.

(b) Design the interface circuit in Fig. P2-39 so the voltage across the 2-kΩ load is 10 V.

(c) Design the interface circuit in Fig. P2-39 so the voltage across the 2-kΩ load is 10 ± 0.5 V using only 1-kΩ ± 5% resistors.

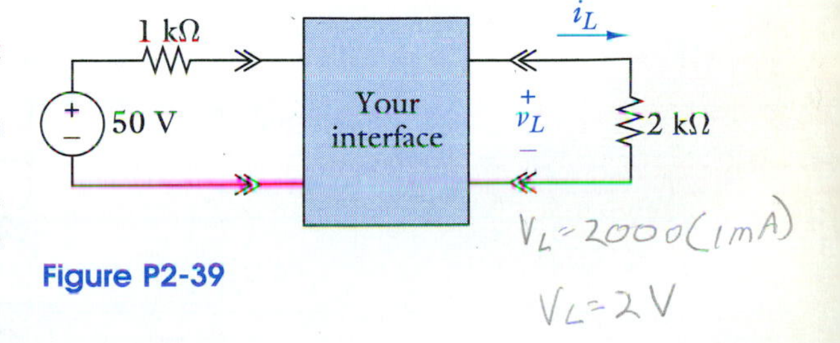

Figure P2-39

2-40 The 1-MΩ potentiometer in Fig. P2-40 is adjusted until the voltage across the 1-MΩ load is 5 V. What is the position of the wiper arm in percent?

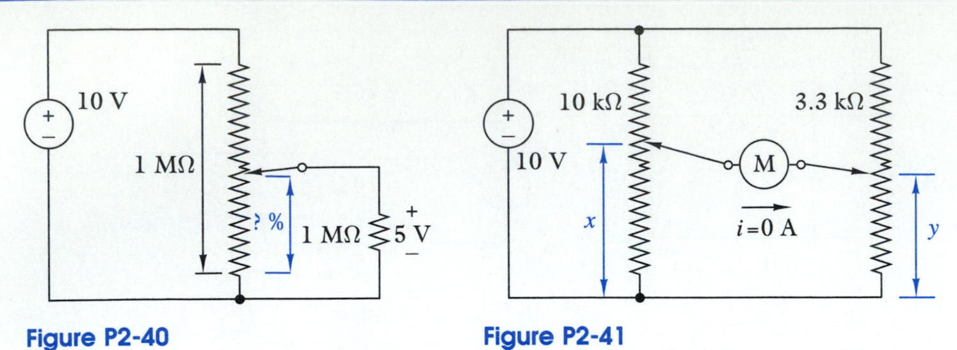

Figure P2-40

Figure P2-41

2-41 A meter with 10 nA full scale and $R_M = 10\ \Omega$ is connected to the wiper arms of two potentiometers as shown in Fig. P2-41. At what percent of the potentiometer should each wiper be set so the current through the meter is zero?

ERO 2-6 Circuit Reduction (Sect. 2-6)

Given a circuit consisting of linear resistors and a constant signal source, find selected signal variables using successive application of series and parallel equivalence, source transformations, and voltage/current division.

2-42 Use circuit reduction techniques to determine the indicated signal variables in circuit C1 of Fig. P2-42.

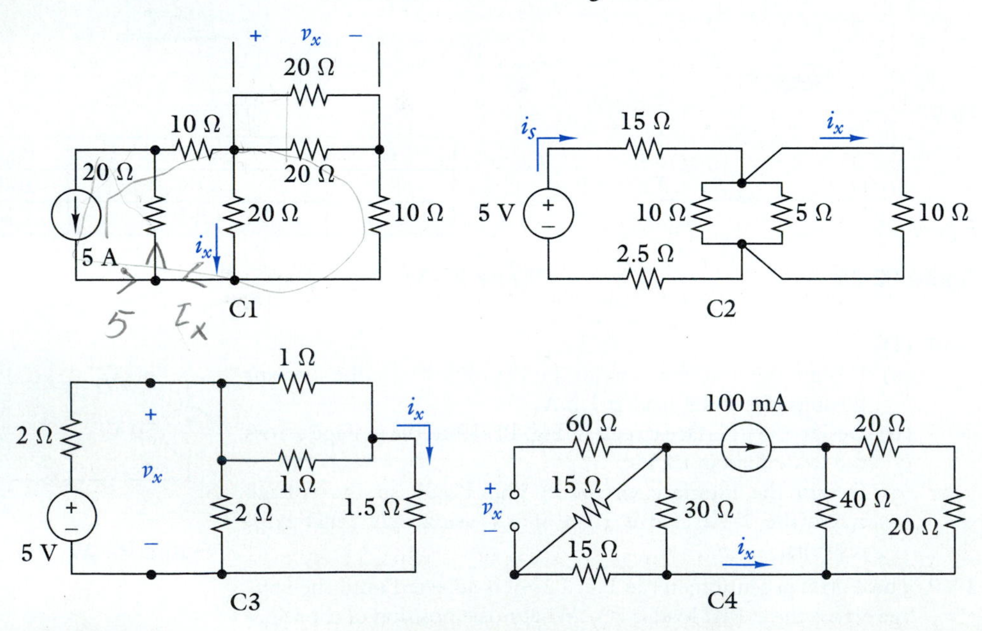

Figure P2-42

2-43 Repeat Problem 2-42 for circuit C2 in Fig. P2-42.

2-44 Repeat Problem 2-42 for circuit C3 in Fig. P2-42.

2-45 Repeat Problem 2-42 for circuit C4 in Fig. P2-42.

2-46 Repeat Problem 2-42 for circuit C1 in Fig. P2-46.

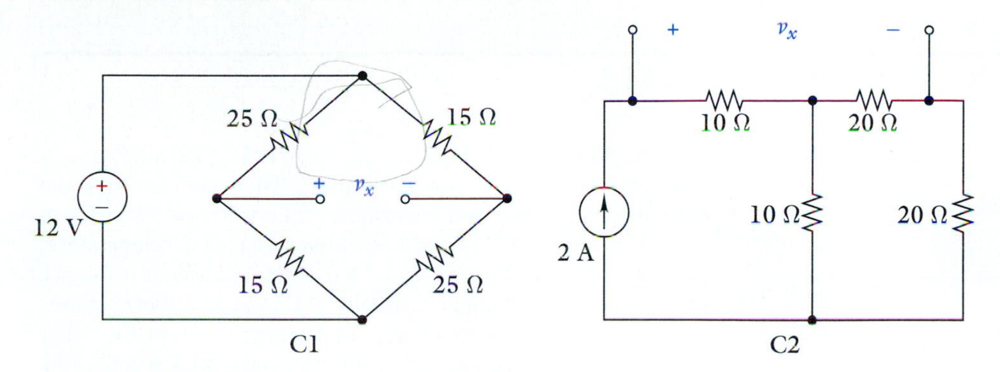

Figure P2-46

2-47 Repeat Problem 2-42 for circuit C2 in Fig. P2-46.

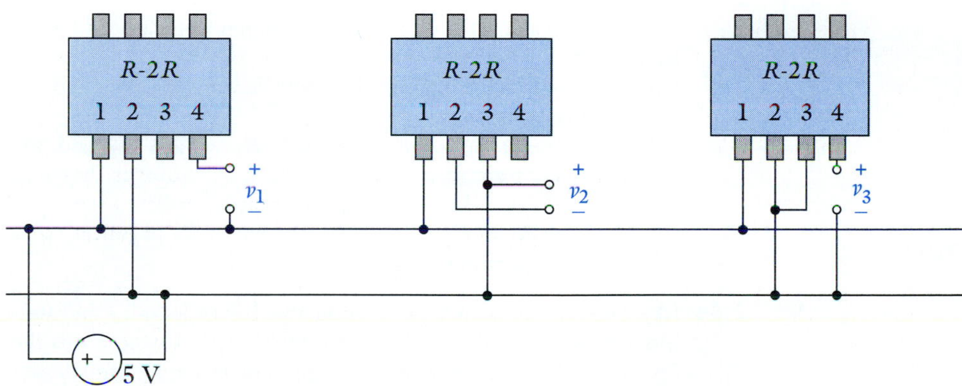

Figure P2-48

2-48 The circuit of Fig. P2-48 contains three of the R-$2R$ resistance arrays described in Problem 2-26 connected to a 5-V bus. Use circuit reduction to determine the voltages v_1, v_2, and v_3.

2-49 Find $v_{D\text{-}B}$, $v_{A\text{-}C}$, and $v_{C\text{-}D}$ in the resistance array circuit in Fig. P2-49.

2-50 Repeat Problem 2-49 when a 50-Ω load resistor is connected between terminals C and D.

2-51 The + terminal of a 5-V source is connected to terminal C of the resistance array in Fig. P2-49 and the − terminal is connected to ground. Terminal B terminal of the array is also connected to ground. Find $v_{A\text{-}D}$.

2-52 A 50-Ω load resistor is connected between terminals A and B of the resistance array in Fig. 2-49. A second 50-Ω load is connected

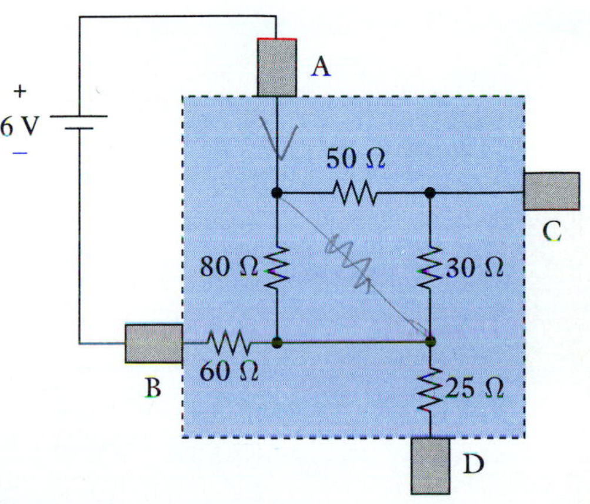

Figure P2-49

between terminals C and D. A 10-V source is connected between terminals B and D with + at terminal D. Find $v_{A\text{-}B}$, $v_{C\text{-}D}$, and the power delivered by the source.

CHAPTER INTEGRATING PROBLEMS

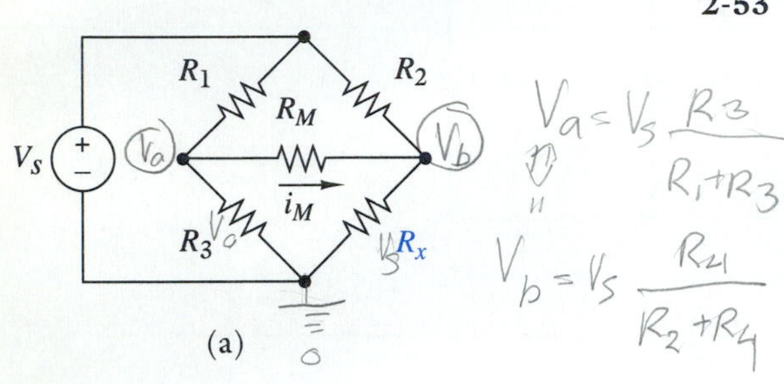

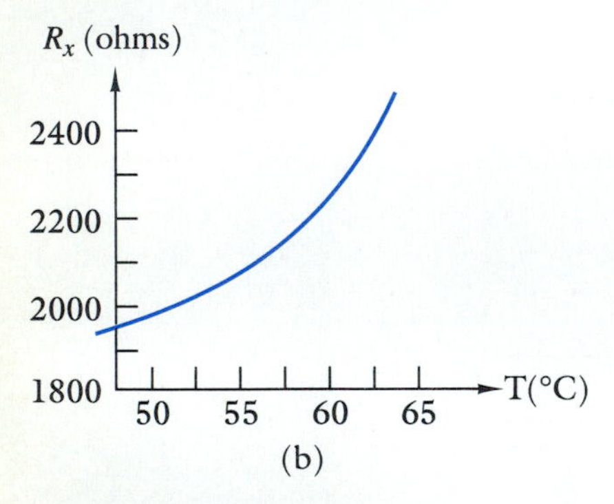

Figure P2-53

2-53 **(A)** The Wheatstone bridge circuit shown in Fig. P2-53(a) occurs extensively in electrical instrumentation. The resistance R_x is the equivalent resistance of a transducer. The value of R_x varies in relation to an external physical phenomenon such as temperature, pressure, or light. The resistance R_M is the equivalent resistance of a measuring instrument, usually a D'Arsonval meter movement. Prior to the start of any measurement one of the other resistors (usually R_3) is adjusted until the current i_M is zero. The resistance of the transducer changes when exposed to the physical phenomenon it is designed to measure. This change causes the bridge to become unbalanced and the meter measures the current through R_M. The deflection of the meter is calibrated to indicate the value of the physical phenomenon measured by the transducer.

(a) **(A)** What is the relationship between R_1, R_2, R_3, and R_x when $i_M = 0$ A?

(b) **(A)** Suppose $R_1 = R_2 = 1$ kΩ, $R_3 = 2.2$ kΩ and the transducer resistance R_x varies with temperature as shown in Fig. P2-53(b). At what temperature is $i_M = 0$?

(c) **(A)** With $R_1 = R_2 = 1$ kΩ, what value of R_3 produces $i_M = 0$ at a temperature of 47.5 °C?

❖ **2-54** **(A), (D)** The resistance of a conductive bar of length l and uniform cross-sectional area A is given by $R = \rho l/A$, where ρ is the resistivity of the material in ohms-cm. For example, the resistivity of Nichrome (a nickel-chromium alloy) is 150×10^{-6} Ω-cm. In hybrid circuits resistors are fabricated by evaporating a thin film of conducting materials through a mask in a vacuum chamber. The conducting film usually has uniform thickness. The film characteristics are specified in ohms per square (Ω/□), since the only dimensions that are easily altered are the length and the width. The designer of a thin-film resistor specifies the thickness (usually between 0.05 and 5 μm) and the material to be evaporated. The material resistivity over the film thickness determines the Ω/□. The designer then lays out a pattern of so many squares to achieve the resistance desired. For example, suppose a certain thickness of a material produces 500 Ω /□. The film with 2 unit lengths and 1 unit width in Fig. P2-54(a) produces a resistance of 1000 Ω between metallic pads A and B. The film with 1 unit length by 2 units width in Fig. P2-54(b) produces a resistance of only 250 Ω between pads C and D.

(a) Using Nichrome and any film thickness between 0.05 and 5 μm, design masks to obtain resistances of 5 Ω, 150 Ω, and 42.5 Ω.

(b) Consider the thin-film circuit in Fig. P2-54(c) with the ohms per square shown. What voltage would be measured between pads B and C when 10 V is applied between A and C? Assume the films are of uniform thickness.

(c) Suppose the voltage measured between B and C is too small. How can the circuit be easily modified without redesigning the masks or reevaporating the film to obtain a larger value of V_{BC}?

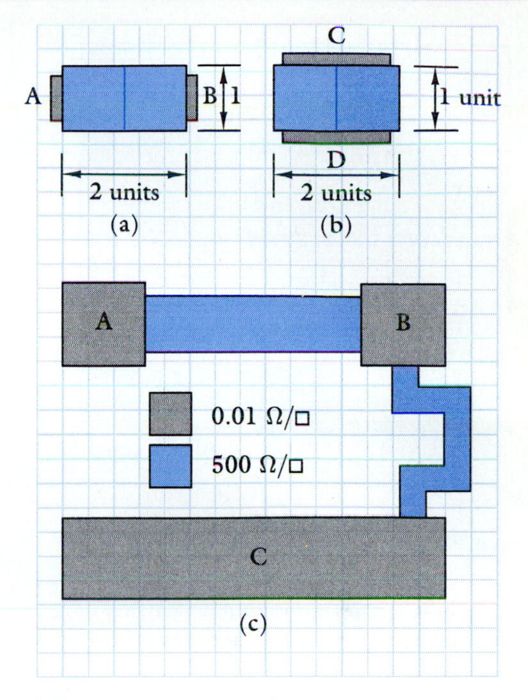

Figure P2-54

2-55 (A) The wheatstone bridge in Fig. P2-55 has $R_1 = R_2 = R_3 = R_{SG} = R_B$. The resistance R_{SG} is a model of a strain gauge, a device whose resistance varies with pressure P.

(a) A small change in pressure ΔP causes a small change in transducer resistance ΔR_{SG}. Derive an express for the resulting small change in the current i_B.

(b) Find Δi_B when $V_s = 10$ V, $R_1 = R_2 = R_2 = R_3 = R_B = R_{SG} = 1$ kΩ, and $\Delta R_{SG} = 1$ Ω.

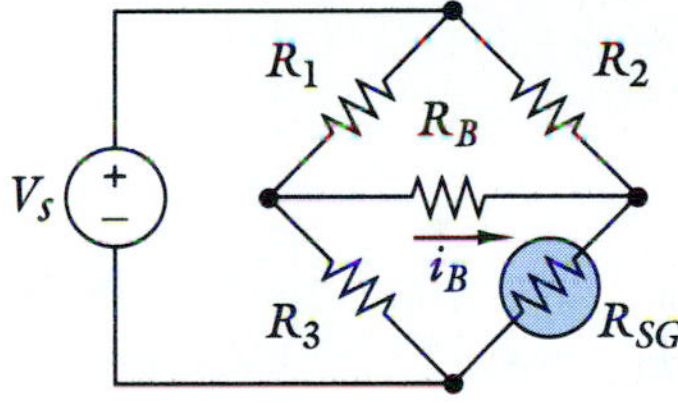

Figure P2-55

2-56 (A) The i-v relationship of a nonlinear resistor is $i = .001v^3$. The device is connected in series with a 10-V voltage source and a 10-Ω resistor. What is the current through and the voltage across each element in the circuit?

2-57 (A) Two resistance arrays are connected as shown in Fig. P2-57. Find the voltage across the 50-Ω load resistor between terminals C and D.

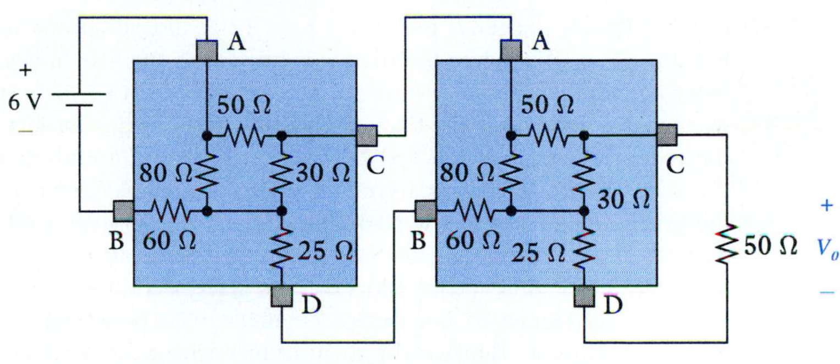

Figure P2-57

2-58 (A) The square grid of equal resistors R shown in Fig. P2-58 extends to infinity in all directions where they are grounded. What will be the equivalent resistance $R_{X\text{-}Y}$ between any two adjacent nodes?

2-59 (A) Incandescent lamps are rated in the number of watts consumed at a specified voltage. If resistance of a 100-W lamp at room temperature is measured and its power consumption computed using v^2/R the answer turns out to be much larger than

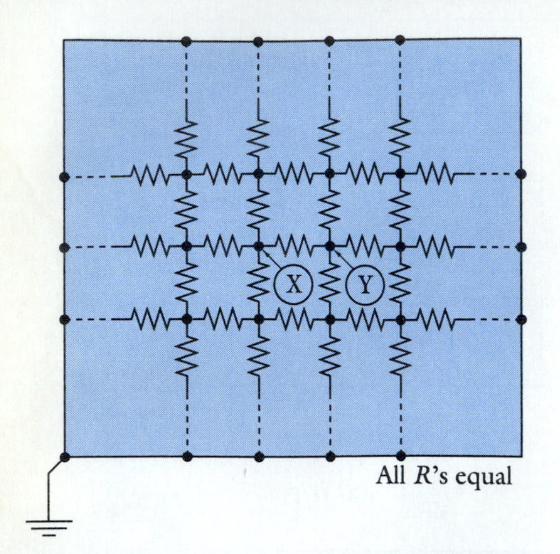

Figure P2-58

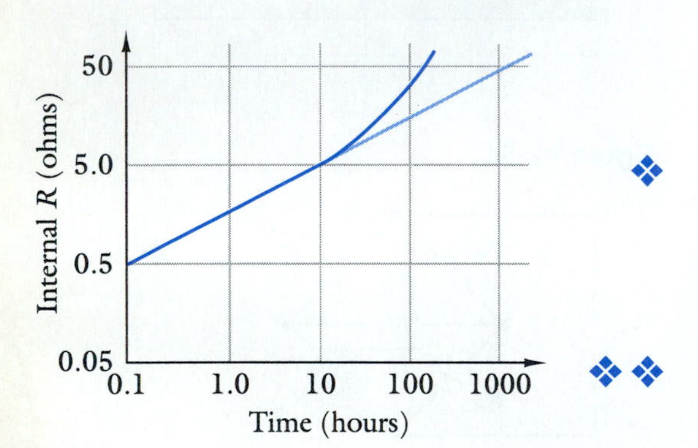

Figure P2-60

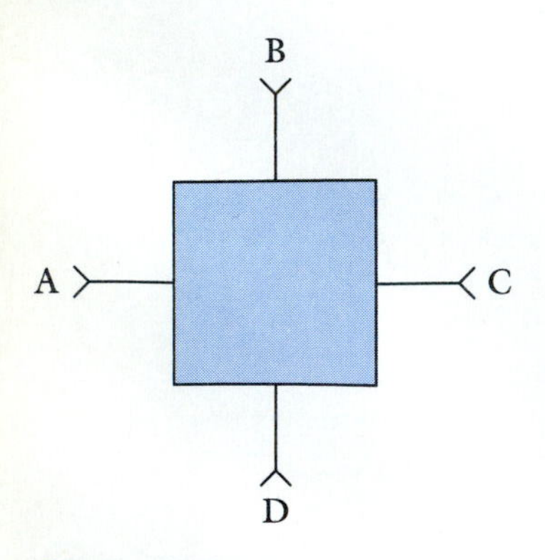

Figure P2-61

100 W. The reason is the resistance of the lamp filament increases dramatically when heated. The temperature coefficient of resistance (α) is defined by

$$\alpha = \frac{\Delta R}{\Delta T R_n}$$

where R_n is the resistance of a material at some known temperature. The resistance at some other temperature is given by $R = R_n + \Delta R$. The resistance of the 100-W tungsten filament at room temperature ($T = 20$ °C) is 12 Ω and $\alpha = 0.00495$ for tungsten. If the lamp consumes 100 W when connected to a 120-V source, how hot is the tungsten filament?

2-60 **(A)** Standard flashlight 1.5-V dry cells contain Zn and MnO_2. As the cell is used, the MnO_2 converts to Mn_2O_3 which has a much higher resistance. The graph in Fig. P2-60 shows the internal resistance of the dry cell versus time.

(a) How much power can the cell deliver to a 100-Ω load when it is new?

(b) How much after 10 hours of use?

(c) How much after 100 hours or use?

❖ **2-61** **(D)** The box in Fig. P2-61 is known to contain only resistors. An ohmmeter is used to measure the resistances between each pair of terminals and the results obtained are $R_{A-B} = 8.33$ kΩ, $R_{A-C} =$ 15 kΩ, $R_{A-D} = 15$ kΩ, $R_{B-C} = 13.3$ kΩ, $R_{B-D} = 13.3$ kΩ, and $R_{C-D} = 0$ kΩ. What is one possible circuit inside the box?

❖❖ **2-62** **(E)** A 1200-W toaster is protected by a 12-A fuse as shown in Fig. P2-62(a). Whenever the toaster is turned on the fuse blows. After some study it was determined the toaster draws more current when it first is turned on than after it has heated up. In fact, the manufacturer, beleaguered by angry customers, found that the current through the toaster varied versus time as shown in the graph of Fig. P2-62(b). Replacing the 12-A fuse with a 30-A or even a 25-A fuse would exceed the safety rating of many home outlet circuits and would not pass safety standards. The manufacturer turned to her design engineer, who proposed the following solution. Replace the existing fuse with a 12-A "slow-blow" fuse whose resistance versus time characteristics are shown in Fig. P2-62(c). Will this solve the problem? Explain.

2-63 **(A)** Starting with the equivalency conditions in Eq. (2-25), derive the Δ-to-Y transformation equations in Eq. (2-26).

2-64 **(A)** Derive the Y-to-Δ transformation equations in Eq. (2-27) using the equivalency conditions in Eq. (2-25) and the Δ-to-Y transformation equations in Eq. (2-26).

❖ **2-65** **(D)** Refer back to Problem 2-6. Design circuits for C1 and C2 that produce the voltages and currents given Fig. P2-6.

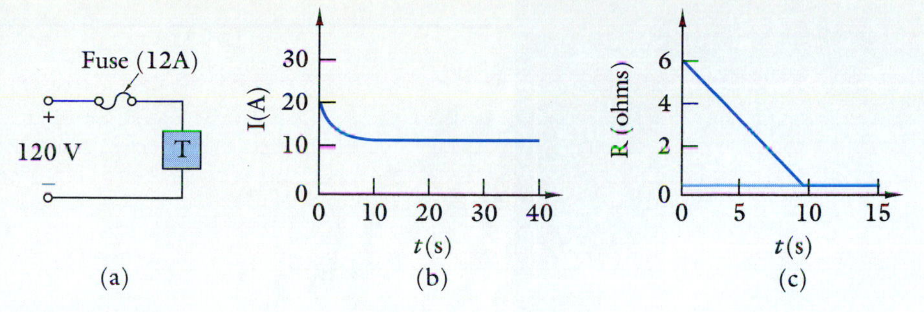

Figure P2-62

2-66 **(E)** Three identical potentiometers are connected as shown in Fig. P2-66. The purpose is to be able to adjust the output voltage v_o with increasing degrees of fineness by sequentially adjusting coarse, medium and fine potentiometers. Derive an expression for v_o/v_i and discuss whether the intended purpose is achieved.

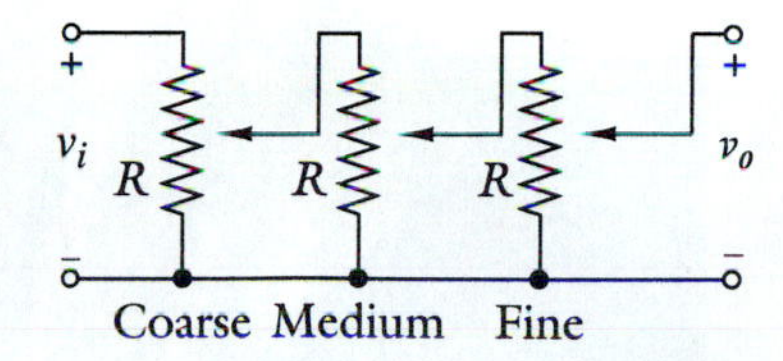

Figure P2-66

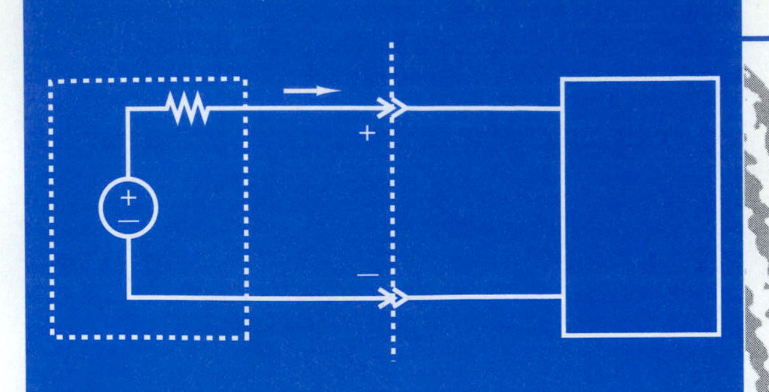

Chapter Three

"Assuming any system of linear conductors connected in such a manner that to the extremities of each one of them there is connected at least one other, a system having electromotive forces $E_1, E_2 \ldots E_3$, no matter how distributed, we consider two points A and A' belonging to the system and having potentials V and V'. If the points A and A' are connected by a wire ABA', which has a resistance r, with no electromotive forces, the potentials of points A and A' assume different values from V and V', but the current i flowing through this wire is given by $i = (V - V')/(r + R)$ in which R represents the resistance of the original wire, this resistance being measured between the points A and A', which are considered to be electrodes."

Leon Charles Thévenin, 1883
French Telegraph Engineer

Circuit Theorems

In this chapter we introduce basic properties of linear circuits that can be stated as theorems, such as Thévenin's theorem. Leon Charles Thévenin (1857–1926) was a distinguished French telegraph engineer and teacher. He was led to his theorem following an extensive study of Kirchhoff's laws. Norton's theorem, which is the dual of Thévenin's theorem, was not proposed until 1926 by Edward L. Norton, an American electrical engineer working on long-distance telephony. Curiously, it turns out that the basic concept had been discovered by Herman von Helmholtz in 1853. Helmholtz's earlier discovery has not been recognized in engineering terminology possibly because Thévenin and Norton both worked in areas of engineering that offered immediate applications for their results, whereas Helmholtz was studying electricity in animal tissue at the time.

Circuit linearity is the foundation of all of the properties and theorems discussed in this chapter. The first two sections present the proportionality and superposition properties, which are the most direct consequences of circuit linearity. These properties are used directly in circuit analysis but, more importantly, provide the basis for developing many other properties of linear circuits. For example, the third section shows superposition leads to Thévenin and Norton's equivalent circuits. These circuits simplify some types of circuit analysis problems and provide a conceptually useful viewpoint for dealing with circuit interfaces. The interface viewpoint leads directly to the maximum power transfer theorem developed in the fourth section. The final section introduces circuit design by showing how these principles are used to create interface circuits that perform in specified ways.

◇ 3-1 Proportionality

The book treats the analysis and design of linear circuits. The hallmark feature of a linear circuit is that signal outputs are linear functions of the inputs. Mathematically a function is said to be linear if it possesses two properties—homogeneity and additivity. In terms of circuit responses, *homogeneity* means the output of a linear circuit is proportional to the input. *Additivity* means the output due to two or more inputs can be found by adding the outputs obtained when each input is applied separately. Mathematically we can write these properties as follows:

$$f(Kx) = Kf(x) \text{ (homogeneity)} \tag{3-1}$$

and

$$f(x_1+x_2) = f(x_1) + f(x_2) \text{ (additivity)} \tag{3-2}$$

where K is a scalar constant. In circuit analysis the homogeneity property is called *proportionality* while the additivity property is called *superposition*. We will study proportionality in this section and superposition in the next section.

The i-v characteristics of a linear resistor, $v = Ri$, shows that if the current (input) is doubled then the voltage (output) is doubled. However, the power delivered is $p = i^2R$. If the current is doubled the power is quadrupled, which is not a proportional change. Only the current and voltage responses of a linear circuit have the proportionality property. Power is equal to the product of current and voltage and is inherently nonlinear even when the circuit is linear.

x
y = Kx
K
input
output

Figure 3-1 Block diagram for scalar multiplier.

For resistive circuits the significance of proportionality is that input-output relationships can be written as

$$y = Kx \tag{3-3}$$

where x is the input (current or voltage), y is an output (current or voltage), and K is a constant. The block diagram shown in Fig. 3-1 describes this relationship. The input x is multiplied by the scalar constant K to produce the output y.

We have already seen several examples of this relationship. For instance, using the voltage division relationship on the circuit in Fig. 3-2(a) produces

$$V_O = \left(\frac{R_2}{R_1 + R_2}\right) V_S$$

which means

$$x = V_S \qquad y = V_O$$

$$K = \frac{R_2}{R_1 + R_2}$$

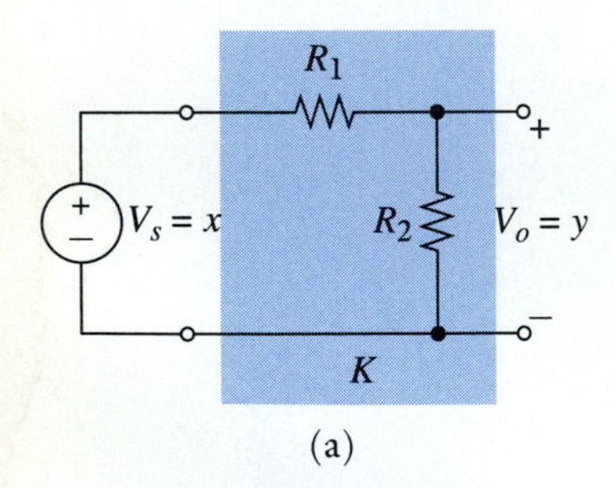

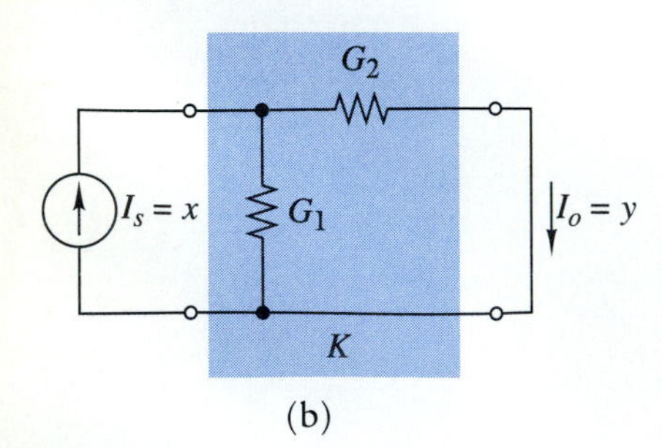

Figure 3-2 Examples of proportionality: (a) Voltage divider. (b) Current divider.

Similarly applying current division to the circuit in Fig. 3-2(b) yields

$$I_O = \left(\frac{G_2}{G_1 + G_2}\right) I_S$$

so that

$$x = I_S \qquad y = I_O$$

$$K = \frac{G_2}{G_1 + G_2}$$

In these two examples the proportionality constant K is dimensionless because the input and output have the same units. In other situations K could carry the units of ohms or siemens when the input or output does not have the same units.

Our first example illustrates finding the input-output relationship for a more complicated circuit and demonstrates that the proportionality constant K can be positive, negative, or even zero.

EXAMPLE 3-1

Given the bridge circuit of Fig. 3-3:

(a) Find the proportionality constant K in the input-output relationship $V_O = KV_s$.

(b) Find the sign of K when $R_2R_3 > R_1R_4$, $R_2R_3 = R_1R_4$, and $R_2R_3 < R_1R_4$.

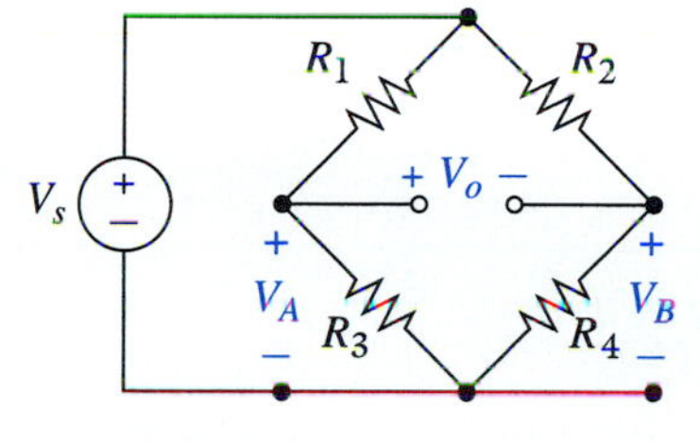

Figure 3-3

SOLUTION:

(a) We observe that the circuit consists of two voltage dividers. Applying voltage division rule to each side of the bridge circuit yields

$$V_A = \frac{R_3}{R_1 + R_3}V_S, \qquad V_B = \frac{R_4}{R_2 + R_4}V_S$$

But KVL allows us to write

$$V_O = V_A - V_B$$

by substituting the equations for V_A and V_B into this KVL equation yields

$$\begin{aligned} V_O &= \left(\frac{R_3}{R_1 + R_3} - \frac{R_4}{R_2 + R_4}\right)V_S \\ &= \left(\frac{R_2R_3 - R_1R_4}{(R_1 + R_3)(R_2 + R_4)}\right)V_S \\ &= (K)V_S \end{aligned}$$

(b) The proportionality constant (K) can be positive, negative, or zero. Specifically

$$\text{If } R_2R_3 > R_1R_4 \text{ then } K > 0$$

$$\text{If } R_2R_3 = R_1R_4 \text{ then } K = 0$$

$$\text{If } R_2R_3 < R_1R_4 \text{ then } K < 0$$

when $K = 0$ the bridge is said to be balanced. This requires the product of the resistances in opposite legs of the two voltage dividers to be equal.

Unit Output Method

The *unit output method* is an analysis technique based on the proportionality property of linear circuits. The method involves finding the input-output proportionality constant K by assuming an output of one unit and determining the input required to produce that unit output. This technique can be applied only to ladder circuits and involves the following steps:

1. A unit output is assumed, that is, $V_O = 1$ V or $I_O = 1$ A.
2. The input required to produce the unit output is then found by successive application of KCL, KVL, and Ohm's law.
3. Because the circuit is linear the proportionality constant relating input and output can be found as

$$K = \frac{\text{Output}}{\text{Input}} = \frac{1}{\text{Input for unit output}}$$

4. The output for any input can then be found as

$$\text{General output} = K \times \text{General input}$$

The unit output method solves the circuit response problem backward, that is, from output to input. The next example illustrates the method.

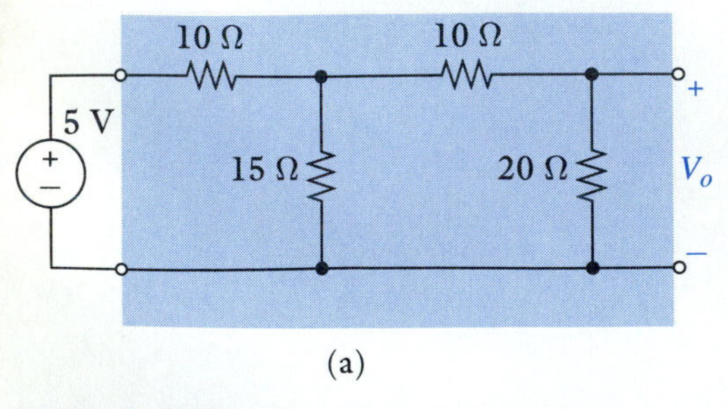

Figure 3-4(a)

EXAMPLE 3-2

Use the unit output method to find V_O in the circuit shown in Fig. 3-4(a).

SOLUTION:

We start by assuming $V_O = 1$ as shown in Fig. 3-4(b). Then using Ohm's law we find I_O

$$I_O = \frac{V_O}{20} = 0.05 \text{ A}$$

Next using KCL at Node B we find I_1

$$I_1 = I_O = 0.05 \text{ A}$$

Again using Ohm's law we find V_1

$$V_1 = 10I_1 = 0.5 \text{ V}$$

Then writing a KVL equation around Loop L2 we can find V_2

$$V_2 = V_1 + V_O = 0.5 + 1.0 = 1.5 \text{ V}$$

Using Ohm's law once more produces

$$I_2 = \frac{V_2}{15} = \frac{1.5}{15} = 0.1 \text{ A}$$

Next writing a KCL equation at Node A yields

$$I_3 = I_1 + I_2 = 0.05 + 0.1 = 0.15 \text{ A}$$

Using Ohm's law one last time

$$V_3 = 10I_3 = 1.5 \text{ V}$$

We can now find the source voltage V_S by applying KVL around Loop L1

$$V_S \Big|_{\text{for } V_O = 1 \text{ V}} = V_3 + V_2 = 1.5 + 1.5 = 3 \text{ V}$$

A source voltage of 3 V is required to produce an output of 1 V. From this result we calculate the proportionality constant K as

$$K = \frac{V_O}{V_S} = \frac{1}{3}$$

Once K has been found we can easily find the output for any given input. For example, the output for a 5-V input is

$$V_O = KV_S = \frac{1}{3} \times 5 = \frac{5}{3} \text{ V}$$

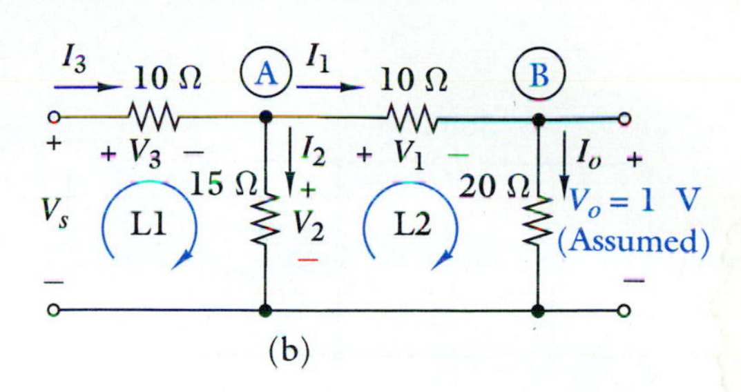

Figure 3-4(b)

■ EXERCISE 3-1

Find V_O in the circuit of Fig. 3-4 when V_S is −5 V, 10 mV, and 3 kV.

ANSWERS:

$V_O = -1.667$ V, 3.333 mV, and 1 kV. ■

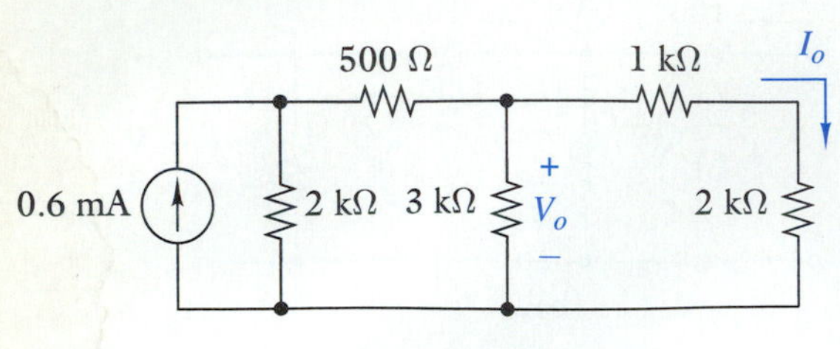

Figure 3-5

EXERCISE 3-2

Use the unit output method to find $K = I_O/I_{IN}$ for the circuit in Fig. 3-5. Then use the proportionality constant K to find I_O for the input current shown in the figure.

ANSWERS:
$K = 1/4$; $I_O = 0.15$ mA.

EXERCISE 3-3

Use the unit output method to find $K = V_O/I_{IN}$ for the circuit in Fig. 3-5. Then use K to find V_O for the input current shown in the figure.

ANSWERS:
$K = 750\ \Omega$; $V_O = 450$ mV.

Note: In this exercise K has the dimensions of ohms because the input is a current and the output a voltage.

3-2 SUPERPOSITION

The superposition principle is a useful computational and conceptual tool used in the analysis of linear circuits with multiple inputs. The input-output relationships of linear circuits possess the additivity property of linear functions. For linear resistance circuits this means we can write any output y as

$$y = K_1x_1 + K_2x_2 + K_3x_3 + \dots \tag{3-4}$$

where $x_1, x_2, x_3, \dots$ are circuit inputs, and $K_1,\ K_2,\ K_3, \dots$ are constants that depend on the circuit. Briefly stated, the output of a linear resistance circuit is a linear combination of the several inputs. Figure 3-6 shows how we can represent this principle using a block diagram. The K's are all scalar multipliers, while the circle (with or without the Σ) represents a summing element.

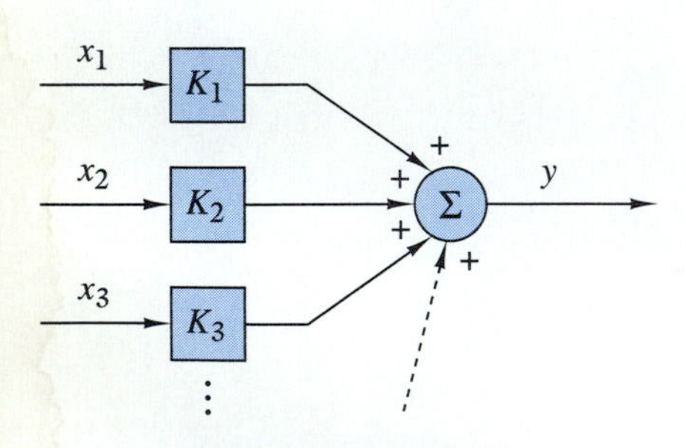

Figure 3-6 Block diagram representing superposition.

To introduce superposition we first determine the output (V_O) of the two-input source circuit in Fig. 3-7(a) using circuit reduction techniques studies in Chapter 2. Then we will use the superposition principle to obtain the same result.

Using circuit reduction we first perform a source transformation to the left of terminals A-B to obtain the circuit shown in Fig. 3-7(b). By redrawing the circuit as shown in Fig. 3-7(c), we observe that the circuit reduces to two current sources and two resistors in parallel. This observation leads to the equivalent circuit in Fig. 3-7(d) where

$$I_{EQ} = \frac{V_S}{R_1} + I_S \qquad \text{and} \qquad R_{EQ} = \frac{R_1R_2}{R_1 + R_2} \tag{3-5}$$

In the final equivalent circuit in Fig. 3-7(d) we see that

$$V_O = I_{EQ} R_{EQ}$$

Using Eq. (3-5) we obtain

$$V_O = \left[\frac{R_2}{R_1 + R_2}\right] V_S + \left[\frac{R_1 R_2}{R_1 + R_2}\right] I_S \qquad (3\text{-}6)$$

$$y = \quad [K_1] \quad x_1 + \quad [K_2] \quad x_2$$

This equation shows that the output is a linear combination of the two inputs. Note that K_1 is dimensionless, since its input and output are voltages and that K_2 has the units of ohms, since its input is a current and output a voltage.

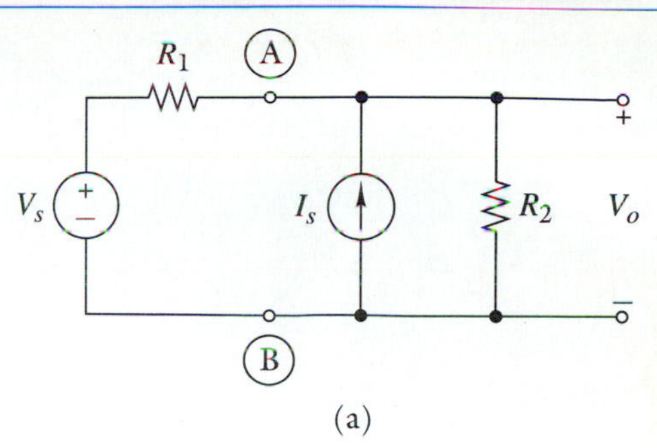

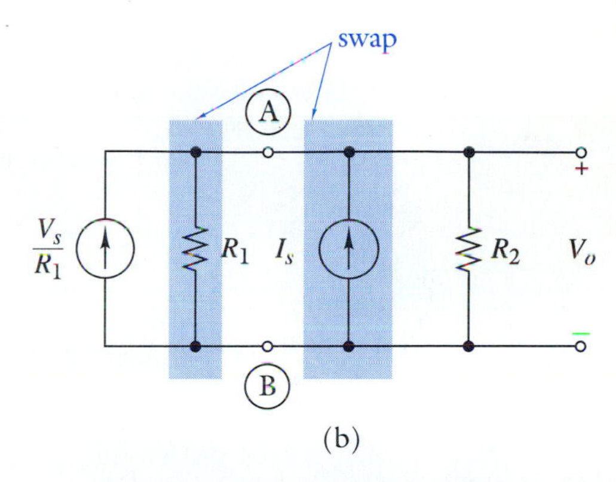

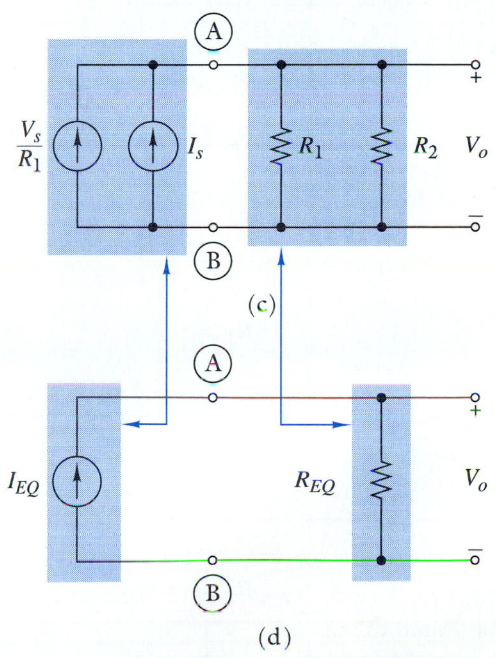

Figure 3-7 Circuit analysis demonstrating superposition: (a) Given circuit. (b) Source transformation. (c) Parallel element reversed. (d) Equivalent circuit.

Superposition Principle

Since the output in Eq. (3-6) is a linear combination, the contribution of each input source is independent of all other inputs. This suggests that the output of a linear circuit with several input sources can be found by the following steps:

STEP 1: "Turn off" all input signal sources except one and find the output of the circuit due to that source acting alone.

STEP 2: Repeat the process in Step 1 until each input source has been turned on and the output due to that source found.

STEP 3: The total output with all sources turned on is then a linear combination (algebraic sum) of the contributions of the individual sources.

These steps outline a circuit analysis technique called *superposition*. The superposition technique will be applied to the circuit of Fig. 3-7 to duplicate the response found by circuit reduction in Eq. (3-6). However, before the technique can be used we must discuss what happens when a voltage or current source is "turned off."

The *i-v* characteristics of voltage and current sources are shown in Fig. 3-8. A voltage source is "turned off" by setting its voltage to zero ($V_S = 0$). This step translates the voltage source *i-v* characteristic to the *i*-axis as shown in Fig. 3-8(a). In Chapter 2 we found that a vertical line on the *i*-axis is *i-v* characteristics of a short circuit. Similarly, "turning off" a current source ($I_S = 0$) translates its *i-v* characteristics to the horizontal axis, which are the *i-v* characteristics of an open circuit. Therefore, when a voltage source is "turned off" we replace it by a short

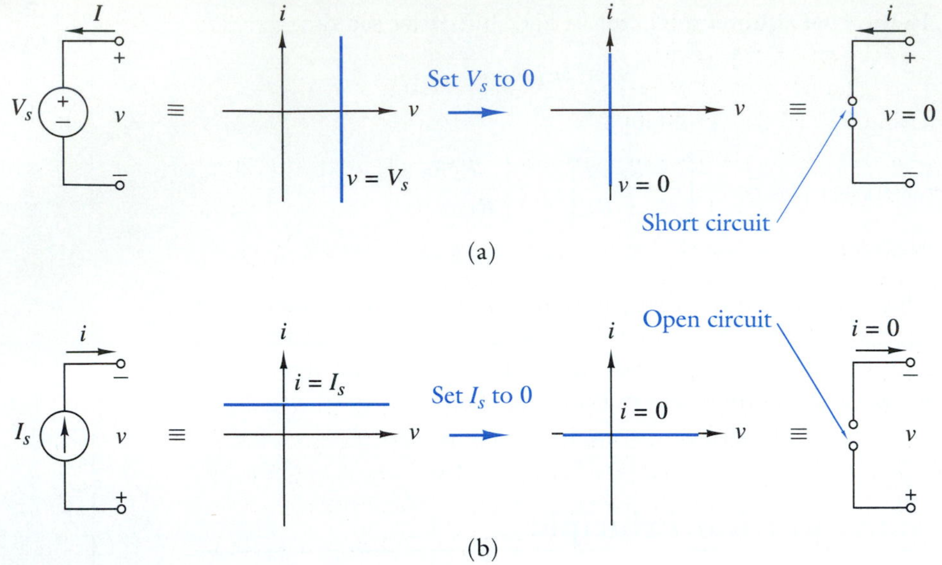

Figure 3-8 Turning off an input signal source: (a) Voltage source. (b) Current source.

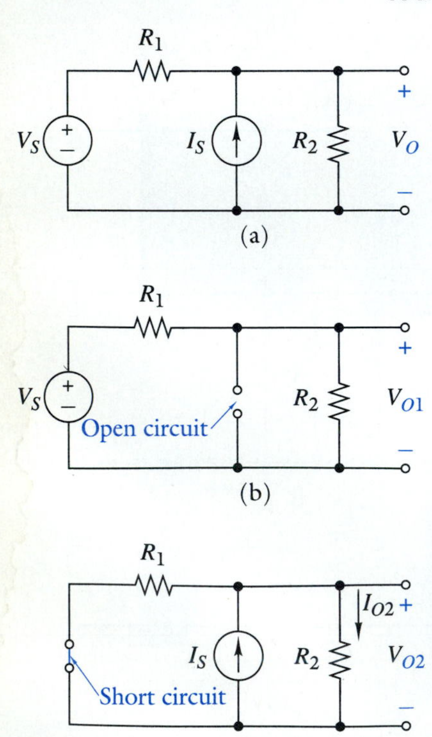

Figure 3-9 Circuit analysis using superposition: (a) Current source off. (b) Voltage source off.

circuit, and when a current source is "turned off" we replace it by an open circuit.

Figure 3-9 shows the steps involved in applying superposition to the circuit in Fig. 3-7. Figure 3-9(a) shows that the circuit has two input sources. We will first "turn off" I_S and replace it with an open circuit as shown in Fig. 3-9(b). It is important to realize that the circuit of Fig. 3-9(b) is not the same circuit as Fig. 3-9(a). The output of the circuit in Fig. 3-9(b) is called V_{O1} and represents that part of the total output caused by the voltage source. Using voltage division in Fig. 3-9(b) yields V_{O1} as

$$V_{O1} = \frac{R_2}{R_1 + R_2} V_S$$

Next the voltage source is "turned off" and the current source is "turned on" as shown in Fig. 3-9(c). Using Ohm's law we get $V_{O2} = I_{O2}R_2$. We use current division to express I_{O2} in terms of I_S to yield V_{O2} as

$$V_{O2} = I_{O2}R_2 = \left[\frac{R_1}{R_1 + R_2} I_S\right] R_2 = \frac{R_1 R_2}{R_1 + R_2} I_S$$

Applying the superposition principle we find the response with both sources "turned on" by adding the two responses V_{O1} and V_{O2}.

$$V_O = V_{O1} + V_{O2} = \left[\frac{R_2}{R_1 + R_2}\right] V_S + \left[\frac{R_1 R_2}{R_1 + R_2}\right] I_S$$

The superposition result above is the same as the circuit reduction result given in Eq. (3-6).

The next example serves to further illustrate superposition.

EXAMPLE 3-3

Figure 3-10(a) shows a resistance circuit used to implement a signal-summing function. Use superposition to show that the output V_O is a weighted sum of the inputs V_{S1}, V_{S2}, and V_{S3}.

SOLUTION:

To determine V_O using superposition we first turn off sources 1 and 2($V_{S1} = 0$ and $V_{S2} = 0$) to obtain the circuit in Fig. 3-10(b). This circuit is a voltage divider in which the output leg consists of two equal resistors in parallel. The equivalent resistance of the output leg is $R/2$ so the voltage division rule yields

$$V_{O3} = \frac{R/2}{R + R/2} V_{S3} = \frac{V_{S3}}{3}$$

Because of the symmetry of the circuit it can be seen that the same technique applies to all three inputs; therefore:

$$V_{O2} = \frac{V_{S2}}{3} \text{ and } V_{O1} = \frac{V_{S1}}{3}$$

Applying the superposition principle the output with all sources "turned on" is

$$V_O = V_{O1} + V_{O2} + V_{O3}$$

$$= \frac{1}{3}[V_{S1} + V_{S2} + V_{S3}]$$

That is, the output is proportional to the sum of the three input signals. Figure 3-11 shows this relationship in block diagram form.

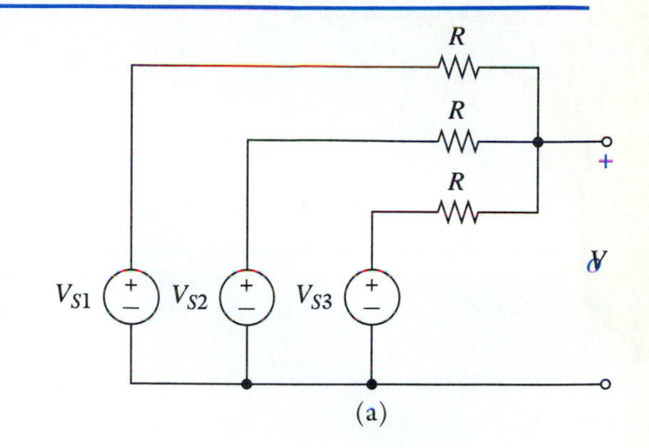

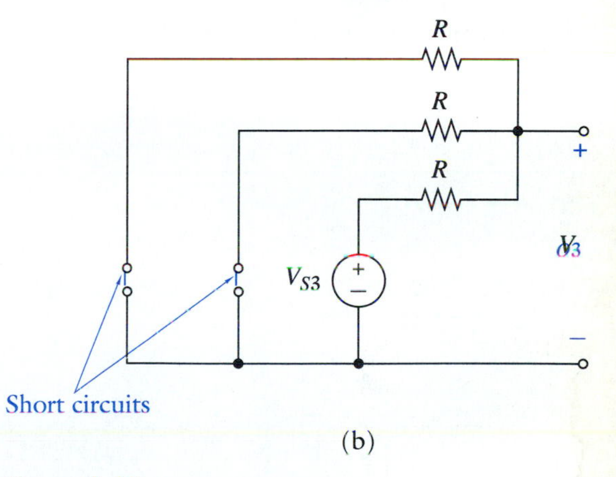

Figure 3-10

EXERCISE 3-4

The circuit of Fig. 3-12 contains two of the R-$2R$ modules discussed in Problem 2-26. Use superposition to find V_O.

ANSWER:
$V_O = 1/2V_{S1} + 1/4V_{S2}$.

EXERCISE 3-5

Repeat Exercise 3-4 with the voltage source V_{S2} replaced by a current source I_{S2} with the current reference arrow directed toward ground.

ANSWER:
$V_O = 3V_{S1}/5 - 4I_{S2}R/5$.

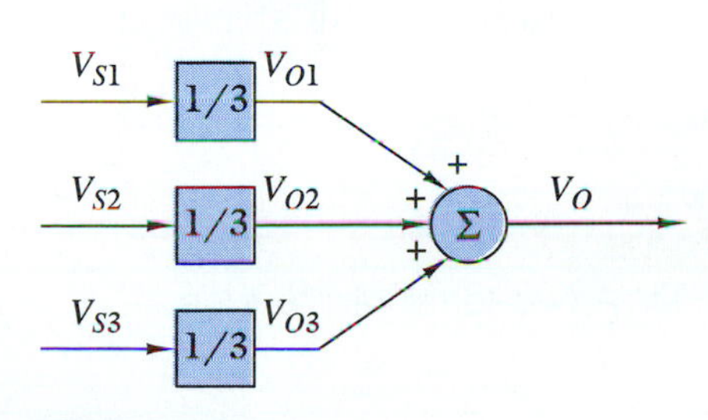

Figure 3-11

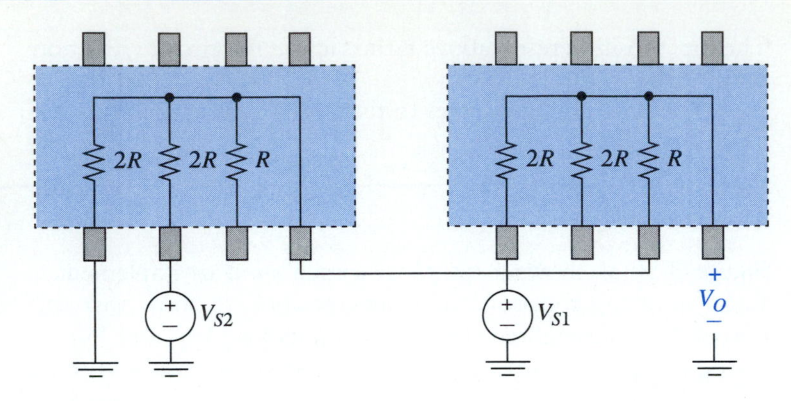

Figure 3-12

The examples and exercises above illustrate applying the superposition principle to analyze linear circuits containing several input signal sources. The reader should not conclude that this is the primary application of the concept. In fact, superposition is not a particularly attractive method, since a circuit with N signal sources requires N different circuit analyses to obtain the final result. Unless the circuit is relatively simple, superposition may not reduce the analysis effort compared with, say, circuit reduction. Superposition is used primarily as a conceptual tool in the development of other circuit analysis and design techniques. For example, superposition is used in the next section to prove Thévenin's theorem.

3-3 THÉVENIN AND NORTON EQUIVALENT CIRCUITS

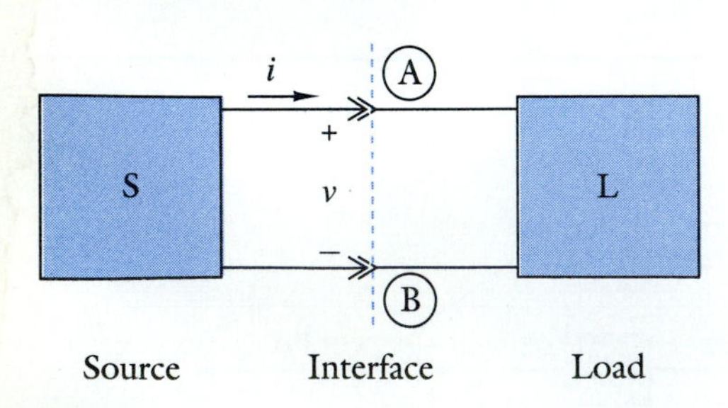

Figure 3-13 A two-terminal interface.

An interface is a connection between two or more circuits that perform different functions. Circuit interfaces occur frequently in electrical and electronic systems so special analysis methods are used to handle them. For the two-terminal interface shown in Fig. 3-13, we normally think of one circuit as the source S and the other as the load L. We think of signals as being produced by the source circuit and delivered to the load. The source-load interaction at an interface is one of the central problems of circuit analysis and design.

The Thévenin and Norton equivalent circuits shown in Fig. 3-14 are valuable tools for dealing with interfaces. The conditions under which these equivalent circuits exist can be stated as a formal theorem:

If the source circuit in a two-terminal interface is linear, then the interface signals v and i do not change when the source circuit is replaced by its Thévenin or Norton equivalent circuit.

The equivalences require the source circuit to be linear, but places no restriction on the linearity of the load circuit. Later in this section we will treat cases in which the load is nonlinear. In subsequent chapters we will study circuits in which the load consists of elements called capacitance and inductance, which are linear models of devices that store energy.

The Thévenin equivalent circuit consists of a voltage source (V_T) in series with a resistance (R_T). The Norton equivalent circuit is a current source (I_N) in parallel with a resistance (R_N). The Thévenin and Norton equivalent circuits are equivalent to each other, since replacing one by the other leaves the interface signals unchanged. Therefore, they must have the same i-v characteristics.

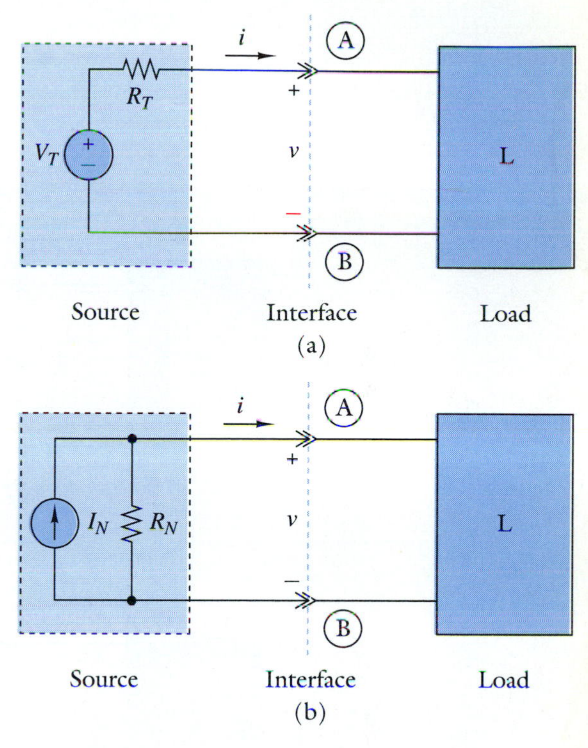

Figure 3-14 (a) Thévenin equivalent circuit. (b) Norton equivalent circuit.

To derive the equivalency conditions we apply KVL and Ohm's law to the Thévenin equivalent in Fig. 3-14(a) to obtain its i-v characteristics at the terminals A-B as

$$v = V_T - iR_T \tag{3-7}$$

Next applying KCL and Ohm's law to the Norton equivalent in Fig. 3-14(b) yields its i-v characteristics at terminals A-B as

$$i = I_N - \frac{v}{R_N} \tag{3-8}$$

Solving Eq. (3-8) for v yields

$$v = I_N R_N - iR_N \tag{3-9}$$

The Thévenin and Norton circuits must have identical i-v characteristics. By comparing Eqs. (3-7) and (3-9) we conclude that

$$\begin{aligned} R_N &= R_T \\ I_N R_N &= V_T \end{aligned} \tag{3-10}$$

In essence the Thévenin and Norton equivalent circuits are related by the source transformation studied in Chapter 2. We do not need to independently find both equivalent circuits. Once one of them is found the other can be determined by a source transformation. In other words, the two equivalent circuits involve four parameters (V_T, R_T, I_N, R_N) and Eq. (3-10) provides two relations between the four parameters. Therefore, only two parameters are needed to determine either equivalent circuit.

Conceptually, the two parameter conditions are easily obtained using open-circuit and short-circuit loads. That is, if the actual load is disconnected from the source as shown in Fig. 3-15(a), then an open-circuit voltage v_{OC} appears between terminals A and B. Connecting an open-circuit load to the Thévenin equivalent reveals that $V_T = v_{OC}$, since current is zero and there is no voltage across R_T. Similarly, disconnecting the load and connecting a short-circuit as shown in Fig. 3-15(b) produce a current i_{SC}. Connecting a short-circuit load to the Norton equivalent requires $I_N = i_{SC}$, since by current division all of the source current I_N is diverted through the short-circuit load.

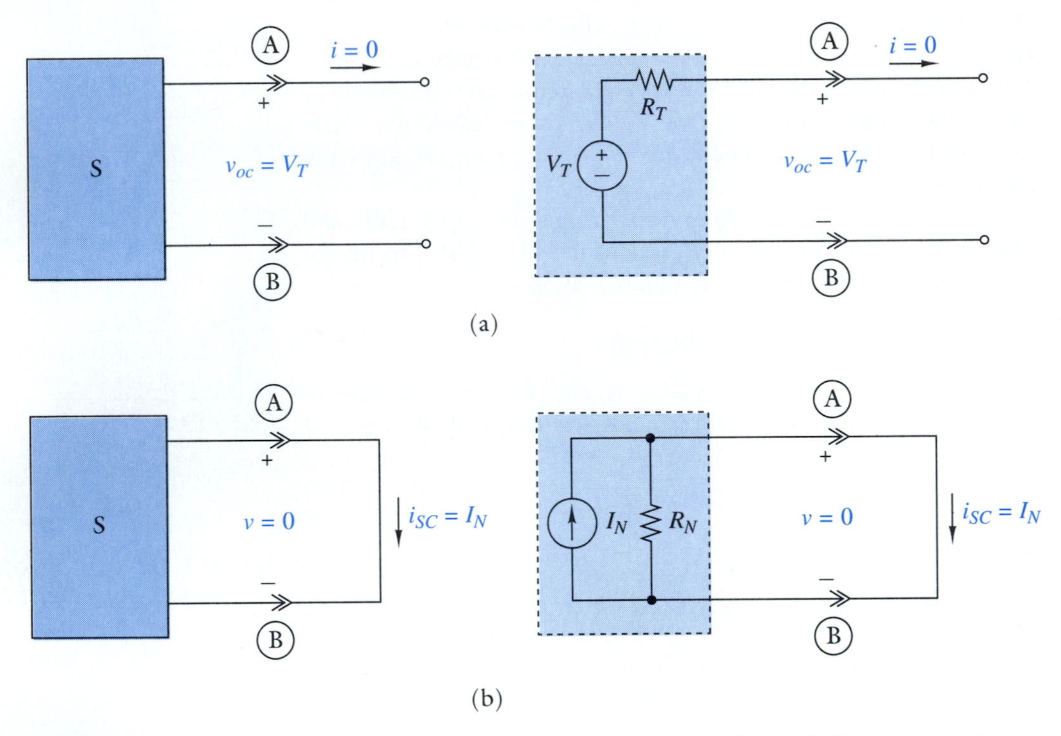

Figure 3-15 Finding Thévenin and Norton equivalent circuits: (a) Thévenin voltage. (b) Norton current.

In summary, if we find the open-circuit voltage and the short-circuit current, we can determine the Thévenin and Norton equivalent circuit parameters as

$$
\begin{aligned}
V_T &= v_{OC} \\
I_N &= i_{SC} \\
R_N = R_T &= \frac{v_{OC}}{i_{SC}}
\end{aligned}
\tag{3-11}
$$

Applications of Thévenin and Norton Equivalent Circuits

Consider the circuit of Fig. 3-16(a). Suppose we need to select a load resistance so the source circuit to the left of the interface A-B delivers 4 volts to the load. This task is easily handled once we have the Thévenin or Norton equivalent for the circuit to the left of interface A-B.

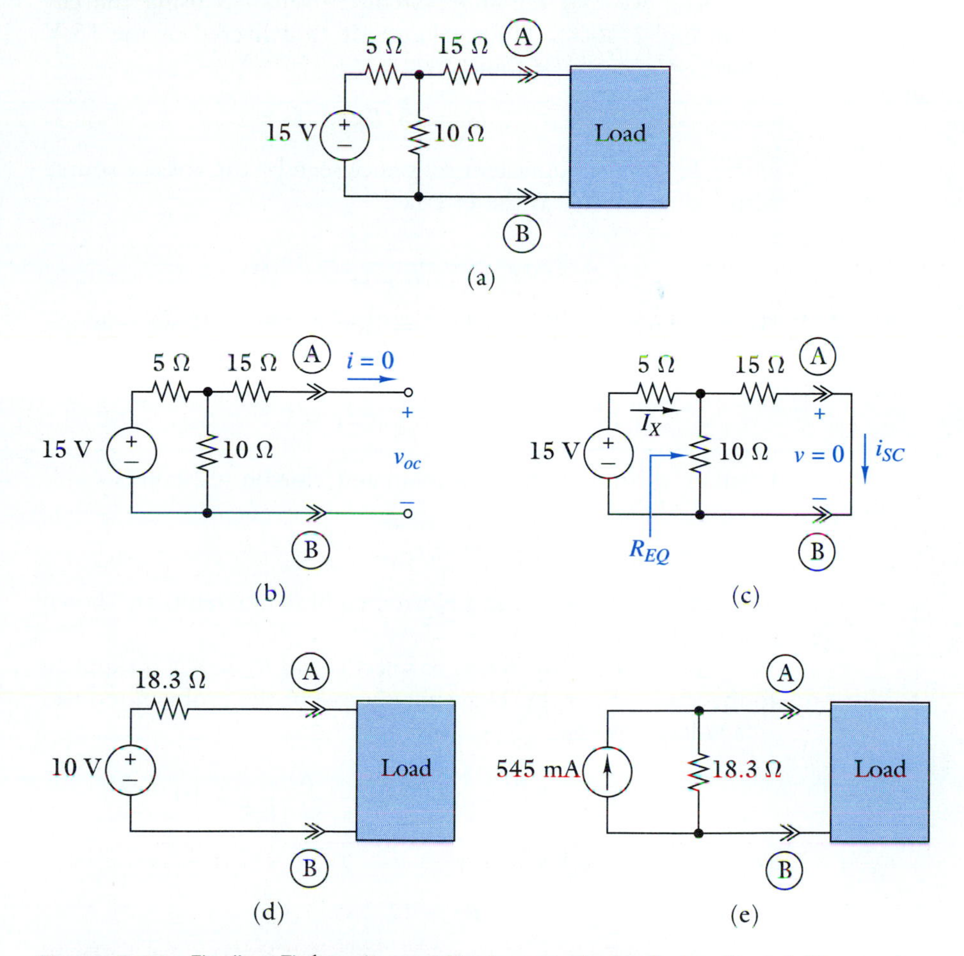

Figure 3-16 Finding Thévenin and Norton equivalent circuits: (a) Given circuit. (b) Open-circuit voltage. (c) Short-circuit current. (d) Thévenin equivalent. (e) Norton equivalent.

To obtain the Thévenin and Norton equivalents we need v_{OC} and i_{SC}. The open-circuit voltage v_{OC} is found by disconnecting the load at the terminals A-B as shown in Fig. 3-16(b). The voltage across the 15-Ω resistor is zero because the current through it is zero due to the open circuit. The open-circuit voltage at the interface is the same as the voltage across the 10-Ω resistor. Using voltage division this voltage is

$$\begin{aligned} V_T &= v_{OC} \\ &= \frac{10}{10+5} \times 15 \\ &= 10\,\text{V} \end{aligned}$$

Next we find the short-circuit current i_{SC} using the circuit in Fig. 3-16(c). The total current I_X delivered by the 15-V voltage source can be found as

$$I_X = 15/R_{EQ}$$

where R_{EQ} is the equivalent resistance seen by the voltage source with a short-circuit at the interface.

$$R_{EQ} = 5 + \frac{10 \times 15}{10+15} = 11\,\Omega$$

We now find I_X as $I_X = 15/11 = 1.36$ A. Given I_X we now use current division to obtain the short-circuit current as

$$I_N = i_{SC} = \frac{10}{10+15} \times I_X = 0.545\ \text{A}$$

Finally, we compute the Thévenin and Norton resistances

$$R_T = R_N = \frac{v_{OC}}{i_{SC}} = 18.3\,\Omega$$

The resulting Thévenin and Norton equivalent circuits are shown in Figs. 3-16(d) and (e).

It now is an easy matter to select a load R_L so 4 V is supplied to the load. Using the Thévenin equivalent the problem reduces to a voltage divider.

$$4\ V = \frac{R_L}{R_L + R_T} \times V_T = \frac{R_L}{R_L + 18.3} \times 10$$

Solving for R_L yields

$$R_L = 12.2\,\Omega$$

The following example shows another application of Thévenin's theorem.

EXAMPLE 3-4

Find the voltage delivered to the load resistor in Fig. 3-17 using Thévenin's theorem.

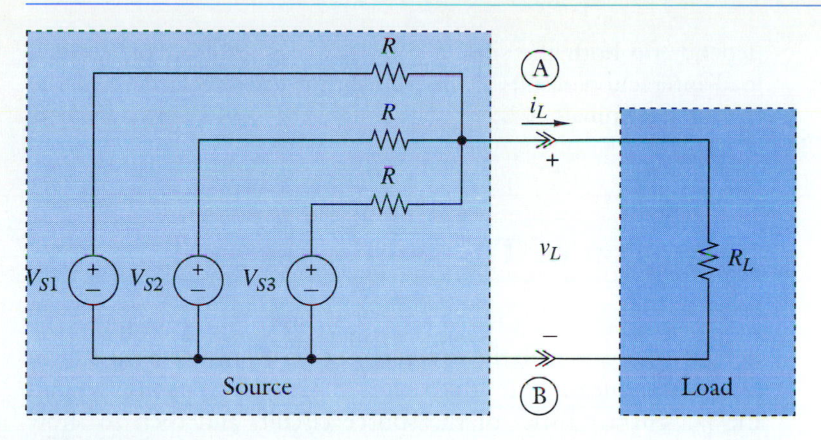

Figure 3-17

SOLUTION:

A source-load interface is first defined at the terminals A-B in Fig. 3-17. In Example 3-3 the open-circuit voltage at terminals A-B (with R_L disconnected) was found using superposition.

$$V_T = v_{OC} = \frac{1}{3}(V_{S1} + V_{S2} + V_{S3})$$

With a short circuit connected between terminals A-B, the short-circuit current is easily found by superposition. With Sources 2 and 3 turned off, the short-circuit current due to Source 1 is seen to be

$$i_{SC1} = \frac{V_{S1}}{R}$$

Because of the symmetry of the circuit we see that same approach applies to all three inputs. By superposition then

$$\begin{aligned} i_{SC} &= i_{SC1} + i_{SC2} + i_{SC3} \\ &= \frac{V_{S1} + V_{S2} + V_{S3}}{R} \end{aligned}$$

We now find R_T as

$$R_T = \frac{v_{OC}}{i_{SC}} = \frac{(1/3)\,(V_{S1} + V_{S2} + V_{S3})}{(1/R)\,(V_{S1} + V_{S2} + V_{S3})} = \frac{R}{3}$$

The resulting Thévenin equivalent circuit is shown in Fig. 3-18.

To calculate the voltage delivered to the load in Fig. 3-18 we use voltage division to write

$$v_L = \left(\frac{R_L}{R_L + R/3}\right) \times V_T = \left(\frac{R_L}{3R_L + R}\right)(V_{S1} + V_{S2} + V_{S3})$$

The voltage across the load is proportional to the sum of the input voltages. The proportionality constant $K = R_L/(3R_L + R_T)$

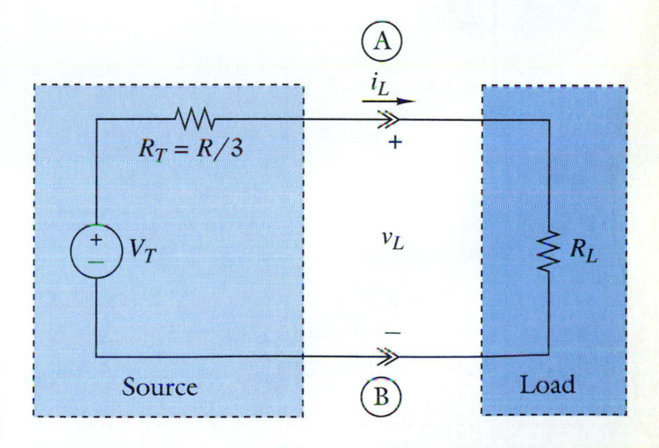

Figure 3-18

depends on both the source and load resistances. This source-load interaction results from connecting the two circuits at the interface terminals A-B.

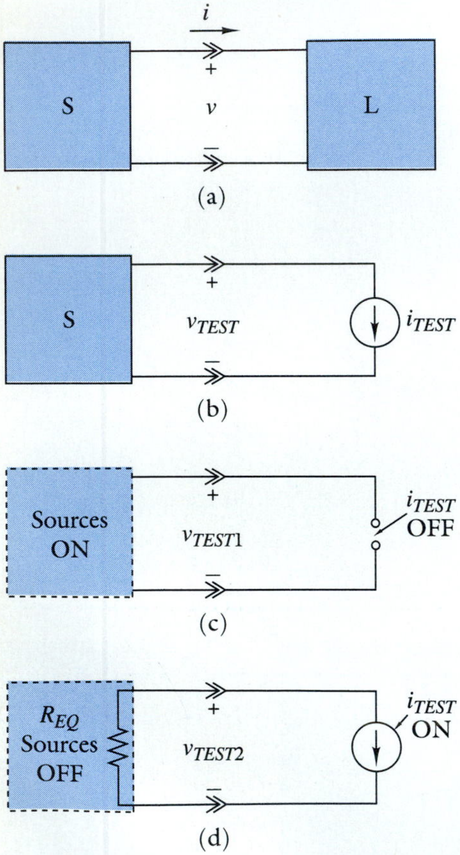

Derivation of Thévenin's Theorem

Thévenin's theorem is based on the superposition principle. The derivation begins with the circuit in Fig. 3-19(a) where the source circuit S is linear. Our approach is to use superposition to find the i-v characteristics of the source circuit, and then to show they are the same as the i-v characteristics of the Thévenin circuit. We first disconnect the load and apply a current source i_{TEST} as shown in Fig. 3-19(b). To find v_{TEST} using superposition we turn i_{TEST} off and leave all of the sources inside S on as in Fig. 3-19(c). Turning a current source off leaves an open circuit, so that

$$v_{TEST1} = v_{OC}$$

Next we turn i_{TEST} back on and all of the sources inside S off. But since S is linear when all internal sources are turned off the source circuit reduces to an equivalent resistance R_{EQ} as shown in Fig. 3-19(d). Using Ohm's law we write

$$v_{TEST2} = -i_{TEST} R_{EQ}$$

The minus sign in this equation results from the reference directions originally assigned to i_{TEST} and v_{TEST} in Fig. 3-19(b). Now using superposition we obtain the i-v characteristics of the source circuit as

$$\begin{aligned} v_{TEST} &= v_{TEST1} + v_{TEST2} \\ &= v_{OC} - i_{TEST} R_{EQ} \end{aligned}$$

This result has the same form as the i-v characteristics of the Thévenin equivalent in Eq. (3-7) with $v_{OC} = V_T$ and $R_T = R_{EQ}$. *QED*.

The derivation points out an alternative and often easier method of determining the Thévenin resistance. As indicated in Fig. 3-20, the source circuit reduces to an interconnection of linear resistors when voltage sources are replaced by short circuits and current sources by open circuits. Under these conditions the source circuit has some equivalent resistance (R_{EQ}) between the terminal A-B. Applying the same condition ($V_T = 0$) to the Thévenin equivalent shows the two circuits are equivalent if

$$R_T = R_{EQ} \tag{3-12}$$

Figure 3-19 Using superposition to prove Thévenin's theorem.

In other words, we can find R_T by determining the resistance seen looking back into the source circuit with all sources turned off. For this reason R_T is sometimes called the *lookback resistance*

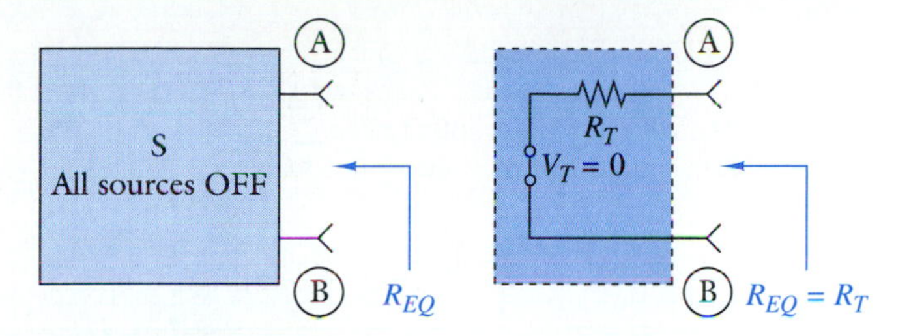

Figure 3-20 Lookback method of finding R_T.

In summary, any two of the following parameters can be used to determine Thévenin/Norton equivalent circuits:

- The open-circuit voltage at the interface
- The short-circuit current at the interface
- The equivalent source circuit resistance at the interface with all sources turned off

All three of these quantities are determined with the load at the interface disconnected. That is, the Thévenin and Norton equivalent circuits characterize the source circuit and do not depend on the load.

EXAMPLE 3-5

Find the Thévenin resistance of the source circuit in Fig. 3-17 using the lookback resistance method.

SOLUTION:

With all the voltage sources turned off in the given circuit we obtain the circuit in Fig. 3-21. Looking back into the circuit at terminals A-B we see three equal resistances connected in parallel. Since the equivalent resistance of three equal resistors in parallel is $R/3$, we obtain the Thévenin resistance as

$$R_T = R_{EQ} = \frac{R}{3}$$

This expression for R_T is the same as we obtained in Example 3-4 using v_{OC} and i_{SC}, but the lookback resistance method is somewhat easier in this circuit.

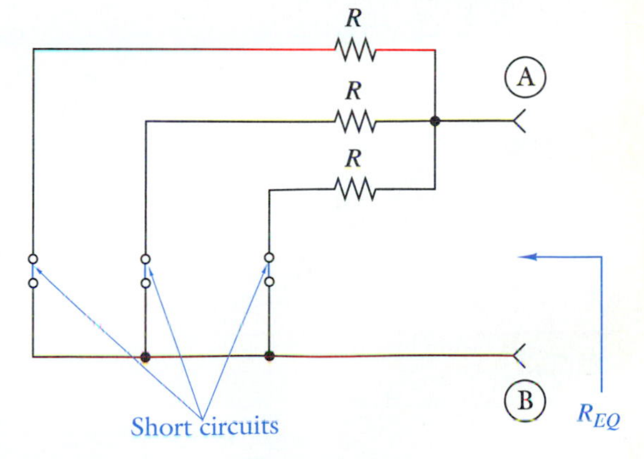

Figure 3-21

EXERCISE 3-6

(a) Find the Thévenin and Norton equivalent circuits seen by the load in Fig. 3-22.

(b) Find the voltage, current, and power delivered to a 50-Ω load resistor.

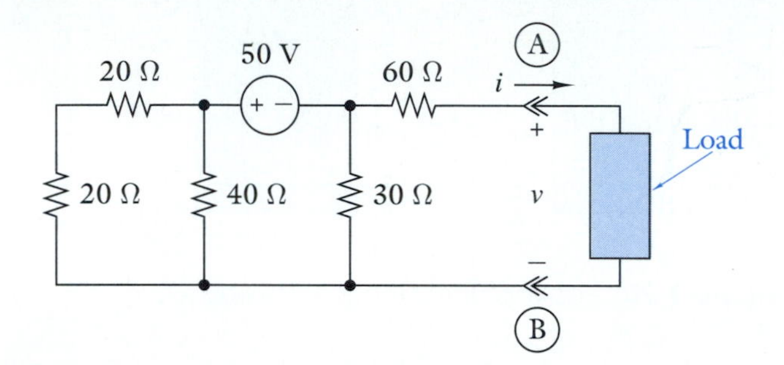

Figure 3-22

ANSWERS:

(a) $V_T = -30$ V, $I_N = -417$ mA, $R_N = R_T = 72$ Ω.

(b) $v = -12.3$ V, $i = -246$ mA, $p = 3.03$ W.

EXERCISE 3-7

Find the current and power delivered to an unknown load in Fig. 3-22 when $v = +6$ V.

ANSWERS:

$i = -\frac{1}{2}$ A; $p = -3$ W.

Application to Nonlinear Loads

An important use of Thévenin and Norton equivalent circuits is finding the voltage across, current through, and power dissipated in a two-terminal nonlinear element (NLE). The approach is a straightforward application of i-v characteristics and proceeds as follows. An interface is defined at the terminals of the nonlinear element and the linear part of the circuit is reduced to a Thévenin equivalent as indicated in Fig. 3-23(a). The i-v characteristics of the Thévenin equivalent are

$$i = -\left(\frac{1}{R_T}\right)v + \left(\frac{V_T}{R_T}\right) \tag{3-13}$$

This is the equation of a straight line in the i-v plane as shown in Fig. 3-23(b). The straight line intersects the i-axis ($v = 0$) at $i = i_{SC} = I_N$ and the v-axis ($i = 0$) at $v = v_{OC} = V_T$. This line could logically be called the source line, since it is determined by the source parameters. Logic notwithstanding, electrical engineers call this the *load line* for reasons that have blurred with the passage of time.

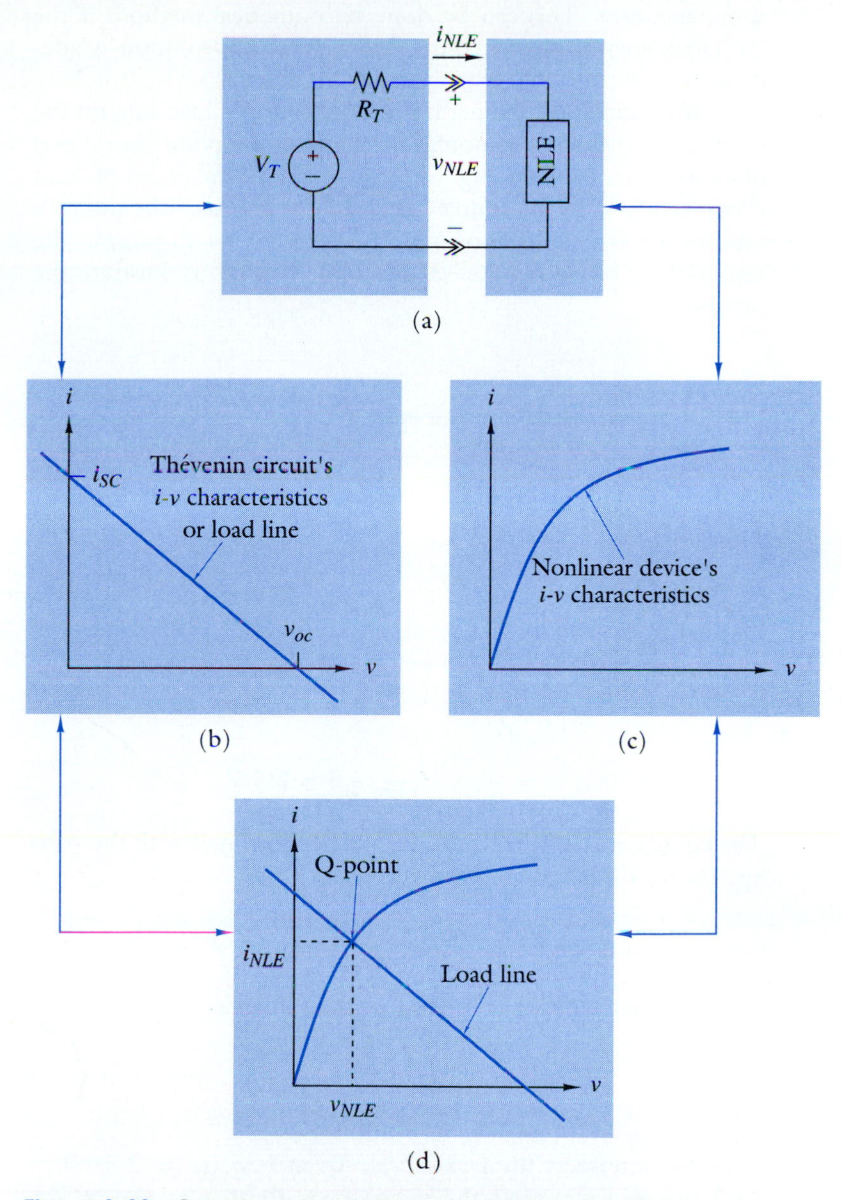

Figure 3-23 Graphical solution of a nonlinear circuit: (a) Given circuit. (b) Load line. (c) NLE *i-v* characteristics. (d) Q-point.

As shown in Fig. 3-23(c), the NLE has nonlinear *i-v* characteristics determined by its physical makeup. Mathematically its characteristics can be written as a function

$$i = f(v) \tag{3-14}$$

Circuit analysis requires that we solve Eqs. (3-13) and (3-14) simultaneously. This can be done by numerical methods if the $f(v)$ is known explicitly, but often a graphical solution is adequate.

If we graphically superimpose the source's load line on the i-v characteristics of the nonlinear element, the point (or points) of intersection shown in Fig. 3-23(d) represent the values of i and v that satisfy both the source and load constraints. The point of intersection is called the operating, quiescent, or *Q-point* in the terminology of electronics. An example is used to illustrate the process.

EXAMPLE 3-6

Find the voltage, current, and power delivered to the diode in Fig. 3-24(a).

SOLUTION:

We first must find the Thévenin equivalent of the circuit to the left of the terminals A-B. By voltage division we see that

$$V_T = v_{OC}$$

$$= \frac{100}{100 + 100} \times 5 = 2.5 \text{ V}$$

The equivalent "lookback" resistance between A-B with the voltage source off is

$$R_T = R_{EQ} = 10 + \frac{100 \times 100}{100 + 100} = 60\,\Omega$$

The i-v characteristics of the source are constrained to a load line given by

$$i = -\frac{1}{60}v + \frac{1}{60} \times 2.5$$

This line intersects the i-axis ($v = 0$) at $i = i_{SC} = 2.5/60 = 41.7$ mA and intersects the v-axis ($i = 0$) at $v = v_{OC} = 2.5$ V. Superimposing this line on the diode's i-v curve yields an intersection, or Q-point, at $i_D = 15$ mA and $v_D = 1.6$ V. This is the point (i_D, v_D) at which both the source and load device constraints are satisfied. Finally, the power delivered to the diode is the product of the voltage across the diode times the current

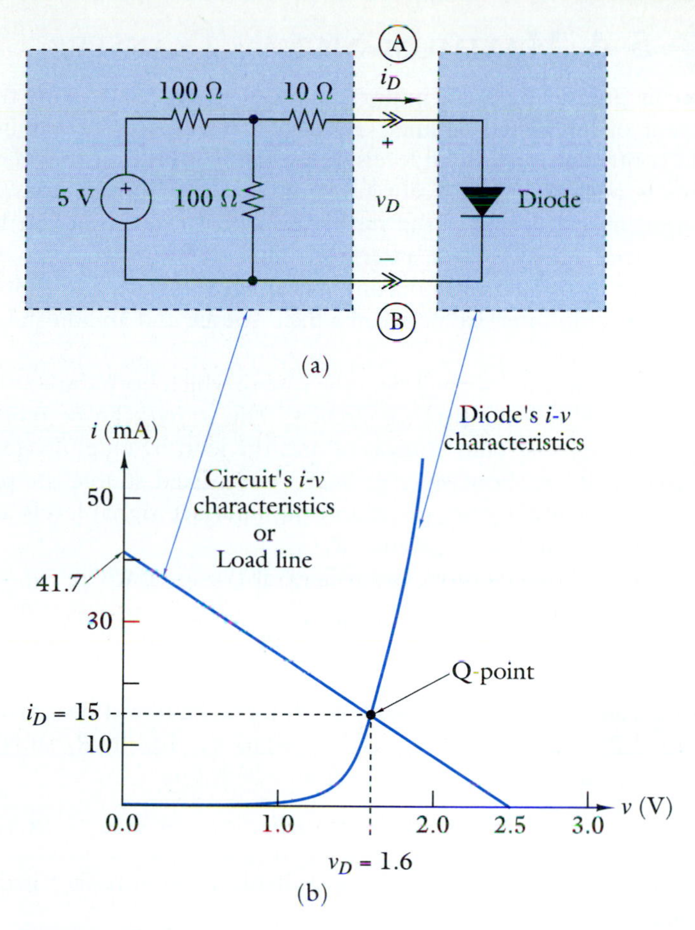

Figure 3-24

through the diode:

$$p_D = i_D v_D = (15 \times 10^{-3})(1.6)$$
$$= 24 \text{ mW}$$

This circuit contains a nonlinear element so the proportionality and superposition properties do not apply. For example, if the source voltage in Fig. 3-24 is increased from 5 V to 10 V, the diode current and voltage do not simply double. Try it.

EXERCISE 3-8

Find the voltage, current, and power delivered to the diode in Fig. 3-24(a) when the 10-Ω resistor is replaced by a short circuit.

ANSWERS:
$v_D = 1.7$ V, $i_D = 18$ mA, and $p_D = 30.6$ mW.

3-4 MAXIMUM SIGNAL TRANSFER

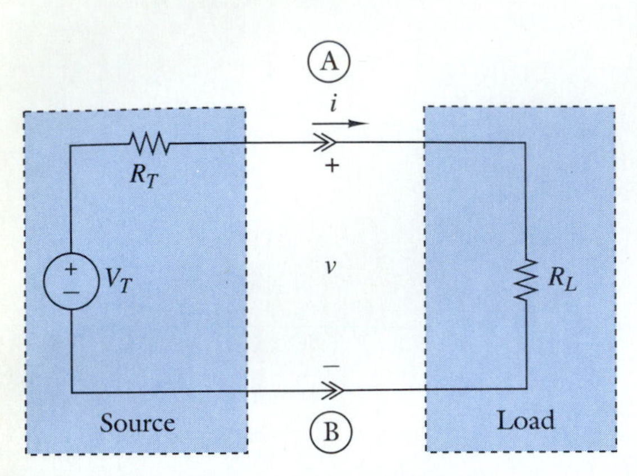

Figure 3-25 Two-terminal interface for deriving the maximum signal transfer conditions.

Circuit interfacing is an important problem especially with the advent of integrated circuits. By interfacing we mean defining and controlling the signal levels at the interconnections between circuits performing different functions. An important consideration in interfacing is the maximum signal levels that can be transferred across a given interface. This section introduces interfacing by defining the maximum voltage, current, and power available at an interface between a fixed source and an adjustable load.

For simplicity we will treat the case in which both the source and load are linear resistance circuits. The source can be represented by its Thévenin equivalent and the load by an equivalent resistance R_L as shown in Fig. 3-25. For a fixed source the parameters V_T and R_T are given and the interface signal levels are functions of the load resistance R_L.

By voltage division, the voltage at the interface is

$$v = \frac{R_L}{R_L + R_T} V_T \tag{3-15}$$

For a fixed source and a variable load, the voltage will be a maximum if R_L is made very large compared to R_T. Ideally R_L should be made infinite (an open circuit), in which case

$$v_{MAX} = V_T = v_{OC} \tag{3-16}$$

Therefore, the maximum voltage available at the interface is the source open-circuit voltage v_{OC}.

The current delivered at the interface is

$$i = \frac{V_T}{R_L + R_T} \tag{3-17}$$

Again, for a fixed source and a variable load, the current will be a maximum if R_L is made very small compared to R_T. Ideally R_L should be zero (a short circuit), in which case

$$i_{MAX} = \frac{V_T}{R_T} = I_N = i_{SC} \tag{3-18}$$

Therefore, the maximum current available at the interface is the source short-circuit current i_{SC}.

The power delivered at the interface is equal to the product $v \times i$. Using Eqs. (3-15) and (3-17) we can write the power as

$$\begin{aligned} p &= v \times i \\ &= \frac{R_L V_T^2}{(R_L + R_T)^2} \end{aligned} \tag{3-19}$$

For a given source, the parameters V_T and R_T are fixed and the delivered power is a function of a single variable R_L. The conditions for obtaining maximum voltage ($R_L \to \infty$) or maximum current ($R_L = 0$) both produce zero power. The value of R_L that maximizes the power lies somewhere between these two extremes. To find this value we differentiate Eq. (3-19) with respect to R_L and solve for the value of R_L that makes $dp/dR_L = 0$.

$$\begin{aligned}\frac{dp}{dR_L} &= \frac{\left[(R_L + R_T)^2 - 2R_L\,(R_L + R_T)\right] V_T^2}{(R_L + R_T)^4} \\ &= \frac{R_T - R_L}{(R_L + R_T)^3} V_T^2 = 0\end{aligned} \tag{3-20}$$

Clearly, the derivative is zero when $R_L = R_T$. Therefore, *maximum power transfer* occurs when the load resistance equals the source Thévenin resistance. When $R_L = R_T$ the source and load are said to be *matched*

Substituting the condition $R_L = R_T$ back into Eq. (3-19) shows the maximum power to be

$$p_{MAX} = \frac{V_T^2}{4R_T} \tag{3-21}$$

But since $V_T = I_N R_T$, this result can also be written as

$$p_{MAX} = \frac{I_N^2 R_T}{4} \tag{3-22}$$

or

$$p_{MAX} = \frac{V_T I_N}{4} = \left[\frac{v_{OC}}{2}\right]\left[\frac{i_{SC}}{2}\right] \tag{3-23}$$

These equations are consequences of what is known as the *maximum power transfer theorem*

> A given resistive source circuit with a Thévenin resistance R_T will deliver maximum power to a resistive load R_L if $R_L = R_T$.

To summarize, for resistance circuits the open-circuit voltage is the maximum voltage available at an interface. The maximum current is the short-circuit current and the maximum power is the product of one-half the open-circuit voltage times one-half the short-circuit current.

Figure 3-26 shows plots of the interface voltage, current, and power as functions of R_L/R_T. The plots of v/v_{OC}, i/i_{SC} and p/p_{MAX} are normalized to the maximum available signal levels so the ordinates in Fig. 3-26 range from 0 to 1. The

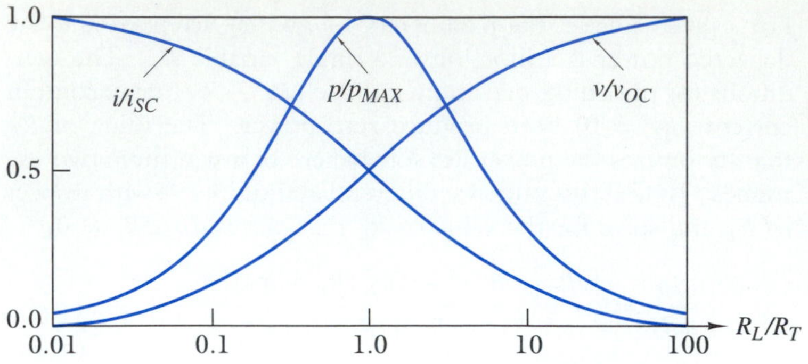

Figure 3-26 Normalized plots of current, voltage, and power versus R_L/R_T.

plot of the normalized power p/p_{MAX} in the neighborhood of the maximum at $R_L/R_T = 1$ is not a particularly strong function of R_L/R_L. Changing R_L/R_L by a factor of two in either direction from the maximum reduces p/p_{MAX} by less than 20%. The normalized voltage v/v_{OC} is within 20% of its maximum as $R_L/R_T \to \infty$ when $R_L/R_T = 4$. Similarly, the normalized current is within 20% of this maximum when $R_L/R_T = 1/4$. In other words, for engineering purposes we can get close to the maximum signal levels with load resistances that only approximate the theoretical requirements.

The reader should remember that these maximum signal levels were all derived for the condition of **a fixed source** and an **adjustable load**. Other possible source-load constraints can change these conclusions, we will see in the next section.

EXAMPLE 3-7

Determine the maximum signals available from the source in Example 3-6.

SOLUTION:

In Example 3-6 we found $V_T = 2.5$ V and $R_T = 60\ \Omega$. The maximum available voltage and current are

$$v_{MAX} = v_{OC} = V_T = 2.5 \text{ V} \quad (R_L \to \infty)$$

$$i_{MAX} = i_{SC} = \frac{V_T}{R_T} = 41.7 \text{ mA} \quad (R_L = 0)$$

Using Eq. (3-23) we obtain the maximum available power as

$$p_{MAX} = \left[\frac{v_{OC}}{2}\right]\left[\frac{I_{SC}}{2}\right] = 26.0 \text{ mW} \quad (R_L = R_T = 60\ \Omega)$$

In the same example we found that a nonlinear diode actually drew the following signal levels from the source:

$$v_D = 1.6 \text{ V}$$

$$i_D = 15 \text{ mA}$$

$$p_D = 24 \text{ mW}$$

These levels are less than the maximum available values. However, the power delivered to the diode is rather close to the maximum available even though the nonlinear diode clearly does not "match" the source resistance.

EXERCISE 3-9

A resistive source circuit delivers 4 V when a 50-Ω resistor is connected across its output and 5 V when 75 Ω is connected. Find the maximum voltage, current, and power available from the source.

ANSWERS:
10 V, 133 mA, and 333 mW.

Tellegen's Theorem

We turn from the question of interface signal transfer to the global power balance in a circuit consisting of two terminal elements. To describe element power we assign an element voltage v_k and current i_k to each element using the passive sign convention. The product $v_k \times i_k$ is positive when the element absorbs power from the circuit and is negative when it delivers power to the circuit.

Tellegen's theorem states that if the voltages v_k satisfy KVL and if the currents i_k satisfy KCL, then

$$\sum_{k=1}^{E} v_k \times i_k = 0 \qquad (3\text{-}24)$$

where the summation extends over all elements and E represents the total number of elements. The theorem requires compliance only with the connection constraints derived from Kirchhoff's laws. The theorem places no restrictions on the i-v characteristics of the elements, so it applies to linear and nonlinear circuits.

The theorem states that in any circuit there is an exact balance between the power absorbed by passive elements and the power delivered by active elements. This conclusion is more than just conservation of energy. It means that the only way energy can be extracted from or injected into a circuit is via the voltage and current at the interface between an element and the rest of the circuit.

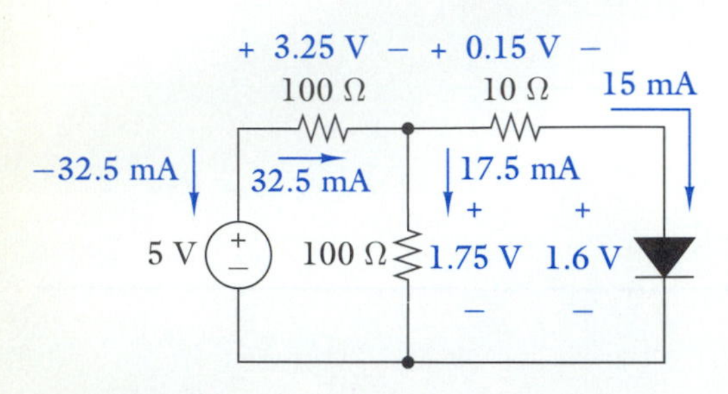

Figure 3-27

EXAMPLE 3-8

Apply Tellegen's theorem to the circuit of Fig. 3-27.

SOLUTION:

The circuit in Fig. 3-27 is taken from Example 3-6 where we used Thévenin's theorem and graphical analysis to show that the diode voltage and current are 1.6 V and 15 mA. The other voltages and currents shown in Fig. 3-27 are easily obtained using KVL, KCL, and Ohm's law. The current through the 10-Ω resistor is 15 mA, since it is in series with the diode. By Ohm's law the voltage across the 10-Ω resistor is $10 \times 0.015 = 0.15$ V. By KVL the voltage across the rightmost 100-Ω resistor is $0.015+1.6 = 1.75$ V, and by Ohm's law its current is $1.75/100 = 17.5$ mA. By KCL the current through the leftmost 100-Ω resistor is $15 + 17.5 = 32.5$ mA and by Ohm's law its voltage is $100 \times 0.0325 = 3.25$ V. The source voltage is 5 V, and using the passive sign convention the current through the source −32.5 mA. We have now found the voltage across and current through every element in the circuit. Applying Tellegen's theorem we find

$$\begin{aligned}\sum_{k=1}^{5} v_k \times i_k &= (1.6 \times .015) + (.15 \times .015) + (1.75 \times .0175) \\ &\quad + (3.25 \times .0325) + (5.0 \times -.0325) \\ &= 0.024 + 0.00225 + 0.030625 \\ &\quad + 0.105625 - 0.1625 \\ &= 0\end{aligned}$$

As expected the 5-volt source delivers the exact amount of power absorbed by all the passive elements in the circuit.

3-5 INTERFACE CIRCUIT DESIGN

The maximum signal levels discussed in the previous section place bounds on what is achievable at an interface. However, the fixed source and adjustable load conditions do not describe all of the situations that arise in practice. There may be circumstances in which the source or the load, or both, can be adjusted to produce prescribed interface signal levels. Sometimes it is necessary to insert an interface circuit between the source and load as indicated in Fig. 3-28. By its very nature the inserted circuit has two terminal pairs, or interfaces, at which voltage and current can be observed or specified. These terminal pairs are also called *ports*, and the interface circuit referred to as a *two-port network*. The port connected to the source is called the input, and the port connected to the load the output. The purpose of this two-port network is to ensure that the source and load interact in a prescribed way.

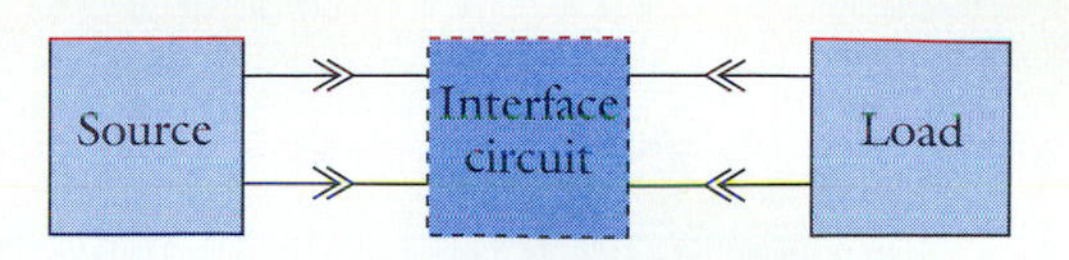

Figure 3-28 An interface circuit.

Four types of interface constraints are listed in Table 3-1. The Type I condition is fairly common, since sources such as transmission lines and laboratory signal generators have standard source resistances such as 50, 75, 300, or 600 ohms. In this case the load resistance must be selected to achieve the desired interface conditions, which may or may not involve matching. The Type II situation arises when we need to select a source to deliver signals to an existing load or receiver. In this case it is often desirable to reduce the source resistance to a minimum, ideally to zero. There are of course limits on achieving this goal, but in the next chapter we will see that OP AMP circuits approach this ideal. The Type III condition requires that we select both the source and the load resistance to achieve specified conditions at their interface. This type problem may involve several trials or iterations, since the source and load interaction affects the interface signal levels.

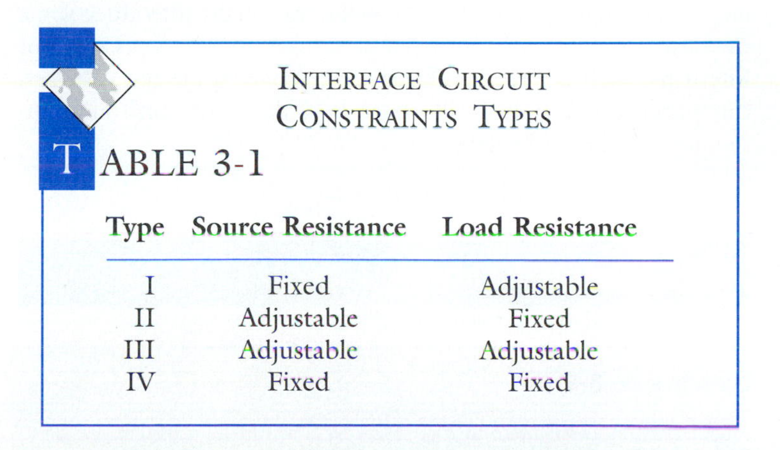

TABLE 3-1 INTERFACE CIRCUIT CONSTRAINTS TYPES

Type	Source Resistance	Load Resistance
I	Fixed	Adjustable
II	Adjustable	Fixed
III	Adjustable	Adjustable
IV	Fixed	Fixed

The Type IV situation in Table 3-1 occurs all too frequently. Here we are faced with the problem of controlling interface conditions at the interconnection between standard building blocks such as integrated circuits or other standard modules that cannot be easily modified. This problem usually requires an interface circuit that will allow the fixed source and fixed load to interact in a prescribed way.

Basic Circuit Design Concepts

Before we treat examples of different interface situations, the reader should recognize that we are now discussing a limited form of circuit design, as contrasted with circuit analysis. Although we use circuit analysis tools in design, there are important differences. A linear circuit analysis problem generally has a unique solution. A circuit design problem may have many solutions or even no solution. The maximum available signal levels found in the preceding section provide bounds that help us test for the existence of a solution. Generally there will be several ways to meet the interface constraints, and it then becomes necessary to evaluate the alternatives using other factors such as cost, power consumption, or reliability.

The only device we now have available for interface circuit design is the linear resistor. In subsequent chapters we will introduce other useful devices such as OP AMPs (Chapter 4), and capacitors and inductors (Chapter 7). But for the moment we are limited to resistors. In a design situation the engineer must choose the resistance values in a proposed circuit. This decision is influenced by a host of practical considerations some of which are discussed in Appendix A. We will occasionally introduce some of these considerations into our design examples. It is not our intent to teach the reader about these practical matters. Rather, our purpose is simply to illustrate how different constraints influence the design process.

Design Example

EXAMPLE 3-9

Select the load resistance in Fig. 3-29 so the interface signals are in the range defined by $v_L \geq 4$ V and $i_L \geq 30$ mA.

SOLUTION:

In this design problem the source circuit is given and we are free to select the load. For a fixed source the maximum signal levels available at the interface are

$$v_{MAX} = V_T = 10 \text{ V}$$

$$i_{MAX} = \frac{V_T}{R_T} = 100 \text{ mA}$$

The bounds given in the design requirements are below the maximum available signal level, so we should be able to find a suitable resistor. Using voltage division the interface voltage constraint requires

$$\frac{R_L}{100 + R_L} \times 10 \geq 4$$

or

$$10R_L \geq 4R_L + 400$$

This condition yields $R_L \geq 400/6 = 66.7\ \Omega$. The interface current constraint can be written as

$$\frac{10}{100 + R_L} \geq 0.03$$

or

$$10 \geq 3 + 0.03R_L$$

which requires $R_L \leq 7/0.03 = 233\ \Omega$. In principle any value of R_L between $66.7\ \Omega$ and $233\ \Omega$ will work. However, to allow for circuit parameter variations we select $R_L = 150\ \Omega$ because it lies at the arithmetic midpoint of allowable range.

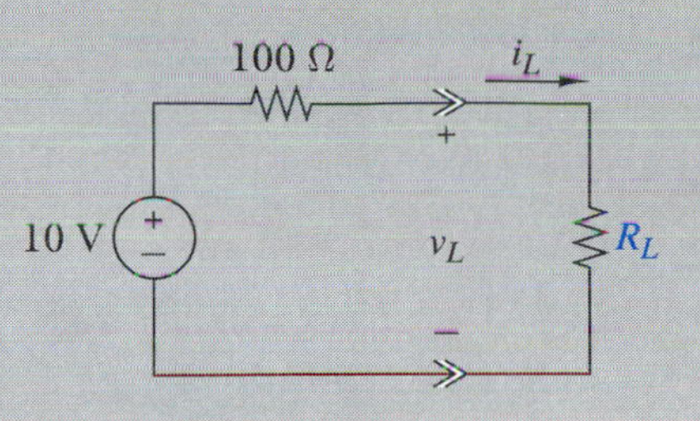

Figure 3-29

❖ Design Example

EXAMPLE 3-10

Select the load resistor in Fig. 3-29 so the voltage delivered to the load is 3 V ±10%. Use only the standard resistance values of ±10% tolerance given in Appendix A.

SOLUTION:

The ±10% voltage tolerance means the interface voltage must fall between 2.7 V and 3.3 V. The required range can be achieved, since the open-circuit voltage of the source circuit is 10 V. Using voltage division we can write the constraints on the value of R_L as

$$2.7 \leq \frac{R_L}{100 + R_L} \times 10 \leq 3.3$$

The left inequality requires

$$270 + 2.7R_L \leq 10R_L$$

or $R_L \geq 270/7.3 = 37.0\ \Omega$. The right inequality requires

$$10R_L \leq 330 + 3.3R_L$$

or $R_L \leq 330/6.7 = 49.2\ \Omega$. Thus, R_L must lie in the range from 37.0 Ω to 49.2 Ω, which has a midpoint at 43.1 Ω.

In practice, resistors do not come in infinitely many values. To limit the inventory costs of part suppliers and equipment manufacturers, the industry has agreed on a finite set of standard resistance values. For resistors with ±10% tolerance the standard values are

10 12 15 18 22 27 33 39 47 56 68 82 Ω

and multiples of 10 times these values. The standard values of 39 Ω and 47 Ω fall inside the required 37.0 to 49.2 Ω range. However, these nominal values and the ±10% tolerance could carry the actual resistance value outside the range. Various series and parallel combinations of standard ±10% resistors produce values that fall in the desired range even with 10% tolerance. For example, two 22-Ω resistors in series produces a nominal equivalent resistance of 44 Ω with a ±10% range from 39.6 Ω to 48.4 Ω. Finally, it turns out that 43 Ω is a standard value for resistors with ±5% tolerance (see Appendix A, Table A-1). This offers an alternative approach, since 43 Ω falls almost exactly at the midpoint of the desired resistance range and the 5% tolerance easily meets the interface voltage range requirement.

In summary, a single standard 39-Ω or 47-Ω resistor with 10% tolerance produces nominal designs that meet the voltage requirements, but runs a risk of being out of tolerance in some cases. Series and parallel combinations of two 10% resistors meet all requirements but lead to a more complicated design. A single 43-Ω 5% resistor meets all requirements, but its tighter tolerance could mean higher parts costs. The final design decision must consider many factors, particularly the sensitivity of the system to changes in the interface voltage.

❖ Design Example

EXAMPLE 3-11

The resistance array package shown in Fig. 3-30 includes seven nearly identical resistors with nominal resistances of 1 kΩ. Use this resistance array to design source and load resistances so a 10-V dc power supply delivers 4.3 V to the load.

SOLUTION:

The required interface voltage is well within the capability of a 10-V source. Using voltage division we can express the design requirement as

$$\frac{R_L}{R_L + R_S} \times 10 = 4.3$$

where R_S and R_L are the source and load resistance, respectively. This design constraint can be rearranged as

$$\frac{R_S}{R_L} = 1.33$$

In other words, the design requirement places a constraint on the ratio of R_S and R_L.

Resistance array packages are especially useful when resistance ratios control a design. The nominal resistance can change by several percentage points from one array to the next. Within a given array the resistances are all nearly equal, so their ratios are nearly unity. In effect the resistance array in Fig. 3-30 contains seven identical resistors, although we do not know their exact resistance. Constructing the source resistance using three resistors in parallel and the load using four resistors in parallel produces $R_S = R/3$ and $R_L = R/4$, where R is nominally 1 kΩ. But regardless of the actual value of R, the ratio $R_S/R_L = 1.33$ as required.

The final design shown in Fig. 3-31 may seem wasteful of resistors. However, in large-volume production a resistance array can cost less than a discrete resistor design. Resistance array packages are better suited to automated manufacturing than discrete resistors. The point is that estimating the cost of a design is not simply a matter of counting the number of resistors.

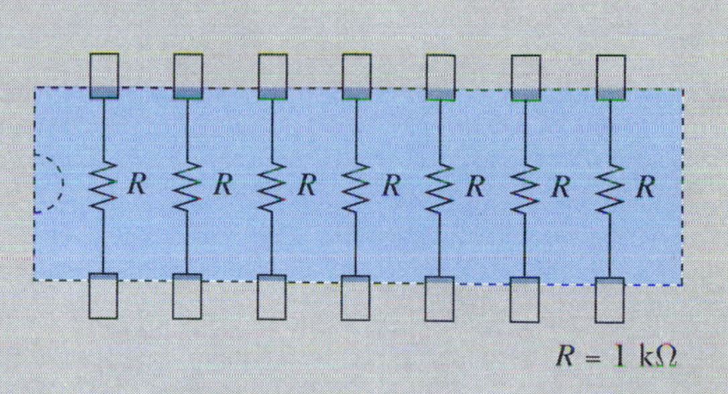

Figure 3-30

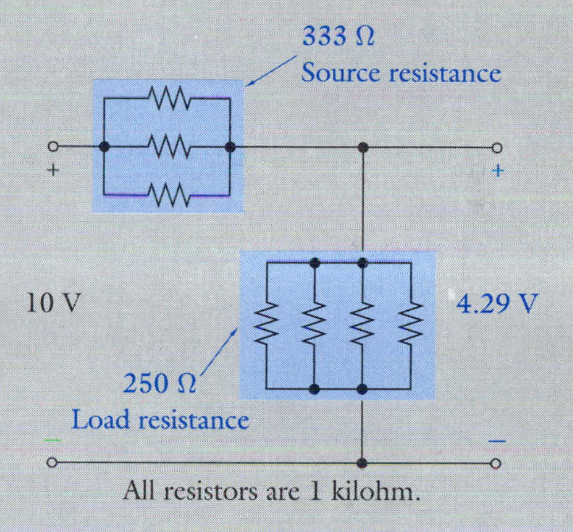

Figure 3-31

❖ Design Example

EXAMPLE 3-12

Design a two-port interface so the circuit in Fig. 3-32 delivers 2 A to the load.

SOLUTION:

The design problem involves a fixed source and fixed load. If the source is connected directly to the load the source current divides equally between the two 50-Ω resistors producing 5 A into the load.

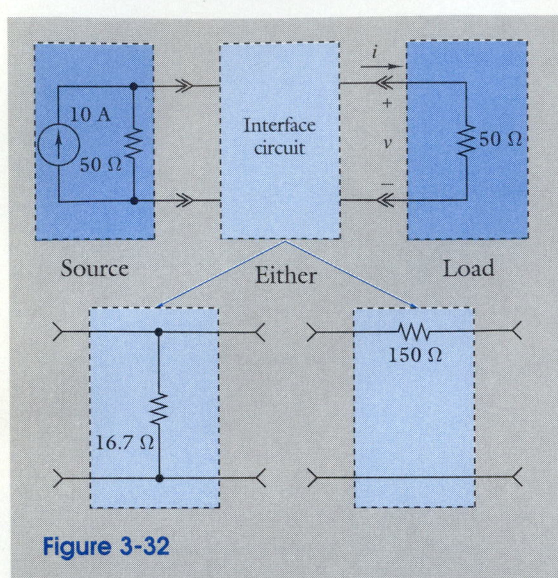

Figure 3-32

Therefore, an interface circuit is needed to reduce the load current to the 2-A level specified. Two possible design solutions are shown in Fig. 3-32. For the parallel resistor case current division yields

$$i = \frac{1/50}{1/50 + 1/50 + 1/R_{PAR}} \times 10$$

For the $i = 2$ A design condition this equation becomes

$$\frac{10}{2 + 50/R_{PAR}} = 2 \text{ A}$$

Solving for R_{PAR} yields

$$R_{PAR} = \frac{50}{3} = 16.7 \ \Omega$$

In the series resistor case a source transformation replaces the 10-A current source in parallel with 50 Ω by a 500-V voltage source in series with 50 Ω. Using series equivalence on the transformed circuit yields the load current as

$$i = \frac{500}{50 + 50 + R_{SER}} = 2 \text{ A}$$

Solving for R_{SER} yields

$$R_{SER} = 150 \ \Omega$$

We have two designs that meet the stated requirement. In practice, an engineer evaluates alternative design using criteria such as power dissipation, cost, and reliability. The interested reader may wish to continue the design process by calculating the power dissipated in the interface circuit in each case and consider the commercially available standard resistance values (see Appendix A, Table A-1).

❖ Design Example

EXAMPLE 3-13

An application calls for a nonlinear element to be driven by the source circuit shown in Fig. 3-33(a). The manufacturer's specifications provide the NLE's i-v characteristics shown as the heavy line in Fig. 3-33(b). The maximum allowable power dissipation in the device is listed as 20 mW. The specification also states that at least 1 V must appear across the device for optimum operation. Design the two-port interface circuit in Fig. 3-33(a) to produce an operating point for the NLE and $v > 1$ V and $p = vi < 20$ mW.

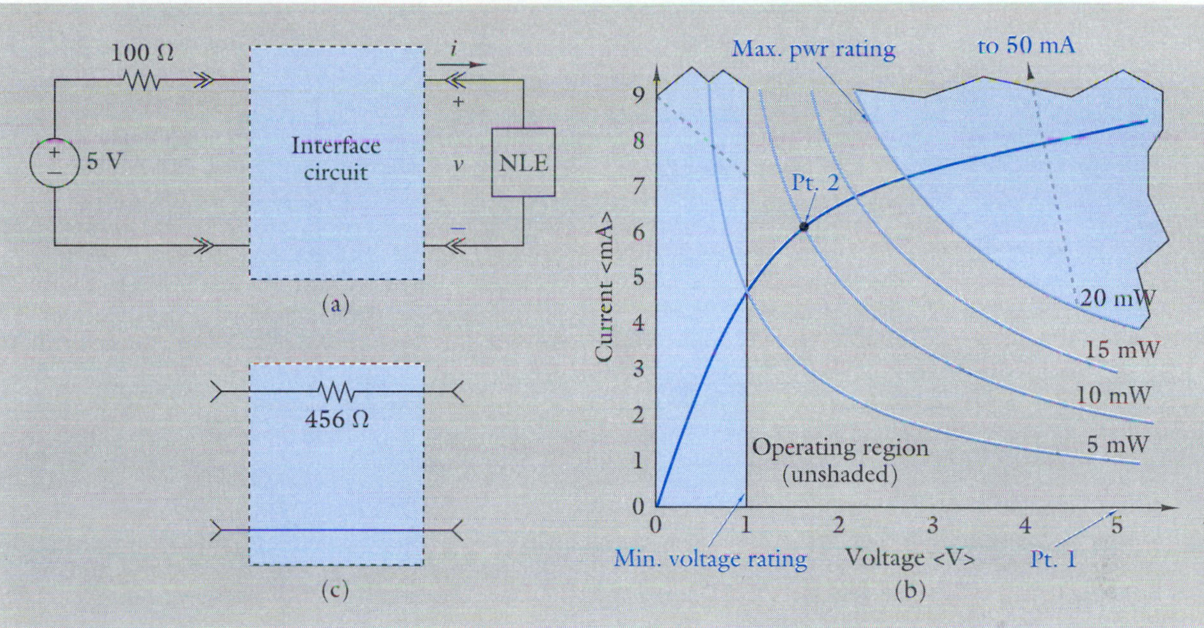

Figure 3-33

SOLUTION:

The operating point must lie in the unshaded region shown in Fig. 3-33(c). The left boundary of the unshaded region is determined by the $v > 1$ V requirement. Curves of constant power dissipation are shown Fig. 3-33(b). The curve corresponding to $p = 20$ mW forms the right boundary of the unshaded region.

We first check to see if an interface circuit is really required. This is accomplished by finding the operating point in Fig. 3-33(b) when the 5-V source is connected directly to the NLE. With no interface circuit the NLE sees a source with $v_{OC} = 5$ V and $i_{SC} = 5/100 = 50$ mA. The corresponding load line intercepts the abscissa in Fig. 3-33(b) at $v = v_{OC} = 5$ V and the ordinate axis at $i = i_{SC} = 50$ mA. The 50-mA intercept is well off the graph in Fig. 3-33(b). Nevertheless, we see that the intersection of this load line and the NLE i-v characteristics falls in the shaded region in Fig. 3-33(b). In other words, with no interface circuit the source delivers more than 20 mW to the NLE.

We can reduce the power dissipated in the NLE by inserting a series resistor interface circuit. The series resistor does not change the open-circuit voltage at the terminals of the NLE but reduces the available short-circuit current. In effect, a series resistance causes the load line to pivot about Point 1 in Fig. 3-33(b). For example, 900-Ω of series resistance rotates the load line so it intercepts the ordinate at $i = i_{SC} = 5/1000 = 5$ mA. The design problem reduces to selecting a series resistance that pivots the load line about Point 1 until it goes through the desired operating point. The question is, where do we want the device to operate?

The manufacturer stated that the device cannot dissipate more than 20 mW. It is common practice to "derate" devices, that is, to operate them well below their maximum power rating. Curves of constant power dissipation are shown on the NLE characteristics in Fig. 3-33(b). The curve corresponding to 10 mW dissipation crosses the device's i-v characteristics at Point 2 in the figure. This point offers a reasonable operating point, since the device voltage is greater than 1 V and the power dissipation is "derated" by a factor of two. When the load line is pivoted until it intersects the NLE i-v characteristics at Point 2, it intersects the ordinate axis at 9 mA. The load line required to produce an operating point at Point 2 in Fig. 3-33(b) is defined by $v_{OC} = 5$ V and $i_{SC} = 9$ mA. The total series resistance required to produce $v_{OC} = 5$ V and $i_{SC} = 9$ mA at the terminal of the NLE is $R_{SER} = 5/0.009 = 556$ Ω. Since the source circuit provides 100 Ω, we add a 456-Ω series resistor in the interface two-port circuit as shown in Fig. 3-33(c).

EXERCISE 3-10

Find the value of series resistance so the NLE in Example 3-13 operates with 5 mW of power dissipation.

ANSWER:
733 Ω.

Design Example

EXAMPLE 3-14

Design the two-port interface circuit in Fig. 3-34 so the load "sees" a Thévenin resistance of 50 Ω between terminals C-D, while simultaneously, the source "sees" a load resistance of 300 Ω between A-B.

SOLUTION:

We first try a single resistor in the interface circuit. A 60-Ω parallel resistor in the interface circuit would make the load see 50-Ω Thévenin resistance, but then the source would see a load resistance much less than 300 Ω. A 250-Ω series resistor in the interface circuit

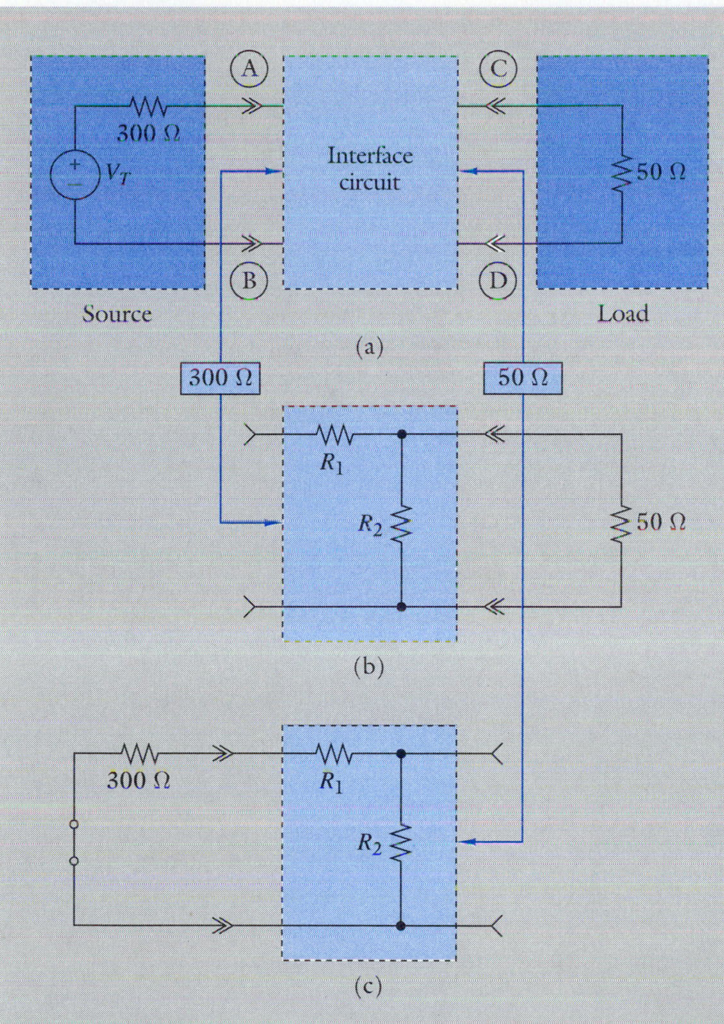

Figure 3-34

resistance than the load, it should look into a series resistor. A configuration that meets these conditions is the "L" circuit shown in Figs. 3-34(b) and 3-34(c).

The above discussion can be summarized mathematically. Using the L-circuit the design requirement at terminals C-D is

$$\frac{(R_1 + 300)\,R_2}{R_1 + 300 + R_2} = 50\ \Omega$$

At terminals A-B the requirement is

$$R_1 + \frac{50R_2}{R_2 + 50} = 300\ \Omega$$

At terminals A-B the requirement is

$$R_1 + \frac{50R_2}{R_2 + 50} = 300\ \Omega$$

The design requirements yield two equations in two unknowns. What could be simpler? It turns out that solving these equations by hand analysis is a bit of a chore. They can easily be solved using PC-based tools such as spreadsheets or prepackaged math analysis programs. But at this point in our development we would encourage the reader to think about the problem in physical terms.

One approach based on the L-circuits in Fig. 3-34(b) might go as follows. If we let $R_2 = 50$ ohms, then the requirement at terminals C—D will be met, at least approximately. Similarly, if $R_1 + R_2 = 300\ \Omega$, the requirements at terminals A-B will be approximately satisfied. In other words, try $R_1 = 250\ \Omega$ and $R_2 = 50\ \Omega$ as a first cut. These values yield equivalent resistances of $R_{CD} = 50 \parallel 550 = 45.8\ \Omega$ and $R_{AB} = 250 + 50 \parallel 50 = 275\ \Omega$. These equivalent resistances are not the exact values specified but are within $\pm 10\%$. Since the tolerance on electrical components may be at least this high a design using these values could be adequate. In this example we used physical reasoning to develop an approximation solution of a problem. Using physical reasoning to finding practical solutions is what engineering is all about.

EXERCISE 3-11

Find precise values of R_1 and R_2 that satisfy the design condition equations in Example 3-14.

ANSWERS:

$R_1 = 273.8613\ \Omega$; $R_2 = 54.77225\ \Omega$.

DESIGN EXERCISE: EXERCISE 3-12

Design an interface two-port using standard $\pm 5\%$ resistors so

(a) The power delivered to the 50-Ω load in Fig. 3-35 is 0.2 W $\pm 5\%$.

(b) Repeat (a) for $p = 0.6$ W $\pm 5\%$.

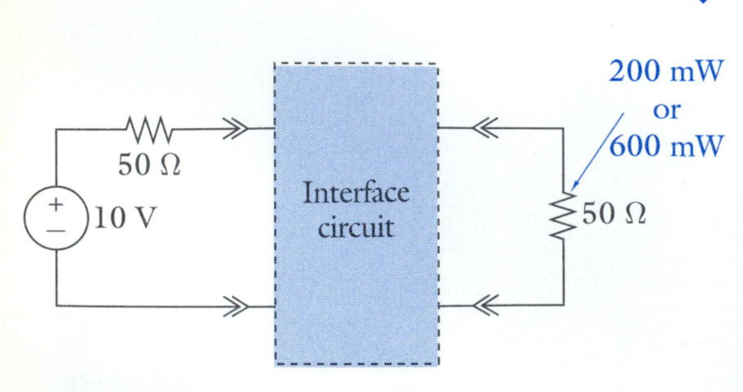

Figure 3-35

ANSWERS:

There are no unique answers to this problem. Some possibilities are discussed below.

DISCUSSION:

(a) Add a standard 56-Ω resistor in series (ideally 58.1 Ω) to obtain 200 mW $\pm 5\%$ dissipation in the 50-Ω load.

(b) No resistive interface will work, since the maximum power available to the load is only 0.5 W. To deliver more power than the signal source alone can provide requires an active device such as an OP AMP or transistor. We will treat problems of this kind in the next chapter.

SUMMARY

- A linear function has two properties: homogeneity and additivity. In a linear circuit homogeneity property is called proportionality and the additivity property is called superposition.
- The Unit Output Method is based on the proportionality property of linear circuits. This method assumes a unit output exists. The input necessary to produce the unit output is found using KVL, KCL, Ohm's law. The proportionality constant is then computed as $K = 1/\text{Input}$.
- Superposition can be used to analyze linear circuits with multiple sources. Each source is used once with all other voltage sources replaced by short circuits and current sources by open circuits. The response with all sources operating is the sum of the individual responses.
- Thévenin's and Norton's theorems simplify the analysis of interface circuits. A Thévenin equivalent circuit consists of a voltage source in series with a resistance. A Norton equivalent circuit consists of a current source in parallel with a resistance. The Thévenin and Norton equivalent circuits are related by a source transformation.
- The parameters of the Thévenin and Norton equivalent circuits can be determined by using any two of the following quantities: (1) the open-circuit voltage at the interface, (2) the short-circuit current at the interface, and (3) the equivalent resistance of the source circuit with all sources turned off.
- For a fixed source and adjustable load interface the maximum interface signal levels are $v_{MAX} = v_{OC}(R_L = \infty)$, $i_{MAX} = i_{SC}(R_L = 0)$, and $p_{MAX} = v_{OC}i_{SC}/4(R_L = R_S)$. The condition $R_L = R_S$ for maximum power transfer is called matching.
- Tellegen's theorem states that the power dissipated by passive elements in a circuit must equal the power supplied by active devices.
- Interface signal transfer conditions are specified in terms of the voltage, current, or power delivered to the load. The design constraints depend on the signal conditions specified and the circuit parameters that are adjustable. Some design requirements may require a two-port interface circuit. An interface design problem may have one, many, or no solutions.

EN ROUTE OBJECTIVES AND ASSOCIATED PROBLEMS

ERO 3-1 Proportionality (Sect. 3-1)

Given a circuit containing linear resistors and one input source, use the response determined at one input signal level to determine the response at other input signal levels.

3-1 Find the proportionality constant $K = i_O/I_S$ for circuit C1 in Fig. P3-1.

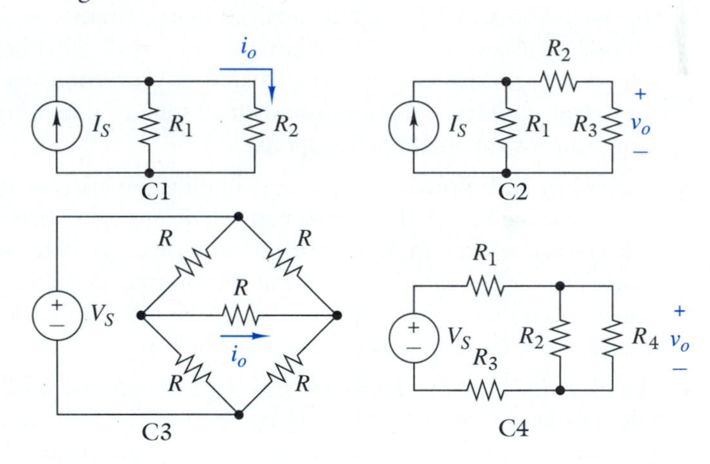

Figure P3-1

3-2 Find the proportionality constant $K = v_O/I_S$ for circuit C2 in Fig. P3-1.

3-3 Find the proportionality constant $K = i_O/V_S$ for circuit C3 in Fig. P3-1.

3-4 Find the proportionality constant $K = v_O/V_S$ for circuit C4 in Fig. P3-1.

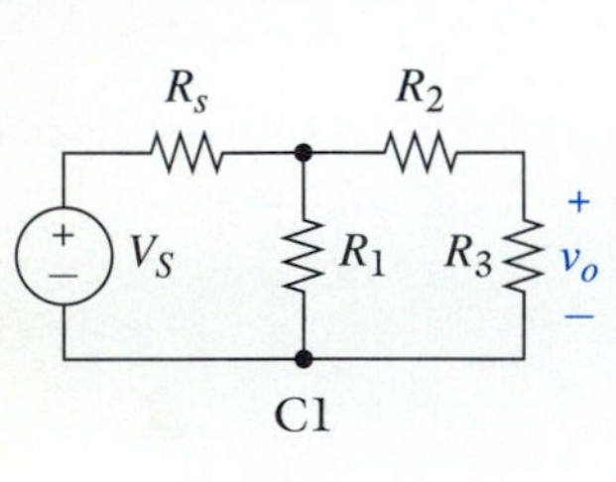

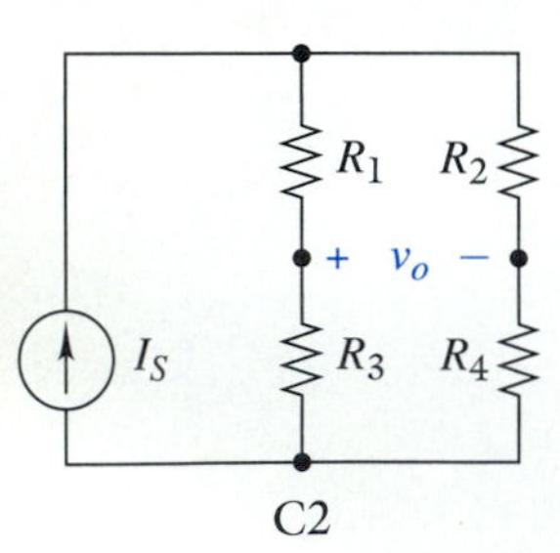

Figure P3-5

3-5 Find the proportionality constant $K = v_O/V_S$ for circuit C1 in Fig. P3-5.

3-6 Find the proportionality constant $K = v_O/I_S$ for circuit C2 in Fig. P3-5.

3-7 Listed below are the i-v relationships for circuits elements. Which elements are linear?

(a) $i = Gv$

(b) $v = L\left(\frac{di}{dt}\right)$

(c) $v = \frac{1}{L}\int_{-\infty}^{t} i\,dt$

(d) $i = I_O\left(e^{v/V_{th}} - 1\right)$

(e) $i = K\sqrt{v}$

3-8 Use the unit output method to find v_O in circuit C1 in Fig. P3-8.

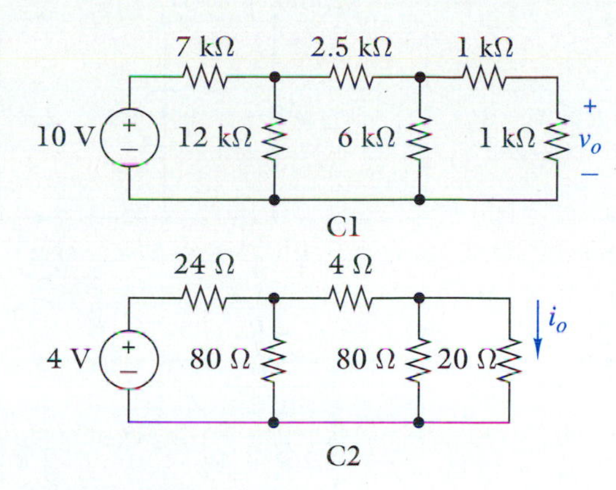

Figure P3-8

3-9 Use the unit output method to find i_O in circuit C2 in Fig. P3- 8.

3-10 Find the input voltage v_S required to produce the 6-volt output shown in Fig. P3-10.

3-11 Use the unit output method to select R in the circuit in Fig. P3-11 so that the proportionality constant $K = V_O/V_S = 1/30$. Find the largest possible value of K for any value of R.

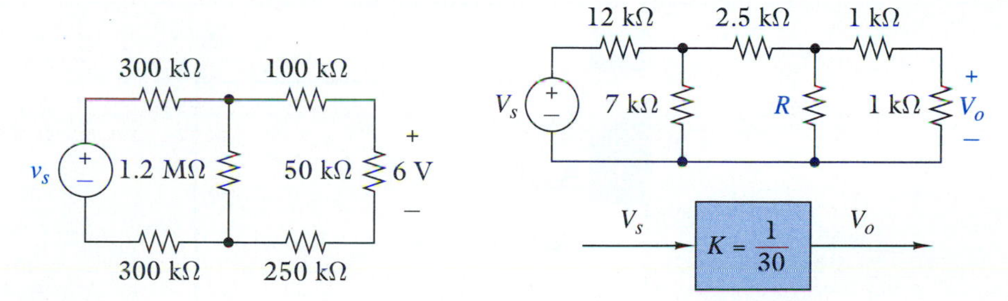

Figure P3-10

Figure P3-11

ERO 3-2 Superposition (Sect. 3-2)

Given a circuit containing linear resistors and two or more input sources, find selected signal variables using the superposition principle.

3-12 Find the output voltage in circuit C1 of Fig. P3-12 using superposition.

3-13 Repeat Problem 3-12 for circuit C2 in Fig. P3-12.

3-14 Repeat Problem 3-12 for circuit C3 in Fig. P3-12.

3-15 Repeat Problem 3-12 for circuit C4 in Fig. P3-12.

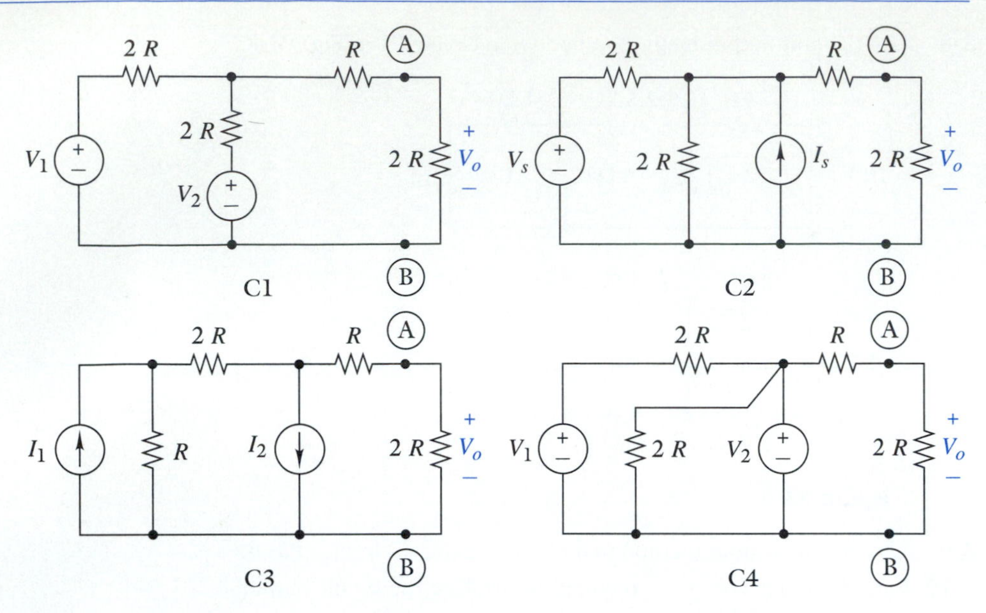

Figure P3-12

3-16 Table P3-16 lists test data showing the output of the circuit in Fig. P3-16 for different values of the three input voltages. Find the input-output relationship for the circuit.

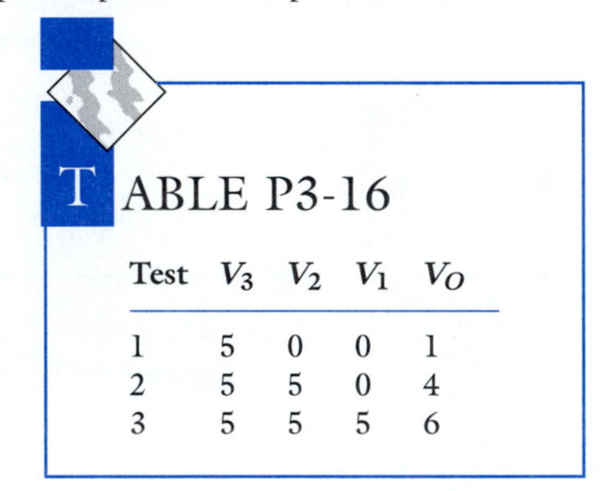

TABLE P3-16

Test	V_3	V_2	V_1	V_O
1	5	0	0	1
2	5	5	0	4
3	5	5	5	6

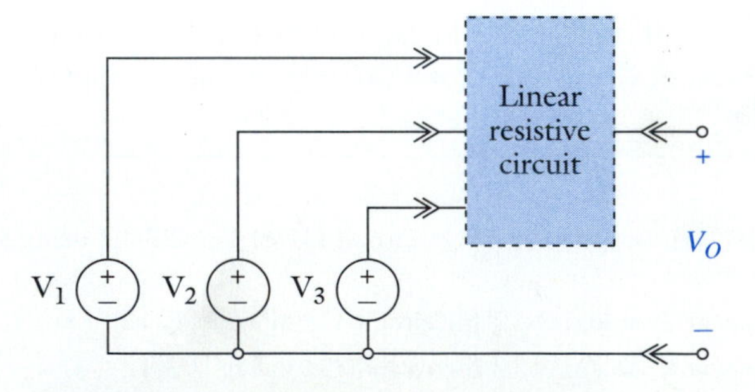

Figure P3-16

3-17 Use superposition to find I_1 and I_2 in the circuit of Fig. P3-17.

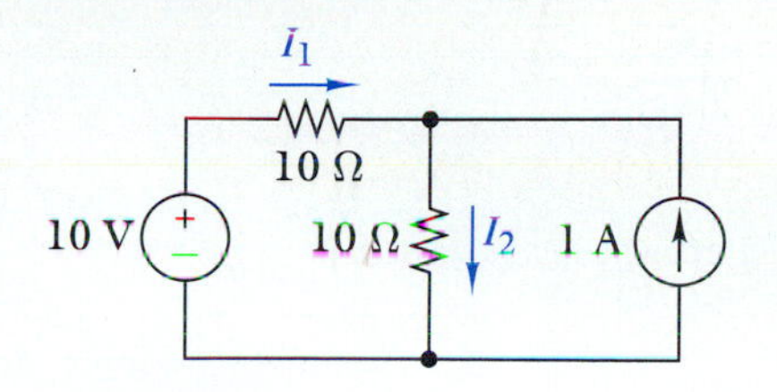

Figure P3-17

3-18 Show that input-output relationship of the R-$2R$ ladder circuit of Fig. P3-18 is

$$V_O = \frac{V_1}{8} + \frac{V_2}{4} + \frac{V_3}{2}$$

Use the relationship to find the output for each input combination given in Table P3-18.

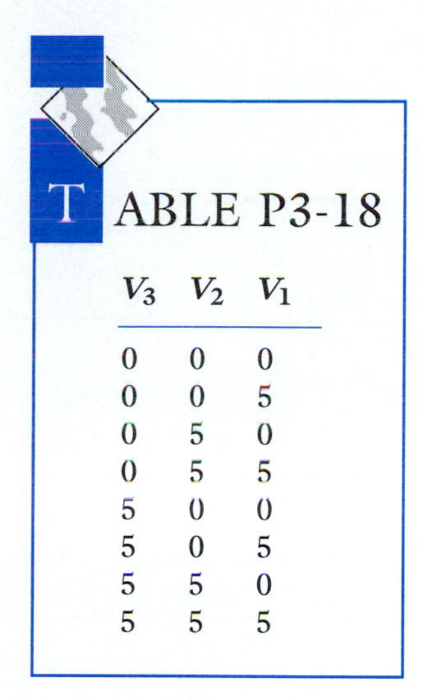

TABLE P3-18

V_3	V_2	V_1
0	0	0
0	0	5
0	5	0
0	5	5
5	0	0
5	0	5
5	5	0
5	5	5

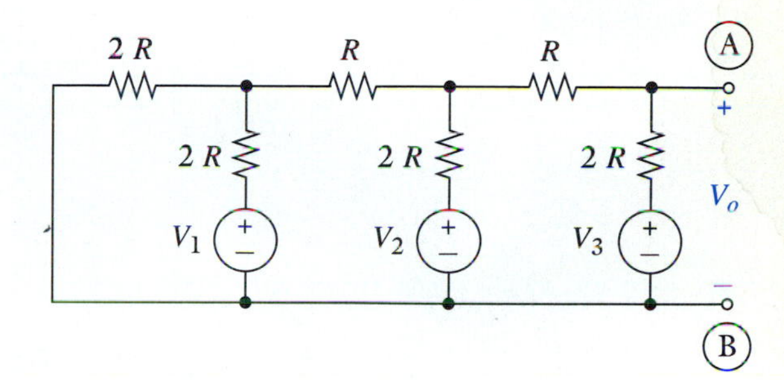

Figure P3-18

3-19 A resistor load in Fig. P3-19 absorbs 0.6 W. What is the value of the load resistance?

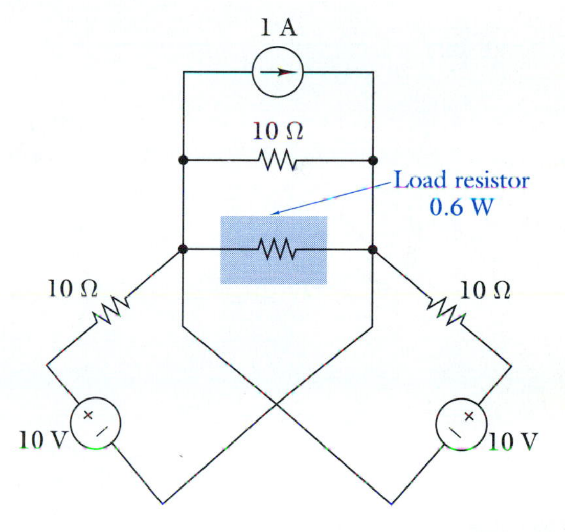

Figure P3-19

ERO 3-3 Thévenin and Norton Equivalent Circuits (Sect. 3-3)

Given a circuit containing linear resistors and constant input sources:

(a) Find the Thévenin or Norton equivalent at a specified pair of terminals.

(b) Use the Thévenin or Norton equivalent to find the signals delivered to linear or nonlinear loads.

3-20 Find the Thévenin and Norton equivalent circuits seen by the load R_L in the circuit C1 in Fig. P3-20.

3-21 Repeat Problem 3-20 for circuit C2 in Fig. P3-20.

3-22 Repeat Problem 3-20 for circuit C3 in Fig. P3-20.

3-23 Repeat Problem 3-20 for circuit C4 in Fig. P3-20.

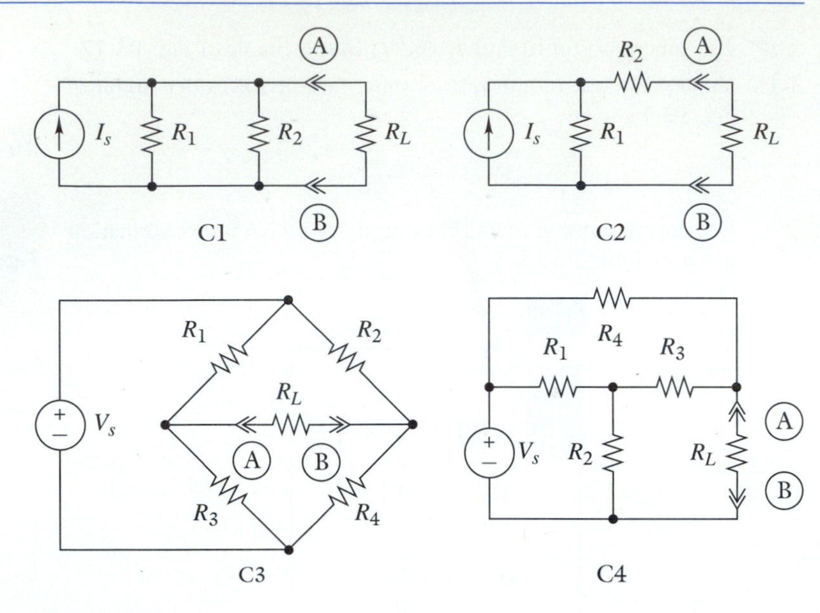

Figure P3-20

3-24 Find the Thévenin equivalent circuit at terminals A-B in circuit C1 of Fig. P3-24 with the switch open and with the switch closed.

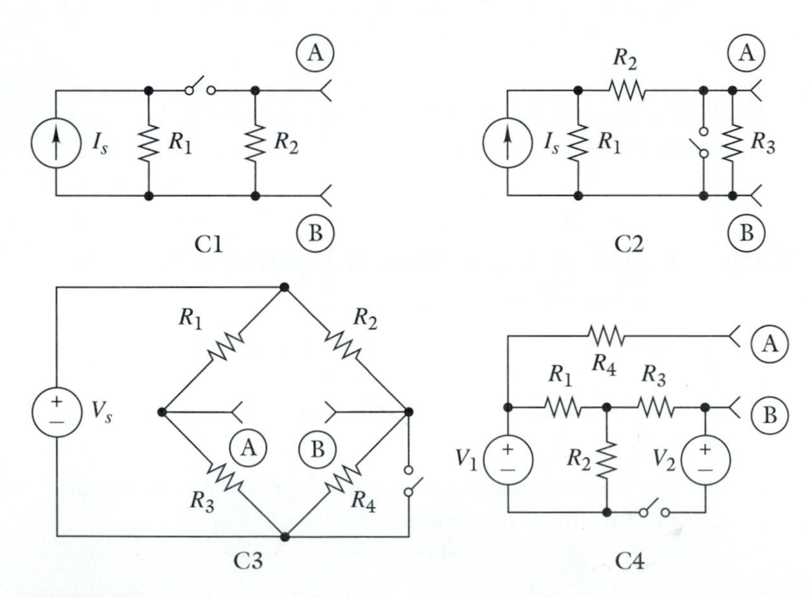

Figure P3-24

3-25 Repeat Problem 3-24 for circuit C2 in Fig. P3-24.

3-26 Repeat Problem 3-24 for circuit C3 in Fig. P3-24.

3-27 Repeat Problem 3-24 for circuit C4 in Fig. P3-24.

3-28 Find the Thévenin equivalent circuit at terminals A-B in circuit C1 of Fig. P3-12.

3-29 Find the Thévenin equivalent circuit at terminals A-B in circuit C2 of Fig. P3-12.

3-30 Find the Thévenin equivalent circuit at terminals A-B in circuit C3 of Fig. P3-12.

3-31 Find the Thévenin equivalent circuit at terminals A-B in circuit C4 of Fig. P3-12.

3-32 The open-circuit voltage of a source circuit is 10 V. When a 2-kΩ resistor is connected across the output the voltage drops to 8 V. Find the Thévenin equivalent of the source circuit and predict the current, voltage, and power delivered to loads of 250 Ω, 500 Ω, and 1000 Ω.

3-33 Find the Thévenin and Norton equivalent circuits at terminals A-B for the circuit in Fig. P3-33.

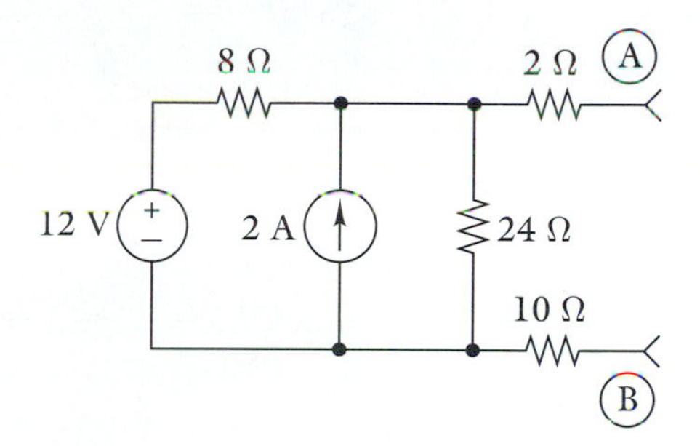

Figure P3-33

3-34 Find the Thévenin equivalent circuit at terminals A-B of the R-$2R$ ladder in Fig. P3-18. (Hint: Use successive source transformations.)

3-35 The diode in the circuit of Fig. P3-35(a) has the i-v characteristics shown in Fig. P3-35(b). Find the power dissipated in the diode. Repeat if the source voltage is increased to 2.0 V.

3-36 A nonlinear resistor with the i-v characteristics shown in Fig. P3-36 is connected to a circuit with the Thévenin equivalent shown in Fig. P3-36. With the nonlinear resistor connected to the source the interface voltage is $v = 30$ V. With the nonlinear resistor disconnected the interface voltage is $v = 40$ V. Find v_S and R_S.

ERO 3-4 Maximum Signal Transfer (Sect. 3-4)

Given a circuit containing linear resistors and constant input sources, find the maximum voltage, current, and power available at a specified pair of terminals. Determine the resistive loads required to obtain the maximum available signal levels.

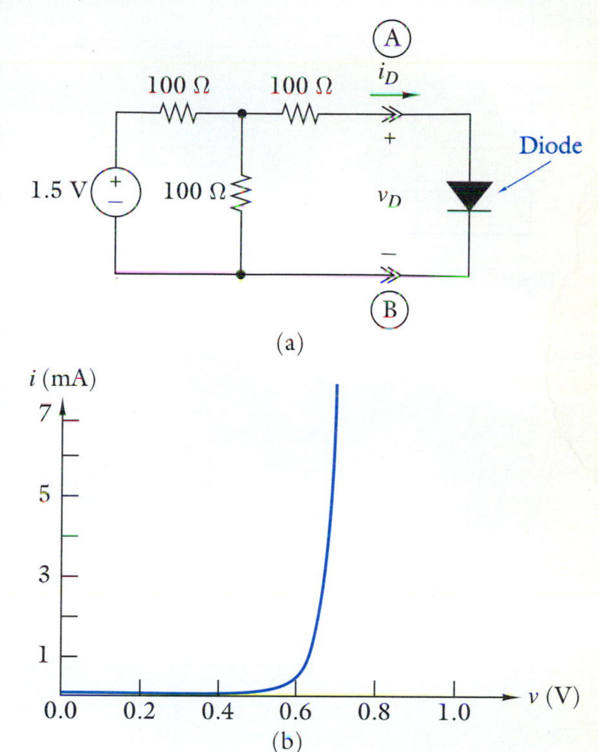

Figure P3-35

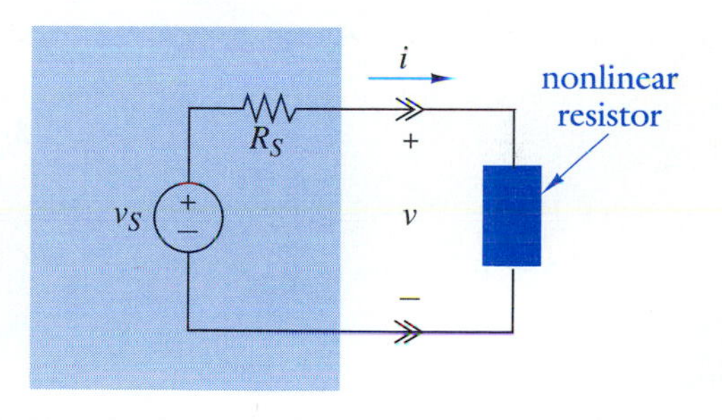

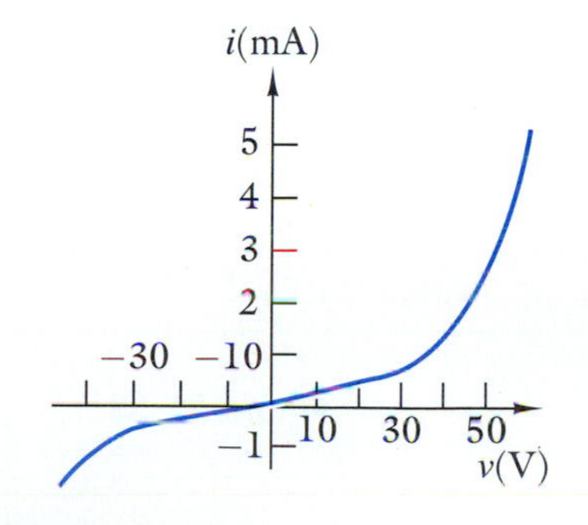

Figure P3-36

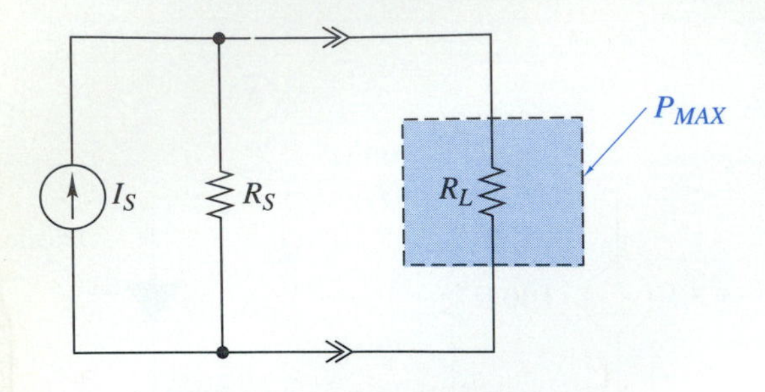

Figure P3-37

3-37 For the circuit of Fig. P3-37 derive an expression for the power delivered to R_L in terms of I_S, R_S, and R_L. When I_S and R_S are fixed, show that the condition for maximum power transfer is $R_L = R_S$. Do **not** use a source transformation.

3-38 For circuits C1 and C2 in Fig. P3-38 find:

(a) The value of the load resistance that causes maximum power to be delivered to R_L and the value of the maximum power.

(b) The value of the load resistance that causes maximum current to be delivered to R_L and the value of the maximum current.

(c) The value of the load resistance that causes maximum voltage to be delivered to R_L and the value of the maximum voltage.

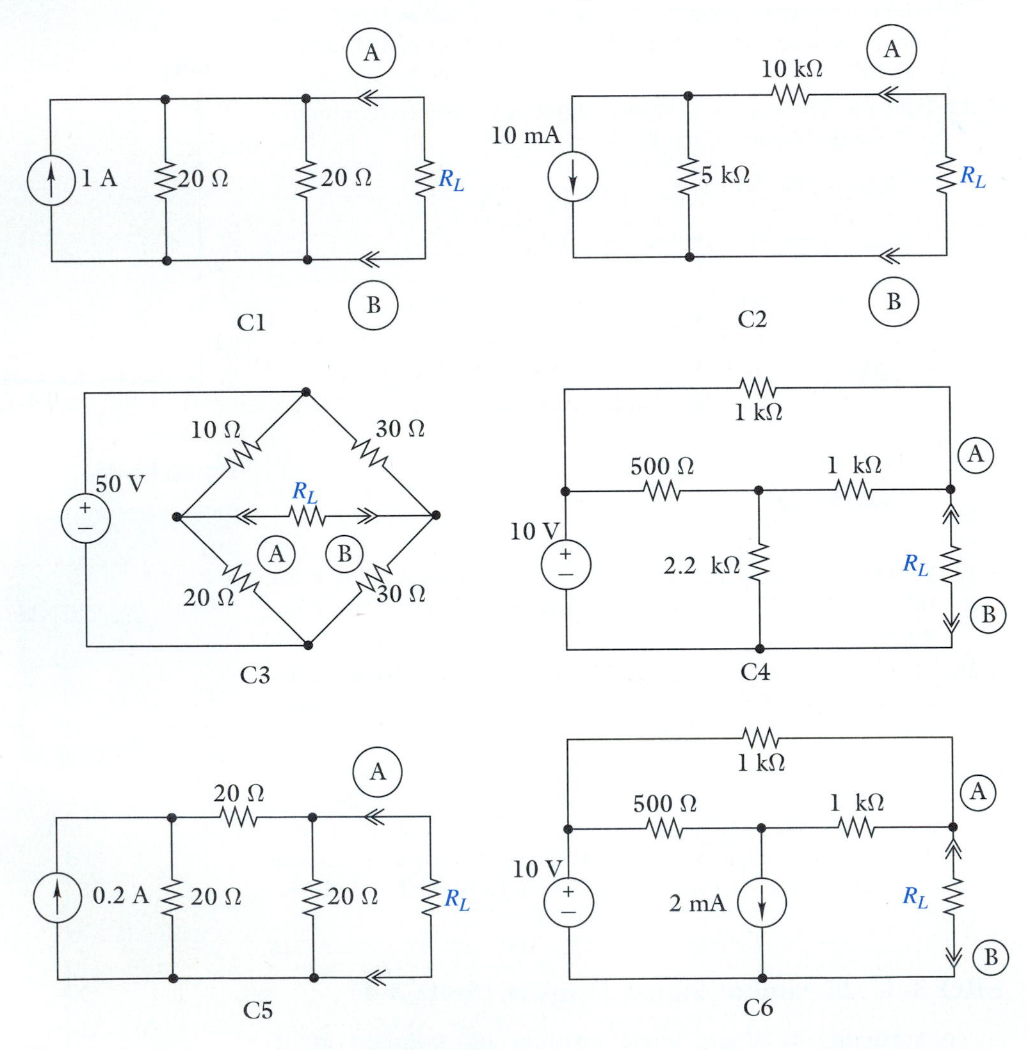

Figure P3-38

3-39 Repeat Problem 3-38 for circuits C3 and C4 in Fig. P3-38.

3-40 Repeat Problem 3-38 for circuits C5 and C6 in Fig. P3-38.

3-41 A stereo speaker can be modeled as an 8-Ω linear resistance. Figure P3-41 shows the Thévenin or Norton equivalent circuits of five stereo amplifiers. Show how to connect each of the stereo amplifiers to one or more 8-Ω speakers so maximum power is delivered to the speaker(s).

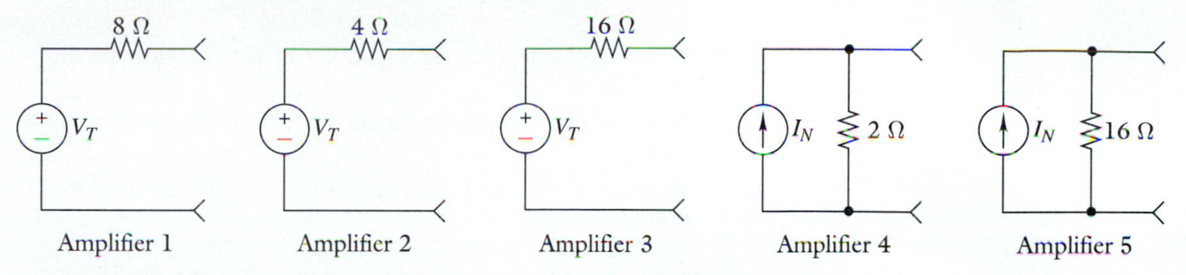

Figure P3-41

3-42 Figure P3-42 shows a two-port interface circuit whose output is connected to a load resistance R_L. The input of the two-port interface can be connected to any of the six sources shown in the figure. When each of the sources in Fig. P3-42 is connected to the input port of the interface circuit find:

(a) The value of the load resistance that causes maximum power to be delivered to R_L and the value of the maximum power.

(b) The value of the load resistance that causes maximum current to be delivered to R_L and the value of the maximum current.

(c) The value of the load resistance that causes maximum voltage to be delivered to R_L and the value of the maximum voltage.

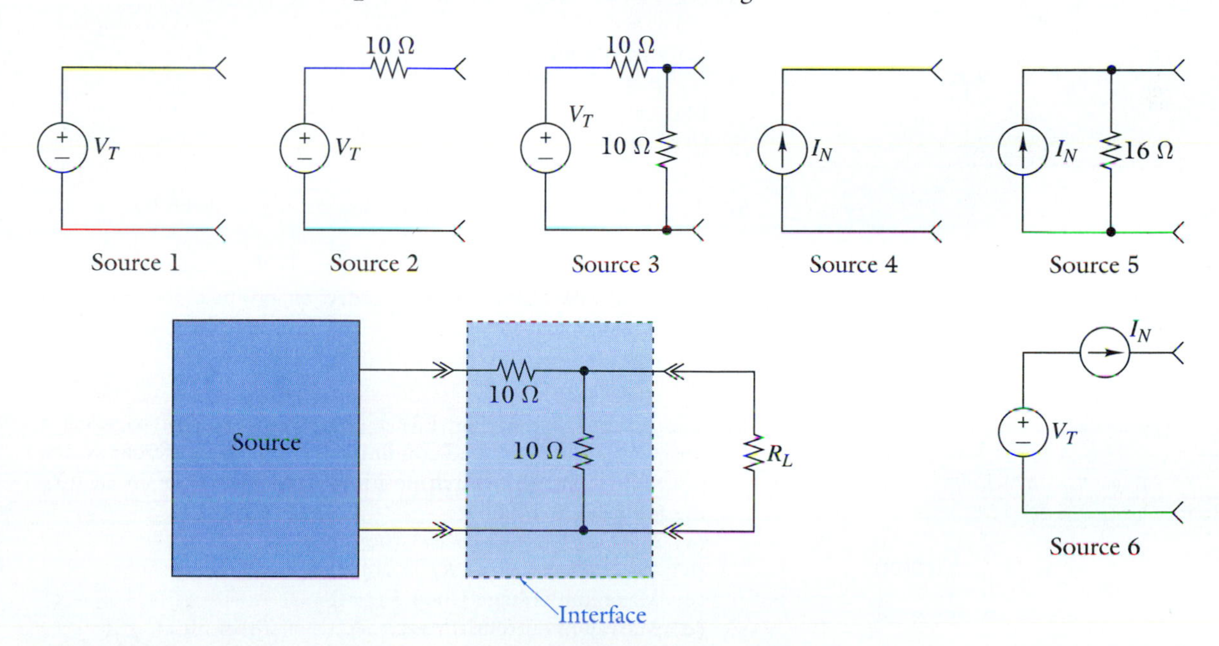

Figure P3-42

3-43 Show that Tellegen's Theorem applies to circuit C1 in Fig. P3-38 with $R_L = 10\ \Omega$.

3-44 Repeat Problem 3-43 for circuit C2 in Fig. P3-38 with $R_L = 15\ \text{k}\Omega$.

3-45 Repeat Problem 3-43 for circuit C3 in Fig. P3-38 with $R_L = 27.7\ \Omega$.

3-46 Repeat Problem 3-43 for circuit C4 in Fig. P3-38 with $R_L = 584.7\ \Omega$.

3-47 Repeat Problem 3-43 for circuit C5 in Fig. P3-38 with $R_L = 13.3\ \Omega$.

3-48 Repeat Problem 3-43 for circuit C6 in Fig. P3-38 with $R_L = 600\ \text{k}\Omega$.

ERO 3-5 Interface Circuit Design (Sect. 3-5)

Given a source-load interface with specified constraints, adjust the circuit parameters or design a two-port interface circuit to achieve given signal transfer objectives.

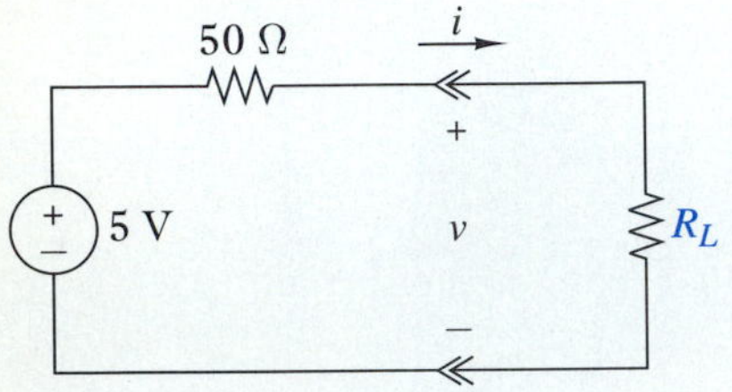

Figure P3-49

3-49 For the source-load interface in Fig. P3-49, find the values of R_L needed to meet each of the following conditions:
(a) $v = 1$ V
(b) $v = 5$ V
(c) $p = 100$ mW
(d) $v = 2.5$ V
(e) $i = 100$ mA
(f) $p = 125$ mW
(g) $v = 5$ V
(h) $i = 0$ A
(i) $p = 200$ mW

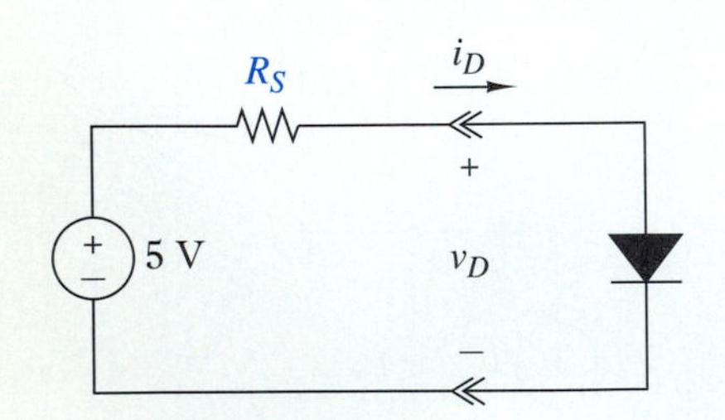

Figure P3-50

3-50 The diode in Fig. P3-50 has i-v characteristics given by

$$i_D = 10^{-15}\left(e^{40v_D} - 1\right)$$

Determine the values of R_S needed to produce the following interface signals:
(a) $v_D = 0.7$ V
(b) $i_D = 0.5$ mA

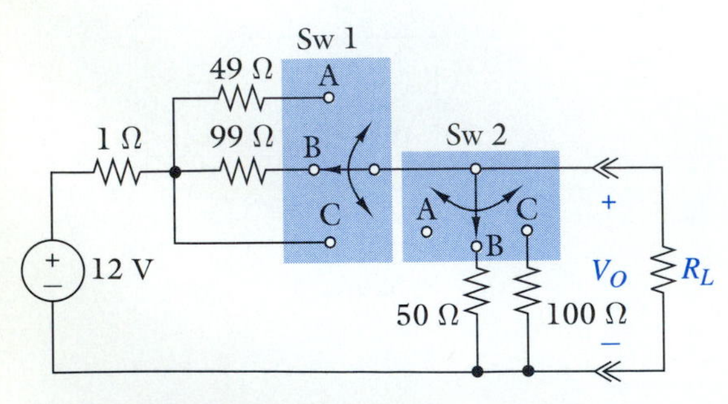

Figure P3-51

3-51 The novel device in Fig. P3-51 permits an operator to select a variety of source and load conditions by selecting different switch positions. Find the switch positions required to obtain each of the following conditions:
(a) Maximum power to R_L when $R_L = 100\ \Omega$.
(b) $V_O = 6$ volts when $R_L = 100\ \Omega$.
(c) Maximum voltage when $R_L = 1\ \text{k}\Omega$.
(d) Maximum current through R_L when $R_L = 50\ \Omega$.
(e) 40-mA current through R_L when $R_L = 100\ \Omega$.

(f) 60 mA current through R_L when $R_L = 100\ \Omega$.
(g) $V_O = 4$ volts when $R_L = 50\ \Omega$.

❖ **3-52** **(D)**
(a) Design an interface circuit that delivers 2 V to the 50-Ω load in Fig. P3-52.
(b) Design another interface circuit that delivers 50 mA to the load.

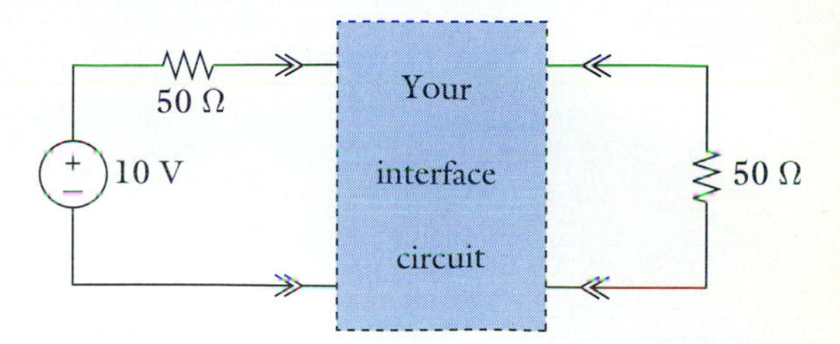

Figure P3-52

❖ **3-53** **(D)** The source circuit in Fig. P3-53 must always "see" an equivalent resistance of 50 Ω to avoid damage. Design an interface circuit using only resistors that ensures that the source "sees" about 50 Ω when the electronic switch is on or off. The interface circuit must transfer at least 1% of the source voltage to the load. The switch characteristics are $R_{switch} = 100\ \Omega$ when $V_G = 5$ V (on) and $R_{switch} = 10^{10}\ \Omega$ when $V_G = 0$ V (off).

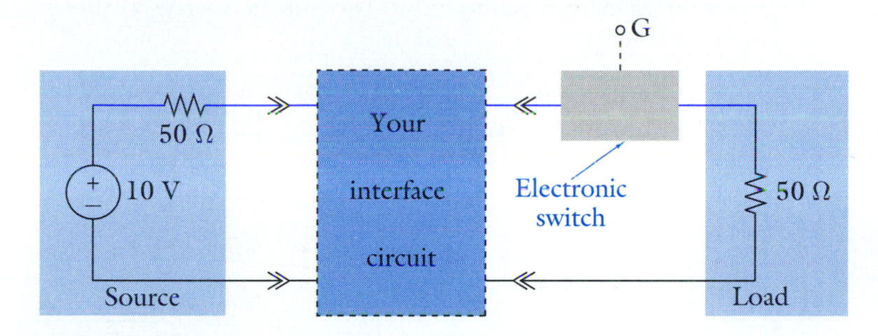

Figure P3-53

3-54 The 1-kΩ resistor in Fig. P3-54 is normally powered by the 110-V source. In case of 110-V power failure, a 12-V backup source is provided. Design an interface that delivers $v_L = 10$ V and $i_L \geq$

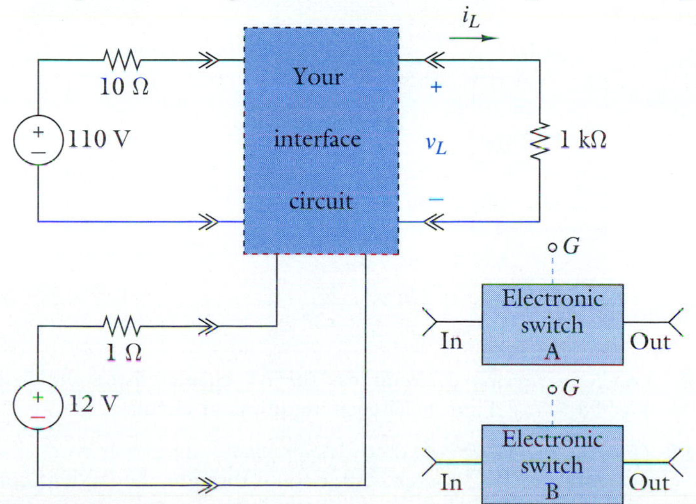

Figure P3-54

2 mA to the 1-kΩ resistor under normal and backup power conditions. Use the electronic switches shown in the figure. Switch A characteristics are $R_{switch} = 100\ \Omega$ when $V_G = 0$ V and $R_{switch} = 10^{10}\ \Omega$ when $V_G = 10$ V. Switch B characteristics are $R_{switch} = 100\ \Omega$ when $v_G = 10$ V and $R_{switch} = 10^{10}\ \Omega$ when $V_G = 0$ V. The switch gates draw no current but you must show how to connect the gates to the main source.

CHAPTER INTEGRATING PROBLEMS

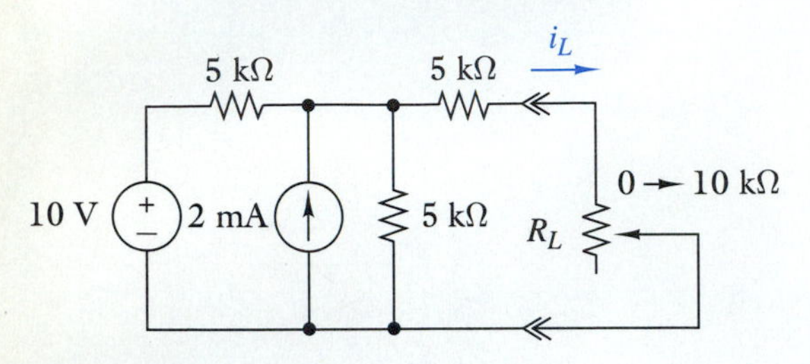

Figure P3-55

3-55 **(A)** How much does i_L vary when R_L in Fig. P3-55 varies from 0 to 10 kΩ?

3-56 **(A)** A three-terminal nonlinear element has the i-v characteristics shown in Fig. P3-56(a) and (b). Find V_{AB}, V_{CB}, I_A, and I_C when the element is connected to two voltage sources as shown in Fig. P3-56(c).

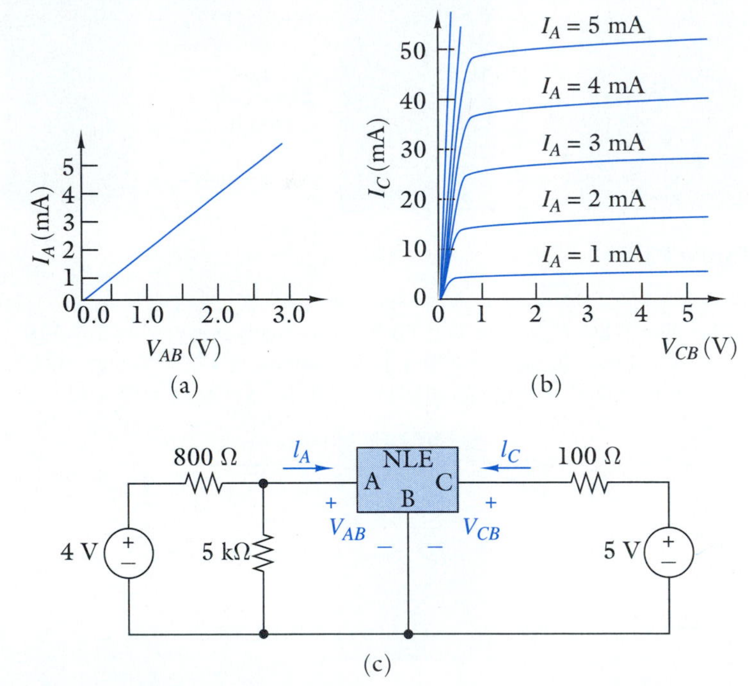

Figure P3-56

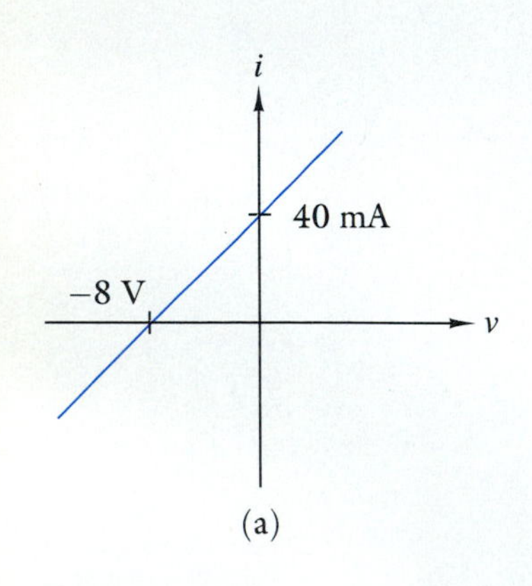

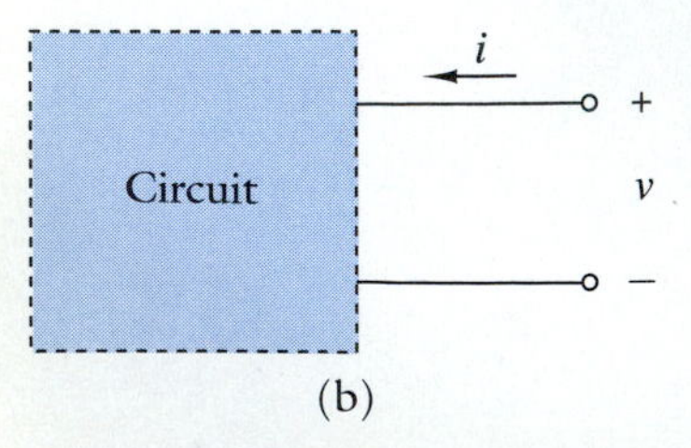

Figure P3-57

3-57 **(A)** A two-terminal circuit has the i-v characteristics shown in Fig. P3-57(a). Find its Thévenin equivalent circuit.

❖❖ **3-58** **(E)** A student was asked to deliver maximum power to a 75-Ω load from a 50-Ω source. After some thought the student connected a 150-Ω resistor in parallel with the 75-Ω load. Does this connection supply maximum power to the 75-Ω load? Explain.

❖ **3-59** **(D)** Design an interface circuit so the 200-Ω source in Fig. P3-59 delivers $V_T/8$ volts to each of the systems in the figure.

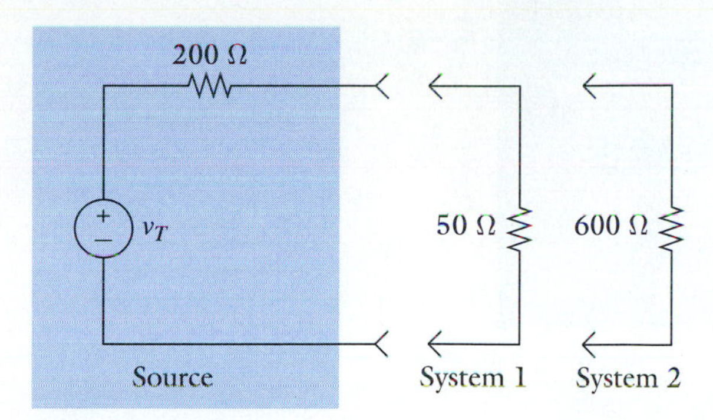

Figure P3-59

3-60 **(A)** The bridge circuit in Fig. P3-60 contains a two-terminal nonlinear element whose i-v characteristics are $i = v^2$. Derive expressions for i_x and v_x in terms of R_x. Check your derivation by showing that $i_x = 0$ and $v_x = 0$ when $R_x = 1$ Ω.

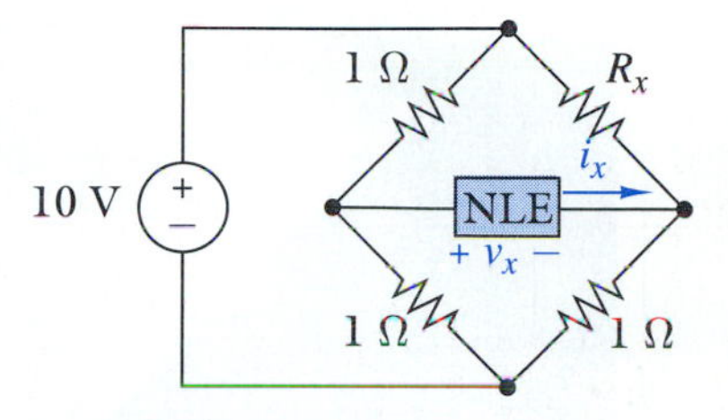

Figure P3-60

3-61 **(A)** Select the value of R_L in Fig. P3-61 so that maximum power is delivered to R_L. What is the value of the maximum power delivered to R_L? How much power does the source supply when maximum power is delivered to R_L?

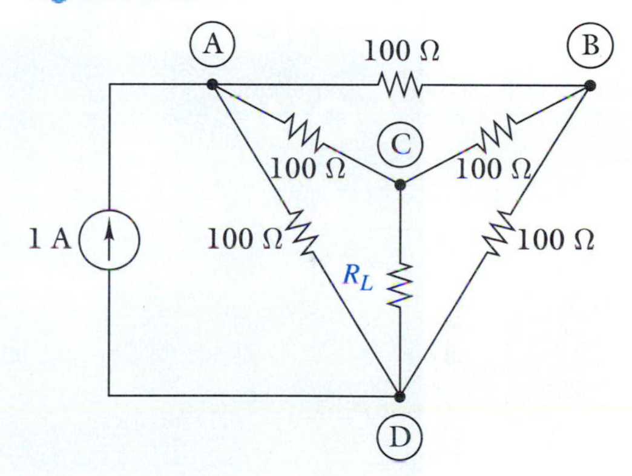

Figure P3-61

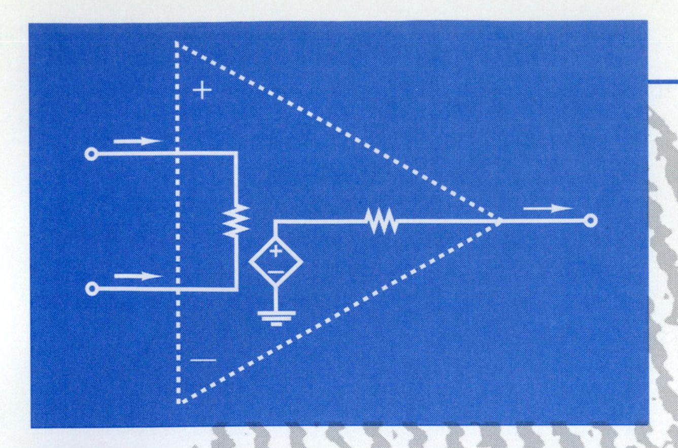

Chapter Four

"Then came the morning of Tuesday, August 2, 1927, when the concept of the negative feedback amplifier came to me in a flash while I was crossing the Hudson River on the Lackawanna Ferry, on my way to work."

Harold S. Black, 1927
American Electrical Engineer

Active Circuits

The integrated circuit (ic) operational amplifier (OP AMP) is the workhorse of present day linear electronic circuits. However, to operate as a linear amplifier the OP AMP must be provided with "negative feedback". The concept of the negative feedback amplifier is one of the key inventions of all time. During the 1920s Harold S. Black (1898–1983) had been working for several years without much success on the problem of improving the performance of vacuum tube amplifiers in telephone systems. The feedback amplifier solution came to him suddenly on the way to work. To document his invention he promptly sketched and signed the canonic feedback amplifier diagram, plus a derivation of the feedback gain equation on his morning copy of the *New York Times*. His invention paved the way for the development of worldwide communication systems and spawned whole new areas of technology such as feedback control systems and robotics.

The four dependent sources introduced in this chapter are the circuit elements needed to model linear active circuits. The analysis techniques used on circuits containing these new elements are illustrated in the second section. Dependent sources and resistors are used to model many electrical and electronic components, including transistor and OP AMP devices discussed in the third and fourth sections. The next two sections are devoted to using the OP AMP model to develop circuit realizations of basic analog signal-processing functions such as summers, subtractors, and inverters. These building blocks are used to design interface circuits in the seventh section. The final section presents the comparator, an application in which the OP AMP intentionally operates in the nonlinear regions of its characteristics.

4-1 LINEAR-DEPENDENT SOURCES

This chapter treats the modeling and analysis of circuits containing active devices such as transistors or operational amplifiers. Active devices are required to achieve signal amplification, one of the most important signal-processing functions in electrical engineering. The amplification concept can be described using the block diagram shown in Fig. 4-1. *Signal amplification* occurs when the input x and output y have the same dimensions and the proportionality factor K is greater than one. The resistance circuits studied thus far cannot produce dimensionless K's greater than one.

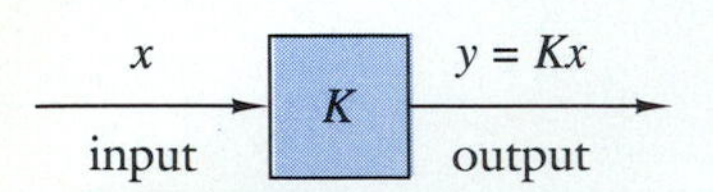

Figure 4-1 Block diagram for a scalar multiplier.

When active devices operate in a linear mode they can be modeled using resistors and one or more of the four dependent source elements shown in Fig. 4-2. The dominant feature of a dependent source is that the strength or magnitude of the voltage source (VS) or current source (CS) is proportional to—that is, controlled by—a voltage (VC) or current (CC) appearing elsewhere in the circuit. For example, the dependent source model for a current-controlled current source (CCCS) is shown in Fig. 4-2(c). The output current βi_1 is dependent on the input current i_1 and the dimensionless factor β. This dependency should be contrasted with the characteristics of the independent voltage and current sources we studied earlier. The current delivered by an ideal independent current source does not depend on the circuit to which it is connected. Likewise, the voltage produced by an ideal independent voltage source is not affected by the circuit. To emphasize this difference, dependent sources are represented by the diamond symbol shown in Fig. 4-2, in contrast to the circle symbol used for independent sources.

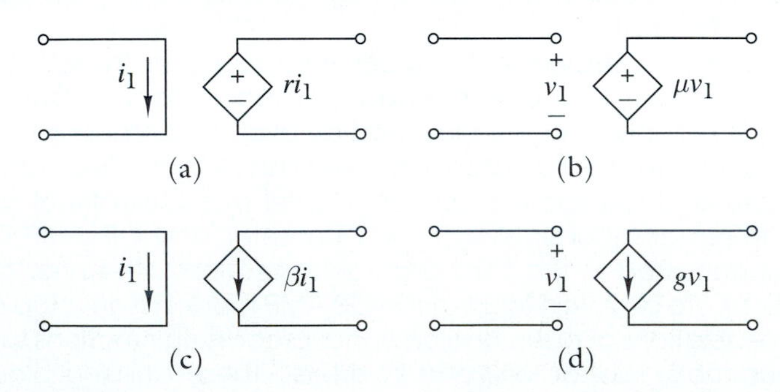

Figure 4-2 Dependent source circuit symbols: (a) Current-controlled voltage source. (b) Voltage-controlled voltage source. (c) Current-controlled current source. (d) Voltage-controlled current source.

Caution: This book uses the diamond symbol to indicate a dependent source and the circle to show an independent source. However, some books use the circle symbol for both dependent and independent sources.

Several matters of notation and symbols should be mentioned. Each dependent source is characterized by a single parameter, either μ, β, r, or g. These parameters are often somewhat loosely called the *gain* of the controlled source. Strictly speaking, the parameters μ and β are dimensionless quantities called the *voltage gain* and *current gain*, respectively. The parameter r has the dimensions of ohms, and is called the *transresistance*, a contraction of transfer resistance. The parameter g is then called *transconductance* and has the dimensions of siemens.

Dependent sources are elements used in circuit analysis. But in some respects they are conceptually different from the other circuit elements we have used. The linear resistor and ideal switch are models of devices called resistors and switches. But you will not find controlled sources in parts lists and catalogs. For this reason dependent sources are more abstract, since they are not models of identifiable physical devices. Dependent sources are used in combination with other elements to model active devices such as transistors and OP AMPs.

In Chapter 3 we learned that a voltage source acts like a short circuit when it is turned off. Likewise, a current source behaves like an open circuit when it is turned off. The same results apply to controlled sources, with one important difference. Controlled sources cannot be turned on and off independently because they depend on excitation supplied by independent sources.

Some consequences of this dependency are illustrated in Fig. 4-3. When the independent current source is on it forces the condition $i_1 = i_S$. Through controlled source action the current controlled voltage source is on and its output is

$$v_O = ri_1 = ri_S$$

When the independent current source is off ($i_S = 0$), it acts like an open circuit forcing the condition $i_1 = 0$. The dependent source is now off and its output is

$$v_O = ri_1 = 0$$

When the independent source is off the dependent source now acts like a short circuit.

In other words, turning the independent source on and off turns the dependent source on and off as well. We must be careful when applying the superposition principle to active circuits, since the state of a dependent source depends on excitation supplied by independent sources. In a linear circuit containing dependent sources, the superposition principle states that the

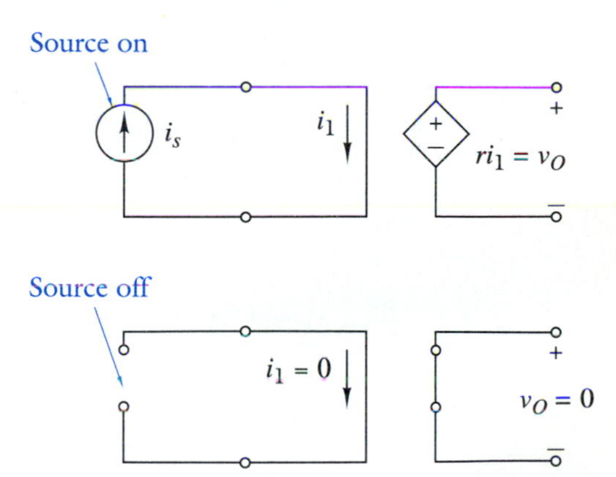

Figure 4-3 Effect on a dependent source of turning off and independent source.

response due to all independent sources acting simultaneously is equal to the sum of the responses due to each independent source acting one at a time.

4-2 Analysis of Circuits with Dependent Sources

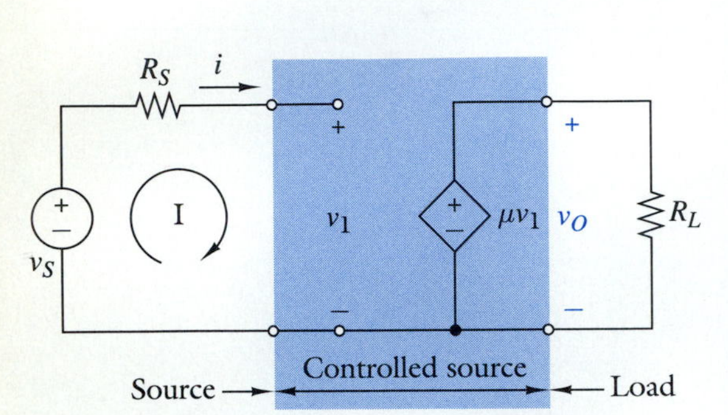

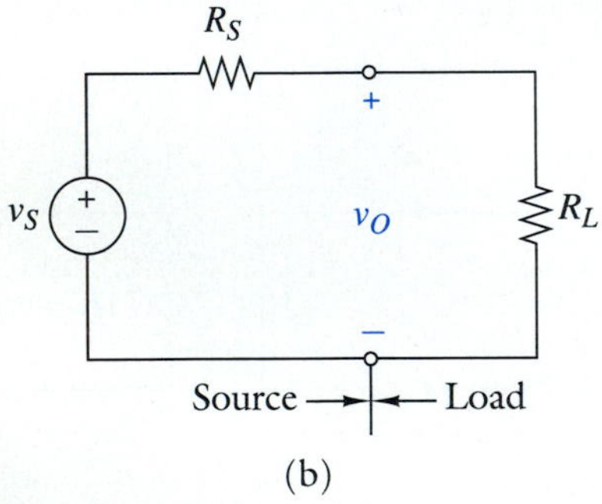

Figure 4-4 (a) Active circuit. (b) Passive circuit.

The analysis of circuits containing dependent sources is basically the same as the resistive circuits we have previously studied. Kirchhoff's laws still apply, and so the connection constraints are the same. The element constraints are still independent of the connections. But we have a new set of elements, so the combined element and connection constraints yield circuit behavior that is significantly different from passive circuits. Our analysis examples are chosen to highlight these differences.

An example of a circuit with a voltage controlled source is shown in Fig. 4-4(a). Applying Kirchhoff's voltage law around Loop I yields

$$v_S = R_S i + v_1$$

but $i = 0$, since the loop contains an open circuit. Therefore, $v_1 = v_S$ and the controlled voltage source output is equal to μv_S. Since v_O defined across a resistor in parallel with the dependent source we have

$$v_O = \mu v_S \tag{4-1}$$

The output voltage is directly proportional to the input voltage. When the magnitude of μ is greater than one the circuit boosts the input and is called an ***amplifier***. When the magnitude of μ is less than one the circuit decreases the input and is called an ***attenuator***. When the magnitude of μ is equal to one the circuit is called a ***buffer*** or ***follower***.

The advantage of linear active circuits can be demonstrated by performing an analysis of the passive circuit in Fig. 4-4(b). Applying voltage division yields

$$v_O = \left(\frac{R_L}{R_S + R_L}\right) v_S \tag{4-2}$$

When we compare the analysis results in Eqs. (4-1) and (4-2), we can make the following observations:

1. The output voltage of the passive circuit [Eq. (4-2)] depends on the source and load resistance, while the output of the active circuit [Eq. (4-1)] depends only on the gain μ.

2. The output of the passive circuit is limited by the maximum signal transfer conditions derived in Chapter 3, while the output of the active circuit is limited by the gain μ.

The controlled source in the active circuit provides unilateral signal transfer from source to load. The coupling is unilateral because the independent source sees an open circuit and the load resistance sees an ideal voltage source. This eliminates the source-load interaction that lead to the maximum signal transfer conditions.

To pursue the difference further, consider the power delivered to the load by the controlled source circuit. Using Eq. (4-1)

$$P_L = \frac{(v_O)^2}{R_L} = \frac{(\mu v_S)^2}{R_L} \tag{4-3}$$

The load power does **not** depend on the source resistance and therefore is not limited by the maximum power transfer condition ($R_S = R_L$). In fact, the power delivered by the independent source is zero, since it is connected to an open circuit. This appears to be a contradiction, since the active circuit produces finite output power with zero input power.

The contradiction is resolved by noting that a dependent source is a model of active devices that require an external power sources to operate in their linear range. In general, we do not show the external power supply in circuit diagrams. We assume that the external supply can provide whatever power the dependent source requires. With real devices this is not the case, and engineers must ensure that the power limits device and external supply are not exceeded. In our discussion of OP AMPs we will consider some limitations of the external supply.

To continue the analysis of dependent source circuits, we find the input-output relationship (v_O/i_S) of the circuit in Fig. 4-5. To determine the dependent source output we must find i_1. Using current division in the input circuit we have

$$i_1 = \left(\frac{R_S}{R_S + R_1}\right) i_S \tag{4-4}$$

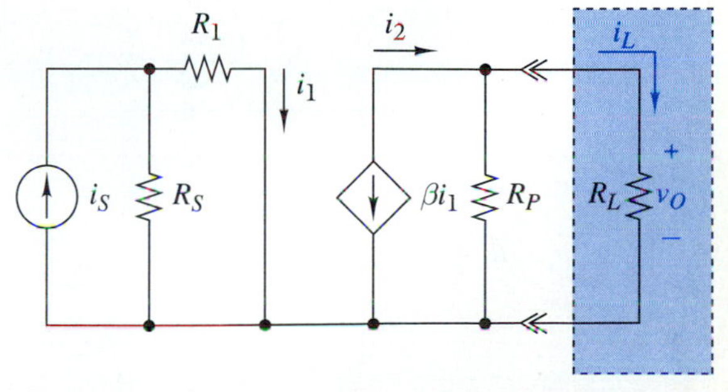

Figure 4-5 A circuit with a dependent source.

To obtain $v_O = i_L R_L$ we apply current division in the output circuit:

$$i_L = \left(\frac{R_P}{R_P + R_L}\right) i_2 \tag{4-5}$$

By KCL $i_2 = -\beta i_1$. Inserting $i_2 = -\beta i_1$ into Eq. (4-5) and then using Eq. (4-4) to eliminate i_1 yields

$$i_L = \left[\frac{R_P}{R_P + R_L}\right](-\beta i_1) = \left[\frac{R_P}{R_P + R_L}\right]\left[\frac{-\beta R_S}{R_S + R_1}\right] i_S \tag{4-6}$$

Finally, we determine v_O using Ohm's law:

$$v_O = i_L R_L = \left[\frac{R_L R_P}{R_P + R_L}\right]\left[\frac{-\beta R_S}{R_S + R_1}\right] i_S \quad (4\text{-}7)$$

and the input-output relationship is

$$\frac{v_O}{i_S} = \frac{-R_L R_P \beta R_S}{(R_P + R_L)(R_S + R_1)} \quad (4\text{-}8)$$

The negative sign in this equation occurs because of the orientation of the dependent source in the problem. Many active devices produce a negative output for a positive input, and vice versa. This common property of active devices is called ***signal inversion***. During circuit analysis it is important that we keep track of signal inversions.

To analyze and design active circuit interfaces it is useful to determine their Thévenin or Norton equivalent circuit. The following example illustrates finding the Thévenin equivalent of a dependent source circuit.

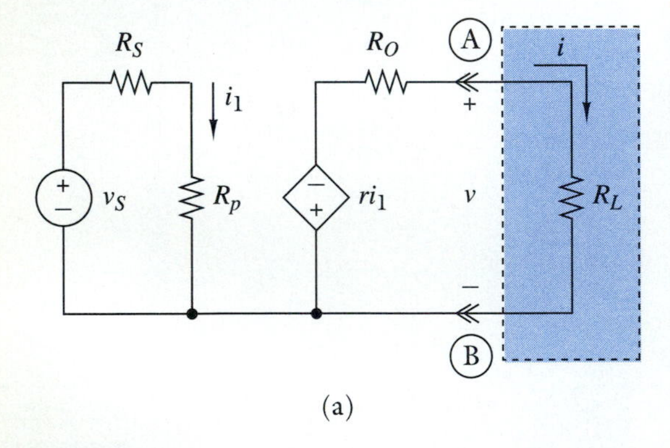

(a)

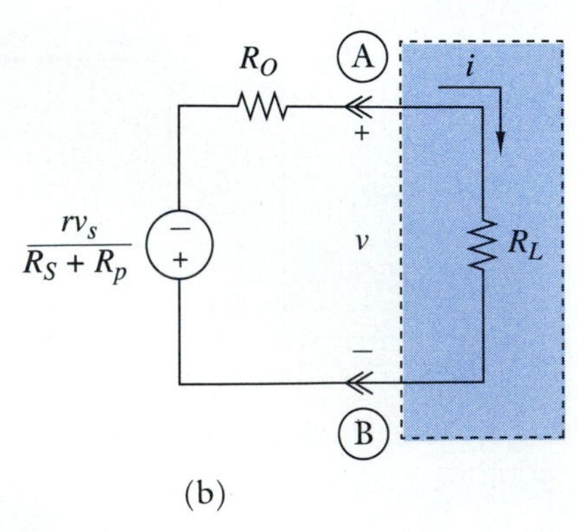

(b)

Figure 4-6

EXAMPLE 4-1

Find the Thévenin equivalent at the A-B interface in the circuit of Fig. 4-6(a).

SOLUTION:

The Thévenin equivalent circuit seen by the load is found by disconnecting the load resistor and determining the open-circuit voltage v_{OC} and short-circuit current i_{SC}. To obtain v_{OC} we need the current i_1, since this current controls the dependent voltage source in the output circuit. Using series equivalence in the input circuit yields i_1:

$$i_1 = \frac{v_s}{R_S + R_P}$$

With an open circuit connected at the output interface the current $i = 0$ and the voltage across R_O is zero. Therefore, the open-circuit voltage is

$$v_{OC} = -ri_1 = -r\left(\frac{v_S}{R_S + R_P}\right)$$

The minus sign in this equation results from the reference directions assigned to the interface voltage v and the orientation of the current-controlled voltage source.

To find i_{SC} we connect a short circuit at the interface and calculate the current through the short circuit.

$$i_{SC} = -\frac{ri_1}{R_O} = \frac{-rv_S}{R_O\,(R_S + R_P)}$$

The minus sign in this equation results from the reference directions assigned to the interface current i and the orientation of the current-controlled voltage source. The Thévenin resistance is then obtained as follows

$$R_T = \frac{V_{oc}}{i_{SC}} = R_O$$

The Thévenin equivalent circuit seen by the load R_L is shown in Fig. 4-6(b).

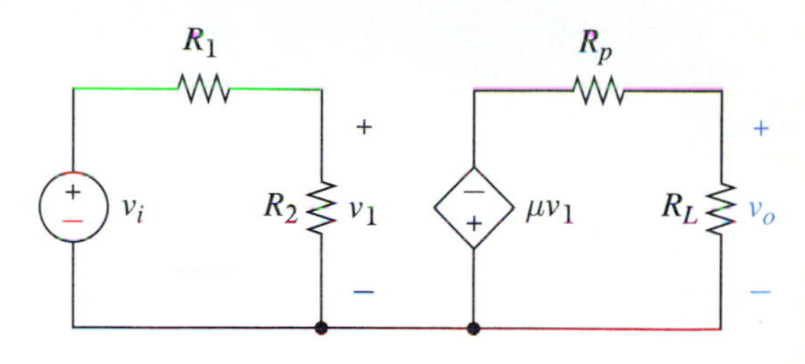

Figure 4-7

EXERCISE 4-1

Find v_o/v_i for the circuit in Fig. 4-7.

ANSWER:

$$\frac{v_o}{v_i} = \left(\frac{-R_L}{R_P + R_L}\right)\left(\frac{\mu R_2}{R_1 + R_2}\right)$$

The Effect of Feedback

The results of Example 4-1 show that finding the Thévenin or Norton equivalent of an active circuit is a straightforward process as long as there is no feedback in the circuit. *Feedback* occurs when there is a path from the output of a dependent source back to the input control circuit.

To illustrate the effect of feedback we determine the equivalent input resistance R_{EQ} between terminals A and B in Fig. 4-8(a). The circuit in Fig. 4-8(a) has no excitation from independent sources, so the dependent current source is inactive and acts like an open circuit. Intuitively it appears that the equivalent input resistance between A and B should be R_L. However, the effect of the dependent source has not been accounted for, since the circuit has no excitation. What is the input resistance when an independent source activates the controlled source?

To find the effect of the dependent source, we must determine the equivalent input resistance under excitation. We provide the excitation by connecting a test current source i_{TEST} as shown in Fig. 4-8(b). To obtain the input resistance we need to compute the voltage produced across terminals A and B. Once we have the voltage v_{TEST} in Fig. 4-8(b), we obtain R_{EQ} as

$$R_{EQ} = \frac{v_{TEST}}{i_{TEST}} \tag{4-9}$$

Using KCL at Node C we can relate i_1 and i_L:

$$i_L = i_1 + \beta i_1$$
$$= (1 + \beta)i_1$$

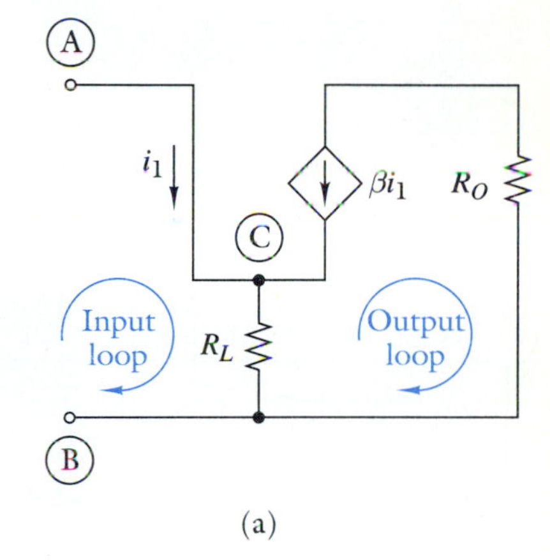

(a)

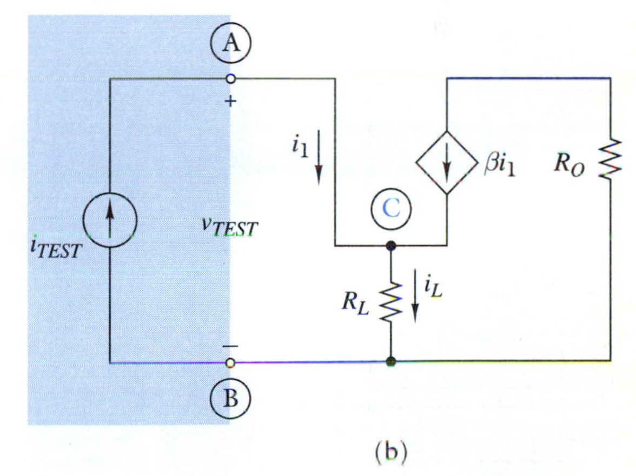

(b)

Figure 4-8 A dependent source circuit with feedback.

A KCL equation at Node A yields $i_1 = i_{TEST}$, so by substitution

$$i_L = (1+\beta)i_{TEST}$$

By Ohm's law

$$v_{TEST} = R_L i_L$$

and v_{TEST} is found as

$$v_{TEST} = R_L(1+\beta)i_{TEST} \qquad (4\text{-}10)$$

Comparing Eqs. (4-9) and (4-10) we conclude that

$$R_{EQ} = (1+\beta)R_L \qquad (4\text{-}11)$$

This value of R_{EQ} is significantly different from our initial intuitive estimate of R_L. Moreover, the circuit in question is a model of a transistor circuit and the gain parameter β would typically be on the order of 100. So not only is our intuition wrong; it is wrong by two orders of magnitude!

Active circuits with feedback can have Thévenin input or output resistances many orders of magnitude larger or smaller than the "lookback" resistance with no excitation. We can supply the required excitation using a test source as shown by the example in Fig. 4-8. We can also supply the excitation by leaving the independent sources on and obtaining the Thévenin equivalent by finding the open-circuit voltage and short-circuit current. The next example shows the importance of leaving independent sources on when calculating a Thévenin equivalent.

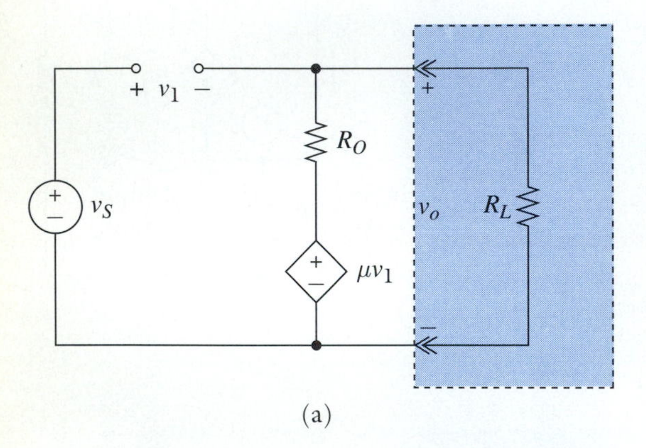

(a)

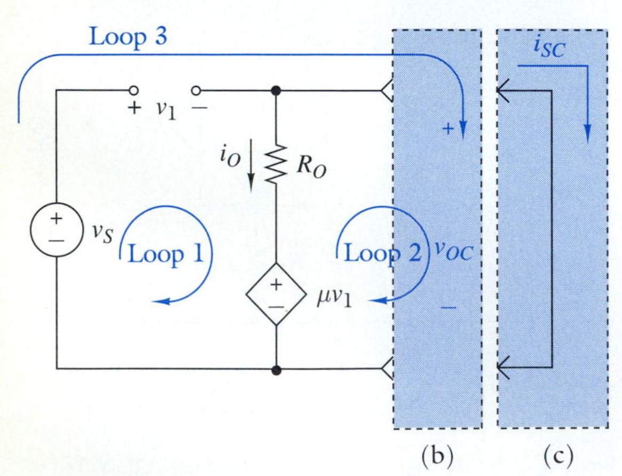

(b) (c)

Figure 4-9

EXAMPLE 4-2

Find the Thévenin equivalent at the output interface of the circuit in Fig. 4-9(a).

SOLUTION:

In this circuit the control voltage v_1 is the open-circuit voltage between the indicated terminals. If we turn v_S off the control voltage is zero and the dependent voltage source becomes a short circuit. Disconnecting the load [Fig. 4-9(b)] and looking back into the circuit we see a resistance of R_O. To show that the Thévenin resistance is actually much smaller than R_O, we turn the dependent source on and find the open-circuit voltage. Applying KVL around Loop 1 yields

$$-v_S + v_1 + R_O i_O + \mu v_1 = 0$$

But $i_O = 0$ because the load is an open circuit. Solving the above KVL equation for v_S gives

$$v_S = (\mu + 1)v_1$$

Writing a KVL equation around Loop 2 we obtain

$$v_{OC} = R_O i_O + \mu v_1$$

Again we note that $i_O = 0$ and we can solve for v_1.

$$v_1 = \frac{v_{OC}}{\mu}$$

Now we can substitute for v_1 into the Loop 1 equation and solve for v_{OC}:

$$v_{OC} = \frac{\mu}{\mu + 1} v_S$$

To solve for i_{SC} we connect a short circuit across the output terminal. Applying KVL around Loop 1 produces the following result:

$$-v_S + v_1 + i_O R_O + \mu v_1 = 0$$

With a short circuit connected at the output interface, the KVL equation around Loop 2 takes the form

$$i_O R_O + \mu v_1 = 0$$

Substituting this result in the Loop 1 equation and noting that $i_O = -i_{SC}$ yields the short-circuit current.

$$i_{SC} = \frac{\mu v_1}{R_O} = \frac{\mu v_S}{R_O}$$

Given v_{OC} and i_{SC} we obtain the Thévenin resistance R_T as

$$R_T = \frac{v_{OC}}{i_{SC}} = \frac{R_O}{1 + \mu}$$

The Thévenin equivalent circuit seen by the output is shown in Fig. 4-10.

$$R_T = \frac{R_O}{1 + \mu}$$

$$v_T = \frac{\mu}{1 + \mu} v_S$$

Figure 4-10

The circuit in Fig. 4-9 is a model of an OP AMP circuit called a voltage follower. The resistance R_O for a general purpose OP AMP is in the order of 100 Ω, while the gain μ is about 10^5. Thus, the Thévenin resistance of the voltage follower with feedback is not 100 Ω as we initially thought, but less than a milliohm!

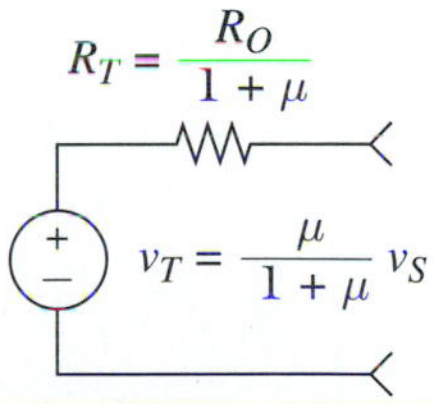

EXERCISE 4-2

Find the Thévenin equivalent circuit at the interface in Fig 4-11.

ANSWER:

$$V_T = \frac{\mu}{\mu - 1} v_s$$

$$R_T = R_O$$

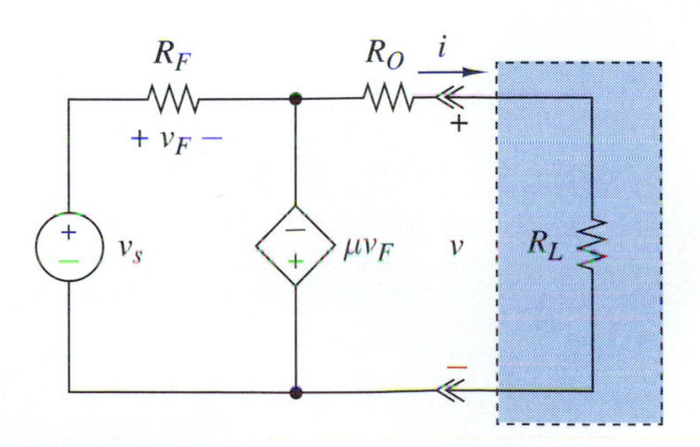

Figure 4-11

4-3 THE TRANSISTOR

The *bipolar junction transistor* (BJT) was developed in 1947 and is still in common use. The term *bipolar* means that positive (holes) and negative (electrons) charge carriers are needed to produce the *i*-*v* characteristics of the BJT. Prior to 1960 all transistors were manufactured and utilized as discrete devices. With the advent of integrated circuits in the 1960s, the predominant technology became the integrated fabrication of transistors along with other devices on a tiny silicon chip. Most of today's integrated circuits use field effect transistors (FETs), which rely on electric fields to control a single type of charge carrier. However, it is not the purpose of this section to describe the detailed operation of either the BJT or the FET. That development is left to subsequent courses in semiconductor electronics.

The purpose of this section is to show how dependent sources are used to create a model of an active device. The BJT is the device we will study. The BJT has several models for different applications, but we limit our study to the *large-signal model*. This relatively simple model has a limited domain of application. Nevertheless, it serves to bridge the gap between the controlled source elements defined in the previous section and the OP AMP device studied in the next section.

Large-Signal Model of a BJT

The circuit symbol of the BJT is shown in Fig. 4-12. The device has three terminals called the *emitter (E)*, *base (B)*, and *collector (C)*. The two voltages V_{BE} and V_{CE} shown in the figure are called the base-emitter voltage and the collector-emitter voltage. The three currents I_C, I_E, and I_B shown in Fig. 4-12 are called the collector, emitter and base currents. Writing a global KCL equation at the circle in the transistor circuit symbol yields

$$I_C + I_E + I_B = 0 \qquad (4\text{-}12)$$

This KCL constraint means that only two of the three transistor currents shown in Fig. 4-12 can be independently specified.

Figure 4-12 also shows the large-signal model of the BJT. The model divides the nonlinear *i*-*v* characteristics of the BJT into three regions, or modes, called *cutoff*, *active*, and *saturation*. Each mode is represented by a circuit consisting of circuit elements we have previously studied. In the cutoff mode the circuit consists of two open circuits; therefore, the device currents I_B, I_E, and I_C are zero. In the saturation mode the open circuits

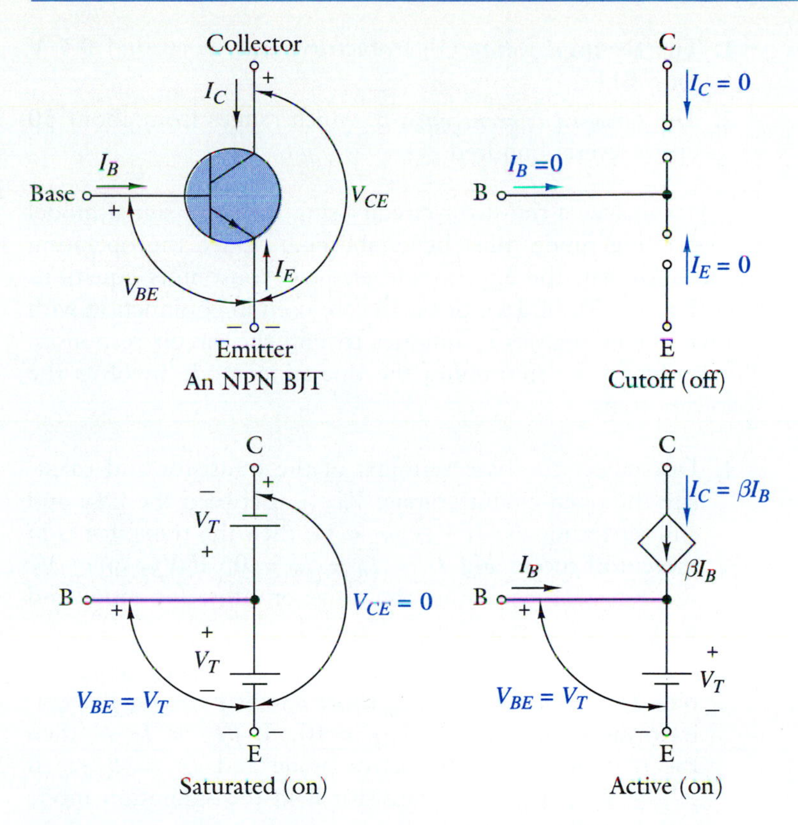

Figure 4-12 Large signal model of a bipolar NPN transistor.

are replaced by two independent voltage sources.[1] Because of the orientation of the voltage sources the device voltages $V_{BE} = V_T$ and $V_{CE} = V_T - V_T = 0$ when the transistor is saturated. In the active mode the circuit contains an independent voltage source V_T and a current-controlled current source whose gain is β. Because of the orientation of these sources we have $V_{BE} = V_T$ and $I_C = \beta I_B$ when the transistor is in the active mode.

In summary, the large-signal model produces three sets of element equations:

Cutoff Mode:	$I_B = I_E = I_C = 0$	(4-13)
Active Mode:	$V_{BE} = V_T$ and $I_C = \beta I_B$	(4-14)
Saturation Mode:	$V_{BE} = V_T$ and $V_{CE} = 0$	(4-15)

The model has two parameters:

[1]In this section the symbol V_T represents the transistor threshold voltage, not the Thévenin voltage.

1. The *threshold voltage* V_T, which typically is around 0.7 V for a BJT
2. The *forward current gain* β, which ranges from about 50 up to several hundred

To analyze a transistor circuit using the large-signal model the operating mode must be established. Once the operating mode is known, the appropriate element constraints equations from Eq. (4-13), (4-14), or (4-15) are used in conjunction with normal circuit analysis techniques to find the circuit responses. The problem of determining the operating mode involves the following steps:

1. Disconnect the base terminal of the transistor and calculate the open-circuit voltage $V_{BE,OC}$ between the base and emitter terminals. If $V_{BE,OC} < V_T$ then the transistor is in the cutoff mode and $I_C = I_B = I_E = 0$. If $V_{BE,OC} \geq V_T$ the transistor is in either the active or saturation mode and we proceed to step 2 with $V_{BE} = V_T$.
2. With $V_{BE} = V_T$ we calculate I_B and the short-circuit current $I_{C,SC}$ obtained with a short circuit between the collector and emitter (i.e., $V_{CE} = 0$). If $\beta I_B < I_{C,SC}$, then the transistor is in the active mode and $I_C = \beta I_B$. If $\beta I_B \geq I_{C,SC}$ then the transistor is in the saturation mode and $I_C = I_{C,SC}$.

The next two examples illustrate the analysis of transistor circuits using the large-signal model.

EXAMPLE 4-3

Find I_C and V_{CE} in the transistor circuit of Fig. 4-13 when $V_T = 0.7$ V, $\beta = 100$, $R_B = 100$ kΩ, $R_C = 1$ kΩ, $V_{CC} = 5$ V, and $V_S = 5$ V.

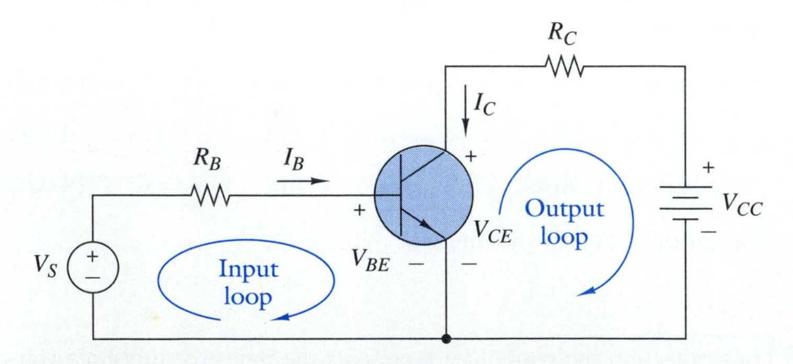

Figure 4-13

SOLUTION:

To find the required responses we determine the transistor operating mode using the two steps listed above.

STEP 1: With the transistor disconnected we see that $V_{BE,OC} = V_S = 5$ V. Since $V_T = 0.7$ V we have $V_{BE,OC} > V_T$ and the transistor is either in the saturation or active modes with $V_{BE} = V_T = 0.7$ V.

STEP 2: Since $V_{BE} = 0.7$ V the base current I_B is

$$I_B = \frac{V_S - V_T}{R_B} = \frac{5 - 0.7}{10^5} = 43\ \mu\text{A}$$

The short-circuit current at the collector terminal is

$$I_{C,SC} = \frac{V_{CC}}{R_C} = \frac{5}{10^3} = 5\ \text{mA}$$

Since $\beta I_B = 100 \times 43 \times 10^{-6} = 4.3$ mA is less than $I_{C,SC} = 5$ mA, we know the transistor is in the active mode with $I_C = \beta I_B = 4.3$ mA.

Writing a KVL equation around the output loop in Fig. 4-13 yields

$$-V_{CE} - I_C R_C + V_{CC} = 0$$

from which we obtain V_{CE} as

$$V_{CE} = V_{CC} - I_C R_C = 5 - 4.3 = 0.7\ V$$

EXAMPLE 4-4

Suppose the input signal in Fig. 4-13 increases linearly as a function of time as $v_S(t) = t$ V for $t > 0$. Plot the output voltage $v_{CE}(t)$ versus t when $V_T = 0.7$ V, $\beta = 100$, $R_B = 100$ kΩ, $R_C = 1$ kΩ, and $V_{CC} = 5$ V.

SOLUTION:

In Example 4-3 we analyzed this circuit and found that $V_{BE,OC} = v_S$ and $I_{C,SC} = 5$ mA. Using the first of these results we see that for $0 \leq t < 0.7$s the transistor is in cutoff, since $V_{BE,OC} = v_S(t) < 0.7$ V. In the cutoff mode $I_C = 0$, so the voltage across R_C is zero and $v_{CE} = V_{CC} = 5$ V.

At $t = 0.7$ s the transistor switches to the active mode since $V_{BE,OC} = 0.7$ V. In the active mode $V_{BE} = 0.7$ V and the collector current is

$$I_C = \beta I_B = 100 \frac{v_S(t) - 0.7}{10^5} \qquad t \geq 0.7\ \text{s} \qquad \text{(4-16)}$$

Combining Eq. (4-16) and a KVL equation around the output loop in Fig. 4-13 shows that as long as the transistor is in the active mode the output voltage is

$$v_{CE}(t) = V_{CC} - I_C R_C$$
$$= 5 - 100 \times \left[\frac{v_S(t) - 0.7}{10^5}\right] \times 10^3$$
$$= 5.7 - v_S(t) \quad (4\text{-}17)$$

The last line in Eq. (4-17) says that $v_{CE} = 5.7 - t$, which is the equation of a straight line in the v_{CE} versus t plane with a slope of -1.

At $t = 5.7$ s Eq. (4-16) shows that

$$I_C = 100 \times \left[\frac{5.7 - 0.7}{10^5}\right] = 5 \text{ mA} = I_{C,SC}$$

and the transistor switches to the saturation mode where $v_{CE} =$ 0 V. The transistor remains in saturation, since for $t \geq 5.7$ Eq. (4-16) shows that $\beta I_B \geq I_{C,SC}$.

The above analysis can be summarized as follows:

1. For $0 \leq t < 0.7$ s the transistor is cutoff and $v_{CE} = 5$ V.
2. For $0.7 \leq t < 5.7$ the transistor is active and $v_{CE} = 5.7 - t$ V.
3. For $5.7 \leq t$ the transistor is saturated and $v_{CE} = 0$ V.

Figure 4-14 shows a plot of these results. The abscissa in this plot is labeled in volts as well as seconds and the different operating modes are indicated.

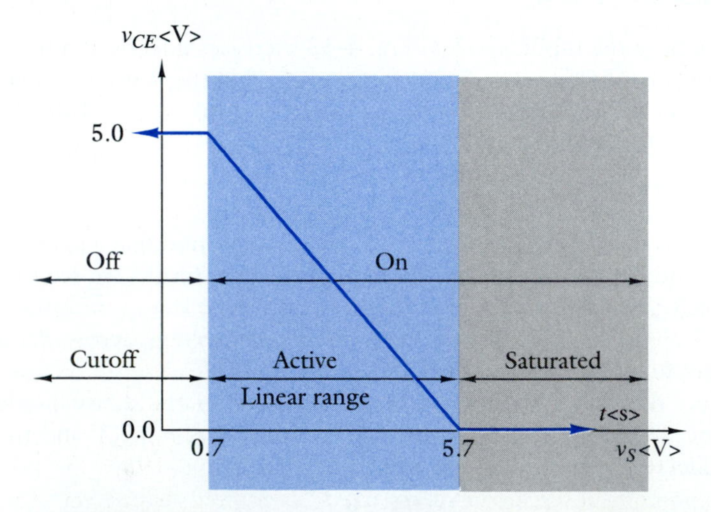

Figure 4-14

COMMENT:

Figure 4-14 is called the transfer characteristics of the transistor circuit in Fig. 4-13. In digital applications the transistor is switched between the cutoff and saturation modes passing quickly through the active mode. In analog application the transistor remains in the active mode and serves as an amplifier. In the next section we will see that the OP AMP also has three operating modes and transfer characteristics similar to those in Fig. 4-14.

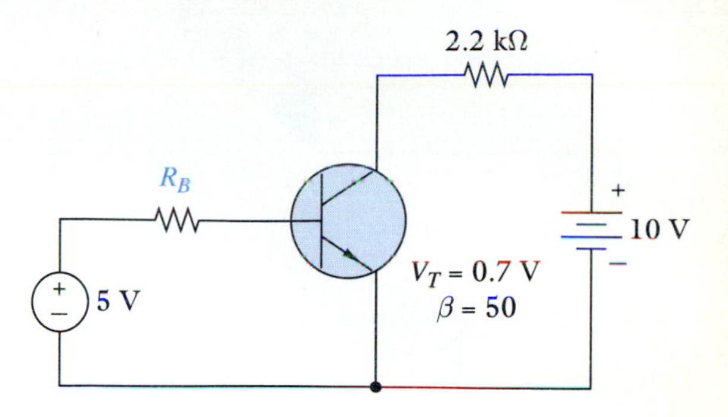

Figure 4-15

EXERCISE 4-3

Find the value of R_B in the transistor circuit of Fig. 4-15 for which $\beta I_B = I_{C,SC}$. That is, find the value of base resistance that causes the transistor to switch from the active to the saturation mode. Is this value a maximum or minimum? Explain.

ANSWER:

$R_B = 47.3$ kΩ. This value is a minimum because further decreases in R_B would increase I_B driving the transistor further into saturation.

EXERCISE 4-4

Find the Norton equivalent circuit at the indicated terminals of the transistor circuit in Fig. 4-16. Assume the transistor is operating in the active mode.

ANSWER:

$$R_N = R_C \qquad I_N = \frac{V_{CC} - \dfrac{\beta(V_S - V_T)}{R_B}}{R_C}$$

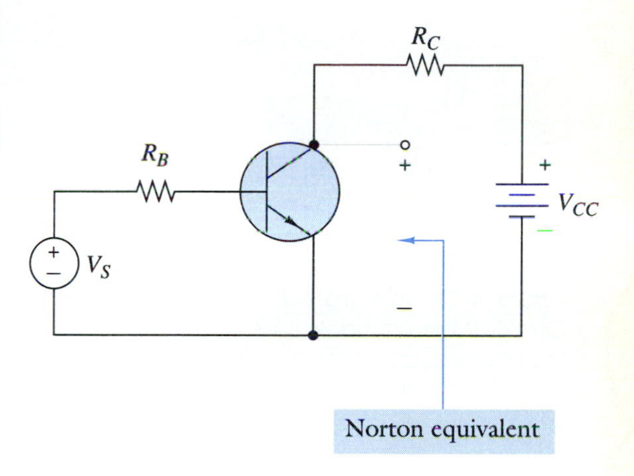

Figure 4-16

4-4 THE OPERATIONAL AMPLIFIER

The operational amplifier is the premier linear active device made available by integrated circuit technology. The term *operational amplifier* was apparently first used in a 1947 paper by John R. Ragazzini and his colleagues, who reported on work carried out for the National Defense Research Council during World War II. The paper described the use of high-gain direct current (dc) amplifiers in circuits that perform mathematical operations (addition, subtraction, multiplication, division, integration, etc.); hence the name "operational" amplifier. For more than a decade the most important applications were general- and special-purpose analog computers using vacuum tube amplifiers. In the early 1960s general-purpose, discrete-transistor, operational amplifiers became readily available and by the mid-1960s the first commercial integrated circuit OP AMPs entered the market. The transition from vacuum tubes to integrated circuits resulted in a

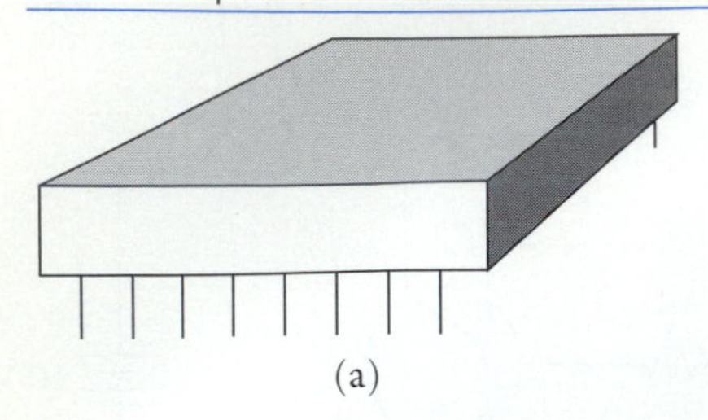

(a)

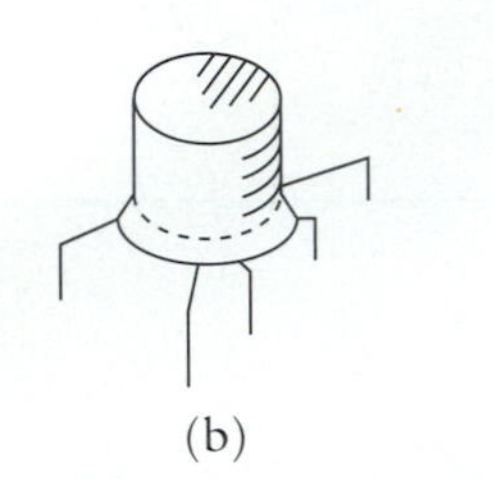

(b)

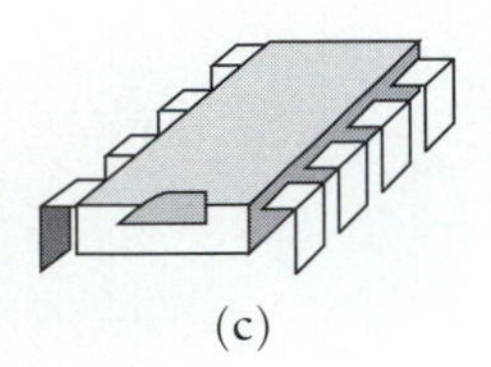

(c)

Figure 4-17 Integrated OP AMP circuit packages: (a) Encapsulated hybrid. (b) TO can. (c) Dual in-line package (DIP).

decrease in size, power consumption, and cost of OP AMPs by nearly three orders of magnitude. By the early 1970s the integrated circuit version became the dominant active device in analog circuits.

The device itself is a complex array of transistors, resistors, diodes, and capacitors all fabricated and interconnnected on a tiny piece of silicon commonly called a chip. Figure 4-17 shows examples of ways the completed device is packaged[2] for use in circuits. In spite of its complexity, the OP AMP can be modeled by rather simple i-v characteristics. We do not need to concern outselves with what is going on inside the package; rather, we can treat the OP AMP as a circuit element with a particular set of constraints between the voltages and currents as its external terminals.

OP AMP Notation

There are certain matters of notation and nomenclature that must be discussed before we develop a circuit model for the OP AMP. The OP AMP is a five-terminal device as shown in Fig. 4-18(a). The "+" and "−" symbols identify the input terminals and are a shorthand notation for the noninverting and inverting input terminals, respectively. These "+" and "−" symbols identify the two input terminals and have nothing to do with the polarity of the voltages applied. The other terminals are the output and the positive and negative supply voltage, usually labeled $+V_{CC}$ and $-V_{CC}$. While some OP AMPs have more than five terminals, these five are always present and are the only ones we will use in this text. Figure 4-18(b) shows how these terminals are arranged in a very common package, the 8-pin miniDIP.

The two power supply terminals in Fig. 4-18 are not usually shown in circuit diagrams. Be assured that they are always there. The power for signal amplification comes through these terminals from an external power source. The $+V_{CC}$ and $-V_{CC}$ voltages applied to these terminals also determine the upper and lower limits on the OP AMP output voltage.

Figure 4-19(a) shows a complete set of voltage and current variables for the OP AMP, while Fig. 4-19(b) shows the abbreviated set of signal variables we will use. All voltages are defined with respect to a common reference node, usually ground. Volt-

[2]The most common packages are the 8-pin metal, the 10-pin flatpack, and the 8-pin MINIDIP, or 14-pin DIP, configurations. DIP stands for dual in-line package and refers to the alignment of the pins on the package.

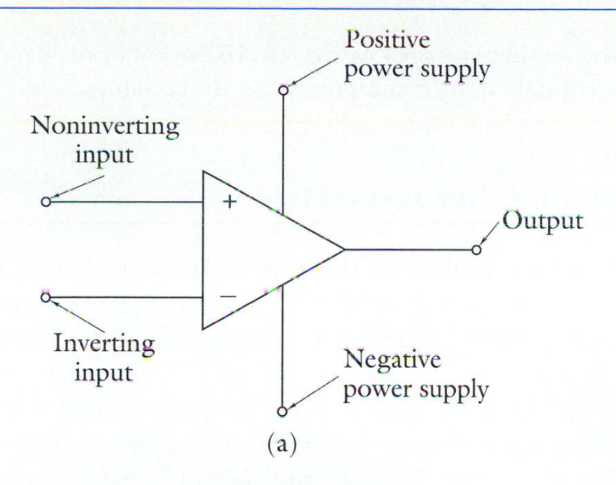

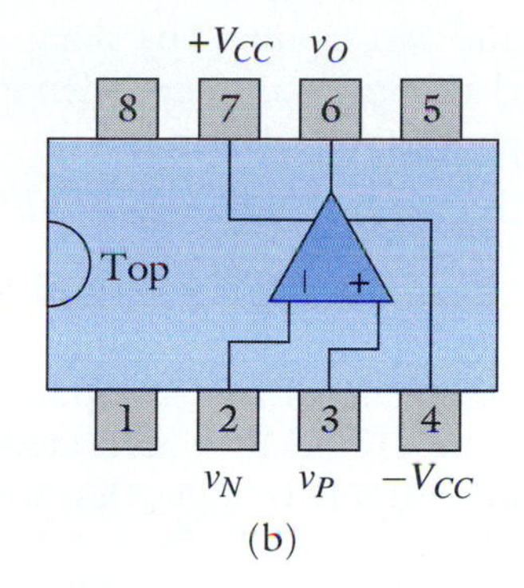

Figure 4-18 The OP AMP: (a) Circuit symbol. (b) Pin out diagram for an 8-pin DIP.

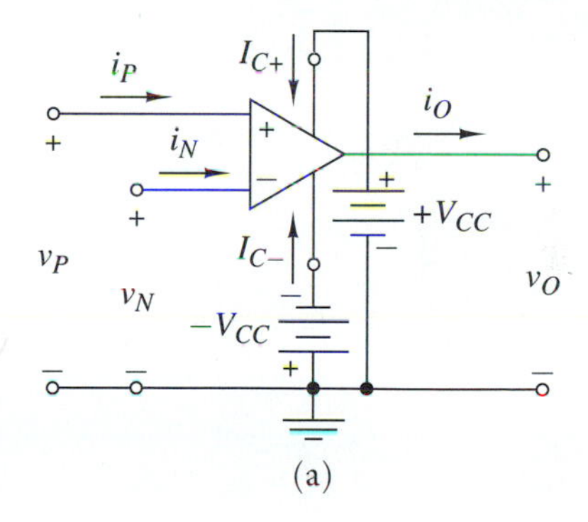

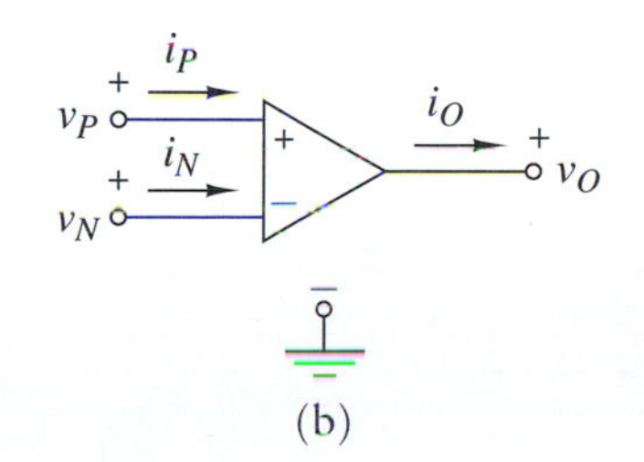

Figure 4-19 OP AMP voltage and current definitions: (a) Complete set. (b) Shorthand set.

age variables v_P, v_N, and v_O are defined by writing a voltage symbol beside the corresponding terminals. This notation means the "+" reference mark is at the terminal in question and the "−" reference mark is at the reference or ground terminal. The reference directions for the currents are directed in at input terminals and out at the output. At times the abbreviated set of current variables may appear to violate KCL. For example, a global KCL equation for the complete set of variable in Fig. 4-19(a) is

$$i_O = I_{C+} + I_{C-} + i_P + i_N \qquad \text{(correct equation)} \qquad (4\text{-}18)$$

A similar equation using the shorthand set of current variables in Fig. 4-19(b) reads

$$i_O = i_N + i_P \qquad \text{(incorrect equation)} \qquad (4\text{-}19)$$

This latter equation is *not* correct, since it does not include all of the currents. More importantly, it implies that the output current comes from the inputs. In fact, this is wrong. The input currents are very small, ideally zero. The output current comes from the supply voltages as Eq. (4-18) points out, even though these terminals are not shown on the abbreviated circuit diagram.

Transfer Characteristics

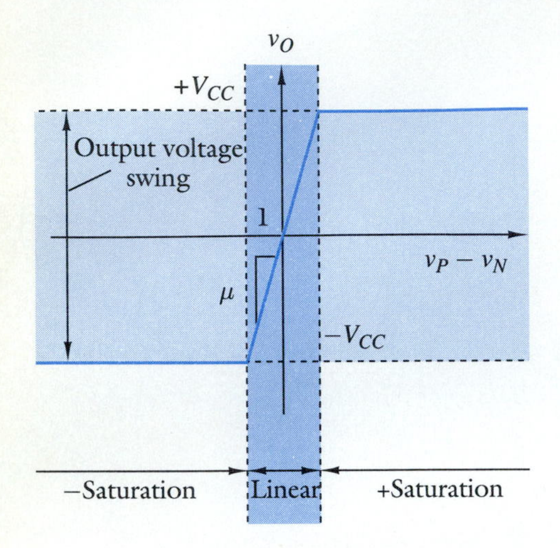

Figure 4-20 OP AMP transfer characteristics.

The dominant feature of the OP AMP is the transfer characteristics shown in Fig. 4-20. These characteristics provide the relationships between *the noninverting input* v_P, the *inverting input* v_N, and the *output voltage* v_O. The transfer characteristics are divided into three regions or modes called *+saturation*, *−saturation*, and *linear*. In the linear region the OP AMP is a *differential amplifier* because the output is proportional to the difference between the two inputs. The slope of the line in the linear range is called the *open-loop gain*, denoted as μ. In the linear region the input-output relation is

$$v_O = \mu\,(v_P - v_N) \tag{4-20}$$

The open-loop gain of an OP AMP is very large, usually greater than 10^5. As long as the net input $(v_P - v_N)$ is very small, the output will be proportional to the input. However, when $\mu|v_P - v_N| > V_{CC}$, the OP AMP is saturated and the output voltage is limited by the supply voltages (less some small internal losses).

Like the BJT described in the preceding section, the OP AMP has three operating modes:

1. +Saturation when $\mu(v_P - v_N) > V_{CC}$ and $v_O = +V_{CC}$.
2. −Saturation when $\mu(v_P - v_N) < -V_{CC}$ and $v_O = -V_{CC}$.
3. Linear when $\mu|v_P - v_N| < V_{CC}$ and $v_O = \mu(v_P - v_N)$.

In the last section of this chapter we will discuss an application in which the OP AMP operates in the saturation modes. However, everywhere else in this book we assume that the OP AMP is operating in the linear mode. In general, we want to analyze and design OP and AMP circuits that do not saturate for the given inputs.

Ideal OP AMP Model

A controlled source model of an OP AMP operating in its linear range is shown in Fig. 4-21. This model incudes an input resistance (R_I), an output resistance (R_O), and a voltage-controlled voltage source whose gain is the open-loop gain μ. Some typical ranges for these OP AMP parameters are given in Table 4-1,

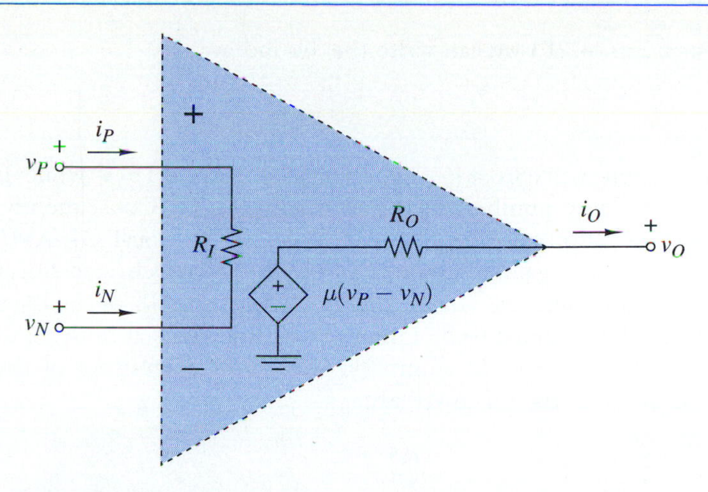

Figure 4-21 Dependent source model of an OP AMP operating in the linear mode.

along with the values for the ideal OP AMP.[3] The high input and low output resistances and high open-loop gain are the key attributes of an OP AMP. The ideal model carries these traits to the extreme limiting values.

TABLE 4-1

Name	Parameter	Range	Ideal Values
Open-loop gain	μ	10^5 to 10^8	∞
Input resistance	R_I	10^6 to 10^{13} Ω	$\infty\Omega$
Output resistance	R_O	10 to 100 Ω	0Ω
Supply voltages	$\pm V_{CC}$	±12 to ±24 V	

The controlled source model can be used to develop the i-v characteristics of the ideal model. We restricted our treatment to the linear region of operation. This means the output voltage is bounded as

$$-V_{CC} \le v_O \le +V_{CC}$$

[3]More detailed characteristics of OP AMP devices are given in Appendix A.

Using Eq. (4-20) we can write this bound as

$$-\frac{V_{CC}}{\mu} \leq (v_P - v_N) \leq \frac{V_{CC}}{\mu}$$

The supply voltages $\pm V_{CC}$ are typically about ± 15 V while μ is a very large number, usually 10^5 or greater. Consequently, linear operation requires that $v_P \approx v_N$. For the ideal OP AMP the open-loop gain is infinite ($\mu \rightarrow \infty$) in which case linear operation requires $v_P = v_N$. The input resistance R_I of the ideal OP AMP is assumed to be infinite, so the currents at both input terminals are zero. In summary, the i-v characteristics of the ***ideal model*** of the OP AMP are

$$\begin{aligned} v_P &= v_N \\ i_P &= i_N = 0 \end{aligned} \tag{4-21}$$

The implications of these element equations are illustrated on the OP AMP circuit symbol in Fig. 4-22.

Figure 4-22 Ideal OP AMP characteristics.

At first glance the element constraints of the ideal OP AMP appear to be fairly useless. They actually look more like connection constraints and are totally silent about the output quantities (v_O and i_O), which are usually the signals of interest. In fact, they seem to say that the OP AMP input terminals are simultaneously a short circuit ($v_P = v_N$) and an open circuit ($i_P = i_N = 0$). The ideal model of the OP AMP is useful because in linear applications feedback is always present. That is, in order for the OP AMP to operate in a linear mode it is necessary for there to be feedback paths from the output to one or both of the inputs. These feedback paths ensure that $v_P \approx v_N$ and allow us to analyze OP AMP circuits using the ideal OP AMP constraints.

Noninverting OP AMP

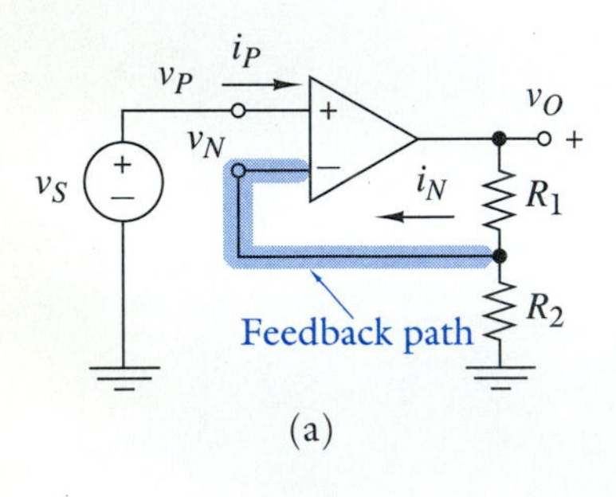

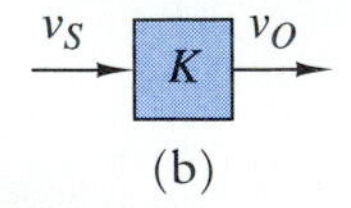

Figure 4-23 The noninverting OP AMP circuit.

To illustrate the effects of feedback let us determine the input-output characteristics of the circuit in Fig. 4-23. This circuit has a feedback path from the output to the inverting input via a voltage divider. Since the ideal OP AMP draws no current at either input ($i_P = i_N = 0$), we can use voltage division to determine the voltage at the inverting input:

$$v_N = \frac{R_2}{R_1 + R_2} v_O \tag{4-22}$$

The input source connection at the noninverting input requires the condition

$$v_P = v_S \tag{4-23}$$

But the ideal OP AMP element constraint demands that $v_P = v_N$; therefore, we can equate the right sides of Eqs. (4-22) and (4-23) to obtain the input-output relationship of the overall circuit.

$$v_O = \frac{R_1 + R_2}{R_2} v_S \tag{4-24}$$

The above analysis strategy uses the input source constraint together with the feedback path to determine the two OP AMP input voltages. The ideal OP AMP constraint requires the two input voltages to be equal and this equality determined the overall circuit input-output relationship.

The circuit in Fig. 4-23(a) is called a ***noninverting amplifier.*** The input-output relationship is of the form $v_O = K v_S$, which reminds us that the circuit is linear. Figure 4-23(b) shows the functional building block for this circuit, where the proportionality constant K is

$$K = \frac{R_1 + R_2}{R_2} \tag{4-25}$$

The constant K is called the ***closed-loop*** gain, since it includes the effect of the feedback path. When discussing OP AMP circuits it is necessary to distinguish between two types of gains. The first is the open-loop gain μ provided by the OP AMP device. The gain μ is a large number with a large uncertainty tolerance. The second type is the closed-loop gain K of the OP AMP circuit with a feedback path. The gain K must be much smaller than μ, and its value is determined by the resistance elements in the feedback path.

For example, the closed-loop gain in Eq. (4-25) is really the voltage division rule upside down. The uncertainty tolerance assigned to K is determined by the quality of the resistors in the feedback path, and not the uncertainty in the actual value of the closed-loop gain. In effect, feedback converts a very large but imprecisely known open-loop gain into a much smaller but precisely controllable closed-loop gain.

❖ Design Example

EXAMPLE 4-5

Design an amplifier with a closed-loop gain K of 10.

SOLUTION:

Using noninverting OP AMP circuit, the design problem is to select the values of the resistors in the feedback path. From Eq. (4-25) the design constraint is

$$10 = \frac{(R_1 + R_2)}{R_2}$$

We have one constraint with two unknowns. Arbitrarily selecting R_2 =10 kΩ we find R_1 =90 kΩ. These resistors would normally have high precision (±1% or less) to produce a precisely controlled closed-loop gain.

COMMENT:

The problem of choosing resistance values in OP AMP circuit design problems deserves some discussion. Although resistances from a few ohms to several hundred megohms are commercially available, we generally limit ourselves to the range from about 1 kΩ to perhaps 1 MΩ. The lower limit of 1 kΩ exists in part because of power dissipation in the resistors. Typically we use resistors with 1/4 watt power ratings. The maximum voltage in OP AMP circuits is usually 15 V. The smallest 1/4 W resistance we can use is $R_{MIN} \geq (15)^2/0.25 = 900\ \Omega$, or about 1 kΩ. The upper bound of 1 MΩ exists because it is difficult to maintain precision in a high-value resistance because of variation in surface leakage caused by humidity. High-value resistors are also noisy, which leads to problems when they are in the feedback path. The 1-kΩ to 1-MΩ range should be used as a guideline and not an inviolate design rule. Actual design choices are influenced by system-specific factors and changes in technology.

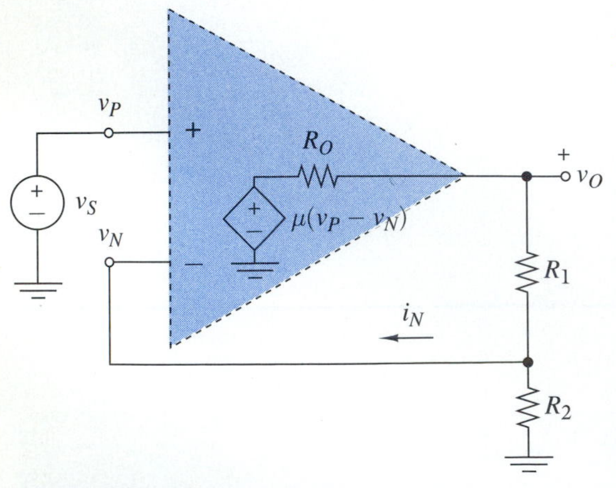

Figure 4-24 The noninverting OP AMP circuit with a dependent source model.

Effects of Finite Open-Loop Gain

We should pause here and reflect for a moment on the rather dramatic properties of the ideal OP AMP. The model has an infinite open-loop gain that is converted into a finite closed-loop gain through feedback. Real OP AMPs have very large, but finite open-loop gains. We now want to address the effect of large but finite gain on the closed-loop response of OP AMP circuits.

The circuit in Fig. 4-24 shows a finite gain OP AMP circuit model in which the input resistance R_I is infinite. As Table 4-1 shows, the actual values of OP AMP input resistance range from 10^6 to 10^{13} Ω, so no important effect is left out by ignoring this resistance. Examining the circuit we see that the noninverting input voltage is determined by the independent voltage source. The inverting input can be found by voltage division, since the current i_N is zero. In other words, Eqs. (4-22) and (4-23) apply to this circuit as well.

We next determine the output voltage in terms of the controlled source voltage using voltage division on the series connection of the three resistors R_O, R_1, and R_2:

$$v_O = \frac{R_1 + R_2}{R_O + R_1 + R_2}\mu\,(v_P - v_N)$$

Substituting v_P and v_N from Eqs. (4-22) and (4-23) yields

$$v_O = \left[\frac{R_1 + R_2}{R_O + R_1 + R_2}\right]\mu\left[v_S - \frac{R_2}{R_1 + R_2}v_O\right] \quad (4\text{-}26)$$

The intermediate result in Eq. (4-26) shows that feedback is present, since v_O appears on both sides of the equation. Solving for v_O yields

$$v_O = \frac{\mu\,(R_1 + R_2)}{R_O + R_1 + R_2\,(1 + \mu)}v_S \quad (4\text{-}27)$$

In the limit as $\mu \to \infty$ Eq. (4-27) reduces to

$$v_O = \frac{R_1 + R_2}{R_2}v_S = K v_S$$

where K is the closed-loop gain we previously found using the ideal OP AMP model.

To see the effect of a finite μ we ignore R_O in Eq. (4-27), since it is generally quite small compared with $R_1 + R_2$. With this approximation Eq. (4-27) can be written in the following form:

$$v_O = \frac{K}{1 + (K/\mu)}v_S \quad (4\text{-}28)$$

In this form it is clear that the closed-loop gain reduces to K as $\mu \to \infty$. Moreover, we see that the finite gain model yields a good approximation to the ideal results as long as $K \ll \mu$. In other words, the ideal model yields good results as long as the closed-loop gain is much less than the open-loop gain of the OP AMP device.

The feedback path also affects the active output resistance. To see this we construct a Thévenin equivalent circuit using the open-circuit voltage and the short-circuit current. Equation (4-28) is the open-circuit voltage and we need only find the short-circuit current. Connecting a short-circuit at the output in Fig. 4-24 forces $v_N = 0$ but leaves $v_P = v_S$. Therefore, the short-circuit current is

$$i_{SC} = \mu\,(v_S/R_O)$$

As a result the Thévenin resistance is

$$R_T = \frac{v_{OC}}{i_{SC}} = \frac{K/\mu}{1 + K/\mu}R_O$$

If the closed-loop gain K is much smaller than the open-loop gain μ, this expression reduces to

$$R_T = \frac{K}{\mu} R_O \approx 0 \ \Omega$$

The OP AMP circuit with feedback has an output resistance that is much smaller than the output resistance of the OP AMP device. In fact, it is essentially zero, since R_O is around 100 Ω and μ is 10^5 or more.

At this point we can summarize our discussion. We introduced the OP AMP as a five-terminal device including two supply terminals not normally shown on circuit diagrams. We then developed an ideal model of this device and used the model to analyze and design circuits that have feedback. Feedback must be present for the device to operate in the linear mode. The most dramatic feature of the ideal model is the assumption of infinite gain. Using a finite gain model we found that the ideal model predicts the closed-loop response rather well as long as the closed-loop gain K is much less than the open-loop gain μ. We also discovered that the Thévenin output resistance of the feedback amplifier is essentially zero.

The ideal OP AMP i-v constraints in Eq. (4-21) will be used throughout the rest of the book to analyze OP AMP circuits. Such circuits have essentially zero output resistance, which means the output voltage is unaffected by any reasonable load. The loads must be within the power-handling capability of the actual device. But the same restrictions apply to all of the models we use, including the linear resistance model of the load itself. Unless otherwise stated, from now on OP AMP means the ideal model.

Voltage Follower

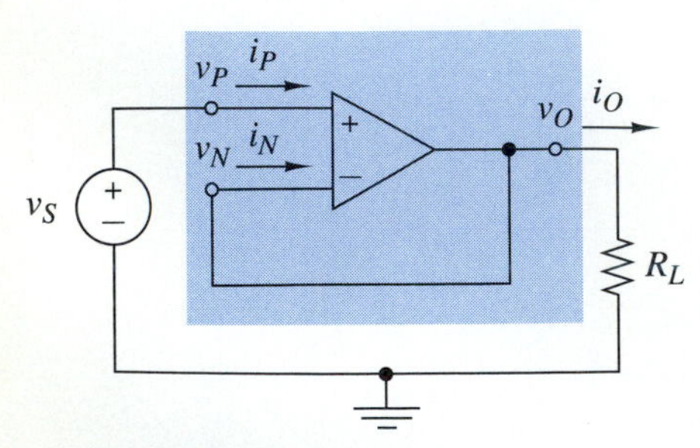

Figure 4-25 The OP AMP voltage follower circuit.

The circuit in Fig. 4-25 is called a *voltage follower* or *buffer*. In this case the feedback path is a direct connection from the output to the inverting input. The feedback connection forces the condition $v_N = v_O$. The input signal source is directly connected to the noninverting input causing the condition $v_P = v_S$. The ideal OP AMP model requires $v_P = v_N$, so we conclude that $v_O = v_S$. By inspection the closed-loop gain is $K = 1$. Since the output identically equals the input, we say the output follows the input, hence the name voltage follower.

The voltage follower is used in interface circuits because it isolates the source and load. By Ohm's law the current delivered to the load is $i_O = v_O/R_L$. But since $v_O = v_S$ the output current can be written as

$$i_O = v_S/R_L$$

If we now write a KCL equation at the reference node we discover an apparent dilemma:

$$i_P = i_O$$

For the ideal model $i_P = 0$, but the above equation says that i_O cannot be zero unless v_S is zero. It appears that KCL is violated.

The dilemma is resolved by noting that the circuit diagram does not include the supply terminals. The output current actually comes from the power supply and not from the input. This dilemma arises only at the reference node (the ground terminal). In OP AMP circuits as in all circuits, KCL must be satisfied. However, we must be alert to the fact that a KCL equation at the reference node could yield misleading results because the power supply terminals are not usually shown in circuit diagrams.

EXERCISE 4-5

The voltage follower circuit in Fig. 4-25 has $v_S = 1.5$ V and a load $R_L = 1$ kΩ. Compute the power absorbed by the load resistor and the power supplied by the voltage source v_S. Explain any discrepancies.

ANSWERS:
$p_L = 2.25$ mW; $p_S = 0.0$ mW.

DISCUSSION:
The answers appear to violate conservation of energy. The input current $i_P = 0$ so the voltage source supplies zero power. The load current is calculated from Ohm's law to be 1.5 mA, resulting in a power consumption of 2.25 mW. The load current comes from the external power supply not shown in the circuit diagram.

4-5 BASIC LINEAR OP AMP CIRCUITS

Analog signal processing systems are often constructed by connecting relatively simple OP AMP circuits in a building block fashion. This section introduces a basic set of circuits that can be used as building blocks. We have already introduced one of these circuits—the noninverting amplifier discussed in the preceding section. The other circuits are the inverting amplifier, the summer, and the subtractor. The key to using the *building block approach* is to recognize the feedback pattern and to isolate the basic circuit as a building block. The first example illustrates this process.

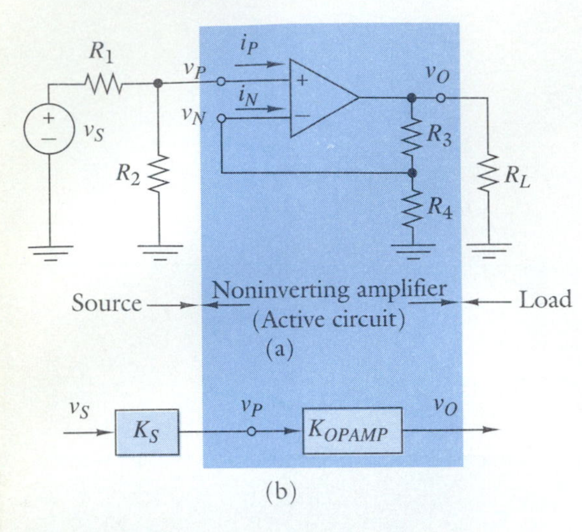

Figure 4-26

EXAMPLE 4-6

Find the input-output relationship of the circuit in Fig. 4-26(a).

SOLUTION:

When the circuit is partitioned as shown in Fig. 4-26(a), we recognize two building block gains: (1) K_S the proportionality constant of the voltage divider circuit, and (2) $K_{OP\ AMP}$ the gain of the noninverting amplifier. Since the input current to the noninverting OP AMP is zero, we can use voltage division to find K_S:

$$K_S = \frac{v_P}{v_S} = \frac{R_2}{R_1 + R_2}$$

Since the noninverting amplifier has zero output resistance, the load R_L has no effect on the output voltage v_O. Using the results in Eq. (4-24) the gain of the noninverting OP AMP circuit is

$$K_{OP\ AMP} = \frac{v_O}{v_P} = \frac{R_3 + R_4}{R_4}$$

The overall circuit gain found as

$$K_{CIRCUIT} = \frac{v_O}{v_S} = \left[\frac{v_O}{v_P}\right]\left[\frac{v_P}{v_S}\right]$$
$$= K_{OP\ AMP} \times K_S$$
$$= \left[\frac{R_3 + R_4}{R_4}\right]\left[\frac{R_2}{R_1 + R_2}\right]$$

Note that the overall gain is the product of the two building block gains because (1) the gain of the voltage divider is not affected by the OP AMP, since the current at the noninverting input is zero, and (2) the gain of the OP AMP circuit is not affected by the load resistance, since the OP AMP has zero output resistance.

The Inverting Amplifier

The circuit in Fig. 4-27 is called an *inverting amplifier*. The key feature of this circuit is that the input signal and the feedback are both applied at the inverting input. Since the noninverting input is grounded, we have $v_P = 0$, an observation we will use shortly. We analyze this circuit by applying KCL at Node A

$$i_1 + i_2 = i_N \tag{4-29}$$

The element constraints for this circuit can then be written as

$$i_1 = \frac{(v_S - v_N)}{R_1}$$

$$i_2 = \frac{(v_O - v_N)}{R_2} \tag{4-30}$$

$$i_N = 0$$

The first two element constraints in Eq. (4-30) are derived from Ohm's law and KVL, while the third constraint is one of the OP AMP i-v characteristics. Substituting Eq. (4-30) into (4-29) yields

$$\frac{v_S - v_N}{R_1} + \frac{v_O - v_N}{R_2} = 0 \tag{4-31}$$

We are now ready to use the fact that the noninverting input is grounded. The OP AMP voltage constraint requires $v_P = v_N$, but since $v_P = 0$ it follows that $v_N = 0$. We solve Eq. (4-31) for the input-output relationship as

$$v_O = -\left(\frac{R_2}{R_1}\right) v_s \tag{4-32}$$

This result is of the form $v_O = K v_S$, where K is the closed-loop gain. However, in this case the closed-loop gain $K = -R_2/R_1$ is negative indicating a signal inversion, hence the name inverting amplifier. We can use the same block diagram symbol in Fig. 4-27(b) to indicate either the inverting or non-inverting OP AMP configuration, since both circuits provide a gain of K. We observe that the closed-loop gain of the inverting amplifier depends only on the resistors in the feedback path.

The next example shows that more complex circuits can often be reduced to the inverting amplifier configuration.

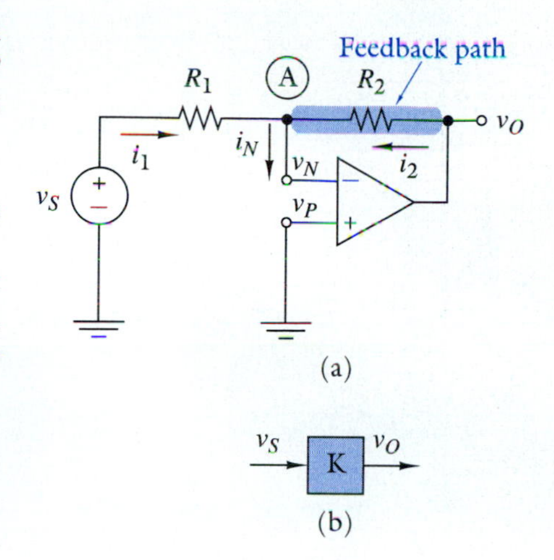

Figure 4-27 The inverting OP AMP circuit.

EXAMPLE 4-7

Find the input-output relationship of the circuit in Fig. 4-28(a).

SOLUTION:

The circuit to the right of Node B is an inverting amplifier. The load resistance R_L has no effect on the circuit transfer characteristics, since the OP AMP has zero output resistance. However, the voltage divider circuit to the left of Node B effects the input resistance of the inverting amplifier. The effect can be seen by constructing a Thévenin equivalent of the circuit to the left of Node B as shown in Fig. 4-28(b). By inspection of Fig. 4-28(a)

$$v_T = \frac{R_2}{R_1 + R_2} v_S$$

$$R_T = \frac{R_1 R_2}{R_1 + R_2}$$

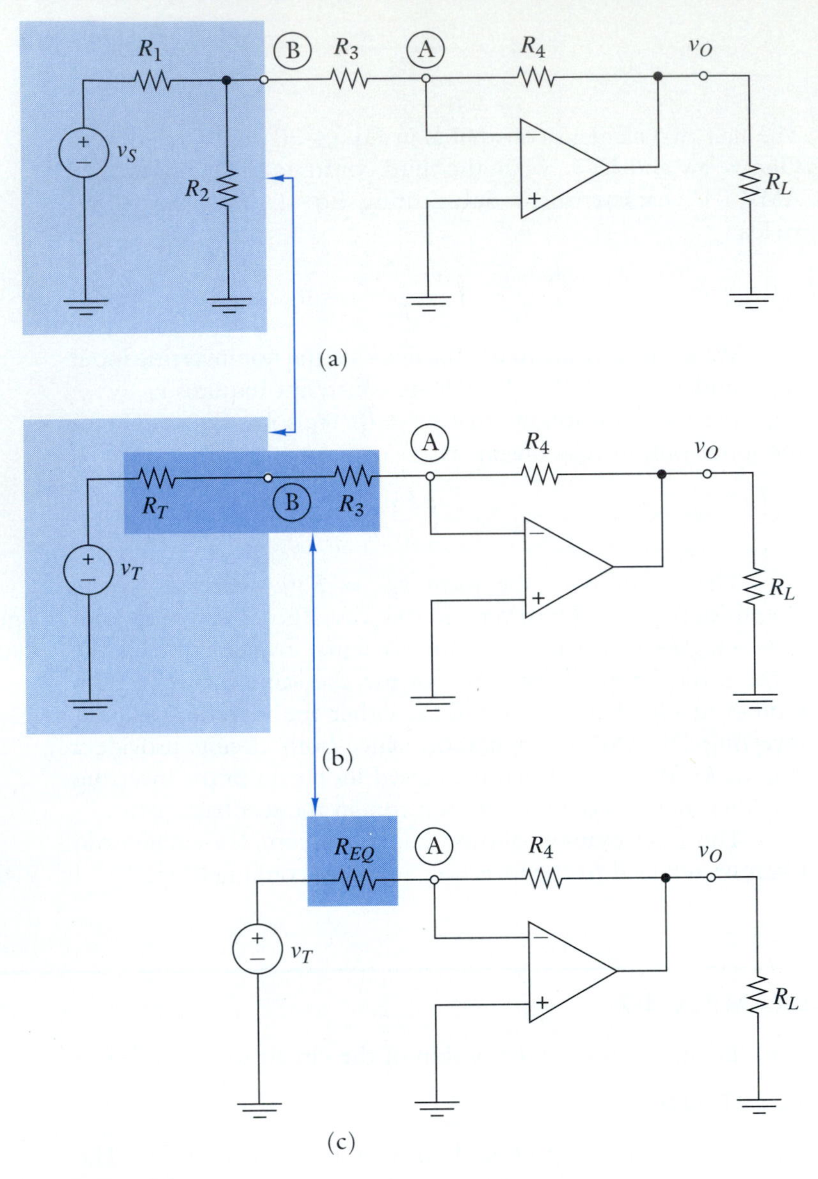

Figure 4-28

In Fig. 4-28(b) the Thévenin resistance is connected in series with the input resistor R_3. Using series equivalence the circuit can be reduced as shown in Fig. 4-28(c), where

$$R_{EQ} = R_3 + R_T$$

This reduced circuit is in the form of the basic inverting amplifier configuration, so we can write the input-output relationship between v_O and v_T as

$$K_1 = \frac{v_O}{v_T} = -\frac{R_4}{R_{EQ}} = -\frac{R_4\,(R_1 + R_2)}{R_1 R_2 + R_1 R_3 + R_2 R_3}$$

The overall input-output relationship from the input source v_S to the OP AMP output v_O is obtained by writing

$$K_{CIRCUIT} = \frac{v_O}{v_S} = \left[\frac{v_O}{v_T}\right]\left[\frac{v_T}{v_S}\right]$$
$$= -\left[\frac{R_4\,(R_1 + R_2)}{R_1 R_2 + R_1 R_3 + R_2 R_3}\right]\left[\frac{R_2}{R_1 + R_2}\right]$$
$$= -\left[\frac{R_2 R_4}{R_1 R_2 + R_1 R_3 + R_2 R_3}\right]$$

It is important to note that the overall gain is not the product of the voltage divider gain $R_2/(R_1 + R_2)$ and the inverting amplifier gain $-R_4/R_3$. In this circuit the two building blocks interact because connecting the inverting OP AMP circuit loads the voltage divider circuit.

EXERCISE 4-6

The OP AMP circuit of Fig. 4-29(a) is driven by a voltage source whose output increases linearly as a function of time as $v_S = t$. The OP AMP is supplied by a ±15 V power supply. Plot the output v_O versus time.

ANSWER:

The solution is shown in Fig. 4-29(b). For $t \geq 0$ the output is $-3.3t$, since the gain of this inverting amplifier is $K = -3.3$. At $t = 4.54$ s the output equals -15 V and the OP AMP switches from the linear to the negative saturation mode. Thereafter, the output remains constant at -15 V, since the OP AMP is saturated.

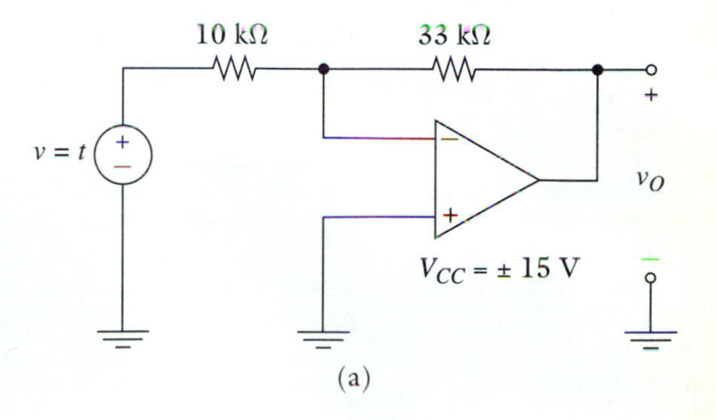

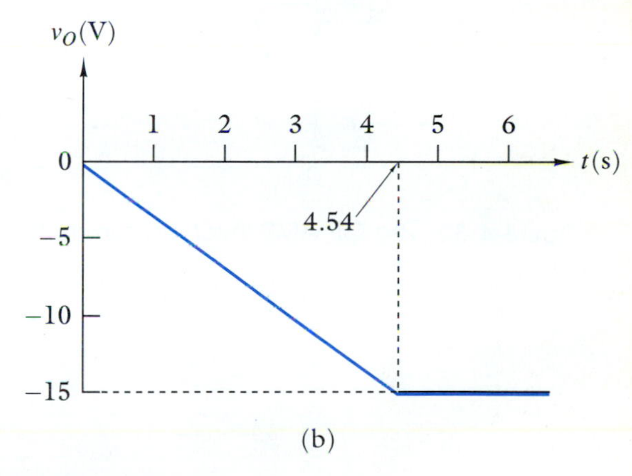

Figure 4-29

The Summing Amplifier

The *summing amplifier* or *adder* circuit is shown in Fig. 4-30(a). This circuit has two inputs connected at Node A, which is called the *summing point*. Since the noninverting input is again grounded we have $v_P = 0$. This configuration is quite similar to the inverting amplifier, so we start by applying KCL at the summing point (Node A):

$$i_1 + i_2 + i_F = i_N \qquad (4\text{-}33)$$

Making use of the fact that the OP AMP element constraint

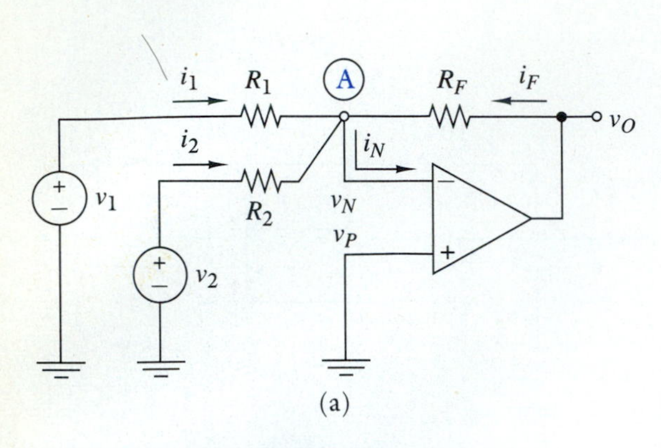

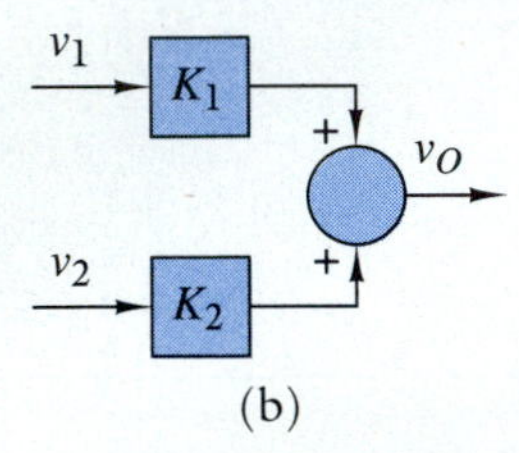

Figure 4-30 The OP AMP inverting summer.

requires $v_N = v_P = 0$, we write the remaining element constraints as

$$\begin{aligned} i_1 &= \frac{v_1}{R_1} \qquad i_2 = \frac{v_2}{R_2} \\ i_F &= \frac{v_O}{R_F} \qquad i_N = 0 \end{aligned} \tag{4-34}$$

Substituting Eq. (4-34) into (4-33) and solving for the output voltage v_O yields

$$\begin{aligned} v_O &= \left(-\frac{R_F}{R_1}\right) v_1 + \left(-\frac{R_F}{R_2}\right) v_2 \\ &= (K_1)\, v_1 + (K_2)\, v_2 \end{aligned} \tag{4-35}$$

The output is a weighted sum of the two inputs. The scale factors, or gains as they are called, are determined by the ratio of the feedback resistor R_F to the input resistor for each input; that is, $K_1 = -R_F/R_1$ and $K_2 = -R_F/R_2$. In the special case $R_F = R_1 = R_2$, Eq. (4-35) reduces to

$$v_O = -(v_1 + v_2)$$

In this case the output is the inverted sum of the two inputs, hence the name summing amplifier or adder. A block diagram representation of this circuit is shown in Fig. 4-30(b).

The summing amplifier in Fig. 4-30 has two inputs, so there are two gains to contend with, one for each input. The input-output relationship in Eq. (4-35) is easily generalized to the case of n inputs as

$$\begin{aligned} v_O &= \left(-\frac{R_F}{R_1}\right) v_1 + \left(-\frac{R_F}{R_2}\right) v_2 + \cdots + \left(-\frac{R_F}{R_n}\right) v_n \\ &= K_1 v_1 + K_2 v_2 + \cdots + K_n v_n \end{aligned}$$

where R_F is the feedback resistor and $R_1,\ R_2, \ldots R_n$ are the input resistors for the n input voltages $v_1,\ v_2, \ldots v_n$. The reader can easily verify this result by expanding the KCL constraint in Eq. (4-33) and the element constraints in Eq. (4-34) to cover n inputs, and then solving for v_O.

❖ DESIGN EXERCISE: EXERCISE 4-7

Design a circuit that combines the following two signals and produce a constant output of 0 V, $v_1(t) = 3t$, $v_2(t) = -5t$.

ANSWER:

There are many solutions. One approach is to use an inverting summer with a feedback resistor equal to 150 kΩ and a v_1 input resistor equal to 30 kΩ and a v_2 input resistor equal to 50 kΩ. ■

❖ Design Example

EXAMPLE 4-8

Design an inverting summer which implements the input-output relationship

$$v_O = -(5v_1 + 13v_2)$$

SOLUTION:

The design problem involves selecting the input and feedback resistors so that

$$\frac{R_F}{R_1} = 5 \quad \text{and} \quad \frac{R_F}{R_2} = 13$$

One solution is to arbitrarily select $R_F = 65\ \text{k}\Omega$, which yields $R_1 = 13\ \text{k}\Omega$ and $R_2 = 5\ \text{k}\Omega$. The resulting circuit is shown in Fig. 4-31(a) The design can be modified to use standard resistance values for resistors with ±5% tolerance (See Appendix A, Table A-1). Selecting the standard value $R_F = 56\ \text{k}\Omega$ requires $R_1 = 11.2\ \text{k}\Omega$ and $R_2 = 4.31\ \text{k}\Omega$. The nearest standard values are 11 kΩ and 4.3 kΩ. The resulting circuit shown in Fig. 4-31(b) uses only standard value resistors and produces gains of $K_1 = 56/11 = 5.09$ and $K_2 = 56/11 = 13.02$. These nominal gains are within 2% of the values in the specified input-output relationship.

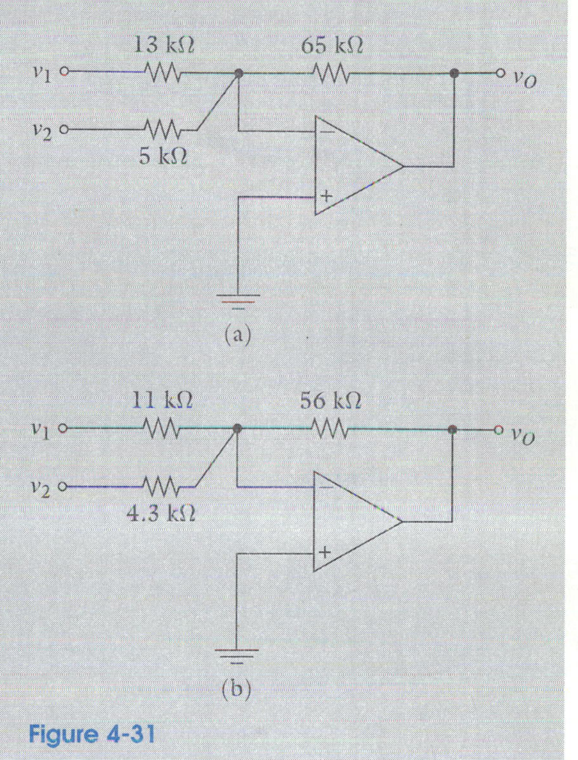

Figure 4-31

The Differential Amplifier

The circuit in Fig. 4-32(a) is called a *differential amplifier* or *subtractor*. Like the summer, this circuit has two inputs, one applied at the inverting input and one at the noninverting input of the OP AMP. The input-output relationship can be obtained using superposition.

First, we turn off source v_2 in which case there is no excitation at the noninverting input and $v_P = 0$. In effect, the noninverting input is grounded and the circuit acts like an inverting amplifier with the result that

$$v_{O1} = -\frac{R_2}{R_1}v_1 \tag{4-36}$$

Now turning v_2 back on and turning v_1 off we see the circuit looks like a noninverting amplifier with a voltage divider connected at its input. This case was treated in Example 4-6, so we

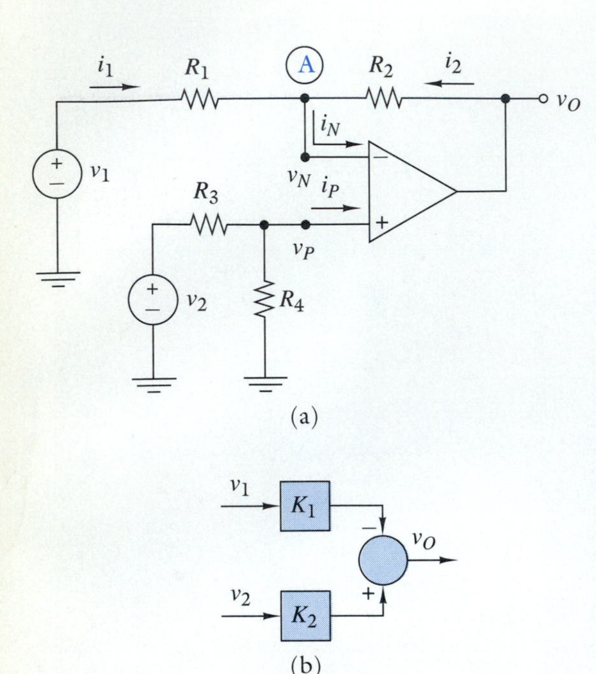

Figure 4-32 The OP AMP subtractor.

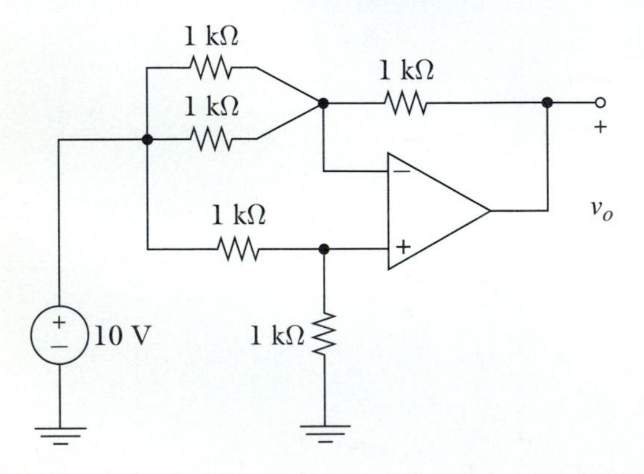

Figure 4-33

can write

$$v_{O2} = \left[\frac{R_4}{R_3 + R_4}\right]\left[\frac{R_1 + R_2}{R_1}\right] v_2 \tag{4-37}$$

Using superposition we can add Eqs. (4-36) and (4-37) to obtain the output with both sources on as

$$\begin{aligned} v_O &= v_{O1} + v_{O2} \\ &= -\left[\frac{R_2}{R_1}\right] v_1 + \left[\frac{R_4}{R_3 + R_4}\right]\left[\frac{R_1 + R_2}{R_1}\right] v_2 \\ &= -[K_1] v_1 + [K_2] v_2 \end{aligned} \tag{4-38}$$

where K_1 and K_2 are the inverting and noninverting gains. Figure 4-32(b) shows how the differential amplifier is represented in a block diagram.

In the special case of $R_1 = R_2 = R_3 = R_4$ Eq. (4-38) reduces to

$$v_O = v_2 - v_1$$

hence the name subtractor.

Exercise 4-8

Find the output of the circuit in Fig. 4-33.

ANSWER:

−5 V

Basic OP AMP Building Blocks

We have introduced four basic OP AMP circuit building blocks. It is often convenient to represent these circuits using their block diagrams. The reason is that analog system analysis and design often takes place at the functional or block diagram level. A major advantage of OP AMP circuits is the near one-to-one correspondence between the elements in a block diagram and the circuits that implement the block diagram.

Only two block diagram symbols have been introduced thus far: *the gain block* and *the summing symbol.* The gain block indicates the output is proportional to the input $v_O = K v_S$, where the gain K can be either positive or negative. The summing symbol indicates that its output is

$$v_O = v_1 + v_2 + \ldots + v_n$$

The block diagram representation of the four basic OP AMP circuits are shown in Fig. 4-34. The noninverting and inverting amplifiers are represented as gain blocks. The summing amplifier

Circuit	Block diagram	Gains
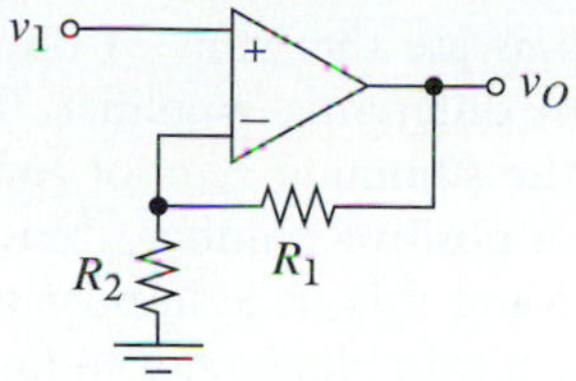 v_1, v_O, R_1, R_2, +, −	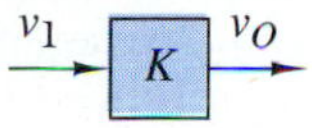 v_1, K, v_O	$K = \dfrac{R_1 + R_2}{R_2}$
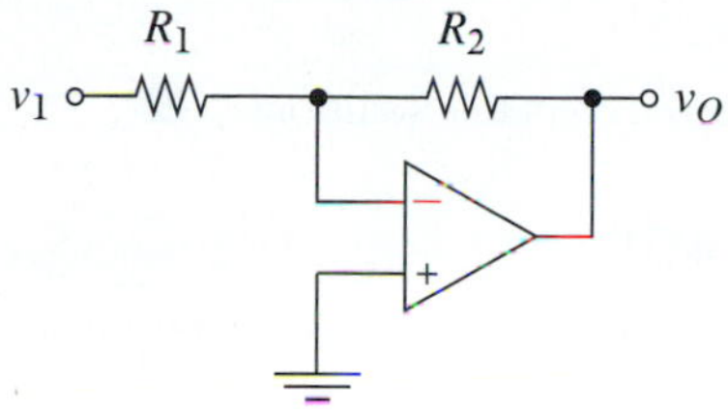 R_1, R_2, v_1, v_O, −, +	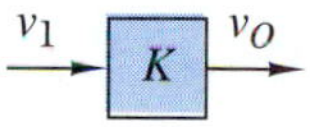 v_1, K, v_O	$K = -\dfrac{R_2}{R_1}$
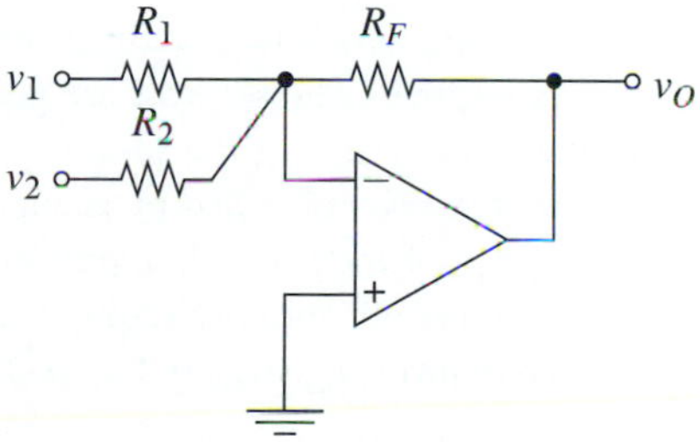 R_1, R_F, v_1, v_O, R_2, v_2, −, +	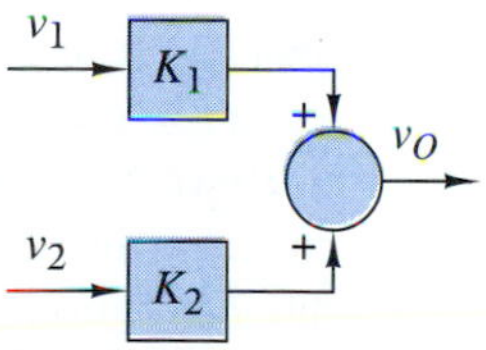 v_1, K_1, +, v_O, v_2, K_2, +	$K_1 = -\dfrac{R_F}{R_1}$ $K_2 = -\dfrac{R_F}{R_2}$
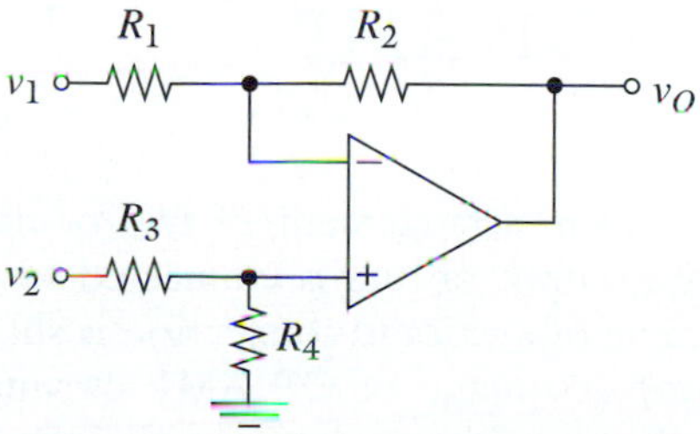 R_1, R_2, v_1, v_O, R_3, v_2, R_4, −, +	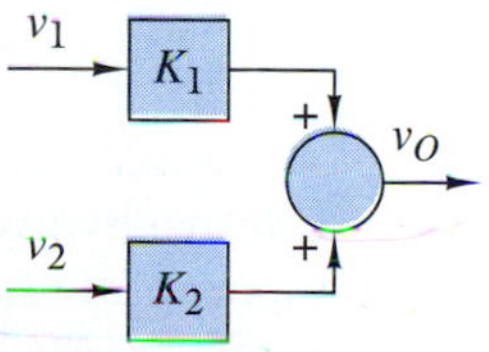 v_1, K_1, +, v_O, v_2, K_2, +	$K_1 = -\dfrac{R_2}{R_1}$ $K_2 = \left(\dfrac{R_1 + R_2}{R_1}\right)\left(\dfrac{R_4}{R_3 + R_4}\right)$

Figure 4-34 Summary of basic OP AMP signal processing circuits.

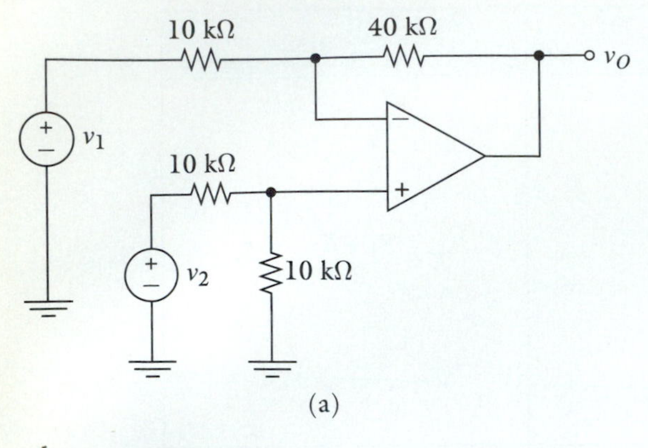

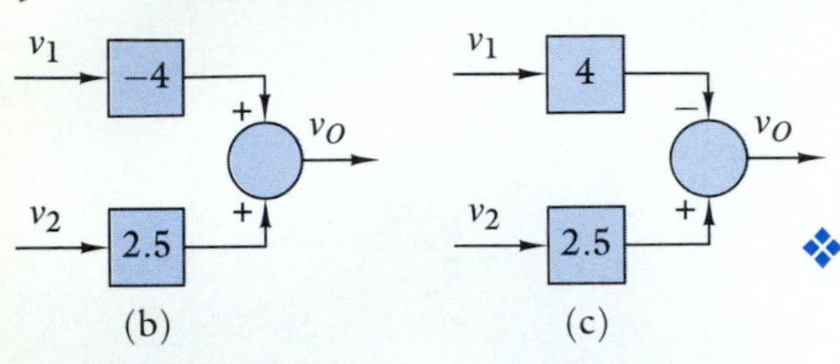

Figure 4-35

and differential amplifier require both gain blocks and the summing symbol. Considerable care must be used when translating from a block diagram to a circuit, and vice versa, since some gain blocks may involve negative gains. For example, the gain of the inverting amplifier is negative as are the gains of the summing amplifier and the K_1 gain of the differential amplifier. The minus sign is sometimes moved to the summing symbol and the gain within the block changed to a positive number. Since there is no standard convention for doing this, it is important to keep track of the signs associated with gain blocks and summing point symbol.

EXERCISE 4-9

Construct a block diagram for the circuit in Fig. 4-35(a).

ANSWER:
Either Fig. 4-35(b) or 4-35(c) are acceptable solutions.

DESIGN EXERCISE: EXERCISE 4-10

Design an OP AMP circuit to implement the block diagram of Fig. 4-36(a).

ANSWER:
Two possible answers are shown in Fig. 4-36.

DISCUSSION:
The circuit in Fig. 4-36(b) uses a summing amplifier and an inverting amplifier, since the specified gains are positive. The resistors are greater than 1 kΩ to reduce the load on the sources driving the summer. The circuit in Fig. 4-36(c) is a *noninverting summer* obtained by grounding the inverting input to a subtractor and applying multiple inputs at the noninverting input. The resulting noninverting summer requires fewer OP AMPs in this case because the second inverting stage in Fig. 4-36(b) is not required.

4-6 CASCADED OP AMP CIRCUITS

A *cascade connection* is a tandem arrangement of two or more similar circuits in which the output of one is connected to the input of the next. An example of a cascade connection is shown in Fig. 4-37. An important advantage of OP AMP circuits is that they can connect in cascade without changing their input-output relationships. As a result, the overall gain of the circuit in Fig. 4-37 is $K_1 \times K_2 \times K_3$, where K_1, K_2, and K_3 are the gains calculated under no-load conditions. The input-output equivalent circuit of the cascade connection in Fig. 4-37 shows

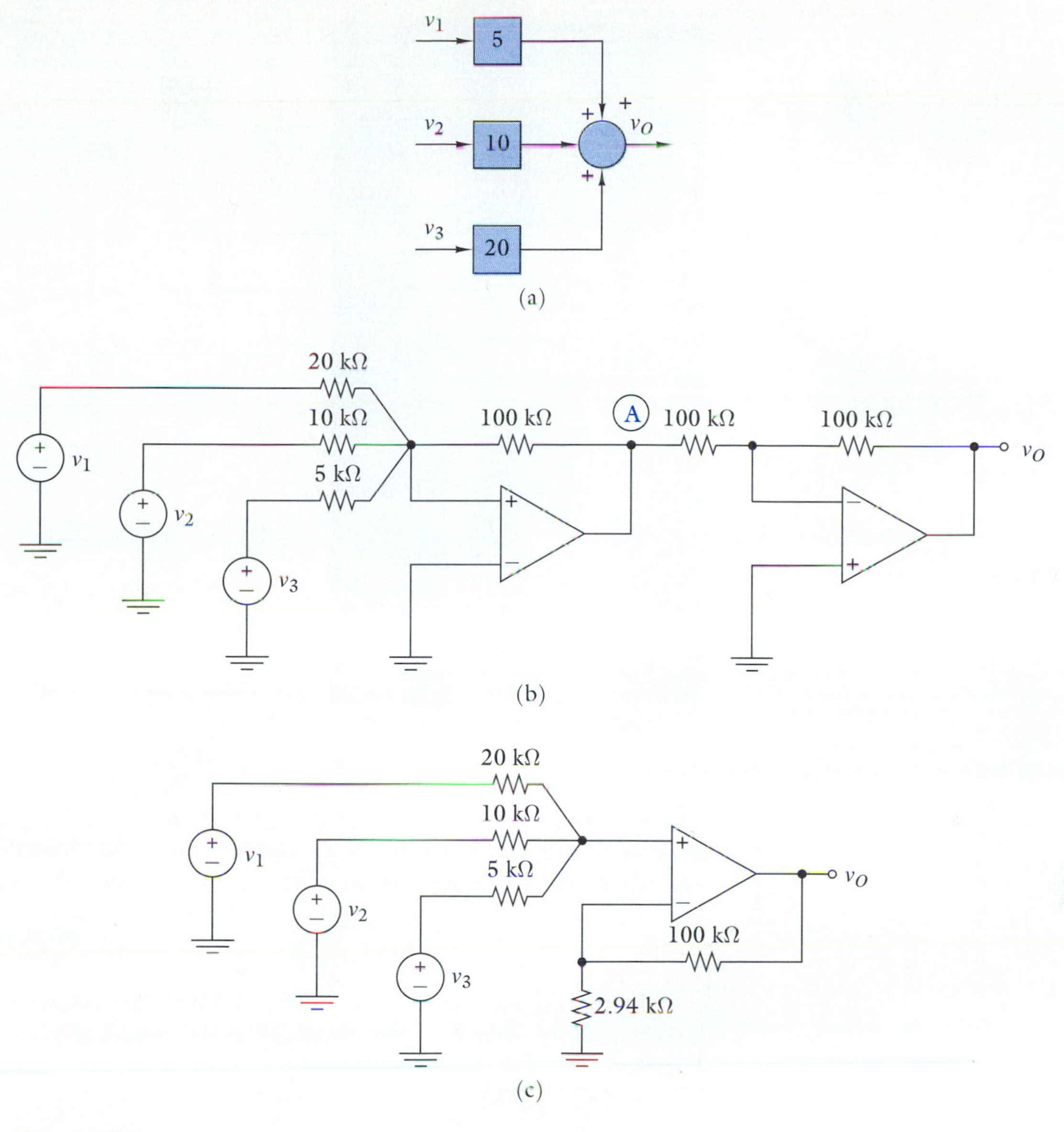

Figure 4-36

that the OP AMP circuits have zero output resistance. Consequently the cascade connection does not produce loading effects to change the circuit input-output relationships. However, the load presented by the next stage in the cascade must be within the power-handling capability of the OP AMP. For this reason it is important to know the input resistance of an OP AMP circuit.

Figure 4-38 shows the method of determining the input resistance for the two basic amplifier configurations. A test volt-

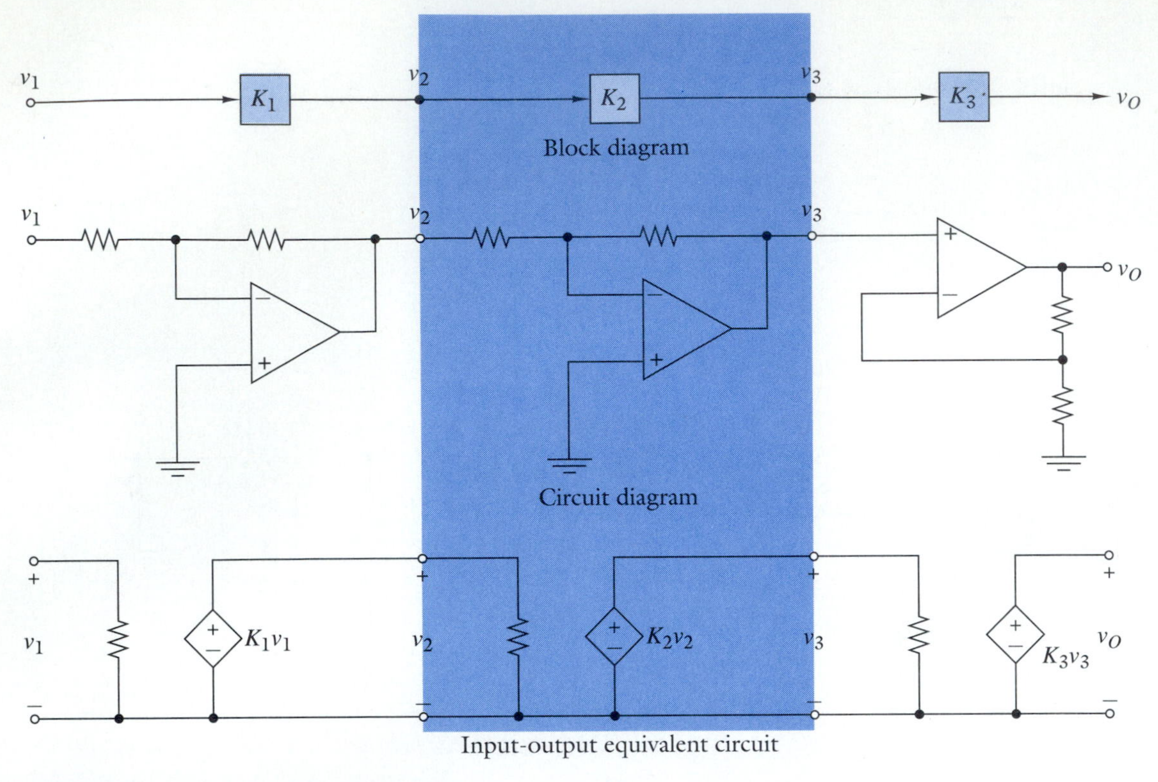

Figure 4-37 The cascade connection.

age is applied at the circuit input, the response i_{TEST} determined by analysis, and R_{IN} calculated as

$$R_{IN} = \frac{v_{TEST}}{i_{TEST}} \tag{4-39}$$

In the noninverting circuit in Fig. 4-38(a), the response i_{TEST} equals i_P, which in the ideal OP AMP model is zero. Therefore,

$$R_{IN(noninverting)} = \infty\ \Omega \tag{4-40}$$

When the finite gain-controlled source model of the OP AMP from Section 4-4 is used, the actual input resistance is found to be approximately $\mu R_I/K$, where μ is the open-loop gain, K the closed-loop gain, and R_I the OP AMP input resistance. For any reasonable closed-loop gain this input resistance is on the order of 10^{10} to 10^{15} Ω. For all practical purposes the noninverting amplifier has an infinite input resistance, as the ideal OP AMP model indicates.

For the inverting amplifier in Fig. 4-38(b) the response i_{TEST} is

$$i_{TEST} = \frac{v_{TEST} - v_N}{R_I}$$

For the ideal OP AMP model $v_N = v_P = 0$, since the noninverting input is grounded. Therefore, the input resistance is

$$R_{IN(inverting)} = R_I \quad (4\text{-}41)$$

When the finite gain-controlled source model of the OP AMP is used the input resistance is found to be approximately $R_1(1 + K/\mu)$, where K is the closed-loop gain and μ the open-loop gain. Since $K/\mu \ll 1$ in practical applications, the input resistance of the inverting amplifier is essentially R_1 as the ideal model indicates.

The fact that the inverting amplifier has a finite input resistance means selecting R_1 involves a tradeoff. The closed-loop gain of the inverting amplifier is inversely proportional to R_1, which suggests that R_1 should be made small. On the other hand, R_1 should be large to reduce the load on the preceding stage. Loading is important when the preceding stage is not the output of an OP AMP. It is particularly important to control the input resistance of an inverting amplifier when interfacing with a source with a high Thévenin resistance.

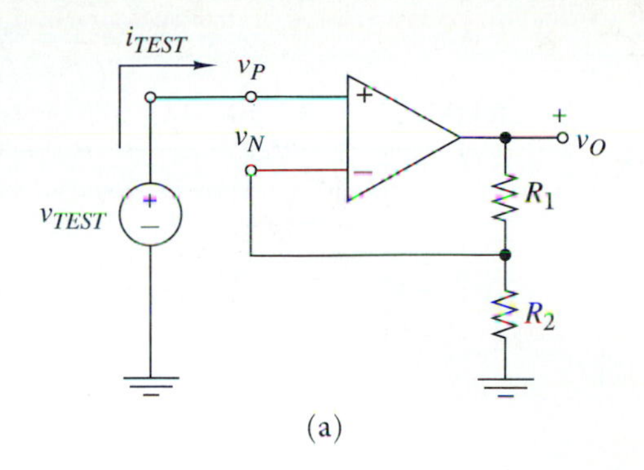

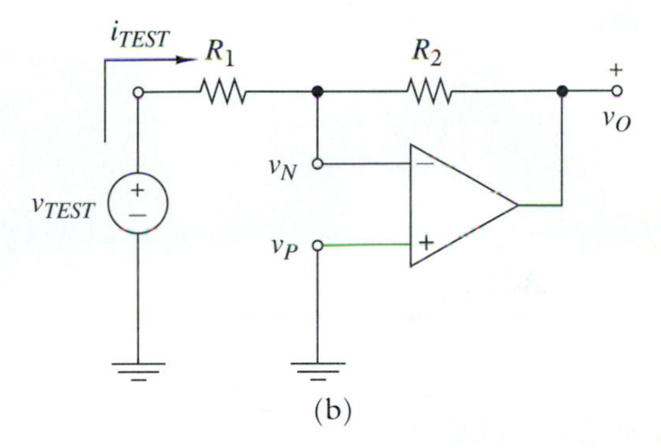

Figure 4-38 Test inputs for finding input resistance. (a) Noninverting circuit. (b) Inverting circuit.

EXAMPLE 4-9

Find the input-output relationship of the circuit in Fig. 4-39(a).

SOLUTION:

This circuit is a cascade connection of three OP AMP circuits. The first-stage is an inverting amplifier with $K = -0.33$, the second stage is a unity gain inverting summer, and the final stage is another inverting amplifier with $K = -5/9$. Given these observations, we construct the block diagram shown in Fig. 4-39(b). In tracing the signal from input to output, we begin with the first stage, whose input is 9.7 V, and find its output as $(-0.33) \times (9.7) = -3.2$ V. This voltage is added to the variable input v_F in the second stage to produce $(-1) \times (-3.2) + (-1) \times (v_F) = 3.2 - v_F$. This result is then multiplied by the gain of the final stage $(-5/9)$ to obtain the overall circuit output as

$$v_C = \frac{5}{9}(v_F - 3.2)$$

The physical significance of this relationship is this: When $v_F = \theta_F/10$, where θ_F is a temperature in degrees Fahrenheit, then $v_C = \theta_C/10$, where θ_C is the equivalent temperature in degrees Celsius.

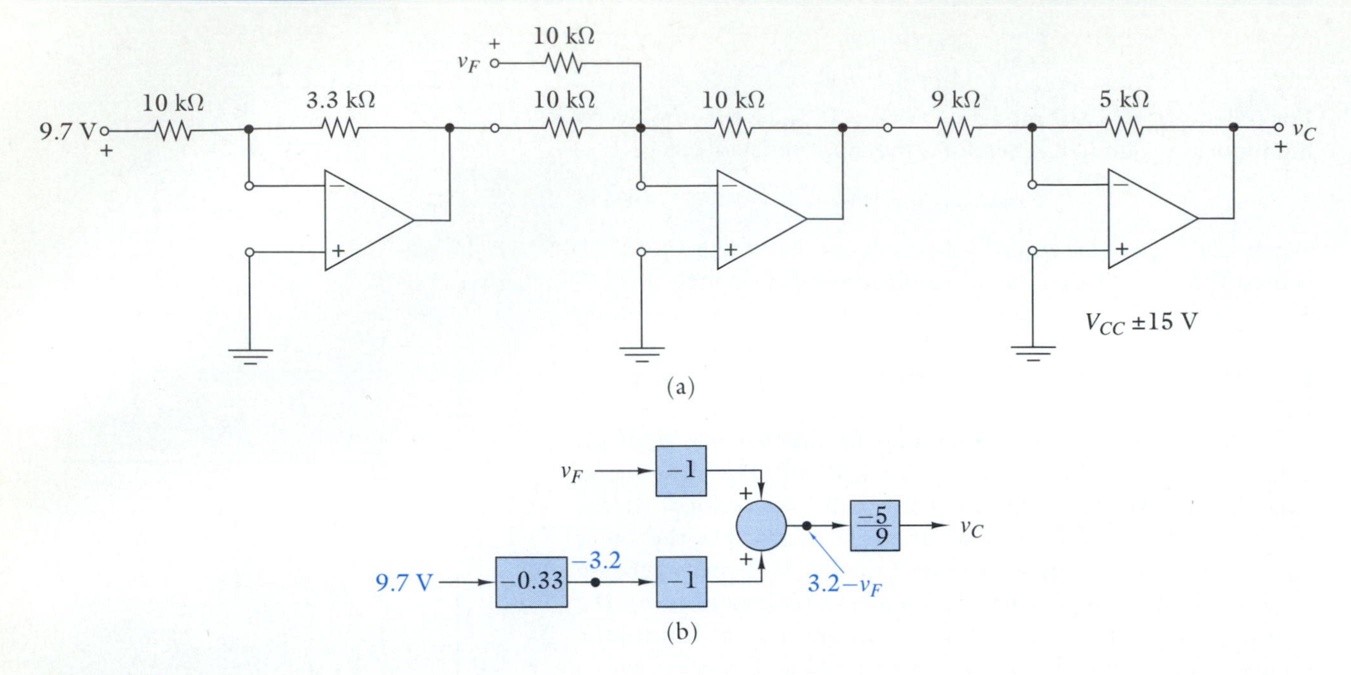

Figure 4-39

Each stage in Fig. 4-39 is an inverting amplifier with input resistances of about 10 kΩ. Since V_{CC} = 15 V, this means the maximum current drawn at the input of any stage is about 15 ÷ 10^4 =1.5 mA—well within the power capabilities of general-purpose OP AMP devices. The reader is now invited to show that the output of any stage will not exceed ±15 V when the input temperature is in the range from −128°F to +182°F.

❖ Design Example

EXAMPLE 4-10

Design an amplifier with a gain of +10, 000 ± 5% and an input resistance of at least 10 kΩ. The closed-loop gain of any stage must not exceed 5% of the open-loop gain of the OP AMP. Assume the open-loop gain is 10^5.

SOLUTION:

Several possible solutions to this design problem are shown in Fig. 4-40.

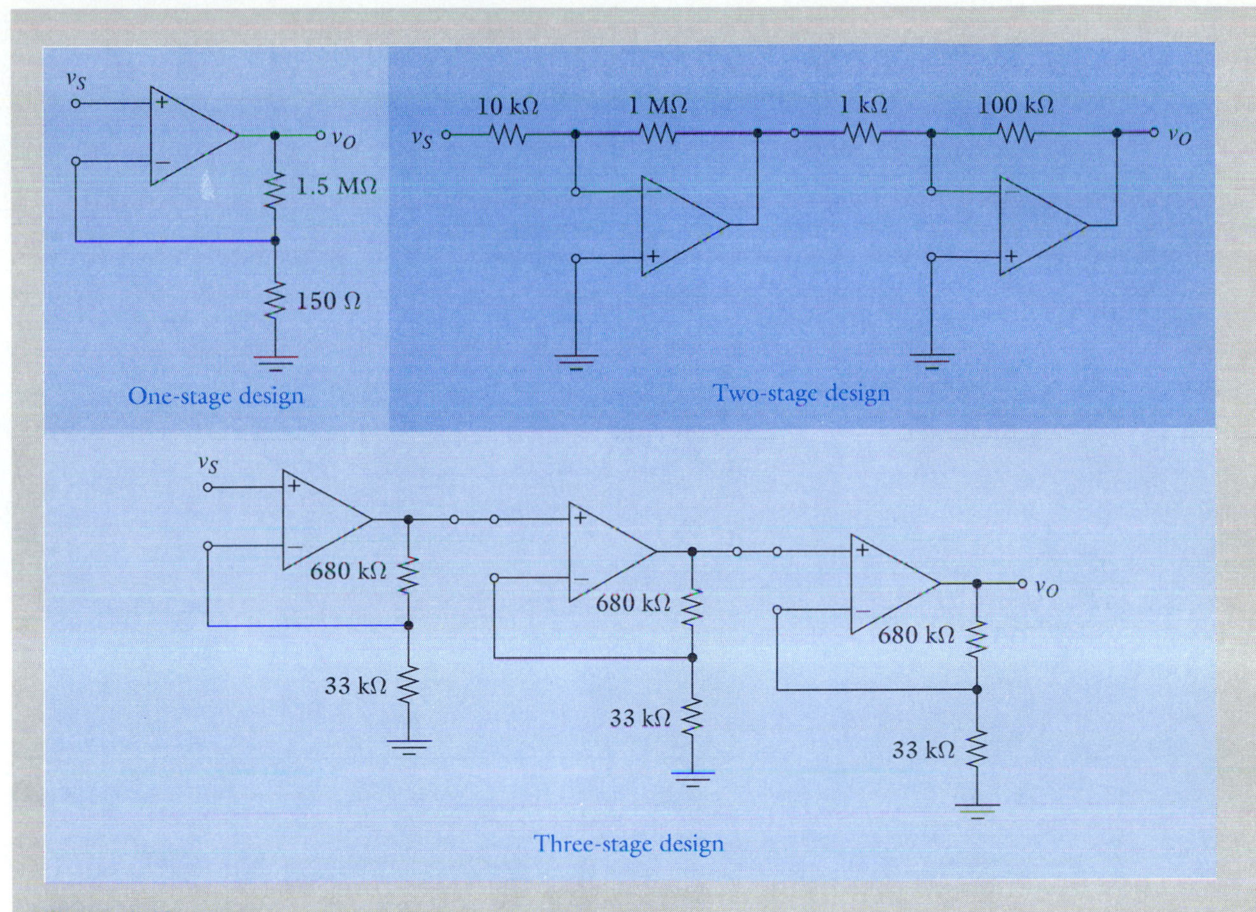

Figure 4-40

The single-stage design uses only one OP AMP in the noninverting configuration, so its input resistance is very high. However, a closed-loop gain of $K = 10^4$ is 10% of the available open-loop gain, which exceeds the prescribed 5% value. The closed-loop gain of the single-stage design is not small compared with the OP AMP open-loop gain, so the gain error could exceed the allowed $\pm 5\%$ tolerance.

The two-stage design involves a cascade of two inverting amplifiers. Each stage has $K = -100$, so the overall gain is $(-100) \cdot (-100) = +10{,}000$, as required. The circuit input resistance is 10 kΩ, which is the minimum value allowed. Although the order in which the two stages are connected does not affect the gain, since $K_1 \times K_2 = K_2 \times K_1$, reversing the order does change the input resistance to 1 kΩ, well below the design requirement.

The three-stage design is a cascade of three identical noninverting amplifiers. The use of identical stages simplifies manufacturing, maintenance, and logistics. Each stage has a gain of $K = 21.606$, so the overall gain is $(21.606)^3 = +10{,}086$, which is within the allowed ±5% tolerance. The noninverting amplifiers in the three-stage design has a very high input resistance. The power consumption and parts count are greater for this design than the others, since three OP AMPs are required.

EXERCISE 4-11

Draw a block diagram of the OP AMP circuit in Fig. 4-41(a). Use the block diagram to find v_1, v_2, and v_O in terms of v_S.

ANSWERS:

(a) Block diagram is shown in Fig. 4-41(b).
(b) $v_1 = (-2.2\ v_S) + (-2.2\ v_O)$.
(c) $v_2 = -10(-2.2\ v_S - 2.2\ v_O) + (-2.2)(-1.0)$.
(d) $v_O = -(44/43)(v_S + 0.1)$.

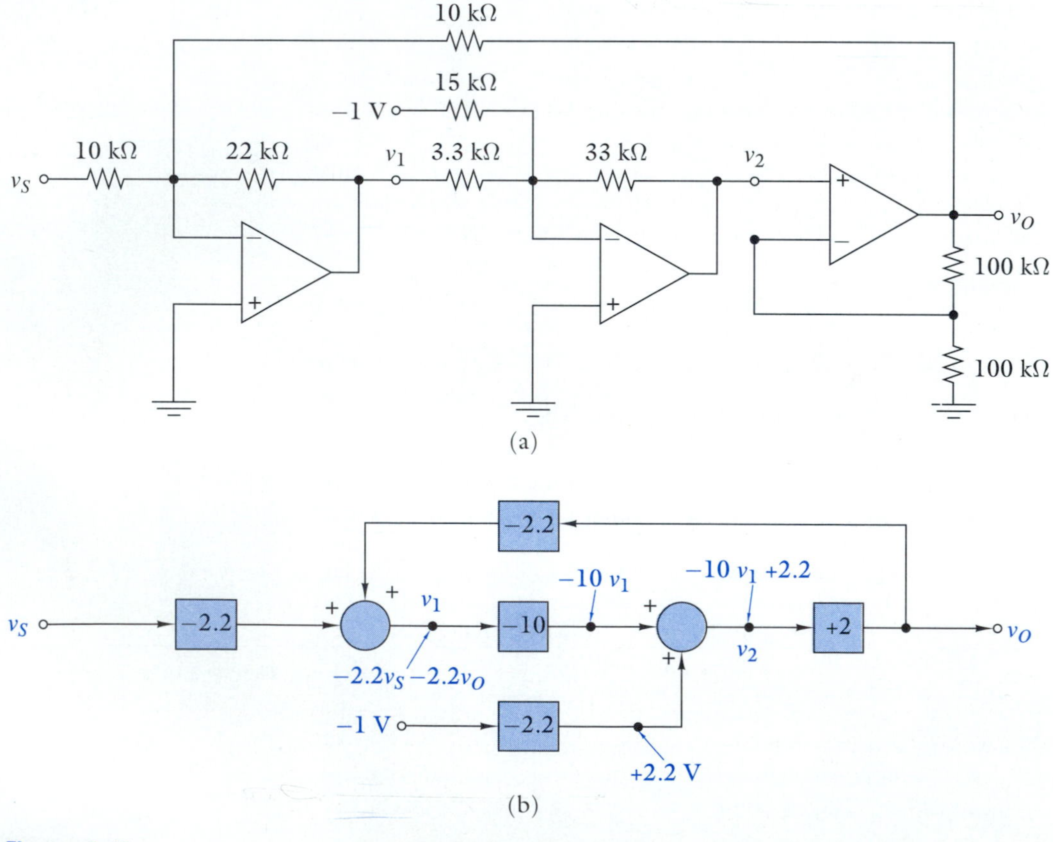

Figure 4-41

4-7 INTERFACE DESIGN WITH OP AMPS

OP AMPs simplify the design of interface circuits because they isolate the source and load. Figure 4-42 shows an elementary example of this isolation feature. The source-load interface in Fig. 4-42(a) is a voltage divider, and so the voltage delivered to the load depends on R_S and R_L. The interaction of the source and load resistances limits the signal levels the source can transfer across the interface. In Fig. 4-42(b) a voltage follower has been inserted between the source and load. The voltage across R_S is zero, since the current into the noninverting input of the follower is zero. Therefore, the voltage at the input of the voltage follower is v_S. The voltage delivered to the load is v_S regardless of the value of R_L, since the voltage follower has zero output resistance. In Fig. 4-42(b) the voltage delivered to the load is v_S regardless of the values of R_S and R_L. In effect, the voltage source v_S in Fig. 4-42 sees an open-circuit and the load sees an ideal voltage source v_S.

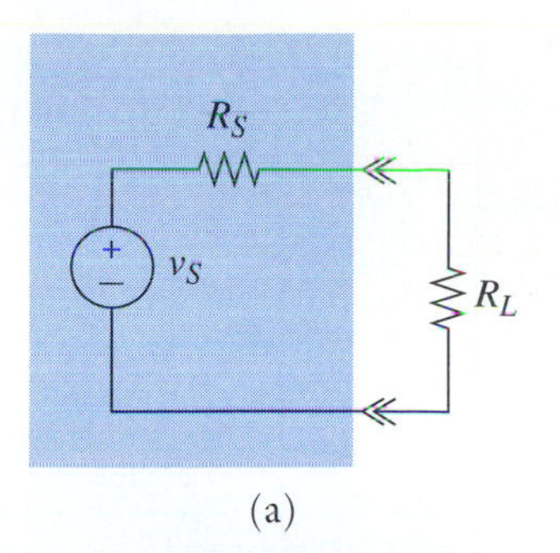

(a)

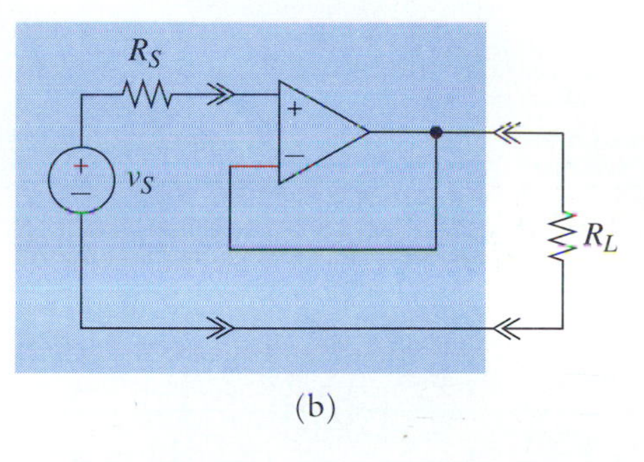

(b)

Figure 4-42 Source-load interface. (a) Directly connection. (b) With OP AMP voltage follower.

There are practical limitations on the ideal situation described above. The OP AMP is an active device requiring a power supply. A power supply will usually exist in a modern electronic system, but voltage followers represent an additional power demand. Theoretically, the voltage follower will deliver v_S to any load resistance. In practice, the current drain must be within the capability of the OP AMP device. For example, the maximum output current for general-purpose OP AMPs is about 10 mA. For a 15-V output the load resistance R_L must be greater than about 1.5 kΩ to avoid exceeding the capability of a general-purpose OP AMP device.

Digital-to-Analog Converters

An important application of OP AMPs is interfacing digital systems with analog systems. The parallel four-bit output in Fig. 4-43 is a digital representation of a signal. Each bit is weighted so that v_1 is worth eight times v_4, v_2 is worth four times v_4, and v_3 is worth two times v_4. We call v_4 the least significant bit (LSB) and v_1 the most significant bit (MSB). Each bit can only have two values: (1) a high or "1" (typically +5 V) and (2) a low or "0" (typically 0 V). To convert the digital representation of the signal to analog form, we must weight the bits so that the analog output v_O is

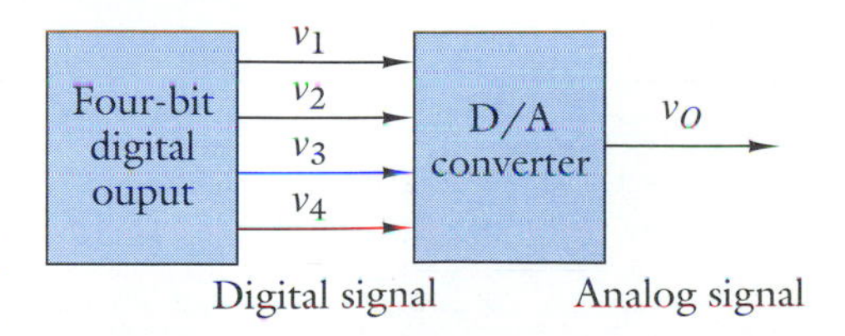

Figure 4-43 A digital-to-analog (D/A) converter.

$$v_O = \pm K\,(8v_1 + 4v_2 + 2v_3 + v_4) \qquad (4\text{-}42)$$

where K is a scale factor or gain applied to the analog signal.

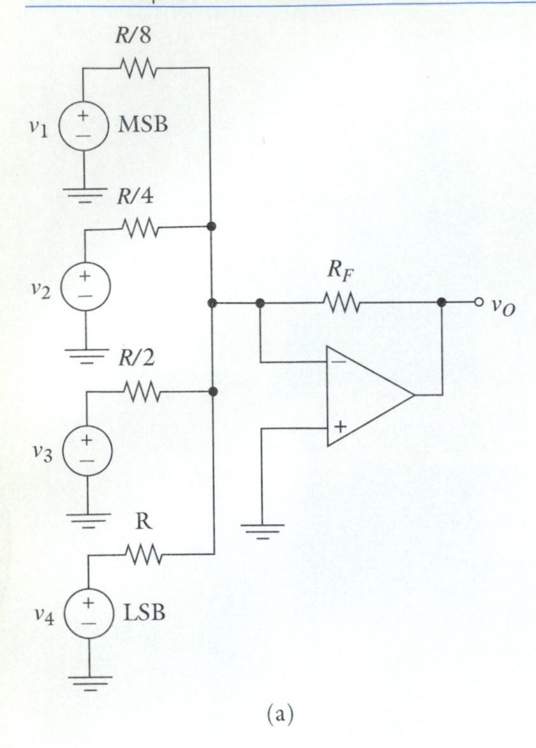

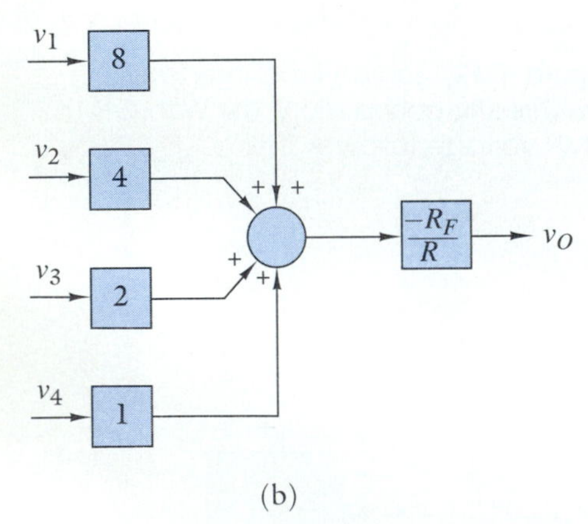

Figure 4-44 A binary-weighted resistance D/A converter. (a) Circuit. (b) Block diagram.

One way to implement Eq. (4-42) is to use an inverting adder with binary-weighted input resistors as shown in Fig. 4-44(a). Figure 4-44(b) shows the four-bit digital to analog (D/A) converter input-output relationship in block diagram form. In either circuit or block diagram form, the output is seen to be a binary-weighted sum of the digital input and then scaled by $-R_F/R$. That is, the output voltage is

$$v_O = \frac{-R_F}{R}(8v_1 + 4v_2 + 2v_3 + v_4) \tag{4-43}$$

The R-$2R$ ladder in Fig. 4-45(a) is another example of a four-bit D/A converter. The Thévenin equivalent circuit of the R-$2R$ network between Node A and ground is shown in Fig. 4-45(b), where

$$v_T = \frac{v_1}{2} + \frac{v_2}{4} + \frac{v_3}{8} + \frac{v_4}{16}$$

The output voltage is found using the inverting amplifier gain relationship:

$$v_O = \frac{-R_F}{R}v_T = \frac{-R_F}{R}\left(\frac{v_1}{2} + \frac{v_2}{4} + \frac{v_3}{8} + \frac{v_4}{16}\right) \tag{4-44}$$

Using $R_F = R$ yields

$$v_O = -\frac{1}{16}(8v_1 + 4v_2 + 2v_3 + v_4) \tag{4-45}$$

which shows the binary weights assigned to the digital inputs.

The circuits in Fig. 4-44(a) and Fig. 4-45(a) perform the same signal-processing function—a four-bit D/A conversion. However, there are important differences between the two circuits. The inverting summer method in Fig. 4-44(a) requires precision resistors with four different values spanning an 8:1 range. An eight-bit converter would require eight precision resistors spanning a 256:1 range. Moreover, the voltage sources in Fig. 4-44(a) see input resistances that span an 8:1 range; therefore, source-load interface is not the same for each digital bit. On the other hand, the resistances in the R-$2R$ ladder converter in Fig. 4-45(a) span only a 2 : 1 range regardless of the number of digital bits. The R-$2R$ ladder also presents the same input resistance to each binary input.

Exercise 4-12

Figure 4-46(a) is a timing diagram showing time variation of the voltages representing the four bits of a binary-weighted representation of a signal. These voltages are the inputs to the R-$2R$ ladder D/A converter in Fig. 4-45(a) whose input-output relationship is given in Eq. (4-44). Plot the output voltage of the converter versus time when $R_F = 16R/5$.

ANSWER:
The output is shown in Fig. 4-46(b).

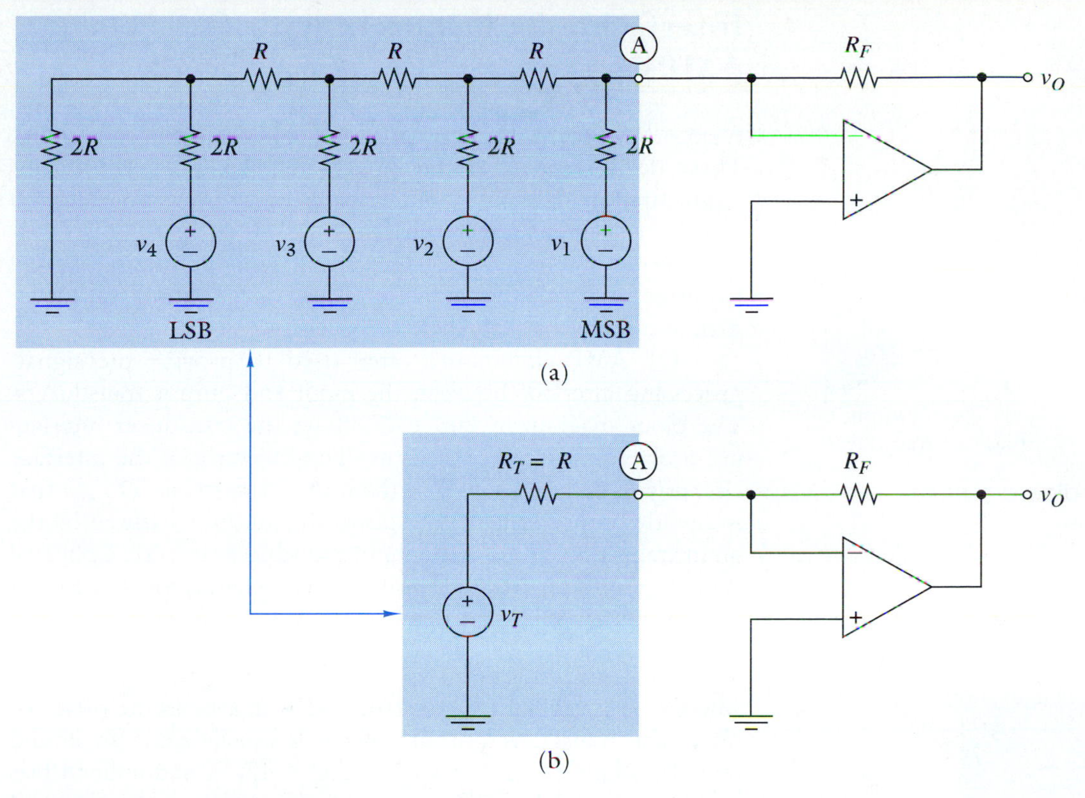

Figure 4-45 An R-$2R$ ladder D/A converter. (a) Circuit. (b) Thévenin equivalent.

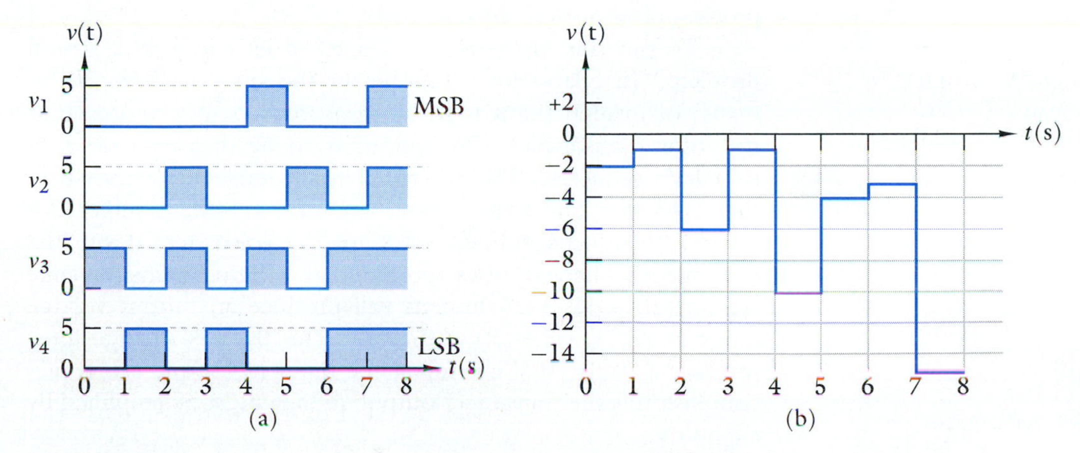

Figure 4-46

Interfacing Transducers with OP AMPs

A *transducer* converts energy from one physical form to another. These devices appear at the input and output of electrical instrumentation systems. Input transducers convert nonelectrical quantities such as temperature, pressure, strain, or angular position into an electrical signal. Electrical signal processing is used to correct and scale the signal for display on an output transducer such as a meter or strip chart recorder.

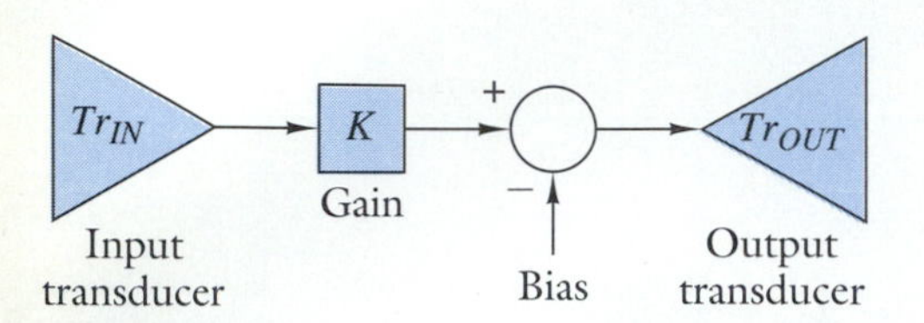

Figure 4-47 Block diagram of a transducer interface.

OP AMP circuits are often used to provide the signal-processing interface between the input and output transducers. The block diagram in Fig. 4-47 shows the transducer interfacing problem in its simplest form. The objective of the interface circuit is to deliver a signal to the output transducer (Tr_{out}) that is directly proportional to the physical quantity measured by the input transducer (Tr_{in}). The input transducer converts a physical variable x into an electrical voltage. The characteristics of most transducers are such that this voltage is of the form

$$v = kx + b$$

where k is a calibration constant and b is a constant offset or bias. The transducer output voltage is usually quite small and must be amplified as indicated in Fig. 4-47. The amplified bias is then removed by adding or subtracting a constant electrical voltage. The resulting output voltage is directly proportional to the quantity measured and goes directly to an output transducer, such as a meter or chart recorder. In other cases the output voltage may go to an analog-to-digital (A/D) converter for further processing or analysis by computer.

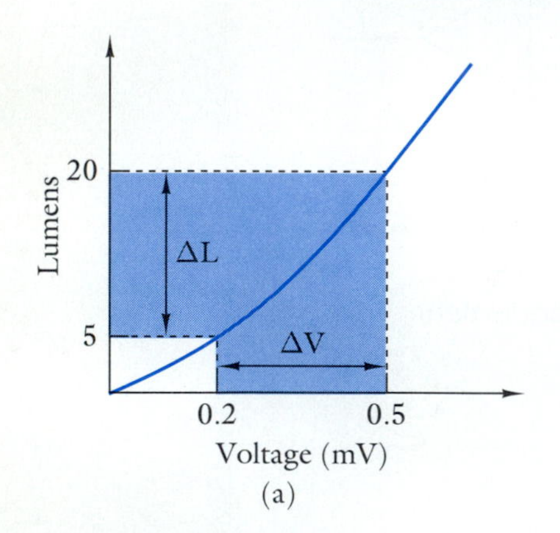

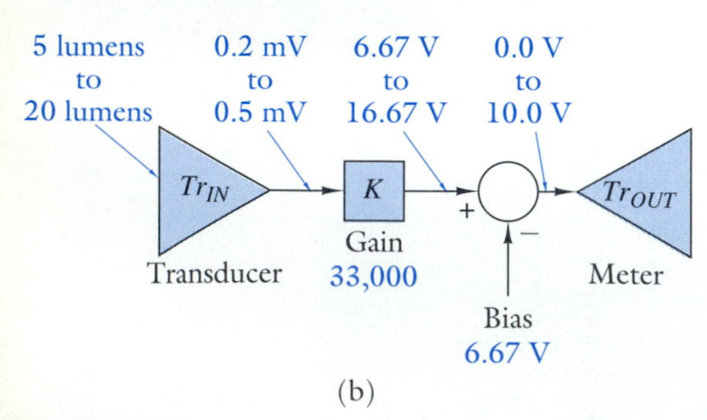

Figure 4-48 Photocell transducer interface design example: (a) Transducer characteristics. (b) Block diagram.

To put this discussion into practice let's look at a typical problem. In a laboratory experiment the amount (5 to 20 lumens) of incident light is to be measured using a photocell as the input transducer. The output is to be displayed on a 0- to 10-V voltmeter. The photocell characteristics are shown in Fig. 4-48(a). The requirements are that 5 lumens indicate 0 V, and 20 lumens indicate 10 V on the voltmeter. From the transducer's characteristics we see that a light intensity range $\Delta L = (20 - 5) = 15$ lumens will produce an output voltage range of $\Delta V = (.5 - .2) = 0.3$ mV. This 0.3-mV change must produce a 0- to 10-V change at the input to the meter. To accomplish this the transducer output voltage must be amplified by a gain of

$$K = \frac{\text{output range}}{\text{input range}} = \frac{(10 - 0)}{(0.5 - 0.2) \times 10^{-3}} = 3.33 \times 10^4 \quad (4\text{-}46)$$

When the transducer's output voltage range (0.2 to 0.5 mV)

is multiplied by the gain K found above, we obtain an output voltage range of 6.67 V to 16.67 V. The output range is shifted to the required 0- to 10-V range by subtracting a constant 6.67 V of bias from the amplified transducer output. A block diagram of the required signal-processing functions is shown in Fig. 4-48(b).

We use a cascade connection of OP AMP circuits to realize the signal-processing functions in the block diagram. Figure 4-49 shows one possible design using an inverting amplifier and an inverting adder. This design uses two circuits in cascade so the two signal inversions cancel. Part of the overall gain of $K = 33.3 \times 10^3$ gain was realized in the inverting amplifier ($K_1 = -330$) and the remainder by the inverting summer ($K_2 = -100$). Dividing the overall gain between the two stages avoids the problem of producing a high gain in a single stage. A single-stage closed-loop gain of 333,000 would be too close to the open-loop gain of the OP AMP to be practical. The high gain would also require a very low input resistance and an uncommonly large feedback resistance, for example, 100 Ω and 3.3 MΩ.

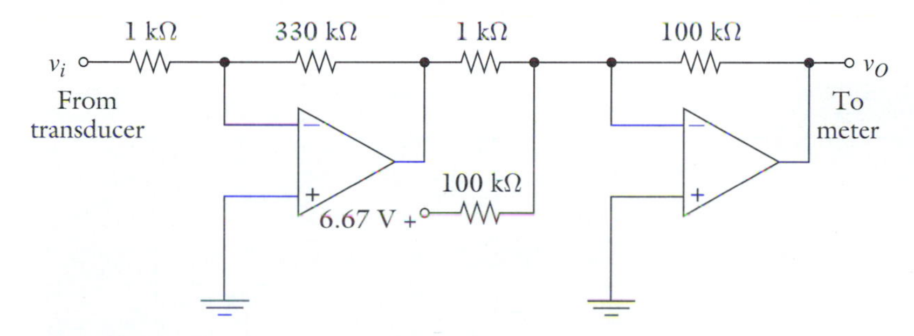

Figure 4-49 Interface circuit design for the photocell transducer problem.

❖ Design Example

EXAMPLE 4-11

A thermocouple is a transducer that converts heat energy into electrical voltage through a phenomenon known as the Seebeck effect. The device consists of a welded junction of two dissimilar metals, such as copper and an alloy called constantan. The output from this junction is a voltage that varies linearly with the temperature of the junction. Figure 4-50(a) shows the characteristic curve of a certain thermocouple. The Thévenin resistance of the thermocouple varies with temperature but never exceeds 100 Ω.

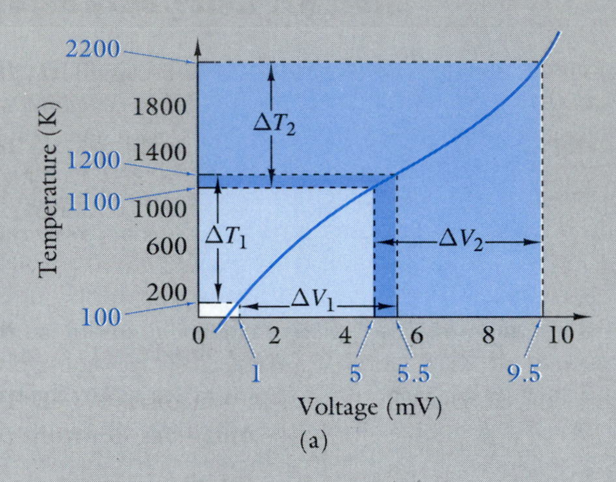

(a)

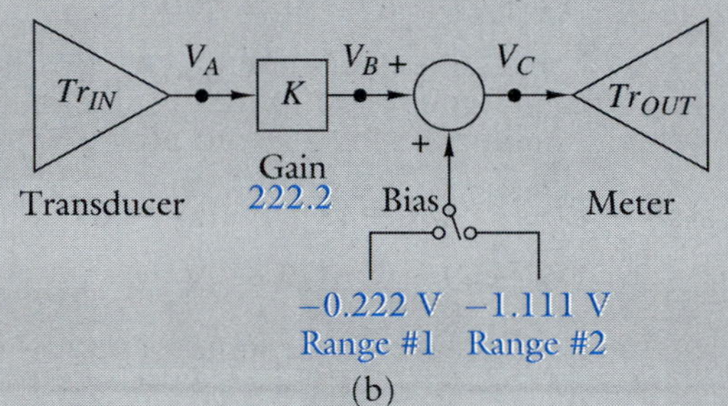

(b)

Range	V_A	Gain	V_B	Bias	V_C
100 to 1200	.001 to .0055	222.2 222.2	0.222 to 1.222	−0.222 −0.222	0.000 to 1.000
1100 to 2200	.005 to .0095	222.2 222.2	1.111 to 2.111	−1.111 −1.111	0.000 to 1.000

(c)

Figure 4-50

Design an interface for this transducer that will provide a 0 to 1.0 V output for each of two temperature ranges: (1) a low range 100° K to 1200° K and (2) a high range 1100° K to 2200° K. The thermocouple will be used in a location with high ambient electromagnetic noise.

SOLUTION:

This design deals with two related problems: (1) interfacing the input transducer with the meter and (2) protecting the signal against the ambient noise environment.

To solve the first problem we need to compute the gain and bias needed in each range. The gain for the low range is found as

$$K_{LOW} = \frac{\text{output range}}{\text{input range}} = \frac{1.0 - 0.0}{0.0055 - 0.001} = \frac{1}{0.0045} = 222.2$$

and the gain for the upper end is found as

$$K_{HIGH} = \frac{1.0 - 0.0}{.0095 - .005} = \frac{1}{0.0045} = 222.2$$

The gains are equal because the input and output ranges are the same in each of the temperature ranges. However, the table in Fig. 4-50(c) shows that multiplying the extreme values of the transducer output voltage by the required gain produces a different value of bias in each range. Figure 4-50(b) shows the block diagram implementing these interfacing requirements. The input transducer output is always amplified by a gain of $K = 222.2$, but different bias inputs are applied depending on the temperature range displayed on the output transducer.

Figure 4-51 outlines the noise problem. The thermocouple output voltage V(T) is in the mV range, and the unwanted electromagnetic signals of nearly equal amplitude can couple to the thermocouple wires generating a noise voltage V_{EM}. Amplifying the output of the single-ended arrangement in Fig. 4-51(a) does not solve the problem because it treats signal and noise equally. The solution is to use the differential mode configuration shown in Fig. 4-51(b). In the differential configuration the signal voltage exists between the two wires *(differential mode)* while the noise V_{EM} couples to both wires equally *(common-mode)*. Connecting the wires to a differential amplifier amplifies the transducer signal but rejects[4] the common-mode noise. To ensure that the noise couples equally, the thermocouple wires are insulated and twisted together, and if necessary, shielded as shown in Fig. 4-51(c). In addition, the differential amplifier is placed as close to the thermocouple as practical to reduce the exposure to noise coupling.

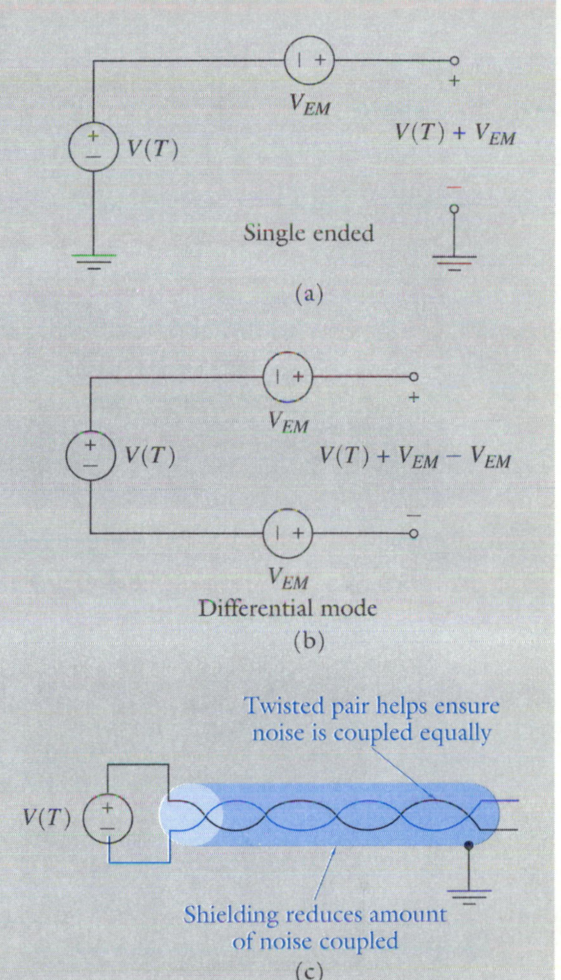

Figure 4-51

We can now design the interface circuit. We can model the thermocouple as a voltage source $V(T)$ in series with a temperature-dependent resistor $R(T)$. Since $R(T)$ never exceeds 100 Ω, we will used 100 Ω as the worst case. The input resistances of the differential amplifier must be much larger than 100 Ω so that $R(T)$ will have minimal effect on the circuit gain.

Figure 4-52 shows one possible design. The input resistances in the differential amplifier are 1 kΩ and 10 kΩ, at least an order of magnitude greater than worst-case $R(T)$. The differential amplifier

[4] A feature of the OP AMP device is its ability to reject common noise appearing at its input terminals. This noise rejection ability is measured by a figure of merit called *common-mode rejection ratio* (CMRR).

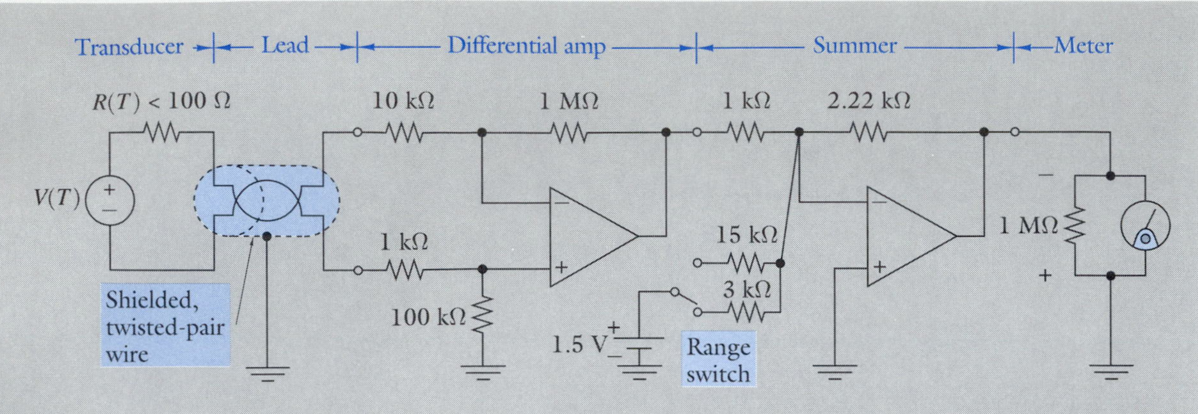

Figure 4-52

has gain of 100 to increase the differential signal above the common-mode noise. Since an overall voltage gain of 222.2 is required, an additional gain of 2.222 is included in the summer. The summer also provides two selectable bias input for the two temperature ranges specified. Using a standard 1.5-V battery, the two input resistances for the bias inputs are selected so that −0.222 V of bias is added in the low range and −1.111 volts in the high range.

❖ DESIGN EXERCISE: EXERCISE 4-13

Design an OP AMP circuit to interface the transducer whose characteristics are shown in Fig. 4-53(a) with a 0- to 1-V meter. The meter should indicate 0 V at −20° C and 1 V at +120° C. The transducer has an output resistance of 880 Ω.

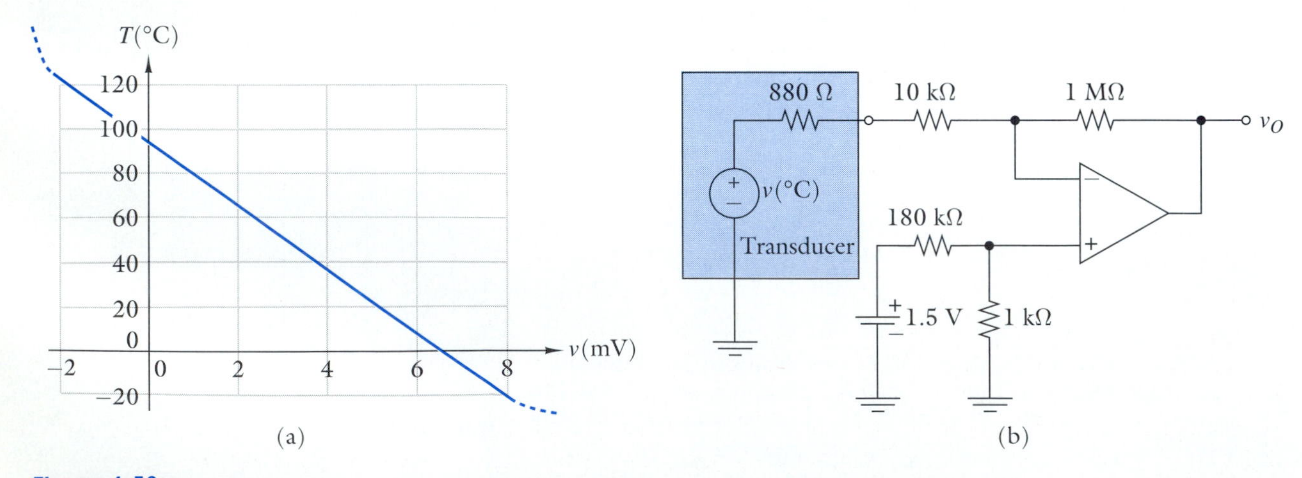

Figure 4-53

ANSWER:
One possible design is shown in Fig. 4-53(b).

DISCUSSION:
One important issue is to select the input resistor of the inverting amplifier with at least a factor of 10 greater than the Thévenin resistance of the transducer. We have used a value of 10 kΩ. ■

4-8 THE COMPARATOR

In the introduction to the OP AMP we found that feedback is required to ensure that the device operates in the linear mode with $v_P = v_N$. When v_P and v_N differ by more than a few millivolts the device switches to one of two saturation modes:

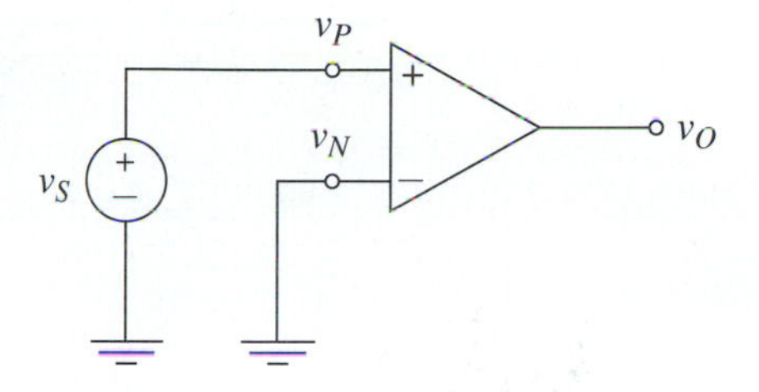

Figure 4-54 An OP AMP comparator.

1. +Saturation with $v_O = +V_{CC}$ when $(v_P - v_N) > 0$.
2. −Saturation with $v_O = -V_{CC}$ when $(v_P - v_N) < 0$.

When no feedback is applied the OP AMP operates in one of the two saturation modes. Under saturation conditions the output voltage indicates whether $v_P > v_N$ or $v_P < v_N$. A device that discriminates between two unequal voltages is called a ***comparator***.

Figure 4-54 shows an example of an OP AMP operating as a comparator. First note there is no feedback and $v_N = 0$, since the inverting input is grounded. The voltage source connected to the noninverting input forces the condition $v_P = v_S$. When $v_S > 0$ it follows that $v_P - v_N > 0$ and $v_O = +V_{CC}$. Conversely, when $v_S < 0$ the output is $v_O = -V_{CC}$. Figure 4-55(a) shows an example plot of $v_S(t)$ versus time. The OP AMP output changes every time v_S crossed zero. For this reason the circuit in Fig. 4-54 is called a ***zero crossing detector***. Notice that the comparator output is not proportional to the input, which shows the element is nonlinear.

A modification of the zero crossing detector is shown in Fig. 4-56(a). In this circuit a constant 2-V source is applied at the inverting input and the input signal v_S in Fig. 4-55 applied to the noninverting input. The input voltage is now compared with 2 V rather than zero. When $v_S > 2$ V the OP AMP mode is +saturation with $v_O = +V_{CC}$ and when $v_S < 2$ V the OP AMP is in −saturation with $v_O = -V_{CC}$. A plot of the resulting output voltage is shown in Fig. 4-56(c). The value of the fixed source determines the input signal level at which the comparator switches from one state to another. With a 10-V fixed source the output of the comparator is always $-V_{CC}$, since v_S never exceeds

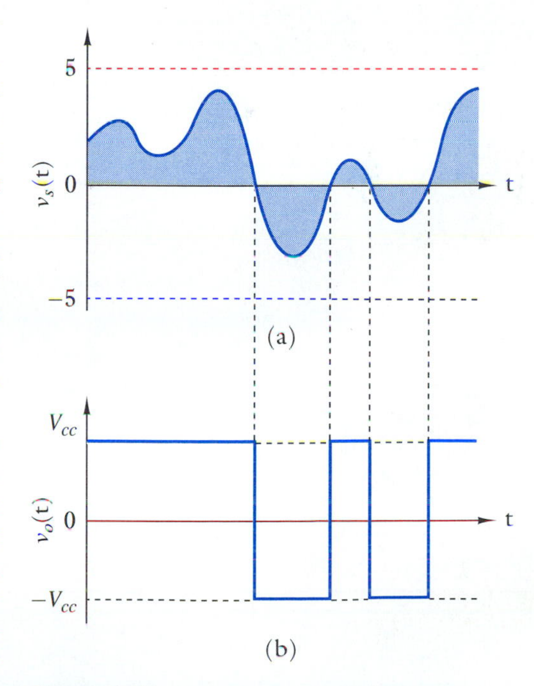

Figure 4-55 Input and output signals of an OP AMP comparator.

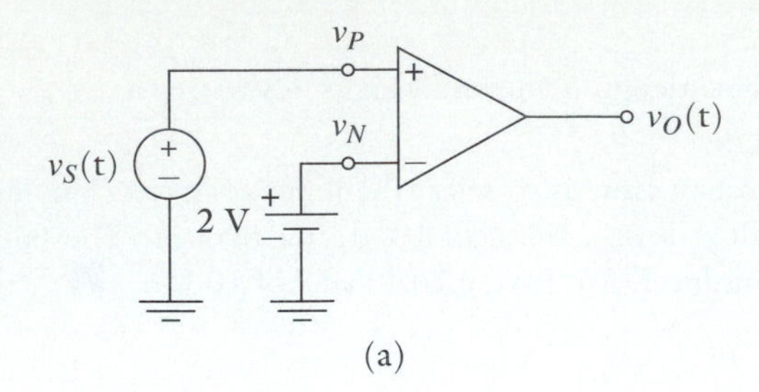

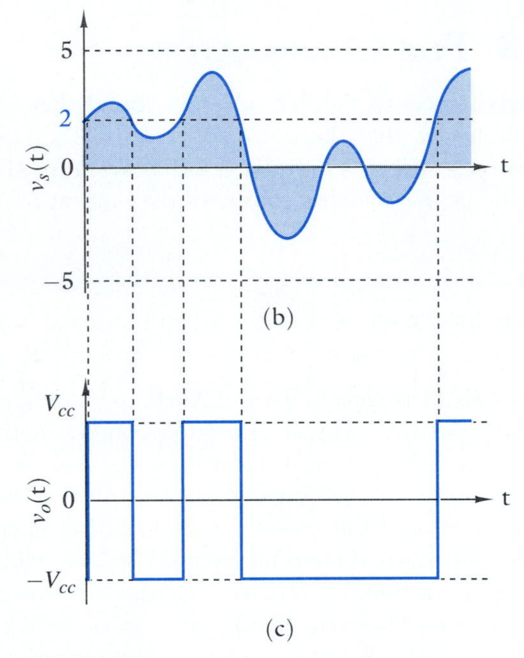

Figure 4-56 Result of applying a 2-V source to a comparator. (a) Circuit diagram. (b) Input voltage. (c) Output voltage.

10 V. With a −10-V source the output is always $+V_{CC}$ since v_S is always greater than −10 V.

In theory any OP AMP operating open loop will function as a comparator. However, OP AMPs designed for linear operation normally saturate at $\pm V_{CC}$, usually ±15 V. These signal levels are not compatible with those used in digital systems. Comparator devices are designed to produce output saturation levels called "high" (V_H) and "low" (V_L) that are compatible with specific types of digital circuits. For example, a comparator with $V_H = +5V$ and $V_L = 0$ V can be directly interfaced with TTL (transistor-transistor logic) digital circuits.

Using the high and low terminology, we write the input-output characteristics of an *ideal comparator* as

$$\begin{aligned} &\text{If } v_P > v_N \text{ then } v_O = V_H \quad \text{else} \\ &\text{If } v_P < v_N \text{ then } v_O = V_L \end{aligned} \tag{4-47}$$

where V_H and V_L are the high and low saturation levels of the element. Note that the equal sign is omitted in the conditional part of the statement because we cannot be sure what the output will be when $v_P = v_N$.

EXAMPLE 4-12

Figure 4-57 shows two transducers that detect temperatures in a rocket engine. For safe operation the temperature T_1 must always be less than T_2. An alarm sounds whenever T_1 is greater than T_2. Show that the comparator in Fig. 4-57 detects alarm and no alarm conditions. Assume that $V_H = 5$ V and $V_L = 0$ V.

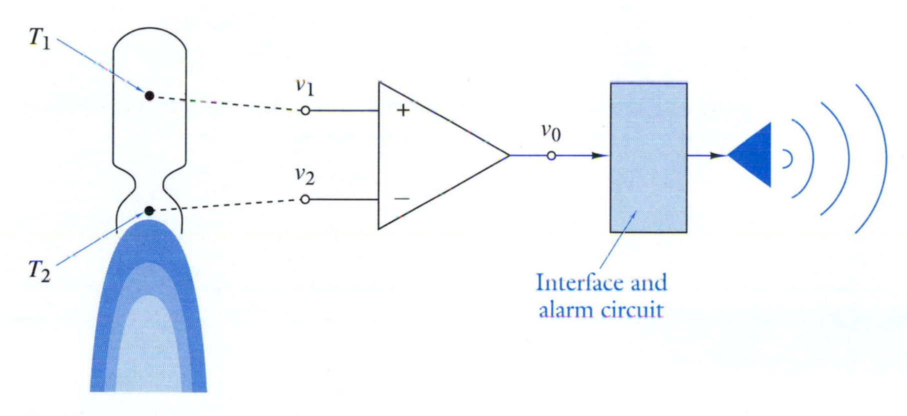

Figure 4-57

SOLUTION:

It is assumed that the two temperature transducers are interfaced with OP AMP circuits whose outputs v_1 and v_2 are proportional to the temperatures T_1 and T_2. These voltages are connected to a comparator as shown in Fig. 4-57. The output of the comparator will be $V_H = 5$ V when $v_1 > v_2$, and $V_L = 0$ V when $v_2 > v_1$. The comparator output is connected to a circuit that generates

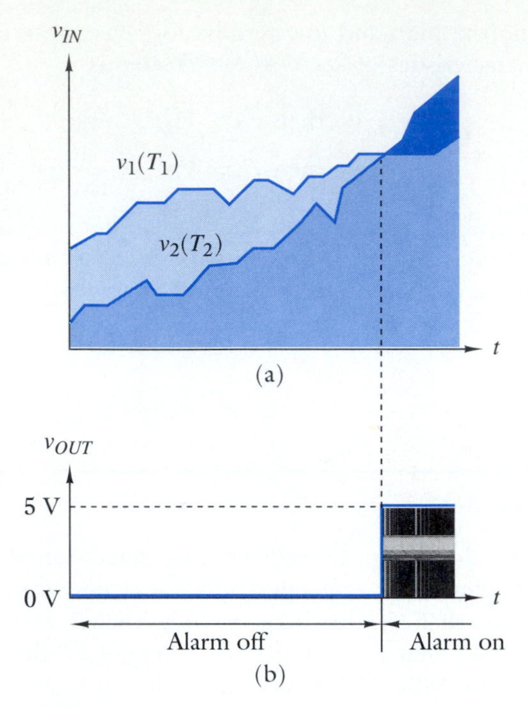

Figure 4-58

an alarm when 5 V is applied at its input. Figure 4-58(a) and (b) show a possible time history of the two temperatures, illustrating both the alarm and no alarm conditions.

Application Note

EXAMPLE 4-13

Figure 4-59 shows a digital application of the comparator. The inputs V_1 and V_2 are digital signals with two possible values: +5 V (high) and 0 V (low). Likewise, the comparator output levels are $V_H = +5$ V and $V_L = 0$ V. The input circuit is a linear resistance summer of the type studied in Chapter 3. Using superposition the voltage v_P is found to be

$$v_P = \frac{1}{3}V_1 + \frac{1}{3}V_2$$

Likewise, voltage division shows that $v_N = 2.5$ V. Using these results and the ideal comparator characteristics in Eq. (4-47), we can write all possible inputs and outputs for this circuit as shown in Table 4-3.

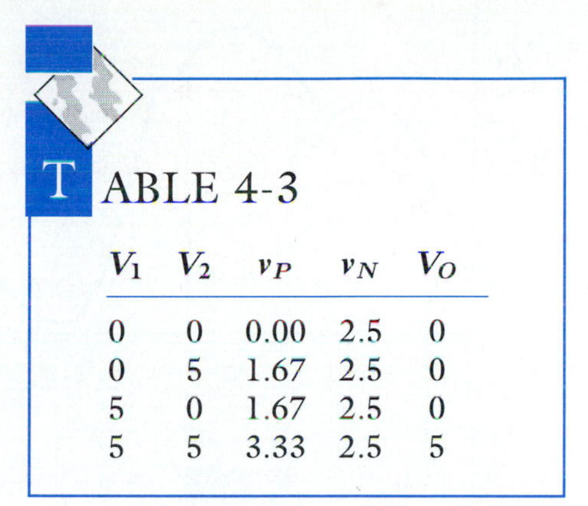

TABLE 4-3

V_1	V_2	v_P	v_N	V_O
0	0	0.00	2.5	0
0	5	1.67	2.5	0
5	0	1.67	2.5	0
5	5	3.33	2.5	5

Notice the pattern in the first two and the last column of Table 4-3. Only when both input voltages are high (5 V) is the output high. The circuit implements the logical statement that V_O is high if and only v_1 and v_2 are high. This statement is called the logical AND operation. A circuit that implements the state is called an *AND gate*—one of the basic building blocks of digital circuits.

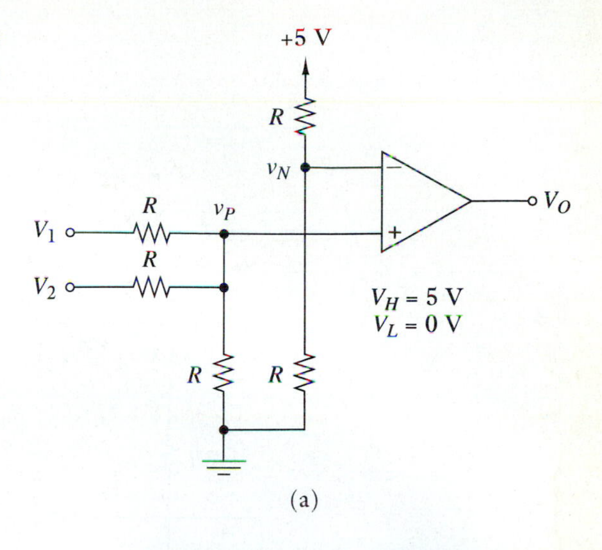

(a)

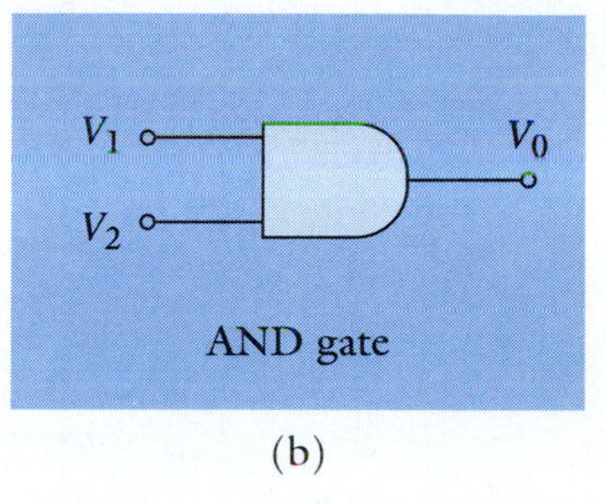

(b)

Figure 4-59

Application Note

EXAMPLE 4-14

The circuit in Fig. 4-60(a) is a *flash converter*, which changes an analog input into a three-bit digital output. The inverting inputs to the comparators are connected to a special voltage divider. Using voltage division we see that $v_{N1} = 1$ V, $v_{N2} = 3$ V and $v_{N3} = 5$ V. The analog input v_{IN} is applied simultaneously to the noninverting inputs of all three comparators. Using Fig. 4-60(b) we see that at time t_1 the analog input is 4 V. At time t_1 $v_P > v_N$ on comparators 1 and 2, and $v_P < v_N$ on Comparator 3. The resulting comparator outputs are $v_{O1} = v_{O2} = 5$ V and $v_{O3} = 0$ V. At time t_2 the analog input increases to 5.3 V. As a result $v_P = 5.3$ V $> v_N$ on all comparators and their outputs are all 5 V. At t_3 the analog input is less than 1 V and all three comparators have outputs of $V_L = 0$ V.

The comparators encode analog inputs in the range 0 to 1 V as a three-bit digital word 000. Inputs in the range from

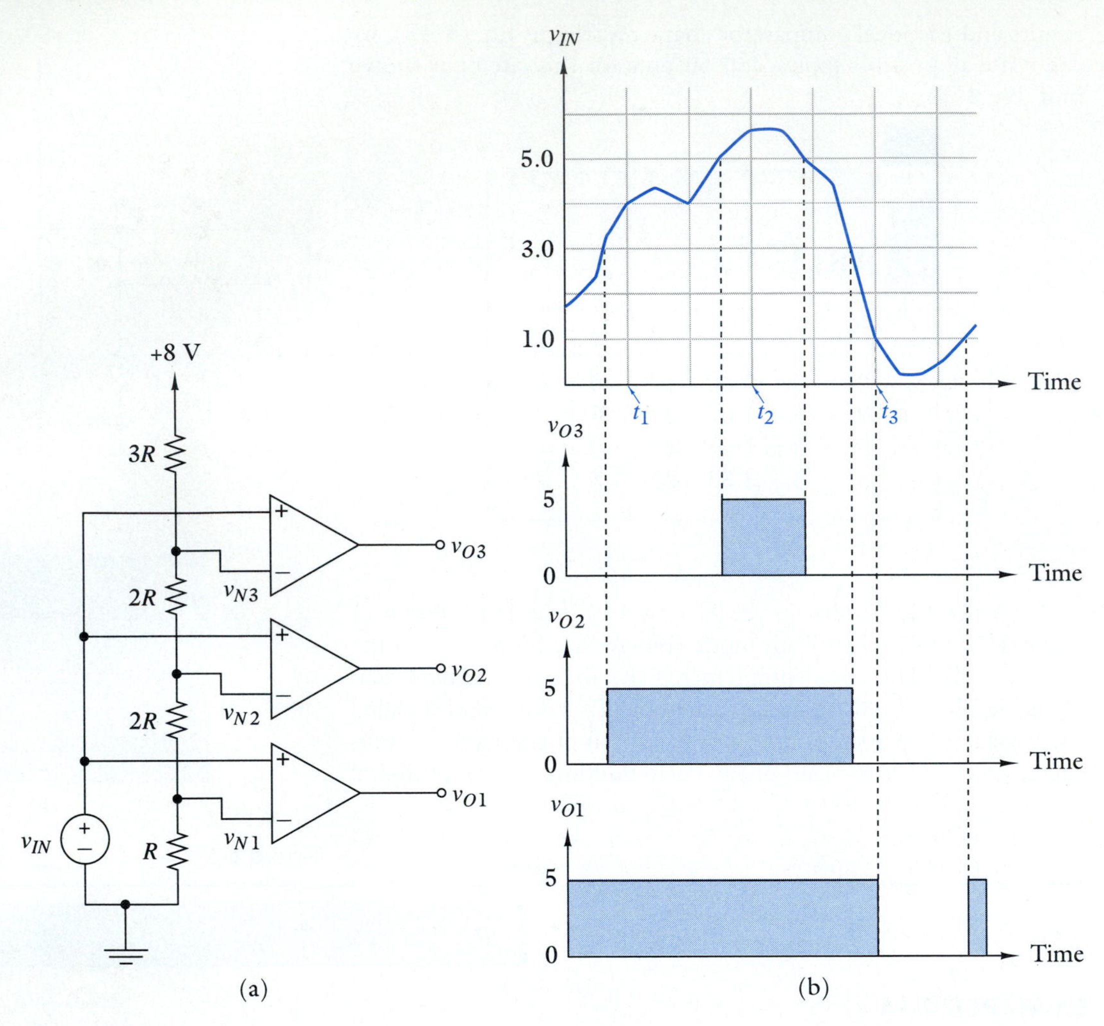

Figure 4-60

1 to 3 V are encoded as 001 and those in the range from 3 to 5 V as 011. In other words, the voltage ranges established by the voltage divider are each converted into a unique digital word. The three-bit digital output is sent to an encoder, which changes this code into a standard two-bit binary word. Since the comparators in Fig. 4-60(a) operate simultaneously the analog-to-digital conversion takes place almost instantaneously, hence the name flash converter.

SUMMARY

- The output of a dependent source is controlled by a signal appearing in a different part of the circuit. Dependent sources are linear circuit elements used to model active devices and are represented by a diamond-shaped source symbol. Each type of controlled source is characterized by a single-gain parameter μ, β, r, or g.

- The Thévenin resistance of a circuit containing dependent sources can be found using open-circuit voltage and short-circuit current, or by applying a test source at the terminals in question. The active Thévenin resistance may be significantly different from the passive lookback resistance.
- An active circuit is capable of delivering more signal power to a load than it receives from the input signal source. The additional power comes from supply voltage sources that are not usually shown in circuit diagrams.
- Active devices can be modeled using linear elements by limiting their operation to a linear region of their nonlinear characteristics.
- The large-signal model can be used to represent the transistor. The nonlinear characteristics of the transistor are divided into three regions or modes. Each mode is represented by a separate circuit and i-v characteristics. Once the operating mode is determined, transistor circuit analysis uses standard methods.
- The OP AMP is an active device with at least five terminals: the inverting input, the noninverting input, the output, and two power supply terminals. The power supply terminals are not usually shown in circuit diagrams. In integrated circuit form the OP AMP is a differential amplifier with a very high open-loop gain μ.
- The OP AMP can be made to operate in a linear mode by providing a feedback circuit from the output to the inverting input. To remain in the linear mode the output voltage is limited to the range $-V_{CC} \leq v_O \leq +V_{CC}$, where $\pm V_{CC}$ are the supply voltages.
- The i-v characteristics of the ideal OP AMP are $i_P = i_N = 0$ and $v_P = v_N$. The ideal OP AMP is assumed to have infinite open-loop gain and input resistance and zero output resistance. The ideal model applies to devices with finite open-loop gain, provided the closed-loop gain is much smaller than the open-loop gain.
- Four basic signal-processing functions performed by OP AMP circuits are the inverting amplifier, noninverting amplifier, inverting summer, and differential amplifier. These arithmetic operations can also be represented in block diagram form.

- OP AMP circuits can be connected in cascade to obtain more complicated signal-processing functions. The analysis and design of the individual stages in the cascade can be treated separately, provided input resistance of the following stage is kept sufficiently high.
- The comparator is a nonlinear signal-processing device obtained by operating an OP AMP device without feedback. The comparator has two analog inputs and a digital output.

EN ROUTE OBJECTIVES AND ASSOCIATED PROBLEMS

ERO 4-1 Linear-Dependent Sources (Sects. 4-1, 4-2)

(a) Given a circuit consisting of linear resistors, controlled sources, and independent sources, find selected signal variables, input-output relationships, or equivalent circuits at specified terminals.

(b) Given an independent source and a specified controlled source model, select circuit parameters so that a specified voltage, current, or power is delivered to a load.

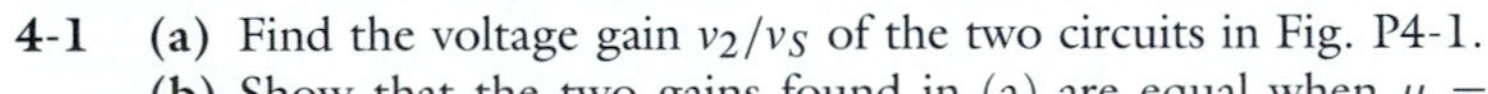

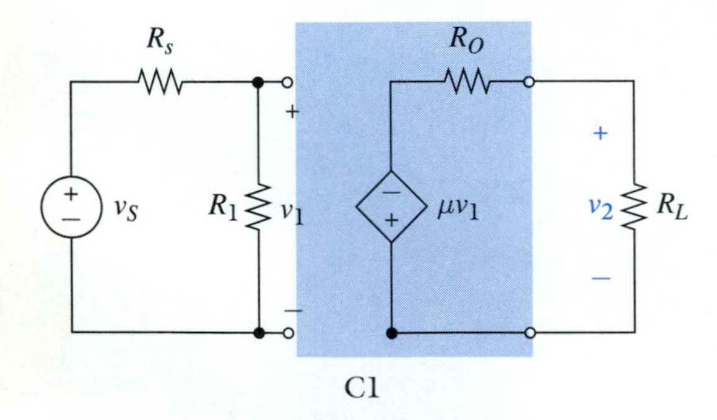

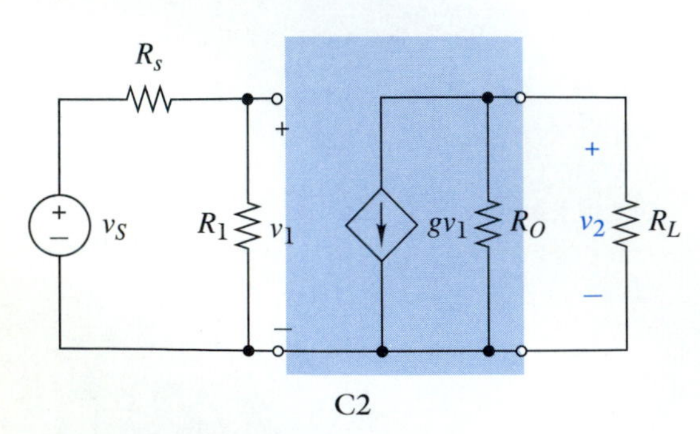

Figure P4-1

4-1 **(a)** Find the voltage gain v_2/v_S of the two circuits in Fig. P4-1.
(b) Show that the two gains found in (a) are equal when $\mu = gR_O$.

4-2 Find the Thévenin equivalent circuit seen by the load resistor R_L in each of the circuits in Fig. P4-1.

4-3 **(a)** Find the current gain i_2/i_S of the two circuits in Fig. P4-3.
(b) Show that the two gains found in (a) are equal when $\beta = gR_1$.

4-4 Find the Norton equivalent circuit seen by load resistor R_L in each of the circuits in Fig. P4-3.

4-5 Determine the voltage gain v_2/v_S for both circuits in Fig. P4-5.

4-6 Show that the input-output relationship of the circuit in Fig. P4-6 is of the form $v_2 = \beta(k_1 v_{S1} + k_2 v_{S2})$.

4-7 The circuit in Fig. P4-7 is the hybrid-π model of a transistor. Find the equivalent resistance seen at terminals A-B.

4-8 Find the Norton equivalent seen by the load resistor R_L in each of the circuits in Fig. P4-5.

4-9 The values of the parameters of circuit C1 in Fig. P4-5 are: $v_S = 5$ V; $R_S = 10$ kΩ; $\beta = 100$; $R_L = 1$ kΩ. Find the value of feedback resistor R_E that produces -10 V across the load R_L.

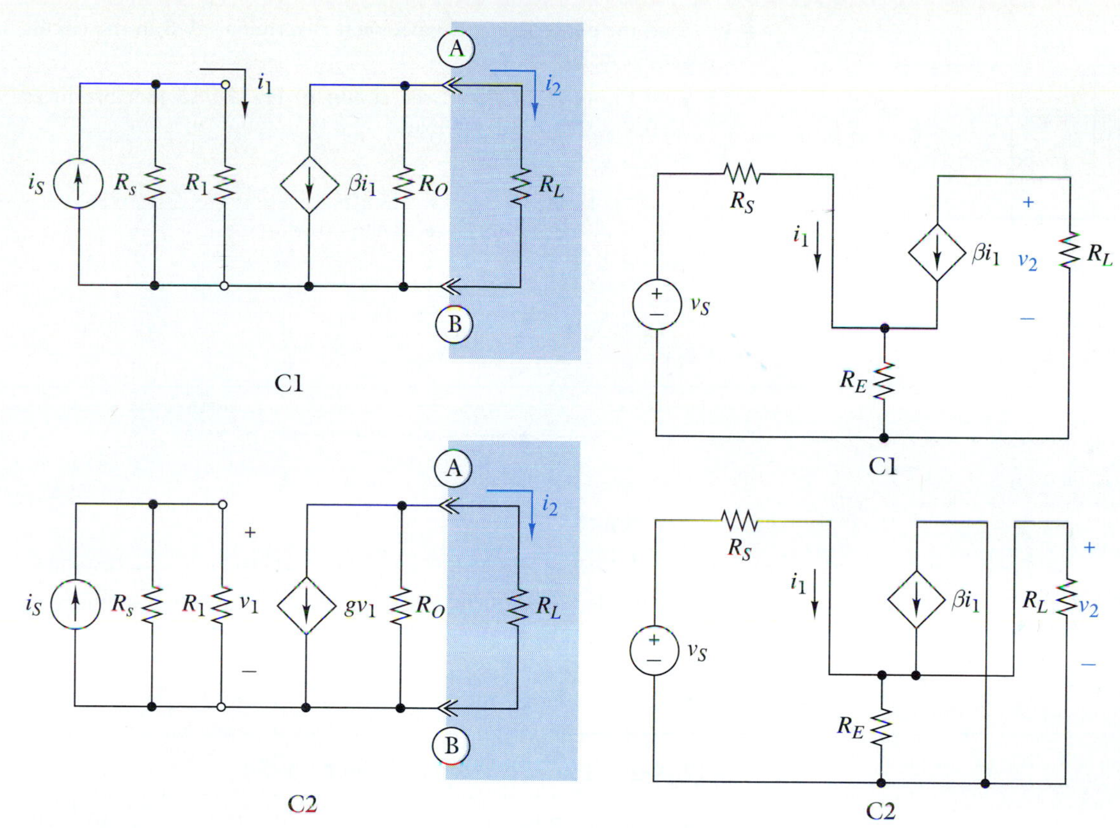

Figure P4-3

Figure P4-5

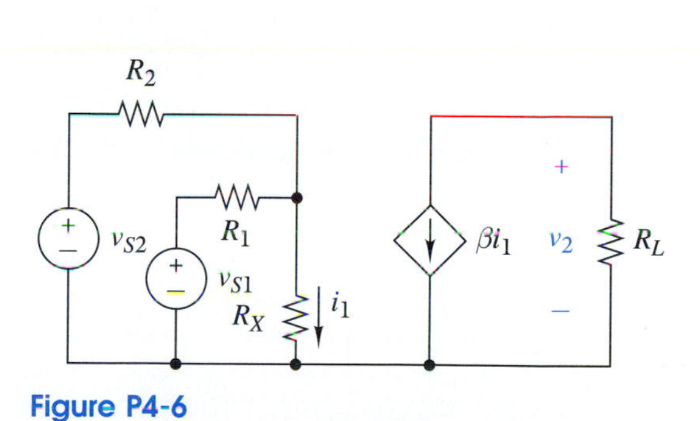

Figure P4-6

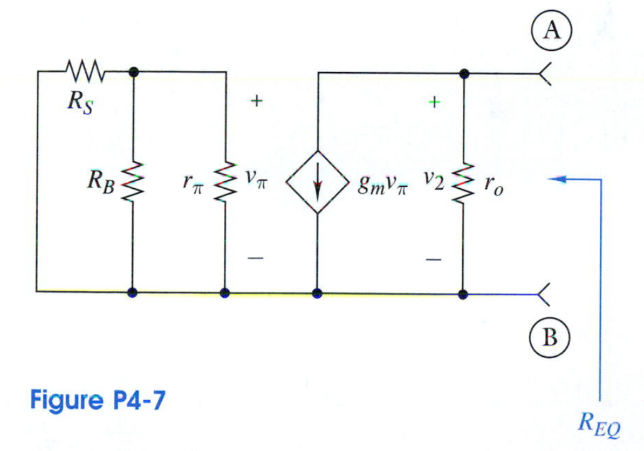

Figure P4-7

4-10 The value of the parameters of circuit C1 in Fig. P4-1 are: $v_S = 5$ V; $R_S = 100\ \Omega$; $R_1 = 200\ \Omega$; $R_O = R_L = 1\ \text{k}\Omega$. Find the values of gain μ that produces $v_2 = -10$ V.

4-11 The value of the parameters of circuit C1 in Fig. P4-3 are: $i_S = 3$ mA; $R_S = 100\ \Omega$; $R_1 = 50\ \Omega$; $R_O = R_L = 1\ \text{k}\Omega$. Find the value of gain β that produces -10 V across the load R_L.

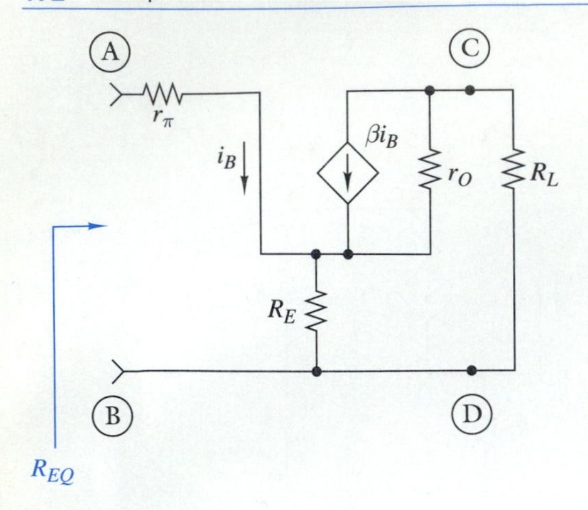

Figure P4-12

4-12 Find the equivalent resistance seen at terminals A-B in the circuit shown in Fig. P4-12.

4-13 Find the value of R_F in the circuit of Fig. P4-13 that produces -10 V across the 1-kΩ load.

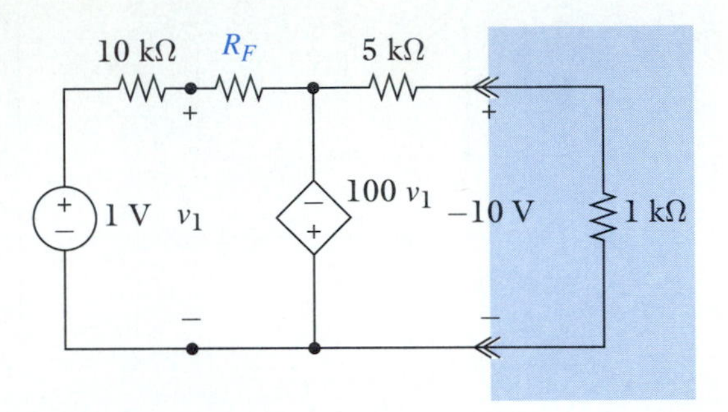

Figure P4-13

4-14 What value of R_F in Fig. P4-13 causes the maximum voltage to be delivered to the load? What is the maximum voltage?

ERO 4-2 The Transistor (Sect. 4-3)

Given a circuit containing resistors, constant sources, and one transistor, use the large signal model to find the operating mode of the transistor or adjust circuit parameters to obtain a specified operating mode.

4-15 Select a value of R_B in the circuit of Fig. P4-15 so that the transistor will switch from the cutoff to active mode when $v_S = V_T$ and from the active to saturation mode when $v_S = 5$ V.

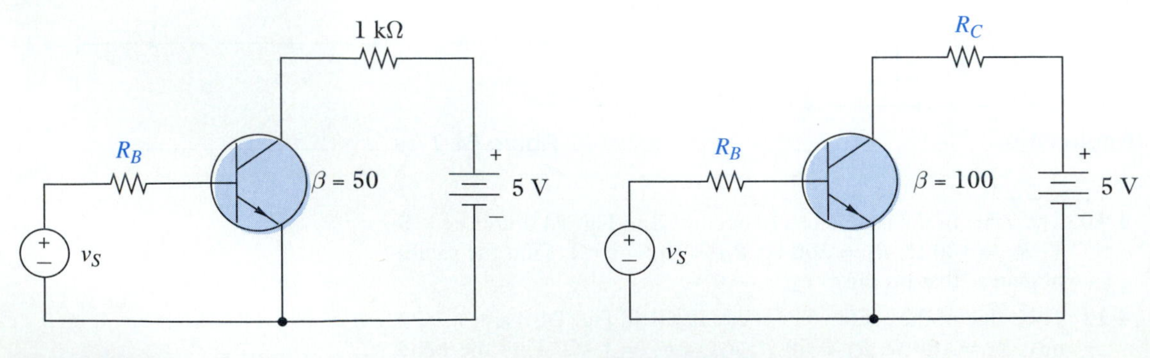

Figure P4-15

Figure P4-16

4-16 The input signal v_S in Fig. P4-16 never exceeds +10 V and is never less than +1 V. Select the values of R_B and R_C to ensure the transistor is always in the active mode when $\beta = 100$ and $V_T = 0.7$ V.

4-17 A 5-V Thévenin voltage in the digital circuit in Fig. P4-17 indicates the incandescent lamp should be turned on. The 5-V output cannot provide the necessary power to light the lamp. The transistor shown in Fig. P4-17 acts as a switch when driven into saturation. Select the value of R_B so the transistor is in saturation when the Thévenin voltage of the digital circuit is 5 V. Assume $V_T = 0.7$ V.

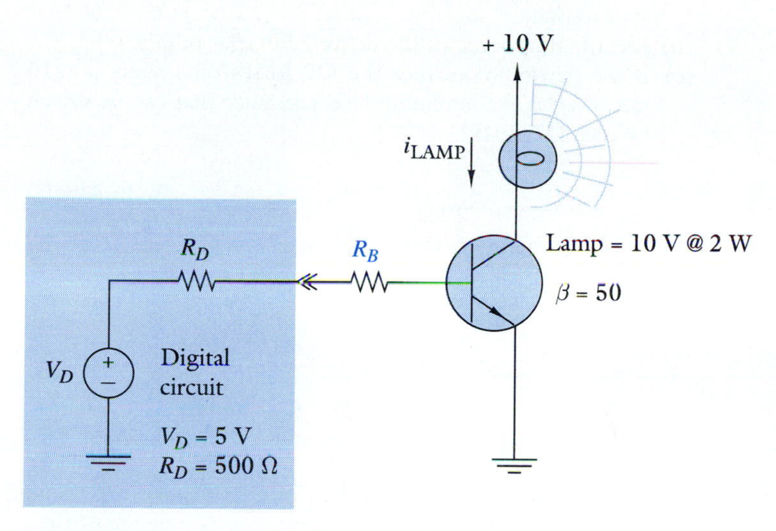

Figure P4-17

4-18 The transistor circuit of Fig. P4-18 uses a 100-Ω feedback resistor in series with the emitter. For $\beta = 50$ and $V_T = 0.7$ V, find V_O when $V_S = 10, 5, 1$, and -1 V.

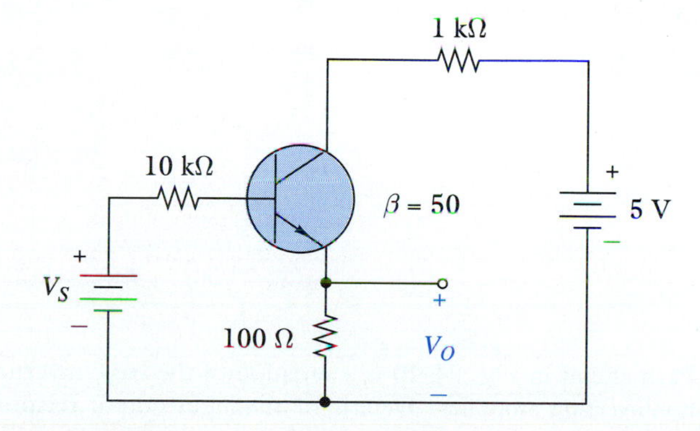

Figure P4-18

ERO 4-3 OP AMP Circuit Analysis (Sects. 4-4, 4-5, 4-6, 4-7)

Given a circuit consisting of linear resistors, OP AMPs, and constant input sources; find selected output signals or input-output relationships in equation or block diagram form.

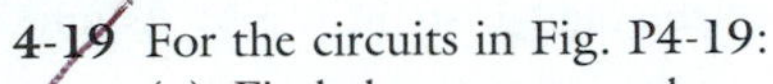

4-19 For the circuits in Fig. P4-19:

(a) Find the current, voltage and power delivered to the 1-kΩ load resistance.

(b) Identify the source of the delivered output power.

(c) If the maximum current the OP AMPs can supply is ±10 mA, what is the minimum load resistance that can be driven by each OP AMP?

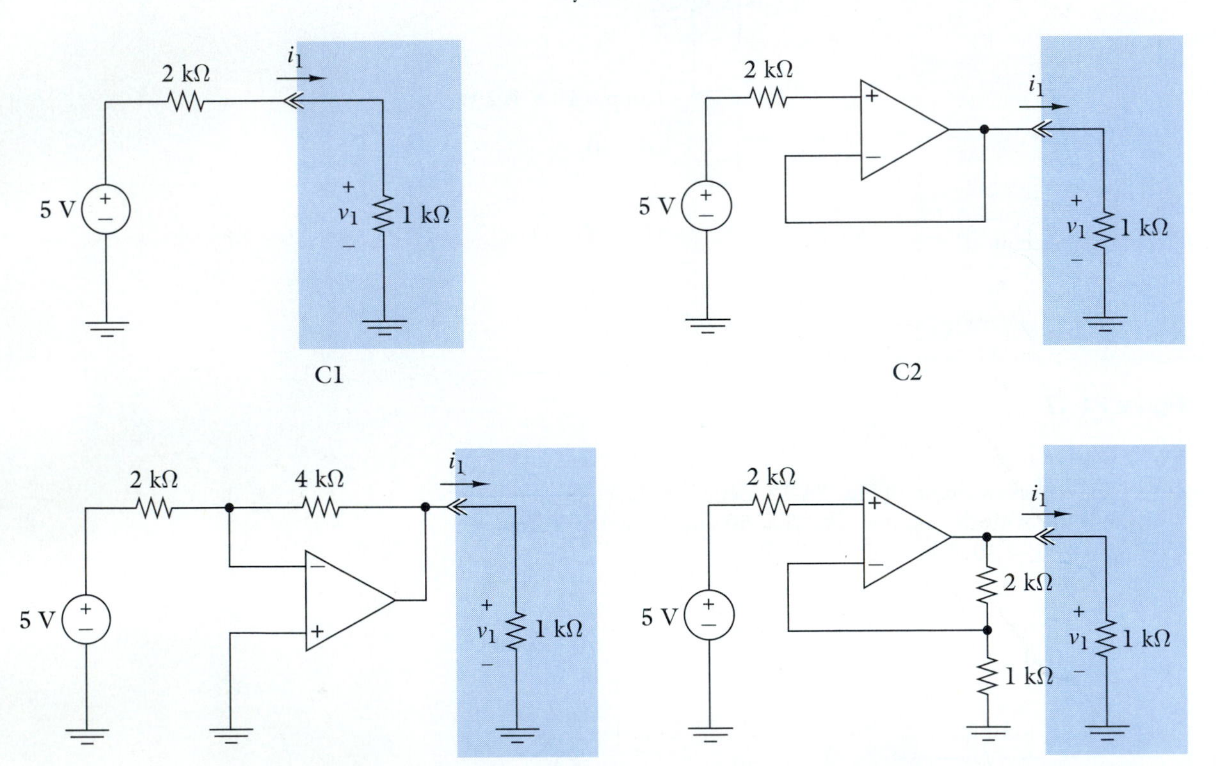

Figure P4-19

4-20 Each circuit in Fig. P4-20 is a variation of the basic inverting or noninverting amplifier. Determine the input-output relationship v_O/v_S, and the input resistance seen by the voltage source in each circuit.

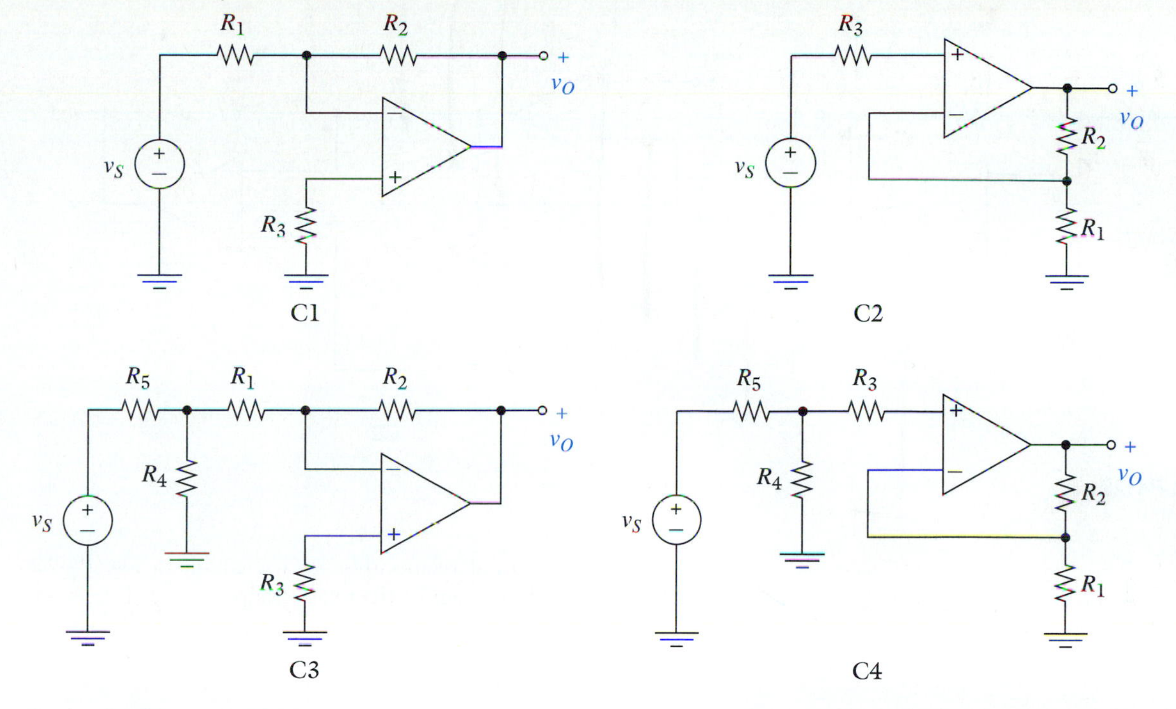

Figure P4-20

4-21 **(E)** Show that the OP AMP circuits in Fig. P4-21 have the same input-output relationship. Draw a block diagram of this relationship. Discuss the advantages and disadvantages of each circuit in terms of parts count, power consumption, input resistance, and input source loading.

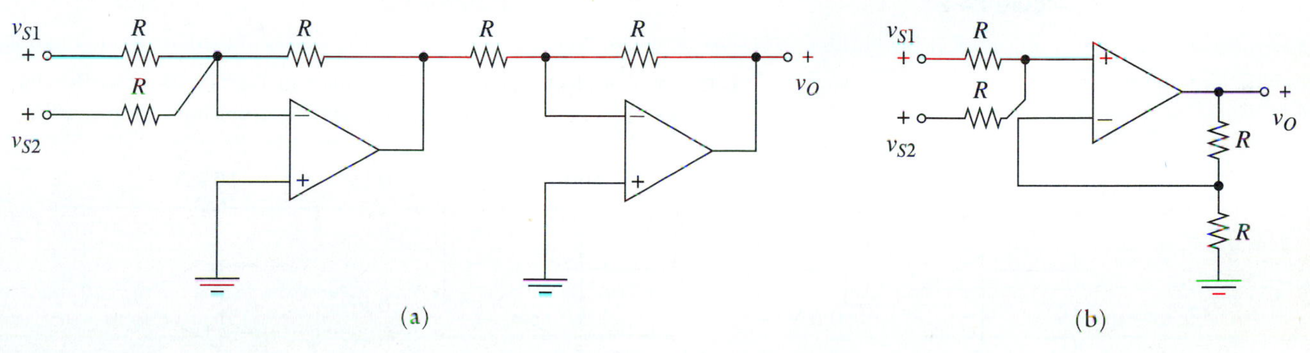

Figure P4-21

4-22 **(E)** Repeat Problem 4-21 for the circuits in Fig. P4-22.

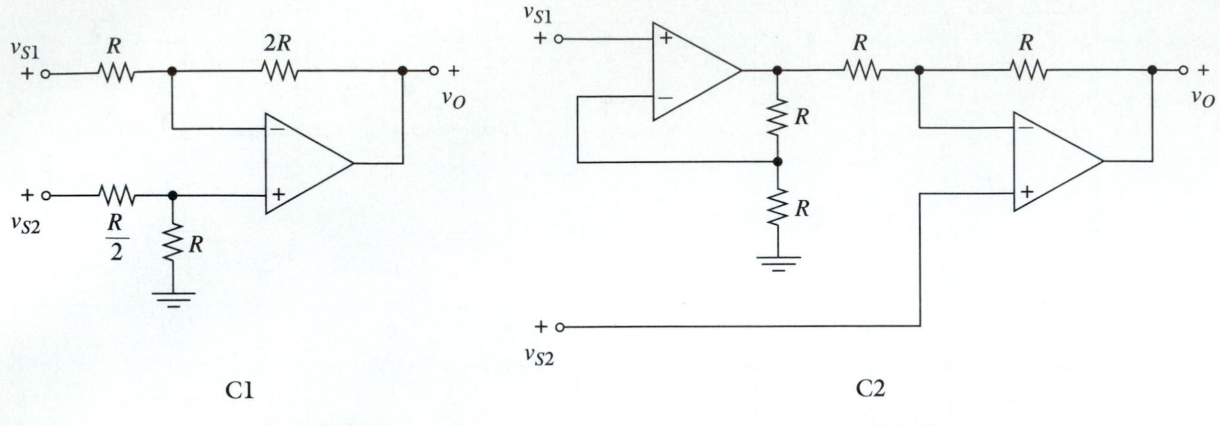

Figure P4-22

4-23 Find the input-output relationship for the circuit in Fig. P4-23 and draw a block diagram of the relationship.

4-24 For the circuit in Fig. P4-24, show that the current $i_L = -2v_S/R$ regardless of the value of R_L.

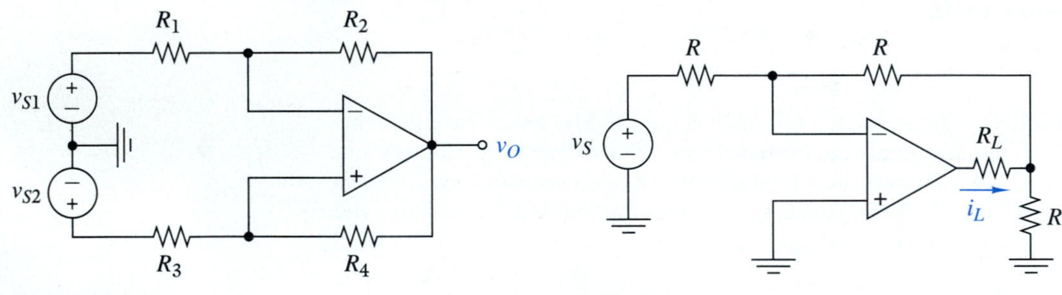

Figure P4-23

Figure P4-24

4-25 Draw a block diagram of the circuit in Fig. P4-25. Use the diagram to determine the input-output relationship of the circuit.

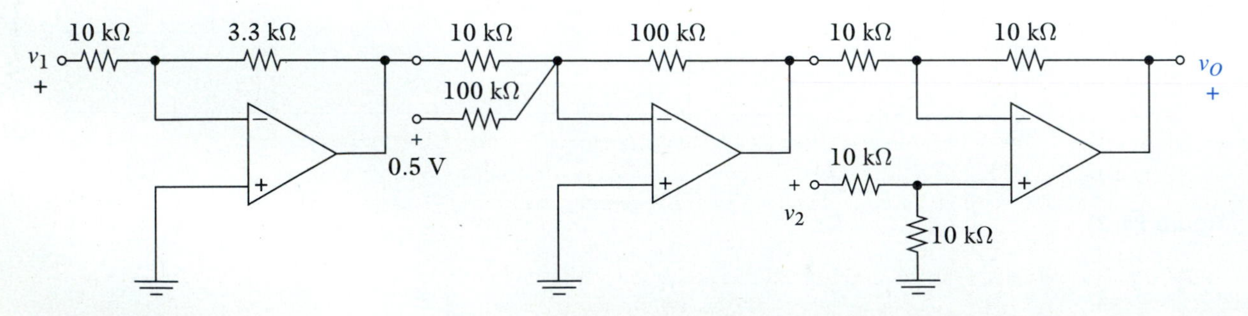

Figure P4-25

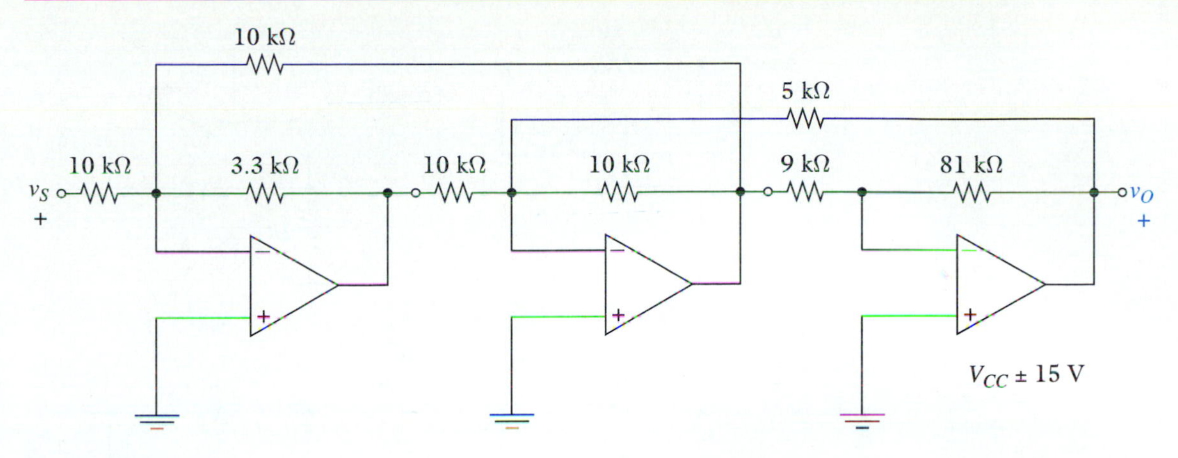

Figure P4-26

4-26 The circuit in Fig. P4-26 is called "leap frog" circuit. Find the input-output relationship v_O/v_S for the circuit.

4-27 For each circuit in Fig. P4-27, find the relationship between i_S and v_O. Then determine the input resistance seen by the current source.

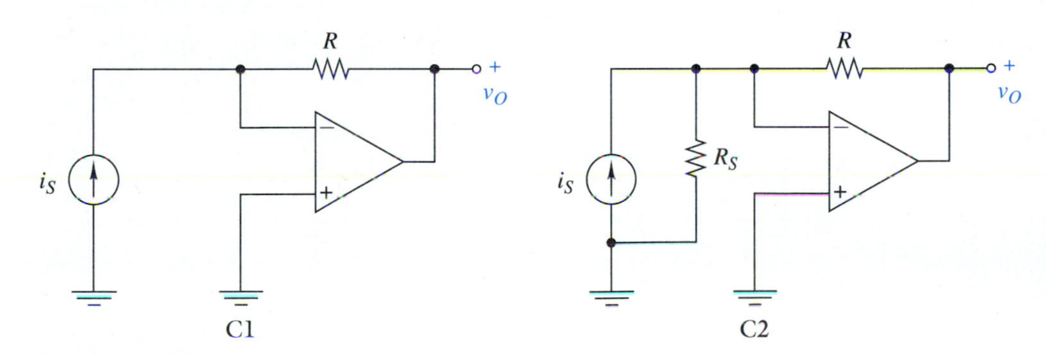

Figure P4-27

4-28 For each of the circuits in Fig. P4-28, find the range of input voltage that will not saturate any of the OP AMPs for $V_{CC} = \pm 15$ V. Then plot the input-output characteristics for the input voltage on the range of ±15 V.

4-29 Plot the output of the three-bit digital-to-analog converter shown in Fig. P4-29 for the digital inputs shown in the figure using $R_F = R$ and $R_F = 2R$.

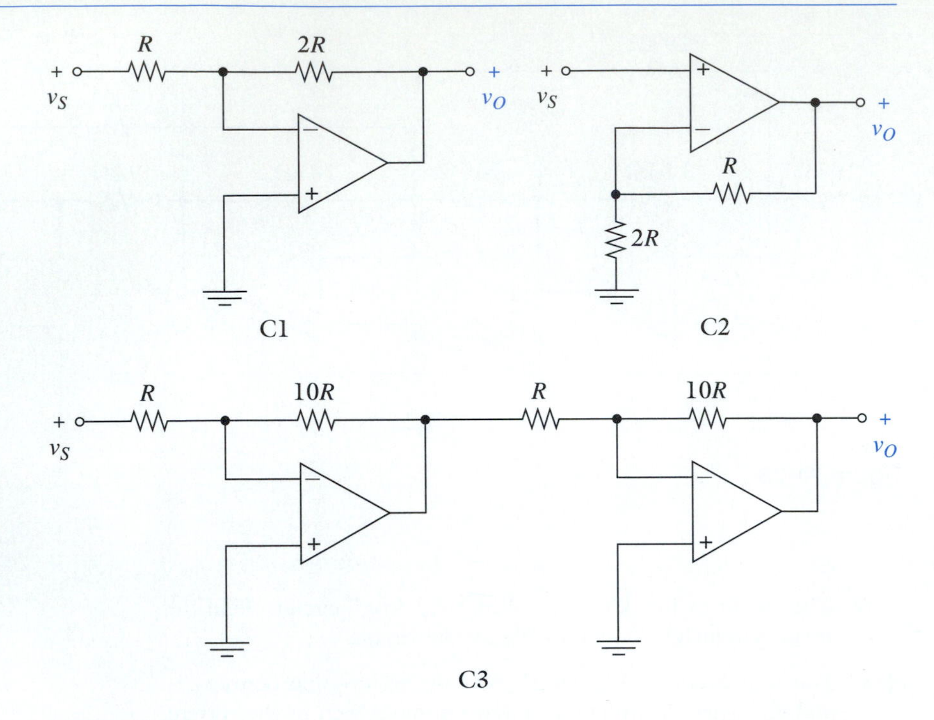

Figure P4-28

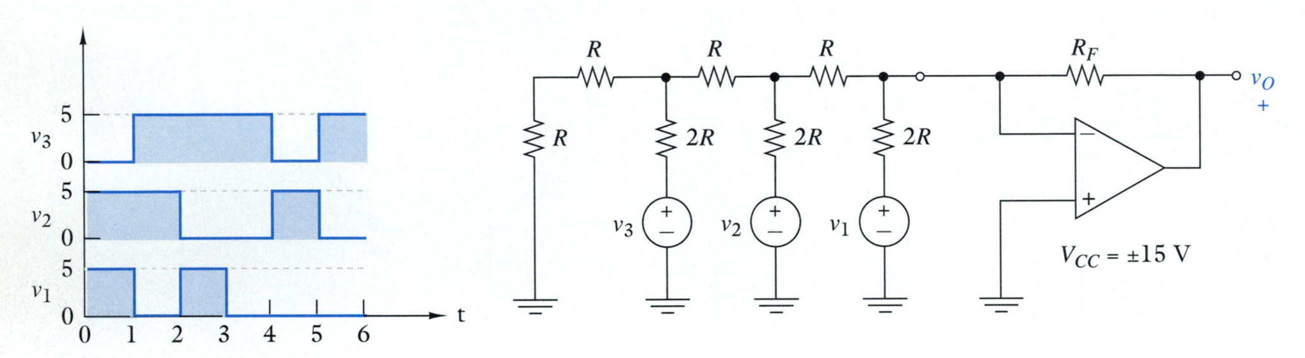

Figure P4-29

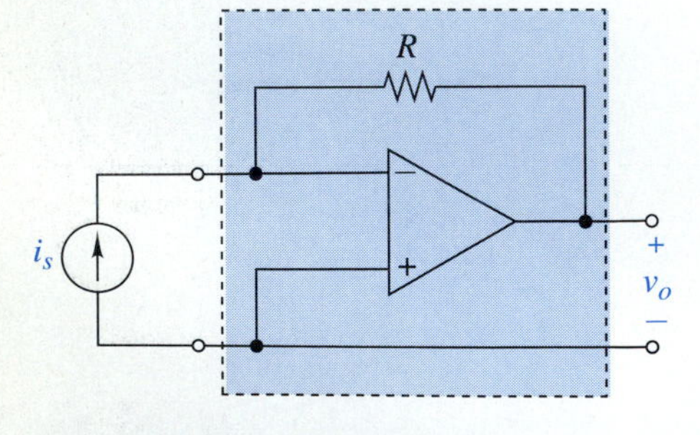

Figure P4-30

4-30 Show that the circuit of Fig. P4-30 is a realization of a current to voltage converter.

4-31 Figure P4-31 shows the controlled source model of an OP AMP voltage follower. If μ is large and $R_I \ll R_O$, show that the voltage follower closed-loop gain, input resistance, and output resistance are approximately $K = \mu/(1+\mu)$, $R_{IN} = \mu R_I$, and $R_{OUT} = R_O/\mu$.

4-32 Analyze the circuit of Fig. P4-32 and show that its input resistance is equal to $-R$ when $R_1 = R_2$.

4-33 A logamp is an amplifier with an output voltage that is proportional to the natural logarithm of its input voltage. Show that

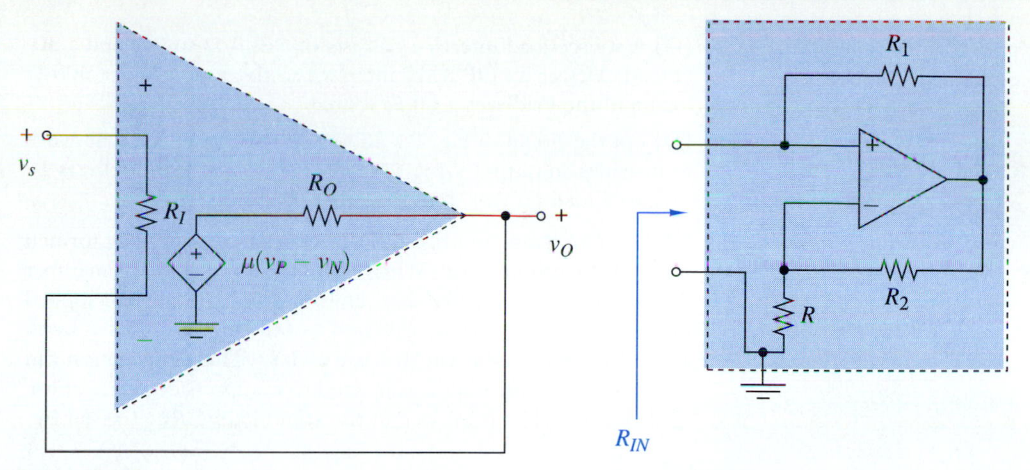

Figure P4-31

Figure P4-32

using a diode in place of the feedback resistor of an inverting OP AMP produces $v_O = K_1 \ln(K_2 v_S)$ when the diode i-v relationship is

$$i_D = 10^{-14} e^{38.5 v_D} \text{ A}$$

for $v_D > 0$. Find the values of K_1 and K_2 in terms of diode parameters.

ERO 4-4 OP AMP Circuit Design (Sects. 4-6, 4-7)

Given an input-output relationship in equation or block diagram form, devise a circuit consisting of linear resistors and OP AMPs that implements the relationship.

❖ **4-34** **(D)** Show how to interconnect a single OP AMP and the R-$2R$ resistor array shown in Fig. P4-34 to obtain voltage gains of ± 3, ± 2, ± 1, and $\pm 1/2$.

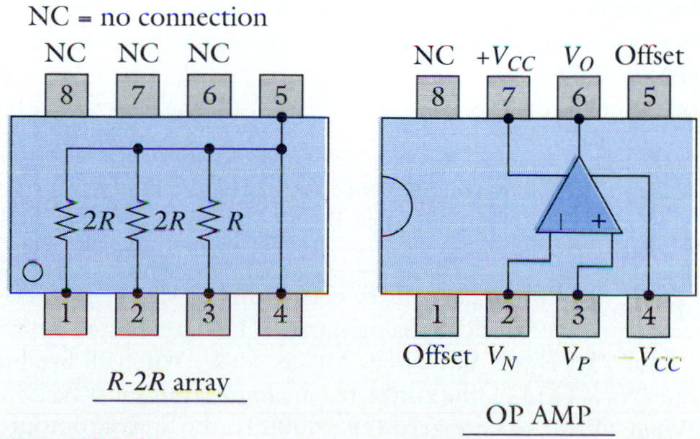

Figure P4-34

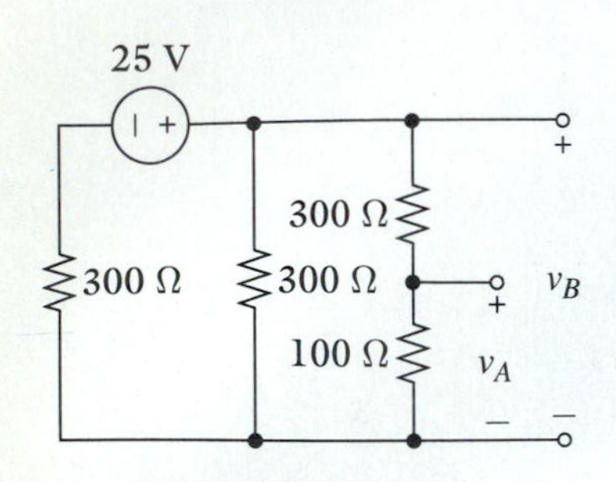

Figure P4-36

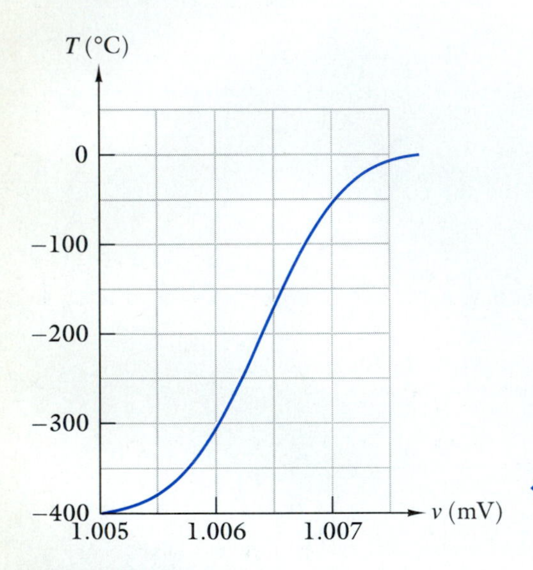

Figure P4-38

❖ **4-35** **(D)** A source-load interface consists of a 300-Ω source and a 50-Ω load. Design an OP AMP interface so the source sees a 300-Ω load and the load sees a 50-Ω source.

❖ **4-36** **(D)** For the circuit of Fig. P4-36, (a) design an OP AMP network to provides an output $v_O = 1/4v_A + 1/2v_B$. (b) If v_O delivers 10 mW to a load resistor, find the value of the resistor.

❖ **4-37** **(D)** A strain gauge is a resistive device such as a specially formed piece of Nichrome wire. When mounted on a physical member the device has a resistance that varies in proportion to the applied mechanical strain. The deformation δ of the member caused by an applied force per unit length ϵ is $\epsilon = \delta/L$. This deformation can occur when the member is subjected to compression, elongation, or deflection. The resistance of the strain gauge R_{St} is given by

$$R_{St} = \frac{\rho L_R}{A_R}$$

where ρ is the resistivity, L_R is the length, and A_R the cross-sectional area of the wire. The strain produced in the strain gauge causes the length to change slightly, ΔL_R, so that

$$\epsilon = \frac{\Delta L_R}{L_R}$$

The total resistance then changes under strain as

$$\Delta R_{St} = 2R_{St}\epsilon$$

ΔR_{St} is generally less than 0.005 Ω per 1000 Ω of R_{st}, far too little to measure with an ohmmeter. Design an OP AMP circuit to detect changes in R_{St} as small as 0.0005 Ω for a 1000-Ω strain gauge, and display the change as a voltage on a 0- to 1-V voltmeter.

❖ **4-38** **(D)** Design an interface circuit to process the output of the temperature transducer whose characteristics are shown in Fig. P4-38 and deliver an output voltage between −1 and +1 V. The output must be −1 V at −300° C and +1 V at −100° C.

❖ **4-39** **(D)** Design a circuit using resistors and OP AMPs to implement the following input-output relationships:

(a) $v_O = 3v_1 - 3v_2$
(b) $v_O = 2v_1 - v_2$
(c) $v_O = 2v_1 + 4v_2$

❖ **4-40** **(D)** Design a circuit to convert a voltage proportional to temperature in °C into a voltage proportional to the temperature in °F. The circuit must operate over the temperature range from −50° C to +150° C.

❖ **4-41** **(D)** A five-bit parallel binary-coded digital signal is to be converted to an equivalent analog signal. The digital signal represents a logic "1" as 0 V and a logic "0" as +5 V. When all five bits are ones (i.e., 11111), the converted analog output must be 23.25 V. When all five bits are zero (i.e., 00000), the analog output must be zero. The digital signals are written with the most significant bit first. You are to use OP AMPs, with $\pm V_{CC} = \pm 25$ V, and resistors with $\pm 1\%$ tolerance.

(a) Design an OP AMP interface circuit to perform the required D/A conversion.

(b) Check your design by inserting all 2^5 possible digital inputs (00001, 00010, 00100, etc.) and checking the linearity of the output. Finally, comment on the adequacy of $\pm 1\%$ tolerance on the resistors.

4-42 **(D)** Design an OP AMP circuit that implements each of the block diagrams in Fig. P4-42 using only the standard resistance values for $\pm 5\%$ resistors.

4-43 **(D)** Repeat Problem 4-42 for the block diagrams in Fig. P4-43.

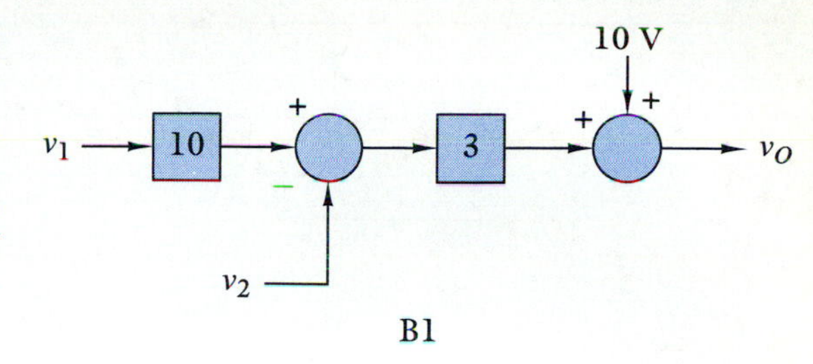

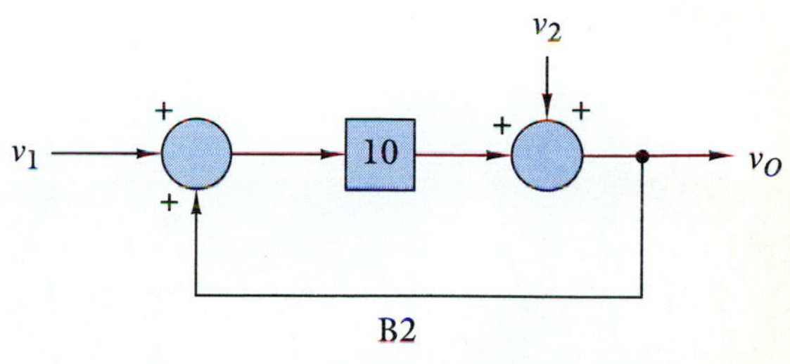

Figure P4-42

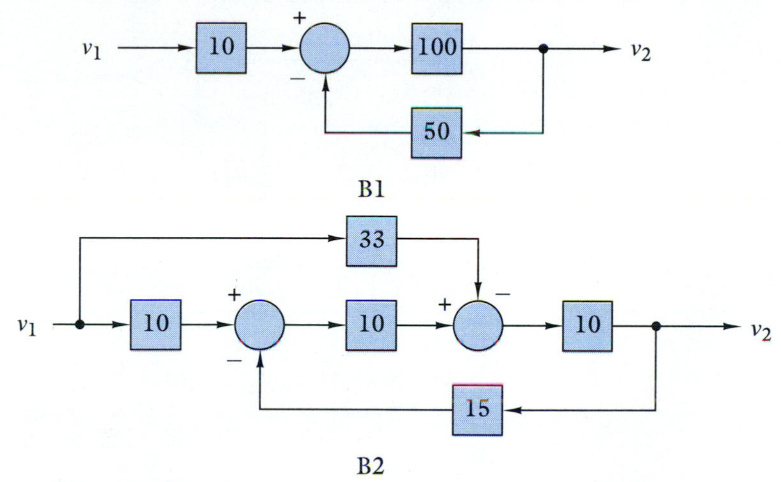

Figure P4-43

4-44 **(D)** The circuit in Fig. P4-44 has been designed to implement a certain input-output relationship. Find the relationship and develop an alternative design using only one OP AMP.

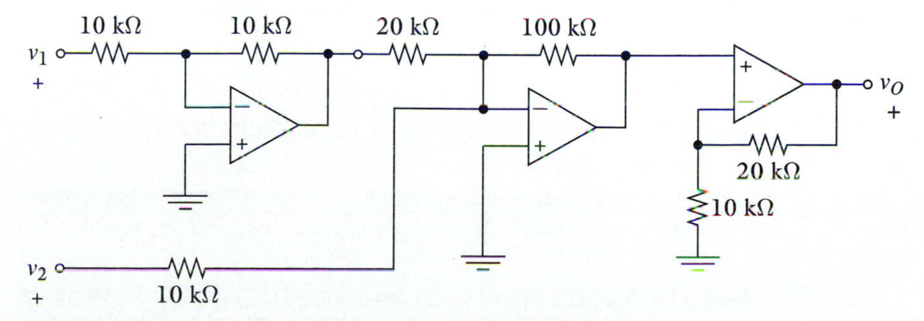

Figure P4-44

ERO 4-5 The Comparator (Sect. 4-8)

(a) Given a circuit with one or more comparators, find the output for given input(s) or find the input required to produce a given output.

(b) Given an input-output relationship, design a circuit consisting of linear resistors, constant voltage sources, and comparators that produces the relationship.

4-45 For each of the circuits in Fig. P4-45 determine the condition for the input to produce $v_O = V_H$. Repeat for $v_O = V_L$.

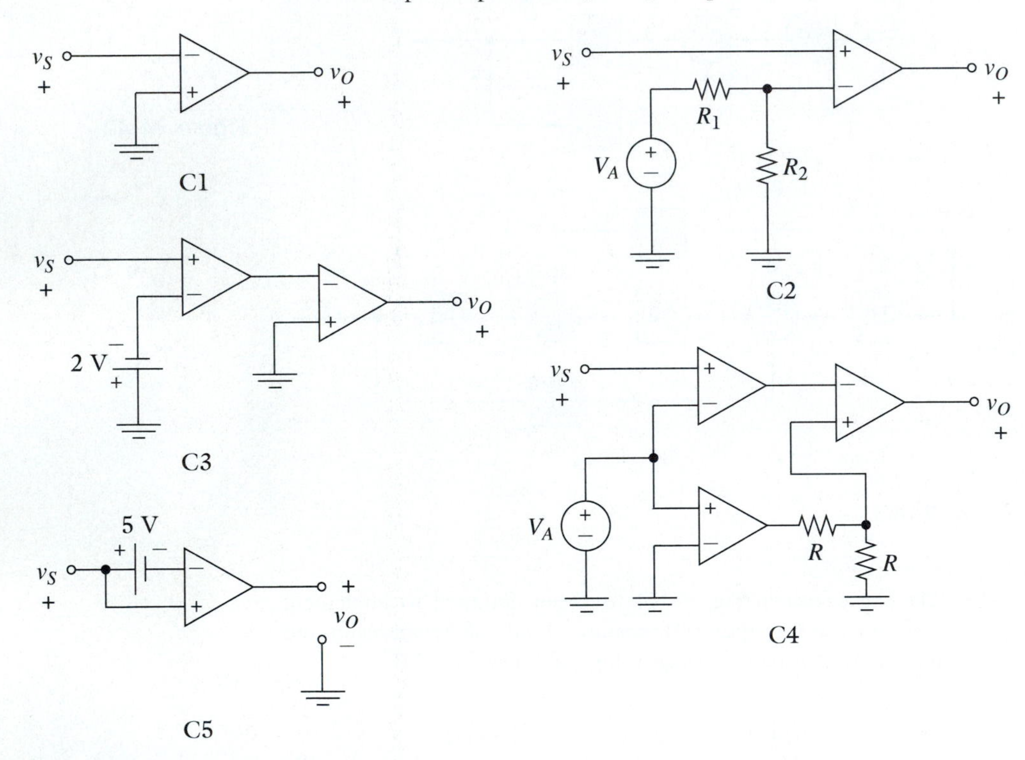

Figure P4-45

4-46 Determine the output of the circuit of Fig. P4-46(a) for the input shown in Fig. P4-46(b).

4-47 A four-bit flash analog-to-digital converter is shown in Fig. P4-47. If $v_S = t$, sketch the outputs v_A, v_B, v_C, and v_D versus t for $0 \leq t \leq 16$.

❖ **4-48** **(D)** A certain signal varies between −10 V to +10 V. Design a circuit that will light a red light whenever the signal is outside the range ±7.5 V.

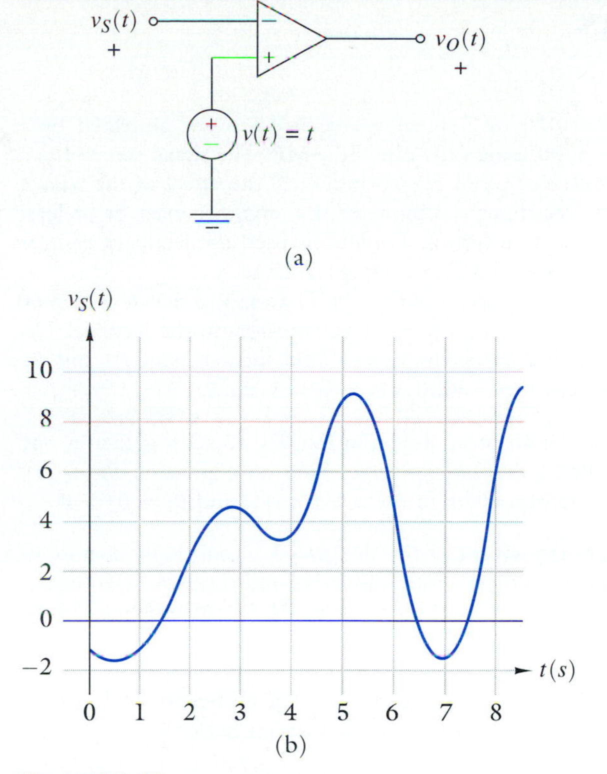

Figure P4-46

Figure P4-47

❖ **4-49** **(D)** A signal varies as a function of time as $v = 2t - 12$. Design a circuit that will detect when the signal crosses 0 volts. At what time does this occur?

❖ **4-50** **(D)** A temperature transducer and interface circuit produce the characteristics shown in Fig. P4-50. Design a circuit that lights a green light whenever the temperature is less than 150°C and a red light whenever it is greater than 120°C.

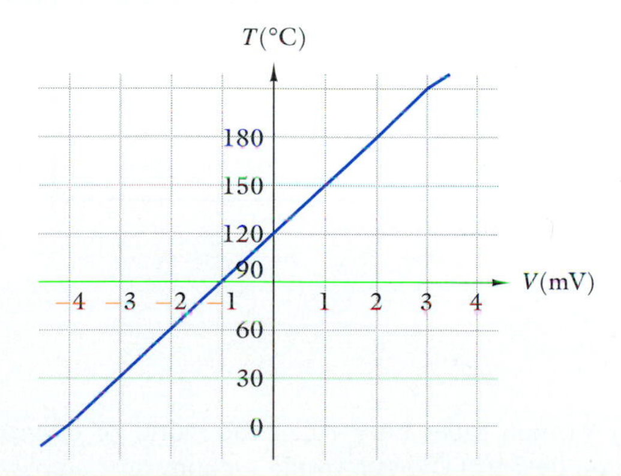

Figure P4-50

CHAPTER INTEGRATING PROBLEMS

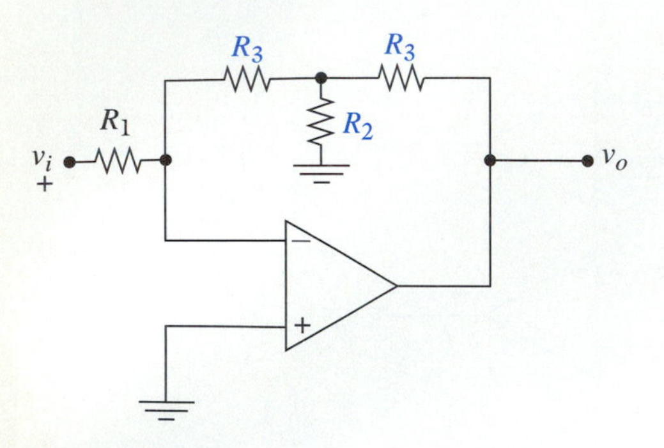

Figure P4-51

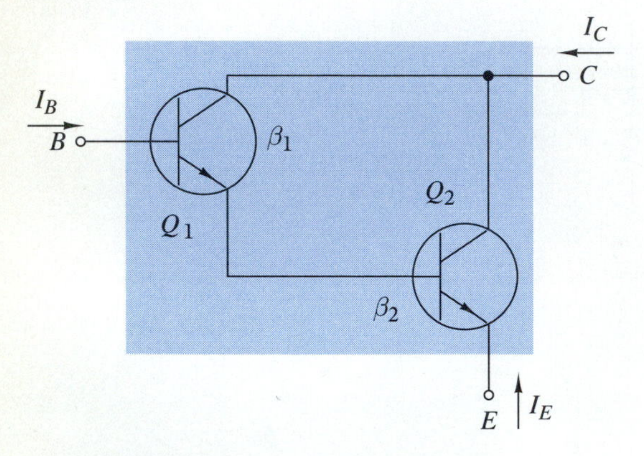

Figure P4-52

4-51 **(A)** The output of a signal source needs to be amplified by a gain of -3000 using only one OP AMP. The signal source has a source resistance of 75 Ω. To minimize the effect of the source resistance, the input resistance of the amplifier must be at least 1 kΩ. If a basic inverting amplifier is used the feedback resistor would be at least 3 MΩ, which is too large.

Analyze the circuit of Fig. P4-51 and show that $K = -3000$ can be achieved using three smaller resistors in the feedback circuit. Note that R_3 occurs twice. Find the values of R_2 and R_3 that produces $K = -3000$ when $R_1 = 1$ kΩ.

4-52 **(A)** The two transistors shown in Fig. P4-52 act together as one device called a *Darlington pair*. Find an expression for I_C/I_B when the transistors are in the active mode and $\beta_1 = \beta_2 = \beta$.

4-53 **(A)** Before the advent of the integrated circuit, logic operations were realized using discrete transistors and resistors. Two examples of such circuits are shown in Fig. P4-53. Assume the transistors switch from cutoff to the saturation mode when a 5-V input is applied and from saturation to the cutoff mode when a 0-V input is applied. Complete the missing entries in Table P4-53 and identify the logic function each circuit performs.

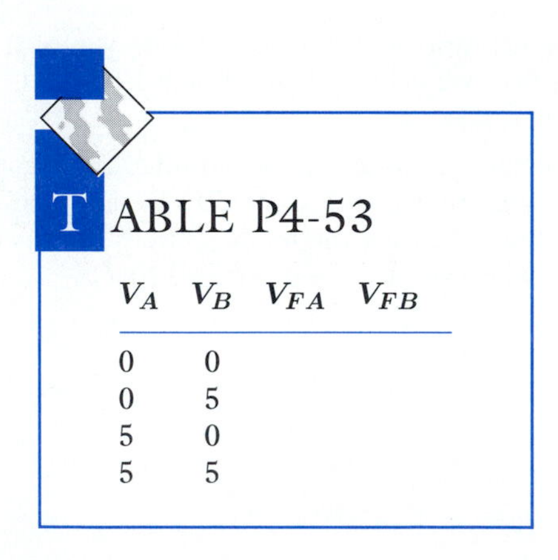
TABLE P4-53

V_A	V_B	V_{FA}	V_{FB}
0	0		
0	5		
5	0		
5	5		

4-54 **(A)** Vacuum tubes once ruled the world of electronic circuits. Figure P4-54(a) shows a triode vacuum tube amplifier. Find the gain v_o/v_i using the dependent source model of Fig. P4-54(b).

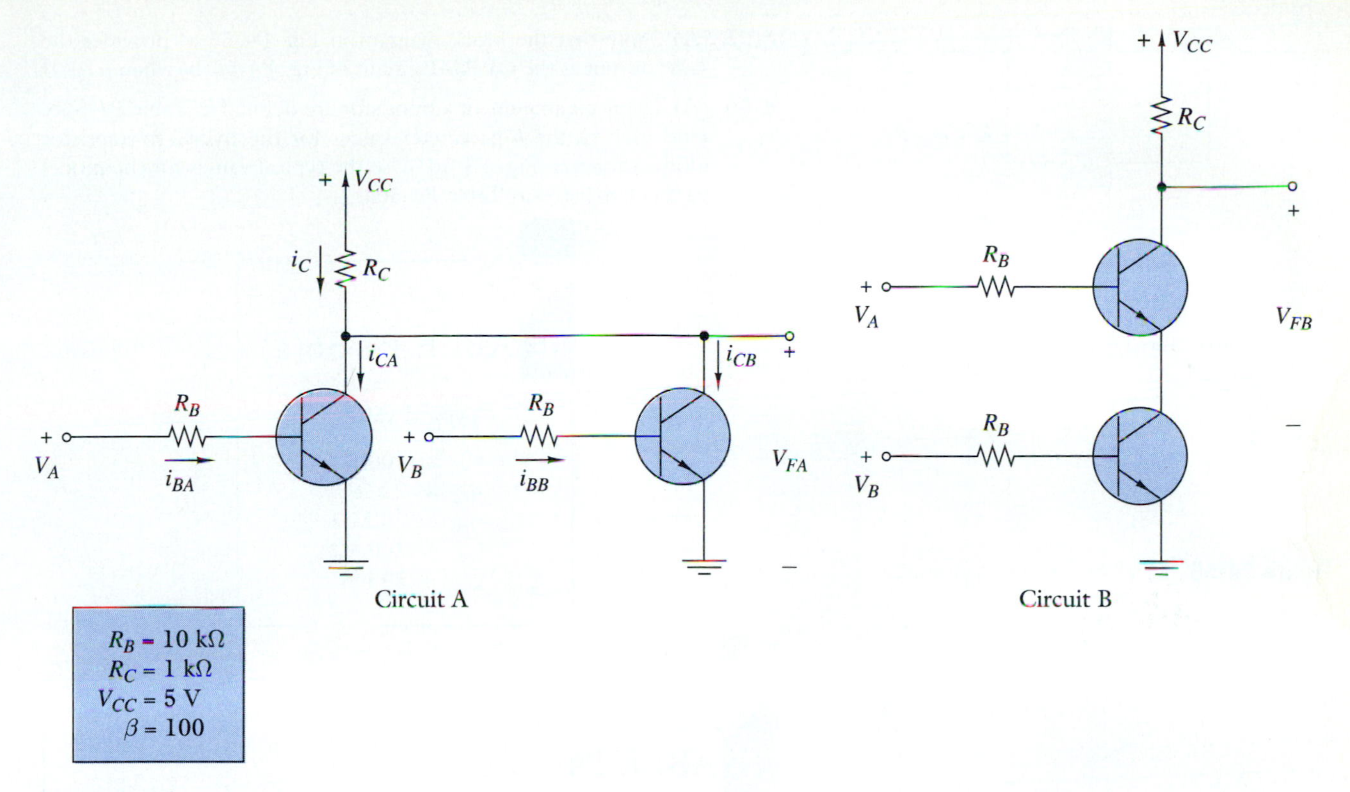

Figure P4-53

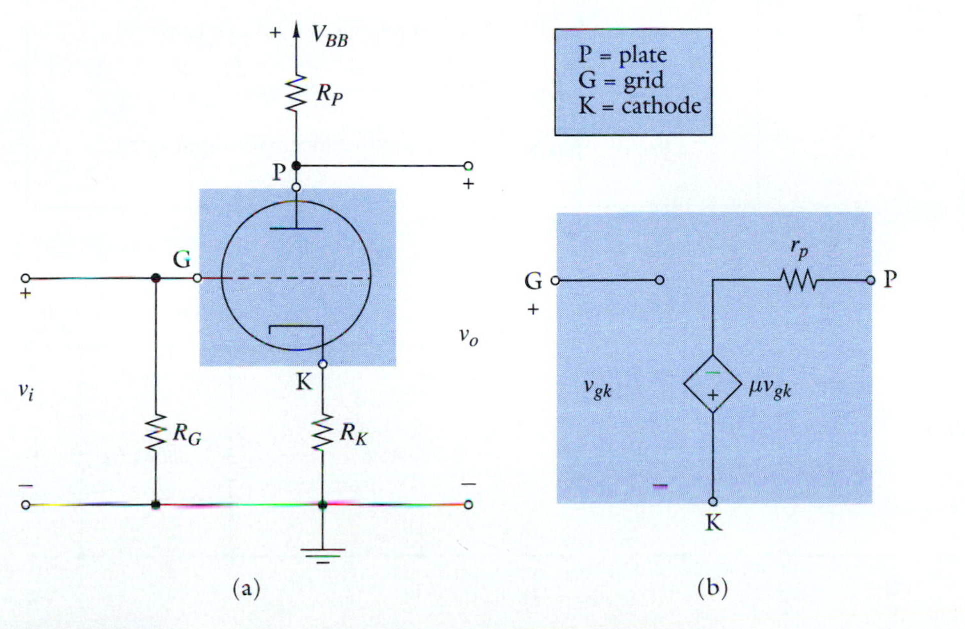

Figure P4-54

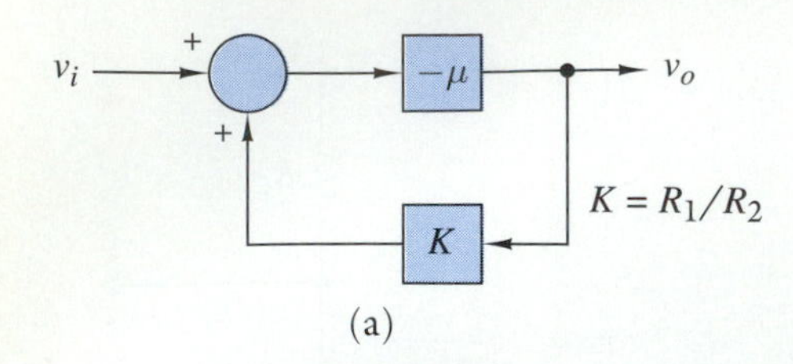

(a)

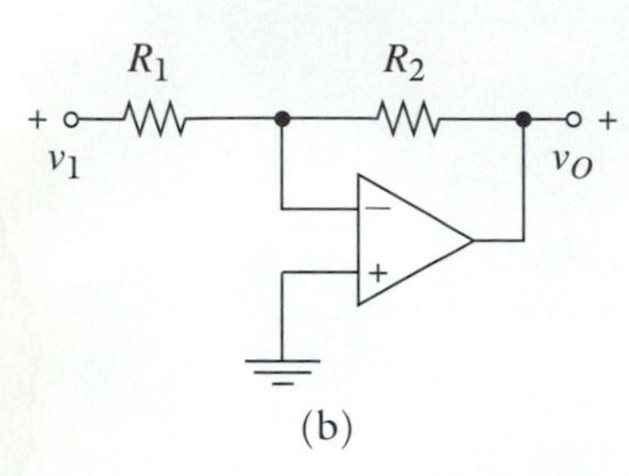

(b)

Figure P4-55

4-55 **(A)** Show that the block diagram in Fig. P4-55(a) provides the same output as the OP AMP circuit of Fig. P4-55(b) when $\mu \to \infty$.

4-56 **(A)** The h-parameters of a transistor are defined in Table P4-56b. Find each of the h-parameter values for the hybrid-π transistor model shown in Fig. P4-56. Use the typical values for the model parameters listed in Table P4-56a.

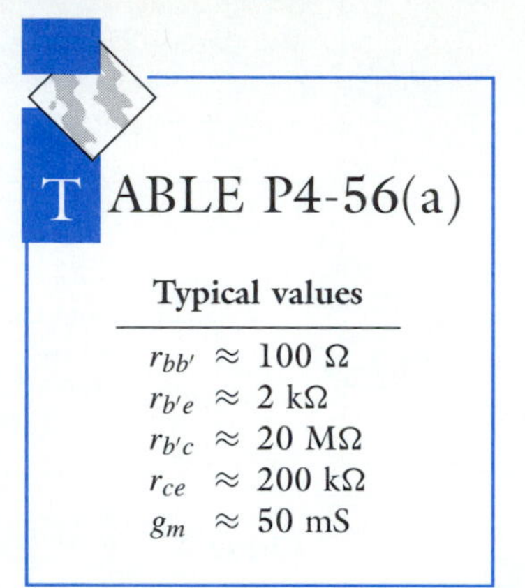

TABLE P4-56(a)

Typical values
$r_{bb'} \approx 100\ \Omega$
$r_{b'e} \approx 2\ \text{k}\Omega$
$r_{b'c} \approx 20\ \text{M}\Omega$
$r_{ce} \approx 200\ \text{k}\Omega$
$g_m \approx 50\ \text{mS}$

TABLE P4-56(b)

$$h_{ie} = \frac{v_b}{i_b}\Big|_{v_c=0} = \text{input resistance with output shorted}$$

$$h_{re} = \frac{v_b}{v_c}\Big|_{i_b=0} = \text{reverse open-circuit voltage gain}$$

$$h_{fe} = \frac{i_c}{i_b}\Big|_{v_c=0} = \text{short-circuit current gain}$$

$$h_{oe} = \frac{i_e}{v_c}\Big|_{i_b=0} = \text{output conductance with input open}$$

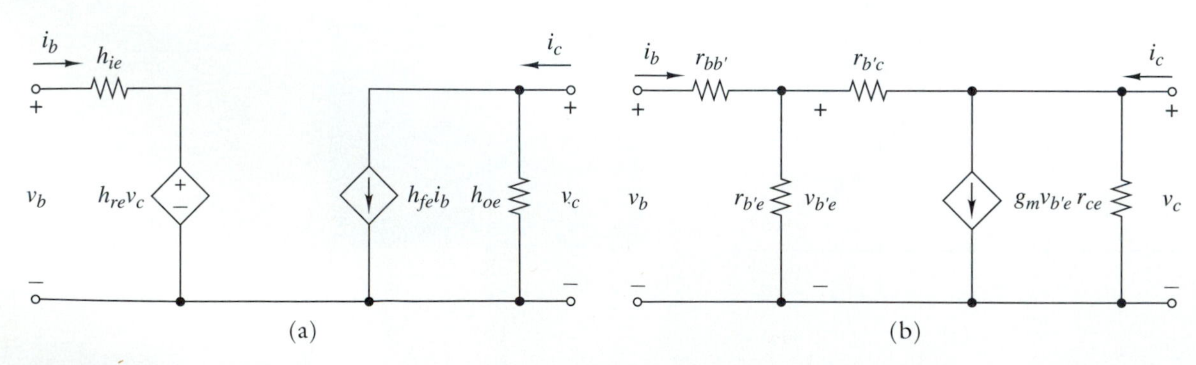

Figure P4-56

4-57 **(A)** A transistor differential amplifier can be modeled as shown in Fig. P4-57. Find v_O as a function of v_1 and v_2.

❖ **4-58** **(D)** A thermistor is used to measure the temperature of a motor from the lowest expected temperature (−10°C) to the highest motor operating temperature (+150°C). Data on the available thermistor are given in Table P4-58. The temperature is to be displayed on a 0 to 5 V voltmeter. Design a suitable interface circuit. Assume you have general-purpose OP AMPs available with $V_{CC} = \pm 15$ V. Use only the standard resistance values for resistors with ±5% tolerance.

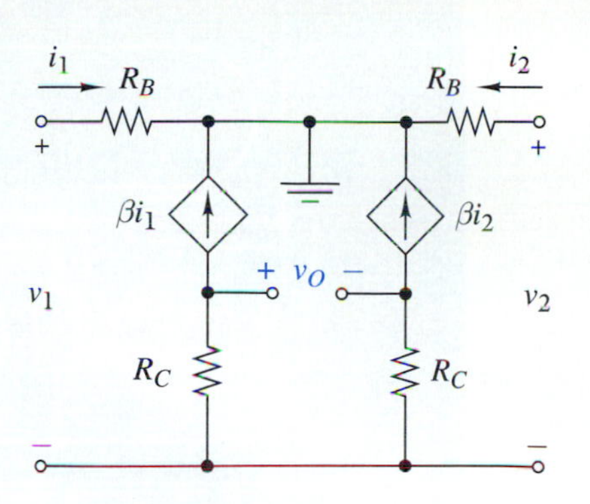

Figure P4-57

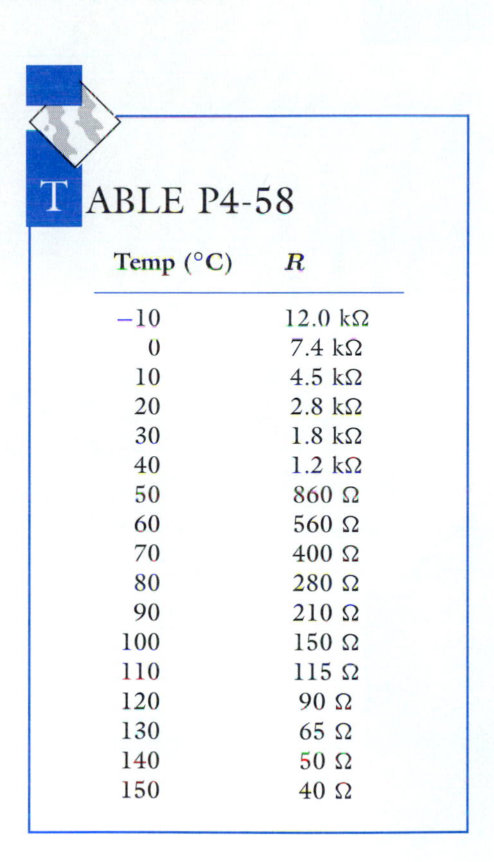

TABLE P4-58

Temp (°C)	*R*
−10	12.0 kΩ
0	7.4 kΩ
10	4.5 kΩ
20	2.8 kΩ
30	1.8 kΩ
40	1.2 kΩ
50	860 Ω
60	560 Ω
70	400 Ω
80	280 Ω
90	210 Ω
100	150 Ω
110	115 Ω
120	90 Ω
130	65 Ω
140	50 Ω
150	40 Ω

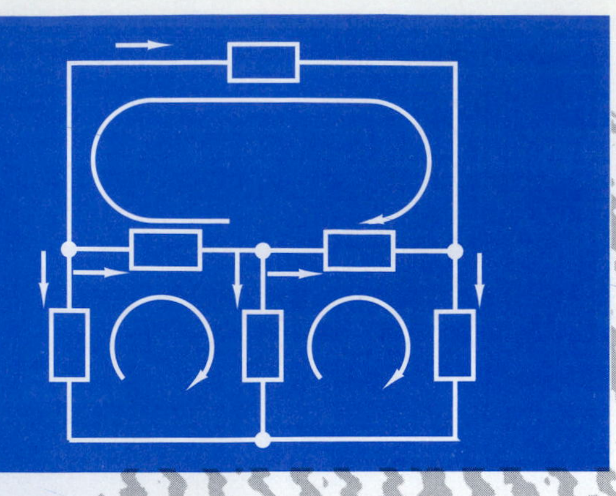

Chapter Five

"(The) advantage of the nodal formulation results from the fact that the equations can be more directly correlated with the physical structure of the network than is possible with the mesh formulation."

Hendrik W. Bode, 1945
American Electrical Engineer

GENERAL CIRCUIT ANALYSIS

The analysis methods studied so far help develop insight into circuit analysis and design because we manipulate the circuit model to find voltage and current responses. With practice and experience we learn which tools to use and in what order to avoid going down blind alleys. This ad hoc approach is practical as long as circuits are fairly simple. As circuit complexity increases, however, a more systematic approach is needed to deal with analysis problems in a more structured way.

The two primary systematic methods are called mesh-current analysis and node-voltage analysis. Although in principle either method will work on any circuit, the node-voltage method generally works best in electronic circuits. Among the first to capitalize on this was Hendrik W. Bode, who spent most of his professional career as a member of the technical staff of the Bell Telephone Laboratories. A central technological problem facing Bell Labs in the 1930s was to provide high-quality long-distance telephone service on a wide scale. Bode, along with colleagues such as Harry Nyquist and R. L. Blackman, pioneered the development of a design theory for feedback amplifiers. Their contributions not only solved the design problems of that era but continue to be useful today.

The node-voltage and mesh-current methods both involve formulating linear equations that completely describe the circuit and then solving these equations using the mathematics of linear algebra. Such methods are necessarily more abstract because we manipulate sets of equations describing the circuit, rather than the circuit itself. In doing so we lose some contact with the more intuitive circuit analysis techniques developed in previous chapters. As compensation, however, we gain access to methods that treat a much wider range of applications. In particular, we develop a logical framework for intelligently using computer-aided circuit analysis tools such as SPICE and MICRO-CAP.

5-1 DEVICE AND CONNECTION ANALYSIS

Before describing the node and mesh methods of circuit analysis, we must first understand the foundation on which all methods are based. We have seen that circuit behavior is fundamentally rooted in constraints of two types. First are the connection constraints derived using Kirchhoff's laws. These constraints depend only on the way the devices are interconnected in the circuit and do not depend on the nature of the devices themselves. Device individuality is described by its i-v characteristic which, in turn, do not depend on the nature of the circuit connections. In sum, the operation of a circuit results from balancing two independent types of constraints: (1) connection constraints (Kirchhoff's laws) and (2) device constraints (i-v characteristics).

These observations are nothing new. In Chapter 2 we showed that the combined connection and device constraints generated enough equations to solve for every voltage and every current in a circuit. When we analysis techniques such as voltage division, equivalence, and reduction, we used a combination of device and connection equations. Given this background, all we need to do at this point is to describe a systematic method of writing device and connection equations for any circuit.

We can formulate a set of device and connection equations that completely describe a circuit with N nodes and E two-terminal elements using the following four steps that make up *device and connection analysis*:

STEP 1: Identify a current and a voltage variable with every element in the circuit.

STEP 2: Identify a reference node and write KCL connection constraints at the remaining $N - 1$ nodes using the element currents defined in Step 1.

STEP 3: Write KVL connection constraints around $E - N + 1$ different loops using the element voltages defined in Step 1.

STEP 4: Use the element i-v characteristics to write the device constraints in terms of the element currents and voltages defined in Step 1.

Step 1 produces $N - 1$ independent KCL connection equations, Step 2 produces $E - N + 1$ independent KVL connection equations, and Step 4 leads to E element equations. Altogether this is a total of $(N - 1) + (E - N + 1) + E = 2E$ independent equations, which is sufficient to solve for all of the $2E$ voltage and current variables identified in Step 1.

For example, the circuit in Fig. 5-1 contains four elements, three resistors, and an independent current source.

STEP 1: The figure shows the reference direction for the current and voltage associated with every element. In defining these variables we have been careful to adhere to the passive sign convention, even for the current source. This completes Step 1 of the procedure.

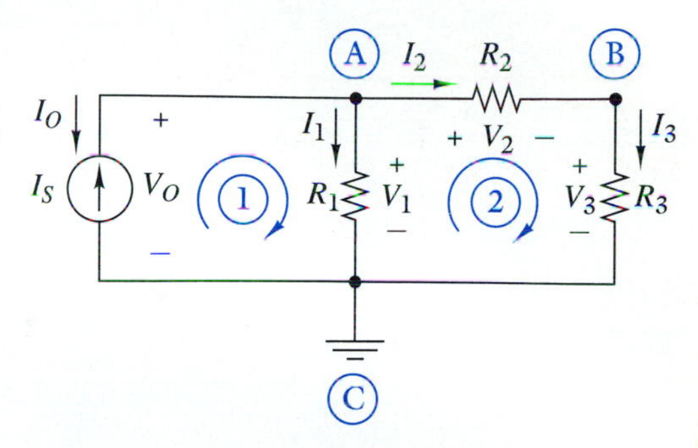

Figure 5-1 Circuit for demonstrating device and connection analysis.

STEP 2: There are three nodes in the circuit. Selecting Node C as the reference node, the KCL equations at the remaining nodes are

$$
\begin{aligned}
&\text{Node A:} \quad -I_O - I_1 - I_2 = 0 \\
&\text{Node B:} \quad +I_2 - I_3 = 0
\end{aligned}
\tag{5-1}
$$

This completes Step 2 of the procedure.

STEP 3: There are $E - N + 1 = 4 - 3 + 1 = 2$ independent loops in the circuit. Step 3 is accomplished by writing a KVL equation around the two loops shown in the figure.

$$
\begin{aligned}
&\text{Loop 1:} \quad -V_O + V_1 = 0 \\
&\text{Loop 2:} \quad -V_1 + V_2 + V_3 = 0
\end{aligned}
\tag{5-2}
$$

Equations (5-1) and (5-2) only describe the way the circuit is connected. They do not depend on the actual devices in the circuit.

STEP 4: Step 4 brings in the contribution of the devices to circuit behavior. We write their i-v characteristics in terms of the element voltages and currents as

$$
\begin{aligned}
&\text{Current source: } I_O = -I_S \qquad &&\text{Resistor } R_2\text{: } V_2 = R_2 I_2 \\
&\text{Resistor } R_1\text{: } V_1 = R_1 I_1 \qquad &&\text{Resistor } R_3\text{: } V_3 = R_3 I_3
\end{aligned}
\tag{5-3}
$$

We have written 2 KCL constraints, 2 KVL constraints, and four device i-v constraints for a total of eight equations. The equations are sufficient to solve for the four element currents (I_0, I_1, I_2, and I_3) and four voltages (V_0, V_1, V_2, and V_3) identified in Step 1. That is, these eight equations completely describe the circuit because we can now solve for every element current and voltage.

EXAMPLE 5-1

Formulate a complete set of device and connection equations for the circuit in Fig. 5-2.

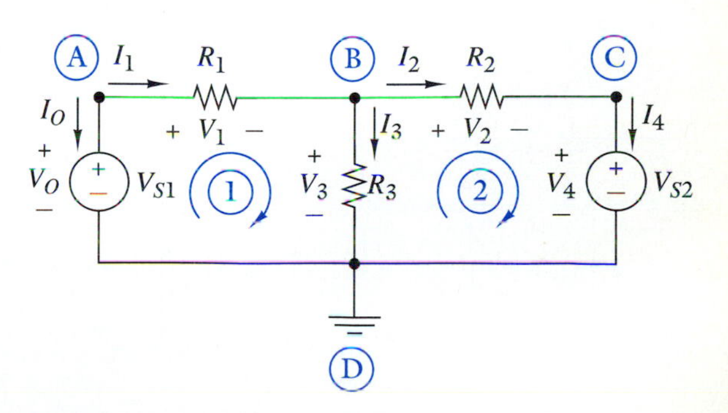

Figure 5-2

SOLUTION:

STEP 1: Figure 5-2 identifies $E = 5$ element currents, 5 element voltages, and $N = 4$ nodes.

STEP 2: We identify Node D as the reference (ground) and write $N - 1 = 3$ KCL equations at the three remaining nonreference nodes.

$$\begin{aligned} \text{Node A:} \quad & -I_O - I_1 &= 0 \\ \text{Node B:} \quad & +I_1 - I_2 - I_3 = 0 \\ \text{Node C:} \quad & +I_2 - I_4 &= 0 \end{aligned}$$

STEP 3: We write $E - N + 1 = 2$ KVL equations around the loops shown in the figure.

$$\begin{aligned} \text{Loop 1:} \quad & -V_O + V_1 + V_3 = 0 \\ \text{Loop 2:} \quad & -V_3 + V_2 + V_4 = 0 \end{aligned}$$

STEP 4: We write the element constraints as

$$V_O = V_{S1} \qquad V_4 = V_{S2} \qquad V_3 = R_3 I_3$$

$$V_1 = R_1 I_1 \qquad V_2 = R_2 I_2$$

We have written a complete set of 10 equations in 10 unknowns.

These examples illustrate the formulation of a set of linear equations that describe circuit behavior by systematically writing a total $2E$ device and connection constraints. As a practical matter device and connection constraints lead to a large number of equations that must be manipulated simultaneously to obtain solutions. To completely solve the circuit in Example 5-1 we would need to treat 10 equations in 10 unknowns. Although this is not an impossible task, it is one that ordinary mortals are unlikely to carry out in an error-free manner. Put differently, the number of equations that must be solved simultaneously is a measure of the practicality of a particular circuit analysis method.

We will shortly develop methods that greatly reduce the number of equations that must be treated simultaneously. However, the reader should not completely abandon the device and connection method. This method is important because it provides the foundation for all other systematic methods of circuit analysis. In subsequent chapters we use the concept of device and connection constraints many times. The device and connection approach provides the theoretical framework for developing many important ideas in circuit analysis. The method does not, however, provide a practical way to calculate numerical answers to specific circuit analysis problems.

5-2 NODE-VOLTAGE ANALYSIS

Node voltages are a convenient set of solution variables for most electronic circuits, particularly those with operational amplifiers (OP AMPs).[1] To define a set of node voltages we must first select a reference node. The *node voltages* are then defined as the voltages between the selected reference node and the remaining nodes. Figure 5-3 shows the notation and interpretation of this definition. The reference node is indicated by the ground symbol. The node voltages are identified by a voltage symbol adjacent to the remaining nodes. This notation means that the positive reference mark for the node voltage is located at the node in question while the negative mark is at the reference node. Obviously any circuit with N nodes will have $N - 1$ node voltages.

A fundamental property of node voltages needs to be covered at the outset. Suppose we are given a two-terminal device whose element voltage is labeled V_1. Suppose further that the terminal with the plus reference mark is connected to a node, say, Node A. The two cases shown in Fig. 5-4 are the only two possible ways the other device terminal can be connected. In case A the other terminal is connected to the reference node, in which case KVL requires $V_1 = V_A$. In case B the other terminal is connected to a nonreference node, say, Node B, in which case KVL requires $V_1 = V_A - V_B$. This example illustrates a fundamental property of node voltages, which can be stated as follows:

> If the kth two-terminal element is connected between nodes X and Y, then the element voltage can be expressed in terms of the two node voltages as
>
> $$V_k = V_X - V_Y \tag{5-4}$$
>
> where X is the node connected to the positive reference for element voltage V_k.

Equation (5-4) is a KVL constraint at the element level. If Node Y is the reference node, then by definition $V_Y = 0$ and Eq. (5-4) reduces to $V_k = V_X$. On the other hand, if Node X is the reference node, then $V_X = 0$ and therefore $V_k = -V_Y$. The minus sign here comes from the fact that the positive reference for the element is connected to the reference node. In any case, the important fact is that the voltage across any two-terminal

[1] In most circuits it is easier to measure voltage than current. Current measurements usually require that a circuit connection be broken to insert an ammeter. Voltage measurements generally do not interrupt the circuit, since the voltmeter is connected between two points. Circuit schematics often show the operating voltages relative to ground at various nodes rather than branch currents.

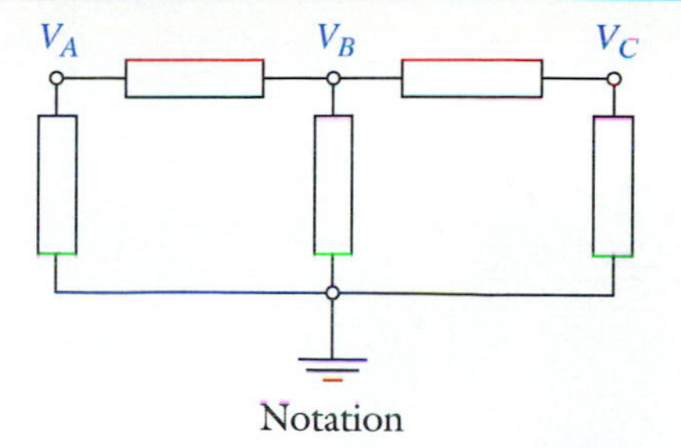

Notation

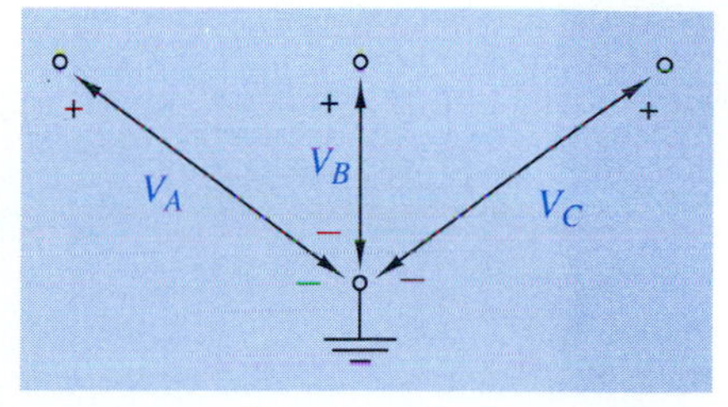

Interpretation

Figure 5-3 Node voltage definition.

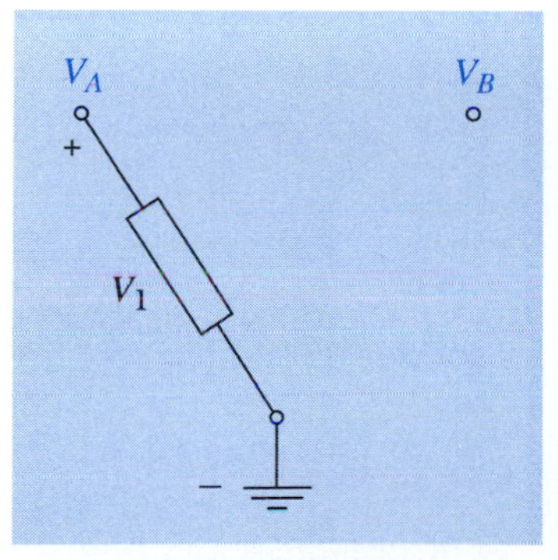

Case A

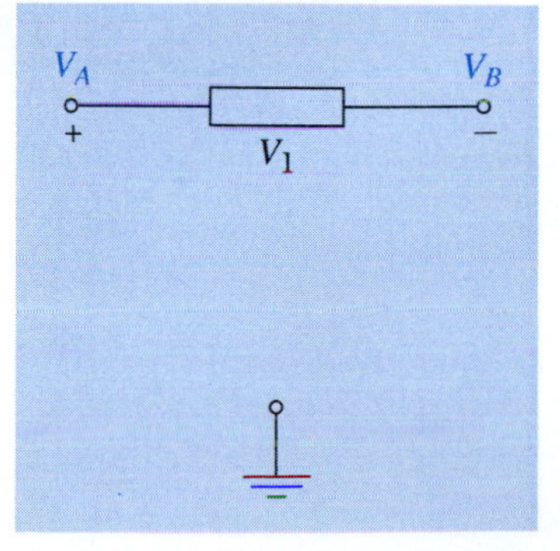

Case B

Figure 5-4 Two possible connections.

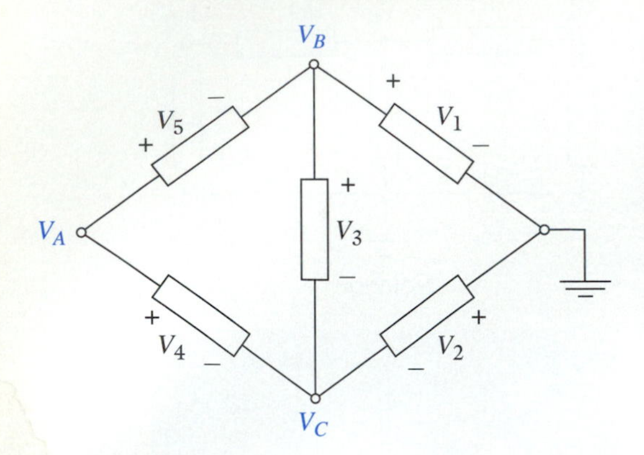

Figure 5-5

element can be expressed as the difference of two node voltages, one of which may be zero.

EXERCISE 5-1

The reference node is as shown and node voltages in the bridge circuit of Fig. 5-5 are $V_A = 5$ V, $V_B = 10$ V, and $V_C = -3$ V. Find the element voltages.

ANSWERS:

$V_1 = 10$ V, $V_2 = 3$ V, $V_3 = 13$ V, $V_4 = 8$ V, and $V_5 = -5$ V.

Formulating Node-Voltage Equations

To formulate a circuit description using node voltages we use device and connection analysis, except that the KVL connection equations are not explicitly written down. Instead, we use the fundamental property of node analysis to express the element voltages in terms of the node voltages.

The circuit in Fig. 5-6 was used in the previous section to illustrate the formulation of device and connection equations. Here we will use the circuit to demonstrate the formulation of node-voltage equations. In Fig. 5-6 we have identified a reference node (indicated by the ground symbol) and two node voltages (V_A and V_B).

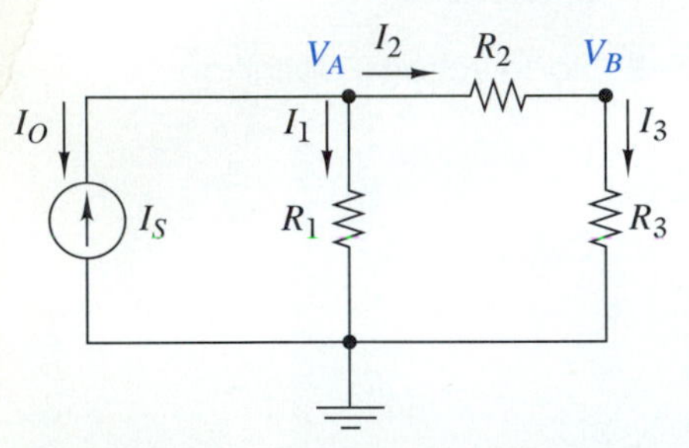

Figure 5-6 Circuit for demonstrating node-voltage analysis.

The KCL constraints at the two nonreference nodes are

$$\begin{aligned} \text{Node A:} \quad & -I_O - I_1 - I_2 = 0 \\ \text{Node B:} \quad & I_2 - I_3 = 0 \end{aligned} \tag{5-5}$$

Using the fundamental property of node analysis we use the device equations to relate the element currents to the node voltages:

$$\begin{aligned} \text{Resistor } R_1: \quad & I_1 = G_1 V_A \\ \text{Resistor } R_2: \quad & I_2 = G_2 (V_A - V_B) \\ \text{Resistor } R_3: \quad & I_3 = G_3 V_B \\ \text{Current source:} \quad & I_O = -I_S \end{aligned} \tag{5-6}$$

We have written six equations in six unknowns—four element currents (I_0, I_1, I_2, I_3) and two node voltages. The right side of the device equations in Eq. (5-6) involve unknown node voltages and the input signal I_S. Substituting the device constraints in Eq. (5-6) into the KCL connection constraints in Eq. (5-5) yields

$$\begin{aligned} \text{Node A:} \quad & I_S - G_1 V_A - G_2 (V_A - V_B) = 0 \\ \text{Node B:} \quad & G_2 (V_A - V_B) - G_3 V_B \quad = 0 \end{aligned}$$

which can be arranged in the following standard form:

$$\begin{aligned} &\text{Node A:} \quad (G_1 + G_2)V_A - G_2 V_B = I_S \\ &\text{Node B:} \quad -G_2 V_A + (G_2 + G_3)V_B = 0 \end{aligned} \tag{5-7}$$

In this standard form all of the unknown node voltages are grouped on one side and the independent sources on the other.

By systematically eliminating the element currents, we have reduced the circuit description to two linear equations in the two unknown node voltages. The coefficients in the equations on the left side ($G_1 + G_2$, G_2, $G_2 + G_3$) depend only on circuit parameters, while the right side contains the known input driving force I_S.

In developing the node-voltage equations in Eq. (5-7) it may appear that we have not used KVL. As noted previously, every method of circuit analysis must satisfy KVL, KCL, and the device i-v characteristics. But KVL is satisfied because the equations $V_1 = V_A$, $V_2 = V_A - V_B$, and $V_3 = V_B$ were implicitly used to write Eq. (5-6). The KVL constraints do not appear explicitly in the formulation of node equations, but are implicitly used when the fundamental property of node analysis is used to write the element voltages in terms of the node voltages.

In summary, four steps are needed to develop node-voltage equations:

STEP 1: Select a reference node. Identify a node voltage at each of the remaining $N - 1$ nodes and a current with every element in the circuit.

STEP 2: Write KCL connection constraints in terms of the element currents at the $N - 1$ nonreference nodes.

STEP 3: Use element i-v characteristics and the fundamental property of mode analysis to express the element currents in terms of the mode voltages.

STEP 4: Substitute the device constraints from Step 3 into the KCL connection constraints from Step 2 and arrange the resulting $N - 1$ equations in a standard form.

Device and connection analysis requires the solution of $2E$ simultaneous equations. The node-voltage method reduces the number of linear equations that must be solved simultaneously to $N - 1$. The reduction from $2E$ to $N - 1$ is especially striking in circuits with a large number of elements (large E) connected in parallel (small N).

EXAMPLE 5-2

Formulate node-voltage equations for the bridge circuit in Fig. 5-7.

SOLUTION:

STEP 1: The reference node, node voltages, and element currents are shown in Fig. 5-7.

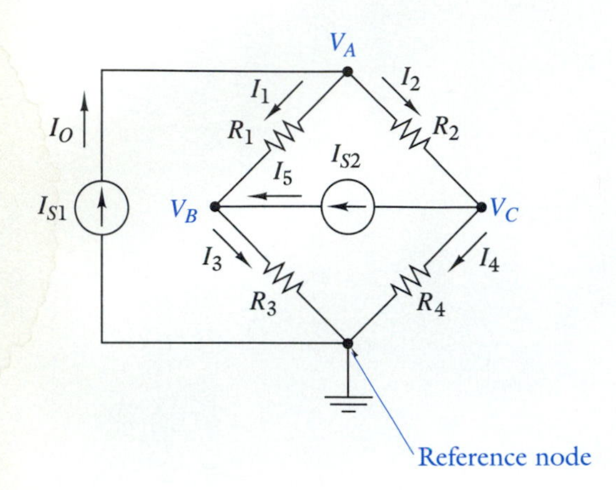

Figure 5-7

STEP 2: The KCL constraints at the three nonreference nodes are

$$\text{Node A:} \quad I_0 - I_1 - I_2 = 0$$
$$\text{Node B:} \quad I_1 - I_3 + I_5 = 0$$
$$\text{Node C:} \quad I_2 - I_4 - I_5 = 0$$

STEP 3: We write the device equations in terms of the node voltages and input signal sources:

$$I_0 = I_{S1} \qquad I_3 = G_3 V_B$$
$$I_1 = G_1 (V_A - V_B) \qquad I_4 = G_4 V_C$$
$$I_2 = G_2 (V_A - V_C) \qquad I_5 = I_{S2}$$

STEP 4: Substituting the device equations into the KCL constraints and arranging the result in standard form yields three equations in the three unknown node voltages:

$$\text{Node A:} \quad (G_1 + G_2) V_A \quad -G_1 V_B \quad -G_2 V_C = I_{S1}$$
$$\text{Node B:} \quad -G_1 V_A + (G_1 + G_3) V_B \quad = I_{S2}$$
$$\text{Node C:} \quad -G_2 V_A \quad + (G_2 + G_4) V_C = -I_{S2}$$

Writing Node-Voltage Equations by Inspection

The node-voltage equations derived in Example 5-2 have a symmetrical pattern. The coefficient of V_B in the Node A equation and the coefficient of V_A in the Node B equation are both the negative of the conductance connected between the nodes $(-G_1)$. Likewise, the coefficients of V_A in the Node C equation and V_C in the Node A equation are both $-G_2$. Finally, coefficients of V_A in the Node A equation, V_B in the Node B equation, and the V_C in the Node C equation are the sum of the conductances connected to the node in question.

The symmetrical pattern always occurs in circuits containing only resistors and independent current sources. To see why, consider any general two-terminal conductance G with one terminal connected to, say, Node A. Then according to the fundamental property of node analysis there are only two possibilities. Either the other terminal of G is connected to the reference node, in which case the current **leaving** Node A via conductance G is

$$I = G(V_A - 0) = GV_A$$

or else it is connected to another nonreference node, say, Node B, in which case the current **leaving** Node A via G is

$$I = G(V_A - V_B)$$

The pattern for node equations follows from these observations. The sum of the currents **leaving** any Node A via conductances is

1. V_A times the sum of conductances connected to Node A.
2. Minus V_B times the sum of conductances connected between Nodes A and B.
3. Minus similar terms for all other nodes connected to Node A by conductances.

The sum of currents leaving Node A via conductances plus the sum of currents directed away from Node A by independent current sources must equal zero because of KCL.

The process outlined above allows us to write node-voltage equations by inspection without going through the intermediate steps involving the KCL constraints and the device equations. We have described the process assuming the circuit contains only resistors and independent current sources. The procedure can be extended to circuits containing dependent current sources as shown in the following discussion.

The circuit in Fig. 5-8 contains an independent current source and voltage-controlled current source. To formulate node equations by inspection, we temporarily treat the dependent source as an independent source.

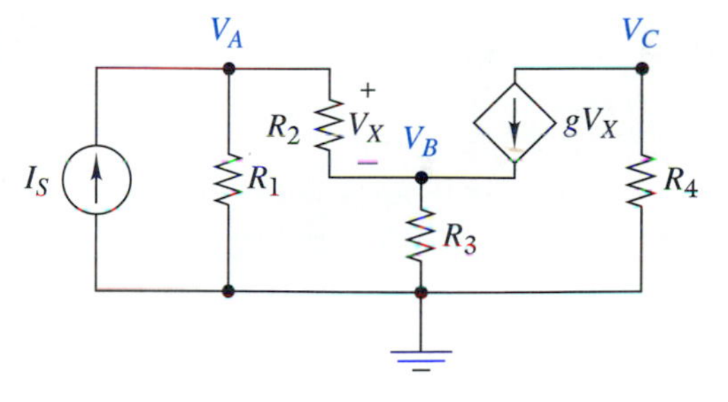

Figure 5-8

Starting with Node A, the sum of conductances connected to Node A is $G_1 + G_2$. The conductances between Nodes A and B is G_2. There is no conductance between Nodes A and C. The reference direction for the source current I_S is into Node A. Pulling all of the observations together we write the sum of currents directed out of Node A as

$$\text{Node A:} \quad (G_1 + G_2)V_A - G_2V_B - 0V_C - I_S = 0 \qquad (5\text{-}8)$$

Similarly, at Node B the sum of conductances is $G_2 + G_3$. Between Nodes B and A the conductance is again G_2, while there is no conductance between Nodes B and C. Treating the dependent current source as if it were independent yields the Node B equation as

$$\text{Node B:} \quad (G_2 + G_3)\, V_B - G_2 V_A - 0 V_C - g V_X = 0 \qquad (5\text{-}9)$$

By similar reasoning the equation at Node C is

$$\text{Node C:} \quad G_4 V_C - 0 V_A - 0 V_B + g V_X = 0 \qquad (5\text{-}10)$$

We have derived a set of three symmetrical node equations by treating the dependent source as an independent source. We are now ready for the final step. Using the fundamental property of node analysis we write the control voltage V_X as

$$V_X = V_A - V_B \qquad (5\text{-}11)$$

Equation (5-11) is now used to eliminate V_X from the node equations. When the results are put in standard form (unknown node voltages on the left and independent sources on the right) we obtain

$$\begin{aligned} &\text{Node A:} \quad (G_1 + G_2)\, V_A \quad -G_2 V_B \quad = I_S \\ &\text{Node B:} \quad -(G_2 + g) V_A \; +(G_2 + G_3 + g) V_B \quad = 0 \\ &\text{Node C:} \quad g V_A \quad -g V_B \; +G_4 V_C \; = 0 \end{aligned} \qquad (5\text{-}12)$$

We now have three equations in three unknowns, including the effect of the dependent current source.

These node-voltage equations do not have coefficient symmetry because of the controlled source in the circuit (see Fig. 5-8). For example, the coefficient of V_A in the Node B equation is not the same as the coefficient of V_B in the Node A equation. More importantly, V_C does not appear in either the Node A or Node B equations, but the Node C equation contains the terms gV_A and $-gV_B$. The symmetry is lost because the controlled source provides one-way coupling. The voltage V_X directly controls the current gV_X, but there is no reverse coupling via the controlled source. *Unidirectional coupling* is an important property of control source models of active devices. In the present context, one-way coupling shows up as a lack of symmetry in the node-voltage equations.

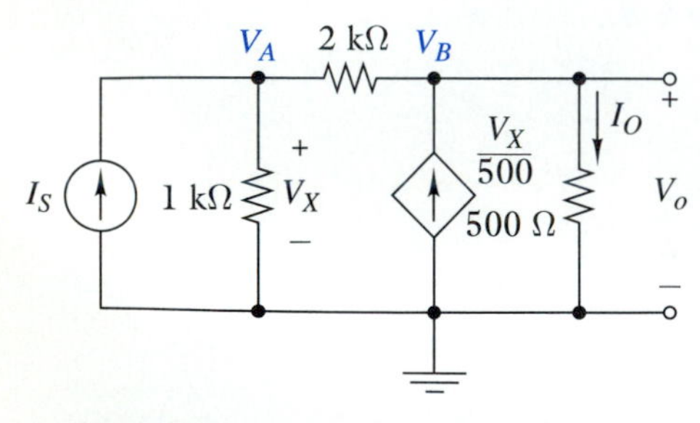

Figure 5-9

EXERCISE 5-2

Formulate node-voltage equations for the circuit in Fig. 5-9.

ANSWER:

$$1.5 \times 10^{-3} V_A - 0.5 \times 10^{-3} V_B = I_S$$

$$-2.5 \times 10^{-3} V_A + 2.5 \times 10^{-3} V_B = 0$$

Solving Linear Algebraic Equations

So far we have only dealt with the problem of formulating a set of linear equations in the node voltages. To complete a circuit analysis problem we must solve this set of linear equations for selected responses. Cramer's rule and Gaussian elimination are standard mathematical tools commonly used for hand-solution of circuit equations. These tools are assumed to be part of the reader's mathematical background. Those needing a review of these matters are referred to Appendix B.

Cramer's rule and Gaussian elimination are suitable for hand calculations involving three or four simultaneous equations. Cramer's rule is generally easier when the circuit parameters are left in symbolic form, while the Gaussian method is more efficient when numerical values of circuit parameters are given. The efficiency advantage of the Gaussian method is not terribly important for two or three equations, but increases dramatically for four or more equations.

At about four or five simultaneous equations numerical solutions are best obtained using some computer-based procedures. Many scientific hand-held calculators have a built-in capability to solve up to five linear equations. Virtually all PC-based mathematical software packages have the capability to solve systems of linear equations, or what is equivalent, to perform matrix manipulations. Appendix B describes how to use a package called MATLAB™ to solve linear equations.[2]

In the rest of the book we generally use Cramer's rule because our examples often leave the circuit parameters in symbolic form and usually involve no more than three equations. This does not mean Cramer's rule is the optimum method, but only that it can easily handle the class of problems treated in this book. The argument for Gaussian elimination is only compelling for four or more equations. The ready availability of computer-aided tools for this class of problem makes the hand solution of linear equations by Gaussian elimination an obsolete skill.

Earlier in this section we formulated node-voltage equations for the circuit in Fig. 5-6 (see Eq. (5-7)):

$$\begin{aligned} \text{Node A:} \quad & (G_1 + G_2)\, V_A - G_2 V_B = I_S \\ \text{Node B:} \quad & -G_2 V_A + (G_2 + G_3)\, V_B = 0 \end{aligned}$$

To solve these equations we use Cramer's rule because it easily

[2]MATLAB is a registered trademark of The MathWorks, Inc.

handles the case where the circuit parameters are left in symbolic form:

$$V_A = \frac{\Delta_A}{\Delta} = \frac{\begin{vmatrix} I_S & -G_2 \\ 0 & G_2 + G_3 \end{vmatrix}}{\begin{vmatrix} G_1 + G_2 & -G_2 \\ -G_2 & G_2 + G_3 \end{vmatrix}}$$

$$= \left(\frac{G_2 + G_3}{G_1 G_2 + G_1 G_3 + G_2 G_3} \right) I_S \quad (5\text{-}13)$$

$$V_B = \frac{\Delta_B}{\Delta} = \frac{\begin{vmatrix} G_1 + G_2 & I_S \\ -G_2 & 0 \end{vmatrix}}{D} = \left(\frac{G_2}{G_1 G_2 + G_1 G_3 + G_2 G_3} \right) I_S \quad (5\text{-}14)$$

These results express the two node voltages in terms of the circuit parameters and the input signal. Note that the node voltages are proportional to the input, since the circuit is linear. Given the two node voltages V_A and V_B we can now determine every element voltage and every current using Ohm's law and the fundamental property of node voltages.

$$V_1 = V_A \qquad V_2 = V_A - V_B \qquad V_3 = V_B$$

$$I_1 = G_1 V_1 \qquad I_2 = G_2 V_2 \qquad I_3 = G_3 V_3$$

In solving the node equations, we left everything in symbolic form to emphasize that responses depend on the values of the circuit parameters (G_1, G_2, G_3) and the input signal (I_S). Even when numerical values are given, some parameters can be left in symbolic form to obtain input-output relationships or to reveal the effect of specific parameters on circuit response.

EXAMPLE 5-3

Given the circuit in Fig. 5-10 with $R_1 = 10$ kΩ, $R_2 = 1$ kΩ, $R_3 = 100$ kΩ, $R_4 = 5$ kΩ, and $g = 100$ mS, find the outputs V_O and I_O in terms of the input I_S.

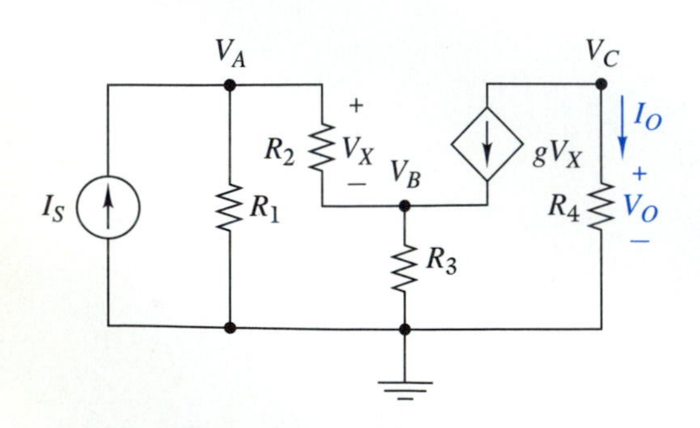

Figure 5-10

SOLUTION:

Earlier in this section we formulated node-voltage equations for this circuit [see Eq. (5-12)] as follows:

$$\text{Node A:} \quad (G_1 + G_2)V_A \qquad -G_2 V_B \qquad = I_S$$

$$\text{Node B:} \quad -(G_2 + g)V_A + (G_2 + G_3 + g)V_B \qquad = 0$$

$$\text{Node C:} \quad gV_A \qquad -gV_B + G_4 V_C = 0$$

Inserting the given numerical values results in

$$\begin{aligned} 1.1\times10^{-3}V_A - 10^{-3}V_B &= I_S \\ -0.101V_A + 0.111V_B &= 0 \\ 0.1V_A - 0.1V_B \quad + 2\times10^{-4}V_C &= 0 \end{aligned}$$

The output voltage corresponds to the node-voltage V_C. Using Cramer's rule to solve for V_C yields

$$V_C = \frac{\Delta_C}{\Delta} = \frac{\begin{vmatrix} 1.1\times10^{-3} & -10^{-3} & I_S \\ -0.101 & 0.111 & 0 \\ 0.1 & -0.1 & 0 \end{vmatrix}}{\begin{vmatrix} 1.1\times10^{-3} & -10^{-3} & 0 \\ -0.101 & 0.111 & 0 \\ 0.1 & -0.1 & 2\times10^{-4} \end{vmatrix}}$$

$$= \frac{I_S(0.101\times0.1 - 0.111\times0.1)}{2\times10^{-4}(1.1\times10^{-3}\times0.111 - 10^{-3}\times0.101)}$$

$$= -2.37\times10^5\, I_S$$

Now using Ohm's law we find the output current:

$$I_O = G_4\, V_O = -47.4\, I_S$$

The negative sign in these results means that the polarity of the output signals are the opposite of the input signal I_S. The signal inversion results from the relationship between the reference directions for the controlled source and the output signals.

EXERCISE 5-3

Solve for the node-voltage equations in Exercise 5-2 for V_O and I_O in terms of I_S.

ANSWERS:

$V_O = 1000I_S$; $I_O = 2I_S$

EXERCISE 5-4

Use node-voltage equations to solve for V_1, V_2, and I_3 in Fig. 5-11.

ANSWERS:

$V_1 = 12$ V, $V_2 = 32$ V, and $I_3 = -10$ mA

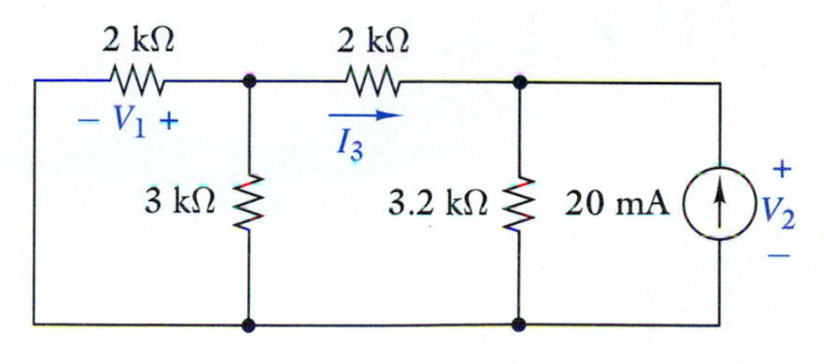

Figure 5-11

5-3 NODE ANALYSIS WITH VOLTAGE SOURCES

Up to this point we have assumed that circuits contain only resistors and current sources. This assumption simplifies node-voltage analysis because applying KCL at a node only involves current sources or resistor currents expressed in terms of the node voltages. Adding voltage sources (dependent and independent) to the circuits modifies node analysis procedures because the current through a voltage source is not directly related to the voltage across it. While initially it may appear that voltage sources complicate the situation, they actually simplify node analysis by reducing the number of equations required.

Figure 5-12 shows three ways to deal with voltage sources in node analysis. Method 1 eliminates the problem by using a source transformation to replace the voltage source and series resistance with an equivalent current source and parallel resistance. We can then formulate node equations at the remaining nonreference nodes in the usual way. The source transformation eliminates Node C, so there are only $N - 2$ nonreference nodes left in the circuit. Obviously, Method 1 applies only if there is a resistance in series with the voltage source.

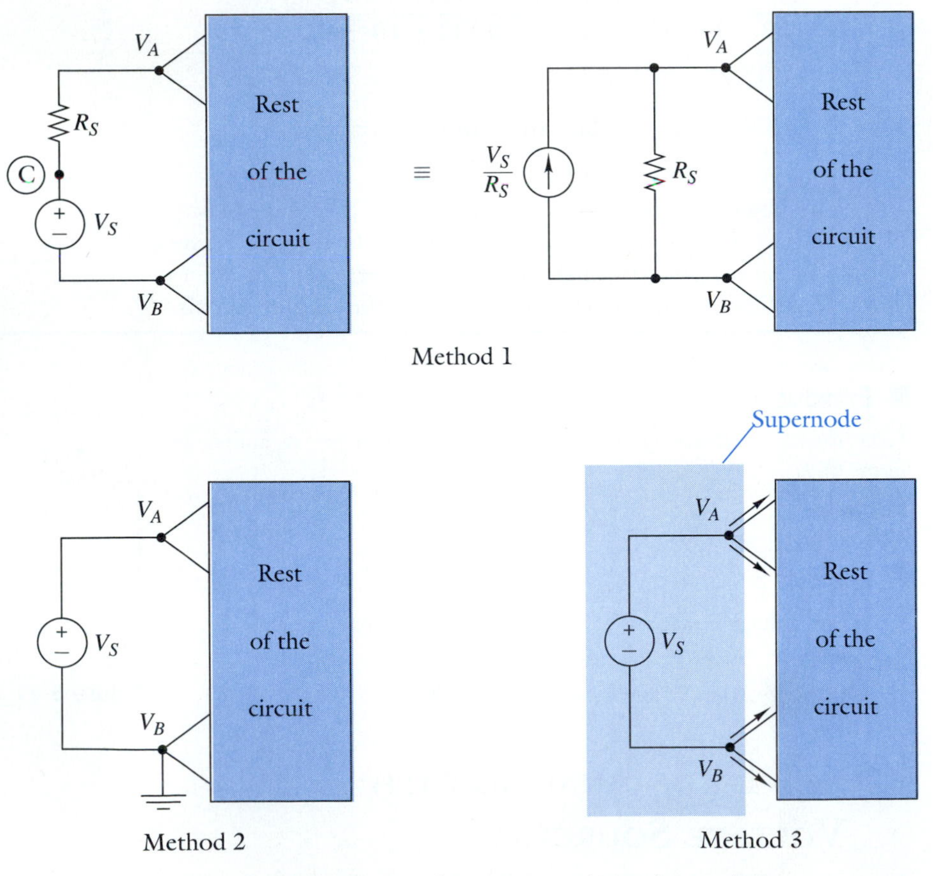

Figure 5-12 Methods of treating voltage source in node analysis.

Method 2 in Fig. 5-12 can be used whether or not there is a resistance in series with the voltage source. When Node B is selected as the reference node, by definition $V_B = 0$ and the fundamental property of node voltages says that $V_A = V_S$. We do not need a node-voltage equation at Node A because its voltage is known to be the same as the source voltage. We write the node equations at the remaining $N - 2$ nonreference nodes in the usual way. In the final step we move all terms involving V_A to the right side, since it is a known input and not an unknown response. Method 2 reduces the number of node equations by one, since no equation is needed at Node A.

The third method in Fig. 5-12 is needed when neither Node A or B can be selected as the reference and the source is not connected in series with a resistance. In this case we combine Nodes A and B into a *supernode* indicated by the boundary in Fig. 5-12. We use the fact that KCL applies to the currents penetrating this boundary to write a node equation at the supernode. We then write node equations at the remaining $N - 3$ nonreference nodes in the usual way. We now have $N - 3$ node equations plus one supernode equation, leaving us one equation short of the $N - 1$ required. Using the fundamental property of moded voltages we can write

$$V_A - V_B = V_S \tag{5-15}$$

The voltage source inside the supernode constrains the difference between the node voltages at Nodes A and B. The voltage source constraint provides the additional relationship needed to write $N - 1$ independent equations in the $N - 1$ node voltages.

For reference purposes we will call these modified node equations, since we either modify the circuit (Method 1), do not write node equations at some nonreference nodes (Method 2), or combine nodes to produce a supernode (Method 3). The three methods are not mutually exclusive. We frequently use a combination of methods as illustrated in the following examples.

EXAMPLE 5-4

The circuit in Fig. 5-13(a) is a model of an inverting OP AMP circuit. Use node-voltage analysis to find the output V_O and I_O.

SOLUTION:

The given circuit in Fig. 5-13(a) has four nodes so it may appear that we need $N - 1 = 3$ node equations. However, a source transformation (Method 1) applied to the independent source leads to the modified three-node circuit shown in Fig. 5-13(b). With the selected reference node the dependent voltage source constrains the voltage at Node B (Method 2). The control voltage is $V_X = V_A$, and the controlled source forces the Node B

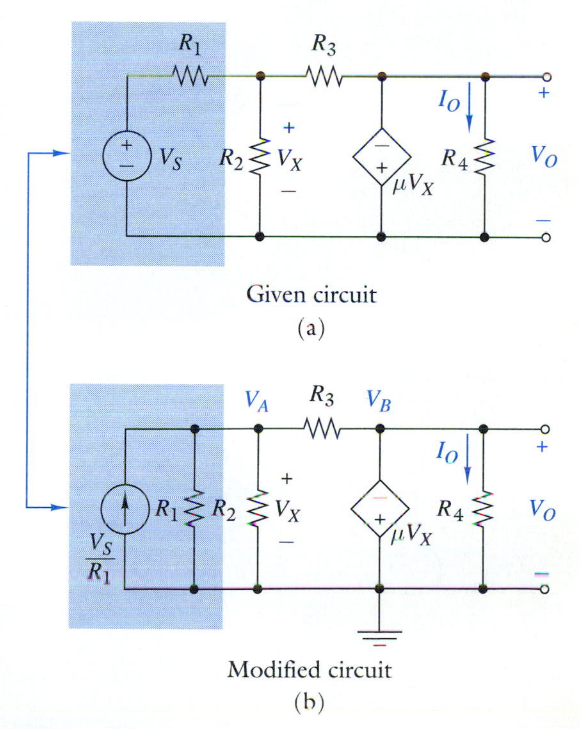

Figure 5-13

voltage to be

$$V_B = -\mu V_X = -\mu V_A$$

Node A is the only independent node in the circuit. We can write the Node A equation by inspection as

$$(G_1 + G_2 + G_3)\, V_A - G_3 V_B = G_1 V_S$$

Substituting in the control source constraint yields the standard form for this equation as

$$(G_1 + G_2 + G_3 + \mu G_4)\, V_A = G_1 V_S$$

We end up with only one node equation even though at first glance the given circuit appeared to need three node equations. The reason is that there are two voltage sources in the original circuit in Fig. 5-13(a). Since the two sources share the reference node, the number of unknown node voltages is reduced from three to one. The general principle illustrated is that the number of independent KCL constraints in a circuit containing N nodes and N_V voltage sources (dependent or independent) is $N-1-N_V$.

The one node equation can easily be solved for V_A. Since the output voltage is $V_O = V_B$ we can write

$$V_O = V_B = -\mu V_A = \left(\frac{-\mu G_1}{G_1 + G_2 + G_3 + \mu G_4}\right) V_S$$

We find the output current using Ohm's law as

$$I_O = G_4 V_O = \left(\frac{-\mu G_1 G_4}{G_1 + G_2 + G_3 + \mu G_4}\right) V_S$$

The minus signs means the circuit provides signal inversion. The signal inversion results from the reference polarity of the controlled source. The output voltage does not depend on the output load resistor R_4, since the load is connected across ideal (though dependent) voltage sources.

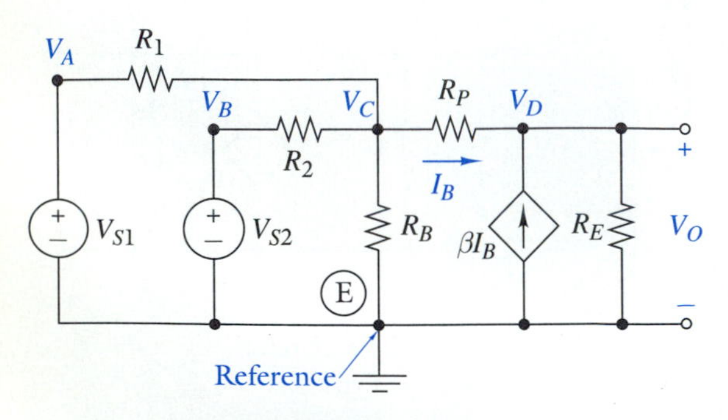

Figure 5-14

EXAMPLE 5-5

The circuit in Fig. 5-14 is a model of a transistor circuit with two input voltages V_{S1} and V_{S2}. Use node-voltage analysis to find the output V_O. Use $R_1 = 1$ kΩ, $R_2 = 3$ kΩ, $R_B = 100$ kΩ, $R_P = 1.3$ kΩ, $R_E = 3.3$ kΩ, and $\beta = 50$.

SOLUTION:

The given circuit has five nodes. We use Method 2 and select Node E as the reference because both voltage sources are connected to this node. With Node E as the reference node the two input sources force the voltages at Node A and B to be $V_A = V_{S1}$

and $V_B = V_{S2}$. Therefore, we only need to write node equations at Nodes C and D because voltages at Nodes A and B are known. As a preliminary step we will treat the dependent current source βI_B as if it were an independent source. Treating βI_B as an independent source allows us to write a set of symmetrical node equations by inspection.

$$\text{Node C:}\quad -G_1V_A - G_2V_B + (G_1 + G_2 + G_B + G_P)V_C - G_PV_D = 0$$

$$\text{Node D:}\quad -G_PV_C + (G_P + G_E)V_D = \beta I_B$$

Note that the term βI_B is on the right, or input side of the Node D equation. However, the current I_B can be written in terms of the node voltages as

$$I_B = G_P(V_C - V_D)$$

This expression for I_B together with the constraints $V_A = V_{S1}$ and $V_B = V_{S2}$ are now substituted into the Node C and D equations and the result put in standard form.

$$\text{Node C:}\quad (G_1 + G_2 + G_B + G_P)\,V_C - G_PV_D = G_1V_{S1} + G_2V_{S2}$$

$$\text{Node D:}\quad -(\beta + 1)G_PV_C + [(\beta + 1)\,G_P + G_E]\,V_D = 0$$

The final result is two equations in the two unknown node voltages. These equations do not have coefficient symmetry because of the current-controlled current source.

Substituting the given numerical values

$$2.11 \times 10^{-3}V_C - 7.69 \times 10^{-4}V_D = 10^{-3}V_{S1} + 33.3 \times 10^{-4}V_{S2}$$

$$-3.92 \times 10^{-2}V_C + 3.95 \times 10^{-2}V_D = 0$$

Since $V_O = V_D$ we solve the second equation for $V_C = 1.008V_D$. When this equation for V_C is substituted into the first equation we obtain

$$V_O = V_D = 0.736\,V_{S1} + 0.245V_{S2}$$

The circuit acts like a noninverting summer. The output is a linear combination of the two inputs because the circuit is linear.

EXAMPLE 5-6

For the circuit in Fig. 5-15:

(a) Formulate node-voltage equations.

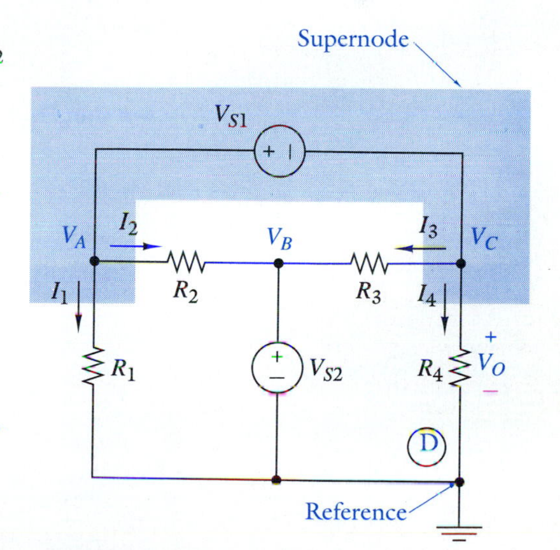

Figure 5-15

(b) Solve for the output voltage V_O using $R_1 = R_4 = 2\ \text{k}\Omega$ and $R_2 = R_3 = 4\ \text{k}\Omega$.

SOLUTION:

(a) The voltage sources in Fig. 5-15 do not have a common node and we cannot select a reference node that includes both sources. Selecting Node D as the reference forces the condition $V_B = V_{S2}$ (Method 2) but leaves the other source V_{S1} ungrounded. We surround the ungrounded source, and all wires leading to it, by the supernode boundary shown in the figure (Method 3). KCL applies to the four element currents that penetrate the supernode boundary and we can write

$$I_1 + I_2 + I_3 + I_4 = 0$$

These currents can easily be expressed in terms of the node voltages.

$$G_1 V_A + G_2(V_A - V_B) + G_3(V_C - V_B) + G_4 V_C = 0$$

But since $V_B = V_{S2}$ the standard form of this equation is

$$(G_1 + G_2)V_A + (G_3 + G_4)V_C = (G_2 + G_3)V_{S2}$$

We have one equation in the two unknown node voltages V_A and V_C. Applying the fundamental property of node voltages inside the supernode we can write

$$V_A - V_C = V_{S1}$$

That is, the ungrounded voltage source constrains the difference between the two unknown node voltages inside the supernode and thereby supplies the relationship needed to obtain two equations in two unknowns.

(b) Substituting in the given numerical values yields

$$7.5 \times 10^{-4} V_A + 7.5 \times 10^{-4} V_C = 5 \times 10^{-4} V_{S2}$$

$$V_A - V_C = V_{S1}$$

To find the output V_O we need to solve these equations for V_C. The second equation yields $V_A = V_C + V_{S1}$, which when substituted into the first equation yields the required output as

$$V_O = V_C = \frac{V_{S2}}{3} - \frac{V_{S1}}{2}$$

The output is a linear combination of the two inputs, since the circuit is linear.

EXAMPLE 5-7

The circuit in Fig. 5-16 is a large signal model of a bipolar junction transistor (BJT) operating in the active mode. Use node-voltage analysis to find the collector current I_C.

SOLUTION:

This problem could be solved by formulating a set of node-voltage equations and formally solving them for the voltages needed to determine the collector current I_C. However, we will use this example to illustrate an informal approach that uses node analysis concepts and principles, but that never formally writes down the requisite number of node equations. This type of node analysis is less structured but more representative of the ad hoc methods used in books on electronics.

The source V_{CC} represents a power supply, and the source V_T is part of the transistor model. These two voltage sources do not have a common node, so we select Node E as the reference and create a supernode around the voltage source V_T as shown in the figure. The controlling current I_B crosses the supernode boundary from the left, so the dependent source in the transistor model forces a current βI_B across the boundary from the top. Applying KCL to the supernode boundary shows that a current $(\beta + 1)I_B$ must *exit* the boundary at the bottom. Using KCL at Node D shows that $(\beta + 1)I_B$ is the current through the resistor R_E. Because one end of R_E is grounded we find the voltage at Node D using Ohm's law:

$$V_D = (\beta + 1)I_B R_E$$

Using this result for V_D together with the fundamental property of node voltages gives the voltage at Node C as

$$V_C = V_T + V_D = V_T + (\beta + 1)I_B R_E$$

Applying the fundamental property of node voltages to resistance R_1 allows us to write $I_B = (V_A - V_C)/R_1$. For the selected reference $V_A = V_{CC}$. Using the expression for V_C above yields

$$I_B = \frac{V_{CC} - V_T - (\beta + 1)I_B R_E}{R_1}$$

which can be solved for I_B as

$$I_B = \frac{V_{CC} - V_T}{R_1 + (\beta + 1)R_E}$$

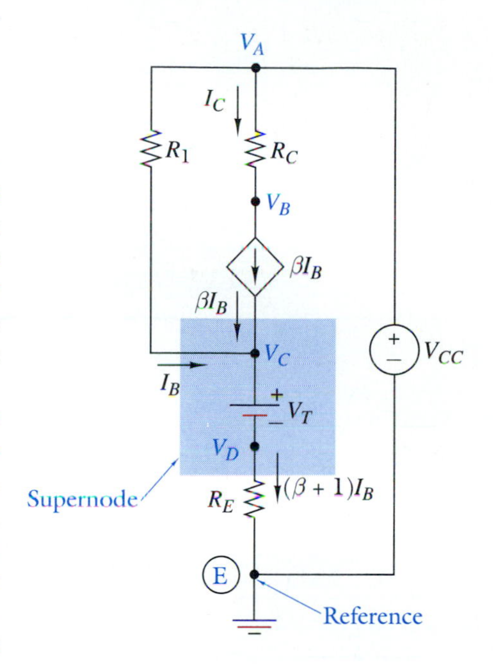

Figure 5-16

Finally, applying KCL at Node B yields the collector current as

$$I_C = \beta I_B = \frac{\beta(V_{CC} - V_T)}{R_1 + (\beta + 1)R_E}$$

The narrative discussion above provides a detailed account of all of the node analysis concepts used to find the collector current. With practice and experience many of these ideas become second nature, and this type of analysis can be carried out by writing only one or two equations. Node analysis provides a systematic method of solving circuit problems. But it also offers a viewpoint that allows us to quickly analyze electronic circuits in ways that are not too different from the methods studied in previous chapters.

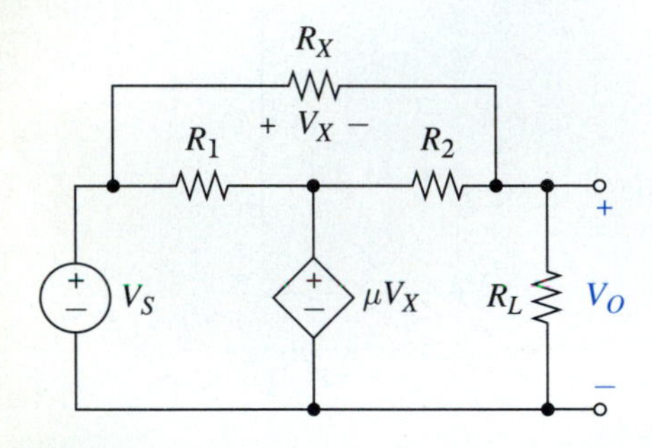

Figure 5-17

EXERCISE 5-5

Use node-voltage analysis to find V_O in Fig. 5-17.

ANSWER:

$$V_O = \frac{G_X + \mu G_2}{G_X + G_L + (\mu + 1)G_2} V_S$$

EXERCISE 5-6

Find V_O in Fig. 5-18 when the element E is

(a) A 10 kΩ resistance.

(b) A 4-mA independent current source with reference arrow pointing left.

(c) A dependent voltage source $10V_X$ with the plus reference on the right.

ANSWERS:

(a) 2.53 V

(b) −17.3 V

(c) 23.4 V

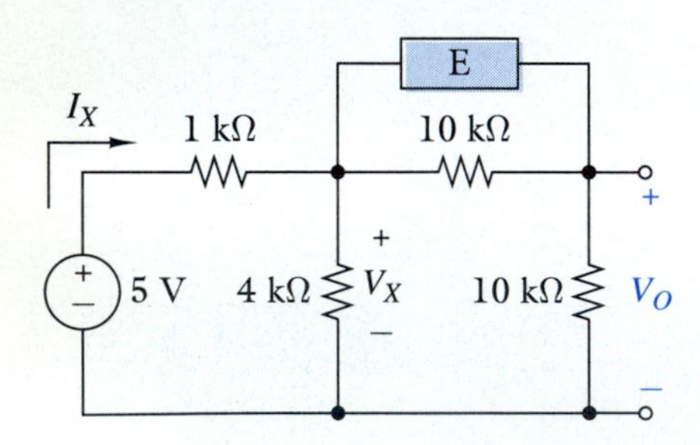

Figure 5-18

EXERCISE 5-7

Find V_O in Fig. 5-18 when the element E is

(a) An open-circuit.

(b) A 10-V independent voltage source with the plus reference on the right.

(c) A dependent current source $10I_X$ with the reference arrow pointing left.

ANSWERS:

(a) 1.92 V

(b) −5.56 V

(c) −9.52 V

Node Analysis of OP AMP Circuits

Node analysis can be extended to the general OP AMP circuit situation shown in Fig. 5-19. Normally we want to find the OP AMP output voltage (V_O) relative to ground—the reference node in the present context. We must assign a node voltage variable to the OP AMP output. However, as we saw in Chapter 4, and again in Example 5-4, the output of an ideal OP AMP acts like a controlled voltage source connected between the output terminal and ground. According to Method 2 of modified node analysis, we do not need to formulate a node equation at such a node. In other words, we do not need to write a node equation at the output terminal of the OP AMP.

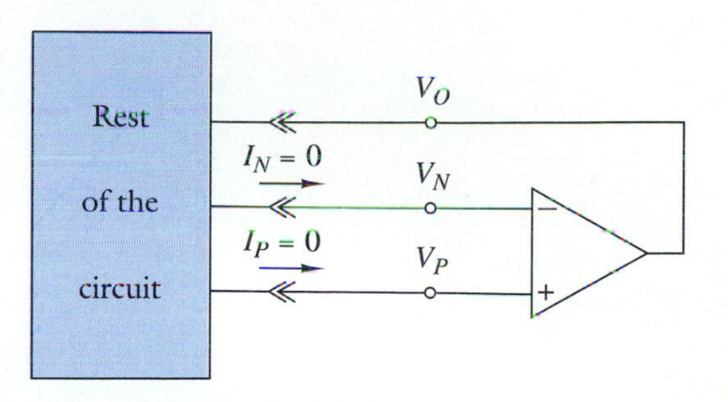

Figure 5-19 General OP AMP circuit analysis.

We then formulate node equations at the remaining $N-2$ nonreference nodes in the usual way. Since we have $N-1$ nodes it would appear that we fall one short of the number of equations needed. However, the ideal OP AMP model forces the net input voltage to zero, so $V_P = V_N$ in Fig. 5-19. The OP AMP forces these two nodes to have identical voltages, thereby eliminating one unknown node voltage. Finally, we recall from Chapter 4 that the ideal OP AMP draws no current at its inputs, $I_P = I_N = 0$ in Fig. 5-19. These currents can be ignored in formulating node equations.

The following steps outline an approach to formulating of node equations for OP AMP circuits:

STEP 1: Identify a node voltage at all OP AMP outputs, but do *not* formulate a node equations at these nodes.

STEP 2: Formulate node equations at the remaining nonreference nodes and use the ideal OP AMP voltage constraint $V_P = V_N$ to reduce the number of unknowns.

EXAMPLE 5-8

Find the input-output relationship for the circuit in Fig. 5-20.

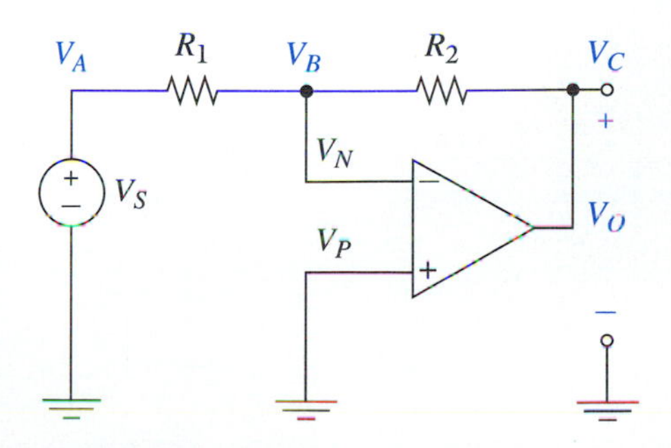

Figure 5-20

SOLUTION:

The circuit is the familiar inverting amplifier configuration. As drawn the circuit has three nonreference nodes. We do not need to formulate a node equation at Node A because it is connected to a grounded input voltage source. As discussed above, we also do not need to write an equation at Node C, the OP AMP output. A node equation is needed only at Node B. By inspection

$$\text{Node B:}\quad (G_1 + G_2)\,V_B - G_1 V_A - G_2 V_C = 0$$

Now using the OP AMP voltage constraint ($V_P = V_N$), we find that $V_B = V_N = V_P = 0$, since the noninverting input is connected to the reference node. The single node equation reduces to

$$-G_1 V_A - G_2 V_C = 0$$

But since $V_A = V_S$ and $V_C = V_O$ the equation above can be rewritten as

$$V_O = -\frac{R_2}{R_1} V_S$$

This answer is the same as the input-output relationship found in Chapter 4.

EXAMPLE 5-9

Find the input-output relationship for the circuit in Fig. 5-21.

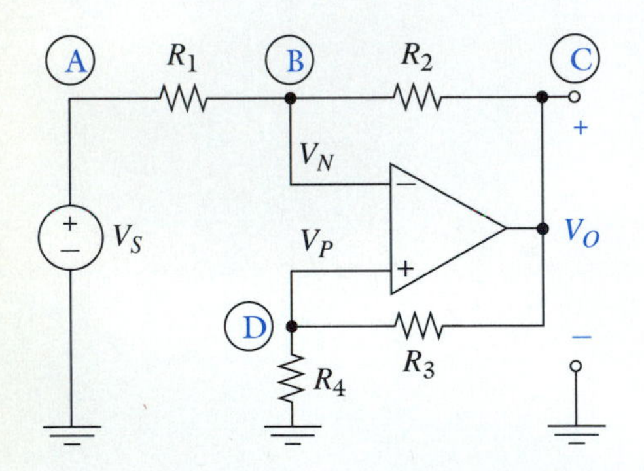

Figure 5-21

SOLUTION:

Neither OP AMP input is grounded in Fig. 5-21, so we need to formulate node equations at Nodes B and D. We do not need a node equations at Node A, since its voltage is $V_A = V_S$. By inspection the Node B and Node D equations are

$$\text{Node B:} \quad (G_1 + G_2)\, V_B - G_1 V_A - G_2 V_C = 0$$

$$\text{Node D:} \quad (G_3 + G_4)\, V_D - G_3 V_C = 0$$

The OP AMP voltage constraint ($V_P = V_N$) requires $V_B = V_D$. Substituting the conditions $V_A = V_S$ and $V_D = V_B$ into the Node B and D equations yields two equations in two unknown node voltages V_B and V_C.

$$(G_1 + G_2)\, V_B - G_2 V_C = G_1 V_S$$

$$(G_3 + G_4)\, V_B - G_3 V_C = 0$$

Since the output voltage is related to the node voltages as $V_O = V_C$, we can solve for the output using Cramer's rule:

$$V_O = V_C = \frac{\Delta_C}{\Delta} = \frac{\begin{vmatrix} G_1 + G_2 & G_1 V_S \\ G_3 + G_4 & 0 \end{vmatrix}}{\begin{vmatrix} G_1 + G_2 & -G_2 \\ G_3 + G_4 & -G_3 \end{vmatrix}} = \frac{G_1 G_4 + G_1 G_3}{G_1 G_3 - G_2 G_4} V_S$$

After a bit of algebraic manipulation, the equation above reduces to

$$V_O = \frac{R_2 R_3 + R_2 R_4}{R_2 R_4 - R_1 R_3} V_S$$

EXAMPLE 5-10

Find the input-output relationship of the circuit in Fig. 5-22.

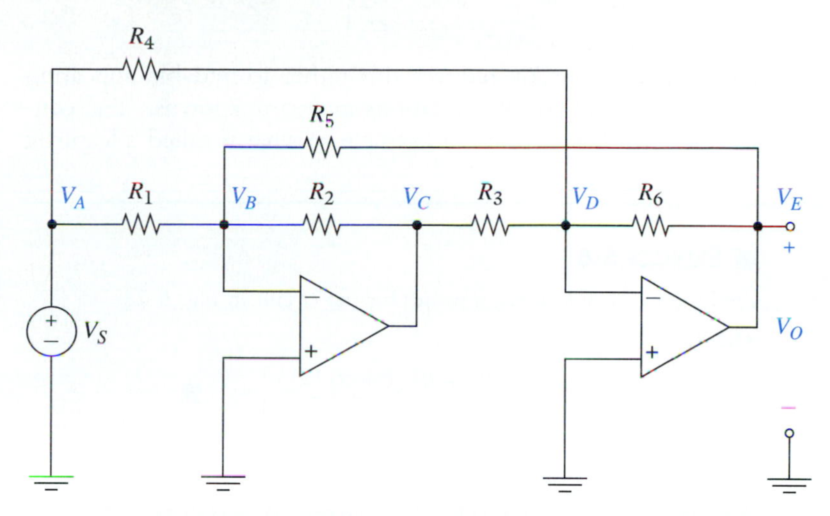

Figure 5-22

SOLUTION:

The circuit contains two OP AMPs and a total of five nonreference nodes. Nodes C and E are connected to OP AMP outputs, and Node A is connected to the grounded input voltage source. We only need node equations at Nodes B and D. By inspection

Node B: $(G_1 + G_2 + G_5)\,V_B - G_1 V_A - G_2 V_C - G_5 V_E = 0$

Node C: $(G_3 + G_4 + G_6)\,V_D - G_4 V_A - G_3 V_C - G_6 V_E = 0$

But both noninverting inputs are grounded and the OP AMP voltage constraint ($V_P = V_N$) means $V_B = 0$ and $V_D = 0$. Finally, the input source constraint means $V_A = V_S$. The two node equations reduce to

$$G_2 V_C + G_5 V_E = -G_1 V_S$$

$$G_3 V_C + G_6 V_E = -G_4 V_S$$

We end up with two equations in two unknowns—the two OP AMP outputs as it happens. These equations are not symmetrical, which points out that the OP AMP is a unilateral device.

Solving for the circuit output voltage yields the input-output relationship as

$$V_O = V_E = \frac{\Delta_E}{\Delta} = \frac{\begin{vmatrix} G_2 & -G_1 V_S \\ G_3 & -G_4 V_S \end{vmatrix}}{\begin{vmatrix} G_2 & G_5 \\ G_3 & G_6 \end{vmatrix}} = \frac{G_1 G_3 - G_2 G_4}{G_2 G_6 - G_3 G_5} V_S$$

Node analysis quickly reduces this rather formidable appearing OP AMP circuit to two equations in two unknowns. The configuration in Fig. 5-22 is an example of what is called a leapfrog circuit.

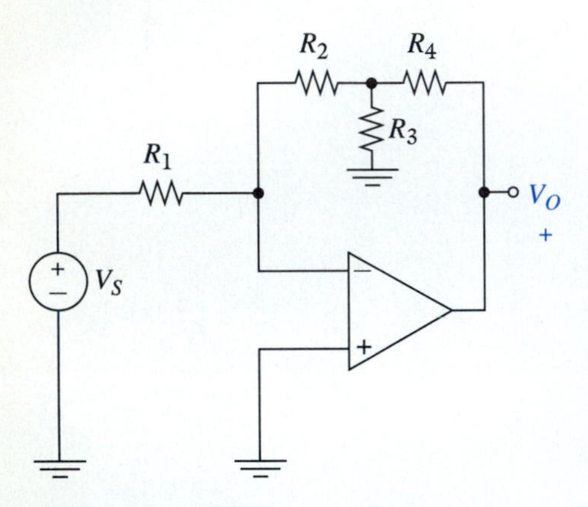

Figure 5-23

EXERCISE 5-8

Find the input-output relationship for the circuit in Fig. 5-23.

ANSWER:

$$v_O = -\left[\frac{R_4 + R_2\,(1 + R_4/R_3)}{R_1}\right] v_S$$

Summary of Node-Voltage Analysis

We have seen that node-voltage equations are very useful in the analysis of a variety of electronic circuits. These equations can always be formulated using KCL, the element constraints, and the fundamental property of node voltages. When in doubt the reader should always fall back on these principles to formulate node equations in new situations. With practice and experience, however, we eventually develop an analysis approach that allows us to recognize shortcuts in the formulation process. The following guidelines or steps summarize our approach and may help readers develop their own analysis style:

1. Simplify the circuit by combining elements in series and parallel wherever possible.
2. If not specified, select a reference node so that as many dependent and independent voltage sources as possible are directly connected to the reference.
3. Label a node voltage adjacent to each of the nonreference node.
4. Create supernodes for dependent and independent voltage sources that are not directly connected to the reference node.
5. Node equations are required at supernodes and all other nonreference nodes except OP AMP outputs and nodes

directly connected to the reference by a voltage source (dependent or independent).

6. Write symmetrical node equations by treating dependent sources as independent sources and using the inspection method.
7. Write expressions relating the node voltages to the controlling current or voltage of the dependent sources.
8. Write expressions equating the node voltages at the input terminals of OP AMPs and nodes directly connected to the reference node by a voltage source.
9. Write expressions relating the node and source voltages for voltage sources included in supernodes.
10. Substitute the expressions from Steps 7, 8, and 9 into the node equations from Step 6 and place the result in standard form.

5-4 MESH-CURRENT ANALYSIS

Mesh currents are an alternative set of analysis variables that are useful in circuits containing many elements connected in series. To review terminology, a loop is a sequence of circuit elements that forms a closed path that passes through each element just once. A mesh is a special type of loop that does not enclose any elements. For example, Loops A and B in Fig. 5-24 are meshes, while the Loop X is not a mesh because it encloses an element.

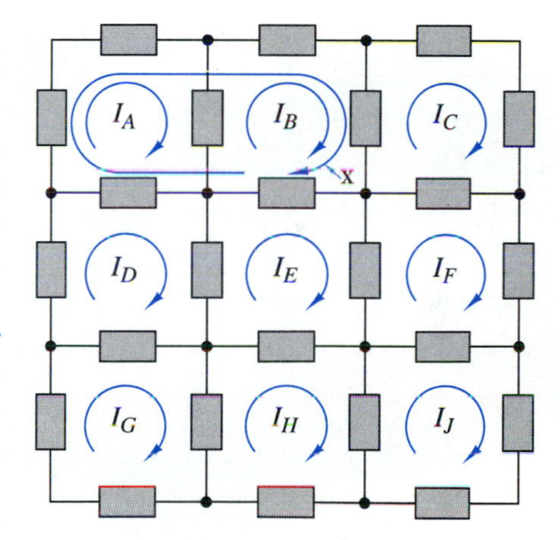

Figure 5-24 Meshes in a planar circuit.

We restrict our development of mesh analysis to *planar circuits*. Planar circuits can be drawn on a flat surface without crossovers in the "window pane" fashion shown in Fig. 5-24. To define a set of variables we associate a *mesh current* (I_A, I_B, I_C, etc.) with each window pane and assign a reference direction. The reference directions for all mesh currents are customarily taken in a clockwise sense. There is no momentous reason for this except perhaps tradition.

We think of these mesh currents as circulating through the elements in their respective meshes, as suggested in Fig. 5-24. We should emphasize that this viewpoint is not based on the physics of circuit behavior. There are not red and blue electrons running around that somehow get assigned to mesh currents I_A or I_B. Mesh currents are variables used in circuit analysis. They are only somewhat abstractly related to the physical operation of a circuit and may be impossible to measure directly. For example, there is no way to cut the circuit in Fig. 5-24 to insert an ammeter that only measures I_E.

Mesh currents have a unique feature that is the dual of the fundamental property of node voltages. If we reexamine Fig. 5-24, we see the elements around the perimeter are contained in only one mesh, while those in the interior are in two meshes. In a planar circuit any given element is contained in at most two meshes. When an element is in two meshes the two mesh currents circulate through the element in opposite directions. In such cases KCL declares that the net element current through the element is the difference of the two mesh currents.

These observations lead us to the fundamental property of mesh currents.

If the kth two-terminal element is contained in Meshes X and Y, then the element current can be expressed in terms of the two mesh currents as

$$I_k = I_X - I_Y \tag{5-16}$$

where X is the mesh whose reference direction agrees with the reference direction of I_k.

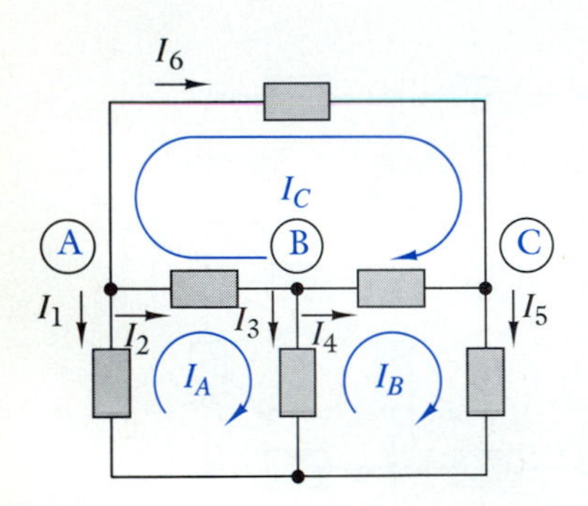

Figure 5-25

Equation (5-16) is a KCL constraint at the element level. If the element is contained in only one mesh, then $I_k = I_X$ or $I_k = -I_Y$ depending on whether the reference direction for the element current agrees or disagrees with the mesh current. The key fact is that the current through every two-terminal element in a planar circuit can be expressed as the difference of at most two mesh currents.

EXERCISE 5-9

In Fig. 5-25 the mesh currents are $I_A = 10$ A, $I_B = 5$ A, and $I_C = -3$ A. Find the element currents I_1 through I_6 and show that KCL is satisfied at Nodes A, B, and C.

ANSWERS:

$I_1 = -10$ A, $I_2 = 13$ A, $I_3 = 5$ A, $I_4 = 8$ A, $I_5 = 5$ A, and $I_6 = -3$ A.

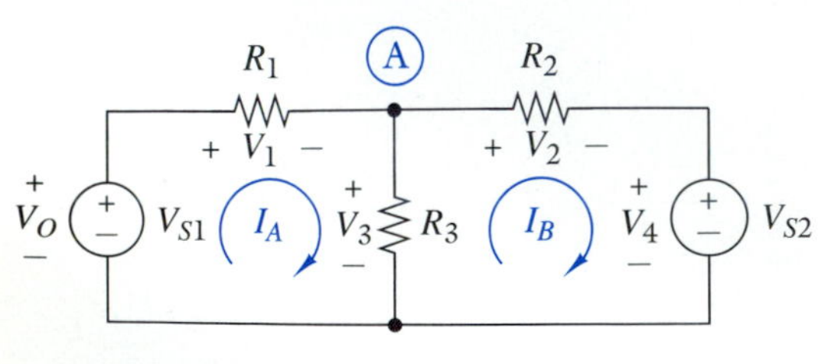

Figure 5-26

To use mesh currents to formulate circuit equations we proceed as in device and connection analysis, except that the KCL constraints are not explicitly written down. Instead, we use the fundamental property of mesh currents to express the device constraints in terms of the mesh currents. By doing so we avoid using the element currents and work only with the element voltages and mesh currents.

For example, the planar circuit in Fig. 5-26 can be analyzed using the mesh-current method. In the figure, we have defined two mesh currents and as well the voltages across each of the five elements. We write KVL constraints around each mesh using the element voltages:

$$\begin{aligned} \text{Mesh A:} \quad & -V_O + V_1 + V_3 = 0 \\ \text{Mesh B:} \quad & -V_3 + V_2 + V_4 = 0 \end{aligned} \tag{5-17}$$

Using the fundamental property of mesh currents we write the element voltages in terms of the mesh currents and input voltages:

$$\begin{aligned} V_1 &= R_1 I_A & V_O &= V_{S1} \\ V_2 &= R_2 I_B & V_4 &= V_{S2} \\ V_3 &= R_3 (I_A - I_B) \end{aligned} \tag{5-18}$$

We substitute these element equations into the KVL connection equations and arrange the result in standard form:

$$\begin{aligned} (R_1 + R_3)\, I_A \quad & - R_3 I_B &= V_{S1} \\ -R_3 I_A + & (R_2 + R_3)\, I_B &= -V_{S2} \end{aligned} \tag{5-19}$$

We have completed the formulation process with two equations in two unknown mesh currents.

While formulating mesh equations it may appear that we have not used KCL. In fact, writing the device constraints in the form in Eq. (5-18) implicitly requires the KCL equations $I_1 = I_A$, $I_2 = I_B$, and $I_3 = I_A - I_B$. In effect, the fundamental property of mesh currents ensures that the KCL constraints are satisfied. As we have previously noted, any general method of circuit analysis must satisfy KCL, KVL, and the device i-v relationships. Mesh-current analysis appears to focus on the latter two, but implicitly satisfies KCL when the device constraints are expressed in terms of the mesh currents.

We use Cramer's rule to solve for the mesh currents:

$$I_A = \frac{\Delta_A}{\Delta} = \frac{\begin{vmatrix} V_{S1} & -R_3 \\ -V_{S2} & R_2 + R_3 \end{vmatrix}}{\begin{vmatrix} R_1 + R_3 & -R_3 \\ -R_3 & R_2 + R_3 \end{vmatrix}} = \frac{(R_2 + R_3)V_{S1} - R_3 V_{S2}}{R_1 R_2 + R_1 R_3 + R_2 R_3} \tag{5-20}$$

and

$$I_B = \frac{\Delta_B}{\Delta} = \frac{\begin{vmatrix} R_1 + R_3 & V_{S1} \\ -R_3 & -V_{S2} \end{vmatrix}}{\begin{vmatrix} R_1 + R_3 & -R_3 \\ -R_3 & R_2 + R_3 \end{vmatrix}} = \frac{R_3 V_{S1} - (R_1 + R_3)\, V_{S2}}{R_1 R_2 + R_1 R_3 + R_2 R_3} \tag{5-21}$$

Equations (5-20) and (5-21) can now be substituted into the device constraints in Eqs. (5-18) to solve for every voltage in the circuit. For instance, the voltage across R_3 is

$$V_A = V_3 = R_3\,(I_A - I_B) = \frac{R_2 R_3 V_{S1} + R_1 R_3 V_{S2}}{R_1 R_2 + R_1 R_2 + R_2 R_3} \qquad (5\text{-}22)$$

The reader is invited to show that the node-voltage analysis yields the same result.

The mesh-current analysis approach illustrated above can be summarized in four steps:

STEP 1: Identify a mesh current with every mesh and a voltage across every circuit element.

STEP 2: Write KVL connection constraints in terms of the element voltages around every mesh.

STEP 3: Use KCL and the element i-v characteristics to express the element voltages in terms of the mesh currents.

STEP 4: Substitute the device constraints from Step 3 into the connection constraints from Step 2 and arrange the resulting equations in standard form.

The number of mesh-current equations derived in this way equals the number of KVL connection constraints in Step 2. In the study of device and connection analysis we noted that there are $E - N + 1$ independent KVL constraints in any circuit. Using the window panes in a planar circuit generates $E - N + 1$ independent mesh currents. Mesh analysis works best when the circuit has many elements (E large) connected in series (N large).

Writing Mesh-Current Equations by Inspection

The mesh equations in Eq. (5-19) have a symmetrical pattern that is similar to the coefficient symmetry observed in node equations. The coefficients of I_B in the first equation and I_A in the second equation are the negative of the resistance common to Meshes A and B. The coefficients of I_A in the first equation and I_B in the second are the sum of the resistances in Meshes A and B, respectively.

This pattern will always occur in planar circuits containing resistors and independent voltage sources when the mesh currents are defined as shown in Fig. 5-26. To see why, consider a general resistance R that is contained in, say, Mesh A. There are only two possibilities. Either R is not contained in any other mesh, in which case the voltage across it is

$$V = R\,(I_A - 0) = R I_A$$

or else it is also contained in only one adjacent mesh, say Mesh B, in which case the voltage across it is

$$V = R\,(I_A - I_B)$$

These observations lead to the following conclusions. The sum of the voltages across resistance in Mesh A is the following:

1. I_A times the sum of the resistances in Mesh A
2. $-I_B$ times the sum of resistances common to Mesh A and Mesh B
3. Minus similar terms for any other mesh adjacent to Mesh A.

The sum of the voltages across resistance plus the sum of the input voltages around Mesh A due to independent voltage sources must equal zero. The same procedure works for the other meshes as well.

The process outlined above allows us to write mesh-current equations by inspection without going through the intermediate steps involving the KVL connection constraints and the device constraints.

EXAMPLE 5-11

For the circuit of Fig. 5-27:

(a) Formulate mesh-current equations.

(b) Find the output V_O using $R_1 = R_4 = 2\ \text{k}\Omega$ and $R_2 = R_3 = 4\ \text{k}\Omega$.

SOLUTION:

(a) To write mesh-current equations by inspection we note that the total resistances in Meshes A, B, and C are $R_1 + R_2$, $R_3 + R_4$, and $R_2 + R_3$, respectively. The resistance common to Meshes A and C is R_2. The resistance common to Meshes B and C is R_3. There is no resistance common to Meshes A and B. Using these observations we write the mesh equations as

$$\text{Mesh A:} \quad (R_1 + R_2)\,I_A - 0I_B - R_2 I_C + V_{S2} = 0$$

$$\text{Mesh B:} \quad (R_3 + R_4)\,I_B - 0I_A - R_3 I_C - V_{S2} = 0$$

$$\text{Mesh C:} \quad (R_2 + R_3)\,I_C - R_2 I_A - R_3 I_B + V_{S1} = 0$$

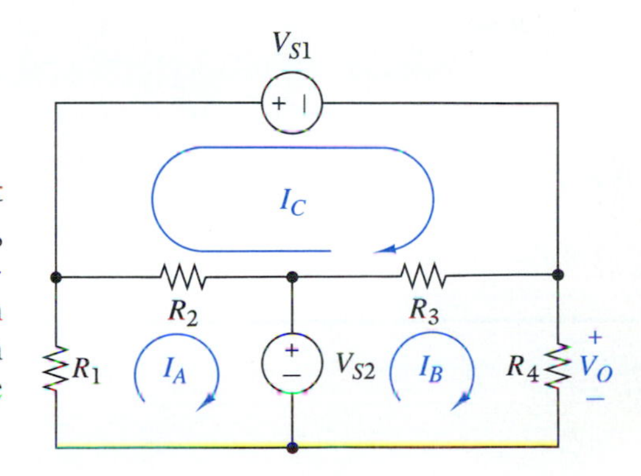

Figure 5-27

The signs assigned to voltage source terms follow the passive convention for the mesh current in question. Arranged in standard form these equations become

$$\begin{aligned}(R_1 + R_2)\,I_A \qquad\qquad\qquad & -R_2 I_C = -V_{S2}\\ +(R_3+R_4)\,I_B \qquad & -R_3 I_C = +V_{S2}\\ -R_2 I_A \qquad\qquad -R_3 I_B & + (R_2+R_3)\,I_C = -V_{S1}\end{aligned}$$

Coefficient symmetry greatly simplifies the formulation of these equations compared with the more fundamental, but time-consuming process of writing device and connection constraints.

(b) Substituting the numerical values into these equations yields

$$\begin{aligned}6\times10^3 I_A \qquad\qquad\qquad & - 4\times10^3 I_C = -V_{S2}\\ 6\times10^3 I_B & - 4\times10^3 I_C = +V_{S2}\\ -4\times10^3 I_A - 4\times10^3 I_B & + 8\times10^3 I_C = -V_{S1}\end{aligned}$$

The output voltage V_O is the voltage across R_4. In terms of mesh currents this voltage is $I_B R_4$. Using Cramer's rule to solve for I_B

$$I_B = \frac{\Delta_B}{\Delta} = \frac{\begin{vmatrix} 6\times10^3 & -V_{S2} & -4\times10^3 \\ 0 & V_{S2} & -4\times10^3 \\ -4\times10^3 & -V_{S1} & 8\times10^3 \end{vmatrix}}{\begin{vmatrix} 6\times10^3 & 0 & -4\times10^3 \\ 0 & 6\times10^3 & -4\times10^3 \\ -4\times10^3 & -4\times10^3 & 8\times10^3 \end{vmatrix}}$$

$$= \frac{16\times10^6 V_{S2} - 24\times10^6 V_{S1}}{96\times10^9}$$

$$= 10^{-3}\left(\frac{V_{S2}}{6} - \frac{V_{S1}}{4}\right)$$

Using Ohm's law the output voltage is

$$V_O = R_4 I_B = \frac{V_{S2}}{3} - \frac{V_{S1}}{2}$$

The same result was obtained in Example 5-6 using node equations. Either approach produces the same answer, but which method do you think is easier?

Mesh Equations with Current Sources

In developing mesh analysis we have assumed that circuits contain only voltage sources and resistors. This assumption simplifies the formulation process because the sum of voltages around a mesh is determined by voltage sources and the mesh currents through resistors. Current sources (independent or dependent) complicate the picture because the voltage across them is not directly related to their current. We need to adapt mesh analysis to accommodate current sources in the same way that we revised node analysis to deal with voltage sources. In fact, the three approaches used here are the duals of the three methods used in node analysis with voltage sources.

There are three ways to handle current sources in mesh analysis:

1. If the current source is connected in parallel with a resistor, then it can be converted to an equivalent voltage source by source conversion. Each source conversion eliminates a mesh and reduces the number of equations required by one. This method is the dual of Method 1 for node analysis.
2. If a current source is contained in only one mesh, then that mesh current is determined by the source current and is no longer an unknown. We write mesh equations around the remaining meshes in the usual way and move the known mesh current to the source side of the equations in the final step. The number of equations obtained is one less than the number of meshes. This method is the dual of Method 2 for node analysis.
3. If a current source is contained in two meshes and is not connected in parallel with a resistance, then neither of the first two approaches will work. In this case we create a *supermesh* by excluding the current source and any elements connected in series with it as shown in Fig. 5-28. We write one mesh equation around the supermesh using the currents I_A and I_B. We then write mesh equations of the remaining meshes in the usual way. This leaves us one equation short because parts of Meshes A and B are included in the supermesh. However, the fundamental property of mesh currents relates the currents I_S, I_A, and I_B as

$$I_A - I_B = I_S$$

This equation supplies the one additional relationship needed to get the requisite number of equations in the unknown mesh currents. This approach is obviously the dual of the supernode (Method 3) method for modified node analysis.

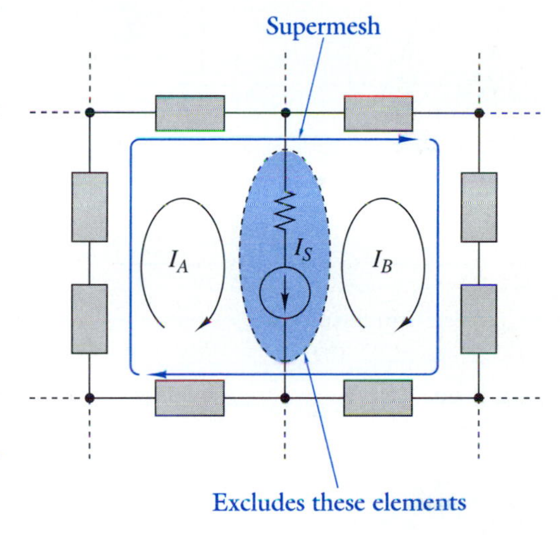

Figure 5-28 Example of a supermesh.

The three methods listed above are not mutually exclusive. We often use more than one method in a circuit as the following examples illustrate.

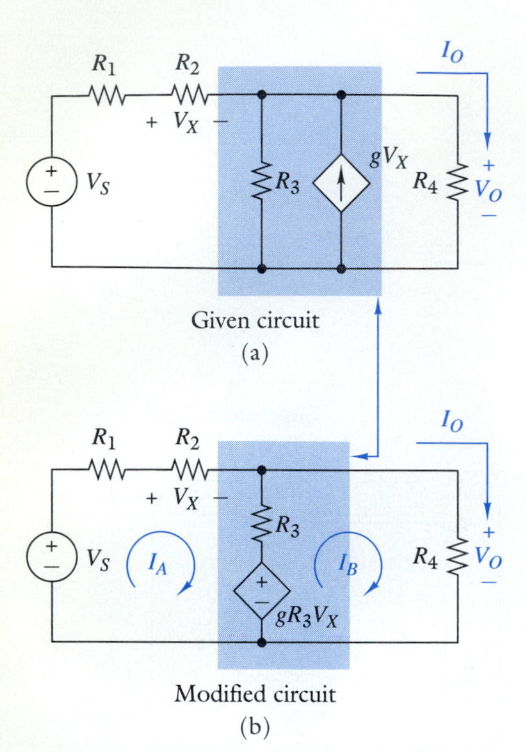

Figure 5-29

EXAMPLE 5-12

For the circuit in Fig. 5-29(a):

(a) Write mesh-current equations.

(b) Find I_O and V_O using $R_1 = 50\ \Omega$, $R_2 = 1\ \text{k}\Omega$, $R_3 = 100\ \Omega$, $R_4 = 5\ \text{k}\Omega$, and $g = 100$ mS.

SOLUTION:

(a) The circuit in Fig. 5-29(a) contains a voltage-controlled current source that can be converted to a voltage-controlled voltage source. After a source conversion the modified circuit is shown in Fig. 5-29(b). We initially treat the dependent voltage source gR_3V_X as an independent source, and then write two symmetrical mesh equations by inspection:

$$\text{Mesh 1:}\quad (R_1 + R_2 + R_3)\, I_A - R_3 I_B - V_S + gR_3V_X = 0$$

$$\text{Mesh 2:}\quad (R_3 + R_4)\, I_B - R_3 I_A - gR_3V_X = 0$$

The control voltage V_X can be written in terms of mesh currents:

$$V_X = R_2 I_A$$

Substituting this equation for V_X into the mesh equations and putting the equations in standard form yields

$$(R_1 + R_2 + R_3 + gR_2R_3)\, I_A - R_3 I_B = V_S$$

$$-(R_3 + gR_2R_3)\, I_A + (R_3 + R_4)\, I_B = 0$$

The resulting mesh equations are not symmetrical because of the controlled source.

(b) Substituting the numerical values into the mesh equations gives

$$1.115 \times 10^4 I_A - 10^2 I_B = V_S$$

$$-1.01 \times 10^4 I_A + 5.1 \times 10^3 I_B = 0$$

Using Cramer's rule the output current is found to be

$$I_O = I_B = \frac{\Delta_B}{\Delta} = \frac{\begin{vmatrix} 1.115 \times 10^4 & V_S \\ -1.01 \times 10^4 & 0 \end{vmatrix}}{\begin{vmatrix} 1.115 \times 10^4 & -10^2 \\ -1.01 \times 10^4 & 5.1 \times 10^3 \end{vmatrix}}$$

$$= \frac{1.01 \times 10^4 V_S}{5.6865 \times 10^7 - 1.01 \times 10^6}$$

$$= 1.808 \times 10^{-4} V_S$$

The output voltage is found from Ohm's law:

$$V_O = R_4 I_O = 0.904\, V_S$$

In this case the input-output relationships do not involve a minus sign indicating that the circuit does not involve signal inversion. The circuit is a small signal model of a transistor emitter follower circuit.

EXAMPLE 5-13

Use mesh analysis to find the current I_C in Fig. 5-30.

SOLUTION:

This circuit is a large signal model of a BJT operating in the active mode. The same circuit was used in Example 5-7 to illustrate node analysis with a supernode. Here the circuit is used to demonstrate using a supermesh. The two mesh currents in Fig. 5-30 are labeled I_1 and I_2 to avoid possible confusion with the transistor base current I_B. The dependent current source βI_B is included in both meshes and is not connected in parallel with a resistance. Therefore, we can use Method 3, which calls for a supermesh. We combine Meshes 1 and 2 into a supermesh by excluding the series subcircuit consisting of the dependent current source and the resistance R_C. Beginning at the bottom of the circuit, we write a KVL mesh equation around the supermesh using unknowns I_1 and I_2.

$$I_2 R_E - V_T + I_1 R_1 + V_{CC} = 0$$

This KVL equation provides one equation in two unknowns. Since the two mesh currents have opposite directions through the current source βI_B, the currents I_1, I_2, and βI_B are related as

$$I_1 - I_2 = \beta I_B$$

This constraint supplies the additional relationship needed to obtain two equations in the two unknown mesh currents. Since $I_B = -I_1$ the constraint means that $I_2 = (\beta + 1)I_1$. When $I_2 = (\beta + 1)I_1$ is substituted into the supermesh KVL equation, we can solve for I_1 as

$$I_1 = -\frac{V_{CC} - V_T}{R_1 + (\beta + 1)R_E}$$

Since $I_B = -I_1$ the collector current is

$$I_C = \beta I_B = \beta(-I_1) = \frac{\beta(V_{CC} - V_T)}{R_1 + (\beta + 1)R_E}$$

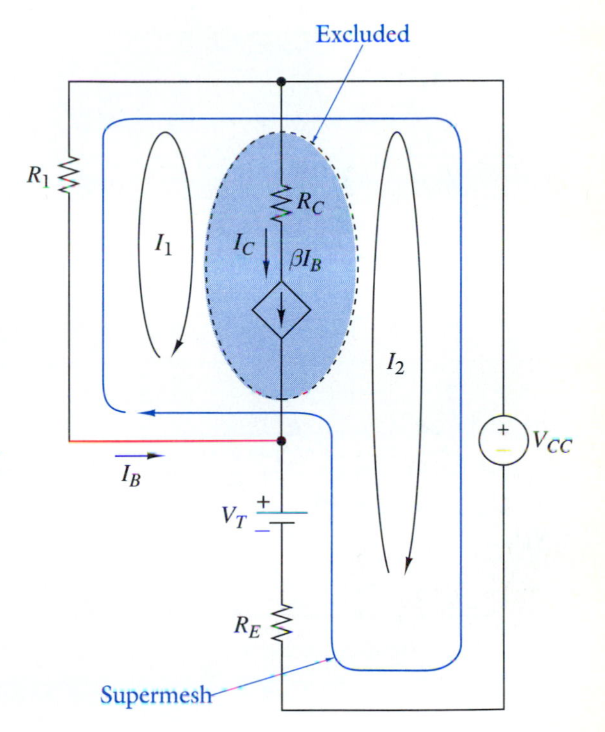

Figure 5-30

This equation for the collector current I_C is the same as was obtained in Example 5-7 using node analysis. Again we see that mesh and node analysis yield identical results.

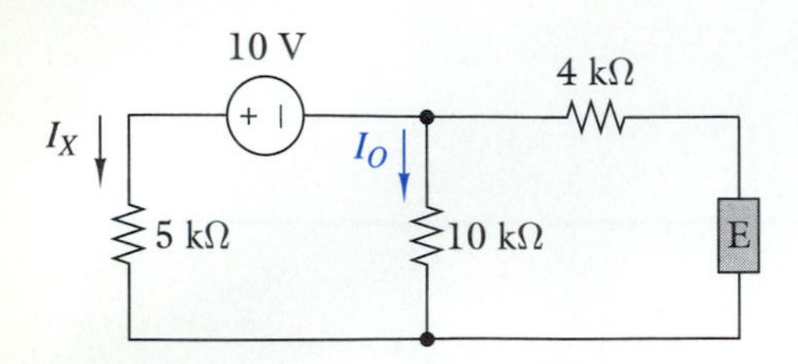

Figure 5-31

EXERCISE 5-10

Use mesh analysis to find the current I_O in Fig. 5-31 when the element E is:

(a) A 5-V voltage source with the positive reference at the top.
(b) A 10-kΩ resistor.
(c) A dependent current source $2I_X$ with the reference arrow directed down.

ANSWERS:
(a) −0.136 mA
(b) −0.538 mA
(c) −0.857 mA ■

EXERCISE 5-11

Use mesh analysis to find the current I_O in Fig. 5-31 when the element E is:

(a) A 1-mA current source with the reference arrow directed down.
(b) Two 20-kΩ resistors in parallel.
(c) A dependent voltage source $2000I_X$ with the plus reference at the top.

ANSWERS:
(a) −1 mA
(b) −0.538 mA
(c) −0.222 mA ■

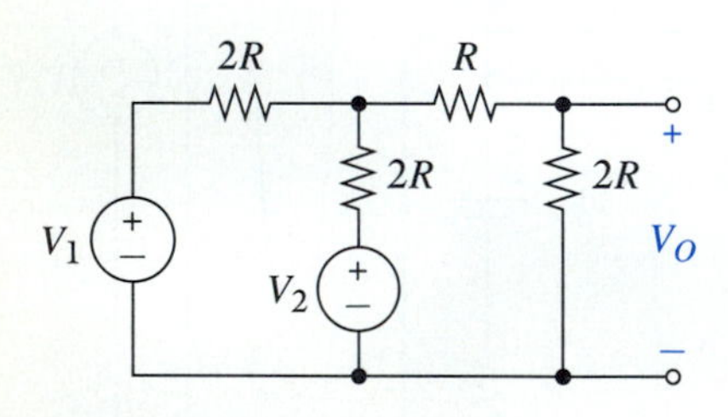

Figure 5-32

EXERCISE 5-12

Use mesh-current equations to find V_O in Fig. 5-32.

ANSWER:
$V_O = (V_1 + V_2)/4$. ■

Summary of Mesh-Current Analysis

Mesh-current equations can always be formulated from KVL, the element constraints, and the fundamental property of mesh currents. When in doubt the reader should always fall back on these principles to formulate mesh equations in new situations. The following guidelines or steps summarize an approach to formulating mesh equations for resistance circuits:

1. Simplify the circuit by combining elements in series or parallel wherever possible.
2. Assign a clockwise mesh current to each mesh.

3. Create a supermesh for dependent and independent current sources that are contained in two meshes.
4. Write symmetrical mesh equations for all meshes by treating dependent sources as independent sources and using the inspection method.
5. Write expressions relating the mesh currents to the controlling current or voltage of the dependent sources.
6. Write expressions relating the mesh and source currents for current sources contained in only one mesh.
7. Write expressions relating the mesh and source currents for current sources included in supermeshes.
8. Substitute the expressions from Steps 5, 6, and 7 into the mesh equations from Step 4 and place the result in standard form.

5-5 COMPUTER-AIDED CIRCUIT ANALYSIS

In this section we introduce computer-aided circuit analysis using a program called SPICE. SPICE is an acronym for **S**imulation **P**rogram with **I**ntegrated **C**ircuit **E**mphasis. The original SPICE program was developed at the University of California under government contract during the 1970s. Various companies have added proprietary features to the basic SPICE program to produce SPICE-based products for personal and mainframe computers. Although we will use SPICE in our study, the underlying concepts apply to other circuit analysis programs such as MICRO-CAP.[3]

Our purpose in introducing SPICE is not simply to teach the reader how to use SPICE to solve circuits and to get numbers called answers. Our objectives are the following:

1. To identify the data and instructions that must be communicated to a circuit analysis program to define a problem
2. To identify the circuit analysis conventions that must be understood to interpret the answers generated by a circuit analysis program
3. To show how to use computer-aided analysis in conjunction with basic principles to calculate circuit parameters or to adjust element values to achieve specified parameters

[3]MICRO-CAP is a registered trademark of Spectrum Software, Inc.

Circuit Analysis Programs

Figure 5-33 presents an overview of a circuit analysis program such as SPICE. The input is a *circuit diagram* that may only be a rough sketch or it may actually be created "on screen" using a graphical schematic editor. The preprocessor in Fig. 5-33 has two main functions: (1) **schematic capture** and (2) **circuit file generation**. Schematic capture documents the circuit in terms of part lists, physical interconnections (wire lists), or functional interconnections (net lists). The preprocessor also creates a circuit file by combining schematic data with component models from the device library. The *circuit file* contains a complete and unambiguous description of the device models and circuit connections in the syntax used by the analysis processor.

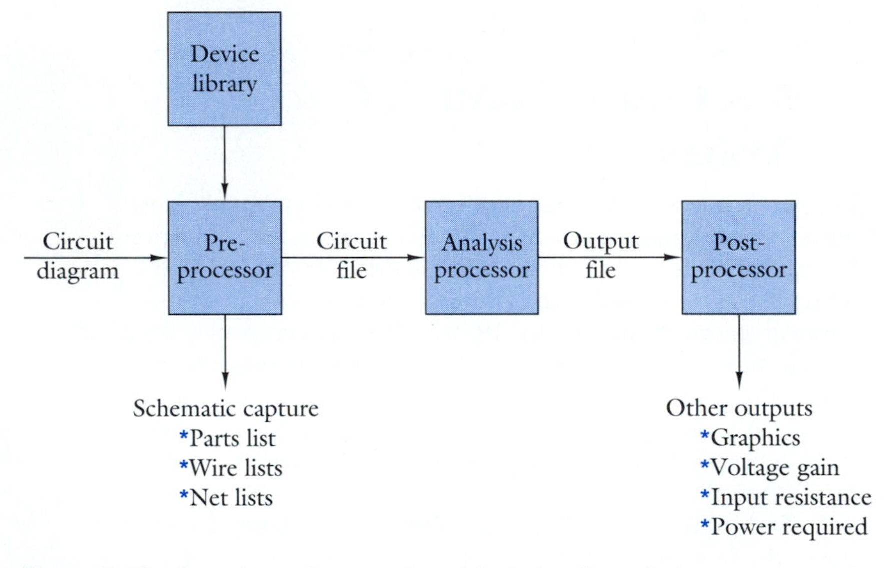

Figure 5-33 Overview of computer-aided circuit analysis.

The circuit analysis processor uses the data in the circuit file to formulate a set of circuit equations. The processor then solves the equations and transfers the calculated responses to the postprocessor via an *output file*. The postprocessor generates other outputs such as graphical plots and responses or parameters derivable from the calculated responses. For example, voltage gain or power dissipated in a load are not normally part of the processor output file. These secondary calculations are normally carried out under user control in the postprocessor.

The preprocessor and postprocessor are dependent on the computer system used and involve rapidly changing technology. On other hand, the basic functions of an analysis processor cannot change because they involve the underlying physical principles of device and connection constraints. These constraints are communicated to the analysis processor via the syntax used to create the circuit file.

The SPICE Syntax

The *SPICE syntax* is a language used in the circuit file to describe device models, circuit interconnections, analysis tasks, and output formats. One of the reasons SPICE is widely used is that standard SPICE circuit files are transportable across many different computer platforms.[4] In addition, the SPICE semiconductor device models are widely accepted in the electronics industry. Semiconductor device manufacturers provided the model parameters for their new products via computer files written in the SPICE syntax.

This chapter introduces the features of SPICE syntax needed to treat the circuit analysis concepts we have studied to this point. As new concepts are introduced in later chapters, we will add the corresponding features of the SPICE syntax. For reference purposes all of the syntax features covered in this book are summarized in Appendix D.

Organization of a SPICE Circuit File

A SPICE circuit file consists of a sequence of statements that do three things:

1. List the elements in the circuit and how they are interconnected.
2. Identify the type of analysis to be performed on the circuit.
3. Control the number and type of calculated responses in the output file.

[4]MICRO-CAP IV can read and execute externally generated circuit files written in the SPICE syntax. Conversely, a MICRO-CAP IV file can be converted into a SPICE file using a built-in TOSPICE.EXE program. The computer-aided circuit analysis problems in this book can be solved using either SPICE or MICRO-CAP IV.

The first statement in a file is a title and the last is an .END statement. The statements in between can be listed in almost any order. However, in this book we will always list the statements in the following order.

```
TITLE STATEMENT
ELEMENT STATEMENTS
COMMAND STATEMENTS
OUTPUT STATEMENTS
.END
```

The SPICE analysis processor assumes that the first statement in a circuit file is a title statement. The first statement has no effect on the circuit or the analysis and can be anything including a blank. SPICE uses the title statement to label the output listing, so for documentation purposes we should include some meaningful description in the title statement. The last statement must be .END. ***Caution:*** The period in the .END is part of the statement and must be included.

Collectively the element, command, and output statements define the elements in the circuit and their interconnections, analysis tasks to be performed, and responses to be included in the output file. The rest of this section introduces the statements we will use in this chapter. The syntax we will use is a subset of the student version of PSpice®,[5] a readily available commercial derivative of the original SPICE program. We begin our syntax discussion with the linear resistor element statement.

Resistor Statements

A SPICE circuit file contains one line for each two-terminal element in the circuit. The syntax for describing a linear resistor is

```
<name> <1st node> <2nd node> <value>
```

The <name> entry is a collection of letters and numbers that begins with the letter R. That is, the name Rxxx. . .xx means linear resistor.

The <1st node> <2nd node> entries define the two nodes to which the resistor is connected. In the SPICE syntax the nodes are numbered using positive integers. The numbers do not need to be sequential, but zero (0) is reserved for the reference or "ground" node.

[5] PSpice is a registered trademark of MicroSim Corporation.

The <value> entry specifies the numerical value of the resistance of Rxxx...xx in ohms. Values can be expressed as decimal number (4300.) or a floating-point number (4.3E3). SPICE uses the following letters to indicate the standard decimal scale factors:

Letter Suffix	Multiplier	
T	tera	10^{+12}
G	giga	10^{+9}
MEG	mega	10^{+6}
K	kilo	10^{+3}
M	milli	10^{-3}
U	micro	10^{-6}
N	nano	10^{-9}
P	pico	10^{-12}
F	femto	10^{-15}

The suffix letters are all capitalized because the computers originally used to develop SPICE did not distinguish between uppercase and lowercase. The only place this causes a major difference is distinguishing between the milli and mega multipliers.

Some examples of resistor statements are

```
RLOAD 3 7 1.2MEG
RS 1 0 40.1
RIN 3 2 10K
```

The first statement defines a 1.2-MΩ resistor named RLOAD connected between Nodes 3 and 7. The second specifies a 40.1-Ω resistor RS connected between Node 1 and ground. The third statement defines a 10-kΩ resistor RIN connected between Nodes 3 and 2.

Independent Source Statements

The syntax for describing independent voltage and current source is

```
<name> <1st node> <2nd node> <type> <value>
```

The <name> entry is a collection of letters and numbers that begins with the letter V for a voltage source and I for a current source. That is, the name Vxxx...xx means independent voltage source and Ixx...xx means independent current source.

The <1st node> <2nd node> entries define the two nodes to which the sources are connected. The <type> entry defines whether the source voltage or current is constant (dc) or time varying. In this chapter we only use dc sources. When the <type> = DC the sources produce constant outputs whose amplitudes are specified by <value> entry in volts for source Vxxxx and amperes for source Ixxxx.

Some examples of independent source statements are

```
VHIGH 12 8 DC -40K
VS 1 0 DC 15
ISINK 3 2 DC 10M
```

The first statement defines a dc voltage source VHIGH connected between Nodes 12 and 8 whose output is −40 kV. The second statement defines a 15-V dc voltage source connected between Node 1 and ground. The third statement defines a dc current source connected between Nodes 3 and 2 whose output is 10 mA. Note that the 10M value in the last statement means "milli" not "mega."

RLOAD 3 7 1.2 MEG

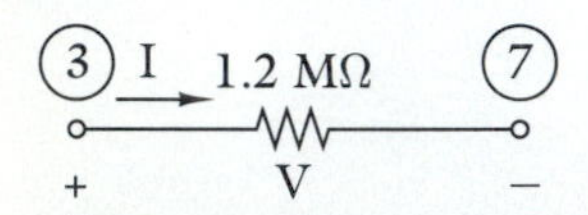

VS 1 0 DC 15

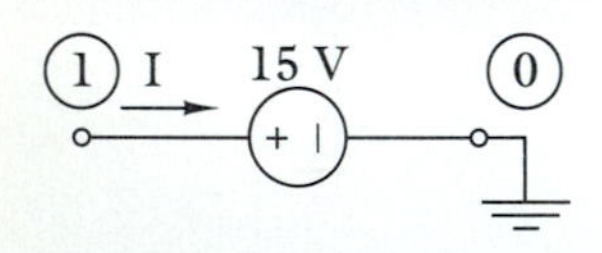

ISINK 3 2 DC 10 M

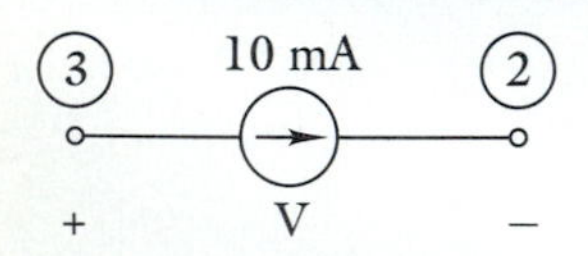

Figure 5-34 SPICE descriptions of two-terminal elements.

Voltage and Current Conventions

There are two conventions that must be understood to use SPICE effectively. The first has to do with sign conventions.

> SPICE assigns the positive voltage reference mark to the <1st node> in the element description and uses the passive sign convention to assign the current reference.

The interpretation of this convention is illustrated in Fig. 5-34 for three of the element descriptions discussed above. Under the SPICE syntax the <1st node> in the element description is assigned the positive voltage reference mark. Under the passive sign convention the reference direction for current is directed from <1st node> toward <2nd node>. In other words, the reference directions for both voltage *and* current are established by the order in which the nodes are listed in the element statement.

The second important feature of SPICE relates to circuit equation formulation and solution variables.

> SPICE uses a modified form of node analysis in which an unknown current variable is assigned to all independent voltage sources.

The basic calculated responses are the N−1 node voltages identified in the SPICE output file as V(1), V(2), . . . V(N−1). SPICE also assigns an unknown current variable to each independent voltage source. The result is that the calculated responses in the output file are the node voltages *plus* the currents through all independent voltage sources.

Although SPICE is node-voltage-oriented we can cause it to solve for a current. Figure 5-35 shows how to insert a current monitoring zero-volt source called VAMP. The element statement for VAMP is

```
VAMP 3 4 DC 0
```

Since Type = DC and value = 0 this statement defines a zero-volt voltage source connected between Nodes 3 and 4. SPICE calculates the current through all independent voltage sources including I(VAMP). But since voltage output of VAMP is zero it acts like an short circuit and does not disturb the operation of the circuit. In effect the voltage source VAMP acts like an ideal ammeter. The reference direction for I(VAMP) is directed from Node 3 to Node 4 because Node 3 is the first node listed in the VAMP element statement. Remember that SPICE assigns the plus voltage reference mark to the first node in an element statement and uses the passive sign convention to assign the current reference direction.

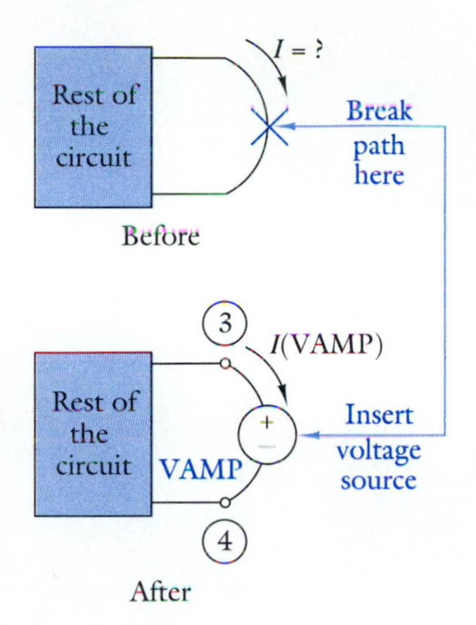

Figure 5-35 Using a voltage source to measure current.

EXERCISE 5-13

A SPICE element listing for the circuit in Fig. 5-36 is given below:

```
EXERCISE 5-13
VS   1   0   DC   10
R1   1   2   1K
R2   2   0   2K
R3   2   3   3K
R4   3   4   4K
VA   4   0   DC   0
.END
```

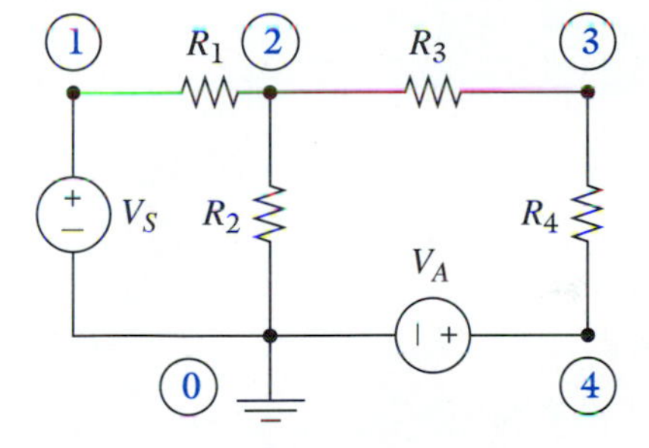

Figure 5-36

One of the calculated responses in the output file is V(2) = 6.0870 V. What other responses are in the output file and what are their values?

ANSWERS:
V(1) = 10 V, V(3) = 3.4783 V, V(4) = 0, I(VS) = −3.913 mA, and I(VA) = 0.869 mA

Dependent-Source Statements

The syntax statement describing a linear dependent source is

```
<name> <1st node> <2nd node> <control variable> <value>
```

The <value> entry gives the numerical value of the dependent source gain. The <name> entry is a collection of letters and numbers. The first letter specifies the dependency and source type using the following codes:

```
Exx...xx
   means voltage-controlled voltage source.
Fxx...xx
   means current-controlled current source.
Gxx...xx
   means voltage-controlled current source.
Hxx...xx
   means current-controlled voltage source.
```

The <1st node> and <2nd node> entries in the element description define the two nodes that the dependent source is connected between. The form of the <control variable> depends on the nature of the controlling signal. If the source is voltage-controlled (Exx...xx and Gxx...xx sources) then the entry is

```
<control variable> := <3rd node> <4th node>
```

The control variable entry specifies the controlling voltage by listing the two nodes across which it is defined. If the source is current-controlled (Fxx...xx and Hxx...xx sources) then the entry is

```
<control variable> := <voltage source>
```

In this case the controlling current is defined by inserting a separate zero-volt independent voltage source. SPICE calculates the current through the zero-volt source and uses it to control the dependent source. The current-monitoring voltage source must be defined in a separate element statement.

Figure 5-37 summarizes the element statements for dependent sources. Note that every controlled source involves four nodes. For voltage-controlled devices the four nodes are all in the element statement. For current-controlled devices, Nodes N1 and N2 are in the element statement while Nodes N3 and N4 are in the accompanying VXX statement. It is particularly important to understand the various reference marks for dependent sources. They can be derived by remembering that SPICE always assigns the positive voltage reference marks to Nodes N1 (and N3 if present). The current reference directions then follow from the passive sign convention.

EXAMPLE 5-14

Create a SPICE element statement listing for the dependent source model of the OP AMP shown in Fig. 5-38.

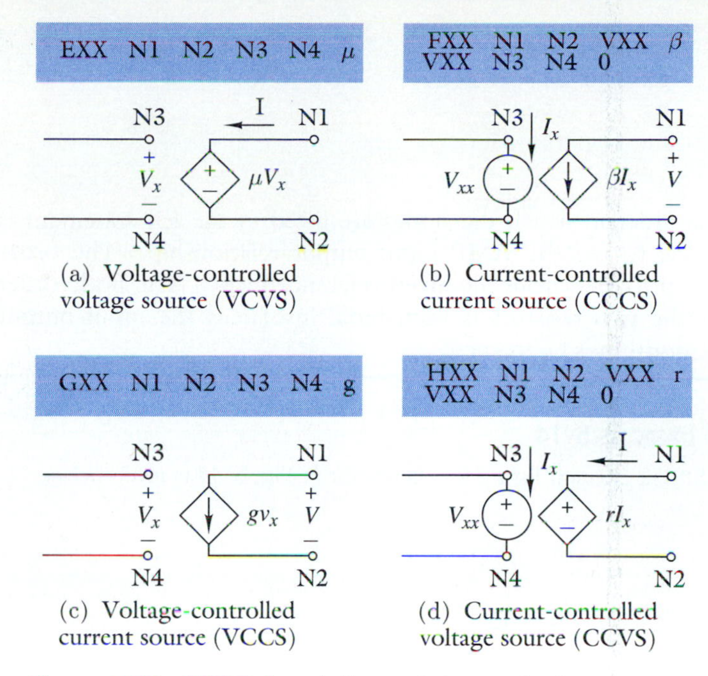

Figure 5-37 SPICE descriptions of dependent sources.

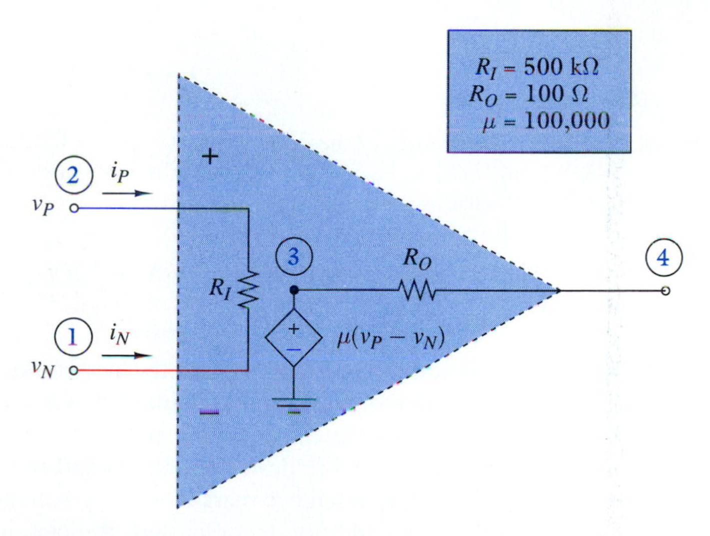

Figure 5-38

SOLUTION:

The model contains two resistors and a voltage-controlled voltage source. The positive reference mark of the dependent source is at Node 3. The positive reference mark for the controlling voltage $v_P - v_N$ is at Node 2. The appropriate SPICE element statements for the circuit Fig. 5-38 are the following:

```
EXAMPLE 5-14
EX 3 0 2 1 100K
RI 2 1 500K
RO 3 4 100
.END
```

The order in which the nodes are listed in the EX statement is critical to the OP AMP input-output relationship. The order shown follows from the given reference marks. The node order for the two resistors is immaterial insofar as the input-output relationship is concerned.

Figure 5-39

EXERCISE 5-14

A SPICE element listing for the circuit in Fig. 5-39 is given below:

```
EXERCISE 5-14
VS  1  0  DC  1
R1  1  2  8K
R2  2  5  1K
VX  5  0  DC  0
R3  2  3  40K
FX  3  0  VX  50
R4  3  4  4K
VA  4  0  DC  0
.END
```

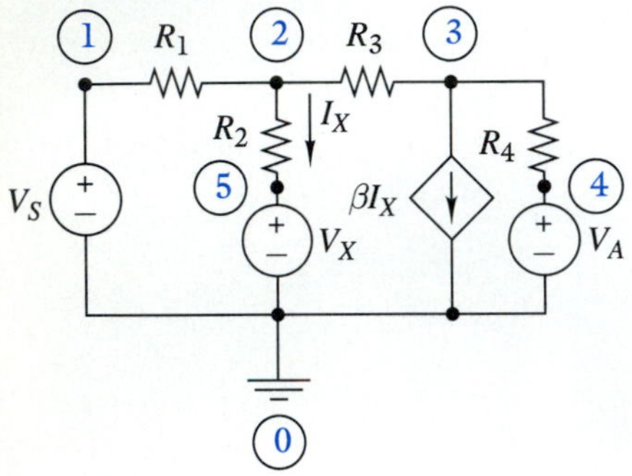

Some of the calculated responses in the output files are V(2) = 0.0220 V, V(3) = −3.99, and I(VS) = −0.1223 mA. What other responses are in the output file and what are their values?

ANSWERS:

V(1) = 1 V, V(4) = 0 V, V(5) = 0, I(VX) = 22 μA, and I(VA) = −0.9975 mA

A SPICE element statement listing defines the devices in a circuit and how they are interconnected. The element statements give a complete description of the circuit in the sense that they contain enough information to write device equations and draw a schematic diagram, including reference marks for every voltage and current in the circuit. In addition to describing the circuit, the SPICE input file must tell the analysis processor what type of analysis to perform.

Command Statements

The SPICE syntax includes command statements to specify the type of analysis to be performed. Only commands for dc analysis will be treated here because at this point in our study voltages and currents are constant. In later chapters we will study alternating current (ac) analysis where all voltages and currents are sinusoids, and transient analysis where they are general functions of time. We will introduce the appropriate SPICE commands for these cases as the need arises.

Three SPICE commands used in dc analysis are the .OP, .TF, and .DC statements. The *.OP command* statement has no parameters so its syntax is simply

```
.OP
```

The .OP statement causes SPICE to calculate the dc operating point (OP) of the circuit and is the *default command* when no command statements are included in the input file.

The syntax of the *.TF (transfer function) command* is

```
.TF <Output Variable> <Input Source>
```

The .TF statement causes SPICE to calculate three numbers: (1) the circuit **gain** from input source to output variable, (2) the **input resistance** seen by the input source, and (3) the **output resistance** at the terminals defining the output variable. The numerical values of the input and output resistances are in ohms. The gain or transfer function is a number defined as

```
GAIN=(Output Variable/Input Source)
```

The units of the gain depend on the units of the input and output variables.

The syntax of the *.DC command* statement is

```
.DC <Source Name> <Start Value> <Stop Value> <Step Value>
```

The statement causes SPICE to perform dc analysis as the value of the source identified by <Source Name> is incrementally stepped across a range beginning at <Start Value> and ending at <Stop Value>. The increments are defined by <Step Value>, which must be greater than zero.

For example, the statement

```
.DC VS -1 3 0.5
```

causes the voltage source named VS to be stepped across the range from -1 V to $+3$ V in increments of 0.5 V. The .DC command listed above initiates a sequence of nine dc analyses for VS $= -1.0, -0.5, 0.0, 0.5, 1.0, 1.5, 2.0, 2.5$, and 3.0 V. These values override the <value> assigned to source VS in its element statement. That is, the element statement for the source VS might be

```
VS 1 0 DC 15
```

The element statement defines a 15-V dc voltage source. However, the .DC command given above overrides the 15-V value and causes the source VS to assume the sequence of values listed above.

The .DC command produces only one source value when the start and stop values are equal. For example, the .DC command statement

```
.DC IDC 5M 5M 1M
```

produces a single dc analysis with the independent current source name IDC set at 5 mA. Again, the value IDC = 5 mA overrides the value assigned to IDC in its element statement.

Caution: The periods in the .OP, .TF, and .DC commands must be included.

Output Statements

Output statements control the types and number of calculated responses included in the output file. If there is **no** output statement in the input file, then by **default** the output file will contain all of the node voltages and currents through independent voltage sources. If the .TF command is included in the input file, then the output file will also contain the numerical values of the gain, input resistance, and output resistance.

Some circuits can be handled with no output control statements because the node voltages and source currents allow us to calculate responses not included in the default output file. However, the .PRINT statement in conjunction with the .DC command statement can be used to specify the calculated responses in the output file. The syntax of the .PRINT command statement is

```
.PRINT <Analysis Type> <Output Variable List>
```

When the <Analysis Type> is DC there must be a .DC command statement in the input file for the .PRINT statement to be executed. The .PRINT statement restricts the calculated responses in the output file to those given in the output variable list. The output variable list identifies responses using the notation V(N), V(N1,N2), and I(Vxxxx). V(N) is the node voltage at Node N. V(N1,N2) is the voltage between Nodes N1 and N2 with the plus reference mark at N1. I(Vxxx) is the current through the element Vxxxx. The reference direction for I(Vxxxx) is determined by the first node listed in the element statement defining Vxx. For example, the statement

```
.PRINT DC V(1) V(3,6) I(VS) I(RE)
```

produces an output file containing four responses: V(1) is the voltage between Node 1 and ground, V(3,6) is the voltage between Nodes 3 and 6, I(VS) is the current through voltage source VS, and I(RE) is the current through resistor RE.

SPICE Analysis Examples

In the following examples we will list examples of SPICE input files needed to calculate circuit characteristics such as voltage gains, Thévenin equivalents, and power dissipation. In addition to defining the circuit, the element statements in these files must do three things:

1. They must define a reference node (Node 0 or ground) otherwise SPICE cannot define node voltages.
2. They must ensure that at least two elements are connected to every node (including the reference node) because the SPICE algorithms are not designed to handle nodes that dangle in the air.
3. They must ensure that there is a dc path from every node to the reference node. A dc path means a route via resistors and independent voltage sources. The route can be very circuitous, passing through many nodes along the way, but it must ultimately lead to Node 0. If a node does not have a dc path to ground, then its dc voltage is undefined and the SPICE algorithms cannot converge.

The SPICE processor always checks these conditions and aborts an analysis run with an error message if the conditions are not satisfied.

EXAMPLE 5-15

The circuit in Fig. 5-40 is an inverting amplifier using the finite gain model of the OP AMP discussed in Example 5-14. Use SPICE to find the voltage gain V_O/V_S, the input resistance seen by V_S, and output resistance seen between Node 3 and ground.

SOLUTION:

For the ideal OP AMP model the circuit gain is $-R_2/R_1 = -5$, the input resistance is $R_1 = 10\ \text{k}\Omega$, and the output resistance is zero. These quantities can be calculated using the .TF command.

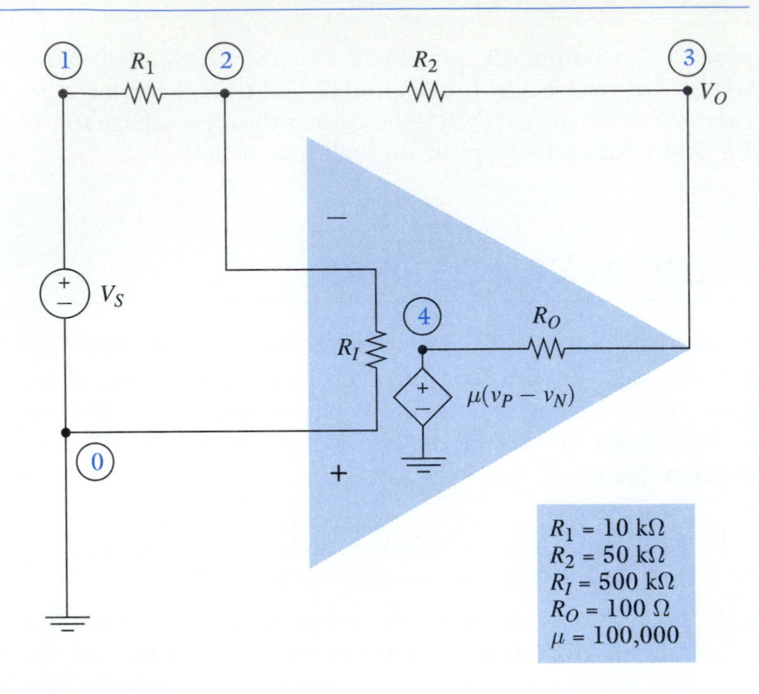

Figure 5-40

A suitable SPICE input file for the circuit in Fig. 5-40 is the following:

```
EXAMPLE 5-15
VS 1 0 DC 1
R1 1 2 10K
RI 2 0 500K
R2 2 3 50K
RO 3 4 100
EX 4 0 0 2 100K
.TF VS V(3)
.END
```

SPICE performs a dc analysis and lists the following quantities in the output file:

```
V(3)/VS = -5.000E+00
INPUT RESISTANCE AT VS = 1.000E+04
OUTPUT RESISTANCE AT V(3) = 6.100E-03
```

The gain and input resistance agree with the ideal model to within the four significant figures given by SPICE. The output resistance is not zero but is very small compared with other circuit resistances.

EXAMPLE 5-16

Use SPICE to solve for the input-output relationship of the circuit in Fig. 5-41.

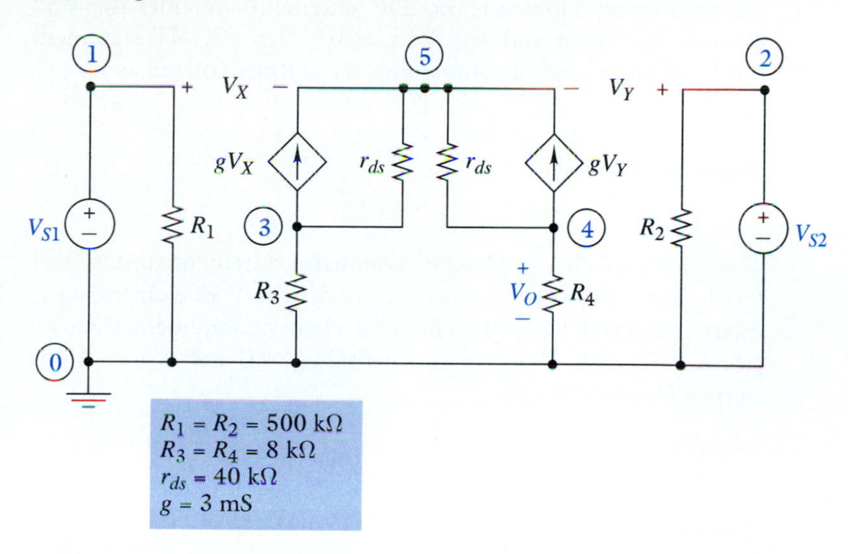

Figure 5-41

SOLUTION:

The circuit is a linear model of a field effect transistor (FET) amplifier with two inputs V_{S1} and V_{S2}. By the superposition principle the input-output relationship is of the form

$$V_O = K_1 V_{S1} + K_2 V_{S2}$$

We can use superposition to find the gain K_1 by setting $V_{S1} = 1$ and $V_{S2} = 0$ and solving for V(4). The gain K_2 is then found by setting $V_{S1} = 0$ and $V_{S2} = 1$ and solving for V(4). A suitable SPICE circuit file for the condition $V_{S1} = 1$ and $V_{S2} = 0$ is the following:

```
Example 5-16
VS1 1 0 DC 1
R1  1 0 500K
VS2 2 0 DC 1
R2 2 0 500K
R3 3 0 8K
R4 4 0 8K
RDS1 5 3 40K
RDS2 5 4 40K
G1 3 5 1 5 3M
G2 4 5 2 5 3M
```

```
.DC VS2 0 0 1
.PRINT DC V(4)
.END
```

The values assigned to $VS1$ and $VS2$ in their element statements are both unity. However, the .DC statement overrides the $VS2$ element statement and sets $VS2 = 0$. The .PRINT statement produces an output file containing the output voltage as

```
V(4) = 1.000E+01
```

Therefore, the gain K_1 is

```
K1 = V(4)/VS1 = 10
```

To find K_2 we change the .DC command statement to .DC VS1 0 0 1. This new .DC statement overrides the $VS1$ element statement and sets $VS1 = 0$. The $VS2$ element statement then restores $VS2 = 1$. For the run with $V_{S1} = 0$ and $V_{S2} = 1$ the output file reports

```
V(4)= -1.000E+01
```

Therefore, $K_2 = \mathrm{V}(4)/\mathrm{V}S2 = -10$ and the input-output relationship for the circuit is

$$V_O = 10\,(V_{S1} - V_{S2})$$

The circuit is a differential amplifier of a type used as the input stage of an OP AMP.

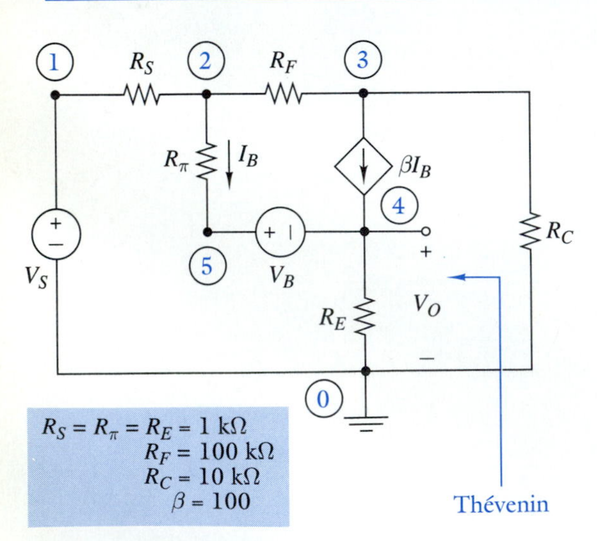

Figure 5-42

EXAMPLE 5-17

(a) Use SPICE to find the Thévenin equivalent at the output between Nodes 4 and ground of the circuit in Fig. 5-42.

(b) Find the value of R_E that produces a Thévenin resistance of 75 Ω.

SOLUTION:

(a) The .TF command statement can be used to calculate the open-circuit voltage and output resistance seen at Node 4. Since the input voltage is not specified we use $VS = 1$. A suitable SPICE input file for $VS = 1$ is the following:

```
Example 5-17
VS 1 0 DC 1
RS 1 2 1K
RP 2 5 1K
RF 2 3 100K
VB 5 4 DC 0
```

```
FB 3 4 VB 100
RE 4 0 1K
RC 3 0 10K
.TF V(4) VS
.END
```

SPICE performs a DC analysis and lists the following quantities in the output file:

```
V(4)/VS = 8.937E-01
INPUT RESISTANCE AT VS = 1.026E+4
OUTPUT RESISTANCE AT V(4) = 98.21E+01
```

Since $VS = 1$ the open-circuit voltage is $V(4) = V_{OC} = 0.8937$ V. The output resistance of 98.21 Ω calculated by the .TF command is the Thèvenin resistance seen at Node 4. The parameters of the Thèvenin equivalent circuit for any input voltage V_S are

$$V_T = 0.8937V_S \qquad \text{and} \qquad R_T = R_{OUT} = 98.21\ \Omega$$

(b) A sequence of SPICE runs using different values of R_E in the input file yields the results given below:

R_E (Ω)	V(4) (V)	R_T (Ω)
1000	0.8937	98.21
500	0.8137	89.43
250	0.6903	75.86
240	0.6817	74.91
241	0.6825	75.01

The series of trial and error runs indicates that $R_T \approx 75\ \Omega$ when $R_E = 241\ \Omega$. The mainframe version of PSpice supports a .STEP command that automatically increments an element value across a specified range. The circuit in 5-42 is a small signal model of a transistor feedback amplifier.

EXAMPLE 5-18

The transistor in Fig. 5-43 is operating in the active mode. Use SPICE and the large signal model of the BJT from Chapter 4 to find the power delivered by the 24-V source and the power absorbed by the transistor and resistors.

SOLUTION:

Figure 5-43 shows the SPICE circuit model, where the active mode large-signal model has replaced the transistor. A SPICE

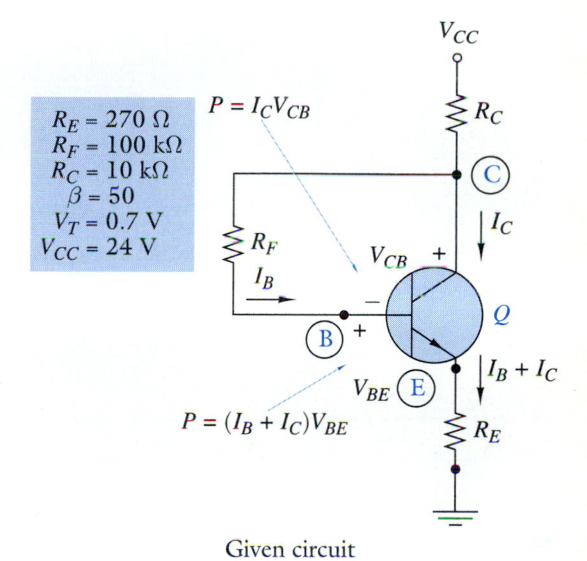

Given circuit

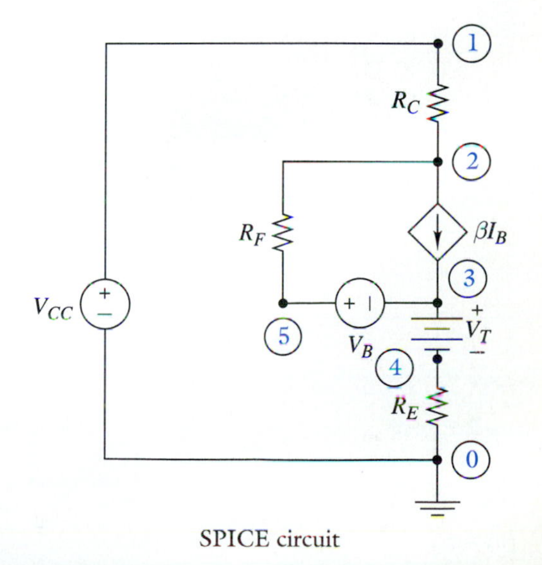

SPICE circuit

Figure 5-43

input file for this circuit is the following:

```
Example 5-18
VCC 1 0 DC 24
RC 1 2 10K
RF 2 5 100K
VB 5 3 DC 0
FB 2 3 VB 50
VT 3 4 DC 0.7
RE 4 0 270
.END
```

The input file does not contain command or output control statements. SPICE defaults to the .OP command, performs a dc analysis, and writes all node voltages and all independent voltage source currents to the output file. The output file lists the following responses:

```
NODE VOLTAGES

V(1) 24.000  V(2) 4.9497  V(3) 1.2144
V(4) 0.5144  V(5) 1.2144

VOLTAGE SOURCE CURRENTS

I(VCC) -1.905E-03
I(VB)   3.735E-05
I(VT)   1.905E-03
```

The power delivered by the 24-V source is

$$P_S = [V_{CC} I\,(VCC)] = [24 \times (-1.905)] = -45.72 \text{ mW}$$

The source power is negative because SPICE follows the passive sign convention under which delivered power is negative. The given circuit in Fig. 5-43 shows that the power absorbed by the transistor consists of two components:

$$P_Q = V_{CB} I_C + V_{BE}\,(I_C + I_B)$$

Using the data in the output the element voltages and currents are

$$V_{CB} = V(2) - V(3) = 3.7353 \text{ V}$$

$$I_C = I(VT) - I(VB) = 1.8676 \text{ mA}$$

$$V_{BE} = V(3) - V(4) = 0.7 \text{ V}$$

$$I_B = I(VB) = 37.35\ \mu\text{A}$$

The powers absorbed by the transistor and resistors are

$$P_Q = 8.309 \text{ mW}$$

$$P_{RC} = (I_C + I_B)^2 R_C = 36.29 \text{ mW}$$

$$P_{RF} = I_B^2 R_F = 0.1395 \text{ mW}$$

$$P_{RE} = (I_C + I_B)^2 R_E = 0.9798 \text{ mW}$$

The sum of element powers is

$$P_S + P_Q + P_{RC} + P_{RF} + P_{RE} = -46.72 + 45.7183 = -0.0017 \text{ mW}$$

which matches the requirement of Tellegen's theorem within roundoff error.

5-6 COMPARISON OF ANALYSIS METHODS

We have completed our study of resistance circuits, so it is important to reflect on what we have learned. We began by describing how the voltages and currents in a circuit are constrained by Kirchhoff's laws and device characteristics. Using the device and connection method as a foundation, we studied a variety of circuit analysis techniques, including circuit reduction, unit output, superposition, Thévenin's and Norton's theorems, node analysis, mesh analysis, and finally computer-aided analysis. Are all of these methods really necessary? Why don't we learn one method and apply it to every situation?

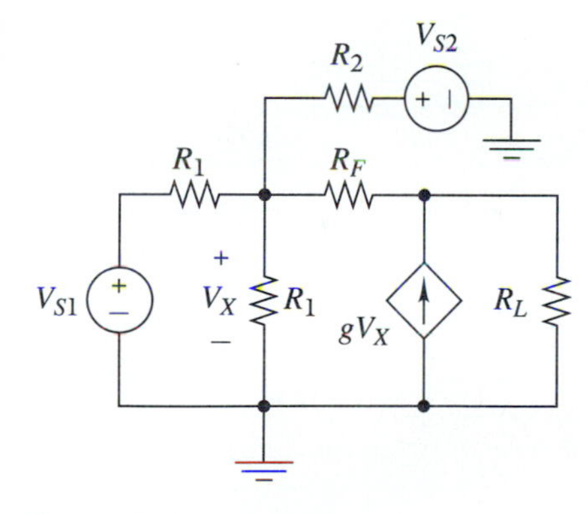

Figure 5-44 An example circuit.

The short answer is that we cannot. All of these methods are useful because each offers a different perspective of circuit behavior. To illustrate the point consider the circuit in Fig. 5.44. If we need a general characterization, then writing two node equations is preferable to three mesh equations. On the other hand, mesh currents might be better if the purpose of the analysis is to determine the current through R_F. Similarly, if the purpose is to determine R_L for maximum power transfer, then either Thévenin's or Norton's equivalent is needed. Superposition would be used if the objective is to isolate the effects of V_{S1} and V_{S2} on the circuit responses. Still another approach is to use successive source conversions and circuit reductions to reduce the circuit to only three elements connected in parallel—a resistor, a dependent, and an independent current source (try it). An engineer must know how to use different tools because some tools may work better than others on any given problem.

Conversely, all of these methods are needed because each has some limitation that makes it difficult or impossible to apply in every situation. Some of these limitations are fundamental and some arise from practical considerations. An example of a fundamental limitation is the fact that the unit output method only works on ladder circuits. A very practical limitation is that the algebraic burden of hand computations begins to get out of hand for circuits with four or more nodes or meshes, making computer-aided analysis a practical necessity.

Of what use are hand analysis methods that become impractical at such a model level of complexity? Why not just use SPICE or MICRO-CAP and forget about everything else? It is certainly true that large-scale circuits are best handled by computer-aided techniques. Programs like SPICE are probably the right tool for circuits of even modest complexity when numerical values of all parameters are known, and a numerical value of the response is the desired end product.

But circuit analysis is not an end product, but a means to an end. What hand analysis does that SPICE cannot do is generate an analytical solution with the circuit parameters left in symbolic form. A symbolic solution often gives us greater insight into circuit operation because we can see how parameters affect the response.

In other words, hand analysis and computer-aided analysis are not simply alternative ways to solve the same old problems. Computer-aided analysis is appropriate when we have a basic understanding of how a circuit works and need to examine the numerical details of its operation. The basic understanding needed to use programs like SPICE intelligently is gained through hand analysis of the relatively simple circuits that form the building blocks of large-scale circuits.

A few general guidelines for using these analysis tools are the following:

1. Simplify the circuit by combining elements in series or parallel wherever possible.
2. Single-input ladder circuits are easily treated using circuit reduction or the unit output method.
3. Superposition is useful in multiple-input circuits to isolate the effect of individual inputs. It is not always the best way to find the combined effects of several inputs.
4. Thévenin's or Norton's theorems are useful in interface situations to examine the effect of different loads or nonlinear loads. They are not particularly useful for a single, fixed linear load.
5. OP AMP circuits can be treated using functional block diagrams for the standard configurations and node analysis for more complex configurations.

6. Node analysis works best in circuits with many elements connected in parallel and all voltage sources connected to the reference node.
7. Mesh analysis works best in circuits with many elements connected in series and only one mesh current through each current source.
8. Computer-aided analysis is appropriate when you have a basic understanding of how a circuit works and need to check the numerical details of its operation.

No single technique fits every task. In most circuit problems several different techniques will be needed. Only practice and experience will give you the insight needed to select the best set of tools in each new situation. Table 5-1 summarizes the major advantages and disadvantages of various methods of circuit analysis.

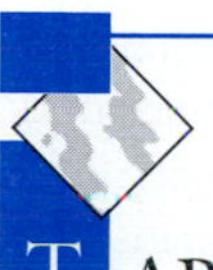

TABLE 5-1 A SUMMARY OF CIRCUIT ANALYSIS TECHNIQUES

Technique	Advantage	Disadvantage
Circuit reduction	Involves working directly with the circuit model	Nonseries/parallel circuits not easily handled
Unit output	Simple and direct application of KCL, KVL, and Ohm's law	Only works for ladder circuits
Superposition	Isolates the effect of different sources on circuit responses	Requires repeated analysis
Thévenin/Norton	Useful for interface situations with many different loads	Requires two analyses to find V_{OC} and I_{SC}
Device and connection analysis	Equations are easy to formulate; useful as a conceptual foundation	Generates many equations
Node analysis	Equations are easy to formulate and apply to electronic circuits	Transformers (Ch. 7) not easily handled
Mesh analysis	Equations are easy to formulate and apply to planar circuits	OP AMPs not easily handled
Computer-aided circuit analysis	Can easily handle simple or large-scale electronic circuits	Requires numerical values for all circuit parameters.

SUMMARY

- General systematic methods of circuit analysis involve formulating and solving linear algebraic equations. Formulation methods include device and connection analysis, node analysis, and mesh analysis. The solution methods include Cramer's rule, Gauss's elimination, and computer methods.
- Device and connection analysis forms the basis of all systematic methods of formulating circuit equations. For a circuit containing E elements and N nodes this method produces a total of $2E$ equations including: $N - 1$ independent KCL connection equations, $E - N + 1$ independent KVL connection equations, and E element equations.
- Node analysis involves identifying a reference node and the node to datum voltages at the remaining $N - 1$ nodes. The KCL connection constraints at the $N - 1$ nonreference nodes are combined with the device constraints written in terms of the node voltages to produce $N - 1$ linear equations in the unknown node voltages.
- Mesh analysis involves identifying mesh currents that circulate around the perimeter of each mesh in a planar circuit. The KVL connection constraints written around each of the $E - N + 1$ meshes are combined with the device constraints written in terms of the mesh currents to produce $E - N + 1$ linear equations in the unknown mesh currents.
- Node and mesh analysis can be adapted to handle both types of sources using a combination of three methods: (1) source transformations, (2) selecting the reference node or mesh currents so that some unknowns are determined by independent sources, and (3) using supernodes and supermeshes.
- OP AMP circuits are best treated using node analysis. A node voltage is identified at each OP AMP output, but a node equation is not written at these nodes. Node equations are then written at the remaining nodes and the ideal OP AMP input voltage constraint ($V_N = V_P$) used to reduce the number of unknowns.
- Computer-aided circuit analysis programs often use the SPICE analysis processor. SPICE formulates node equations using the passive sign convention. The calculated responses in a SPICE output file are the node voltages plus the currents through independent voltage sources.
- The SPICE syntax is a standardized way to describe device models, circuit connections, and analysis tasks. A SPICE

circuit file contains element statements describing the devices and connections, and command statements prescribing the desired analysis method.

- Table 5-1 summarizes the major advantages and disadvantages of the circuit analysis techniques studied in the first five chapters of this book. Each technique is useful because it provides a different perspective on circuit analysis and design problems. Some general guidelines exist, but only experience and practice develop the ability to select the best technique in a given situation.

EN ROUTE OBJECTIVES AND ASSOCIATED PROBLEMS

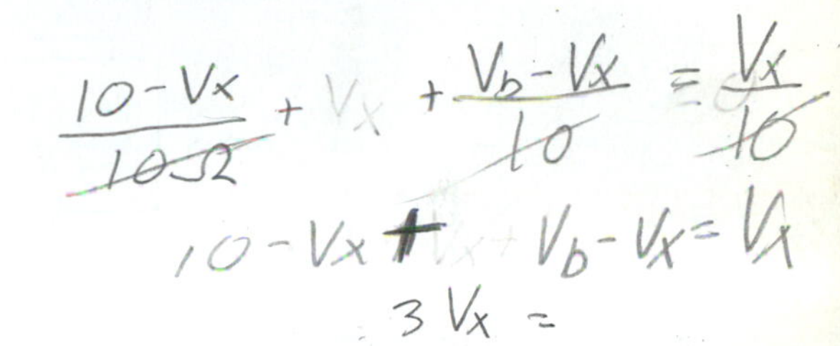

ERO 5-1 General Circuit Analysis (Sects. 5-1 to 5-4)

Given a circuit consisting of linear resistors, controlled sources, OP AMPs, and input signal sources:

(a) (Formulation) Write a complete set of device/connection or node-voltage or mesh-current equations for the circuit.

(b) (Solution) Solve the system of linear equations for selected signal variables or input-output relationships.

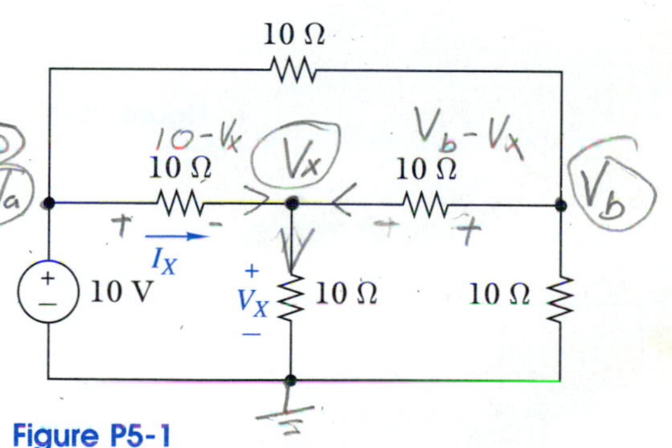

Figure P5-1

5-1 How many device and connection equations are needed to describe the circuit in Fig. P5-1? Formulate (but do not solve) a complete set of device and connection equations for this circuit. Given $V_X = 5$ V and $I_X = 0.5$ A, use these equations to find all of the other currents and voltages in the circuit.

5-2 How many device and connection equations are needed to describe the circuit in Fig. P5-2? Formulate (but do not solve) a complete set of device and connection equations for this circuit. Given $V_X = 30$ V and $I_X = -1$ A, find all of the other currents and voltages in the circuit.

5-3 How many node-voltage equations are needed to describe the circuit in Fig. P5-1? Formulate (but do not solve) a complete set of node-voltage equations for this circuit. Given $V_X = 5$ V and $I_X = 0.5$ A, use these equations to find all of the other currents and voltages in the circuit.

5-4 How many node-voltage equations are needed to describe the circuit in Fig. P5-2? Formulate (but do not solve) a complete set of node-voltage equations for this circuit. Given $V_X = 30$ V and $I_X = -1$ A, find all of the other currents and voltages in the circuit.

Figure P5-2

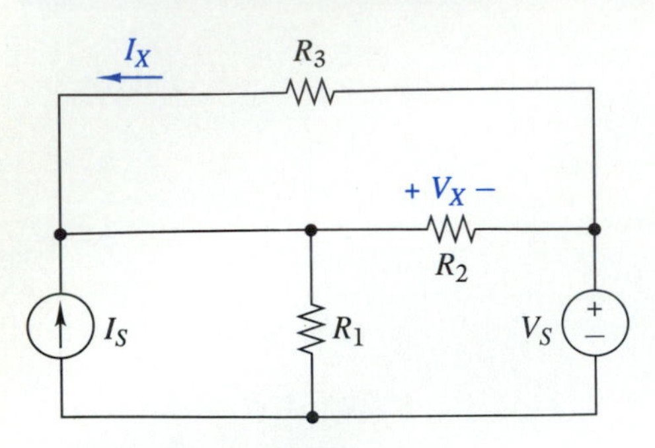

Figure P5-5

5-5 Formulate node-voltage equations for the circuit in Fig. P5-5 and solve for V_X and I_X.

5-6 Formulate node-voltage equations for the circuit in Fig. P5-6. Solve for V_X and I_X using $R_1 = R_2 = R_3 = R_4 = 1$ kΩ, $R_5 = R_6 = 2$ kΩ, and $I_{S1} = I_{S2} = 100$ mA. Find the total power delivered to the circuit by the current sources.

5-7 Formulate node-voltage equations for the circuit in Fig. P5-7. Solve for V_X and I_X using $R_1 = 10$ kΩ, $R_2 = 20$ kΩ, $R_3 = 30$ kΩ, $R_4 = 40$ kΩ, $R_X = 3$ kΩ, and $V_S = 10$ V. What is the power absorbed by resistor R_2?

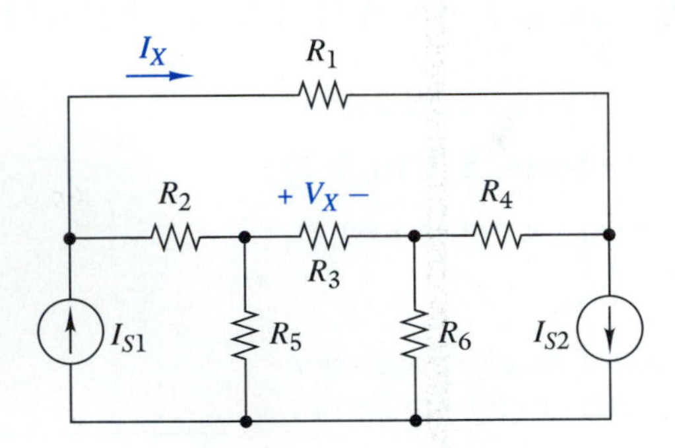

Figure P5-6

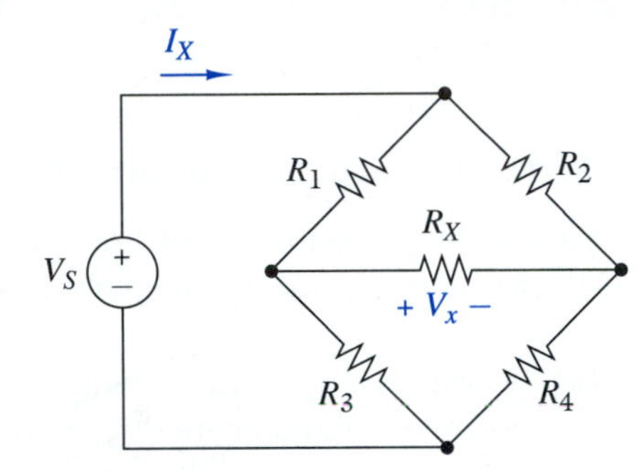

Figure P5-7

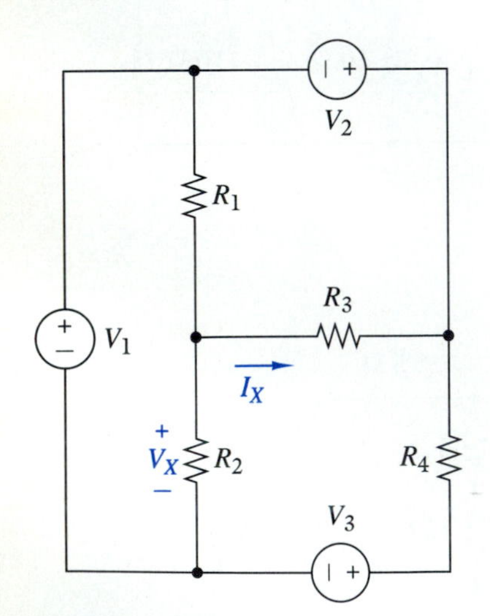

Figure P5-8

5-8 Formulate node-voltage equations for the circuit in Fig. P5-8. Solve for V_X and I_X using $R_1 = 10$ kΩ, $R_2 = 20$ kΩ, $R_3 = 30$ kΩ, $R_4 = 40$ kΩ and $V_1 = V_2 = V_3 = 10$ V. Find the power delivered to R_1.

5-9 Formulate node-voltage equations for the circuit in Fig. P5-9. Solve for V_X and I_X using $R_1 = 1$ kΩ, $R_2 = 2$ kΩ, $R_3 = 1$ kΩ, $R_4 = 3$ kΩ, $R_5 = 1$ kΩ, $R_X = 3$ kΩ, $I_S = 275$ mA and $V_S = 100$ V. Find the power delivered by the current source.

5-10 How many mesh-current equations are needed to describe the circuit in Fig. P5-1? Formulate (but do not solve) a complete set of mesh-current equations for this circuit. Given $V_X = 5$ V and $I_X = 0.5$ A, find all of the other currents and voltages in the circuit.

5-11 How many mesh-current equations are needed to describe the circuit in Fig. P5-2? Formulate (but do not solve) a complete set of mesh equations for this circuit. Given $V_X = 30$ V and $I_X = -1$ A, find all of the other currents and voltages in the circuit.

5-12 Formulate mesh-current equations for the circuit in Fig. P5-5 and solve for V_X and I_X.

5-13 Formulate mesh-current equations for the circuit in Fig. P5-6. Use the mesh equations to find V_X and I_X using $R_1 = R_2 = R_3 = R_4 = 1$ kΩ, $R_5 = R_6 = 2$ kΩ, and $I_{S1} = I_{S2} = 100$ mA. Find the total power delivered by the two current sources.

5-14 Formulate mesh-current equations for the circuit in Fig. P5-7. Use the mesh equations to find V_X and I_X using $R_1 = 10$ kΩ, $R_2 = 20$ kΩ, $R_3 = 30$ kΩ, $R_4 = 40$ kΩ, $R_X = 3$ kΩ, and $V_S = 10$ V. Find the power delivered to the resistor R_2.

5-15 Formulate mesh-current equations for the circuit in Fig. P5-8. Use the mesh current to find V_X and I_X using $R_1 = 10$ kΩ, $R_2 = 20$ kΩ, $R_3 = 30$ kΩ, $R_4 = 40$ kΩ and $V_1 = V_2 = V_3 = 10$ V. Find the total power deliver by the three voltage sources.

5-16 Formulate mesh-current equations for the circuit in Fig. P5-9. Use the mesh-current equations to find V_X and I_X using $R_1 = 1$ kΩ, $R_2 = 2$ kΩ, $R_3 = 1$ kΩ, $R_4 = 3$ kΩ, $R_5 = 1$ kΩ, $R_X = 3$ kΩ, $I_S = 275$ mA and $V_S = 100$ V. Find the power delivered by the voltage source.

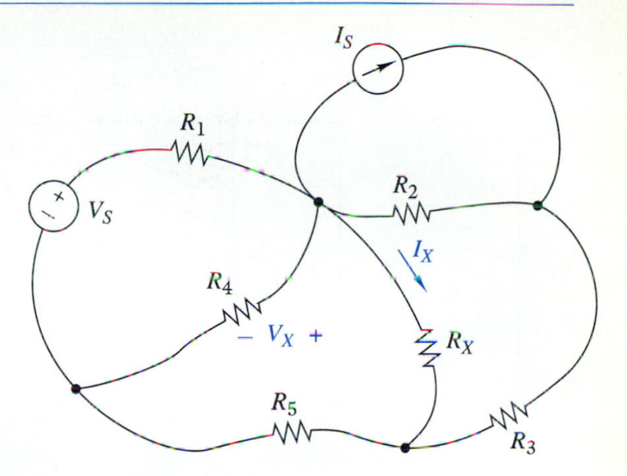

Figure P5-9

5-17 The circuit in Fig. P5-17 is a model of a two-stage amplifier using identical transistors. Formulate node-voltage or mesh-current equations for this circuit. Use these equations to solve for the input-output relationship $V_O = KV_S$ using $R_S = 50$ Ω, $r_\pi = 1$ kΩ, $\beta = 100$ and $R_L = 1$ kΩ.

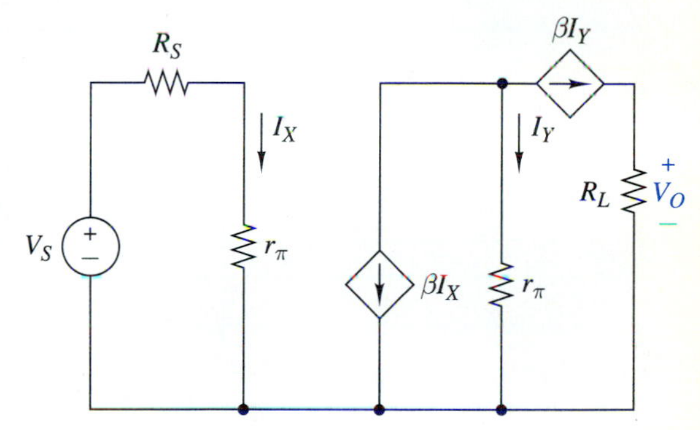

Figure P5-17

5-18 The circuit in Fig. P5-18 is a model of a feedback amplifier using two identical transistors. Formulate either node-voltage or mesh-current equations for this circuit. Use these equations to solve for the input-output relationship $V_O = KV_S$ using $r_\pi = 1$ kΩ, $R_E = 100$ Ω, $R_C = 10$ kΩ, $R_L = 5$ kΩ, $R_F = 5$ kΩ, and $\beta = 50$.

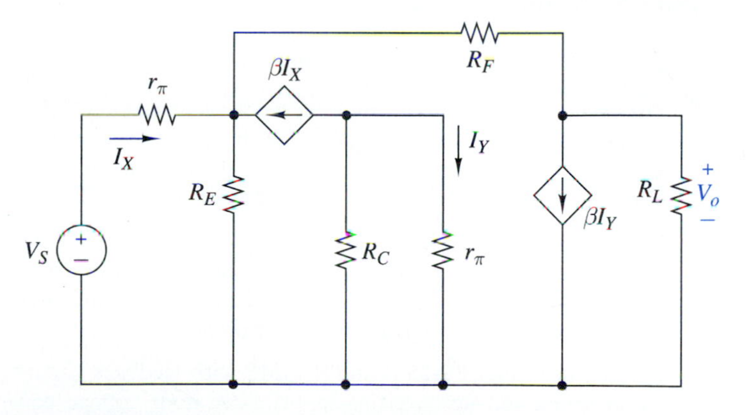

Figure P5-18

5-19 The circuit in Fig. P5-19 is a model of two identical transistors connected as Darlington pair. Formulate either node-voltage or mesh-current equations for this circuit. Use these equations to solve for the input-output relationship $V_O = KV_S$ and input resistance V_S/I_X in terms of the circuit parameters r_π, β, and R_L.

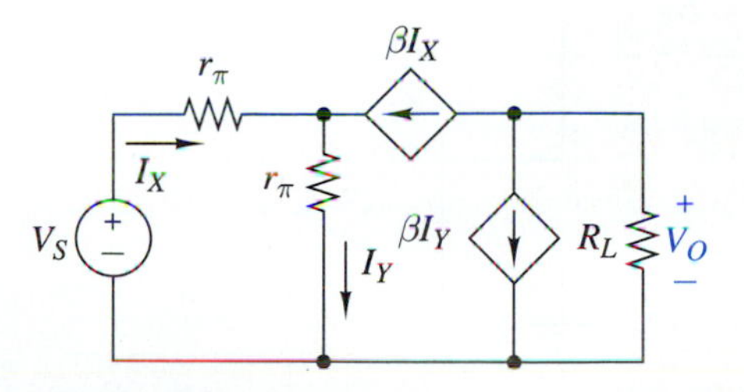

Figure P5-19

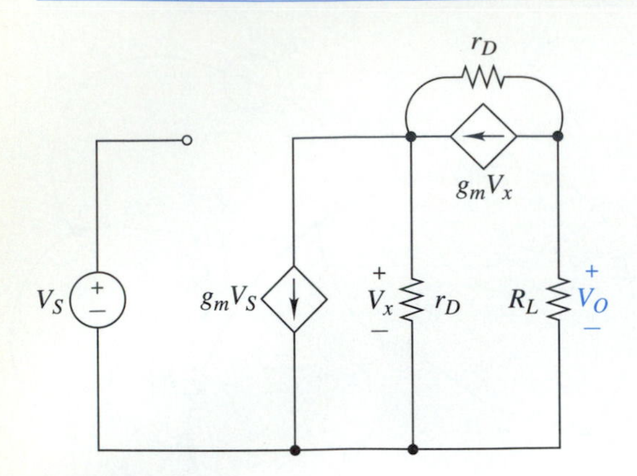

Figure P5-20

5-20 The circuit in Fig. P5-20 is a model of two identical MOS (metal-oxide semiconductor) transistors connected in a circuit called a cascode amplifier. Use node-voltage or mesh-current equations to solve for the amplifier gain V_O/V_S in terms of the circuit parameters r_D, g_M, and R_L.

5-21 The circuit in Fig. P5-21 is a model of a noninverting OP AMP circuit. Use node-voltage equations to solve for the voltage gain V_O/V_S in terms of R_1, R_2, R_X, R_O, and μ. Show that the gain approaches the ideal value $(R_2 + R_1)/R_2$ as μ becomes very large.

5-22 The circuit in Fig. P5-21 is a model of a noninverting OP AMP circuit. Turn the input voltage source V_S off and connect a current source I_{TEST} across the output. Use node-voltage equations to solve for the circuit output resistance V_{TEST}/I_{TEST} in terms of R_1, R_2, R_X, R_O, and μ. Show that the output resistance approaches R_O/μ as μ becomes very large.

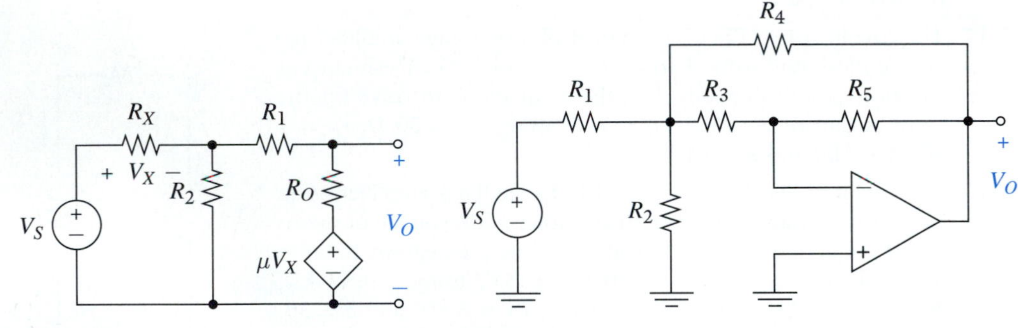

Figure P5-21

Figure P5-23

5-23 The circuit in Fig. P5-23 is a multiple feedback inverting OP AMP circuit. Use node-voltage equations and the ideal OP AMP model to solve for the voltage gain V_O/V_S in terms of circuit parameters.

5-24 The circuit in Fig. P5-24 is the inverting OP AMP configuration with a bridged-T circuit in the feedback path. Use node-voltage equations and the ideal OP AMP model to solve for the circuit voltage gain V_O/V_S in terms of circuit parameters.

5-25 The circuit in Fig. P5-25 is an OP AMP with feedback paths to the inverting and noninverting inputs. Use node-voltage equations and the ideal OP AMP model to solve for the circuit voltage gain V_O/V_S in terms of circuit parameters.

5-26 Use node-voltage equations and the ideal OP AMP model to show that the input resistance of the OP AMP circuit in Fig. P5-26 is $-R_L$. Connect a constant voltage source at the input and show that the circuit delivers power to the source. Where does this power come from?

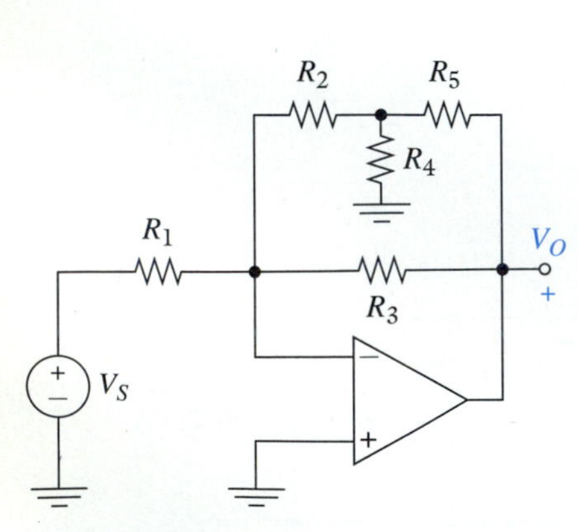

Figure P5-24

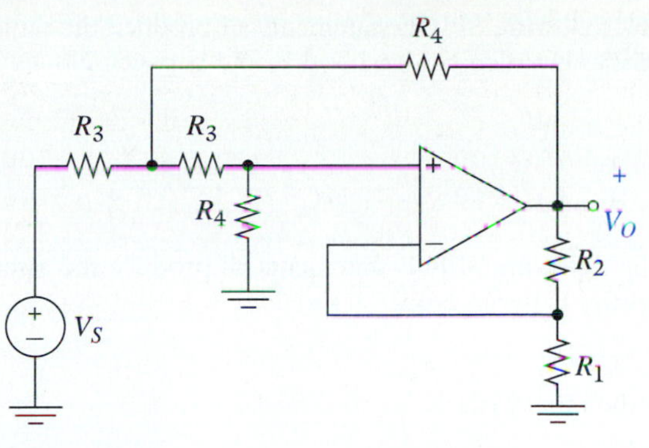

Figure P5-25

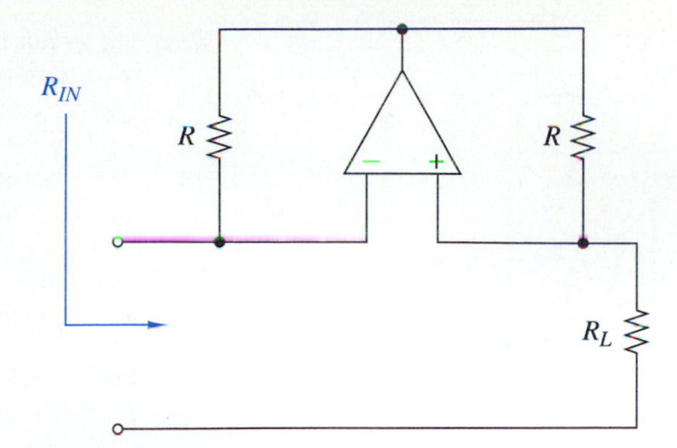

Figure P5-26

5-27 Formulate a complete set of either mesh or node-voltage equations for the circuit in Fig. P5-27. Write the equations in a standard form. Do not solve the equations.

5-28 Formulate a complete set of mesh or node-voltage equations for the circuit in Fig. P5-28. Write the equations in a standard form. Do not solve the equations.

5-29 The node-voltage equations for the circuit in Fig. P5-29 with switch open are

$$(G_1 + G_3 + G_5)\,V_A - G_5 V_C = G_3 V_S$$

$$(G_2 + G_4 + G_5)\,V_C - G_5 V_A = G_4 V_S$$

What are the node-voltage equations with the switch closed?

5-30 The mesh-current equations for the circuit in Fig. P5-29 with the switch open are

$$(R_1 + R_3)\,I_1 - R_3 I_3 = -V_S$$

$$(R_2 + R_4)\,I_2 - R_4 I_3 = V_S$$

$$(R_3 + R_4 + R_5)\,I_3 - R_3 I_1 - R_4 I_2 = 0$$

What are the mesh-current equations with the switch closed?

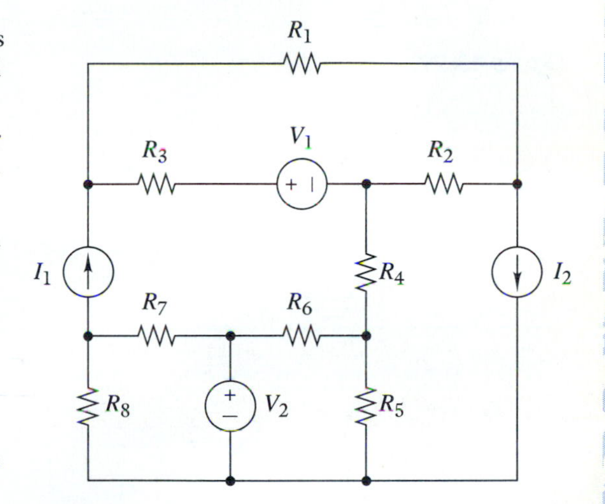

Figure P5-27

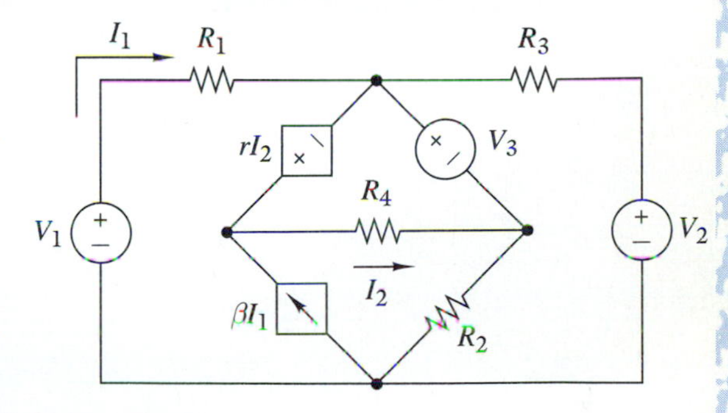

Figure P5-28

ERO 5-2 Computer-Aided Circuit Analysis (Sect. 5-5)

(a) Given a linear resistance circuit, write a SPICE input file for the circuit, or conversely, draw the circuit diagram for a given SPICE input file.

(b) Use a computer-aided analysis program in conjunction with basic circuit principles to calculate circuit parameters or to adjust parameters to achieve specified performance.

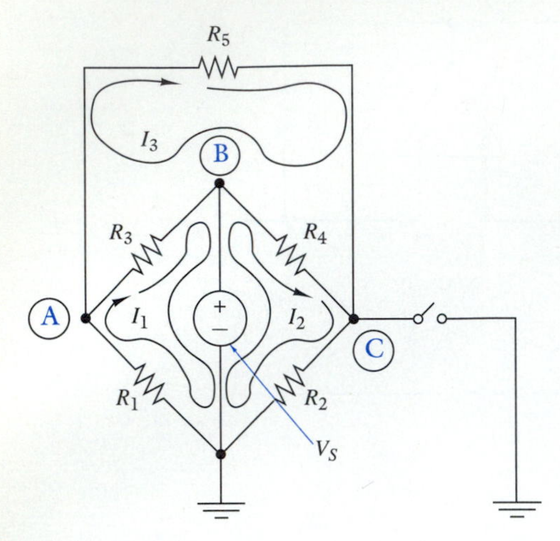

Figure P5-29

5-31 Show that the following SPICE statements all produce the same i-v characteristics between Nodes 1 and 2, and between Nodes 3 and 4:

```
EX  1  2  3  4  +100        EY  2  1  4  3  +100
EZ  1  2  4  3  -100        EW  2  1  3  4  -100
```

5-32 Show that the following SPICE statements all produce the same i-v characteristics between Nodes 1 and 2:

```
R1  1  2  100
G1  1  2  1  2  0.01
```

5-33 Explain why SPICE cannot analyze the circuits described in the following input files. Draw a circuit diagram of the circuit and solve for the voltage across R_L using hand analysis.

```
EXAMPLE 1                   EXAMPLE 2
VS  1  3DC1                 VS  1  0DC1
RS  1  2  10K               RS  1  2  10K
EX  2  3  1  3  100         EX  3  0  2  0  100
RL  2  3  1K                RL  3  0  1K
.END                        .END
```

5-34 Draw the circuit diagram for the SPICE input file given below. An analysis run reports that V(2) = 5.9933 V and I(VA) = 0.2694 mA. What other variables are in the output file and what are their values? Use these results to calculate the input resistance seen by V_S.

```
EXAMPLE CIRCUIT
VS  1  0DC10
R1  1  2  1K
R2  2  0  2K
R3  2  3  3K
R4  3  0  4K
R5  3  4  5K
R6  4  5  6K
VA  5  0DC0
.END
```

5-35 Draw the circuit diagram for the SPICE input file below. An analysis run reports that V(2) = −0.0020 V. What other variables are in the output file and what are their values? What is the input resistance seen by VS and the voltage gain V(3)/VS?

```
EXAMPLE CIRCUIT
VS  1  0DC10
R1  1  2  1K
R2  2  3  2K
```

```
R4  3  0  4K
E1  3  0  2  0  10K
.END
```

5-36 Draw the circuit diagram for the SPICE input file below. An analysis run reports that V(1) = 0.9436 V and I(VB) = 9.251 μA. What other variables are in the output file and what are their values? What is the voltage input resistance seen by IS and the current gain I(VB)/IS?

```
EXAMPLE CIRCUIT
IS  1  0DC1M
R1  1  0  1K
R2  1  0  20K
R4  1  2  1K
VB  2  3DC0
FB  0  3  VB  100
R5  3  4  1K
VA  4  0DC0
.END
```

5-37 Use a computer-aided circuit analysis program to find the value of R_F that produces a voltage gain of −5 in the circuit in P5-37. Use the OP AMP model shown.

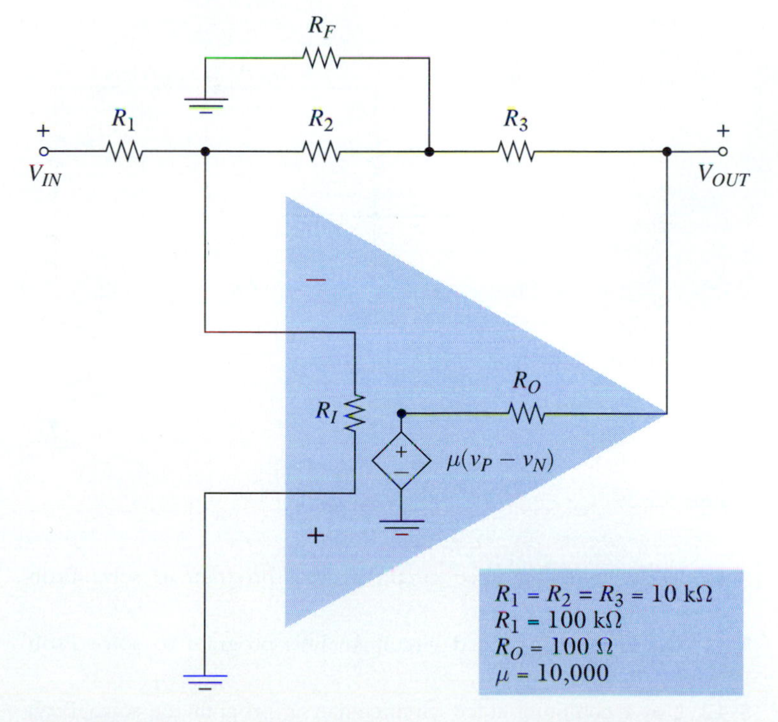

Figure P5-37

5-38 Use a computer-aided circuit analysis program to find the value of R_F in Fig. P5-38 that causes the input resistance seen by I_S to be 50 Ω. Find the current gain I_O/I_S for this value of R_F. Use $\beta = 50$, $r_\pi = 1.1$ kΩ, $R_C = 10$ kΩ, R_E, and $R_L = 100$ Ω.

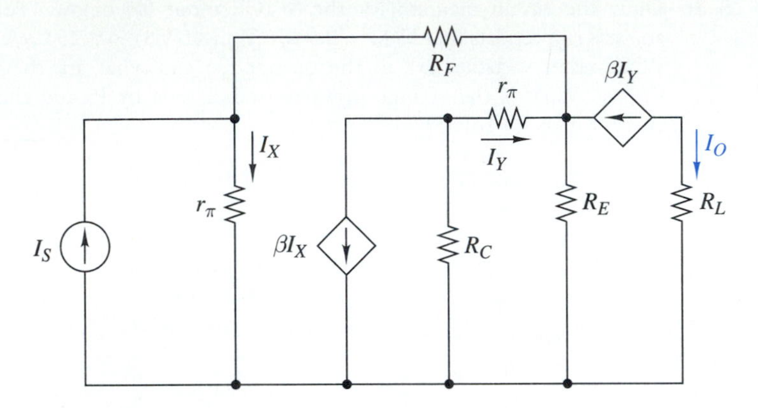

Figure P5-38

5-39 The two identical transistors in Fig. P5-39 are operating in the active mode. Use the large-signal model from Chapter 4 and a computer-aided circuit analysis program to calculate the power dissipated in Q1 and Q2.

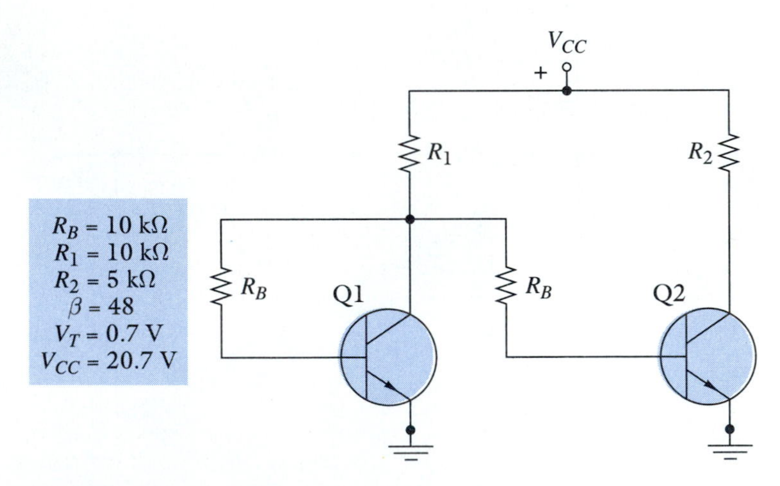

Figure P5-39

5-40 Use a computer-aided circuit analysis program to solve Problem 5-6.

5-41 Use a computer-aided circuit analysis program to solve Problem 5-7.

5-42 Use a computer-aided circuit analysis program to solve Problem 5-8.

5-43 Use a computer-aided circuit analysis program to solve Problem 5-9.

5-44 Use a computer-aided circuit analysis program to solve Problem 5-17. Find the input resistance and output resistance of the circuit.

5-45 Use a computer-aided circuit analysis program to solve Problem 5-18. Find the input and output resistance of the circuit.

ERO 5-3 Comparison of Analysis Methods (Sect. 5-6)

Given a circuit consisting of linear resistors, controlled sources, OP AMPs, and input signal sources, identify and compare analysis techniques for determining specified circuit variables or conditions.

5-46 Discuss the analysis techniques you would use on the circuit in Fig. P5-27 to solve for the following quantities: (a) The voltage across R_4, (b) The current through R_2, (c) The value of R_3 that will extract maximum power from the circuit, (d) The contribution of V_2 to the current through R_2 and (e) The value of V_1 that makes the voltage across R_2 zero.

5-47 Repeat Problem 5-46 using the circuit in Fig. P5-28.

5-48 Discuss the analysis techniques you would use on the circuit in Fig. P5-48 to find the voltage across R_L when the element E_2 is a resistor and element E_1 is (a) an open circuit, (b) an independent current source and (c) a resistor.

5-49 Discuss the analysis techniques you would use on the circuit in Fig. P5-48 to find the current through R_2 when the element E_1 is an independent voltage source and element E_2 is (a) an open circuit, (b) an independent current source, and (c) a resistor.

5-50 Repeat Problem 5-49 when element E_1 is an independent current source.

5-51 Discuss the analysis techniques you would use on the circuit in Fig. P5-48 to find the voltage across R_1 when the element E_1 is an independent voltage source and element E_2 is (a) an independent voltage source, (b) an independent current source, and (c) a short circuit.

5-52 In Fig. P5-48 $R_1 = R_2 = 10\ \Omega$, $V_S = 10$ V and element E_2 is a 1/4 A fuse. Discuss what analysis techniques you would use to find the value of R_L that will blow the fuse when element E_1 is (a) an open circuit, (b) a 10-Ω resistor, and (c) a short circuit.

5-53 Discuss the analysis techniques you would use on the circuit in Fig. P5-48 to find the Thévenin voltage and resistance seen by resistance R_2 when element E_1 is an open circuit and E_2 is (a) an

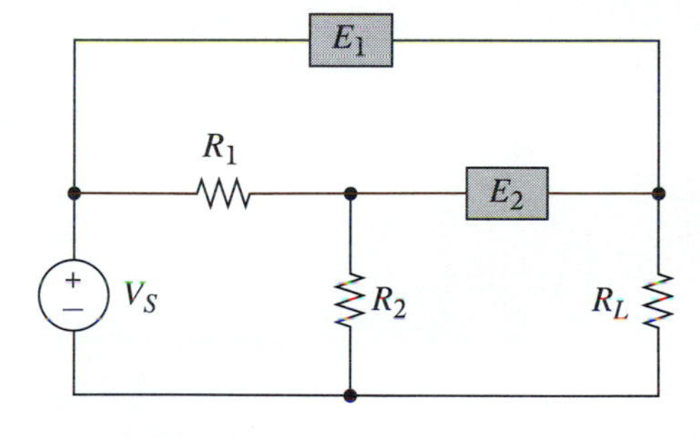

Figure P5-48

independent voltage source, (b) an independent current source, and (c) a resistor.

5-54 Discuss the analysis techniques you would use to find the input resistance in Fig. P5-17.

5-55 Discuss the analysis techniques you would use to find the value of R_F in Fig. P5-18 that produces a voltage gain $V_O/V_S = 50$.

5-56 Discuss the analysis techniques you would use to find the input resistance of the circuit in Fig. P5-19.

CHAPTER INTEGRATING PROBLEMS

5-57 **(A)** Under the principle of duality the dual of a circuit can be constructed by interchanging the following quantities: voltage and current, resistance and conductance, mesh and node. Draw a circuit diagram corresponding to the following set of mesh equations. Then draw the circuit diagram of its dual.

$$4I_1 - I_2 - 2I_3 = -3$$

$$-I_1 + 3I_2 - I_3 = 0$$

$$-2I_1 - I_2 + 3I_3 = 3$$

5-58 **(A)** Figure P5-58 is the large-signal equivalent circuit for an amplifier containing two transistors in the active mode. Derive expressions for V_{CE1}, V_{CE2}, I_{C1}, and I_{C2}. Assume that β is very large.

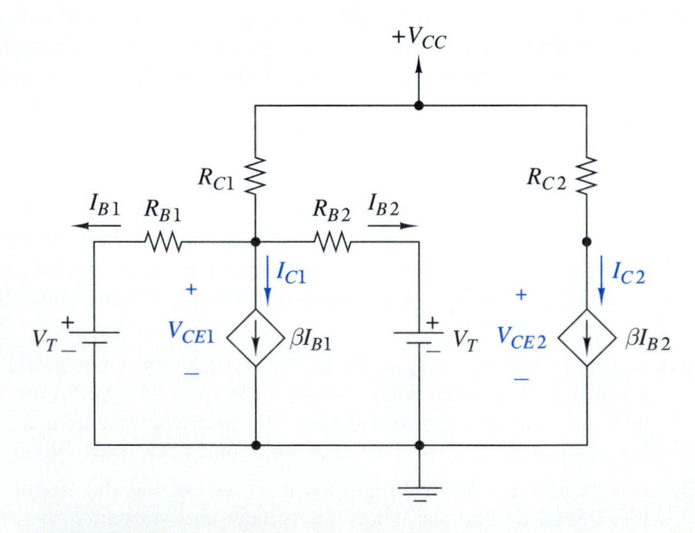

Figure P5-58

5-59 **(A)** The circuit in Fig. P5-59 is an idealized model of a dc power system. (a) Create a supernode enclosing all three sources and use the node-voltage method to show that the three currents are $I_A = 1$ A, $I_B = -1$ A, and $I_C = 0$ A. (b) Use these results to show that all of the device and connection constrains are satisfied. (c) There is no unique way to determine contribution of each voltage source to these currents. To convince yourself of this try using superposition to find the contribution of the 10-V source to each current. (d) Try using SPICE to solve for the currents and see what error message you get.

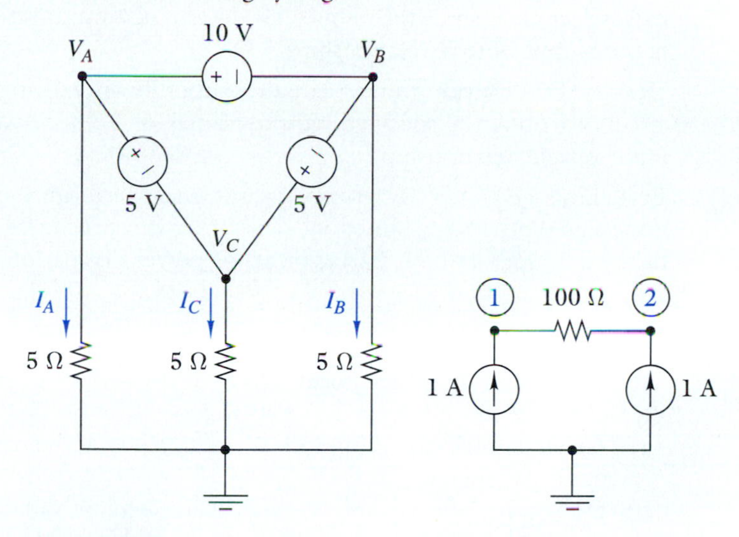

Figure P5-59

Figure P5-60

5-60 **(A)** There are pathological circuits that have irreconcilable conflicts between the device and connection constraints. An example is shown in Fig. P5-60. Try using node analysis to solve for the voltage at Node 1. Try using SPICE and see what error message is generated. Insolvable circuits such as this one are easy to draw on paper but impossible to actually construct in the laboratory. In the laboratory the response is controlled by the parasitic devices not included in our ideal model and they resolve the irreconcilable conflict in a blinding flash. For example, connect a 1000-MΩ resistor (almost an open-circuit) between Node 1 and ground, and then solve for the Node 1 voltage.

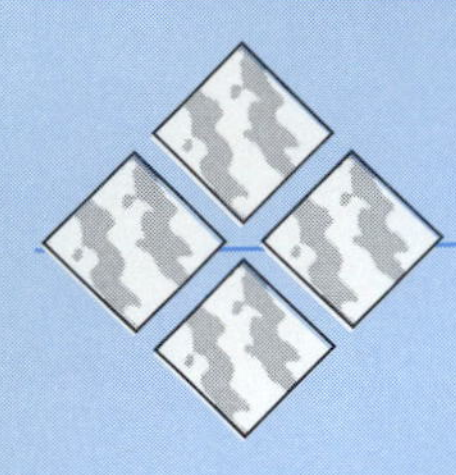

PART ONE

Resistance Circuits COURSE INTEGRATING PROBLEMS

Chapters 1 to 5 Course Objectives

Analysis **(A)**—Given a linear circuit with known input signals, select an analysis technique and determine output signals or input-output relationships.

Design **(D)**—Devise a linear circuit or modify an existing circuit to obtain a specified output signal or a specified input-output relationship.

Evaluation **(E)**—Given several circuits that perform the same signal processing function, rank order the circuits using criteria such as cost, parts count, or power dissipation.

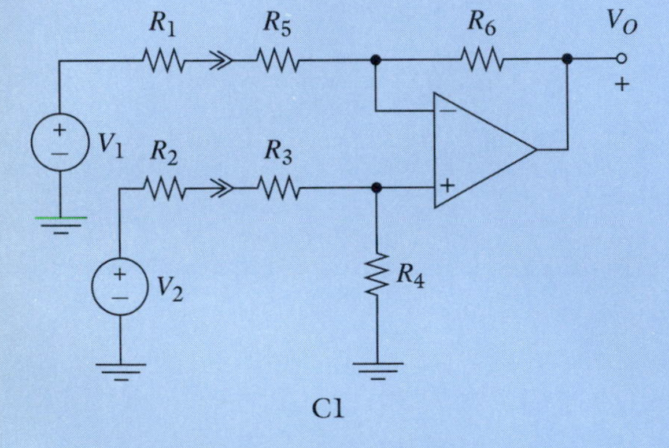

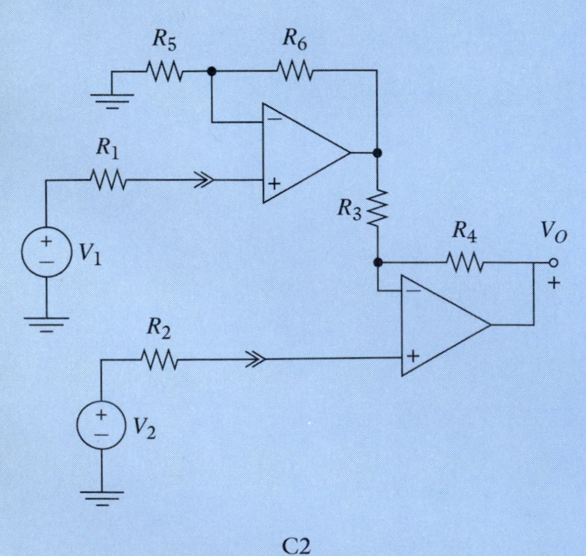

Figure PI-1

I-1 **(A, D, E)** The input-output relationship for the circuit C1 in Fig. PI-1 is of the form $V_O = K_2V_2 + K_1V_1$. If $R_3 = R_6 = 10\ \Omega$ and $R_4 = R_5 = 90\ \Omega$,

(a) **(A)** Determine the constants K_1 and K_2 in terms of circuit parameters.

(b) **(D)** For $R_1 = 1\ \text{k}\Omega$ and $R_2 = 2\ \text{k}\Omega$, select the values of the remaining circuit resistances to achieve $V_0 = 10(V_2 - V_1)$

(c) **(E)** Show that the circuit C2 in the figure meets the design requirement listed in (b) above. Compare these two designs on the basis of the number of devices required and the loading of the two input signal sources.

I-2 **(A, D, E)** Circuit C1 in Fig. PI-2 contains a photoresistor whose resistance varies inversely with the intensity of incident light. In complete darkness its resistance is 10 kΩ. In bright sunlight its resistance is 1 kΩ.

(a) **(A)** At any given light intensity the circuit is linear so its input-output relationship is of the form $V_0 = KV_1$. Determine the constant K in terms of circuit resistances.

(b) **(D)** For $V_1 = +15$ V, select the values of R and R_F so that V_0 is -10 V in bright sunlight and $+10$ V in complete darkness.

(c) **(E)** Show that circuit C2 in the figure meets the design requirement given in (b) above when $R = 4.71\ \text{k}\Omega$ and $R_F = 2.96\ \text{k}\Omega$. Compare these two designs on the basis of the number of devices required and the total power dissipated.

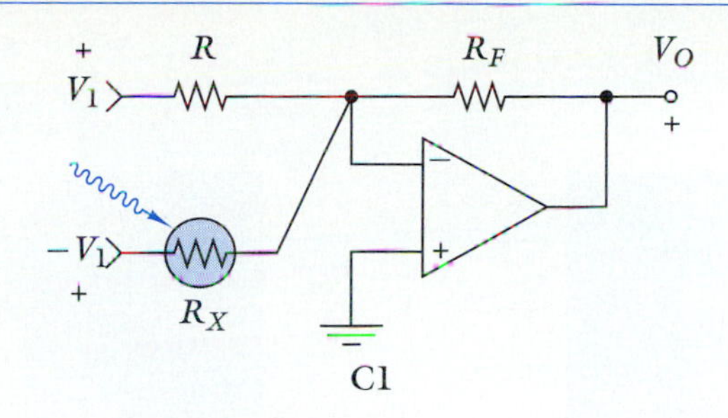

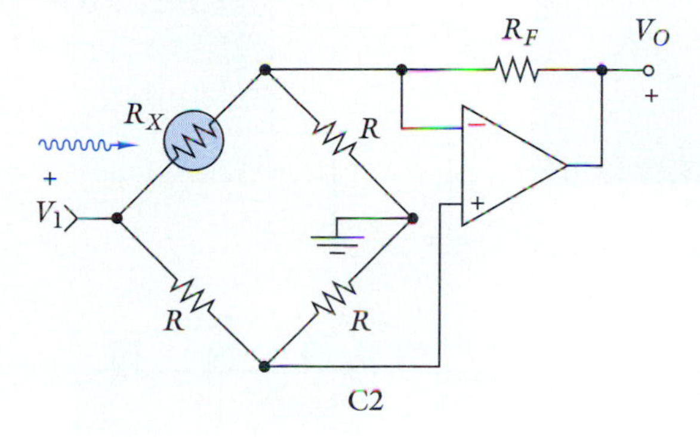

Figure PI-2

I-3 **(A, D)** The source circuit in Fig. PI-3 is to be connected to R_L.

(a) **(A)** Select the value of R_L so that maximum power is delivered to the load.

(b) **(A)** Select the value of R_L so that maximum voltage is delivered to the load.

(c) **(A)** Select the value of R_L so that the load current is 10 mA.

(d) **(D)** For $R_L = 1\ \text{k}\Omega$, design an interface circuit so the load current is 5 mA.

(e) **(D)** Repeat (d) above for a load current of 10 mA.

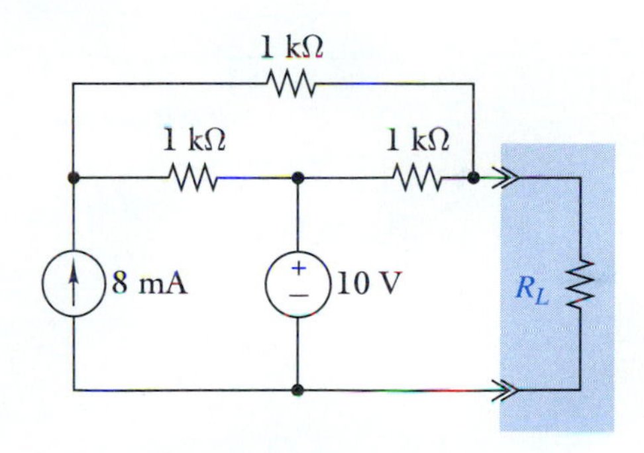

Figure PI-3

I-4 **(A, D)** The source circuit in Fig. PI-4 contains an adjustable potentiometer. The load resistance is fixed. An interface circuit is to be designed so that the power delivered to the load varies between 0 and 20 mW as the potentiometer is adjusted over its full range.

(a) **(A)** Show that the interface circuit must contain an amplifier with a voltage gain of at least 4.

(b) **(D)** Design an interface circuit to meet the objective given above.

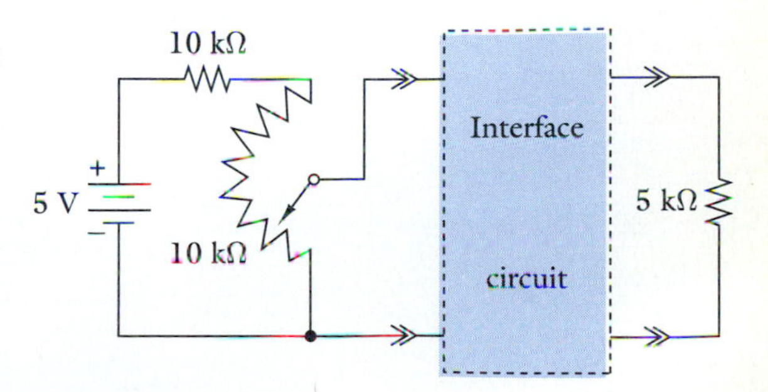

Figure PI-4

I-5 **(D, E)** A system in Fig. PI-5 contains a sensor that requires an excitation voltage of 10 ± 0.1 V. The input

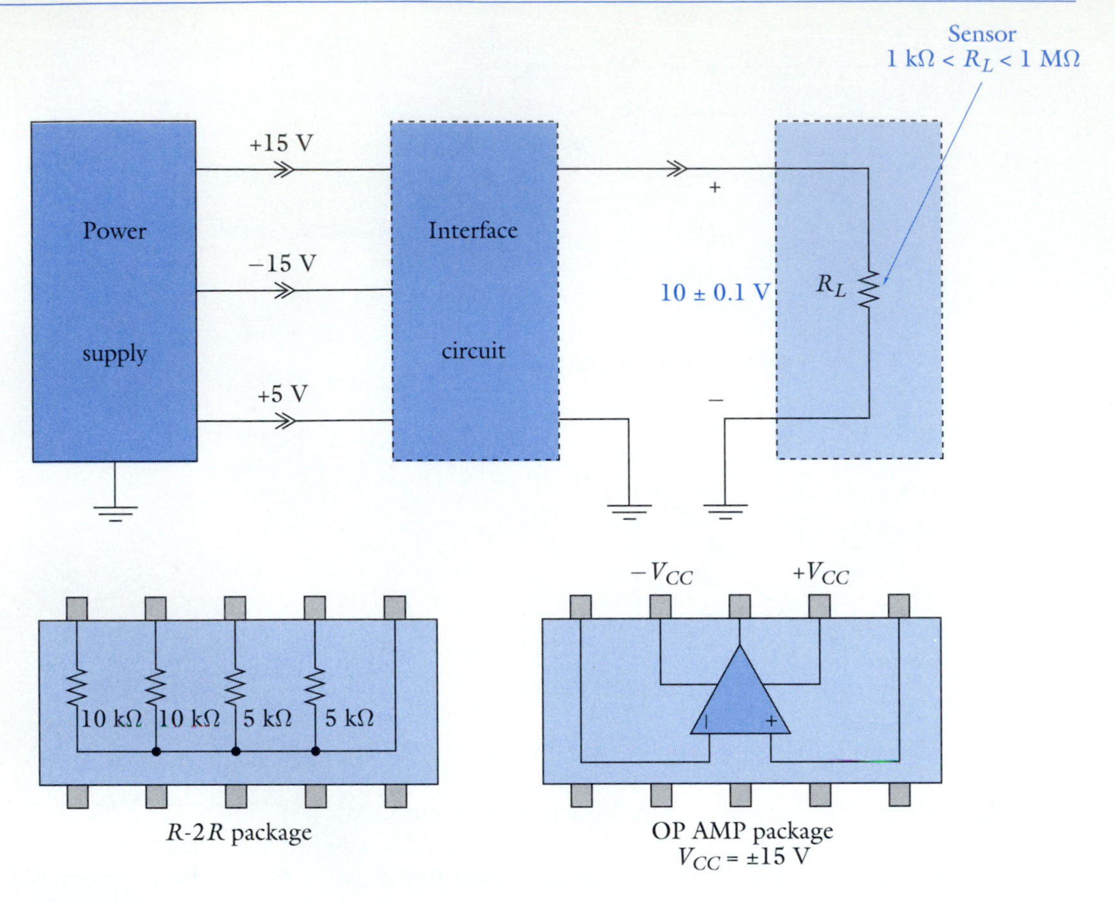

Figure PI-5

resistance of the sensor varies from 1 kΩ to 1 MΩ. The system contains a power supply that provides ±15 V and +5 V. Figure PI-5 shows two integrated circuit packages on the approved parts list for the system. These parts must be used in the design.

(a) **(D)** Design at least two interface circuits using only the two IC packages in the figure. You may use any number of packages but the best design uses the fewest total package count. There are at least a half dozen two-package designs.

(b) **(E)** Compare your two designs in terms of the power dissipated in the *R*-2*R* package. Two-package designs range from 5 to 100 mW.

I-6 **(A)** Figure PI-6 shows a simplified version of a system called a logic analyzer. When the probe is connected to a test point the analyzer detects whether the voltage (relative to

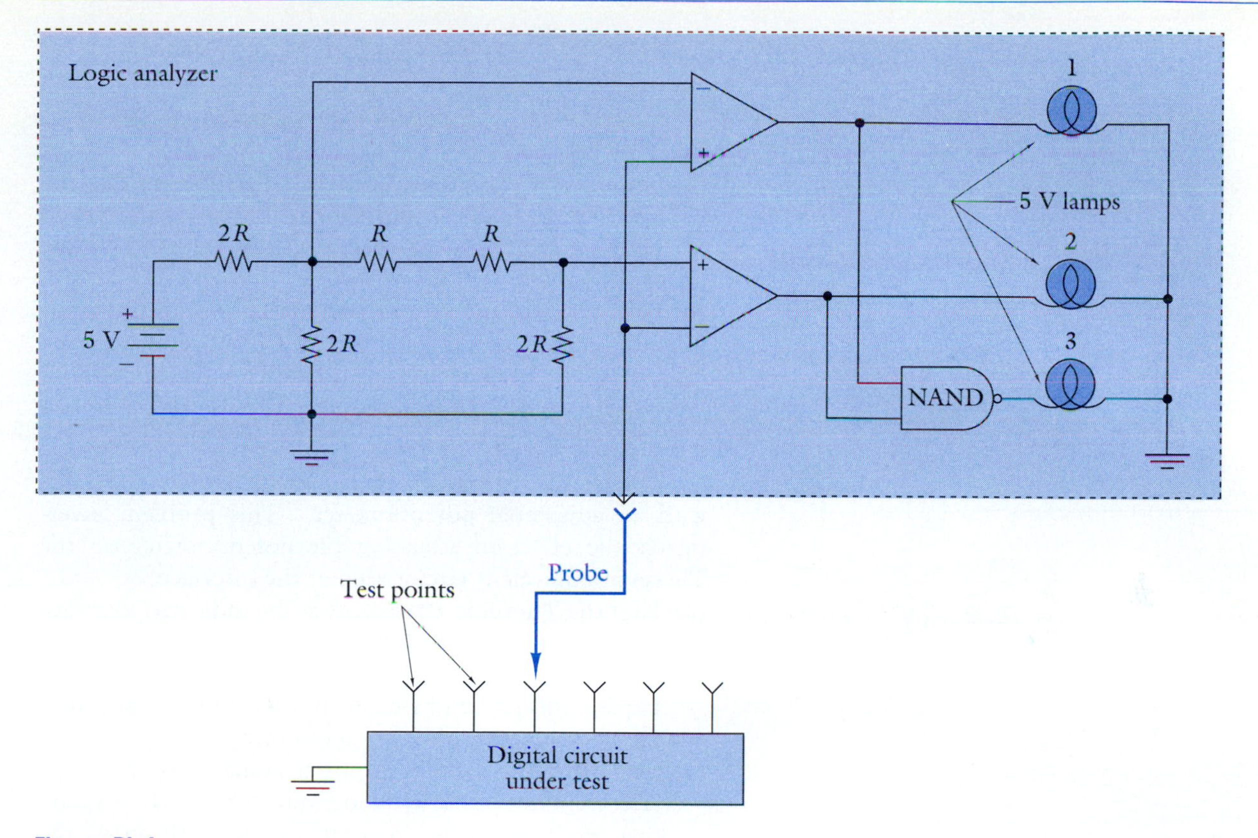

Figure PI-6

ground) is greater than 2 V (logic "one"), less than 1 V (logic "zero"), or between 1 V and 2 V (an ambiguous case indicating a circuit fault). In this circuit the comparator outputs are 5 V if $v_P > v_N$ and 0 V otherwise. The NAND gate is a digital device whose output is 5 V if and only if both of its inputs are zero.

(a) Show that if the test point voltage is greater than 2 V, then light number 1 is on and lights 2 and 3 are off.

(b) Show that if the test point voltage is less than 1 V, then light number 2 is on and lights 1 and 3 are off.

(c) Show that if the test point voltage is between 1 V and 2 V, then light number 3 is on and lights 1 and 2 are off.

(d) Identify three points within the logic analyzer that could be used to perform a self-test of the analyzer.

I-7 **(A)** Circuit C1 in Fig. PI-7 shows a three-terminal source configuration. Circuit C2 is a related three-terminal source with

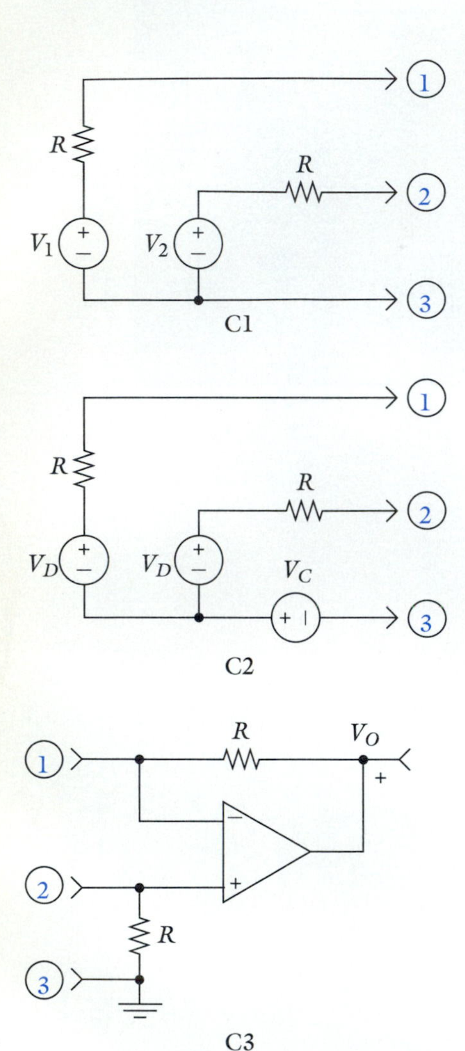

Figure PI-7

$$V_C = \frac{V_1 + V_2}{2} = \text{Common Mode Voltage}$$

$$V_D + \frac{V_1 - V_2}{2} = \text{Differential Mode Voltage}$$

(a) Show that C2 is equivalent to C1 by showing that the open-circuit voltage and short-circuit current "seen" between any two pairs of terminals in C2 are the same as those "seen" in C1.

(b) Now connect the output terminals of C2 to the input terminals of the OP AMP circuit C3 and show that the OP AMP only responds to the differential mode input and rejects the common mode. The use of superposition is suggested.

I-8 **(A)** Figure PI-8 shows an ideal voltage source in parallel with an adjustable potentiometer. This problem investigates the effect of adjusting the potentiometer on the Thévenin equivalent circuit seen at the interface.

(a) Find the Thévenin equivalent at the indicated interface in terms of k, V_O, and R.

(b) Show that $R_{TH} = 0$ when either $k = 0$ or $k = 1$. Explain this result physically in terms of the position of the movable arm of the potentiometer.

(c) Show that the maximum power available at the interface is infinity for $k = 0$ and zero for $k = 1$. Explain this result physically in terms of the position of the movable arm of the potentiometer.

(d) Show that the short-circuit current is $V_O/(kR)$ for $k \neq 1$, but is physically difficult to determine for $k \equiv 1$.

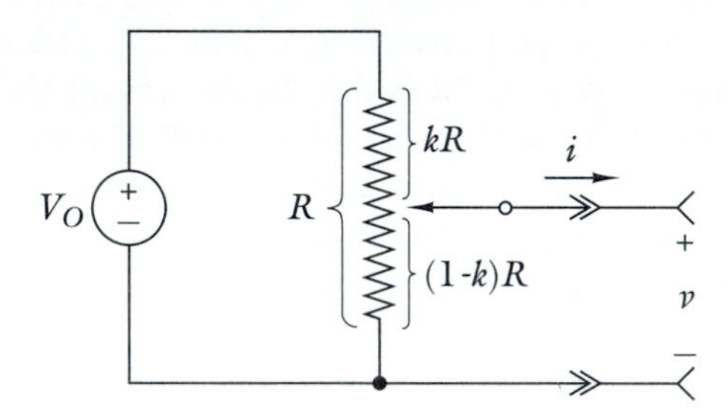

Figure PI-8

I-9 **(A)** Circuit C1 in Fig. PI-9 shows a constant voltage source (V) in series with a source resistance (R) and connected at an interface to a load resistance (R_L). The maximum power transfer theorem states that the maximum power is delivered to the load when the two resistors are matched ($R_L = R$), and that the maximum power is $P_{MAX} = V^2/(4R)$.

(a) Circuit C2 shows the same source connected to a battery whose voltage is V/2. Show that the power delivered to the battery is equal to P_{MAX}.

(b) Since the battery clearly does not "match" the source resistance, can we conclude that circuit C2 disproves the maximum power transfer theorem? Before answering, review the statement of the theorem in Chapter 3, noting carefully what it says and does not say.

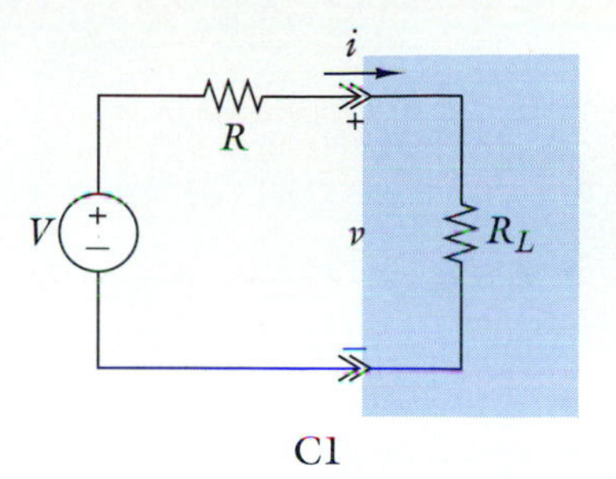

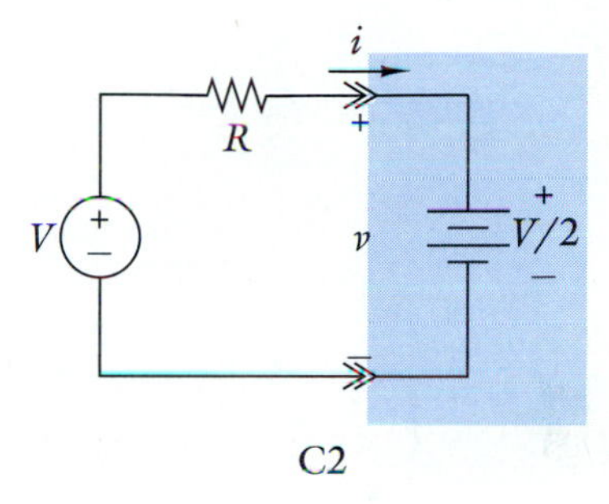

Figure PI-9

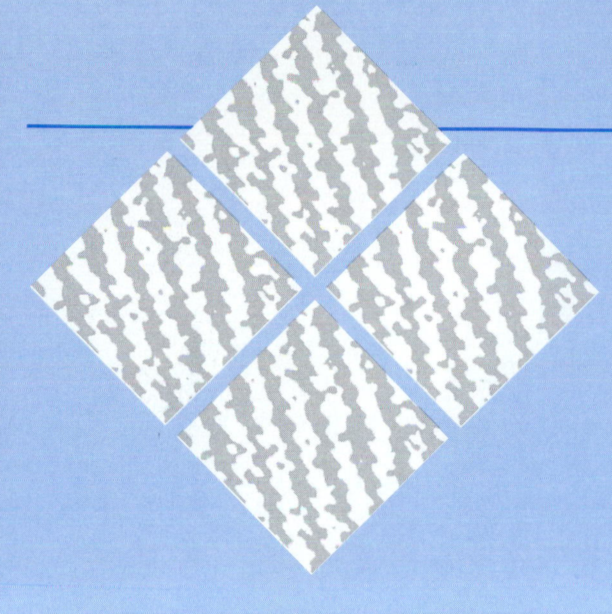

PART TWO

Dynamic Circuits

Course Objectives

- **Analysis (A)**—Given a linear circuit with known input signals, select an analysis technique and determine output signals or input-output relationships.
- **Design (D)**—Devise a linear circuit or modify an existing circuit to obtain a specified output signal or a specified input-output relationship.
- **Evaluation (E)**—Given several circuits that perform the same signal-processing function, rank-order the circuits using criteria such as cost, parts count, or power dissipation.

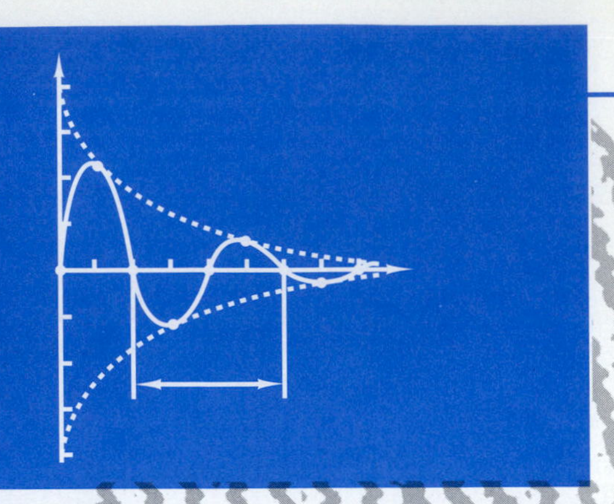

Chapter Six

"Under the sea, under the sea
mark how the telegraph
motions to me.
Under the sea, under the sea
signals are coming along."

James Clerk Maxwell, 1873
Scottish Physicist and Occasional
Humorous Poet

Signal Waveforms

James Clerk Maxwell (1831–1879) is considered the unifying founder of the mathematical theory of electromagnetics. This genial Scotsman often communicated his thoughts to friends and colleagues via whimsical poetry. In the short excerpt given on the opposite page, Maxwell reminds us that the purpose of a communication system (the submarine cable telegraph in this case) is to transmit signals and that those signals must be changing, or *in motion* as he put it.

This chapter marks the beginning of a new phase of our study of circuits. Up to this point we have dealt with resistive circuits in which voltages and currents are constant, for example, +15 V or −3 mA. From this point forward we will be dealing with dynamic circuits in which voltages and currents vary as functions of time. To analyze dynamic circuits we need models for the time-varying signals and models for devices that describe the effects of time-varying signals in circuits. Signal models are introduced in this chapter. The dynamic circuit elements called capacitance and inductance are introduced in the next chapter. Chapters 8, 9, and 10 then show how we combine signal models and device models to analyze dynamic circuits.

This chapter introduces the basic signal models used in the remainder of the book. We do this in a separate chapter so the reader can master these models prior to launching into the complexities of dynamic circuits. We first introduce three key waveforms: the step, exponential, and sinusoid functions. By combining these three models we can build composite waveforms for all signals encountered in this book. Descriptors used to classify and describe waveforms are introduced because they highlight important signal attributes. The concept of a signal spectrum is briefly introduced to begin developing an understanding of the dual nature of signals and circuits. Chapter 6 concludes with a summary of the SPICE syntax that defines the signal models available in this computer-aided circuit analysis program.

6-1 INTRODUCTION

Electrical engineers normally think of a signal as an electrical current $i(t)$, voltage $v(t)$ or power $p(t)$. In any case, the time-variation of the signal is called a waveform. More formally:

A *waveform* is an equation or graph that defines the signal as a function of time.

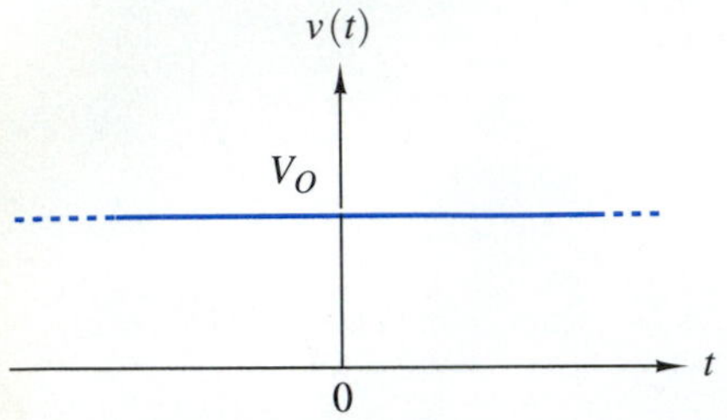

Up to this point our study has been limited to the type of waveform shown in Fig. 6-1. Waveforms that are constant for all time are called *dc signals*. The abbreviation *dc* stands for direct current, but it applies to either voltage or current. Mathematical expressions for a dc voltage $v(t)$ or current $i(t)$ take the form

$$\left.\begin{aligned} v(t) &= V_O \\ i(t) &= I_O \end{aligned}\right\} \text{for } -\infty < t < \infty \tag{6-1}$$

Figure 6-1 A constant or dc waveform.

This equation is only a model. No physical signal can remain constant forever. It is a useful model, however, because it approximates the signals produced by physical devices such as batteries.

There are two matters of notation and convention that must be discussed before continuing. First, quantities that are constant (nontime varying) are usually represented by uppercase letters (V_A, I, T_O) or lowercase letters in the early part of the alphabet (a, b_7, f_O). Time-varying quantities are represented by lowercase letters that are from the end of the alphabet. The time variation is expressly indicated when we write waveforms as $v_1(t)$, $i_A(t)$, or $u(t)$. Time variation is implicit when they are written as v_1, i_A, or u.

Second, in a circuit diagram signal variables are normally accompanied by reference marks ($+$, $-$, $\rightarrow$, or $\leftarrow$). It is important to remember that these reference marks *do not* indicate the polarity of a voltage or the direction of current. The marks provide a baseline for determining the sign of the numerical value of the actual waveform. When the actual voltage polarity or current direction coincides with the reference directions, then the signal has a positive value. When the opposite occurs, the value is negative. Figure 6-2 shows examples of waveforms that assume both positive and negative values, indicating the actual voltage polarity is changing as a function of time.

The waveforms in Fig. 6-2 are examples of signals used in electrical engineering. Since there are many such signals, it may seem that the study of signals involves the uninviting task of compiling a lengthy catalog of waveforms. However, it turns out that a long list is not needed. In fact, we can derive most of the waveforms of interest using just three basic signal models: the

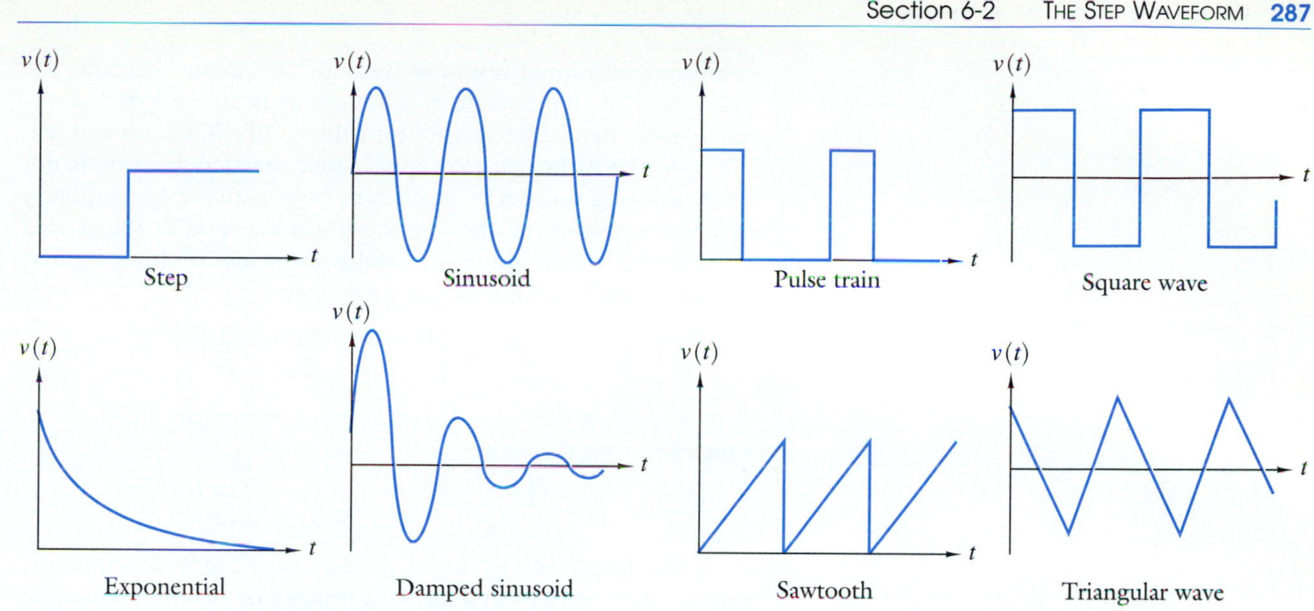

Figure 6-2 Example waveforms.

step, exponential, and sinusoidal functions. The small number of basic signals illustrates why models are so useful to engineers. In reality, waveforms are very complex, but their time variation can be approximated adequately using only a few basic building blocks.

Finally, in this chapter we will generally use voltage $v(t)$ to represent a signal waveform. The reader should remember, however, a signal can be either a voltage $v(t)$ or current $i(t)$.

6-2 THE STEP WAVEFORM

The first basic signal in our catalog is the step waveform. The general step function is based on the *unit step function* defined as

$$u(t) = \begin{cases} 0 & \text{for } t < 0 \\ 1 & \text{for } t \geq 0 \end{cases} \tag{6-2}$$

The step function waveform is equal to zero when its argument t is negative, and is equal to unity when its argument is zero or positive. Mathematically, the function $u(t)$ has a jump discontinuity at $t = 0$.

Strictly speaking, it is impossible to generate a true step function since signal variables like current and voltage cannot transition from one value to another in zero time. Practically speaking, we can generate very good approximations to the step function. What is required is that the transition time be short compared with other response times in the circuit. Actually, the generation of approximate step functions is an everyday occurrence, since people frequently turn things like lights on and off.

On the surface, it may appear that the step function is not a very exciting waveform or, at best, only a source of temporary excitement. However, the step waveform is a versatile signal used to construct a wide range of useful waveforms. Multiplying $u(t)$ by a constant V_A produces the waveform

$$V_A u(t) = \begin{cases} 0 & \text{for } t < 0 \\ V_A & \text{for } t \geq 0 \end{cases} \tag{6-3}$$

Replacing t by $(t - T_S)$ produces a waveform $V_A u(t - T_S)$, which takes on the values

$$V_A u(t - T_S) = \begin{cases} 0 & \text{for } t - T_S < 0 \text{ or } t < T_S \\ V_A & \text{for } t - T_S \geq 0 \text{ or } t \geq T_S \end{cases} \tag{6-4}$$

The ***amplitude*** V_A scales the size of the step discontinuity and the ***time-shift*** parameter T_S advances or delays the time at which the step occurs.

Amplitude and time-shift parameters are required to define the general step function. The amplitude V_A carries the units of volts. The amplitude of step function in electric current is I_A and carries the units of amperes. The constant T_S carries the units of time, usually seconds. The parameters V_A (or I_A) and T_S can be positive, negative, or zero as shown in Fig. 6-3. By combining

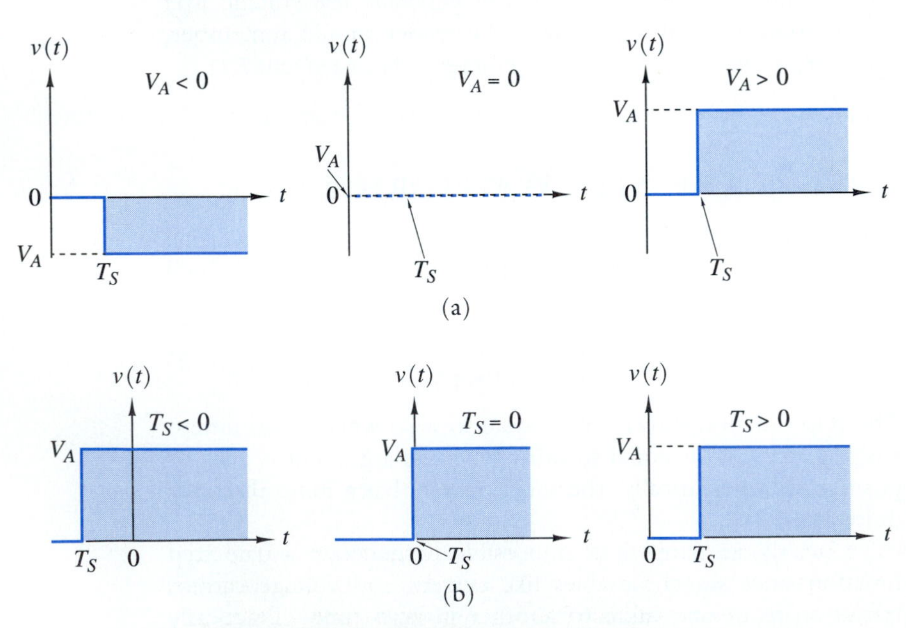

Figure 6-3 Effect of amplitude and time shift. (a) Amplitude. (b) Time shift.

several step functions we can represent a number of important waveforms. One possibility is illustrated in the following example.

EXAMPLE 6-1

Express the waveform in Fig. 6-4(a) in terms of step functions.

SOLUTION:

The amplitude of the pulse jumps to a value of 3 V at $t = 1$ s; therefore, $3\ u(t-1)$ is part of the equation for the waveform. The pulse returns to zero at $t = 3$ s, so an equal and opposite step must occur at $t = 3$ s. Putting these observations together we express the rectangular pulse as

$$v(t) = 3u(t-1) - 3u(t-3)$$

Figure 6-5(b) shows how the two step functions combine to produce the given rectangular pulse.

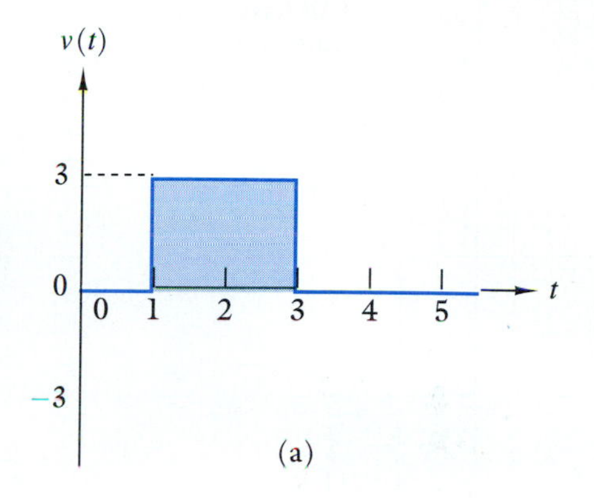

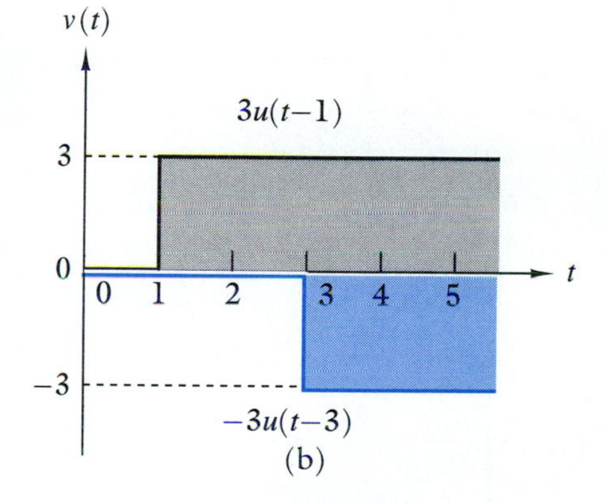

Figure 6-4

The Impulse Function

The generalization of Example 6-1 is the waveform

$$v(t) = V_A\left[u(t-T_1) - u(t-T_2)\right]$$

This waveform is a rectangular pulse of amplitude V_A, that turns on at $t = T_1$ and off at $t = T_2$. The pulse train and square wave signals in Fig. 6-2 can be generated by a series of these pulses. Pulses that turn on at some time T_1 and off at some later time T_2 are sometimes called *gating functions* because they are used in conjunction with electronic switches to enable or inhibit the passage of another signal.

A rectangular pulse centered on $t = 0$ is written in terms of step functions as

$$v_1(t) = \frac{1}{T}\left[u\left(t+\frac{T}{2}\right) - u\left(t-\frac{T}{2}\right)\right] \qquad (6\text{-}5)$$

The pulse in Eq. (6-5) is zero everywhere except in the range $-T/2 \leq t \leq T/2$ where its amplitude is $1/T$. The area under the pulse is one because its amplitude is inversely proportional to its duration. As shown in Fig. 6-5(a), the pulse becomes narrower and higher as T decreases but maintains its unit area. In the

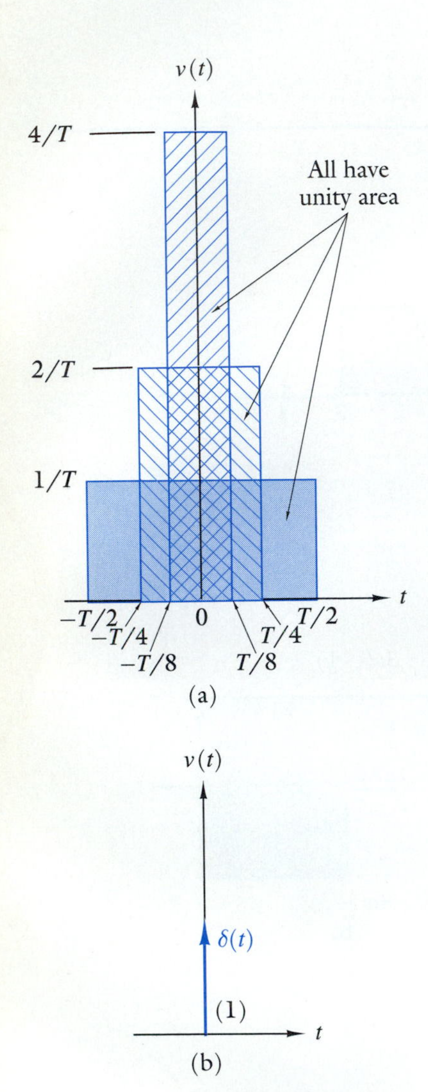

Figure 6-5 Rectangular pulse and impulse waveforms.

limit as $T \to 0$ the amplitude approaches infinity but the area remains unity. The function obtained in the limit is called a ***unit impulse***, symbolized as $\delta(t)$. The graphical representation of $\delta(t)$ is shown in Fig. 6-5(b). The impulse is an idealized model of a large-amplitude, short-duration pulse.

A formal definition of the unit impulse is

$$\delta(t) = 0 \text{ for } t \neq 0 \qquad \text{and} \qquad \int_{-\infty}^{t} \delta(x)dx = u(t) \tag{6-6}$$

The first condition says the impulse is zero everywhere except at $t = 0$. The second condition suggests that the impulse is the derivative of a step function.

$$\delta(t) = \frac{du(t)}{dt} \tag{6-7}$$

The conclusion in Eq. (6-7) cannot be justified using elementary mathematics, since the function $u(t)$ has a discontinuity at $t = 0$ and its derivative at that point does not exist in the usual sense. However, the concept can be justified using limiting conditions on continuous functions as discussed in texts on signals and systems.[1] Accordingly, engineers usually ignore mathematical rigor at this point and think of the impulse as the derivative of a step function.

The amplitude of an impulse is infinite, so its strength is defined in terms of its area. An impulse of strength K is denoted $K\delta(t)$, where K is the **area** under the impulse. In the graphical representation of the impulse the value of K is written in parentheses beside the arrow as shown in Fig. 6-5(b).

EXAMPLE 6-2

Calculate and sketch the derivative of the pulse in Fig. 6-6(a).

SOLUTION:

In Example 6-1 the pulse waveform was written as

$$v(t) = 3u(t-1) - 3u(t-3) \text{ V}$$

Using the derivative property of the step function we write

$$\frac{dv(t)}{dt} = 3\delta(t-1) - 3\delta(t-3)$$

[1]For example, see Alan V. Oppenheim and Allan S. Willsky, *Signals and Systems Analysis*, Prentice Hall, Englewood Cliffs, NJ, 1983, pp. 22–23.

The derivative waveform consists of a positive going impulse at $t = 1$ s and a negative going impulse at $t = 3$ s. Figure 6-6(b) shows how the impulse train is represented graphically. The waveform $v(t) = 3u(t-1) - 3u(t-3)$ has the units of volts. Consequently the scale factors $K = \pm 3$ have the units of volts and the impulse $3\delta(t-1)$ has an area (not amplitude) of 3 V. The area of the impulse $-3\delta(t)$ is -3 V.

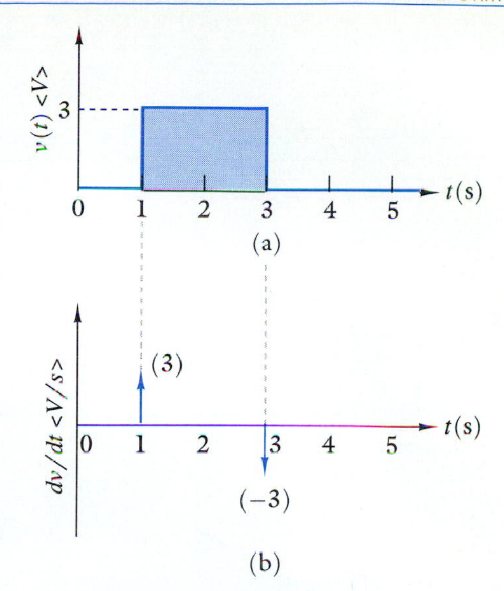

Figure 6-6

The Ramp Function

The *unit ramp* is defined as the integral of a step function.

$$r(t) = \int_{-\infty}^{t} u(x)dx = tu(t) \tag{6-8}$$

The unit ramp waveform $r(t)$ in Fig. 6-7(a) is zero for $T \leq 0$ and is equal to t for $t > 0$. Notice that the slope of $r(t)$ is one. The general ramp waveform shown in Fig. 6-7(b) is written $Kr(t - T_S)$. The general ramp is zero for $t \leq T_S$ and equal to $K(t - T_S)$ for $t > 0$. The scale factor K defines the slope of the ramp for $t > 0$. By adding a series of ramps we can create the triangular and sawtooth waveforms shown in Fig. 6-2.

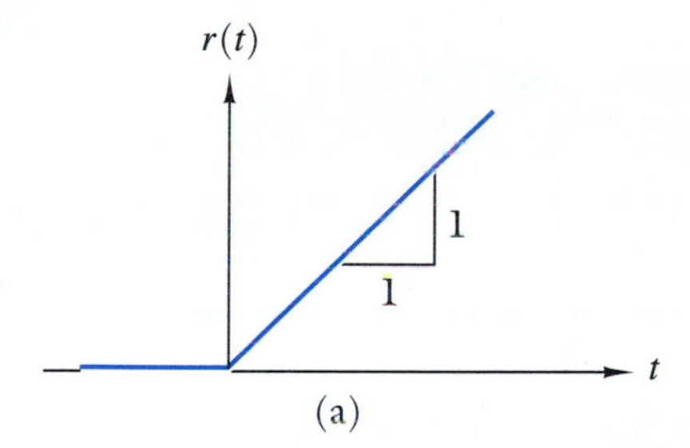

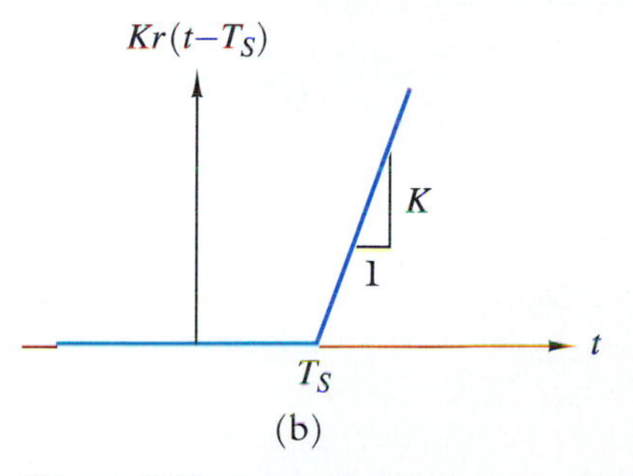

Figure 6-7 (a) Unit ramp waveform. (b) General ramp waveform.

Singularity Functions

The impulse, step, and ramp form a triad of related signals that are referred to as *singularity functions*. They are related by integration as

$$\begin{aligned} u(t) &= \int_{-\infty}^{t} \delta(x)dx \\ r(t) &= \int_{-\infty}^{t} u(x)dx \end{aligned} \tag{6-9}$$

or by differentiation as

$$\begin{aligned} \delta(t) &= \frac{du(t)}{dt} \\ u(t) &= \frac{dr(t)}{dt} \end{aligned} \tag{6-10}$$

These signals are used to generate other waveforms and as test inputs to linear systems to characterize their responses. When applying the singularity functions in circuit analysis it is important to remember that $u(t)$ is a dimensionless function. But Eqs. (6-9) and (6-10) point out that $\delta(t)$ carries the units of seconds^{-1} and $r(t)$ carries units of seconds.

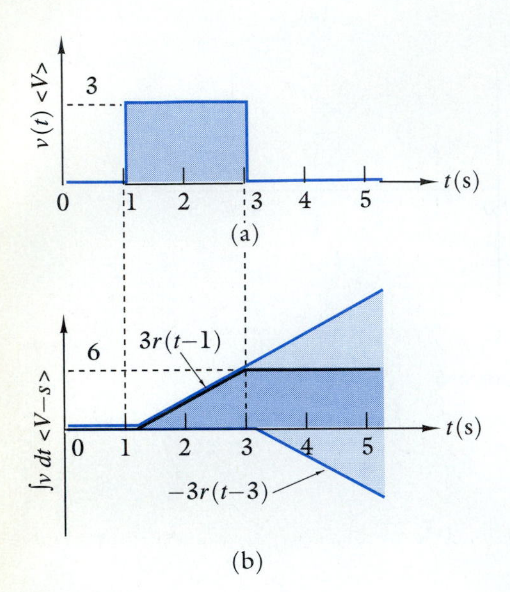

Figure 6-8

EXAMPLE 6-3

Derive the waveform for the integral of the pulse in Fig. 6-8(a).

SOLUTION:

In Example 6-1 the pulse waveform was written as

$$v(t) = 3u(t-1) - 3u(t-3) \text{ V}$$

Using the integration property of the step function we write

$$\int_{-\infty}^{t} v(x)dx = 3(t-1)u(t-1) - 3(t-3)u(t-3)$$

$$= 3r(t-1) - 3r(t-3)$$

The integral is zero for $t < 1$ s. For $1 < t < 3$ the waveform is $3(t-1)$. For $t > 3$ it is $3(t-1) - 3(t-3) = 6$. These two ramps produce the pulse integral as shown in Fig. 6-8(b). The waveform $v(t) = 3u(t-1) - 3u(t-2)$ carries the units of volts. Consequently the scale factor $K = 3$ has the units of volts so the ramp $3r(t-1)$ has a slope (not the amplitude) of 3 V. The ramp $-3r(t-3)$ has a slope of -3 V.

Figure 6-9(a)

EXAMPLE 6-4

Figure 6-9(a) shows an ideal electronic switch whose input is a ramp $2r(t)$, where the amplitude factor $A = 2$ has the units of V/s. Find the switch output $v_O(t)$ when the gate function in Example 6-1 is applied to the control terminal (G) of the switch.

SOLUTION:

In Example 6-1 the gate function was written as

$$v_G(t) = 3u(t-1) - 3u(t-3) \text{ V}$$

The gate function turns the switch turn on at $t = 1$ s and off at $t = 3$ s. The output voltage of the switch is

$$v_O(t) = \begin{cases} 0 & t < 1 \\ 2t & 1 \leq t \leq 3 \\ 0 & 3 < t \end{cases}$$

Only the portion of the input waveform within the gate interval appears at the output. Figures 6-9 (b), (c), and (d) show how the gate function $v_G(t)$ controls the passage of the input signal through the electronic switch.

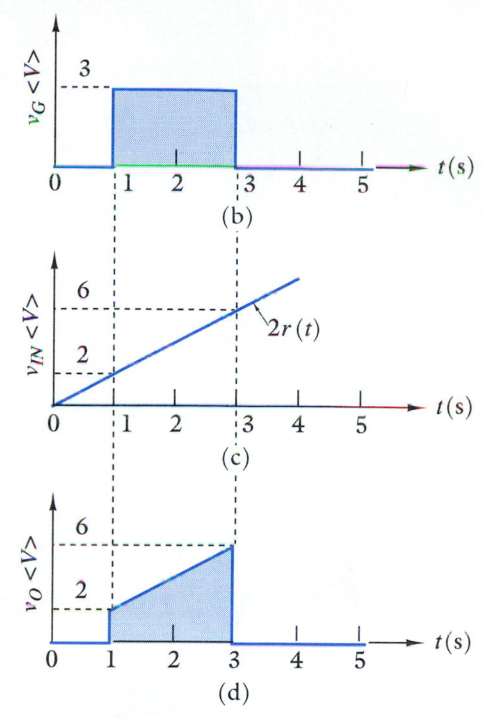

Figure 6-9(b–d)

EXERCISE 6-1

Express the following signals in terms of singularity functions.

(a) $v_1(t) = \begin{cases} 0 & t < 2 \\ 4 & 2 \leq t \leq 4 \\ -4 & 4 < t \end{cases}$

(b) $v_2(t) = \begin{cases} 0 & t < 2 \\ 4 & 2 \leq t \leq 4 \\ -2t + 12 & 4 < t \end{cases}$

(c) $v_3(t) = \int_{-\infty}^{t} v_1(x)dx$

(d) $v_4(t) = \dfrac{dv_2(t)}{dt}$

ANSWERS:

(a) $v_1(t) = 4\ u(t-2) - 8\ u(t-4)$
(b) $v_2(t) = 4\ u(t-2) - 2\ r(t-4)$
(c) $v_3(t) = 4\ r(t-2) - 8\ r(t-4)$
(d) $v_4(t) = 4\ \delta(t-2) - 2\ u(t-4)$

EXERCISE 6-2

(a) Write an expression for a rectangular pulse with an amplitude of 15 V that begins at $t = -5$ s and ends at $t = 10$ s.
(b) Write an expression for the derivative of the pulse defined in (a).
(c) Write an expression for the integral of the pulse in (a).

ANSWERS:

(a) $15[u(t+5) - u(t-10)]$ V
(b) $15[\delta(t+5) - \delta(t-10)]$
(c) $15(t+5)u(t+5) - 15(t-10)u(t-10) = 15[r(t+5) - r(t-10)]$

6-3 THE EXPONENTIAL WAVEFORM

The *exponential signal* is a step function whose amplitude gradually decays to zero. The equation for this waveform is

$$v(t) = \left[V_A e^{-t/T_C}\right] u(t) \tag{6-11}$$

A graph of $v(t)$ versus t/T_c is shown in Fig. 6-10. The exponential starts out like a step function. It is zero for $t < 0$ and jumps to a maximum amplitude of V_A at $t = 0$. Thereafter it monotonically decays toward zero as time marches on. The two parameters that define the waveform are the *amplitude* V_A (in volts) and the *time constant* T_C (in seconds). The amplitude of a current exponential would be written I_A and carry the units

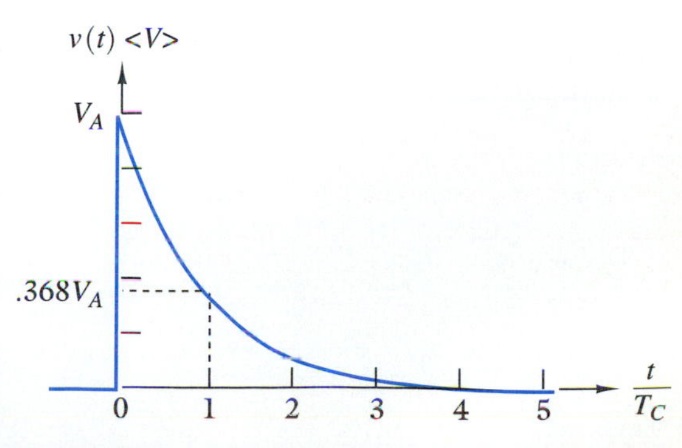

Figure 6-10 The exponential waveform.

of amperes. Figure 6-11 shows how the exponential waveform changes for different values of the amplitude and time constant.

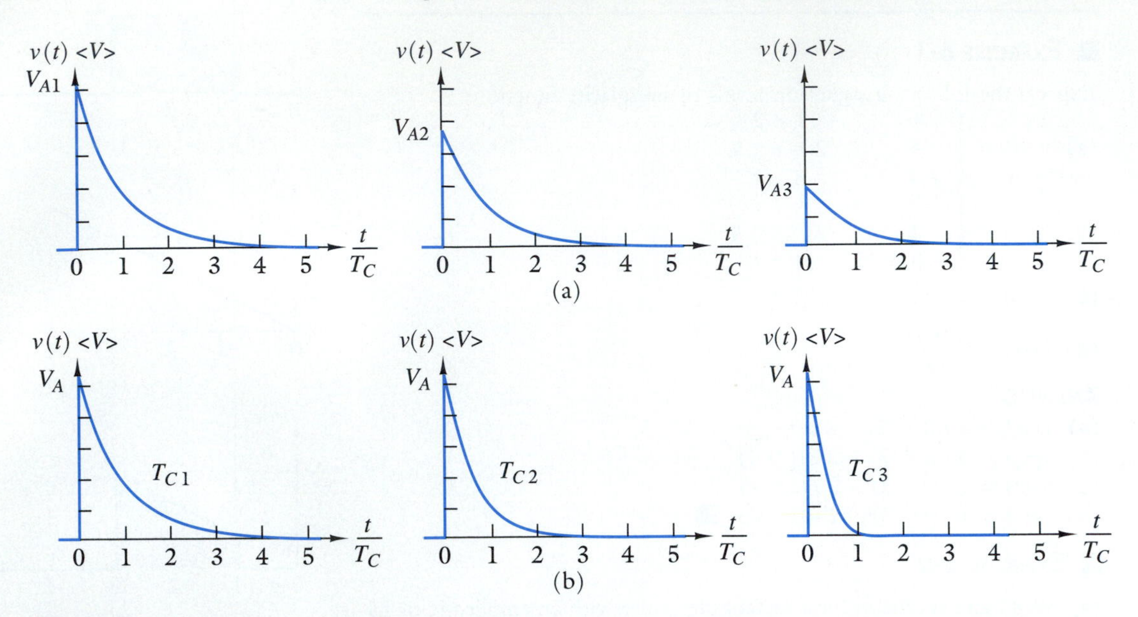

Figure 6-11 Effect of amplitude and time constant. (a) Amplitude. (b) Time constant.

The time constant is of special interest, since it determines the rate at which the waveform decays to zero. An exponential decays to about 37% of its initial amplitude $v(0) = V_A$ in one time constant, because at $t = T_C$, $v(T_C) = V_A e^{-1}$ or approximately $0.368 \times V_A$. At $t = 5T_C$, the value of the waveform is $V_A e^{-5}$ or approximately $0.00674\ V_A$. An exponential signal decays to less than 1% of its initial amplitude in a time span of five time constants. In theory an exponential endures forever, but practically speaking after about $5T_C$ the waveform amplitude becomes negligibly small. For this reason we define the *duration* of an exponential waveform to be $5T_C$.

EXAMPLE 6-5

Sketch the waveform $v(t) = [-17\ e^{-100t}]u(t)$ V.

SOLUTION:

We recognize that $V_A = -17$ V and $T_C = 1/100$ s or 10 ms. The minimum value of $v(t)$ is $v(0) = -17$ V and the maximum value is approximately 0 V as t approaches $5T_C = 50$ ms. These observations define appropriate scales for plotting the waveform. The values of $v(t)$ are calculated and listed below and result in the plot of Fig. 6-12.

$$v(0.00) = -17e^{-0} = -17.0 \text{ V}$$
$$v(0.01) = -17e^{-1} = -6.25 \text{ V}$$
$$v(0.02) = -17e^{-2} = -2.30 \text{ V}$$
$$v(0.03) = -17e^{-3} = -0.846 \text{ V}$$
$$v(0.04) = -17e^{-4} = -0.311 \text{ V}$$
$$v(0.05) = -17e^{-5} = -0.114 \text{ V}$$

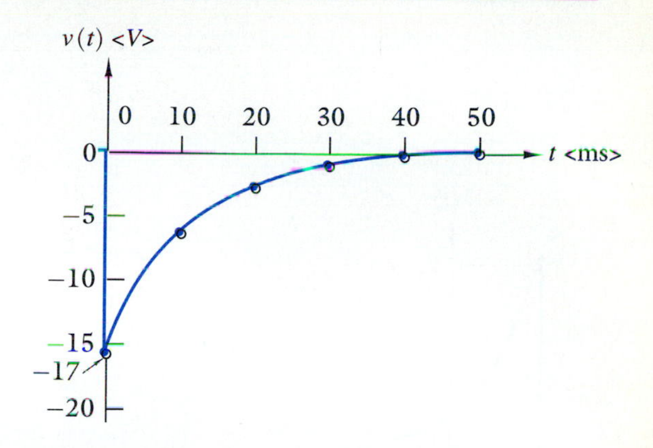

Figure 6-12

Properties of Exponential Waveforms

The *decrement property* describes the decay rate of an exponential signal. For $t > 0$ the exponential waveform is given by

$$v(t) = V_A e^{-t/T_C} \tag{6-12}$$

The step function can be omitted, since it is unity for $t > 0$. At time $t + \Delta t$ the amplitude is

$$v(t + \Delta t) = V_A e^{(t+\Delta t)/T_C} = V_A e^{-t/T_C} e^{-\Delta t/T_C} \tag{6-13}$$

The ratio of these two amplitudes is

$$\frac{v(t + \Delta t)}{v(t)} = \frac{V_A e^{-t/T_C} e^{-\Delta t/T_C}}{V_A e^{-t/T_C}} = e^{-\Delta t/T_C} \tag{6-14}$$

The decrement ratio is independent of amplitude and time. In any fixed time period Δt, the fractional decrease depends only on the time constant. The decrement property states that the same percentage decay occurs in equal time intervals.

The slope of the exponential waveform (for $t > 0$) is found by differentiating Eq. (6-12) with respect to time:

$$\frac{dv(t)}{dt} = -\frac{V_A}{T_C} e^{-t/T_C} = -\frac{v(t)}{T_C} \tag{6-15}$$

The *slope property* states that the time rate of change of the exponential waveform is inversely proportional to the time constant. Small time constants lead to large slopes or rapid decays, while large time constants produce shallow slopes and long decay times.

Equation (6-15) shows that $v(t)$ and dv/dt are both of the form e^{-t/T_C}. Differentiating an exponential signal produces another exponential signal with the same time constant. As a

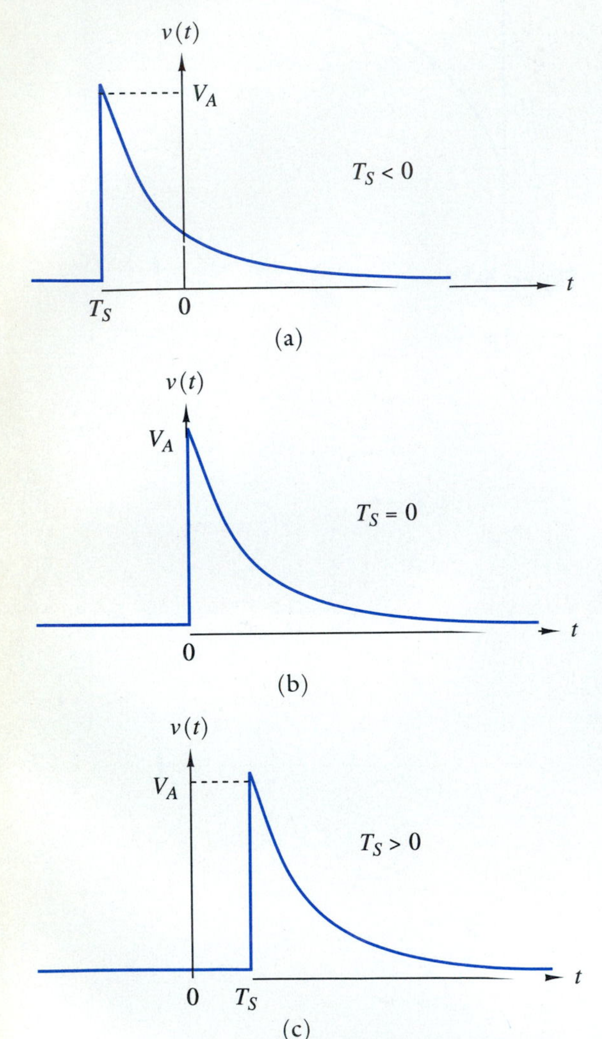

Figure 6-13 Effect of time shift on the exponential waveform.

result, Eq. (6-15) can be rearranged as

$$T_C \frac{dv(t)}{dt} + v(t) = 0 \tag{6-16}$$

That is, $v(t)$ plus T_C times its derivative add to zero. A corollary of the slope property is that an exponential waveform is a solution of the first-order linear differential equation in Eq. (6-16). We make use of this fact in Chapter 8.

The time-shifted exponential waveform is obtained by replacing t in Eq. (6-11) by $t - T_S$. The general exponential waveform is written as

$$v(t) = \left[V_A e^{-(t-T_S)/T_C}\right] u(t - T_S) \tag{6-17}$$

where T_S is the *time-shift* parameter for the waveform. Figure 6-13 shows exponential waveforms with the same amplitude and time constant but different values of T_S. Time shifting translates the waveform to the left or right depending on whether T_S is negative or positive. *Caution:* The factor $t - T_S$ must appear in both the argument of the step function **and** the exponential as shown in Eq. (6-17).

EXERCISE 6-3

(a) An exponential waveform has $v(0) = 1.2$ V and $v(3) = 0.5$ V. What are V_A and T_C for this waveform?
(b) An exponential waveform has $v(0) = 5$ V and $v(2) = 1.25$ V. What are values of $v(t)$ at $t = 1$ and $t = 4$?
(c) An exponential waveform has $v(0) = 5$ and an initial ($t = 0$) slope of -25 V/s. What are V_A and T_C for this waveform?
(d) An exponential waveform decays to 10% of its initial value in 3 ms. What is T_C for this waveform?
(e) A waveform has $v(2) = 4$ V, $v(6) = 1$ V and $v(10) = 0.5$ V. Is it an exponential waveform?

ANSWERS:
(a) $V_A = 1.2$ V, $T_C = 3.43$ s
(b) $v(1) = 2.5$ V, $v(4) = 0.3125$ V
(c) $V_A = 5$ V, $T_c = 200$ ms
(d) $T_C = 1.303$ ms
(e) No, it violates the decrement property.

EXERCISE 6-4

Find the amplitude and time constant of the following exponential signals.
(a) $v_1(t) = [-15e^{-1000t}]u(t)$ V
(b) $v_2(t) = [+12e^{-t/10}]u(t)$ mV
(c) $i_3(t) = [15e^{500t}]u(-t)$ mA
(d) $i_4(t) = [4e^{-200(t-100)}u(t - 100)]$ A

ANSWERS:
(a) $V_A = -15$ V, $T_C = 1$ ms
(b) $V_A = 12$ mV, $T_C = 10$ s
(c) $I_A = 15$ mA, $T_C = 2$ ms
(d) $V_A = 4$ A, $T_C = 5$ ms

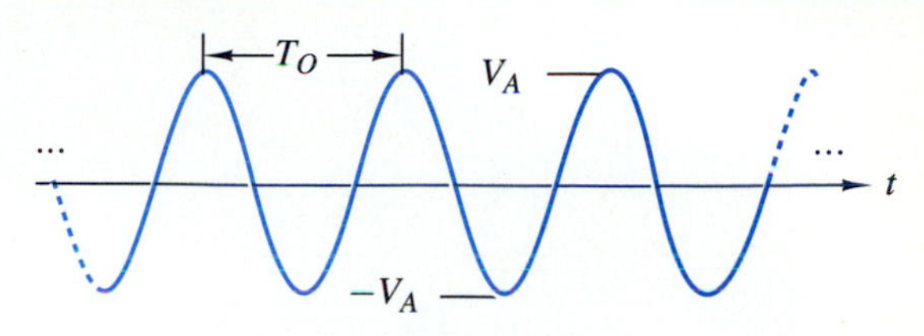

Figure 6-14 The eternal sinusoid.

6-4 THE SINUSOIDAL WAVEFORM

The cosine and sine functions are important in all branches of science and engineering. The corresponding time-varying waveform in Fig. 6-14 plays an especially prominent role in electrical engineering.

In contrast with the step and exponential waveforms studied earlier, the *sinusoid*, like the dc waveform in Fig. 6-1, extends indefinitely in time in both the positive and negative directions. The sinusoid has neither a beginning nor an end! Of course, real signals have finite durations. They were turned on at some finite time in the past and will be turned off at some time in the future. While it may seem unrealistic to have a signal model that lasts forever, it turns out that the eternal sinewave is a very good approximation in many practical applications.

The sinusoid in Fig. 6-14 is an endless repetition of identical oscillations between positive and negative peaks. The *amplitude* V_A defines the maximum and minimum values of the oscillations. The *period* T_O is the time required to complete one cycle of the oscillation. Using these two parameters we can write an equation for the sinusoid in terms of a cosine function:

$$v(t) = V_A \cos(2\pi t/T_O) \tag{6-18}$$

The waveform $v(t)$ carries the units of V_A (volts in this case) and the period T_O carries the units of time t (usually seconds). Equation (6-18) produces the waveform in Fig. 6-15(a) which has a positive peak at $t = 0$, since $v(0) = V_A$.

As in the case of the step and exponential functions, the general sinusoid is obtained by replacing t by $(t - T_S)$. Inserting this change in Eq. (6-18) yields a general expression for the sinusoid as

$$v(t) = V_A \cos[2\pi(t - T_S)/T_O] \tag{6-19}$$

where the constant T_S is the *time-shift* parameter. Figures 6-15(b) and (c) show that the sinusoid shifts to the right when $T_S > 0$ and to the left when $T_S < 0$. In effect, time shifting causes the positive peak **nearest** the origin to occur at $t = T_S$.

The time-shifting parameter can also be represented by an angle:

$$v(t) = V_A \cos[2\pi t/T_O + \phi] \tag{6-20}$$

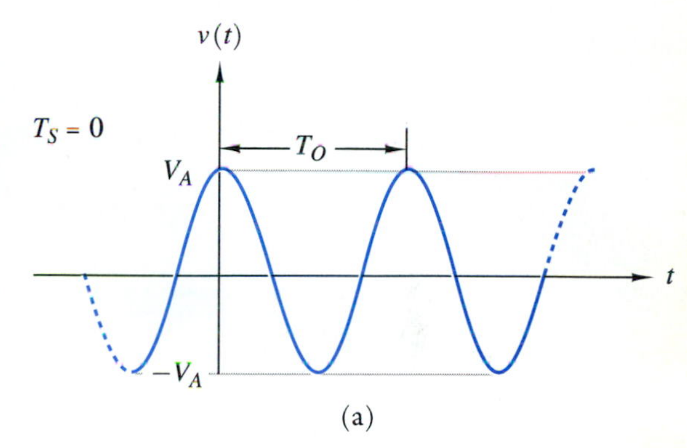

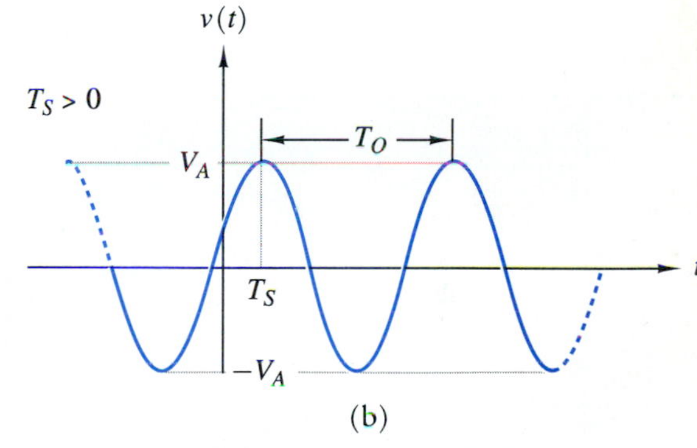

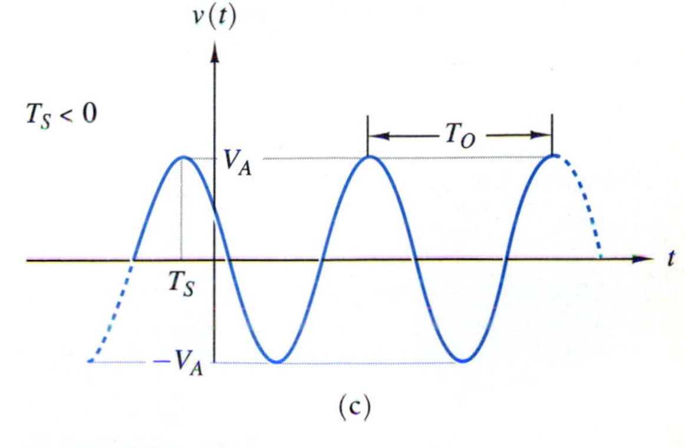

Figure 6-15 Effect of time shift on the sinusoidal waveform.

The parameter ϕ is called the *phase angle*. The term phase angle is based on the circular interpretation of the cosine function. We think of the period as being divided into 2π radians. In this sense the phase angle is the angle between $t = 0$ and the nearest positive peak. Comparing Eqs. (6-19) and (6-20) we find the relation between T_S and ϕ to be

$$\phi = -2\pi \frac{T_S}{T_O} \tag{6-21}$$

Changing the phase angle moves the waveform to the left or right, revealing different phases of the oscillating waveform, hence the name phase angle.

The phase angle should be expressed in radians, but is often reported in degrees. Care must be used when numerically evaluating the expression $(2\pi t/T_O + \phi)$ to ensure that both terms have the same units. The term $2\pi t/T_O$ has the units of radians, so it is necessary to convert ϕ to radians when it is given in degrees.

An alternative form of the general sinusoid is obtained by expanding Eq. (6-20) using the identity $\cos(x+y) = \cos(x)\cos(y) - \sin(x)\sin(y)$.

$$v(t) = [V_A \cos\phi]\cos(2\pi t/T_O) + [-V_A \sin\phi]\sin(2\pi t/T_O)$$

The quantities inside the brackets in this equation are constants; therefore, we can write the general sinusoid in form

$$v(t) = a\ \cos(2\pi t/T_O) + b\ \sin(2\pi t/T_O) \tag{6-22}$$

The two amplitude-like parameters a and b have the same units as the waveform (volts in this case) and are called *Fourier coefficients*. By definition the Fourier coefficients are related to the amplitude and phase parameters by the equations

$$a = V_A \cos\phi$$
$$b = -V_A \sin\phi \tag{6-23}$$

The inverse relationships are obtained by squaring Eq. (6-23):

$$V_A = \sqrt{a^2 + b^2} \tag{6-24}$$

and by dividing the second equation in Eq. (6-23) by the first:

$$\phi = \tan^{-1}\frac{-b}{a} \tag{6-25}$$

It is customary to describe the time variation of the sinusoid in terms of a frequency parameter. *Cyclic frequency* f_O is defined as the number of periods per unit time. By definition the period T_O is the number of seconds per cycle; consequently the number of cycles per second is

$$f_O = \frac{1}{T_O} \tag{6-26}$$

where f_O is the cyclic frequency or simply the frequency. The unit of frequency (cycles per second) is the *hertz* (Hz). The *angular frequency* ω_O in radians per second is related to cyclic frequency by the relationship

$$\omega_O = 2\pi f_O = \frac{2\pi}{T_O} \tag{6-27}$$

because there are 2π radians per cycle.

There are two ways to express the concept of sinusoidal frequency: cyclic frequency (f_O, hertz) and angular frequency (ω_0, radians per second). When working with signals, we tend to use the former. For example, radio stations transmit carrier signals at frequencies specified as 690 kHz (AM band) or 101 MHz (FM band). Radian frequency is more convenient when working with the response of circuits to sinusoidal signals for reasons that will become clear in later chapters.

In summary, there are several equivalent ways to describe the general sinusoid:

$$v(t) = V_A \cos\left[\frac{2\pi f_O(t - T_S)}{T_O}\right] = V_A \cos\left(\frac{2\pi t}{T_O} + \phi\right)$$

$$= a \cos\left(\frac{2\pi t}{T_O}\right) + b \sin\left(\frac{2\pi t}{T_O}\right)$$

$$= V_A \cos[2\pi f_O(t - T_S)] = V_A \cos(2\pi f_O t + \phi)$$

$$= a \cos(2\pi f_O t) + b \sin(2\pi f_O t)$$

$$= V_A \cos[\omega_O(t - T_S)] = V_A \cos(\omega_O t + \phi)$$

$$= a \cos(\omega_O t) + b \sin(\omega_O t)$$

To use any one of these expressions we need three types of parameters:

1. **Amplitude**—either V_A or the Fourier coefficients a and b.
2. **Time Shift**—either T_S or the phase angle ϕ.
3. **Time/Frequency**—either T_O or f_O or ω_O.

In different parts of this book we use different forms of the sinusoid. Therefore, it is important for the reader to thoroughly understand the relationships among the various parameters in Eqs. (6-21) through (6-27).

EXAMPLE 6-6

An oscilloscope is a laboratory instrument that visually displays waveform amplitude versus time. Figure 6-16 shows an oscilloscope display of a sinusoid. The vertical axis in Fig. 6-14 is calibrated at 5 V per division. The horizontal axis is calibrated in at 0.1 ms per division. Derive an expression for the sinusoid displayed in Fig. 6-16.

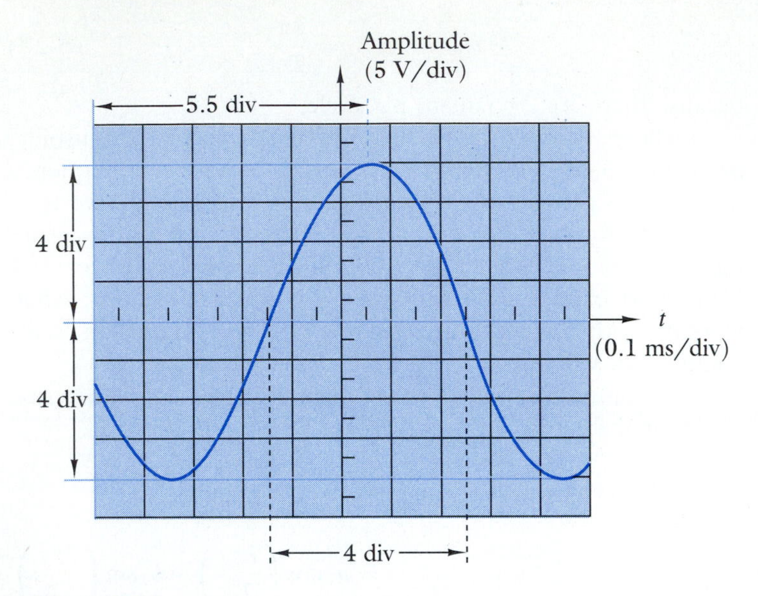

Figure 6-16

SOLUTION:

The maximum amplitude of the waveform is seen to be four vertical divisions, therefore

$$V_A = (4 \text{ div})(5 \text{ V/div}) = 20 \text{ V}$$

There are four horizontal divisions between successive zero crossings, which means there are a total of eight divisions in one cycle. The period of the waveform is

$$T_O = (8 \text{ div})(0.1 \text{ ms/div}) = 0.8 \text{ ms}$$

The two frequency parameters are $f_O = 1/T_O = 1.25$ kHz and $\omega_O = 2\pi f_O = 7854$ rad/s. The parameters V_A, T_O, f_O, and ω_O do not depend on the location of the $t = 0$ axis.

To determine the time shift T_S we need to define a time origin. The $t = 0$ axis is arbitrarily taken at the left edge of the display in Fig. 6-16. The positive peak shown in the display is 5.5 divisions to the right of $t = 0$, which is more than half a cycle (four divisions). The positive peak closest to $t = 0$ is not shown in Fig. 6-16 because it must lie beyond the left edge of the display. However, the positive peak shown in the display is located at $t = T_S + T_O$, since it is one cycle after $t = T_S$. We can write

$$T_S + T_O = (5.5 \text{ div})(0.1 \text{ ms/div}) = 0.55 \text{ ms}$$

which yields $T_S = 0.55 - T_O = -0.25$ ms. As expected, T_S is negative because the nearest positive peak is to the left of $t = 0$.

Given T_S we can calculate the remaining parameters of the sinusoid as

$$\phi = -\frac{2\pi T_S}{T_O} = 1.96 \text{ Rad or } 112.5^\circ$$

$$a = V_A \cos\phi = -7.65 \text{ V}$$

$$b = -V_A \sin\phi = -18.5 \text{ V}$$

So finally the displayed sinusoid can be written as

$$v(t) = 20 \cos[(7854(t + 0.25 \times 10^{-3})]$$

$$= 20 \cos(7854t + 112.5^\circ)$$

$$= -7.65 \cos 7854t - 18.5 \sin 7854t$$

EXERCISE 6-5

Derive an expression for the sinusoid displayed in Fig. 6-16 when $t = 0$ is placed in the middle of the display.

ANSWER:

$v(t) = 20 \cos(7854t - 22.5^\circ)$

Properties of Sinusoids

Waveforms that are endlessly repetitious are called periodic. Stated formally, a waveform is said to be *periodic* if

$$v(t + T_O) = v(t)$$

for all values of t. The constant T_O is called the period of the waveform if it is the smallest nonzero interval for which $v(t + T_O) = v(t)$. The sinusoid is periodic, since

$$v(t + T_O) = V_A \cos[2\pi(t + T_O)/T_O + \phi]$$

$$= V_A \cos[2\pi(t)/T_O + \phi + 2\pi]$$

But $\cos(x + 2\pi) = \cos(x)\cos(2\pi) - \sin(x)\sin(2\pi) = \cos(x)$. Consequently

$$v(t + T_O) = V_A \cos(2\pi t/T_O + \phi) = v(t)$$

for all t.

Other examples of periodic waveforms are the sawtooth, square wave, and triangular wave in Fig. 6-2. Signals that are not periodic are called *aperiodic*.

The *additive property* of sinusoids states that summing two or more sinusoids with the same frequency yields a sinusoid with different amplitude and phase parameters but the same frequency. To illustrate, consider two sinusoids

$$v_1(t) = a_1 \cos(2\pi f_O t) + b_1 \sin(2\pi f_O t)$$

$$v_2(t) = a_2 \cos(2\pi f_O t) + b_2 \sin(2\pi f_O t)$$

The waveform $v_3(t) = v_1(t) + v_2(t)$ can be written as

$$v_3(t) = (a_1 + a_2)\cos(2\pi f_O t) + (b_1 + b_2)\sin(2\pi f_O t)$$

because cosine and sine are linearly independent functions. We obtain the Fourier coefficients of the sum of two sinusoids by adding their Fourier coefficients, provided the two have the same frequency. *Caution:* The summation must take place with the sinusoids in Fourier coefficient form. Sums of sinusoids **cannot** be found by adding amplitudes and phase angles.

The *derivative* and *integral* properties of the sinusoid state that a sinusoid maintains its waveshape when differentiated or integrated:

$$\frac{d(V_A \cos \omega t)}{dt} = -\omega V_A \sin \omega t = \omega V_A \cos(\omega t + \pi/2)$$

$$\int V_A \cos(\omega t)\, dt = \frac{V_A}{\omega} \sin \omega t = \frac{V_A}{\omega} \cos(\omega t - \pi/2)$$

These operations change the amplitude and phase angle but do not change the basic sinusoidal waveshape or frequency. The fact that the waveshape is unchanged by differentiation and integration is a key property of the sinusoid. No other periodic waveform has this shape-preserving property.

EXAMPLE 6-7

(a) Find the periods, cyclic, and radian frequencies of the sinusoids: $v_1(t) = 17 \cos(2000t - 30°)$
$v_2(t) = 12 \cos(2000t + 30°)$.

(b) Find the waveform of $v_3(t) = v_1(t) + v_2(t)$.

SOLUTION:

(a) The two sinusoids have the same frequency $\omega_O = 2000$ rad/s since a term 2000t appears in the arguments of $v_1(t)$ and $v_2(t)$. Therefore, $f_O = \omega_O/2\pi = 318.3$ Hz and $T_O = 1/f_O = 3.14$ ms.

(b) We use the additive property, since the two sinusoids have the same frequency. Beyond this checkpoint the frequency plays no further role in the calculation. The two sinusoids must be converted to the Fourier coefficient form using Eq. (6-23).

$$a_1 = 17\ \cos(-30^\circ) = +14.7\ \text{V}$$

$$b_1 = -17\ \sin(-30^\circ) = +8.50\ \text{V}$$

$$a_2 = 12\ \cos(30^\circ) = +10.4\ \text{V}$$

$$b_2 = -12\ \sin(30^\circ) = -6.00\ \text{V}$$

The Fourier coefficients of the signal $v_3 = v_1 + v_2$ are found as

$$a_3 = a_1 + a_2 = 25.1\ \text{V}$$

$$b_3 = b_1 + b_2 = 2.50\ \text{V}$$

The amplitude and phase angle of $v_3(t)$ are found using Eqs. (6-24) and (6-25):

$$V_A = \sqrt{a_3^2 + b_3^2} = 25.2\ \text{V}$$

$$\phi = \tan^{-1}(2.5/25.1) = 5.69^\circ$$

Two equivalent representations of $v_3(t)$ are

$$v_3(t) = 25.1\ \cos(200t) + 2.5\ \sin(200t)$$

and

$$v_3(t) = 25.2\ \cos(200t + 5.69^\circ)$$

EXAMPLE 6-8

The balanced three-phase voltages used in electrical power systems can be written as

$$v_A(t) = V_O \cos(2\pi f_0 t)$$

$$v_B(t) = V_O \cos(2\pi f_0 t + 120^\circ)$$

$$v_C(t) = V_O \cos(2\pi f_0 t + 240^\circ)$$

Show that the sum of these voltages is zero.

SOLUTION:

The three voltages are given in amplitude/phase angle form. They can be converted to the Fourier coefficient form using Eq. (6-23):

$$v_A(t) = +V_O \cos(2\pi f_O t)$$

$$v_B(t) = -\frac{V_O}{2} \cos(2\pi f_O t) + \frac{\sqrt{3}V_O}{2} \sin(2\pi f_O t)$$

$$v_C(t) = -\frac{V_O}{2} \cos(2\pi f_O t) - \frac{\sqrt{3}V_O}{2} \sin(2\pi f_O t)$$

The sum of the cosine and sine terms is zero, consequently $v_A(t) + v_B(t) + v_C(t) = 0$. If the individual amplitudes were not exactly equal or the phase angle not exactly 0°, 120°, and 240°, then the sum is not zero, in which case the voltages are said to be unbalanced.

EXERCISE 6-6

Write an equation for the waveform obtained by integrating and differentiating the following signals.

(a) $v_1(t) = 30 \cos(10t - 60°)$

(b) $v_2(t) = 3 \cos(4000\pi t) - 4 \sin(4000\pi t)$

ANSWERS:

(a)
$$\frac{dv_1}{dt} = 300 \cos(10t + 30°)$$

$$\int v_1(t)\, dt = 3 \cos(10t - 150°)$$

(b)
$$\frac{dv_2}{dt} = 6.28 \times 10^4 \cos(4000\pi t + 143°)$$

$$\int v_2\, dt = 3.98 \times 10^{-4} \cos(4000\pi t - 36.9°)$$

EXERCISE 6-7

A sinusoid has a period of 5 μs. At $t = 0$ the amplitude is 12 V. The waveform reaches its first positive peak after $t = 0$ at $t = 4$ μs. Find its amplitude, frequency, and phase angle.

ANSWER:

$V_A = 38.8$ V, $f_O = 200$ kHz, $\phi = +72°$

6-5 COMPOSITE WAVEFORMS

In the previous sections we introduced the step, exponential, and sinusoidal waveforms. These waveforms are basic signals because they can can be combined to synthesize all other signals used in this book. Signals generated by combining the three basic waveforms are called *composite signals*. The impulse and ramp waveforms are examples of composite waveforms obtained by differentiating and integrating the step function. This section provides more examples of composite waveforms.

EXAMPLE 6-9

Characterize the composite waveform generated by subtracting an exponential from a step function with the same amplitude.

SOLUTION:

The equation for this composite waveform is

$$v(t) = V_A u(t) - \left[V_A e^{-t/T_C}\right] u(t)$$
$$= V_A \left[1 - e^{-t/T_C}\right] u(t)$$

For $t < 0$ the waveform is zero because of the step function. At $t = 0$ the waveform is still zero since the step and exponential cancel:

$$v(0) = V_A \left[1 - e^{0}\right](1) = 0$$

For $t \gg T_C$ the waveform approaches a constant value V_A because the exponential term decays to zero. For practical purposes $v(t)$ is within less than 1% of its final value V_A when $t = 5T_C$. At $t = T_C$, $v(T_C) = V_A(1 - e^{-1}) = 0.632V_A$. The waveform rises to about 63% of its final value in one time constant. All of the observations are summarized in the plot shown in Fig. 6-17. This waveform is called an *exponential rise*. It is also sometimes referred to as a "charging exponential," since it represents the behavior of signals that occur during the build up of voltage in resistor-capacitor circuits studied in Chapters 7 and 8.

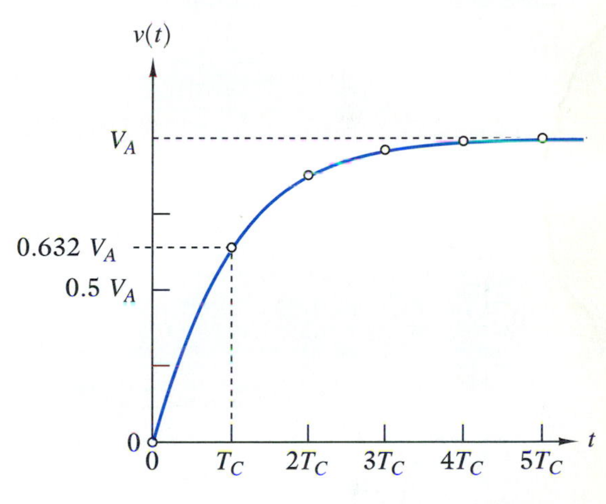

Figure 6-17 The exponential rise waveform.

EXAMPLE 6-10

Characterize the composite waveform obtained by multiplying the ramp $r(t)/T_C$ times an exponential.

SOLUTION:

The equation for this composite waveform is

$$v(t) = \frac{r(t)}{T_C}\left[V_A e^{-t/T_C}\right] u(t)$$
$$= \frac{V_A}{T_C}\left[t e^{-t/T_C}\right] u(t)$$

For $t < 0$ the waveform is zero because of the step function. At $t = 0$ the waveform is zero because $r(0) = 0$. For $t > 0$

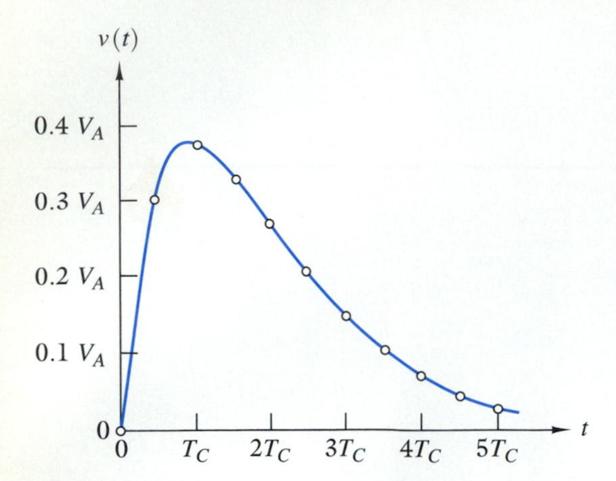

Figure 6-18 The damped ramp waveform.

there is a competition between two effects—the ramp increases linearly with time while the exponential decays to zero. Since the composite waveform is the product of these terms, it is important to determine which effect dominates. In the limit as $t \to \infty$ the product of the ramp and exponential takes on the indeterminant form of infinity times zero. A single application of *l'Hopital's rule* then shows that the exponential dominates forcing the $v(t)$ to zero as t becomes large. That is, the exponential decay overpowers the linearly increasing ramp. The waveform has the shape shown in Fig. 6-18 and is called a *damped ramp*. We will encounter the damped ramp in Chapter 8 as a critically damped response.

EXAMPLE 6-11

Characterize the composite waveform obtained by multiplying $\sin \omega_0 t$ by an exponential.

SOLUTION:

In this case the composite waveform is expressed as

$$v(t) = \sin \omega_O t \left[V_A e^{-t/T_C}\right] u(t)$$
$$= V_A \left[e^{-t/T_C} \sin \omega_O t\right] u(t)$$

The waveform is called a *damped sine* for reasons that are apparent from its graph in Fig. 6-19 (plotted for $T_O = 2T_C$). For $t < 0$ the step function forces the waveform to be zero. At $t = 0$, and periodically thereafter, the waveform passes through zero because $\sin(n\pi) = 0$. The waveform is not periodic because the exponential gradually reduces the amplitude of the oscillation. For all practical purposes the oscillations become negligibly small for $t > 5T_C$. The damped sine waveform occurs in Chapter 8, where it is called an underdamped response because it undershoots and overshoots its final value.

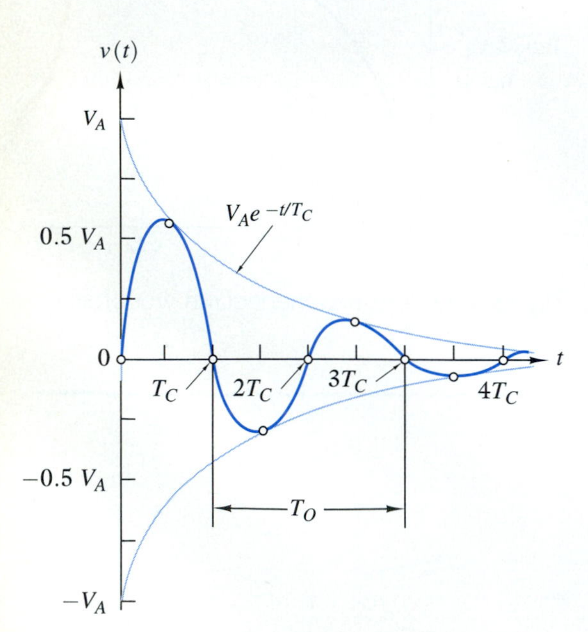

Figure 6-19 The damped sine waveform.

EXAMPLE 6-12

Characterize the composite waveform obtained as the difference of two exponentials with the same amplitude.

SOLUTION:

The equation for this composite waveform is

$$v(t) = \left[V_A e^{-t/T_1}\right] u(t) - \left[V_A e^{-t/T_2}\right] u(t)$$
$$= V_A \left(e^{-t/T_1} - e^{-t/T_2}\right) u(t)$$

For $T_1 > T_2$ the resulting waveform is illustrated in Fig. 6-20 (plotted for $T_1 = 2T_2$).

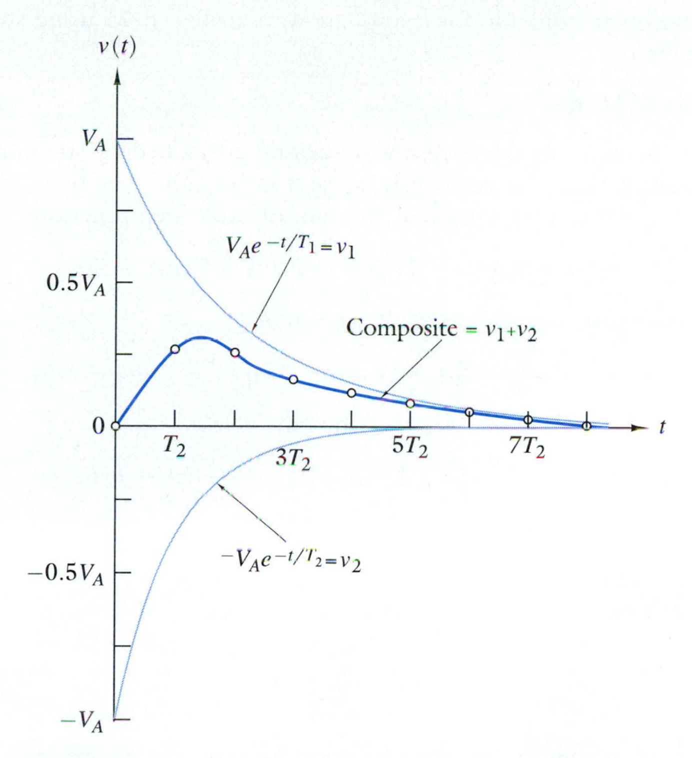

Figure 6-20 The double exponential waveform.

For $t < 0$ the waveform is zero. At $t = 0$ the waveform is still zero, since

$$v(0) = V_A\left(e^{-0/T_1} - e^{-0/T_2}\right)$$
$$= V_A(1 - 1) = 0$$

For $t \gg T_1$ the waveform returns to zero because both exponentials decay to zero. For $5T_1 > t > 5T_2$ the second exponential is negligible and the waveform essentially reduces to the first exponential. Conversely, for $t < T_1$ the first exponential is essentially constant so the second exponential determines the early time variation of the waveform. The waveform is called a *double exponential,* since both exponential components make important contributions to the waveform. The exponential with the longest time constant is said to be the *dominant exponential.* The double exponential waveform occurs in Chapter 8, where it is called an overdamped response because it gradually returns to zero without undershooting its final value.

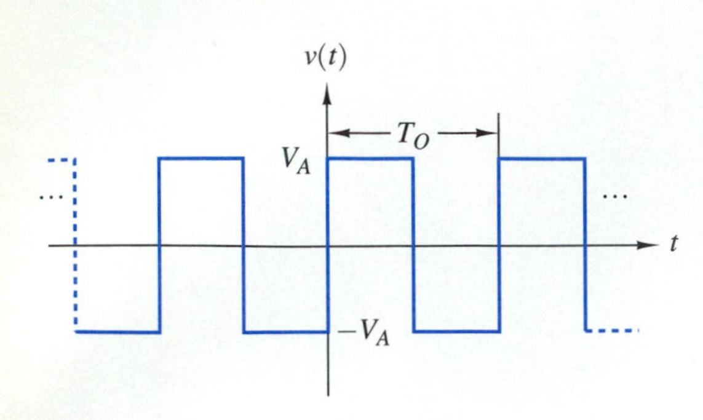

Figure 6-21 The square wave.

EXAMPLE 6-13

Develop an equation for the square wave in Fig. 6-21 using step functions.

SOLUTION:

An equation for the square wave can be developed by summing the expressions for each cycle, since it is periodic. The first cycle after $t = 0$ can be written as the sum of three step functions:

$$v_1(t) = V_A u(t) - 2V_A u(t - T_0/2) + V_A u(t - T_0)$$

Similarly, the second cycle after $t = 0$ is

$$v_2(t) = V_A u(t - T_0) - 2V_A u(t - 3T_0/2) + V_A u(t - 2T_0)$$

From these results we see that the kth cycle is

$$v_k(t) = V_A u(t - [k-1]T_0) - 2V_A u(t - [2k-1]T_0/2) + V_A u(t - kT_0)$$

We can produce the square wave $v(t)$ cycle by cycle, by summing $v_k(t)$ from $k = -\infty$ to $k = \infty$.

$$v(t) = \sum_{k=-\infty}^{k=\infty} v_k(t)$$

An infinite sum is required because the square wave extends to infinity along the positive and negative time axis. The square wave can be represented by an infinite sum of step functions. In Chapter 16 we will see that the square wave and other periodic signals can be constructed using an infinite sum of sinusoids called a **Fourier series.**

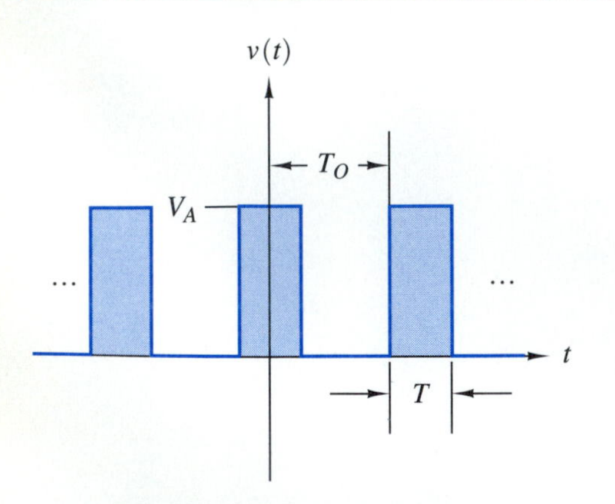

Figure 6-22

EXERCISE 6-8

Express the pulse train in Fig. 6-22 in terms of step functions.

ANSWER:

$$v(t) = V_A \sum_{k=-\infty}^{k=\infty} \left[u\left(t - kT_O + \frac{T}{2}\right) - u\left(t - kT_O - \frac{T}{2}\right) \right]$$

EXERCISE 6-9

Find the maximum amplitude and the approximate duration of the following composite waveforms.

(a) $v_1(t) = [25 \sin 1000t][u(t) - u(t - 10)]$ V
(b) $v_2(t) = [50 \cos 1000t][e^{-200t}]u(t)$ V
(c) $i_3(t) = [3000\, te^{-1000t}]u(t)$ mA
(d) $i_4(t) = [10e^{5000t}]u(-t) + [10e^{-5000t}]u(t)$ A

ANSWERS:
(a) 25 V, 10 s
(b) 50 V, 25 ms
(c) 1.10 mA, 5 ms
(d) 10 A, 2 ms ■

6-6 WAVEFORM PARTIAL DESCRIPTORS

An equation or graph defines a waveform for all time. The value of a waveform $v(t)$, $i(t)$, or $p(t)$ at time t is called the *instantaneous value* of the waveform. We often use numerical values or terminology that characterize a waveform but do not give a complete description. These waveform *partial descriptors* fall into two categories: (1) those that describe temporal features and (2) those that describe amplitude features.

Temporal Descriptors

Temporal descriptors identify waveform attributes relative to the time axis. For example, waveforms that repeat themselves at fixed time intervals are said to be *periodic*. Stated formally

A signal $v(t)$ is periodic if $v(t + T_O) = v(t)$ for all t, where the period T_O is the smallest value that meets this condition. Signals that are not periodic are called aperiodic.

The fact that a waveform is periodic provides important information about the signal, but does not specify all of its characteristics. The period and periodicity of a waveform are partial descriptors. The eternal sinewave is the premier example of a periodic signal. The square wave and triangular wave in Fig. 6-2 are also periodic. Examples of aperiodic waveforms are the step function, exponential, and damped sine.

Waveforms that are identically zero prior to some specified time are said to be *causal*. Stated formally

A signal $v(t)$ is causal if $v(t) \equiv 0$ for $t < T$; otherwise it is noncausal.

It is usually assumed that a causal signal is zero for $t < 0$, since time shifting can always place the starting point of a waveform at $t = 0$. Examples of causal waveforms are the step function, exponential, and damped sine. The eternal sinewave is, of course, noncausal.

Causal waveforms play a central role in circuit analysis. When the input driving force $x(t)$ is causal, the circuit response $y(t)$ must also be causal. That is, a physically realizable circuit cannot anticipate and respond to an input before it is applied. Causality is an important temporal feature, but only a partial description of the waveform.

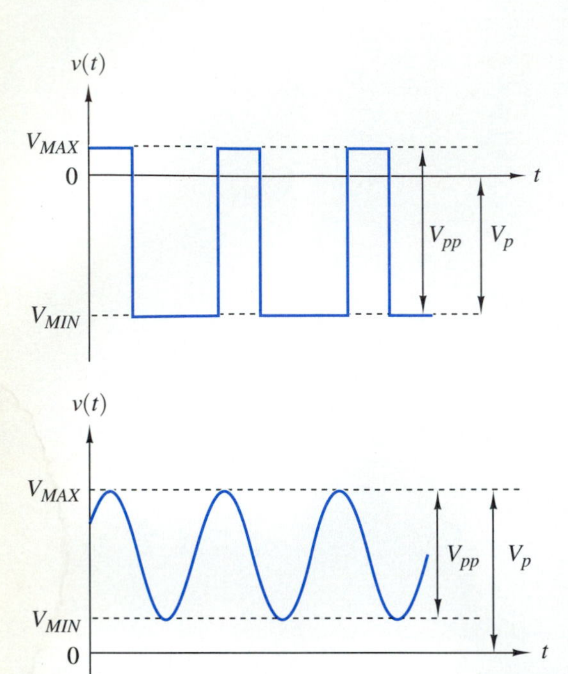

Figure 6-23 Peak values and peak-to-peak values.

Amplitude Descriptors

Amplitude descriptors are positive scalars that identify size features of the waveform. Generally a waveform amplitude varies between two extreme values denoted as V_{MAX} and V_{MIN}. The *peak-to-peak* value (V_{pp}) describes the total excursion of $v(t)$ and is defined as

$$V_{pp} = V_{MAX} - V_{MIN} \tag{6-28}$$

Under this definition V_{pp} is always positive even if V_{MAX} and V_{MIN} are both negative. The *peak value* (V_p) is the maximum of the absolute value of the waveform. That is

$$V_p = Max\,\{|V_{MAX}|,\ |V_{MIN}|\} \tag{6-29}$$

The peak value is a positive number that indicates the maximum absolute excursion of the waveform from zero. Figure 6-23 shows examples of these two amplitude descriptors.

The peak and peak-to-peak values describe waveform variation using the extreme values. The average value smooths things out to reveal the underlying waveform baseline. Average value is the area under the waveform over some period of time T, divided by that time period. Mathematically, we define *average value* (V_{avg}) over the time interval T as

$$V_{avg} = \frac{1}{T}\int_{t_O}^{t_O+T} v(x)\,dx \tag{6-30}$$

For periodic signals the averaging interval T equals the period T_O.

For some periodic waveforms the integral in Eq. (6-30) can be evaluated graphically. The net area under the waveform is the area above the time axis minus the area below the time axis. For example, the sinusoid in Fig. 6-24 has zero average value, since the area above the axis is exactly equal to the area below. The sawtooth in Fig. 6-24 clearly has a positive average value. By geometry the net area under one cycle of the sawtooth waveform is $V_A T_O/2$, so its average value is $(1/T_O)(V_A T_O/2) = V_A/2$.

The average value measures the waveform's baseline with respect to $v = 0$ axis. In other words, it indicates whether the waveform contains a constant, non-time-varying component. The average value is also called the *dc component* of the waveform because dc signals are constant for all t.

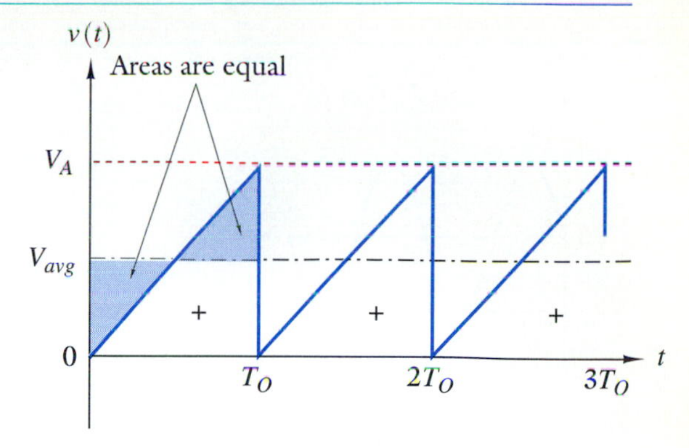

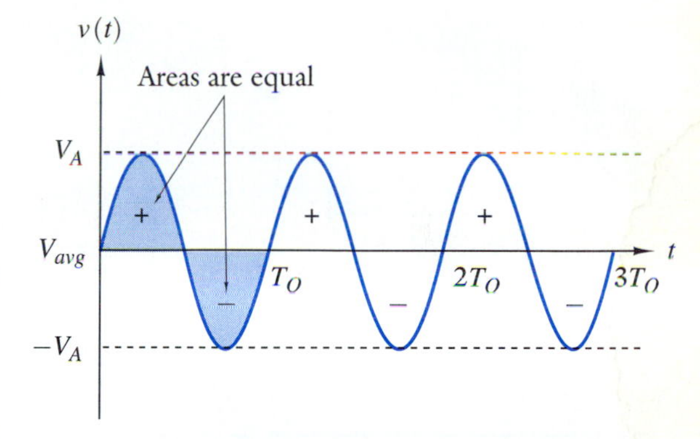

Figure 6-24 Average values of periodic waveform.

EXAMPLE 6-14

Find the peak, peak-to-peak and average values of the periodic input and output waveforms in Fig. 6-25.

SOLUTION:

The input waveform is a sinusoid whose amplitude descriptors are

$$V_{pp} = 2V_A \qquad V_p = V_A \qquad V_{avg} = 0$$

The output waveform is obtained by clipping off the negative half-cycle of the input sinusoid. The amplitude descriptors of the output waveform are

$$V_{pp} = V_p = V_A$$

The output has a nonzero average value, since there is a net positive area under the waveform. The upper limit in Eq. (6-30) can be taken as $T_O/2$, since the waveform is zero from $T_O/2$ to T_O.

$$V_{avg} = \frac{1}{T_O}\int_0^{T_O/2} V_A \sin(2\pi t/T_O)\, dt = -\frac{V_A}{2\pi} \cos(2\pi t/T_O)\Big|_0^{T_O/2}$$

$$= \frac{V_A}{\pi}$$

The signal processor produces an output with a dc value from an input with no dc component. Rectifying circuits described in electronics courses produce waveforms like the output in Fig. 6-25.

Root-Mean-Square Value

The *root-mean-square value* (V_{rms}) of a waveform is a measure of the average power carried by the signal. The instantaneous power delivered to a resistor R by a voltage $v(t)$ is

$$p(t) = \frac{1}{R}[v(t)]^2 \qquad (6\text{-}31)$$

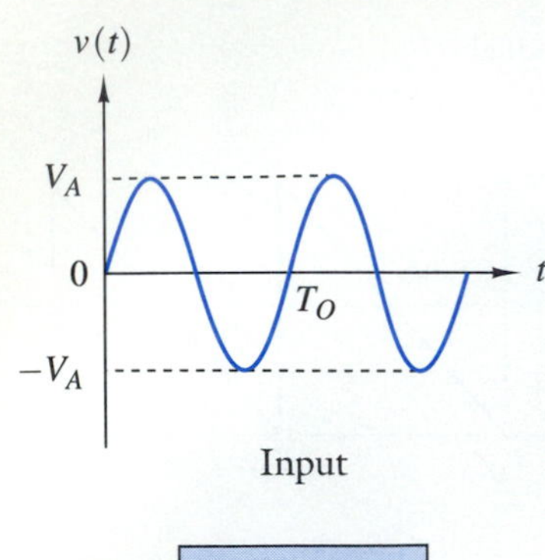

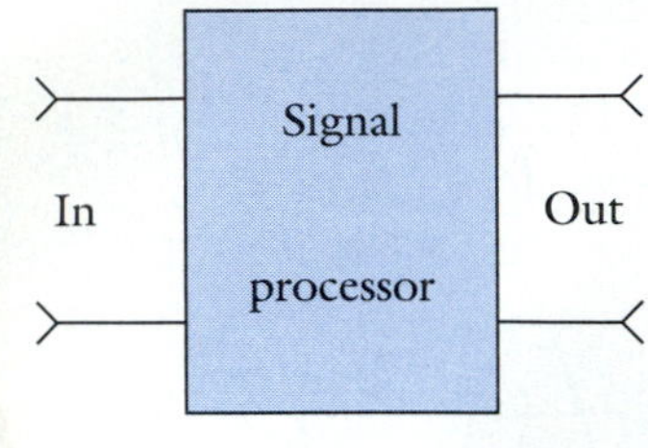

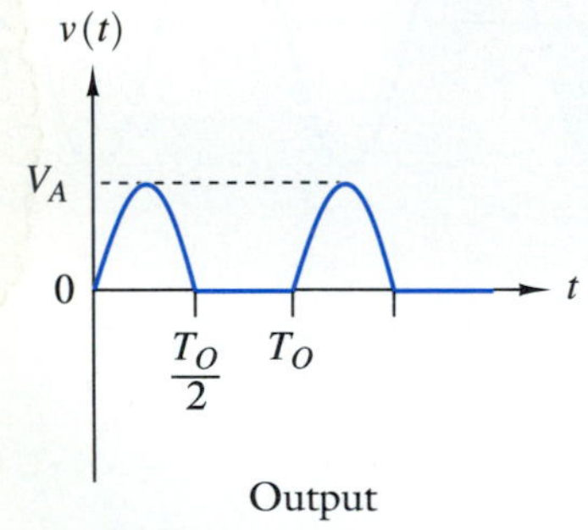

Figure 6-25

The average power delivered to the resistor in time span T is defined as

$$P_{avg} = \frac{1}{T}\int_{t_O}^{t_O+T} p(t)\,dt \tag{6-32}$$

Combining Eqs. (6-31) and (6-32) yields

$$P_{avg} = \frac{1}{R}\left[\frac{1}{T}\int_{t_O}^{t_O+T} [v(t)]^2\,dt\right] \tag{6-33}$$

The quantity inside the large brackets in Eq. (6-33) is the average value of the square of the waveform. The units of the bracketed term are volts squared. The square root of this term to define the amplitude descriptor V_{rms}:

$$V_{rms} = \sqrt{\frac{1}{T}\int_{t_O}^{t_O+T} [v(t)]^2\,dt} \tag{6-34}$$

The amplitude descriptor V_{rms} is called the root-mean-square (rms) value because it is obtained by taking the square **root** of the average (**mean**) of the **square** of the waveform. For periodic signals the averaging interval is one cycle, since such a waveform repeats itself every T_O seconds.

We can express the average power delivered to a resistor in terms of V_{rms} as

$$P_{avg} = \frac{1}{R}V_{rms}^2 \tag{6-35}$$

The equation for average power in terms of V_{rms} has the same form as the instantaneous power in Eq. (6-31). For this reason the rms value is also called the *effective value*, since it determines the average power delivered to a resistor in the same way that a dc waveform $v(t) = V_{dc}$ determines the instantaneous power. If the amplitude of waveform amplitude is doubled, its rms value is doubled, and the average power is quadrupled. Commercial electrical power systems use transmission voltages in the range of several hundred kilovolts (rms) to transfer large blocks of electrical power.

EXAMPLE 6-15

Find the rms value of the sinusoid and sawtooth in Fig. 6-24.

SOLUTION:

Applying Eq. (6-34) the sinusoid yields an rms value of

$$V_{rms} = \sqrt{\frac{V_A^2}{T_0}\int_0^{T_0} \sin^2(2\pi t/T_0)\, dt}$$

$$= \sqrt{\frac{V_A^2}{T_O}\left[\frac{t}{2} - \frac{\sin(4\pi t/T_0)}{8\pi/T_0}\right]_0^{T_0}} = \frac{V_A}{\sqrt{2}}$$

For the sawtooth the rms value is found as

$$V_{rms} = \sqrt{\frac{1}{T_0}\int_0^{T_0} (V_A t/T_0)^2\, dt} = \sqrt{\frac{V_A^2}{T_0^3}\left[\frac{t^3}{3}\right]_0^{T_0}} = \frac{V_A}{\sqrt{3}}$$

Application Note

EXAMPLE 6-16

In the United States residential ac power is supplied at 60 Hz and voltages of 115 and 230 V. These voltages are the rms values of the sinusoidal waveforms supplied by the power company. An oscilloscope displaying the waveform at a 115 V (rms) outlet indicates that the peak value of the sinusoid is larger than 115 V. Example 6-15 showed that the rms value of a sinusoid equals its peak divided by $\sqrt{2}$. If the rms value is 115 V then the peak value is $115\sqrt{2}$ or about 163 V. Most ac voltmeters are calibrated to indicate the rms value of a sinusoid. If we measure a sinusoid with a peak value of 200 V using an ac voltmeter, the meter will indicate $200/\sqrt{2}$ or about 141 V (rms). In general, ac voltmeters are designed to measure pure sinusoids and will not correctly measure the rms value of nonsinusoidal waveforms such as a square wave.

The concept of rms value applies to sinusoidal currents as well. The reader is invited to return to Eq. (6-31) and write the instantaneous power delivered to the resistor in terms of the current through it. Continuing the derivation in terms of current then shows that

$$I_{rms} = \sqrt{\frac{1}{T}\int_{t_0}^{t_0+T} [i(t)]^2\, dt}$$

and

$$P_{avg} = RI_{rms}^2$$

Since this power relationship has the same form as the dc case studied in Chapter 2, we see why the rms value is also called the effective value. Most ac ammeters are calibrated to display the rms value of the current. Similarly the ampere ratings of devices such as fuses and circuit breakers mean rms current.

Since rms values relate to the average power, it should come as no surprise to discover that ac power ratings of devices refers to average power. A 75-W lightbulb will draw an average power of 75 W at its rated (rms) voltage. The instantaneous power delivered to the bulb has a constant average value and an oscillatory component. As a result, its light output fluctuates slightly, although the variation is not noticeable to most people. We will return to the concept of average power again in Chapter 14 when we study maximum power transfer conditions. The consequences of the oscillatory component of the instantaneous power will also be studied in Chapter 14.

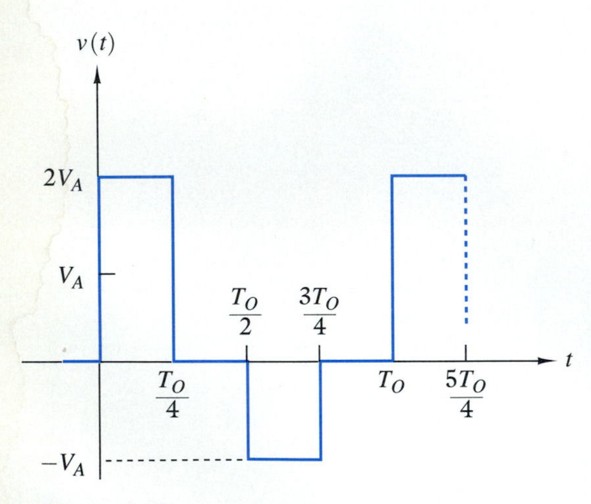

Figure 6-26

EXERCISE 6-10

Find the peak, peak-to-peak, average and rms values of the periodic waveform in Fig. 6-26.

ANSWERS:

$$V_p = 2V_A \qquad V_{pp} = 3V_A \qquad V_{avg} = \frac{V_A}{4} \qquad V_{rms} = \frac{\sqrt{5}}{2}V_A$$

EXERCISE 6-11

Classify each of the signals below as periodic or aperiodic and causal or noncausal. Then calculate the average and rms values of the periodic waveforms, and the peak and peak-to-peak values of the other waveforms.

(a) $v_1(t) = 99 \cos 3000t - 132 \sin 3000t$ V
(b) $v_2(t) = [34 \sin 800\pi t]u(t)$ V
(c) $i_3(t) = 120[u(t+5) - u(t-5)]$ mA
(d) $i_4(t) = 50$ A

ANSWERS:

(a) Periodic, noncausal $V_{avg} = 0$ and $V_{rms} = 117$ V
(b) Aperiodic, causal, $V_p = 34$ V and $V_{pp} = 68$ V
(c) Aperiodic, causal, $V_p = V_{pp} = 120$ mA
(d) Aperiodic, noncasual, $V_p = 50$ A and $V_{pp} = 0$

EXERCISE 6-12

Construct waveforms that have the following characteristics.

(a) Aperiodic and causal with $V_p = 8$ V and $V_{pp} = 15$ V
(b) Periodic and noncasual with $V_{avg} = 10$ V and $V_{pp} = 50$ V
(c) Periodic and noncausal with $V_{avg} = V_p/2$
(d) Aperiodic and causal with $V_p = V_{pp} = 10$ V

ANSWERS:

There are many possible correct answers, since the given parameters are only partial descriptions of the required waveforms. Some examples that meet the requirements are

(a) $v(t) = 8\ u(t) - 15\ u(t-1) + 7\ u(t-2)$
(b) $v(t) = 10 + 25\ \sin\ 1000t$
(c) A sawtooth wave
(d) $v(t) = 10\left[e^{-100t}\right]u(t)$ ■

6-7 SIGNAL SPECTRA

Up to this point we have described signals as waveforms, where a **waveform** is an equation or graph that characterizes a signal as a function of time. The waveform is the appropriate starting place for studying signals because we live in a world in which time is the independent variable. But signals have a dual nature. They can also be described by a *spectrum*, which characterizes the signal as a function of frequency. Accordingly, *a signal can be studied either in the time domain as a waveform or in the frequency domain as a spectrum.*

The dual nature of signals and circuits is one of the most profound and useful tools in electrical engineering. While we cannot fully develop the frequency domain concept at this point in our study, it is important to introduce the concept, since it is the central theme of the rest of the book. As a starting place we examine periodic signals that can be written as sums of sinusoids.

The sinusoid is an important signal because it is the building block for describing the spectra of all signals of interest. One of its most important properties is that periodic signals arising in the engineering applications can be decomposed into a sum of sinusoids with a different amplitudes, phase angles, and frequencies. The decomposition includes a dc component (the signal's average value) and sinusoids at a *fundamental frequency* f_O and at *harmonic frequencies* nf_O, which are integer multiples of the fundamental. The frequency f_O is called the fundamental because it determines the period of the signal $T_O = 1/f_O$ and vice versa. The harmonic frequencies begin with the second harmonic ($2f_O$), followed by the third harmonic ($3f_O$), or, in general, the nth harmonic (nf_O).

For example, the periodic signal

$$v(t) = 10 + 30\ \cos(2\pi f_O t) + 15\ \sin(2\pi 2 f_O t) - 7.5\ \cos(2\pi 4 f_O t) \tag{6-36}$$

is the sum of a dc component plus three sinusoidal ac components. A plot of the waveform in Fig. 6-27 reveals the periodic nature of the signal.

We associated zero frequency with the dc component in Eq. (6-36) because the signal $v(t) = V_A \cos(2\pi f t) = V_A$ when $f = 0$. The frequency of the first ac component is fundamental

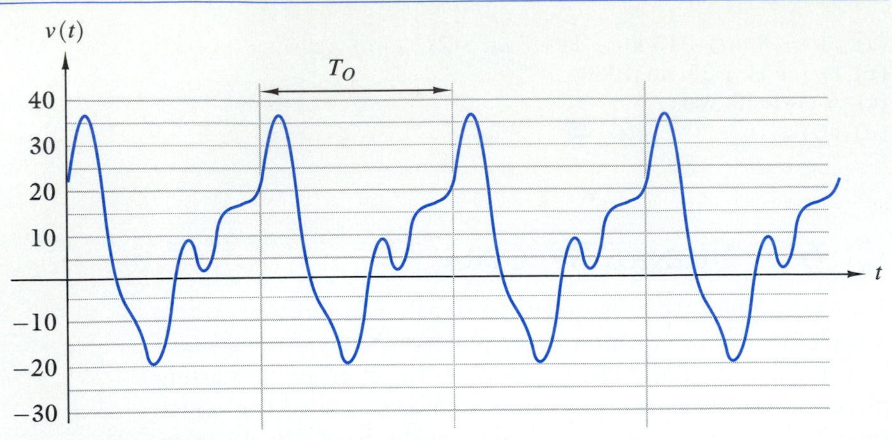

Figure 6-27 Periodic waveform in Eq. (6-36).

frequency, since it is the lowest nonzero frequency. The third and fourth terms in Eq. (6-36) are the second harmonic and fourth harmonic since their frequencies are $2f_O$ and $4f_O$, respectively.

The periodic signal in Eq. (6-36) contains four frequencies, 0, f_O, $2f_O$ and $4f_O$. The *spectrum* of the signal is an equation or graph defining the amplitude and phase angle associated with each frequency. To obtain the correct phase angles we must write each term in Eq. (6-36) in the same form as $V_A \cos(2\pi ft + \phi)$. Using the identities $\sin x = \cos(x - 90°)$ and $-\cos x = \cos(x + 180°)$ we rewrite Eq. (6-36) in the following way:

$$v(t) = 10 + 30\ \cos(2\pi f_O t) + 15\ \cos(2\pi 2 f_O t - 90°) + 7.5\ \cos(2\pi 4 f_O t + 180°) \tag{6-37}$$

Each term is now in a standard form, so we can list the amplitude and phase angle of each cosine component using the format shown below:

Frequency	0	f_O	$2f_O$	$4f_O$
Amplitude(V)	10	30	15	7.5
Phase Angle	–	0°	−90°	180°

(6-38)

The listing above defines the spectrum of the signal in Eq. (6-37) because it gives the amplitude and phase angles of the constituent cosines **as functions of frequency**.

To graphically visualize the spectrum we plot these values as shown in Fig. 6-28. The plot of V_n versus frequency is called an *amplitude spectrum* and a plot of ϕ_n versus frequency is called the *phase spectrum* of the signal. Both plots are called *line spectra* because component amplitudes and phase angles only exist at certain discrete frequencies.

The example above illustrates the dual nature of periodic signals. Equation (6-37) and Fig. 6-27 describe the signal wave-

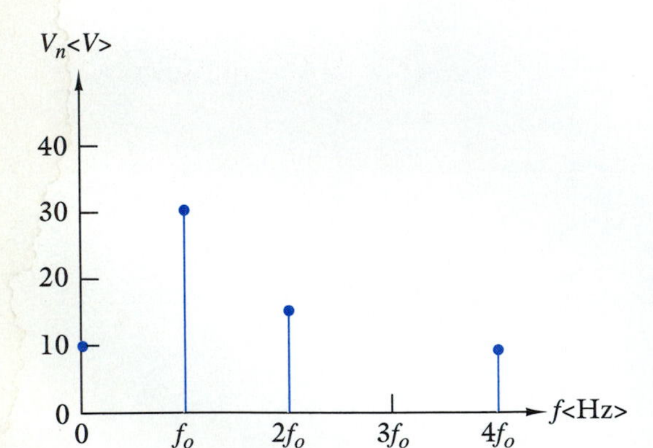

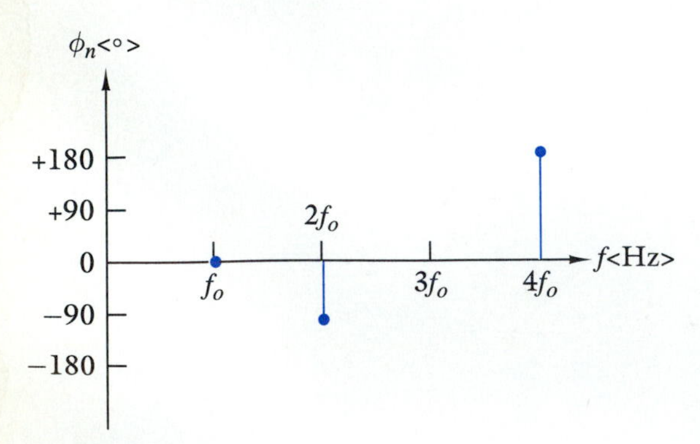

Figure 6-28 Amplitude and phase of the spectrum in Eq. (6-38).

form in the time domain, while Eq. (6-38) and Fig. 6-28 describe the signal spectrum in the frequency domain. In the discussion above we derived the spectrum from the waveform. The process is reversible, since we could use the spectrum in Eq. (6-38) or Fig. 6-28 to construct the waveform in Eq. (6-37). The waveform and the spectrum are seen to be two sides of the same coin. They are two equivalent ways of representing and studying periodic signals.

The decomposition of signals into sums of harmonic sinusoids can be extended to any of the periodic waveforms shown in Fig. 6-2. For example, the first four terms in decomposition for the square wave are

$$v(t) = V_1 \cos(2\pi f_O t - 90°) + \frac{V_1}{3}\cos(2\pi 3 f_O t - 90°) + \frac{V_1}{5}\cos(2\pi 5 f_O t - 90°) + \frac{V_1}{7}\cos(2\pi 7 f_O t - 90°) + \ldots \quad (6\text{-}39)$$

In this case the harmonics all have the same phase angle and their amplitudes decrease with frequency as $1/n$. There is no dc component and no terms at the even harmonics $2f_O$, $4f_O$, $6f_O, \ldots$. Therefore, the spectrum of the square wave contains only odd harmonics beginning with the fundamental.

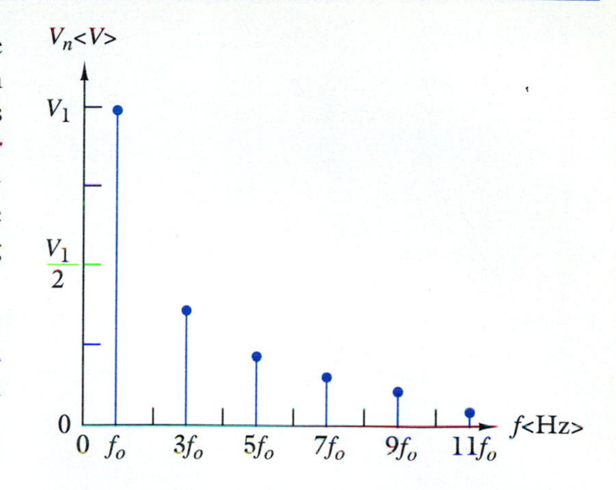

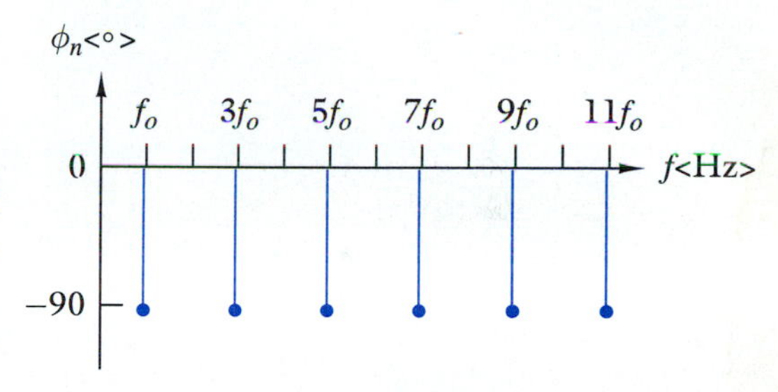

Figure 6-29

Frequency	0	f_O	$2f_O$	$3f_O$	$4f_O$	$5f_O$	⋯	nf_O	⋯
Amplitude	0	V_1	0	$V_1/3$	0	$V_1/5$	⋯	V_1/n	⋯
Phase Angle	–	−90°	–	−90°	–	−90°	⋯	−90°	⋯

(6-40)

Figure 6-29 shows the amplitude and phase plots of the square wave spectrum.

A summation of harmonic sinusoids is called a Fourier series.

The analytical tools needed to determine the amplitude and phase angle for each harmonic components will be provided in Chapter 16 when we study Fourier analysis. At this point we are not concerned with the mathematical details of these tools but simply the fact that such a series decomposition is possible.

Figure 6-30 shows how the harmonics components in Eq. (6-40) combine to produce a square wave. We begin with the fundamental in Fig. 6-30(a) which is just a sine wave. Adding the third harmonic produces the result in Fig. 6-30(b). The result in (b) plus the fifth harmonic produces (c), and so on. After just three terms the composite begins to look like a square wave. As we add more and more harmonics the composite gets closer and closer to a square wave.

The square wave can be thought of as being generated by a set of signal sources, with one source for each harmonic as indicated by the block diagram in Fig. 6-31. The real power of

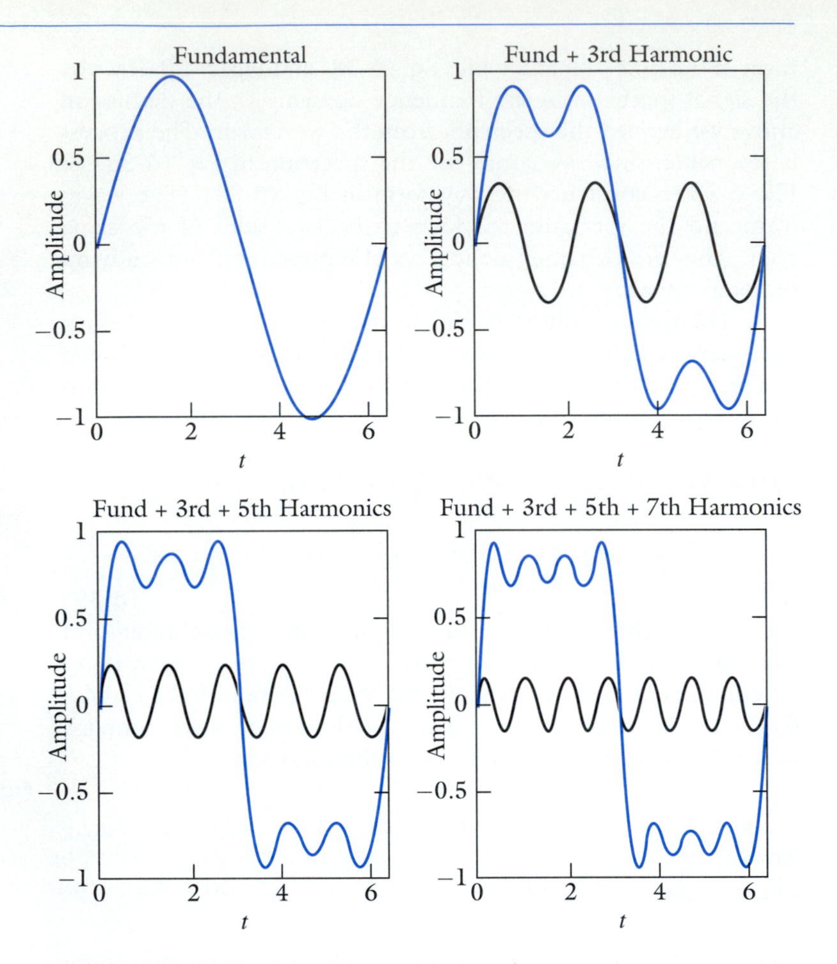

Figure 6-30 Summing harmonics to produce a square wave.

the spectrum viewpoint is revealed once we know how to find circuit responses for a single sinusoidal input (Chapters 12 and 13). We can then handle periodic input signals like the square wave, using superposition. That is, we find the response to each harmonic separately and then sum the individual responses to obtain the total response.

Mathematically the decomposition in Eq. (6-39) is an infinite sum, so theoretically the square wave contains infinitely many discrete frequencies. However, the amplitudes decrease as $1/n$ so at some point the high-frequency components become negligibly small. The fact that higher-order harmonics are negligible leads to the concepts of the highest frequency f_H and lowest frequency f_L contained in a signal. For periodic waveforms, $f_L = f_O$ unless the signal has a dc component in which case $f_L = 0$.

For most periodic waveforms it is not possible to define f_H quite so precisely. One rule of thumb is to compare the ampli-

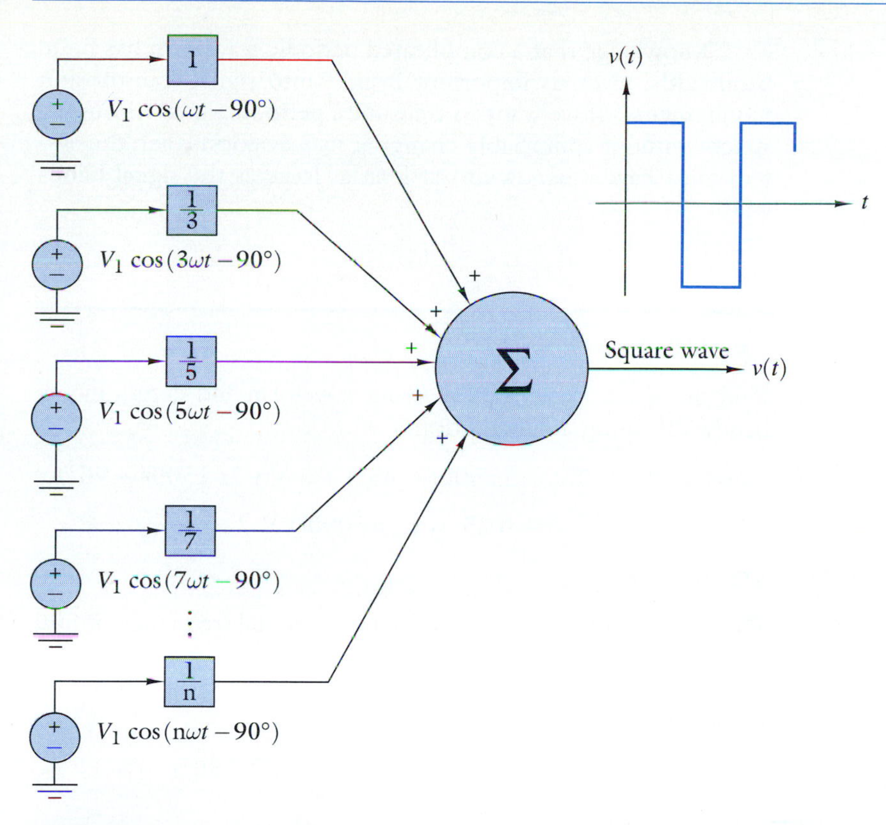

Figure 6-31 Conceptual block diagram for generating a square wave.

tudes of the high-frequency harmonics with the largest ac component amplitude, usually the fundamental. Harmonics whose amplitudes are smaller than some specified fraction of the largest ac amplitude are considered negligible. Using this approach we neglect all harmonics whose amplitudes are smaller than, say, 2% of the fundamental. For engineering purposes the spectrum of the signal contains a finite number of frequencies included within the *bandwidth* (BW).

$$BW = f_H - f_L$$

Bandwidth is a partial descriptor of the spectrum of a signal. Its definition involves an approximation, since we may need to neglect high-frequency components to define f_H. The approximation is of the same type used to define the time duration of the exponential waveform. In principle the exponential waveform endures forever, but in engineering practice it becomes negligibly small after about five time constants. Similarly, in principle a harmonic decomposition contains infinitely many frequencies, but in practice the higher-order harmonics become negligibly small.

Knowledge that a complicated periodic waveform has finite bandwidth, gives us important insight into signal transmission requirements. If we want to transmit a periodic signal through a system without appreciably changing its waveform, then the system must have a bandwidth at least as large as the signal bandwidth.

EXAMPLE 6-17

Find the spectrum of the following waveform and determine its bandwidth using a 5% criterion.

$$v(t) = 10 + 20\ \cos(2\pi 500t - 45°) + 5\ \sin(2\pi 1500t - 90°) + 0.25\ \cos(2\pi 5000t)\ \text{V}$$

SOLUTION:

The signal has a dc component, a fundamental frequency of 500 Hz, a third harmonic at 1500 Hz and a tenth harmonic at 5 kHz. The third harmonic term is not written as a cosine. Using the identities

$$\sin(x - 90°) = -\cos\ x = \cos(x + 180°)$$

we rewrite $v(t)$ as

$$v(t) = 10 + 20\ \cos(2\pi 500t - 45°) + 5\ \cos(2\pi 1500t + 180°) + 0.25\ \cos(2\pi 5000t)$$

The spectrum corresponding to the waveform $v(t)$ is

Frequency (kHz)	0	0.5	1.5	5
Amplitude (V)	10	20	5	0.25
Phase Angle	–	−45°	180°	0°

To find the bandwidth we note that the spectrum has a dc component; therefore, $f_L = 0$. The 5-kHz component can be neglected, since its amplitude is less than 5% of the amplitude of the fundamental ($0.05 \times 20 = 1$ V). Hence $f_H = 1.5$ kHz and $BW = f_H - f_L = 1.5$ kHz.

EXAMPLE 6-18

Determine the bandwidth of a 1-kHz square wave using a 2% criterion.

SOLUTION:

From Eq. (6-40) $V_n = V_1/n$, n odd. The waveform has no dc component, so $f_L = f_O = 1$ kHz. The amplitudes decrease as $1/n$. Using a 2% criteria we ignore harmonics for which $1/n < 0.02$ or $n > 50$. But only odd harmonics are present, so the highest frequency with more than 2% amplitude is the 49th. The 2% bandwidth of a square wave is

$$BW = 49f_O - f_O = 48f_O = 48 \text{ kHz}$$

Figure 6-32 shows one cycle of the approximation obtained by summing the first 49 harmonics of the square wave. The high-frequency oscillations are caused by the absences of harmonics above $n = 49$.

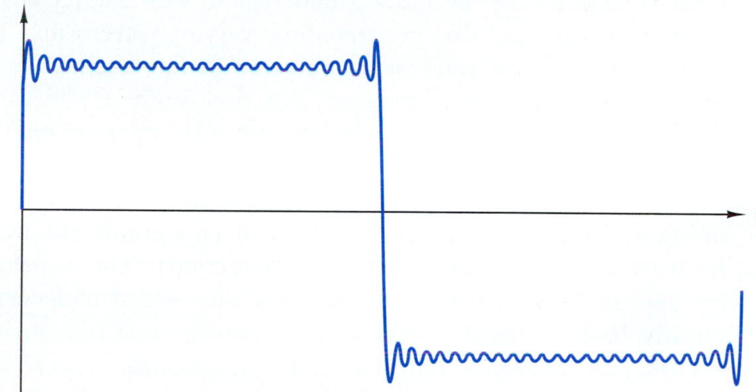

Figure 6-32 Waveform of the first 49 harmonics in a square wave.

EXERCISE 6-13

Estimate the bandwidth of the following signals using a 10% criterion.

(a) $v(t) = 25 \sin 2\pi 20t + 10 \cos(2\pi 40t - 60°) + 5 \sin 2\pi 60t + 2 \cos 2\pi 100t$.

(b) $v(t) = -5 + 10 \cos 1000t + 3 \cos 3000t + 2 \sin 9000t$.

(c) $v(t) = 40 + 20 \cos 200t + 20 \sin 200t + 5 \cos 600t + 2.5 \sin(1200t + 60°)$.

ANSWERS:

(a) 40 Hz

(b) 1.432 kHz

(c) 95.5 Hz ■

6-8 SPICE WAVEFORMS

Computer-based circuit analysis programs can simulate the time-varying waveforms that serve as input signals for circuit analysis. In a circuit analysis context we think of independent sources as generators of the input waveforms. The catalog of available waveforms in analysis programs include composite signals such as pulses, exponentials, and damped sines. To continue the development begun in Chapter 5, we use the SPICE syntax to illustrate the way source waveforms are defined in computer-aided analysis.

In Chapter 5 the SPICE element for an independent voltage or current sources was defined in the following form:

```
<Name> <1st node> <2nd node> <type> <value>
```

The <Name> parameter identifies the source using an alphanumeric string Vxxxxx for voltage source and Ixxxxx for a current source. The <1st> and <2nd> parameters define the two nodes to which the source is connected. In Chapter 5 we only considered dc sources (<type> := DC) that delivered a constant waveform defined by the <value> parameter. In this chapter we introduce source types that generate time-varying waveforms. The parameters for these waveforms are given in the <value> entry of the source element statement. For time-varying waveforms the <value> is not just one number, but a set of numbers that define a time-varying waveform.

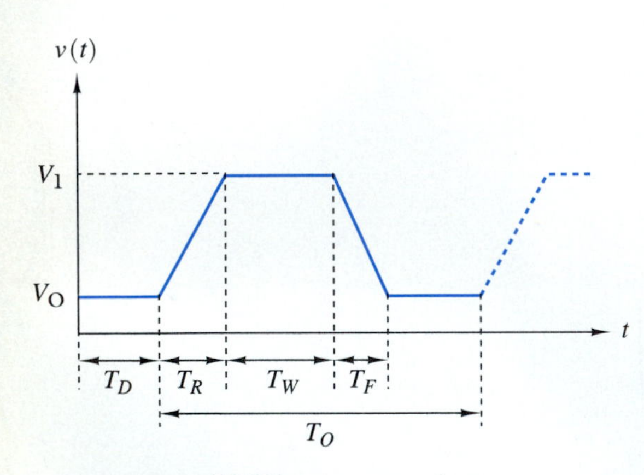

Figure 6-33 SPICE pulse waveform.

The first SPICE waveform is the periodic pulse train shown in Fig. 6-33. After an initial time delay of T_D seconds, the waveform rises linearly from V_O to V_1 in T_R seconds. The waveform remains at the V_1 level for T_W seconds after which it decreases linearly back to the V_O level in T_F seconds. For $t > T_D$ the rise/flat/fall pattern is repeated every T_O seconds. The SPICE calls this type waveform a PULSE. The waveform type and the seven parameters shown in Fig. 6-33 are listed in the value entry in the order shown below:

```
<type>:= PULSE
<value>:=(VO V1 TD TR TF TW TO)
```

For example, the source statement

```
VIN 2 0 PULSE (0 5 0 1M 2M 5M 10M)
```

defines a voltage source VIN connected between Nodes 2 and 0 with a pulse waveform with $V_O = 0$, $V_1 = 5$, $T_D = 0$, $T_R =$ 1 ms, $T_F = 2$ ms, $T_W = 5$ ms, and $T_O = 10$ ms.

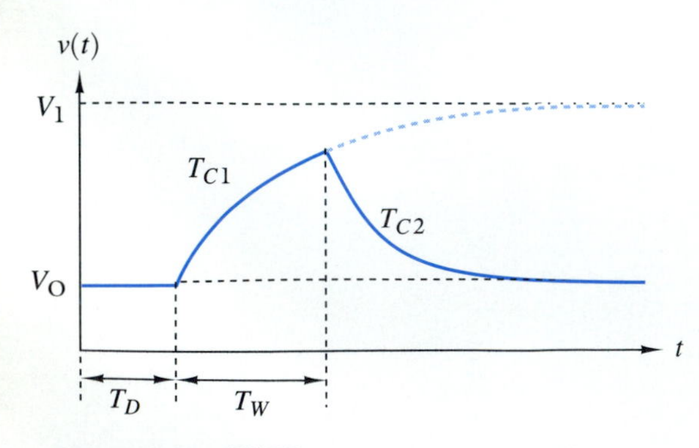

Figure 6-34 SPICE exponential waveform.

The exponential waveform in Fig. 6-34 is similar to the pulse waveform except that rise and fall are exponential rather than linear. After an initial time delay of T_D seconds, the waveform rises exponentially for T_W seconds with a time constant of T_{C1}. During interval $T_D < t < T_D + T_W$ the waveform is described by the equation

$$v(t) = V_O + (V_1 - V_O)\left[1 - e^{-(t-T_D)/T_{C1}}\right]$$

For $t > T_D + T_W$ the waveform decays exponentially back to the V_O level with a time constant of T_{C2} and is governed by the equation

$$v(t) = V_O + (V_O - V_1)(1 - e^{-T_W/T_{C1}})e^{-(t-T_D-T_W)/T_{C2}}$$

The SPICE calls this type of waveform an EXP. The waveform type and the six parameters shown in Fig. 6-33 are listed in the value in the order shown below:

```
<type>:= EXP
<value>:=(VO V1 TD TC1 TW TC2)
```

For example the source statement

```
IGS 12 14 EXP (10M 20M 0 12U 20U 6U)
```

defines a current source IGS connected between Nodes 12 and 14 with an exponential waveform with $I_O = 10$ mA, $I_1 = 20$ mA, $T_D = 0$, $T_{C1} = 12$ μs, $T_W = 20$ μs, and $T_{C2} = 6$ μs.

The SPICE sinusoid in Fig. 6-35 has an initial delay of T_D seconds. Thereafter the waveform is defined by the equation

$$v(t) = V_O + V_1 \left[e^{-\alpha(t-T_D)} \sin[2\pi f_O(t - T_D)]\right] u(t - T_D)$$

When $\alpha > 0$ the waveform is a damped sine oscillating about the V_O level. If $\alpha = 0$ the waveform is undamped and periodically oscillates between the limits $V_O \pm V_1$ for $t > T_D$. It is not, however, an eternal sine wave because the sinusoidal part is zero for $t < T_D$. The SPICE calls this type of waveform a SIN. The waveform type and the five parameters shown in Fig. 6-33 are listed in the value in the order shown below:

```
<type>:= SIN
<value>:=(VO V1 fO TD α )
```

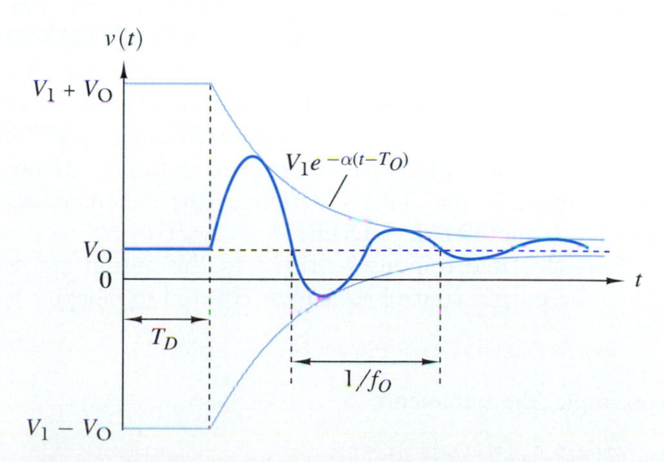

Figure 6-35 SPICE sinusoidal waveform.

For example, the source statement

```
I22 7 0 SIN (0 5 23K 0 140K)
```

defines a current source I22 connected between Nodes 7 and 0 with a damped sine waveform with $I_O = 0$, $I_1 = 5$ A, $f_O = 23$ kHz, $T_D = 0$, and $\alpha = 140$ krad/s.

The SPICE waveform is the piecewise linear waveform (PWL) in Fig. 6-36. The PWL waveform can be used to generate complicated waveforms by connecting straight lines between a sequence of time-voltage pairs. The SPICE type and value description for this waveform is

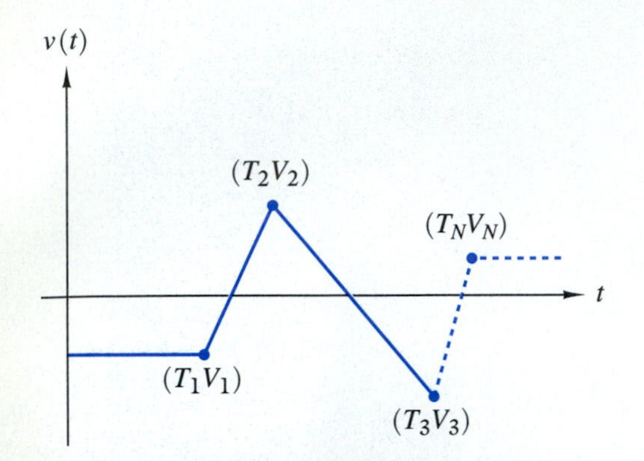

Figure 6-36 SPICE piecewise linear waveform.

```
<type>:= PWL
<value>:=(T1 V1 T2 V2 ... TN VN)
```

For $T_1 < t < T_N$ the waveform amplitude is found by linear interpolation between the adjacent time-voltage pairs. For $< T_1$ the amplitude if V_1 and for $t > T_N$ the amplitude is V_N. The time entries must meet the condition

$$0 \leq T_1 < T_2 < \ldots < T_N$$

That is, the times cannot be negative and must increase as we proceed down the list of time-voltage pairs.

These waveforms serve as inputs for the SPICE transient analysis option. The basic SPICE command statement for transient analysis takes the form

```
.TRAN TSTEP TSTOP TSTART TMAX
```

The .TRAN statement causes SPICE to calculate the circuit responses at a sequence of discrete times beginning at $t = 0$ and ending at t =TSTOP. TSTEP is the time increment at which the calculated responses are written to the output file. The internal step size used by SPICE is never less than TSTEP. TMAX allows the user to specify a smaller maximum internal step size. A transient analysis always starts at $t = 0$. However, writing the calculated responses to the output file is supressed until t =TSTART. The TSTART and TMAX parameters are optional. If no values are assigned in the .TRAN statement the default values are TSTART = 0 and TMAX = TSTEP.

The calculated responses written to the output are designated in the output control statement covered in Chapter 5.

```
.PRINT <analysis type> <variable list>
```

For example, the statements

```
.TRAN 0.1U 10U 1U 50N
.PRINT TRAN V(1) V(7) I(VAMP)
```

cause SPICE to perform a transient analysis beginning at $t = 0$ and continuing to $t = 10\ \mu s$ (TSTOP). The responses V(1), V(7), and I(VAMP) are written to the output file every 0.1 μs (TSTEP) beginning at $t = 1\ \mu s$ (TSTART). The maximum internal step size is TMAX = 50 ns.

Computer circuit analysis programs, including SPICE, have numerical convergence problems if the input waveform has jump discontinuities. Unfortunately, it is all too easy to intentionally or unintentionally create waveform statements that call for jump changes. For example, $T_R = 0$ in a PULSE waveform statement calls for a jump from V_O to V_1 at $t = T_D$. When the SPICE processor detects jump discontinuity in a waveform specified in a source statement, it overrides the user's choices and assigns a **default value** to the offending parameter. In most cases the default value is the TSTEP parameter in .TRAN command. The value of the TSTEP parameter should be carefully selected because it serves two purposes. It defines the time between successive response values in the output file and serves as a default value to used by SPICE to ensure that the input waveforms are all continuous.

EXAMPLE 6-19

Create a SPICE circuit file to produce the waveform in Fig. 6-37.

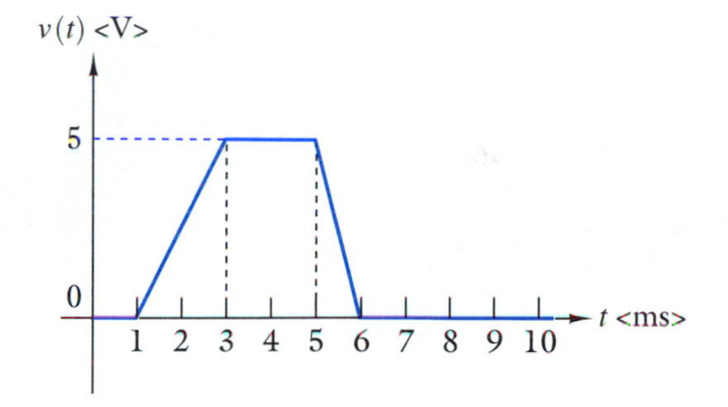

Figure 6-37

SOLUTION:

The waveform could be generated using the pulse or the piecewise linear waveform. Using the pulse waveform a suitable SPICE input file is

```
EXAMPLE 6-19
VS 1 0 PULSE (0 5 1M 2M 1M 2M 20M)
RS 1 0 1K
.TRAN 0.1M 10M
.PRINT TRAN V(1)
.END
```

Note that the period (20 ms) of the PULSE waveform plays no role in this example since the .TRAN command ends the analysis at TSTOP= 10 ms. Note as well that the resistor RS is connected across the source. Although RS does not affect the waveform in this example it must be there; otherwise, Node 1 would be dangling in the air and SPICE would issue an error message. Remember that SPICE requires at last two elements to be connected at every node.

EXERCISE 6-14

Find the peak and peak-to-peak values of the following SPICE waveforms.

(a) `PULSE (1 -5 1 2 1 3 10)`
(b) `EXP (0 5 0 1 2)`
(c) `SIN (3 -5 10K 0)`
(d) `PWL (0 1 0 2 5 3 4 -3)`

ANSWERS:

(a) $V_p = 5$ V, $V_{pp} = 6$ V
(b) $V_p = V_{pp} = 4.3$ V
(c) $V_p = 8$ V, $V_{pp} = 10$ V
(d) $V_p = 5$, $V_{pp} = 8$ V

SUMMARY

- A waveform is an equation or graph that describes a voltage or current as a function of time. Most signals of interest in electrical engineering can be derived using three basic waveforms: the step, exponential, and sinusoid.
- The step function is defined by its amplitude and time-shift parameters. The impulse, step, and ramp are called singularity functions and are often used as test inputs for circuit analysis purposes.
- The exponential waveform is defined by its amplitude, time constant, and time-shift parameter. For practical purposes the duration of the exponential waveform is five time constants.
- A sinusoid can be defined using the two Fourier coefficients or the amplitude and phase angle or time-shift parameter. The time-frequency parameters are defined by the period, frequency, or angular frequency.
- Many composite waveforms can be derived using the three basic waveforms. Some examples are the impulse, ramp, damped ramp, damped sinusoid, exponential rise, and double exponential.
- Partial descriptors are used to classify or describe important attributes of waveforms. Temporal descriptors are periodicity and causality. Periodic waveforms repeat themselves every T_O seconds. Causal signals are zero for $t < 0$. Some important amplitude descriptors are peak value (V_p), peak-to-peak value V_{pp}, average value V_{avg}, and root-mean-square value (V_{rms}).
- Periodic signals can be resolved into a dc component and a sum of ac components at harmonic frequencies. The spec-

trum of a periodic signal is an equation or graphic that defines the amplitudes and phase angles of these components. The signal bandwidth (BW) is a spectrum partial descriptor that defines the range of frequencies within which the component amplitudes exceed specified limits.

- The SPICE syntax includes pulse, exponential, damped sine, and piecewise linear waveforms. These waveforms are generated by independent voltage and current sources and are used as inputs for the transient analysis option.

EN ROUTE OBJECTIVES AND ASSOCIATED PROBLEMS

ERO 6-1 Basic Waveforms (Sects. 6-2, 6-3, 6-4)

Given an equation, graph or word description of a step, exponential or sinusoidal waveform:

(a) Construct an alternative description of the waveform.

(b) Find new waveforms obtained by summing, integrating, or differentiating the given waveforms.

6-1 Graph the following step function waveforms.
(a) $v_1(t) = 5\ u(t)$
(b) $v_2(t) = -5\ u(t-1)$
(c) $v_3(t) = 5\ u(1-t)$
(d) $v_4(t) = -10\ u(t-1)$

6-2 Graph the waveforms obtained by adding the waveforms in Prob. 6-1.
(a) $v_A(t) = v_1(t) + v_2(t)$
(b) $v_B(t) = v_1(t) + v_3(t)$
(c) $v_C(t) = v_1(t) + v_4(t)$
(d) $v_D(t) = v_2(t) + v_3(t)$

6-3 Graph the waveforms obtained by integrating and differentiating the waveforms in Prob. 6-1.

6-4 Write an equation for an exponential waveform that starts at $t = 0$, has an amplitude of 10 V, and a time constant of 20 ms. Graph the waveform.

6-5 Determine the time constants and then graph the following exponential waveforms.
(a) $v_1(t) = [10\ e^{-2t}]u(t)$
(b) $v_2(t) = [10\ e^{-t/2}]u(t)$
(c) $v_3(t) = [-10\ e^{-20t}]u(t)$
(d) $v_4(t) = [-10\ e^{-t/20}]u(t)$

6-6 Graph the waveforms obtained by adding the exponential waveforms in Prob. 6-5.

(a) $v_A(t) = v_1(t) + v_2(t)$
(b) $v_B(t) = v_1(t) + v_3(t)$
(c) $v_C(t) = v_1(t) + v_4(t)$
(d) $v_D(t) = v_2(t) + v_4(t)$

6-7 Graph the waveform obtained by (a) integrating and (b) differentiating the exponential waveforms in Prob. 6-5.

6-8 Show that the exponential waveform $v(t) = V_A e^{-\alpha t} u(t)$ is a solution of the following first order differential equation.

$$\frac{dv(t)}{dt} + \alpha\, v(t) = 0$$

6-9 Write an equation for a sinusoid with an amplitude of 20 V, a period of 10 ms, and its first positive peak at $t = 5$ ms. Graph the waveform.

6-10 A sinusoid has a frequency of 50 Hz, a value of 10 V at $t = 0$, and reaches its first positive peak at $t = 2.5$ ms. Determine its amplitude, phase angle, and Fourier coefficients.

6-11 Determine the period, frequency, amplitude, time shift, and phase angle of the following sinusoids. Graph the waveform.
(a) $v_1(t) = 10\ \cos(2000\pi t) + 10\ \sin(2000\pi t)$
(b) $v_2(t) = -30\ \cos(2000\pi t) - 20\ \sin(2000\pi t)$
(c) $v_3(t) = 10\ \cos(2\pi t/10) - 10\ \sin(2\pi t/10)$
(d) $v_4(t) = -20\ \cos(800\pi t) + 30\ \sin(800\pi t)$

6-12 Determine the frequency, period, amplitude, time shift, and phase angle of the sum of the first two sinusoids in Prob. 6-11.

6-13 Graph the waveforms obtained by (a) integrating and (b) differentiating the sinusoidal waveforms in Prob. 6-11.

6-14 Determine the frequency, period, and Fourier coefficients of the following sinusoids. Graph the waveforms.
(a) $v_1(t) = 20\ \cos(4000\pi t - 180°)$
(b) $v_2(t) = 20\ \cos(4000\pi t - 90°)$
(c) $v_3(t) = 30\ \cos(2\pi t/400 - 45°)$
(d) $v_4(t) = 60\ \sin(2000\pi t + 45°)$

6-15 Determine the frequency, period, phase angle, and amplitude of the sum of the first two sinusoids in Prob. 6-14.

6-16 Graph the waveforms obtained by (a) integrating and (b) differentiating the sinusoids in Prob. 6-14.

6-17 Show that the sinusoid $v(t) = a\ \cos\ \omega t + b\ \sin\ \omega t$ is a solution of the following second-order differential equation:

$$\frac{d^2v(t)}{dt^2} + \omega^2 v(t) = 0$$

ERO 6-2 Composite Waveforms (Sect. 6-5)

Given an equation, graph or word description of a composite waveform:

(a) Construct an alternative description of the waveform.

(b) Find new waveforms obtained by summing, integrating, differentiating, or clipping given waveforms.

6-18 Graph the following waveforms.
(a) $v_1(t) = 20[1 - e^{-100t}]u(t)$
(b) $v_1(t) = [20 - 10e^{-100t}]u(t)$
(c) $v_2(t) = 20[2 - \sin(10\pi t)]u(t)$
(d) $v_3(t) = 20[1 + e^{-t}\sin(10\pi t)]u(t)$

6-19 Graph the following ramp waveforms. Then graph and write an equation for the derivative of each waveform.
(a) $v_1(t) = tu(t) - (t-2)u(t-2)$
(b) $v_2(t) = tu(t) - 2(t-1)u(t-1) + (t-2)u(t-2)$

6-20 A raised cosine pulse is defined as

$$v(t) = [V_O + V_O\cos(2\pi t/T_O)][u(t + T_O/2) - u(t - T_O/2)]$$

What is the purpose of the term $[u(t+T_O/2)-u(t-T_O/2)$? What is the pulse amplitude and duration? Graph the pulse waveform. Derive an equation for and graph the derivative of the pulse.

6-21 A full-wave rectified cosine is defined as

$$v(t) = |V_A\cos(2\pi t/T_O)|$$

Graph the waveform. Then write an equation for and graph the derivative of the waveform.

6-22 Write an equation for the first cycle of the waveform in Fig. P6-22. Graph the waveforms obtained by differentiating the waveform.

6-23 Repeat Prob. 6-22 for the waveform in Fig. P6-23.

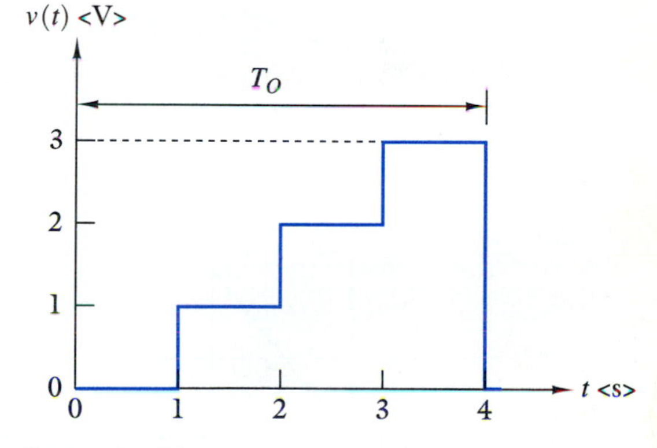

Figure P6-22

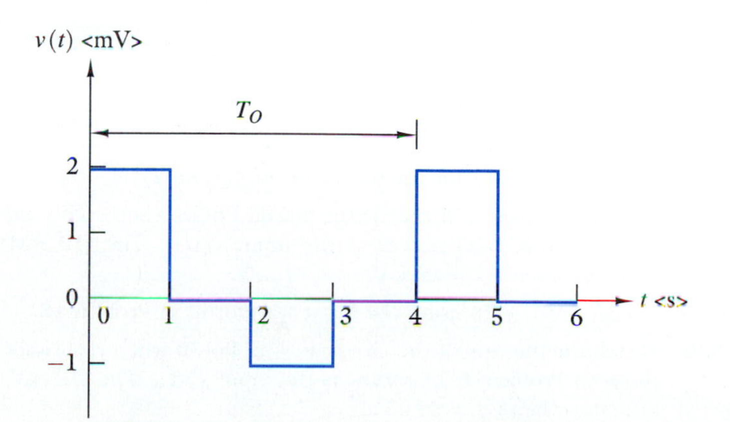

Figure P6-23

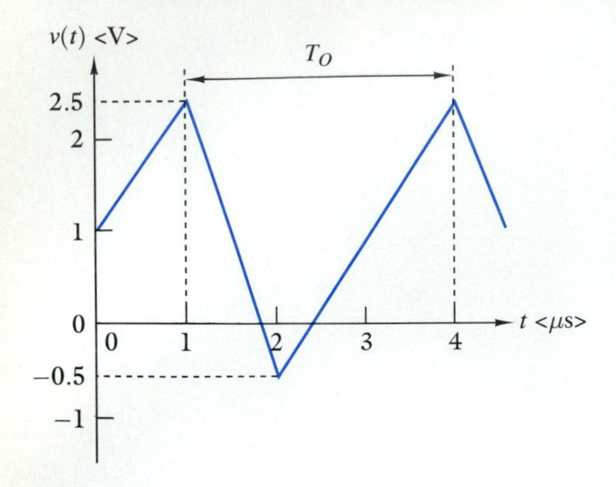

Figure P6-24

6-24 Repeat Prob. 6-22 for the waveform in Fig. P6-24.

6-25 Repeat Prob. 6-22 for the waveform in Fig. P6-25.

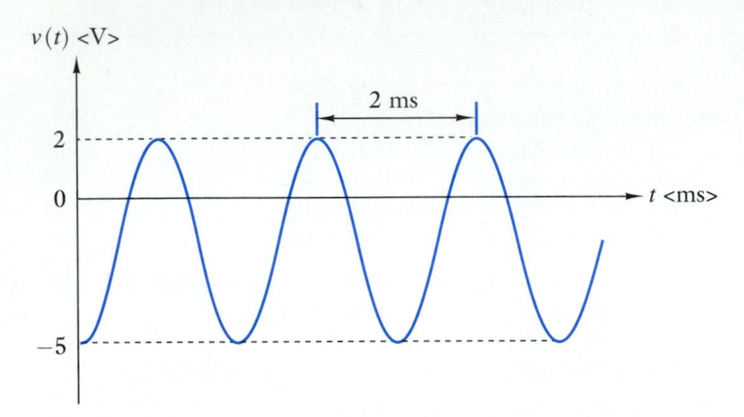

Figure P6-25

6-26 Repeat Prob. 6-22 for the waveform in Fig. P6-26.

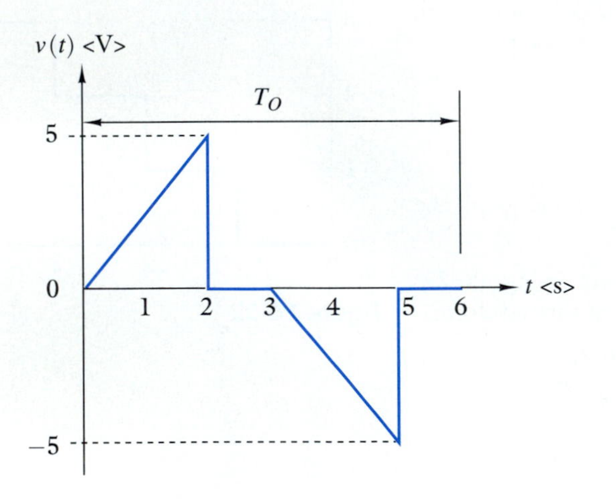

Figure P6-26

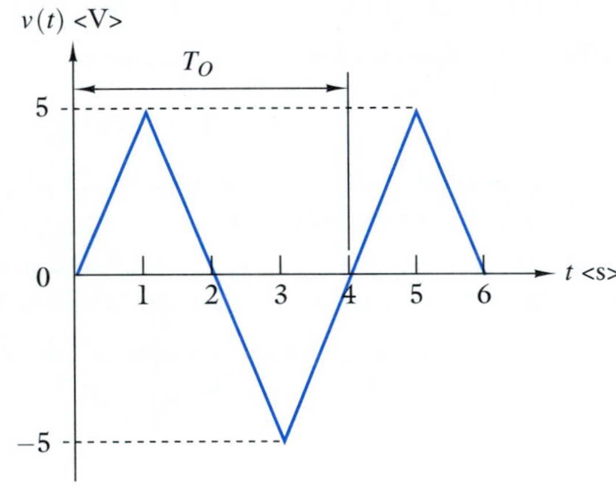

Figure P6-27

6-27 Repeat Prob. 6-22 for the waveform in Fig. P6-27.

6-28 Sketch the outputs of the circuit in Fig. P6-28 when each waveform in Prob. 6-11 serves as the input $v_S(t)$. The OP AMP saturates when $v_O = \pm 15$ V.

6-29 Repeat Prob. 6-28 using the input waveforms in Prob. 6-18.

6-30 Sketch the outputs of the circuit in Fig. P6-30 when each waveforms in Problem 6-11 serves as the input $v_S(t)$. The OP AMP saturates when $v_O = \pm 15$V.

6-31 Repeat Prob. 6-30 using the input waveforms in Figs. P6-26 and P6-27.

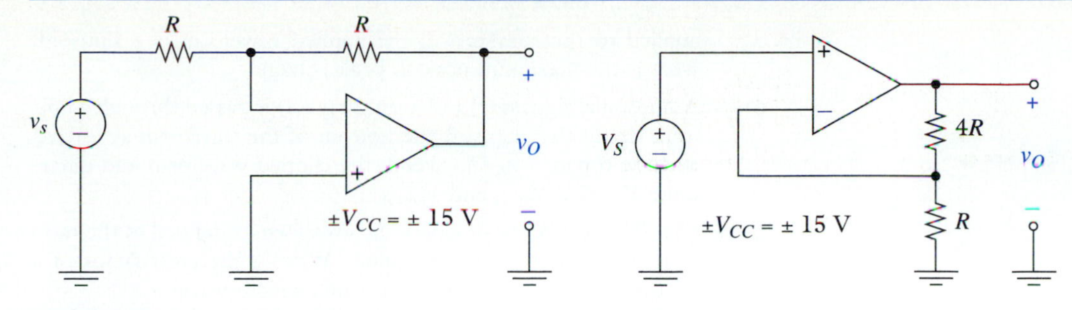

Figure P6-28

Figure P6-30

6-32 Show that the damped sinusoid $v(t) = V_A[e^{-\alpha t}\cos\beta t]u(t)$ is a solution of the following second-order differential equation.

$$\frac{d^2v(t)}{dt^2} + 2\,\alpha\frac{dv(t)}{dt} + (\alpha^2 + \beta^2)v(t) = 0$$

ERO 6-3 Waveform Partial Descriptors (Sect. 6-6)

Given a complete description of a basic or composite waveform:

(a) Find partial waveform descriptors.

(b) Classify the waveform as a periodic or aperiodic and causal or noncausal.

6-33 Determine V_p and V_{pp} for the following waveforms. Determine V_{rms} and V_{avg} for those waveforms that are periodic. Classify the waveforms as causal or noncausal.

(a) $v_1(t) = 10(1 - e^{-100t})u(t)$

(b) $v_2(t) = 10\ \cos(1000\pi t) - 10\ \sin(1000\pi t)$

(c) $v_3(t) = 100[e^{-t}\cos(1000\pi t)]u(t)$

(d) $v_4(t) = 10\ \cos(1000\pi t)[u(t) - u(t-1)]$

6-34 Repeat Prob. 6-33 for the waveforms in Prob. 6-11.

6-35 Repeat Prob. 6-33 for the waveforms in Prob. 6-18.

6-36 Repeat Prob. 6-33 for the waveform in Prob. 6-20.

6-37 Repeat Prob. 6-33 for the waveform in Prob. 6-21.

6-38 Repeat Prob. 6-33 for the waveform in Prob. 6-23.

6-39 Repeat Prob. 6-33 for the waveform in Prob. 6-24.

6-40 Repeat Prob. 6-33 for the waveform in Prob. 6-26.

6-41 The waveform $v(t) = V_O + 10\ \sin\ 2\pi t$ is applied at the input of the OP AMP circuit in Fig. P6-28. What range of values of the dc component V_O ensures that the OP AMP does not saturate?

6-42 A loudspeaker has a average power rating of 100 W and a resistance of 8 Ω. What is the maximum rms voltage that can be applied to the speaker? If the applied waveform is a sinusoid, what is the maximum peak-to-peak voltage?

6-43 A sinusoidal signal $v(t) = V_A \sin(2\pi t/T_O)$ is passed through a limiting circuit that clips off the bottom of the waveform at $-V_A/2$ and the top at $+V_A/2$. Sketch the clipped waveform and determine V_p, V_{pp}, V_{avg}, and V_{rms}.

6-44 **(A)** The crest factor of a periodic waveform is defined as the ratio of its peak value to its rms value. What is the crest factor of a sinusoid? What is the crest factor of a square wave?

6-45 Classify the waveforms in Prob. 6-11 as continuous or discontinuous, periodic or aperiodic, and causal or noncausal signals. If the waveform is periodic specify its period.

6-46 Repeat Prob. 6-45 for the waveforms in Prob. 6-18.

6-47 Repeat Prob. 6-45 for the ramp waveforms in Prob. 6-19.

6-48 Repeat Prob. 6-45 for the rectified waveform in Prob. 6-21.

6-49 Repeat Prob. 6-45 for the clipped sinusoid in Prob. 6-43.

6-50 Repeat Prob. 6-45 for the following waveforms.
(a) $v(t) = e^{-|t|}$
(b) $v(t) = u(t) - u(-t)$
(c) $v(t) = [\sin(t)]/t$
(d) $v(t) = \sin^2(t)$

ERO 6-4 Signal Spectra (Sect. 6-7)

(a) Given a waveform expressed as a sum of harmonic sinusoids, construct plots of its spectrum and determine its bandwidth using a specified criteria.

(b) Given an equation or graph of the spectrum of a periodic signal, determine its bandwidth using a specified criteria and write an expression for its waveform as a sum of harmonic sinusoids.

6-51 Construct the amplitude and phase plots of the spectrum of each of the following waveforms and determine their BWs using a 5% criteria. Repeat using $v(t) = v_1(t) + v_2(t)$.
(a) $v_1(t) = 4 + 5 \sin 2\pi 2500t - 2 \cos 2\pi 5000t + 0.2 \sin 2\pi 10^4 t$ V
(b) $v_2(t) = 3 \cos(2\pi 1250t - 60°) - 2 \sin 2\pi 2500t + \cos 2\pi 5000t$ V

6-52 Construct the amplitude and phase plots of the spectrum of each of the following waveforms and determine their BWs using a 5% criteria. Repeat using $v(t) = v_1(t) + v_2(t)$.
(a) $v_1(t) = -20 + 40 \cos 2\pi 1000t + 10 \cos 2\pi 2000t + 2.5 \cos 2\pi 4000t$ V

(b) $v_2(t) = 20 + 40\ \sin(2\pi 1000t - 90°)$
$+ 30\ \cos\ 2\pi 3000t - 1.6\ \cos\ 2\pi 5000t$ V.

6-53 The inputs to the OP AMP circuit in Fig. P6-53 are $v_1(t) = 5\ \cos(2\pi 3000t)$ V, $v_2(t) = 1.25\ \cos(2\pi 6000t)$ V, and $v_3 = 3$ V.

(a) Write an expression for the output voltage and sketch its waveform.

(b) Construct the amplitude and phase plots of the spectrum of the output voltage.

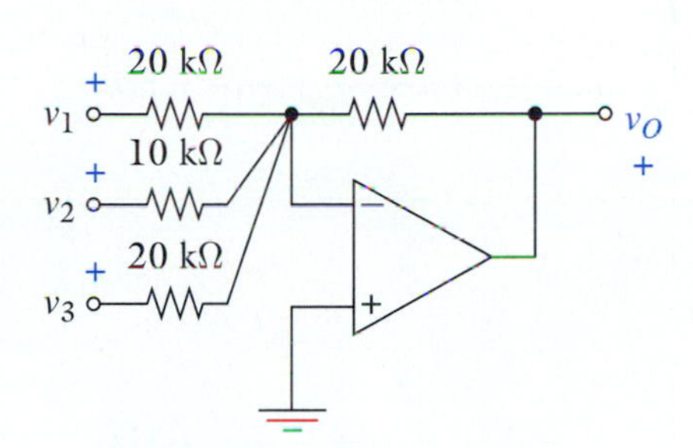

Figure P6-53

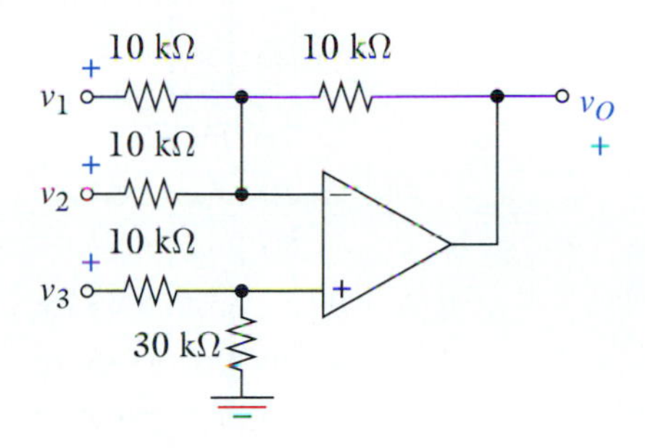

Figure P6-54

6-54 Repeat Prob. 6-53 using the OP AMP circuit in Fig. P6-54.

6-55 The amplitude and phase angles of the nth ($n = 1, 2, 3, \ldots$) harmonic of a period waveform with an average value of $V_A/2$ are

$$V_n = \frac{V_A}{n\pi} \qquad \phi_n = 270°$$

Using $T_O = 0.2$ ms and $V_A = 5$ V, determine the signal bandwidth using a 10% criteria. Write an equation for the waveform using only those frequency components that lie within this bandwidth.

6-56 The amplitude and phase angles of the nth ($n = 1, 2, 3, \ldots$) harmonic of a period waveform with zero average value are

$$V_n = \frac{8V_A}{(n\pi)^2} \qquad \phi_n = 0°,\ n \text{ odd}$$

$$V_n = 0,\ n \text{ even}$$

Using $T_O = 0.2$ ms and $V_A = 5$ V, determine the signal bandwidth using a 10% criteria. Write an equation for the waveform using only those frequency components that lie within this bandwidth.

6-57 A differentiator is a linear signal processor whose output is proportional to the time derivative of its input. For example, the input-output relationship for a certain differentiator is

$$v_{OUT} = \frac{1}{1000}\frac{dv_{IN}(t)}{dt}$$

Using the input voltage

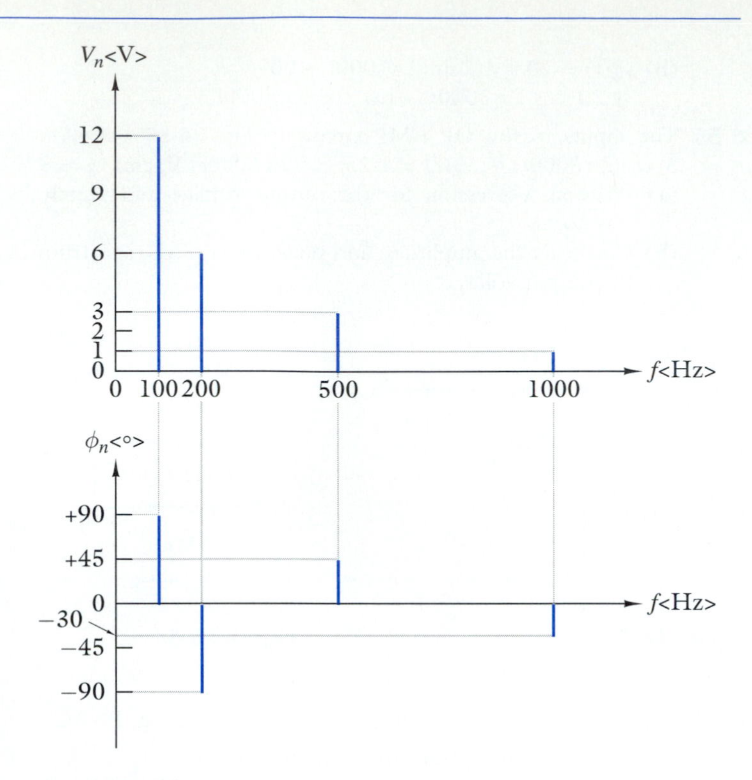

Figure P6-58

$$v_{IN}(t) = 10 \cos 200t + 2 \cos 600t + \cos 1000t + 0.3 \cos 3000t + 0.02 \cos 10^4 t$$

construct the amplitude and phase plots for the input spectrum and output spectrum. Determine the bandwidths of the input and output signals using a 5% criteria and discuss the spectral changes caused by the differentiator.

6-58 Figure P6-58 shows the amplitude and phase plots of the spectrum of a periodic signal. Write an equation for the signal waveform and sketch it time variation. Determine the signal bandwidth using a 10% criteria.

6-59 An ideal lowpass filter acts like an amplifier with a gain $K = 1$ for sinusoidal inputs with $\omega \leq \omega_C$, and a gain $K = 0$ for input with $\omega > \omega_C$. For a filter with $\omega_C = 2$ krad/s and an input

$$v_{IN}(t) = 10 + 10 \cos(2\pi 500t) + 3 \cos(2\pi 1000t) + 2 \cos(2\pi 4000t)$$

construct the amplitude plots of the input spectrum and output spectrum. Determine the bandwidths of the input and output signals using a 5% criteria and discuss the spectral changes caused by the lowpass filter.

6-60 An ideal time delay is a signal processor whose output at any time

t is equal to the earlier input at time $t = T_D$, that is, $v_{OUT}(t) = v_{IN}(t - T_D)$. For time delay of $T_D = 1$ ms and an input

$$v_{IN}(t) = 10 + 10\cos(2\pi 500t) + 3\cos(2\pi 1000t) + 2\cos(2\pi 4000t)$$

construct the amplitude and phase plots of the input spectrum and output spectrum. Determine the bandwidths of the input and output signals using a 5% criteria and discuss the spectral changes caused by the time delay.

6-61 An ideal squarer is a nonlinear signal processor whose output is proportional to the square of its input, that is, $v_{out} = K[v_{IN}(t)]^2$. For $K = 0.01$ and an input $v_{IN} = 20\cos(2\pi 500t)$, construct the amplitude and phase plots of the input spectrum and output spectrum, and discuss the spectral changes caused by the squarer. [Hint: Use the identity $\cos^2 x = \frac{1}{2}(1 + \cos 2x)$].

6-62 Use a PC-based spreadsheet or math analysis package to create a plot of the sum of the first 10 harmonics of the periodic signal defined in Prob. 6-55. Use a time interval $0 \leq t \leq 2T_O$. Can you recognize this waveform as one of the waveforms shown in Fig. 6-2?

6-63 Repeat Prob. 6-62 using the signal defined in Prob. 6-56.

ERO 6-5 SPICE Waveforms (Sect. 6-8)

(a) Given an equation, graph or word description of a composite waveform, construct a SPICE circuit file that will produce the waveform.

(b) Given a SPICE circuit file, construct an equation or graph of the waveform produced.

6-64 Write a SPICE circuit file that produces the signals in Prob. 6-18.

6-65 Write a SPICE circuit file that produces the signal in Prob. 6-23.

6-66 Write a SPICE circuit file that produces the signal in Prob. 6-26.

6-67 Construct a graph of the waveform generated by the following SPICE circuit file.

```
EXAMPLE CIRCUIT
VS 1 0 PULSE 0 15 1M 2M 2M 3M 10M
RS 1 0 10K
.TRAN 1M 20M
.PRINT TRAN V(1)
.END
```

6-68 Construct a graph of the waveform generated by the following SPICE circuit file.

```
EXAMPLE CIRCUIT
VS 1 0 SIN 5 -15 1K 0 0
RS 1 0 10K
.TRAN 0.1M 2M
.PRINT TRAN V(1)
.END
```

6-69 Write an expression for and construct a graph of the waveform generated by the following SPICE circuit file.

```
EXAMPLE CIRCUIT
VS 1 0 EXP 5 10 1M 2M 10M 1M
RS 1 0 10K
.TRAN 0.1M 6M
.PRINT TRAN V(1)
.END
```

6-70 Write an expression for and construct a graph of the waveform generated by the following SPICE circuit file.

```
EXAMPLE CIRCUIT
VS 1 0 EXP 5 10 1M 1U 5U 1M
RS 1 0 10K
.TRAN 0.1M 6M
.PRINT TRAN V(1)
.END
```

CHAPTER INTEGRATING PROBLEMS

6-71 **(A)** The output of a comparator is +15 V if $v_1(t) > v_2(t)$ and −15 V if $v_1(t) < v_2(t)$. Plot the comparator output voltage for $v_1(t) = 1$ and $v_2(t) = 2\cos(2\pi t)$. What is the period of the output? What are $V_p V_{pp}$, V_{avg}, and V_{rms} for the output? What frequencies are contained in the input waveforms? What frequencies are contained in the output waveform?

6-72 **(A)** Derive an expression for and sketch the waveform of the instantaneous power $p(t)$ delivered to a resistor R by a sinusoidal current $v(t) = I_A \sin(2\pi t/T_0)$. Show that $p(t)$ is periodic with a period of $T_0/2$. What are the peak, peak-to-peak, and average values of $p(t)$?

6-73 **(A)** Determine V_p, V_{pp}, V_{avg}, and V_{rms} for the SPICE waveform

PULSE V_0 V_1 0 T_R T_F T_W T_0.

6-74 **(A)** A full wave rectified sinusoid $|V_A \sin \omega t|$ is applied to a dc voltmeter, which measures average values. The meter indicates 56 volts. What is the rms value of the sinusoid?

6-75 **(A)** The two-sided exponential is a composite waveform defined as

$$v_1(t) = V_A[e^{-\alpha t}u(t) + e^{\alpha t}u(-t)]$$

(a) Sketch the waveform.
(b) Is the signal causal or noncausal? Periodic or aperiodic?
(c) What is the practical duration of the waveform?
(d) What are the peak and peak-to-peak values for this waveform?

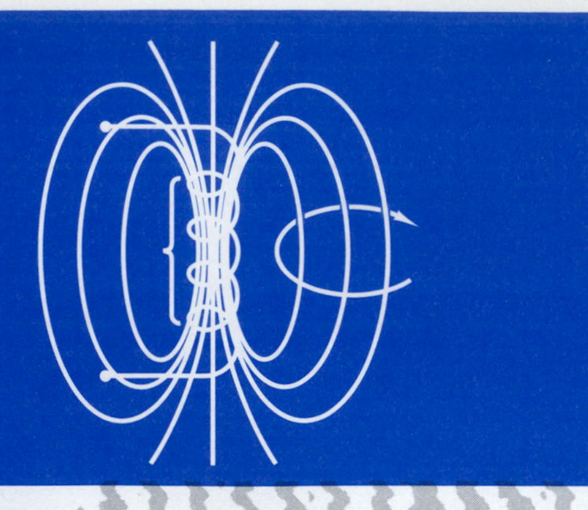

Chapter Seven

"From the foregoing facts, it appears that a current of electricity is produced, for an instant, in a helix of copper wire surrounding a piece of soft iron whenever magnetism is induced in the iron; also that an instantaneous current in one or the other direction accompanies every change in the magnetic intensity of the iron."

Joseph Henry, 1831
American Physicist

CAPACITANCE AND INDUCTANCE

Joseph Henry (1797–1878) and the British physicist Michael Faraday (1791–1867) independently discovered magnetic induction almost simultaneously. The quotation on the opposite page is Henry's summary of the experiments leading to his discovery of magnetic induction. Although Henry and Faraday used similar apparatus and observed almost the same results, Henry was the first to fully recognize the importance of the discovery. The units of circuit inductance (henrys) honors Henry, while the mathematical generalization of magnetic induction is called Faraday's law. Michael Faraday had wide-ranging interests and performed many fundamental experiments in chemistry and electricity. His electrical experiments often used capacitors or Leyden jars as they were called in those days. Faraday was a meticulous experimenter, and his careful characterization of these devices may be the reason that the unit of capacitance (the farad) honors Faraday.

The dynamic circuit responses involve memory effects that cannot be explained by resistance circuits. Two new circuit elements required to explain these effects are called capacitance and inductance. The purpose of this chapter is to introduce the *i-v* characteristics of these dynamic elements and to explore circuit applications that illustrate dynamic behavior.

The first two sections of this chapter develop the device constraints for the linear capacitor and inductor. The third section is a circuit application in which resistors, capacitors, and operational amplifiers (OP AMPs) perform waveform integration and differentiation. The fourth section shows that capacitors or inductors connected in series or parallel can be replaced by a single equivalent capacitance or inductance.

The fifth section covers mutual inductance effects in which a time-varying current in one inductor induces a voltage in another inductor. Mutual inductance is the basis for a widely used electrical device called a transformer, whose ideal model is developed in the sixth section. The last section presents the SPICE syntax that describes capacitance and inductance in computer-aided circuit analysis.

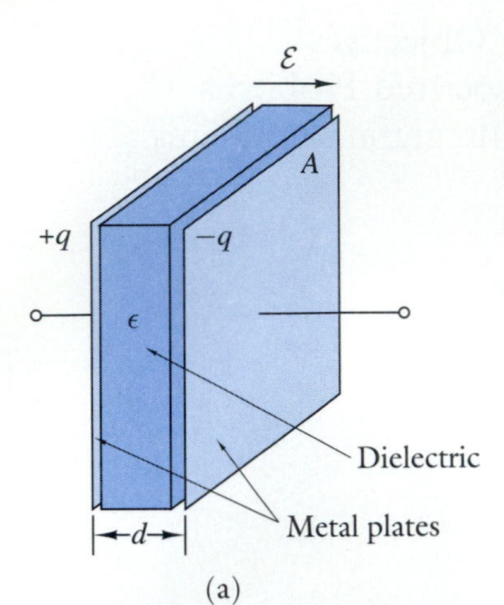

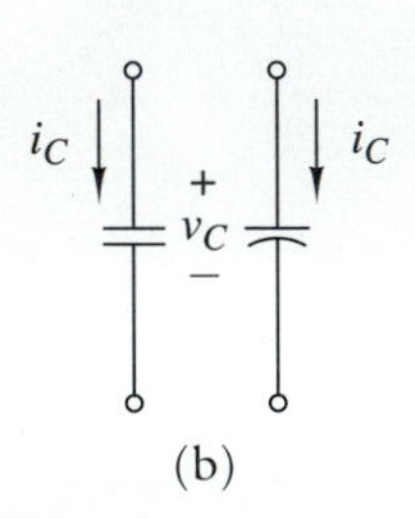

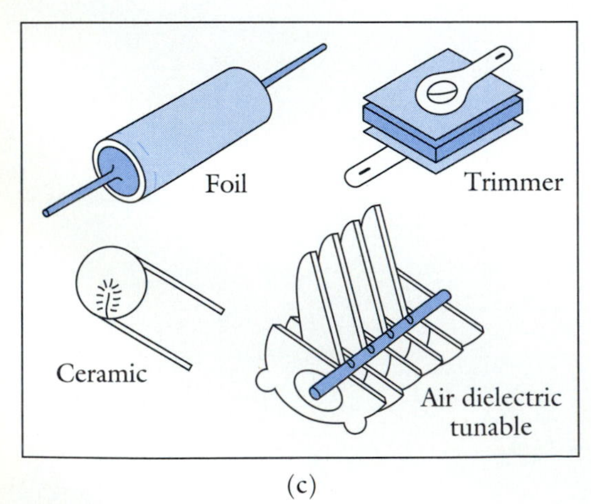

Figure 7-1 The capacitor: (a) Parallel plate device. (b) Circuit symbol. (c) Example devices.

7-1 THE CAPACITOR

A *capacitor* is a dynamic element based upon the time variation of an electric field produced by a voltage. Figure 7-1(a) shows the parallel plate capacitor, which is the simplest physical form of a capacitive device. Figure 7-1 also shows two alternative circuit symbols and sketches of actual devices. Some of the physical features of commercially available devices are discussed in Appendix A.

Electrostatics shows that a uniform electric field $\mathcal{E}(t)$ exists between the metal plates in Fig. 7-1(a) when a voltage exists across the capacitor.[1] The electric field produces charge separation with equal and opposite charges appearing on the capacitor plates. When the separation d is small compared with the dimension of the plates, the electric field between the plates is

$$\mathcal{E}(t) = \frac{q(t)}{\epsilon A} \tag{7-1}$$

where ϵ is the permitivity of the dielectric, A is the area of the plates, and $q(t)$ is the magnitude of the electric charge on each plate. The electric field is related to the voltage across the capacitor $v_C(t)$ by

$$\mathcal{E}(t) = \frac{v_C(t)}{d} \tag{7-2}$$

Substituting Eq. (7-2) into Eq. (7-1) and solving for the charge $q(t)$ yields

$$q(t) = \left[\frac{\epsilon A}{d}\right] v_C(t) \tag{7-3}$$

The proportionality constant inside the bracket in this equation is the *capacitance* C of the capacitor. That is, by definition

$$C \equiv \frac{\epsilon A}{d} \tag{7-4}$$

so Eq. (7-3) becomes

$$q(t) = Cv_C(t) \tag{7-5}$$

The unit of capacitance is the *farad* (F), a term that honors the British physicist Michael Faraday. Values of capacitance range from a few pF (10^{-12} F) in semiconductor devices to tens of mF (10^{-3} F) in industrial capacitor banks.

[1] An electrical field is a vector quantity. In Fig. 7-1(a) the field is confined to the space between the two plates and is perpendicular to the plates.

i-v Relationships

Equation (7-5) is the capacitor element constraint in terms of voltage and electric charge. To express the element relationship in terms of voltage and current, we differentiate Eq. (7-5) with respect to time t:

$$\frac{dq(t)}{dt} = \frac{d[Cv_C(t)]}{dt}$$

Since C is constant and $i_C(t)$ is the time derivative of $q(t)$, we obtain a capacitor i-v relationship in the form

$$i_C(t) = C\frac{dv_C(t)}{dt} \qquad (7\text{-}6)$$

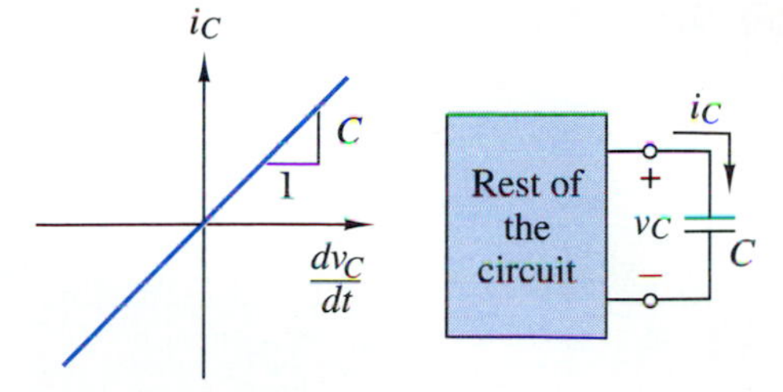

Figure 7-2 Capacitor *i-v* characteristics.

The relationship in Eq. (7-6) is shown graphically in Fig. 7-2. The relationship assumes the reference marks for the current and voltage follow the passive sign convention.

The time derivative in Eq. (7-6) means the current is zero when the voltage across the capacitor is constant, and vice versa. In other words, the capacitor acts like an open circuit ($i_C = 0$) when dc excitations are applied. The capacitor is a dynamic element because the current is zero unless the voltage is changing. However, a discontinuous change in voltage requires an infinite current, which is physically impossible. Therefore, the capacitor voltage must be a continuous function of time.

Equation (7-6) relates the capacitor current to the rate of change of the capacitor voltage. To express the voltage in terms of the current we multiply both sides of Eq. (7-6) by dt, solve for the differential dv_C, and integrate:

$$\int dv_C = \frac{1}{C}\int i_C(t)dt$$

Selecting the integration limits requires some discussion. We assume that at some time t_O the voltage across the capacitor $v_C(t_O)$ is known and we want to determine the voltage at some later time $t > t_O$. Therefore, the integration limits are

$$\int_{v_C(t_O)}^{v_C(t)} dv_C = \int_{t_O}^{t} i_C(x)dx$$

where x is a dummy integration variable. The left side of this equation can be integrated directly to obtain

$$v_C(t) = v_C(t_O) + \frac{1}{C}\int_{t_O}^{t} i_C(x)dx \qquad (7\text{-}7)$$

In practice the time t_O is established by a physical event such as closing a switch or the start of a particular clock pulse. Nothing is lost in the integration in Eq. (7-7) if we arbitrarily define t_O

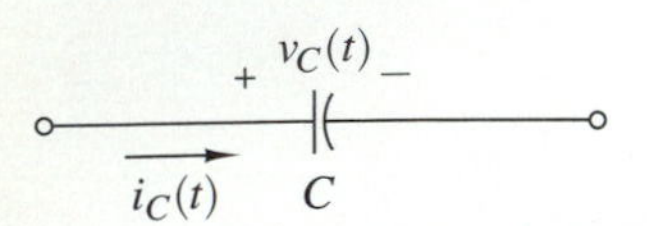

Figure 7-3 Capacitor current and voltage.

to be zero. Using $t_O = 0$ in Eq. (7-7) yields

$$v_C(t) = v_C(0) + \frac{1}{C}\int_0^t i_C(x)dx \tag{7-8}$$

Equation (7-8) is the integral form of the capacitor i-v constraint. Both the integral form and the derivative form in Eq. (7-6) assume that the reference marks for current and voltage follow the passive sign convention in Fig. 7-3.

Power and Energy

With the passive convention the power associated with the capacitor is

$$p_C(t) = i_C(t)v_C(t) \tag{7-9}$$

Using Eq. (7-6) to eliminate $i_C(t)$ from Eq. (7-9) yields the capacitor power in the form

$$p_C(t) = Cv_C(t)\frac{dv_C(t)}{dt} = \frac{d}{dt}\left[\frac{1}{2}Cv_C^2(t)\right] \tag{7-10}$$

This equation shows that the power can be either positive or negative because the capacitor voltage and its time rate of change can have opposite signs. With the passive sign convention, a positive sign means the element absorbs power, while a negative sign means the element delivers power. The ability to deliver power implies that the capacitor can store energy.

To determine the stored energy, we note that the expression for power in Eq. (7-10) is a perfect derivative. Since power is the time rate of change of energy, the quantity inside the bracket must be the energy stored in the capacitor. Mathematically what we can infer from Eq. (7-10) is that the energy at time t is

$$w_C(t) = \frac{1}{2}Cv_C^2(t) + \text{constant}$$

The constant in this equation is the value of stored energy at some reference point in the past. We assume that the zero energy is stored at the reference point and write the capacitor energy as

$$w_C(t) = \frac{1}{2}Cv_C^2(t) \tag{7-11}$$

The stored energy is never negative, since it is proportional to the square of the voltage. The capacitor absorbs power from the circuit when storing energy and returns previously stored energy when delivering power to the circuit.

The relationship in Eq. (7-11) also implies that voltage is a continuous function of time, since an abrupt change in the voltage implies a discontinuous change in energy. Since power is the time derivative of energy, a discontinuous change in energy implies infinite power, which is physically impossible. The capacitor voltage is called a *state variable* because it determines the energy state of the element.

To summarize, the capacitor is a dynamic circuit element with the following properties:

1. The current through the capacitor is zero unless the voltage is changing. The capacitor acts like an open circuit to dc excitations.
2. The voltage across the capacitor is a continuous function of time. Discontinuous changes in capacitor voltage require infinite current and power, which is physically impossible.
3. The capacitor absorbs power from the circuit when storing energy and returns previously stored energy when delivering power. The net energy transfer is nonnegative, indicating that the capacitor is a passive element.

The following examples illustrate these properties.

EXAMPLE 7-1

The voltage in Fig. 7-4(a) appears across a $\frac{1}{2}$-μF capacitor. Find the current through the capacitor.

SOLUTION:

The capacitor current is proportional to the time rate of change of the voltage. For $0 < t < 2$ ms the slope of the voltage waveform has a constant value

$$\frac{dv_C}{dt} = \frac{10}{2 \times 10^{-3}} = 5000 \text{ V/s}$$

The capacitor current during this interval is

$$i_C(t) = C\frac{dv_C}{dt} = (0.5 \times 10^{-6}) \times (5 \times 10^3) = 2.5 \text{ mA}$$

For $2 < t < 3$ ms the rate of change of the voltage is -5000 V/s. Since the rate of change of voltage is negative, the current changes direction and takes on the value $i_C(t) = -2.5$ mA. For $t > 3$ ms the voltage is constant so the slope and current are both zero. The resulting current waveform is shown in Fig. 7-4(b). Note that the voltage across the capacitor is continuous, but the current is discontinuous.

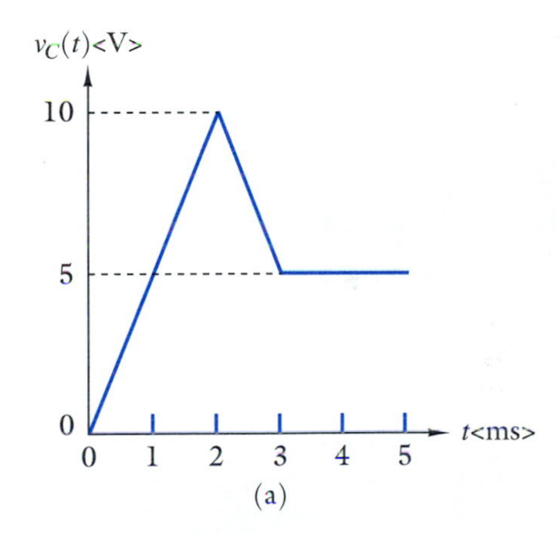

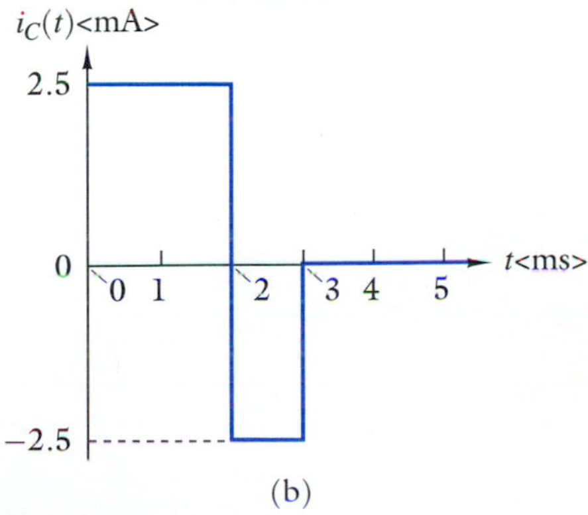

Figure 7-4

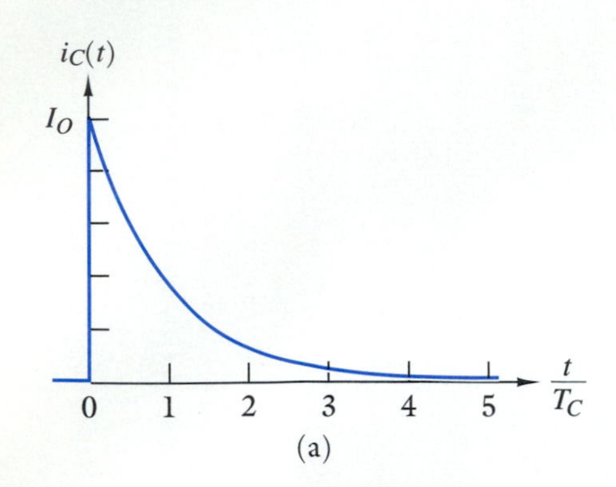

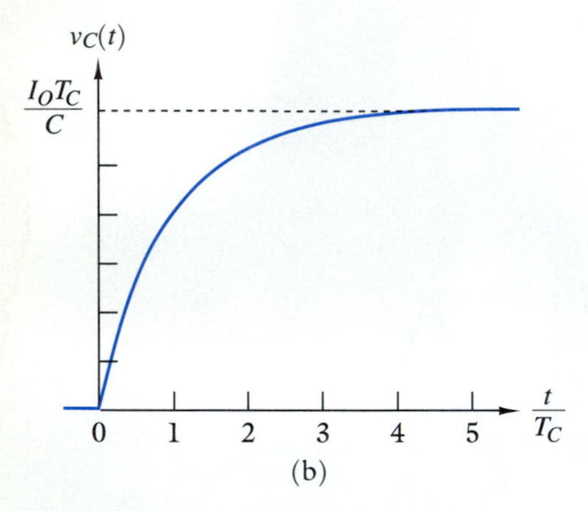

Figure 7-5

EXAMPLE 7-2

The $i_C(t)$ in Fig. 7-5(a) is given by

$$i_C(t) = I_0 \left[e^{-t/T_C}\right] u(t)$$

Find the voltage across the capacitor if $v_C(0) = 0$ V.

SOLUTION:

Using the capacitor i-v relationship in integral form

$$\begin{aligned} v_C(t) &= v_C(0) + \frac{1}{C}\int_0^t i_C(x)dx \\ &= 0 + \frac{1}{C}\int_0^t I_0 e^{-x/T_C}dx = \frac{I_0 T_C}{C}\left(-e^{-x/T_C}\right)\Big|_0^t \\ &= \frac{I_0 T_C}{C}\left(1 - e^{-t/T_C}\right) \end{aligned}$$

The graphs in Fig. 7-5(b) shows that the voltage is continuous.

EXERCISE 7-1

(a) The voltage across a 10-μF capacitor is $25[\sin 2000t]u(t)$. Derive an expression for the current through the capacitor.

(b) At $t = 0$ the voltage across a 100-pF capacitor is -5 V. The current through the capacitor is $10[u(t) - u(t - 10^{-4})]\mu$A. What is the voltage across the capacitor for $t > 0$?

ANSWERS:

(a) $i_C(t) = 0.5[\cos 2000t]u(t)$ A

(b) $v_C(t) = -5 + 10^5 t$ V for $0 < t < 0.1$ ms and $v_C(t) = 5$ V for $t > 0.1$ ms ■

EXERCISE 7-2

For $t \geq 0$ the voltage across a 200-pF capacitor is $5e^{-4000t}$.

(a) What is the charge on the capacitor at $t = 0$ and $t = +\infty$?

(b) Derive an expression for the current through the capacitor for $t \geq 0$.

(c) For $t > 0$ is the device absorbing or delivering power?

ANSWERS:

(a) 1 nC and 0 C

(b) $i_C(t) = -4e^{-4000t}\,\mu$A

(c) delivering ■

EXAMPLE 7-3

Figure 7-6(a) shows the voltage across 0.5-μF capacitor. Find the capacitor's energy and power.

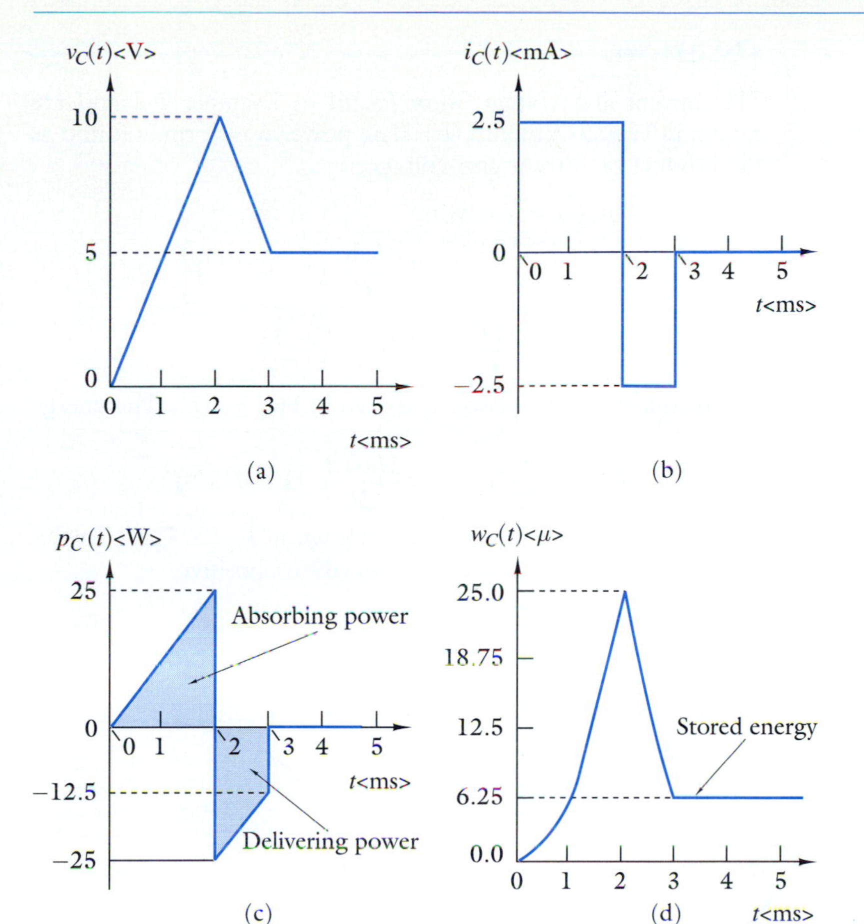

Figure 7-6

SOLUTION:

The current through the capacitor was found in Example 7-1. The power waveform is the point-by-point product of the voltage and current waveforms. The energy is found by either integrating the power waveform or by calculating $\frac{1}{2}C[v_C(t)]^2$ point by point. The current, power, and energy are shown in Figs. 7-6(b), (c), and (d). Note that the capacitor energy increases when it is absorbing power ($p_C > 0$) and decreases when delivering power ($p_C < 0$).

EXAMPLE 7-4

The current through a capacitor is given by

$$i_C(t) = I_0\left[e^{-t/T_C}\right]u(t)$$

Find the capacitor's energy and power.

SOLUTION:

The current and voltage were found in Example 7-2 and are shown in Figs. 7-7(a) and (b). The power waveform is found as the product of current and voltage.

$$\begin{aligned} p_C(t) &= i_C(t)v_C(t) \\ &= \left[I_0 e^{-t/T_C}\right]\left[\frac{I_0 T_C}{C}\left(1 - e^{-t/T_C}\right)\right] \\ &= \frac{I_0^2 T_C}{C}\left(e^{-t/T_C} - e^{-2t/T_C}\right) \end{aligned}$$

The waveform of the power is shown in Fig. 7-7(c). The energy is

$$w_C(t) = \frac{1}{2}Cv_C^2(t) = \frac{(I_0 T_C)^2}{2C}\left(1 - e^{-t/T_C}\right)^2$$

The time history of the energy is shown in Fig. 7-7(d). In this example both power and energy are always positive.

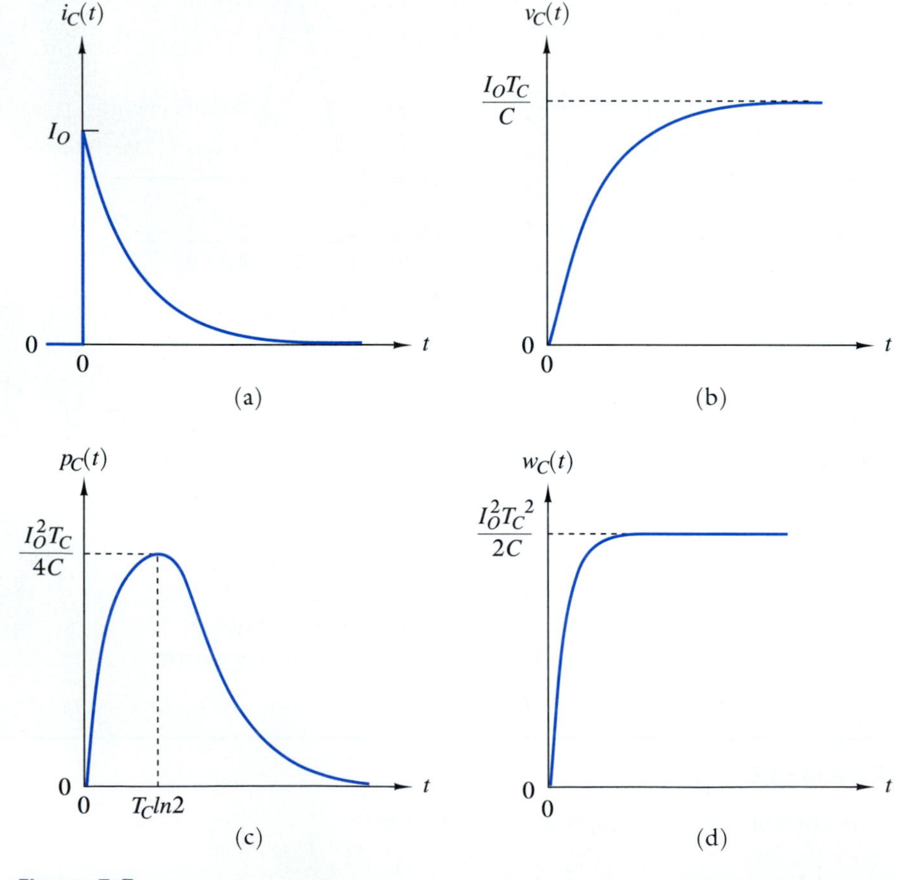

Figure 7-7

EXERCISE 7-3

Find the power and energy for the capacitors in Exercise 7-1.

ANSWERS:

(a) $p_C(t) = 6.25[\sin\ 4000t]u(t)$ W
$w_C(t) = 3.125 \sin^2 2000t$ mJ

(b) $p_C(t) = -0.05 + 10^3 t$ mW for $0 < t < 0.1$ ms
$p_C(t) = 0$ for $t > 0.1$ ms
$w_C(t) = 1.25 - 5 \times 10^4 t + 5 \times 10^8 t^2$ nJ for $0 < t < 0.1$ ms
$w_C(t) = 1.25$ nJ for $t > 0.1$ ms

EXERCISE 7-4

Find the power and energy for the capacitor in Exercise 7-2.

ANSWERS:

$p_C(t) = -20e^{-8000t}\ \mu$W $\quad w_C(t) = 2.5e^{-8000t}$ nJ

Application Note

EXAMPLE 7-5

A track-hold circuit is usually required at the input to an analog-to-digital (A/D) converter. The purpose of the circuit is to continuously follow a time-varying input voltage and then to hold the voltage value during the time required to perform the A/D conversion. This example shows that a capacitor can perform the track-hold function.

Figure 7-8(a) shows the simplest form of a capacitor track-hold circuit. When the switch is closed, the circuit is in the track mode and the voltage across the capacitor follows or tracks the input source voltage. At some time $t = t_O$ the switch is opened and the circuit changes to the hold mode shown in Fig. 7-8(b). For $t > t_O$ the source voltage continues to change, producing the waveform $v_S(t)$ shown in blue in Fig. 7-8(c). However, Fig. 7-8(c) shows in black that the capacitor voltage $v_C(t)$ is constant for $t > t_O$. Opening the switch at $t = t_O$ forces the capacitor current to zero, and zero current means the voltage is constant. The capacitor holds or remembers the voltage existing across it at the instant the switch was opened, namely, $v_S(t_O)$.

Theoretically, the capacitor holds $v_S(t_O)$ forever. In reality some leakage occurs through the dielectric, so the capacitor gradually loses stored energy and the voltage decreases. The stored value in a track-hold circuit must be occasionally refreshed for long-term storage. The capacitor can hold or remember past events because it can store energy.

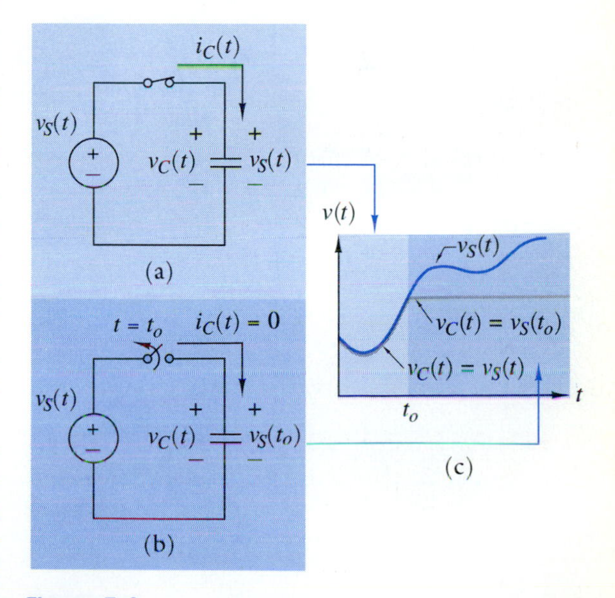

Figure 7-8

7-2 THE INDUCTOR

The *inductor* is a dynamic circuit element based upon the time variation of the magnetic field produced by a current. Magnetostatics shows that a *magnetic flux* ϕ surrounds a wire carrying an electric current. When the wire is wound into a coil the lines of flux concentrate along the axis of the coil as indicated in Fig. 7-9. In a linear magnetic medium the flux is proportional to both the current and the number of turns in the coil. Therefore, the total flux is

$$\phi(t) = k_1 N i_L(t) \tag{7-12}$$

where k_1 is a constant of proportionality.

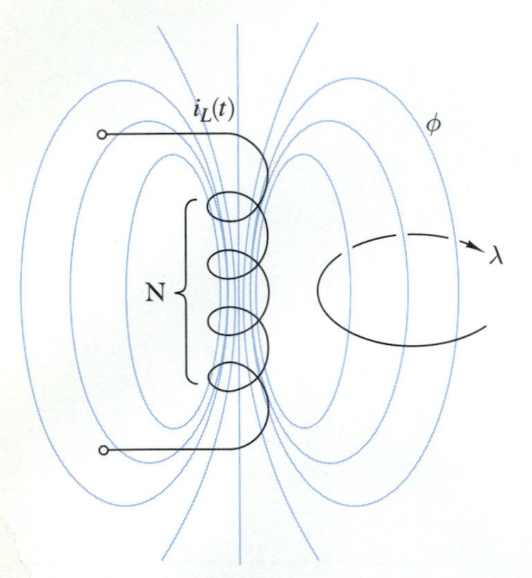

Figure 7-9 Magnetic flux surrounding a current carrying coil.

The magnetic flux intercepts or links the turns of the coil. The *flux linkages* in a coil are represented by the symbol λ, with units of webers (Wb), named after the German scientist Wilhelm Weber (1804–1891). The number of flux linkages is proportional to the number of turns in the coil and to the total magnetic flux, so λ is

$$\lambda(t) = N\phi(t) \tag{7-13}$$

Substituting Eq. (7-12) into Eq. (7-13) gives

$$\lambda(t) = \left[k_1 N^2\right] i_L(t) \tag{7-14}$$

The proportionality constant inside the brackets in this equation is the *inductance* L of the coil. That is, by definition

$$L \equiv k_1 N^2 \tag{7-15}$$

so Eq. (7-14) becomes

$$\lambda(t) = L i_L(t) \tag{7-16}$$

The unit of inductance is the henry (H) (plural henrys), a name that honors American scientist Joseph Henry. Figure 7-10 shows the circuit symbol for an inductor and some examples of actual devices. Some physical features of commercially available inductors are discussed in Appendix A.

i-v Relationship

Equation (7-16) is the inductor element constraint in terms of current and flux linkages. To obtain the element characteristic in terms of voltage and current we differentiate Eq. (7-16) with respect to time:

$$\frac{d[\lambda(t)]}{dt} = \frac{d\,[L i_L(t)]}{dt} \tag{7-17}$$

The inductance L is a constant. According to Faraday's law the voltage across the inductor is equal to the time rate of change of

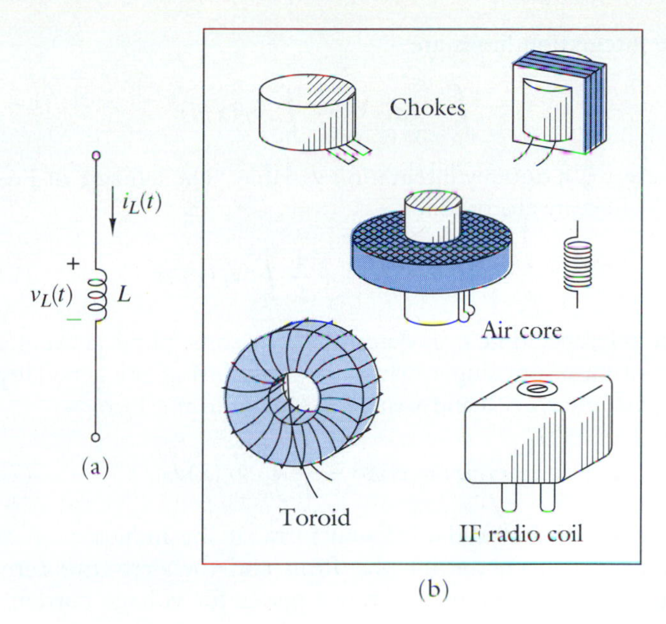

Figure 7-10 The inductor: (a) Circuit symbol. (b) Example devices.

flux linkages. Therefore, we obtain an inductor i-v relationship in the form

$$v_L(t) = L\frac{di_L(t)}{dt} \tag{7-18}$$

The i-v characteristics of the linear inductor are shown in Fig. 7-11.

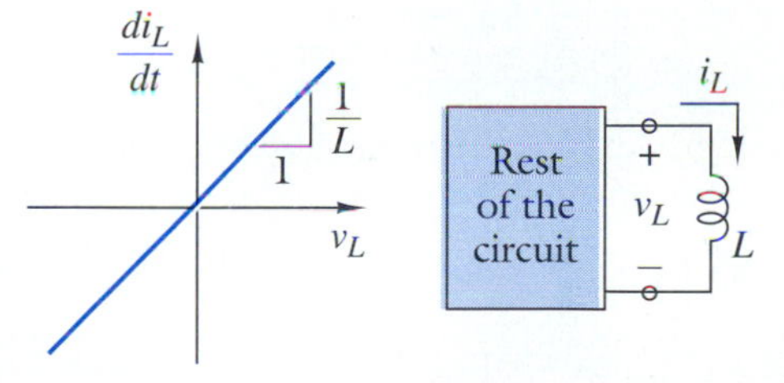

Figure 7-11 Inductor i-v characteristics.

The time derivative in Eq. (7-18) means that the voltage across the inductor is zero unless the current is time varying. Under dc excitation the current is constant and $v_L = 0$, so the inductor acts like a short circuit. The inductor is a dynamic element because only a changing current produces a nonzero voltage. However, a discontinuous change in current produces an infinite voltage, which is physically impossible. Therefore, the current $i_L(t)$ must be a continuous function of time t.

Equation (7-18) relates the inductor voltage to the rate of change of the inductor current. To express the inductor current in terms of the voltage, we multiply both sides of Eq. (7-18) by dt, solve for the differential di_L, and integrate:

$$\int di_L = \frac{1}{L}\int v_L(t)dt \tag{7-19}$$

To set the limits of integration, we assume that the inductor current $i_L(t_O)$ is known at some time t_O. Under this assumption the integration limits are

$$\int_{i_L(t_O)}^{i_L(t)} di_L = \frac{1}{L}\int_{t_O}^{t} v_L(x)dx \tag{7-20}$$

where x is a dummy integration variable. The left side of Eq. (7-20) integrates to produce

$$i_L(t) = i_L(t_O) + \frac{1}{L}\int_{t_O}^{t} v_L(x)dx \tag{7-21}$$

The reference time t_O is determined by some physical event such as closing or opening a switch. Without losing any generality we can assume $t_O = 0$ and write Eq. (7-21) in the form

$$i_L(t) = i_L(0) + \frac{1}{L}\int_{0}^{t} v_L(x)dx \tag{7-22}$$

Equation (7-22) is the integral form of the inductor i-v characteristics. Both the integral form and the derivative form in Eq. (7-18) assume the reference marks for voltage current follow the passive sign convention.

With the passive sign convention the inductor power is

$$p_L(t) = i_L(t)v_L(t) \tag{7-23}$$

Using Eq. (7-18) to eliminate v_L from this equation puts the inductor power in the form:

$$p_L(t) = [i_L(t)]\left[L\frac{di_L(t)}{dt}\right] = \frac{d}{dt}\left[\frac{1}{2}Li_L^2(t)\right] \tag{7-24}$$

This expression shows that power can be positive or negative because the inductor current and its time derivative can have opposite signs. With the passive sign convention a positive means the element absorbs power, while a negative sign means the element delivers power. The ability to deliver power indicates that the inductor can store energy. To find the stored energy we note that the power relation in Eq. (7-24) is a perfect derivative. Since power is the time rate of change of energy, the quantity inside the bracket must represent the energy stored in the magnetic field of the inductor:

$$w_L(t) = \frac{1}{2}Li_L^2(t) \tag{7-25}$$

The derivation of this equation assumes the stored energy is zero at some reference time in the past. The energy stored in an inductor is never negative because it is proportional to the square of the current. The inductor stores energy when absorbing power and returns previously stored energy when delivering power.

Equation (7-25) implies that inductor current is a continuous function of time because an abrupt change in current causes a discontinuity in the energy. Since power is the time derivative of energy, an energy discontinuity implies infinite power, which is physically impossible. Current is called the *state variable* of the inductor because it determines the energy state of the element.

In summary, the inductor is a dynamic circuit element with the following properties:

1. The voltage across the inductor is zero unless the current through the inductor is changing. The inductor acts like a short circuit for dc excitations.
2. The current through the inductor is a continuous function of time. Discontinuous changes in inductor current require infinite voltage and power, which is physically impossible.
3. The inductor absorbs power from the circuit when storing energy and delivers power to the circuit when returning previously stored energy. The net energy transferred is nonnegative, indicating that the inductor is a passive element.

EXAMPLE 7-6

The current through a 2-mH inductor is $i_L = 2 \sin 1000t$ A. Find waveforms of the resulting voltage, power, and energy.

SOLUTION:

The voltage is found from the derivative form of the i-v relationship:

$$\begin{aligned} v_L(t) &= L\frac{di_L(t)}{dt} = 0.002[2 \times 1000 \ \cos 1000t] \\ &= 4 \cos\ 1000t \text{ V} \end{aligned}$$

The inductor power is

$$\begin{aligned} p_L(t) &= i_L(t)v_L(t) = (2 \sin 1000t)(4 \cos 1000t) \\ &= 4 \sin 2000t \text{ W} \end{aligned}$$

The energy is given by

$$w_L(t) = \frac{1}{2}Li_L^2(t) = 4 \sin^2 1000t \text{ mJ}$$

These waveforms are shown in Fig. 7-12. Note that the power is alternately positive and negative, whereas the energy is nonnegative.

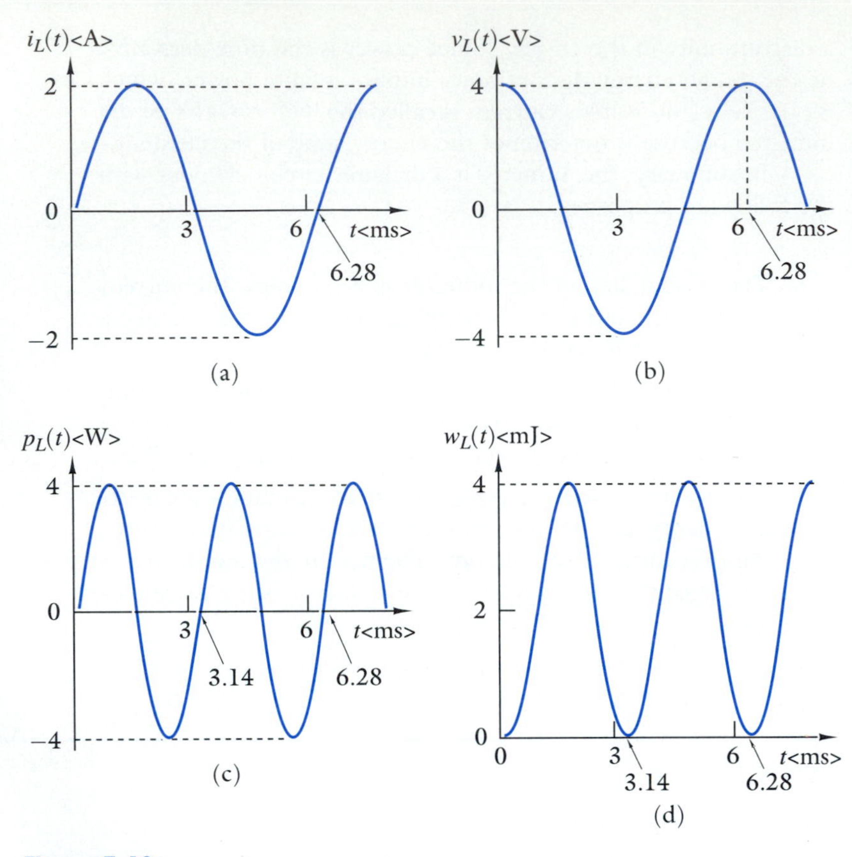

Figure 7-12

EXAMPLE 7-7

Figure 7-13 shows the current through and voltage across an unknown energy storage element.

(a) What is the element and what is its numerical value?

(b) If the energy stored in the element at $t = 0$ is zero, how much energy is stored in the element at $t = 1$ sec?

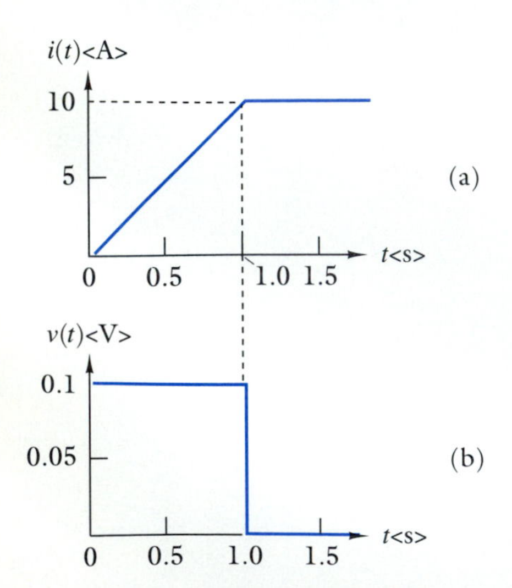

Figure 7-13

SOLUTION:

(a) By inspection the voltage across the device is proportional to the derivative of the current so the element is a linear inductor. During the interval $0 < t < 1$ s the slope of the current waveform is 10 A/s. During the same interval the voltage is a constant 100 mV. Therefore, the inductance is

$$L = \frac{v}{di/dt} = \frac{0.1 \text{ V}}{10 \text{ A/s}} = 10 \text{ mH}$$

(b) The energy stored at $t = 1$ sec is

$$w_L(1) = \frac{1}{2}Li_L^2(1) = 0.5(0.01)(10)^2 = 0.5 \text{ J}$$

EXERCISE 7-5

For $t > 0$ the voltage across a 4-mH inductor is $v_L(t) = 20e^{-2000t}$ V. The initial current $i_L(0) = 0$.

(a) What is the current through the inductor for $t > 0$?
(b) What are the power and
(c) energy for $t > 0$?

ANSWERS:

(a) $i_L(t) = 2.5\left(1 - e^{-2000t}\right)$ A
(b) $p_L(t) = 50\left(e^{-2000t} - e^{-4000t}\right)$ W
(c) $w_L(t) = 12.5\left(1 - 2e^{-2000t} = e^{-4000t}\right)$ mJ

EXERCISE 7-6

For $t < 0$ the current through a 100-mH inductor is zero. For $t \geq 0$ the current is $i_L(t) = 20e^{-2000t} - 20e^{-4000t}$ mA.

(a) Derive an expression for the voltage across the inductor for $t > 0$.
(b) Find the time $t > 0$ at which the inductor voltage passes through zero.
(c) Derive an expression for the inductor power for $t > 0$.
(d) Find the time interval over which the inductor absorbs power and the interval over which it delivers power.

ANSWERS:

(a) $v_L(t) = -4e^{-2000t} + 8e^{-4000t}$ V
(b) $t = 0.347$ ms
(c) $p_L(t) = -80e^{-4000t} + 240e^{-6000t} - 160e^{-8000t}$ mW
(d) Absorbing for $0 < t < 0.347$ ms, delivering for $t > 0.347$ ms

More about Duality

The capacitor and inductor characteristics are quite similar. Replacing C by L, i by v, and v by i in the capacitor equations gives the inductor equations, and vice versa. The interchangeability of capacitor and inductor relations falls under the *principle of duality*. Some dual concepts seen so far are the following:

KVL	KCL
resistance	conductance
voltage source	current source
Thévenin	Norton
short circuit	open circuit
series	parallel
capacitance	inductance
flux linkages	charge

The concept or quantity in one column is the dual of the quantity in the other. Duality is extremely useful in gaining confidence in working circuit analysis problems. When the concept in one column is understood, the dual concept in the other column is easier to remember and apply.

7-3 DYNAMIC OP AMP CIRCUITS

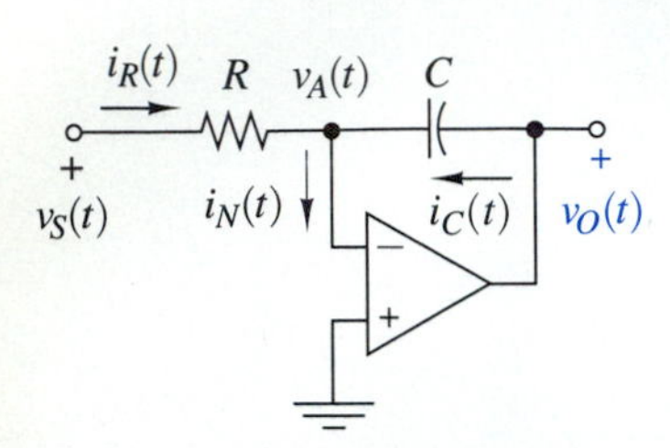

Figure 7-14 The inverting OP AMP integrator.

The dynamic characteristics of capacitors and inductors produce signal-processing functions that cannot be obtained using resistors. The OP AMP circuit in Fig. 7-14 is similar to the noninverting amplifier circuit except for the capacitor in the feedback path. To determine the signal-processing function of the circuit, we need to write the device and connection constraints and solve for the input-output relationship.

We begin by writing a KCL equation at Node A:

$$i_R(t) + i_C(t) = i_N(t)$$

The resistor and capacitor device equations are written using their i-v relationships and the fundamental property of node voltages:

$$i_C(t) = C\frac{d[v_O(t) - v_A(t)]}{dt}$$

$$i_R(t) = \frac{1}{R}[v_S(t) - v_A(t)]$$

The ideal OP AMP device equations are $i_N(t) = 0$ and $v_A(t) = 0$. Substituting all of the element constraints into the KCL connection constraint produces

$$\frac{v_S(t)}{R} + C\frac{dv_O(t)}{dt} = 0$$

To solve for the output $v_O(t)$ we multiply this equation by dt, solve for the differential dv_O, and integrate:

$$\int dv_O = -\frac{1}{RC}\int v_S(t)dt$$

Assuming the output voltage is known at time $t_O = 0$, the integration limits are

$$\int_{v_O(0)}^{v_O(t)} dv_O = -\frac{1}{RC}\int_0^t v_S(x)dx$$

which yields

$$v_O(t) = v_O(0) - \frac{1}{RC}\int_0^t v_S(x)dx$$

The output voltage is equal to the voltage across the capacitor. To show that $v_O = v_C$ we apply the fundamental property of node voltages to the feedback capacitor in Fig. 7-14 and obtain the KVL constraint $v_C(t) = v_O(t) - v_A(t)$. But $v_A = 0$, so in general $v_O(t) = v_C(t)$. When the voltage on the capacitor is zero at $t = 0$, the circuit input-output relationship reduces to

$$v_O(t) = -\frac{1}{RC}\int_0^t v_S(x)dx \tag{7-26}$$

The output voltage is proportional to the integral of the input voltage when the initial capacitor voltage is zero. The circuit in Fig. 7-14 is an *inverting integrator* since the proportionality constant is negative. The constant $1/RC$ has the units of reciprocal seconds (s^{-1}) so that both sides of Eq. (7-26) have the units of volts.

Interchanging the resistor and capacitor in Fig. 7-14 produces the OP AMP differentiator in Fig. 7-15. To show that this circuit performs time differentiation we must write the device and connection equations and solve for the relationship between input and output.

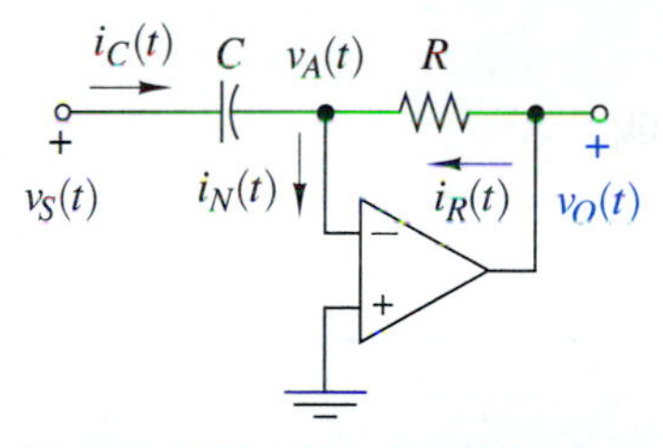

Figure 7-15 The inverting OP AMP differentiator.

The KCL connection constraint at Node A is

$$i_R(t) + i_C(t) = i_N(t)$$

The device equations for the input capacitor and feedback resistor are

$$i_C(t) = C\frac{d[v_S(t) - v_A(t)]}{dt}$$

$$i_R(t) = \frac{1}{R}[v_O(t) - v_A(t)]$$

The device equations for the OP AMP are $i_N(t) = 0$ and $v_A(t) = 0$. Substituting all of these element constraints into the KCL connection constraint produces

$$\frac{v_O(t)}{R} + C\frac{dv_S(t)}{dt} = 0$$

Solving this equation for $v_O(t)$ produces the circuit input-output relationship:

$$v_O(t) = -RC\frac{dv_S(t)}{dt} \tag{7-27}$$

The output voltage is proportional to the derivative of the input voltage. The circuit in Fig. 7-15 is an *inverting differentiator*, since the proportionality constant $(-RC)$ is negative. The units of the constant RC are seconds, so that both sides of Eq. (7-27) have the units of volts.

Because of duality, there are OP AMP inductor circuits that produce the inverting integrator and differentiator functions. These circuits are of little practical interest because of the size and loss characteristic of inductor devices. For completeness the dual inductor circuits are treated in Problem 7-18.

Figure 7-16 shows OP AMP circuits and block diagrams for the inverting integrator and differentiator, together with signal-processing functions introduced in Chap. 4. The term *"operational" amplifier* results from the ability of these circuits to perform a variety of mathematical operations on the input signals. The following examples illustrate using the collection of circuits in Fig. 7-16 in the analysis and design of signal-processing functions.

EXAMPLE 7-8

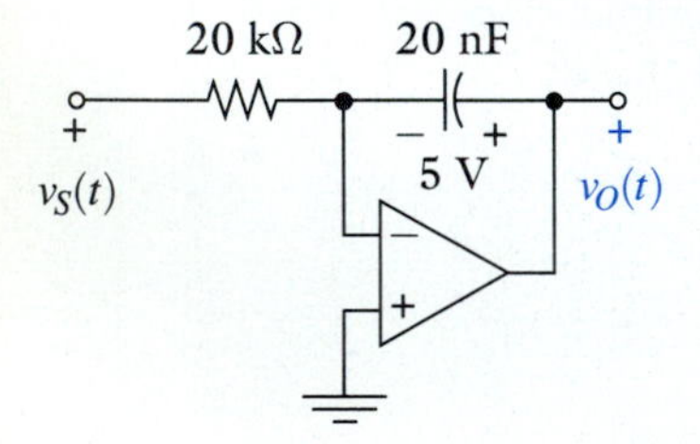

Figure 7-17

The input to the circuit in Fig. 7-17 is $v_S(t) = 10u(t)$. Derive an expression for the output voltage. The OP AMP saturates when $v_O(t) = \pm 15$ V.

SOLUTION:

The circuit is the inverting integrator with an initial voltage of 5 V across the capacitor. For the reference marks shown in Fig. 7-17 this means that $v_O(0) = +5$ V. Assuming the OP AMP is operating in the linear mode, the output voltage is

$$\begin{aligned} v_O(t) &= v_O(0) - \frac{1}{RC}\int_0^t v_S dt \\ &= 5 - 2500\int_0^t 10\, dt \\ &= 5 - 25000t \qquad t > 0 \end{aligned}$$

The output contains a negative going ramp because the circuit is an inverting integrator. The ramp output response is valid only as long as the OP AMP remains in its linear range. Negative saturation will occur when $5 - 25000t = -15$, or at $t = 0.8$ ms. For $t > 0.8$ ms the OP AMP is in the negative saturation mode with $v_O = -15$ V.

Circuit	Block diagram	Gains
v_1, v_O, R_1, R_2	v_1 → K → v_O	$K = \dfrac{R_1 + R_2}{R_2}$
v_1, R_1, R_2, v_O	v_1 → K → v_O	$K = -\dfrac{R_2}{R_1}$
R_1, R_F, v_O, v_1, R_2, v_2	v_1 → K_1, v_2 → K_2, + + → v_O	$K_1 = -\dfrac{R_F}{R_1}$ $K_2 = -\dfrac{R_F}{R_2}$
R_1, R_2, v_O, v_1, R_3, v_2, R_4	v_1 → K_1, v_2 → K_2, + + → v_O	$K_1 = -\dfrac{R_2}{R_1}$ $K_2 = \left(\dfrac{R_1 + R_2}{R_1}\right)\left(\dfrac{R_4}{R_3 + R_4}\right)$
v_1, R, C, v_O	v_1 → K → $\int$ → v_O	$K = -\dfrac{1}{RC}$
v_1, C, R, v_O	v_1 → K → $\dfrac{d}{dt}$ → v_O	$K = -RC$

Figure 7-16 Summary of basic OP AMP signal processing circuits.

EXERCISE 7-7

The input to the circuit in Fig. 7-17 is $v_S(t) = 10[e^{-5000t}]u(t)$ V.

(a) For $v_C(0) = 0$, derive an expression for the output voltage assuming the OP AMP is in its linear range.

(b) Does the OP AMP saturate with the given input?

ANSWERS:

(a) $v_O(t) = 5(e^{-5000t} - 1)u(t)$

(b) Does not saturate

EXAMPLE 7-9

The input to the circuit in Fig. 7-18(a) is a trapeziodal waveform shown in (b). Find the output waveform. The OP AMP saturates when $v_O(t) = \pm 15$ V.

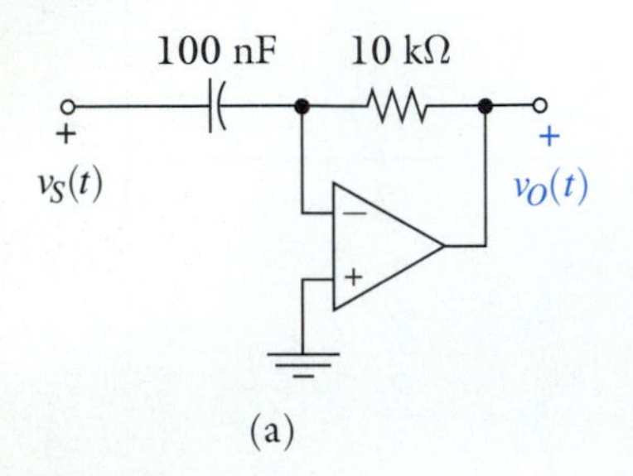

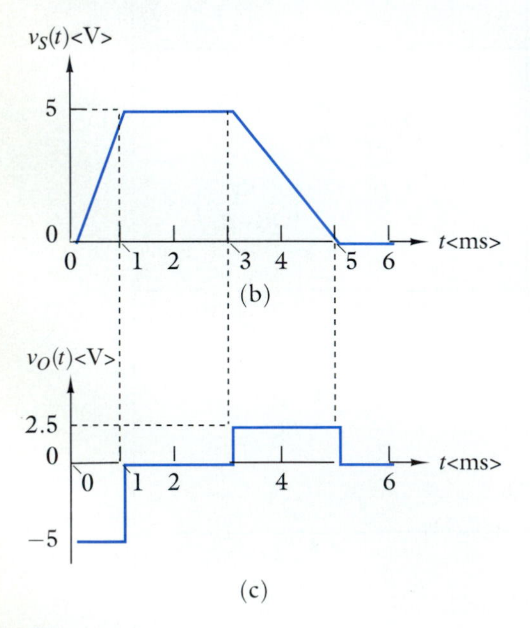

Figure 7-18

SOLUTION:

The circuit is the inverting differentiator with the following input-output relationship:

$$v_O(t) = -RC\frac{dv_S(t)}{dt} = -\frac{1}{1000}\frac{dv_S(t)}{dt}$$

The output voltage is constant over each of the following three time intervals:

1. For $0 < t < 1$ ms the input slope is 5000 V/s and the output is $v_O = -5$ V.
2. For $1 < t < 3$ ms the input slope is zero so the output is zero as well.
3. For $3 < t < 5$ ms the input slope is -2500 V/s and the output is $+2.5$ V.

The resulting output waveform is shown in Fig. 7-18(c).

The output voltage remains within ± 15 V limits, so the OP AMP operates in its linear range.

EXERCISE 7-8

The input to the circuit in Fig. 7-18 is $v_s(t) = V_A \cos 2000t$. The OP AMP saturates when $v_O = \pm 15$ V.

(a) Derive an expression for the output assuming the OP AMP is in the linear mode.

(b) What is the maximum value of V_A for linear operation?

ANSWERS:

(a) $v_O(t) = 2V_A \sin 2000t$

(b) $|V_A| \leq 7.5$ V

EXAMPLE 7-10

Determine the input-output relationship of the circuit in Fig. 7-19(a).

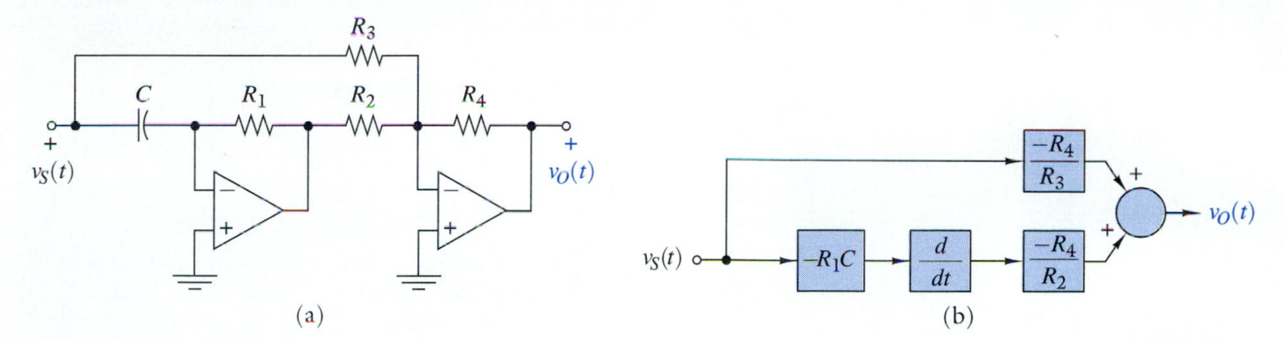

Figure 7-19

SOLUTION:

The circuit contains an inverting differentiator and an inverting summer. To find the input-output relationship it is helpful to develop a block diagram for the circuit. Figure 7-19(b) shows block diagram using the functional blocks in Fig. 7-16. The product of the gains along the lower path yields its contribution to the output as

$$(-R_1C)\left[\frac{d}{dt}\left(-\frac{R_4}{R_2}\right)v_S(t)\right]$$

The upper path contributes $(-R_4/R_3)v_S(t)$ to the output. The total output is the sum of the contributions from each path:

$$v_O(t) = \left(\frac{R_1CR_4}{R_2}\right)\frac{dv_S(t)}{dt} - \left(\frac{R_4}{R_3}\right)v_S(t)$$

This equation assumes that both OP AMPs remain within their linear range.

EXERCISE 7-9

Find the input-output relationship of the circuit in Fig. 7-20.

ANSWER:

$$v_O(t) = v_O(0) + \frac{1}{RC}\int_0^t (v_{S1} - v_{S2})dt$$

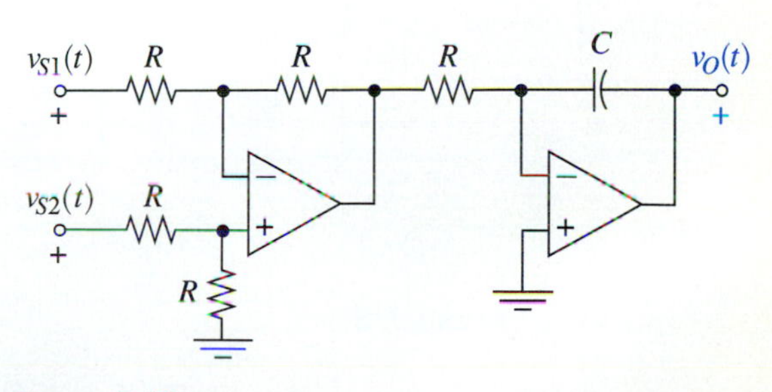

Figure 7-20

❖ Design Example

EXAMPLE 7-11

Use the functional blocks in Fig. 7-16 to design an OP AMP circuit to implement the following input-output relationship:

$$v_O(t) = 10v_S(t) + \int_0^t v_S dt$$

SOLUTION:

There is no unique solution to this design problem. We begin by drawing the block diagram in Fig. 7-21(a), which shows that we need a gain block, an integrator, and a summer. However, the integrator and summer in Fig. 7-16 are inverting circuits. Figure 7-21(b) shows how to overcome this problem. The inverting gains are partitioned so that there are an even number of inverting operations in each path. The overall transfer characteristic is noninverting as required and the inverting building blocks are realizable using OP AMP circuits. One (of many) possible circuit realization of these processors is shown in Fig. 7-21(c). The element parameter constraint on this circuit is $RC = 1$. Selecting the OP AMPs and the values of R and C depend on many additional factors such as the required accuracy, internal resistance of the input source, and output load.

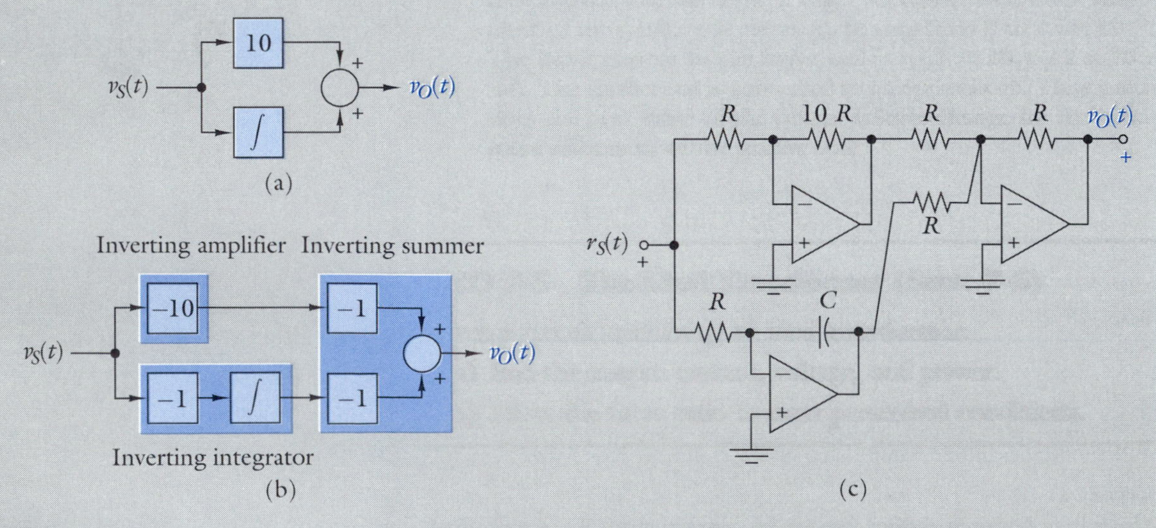

Figure 7-21

EXAMPLE 7-12

Draw a block diagram that implements the following input-output relationship:

$$100\frac{d^2 v_O}{dt^2} + 20\frac{dv_O}{dt} + v_O = v_S$$

SOLUTION:

The relationship is a linear second-order differential equation of the type we will study in the next chapter. A block diagram for linear differential equations can be created using summers, integrators, and gain blocks. We begin by solving for the highest-order derivative in the equation:

$$\frac{d^2 v_O}{dt^2} = \frac{1}{100}v_S - \frac{1}{5}\frac{dv_O}{dt} - \frac{1}{100}v_O$$

The sum on the right side of this equation is shown in block diagram form in Fig. 7-22(a). The summing operation generates the second derivative of the output, which is then integrated twice to generate the first derivative and the output itself as shown in Fig. 7-22(b). The two integrations generate the signals used in the original summing operation. Figure 7-22(c) shows how these signals are feedback to the summer via gain blocks, which provide the required scale factors and inversions. The net result is that the summer, integrators, and feedback gains produce the required input-output relationship.

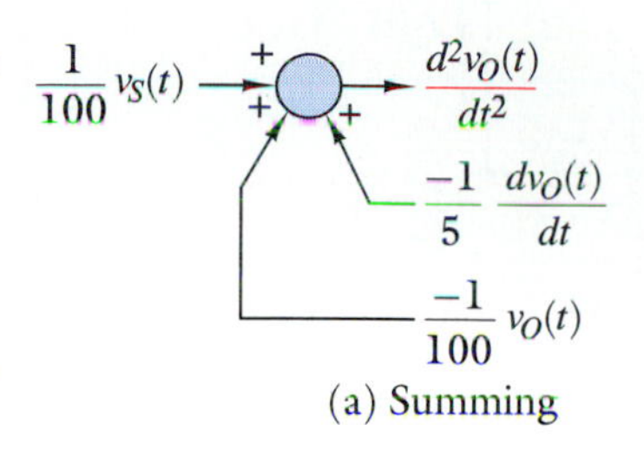

(a) Summing

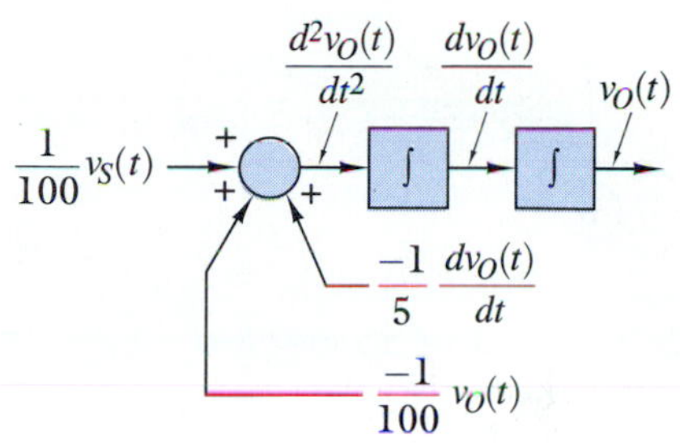

(b) Summing plus integration

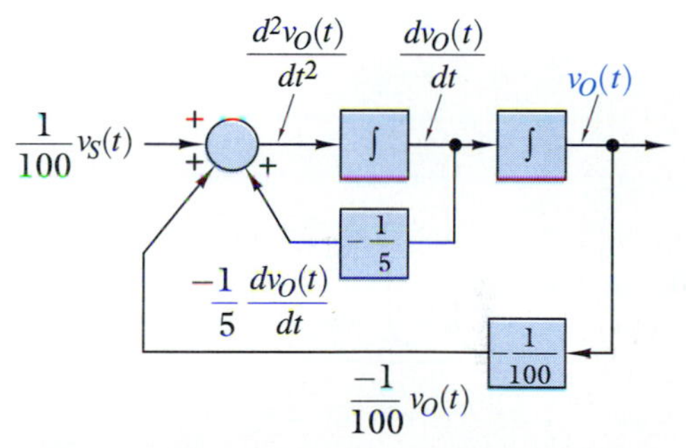

(c) Summing plus integration plus feedback

Figure 7-22

7-4 EQUIVALENT CAPACITANCE AND INDUCTANCE

In Chapter 2 we found that resistors connected in series or parallel can be replaced by equivalent resistances. The same principle applies to series and parallel connections of capacitors and inductors. The parallel connection of capacitors in Fig. 7-23(a) is an example.

Applying KCL at Node A in Fig. 7-23(a) yields

$$i(t) = i_1(t) + i_2(t) + \ldots + i_N(t)$$

Since the elements are connected in parallel KVL requires

$$v_1(t) = v_2(t) = \ldots = v_N(t) = v(t)$$

Because the capacitors all have the same voltage, their i-v relationships are all of the form $i_k(t) = C_k dv(t)/dt$. Substituting the

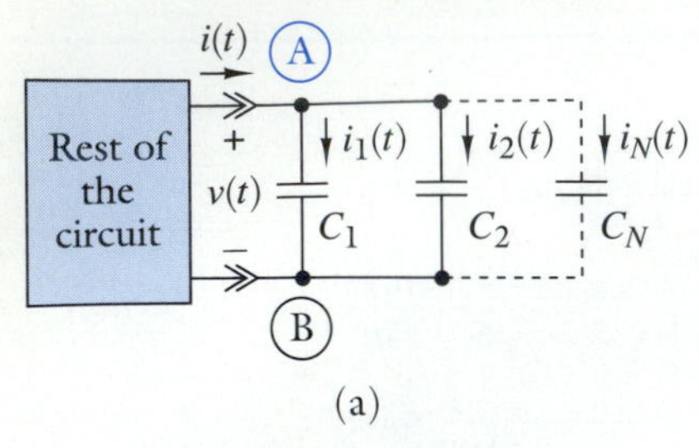

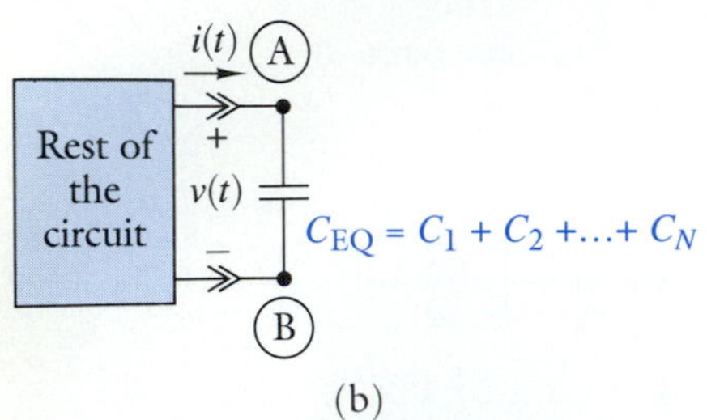

Figure 7-23 Capacitors connected in parallel. (a) Given circuit. (b) Equivalent circuit.

i-v relationships into the KCL equation yields

$$i(t) = C_1\frac{dv(t)}{dt} + C_2\frac{dv(t)}{dt} + \ldots + C_N\frac{dv(t)}{dt}$$

Factoring the derivative out of each term produces

$$i(t) = (C_1 + C_2 + \ldots + C_N)\frac{dv(t)}{dt}$$

This equation states that the responses $v(t)$ and $i(t)$ in Fig. 7-23(a) are unchanged when the N parallel capacitors are replaced by a single equivalent capacitance:

$$C_{EQ} = C_1 + C_2 + \ldots + C_N \qquad \text{(Parallel connection)} \quad (7\text{-}28)$$

The equivalent capacitance simplification is shown in Fig. 7-23(b). The initial voltage, if any, on the equivalent capacitance is $v(0)$, the common voltage across each of the original N capacitors at $t = 0$.

Next consider the series connection of N capacitors in Fig. 7-24(a). Applying KVL around the Loop 1 in Fig. 7-24(a) yields the equation

$$v(t) = v_1(t) + v_2(t) + \ldots + v_N(t)$$

Since the elements are connected in series, KCL requires

$$i_1(t) = i_2(t) = \ldots = i_N(t) = i(t)$$

Since the same current exists in all capacitors, their i-v relationships are all of the form

$$v_k(t) = v_k(0) + \frac{1}{C}\int_0^t i(x)dx$$

Substituting the i-v relationships into the Loop 1 KVL equation yields

$$v(t) = v_1(0) + \frac{1}{C_1}\int_0^t i(x)dx + v_2(0) + \frac{1}{C_2}\int_0^t i(x)dx + \ldots + v_N(0) + \frac{1}{C_N}\int_0^t i(x)dx$$

We can factor the integral out of each term to obtain

$$v(t) = [v_1(0) + v_2(0) + \ldots + v_N(0)] + \left(\frac{1}{C_1} + \frac{1}{C_2} + \ldots + \frac{1}{C_N}\right)\int_0^t i(x)dx$$

This equation indicates that the responses $v(t)$ and $i(t)$ in Fig. 7-24(a) will be unchanged if the N series capacitors are replaced

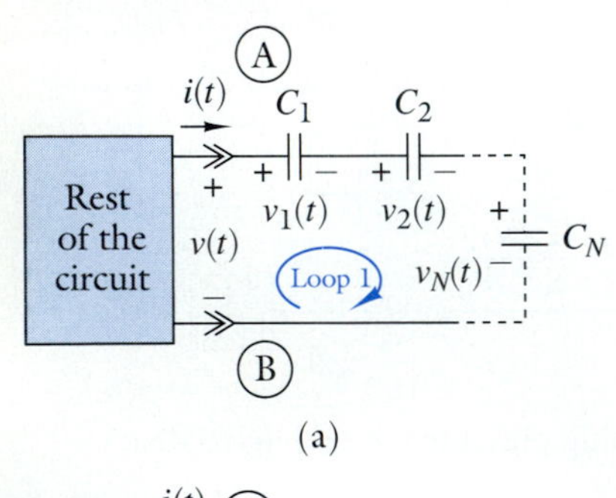

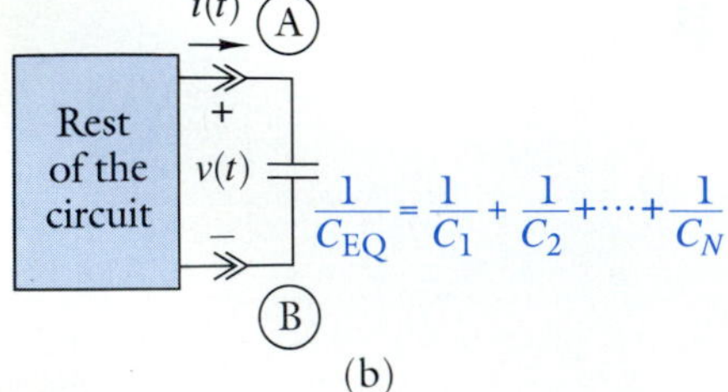

Figure 7-24 Capacitors connected in series. (a) Given circuit. (b) Equivalent circuit.

by a single equivalent capacitance:

$$\frac{1}{C_{EQ}} = \frac{1}{C_1} + \frac{1}{C_2} + \ldots + \frac{1}{C_N} \qquad \text{(Series connection)} \quad (7\text{-}29)$$

The equivalent capacitance is shown graphically in Fig. 7-24(b). The initial voltage on the equivalent capacitance is the sum of the initial voltages on each of the original N capacitors.

A parallel connection of capacitors produces an equivalent capacitance, which is the sum of the individual capacitances. In a series connection the reciprocals of the capacitances add to produce the reciprocal of the equivalent capacitance. Since the capacitor and inductor are dual elements, the corresponding results for inductors are obtained by interchanging the series and parallel equivalence rules for the capacitor. That is, in a series connection the equivalent inductance is the sum of the individual inductances:

$$L_{EQ} = L_1 + L_2 + \ldots + L_N \qquad \text{(Series connection)} \quad (7\text{-}30)$$

For the parallel connection the reciprocals add to obtain the equivalent inductance:

$$\frac{1}{L_{EQ}} = \frac{1}{L_1} + \frac{1}{L_2} + \ldots + \frac{1}{L_N} \qquad \text{(Parallel connection)} \quad (7\text{-}31)$$

Derivation of Eqs. (7-30) and (7-31) uses the approach given above for the capacitor except that the roles of voltage and current are interchanged. Completion of the derivation is left as a problem for the reader (see Problems 7-46 and 7-47).

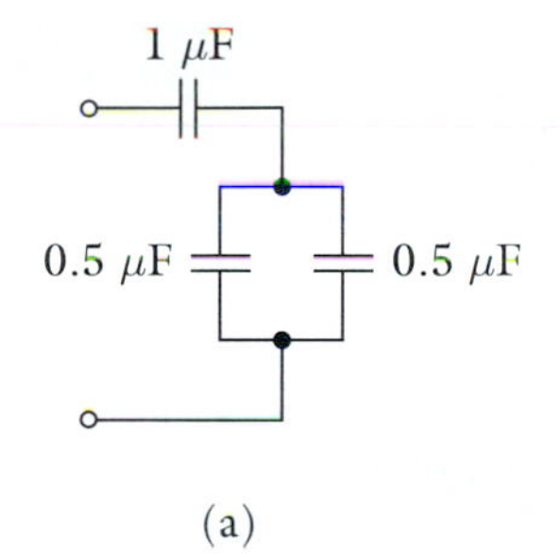

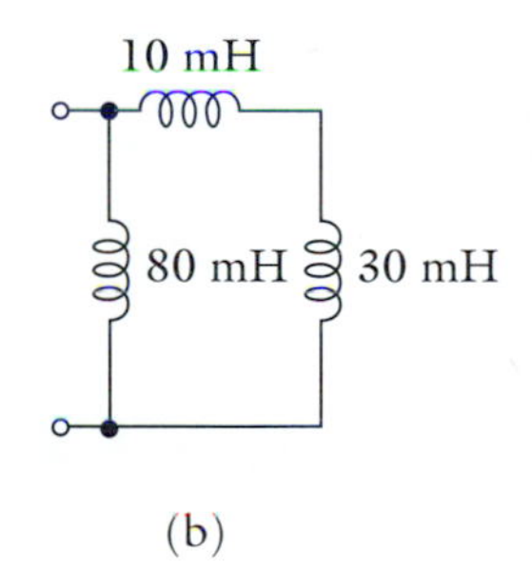

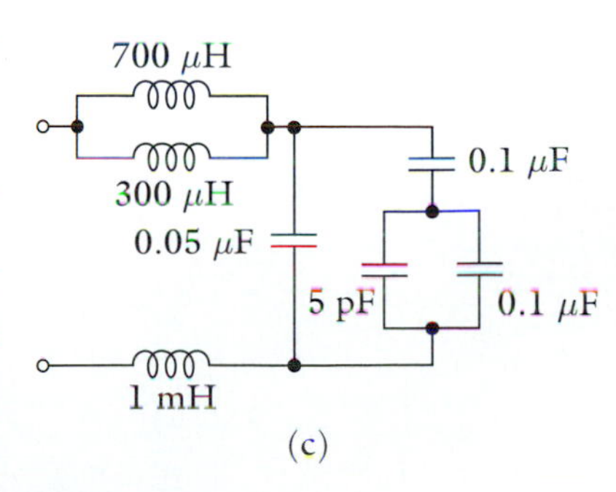

Figure 7-25

EXAMPLE 7-13

Find the equivalent capacitance and inductance of the circuits in Fig. 7-25.

SOLUTION:

(a) For the circuit in Fig. 7-25(a) the two 0.5-μF capacitors in parallel combine to yield an equivalent $0.5 + 0.5 = 1$-μF capacitance. This 1-μF equivalent capacitance is in series with a 1-μF capacitor yielding an overall equivalent of $V_{EQ} = 1/(1/1 + 1/1) = 0.5\ \mu$F.

(b) For the circuit of Fig. 7-25(b), the 10-mH and the 30-mH inductors are in series and add to produce an equivalent inductance of 40 mH. This 40-mH equivalent inductance is in parallel with the 80-mH inductor. The equivalent inductance of the parallel combination is $L_{EQ} = 1/(1/40 + 1/80) = 26.67$ mH.

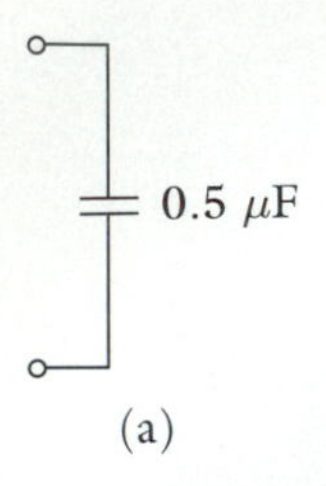

(a)

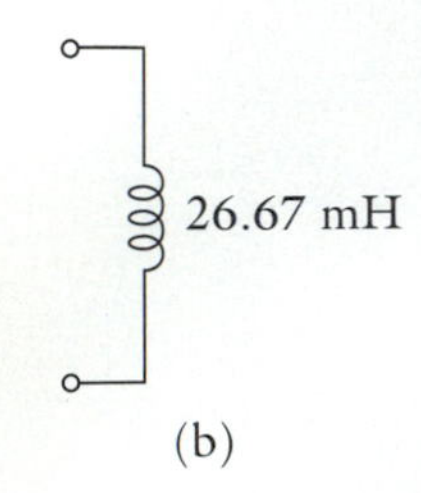

(b)

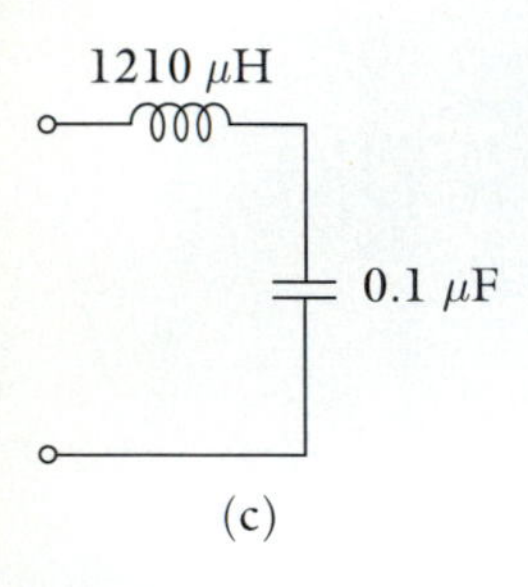

(c)

Figure 7-26

(c) The circuit of Fig. 7-25(c) contains both inductors and capacitors. In later chapters we will learn how to combine all of these into a single equivalent element. For now we combine the inductors and the capacitors separately. The 5-pF capacitor in parallel with the 0.1-μF capacitor yields an equivalent capacitance of 0.100005 μF. For all practical purposes the 5-pF capacitor can be ignored, leaving two 0.1-μF capacitors in series with equivalent capacitance of 0.05 μF. Combining this equivalent capacitance in parallel with the remaining 0.05-μF capacitor yields an overall equivalent capacitance of 0.1 μF. The parallel 700-μH and 300-μH inductors yield an equivalent inductance of $1/(1/700 + 1/300) = 210$ μH. This equivalent inductance is effectively in series with the 1-mH inductor at the bottom, yielding $1000 + 210 = 1210$ μH as the overall equivalent inductance.

Figure 7-26 shows the simplified equivalent circuits for each of the circuits of Fig. 7-25.

EXERCISE 7-10

The current through a series connection of two 1-μF capacitors is a rectangular pulse with an amplitude of 2 mA and a duration of 10 ms. At $t = 0$ the voltage across the first capacitor is +10 V and across the second is zero.

(a) What is the voltage across the series combination at $t = 10$ ms?

(b) What is the maximum instantaneous power delivered to the series combination?

(c) What is the energy stored on the first capacitor at $t = 0$ and $t = 10$ ms?

ANSWERS:

(a) 50 V

(b) 100 mW at $t = 10$ ms

(c) 50 μJ and 450 μJ

DC Equivalent Circuits

Sometimes it is necessary to determine the dc response of circuits containing capacitors and inductors. In the first two sections of this chapter we found that under dc conditions a capacitor acts like an open circuit and an inductor acts like a short circuit. In other words, under dc conditions an equivalent circuit for a capacitor is an open circuit and an equivalent circuit of an inductor is a short circuit.

To determine dc responses we replace capacitors by open circuits and inductors by short circuits and analyze the resulting resistance circuit using any of the methods in Chapters 2 through 5. The circuit analysis involves only resistance circuits and yields capacitor voltages and inductor currents along with any other variables of interest. The dc capacitor voltages and inductor currents are initial conditions for a transient response that begins at $t = 0$ when something in the circuit changes, such as the position of a switch. Programs like SPICE and MICRO-CAP use dc analysis to find the initial operating point of a circuit to be analyzed.

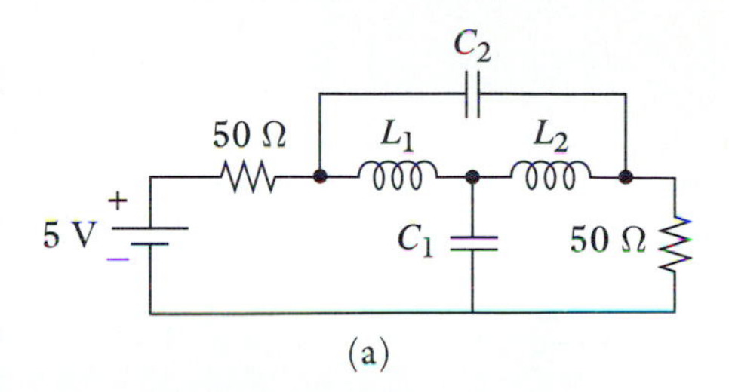

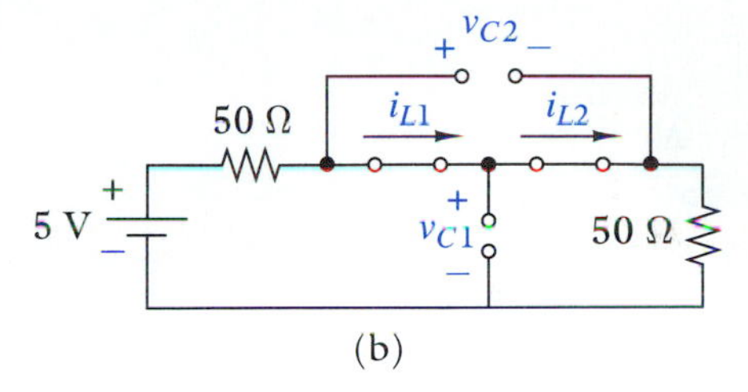

Figure 7-27

EXAMPLE 7-14

Determine the voltage across the capacitors and current through the inductors in Fig. 7-27(a).

SOLUTION:

The circuit is driven by a 5-V dc source. Figure 7-27(b) shows the equivalent circuit under dc conditions. The current in the resulting series circuit is $5/(50 + 50) = 50$ mA. This dc current exists in both inductors so $i_{L1} = i_{L2} = 50$ mA. By voltage division the voltage across the 50-Ω output resistor is $v = 5 \times 50/(50 + 50) = 2.5$ V; therefore $v_{C1}(0) = 2.5$ V. The voltage across C_2 is zero because the two inductors short out the capacitor.

EXERCISE 7-11

Find the OP AMP output voltage in Fig. 7-28.

ANSWER:

$$v_O = \frac{R_2 + R_1}{R_1} V_{dc}$$

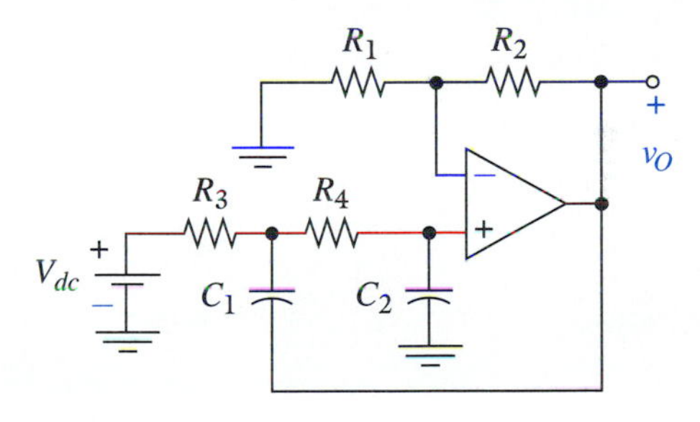

Figure 7-28

7-5 MUTUAL INDUCTANCE

The i-v characteristics of the inductor result from the magnetic field surrounding a coil of wire-carrying current. The magnetic flux spreads out around the coil forming closed loops that cut or link with the turns in the coil. If the current is changing, then Faraday's law states that voltage across the coil is equal to the time rate of change of the total flux linkages.

Now suppose that a second coil is brought close to the first coil. The flux from the first coil will link with the turns of the second coil. If the current in the first coil is changing, then these flux linkages will generate a voltage in the second coil. The coupling between a changing current in one coil and a voltage across a second coil results in ***mutual inductance***.

i-v Characteristics

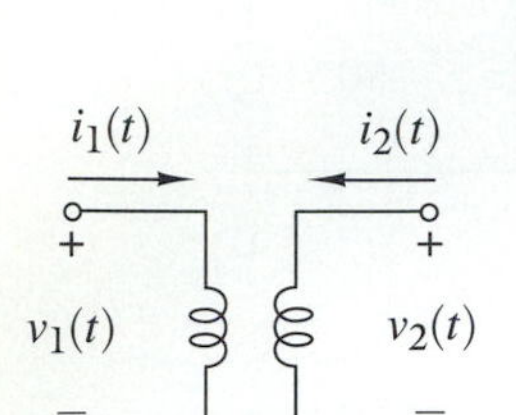

Figure 7-29
Coupled coils.

Let us assume there is coupling between the two coils in Fig. 7-29. Our objective is to develop the i-v characteristics of these coils. Such a derivation unavoidably involves describing the effects observed in one coil due to causes occurring in the other. We will use a double subscript notation because it clearly identifies the various cause and effect relationships. The first subscript indicates the coil in which the effect takes place and the second identifies the coil in which the cause occurs. For example, $v_{11}(t)$ is the voltage across Coil 1 due to causes occurring in Coil 1 itself, while $v_{12}(t)$ is the voltage across Coil 1 due to causes occurring in Coil 2.

We begin by assuming the coils in Fig. 7-29 are far apart so there is no mutual inductance between them. A current $i_1(t)$ passes through the N_1 turns of the first coil in Fig. 7-29 and $i_2(t)$ through N_2 turns in the second. Each coil produces a flux

$$\begin{array}{cc} \text{Coil 1} & \text{Coil 2} \\ \phi_1(t) = k_1 N_1 i_1(t) & \phi_2(t) = k_2 N_2 i_2(t) \end{array} \qquad (7\text{-}32)$$

where k_1 and k_2 are proportionality constants. The flux linkages in each coil are proportional to the number of turns.

$$\begin{array}{cc} \text{Coil 1} & \text{Coil 2} \\ \lambda_{11}(t) = N_1 \phi_1(t) & \lambda_{22}(t) = N_2 \phi_2(t) \end{array} \qquad (7\text{-}33)$$

By Faraday's law the voltage across a coil is equal to the time rate of change of the flux linkages. Using Eqs. (7-32) and (7-33) and the derivative relationship between voltage and flux linkages gives

$$\begin{aligned} \text{Coil 1: } v_{11}(t) &= \frac{d\lambda_{11}(t)}{dt} = N_1 \frac{d\phi_1(t)}{dt} = [k_1 N_1^2] \frac{di_1(t)}{dt} \\ \text{Coil 2: } v_{22}(t) &= \frac{d\lambda_{22}(t)}{dt} = N_2 \frac{d\phi_2(t)}{dt} = [k_2 N_2^2] \frac{di_2(t)}{dt} \end{aligned} \qquad (7\text{-}34)$$

Equations (7-34) provide the i-v relationships for the coils when there is no mutual coupling. These results are the same as previously found in Sect. 7-2.

Now suppose the coils are brought close together so that part of the flux produced by each coil intercepts the other. That is, part but not all of the fluxes $\phi_1(t)$ and $\phi_2(t)$ in Eq. (7-32) intercept the opposite coil. We describe the cross coupling using the double subscript notation:

$$\begin{array}{cc} \text{Coil 1} & \text{Coil 2} \\ \phi_{12}(t) = k_{12}N_2 i_2(t) & \phi_{21}(t) = k_{21}N_1 i_1(t) \end{array} \quad (7\text{-}35)$$

The quantity $\phi_{12}(t)$ is the flux intercepting Coil 1 due to the current in Coil 2, and $\phi_{21}(t)$ is the flux intercepting Coil 2 due to the current in Coil 1. The total flux linkages in each coil are proportional to the number of turns:

$$\begin{array}{cc} \text{Coil 1} & \text{Coil 2} \\ \lambda_{12}(t) = N_1\phi_{12}(t) & \lambda_{21}(t) = N_2\phi_{21}(t) \end{array} \quad (7\text{-}36)$$

By Faraday's law the voltage across a coil is equal to the time rate of change of the flux linkages. Using Eqs. (7-35) and (7-36) and the derivative relationship between flux linkages and voltages gives

$$\text{Coil 1:} \quad v_{12}(t) = \frac{d\lambda_{12}(t)}{dt} = N_1\frac{d\phi_{12}(t)}{dt} = [k_{12}N_1N_2]\frac{di_2(t)}{dt}$$

$$\text{Coil 2:} \quad v_{21}(t) = \frac{d\lambda_{21}(t)}{dt} = N_2\frac{d\phi_{21}(t)}{dt} = [k_{21}N_1N_2]\frac{di_1(t)}{dt} \quad (7\text{-}37)$$

Equation (7-37) are the i-v relationships describing the cross coupling between coils when they are sufficiently close together that there is mutual coupling.

When the magnetic medium supporting the fluxes is linear the results in Eqs. (7-34) and (7-37) can be superimposed to obtain the total voltage across the coil:

$$\begin{aligned} \text{Coil 1:} \quad v_1(t) &= v_{11}(t) + v_{12}(t) \\ &= [k_1N_1^2]\frac{di_1(t)}{dt} + [k_{12}N_1N_2]\frac{di_2(t)}{dt} \\ \text{Coil 2:} \quad v_2(t) &= v_{21}(t) + v_{22}(t) \\ &= [k_{21}N_1N_2]\frac{di_1(t)}{dt} + [k_2N_2^2]\frac{di_2(t)}{dt} \end{aligned} \quad (7\text{-}38)$$

We can identify four inductance parameters in these equations:

$$L_1 = k_1N_1^2 \qquad L_2 = k_2N_2^2 \quad (7\text{-}39)$$

and

$$M_{12} = k_{12}N_1N_2 \qquad M_{21} = k_{21}N_2N_1 \quad (7\text{-}40)$$

The two inductance parameters in Eq. (7-39) are the *self-inductance* of the coils. The two parameters in Eq. (7-40) are the *mutual inductances* between the two coils. In a linear magnetic medium $k_{12} = k_{21} = k_M$, so it is not necessary to use double subscripts to identify the mutual inductances. For a linear medium there is a single mutual inductance parameter M defined as

$$M = M_{12} = M_{21} = k_M N_1 N_2 \tag{7-41}$$

Using the definitions in Eqs. (7-39) and (7-41) the i-v characteristics of two coupled coils are

$$\begin{aligned} &\text{Coil 1:} \quad v_1(t) = L_1 \frac{di_1(t)}{dt} + M \frac{di_2(t)}{dt} \\ &\text{Coil 2:} \quad v_2(t) = M \frac{di_1(t)}{dt} + L_2 \frac{di_2(t)}{dt} \end{aligned} \tag{7-42}$$

Coupled coils involve three inductance parameters, the two self-inductances L_1 and L_2 and the mutual inductance M.

Our development has assumed that the cross coupling is additive. Additive coupling means that a positive rate of change of current in Coil 2 induces a positive voltage in Coil 1, and vice versa. The additive assumption produces the positive sign on the mutual inductance terms in Eq. (7-42). Unhappily it is possible for a positive rate of change of current in Coil 2 to induce a negative voltage in Coil 1, and vice versa. To account for additive and subtractive coupling, the general form of the coupled coil i-v characteristics include a $\pm$ sign on the mutual enductance terms:

$$\begin{aligned} &\text{Coil 1:} \quad v_1(t) = L_1 \frac{di_1(t)}{dt} \pm M \frac{di_2(t)}{dt} \\ &\text{Coil 2:} \quad v_2(t) = \pm M \frac{di_1(t)}{dt} + L_2 \frac{di_2(t)}{dt} \end{aligned} \tag{7-43}$$

When applying these element equations it is necessary to know when to use a plus sign and when to use a minus sign.

The Dot Convention

First of all the mutual inductance parameter M is **always** positive. The question is, What sign should be put in front of this positive parameter in the i-v relationships in Eq. (7-43)? The correct sign depends on two things:

1. The spatial orientation of the two coils
2. The reference marks assigned to the coil currents and voltages

Figure 7-30 shows the additive and subtractive spatial orientation of two coupled coils. In either case, the right-hand rule determines the direction of the flux produced by a current. In the additive case currents i_1 and i_2 produce additive clockwise fluxes ϕ_1 and ϕ_2. In the subtractive case the currents produce opposing fluxes, since ϕ_1 is clockwise and ϕ_2 is counterclockwise. The sign for the mutual inductance term is positive for the additive orientation and negative for the subtractive case. In general, it is awkward to show the spacial features of the coils in circuit diagrams. The dots shown near one terminal of each coil in Fig. 7-30 are used as special reference marks to indicate the relative orientation of the coils. The reference directions for the coil currents and voltages are arbitrary. They can be changed as long as we use the passive sign convention. However, the dots indicate physical attributes of the coils. **They are not arbitrary. They cannot be changed.**

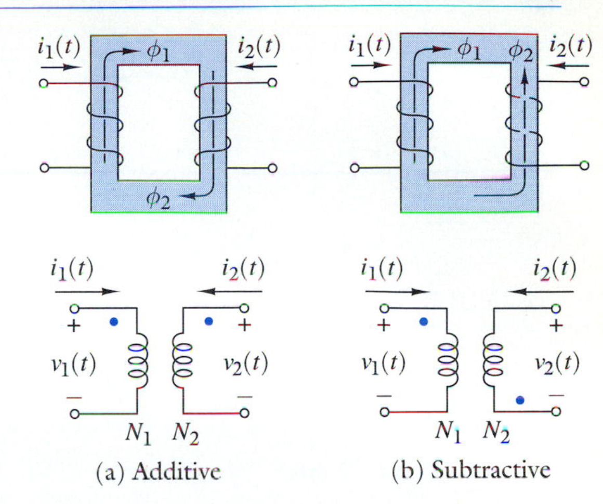

Figure 7-30 Coil orientations and corresponding reference dots. (a) Additive. (b) Subtractive.

The correct sign for the mutual inductance term hinges on how the reference marks for currents and voltages are assigned relative to the coil dots. Figure 7-31 shows all four possible current and voltage reference assignments under the passive sign convention for a given coil orientation. In cases A and B the fluxes are additive and the mutual inductance term is positive. In cases C and D the fluxes are subtractive and the mutual inductance term is negative. From these results we derive the following rule:

> Mutual inductance is additive when both current reference directions point toward or away from dotted terminals; otherwise, it is subtractive.

Since the current reference directions can be changed, a corollary of this rule is that we can always assign reference directions so that the positive sign applies to the mutual inductance. This corollary is important because a positive sign is built into the mutual inductance models used by computer-aided analysis programs like SPICE and MICRO-CAP.

The reader may feel that the discussion above is much ado about nothing. Not so. First of all, selecting the wrong sign can have nontrivial consequences because the sign of the output signal is wrong. If the signal is a command to your car's autopilot, then you really need to know whether stepping on the brake pedal will slow the car down or speed it up. The other problem is that coupled coils may appear in different parts of a circuit diagram and may be assigned voltage and current reference marks for other reasons. In such circumstances it is important to understand the underlying principle to select the correct sign for the mutual inductance term.

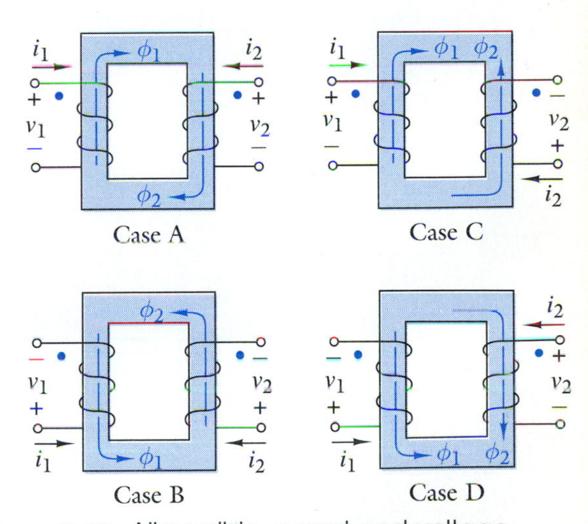

Figure 7-31 All possible current and voltage reference mark assignments using the passive sign convention.

The following examples and exercises illustrate selecting the correct sign and applying the i-v characteristics in Eq. (7-43).

EXAMPLE 7-15

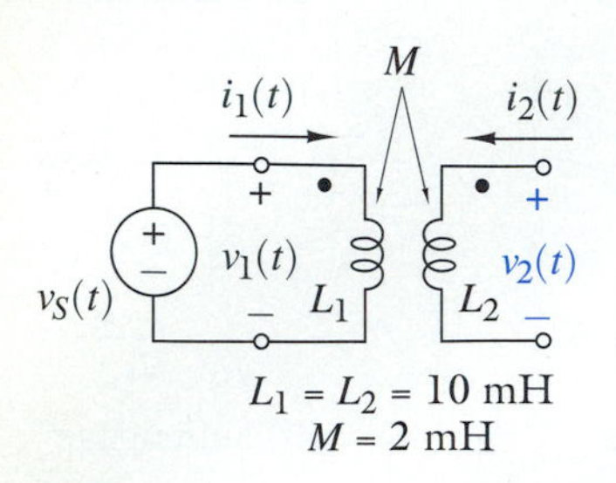

Figure 7-32

In Fig. 7-32 the source voltage is $v_S(t) = 10 \cos 100t$. Find the output voltage $v_2(t)$.

SOLUTION:

The sign of the mutual inductance term is positive because both current reference directions are toward the coil dots. Since the load on Coil 2 is an open circuit $i_2(t) = 0$ and the i-v equations of the coupled coils reduce to

$$\text{Coil 1: } 10 \cos 100t = 0.01 \frac{di_1(t)}{dt} + 0$$

$$\text{Coil 2: } v_2(t) = 0.002 \frac{di_1(t)}{dt} + 0$$

Solving the Coil 1 equation for di_1/dt yields

$$\frac{di_1(t)}{dt} = 1000 \cos 100t$$

Substituting this equation for $i_1(t)$ into the Coil 2 equation yields

$$v_2(t) = 2 \cos 100t \text{ (Correct)}$$

Incorrectly choosing a minus sign for the mutual inductance term produces a Coil 2 voltage of

$$v_2(t) = -2 \cos 100t \text{ (Incorrect)}$$

Note that the incorrect and correct solutions differ by a signal inversion.

EXAMPLE 7-16

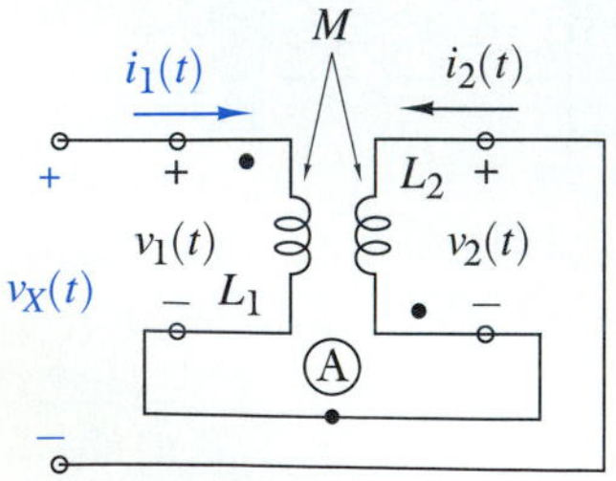

Figure 7-33

Solve for $v_X(t)$ in terms of $i_1(t)$ for the coupled coils in Fig. 7-33.

SOLUTION:

In this case the signs of mutual inductance terms are negative because the reference direction for $i_1(t)$ points toward the Coil 1 dot and the reference direction for $i_2(t)$ points away from the Coil 2 dot. The coupled coil i-v equations are

$$\text{Coil 1: } v_1(t) = L_1 \frac{di_1(t)}{dt} - M \frac{di_2(t)}{dt}$$

$$\text{Coil 2: } v_2(t) = -M \frac{di_1(t)}{dt} + L_2 \frac{di_2(t)}{dt}$$

A KCL constraint at Node A requires that $i_1(t) = -i_2(t)$. Therefore, the coil i-v equations can be written in the form

$$\text{Coil 1: } v_1(t) = L_1\frac{di_1(t)}{dt} - M\frac{d[-i_1(t)]}{dt}$$

$$\text{Coil 2: } v_2(t) = -M\frac{di_1(t)}{dt} + L_2\frac{d[-i_1(t)]}{dt}$$

A KVL constraint around the loop requires $v_X(t) = v_1(t) - v_2(t)$. Subtracting the second equation above from the first yields

$$v_X(t) = L_1\frac{di_1(t)}{dt} + M\frac{di_1(t)}{dt} + M\frac{di_1(t)}{dt} + L_2\frac{di_1(t)}{dt}$$

$$= (L_1 + L_2 + 2M)\frac{di_1(t)}{dt}$$

The two mutual inductance terms add producing an equivalent inductance of $L_1 + L_2 + 2M$. With a plus sign in the coil i-v relations the mutual inductance cancel leaving an equivalent inductance of $L_1 + L_2$. Thus, it is important to have the right sign in the i-v relations.

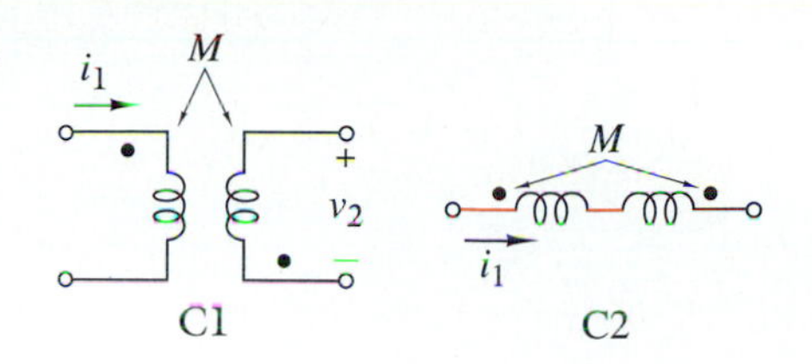

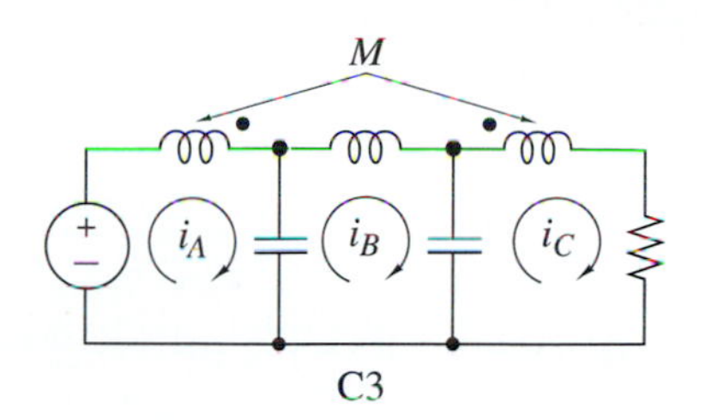

Figure 7-34

EXERCISE 7-12

Select the appropriate sign for the mutual inductance terms for each pair of coupled coils shown in Fig. 7-34.

ANSWER:
All are negative.

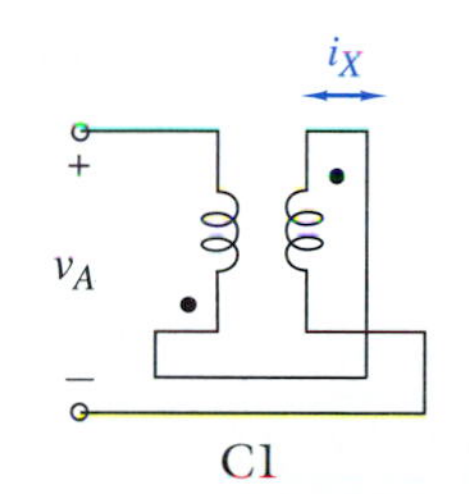

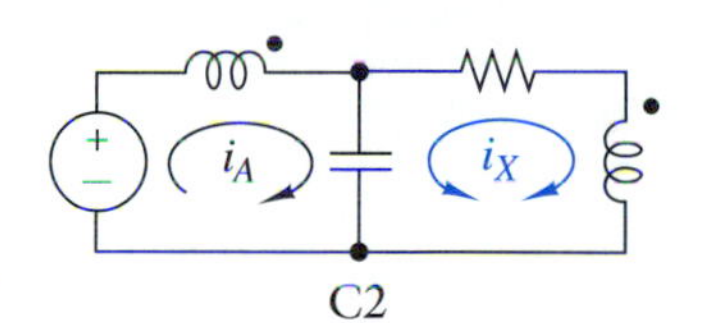

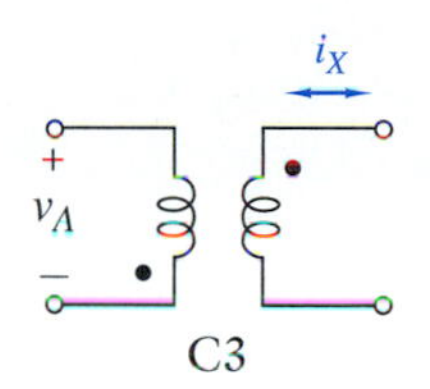

Figure 7-35

EXERCISE 7-13

What reference direction should be assigned to i_X so the mutual inductance terms have a positive sign for each pair of coupled coils in Fig. 7-35?

ANSWER:
C1: To the right. C2: Counterclockwise. C3: To the right.

EXERCISE 7-14

Find $v_1(t)$ and $v_2(t)$ for the circuit in Fig. 7-36.

ANSWERS:
$v_1(t) = -10 \sin 10^4 t - 3 \cos 5 \times 10^3 t$
$v_2(t) = -15 \sin 10^4 t - 5 \cos 5 \times 10^3 t$

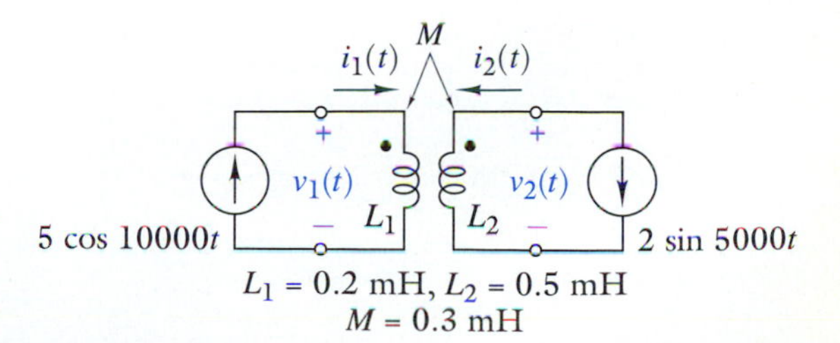

Figure 7-36

Energy Analysis

Calculating the energy stored in a pair of coupled coils reveals a fundamental limitation on allowable values of the self and mutual inductances. We determine the total energy stored in a pair of coupled coils by first calculating the total power absorbed. Multiplying the Coil 1 equation in Eq. (7-43) by $i_1(t)$ and the Coil 2 equation by $i_2(t)$ produces

$$\begin{aligned} p_1(t) &= v_1(t)i_1(t) = L_1 i_1(t)\frac{di_1(t)}{dt} \pm M i_1(t)\frac{di_2(t)}{dt} \\ p_2(t) &= v_2(t)i_2(t) = \pm M i_2(t)\frac{di_1(t)}{dt} + L_2 i_2(t)\frac{di_2(t)}{dt} \end{aligned} \quad (7\text{-}44)$$

The quantities $p_1(t)$ and $p_2(t)$ are the power associated with Coils 1 and 2. The total power is the sum of individual coil powers:

$$\begin{aligned} p(t) &= p_1(t) + p_2(t) \\ &= L_1\left[i_1(t)\frac{di_1(t)}{dt}\right] \pm M\left[i_1(t)\frac{di_2(t)}{dt} + i_2(t)\frac{di_1(t)}{dt}\right] \\ &\quad + L_2\left[i_2(t)\frac{di_2(t)}{dt}\right] \end{aligned} \quad 7\text{-}45)$$

Each of the bracketed terms in Eq. (7-45) is a perfect derivative. Specifically

$$\begin{aligned} i_1(t)\frac{di_1(t)}{dt} &= \frac{1}{2}\frac{di_1^2(t)}{dt} \\ i_2(t)\frac{di_2(t)}{dt} &= \frac{1}{2}\frac{di_2^2(t)}{dt} \\ i_1(t)\frac{di_2(t)}{dt} + i_2(t)\frac{di_1(t)}{dt} &= \frac{di_1(t)i_2(t)}{dt} \end{aligned} \quad (7\text{-}46)$$

Therefore, the total power in Eq. (7-45) is

$$p(t) = \frac{d}{dt}\left[\frac{1}{2}L_1 i_1^2(t) \pm M i_1(t)i_2(t) + \frac{1}{2}L_2 i_2^2(t)\right] \quad (7\text{-}47)$$

Since power is the time rate of change of energy, the quantity inside the bracket in Eq. (7-47) is the total energy stored in the two coils

$$w(t) = \frac{1}{2}L_1 i_1^2(t) \pm M i_1(t)i_2(t) + \frac{1}{2}L_2 i_2^2(t) \quad (7\text{-}48)$$

In Eq. (7-48) the self-inductance terms are always positive. However, the mutual inductance term can be either positive or negative. At first glance it appears that the total energy could be negative. But the total energy must be positive; otherwise, the coils would deliver net energy to the outside world.

The condition $w(t) \geq 0$ places a constraint on the values of the self and mutual inductances. To see why, we divide $w(t)$ by $[i_2(t)]^2$ and defined a variable $x = i_1/i_2$. With these changes the energy constraint $w(t) > 0$ becomes

$$\frac{w(t)}{i_{22}^2(t)} = f(x) = \frac{1}{2}L_1 x^2 \pm Mx + \frac{1}{2}L_2 \geq 0 \quad (7\text{-}49)$$

The minimum value of $f(x)$ occurs when

$$\frac{df(x)}{dx} = L_1 x \pm M = 0 \qquad \text{hence} \qquad x_{MIN} = \mp\frac{M}{L_1} \tag{7-50}$$

The value x_{MIN} yields the minimum of $f(x)$ because the second derivative of $f(x)$ is positive. Substituting x_{MIN} back into Eq. (7-49) yields the condition

$$f(x_{MIN}) = \frac{1}{2}L_1\frac{M^2}{L_1^2} - \frac{M^2}{L_1} + \frac{1}{2}L_2 = \frac{1}{2}\left[-\frac{M^2}{L_1} + L_2\right] \geq 0 \tag{7-51}$$

The constraint in Eq. (7-51) means the stored energy in a pair of coils is positive if

$$L_1 L_2 \geq M^2 \tag{7-52}$$

Energy considerations dictate that the product of the self-inductances must exceed the square of the mutual inductance in any pair of coupled coils.

The constraint in Eq. (7-52) is usually written in terms of a new parameter called the *coupling coefficient k*.

$$k = \frac{M}{\sqrt{L_1 L_2}} \leq 1 \tag{7-53}$$

The parameter k ranges from 0 to 1. If $M = 0$ then $k = 0$ and the coupling between the coils is zero. The condition $k = 1$ requires *perfect coupling* in which all of the flux in Coil 1 links Coil 2, and vice versa. Perfect coupling is physically impossible, although careful design can produce coupling coefficients of 0.99 and higher.

The next section discusses a transformer model that assumes perfect coupling ($k = 1$). The coupling coefficient comes up again in the following section because SPICE specifies the parameters of a pair of coupled coils in terms of the self-inductances L_1 and L_2, and coupling coefficient k.

EXERCISE 7-15

What is the coupling coefficient of the coils in Exercise 7-14?

ANSWER:
0.949

7-6 THE IDEAL TRANSFORMER

A *transformer* is a device based on mutual inductance coupling between two coils. Transformers find application in virtually every type of electrical system, but especially in power supplies and

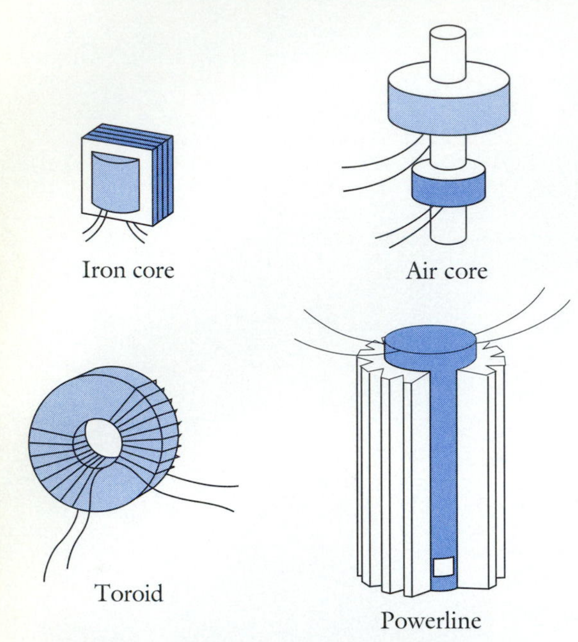

Figure 7-37 Examples of transformer devices.

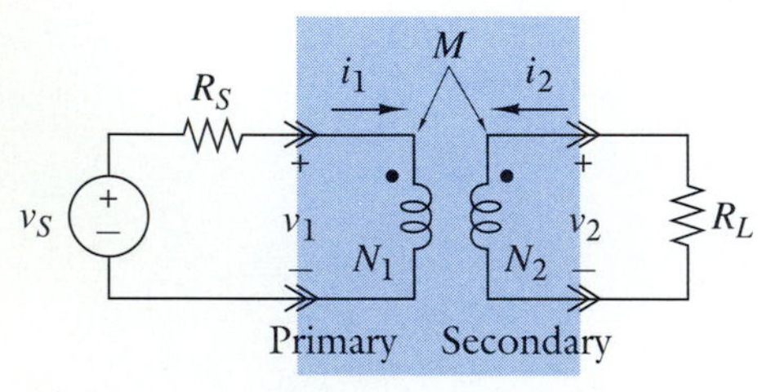

Figure 7-38 The transformer connected as an interface circuit.

commercial power grids. Some example devices from these applications are shown in Fig. 7-37.

In a transformer the first coil is called the *primary winding* and the second coil is the *secondary winding*. In most applications a transformer is an interface device connecting a source on the primary side, with a load on the secondary winding side as shown in Fig. 7-38. In interface applications the transformer is a coupler that transfers signals (especially power) from the source to the load. The basic purpose of the device is to change voltage and current levels so the conditions at the source and load are compatible.

Transformer design involves two goals: (1) to maximize the magnetic coupling between the two windings and (2) to minimize the power loss in the windings. The first goal produces near perfect coupling ($k \approx 1$) so that almost all of the flux in one winding links the other. Minimizing power losses means almost all of the power delivered to the primary winding transfers to the load via the secondary winding. These two features are the basis for the ideal transformer model.

Perfect Coupling

The ideal transformer model assumes perfect coupling between the primary and secondary windings. *Perfect coupling* means all of the flux in the first coil links the second coil, and vice versa. Equation (7-32) defines the total flux in each coil as

$$\begin{array}{cc} \text{Coil 1} & \text{Coil 2} \\ \phi_1(t) = k_1 N_1 i_1(t) & \phi_2(t) = k_2 N_2 i_2(t) \end{array} \tag{7-54}$$

where k_1 and k_2 are proportionality constants. Equation (7-35) defines the cross coupling using the double subscript notation:

$$\begin{array}{cc} \text{Coil 1} & \text{Coil 2} \\ \phi_{12}(t) = k_{12} N_2 i_2(t) & \phi_{21}(t) = k_{21} N_1 i_1(t) \end{array} \tag{7-55}$$

In this equation $\phi_{12}(t)$ is the flux intercepting Coil 1 due to the current in Coil 2 and $\phi_{21}(t)$ is the flux intercepting Coil 2 due to the current in Coil 1. Perfect coupling means that

$$\phi_{21}(t) = \phi_1(t) \quad \text{and} \quad \phi_{12}(t) = \phi_2(t) \tag{7-56}$$

Comparing Eqs. (7-54) and (7-55) shows that perfect coupling requires $k_1 = k_{21}$ and $k_2 = k_{12}$. But in a linear magnetic medium $k_{12} = k_{21} = k_M$, so perfect coupling implies

$$k_1 = k_2 = k_{12} = k_{21} = k_M \tag{7-57}$$

Substituting the perfect coupling conditions in Eq. (7-57) into the coupled-coil i-v characteristics in Eq. (7-38) gives

$$\begin{aligned} v_1(t) &= \left[k_M N_1^2\right]\frac{di_1(t)}{dt} \pm \left[k_M N_1 N_2\right]\frac{di_2(t)}{dt} \\ v_2(t) &= \pm\left[k_M N_1 N_2\right]\frac{di_1(t)}{dt} + \left[k_M N_2^2\right]\frac{di_2(t)}{dt} \end{aligned} \tag{7-58}$$

Factoring N_1 out of the first equation and $\pm N_2$ out of the second produces

$$\begin{aligned} v_1(t) &= N_1\left(k_M N_1\frac{di_1(t)}{dt} \pm k_M N_2\frac{di_2(t)}{dt}\right) \\ v_2(t) &= \pm N_2\left(k_M N_1\frac{di_1(t)}{dt} \pm k_M N_2\frac{di_2(t)}{dt}\right) \end{aligned} \tag{7-59}$$

Dividing the second equation by the first shows that perfect coupling implies

$$\frac{v_2(t)}{v_1(t)} = \pm\frac{N_2}{N_1} = \pm n \tag{7-60}$$

where the parameter n is called the ***turns ratio***

The $\pm$ sign in Eq. (7-60) reminds us that the mutual inductance effects can be additive or subtractive. Selecting the correct sign is important because a signal inversion can take place. The sign depends on the reference marks assigned to the primary and secondary currents relative to the dots indicating the relative coil orientations. The rule for the ideal transformer is a corollary of the rule for selecting the sign of the mutual inductance term in coupled-coil element equations.

The sign in Eq. (7-60) is a positive when the reference directions for both currents point toward or away from a dotted terminal; otherwise it is negative.

With perfect coupling the primary and secondary voltages are proportional so they have the same waveshape. For example, when $v_1(t) = V_A \sin \omega t$ the secondary voltage is $v_2(t) = \pm n V_A \sin \omega t$. When the turns ratio $n > 1$ the secondary voltage amplitude is larger than the primary and the device is called a ***step-up transformer.*** Conversely, when $n < 1$ the secondary voltage is smaller than the primary and the device is called a ***step-down transformer.*** The ability to step voltage levels up and down is a basic feature of transformers. Commercial power systems use transmission voltages of several hundred kilovolts. For residential applications the transmission voltage is reduced to safer levels (typically 220/110 V_{RMS}) using step-down transformers.

EXERCISE 7-16

The transformer in Fig. 7-39 has perfect coupling and a turns ratio of $n = 0.1$. The input voltage is $v_S(t) = 120 \sin 377t$ V.

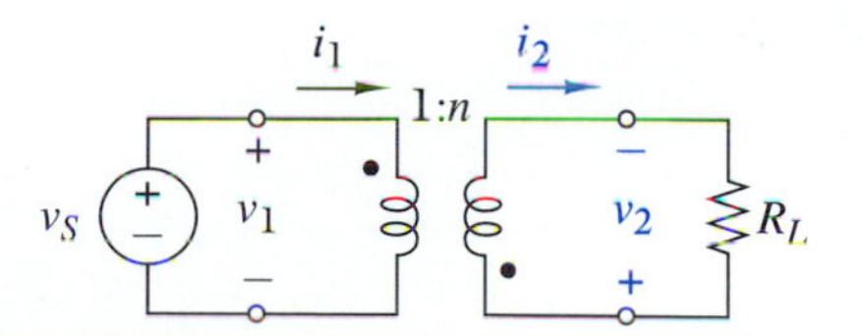

Figure 7-39

(a) What is the secondary voltage?
(b) What is the secondary current for a 50-Ω load?
(c) Is this a step-up or step-down transformer?

ANSWERS:
(a) $+12 \sin 377t$ V
(b) $-0.24 \sin 377t$ A
(c) Step down ■

Zero Power Loss

The ideal transformer model also assumes that there is no power loss in the transformer. With the passive sign convention, the power in the primary winding and secondary windings are $v_1(t)$ $i_1(t)$ and $v_2(t)i_2(t)$, respectively. The zero loss assumption means that the power absorbed (lost) in the two windings is zero. *Zero power loss* requires

$$v_1(t)i_1(t) + v_2(t)i_2(t) = 0 \tag{7-61}$$

which can be rearranged in the form

$$\frac{i_2(t)}{i_1(t)} = -\frac{v_1(t)}{v_2(t)} \tag{7-62}$$

But under the perfect coupling assumption $v_2(t)/v_1(t) = \pm n$. With zero power loss and perfect coupling the primary and secondary currents are related as

$$\frac{i_2(t)}{i_1(t)} = \mp\frac{1}{n} \tag{7-63}$$

The correct sign in this equation depends on the orientation of the assigned current reference directions relative to the dots describing the transformer structure.

With both perfect coupling and zero power loss the secondary current is inversely proportional to the turns ratio. A step-up transformer ($n > 1$) increases the voltage and decreases the current, which improves transmission line efficiency because the i^2R losses in the conductors are smaller.

■ EXERCISE 7-17

The transformer in Fig. 7-39 has perfect coupling, zero power loss, and a turns ratio of $n = 10$. The input is $v_S = 120 \sin 377t$ V.
(a) What is the secondary voltage?
(b) What is the secondary current for a 50-Ω load?
(c) What is the primary current?
(d) What is the power absorbed by the load?
(e) Is this a step-up or step-down transformer?

ANSWERS:

(a) $= 1200 \sin 377t$ V
(b) $-24 \sin 377t$ A
(c) $240 \sin 377t$ A
(d) $28.8 \sin^2 377t$ kW
(e) Step up ■

i-v Characteristics

Equations (7-60) and (7-63) define the i-v characteristics of the ideal transformer circuit element:

$$\begin{aligned} v_2(t) &= \pm n v_1(t) \\ i_2(t) &= \mp \frac{1}{n} i_1(t) \end{aligned} \tag{7-64}$$

where $n = N_2/N_1$ is the turns ratio. The correct sign in these equations depends on the assigned reference directions and transformer dots as previously discussed.

Some *caution* is required when using the ideal transformer model. Equation (7-64) states that the secondary signals are proportional to the primary signals. These element equations appear to apply to dc signals. This is of course wrong. The element equations are an idealization of mutual inductance, which requires time-varying signals to provide the coupling between two coils.

The lesson to remember is that it is up to the user to know and conform to the limitations of a circuit model. The model will not do this for you. If you apply a dc signal to the primary of an ideal transformer element, the model will happily report a proportional dc signal in the secondary. The point is that models don't tell you when they give nonsense for answers. The derivation of a circuit model is not simply an academic exercise. The derivation must be understood in order to know the limitations of a model.

Caution: Do not try to write coupled-coil equations for an ideal transformer because the two assumptions in the model cause the coil inductances to be infinite.

Equivalent Input Resistance

Because a transformer changes the voltage and current levels it effectively changes the load seen by a source in the primary circuit. To derive the equivalent input resistance, we write the device

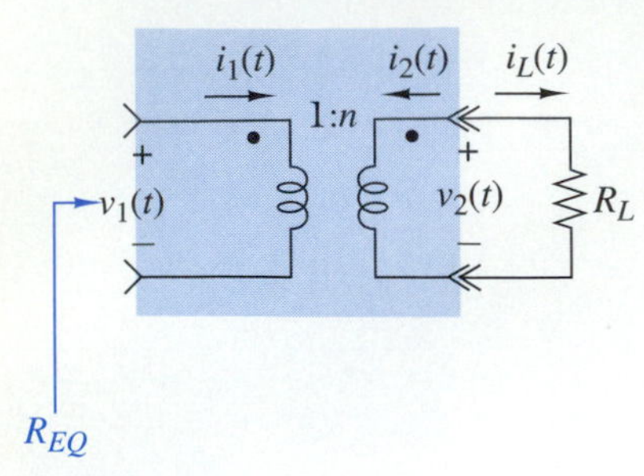

Figure 7-40 Transformer input resistance.

equations for the ideal transformer shown in Fig. 7-40:

$$\text{Resistor:} \quad v_2(t) = R_L i_L(t)$$

$$\text{Transformer:} \quad v_2(t) = n v_1(t)$$

$$i_2(t) = -\frac{1}{n} i_1(t)$$

Dividing the first transformer equation by the second and inserting the load resistance constraint yields

$$\frac{v_2(t)}{i_2(t)} = \frac{i_L(t) R_L}{i_2(t)} = -n^2 \frac{v_1(t)}{i_1(t)}$$

Applying KCL at the output interface tells us $i_L(t) = -i_2(t)$. Therefore, the equivalent resistance seen on the primary side is

$$R_{EQ} = \frac{v_1(t)}{i_1(t)} = \frac{1}{n^2} R_L \tag{7-65}$$

The equivalent load resistance seen on the primary side depends on the turns ratio and the load resistance. Adjusting the turns ratio can make R_{EQ} equal to the source resistance. Transformer coupling can produce the resistance match condition for maximum power transfer when the source and load resistance are not equal.

The following example illustrates this capability.

EXAMPLE 7-17

A stereo amplifier has an output resistance of 600 Ω. The input resistance of the speaker (the load) is 8 Ω. Select the turns ratio of a transformer to obtain maximum power transfer.

SOLUTION:

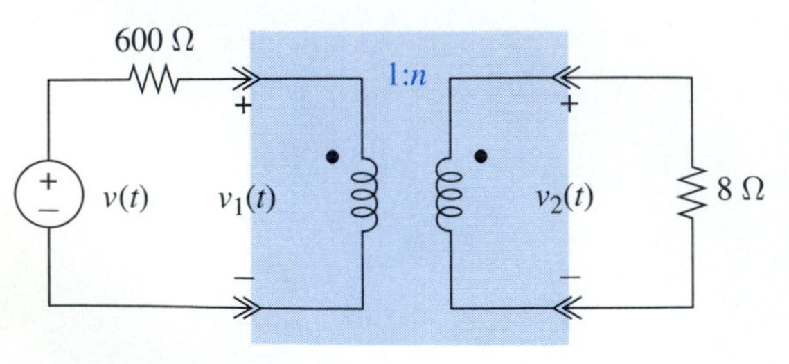

Figure 7-41

The maximum power transfer theorem in Chapter 3 states that the source and load resistance must be matched (equal) to achieve maximum power. In this case directly connecting the amplifier (600 Ω) to the speaker (8 Ω) produces a mismatch. If a transformer is inserted as shown in Fig. 7-41, then the equivalent load resistance seen by the amplifier is

$$R_{EQ} = \frac{1}{n^2} R_L = \frac{1}{n^2} 8$$

To produce a resistance match we need $R_{EQ} = 600 = 8/n^2$ or a turns ratio of $n = 1/8.66$.

7-7 SPICE SYNTAX FOR CAPACITORS AND INDUCTORS

Computer-aided circuit analysis programs simulate linear and nonlinear capacitors, inductors, and mutual inductance. This section shows how the SPICE syntax describes capacitors, inductors and mutual inductance.

The SPICE syntax describes all devices using one or more element statements. The element name is the first entry in an element statement and the first letter in the element name defines the element type. The element statements for capacitors and inductors have the following syntax:

```
Cxx...xx <1st node> <2nd node> <value> [IC=<initial voltage>]
Lxx...xx <1st node> <2nd node> <value> [IC=<initial current>]
```

The first element statement defines a linear capacitor named Cxx ... xx connected between <1st node> and <2nd node> with a capacitance equal to the <value> listed. The 1st node, 2nd node, and value parameters must be present to define a capacitor in the SPICE syntax. SPICE uses dc analysis to determine the initial ($t = 0$) voltage on the capacitor and current through inductors. The optional IC = <initial voltage> entry can be used to set the initial voltage at a user selected level. The sign given to the value of the initial voltage follows from the SPICE convention of assigning the plus reference mark to the 1st node listed in the element statement.

The Lxx...xx statement defines a linear inductor in the same way as the Cxx...xx statement, except that the optional IC feature sets the initial current through the inductor. The sign given to the value of the initial current follows from the passive sign convention and the fact that SPICE assigns the plus voltage reference mark to the 1st node.

The following are sample SPICE element statements:

```
C17 3 8 3U IC = +12
LPAR 2 5 2M IC = -4M
```

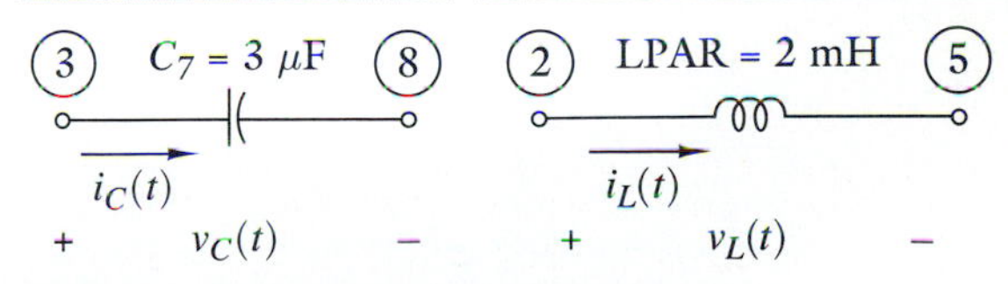

Figure 7-42 SPICE element statements for a capacitor and inductor.

These statements define the element voltages and currents shown in Fig. 7-42. As noted above, SPICE always assigns the positive voltage reference to the first node in an element statement and then assigns the current reference direction using the passive sign convention. The specified initial conditions in these statements mean the $v_C(0) = +12$V and $i_L(0) = -4$ mA.

SPICE handles mutual inductance by defining the coupling coefficient (k) between two inductors using the following syntax:

```
Kxx...xx <1st inductor> <2nd inductor> <value>
```

This statement means the mutual inductance coupling between the first and second named inductors has a coupling coefficient of $k =$ <value>. The Kxx...xx statement must be accompanied by two separate Lxx...xx statements. The coil dots are associated with the first node named in each of the inductor statements. The coupling coefficient must lie on the range $0 < k < 1$. SPICE does not support perfect coupling ($k = 1$), decoupling ($k = 0$), overcoupling ($k > 1$), or negative coupling ($k < 0$) and returns an error message if you try these cases.

An example of SPICE statements defining a pair of coupled coils is

```
L1 7 3 12.5M
L2 5 8 8M
K L1 L2 0.9
```

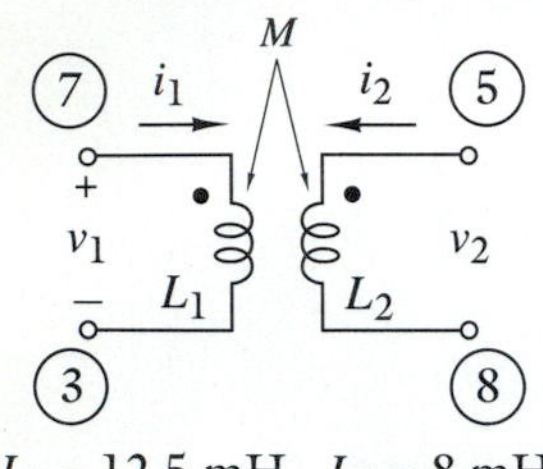

$L_1 = 12.5$ mH, $L_2 = 8$ mH,
$M = k\sqrt{L_1 L_2} = 9$ mH

Figure 7-43 Example coupled coils.

These statements define the coupled coils shown in Fig. 7-43, including the mutual inductance calculated using L_1, L_2, and the coupling coefficient k. The coil dots are associated with the first nodes named in the L1 and L2 statements. In the SPICE syntax the user must ensure that the terminals with the coil dots are the first named nodes in each of the inductor statements; otherwise, an erroneous signal inversion will occur.

SPICE assumes that mutual coupling is additive. The reason is that SPICE always assigns the positive voltage reference mark and the coil dots to the first node in an inductor element statement. SPICE assigns the current reference directions using the passive sign convention. As a result, the reference directions for both coil currents always point toward the coil dots, as illustrated in Fig. 7-43. The net result is that the sign of the mutual inductance term is **always positive**.

SPICE analyzes circuits containing capacitors, inductors, and mutual inductance using the waveforms and transient analysis statement covered in Chapter 6. The SPICE transient analysis command includes an option feature for defining initial conditions:

```
.TRAN TSTEP TSTOP TSTART TMAX [UIC]
```

The role of the parameters TSTEP. TSTOP, TSTART, and TMAX were discussed in Chapter 6. The optimal [UIC] parameter causes SPICE to *u*se the *i*nitial *c*onditions specified in the capacitor or inductor element statements. SPICE ignores the initial conditions in the element statement when [UIC] option is left blank. If you want SPICE to use the initial values given in the element statements the .TRAN command statement *must* contain a UIC entry.

The following examples show applications of SPICE transient analysis for capacitors and inductors.

EXAMPLE 7-18

The voltage across a 0.5-μF capacitor is shown in Fig. 7-44(a). Use SPICE to calculate the current through the capacitor. Assume $v_C(0) = 0$.

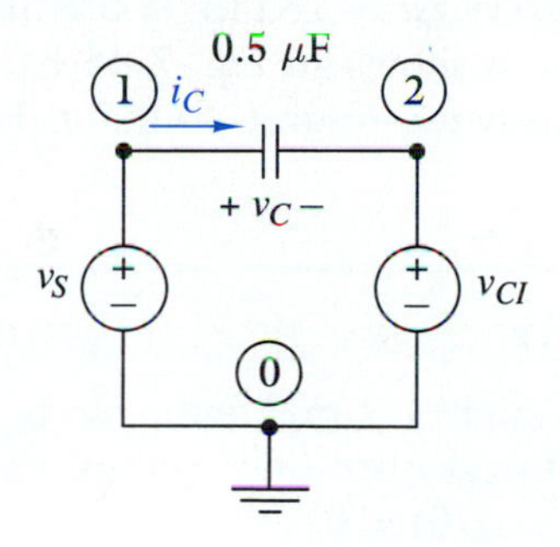

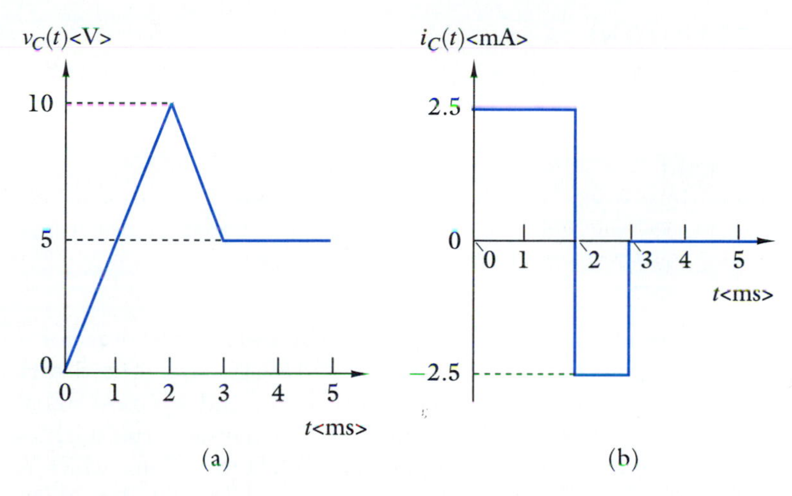

Figure 7-44

SOLUTION:

Figure 7-44 shows a SPICE circuit model for this problem. The source V_{CI} is a current monitoring voltage source inserted in series with the capacitor to measure $i_C(t)$. A suitable SPICE input file for this circuit is

```
EXAMPLE 7-18
VS 1 0 PWL 0 0 2M 10 3M 5 5M 5
C 1 2 0.5U IC = 0
VCI 2 0 DC 0
.TRAN 0.1M 5M UIC
.PRINT TRAN I(VCI)
.END
```

The input voltage source VS generates a piecewise linear (PWL) waveform using the break points shown in Fig. 7-44(a). The .TRAN statement produces a transient analysis run beginning at $t = 0$ and running to t = TSTOP = 5 ms, which is the input duration shown in Fig. 7-44(a). The .PRINT command causes SPICE to write an output file containing time t and the capacitor current I(VCI) every Δt = TSTEP = 0.1 ms. A plot of the data in the output file is shown in Fig. 7-44(b). The SPICE results agree with the analytical result is found in Example 7-1.

EXAMPLE 7-19

The current through a 2-mH inductor is $i(t) = 2 \sin 1000t$ A. Use SPICE to calculate one cycle of the voltage across the inductor. Assume $i_L(0) = 0$.

SOLUTION:

Suitable SPICE input file for this problem is

```
EXAMPLE 7-19
IS 0 1 SIN 0 2 159.15 0 0
L 1 0 2M IC = 0
.TRAN 0.08M 8M UIC
.PRINT TRAN V(1)
.END
```

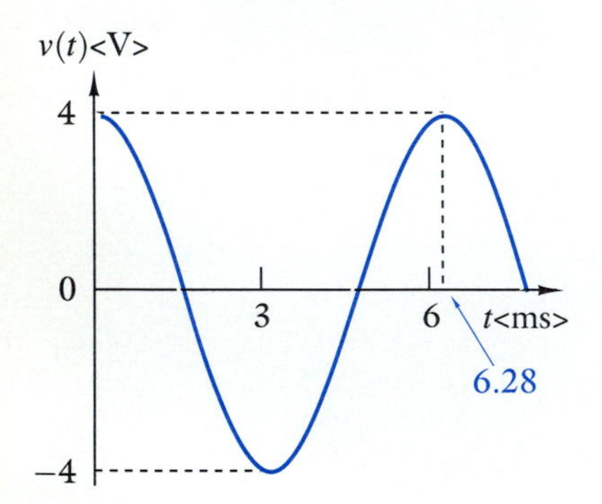

Figure 7-45

The current source IS generates a sinusoidal (SIN) waveform whose frequency is $f = 1000/2\pi \approx 159.15$ Hz. The 2-mH inductor is connected in series with source and has zero initial current. The .TRAN statement causes a transient run analysis beginning at $t = 0$ and running to t = TSTOP = 8 ms, which is slightly more than one cycle of the sinusoid ($T_0 = 1/f \approx 6.28$ ms). The .PRINT statement creates an output file containing time t and the voltage across inductor V(1) every Δt = TSTEP + 0.08 ms, or roughly 100 points per cycle. A plot of this file is shown in Fig. 7-45. The SPICE results agree with the analytical results found in Example 7-6.

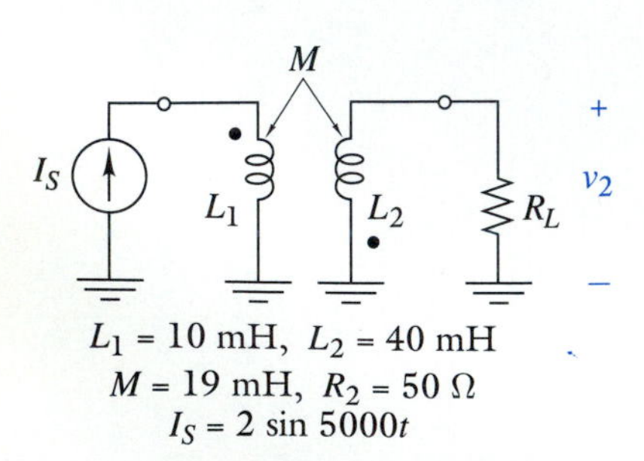

Figure 7-46

EXERCISE 7-18

Write a SPICE input file to calculate two cycles of the voltage across the 50-Ω resistor in Fig. 7-46.

ANSWER:

```
EXERCISE 7-18
IS 0 1 SIN 0 2 795.77 0 0
L1 1 0 10M
```

```
L2 0 2 40M
K L1 L2 0.95
RL 2 0 50
.TRAN .025M 2.5M
.PRINT TRANS V(2)
.END
```

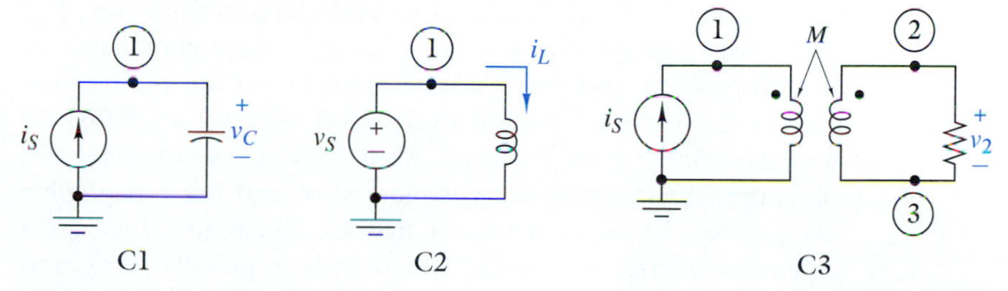

Figure 7-47 Some example circuits.

Circuit Limitations in SPICE

Figure 7-47 shows three circuits that cannot be analyzed using SPICE. Circuit C1 drives a current through a capacitor producing a voltage response. Circuit C2 specifies the voltage across an inductor producing a current response. Circuit C3 drives a current into the primary of a transformer producing a voltage across a load in the secondary circuit. Circuits like these were analyzed earlier in this chapter using analytical methods.

SPICE was designed to analyze semiconductor electronic circuits that always require dc power supplies. To determine the initial operating conditions SPICE first performs a dc analysis prior to a transient analysis. SPICE replaces all capacitors by open circuits and all inductors by short circuits to perform a dc analysis.

Figure 7-48 shows what happens when dc analysis is attempted on the circuits in Fig. 7-47. In circuit C1 the current source tries to drive current into an open circuit, leading to an infinite voltage. In circuit C2 the voltage source is connected to a short circuit, leading to an infinite current. In circuit C3 there is no dc path to ground from Nodes 2 and 3 and SPICE cannot calculate the dc voltages at these nodes.

The generalizations of the examples are:

Circuit C1: SPICE will not analyze circuits with nodes to which only capacitors and current sources are connected.

Circuit C2: SPICE will not analyze circuits with loops containing only voltage sources and inductors.

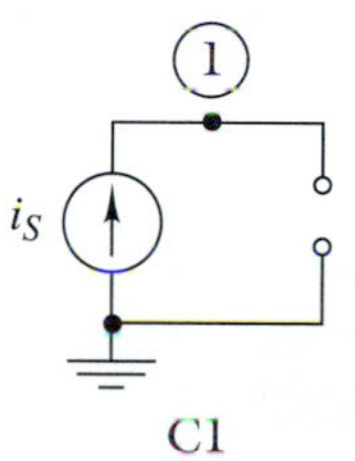

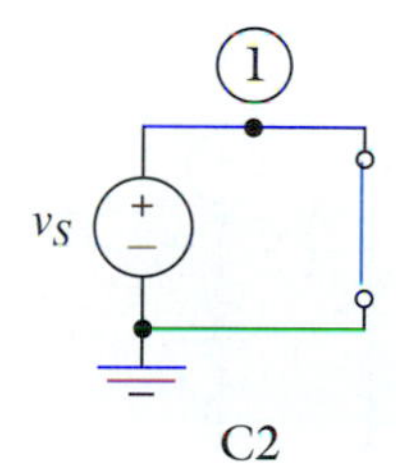

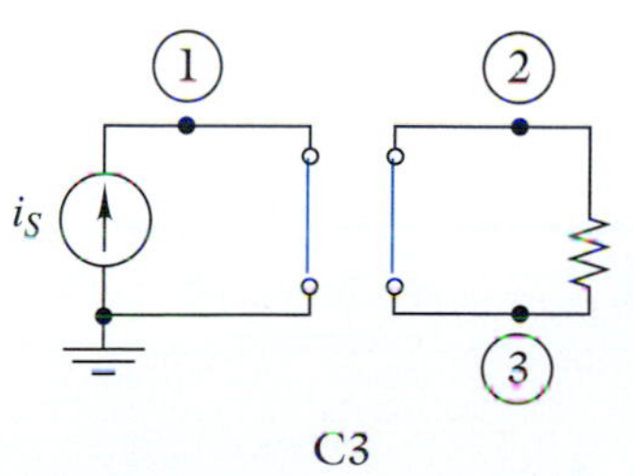

Figure 7-48 Equivalent dc circuits.

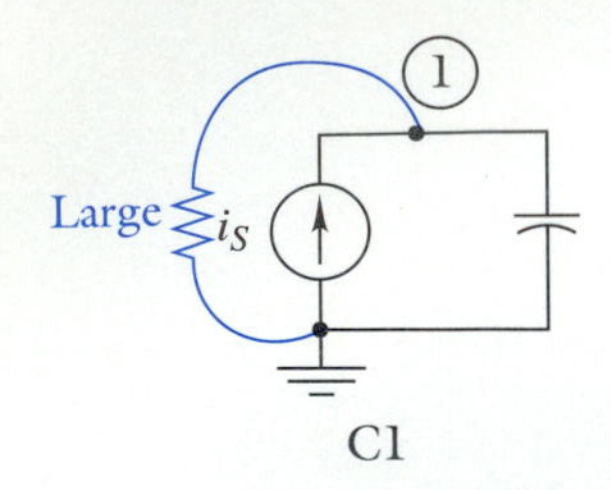

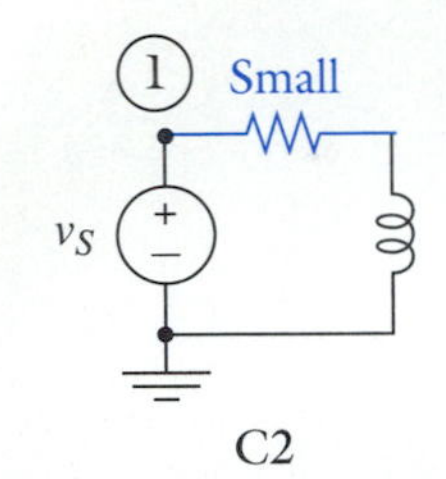

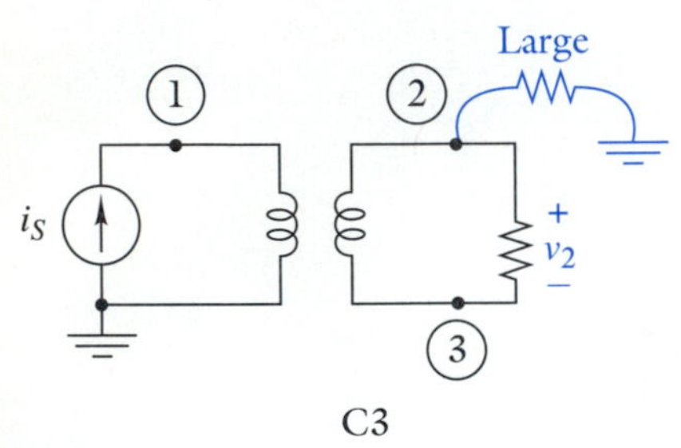

Figure 7-49 Modified SPICE circuits.

Circuit C3: SPICE will not analyze transformer circuits unless both the primary and the secondary circuits have dc paths to ground.

If you create a circuit file with any of these features, SPICE terminates the analysis and returns an error message flagging the problem. These safeguards can be circumvented using the ruses shown in Fig. 7-49. In circuit C1 we add a large resistor (say, 100 MΩ) across the current source. A large resistor eliminates the dc path problem and draws little current in during the transient analysis. In circuit C2 we add a very small resistor (say, 0.001 Ω) in series with the voltage source. A small series resistor eliminates the short circuit across the voltage source and has a negligible voltage drop during the transient analysis. In circuit C3 we add a large resistor from a secondary circuit node to ground. Programs like MICRO-CAP automatically added these features to every voltage and current source to avoid dc analysis problems.

The important message to carry away from this discussion is the following. Computer aided analysis offers powerful tools that can relieve much of the numerical drudgery of hand analysis. These programs handle large-scale circuits that are simply beyond the scope of practical hand analysis. However, using these tools wisely requires some understanding of how the programs work and at least a rough understanding of what the response is expected to be. Without both of these understandings there is a very real danger of falling into the "garbage-in" and "garbage-out" trap.

■ EXERCISE 7-19

Figure 7-50 shows a high frequency model of a transformer. Can SPICE analyze this circuit as drawn?

ANSWER:
No, under dc conditions the inductance of the primary circuit shorts the voltage source and the secondary circuit has no dc path to ground.
■

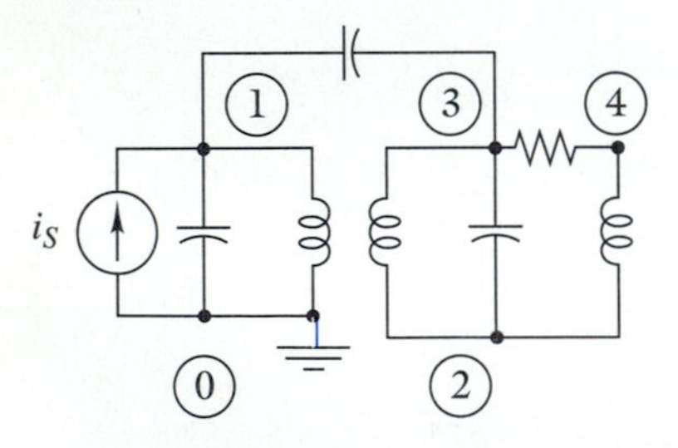

Figure 7-50

SUMMARY

- The linear capacitor and inductor are dynamic circuit elements that can store energy. The instantaneous element power is positive when they are storing energy and negative when they are delivering previously stored energy. The net energy transfer is never negative because inductors and capacitors are passive elements.

- The current through a capacitor is zero unless the voltage is changing. A capacitor acts like an open circuit to dc excitations.
- The voltage across an inductor is zero unless the current is changing. An inductor acts like a short circuit to dc excitations.
- Capacitor voltage and inductor current are called state variables because they define the energy state of a circuit. Circuit state variables are continuous functions of time as long as the circuit-driving forces are finite.
- OP AMP capacitor circuits perform signal integration or differentiation. These operations combined with the summer and gain functions provide the building blocks for designing dynamic input-output characteristics.
- Capacitors or inductors in series or parallel can be replaced with an equivalent element, which is found by adding the individual capacitances or inductances or their reciprocals. The dc response of a dynamic circuit can be found by replacing all capacitors with open circuits and all inductors with short circuits.
- Mutual inductance describes magnetic coupling between two coils. The dot convention describes physical orientation of the two magnetically coupled coils. Mutual inductance coupling is additive when both current reference arrows point toward or away from dotted terminals; otherwise, it is subtractive.
- The ideal transformer is a model that assumes perfect coupling and no power loss. An ideal transformer is characterized by its turns ratio.
- The SPICE syntax includes element statements to describe capacitors, inductors, and mutual inductance. The element statements allow the assignment of initial conditions on the capacitor voltage and inductor current. Mutual inductance is always assumed to be additive.

EN ROUTE OBJECTIVES AND ASSOCIATED PROBLEMS

ERO 7-1 Capacitor and Inductor Responses (Sects. 7-1, 7-2)

Given the waveform of the current through (voltage across) a capacitor or inductor:

(a) determine the voltage across (current through) the element.

(b) determine the power absorbed and energy stored by the element.

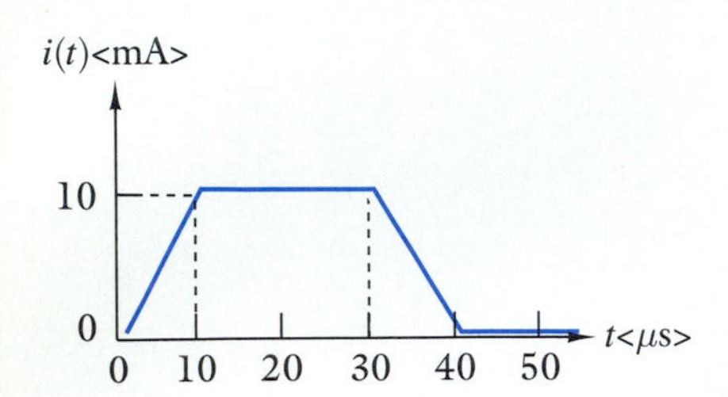

Figure P7-9

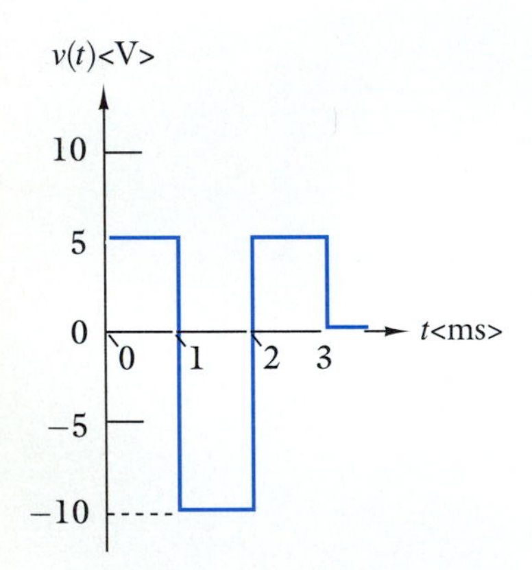

Figure P7-10

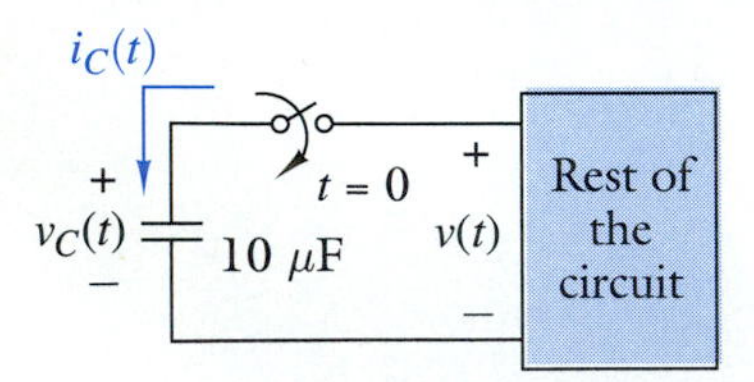

Figure P7-12

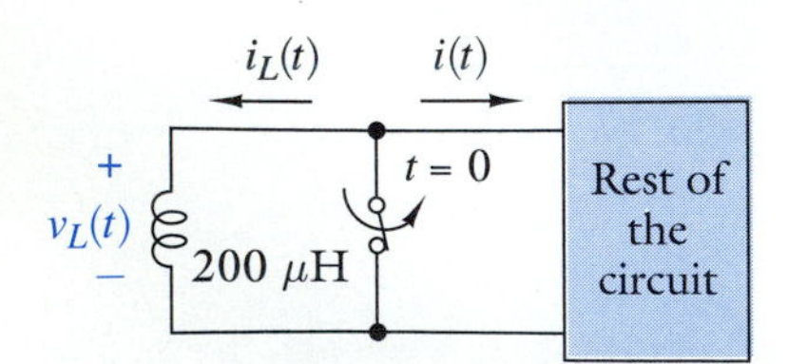

Figure P7-13

7-1 The voltage across a 10-μF capacitor is $v_C(t) = 3[1 - e^{-200t}]u(t)$ V. Find and sketch the waveforms of $i_C(t)$, $p_C(t)$ and $w_C(t)$.

7-2 The voltage across a 10-μF capacitor is $v_C(t) = 3[e^{-200t} \sin 800t]$ $u(t)$ V. Find and sketch the waveforms of $i_C(t)$, $p_C(t)$, and $w_C(t)$.

7-3 The current through a 25-μF capacitor is $i_C(t) = 50[u(t) - u(t - 5 \times 10^{-3})]$ mA. Find and sketch the waveforms of $v_C(t)$, $p_C(t)$, and $w_C(t)$.

7-4 The current through a 25-μF capacitor is $i_C(t) = 50[u(t) - 2u(t - 25 \times 10^{-3})]$ mA. Find and sketch the waveforms of $v_C(t)$, $p_C(t)$, and $w_C(t)$.

7-5 At $t = 0$ the initial voltage across a 200-pF capacitor is 30 V. For $t > 0$ the current through the capacitor is $i_C(t) = 0.4 \cos 10^5 t$ A. Find and sketch the waveforms of $v_C(t)$, $p_C(t)$, and $w_C(t)$.

7-6 The current through a 2-mH inductor is $i_L(t) = [100te^{-1000t}]$ $u(t)$ A. Find and sketch the waveforms of $v_L(t)$, $p_L(t)$, and $w_L(t)$.

7-7 For $t > 0$ the current through a 2-mH inductor is $i_L(t) = 10 \sin 2 \times 10^5 t$ mA. Find and sketch the waveforms of $v_L(t)$, $p_L(t)$, and $w_L(t)$.

7-8 A voltage pulse $v_L(t) = 5[u(t - 10^{-3}) - u(t - 2 \times 10^{-3})]$ appears across a 50-mH inductor. Find and sketch the waveforms of $i_L(t)$, $p_L(t)$, and $w_L(t)$.

7-9 The current through a 20-mH inductor is shown in Fig. P7-9. Find and sketch the waveforms of $v_L(t)$, $p_L(t)$, and $w_L(t)$.

7-10 The voltage pulse in Fig. P7-10 appears across a 25-mH inductor. Find the energy stored in the inductor at $t = 1$, 2, and 3 ms. Assume $i_L(0) = 0$.

7-11 The current through a 30-μF capacitor is shown in Fig. P7-9. Find the energy stored in the capacitor at $t = 10$, 30, and 40 μs. Assume $v_C(0) = 0$.

7-12 The capacitor in Fig. P7-12 carries an initial voltage $v_C(0) = 10$ V. At $t = 0$ the switch is closed and thereafter the voltage across the capacitor is $v_C(t) = 10\ e^{-1000t}$ V. For $t > 0$ find the waveforms of $i_C(t)$ and $p_C(t)$. Is the capacitor absorbing power from or delivering power to the rest of the circuit?

7-13 The inductor in Fig. P7-13 carries an initial current of $i_L(0) = 100$ mA. At $t = 0$ the switch opens and thereafter the current into the rest of the circuit is $i(t) = -100\ e^{-200t}$ mA. For $t > 0$ find the waveforms of $v_L(t)$ and $p_L(t)$. Is the inductor adsorbing power from or delivering power to the rest of the circuit?

7-14 The initial voltage across a 0.1-μF capacitor is $v_C(0) = 15$ V. For $t \geq 0$ the voltage across and current through the capacitor are $v_C(t) = Ke^{-\alpha t}$ V and $i_C(t) = 10^{-2}e^{-\alpha t}$ A. Find the constants K and α.

7-15 Using the passive sign convention the voltage across and current through a linear energy storage element are $v(t) = 100 \cos 5000t$ V and $i(t) = \sin 5000t$ A. Is the element an inductor or a capacitor? What is the value of its inductance or capacitance?

7-16 Using the passive sign convention the voltage across and current through a linear energy storage element are $v(t) = 100(1 - e^{-1000t})u(t)$ V and $i(t) = 0.1[e^{-1000t}]u(t)$ A. Is the element an inductor or a capacitor? What is the value of its inductance or capacitance?

7-17 Using the passive sign convention the voltage across and current through a linear energy storage element are $v(t) = 15[e^{-100t}]u(t)$ V and $i(t) = 0.1(1 - e^{-100t})u(t)$ A. Is the element an inductor or a capacitor? What is the value of its inductance or capacitance?

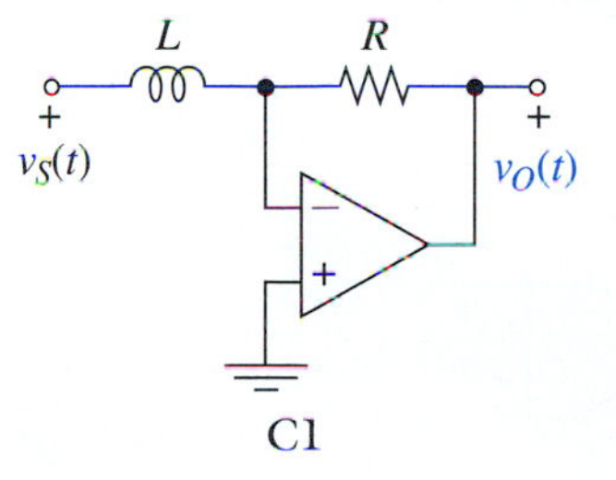

ERO 7-2 Dynamic OP AMP Circuits (Sect. 7-3)

(a) Given a circuit consisting of resistors, capacitors, inductors and OP AMPs, determine its input-output relationship and use the relationship to find output waveforms for specified inputs.

(b) Design an OP AMP circuit to implement a given input-output relationship or a block diagram.

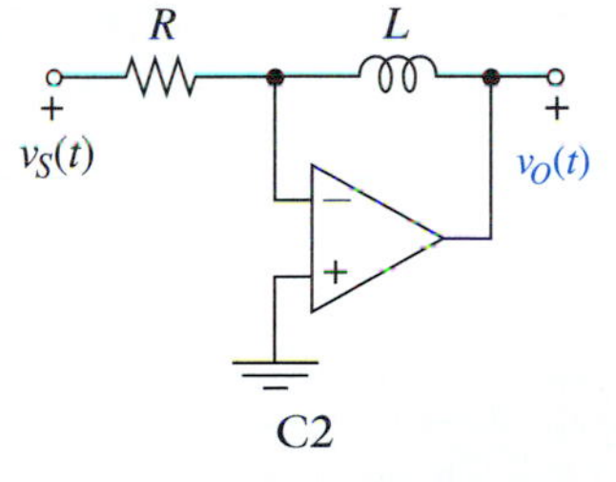

Figure P7-18

7-18 Show that the OP AMP inductor circuits in Fig. P7-18 provide integration and differentiation of the input $v_S(t)$.

7-19 Show that the OP AMP capacitor circuit in Fig. P7-19 is a noninverting integrator whose input-output relationship is

$$v_O(t) = \frac{2}{RC}\int_0^t v_S(x)dx$$

Draw a functional block diagram for this input-output relationship. Hint: By voltage division the voltage at the inverting input is $v_O(t)/2$; hence, for an ideal OP AMP the voltage at the noninverting input is $v_O(t)/2$.

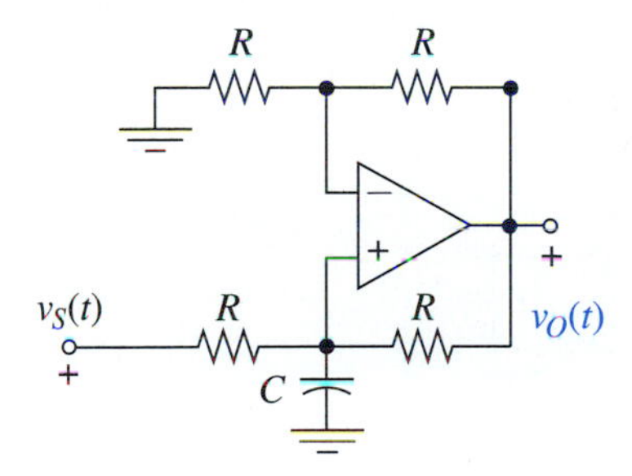

Figure P7-19

7-20 Show that the OP AMP circuit in Fig. P7-20 has the input-output relationship

$$v_O(t) = \frac{v_S(t)}{2} - \frac{RC}{2}\frac{dv_S(t)}{dt}$$

Draw a functional block diagram for this input-output relationship. Hint: By voltage division the voltage at the noninverting input is $v_S(t)/2$; hence for an ideal OP AMP the voltage at the inverting input is $v_S(t)/2$ as well.

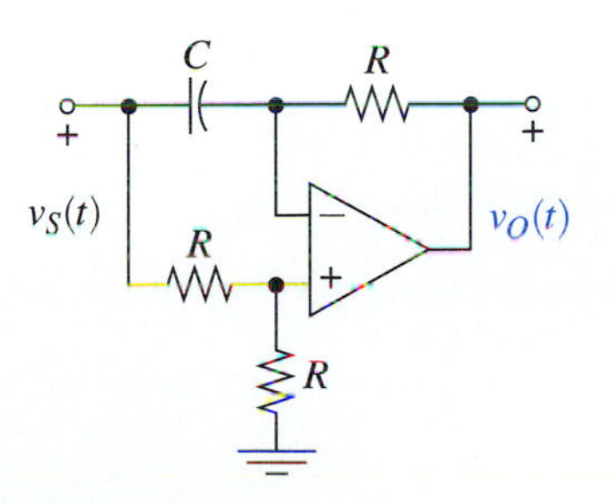

Figure P7-20

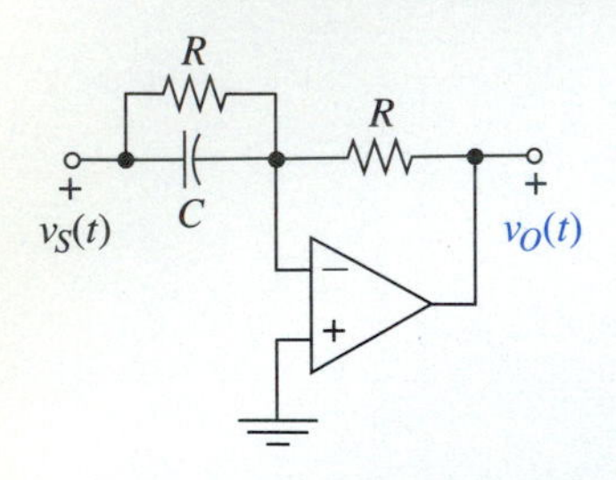

Figure P7-21

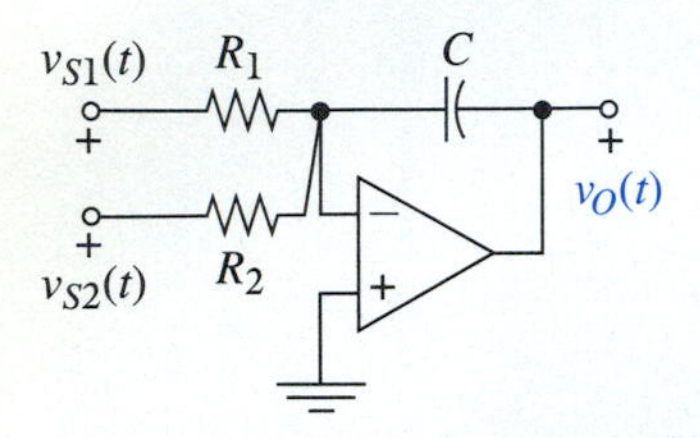

Figure P7-22

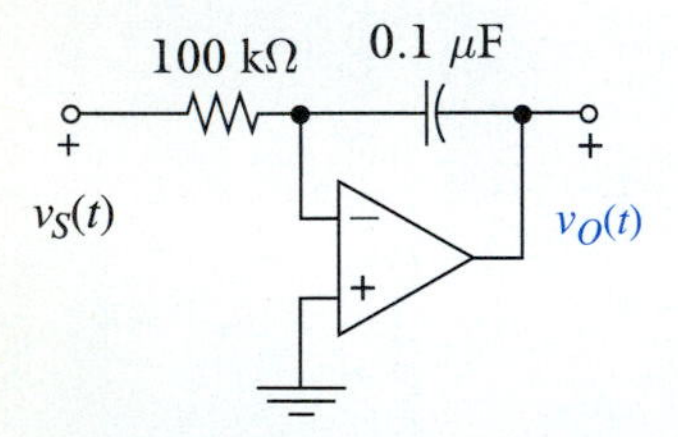

Figure P7-23

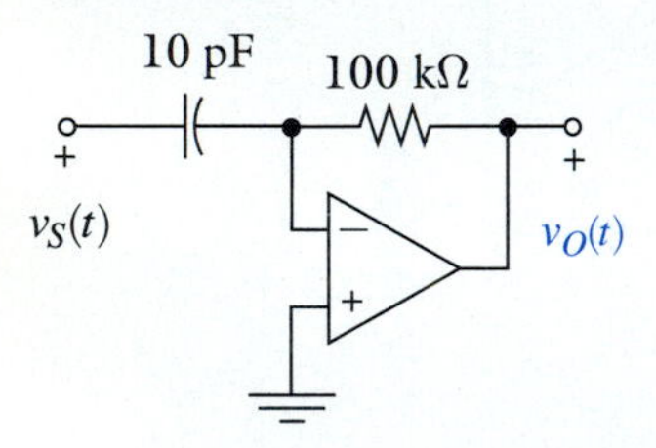

Figure P7-27

7-21 Show that the OP AMP circuit in Fig. 7-21 has the input-output relationship

$$v_O(t) = -v_S(t) - RC\frac{dv_S(t)}{dt}$$

Draw a functional block diagram for this input-output relationship.

7-22 Show that the OP AMP circuit in Fig. P7-22 has the input-output relationship

$$v_O(t) = -\frac{1}{R_1 C}\int_0^t v_{S1}(x)dx - \frac{1}{R_2 C}\int_0^t v_{S2}(x)dx$$

Draw a functional block diagram for this input-output relationship.

7-23 At $t = 0$ the voltage across the capacitor in Fig. P7-23 is zero. The OP AMP saturates at ±15 V. For $v_S(t) = 5u(t)$ V, derive an equation for the output voltage when the OP AMP is in its linear range. At what time $t > 0$ does the OP AMP saturate?

7-24 Repeat Prob. 7-23 when the initial voltage on the capacitor results in $v_O(0) = +10$ V.

7-25 Repeat Prob. 7-23 when $v_S(t) = 5[u(t) - u(t - 0.02)]$.

7-26 At $t = 0$ the voltage across the capacitor in Fig. P7-23 is zero. The OP AMP output saturates at ±15 V. For $v_S(t) = V_A[\sin 1000t]$ $u(t)$ V, derive an equation for the output voltage when the OP AMP is in its linear range. What is the maximum value of V_A for linear operation?

7-27 The input to the circuit in Fig. P7-27 is $v_S(t) = V_A[\sin 1000t]u(t)$ V. Derive an equation for the output voltage when the OP AMP is in its linear range. The OP AMP output saturates at ±15 V. What is the maximum value of V_A for linear operation?

7-28 The input to the circuit in Fig. P7-27 is $v_S(t) = 5[\sin 1000t]u(t)$ V. Derive an equation for the output voltage when the OP AMP is in its linear range. The OP AMP output saturates at ±15 V. What is the maximum value of the feedback resistor for linear operation?

7-29 The input to the circuit in Fig. P7-19 is $v_S(t) = 5[\sin 1000t]u(t)$ V. Derive an equation for the output voltage when the OP AMP is in its linear range. The OP AMP output saturates at ±15 V. What is the minimum value of the RC product for linear operation?

7-30 Repeat Prob. P7-29 using the circuit in Fig. P7-20.

7-31 Find the overall input-output relationship when the circuits in Figs. P7-19 and P7-20 are connected in cascade.

7-32 Find the overall input-output relationship when the circuits in Figs. P7-19 and P7-21 are connected in cascade.

7-33 Find the overall input-output relationship when two identical versions of the circuit in Fig. P7-20 are connected in cascade.

7-34 Find the input-output relationship of the block diagrams in Fig. P7-34.

7-35 Suppose the inputs to the block diagrams in Fig. P7-34 are tied together and their outputs subtracted. Find the overall input-output relationship of the resulting block diagram.

7-36 Find the input-output relationship of the block diagrams in Fig. P7-36.

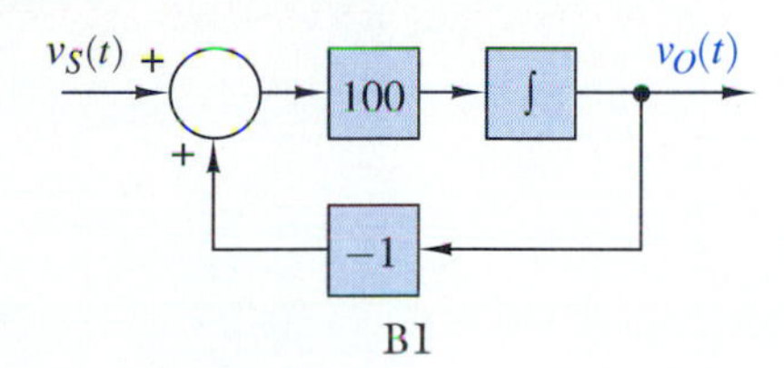

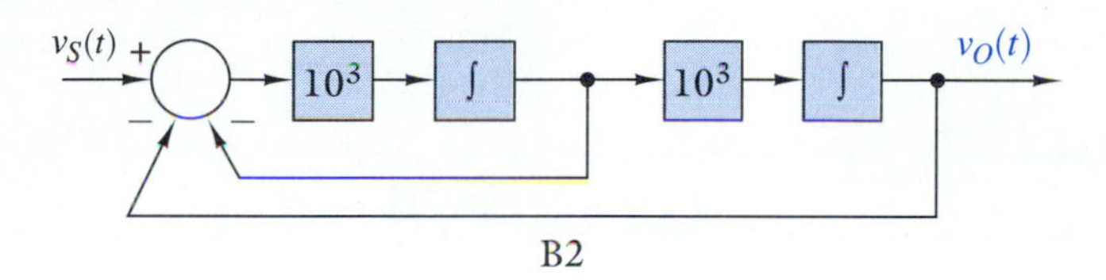

Figure P7-36

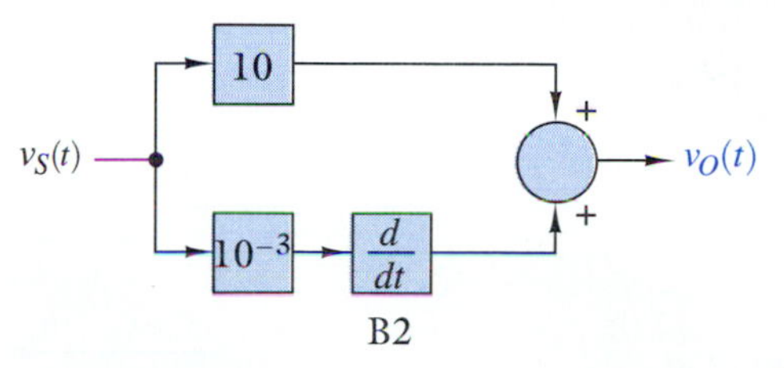

Figure P7-34

7-37 Construct block diagrams for the following input-output relationships:

(a) $v_O(t) = v_S(t) + \dfrac{1}{100}\dfrac{dv_S(t)}{dt}$

(b) $v_O(t) = -3v_S(t) - 100\int_0^t v_S(x)dx$

7-38 Construct block diagrams for the following differential equations:

(a) $200\dfrac{dv_O(t)}{dt} + v_O(t) = v_S(t) + 100\dfrac{dv_S(t)}{dt}$

(b) $\dfrac{d^2v_O(t)}{dt^2} + 100\dfrac{dv_O(t)}{dt} + 800v_O(t) = -300v_S(t)$

❖ **7-39** **(D)** Design OP AMP RC circuits to implement the block diagrams in Fig. P7-34.

❖ **7-40** **(D)** Design OP AMP RC circuits to implement the block diagrams in Fig. P7-36.

❖ **7-41** **(D)** Design OP AMP RC circuits to implement the input-output relationships in Prob. 7-37.

❖ **7-42** **(D)** Design RC OP AMP circuits to implement the input-output relationships in Prob. 7-38.

ERO 7-3 Equivalent Capacitance and Inductance (Sect. 7-4)

(a) Derive equivalence properties of inductors and capacitors or use equivalence properties to simplify LC circuits.

(b) Solve for currents and voltages in RLC circuits with dc input signals.

7-43 Reduce the circuits in Fig. P7-43 to a single equivalent element.

7-44 Reduce the circuit in Fig. P7-44 to as few equivalent elements as possible.

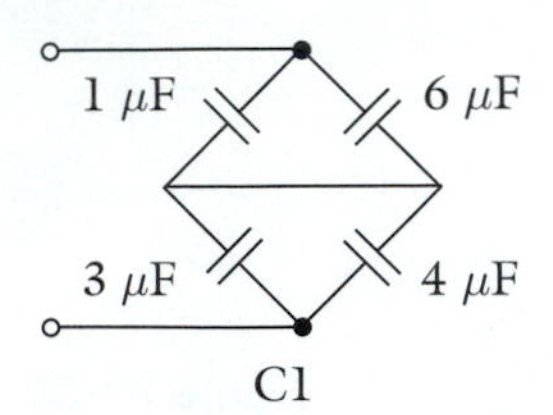

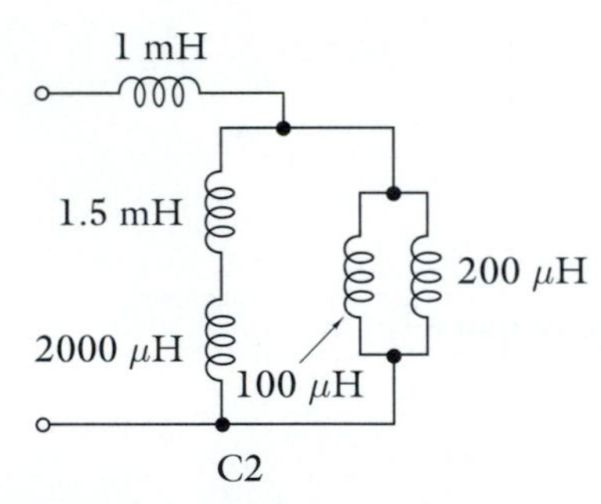

Figure P7-43

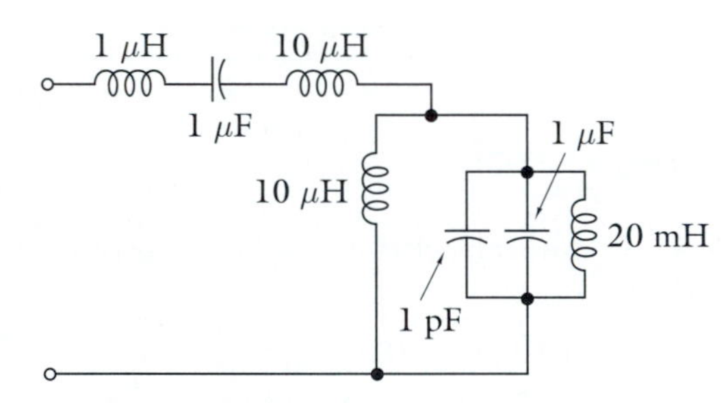

Figure P7-44

7-45 Show that the i-v relationships for the linear inductor have the proportionality and superposition properties if the initial current is zero.

7-46 Derive the expression in Eq. (7-30) for the equivalent inductance of a series connection of inductors.

7-47 Derive the expression in Eq. (7-31) for the equivalent inductance of a parallel connection of inductors.

7-48 A capacitor bank for a large pulse generator consists of 11 capacitor strings connected in parallel. Each string consists of 16 1.5-mF capacitors connected in series.

(a) What is the total equivalent capacitance of the bank?

(b) If each capacitor in a series string is charged to 300 V, what is the total energy stored in the bank?

(c) In the discharge mode the voltage delivered by the pulse generator is $4.8[e^{-500t}]u(t)$ kV. What is the peak power delivered by the pulse generator?

7-49 The circuits in Fig. P7-49 are driven by dc sources. Find the voltage across capacitors and the current through inductors.

1 kΩ, 15 V, 3 kΩ, C — C1

L, 5 V, 10 kΩ, 20 kΩ — C2

L, 1 mA, 600 Ω, C — C3

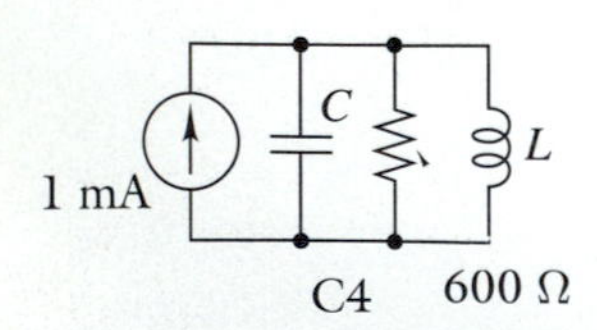

Figure P7-49

7-50 The circuits in Fig. P7-50 are driven by dc sources. Find the voltage across capacitors and the current through inductors.

7-51 The OP AMP circuits in Fig. P7-51 are driven by dc sources. Find the output voltage $v_O(t)$.

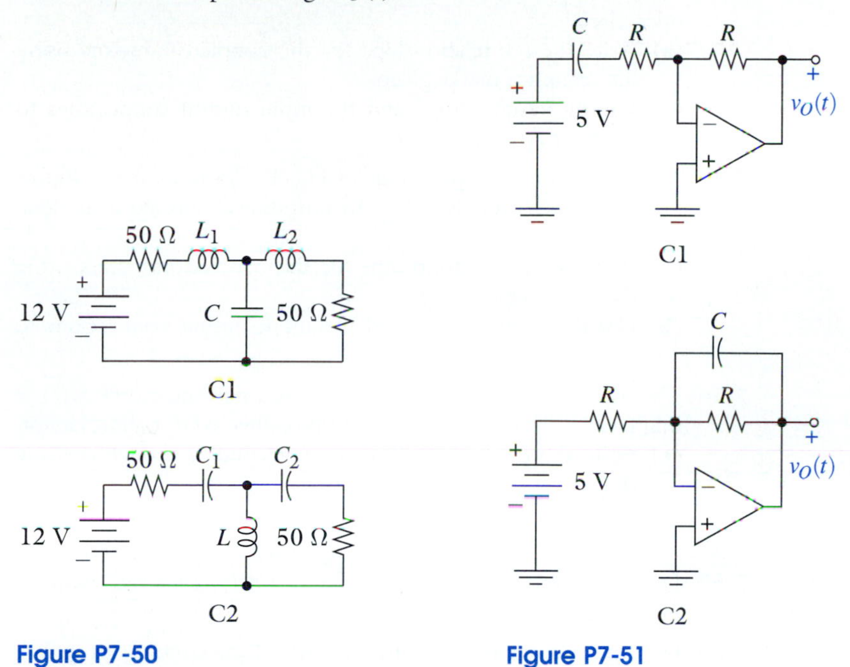

Figure P7-50

Figure P7-51

ERO 7-4 Mutual Inductance (Sect. 7-5)

(a) Given the current through or voltage across two coupled inductors, find the unspecified currents or voltages.

(b) Find the equivalent inductance of series and parallel connections of coupled inductors.

7-52 The input to the coupled coils in Fig. P7-52 is a voltage source $v_S(t) = 25[\sin 1000t]u(t)$ V. The output is connected to an open circuit.

(a) Write the i-v relationships for the coupled inductors using the reference marks in the figure.

(b) Use the results in (a) and the input-output connections to solve for $i_1(t)$ and $v_2(t)$.

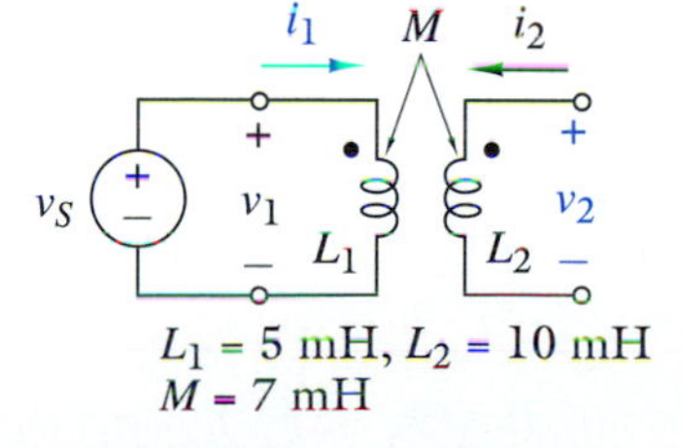

Figure P7-52

7-53 The input to the coupled coils in Fig. P7-52 is a voltage source $v_S(t) = 25[\sin 1000t]u(t)$ V. The output is connected to a short circuit.

(a) Write the i-v relationships for the coupled inductors using the reference marks in the figure.

(b) Use the results in (a) and the input-output connections to solve for $i_1(t)$ and $i_2(t)$.

7-54 The input of the coupled coils in Fig. P7-52 is a voltage source $v_S(t) = 10[1 - e^{-1000t}]u(t)$ V. The output is connected to an open circuit.

(a) Write the i-v relationships for the coupled inductors using the reference marks given.

(b) Use the results in (a) and the input-output connections to solve for $i_1(t)$ and $v_2(t)$.

7-55 The input to the coupled coils in Fig. P7-55 is a current source $i_s(t) = 0.5[\sin 1000t]u(t)$ A. The output is connected to an open circuit.

(a) Write the i-v relationships for the coupled inductors using the reference marks given.

(b) Use the results in (a) and the input-output connections to solve for $v_1(t)$ and $v_2(t)$.

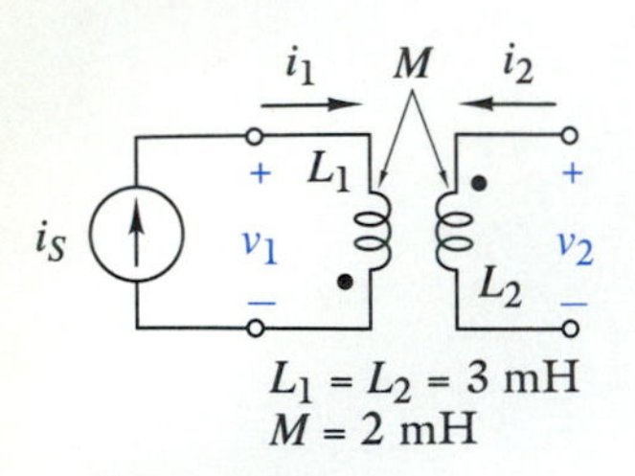

Figure P7-55

7-56 The input to the coils in Fig. P7-55 is a current source $i_S(t) = 0.2[1 - e^{-1000t}]u(t)$ A. The output is connected to a short circuit.

(a) Write the i-v relationships for the coupled inductors using the reference marks given.

(b) Use the results in (a) and the input-output connections to solve for $v_1(t)$ and $i_2(t)$.

7-57 Derive an expression for $v_X(t)$ in terms of $i_1(t)$ for the circuit of Fig. P7-57.

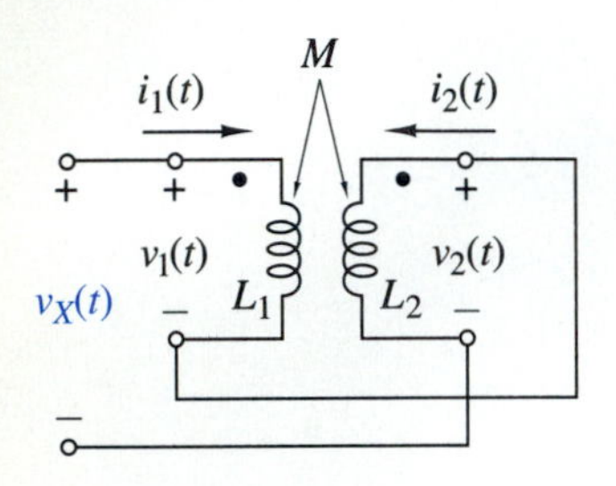

Figure P7-57

7-58 Find the equivalent inductance of each of the circuits of Fig. P7-58.

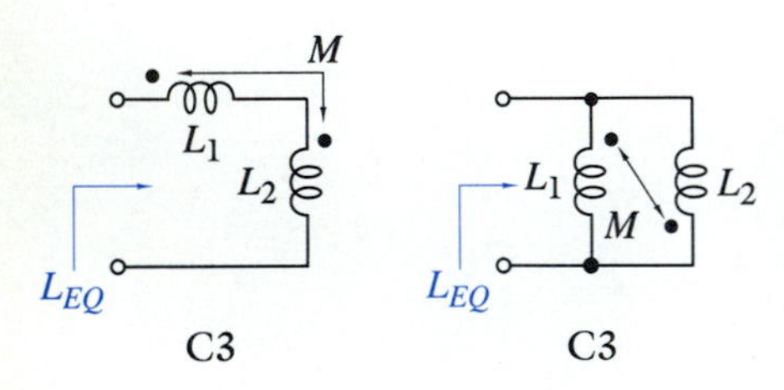

Figure P7-58

7-59 The coupled coils in Fig. P7-59 are constructed so that a smaller movable coil can slip inside a larger fixed coil. Maximum movement of the smaller coil causes k_m to vary from 0 to 0.5×10^{-5}. The input current to the larger coil is $i_1(t) = 10 \cos 2 \times 10^4 t$ mA. The smaller coil is connected to an open circuit. How much does the peak value of the output voltage change for the maximum movement of the smaller coil?

ERO 7-5 The Ideal Transformer (Sect. 7-6)

Given a circuit containing an ideal transformer:

(a) find the output current, voltage, and power.

(b) select the turns ratio to meet prescribed conditions.

7-60 For an ideal transformer, what turns ratio is required to change a 120 V_{rms}, 60 Hz sinusoid into 12 V_{rms} sinusoid? If the output is connected to a 50-Ω resistor, write expressions for the input current, voltage, and power.

7-61 Two amplifier stages are to be coupled using a transformer. The output resistance of the first stage is 300 Ω and the input to the second is 2200 Ω. Select the transformer turns ratio so that maximum power is transferred.

7-62 The turns ratios of a multi-output filament transformer are shown in Fig. P7-62. Find the rms values of $v_2(t)$ and $v_3(t)$ when the input voltage is $v_1(t) = 156 \cos 377t$.

7-63 Select the turns ratio of an ideal transformer in the interface circuit shown in Fig. P7-63 so that the maximum available power is delivered to the 1-kΩ output resistor.

7-64 Show that the controlled source model in Fig. P7-64 has the same i-v characteristics as an ideal transformer.

7-65 Two transformers are connected in cascade as shown in Fig. P7-65. The first is ideal and the second is not. Find $v_O(t)$.

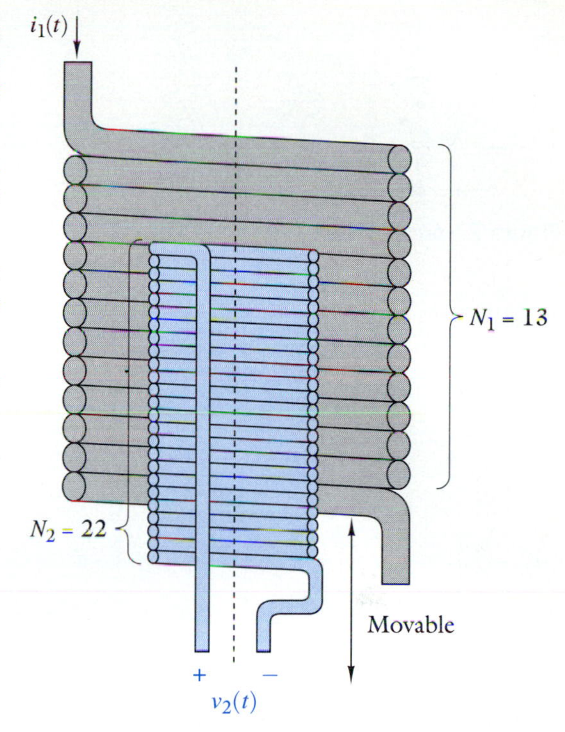

Figure P7-59

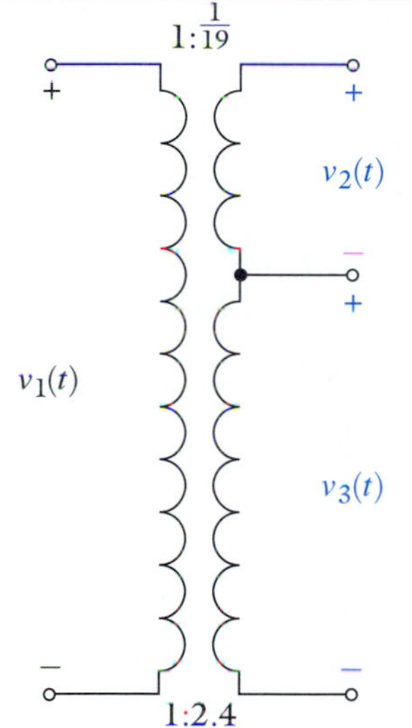

Figure P7-62

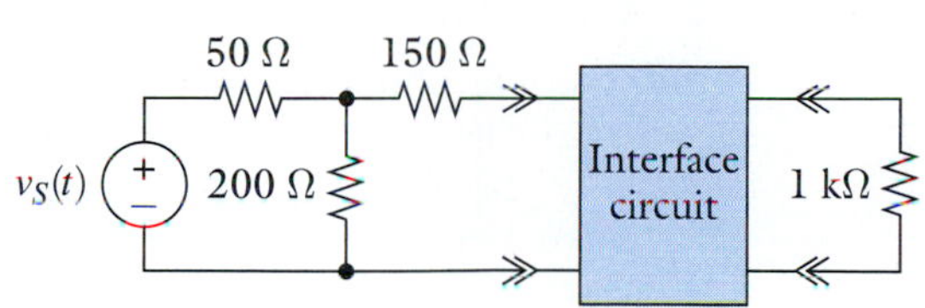

Figure P7-63

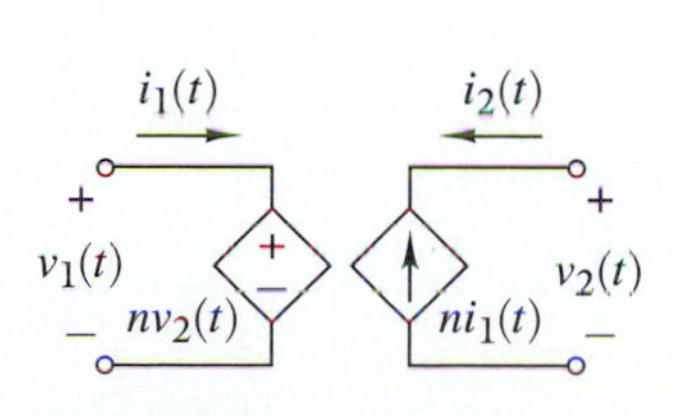

Figure P7-64

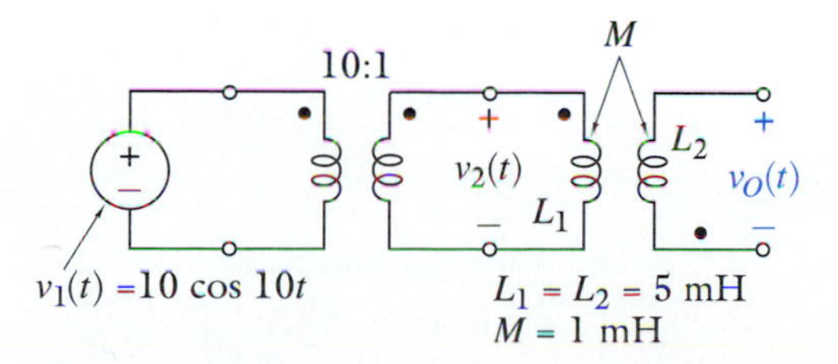

Figure P7-65

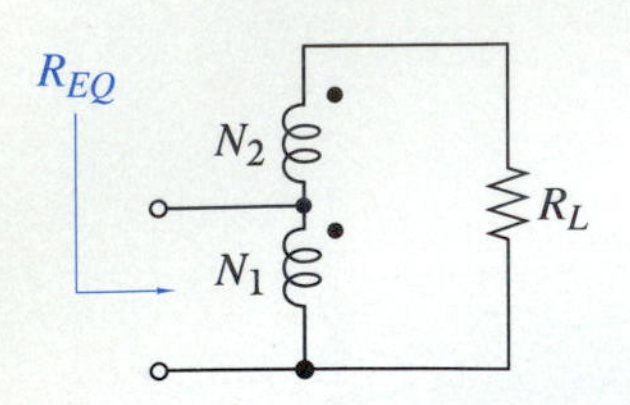

Figure P7-66

7-66 Derive an expression for the equivalent resistance seen at the input of the ideal transformer circuit in Fig. P7-66.

ERO 7-6 Computer Aided Analysis of Capacitors and Inductors (Sect. 7-7)

Use SPICE or MICRO-CAP to:

(a) find the waveform of voltage across or current through a dynamic circuit element.

(b) find the voltage across or current through coupled inductors.

(c) find the dc response of RLC circuits.

7-67 Use SPICE or MICRO-CAP to solve Prob. 7-1.

7-68 Use SPICE or MICRO-CAP to solve Prob. 7-2.

7-69 Use SPICE or MICRO-CAP to solve Prob. 7-7.

7-70 Use SPICE or MICRO-CAP to solve Prob. 7-9.

7-71 Use SPICE or MICRO-CAP to solve Prob. 7-49.

7-72 Use SPICE or MICRO-CAP to solve Prob. 7-50.

7-73 Use SPICE or MICRO-CAP to solve Prob. 7-52.

7-74 Use SPICE or MICRO-CAP to solve Prob. 7-55.

7-75 Use SPICE or MICRO-CAP to solve Prob. 7-56.

7-76 For the SPICE input file given below:

(a) Draw the circuit diagram corresponding to the SPICE file. Be sure to indicate the coil dots.

(b) What is the mutual inductance?

(c) How many data points will be included in the output file?

```
PROBLEM 7-76
IS 0 1 SIN 0 3 60 0 0
RS 1 2 50
L1 2 0 10M
L2 0 3 10M
K L1 L2 0.9999
RL 3 4 5
L3 4 0 100M
.TRAN 1M 100M
.PRINT TRAN V(3)
.END
```

CHAPTER INTEGRATING PROBLEMS

7-77 **(A)** A parallel-plate capacitor is constructed by separating two 6 cm by 6 cm aluminum plates. The separation between the plates is adjustable and is filled with air ($\epsilon = 8.86 \times 10^{-12}$ F/m).

(a) If the applied voltage is 50 V, what must the separation be to store 5×10^{-7} C of charge?

(b) What is the capacitance for the separation found in (a)?

(c) Air breaks down when the electric field exceeds 3×10^{6} V/m. Will the air filled capacitor found in (a) breakdown?

❖ **7-78** **(E)** The two equal capacitors and a switch are connected in series. One of the capacitors carries an initial voltage of V_O volts. At $t = 0$ the switch is closed so that the two capacitors are now connected in parallel. First assume that charge is conserved and show that after the switch is closed the voltage across each capacitor is $V_O/2$. Then assume that energy is conserved and show that the voltage across each capacitor is $V_O/\sqrt{2}$. Ask your instructor for an explanation of this dilemma.

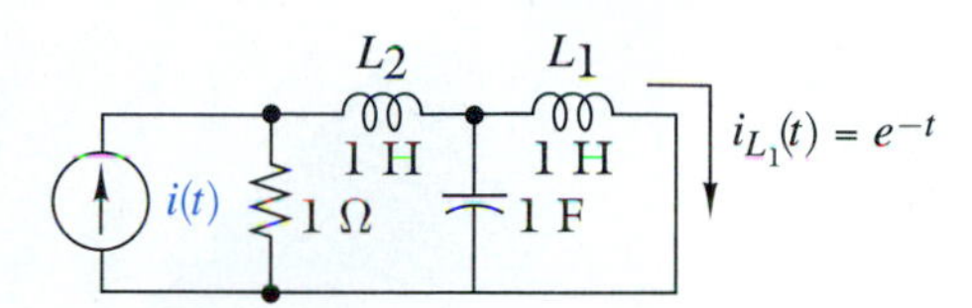

Figure P7-79

7-79 **(A)** Use KVL, KCL, and the element equations to answer the following questions about the circuit in Fig. P7-79.

(a) What input current $i(t)$ is required to produce the output current $i_{L1}(t) = e^{-t}$?

(b) How much energy is stored in the capacitor at $t = 1$ sec?

(c) How much is stored in the inductor L_2 at $t = 1$ sec?

7-80 **(A)** The maximum allowable output current range for the OP AMP in an inverting integrator is $\pm I_{MAX}$. If the input voltage is a step function $V_O u(t)$, show that the input resistor R must equal or exceed V_O/I_{MAX} to keep the OP AMP output current within its allowable range.

7-81 **(A)** Show that the circuits in Fig. P7-81 have identical i-v characteristics.

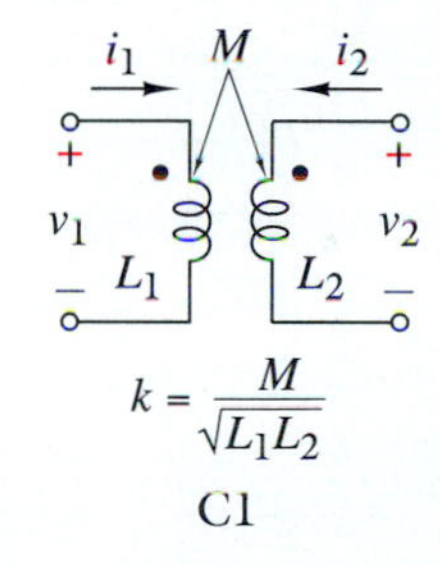

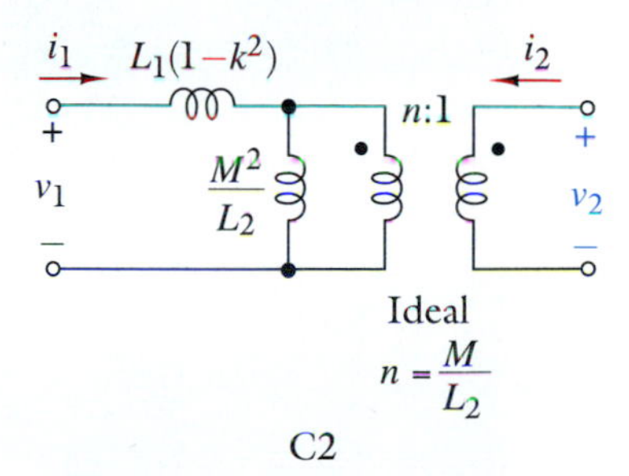

Figure P7-81

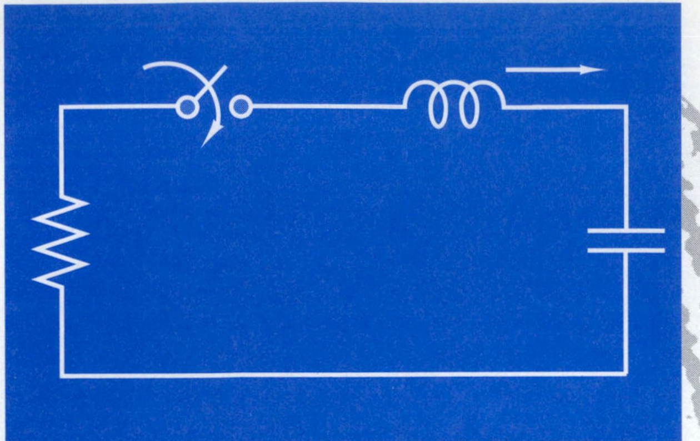

Chapter Eight

"When a mathematician engaged in investigating physical actions and results has arrived at his own conclusions, may they not be expressed in common language as fully, clearly and definitely as in mathematical formula? If so, would it not be a great boon to such as we to express them so—translating them out of their hieroglyphics that we also might work upon them by experiment."

Michael Faraday, 1857
British Physicist

First- and Second-Order Circuits

Michael Faraday (1791–1867) was appointed a Fellow in the Royal Society at age 32 and was a lecturer at the Royal Institution of London for more than 50 years. During this time he published over 150 papers on chemistry and electricity. The most important of these papers was the series *Experimental Researches In Electricity*, which included a description of his discovery of magnetic induction. A gifted experimentalist, Faraday had no formal education in mathematics and apparently felt that mathematics obscured the physical truths he discovered through experimentation. The quotation given on the opposite page is taken from a letter written by Faraday to James Clerk Maxwell in 1857. Faraday's comments are perhaps ironic because Maxwell is best known today for his formulation of the mathematical theory of electromagnetics, which he published in 1873.

Since Faraday's time, tremendous strides have been made in correlating mathematical theory and experiment. Electrical circuit analysis is one of the areas in which there is a close relationship between mathematical theory and laboratory measurements. The near one-to-one correspondence between the predictions of circuit analysis and hardware results enormously simplifies the design and development of electrical and electronic systems.

The *i-v* characteristics of capacitors and inductors were treated in Chapter 7. This chapter combines these device characteristics with the connection constraints to derive the differential equations governing the response of dynamic circuits. Mathematical methods for analytically solving these differential equations are also treated. The first four sections treat the formulation and solution of the first-order differential equations governing circuits with one capacitor or inductor. The next three sections treat circuits with two dynamic elements that lead to second-order differential equations. The last section describes using SPICE to generate numerical solutions for the response of dynamic circuits.

8-1 RC AND RL CIRCUITS

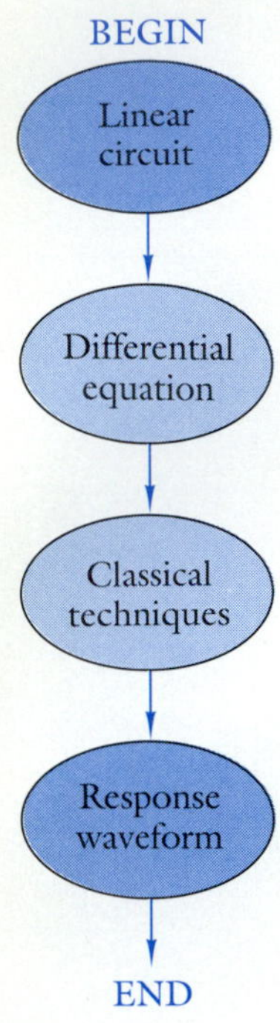

Figure 8-1 Dynamic circuit analysis flow diagram.

The flow diagram in Fig. 8-1 shows the two major steps in the analysis of a dynamic circuit. In the first step we use device and connection equations to formulate the differential equation describing the circuit. In the second step we solve the differential equation for the waveform of the response. In this chapter we examine basic methods of formulating circuit differential equations and the time-honored, classical methods of solving for responses. In solving for the responses of dynamic circuits via classic means, we develop an insight into the physical behavior of circuits that form the basic modules of more complex networks we will study in subsequent chapters. This insight will help us correlate circuit behavior as we study other methods of circuit analysis. The treatment of more powerful techniques using the Laplace transformation begins in the next chapter.

Formulating RC and RL Circuit Equation

RC and RL circuits contain linear resistors and a single capacitor or a single inductor. Figure 8-2 shows how we can divide RC and RL circuits into two parts: (1) the dynamic element and (2) the rest of the circuit containing only linear resistors and sources. To formulate the equation governing either of these circuits we replace the resistors and sources by their Thévenin and Norton equivalent shown in Fig. 8-2.

Dealing first with the RC circuit in Fig. 8-2(a), we note that the Thévenin equivalent source is governed by the constraint

$$R_T i(t) + v(t) = v_T(t) \tag{8-1}$$

The capacitor i-v constraint is

$$i(t) = C\frac{dv(t)}{dt} \tag{8-2}$$

Substituting the i-v constraint into the source constraint yields

$$R_T C\frac{dv(t)}{dt} + v(t) = v_T(t) \tag{8-3}$$

Combining the source and element constraints produces the equation governing the RC series circuit. The unknown in Eq. (8-3) is the capacitor voltage $v(t)$, which is the ***state variable***, because it determines the amount of energy stored in the capacitive element.

Mathematically Eq. (8-3) is a ***first-order linear differential equation with constant coefficients***. The equation is first-order be-

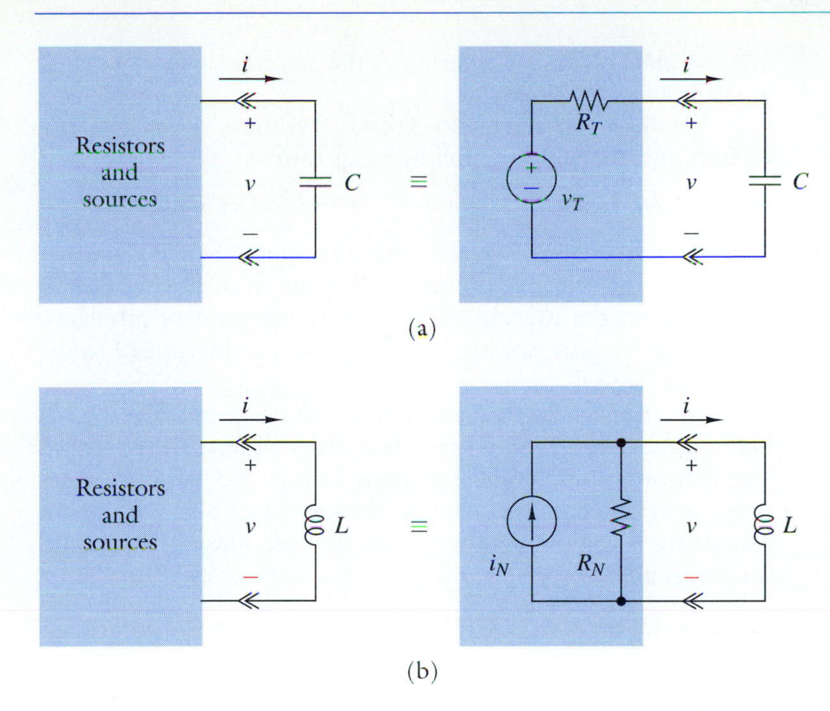

Figure 8-2 First-order circuits: (a) RC circuit. (b) RL circuit.

cause the first derivative of the dependent variable is the highest-order derivative in the equation. The product $R_T C$ is a constant coefficient because it depends on fixed circuit parameters. The signal $v_T(t)$ is the Thévenin equivalent of the independent sources driving the circuit. The voltage $v_T(t)$ is the input and the capacitor voltage $v(t)$ is the circuit response.

The Norton equivalent source in the RL circuit in Fig. 8-2(b) is governed by the constraint

$$G_N v(t) + i(t) = i_N(t) \tag{8-4}$$

The element constraint for the inductor can be written

$$v(t) = L\frac{di(t)}{dt} \tag{8-5}$$

Combining the element and source constraints produces the differential equation for the RL circuit:

$$G_N L\frac{di(t)}{dt} + i(t) = i_N(t) \tag{8-6}$$

The response of the RL circuit is also governed by a first-order linear differential equation with constant coefficients. The dependent variable in Eq. (8-6) is the inductor current. The circuit parameters enter as the constant product $G_N L$ and the driving forces are represented by a Norton equivalent current $i_N(t)$. The unknown in Eq. (8-6) is the inductor current $i(t)$, which is the *state variable* because it determines the amount of energy stored in the inductive element.

We observe that Eqs. (8-3) and (8-6) have the same form. In fact, interchanging the following quantities

$$R_T \leftrightarrow G_N \qquad C \leftrightarrow L \qquad v \leftrightarrow i \qquad v_T \leftrightarrow i_N$$

converts one equation into the other. This interchange is another example of the principle of duality. Because of duality we do not need to study the RC and RL circuits as independent problems. Everything we learn solving the RC circuit can be applied to the RL circuit as well.

We refer to the RC and RL circuits as *first-order circuits* because they are described by a first-order differential equation. The first-order differential equations in Eq. (8-3) and (8-6) describe general RC and RL circuits shown in Fig. 8-2. Any circuit containing a single capacitor or inductor and linear resistors and sources is a first-order circuit.

Zero-Input Response of First-Order Circuits

The response of a first-order circuit is found by solving the circuit differential equation. For the RC circuit the response $v(t)$ must satisfy the differential equation in Eq. (8-3) and the initial condition $v(0)$. By examining Eq. (8-3) we see the response depends on three factors:

1. The inputs driving the circuit $v_T(t)$
2. The values of the circuit parameters R_T and C
3. The value of $v(t)$ at $t = 0$, i.e., the initial condition

The first two factors apply to any linear circuit, including resistance circuits. The third factor relates to the initial energy stored in the circuit. The initial energy can cause the circuit to have a nonzero response even when the input $v_T(t) = 0$ for $t \geq 0$. The existence of a response with no input is new in our study of linear circuits.

To explore this discovery we find the *zero-input response*. Setting all independent sources in Fig. 8-2 to zero makes $v_T = 0$ in Eq. (8-3):

$$R_T C \frac{dv}{dt} + v = 0 \qquad (8\text{-}7)$$

Mathematically Eq. (8-7) is a *homogeneous equation* because the right side is zero. The classical approach to solving a linear homogeneous differential equation is to try a solution in the form of an exponential

$$v(t) = Ke^{st} \tag{8-8}$$

where K and s are constants to be determined.

The form of the homogenous equation suggests an exponential solution for the following reasons. Equation (8-7) requires that $v(t)$ plus $R_T C$ times its derivative must add to zero for **all time** $t \geq 0$. This can only occur if $v(t)$ and its derivative have the same form. In Chapter 6 we saw that an exponential waveform and its derivative are both of the form e^{-t/T_C}. Therefore, the exponential is a reasonable starting place.

If Eq. (8-8) is indeed a solution, then it must satisfy the differential equation in Eq. (8-7). Substituting the trial solution into Eq. (8-7) yields

$$R_T C K s e^{st} + K e^{st} = 0$$

or

$$K e^{st}(R_T C s + 1) = 0$$

The exponential function e^{st} cannot be zero for all t. The condition $K = 0$ is a trivial solution because it implies that $v(t)$ is zero for all time t. The only nontrivial way to satisfy the equation involves the condition

$$R_T C s + 1 = 0 \tag{8-9}$$

Equation (8-9) is the circuit *characteristic equation* because its root determines the attributes of $v(t)$. The characteristic equation has a single root at $s = -1/R_T C$ so the zero-input response of the RC circuit has the form

$$v(t) = K e^{-t/R_T C} \qquad t \geq 0$$

The constant K can be evaluated using the value of $v(t)$ at $t = 0$. Using the notation $V_O = v(0)$ yields

$$v(0) = K e^0 = K = V_O$$

The final form of the zero-input response is

$$v(t) = V_O e^{-t/R_T C} \qquad t \geq 0 \tag{8-10}$$

The zero-input response of the RC circuit is the familiar exponential waveform shown in Fig. 8-3. At $t = 0$ the waveform starts out at $v(0) = V_O$ and then decays to zero at $t \to \infty$. The time constant $T_C = R_T C$ depends only on fixed circuit parameters. From our study of exponential waveforms in Chapter 6, we know that the $v(t)$ decays to about 37% of its initial amplitude in one time constant and to essentially zero after about five time

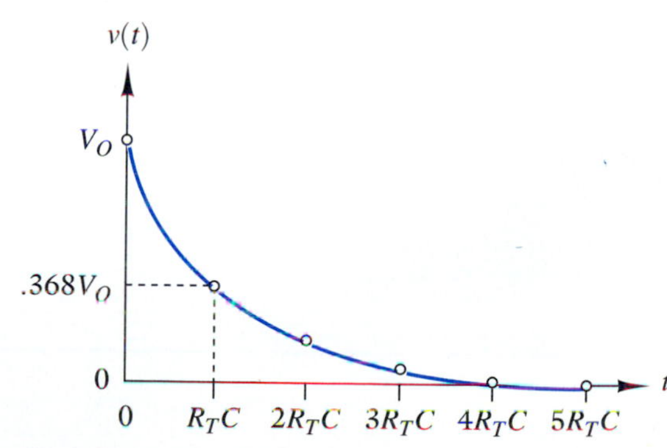

Figure 8-3 First-order RC circuit zero-input response.

constants. The zero-input response of the RC circuit is determined by two quantities: (1) the circuit time constant and (2) the value of the capacitor voltage at $t = 0$.

The zero-input response of the RL circuit in Fig. 8-2(b) is found by setting the Norton current $i_N(t) = 0$ in Eq. (8-6):

$$G_N L \frac{di}{dt} + i = 0 \qquad (8\text{-}11)$$

The unknown in this homogeneous differential equation is the inductor current $i(t)$. Equation (8-11) has the same form as the homogeneous equation for the RC circuit, which suggests a trial solution of the form

$$i(t) = Ke^{st}$$

where K and s are constants to be determined. Substituting the trial solution into Eq. (8-11) yields the RL circuit characteristic equation:

$$G_N Ls + 1 = 0 \qquad (8\text{-}12)$$

The root of this equation is $s = -1/G_N L$. Denoting the initial value of the inductor current by I_O, we evaluate the constant K:

$$i(0) = I_O = Ke^0 = K$$

The final form of the zero-input response of the RL circuit is

$$i(t) = I_O e^{-t/G_N L} \qquad t \geq 0 \qquad (8\text{-}13)$$

The RL circuit zero-input response is a exponential waveform in the state variable $i(t)$. The exponential transient has a time constant of $T_C = G_N L = L/R_T$. The transient connects the initial state $i(0) = I_O$ with the final state $i(\infty) = 0$.

The zero-input responses in Eqs. (8-10) and (8-13) show the duality between first-order RC and RL circuits. These results point out that the zero-input responses in a first-order circuit depends on two quantities: (1) the circuit time constant and (2) the value of the state variable at $t = 0$. Capacitor voltage and inductor current are state variables because they determine the amount of energy stored in the circuit at any time t. The following examples show that the zero-input response of the state variable provides enough information to determine the zero-input response of every other voltage amd current in the circuit.

EXAMPLE 8-1

The switch in Fig. 8-4 is closed at $t = 0$ connecting a capacitor with an initial voltage of 30 V to the resistances shown. Find the responses $v_C(t)$, $i(t)$, $i_1(t)$, and $i_2(t)$ for $t \geq 0$.

SOLUTION:

There is no input source in the resistance circuit, so this problem involves zero-input responses of an RC circuit. To find the required responses we first determine the circuit time constant with the switch closed ($t \geq 0$). The equivalent resistance seen by the capacitor is

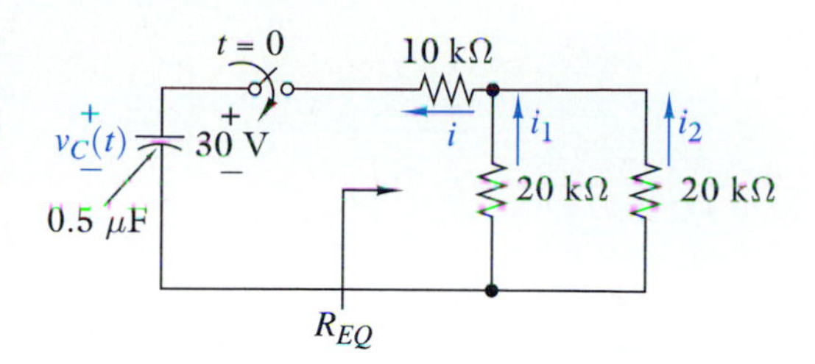

Figure 8-4

$$R_{EQ} = 10 + (20 \parallel 20) = 20 \text{ k}\Omega$$

For $t \geq 0$ the circuit time constant is

$$T_C = R_T C = 20 \times 10^3 \times 0.5 \times 10^{-6} = 10 \text{ ms}$$

The initial capacitor voltage is given by $V_O = 30$ V. Using Eq. (8-10), the zero-input response of the capacitor voltage is

$$v_C(t) = 30e^{-100t} \qquad t \geq 0$$

The capacitor voltage provides the information needed to solve for all other zero-input responses. The current $i(t)$ through the capacitor is

$$i(t) = C\frac{dv_C}{dt} = (0.5 \times 10^{-6})(30)(-100)e^{-100t}$$
$$= -1.5 \times 10^{-3}e^{-100t} \text{ A} \qquad t \geq 0$$

The minus sign means the actual current direction is opposite of the reference direction shown in Fig. 8-4. The minus sign makes physical sense because the initial voltage on the capacitor is positive which forces current into the resistances to the right of the switch. The other current responses are found by current division.

$$i_1(t) = i_2(t) = \frac{20}{20+20}i(t) = -0.75 \times 10^{-3}e^{-100t} \text{ A} \qquad t > 0$$

Notice the analysis pattern. We first determine the zero-input response of the capacitor voltage. The state variable response and resistance circuit analysis techniques were then used to find other voltages and currents. The circuit time constant and the value of the state variable at $t = 0$ provide enough information to determine the zero-input response of every voltage or current in the circuit.

EXAMPLE 8-2

Find the response of the state variable of the RL circuit in Fig. 8-5 using $L_1 = 10$ mH, $L_2 = 30$ mH, $R_1 = 2$ kΩ, $R_2 = 6$ kΩ, and $i_L(0) = 100$ mA.

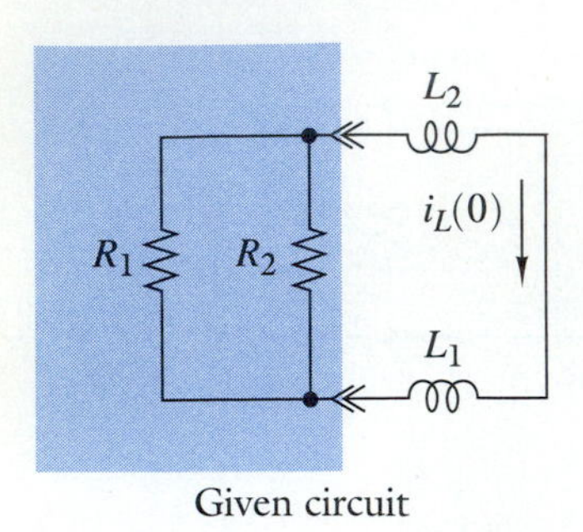

Given circuit

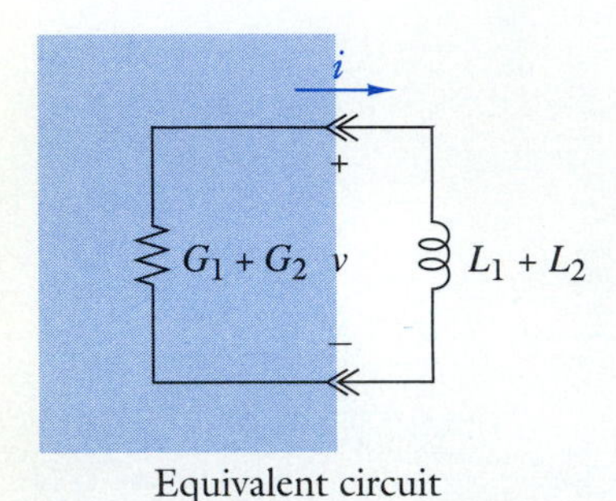

Equivalent circuit

Figure 8-5

SOLUTION:

The inductors are connected in series and can be replaced by an equivalent inductor

$$L_{EQ} = L_1 + L_2 = 10 + 30 = 40 \text{ mH}$$

Likewise, the resistors are connected in parallel and the conductance seen by L_{EQ} is

$$G_{EQ} = G_1 + G_2 = 10^{-3}/2 + 10^{-3}/6 = 2 \times 10^{-3}/3 \text{ S}$$

Figure 8-5 shows the resulting equivalent circuit. The interface signals $v(t)$ and $i(t)$ are the voltage across and current through $L_{EQ} = L_1 + L_2$. The time constant of the equivalent RL circuit is

$$T_C = G_{EQ}L_{EQ} = 8 \times 10^{-5}/3 \text{ s} = 1/37500 \text{ s}$$

The initial current through L_{EQ} is $i_L(0) = 0.1$ A. Using Eq. (8-13) with $I_O = 0.1$ yields the zero-state response of the inductor current.

$$i(t) = 0.1e^{-37500t} \text{ A} \qquad t \geq 0$$

Given the state variable response we can find every other response in the original circuit. For example, by KCL and current division the currents through R_1 and R_2 are

$$i_{R_1}(t) = \frac{R_2}{R_1 + R_2} i(t) = 0.075e^{-37500t} \text{ A} \qquad t \geq 0$$

$$i_{R_2}(t) = \frac{R_1}{R_1 + R_2} i(t) = 0.025e^{-37500t} \text{ A} \qquad t \geq 0$$

Example 8-2 illustrates an important point. The RL circuit in Fig. 8-5 is a first-order circuit even though it contains two inductors. The two inductors are connected in series and can be replaced by a single equivalent inductor. In general, capacitors or inductors in series and parallel can be replaced by a single equivalent element. Thus, any circuit containing the **equivalent** of a single inductor or a single capacitor is a first-order circuit.

In some cases it may be difficult to determine the Thévenin or Norton equivalent seen by the dynamic element in a first-order circuit. In such cases we use the basic concepts of device and connection constraints to derive the differential equation in terms of a more convenient signal variable. For example, the OP AMP RC circuit in Fig. 8-6 is a first-order circuit because it contains a single capacitor. However, it is somewhat awkward to determine the Thévenin equivalent seen by the feedback capacitor. From previous experience we know the key to analyzing an inverting

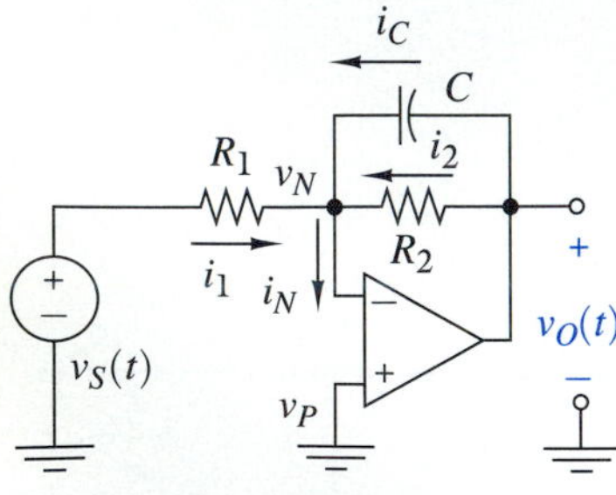

Figure 8-6 First-order OP AMP RC circuit.

OP AMP circuit is to write a KCL equation at the inverting input:

$$\text{KCL: } i_1(t) + i_2(t) + i_C(t) = i_N(t)$$

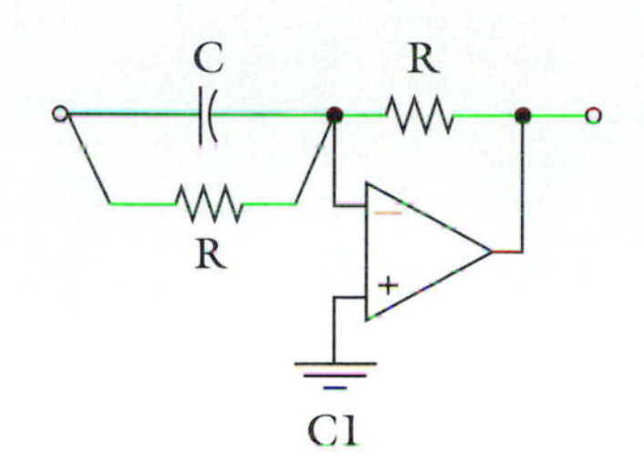

The element equations are

$$C: \quad i_C = C\frac{d(v_O - v_N)}{dt}$$

$$R_1: \quad i_1 = G_1(v_S - v_N) \qquad\qquad R_2: \quad i_2 = G_2(v_O - v_N)$$

$$\text{OP AMP:} \quad i_N = 0 \qquad\qquad v_N = 0$$

The second OP AMP element equation reflects the fact that the noninverting input is grounded.

Substituting the element constraints into the KCL connection constraint yields

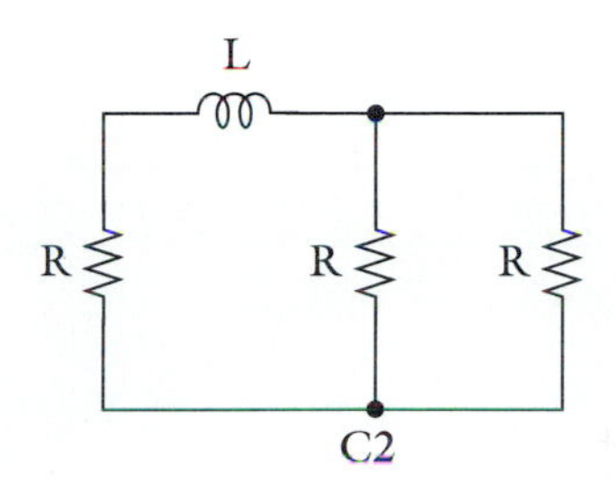

$$G_1 v_S + G_2 v_O + C\frac{dv_O}{dt} = 0$$

which can be rearranged in standard form as

$$R_2 C\frac{dv_O}{dt} + v_O = -\frac{R_2}{R_1}v_S(t) \qquad (8\text{-}14)$$

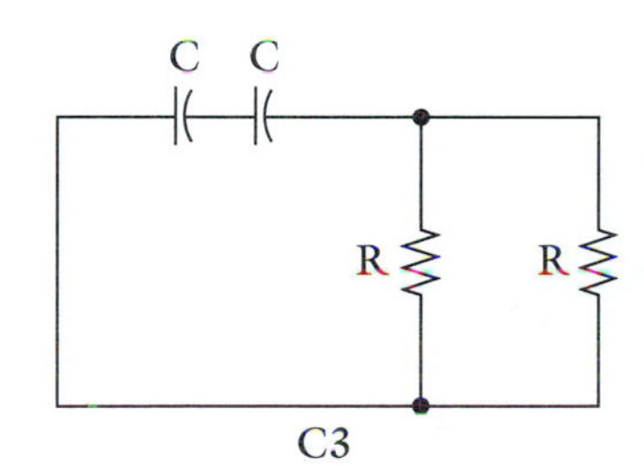

The unknown Eq. (8-14) is the OP AMP output voltage rather than the capacitor voltage. The form of the differential equation indicates that the circuit time constant is $T_C = R_2 C$.

The OP AMP circuit example illustrates that we can formulate first-order differential equations using other circuit variables besides the capacitor voltage and inductor current. However, opening or closing a switch may cause jump discontinuities in these other variables. Usually it is best to use capacitor voltage and inductor current because they are continuous functions of time without jump discontinuities.

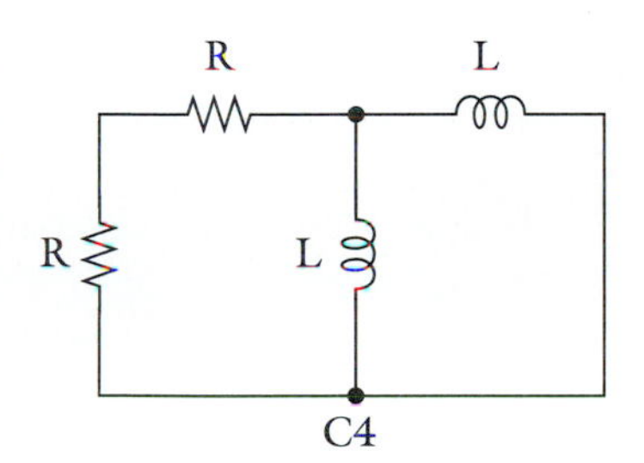

Figure 8-7

EXERCISE 8-1

What are the time constants of the circuits shown in Fig. 8-7?

ANSWERS:

C1: RC C2: 2L/3R C3: RC/4 C4: L/4R

EXERCISE 8-2

The switch in Fig. 8-8 closes at $t = 0$. For $t \geq 0$ the current through the resistor is $i_R(t) = e^{-100t}$ mA.

(a) What is the capacitor voltage at $t = 0$?
(b) Write an equation for $v(t)$ for $t \geq 0$.
(c) Write an equation for the power in the resistor for $t \geq 0$.
(d) How much energy does the resistor dissipate for $t \geq 0$?
(e) How much energy is stored in the capacitor at $t = 0$?

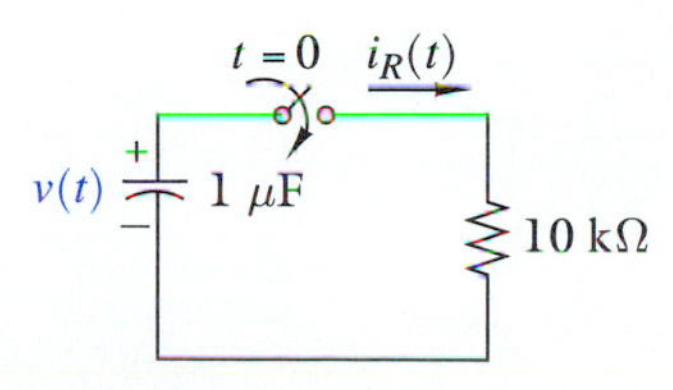

Figure 8-8

ANSWERS:

(a) 10 V
(b) $v(t) = 10e^{-100t}$ V
(c) $p_R(t) = 10e^{-200t}$ mW
(d) 50 μJ
(e) 50 μJ ■

■ EXERCISE 8-3

For $t > 0$ the current through the 40–mH inductor in a first-order circuit is $20e^{-500t}$ mA.

(a) What is the circuit time constant?
(b) How much energy is stored in the inductor at $t = 0$, $t = T_C$, and $t = 5T_C$?
(c) Write an equation for the voltage across the inductor.
(d) What is the equivalent resistance seen by the inductor?

ANSWERS:

(a) 2 ms
(b) 8, 1.08, 0.000363 μJ
(c) $-400e^{-500t}$ mV
(d) 20 Ω ■

8-2 FIRST-ORDER CIRCUIT STEP RESPONSE

Linear circuits are characterized by applying step function and sinusoid inputs. This section introduces the step response of first-order circuits. Later in this chapter we treat the sinusoidal response of first-order circuits and step response of second-order circuits. The step response analysis introduces the concepts of forced, natural, and zero-state responses that appear extensively in later chapters.

The development of the step response of a first-order circuit treats the RC circuit in detail and then summarizes the corresponding results for its dual, the RL circuit. When the input to the RC circuit in Fig. 8-1 is a step function we can write the Thévenin source as $v_T(t) = V_A u(t)$. The circuit differential equation in Eq. (8-3) becomes

$$R_T C \frac{dv}{dt} + v = V_A u(t) \tag{8-15}$$

The step response is a function $v(t)$ that satisfies this differential equation for $t \geq 0$ and meets the initial condition $v(0)$. Since $u(t) = 1$ for $t \geq 0$ we can write Eq. (8-15) as

$$R_T C \frac{dv(t)}{dt} + v(t) = V_A \qquad \text{for} \qquad t \geq 0 \tag{8-16}$$

There are a number of mathematical approaches to solve this equation, including separation of variables and integrating factors. However, because the circuit is linear we chose a method that divides solution $v(t)$ into two components:

$$v(t) = v_N(t) + v_F(t) \tag{8-17}$$

The first component $v_N(t)$ is the ***natural response*** and is the general solution Eq. (8-16) when the input is set to zero. The natural response has its origin in the physical characteristic of the circuit and does not depend on the form of the input. The component $v_F(t)$ is the ***forced response*** and is a particular solution of Eq. (8-16) when the input is the step function. We call this the forced response because it represents what the circuit is compelled to do by the form of the input.

Finding the natural response requires the general solution of Eq. (8-16) with the input set to zero:

$$R_T C \frac{dv_N(t)}{dt} + v_N(t) = 0 \qquad \text{for} \qquad t \geq 0$$

But this is the homogeneous equation that produces the zero-input response in Eq. (8-7). Therefore, we know that the natural response takes the form

$$v_N(t) = Ke^{-t/R_T C} \qquad t \geq 0 \tag{8-18}$$

This is a general solution of the homogeneous equation because it contains an arbitrary constant K. At this point we cannot evaluate K from the initial condition as we did for the zero-input response. The initial condition applies to the total response (natural plus forced) and we have yet to find the forced response.

Turning now to the forced response, we seek a particular solution of the equation

$$R_T C \frac{dv_F(t)}{dt} + v_F(t) = V_A \qquad \text{for} \qquad t \geq 0 \tag{8-19}$$

The equation requires that $v_F(t)$ plus $R_T C$ times its derivative produce a constant V_A for $t \geq 0$. Setting $v_F(t) = K_F$, a constant, meets this condition since $dv_F/dt = dK_F/dt = 0$. Therefore, $v_F = V_A$, since Eq. (8-19) reduces to the identity $K_F = V_A$.

Now combining the forced and natural responses we obtain

$$\begin{aligned} v(t) &= v_N(t) + v_F(t) \\ &= Ke^{-t/R_T C} + V_A \qquad t \geq 0 \end{aligned}$$

This equation is the general solution for the step response because it satisfies Eq. (8-16) and contains an arbitrary constant K. This constant can now be evaluated using the initial condition:

$$v(0) = V_O = Ke^0 + V_A = K + V_A$$

The initial condition requires that $K = (V_O - V_A)$. Substituting this conclusion into the general solution yields the step response of the RC circuit.

$$v(t) = (V_O - V_A)e^{-t/R_TC} + V_A \qquad t \geq 0 \tag{8-20}$$

A typical plot of the waveform of $v(t)$ is shown in Fig. 8-9.

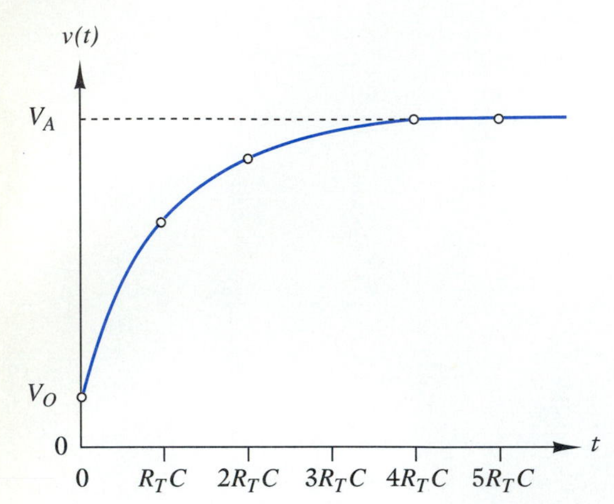

Figure 8-9 Step response of a first-order RC circuit.

The RC circuit step response in Eq. (8-20) starts out at the initial condition V_O and is driven to a final condition V_A, which is determined by the amplitude of the step function input. That is, the initial and final states are

$$\lim_{t \to 0+} v(t) = (V_O - V_A)e^{-0} + V_A = V_O$$

$$\lim_{t \to \infty} v(t) = (V_O - V_A)e^{-\infty} + V_A = V_A$$

The path between the two end points is an exponential waveform whose time constant is the circuit time constant. We know from our study of exponential signals in Chapter 6 that the step response will reach its final value after about five time constants. In other words, after about five time constants the natural response decays to zero and we are left with a constant forced response caused by the step function input.

The RL circuit in Fig. 8-1 is the dual of the RC circuit, so the development of its step responses follows the same pattern discussed above. Briefly sketching the main steps, the Norton equivalent input is a step function $I_A u(t)$ and for $t \geq 0$ the RL circuit differential equation Eq. (8-5) becomes

$$G_N L \frac{di(t)}{dt} + i(t) = I_A \qquad t \geq 0 \tag{8-21}$$

The solution of this equation is found by superimposing the natural and forced components. The natural response is the solution of the homogeneous equation [right side of Eq. (8-21) set to zero], and takes the same form as the zero-input response found in the previous section:

$$i_N(t) = Ke^{-t/G_N L} \qquad t \geq 0$$

where K is a constant to be evaluated from the initial condition once the complete response is known. The forced response is a particular solution of the equation

$$G_N L \frac{di_F(t)}{dt} + i_F(t) = I_A \qquad t \geq 0$$

Setting $i_F = I_A$ satisfies this equation since $dI_A/dt = 0$.

Combining the forced and natural responses we obtain the general solution of Eq. (8-21) in the form

$$\begin{aligned} i(t) &= i_N(t) + i_F(t) \\ &= Ke^{-t/G_N L} + I_A \qquad t \geq 0 \end{aligned}$$

The constant K is now evaluated from the initial condition:

$$i(0) = I_O = Ke^{-0} + I_A = K + I_A$$

The initial condition requires that $K = I_O - I_A$, so the step response of the RL circuit is

$$i(t) = (I_O - I_A)e^{-t/G_N L} + I_A \qquad t \geq 0 \tag{8-22}$$

The RL circuit step response has the same form as the RC circuit step response in Eq. (8-20). At $t = 0$ the starting value of the response is $i(0) = I_O$ as required by the initial condition. The final value is the forced response $i(\infty) = i_F = I_A$, since the natural response decays to zero as time increases.

A step function input to the RC or RL circuit drives the state variable from an initial value determined by what happened prior to $t = 0$, to a final state determined by amplitude of the step function applied at $t = 0$. The time needed to transition from one state to the other is about $5T_C$, where T_C is the circuit time constant. We conclude that the step response of a first-order circuits depends on three quantities:

1. the amplitude of the step input (V_A or I_A)
2. the circuit time constant ($R_T C$ or $G_N L$)
3. the value of the state variable at $t = 0$ (V_O or I_O)

EXAMPLE 8-3

Find the response of the RC circuit in Fig. 8-10.

SOLUTION:

The circuit is first order, since the two capacitors in series can be replaced by a single equivalent capacitor

$$C_{EQ} = \frac{C_1 C_2}{C_1 + C_2} = \frac{(0.1)(0.5)}{0.1 + 0.5} = 0.0833\ \mu\text{F}$$

The initial voltage on C_{EQ} is the sum of the initial voltages on the original capacitors.

$$V_O = V_{O1} + V_{O2} = 5 + 10 = 15\ \text{V}$$

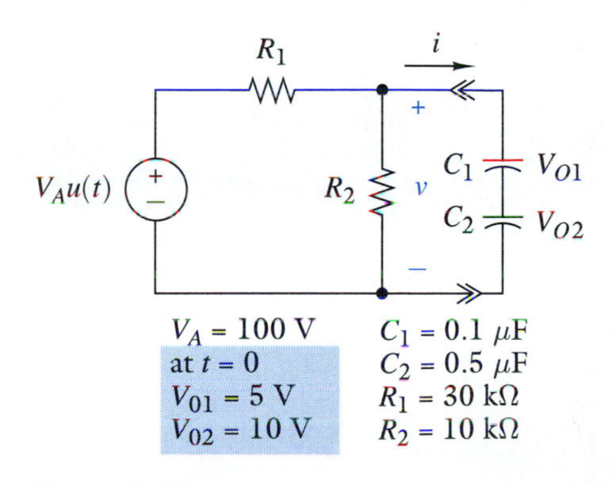

Figure 8-10

The Thévenin equivalent seen by the equivalent capacitor is found by first determining the open-circuit voltage. Disconnecting the capacitors in Fig. 8-10 and using voltage division at the interface yields

$$v_T = v_{OC} = \frac{R_2}{R_1 + R_2} V_A u(t) = \frac{10}{40} 100u(t) = 25u(t)$$

Replacing the voltage source by a short circuit and looking to the left at the interface we see R_1 in parallel with R_2. The Thévenin resistance of this combination is

$$R_T = \frac{R_1 R_2}{R_1 + R_2} = \frac{(30)(10)}{40} = 7.5 \text{ k}\Omega$$

The circuit time constant is

$$T_C = R_T C_{EQ} = (7.5 \times 10^3)(8.33 \times 10^{-8}) = \frac{1}{1600} \text{ s}$$

For the Thévenin equivalent circuit the initial capacitor voltage is $V_O = 15$ V, the step input is $25u(t)$, and the time constant is $1/1600$ s. Using the RC circuit step response in Eq. (8-20) yields

$$v(t) = (15 - 25)e^{-1600t} + 25$$
$$= 25 - 10e^{-1600t} \qquad t \geq 0$$

The initial ($t = 0$) value of $v(t)$ is $25 - 10 = 15$ V as required. The equivalent capacitor voltage is driven to a final value of 25 V by the step input in the Thévenin equivalent circuit. For practical purposes $v(t)$ reaches 25 V after about $5T_C = 3.125$ ms.

EXAMPLE 8-4

Find the step response of the RL circuit in Fig. 8-11(a). The initial condition is $i(0) = I_O$.

SOLUTION:

We first find the Norton equivalent to the left of the interface. By current division the short-circuit current at the interface is

$$i_{SC}(t) = \frac{R_1}{R_1 + R_2} I_A u(t)$$

Looking to the left at the interface with the current source off (replaced by an open circuit) we see R_1 and R_2 in series producing a Thévenin resistance

$$R_T = \frac{1}{G_N} = R_1 + R_2$$

The time constant of the Norton equivalent circuit in Fig. 8-11(b) is

$$T_C = G_N L = \frac{L}{(R_1 + R_2)}$$

The natural response of the Norton equivalent circuit is

$$i_N(t) = Ke^{-(R_1+R_2)t/L}$$

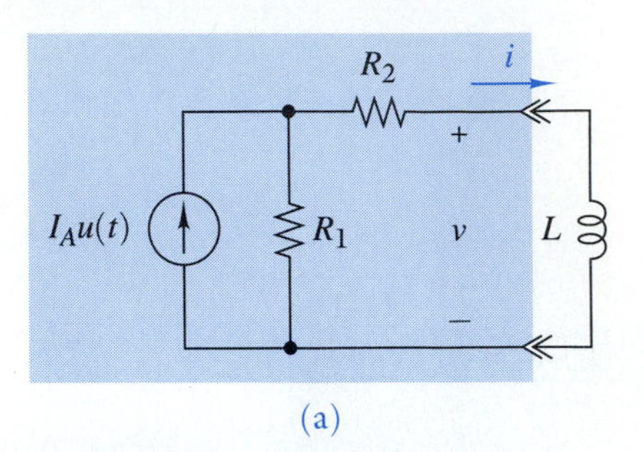

(a)

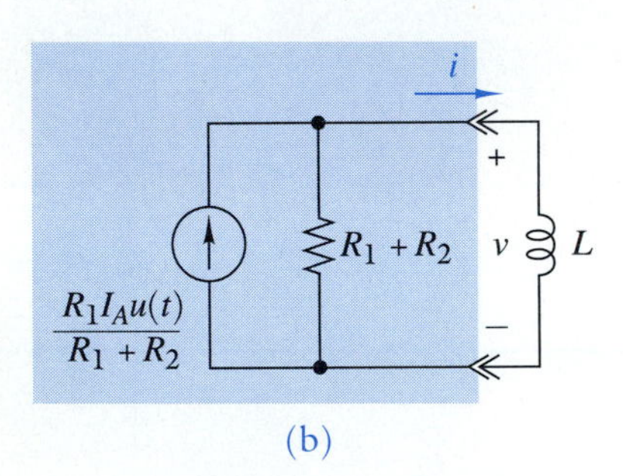

(b)

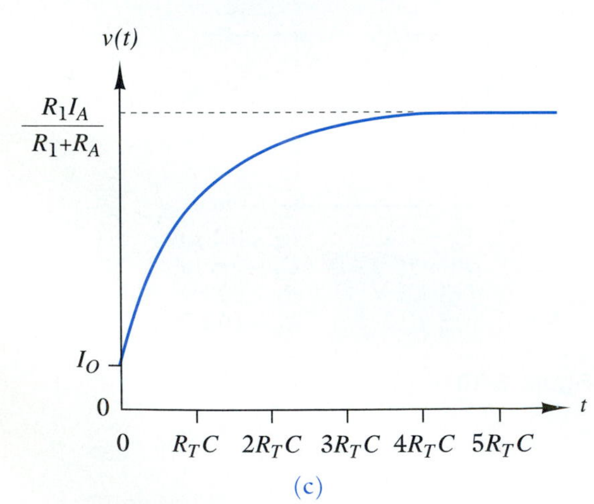

(c)

Figure 8-11

The short-circuit current $i_{SC}(t)$ is the step function input in the Norton circuit. Therefore, the forced response is

$$i_F(t) = i_{SC}(t) = \frac{R_1 I_A}{(R_1 + R_2)} u(t)$$

Superimposing the natural and forced responses yields

$$i(t) = Ke^{-(R_1+R_2)t/L} + \frac{R_1 I_A}{R_1 + R_2} \qquad t \geq 0$$

The constant K can be evaluated from the initial condition.

$$i(0) = I_O = K + \frac{R_1 I_A}{R_1 + R_2}$$

which requires that

$$K = I_O - \frac{R_1 I_A}{R_1 + R_2}$$

So circuit step response is

$$i(t) = \left[I_O - \frac{R_1 I_A}{R_1 + R_2} \right] e^{\frac{-(R_1+R_2)t}{L}} + \frac{R_1 I_A}{R_1 + R_2} \qquad t \geq 0$$

An example of this response waveform is shown in Fig. 8-11(c).

EXAMPLE 8-5

The state variable response of a first-order RC circuit for a step function input is

$$v_C(t) = 20e^{-200t} - 10, \qquad t \geq 0.$$

(a) What is the circuit time constant?

(b) What is the initial voltage across the capacitor?

(c) What is the amplitude of the forced response?

(d) At what time is $v_C(t) = 0$?

SOLUTION:

(a) The natural response of a first-order circuit is of the form Ke^{-t/T_C}. Therefore, the time constant of the given responses is $T_C = 1/200 = 5$ ms.

(b) The initial ($t = 0$) voltage across the capacitor is

$$v_C(0) = 20e^{-0} - 10 = 20 - 10 = 10 \text{ V}$$

(c) The natural response decays to zero so the forced response is the final value $v_C(t)$.

$$v_C(\infty) = 20e^{-\infty} - 10 = 0 - 10 = -10 \text{ V}$$

(d) The capacitor voltage must pass through zero at some intermediate time, since the initial value is positive and the final value negative. This time is found by setting the given step response equal to zero:

$$20e^{-200t} + 10 = 0$$

which yields the condition $e^{200t} = 2$ or $t = \ln 2/200 = 3.47$ ms.

EXERCISE 8-4

Given the following first-order circuit step responses

(1) $v_C(t) = 20 - 20e^{-1000t} \quad t \geq 0$
(2) $i_L(t) = -3 + 3e^{-200t} \quad t \geq 0$
(3) $v_C(t) = -20 + 10e^{-t/10} \quad t \geq 0$
(4) $i_L(t) = 3e^{-10t} \quad t \geq 0.$

- **(a)** What is the amplitude of the step input?
- **(b)** What is the circuit time constant?
- **(c)** What is the initial value of the state variable?
- **(d)** What is the circuit differential equation?

ANSWERS:

(1) **(a)** 20 V
(b) 1 ms
(c) 0 V
(d) $10^{-3} dv_C/dt + v_C = 20u(t)$

(2) **(a)** −3 A
(b) 5 ms
(c) 0 A
(d) $di_L/dt + 200i_L = -600u(t)$

(3) **(a)** −20 V
(b) 10 s
(c) −10 V
(d) $10dv_C/dt + v_C = -20u(t)$

(4) **(a)** 0 A
(b) 100 ms
(c) 3 A
(d) $di_L/dt + 10i_L = 0.$ ■

EXERCISE 8-5

Find the solution of the following first-order differential equations.

(a) $10^{-4}\dfrac{dv_C}{dt} + v_C = -5u(t) \qquad v_C(0) = 5 \text{ V}$

(b) $5 \times 10^{-2}\dfrac{di_L}{dt} + 2000i_L = 10u(t) \qquad i_L(0) = -5 \text{ mA}$

ANSWERS:

(a) $v_C(t) = -5 + 10e^{-10000t}$ V $\quad t \geq 0$
(b) $i_L(t) = 5 - 10e^{-40000t}$ mA $\quad t \geq 0.$ ■

Zero-State Response

Additional properties of dynamic circuit responses are revealed by rearranging the RC and RL circuit step responses in Eqs. (8-20) and (8-22) in the following way:

$$\text{RC Circuit:} \quad v(t) = V_O e^{-t/R_T C} + V_A(1 - e^{-t/R_T C}) \qquad t \geq 0$$

$$\text{RL Circuit:} \quad i(t) = I_O e^{-t/G_N L} + I_A(1 - e^{-t/G_N L}) \qquad t \geq 0$$

We recognize the first term on the right in each equation as the zero-input response discussed in Sect. 8-1. By definition the *zero-input response* occurs when the step function input is zero ($V_A = 0$ or $I_A = 0$). The second term on the right in each equation is called the *zero-state response* because this part occurs when the initial state of the circuit is zero ($V_O = 0$ or $I_O = 0$).

The zero-state response is proportional to the amplitude of the input step function (V_A or I_A). However, the total response (zero-input plus zero-state) is not directly proportional to the input amplitude. When the initial state is not zero, the circuit appears to violate the proportionality property of linear circuits. However, bear in mind that the proportionality property applies only to linear circuits with a single input.

The RC and RL circuits can store energy and have memory. In effect they have two inputs: (1) the input that occurred before $t = 0$ and (2) the step function applied at $t = 0$. The first input produces the initial energy state of the circuit at $t = 0$ and the second causes the zero-state response for $t \geq 0$. In general, for $t \geq 0$ the total response of a dynamic circuit is the superposition two responses: (1) the zero-input response caused by initial conditions produced by inputs applied before $t = 0$ and (2) the zero-state response caused by inputs applied after $t = 0$.

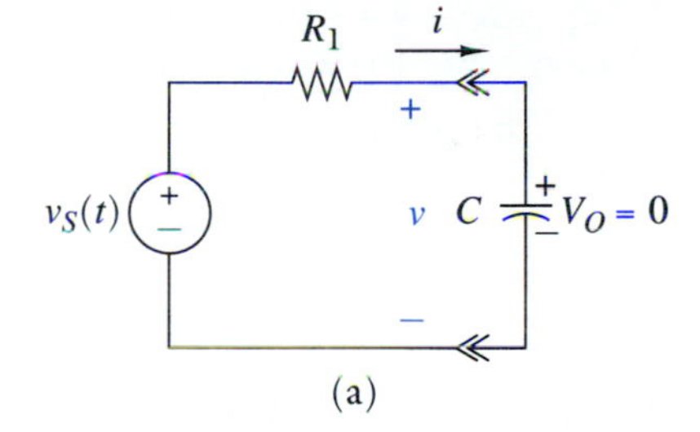

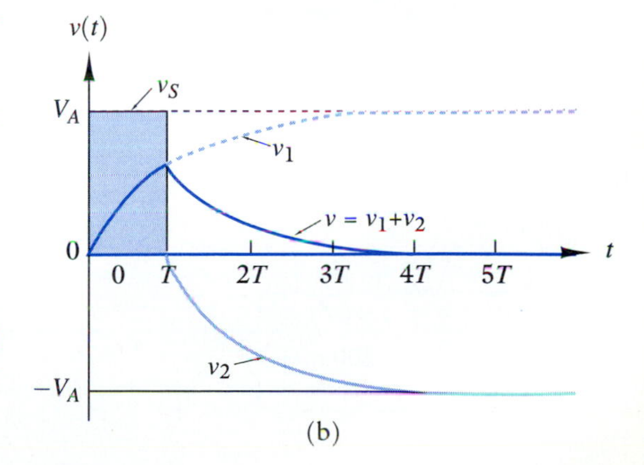

Figure 8-12

EXAMPLE 8-6

Find the zero-state response of the RC circuit of Fig. 8-12(a) for an input

$$v_S(t) = V_A[u(t) - u(t - T)]$$

SOLUTION:

The input is the rectangular pulse shown by the shading in Fig. 8-12(b). Since the initial condition is zero (zero-state), the total response is the superposition of responses caused by two inputs:

1. A step of amplitude V_A applied at $t = 0$
2. A step of amplitude $-V_A$ applied at $t = T$

The first input causes a zero-state response of

$$v_1(t) = V_A(1 - e^{-t/RC})u(t)$$

The second input causes a zero-state response of

$$v_2(t) = -V_A\left(1 - e^{\frac{-(t-T)}{RC}}\right)u(t - T)$$

The total response is the superposition of these two responses.

$$v(t) = v_1(t) + v_2(t)$$

Figure 8-12(b) shows how the two responses combine to produce the overall pulse response of the circuit. The first step function causes a response $v_1(t)$ that begins at zero and would eventually reach an amplitude of $+V_A$ at about t = 5RC. However, at $t = T < 5T_C$ the second step function causes an equal and opposite response $v_2(t)$ to begin. At about $t = T +$ 5RC the second response reaches its final state and cancels the first response, so the total pulse response returns to zero.

Figure 8-13

Exercise 8-6

The switch in Fig. 8-13 closes at $t = 0$.

(a) Find the zero-state response of the circuit state variable.

(b) Find the zero-state response of the current through the 10-V voltage source.

ANSWERS:

(a) $1.818(1 - e^{-200t})$ V

(b) $-0.4545(1 - e^{-200t})$ mA

Exercise 8-7

Given the following first-order circuit zero-state step responses:

(1) $v_C(t) = 10(1 - e^{-2000t})u(t)$ V

(2) $i_L(t) = 5(1 - e^{-5000t})u(t)$ mA

(a) What are the circuit characteristic equations?

(b) What are the circuit time constants?

(c) Design a first-order circuit that realizes these responses.

ANSWERS:

(1) **(a)** $(s + 2000) = 0$

(b) 0.5 ms

(c) There are no unique answers to the (c) part of this exercise because the time constants only specify the products RC and GL. Figure 8-14 shows some possible answers.

(2) **(a)** $(s + 5000) = 0$

(b) 0.2 ms

(c) same as above

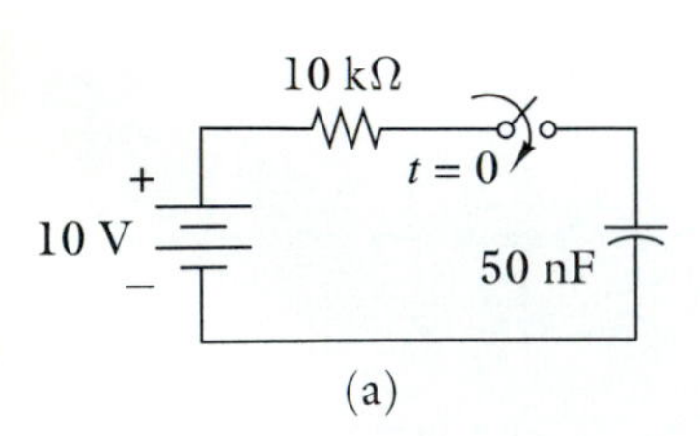

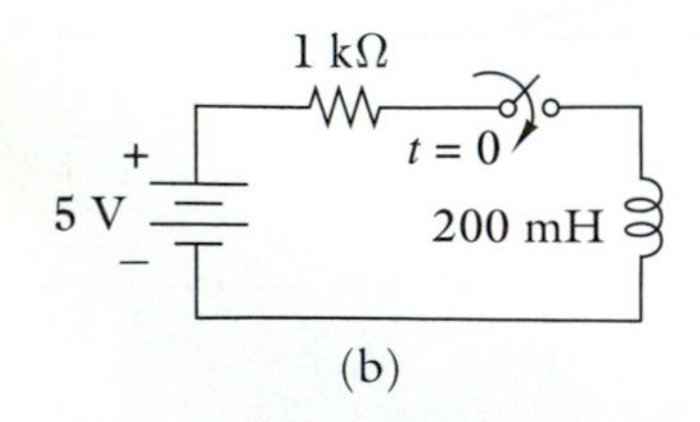

Figure 8-14

DISCUSSION:

The response in (1) is an RC circuit with V_A = 10 V and a time RC = 5×10^{-4} s. Selecting R = 10 kΩ yields C = 50 nF producing the circuit in Fig. 8-14(a). The response in (2) is an RL circuit with $I_A = 5$ mA and a time constant of GL $= 2 \times 10^{-4}$ s. Selecting R $= 1$ kΩ yields L $= 200$ mH. A source conversion changes the 5-mA current source in parallel with R into a $(5 \times 10^{-3})(10^3) = 5$-V voltage source in series with R as shown in Fig. 8-14(b). ■

8-3 INITIAL AND FINAL CONDITIONS

Reviewing the first-order step responses of the last section shows that the state variable responses can be written in the form

$$\text{RC Circuit:} \quad v_C(t) = (IC - FC)e^{-t/T_C} + FC \qquad t \geq 0$$

$$\text{RL Circuit:} \quad i_L(t) = (IC - FC)e^{-t/T_C} + FC \qquad t \geq 0 \tag{8-23}$$

where IC stands for the initial condition ($t = 0$) and FC for the final condition ($t \to \infty$). To determine the step response of any first-order circuit we need three quantities: IC, FC, and T_C. Since we know how to get the time constant directly from the circuit, it also would be useful if we had a direct way to determine IC and FC by inspecting the circuit itself.

The final condition can be calculated directly from the circuit by observing that for time larger than $5T_C$, the step responses approach a constant value or dc value. Under dc condition a capacitor acts like an open circuit and an inductor acts like a short circuit. After roughly five time constants a capacitor acts like an open circuit and an inductor like a short circuit and we can calculate the final value of the state variable using resistance circuit analysis methods.

In addition, we can use the dc analysis method to determine the initial condition in many practical situations. One common situation is a circuit containing a switch that remains in one state for a period of time that is long compared with the circuit time constant. If the switch is closed for a long period of time, then the state variable approaches final values determined by the dc inputs. If the switch is now opened at $t = 0$ a transient occurs in which the state variables are driven to a new final condition.

But note: The initial condition at $t = 0$ is the dc value of the state variable for the circuit configuration that existed before the switch was opened at $t = 0$. The switching action cannot cause an instantaneous change in the initial condition because capacitor voltage and inductor current are continuous functions of time. In other words, opening a switch at $t = 0$ marks the boundary between two eras. The dc condition of the state variables for the $t < 0$ era are the initial conditions for the $t > 0$ era that follows.

The usual way to state a switched circuit problem is to say that a switch has been closed (open) for a "long time" and then is opened (closed) at $t = 0$. In this context, a "long time" means at least five time constants. Time constants rarely exceed a few hundred milliseconds in electrical circuits so a "long time" passes rather quickly.

The parameters IC, FC, and T_C in switched dynamic circuits are found using the following steps:

STEP 1: Find the initial condition IC by applying dc analysis to the circuit configuration for $t < 0$

STEP 2: Find the final conditions FC by applying dc analysis to the circuit configuration for $t \geq 0$.

STEP 3: Find the time constant T_C of the circuit in the configuration for $t \geq 0$.

STEP 4: Write the step response directly using Eq. (8-23) without formulating and solving the circuit differential equation.

(a)

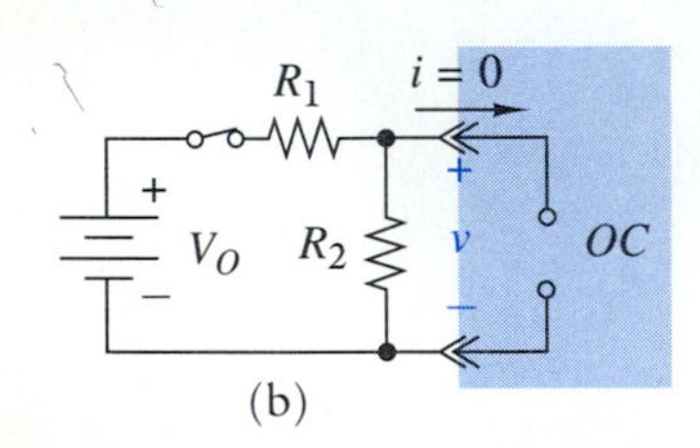

(b)

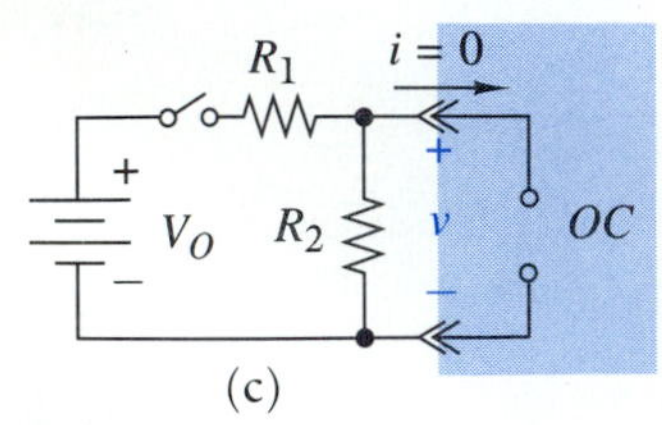

(c)

Figure 8-15 Solving switched dynamic circuit using the initial and final conditions.

For example, the switch in Fig. 8-15(a) has been closed for a "long time" and is opened at $t = 0$. We need IC, FC, and T_C to determine the capacitor voltage $v(t)$ for $t \geq 0$.

STEP 1: The initial condition is found by dc analysis of the circuit in Fig. 8-15(b) in which the switch is closed. Using voltage division the initial capacitor voltage is found to be

$$v_C(0-) = IC = \frac{R_2 V_O}{R_1 + R_2}$$

STEP 2: The final condition is found by dc analysis of the circuit in Fig 8-15(c) in which the switch is open. With the switch open the circuit has no dc excitation, so the final value of the capacitor voltage is zero.

STEP 3: The circuit in Fig. 8-15(c) also gives us the time constant. Looking back at the interface we see an equivalent resistance of R_2, since R_1 is connected in series with an open switch. For $t \geq 0$ the time constant is R_2C.

STEP 4: Using Eq. (8-23) the capacitor voltage for $t \geq 0$ is

$$v_C(t) = (IC - FC)e^{-t/T_C} + FC$$

$$= \frac{R_2 V_O}{R_1 + R_2} e^{-t/R_2 C} \qquad t \geq 0$$

This result is a zero-input response, since there is no excitation for $t \geq 0$. But now we see how the initial condition for the zero-input is physically established by opening a switch that has been closed for a long time.

To complete the analysis we determine the capacitor current by using its element constraint:

$$i_C(t) = C\frac{dv_C}{dt} = -\frac{V_O}{R_1 + R_2}e^{-t/R_2C}$$

This is the capacitor current for $t \geq 0$. For $t < 0$ the initial condition circuit in Fig. 8-15(b) points out that $i_C(0-) = 0$, since the capacitor is an open circuit.

Both the capacitor voltage and current responses are plotted in Fig. 8-16. The capacitor voltage is continuous at $t = 0$, but the capacitor current has a jump discontinuity at $t = 0$. In other words, state variables are continuous, but nonstate variables can have discontinuities at $t = 0$. Since the state variable is continuous, it is desirable to first determine the circuit state variable and then solve for other circuit variables using the element and connection constraints.

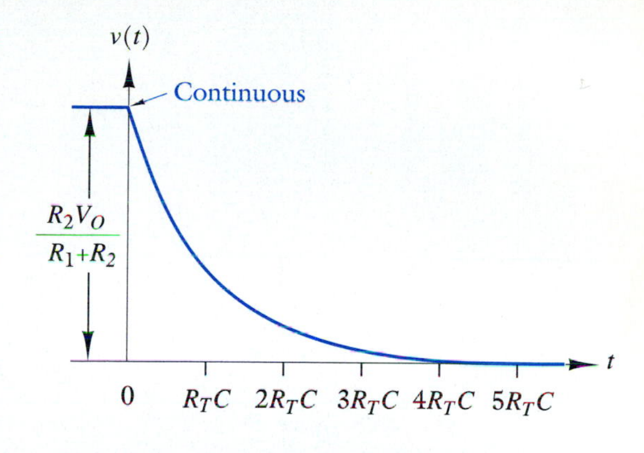

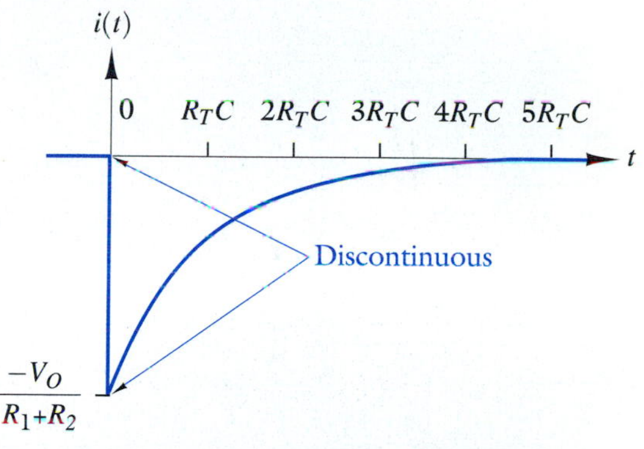

Figure 8-16 Response of the RC circuit in Fig. 8-15(a).

EXAMPLE 8-7

The switch in Fig. 8-17(a) has been open for a "long time" and is closed at $t = 0$. Find the inductor current for $t > 0$.

SOLUTION:

We first find the initial condition using the circuit in Fig. 8-17(b). By series equivalence the initial current is

$$i(0-) = IC = \frac{V_O}{R_1 + R_2}$$

The final condition and the time constant are determined from the circuit in Fig. 8-17(c). Closing the switch shorts out R_2 and the final condition and time constant for $t > 0$ are

$$i(\infty) = FC = \frac{V_O}{R_1} \qquad T_C = G_N L = \frac{L}{R_1}$$

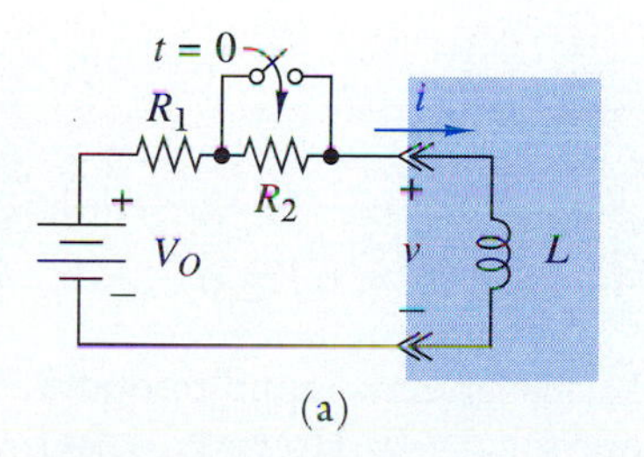

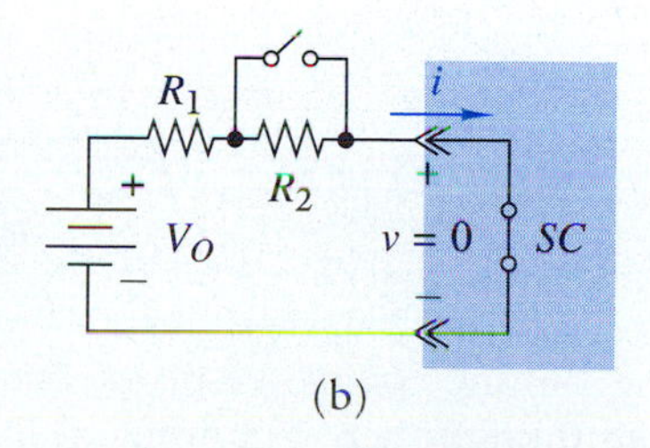

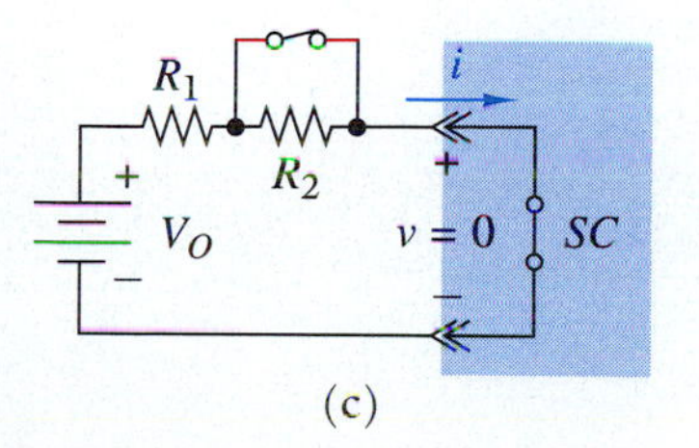

Figure 8-17

Using Eq. (8-23) the inductor current for $t \geq 0$ is

$$i(t) = (IC - FC)e^{-t/T_C} + FC$$

$$= \left[\frac{V_O}{R_1 + R_2} - \frac{V_O}{R_1}\right] e^{-R_1 t/L} + \frac{V_O}{R_1} \qquad t \geq 0$$

EXAMPLE 8-8

The switch in Fig. 8-18(a) has been closed for a "long time" and is opened at $t = 0$. Find the voltage $v_O(t)$.

SOLUTION:

The problem asks for voltage $v_O(t)$, which is not the circuit state variable. Our approach is to first find the state variable response and then use this response to solve for the required nonstate variable.

For $t < 0$ the circuit in Fig. 8-18(b) applies. By voltage division the initial capacitor voltage is

$$v(0-) = IC = \frac{R_1 V_O}{R_1 + R_2}$$

The final value and time constant are found from the circuit in Fig. 8-18(c).

$$v(\infty) = FC = V_O$$

$$T_C = R_T C = R_2 C$$

Using Eq. (8-23) the capacitor voltage for $t > 0$ is

$$v(t) = (IC - FC)e^{-t/T_C} + FC$$

$$= \left[\frac{R_1 V_O}{R_1 + R_2} - V_O\right] e^{-t/R_2 C} + V_O$$

$$= V_O + \frac{R_2 V_O}{R_1 + R_2} e^{-t/R_2 C} \qquad t \geq 0$$

Given the state variable, we can determine the voltage $v_O(t)$ by writing a KVL equation around the perimeter of the original circuit.

$$-V_O + v(t) + v_O(t) = 0$$

or

$$v_O(t) = V_O - v(t) = -\frac{R_2 V_O}{R_1 + R_2} e^{-t/R_2 C} \qquad t \geq 0$$

The output voltage response looks like a zero-input response even though there is a nonzero input for $t \geq 0$. However, $v_O(t)$ is not the state variable but the voltage across the resistor R_2. By

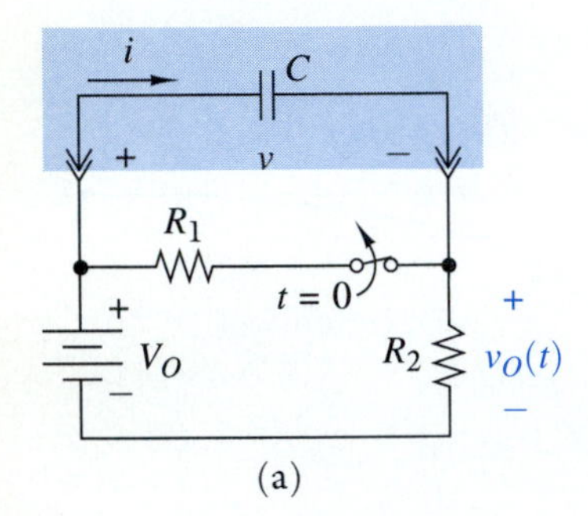

(a)

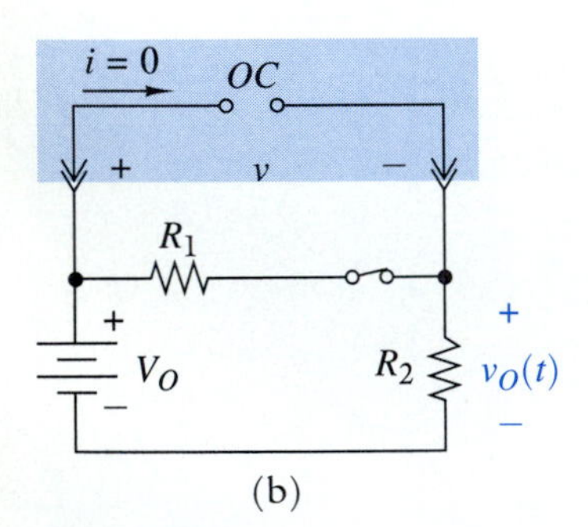

(b)

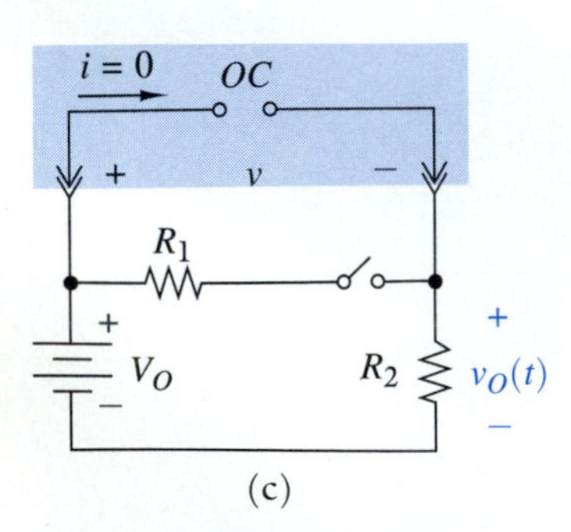

(c)

Figure 8-18

Ohm's law the voltage across R_2 is proportional to the capacitor current, which goes to zero in the final circuit configuration in Fig. 8-18(c) because the capacitor acts like an open-circuit.

EXAMPLE 8-9

The switch in Fig. 8-19 has been in position A for "quite a while" and is moved to position B at $t = 0$. Find the OP AMP output voltage $v_O(t)$.

SOLUTION:

In Sect. 8-1 we showed that this OP AMP circuit is a first-order circuit with a time constant of R_2C. When the capacitor is replaced by an open circuit the configuration is an inverting amplifier with a gain of $-R_2/R_1$. For $t < 0$ the switch is in position A and the initial output voltage is

$$v_O(0-) = IC = -\frac{R_2}{R_1}V_A$$

For $t \geq 0$ the switch is in Position B and the final output voltage is

$$v_O(\infty) = FC = -\frac{R_2}{R_1}(-V_A) = -IC$$

Using Eq. (8-23) the step response is

$$\begin{aligned} v_O(t) &= (IC - FC)e^{-t/T_C} + FC \\ &= -2\frac{R_2}{R_1}V_A e^{t/R_2C} + \frac{R_2}{R_1}V_A \\ &= \frac{R_2}{R_1}V_A\left(1 - 2e^{-t/R_2C}\right) \qquad t \geq 0 \end{aligned}$$

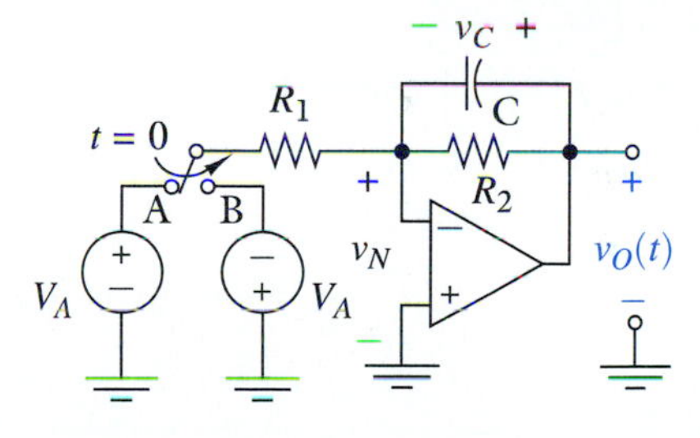

Figure 8-19

Since $v_O(t)$ is not a state variable we may wonder if there is a discontinuity at $t = 0$. Applying the fundamental principle of node voltages to the capacitor voltage yields

$$v_C(t) = v_O(t) - v_N(t)$$

But $v_N = 0$, since the noninverting input is grounded. In other words, the OP AMP output voltage equals the capacitor voltage, which is the state variable, and $v_O(t)$ cannot have a jump change at $t = 0$.

EXERCISE 8-8

In each circuit shown in Fig. 8-20 the switch has been in position A for a "long time" and is moved to position B at $t = 0$. Find the circuit state variable for each circuit for $t \geq 0$.

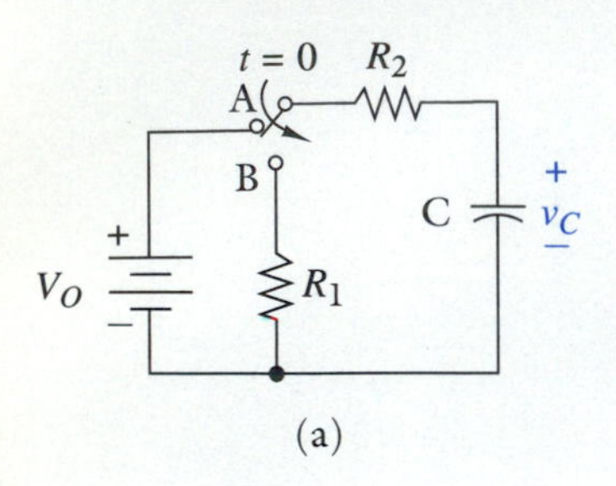

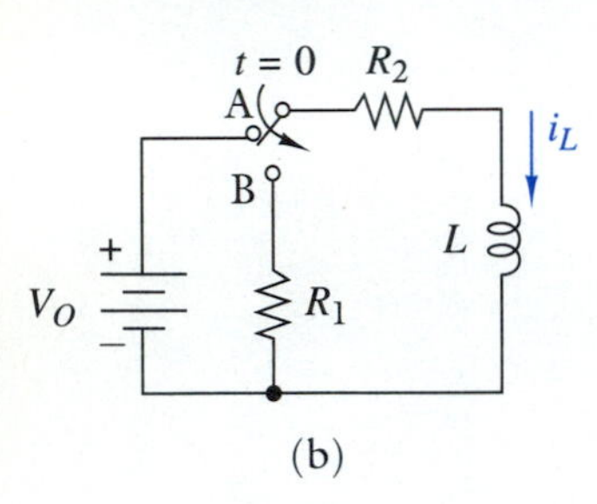

Figure 8-20

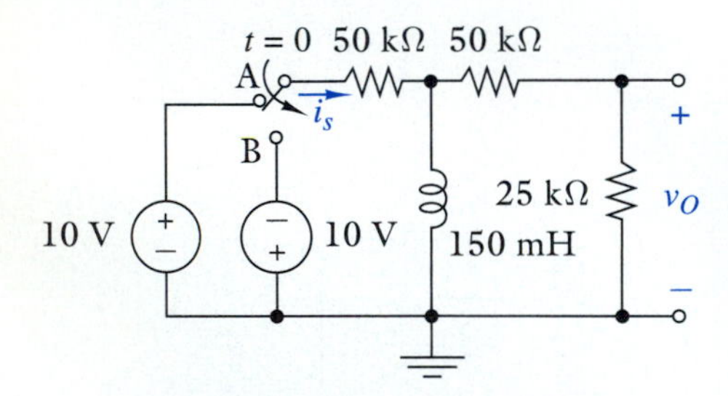

Figure 8-21

ANSWERS:

C1: $v_C(t) = V_O e^{-t/(R_1+R_2)C}$

C2: $i_L(t) = \dfrac{V_O}{R_2} e^{-(R_1+R_2)t/L}$ ■

■ EXERCISE 8-9

In each circuit shown in Fig. 8-20 the switch has been in position B for a "long time" and is moved to position A at $t = 0$. Find the circuit state variable for each circuit for $t \geq 0$.

ANSWERS:

C1: $v_C(t) = V_O\left(1 - e^{-t/R_2C}\right)$

C2: $i_L(t) = \dfrac{V_O}{R_2}\left(1 - e^{-R_2t/L}\right)$ ■

■ EXERCISE 8-10

In the circuit in Fig. 8-21 the switch has been in position A for a "long time" and is moved to position B at $t = 0$. For $t \geq 0$ find:

(a) the output voltage $v_O(t)$

(b) the current $i_S(t)$ through the switch

ANSWERS:

(a) $-4e^{-200t}$ V

(b) $-0.2 + 0.24e^{-200t}$ A ■

8-4 FIRST-ORDER CIRCUIT SINUSOIDAL RESPONSE

The response of linear circuits to sinusoidal inputs is one of the central themes of electrical engineering. In this introduction to the concept we treat sinusoidal response of first-order circuits using differential equations. In later chapters we see that sinusoidal response can be found using the Laplace transformations and ultimately the steady-state response from the circuit itself using the concept of a phasor. But for the moment, we concentrate on the classical method of finding the forced response from the circuit differential equation.

If the input to the RC circuit in Fig. 8-1 is a sinusoid, then the circuit differential equation in Eq. (8-3) is written as

$$R_T C \frac{dv(t)}{dt} + v(t) = V_A [\cos \omega t] u(t) \tag{8-24}$$

The input on the right side of Eq. (8-24) is **not** the eternal sinewave but a sinusoid that starts at $t = 0$, through some action such as the closing of a switch. We seek a solution function $v(t)$ that satisfies Eq. (8-24) for $t \geq 0$, and that meets the prescribed initial condition $v(0) = V_O$.

As with the step response, we find the solution in two parts: natural response and forced response. The natural response is of the form

$$v_N(t) = Ke^{-t/R_TC} \qquad t \geq 0$$

The natural response of a first-order circuit always has this form because it is a general solution of the homogeneous equation with input set to zero. The form of the natural response depends on physical characteristics of the circuit and is independent of the input.

The forced response depends on both the circuit and the nature of the forcing function. The forced response is a particular solution of the equation

$$R_TC\frac{dv_F(t)}{dt} + v_F(t) = V_A \cos \omega t \qquad t \geq 0$$

This equation requires that $v_F(t)$ plus R_TC times its first derivative, add to produce a cosine waveform for $t \geq 0$. The only way this can happen is for $v_F(t)$ and its derivative to be sinusoids. This requirement brings to mind the derivative property of the sinusoid. So we try a solution in the form of a general sinusoid:

$$v_F(t) = a \cos \omega t + b \sin \omega t \qquad (8\text{-}25)$$

In this expression the Fourier coefficients a and b are unknown. The approach we are using is called the method of undetermined coefficients. To evaluate the unknown coefficients we insert proposed forced response in Eq. (8-25) into the differential equation.

$$R_TC\frac{d}{dt}(a \cos \omega t + b \sin \omega t)$$
$$+ (a \cos \omega t + b \sin \omega t) = V_A \cos \omega t \qquad t \geq 0$$

Performing the differentiation gives

$$R_TC(-\omega a \sin \omega t + \omega b \cos \omega t) + (a \cos \omega t + b \sin \omega t) = V_A \cos \omega t$$

We next gather all sine and cosine terms on one side of the equation.

$$[R_TC\omega b + a - V_A] \cos \omega t + [-R_TC\omega a + b] \sin \omega t = 0$$

The left side of this equation is zero **for all** $t \geq 0$ only when the coefficients of the cosine and sine terms are identically zero. This requirement yields two linear equations in the unknown coefficients a and b:

$$a + (R_TC\omega)\, b = V_A$$
$$-(R_TC\omega)\, a + b = 0$$

The solutions of these linear equations are

$$a = \frac{V_A}{1 + (\omega R_T C)^2} \qquad b = \frac{\omega R_T C V_A}{1 + (\omega R_T C)^2}$$

The undetermined Fourier coefficients are now known, since these equations express a and b in terms of known circuit parameters ($R_T C$) and known input signal parameters (ω and V_A).

We combine the forced and natural responses as

$$v(t) = Ke^{-t/R_T C} + \frac{V_A}{1 + (\omega R_T C)^2} \cdot (\cos \omega t + \omega R_T C \sin \omega t) \qquad t \geq 0 \tag{8-26}$$

The initial condition requires

$$v(0) = V_O = K + \frac{V_A}{1 + (\omega R_T C)^2}$$

which means K is

$$K = V_O - \frac{V_A}{1 + (\omega R_T C)^2}$$

We substitute this value of K into Eq. (8-26) to obtain the function $v(t)$ that satisfies the differential equation and the initial conditions.

$$v(t) = \underbrace{\left[V_O - \frac{V_A}{1 + (\omega R_T C)^2}\right] e^{-t/R_T C}}_{\Longleftarrow \text{ Natural Response } \Longrightarrow} + \underbrace{\frac{V_A}{1 + (\omega R_T C)^2} (\cos \omega t + \omega R_T C \sin \omega t)}_{\Longleftarrow \text{ Forced Response } \Longrightarrow}$$

This expression is somewhat less formidable when we convert the forced response to an amplitude and phase angle format.

$$v(t) = \underbrace{\left[V_O - \frac{V_A}{1 + (\omega R_T C)^2}\right] e^{-t/R_T C}}_{\Longleftarrow \text{ Natural Response } \Longrightarrow} + \underbrace{\frac{V_A}{\sqrt{1 + (\omega R_T C)^2}} \cos(\omega t + \theta)}_{\Longleftarrow \text{ Forced Response } \Longrightarrow} \tag{8-27}$$

Where

$$\theta = \tan^{-1}(-b/a) = \tan^{-1}(-\omega R_T C)$$

Equation (8-27) is the complete response of the RC circuit for an initial condition V_O and a sinusoidal input $[V_A \cos \omega t]u(t)$.

Several aspects of the response deserve comment:

1. After roughly five time constants the natural response decays to zero but the sinusoidal forced response persists.
2. The forced response is a sinusoid with the same frequency (ω) as the input but with a different amplitude and phase angle.
3. The forced response is proportional to V_A. This means that amplitude of the forced component has the proportionality property because the circuit is linear.

In the terminology of electrical engineering the forced component is called the ***sinusoidal steady-state response***. The words *steady state* may be misleading, since together they seem to imply a constant or "steady" value, whereas the forced response is a sustained oscillation. To electrical engineers *steady state* means the conditions reached after the natural response has died out. The sinusoidal steady-state response is also called the **ac steady-state response**. Often the words *steady state* are dropped and it is called simply the **ac response**. Hereafter, ac response, sinusoidal steady-state response, and the forced response for a sinusoidal input will be used interchangeably.

Finally, the forced response due to a step function input is called the ***zero-frequency*** or ***dc steady-state response***. The zero-frequency terminology means that we think of a step function as a cosine $V_A[\cos \omega t]u(t)$ with $\omega = 0$. The reader can easily show that inserting $\omega = 0$ reduces Eq. (8-27) to the RC circuit step response in Eq. (8-20).

EXAMPLE 8-10

The switch in Fig. 8-22(a) has been open for a "long time" and is closed at $t = 0$. Find the voltage $v(t)$ for $t \geq 0$ when $v_S(t) = [20 \sin 1000t]u(t)$.

SOLUTION:

We first derive the circuit differential equation. By voltage division the Thévenin voltage seen by the capacitor is

$$v_T(t) = \frac{4}{4+4} v_S(t) = 10 \sin 1000t$$

The Thévenin resistance (switch closed and source off) looking back into the interface is two 4-kΩ resistors in parallel so $R_T = 2$ kΩ. The circuit time constant is

$$T_C = R_T C = (2 \times 10^3)(1 \times 10^{-6}) = 2 \times 10^{-3} = 1/500 \text{ s}$$

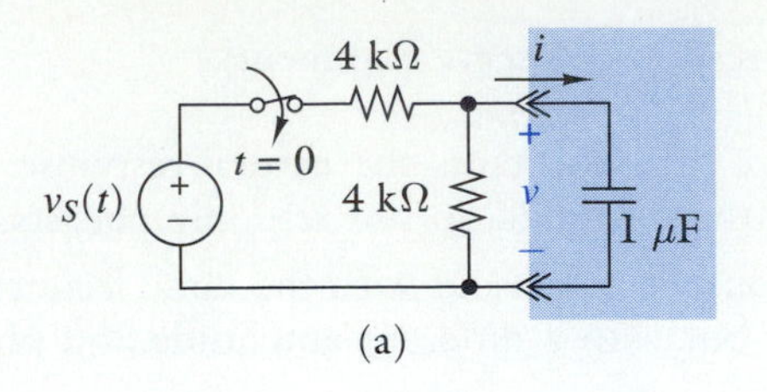

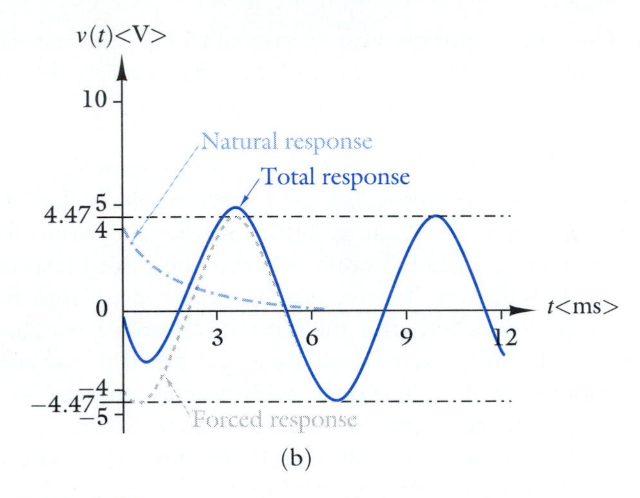

Figure 8-22

Given the Thévenin equivalent seen by the capacitor and the circuit time constant, the circuit differential is

$$2 \times 10^{-3} \frac{dv(t)}{dt} + v(t) = 10 \sin 1000t \qquad t \geq 0$$

Note that the right side of the circuit differential equation is the Thévenin voltage $v_T(t)$ and not the original source input $v_S(t)$. The natural response is of the form

$$v_N(t) = Ke^{-500t} \qquad t \geq 0$$

The forced response with undetermined Fourier coefficients is

$$v_F(t) = a \cos 1000t + b \sin 1000t$$

Substituting the forced response into the differential equation produces

$$2 \times 10^{-3}(-1000a \sin 1000t + 1000b \cos 1000t) + a \cos 1000t + b \sin 1000t = 10 \sin 1000t$$

Collecting all sine and cosine terms on one side of this equation yields

$$(a + 2b) \cos 1000t + (-2a + b - 10) \sin 1000t = 0$$

The left side of this equation is zero for all $t \geq 0$ only when the coefficient of the sine and cosine terms vanish:

$$a + 2b = 0$$

$$-2a + b = 10.$$

The solutions of these two linear equations are $a = -4$ and $b = 2$. We combine the forced and natural responses

$$v(t) = Ke^{-500t} - 4\cos 1000t + 2\sin 1000t \qquad t \geq 0$$

The constant K can be determined from the initial conditions

$$v(0) = V_O = K - 4$$

The initial condition is $V_O = 0$ because with the switch open the capacitor had no input for a "long time" prior to $t = 0$. The initial condition $v(0) = 0$ requires $K = 4$, so we can now write the complete response in the form

$$v(t) = 4e^{-500t} - 4\cos 1000t + 2\sin 1000t \qquad t \geq 0$$

or in an amplitude, phase angle format as

$$v(t) = 4e^{-500t} + 4.47\cos(1000t + 153°) \qquad t \geq 0$$

Plots of the natural response, forced response, and total response are shown in Fig. 8-22(b). The natural response decays to zero, and the total response merges into a sustained oscillation caused by the forced response. After five time constants or so the circuit settles down to the sinusoidal or ac steady-state condition.

EXAMPLE 8-11

Find the sinusoidal steady-state response of the output voltage $v_O(t)$ in Fig. 8-23 when the input current is $i_S(t) = [I_A \cos \omega t]u(t)$.

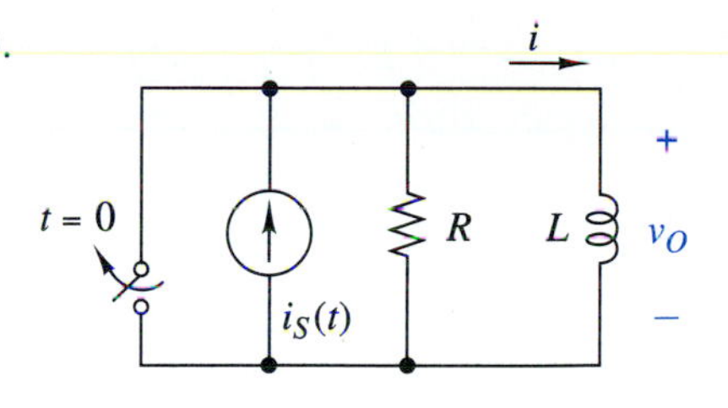

Figure 8-23

SOLUTION:

In keeping with our general analysis approach, we first find the steady-state response of the state variable $i(t)$ and use it to determine the required output voltage. The differential equation of the circuit in terms of the inductor current is

$$GL\frac{di}{dt} + i = I_A \cos \omega t \qquad t \geq 0$$

To find the steady-state response we need to find the undetermined coefficients a and b in the forced component:

$$i_F(t) = a\cos \omega t + b\sin \omega t$$

Substituting this into the differential equation yields

$$GL(-a\omega \sin \omega t + b\omega \cos \omega t) + a\cos \omega t + b\sin \omega t = I_A \cos \omega t$$

Collecting sine and cosine terms produces

$$(a + GLb\omega - I_A)\cos\omega t + (-GLa\omega + b)\sin\omega t = 0$$

The left side of this equation is zero for all $t \geq 0$ only if

$$a + (GL\omega)\,b = I_A$$
$$-(GL\omega)\,a + b = 0$$

The solutions of these linear equations are

$$a = \frac{I_A}{1 + (GL\omega)^2} \qquad b = \frac{\omega GLI_A}{1 + (GL\omega)^2}$$

Therefore, the forced component of the inductor current is

$$i_F(t) = \frac{I_A}{1 + (\omega GL)^2}(\cos\omega t + \omega GL\sin\omega t)$$

The prescribed output is the voltage across the inductor. The steady-state output voltage is found using the inductor element equation.

$$\begin{aligned} v_O = L\frac{di_F}{dt} &= \left[\frac{I_A L}{1 + (\omega GL)^2}\right]\frac{d}{dt}[\cos\omega t + \omega GL\sin\omega t] \\ &= \left[\frac{I_A L}{1 + (\omega GL)^2}\right]\left[-\omega\sin\omega t + \omega^2 GL\cos\omega t\right] \\ &= \frac{I_A\omega L}{\sqrt{1 + (\omega GL)^2}}\cos(\omega t + \theta) \end{aligned}$$

where $\theta = \tan^{-1}(1/\omega GL)$. The output voltage is a sinusoid with the same frequency as the input signal, but with a different amplitude and phase angle. In fact, in the sinusoidal steady state every voltage and current in a linear circuit is sinusoidal with the same frequency.

EXERCISE 8-11

Find the forced component solution of the following differential equation for:

(a) $\omega = 500$
(b) $\omega = 1000$
(c) $\omega = 2000$ rad/s

$$10^{-3}\frac{dv}{dt} + v = 10\cos\omega t$$

ANSWERS:

(a) $v_F(t) = 8\cos 500t + 4\sin 500t$
(b) $v_F(t) = 5\cos 1000t + 5\sin 1000t$
(c) $v_F(t) = 2\cos 2000t + 4\sin 2000t$

EXERCISE 8-12

The circuit in Fig. 8-24 is operating in the sinusoidal steady state with

$$v_O(t) = 10\cos(100t - 45°).$$

Find the source voltage $v_S(t)$.

ANSWER:
$10\sqrt{2}\cos 100t$ V

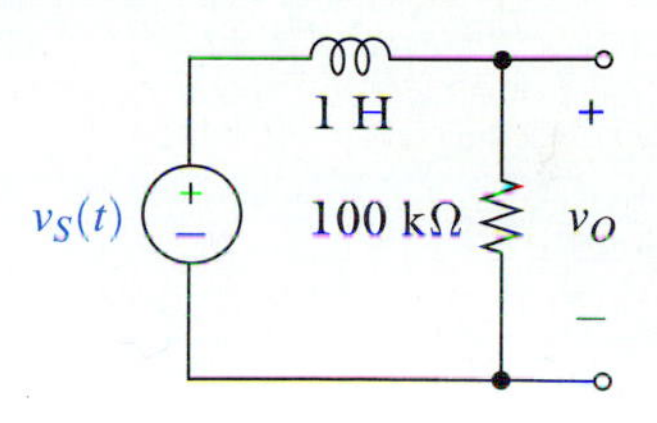

Figure 8-24

8-5 THE SERIES RLC CIRCUIT

Second-order circuits contain two energy storage elements that cannot be replaced by a single equivalent element. They are called *second-order circuits* because the circuit differential equation involves the second derivative of the dependent variable. Although there is an endless number of such circuits, in this chapter we will concentrate on two classical forms: (1) the series RLC circuit and (2) the parallel RLC circuit. These two circuits illustrate almost all of the basic concepts of second-order circuits and serve as vehicles for studying the solution of second-order differential equations. In subsequent chapters we use Laplace transform techniques to analyze any second-order circuit.

Formulating Series RLC Circuit Equations

We begin with the circuit in Fig. 8-25(a) where the inductor and capacitor are connected in series. The source-resistor circuit can be reduced to the Thévenin equivalent shown in Fig. 8-25(b). The result is a circuit in which a voltage source, resistor, inductor, and capacitor are connected in series, hence the name *series RLC circuit.*

The first task is to develop the equations that describe the series RLC circuit. The Thévenin equivalent to the left of the interface in Fig. 8-25(b) produces the KVL constraint

$$v + R_T i = v_T \tag{8-28}$$

Applying KVL around the loop on the right side of the interface yields

$$v = v_L + v_C \tag{8-29}$$

Finally, the i-v characteristics of the inductor and capacitor are

$$v_L = L\frac{di}{dt} \tag{8-30}$$

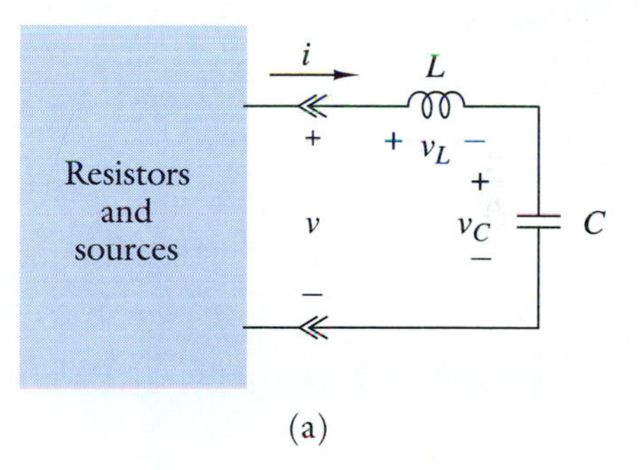

(a)

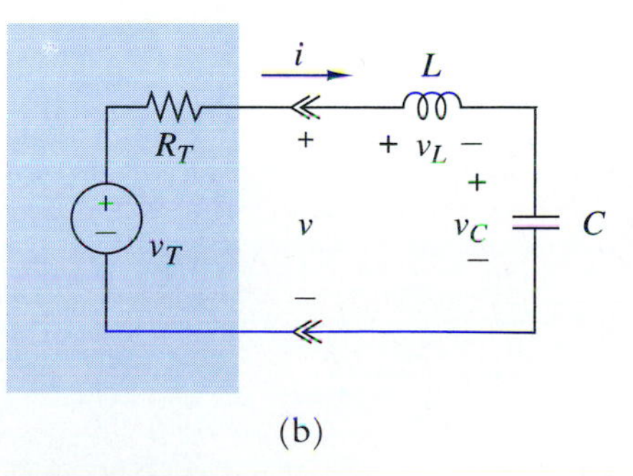

(b)

Figure 8-25 The series RLC circuit.

$$i = C\frac{dv_C}{dt} \tag{8-31}$$

Equations (8-28) through (8-31) are four independent equations in four unknowns (i, v, v_L, v_C). Collectively this set of equations provides a complete description of the dynamics of the series RLC circuit. To find the circuit response using classical methods, we must derive a circuit equation containing only one of these unknowns.

We are using circuit state variables as solution variables because they are continuous functions of time. In the series RLC circuit in Fig. 8-25(b) there are two state variables: (1) the capacitor voltage $v_C(t)$ and (2) the inductor current $i(t)$. We first show how to describe the circuit using the capacitor voltage as the starting place.

To derive a single equation in $v_C(t)$ we substitute Eqs. (8-29) and (8-31) into (8-28).

$$v_L + v_C + R_T C\frac{dv_C}{dt} = v_T \tag{8-32}$$

These substitutions eliminate the unknowns except v_C and v_L. To eliminate the inductor voltage, we substitute Eq. (8-31) into (8-30) to obtain

$$v_L = LC\frac{d^2v_C}{dt^2}$$

Substituting this result into Eq. (8-32) produces

$$\begin{aligned} LC\frac{d^2v_C}{dt^2} + R_T C\frac{dv_C}{dt} + v_C &= v_T \\ v_L \quad + \quad v_R \quad + v_C &= v_T \end{aligned} \tag{8-33}$$

In effect, this is a KVL equation around the loop in Fig. 8-25(b), where the inductor and resistor voltages have been expressed in terms of the capacitor voltage.

Equation (8-33) is a ***second-order linear differential equation with constant coefficients***. It is a second-order equation because the highest-order derivative is the second derivative of the dependent variable $v_C(t)$. The coefficients are constant because the circuit parameters L, C, and R_T do not change. The Thévenin voltage $v_T(t)$ is a known driving force. The initial conditions

$$v_C(0) = V_O \qquad \text{and} \qquad \frac{dv_C}{dt}(0) = \frac{1}{C}i(0) = \frac{I_O}{C} \tag{8-34}$$

are determined by the values of the capacitor voltage and inductor current at $t = 0$, namely V_O and I_O.

In summary, the second-order differential equation in Eq. (8-33) characterizes the response of the series RLC circuit in terms of the capacitor voltage $v_C(t)$. Once the solution $v_C(t)$

is found, we can solve for every other voltage or current, including the inductor current, using the element and connection constraints in Eqs. (8-28) to (8-31).

Alternatively, we can characterize the series RLC circuit using the inductor current. We first write the capacitor *i-v* characteristics in integral form:

$$v_C(t) = \frac{1}{C}\int_0^t i(x)dx + v_C(0) \tag{8-35}$$

Equations (8-35), (8-30), and (8-29) are inserted into the interface constraint of Eq. (8-28) to obtain a single equation in the inductor current $i(t)$:

$$\begin{aligned} L\frac{di}{dt} + \frac{1}{C}\int_0^t i(x)dx + v_C(0) + R_T i &= v_T \\ v_L + \qquad v_C \qquad + v_R &= v_T \end{aligned} \tag{8-36}$$

In effect, this a KVL equation around the loop in Fig. 8-25(b), where the capacitor and resistor voltages have been expressed in terms of the inductor current.

Equation (8-36) is a ***second-order linear integro-differential equation with constant coefficients*** It is second order because it involves the first derivative and the first integral of the dependent variable $i(t)$. The coefficients are constant because the circuit parameters L, C, and R_T do not change. The Thévenin equivalent voltage $v_T(t)$ is a known driving force and the initial conditions are $v_C(0) = V_O$ and $i(0) = I_O$.

Equations (8-33) and (8-36) involve the same basic ingredients: (1) a single-state variable, (2) three circuit parameters (R_T, L, C), (3) a known input $v_T(t)$, and (4) two initial conditions (V_O and I_O). The only difference is that one expresses the sum of voltages around the loop in terms of the capacitor voltage, while the other uses the inductor current. Either equation characterizes the dynamics of the series RLC circuit because once a state variable is found, every other voltage or current can be found from using element and connection constraints.

Zero-Input Response of the Series RLC Circuit

The circuit dynamic rsponse for $t \geq 0$ can be divided into two components: (1) the zero-input response caused by the initial conditions and (2) the zero-state response caused by driving forces applied after $t = 0$. Because the circuit is linear we can solve for these responses separately and superimpose them to get the total response. We first deal with the zero-input response.

With $v_T = 0$ (zero-input) Eq. (8-33) becomes

$$LC\frac{d^2v_C}{dt^2} + R_TC\frac{dv_C}{dt} + v_C = 0 \tag{8-37}$$

This result is a second-order homogeneous differential equation in the capacitor voltage. Alternatively, we set $v_T = 0$ in Eq. (8-36) and differentiate once to obtain the following homogeneous differential equation in the inductor current:

$$LC\frac{d^2i}{dt^2} + R_TC\frac{di}{dt} + i = 0 \tag{8-38}$$

We observe that Eqs. (8-37) and (8-38) have exactly the same form except the dependent variables are different. The zero-input response of the capacitor voltage and inductor current have the same general form. We do not need to study both to understand the dynamics of the series RLC circuit. In other words, in the series RLC circuit we can use either state variable to describe the zero-input response.

In the following we will concentrate on the capacitor voltage response. Equation (8-37) requires the capacitor voltage, plus RC times its first derivative, plus LC times its second derivative to add to zero for all $t \geq 0$. The only way this can happen is for $v_C(t)$, its first derivative, and its second derivative to all have the same form. No matter how many times we differentiate an exponential of the form e^{st}, we are left with a signal of the same form. This observation, plus our experience with first-order circuits, suggests we try a solution of the form

$$v_C(t) = Ke^{st}$$

where the parameters K and s are to be evaluated. When the trial solution is inserted in Eq. (8-37) we obtain the condition

$$Ke^{st}(LCs^2 + R_TCs + 1) = 0$$

The function e^{st} cannot be zero for all $t \geq 0$. The condition $K = 0$ is not allowed because it is a trivial solution declaring $v_C(t)$ is zero for all t. The only useful way to meet the condition is to require

$$LCs^2 + R_TCs + 1 = 0 \tag{8-39}$$

Equation (8-39) is the ***characteristic equation*** of the series RLC circuit. The characteristic equation is a quadratic because the circuit contains two energy storage elements. Inserting Ke^{st} into the homogeneous equation of the inductor current in Eq. (8-38) produces the same characteristic equation. Thus, Eq. (8-39) relates the zero-input response to circuit parameters for both state variables, hence the name characteristic equation.

In general a quadratic characteristic equation has two roots:

$$s_1,\ s_2 = \frac{-R_T C \pm \sqrt{(R_T C)^2 - 4LC}}{2LC} \tag{8-40}$$

From the form of the expression under the radical in Eq. (8-40), we see there are three distinct possibilities:

Case A: If $(R_T C)^2 - 4LC > 0$, the radicand is positive and there will be two real, unequal roots ($s_1 = \alpha_1 \neq s_2 = \alpha_2$).

Case B: If $(R_T C)^2 - 4LC = 0$, the radicand vanishes and there will be two real, equal roots ($s_1 = s_2 = \alpha$).

Case C: If $(R_T C)^2 - 4LC < 0$, the radicand will be negative and there are two complex conjugate roots ($s_1 = -\alpha - j\beta$ and $s_2 = -\alpha + j\beta$).

Before dealing with zero-input response for each case we consider an example.

EXAMPLE 8-12

In a series RLC circuit $C = 0.25\ \mu$F and $L = 1$ H. Find the roots of the characteristic equation for $R_T =$

(a) 8.5 kΩ
(b) 4 kΩ
(c) 1 kΩ

SOLUTION:

The values of the capacitance and inductance are fixed, and we need to find the roots Eq. (8-39) for three different values of the resistance.

(a) For $R_T = 8.5$ kΩ the characteristic equation is

$$0.25 \times 10^{-6} s^2 + 2.125 \times 10^{-3} s + 1 = 0$$

whose roots are

$$s_1,\ s_2 = -4250 \pm \sqrt{(3750)^2} = -500,\ -8000$$

These roots illustrate Case A. The quantity under the radical is positive, and there are two real, unequal roots at $s_1 = -500$ and $s_2 = -8000$.

(b) For $R_T = 4$ kΩ the characteristic equation is

$$0.25 \times 10^{-6} s^2 + 10^{-3} s + 1 = 0$$

whose roots are

$$s_1,\ s_2 = -2000 \pm \sqrt{4 \times 10^6 - 4 \times 10^6} = -2000$$

This is an example of Case B. The quantity under the radical is zero and there are two real, equal roots at $s_1 = s_2 = -2000$.

(c) For $R_T = 1$ kΩ the characteristic equation is

$$0.25 \times 10^{-6} s^2 + 0.25 \times 10^{-3} s + 1 = 0$$

whose roots are

$$s_1,\ s_2 = -500 \pm 500\sqrt{-15}$$

The quantity under the radical is negative illustrating Case C. Defining the symbol $j = \sqrt{-1}$, the two roots are[1]

$$s_1,\ s_2 = -500 \pm j500\sqrt{15}$$

In Case C the two roots are complex conjugates.

EXERCISE 8-13

For a series RLC circuit:

(a) Find the roots of the characteristic equation when $R_T = 2$ kΩ, $L = 100$ mH, and $C = 0.4\ \mu$F.

(b) Find the values of R_T and C when $L = 100$ mH and the roots of the characteristic equation are $s_1,\ s_2 = -1000 \pm j2000$.

(c) Select the values of R_T, L, and C so $s_1 = s_2 = -10^4$.

ANSWERS:

(a) $-1340, -18660$.

(b) $R_T = 200$ Ω, C = 2 μF.

(c) There is no unique answer to part (c), since the requirement

$$(10^{-4}s + 1)^2 = LCs^2 + R_TCs + 1$$
$$= 10^{-8}s^2 + 10^{-4}s + 1$$

gives two equations $R_TC = 10^{-4}$ and $LC = 10^{-8}$ in three unknowns. One solution is to select $C = 1\ \mu$F which yields $L = 10$ mH and $R_T = 200$ Ω. ■

We have not introduced complex numbers simply to make things, well, complex. Complex numbers arise quite naturally in practical physical situations involving nothing more than factoring a quadratic equation. The ability to deal with complex

[1]Mathematicians use the letter i to represent the imaginary number $\sqrt{-1}$. Electrical engineers use j, since the letter i represents electric current.

numbers is essential to our study. For those who need a review of such matters there is a concise discussion in Appendix C.

Form of the Zero-Input Response

Since the characteristic equation has two roots there are two solutions to the homogeneous differential equation:

$$v_{C1}(t) = K_1 e^{s_1 t}$$

$$v_{C2}(t) = K_2 e^{s_2 t}$$

That is,

$$LC \frac{d^2}{dt^2}\left(K_1 e^{s_1 t}\right) + R_T C \frac{d}{dt}\left(K_1 e^{s_1 t}\right) + K_1 e^{s_1 t} = 0$$

and

$$LC \frac{d^2}{dt^2}\left(K_2 e^{s_2 t}\right) + R_T C \frac{d}{dt}\left(K_2 e^{s_2 t}\right) + K_2 e^{s_2 t} = 0$$

The superposition or sum of these two solutions is also a solution, since

$$LC \frac{d^2}{dt^2}\left(K_1 e^{s_1 t} + K_2 e^{s_2 t}\right)$$
$$+ R_T C \frac{d}{dt}\left(K_1 e^{s_1 t} + K_2 e^{s_2 t}\right) + K_1 e^{s_1 t} + K_2 e^{s_2 t} = 0$$

Therefore, the general solution for the zero-input response is of the form

$$v_C(t) = K_1 e^{s_1 t} + K_2 e^{s_2 t} \qquad (8\text{-}41)$$

The constants K_1 and K_2 can be found using the initial conditions given in Eq. (8-34). At $t = 0$ the condition on the capacitor voltage yields

$$v_C(0) = V_O = K_1 + K_2 \qquad (8\text{-}42)$$

To use the initial condition on the inductor current we differentiate Eq. (8-41).

$$\frac{dv_C}{dt} = K_1 s_1 e^{s_1 t} = K_2 s_2 e^{s_2 t}$$

Using Eq. (8-34) to relate the initial value of the derivative of the capacitor voltage to the initial inductor current $i(0)$ yields

$$\frac{dv_C(0)}{dt} = \frac{I_O}{C} = K_1 s_1 + K_2 s_2 \qquad (8\text{-}43)$$

Equations (8-42) and (8-43) provide two equations in the two unknown constants K_1 and K_2:

$$K_1 + K_2 = V_O$$

$$s_1 K_1 + s_2 K_2 = I_O/C$$

The solutions of these equations are

$$K_1 = \frac{s_2 V_O - I_O/C}{s_2 - s_1} \quad \text{and} \quad K_2 = \frac{-s_1 V_O + I_O/C}{s_2 - s_1}$$

Inserting these solutions back into Eq. (8-41) yields

$$v_C(t) = \frac{s_2 V_O - I_O/C}{s_2 - s_1} e^{s_1 t} + \frac{-s_1 V_O + I_O/C}{s_2 - s_1} e^{s_2 t} \qquad t \geq 0 \quad (8\text{-}44)$$

Equation (8-44) is the general zero-input response of the series RLC circuit. The response depends on two initial conditions V_O and I_O, and the circuit parameters R_T, L, and C since s_1 and s_2 are the roots of the characteristic equation $LCs^2 + R_T Cs + 1 = 0$. The response has different waveforms depending on whether the roots s_1 and s_2 fall under Case A, B, or C.

For Case A the two roots are real and distinct. Using the notation $s_1 = -\alpha_1$ and $s_2 = -\alpha_2$, the form zero-input response for $t \geq 0$ is

$$v_C(t) = \left[\frac{\alpha_2 V_O + I_O/C}{\alpha_2 - \alpha_1}\right] e^{-\alpha_1 t} + \left[\frac{-\alpha_1 V_O - I_O/C}{\alpha_2 - \alpha_1}\right] e^{-\alpha_2 t} \quad (8\text{-}45)$$

This form is called the ***overdamped response***. The overdamped response is the sum of two exponential waveforms similar to the double exponential waveform treated in Example 6-12. The waveform has two time constants $1/\alpha_1$ and $1/\alpha_2$. The time constant can be greatly different, or nearly equal, but they cannot be equal because we would have Case B.

With Case B the roots are real and equal. Using the notation $s_1 = s_2 = -\alpha$ the general form in Eq. (8-44) becomes

$$v_C(t) = \frac{(\alpha V_O + I_O/C)e^{-\alpha t} + (-\alpha V_O - I_O/C)e^{-\alpha t}}{\alpha - \alpha}$$

We immediately see a problem here because the denominator vanishes. However, a closer examination reveals that the numerator vanishes as well, so the solution reduces to the indeterminate form 0/0. To investigate the indeterminacy, we let $s_1 = -\alpha$ and $s_2 = -\alpha + x$, and explore the situation as x approaches zero. Inserting s_1 and s_2 in this notation in Eq. (8-44) produces

$$v_C(t) = V_O e^{-\alpha t} + \left[\frac{-\alpha V_O - I_O/C}{x}\right] e^{-\alpha t} + \left[\frac{\alpha V_O + I_O/C}{x}\right] e^{-\alpha t} e^{xt}$$

which can be arranged in the form

$$v_C(t) = e^{-\alpha t}\left[V_O - (\alpha V_O + I_O/C)\frac{1 - e^{xt}}{x}\right]$$

We see that the indeterminacy comes from the term $(1 - e^{xt})/x$, which reduces to $0/0$ as x approaches zero. Application of l'Hopital's rule reveals

$$\lim_{x \to 0} \frac{1 - e^{xt}}{x} = \lim_{x \to 0} \frac{-te^{xt}}{1} = -t$$

This result removes the indeterminancy and as x approaches zero the zero-input response reduces to

$$v_C(t) = V_O e^{-\alpha t} + (\alpha V_O + I_O/C)\, t e^{-\alpha t} \qquad t \geq 0 \qquad (8\text{-}46)$$

This special form is called the ***critically damped reponse***. The critically damped response includes an exponential and the damped ramp waveform studied in Example 6-10. The damped ramp is required, rather than two exponentials, because in Case B the two equal roots produce the same exponential waveform.

Case C produces complex conjugate roots of the form

$$s_1 = -\alpha - j\beta \qquad \text{and} \qquad s_2 = -\alpha + j\beta$$

Inserting these roots into Eq. (8-44) yields

$$v_C(t) = \left[\frac{(-\alpha + j\beta)V_O - I_O/C}{j2\beta}\right] e^{-\alpha t} e^{-j\beta t} + \left[\frac{(\alpha + j\beta)V_O + I_O/C}{j2\beta}\right] e^{-\alpha t} e^{-j\beta t}$$

which can be arranged in the form

$$v_C(t) = V_O e^{-\alpha t}\left[\frac{e^{j\beta t} + e^{-j\beta t}}{2}\right] + \frac{\alpha V_O + I_O/C}{\beta} e^{-\alpha t}\left[\frac{e^{j\beta t} - e^{-j\beta t}}{2j}\right] \qquad (8\text{-}47)$$

The expressions within the brackets have been arranged in a special way. Euler's relationships for an imaginary exponential are written as

$$e^{j\theta} = \cos\theta + j\sin\theta$$

and

$$e^{-j\theta} = \cos\theta - j\sin\theta$$

When we add and subtract these equations we obtain

$$\cos\theta = \frac{e^{j\theta} + e^{-j\theta}}{2} \qquad \text{and} \qquad \sin\theta = \frac{e^{j\theta} - e^{-j\theta}}{2j}$$

Comparing these expressions for $\sin\theta$ and $\cos\theta$ with the complex terms in Eq. (8-47) reveals we can write $v_C(t)$ in the form

$$v_C(t) = V_O e^{-\alpha t} \cos \beta t + \frac{\alpha V_O + I_O/C}{\beta} e^{-\alpha t} \sin \beta t \qquad t \geq 0 \tag{8-48}$$

This form is called the ***underdamped response***. The underdamped response contains the damped sinusoid waveform studies in Example 6-11. The real part of the roots (α) provides the damping term in the exponential, while the imaginary parts (β) defines the frequency of the oscillations.

In summary, the roots of the characteristic equation affect the form of the zero-input response in the following way. When the roots are real and unequal (Case A) the response is the sum of two exponentials. When the roots are real and equal (Case B) the response is the sum of an exponential and a damped ramp. When the roots are complex conjugates the response is the sum of a damped cosine and a damped sine (Case C).

EXAMPLE 8-13

The circuit of Fig. 8-26 has $C = 0.25\ \mu$F and $L = 1$ H. The switch has been open for a long time and is closed at $t = 0$. For $t \geq 0$ find the zero-input capacitor voltage for three different values of R: (a) $R = 8.5$ kΩ, (b) $R = 4$ kΩ, and (c) $R = 1$ kΩ. The initial conditions are $I_O = 0$ and $V_O = 15$ V.

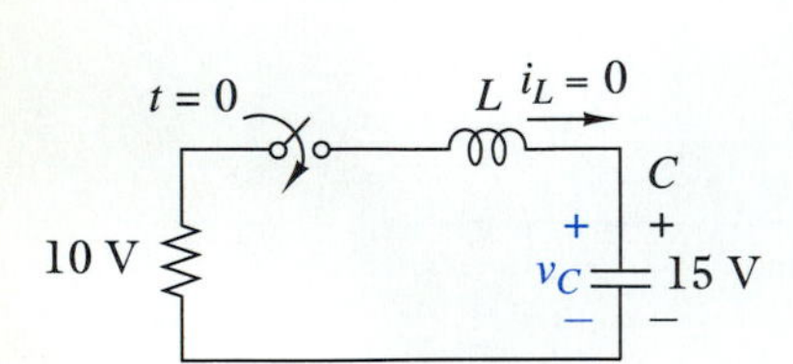

Figure 8-26

SOLUTION:

The roots of the characteristic equation for these three values of resistance were found in Example 8-12. We are now in a position to use those results to find the corresponding zero-input responses.

(a) For $R = 8.5$ kΩ we have Case A. In Example 8-12 the roots were found to be

$$s_1 = -\alpha_1 = -500 \qquad \text{and} \qquad s_2 = -\alpha_2 = -8000$$

Inserting these roots and the initial conditions into Eq. (8-45) yields

$$v_C(t) = 16e^{-500t} - e^{-8000t} \qquad t \geq 0$$

(b) For $R = 4$ kΩ we have Case B. In Example 8-12 the roots were found to be

$$s_1 = s_2 = -\alpha = -2000$$

Inserting these roots and the initial conditions into Eq. (8-46) yields

$$v_C(t) = 15e^{-2000t} + 15(2000t)e^{-2000t} \qquad t \geq 0$$

(c) For $R = 1$ kΩ we have Case C. In Example 8-12 the roots were found to be

$$s_1,\ s_2 = -500 \pm j500\sqrt{15}$$

Inserting these roots and the initial conditions into Eq. (8-48) yields

$$v_C(t) = 15e^{-500t}\cos 500\sqrt{15}t + \frac{15}{\sqrt{15}}e^{-500t}\sin 500\sqrt{15}t \qquad t \geq 0$$

Figure 8-27 shows plots of these responses. All three responses start out at 15 V (the initial condition) and eventually decay to zero. However, the form of the response between the end points changes. The overdamped Case A gradually decays to zero. The critically damped Case B decays more rapidly but does not overshoot the final value. The underdamped Case C passes through zero quickly but then overshoots the mark, eventually decaying to zero in a series of damped oscillations.

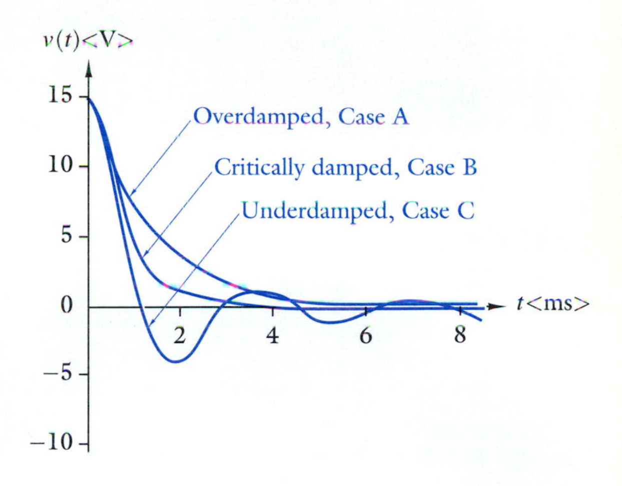

Figure 8-27

EXERCISE 8-14

For a series RLC circuit $R = 5$ kΩ, $C = 0.1$ μF, $L = 100$ mH, $V_O = 0$ V, and $I_O = 25$ mA.

(a) What is the capacitor voltage for $t \geq 0$?
(b) What is the inductor current for $t \geq 0$?
(c) What is the capacitor voltage if R is changed to 100 Ω?

ANSWERS:

(a) $v_C(t) = 5.45(e^{-2087t} - e^{-47913t})$V
(b) $i_L(t) = -1.14e^{-2087t} + 26.14e^{-47913t}$ mA
(c) $v_C(t) = 25.03e^{-500t}\sin 9987t$ V

EXERCISE 8-15

In a series RLC circuit the zero-input responses are:

$$v_C(t) = 10e^{-1000t}\sin 1000t \text{ V}$$

$$i_L(t) = 10e^{-1000t}(\cos 1000t - 1000t) \text{ mA}$$

(a) What is the circuit characteristic equation?
(b) What are the initial values of the state variables?
(c) What are the values of R, L, and C?

ANSWERS:

(a) $s^2 + 2000s + 2 \times 10^6 = 0$
(b) $V_O = 0,\ I_O = 10$ mA
(c) $R = 1$ kΩ, $L = 0.5$ H, $C = 1\mu$F

8-6 THE PARALLEL RLC CIRCUIT

The inductor and capacitor in Fig. 8-28(a) are connected in parallel. The source-resistor circuit can be reduced to the Norton equivalent shown in Fig. 8-28(b). The result is a *parallel RLC circuit* consisting of a current source, resistor, inductor, and capacitor. Our first task is to develop a differential equation for this circuit. We expect to find a second-order differential equation because there are two energy storage elements.

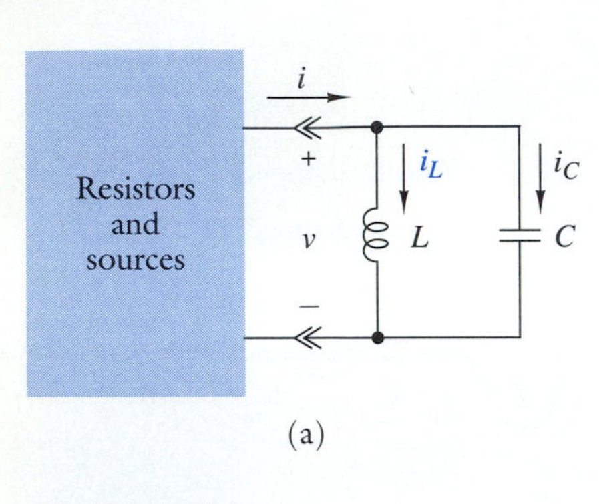

(a)

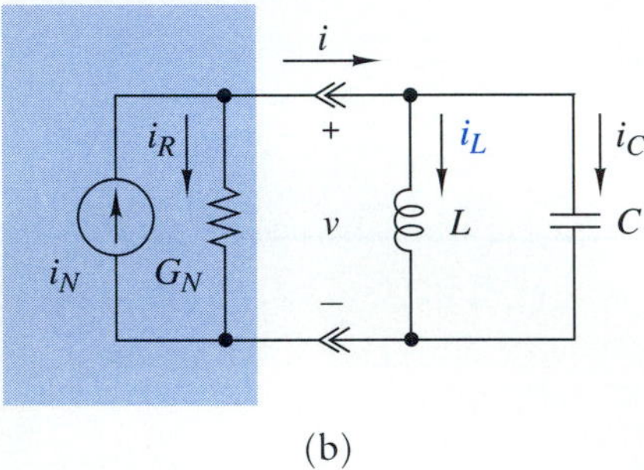

(b)

Figure 8-28 The parallel RLC circuit.

The Norton equivalent to the left of the interface introduces the constraint

$$i + G_N v = i_N \tag{8-49}$$

Writing a KCL equation at the interface yields

$$i = i_L + i_C \tag{8-50}$$

The i-v characteristics of the inductor and capacitor are

$$i_C = C\frac{dv}{dt} \tag{8-51}$$

$$v = L\frac{di_L}{dt} \tag{8-52}$$

Equations (8-49) through (8-52) provide four independent equations in four unknowns (i, v, i_L, i_C). Collectively this set of equations describes the dynamics of the parallel RLC circuit. To solve for the circuit response using classical methods we must derive a circuit equation containing only one of these four variables.

We prefer using state variables because they are continuous. To obtain a single equation in the inductor current we substitute Eqs. (8-50) and (8-52) into Eq. (8-49):

$$i_L + i_C + G_N L\frac{di_L}{dt} = i_N \tag{8-53}$$

The capacitor current can be eliminated from this result by substituting Eq. (8-52) into Eq. (8-51) to obtain

$$i_C = LC\frac{d^2 i_L}{dt^2} \tag{8-54}$$

Inserting this equation into Eq. (8-53) produces

$$\begin{aligned} LC\frac{d^2 i_L}{dt^2} &+ G_N L\frac{di_L}{dt} + i_L = i_N \\ i_C \quad &+ \quad\; i_R \quad\;\; + i_L = i_N \end{aligned} \tag{8-55}$$

This result is a KCL equation in which the resistor and capacitor currents are expressed in terms of the inductor current.

Equation (8-55) is a second-order linear differential equation of the same form as the series RLC circuit equation in Eq. (8-33). In fact, if we interchange the following quantities

$$v_C \leftrightarrow i_L \qquad L \leftrightarrow C \qquad R_T \leftrightarrow G_N \qquad v_T \leftrightarrow i_N$$

we change one equation into the other. The two circuits are duals, which means that the results developed for the series case apply to the parallel circuit with the above duality interchanges.

However, it is still helpful to outline the major features of the analysis of the parallel RLC circuit. The initial conditions in the parallel circuit are the initial inductor current I_O and capacitor voltage V_O. The initial inductor current provides the condition $i_L(0) = I_O$ for the differential equation in Eq. (8-55). By using Eq. (8-52) the initial capacitor voltage specifies the initial rate of change of the inductor current as

$$\frac{di_L(0)}{dt} = \frac{1}{L}v_C(0) = \frac{1}{L}V_O$$

These initial conditions are the dual of those obtained for the series RLC circuit in Eq. (8-34).

To solve for the zero-input response we set $i_N = 0$ in Eq. (8-55) and obtain a homogeneous equation in the inductor current:

$$LC\frac{d^2i_L}{dt^2} + G_N L\frac{di_L}{dt} + i_L = 0$$

A trial solution of the form $i_L = Ke^{st}$ leads to the characteristic equation:

$$LCs^2 + G_N Ls + 1 = 0 \qquad (8\text{-}56)$$

The characteristic equation is quadratic because there are two energy storage elements in the parallel RLC circuit. The characteristic equation has two roots:

$$s_1,\ s_2 = \frac{-G_N L \pm \sqrt{(G_N L)^2 - 4LC}}{2LC}$$

and as in the series case, there are three distinct cases.

Case A (Overdamped): When $(G_N L)^2 > 4LC$, the radicand is positive and there are two unequal real roots $s = -\alpha_1$ and $s_2 = -\alpha_2$. The overdamped zero-input response is

$$i_L(t) = K_1e^{-\alpha_1 t} + K_2e^{-\alpha_2 t} \qquad t \geq 0 \qquad (8\text{-}57)$$

which is the overdamped form.

Case B (Critically Damped): When $(G_N L)^2 = 4LC$, the radicand vanishes leaving two real equal roots $s_1 = s_2 = -\alpha$. The zero-input response is

$$i_L(t) = K_1e^{-\alpha t} + K_2te^{-\alpha t} \qquad t \geq 0 \qquad (8\text{-}58)$$

which is the critically damped form.

Case C (Underdamped): When $(G_N L)^2 < 4LC$, the radicand is negative and there will be a pair of complex, conjugate roots $s_1, s_2 = -\alpha \pm j\beta$. The zero-input response is

$$i_L(t) = K_1 e^{-\alpha t} \cos \beta t + K_2 e^{-\alpha t} \sin \beta t \qquad t \geq 0 \qquad (8\text{-}59)$$

which is the underdamped form.

The analysis results for the series RLC circuit apply to the parallel RLC case with the appropriate duality replacements. In particular, the concept of overdamped, critically damped, and underdamped response applies to both circuits. The form of the responses in Eqs. (8-57), (8-58), and (8-59) have been written with two arbitrary constants K_1 and K_2. The following example shows how to evaluate these constants using the initial conditions for the two state variables.

EXAMPLE 8-14

In a parallel RLC circuit $R_T = 1/G_N = 500\ \Omega$, $C = 1\ \mu\text{F}$, $L = 0.2$ H. The initial conditions are $I_O = 50$ mA and $V_O = 0$. Find the zero-input response of inductor current, resistor current, and capacitor voltage.

SOLUTION:

From Eq. (8-56) the circuit characteristic equation is

$$LCs^2 + G_N Ls + 1 = 2 \times 10^{-7} s^2 + 4 \times 10^{-4} s + 1 = 0$$

The roots of the characteristic equation are

$$s_1,\ s_2 = \frac{-4 \times 10^{-4} \pm \sqrt{16 \times 10^{-8} - 8 \times 10^{-7}}}{4 \times 10^{-7}}$$

$$= -1000 \pm j2000$$

Since roots are complex conjugates, we have the underdamped case. The zero-input response of the inductor current takes the form of Eq. (8-59).

$$i_L(t) = K_1 e^{-1000t} \cos 2000t + K_2 e^{-1000t} \sin 2000t \qquad t \geq 0$$

The constants K_1 and K_2 are evaluated from the initial conditions. At $t = 0$ the inductor current reduces to

$$i_L(0) = I_O = K_1 e^0 \cos 0 + K_2 e^0 \sin 0 = K_1$$

We conclude that $K_1 = I_O = 50$ mA. To find K_2 we use the initial capacitor voltage. In a parallel RLC circuit the capacitor and inductor voltages are equal.

$$L\frac{di_L}{dt} = v_C(t)$$

In this example the initial capacitor voltage is zero so the initial rate of change of inductor current is zero at $t = 0$. Differentiating the zero-input response produces

$$\frac{di_L}{dt} = -2000K_1e^{-1000t}\sin 2000t - 1000K_1e^{-1000t}\cos 2000t$$
$$- 1000K_2e^{-1000t}\sin 2000t + 2000K_2e^{-1000t}\cos 2000t$$

Evaluating this expression at $t = 0$ yields

$$\frac{di_L}{dt}(0) = -2000K_1e^0\sin 0 - 1000K_1e^0\cos 0$$
$$- 1000K_2e^0\sin 0 + 2000K_2e^0\cos 0$$
$$= -1000K_1 + 2000K_2 = 0$$

The derivative initial condition gives condition $K_2 = K_1/2 = 25$ mA. Given the values of K_1 and K_2, the zero-input response of the inductor current is

$$i_L(t) = 50e^{-1000t}\cos 2000t + 25e^{-1000t}\sin 2000t \qquad \text{mA} \quad t \geq 0$$

The zero-input response of the inductor current allows us to solve for every voltage and current in the parallel RLC circuit. For example, the capacitor voltage and resistor current are

$$v_C(t) = v_L = L\frac{di_L}{dt} = -25e^{-1000t}\sin 2000t \qquad \text{V} \quad t \geq 0$$

$$i_R(t) = \frac{v_L}{R_T} = -50e^{-1000t}\sin 2000t \qquad \text{mA} \quad t \geq 0$$

EXAMPLE 8-15

The switch in Fig. 8-29 has been open for a long time and is closed at $t = 0$.

(a) Find the initial conditions at $t = 0$.

(b) Find the inductor current for $t \geq 0$.

(c) Find the capacitor voltage and current through the switch for $t \geq 0$.

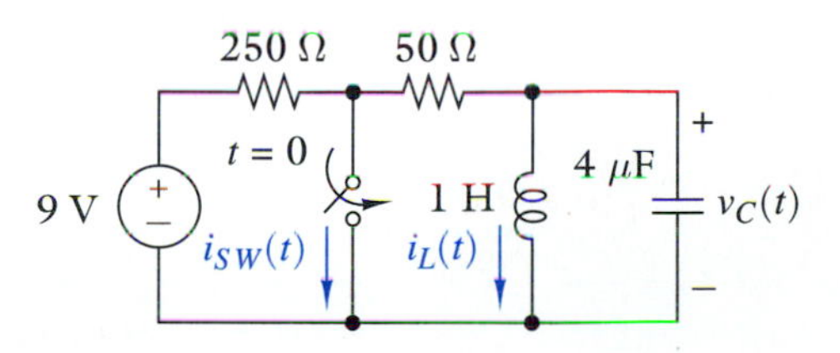

Figure 8-29

SOLUTION:

(a) For $t < 0$ the circuit is in the dc steady state, so the inductor acts like a short circuit and the capacitor like an open circuit. Since the inductor shorts out the capacitor, the initial conditions just prior to closing the switch at $t = 0$ are

$$v_C(0) = 0 \qquad i_L(0) = \frac{9}{250 + 50} = 30 \text{ mA}$$

(b) For $t \geq 0$ the circuit is a zero-input parallel RLC circuit with initial conditions found in (a) above. The circuit characteristic equation is

$$LCs^2 + GLs + 1 = 4 \times 10^{-6}s^2 + 2 \times 10^{-2}s + 1 = 0$$

The roots of this equation are

$$s_1 = -50.51 \qquad \text{and} \qquad s_2 = -4950$$

The circuit is overdamped (Case A), since the roots are real and unequal. The general form of the inductor current zero-input response is

$$i_L(t) = K_1 e^{-50.51t} + K_2 e^{-4950t} \qquad t \geq 0$$

The constants K_1 and K_2 are found using the initial conditions. At $t = 0$ the zero-input response is

$$i_L(0) = K_1 e^0 + K_2 e^0 = K_1 + K_2 = 30 \times 10^{-3}$$

The initial capacitor voltage establishes an initial condition on the derivative of the inductor current, since

$$L\frac{di_L}{dt}(0) = v_C(0) = 0$$

The derivative of the inductor response at $t = 0$ is

$$\frac{di_L}{dt}(0) = (-50.51K_1 e^{-50.51t} - 4950K_2 e^{-4950t})|_{t=0}$$

$$= -50.51K_1 - 4950K_2 = 0$$

The initial conditions on inductor current and capacitor voltage produce two equations in the unknown constants K_1 and K_2:

$$K_1 + K_2 = 30 \times 10^{-3}$$

$$-50.51K_1 - 4950K_2 = 0$$

Solving these equations yields $K_1 = 30.3$ mA and $K_2 = -0.309$ mA. The zero-input response of the inductor current is

$$i_L(t) = 30.3e^{-50.51t} - 0.309e^{-4950t} \text{ mA} \qquad t \geq 0$$

(c) Given the inductor current in (b) above, the capacitor voltage is

$$v_C(t) = L\frac{di_L}{dt} = -1.53e^{-50.51t} + 1.53e^{-4950t} \text{ V} \quad t \geq 0$$

For $t \geq 0$ the current $i_{SW}(t)$ is the current through the 50-Ω resistor **plus** the current through the 250-Ω resistor.

$$i_{SW}(t) = i_{250} + i_{50} = \frac{9}{250} + \frac{v_C(t)}{50}$$
$$= 36 - 30.6e^{-50.51t} + 30.6e^{-4950t} \text{ mA} \qquad t \geq 0$$

EXERCISE 8-16

The zero-input responses of a parallel RLC circuit are observed to be

$$i_L(t) = 10te^{-2000t}$$
$$v_C(t) = 10e^{-2000t} - 20000te^{-2000t} \qquad t \geq 0$$

(a) What is the circuit characteristic equation?
(b) What are the initial values of the state variables?
(c) What are the values of R, L, and C?
(d) Write an expression for the current through the resistor.

ANSWERS:
(a) $s^2 + 4000s + 4 \times 10^6 = 0$.
(b) $i_L(0) = 0$, $v_C(0) = 10$ V.
(c) $L = 1$ H, $C = 0.25\ \mu F$, $R = 1\ k\Omega$.
(d) $i_R(t) = 10e^{-2000t} + 20000te^{-2000t}$ mA, $t \geq 0$.

8-7 SECOND-ORDER CIRCUIT STEP RESPONSE

The step response provides important insights into the response of dynamic circuits in general. So it is natural that we investigate the step response of second-order circuits. In Chapter 11 we will develop general techniques for determining the step response of any linear circuit. However, in this introduction we use classical differential equation methods to find the step response of second-order circuits.

The general second-order linear differential equation with a step function input has the form

$$a_2\frac{d^2y(t)}{dt^2} + a_1\frac{dy(t)}{dt} + a_O y(t) = Au(t) \qquad (8\text{-}60)$$

where $y(t)$ is a voltage or current response, $Au(t)$ is the step function input, and a_2, a_1, and a_O are constant coefficients. The step response is the general solution of this differential equation for $t \geq 0$. The step response can be found by partitioning $y(t)$ into forced and natural components:

$$y(t) = y_N(t) + y_F(t) \qquad (8\text{-}61)$$

The natural response $y_N(t)$ is the general solution of the homogeneous equation (input set to zero), while the forced response $y_F(t)$ is a particular solution of the equation

$$a_2\frac{d^2y_F}{dt^2}+a_1\frac{dy_F}{dt}+a_Oy_F=A \qquad t\geq 0$$

It is readily apparent that $y_F = A/a_O$ is a particular solution of this differential equation, since dA/dt and d^2A/dt^2 are both zero. So much for the forced response.

Turning now to the natural response we seek a general solution of the homogeneous equation. The natural response has the same form as the zero-state response studied in the previous section. In a second-order circuit the zero-state and natural responses take one of the three possible forms: overdamped, critically damped, or underdamped. To describe the three possible forms we introduced two new parameters: ω_O (omega zero) and ζ (zeta). These parameters are defined in terms of the coefficients of the general second-order equation in Eq. (8-60):

$$\omega_O^2=\frac{a_O}{a_2} \qquad \text{and} \qquad 2\zeta\omega_O=\frac{a_1}{a_2} \tag{8-62}$$

The parameter ω_O is called the ***undamped natural frequency***, or, simply, the natural frequency, and ζ is called the ***damping ratio***. Using these two parameters, the general homogeneous equation is written in the form

$$\frac{d^2y_N}{dt^2}+2\zeta\omega_O\frac{dy_N}{dt}+\omega_O^2y_N=0 \tag{8-63}$$

The left side of Eq. (8-63) is called the ***standard form*** of the second-order linear differential equation. When a second-order equation is arranged in this format we can determine its damping ratio and undamped natural frequency by equating its coefficients with those in the standard form. For example, in standard form the homogeneous equation for the series RLC circuit in Eq. (8-37) is

$$\frac{d^2v_C}{dt^2}+\frac{R_T}{L}\frac{dv_C}{dt}+\frac{1}{LC}v_C=0$$

Equating like terms yields

$$\omega_O^2=\frac{1}{LC} \qquad \text{and} \qquad 2\zeta\omega_O=\frac{R_T}{L}$$

for the series RLC circuit. Note that the circuit elements determine the values of the parameters ω_O and ζ.

To determine the form of the natural response using ω_O and ζ, we insert a trial solution $y_N(t) = Ke^{st}$ into the standard form in Eq. (8-63). The trial function Ke^{st} is a solution provided

$$Ke^{st}\left[s^2+2\zeta\omega_Os+\omega_O^2\right]=0$$

Since $K = 0$ is the trivial solution and $e^{st} \neq 0$ for all $t \geq 0$, the only useful way for the right side of this equation to be zero for all t is for the quadratic expression within the brackets to vanish. The quadratic expression is the characteristic equation for the general second-order differential equation:

$$s^2 + 2\zeta\omega_O s + \omega_O^2 = 0$$

The roots of the characteristic equation are

$$s_1,\ s_2 = \omega_O(-\zeta \pm \sqrt{\zeta^2 - 1})$$

We begin to see the advantage of using the parameters ω_O and ζ. The constant ω_O is like a scale factor that designates the "size" of the roots. The radicand defining the form of the roots depends only on the damping ratio ζ. As a result we can express the three possible forms of the natural response in terms of the damping ratio.

Case A: For $\zeta > 1$ the radicand is positive and there are two unequal, real roots

$$s_1,\ s_2 = -\alpha_1,\ -\alpha_2 = \omega_O(-\zeta \pm \sqrt{\zeta^2 - 1})$$

and the natural response is of the form

$$y_N(t) = K_1 e^{-\alpha_1 t} + K_2 e^{-\alpha_2 t} \qquad t \geq 0 \qquad (8\text{-}64)$$

Case B: For $\zeta = 1$ the radicand vanishes and there are two real, equal roots

$$s_1 = s_2 = -\alpha = -\zeta\omega_O$$

and the natural response is of the form

$$y_N(t) = K_1 e^{-\alpha t} + K_2 t e^{-\alpha t} \qquad t \geq 0 \qquad (8\text{-}65)$$

Case C: For $\zeta < 1$, the radicand is negative leading to two complex, conjugate roots $s_1,\ s_2 = -\alpha \pm j\beta$, where

$$\alpha = \zeta\omega_O \qquad \text{and} \qquad \beta = \omega_O\sqrt{1 - \zeta^2}$$

and the natural response is of the form

$$y_N(t) = K_1 e^{-\alpha t} \cos\beta t + K_2 e^{-\alpha t} \sin\beta t \qquad t \geq 0 \qquad (8\text{-}66)$$

In other words, for $\zeta > 1$ the natural response is overdamped, for $\zeta = 1$ the natural response is critically damped, and for $\zeta < 1$ the response is underdamped.

Combining the forced and natural responses yields the step response of the general second-order differential equation in the form

$$y(t) = y_N(t) + A/\alpha_O \qquad t \geq 0 \qquad (8\text{-}67)$$

The factor A/a_O is the forced response. The natural response $y_N(t)$ takes one of the forms in Eq. (8-64) through (8-66) depending on the value of the damping ratio. The constants K_1 and K_2 in natural response can be evaluated from the initial conditions.

In summary, the step response of a second-order circuit is determined by

1. The amplitude of the step function input $Au(t)$
2. The damping ratio ζ and natural frequency ω_O
3. The initial conditions $y(0)$ and $dy/dt(0)$

In this regard the damping ratio and natural frequency play the same role for second-order circuits that the time constant plays for first-order circuits. That is, these circuit parameters determine the basic form of the natural response just as time constant defines the form of the natural response in a first-order circuit. It is not surprising that a second-order circuit takes two parameters, since it contains two energy storage elements.

EXAMPLE 8-16

The series RLC circuit in Fig. 8-30 is driven by a step function and is in the zero-state at $t = 0$. Find the capacitor voltage for $t \geq 0$.

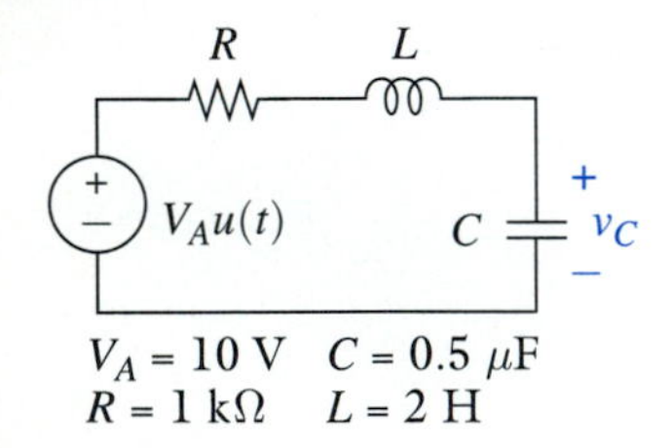

Figure 8-30

SOLUTION:

The differential equation for the capacitor voltage is

$$10^{-6}\frac{d^2v_C}{dt^2} + 0.5 \times 10^{-3}\frac{dv_C}{dt} + v_C = 10 \qquad t \geq 0$$

By inspection the forced response is $v_{CF} = 10$ V. In standard format the homogeneous equation is

$$\frac{d^2v_{CN}}{dt^2} + 500\frac{dv_{CN}}{dt} + 10^6 v_{CN} = 0 \qquad t \geq 0$$

Comparing this format the standard form in Eq. (8-63) yields

$$\omega_O^2 = 10^6 \qquad \text{and} \qquad 2\zeta\omega_O = 500$$

So that $\omega_O = 1000$ and $\zeta = 0.25$. Since the $\zeta < 1$ the natural response is underdamped (Case C) and has the form

$$\alpha = \zeta\omega_O = 250$$

$$\beta = \omega_O\sqrt{1 - \zeta^2} = 968$$

$$v_{CN}(t) = K_1 e^{-250t}\cos 968t + K_2 e^{-250t}\sin 968t$$

The general solution of the circuit differential equation is the sum of the forced and natural responses:

$$v_C(t) = 10 + K_1 e^{-250t}\cos 968t + K_2 e^{-250t}\sin 968t \qquad t \geq 0$$

The constants K_1 and K_2 are determined by the initial conditions. The circuit is in the zero state at $t = 0$, so the initial conditions are $v_C(0) = 0$ and $i_L(0) = 0$. Applying the initial condition constraints to the general solution yields two equations in the constants K_1 and K_2:

$$v_C(0) = 10 + K_1 = 0$$

$$\frac{dv_C}{dt}(0) = -250K_1 + 968K_2 = 0$$

These equations yield $K_1 = -10$ and $K_2 = -2.58$. The step response of the capacitor voltage step response is

$$v_C(t) = 10 - 10e^{-250t}\cos 968t - 2.58e^{-250t}\sin 968t \qquad t \geq 0$$

A plot of $v_C(t)$ versus time is shown in Fig. 8-31. The waveform and its first derivative at $t = 0$ satisfy the initial conditions. The natural response decays to zero, so the forced response determines the final value of $v_C(\infty) = 10$ V. Beginning at $t = 0$ the response climbs rapidly but overshoots the mark several times before eventually settling down to the final value. The damped sinusoidal behavior results from the fact that $\zeta < 1$, producing an underdamped natural response.

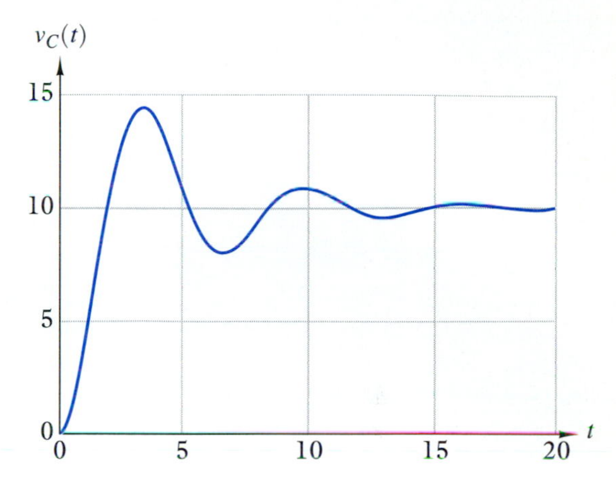

Figure 8-31

EXERCISE 8-17

Find the zero-state solution of the following differential equations.

(a) $10^{-4}\dfrac{d^2v}{dt^2} + 2\times 10^{-2}\dfrac{dv}{dt} + v = 100u(t)$

(b) $\dfrac{d^2i}{dt^2} + 2500\dfrac{di}{dt} + 4\times 10^6 i = 10^5 u(t)$

ANSWERS:

(a) $v(t) = 100 - 100e^{-100t} - 10^4 te^{-100t}$ V $\quad t \geq 0$

(b) $i(t) = \dfrac{1}{40} - \dfrac{1}{40}e^{-1250t}\cos 1561t - \dfrac{1}{50}e^{-1250t}\sin 1561t A$
$t \geq 0$

EXERCISE 8-18

The step response of a series RLC circuit is observed to be

$$v_C(t) = 15 - 15e^{-1000t}\cos 100t \text{ V} \qquad t \geq 0$$

$$i_L(t) = 45e^{-1000t}\cos 1000t + 45e^{-1000t}\sin 1000t \text{ mA} \qquad t \geq 0$$

(a) What is the circuit characteristic equation?
(b) What are the initial values of the state variables?

(c) What is the amplitude of the step input?
(d) What are the values of R, L, and C?
(e) What is the voltage across the resistor?

ANSWERS:
(a) $s^2 + 2000s + 2 \times 10^6 = 0$
(b) $v_C(0) = 0$, $i_L(0) = 45$ mA
(c) $V_A = 15$ V
(d) $R = 333\ \Omega$, $L = 167$ mH, $C = 3\ \mu$F
(e) $v_R(t) = 15e^{-1000t} \cos 1000t + 15e^{-1000t} \sin 1000t$ V ■

Some Other Second-Order Circuits

Up to this point the series and parallel RLC circuits are the only second-order circuits we have considered. Using the classical differential equation methods, we found that the form of the natural (zero-input) response can be overdamped, critically damped, or underdamped depending on the roots of the characteristic equation. We now show that these concepts apply to other second-order circuits besides the series and parallel RLC.

The OP AMP circuit in Fig. 8-32 contains three equal resistors and two unequal capacitors. The circuit is second-order because the two capacitors are not in series or parallel and cannot be replaced by a single equivalent capacitor. We use the basic concept of device and connection equations to formulate the circuit differential equation.

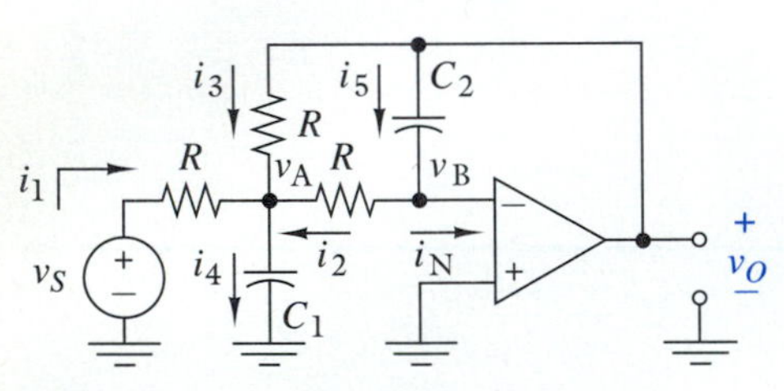

Figure 8-32 A second-order OP AMP RC circuit.

From our experience with OP AMP resistance circuit, we know that two node equations describe an OP AMP circuit of this type. We formulate these equations by writing KCL connection constraints at Nodes A and B:

$$\text{Node A: } i_1 + i_2 + i_3 - i_4 = 0$$

$$\text{Node B: } -i_2 + i_5 - i_N = 0$$

The resistor and capacitor element constraints in terms of the node voltages are

$$i_1 = G(v_S - v_A) \qquad i_4 = C_1 \frac{dv_A}{dt}$$

$$i_2 = G(v_B - v_A) \qquad i_5 = C_2 \frac{d(v_O - v_B)}{dt}$$

$$i_3 = G(v_O - v_A)$$

The ideal OP AMP model requires $i_N = 0$. The model also requires $v_B = 0$ because the noninverting input is connected to ground. Substituting all of these element equations into the KCL connection equations at Nodes A and B results in

$$-3Gv_A - C_1\frac{dv_A}{dt} + Gv_O + Gv_S = 0$$

$$Gv_A + C_2\frac{dv_O}{dt} = 0$$

Solving the second of these equations for v_A yields

$$v_A = -RC_2\frac{dv_O}{dt}$$

Inserting this result into the first equation produces the circuit differential equation:

$$R^2C_1C_2\frac{d^2v_O}{dt^2} + 3RC_2\frac{dv_O}{dt} + v_O = -v_S$$

This expression is a second-order linear differential equation in which the OP AMP output voltage is the dependent variable.

In standard format the homogeneous equation for this circuit is

$$\frac{d^2v_O}{dt^2} + \frac{3}{RC_1}\frac{dv_O}{dt} + \frac{1}{R^2C_1C_2}v_O = 0$$

Comparing this expression with the standard form yields

$$\omega_O^2 = \frac{1}{R^2C_1C_2} \quad \text{and} \quad 2\zeta\omega_O = \frac{3}{RC_1}$$

or

$$\omega_O = \frac{1}{\sqrt{RC_1RC_2}} \quad \text{and} \quad \zeta = \frac{3}{2}\sqrt{\frac{C_2}{C_1}}$$

The natural frequency ω_O of the circuit is determined by RC products and the damping ratio ζ is determined by C_2 and C_1. In particular, we see that the natural response has three forms.

Case A: If $9C_2 > 4C_1$ then $\zeta > 1$ and the natural reponse is overdamped.

Case B: If $9C_2 = 4C_1$ then $\zeta = 1$ and the natural reponse is critically damped.

Case C: If $9C_2 < 4C_1$ then $\zeta < 1$ and the natural reponse is underdamped.

The RC OP AMP circuit in Fig. 8-32 produces the same response forms as the series and parallel RLC circuits without using an inductor. The next example shows that active RC circuits can produce responses that cannot be produced by these RLC circuits.

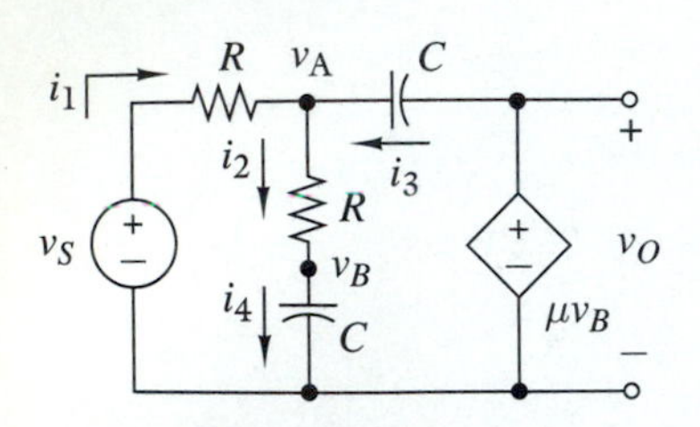

Figure 8-33

EXAMPLE 8-17

Find the damping ratio and natural frequency of the second-order circuit in Fig. 8-33.

SOLUTION:

The circuit differential equation is obtained from the connection and element constraints. Writing KCL equations at Nodes A and B we obtain the following connection constraints.

$$\text{Node A: } i_1 - i_2 + i_3 = 0$$

$$\text{Node B: } i_2 - i_4 = 0$$

The element constraints are

$$i_1 = G(v_S - v_A) \qquad i_3 = C\frac{d(v_O - v_A)}{dt}$$

$$i_2 = G(v_A - v_B) \qquad i_4 = C\frac{dv_B}{dt}$$

$$v_O = \mu v_B$$

Substituting the element equations into the connection equations yields

$$\text{Node A: } Gv_S - 2Gv_A + Gv_B + \mu C\frac{dv_B}{dt} - C\frac{dv_A}{ct} = 0$$

$$\text{Node B: } \qquad Gv_A - Gv_B - C\frac{dv_B}{dt} = 0$$

We have two coupled first-order differential equations in node voltages v_A and v_B. To derive a single second-order equation we need to eliminate one of these node voltages. Solving the Node B equation for v_A produces

$$v_A = v_B + RC\frac{dv_B}{dt}$$

Substituting this result into the Node A equation yields the circuit differential equation:

$$\frac{d^2v_B}{dt^2} + \frac{3-\mu}{RC}\frac{dv_B}{dt} + \frac{1}{(RC)^2}v_B = \frac{1}{(RC)^2}v_S$$

Comparing this equation with the standard form of the second-order differential equation in Eq. (8-63) identities the damping ratio and undamped natural frequency of the circuit:

$$\omega_O = \frac{1}{RC} \qquad \zeta = \frac{3-\mu}{2}$$

The natural frequency is determined by the RC product and the damping ratio by the dependent source gain μ. There are the usual three responses.

Case A: If $\mu < 1$, then $\zeta > 1$ and the circuit is overdamped.

Case B: If $\mu = 1$, then $\zeta = 1$ and the circuit is critically damped.

Case C: If $3 > \mu > 1$, then $0 < \zeta < 1$ and the circuit is underdamped.

In addition, this active circuit has two forms of responses that are not possible in passive RLC circuits. If $\mu = 3$ then $\zeta = 0$ and there is no damping. The natural response has the form

$$v_{ON}(t) = K_1 \cos \omega_O t + K_2 \sin \omega_O t \qquad t \geq 0$$

No damping corresponds to a sustained oscillation where the energy to overcome the losses in the resistors comes from the dependent source. Note that when $\zeta = 0$, the frequency of the undamped sinusoid is ω_O, hence the name undamped natural frequency.

Even more dramatic is the case $\mu > 3$ and $\zeta < 0$. When $\zeta < 0$ there is negative damping, which means that the exponential in the natural response

$$v_{ON}(t) = K_1 e^{-\zeta\omega_O t} \cos \beta t + K_2 e^{-\zeta\omega_O t} \sin \beta t \qquad t \geq 0$$

grow without bound. A circuit whose natural response increases without bound is said to be ***unstable***. We will treat circuit stability again in Chapter 10.

EXERCISE 8-19

Given the following second-order circuit zero-state step responses:

(1) $v_C(t) = (10 - 15e^{-1000t} + 5e^{-3000t})u(t)$ V

(2) $v_C(t) = (10 - 10e^{-1000t} \cos 3000t - 3.33e^{-1000t} \sin 3000t)u(t)$ V

- **(a)** What are the characteristic equations?
- **(b)** What are the damping ratios and undamped natural frequencies?
- **(c)** Design series RLC circuits that realize these responses.

ANSWERS:

For the response in (1):

(a) $s^2 + 4000s + 3 \times 10^6 = 0$

(b) $\omega_O = 1732$ rad/s, $\zeta = 1.55$

(c) $C = 1/3\ \mu$F, $L = 1$ H, $R = 4$ kΩ

For the response in (2):

(a) $s^2 + 2000s + 10^7 = 0$

(b) $\omega_O = 3000$, $\zeta = 0.333$

(c) $C = 100$ nF, $L = 1$ H, $R = 2$ kΩ.

DISCUSSION:
The answers in part (c) are not unique. The standard form of the characteristic equation for a second-order series RLC circuit is $LCs^2 + RCs + 1$. For the response in (1) the results from (a) require $LC = 10^{-6}/3$ and $RC = 4 \times 10^{-3}/3$. Selecting $C = 1/3\ \mu$F yields $L = 1$ H and $R = 4$ kΩ. For the response in (2) part (a) requires $LC = 10^{-7}$ and $RC = 2 \times 10^{-4}$. Selecting $C = 100$ nF yields $L = 1$ H and $R = 2$ kΩ. ■

8-8 TRANSIENT ANALYSIS USING SPICE

Transient analysis is an important application of computer-aided circuit analysis. By transient analysis we mean determining the time-varying waveforms representing responses in circuits containing capacitance and inductance. Analysis processors like SPICE and MICRO-CAP are capable of numerically estimating these responses. The SPICE statements required for transient analysis were introduced in previous chapters. Before presenting examples we need to discuss some features of computer-aided analysis.

Programs like SPICE numerically solve for the node voltages and voltage source currents at a series of discrete times t_K, $K = 1, 2, \ldots N$. The time interval between solution times is called the *internal step size* The step size is not fixed because the analysis processor continually adjusts the step size to achieve accuracy goals with minimum computational effort. When circuit variables are changing slowly, the step size expands to reduce the computational burden. When things change rapidly, the step size shrinks to ensure accuracy.

For the most part the internal step size adjustment takes place under the hood and is beyond the direct control of the beginning user. The numerically generated responses are approximations resulting from a compromise between using small step sizes to increase accuracy and large step sizes to reduce the computational burden. In most cases the computer program's compromises are quite good and the accuracy is more than adequate for engineering purposes.

At the same time the analysis processor writes responses to the output file at discrete times t_J, $J = 1, 2, \ldots M$. These output times are separated by **a constant interval of time**. The user has direct control over the fixed time interval between the successive points and the time range included in the output file. It is up to the user to select these time parameters so that the data in the output file adequately describe the transient response. This is not a question of data accuracy, but rather a question of whether the user-specified sampling of the calculated responses presents an accurate picture of the responses.

The user must have some knowledge of what the response will look like in order to use computer-aided analysis effectively. The user cannot have a complete knowledge because there would be no reasons to use a computer. On the other hand, the user cannot be totally uninformed either, because there would be no way to specify the response sampling for the output file. In other words, computer-aided analysis tools do not eliminate the need for a basic understanding of the response of dynamic circuits.

In Chapter 6 we introduced the SPICE command statement for transient analysis. In its basic form this command is

```
.TRAN TSTEP TSTOP [TSTART] [TMAX] [UIC]
```

This .TRAN statement causes SPICE to calculate the circuit transient response beginning at $t = 0$ and ending at t =TSTOP. SPICE writes the responses designated in a .PRINT statement to the output file every TSTEP seconds starting at t = TSTART and ending at t = TSTOP. The parameter TMAX specifies a maximum internal step size for the computation. The UIC instruction tells SPICE to use the initial conditions specified in the capacitor and inductor element statement.

The TSTEP and TSTOP parameters are mandatory and the others are optional. If TSTART and TMAX are left blank, the default values are TSTART = 0 and TMAX = TSTOP/50. If the UIC instruction is not included, SPICE uses dc analysis to calculate the initial capacitor voltages and inductor currents prior to beginning the transient analysis.

For example, the statements

```
.TRAN 0.1U 10U UIC
.PRINT TRAN V(1) V(7) I(VAMP)
```

direct SPICE to perform a transient analysis using the initial condition specified in the capacitor and inductor element statements. The analysis begins at $t = 0$ (always) and ends at t = TSTOP = 10 μs. The responses V(1), V(7), and I(VAMP) listed in the .PRINT command will be written to the output files every TSTEP = 0.1 μs. No value is assigned to TSTART so by default the output responses are written beginning at $t = 0$. Likewise, no value is given to TMAX, so by default the maximum computational step size is TSTOP/50 = 0.2 μs.

To use SPICE for transient analysis we must decide how long to run the analysis (TSTOP) and how often to write (TSTEP) the output variables into the output file. One way to specify TSTOP is to estimate the duration of the natural response. A useful rule of thumb is to let TSTOP be about five time constants. A rule of thumb for TSTEP is to let TSTEP = TSTOP/100. This rule generates 101 samples (counting $t = 0$) in the output file and is often adequate for graphical purposes. Although TSTEP/100 is a good starting place, TSTEP may need to be modified in subsequent runs if the calculated responses include short rise-time pulses or lightly damped oscillations.

Initial conditions can cause problems when using SPICE to generate transient data. First of all, the user can determine the initial conditions outside of SPICE. The initial conditions are inserted in the capacitor or inductor element statements as shown in the following examples:

```
C17 3 8 3U IC = 12
LPAR 2 5 2M IC = -3M
```

The first statement defines a 3-μF capacitor C17 connected between Nodes 3 and 8 with an initial voltage of +12 V. The second specifies a 2-mH inductor LPAR connected between Nodes 2 and 5 with an initial current of −3 mA. In SPICE the reference directions for the initial conditions are determined by the order in which the nodes are listed in the element statement and the passive sign convention.

When initial conditions are specified in the element statements the UIC instruction must be included in the .TRAN command, even if the initial conditions are zero. The reason is that SPICE always performs a dc analysis before the transient analysis run. When the UIC instruction is included, the dc analysis is performed with all capacitors replaced by voltage sources equal to their initial conditions and all inductors replaced by current sources equal to their initial condition. This guarantees that the dc analysis does not alter the initial conditions prior to the start of the transient analysis.

We will illustrate SPICE transient analysis using example problems we solved earlier in this chapter using analytical methods and then close with a design example.

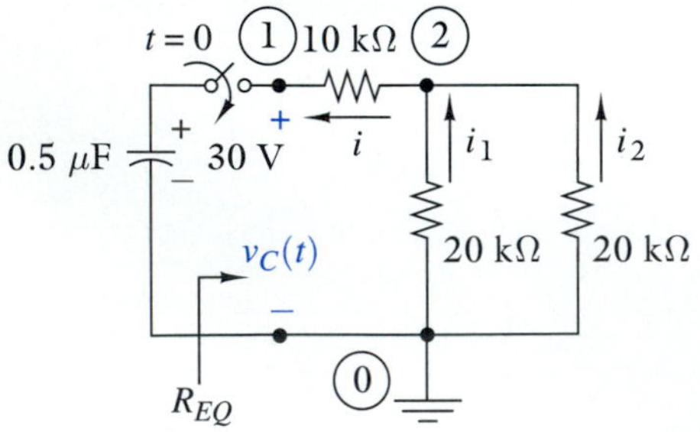

Figure 8-34

EXAMPLE 8-18

Use SPICE to calculate the capacitor voltage in Fig. 8-34 for $t \geq 0$.

SOLUTION:

This problem was solved analytically in Example 8-1. In that example we found $R_{EQ} = 20$ kΩ and the circuit time constant is $T_C = 10$ ms. We use TSTEP $= 5T_C = 50$ ms and TSTEP = TSTOP/100 = 0.5 ms. An appropriate SPICE circuit file for

$t > 0$ (switch closed) is

```
EXAMPLE 8-18
C 1 0 0.5U IC=30
R1 1 2 10K
R2 2 0 20K
R3 2 0 20K
.TRAN 0.5M 50M UIC
.PRINT TRAN V(1)
.END
```

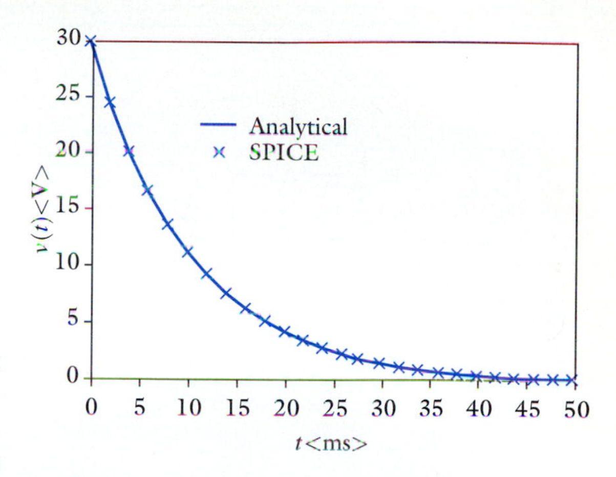

Figure 8-35

The 30-V initial condition is included in the capacitor element statement. The UIC instruction in the .TRAN command ensures that SPICE uses this value at the start of the transient run. Figure 8-35 compares the numerically generated SPICE response with the analytically generated response from Example 8-1. The "X's" in this figure indicate the discrete values of V(1) written in the output file by SPICE. The solid curve is the anaytical solution.

EXAMPLE 8-19

Use SPICE to calculate the zero-state response of the capacitor voltage in Fig. 8-36 for $t \geq 0$.

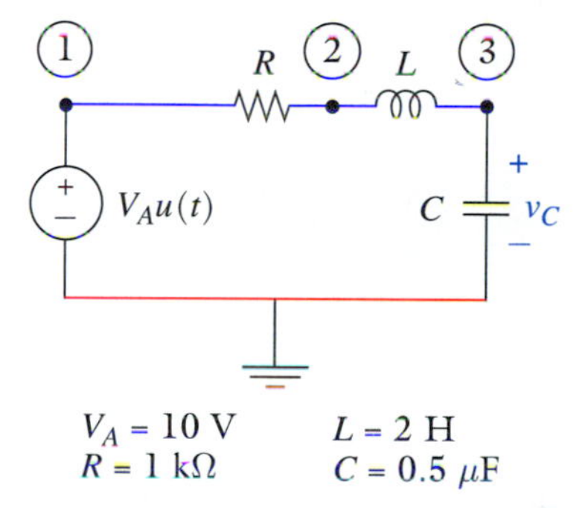

Figure 8-36

SOLUTION:

This problem was solved analytically in Example 8-16. In that example we found $\omega_O = 10^3$ and $\zeta = 0.25$. Since $\zeta < 1$ the response is underdamped and the circuit "time constant" is $1/\zeta\omega_O = 4$ ms. We use TSTOP $= 5T_C = 20$ ms and TSTEP $=$ TSTOP$/100 = 0.2$ ms. An appropriate SPICE circuit file is

```
EXAMPLE 8-19
VS 1 0 DC 10
R 1 2 1K
L 2 3 2 IC=0
C 3 0 0.5U IC=0
.TRAN 0.2M 20M UIC
.PRINT TRAN V(3)
.END
```

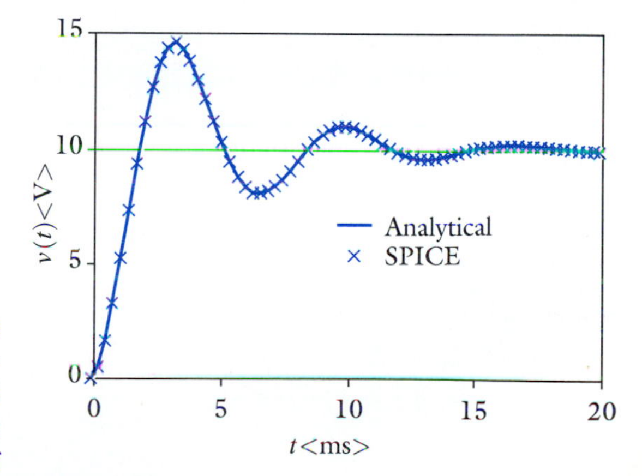

Figure 8-37

The initial condition in the capacitor and inductor statements are zero and the UIC instruction is included in the .TRAN command. Figure 8-37 compares the numerically generated SPICE response with the analytically generated response from Example 8-16. The "X's" in this figure indicate the discrete values of V(3) written in the output file by SPICE. The solid curve is the analytical solution.

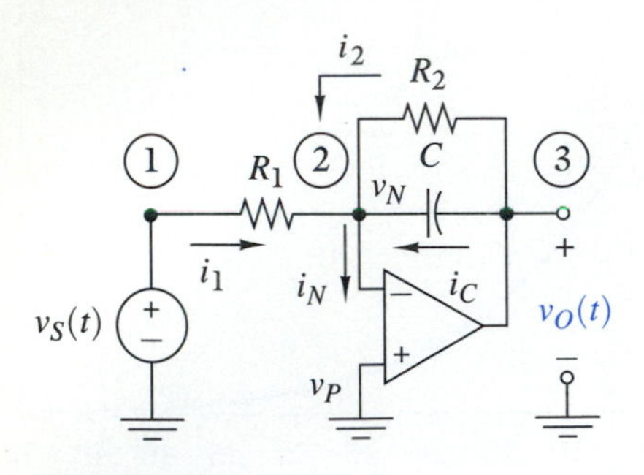

Figure 8-38

EXAMPLE 8-20

Select the element values in Fig. 8-38 so that the circuit time constant is 100 μs and the dc gain is -10. Use SPICE to verify the design for a 1-V step function input.

SOLUTION:

In Section 8-1 we found the time constant of this inverting amplifier to be $T_C = R_2C$. To determine the dc gain we replace the capacitor by an open circuit and find that the dc gain is $-R_2/R_1$. There are three circuit elements and two design constraints:

$$R_2C = 10^{-4} \qquad \text{and} \qquad R_2 = 10R_1$$

Selecting $R_1 = 10$ kΩ requires $R_2 = 100$ kΩ and $C = 1$ nF. To verify the design we create a SPICE circuit file with TSTOP $= 5T_C = 500$ μs and TSTEP = TSTOP/100 = 5 μs.

```
EXAMPLE 8-20
VS 1 0 DC 1
R1 1 2 10K
R2 2 3 100K
C 2 3 1N IC=0
EOPAMP 3 0 0 2 1MEG
.TRAN 5U 500U UIC
.PRINT V(3)
.END
```

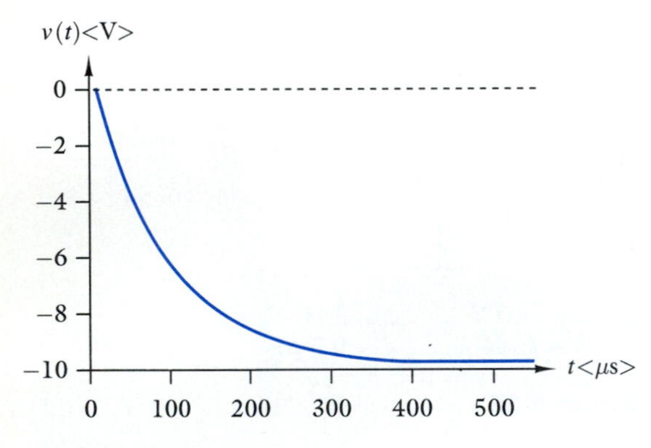

Figure 8-39

The voltage-controlled source EOPAMP represents the OP AMP. Figure 8-39 shows the SPICE-generated response. For a +1-V input, the output goes negative approaching a final value of -10 V, indicating a dc gain of -10 as required. At $t = 100$ μs the output is about $-6.3 \approx -10(1 - e^{-1})$ V, which indicates that the time constant is correct.

SUMMARY

- Circuits containing linear resistors and the equivalent of one capacitor or one inductor are described by first-order differential equations in which the unknown is the circuit state variable.
- The zero-input response in a first-order circuit is an exponential whose time constant depends on circuit parameters. The amplitude of the exponential is equal to the initial value of the state variable.
- The natural response is the general solution of the homogeneous differential equation obtained by setting the input

to zero. The forced response is a particular solution of the differential equation for the given input. For linear circuits the total response is the sum of the forced and natural responses.

- For linear circuits the total response is the sum of the zero-input and zero-state responses. The zero-input response is caused by the initial energy stored in capacitors or inductors. The zero-state response results from the input driving forces.

- The initial and final values of the step response of a first and second-order circuit can be found by replacing capacitors by open circuits and inductors by short circuits and then using resistance circuit analysis methods.

- For a sinusoidal input the forced response is called the sinusoidal steady-state response or the ac response. The ac response is a sinusoid with the same frequency as the input but with a different amplitude and phase angle. The ac response can be found from the circuit differential equation using the method of undetermined coefficients.

- Circuits containing linear resistors and the equivalent of two energy storage elements are described by second-order differential equations in which the dependent variable is one of the state variables. The initial conditions are the values of the two state variables at $t = 0$.

- The zero-input response of a second-order circuit takes different forms depending on the roots of the characteristic equation. Unequal real roots produce the overdamped response, equal real roots produce the critically damped response, and complex conjugate roots produce underdamped responses.

- The circuit damping ratio ζ and undamped natural frequency ω_O determine the form of the zero-input and natural responses of any second-order circuit. The response is overdamped if $\zeta > 1$, critically damped if $\zeta = 1$, and underdamped if $\zeta < 1$. Active circuits can produce undamped ($\zeta = 0$) and unstable ($\zeta < 0$) responses.

- Computer-aided circuit analysis programs such as SPICE or MICRO-CAP can generate numerical approximations to circuit transient responses. Some knowledge of analytical methods and an estimate of the general form of the expected response are necessary to meaningfully use these analysis tools.

EN ROUTE OBJECTIVES AND ASSOCIATED PROBLEMS

ERO 8-1 First-Order Circuits (Sects. 8-1, 8-2, 8-3, 8-4)

Given a first-order RC or RL circuit:

(a) Determine the circuit differential equation, characteristic equation, time constant, and initial conditions (if not given).

(b) Find the zero-input response or the natural and the forced components of the response for step function and sinusoidal inputs.

Conversely, given the response of a first-order circuit:

(a) Determine the circuit characteristic equation, time constant, and initial conditions.

(b) Select values of circuit parameters to produce the given response.

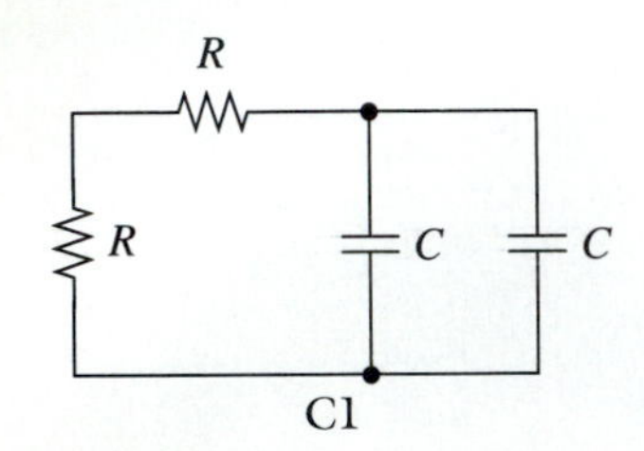

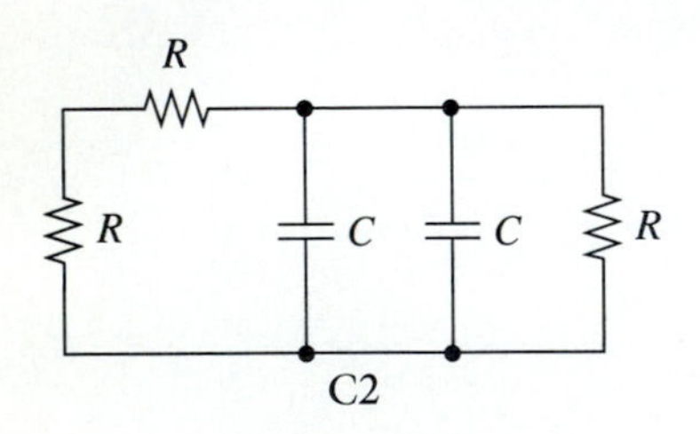

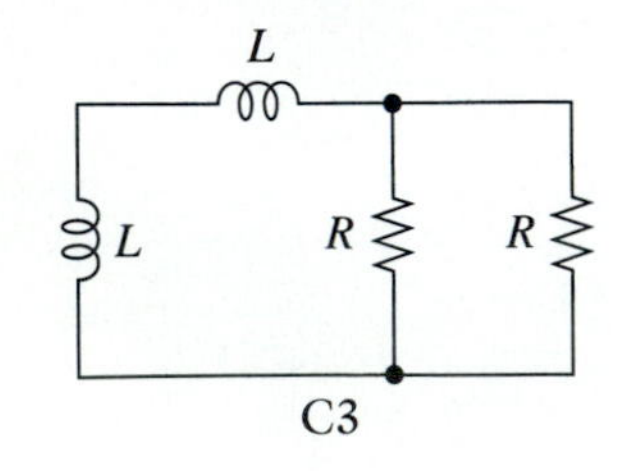

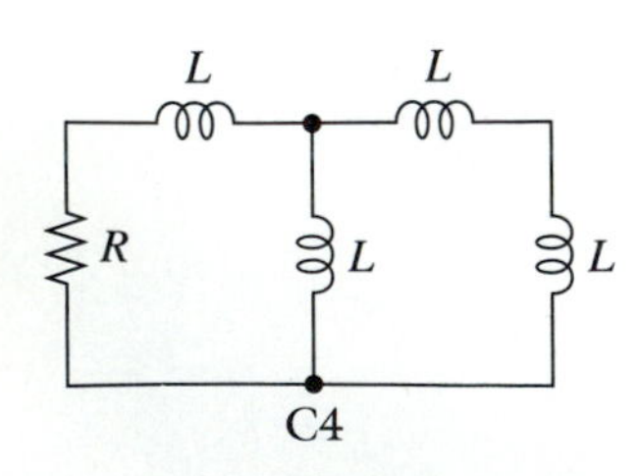

Figure P8-3

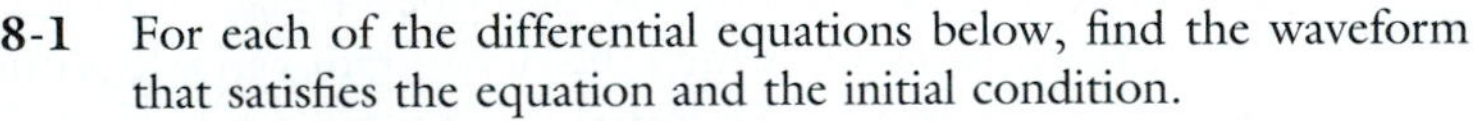

8-1 For each of the differential equations below, find the waveform that satisfies the equation and the initial condition.

(a) $\dfrac{dv(t)}{dt} + 10v(t) = 0,\ v(0) = 5\ \text{V}.$

(b) $10^{-4}\dfrac{di(t)}{dt} + i(t) = 10^{-2}u(t),\ i(0) = -10\ \text{mA}$

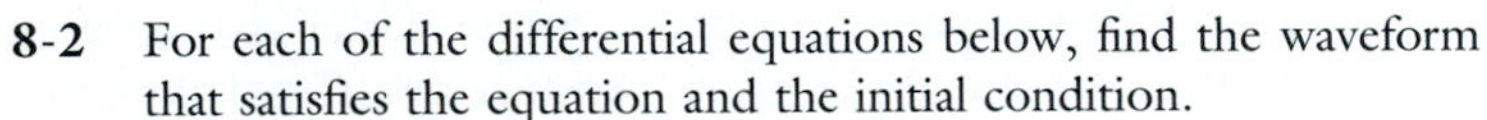

8-2 For each of the differential equations below, find the waveform that satisfies the equation and the initial condition.

(a) $\dfrac{dv(t)}{dt} + 5v(t) = 10(\cos 5t)u(t),\ v(0) = 0$

(b) $10^{-4}\dfrac{di(t)}{dt} + i(t) = 10(\sin 100t)u(t),\ i(0) = 3\ \text{A}$

8-3 For each of the circuits in Fig. P8-3 write an expression for the circuit time constant in terms of R and C or R and L.

8-4 Determine the time constant of each of the circuits in Fig. P8-4.

8-5 The switch in each circuit in Fig. P8-5 has been open for a "long time" and is closed at $t = 0$.

(a) Determine the value of the state variable at $t = 0$.

(b) Determine the value of the state variable as $t \to \infty$.

(c) Determine the circuit time constant.

(d) Write an expression for the state variable for $t \geq 0$. Identify the forced and natural components, and sketch their waveforms.

8-6 Repeat Prob. 8-5 if the switch in each circuit has been closed for a "long time" and then is opened at $t = 0$.

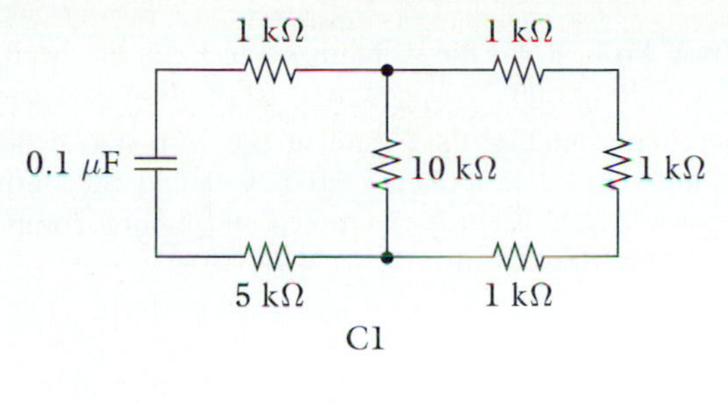

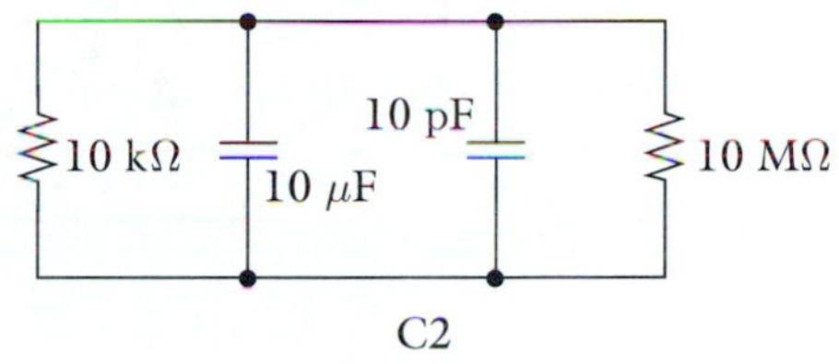

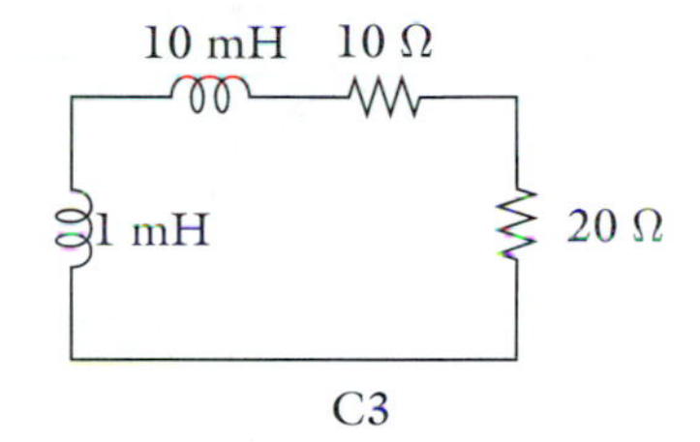

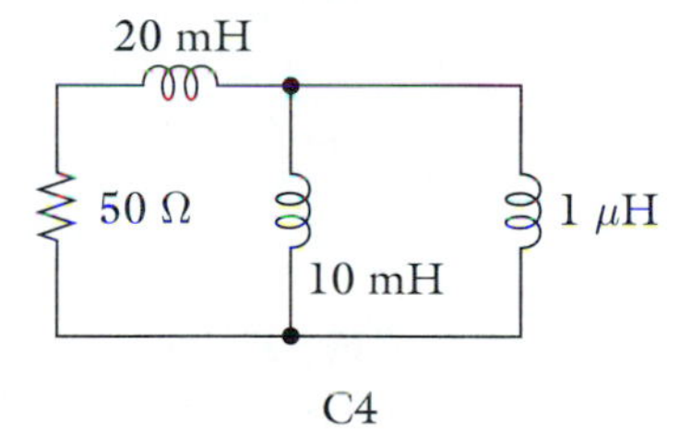

Figure P8-4

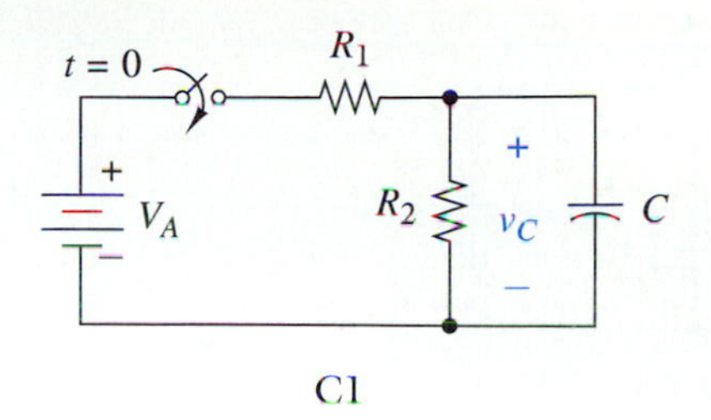

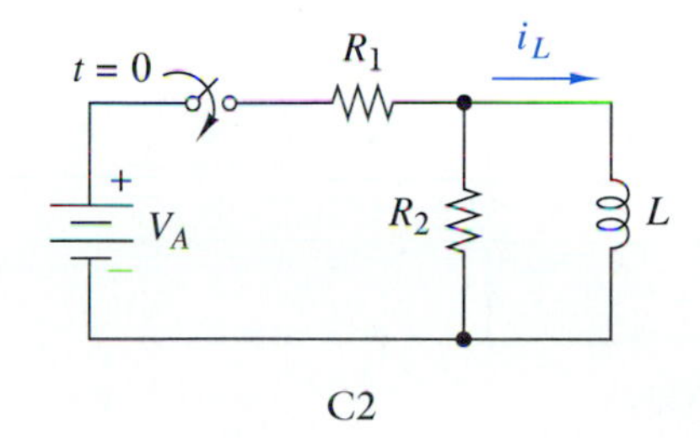

Figure P8-5

8-7 The switch in each circuit in Fig. P8-7 has been open for a "long time" and is closed at $t = 0$.

(a) Determine the value of the state variable at $t = 0$.

(b) Determine the value of the state variable as $t \to \infty$.

(c) Determine the circuit time constant.

(d) Write an expression for the state variable for $t \geq 0$. Identify the forced and natural components and sketch their waveforms.

8-8 Repeat Prob. 8-7 if the switch in each circuit has been closed for a "long time" and then is opened at $t = 0$.

8-9 The switch in each circuit in Fig. P8-9 has been in position A for a "long time" and is moved to position B at $t = 0$. Write an expression for the state variable for $t \geq 0$. Identify the forced and natural components and sketch their waveforms.

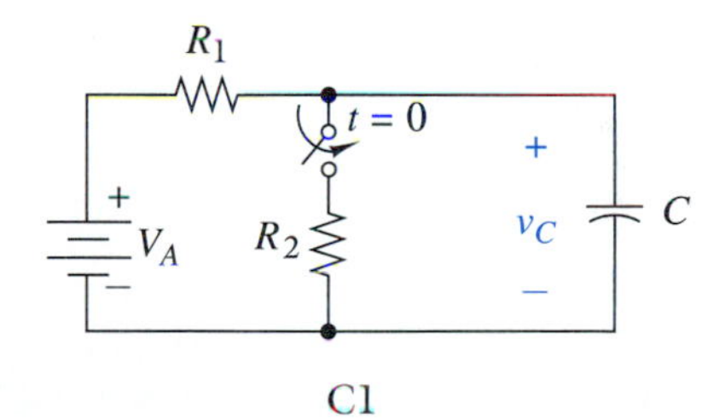

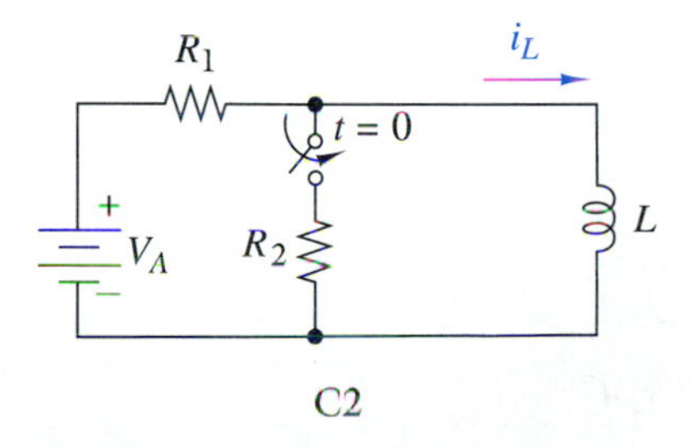

Figure P8-7

8-10 Repeat Prob. 8-9 if the switch in each circuit has been in position B for a "long time" and is moved to position A at $t = 0$.

8-11 The circuits in Fig. P8-11 are in the zero-state when the step function $V_A u(t)$ is applied. For $t \geq 0$ find the output voltage response $v_O(t)$. Identify the forced and natural components and sketch their waveforms.

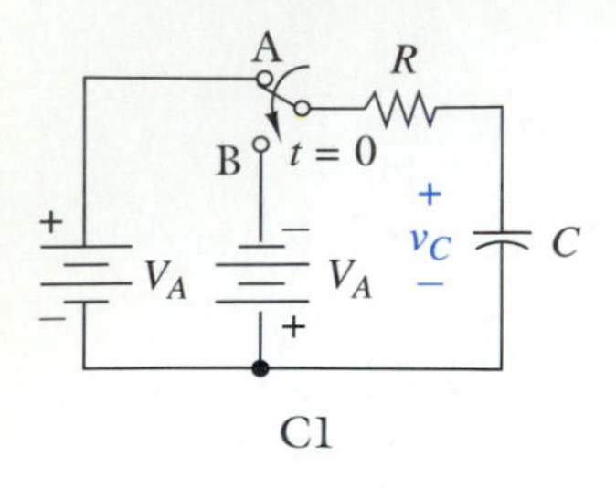

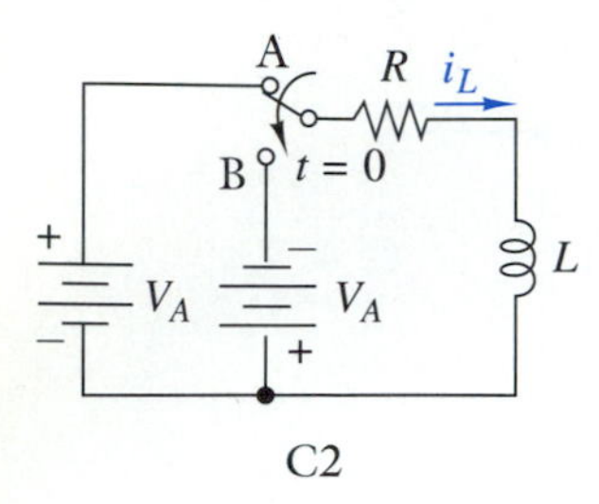

Figure P8-9

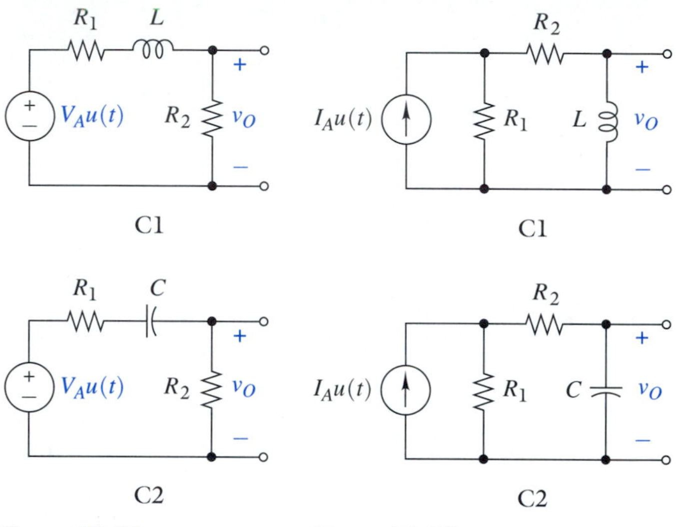

Figure P8-11 **Figure P8-12**

8-12 The circuits in Fig. P8-12 are in the zero-state when the step function $I_A u(t)$ is applied. For $t \geq 0$ find the output voltage response $v_O(t)$. Identify the forced and natural components and sketch their waveforms.

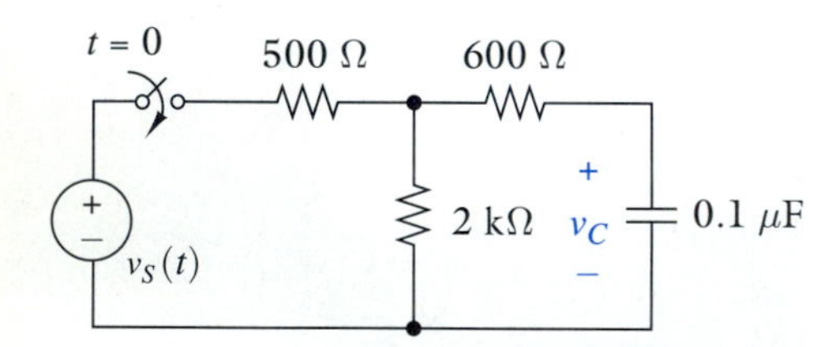

Figure P8-13

8-13 The switch in the circuit of Fig. P8-13 has been open for a "long time" and is closed at $t = 0$. When $v_S(t) = 10$ solve for $v_C(t)$. Identify the forced and natural components and sketch the waveform.

8-14 Repeat Prob. 8-13 for $v_S(t) = 10 \cos 5000t$.

8-15 The switches in Fig. P8-15 have been open for a "long time" and are closed at $t = 0$. When $v_S(t) = 10$ solve for the indicated output variables. Identify the forced and natural components and sketch their waveforms.

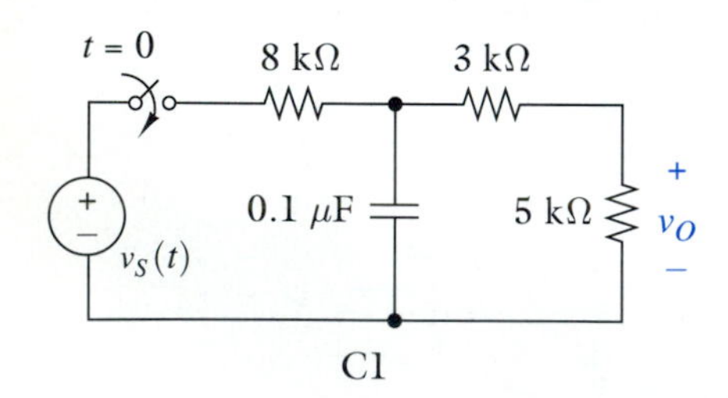

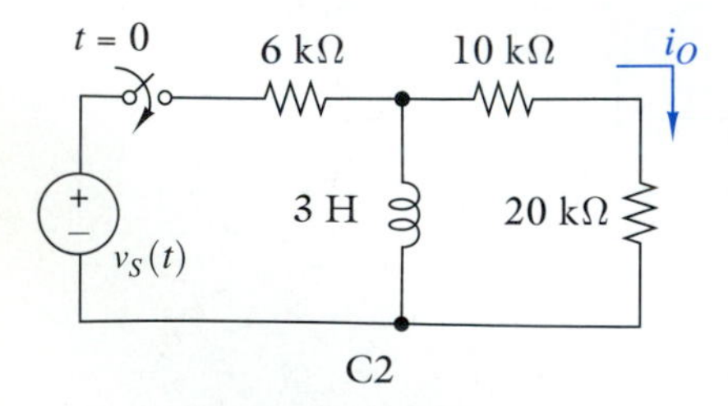

Figure P8-15

8-16 Repeat Prob. 8-15 for $v_S(t) = 10 \cos 5000t$.

8-17 Each of the circuits in Fig. P8-17 is operating in the sinusoidal steady state with $v_S(t) = [V_A \cos \omega t]u(t)$. Determine the forced component of the output voltage.

8-18 Each of the circuits in Fig. P8-18 has a time constant of RC and is in the zero state. An input $v_S(t) = V_A u(t)$ is applied to each circuit.

(a) Find the final value of $v_C(t)$ as $t \to \infty$.

(b) Write an expression for the voltage $v_C(t)$.

(c) Use (b) to derive an expression for the output voltage $v_O(t)$, identify the forced and natural components, and sketch their waveforms.

8-19 Waveforms listed below are the state variable responses of first-order RC and RL circuits.

(1) $v_C(t) = 10 - 5e^{-3000t}$ V for $t \geq 0$

(2) $i_L(t) = 5e^{-6000t}$ mA for $t \geq 0$

For each response:

(a) Determine the circuit characteristic equation, the circuit time constant, and the initial condition of the state variable.

(a) Determine the input waveform and its amplitude.

8-20 Waveforms listed below are the state variable responses of first-order RC and RL circuits.

(1) $v_C(t) = 10\cos 1000t + 10\sin 1000t - 5e^{-1000t}$ V for $t \geq 0$.

(2) $i_L(t) = -6\cos 300t + 2\sin 300t + 6e^{-100t}$ mA for $t \geq 0$.

For each response:

(a) Determine the circuit characteristic equation, the circuit time constant, and the initial condition.

(b) Determine the input waveform and its amplitude.

8-21 Waveforms listed below are zero-input responses of first-order RC and RL circuits. The given responses are *not* state variables, but the initial value of the state variable is given.

(1) $i_C(t) = -10e^{-2000t}$ mA for $t \geq 0$ with $v_C(0) = 10$ V

(2) $v_L(t) = +5e^{-5000t}$ V for $t \geq 0$ with $i_L(0) = 10$ mA.

For each response:

(a) Determine the circuit characteristic equation and the circuit time constant.

(b) Determine the value of the circuit parameters R and C or R and L.

❖ **8-22** **(D)** Design first-order circuits that produce the responses given in Prob. 8-19.

❖ **8-23** **(D)** Design first-order circuits that produce the responses given in Prob. 8-20.

❖ **8-24** **(D)** Design first-order circuits that produce the responses given in Prob. 8-21.

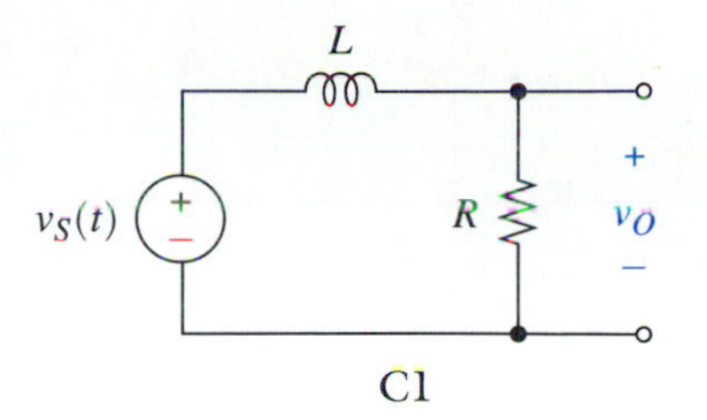

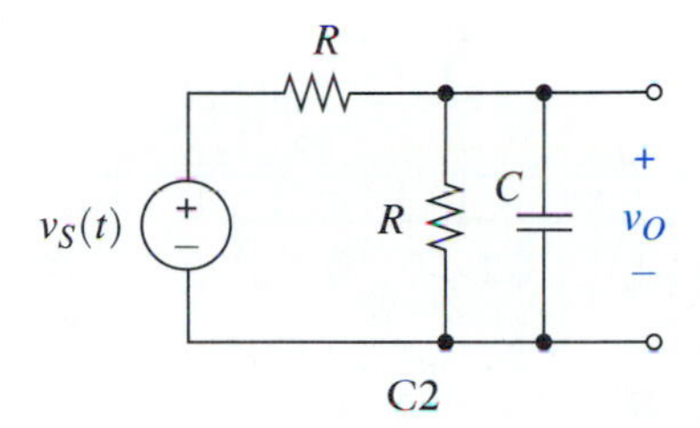

Figure P8-17

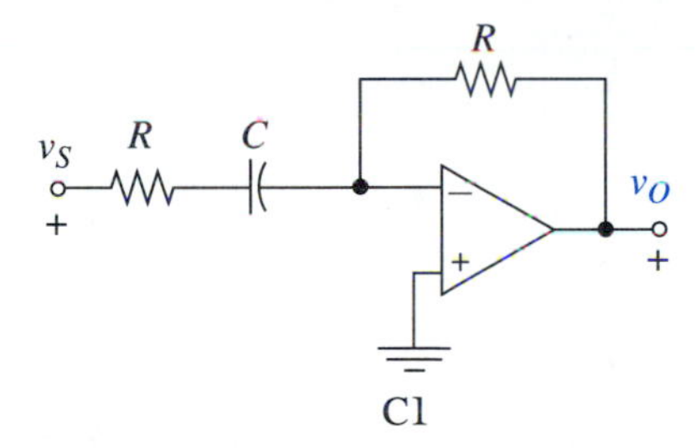

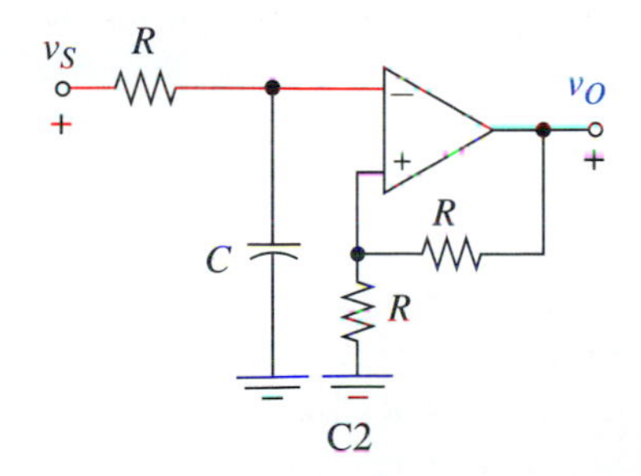

Figure P8-18

ERO 8-2 Second-Order Circuits (Sects. 8-5, 8-6, 8-7)

Given a second-order RLC or OP AMP RC circuit:

(a) Determine the circuit differential equation, characteristic equation, damping ratio and undamped natural frequency, and initial conditions (if not given).

(b) Find the zero-input response or the natural and forced components of circuit response for a step function input.

Conversely, given the step response of a second-order circuit:

(a) Determine the circuit differential equation, characteristic equation, damping ratio and undamped natural frequency, and initial conditions (if not given).

(b) Select the value of circuit parameters to produce the given step response.

8-25 For each of the following differential equations, find the waveform $v(t)$ that satisfies the equation and meets the initial conditions.

(a) $\dfrac{d^2v}{dt^2} + 7\dfrac{dv}{dt} + 10v = 0,\ v(0) = 0,\ \dfrac{dv(0)}{dt} = 15\ \text{V/s}$

(b) $\dfrac{d^2v}{dt^2} + 4\dfrac{dv}{dt} + 4v = 0,\ v_0 = 0,\ \dfrac{dv(0)}{dt} = 2\ \text{V/s}$

8-26 Repeat Prob. 8-25 for the following differential equations.

(a) $\dfrac{d^2v}{dt^2} + 10\dfrac{dv}{dt} + 25v = 100u(t),\ \ v(0) = 5,\ \dfrac{dv(0)}{dt} = 25\ \text{V/s}$

(b) $\dfrac{d^2v}{dt^2} + 2\dfrac{dv}{dt} + 10v = 100u(t),\ \ v(0) = 5\ \text{V},\ \dfrac{dv(0)}{dt} = 5\ \text{V/s}$

8-27 Determine the numerical value of the circuit damping ratio ζ and the undamped natural frequency ω_O for each circuit in Fig. P8-27. Specify whether the circuits are overdamped, underdamped, or critically damped.

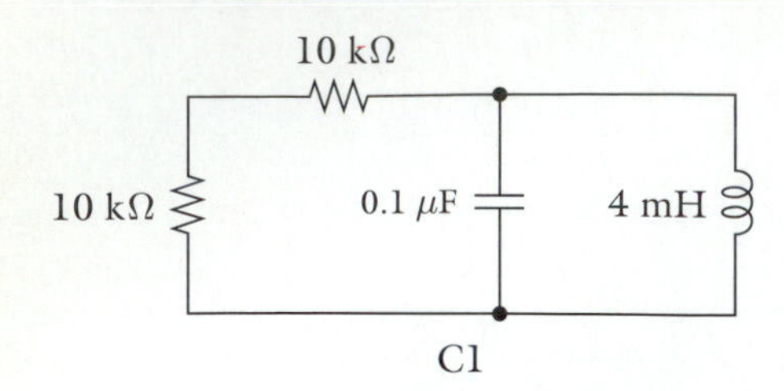

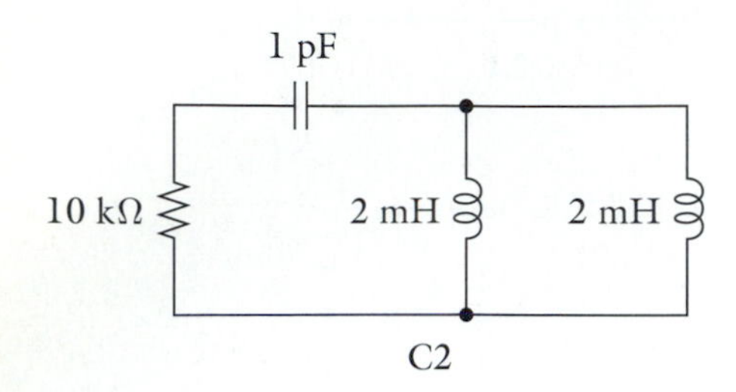

Figure P8-27

8-28 For each circuit in Fig. P8-28 write an expression for the circuit damping ratio ζ and the undamped natural frequency ω_O in terms of the circuit parameters R, L, and C. Then derive the expression for R in terms of L and C that produces critical damping.

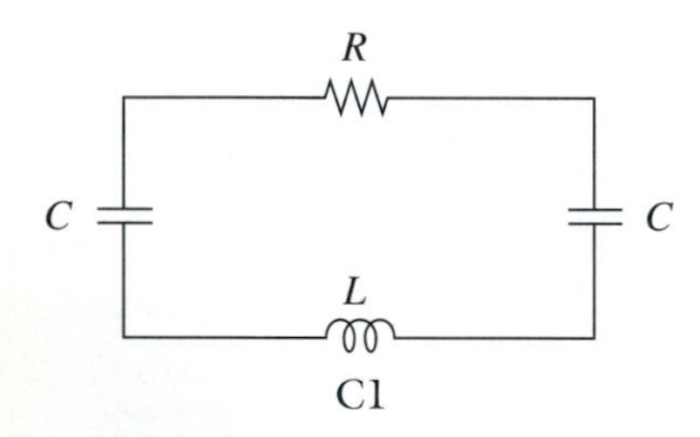

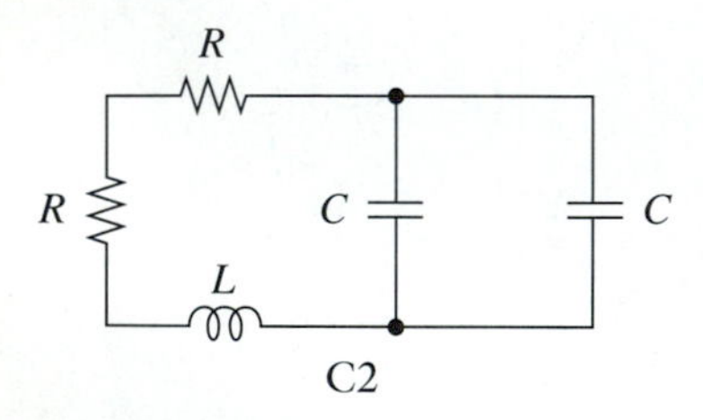

Figure P8-28

8-29 The switch in Fig. P8-29 has been open for a "long time" and is closed at $t = 0$. The circuit parameters are $L = 0.4$ H, $C = 0.1\ \mu$F, $R = 3$ kΩ, and $V_A = 10$ V.

(a) Determine the initial values of v_C and i_L at $t = 0$.

(b) Determine the final values of v_C and i_L as $t \to \infty$.

(c) Develop the differential equation for $v_C(t)$.

(d) Determine $v_C(t)$, identify the forced and natural components of the response, and sketch their waveform.

(e) Is the response overdamped or underdamped?

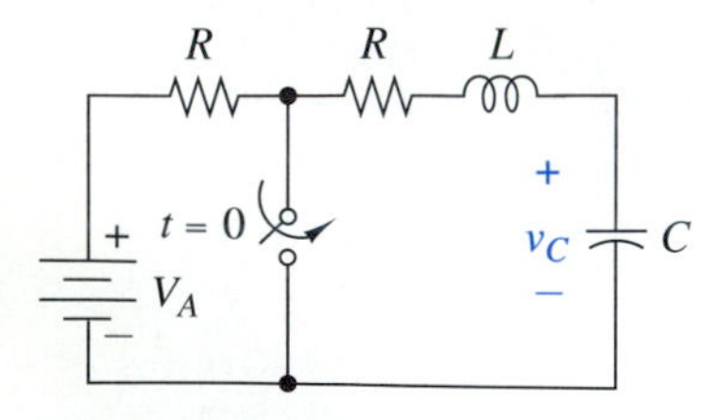

Figure P8-29

8-30 The switch in Fig. P8-30 has been open for a "long time" and is closed at $t = 0$. The circuit parameters are $L = 0.4$ H, $C = 25$ pF, $R = 6$ kΩ, and $V_A = 10$ V.

(a) Determine the initial values of v_C and i_L at $t = 0$.

(b) Determine the final values of v_C and i_L as $t \to \infty$.

(c) Develop the differential equation for $v_C(t)$.

(d) Determine $v_C(t)$, identify the forced and natural components of the response, and sketch their waveform.

(e) Is the response overdamped or underdamped?

8-31 The switch in Fig. P8-31 has been open for a "long time" and is closed at $t = 0$. The circuit parameters are $L = 0.1$ H, $C = 0.1\ \mu$F, $R = 12$ kΩ, and $V_A = 10$ V.

(a) Determine the initial values of v_C and i_L at $t = 0$.
(b) Determine the final values of v_C and i_L as $t \to \infty$.
(c) Develop the differential equation for $i_L(t)$.
(d) Determine $i_L(t)$, identify the forced and natural components of the response, and sketch their waveform.
(e) Is the response overdamped or underdamped?

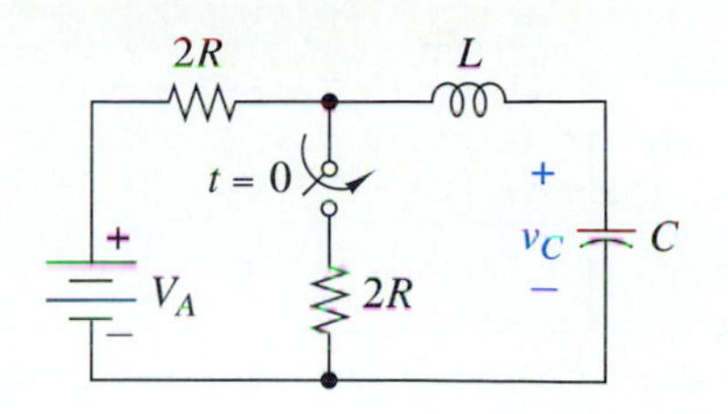

Figure P8-30

8-32 The switch in Fig. P8-32 has been open for a "long time" and is closed at $t = 0$. The circuit parameters are $L = 0.1$ H, $C = 0.1\ \mu$F, $R = 12$ kΩ, and $V_A = 10$ V.
(a) Determine the initial values of v_C and i_L at $t = 0$.
(b) Determine the final values of v_C and i_L as $t \to \infty$.
(c) Develop the differential equation for $i_L(t)$.
(d) Determine $i_L(t)$, identify the forced and natural components of the response, and sketch their waveform.
(e) Is the response overdamped or underdamped?

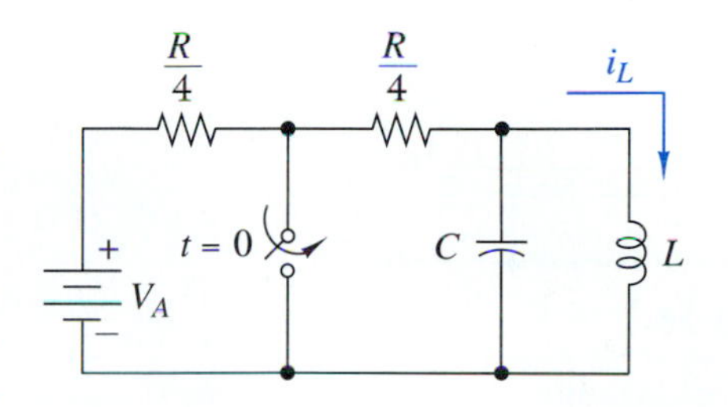

Figure P8-31

8-33 The switch in Fig. P8-33 has been in position A for a "long time" and is moved to position B at $t = 0$. The circuit parameters are $R_1 = 40$ kΩ, $R_2 = 2.5$ kΩ, $L = 2$ H, $C = 0.02\mu$F, and $V_A = 15$ V. Determine $i_L(t)$ for $t > 0$. Identify the forced and natural components of the response and sketch their waveforms. Is the circuit overdamped or underdamped?

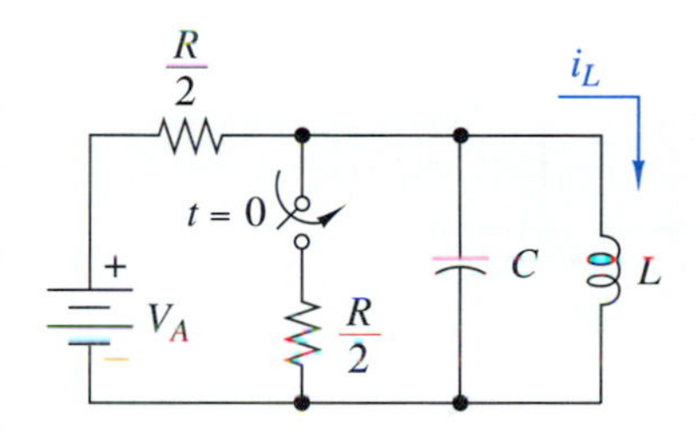

Figure P8-32

8-34 The switch in the circuit shown in Fig. P8-34 has been in position A for a "long time." At $t = 0$ it is moved to position B. The circuit parameters are $R_1 = 400$ kΩ, $R_2 = 25$ kΩ, $L = 10$ mH, $C = 0.01\ \mu$F, and $V_A = 15$ V. Determine $i_L(t)$ for $t > 0$. Identify the forced and natural components of the response and sketch their waveforms. Is the circuit overdamped or underdamped?

8-35 The switch in Fig. P8-35 the switch has been in position A for a "long time" and is moved to position B at $t = 0$. The circuit parameters are $R_1 = 40$ kΩ, $R_2 = 2.5$ kΩ, $L = 2$ H, $C = 0.02\ \mu$F, and $V_A = 15$ V. Determine $v_C(t)$ for $t > 0$. Identify the forced and natural components of the response and sketch their waveforms. Is the circuit overdamped or underdamped?

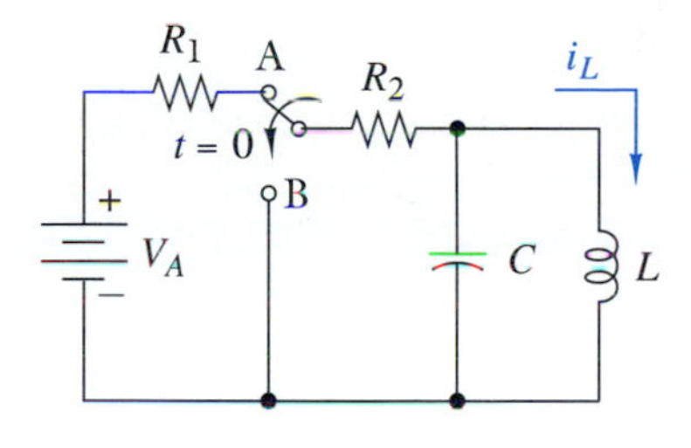

Figure P8-33

8-36 The switch in Fig. P8-36 has been in position A for a "long time" and is moved to position B at $t = 0$. The circuit parameters are $R_1 = 400$ kΩ, $R_2 = 25$ kΩ, $L = 10$ mH, $C = 0.01\ \mu$F, and $V_A = 15$ V. Determine $v_C(t)$ for $t > 0$. Identify the forced and natural components of the response, and sketch their waveforms. Is the circuit overdamped or underdamped?

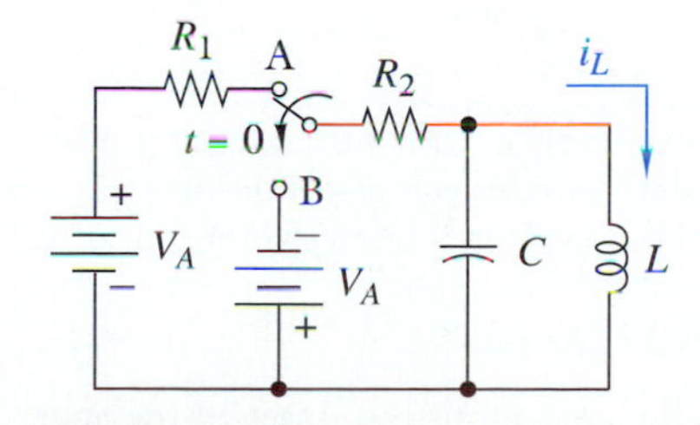

Figure P8-34

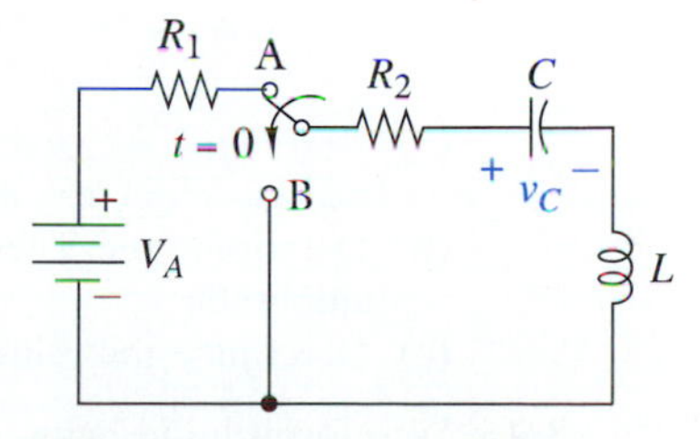

Figure P8-35

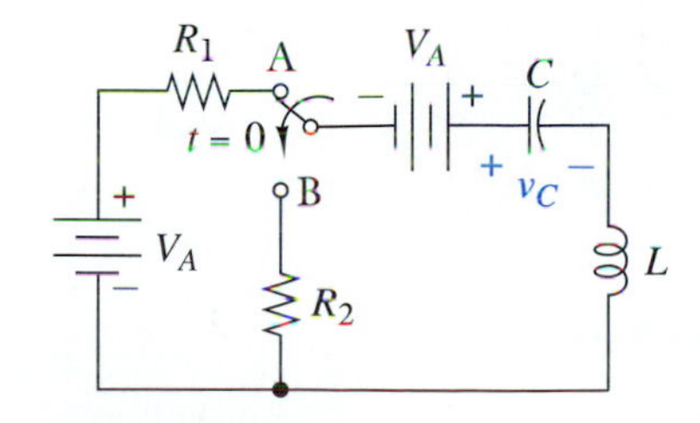

Figure P8-36

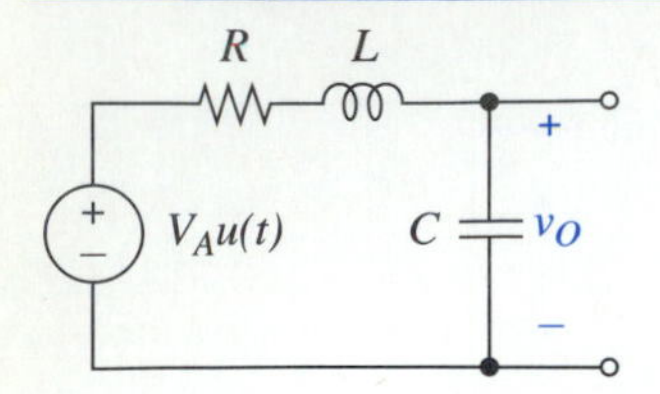

Figure P8-37

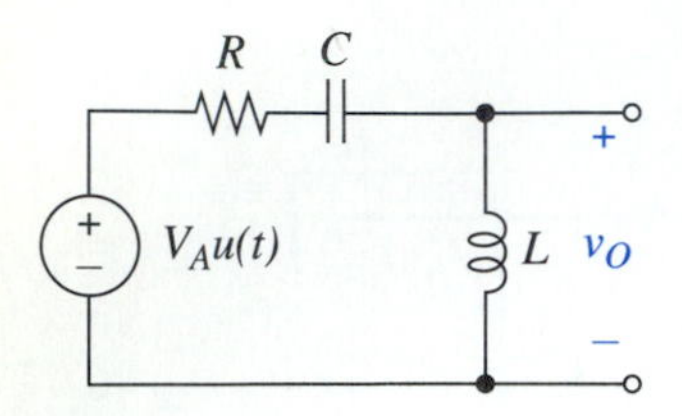

Figure P8-38

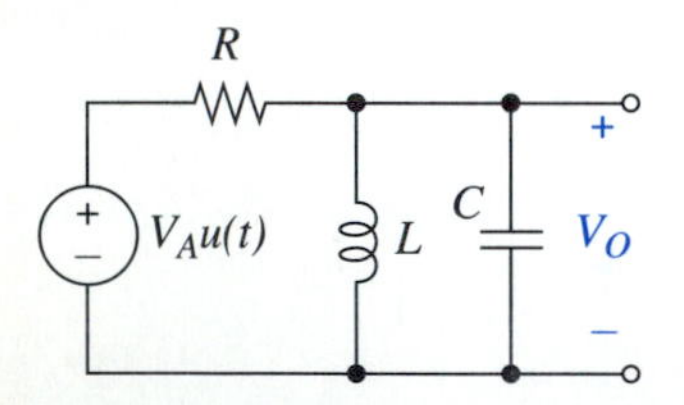

Figure P8-39

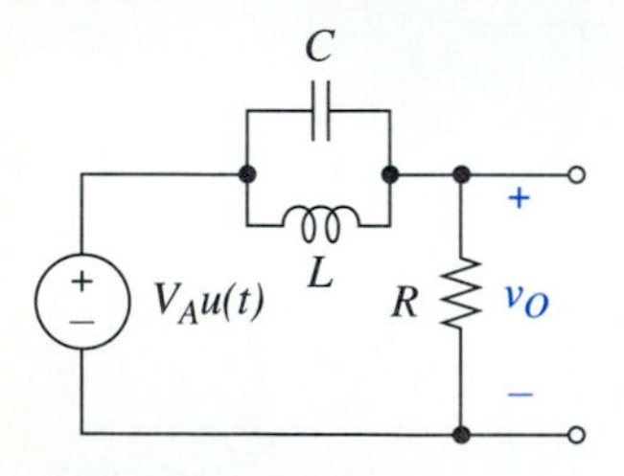

Figure P8-40

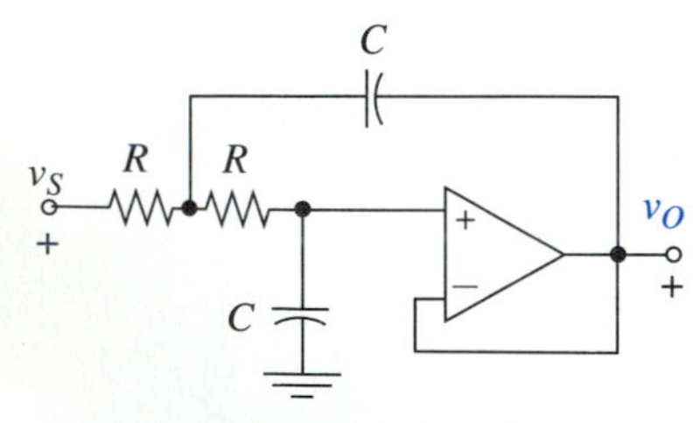

Figure P8-41

8-37 The circuit in Fig. P8-37 is in the zero-state when the step function input is applied. Find the output $v_O(t)$ using $L = 50$ mH, $C = 0.05\ \mu$F, $R = 4$ kΩ, and $V_A = 100$ V.

8-38 The circuit in Fig. P8-38 is in the zero state when the step function input is applied. Find the output $v_O(t)$ using $L = 50$ mH, $C = 0.05\ \mu$F, $R = 4$ kΩ, and $V_A = 100$ V.

8-39 The circuit in Fig. P8-39 is in the zero state when the step function input is applied. Find the output $v_O(t)$ using $L = 50$ mH, $C = 0.05\ \mu$F, $R = 4$ kΩ, and $V_A = 100$ V.

8-40 The circuit in Fig. P8-40 is in the zero state when the step function input is applied. Find the output $v_O(t)$ using $L = 50$ mH, $C = 0.05\ \mu$F, $R = 4$ kΩ, and $V_A = 100$ V.

8-41 Show that the circuit in Fig. P8-41 is critically damped and that $\omega_O = 1/RC$.

8-42 Waveforms listed below are the state variable responses of a series RLC circuit.

$v_C(t) = 10 - 5e^{-3000t} - 5e^{-6000t}$ V for $t > 0$

$i_L(t) = 2e^{-3000t} + 4e^{-6000t}$ mA for $t > 0$

(a) Determine the circuit characteristic equation, damping ratio, natural frequency, and state variable initial conditions.

(b) Determine the Thévenin equivalent input waveform and its amplitude.

(c) Determine the values of R_T, L, and C.

8-43 Waveforms listed below are the state variable responses of a series RLC circuit.

$v_C(t) = 10 - 5e^{-100t} \sin 100t$ V for $t \geq 0$

$i_L(t) = 0.5e^{-100t}(\sin 100t - \cos 100t)$ mA for $t \geq 0$

(a) Determine the circuit characteristic equation, damping ratio, undamped natural frequency, and state variable initial conditions.

(b) Determine the Thévenin equivalent input waveform and its amplitude.

(c) Determine the values of R_T, L and C.

8-44 Waveforms listed below are the state variable responses of a parallel RLC circuit.

$$v_C(t) = 2.5e^{-100t} - 2.5e^{-500t} \text{ V for } t \geq 0$$

$$i_L(t) = 20 - 25e^{-100t} + 5e^{-500t} \text{ mA for } t \geq 0$$

(a) Determine the circuit characteristic equation, damping ratio, natural frequency, and state variable initial conditions.

(b) Determine the Thévenin equivalent input waveform and its amplitude.

(c) Determine the values of R_T, L, and C.

8-45 Waveforms listed below are the zero-input responses of the state variables in an RLC circuit.

$$v_C(t) = 2e^{-1000t}\cos 2000t$$

$$- 4\, e^{-1000t}\sin 2000t \text{ V for } t \geq 0$$

$$i_L(t) = 5e^{-1000t}\cos 2000t \text{ mA for } t \geq 0$$

(a) Is this a series RLC or a parallel RLC circuit?

(b) Determine the circuit characteristic equation, damping ratio, natural frequency, and the state variable initial condition.

(c) Determine the values of R_T, L, and C.

8-46 Waveforms listed below are zero-input responses of a series RLC circuit. The initial value of the state variables are given, but the given responses are **not** state variables.

$$i_C(t) = 2e^{-2000t} - 4000te^{-2000t} \text{ mA}$$

$$\text{for } t \geq 0 \text{ with } v_C(0) = 0 \text{ V}$$

$$v_L(t) = -8e^{-2000t} + 8000te^{-2000t} \text{ V}$$

$$\text{for } t \geq 0 \text{ with } i_L(0) = 2 \text{ mA.}$$

(a) Determine the circuit characteristic equation, damping ratio, and undamped natural frequency.

(b) Determine the value of the circuit parameters R_T and C or R and L.

❖ **8-47** **(D)** Design a second-order series RLC circuit so that an input $v_S(t) = V_A u(t)$ produces a zero-state response of

$$v_C(t) = V_A - 2V_A e^{-200t} + 0.5V_A e^{-400t} \text{ V for } t > 0$$

ERO 8-3 Computer-Aided Transient Analysis (Sect. 8-8)

Use SPICE or MICRO-CAP to generate numerical values of the transient response of first and second-order linear circuits.

8-48 Solve Prob. 8-13 using analytical methods and verify your results using SPICE or MICRO-CAP

8-49 Solve Prob. 8-14 using analytical methods and verify your results using SPICE or MICRO-CAP.

8-50 Solve Prob. 8-15 using analytical methods and verify your results using SPICE or MICRO-CAP.

8-51 Solve Prob. 8-29 using analytical methods and verify your results using SPICE or MICRO-CAP.

8-52 Solve Prob. 8-33 using analytical methods and verify your results using SPICE or MICRO-CAP.

8-53 Solve Prob. 8-40 using analytical methods and verify your results using SPICE or MICRO-CAP.

❖ **8-54** **(D)** Design a circuit to meet the requirements in Prob. 8-47. Use SPICE or MICRO-CAP to verify your design.

❖ **8-55** **(D)** Design a first-order OP AMP RC circuit with a time constant of 20 ms and a dc gain of +100. Verify your design using SPICE or MICRO-CAP.

CHAPTER INTEGRATING PROBLEMS

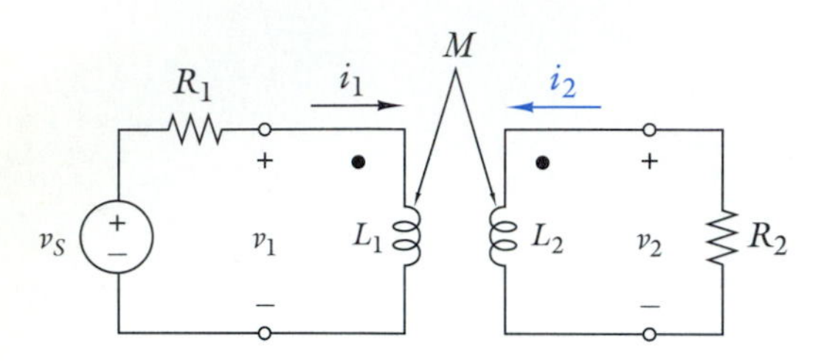

Figure P8-56

8-56 **(A)** For the circuit in Fig. P8-56:

(a) Write the device and connection equations for the coupled coils shown.

(b) Reduce the equations in (a) to a single second-order differential equation in the current $i_2(t)$.

(c) Show that perfect coupling reduces the second-order equation in (b) to a first-order equation with a time constant of $L_1/R_1 + L_2/R_2$.

(d) With perfect coupling find the sinusoidal steady-state component of $i_2(t)$ using $L_1 = L_2 = M = 100$ mH, $R_1 = R_2 = 100\ \Omega$, and $v_S(t) = 10 \sin 500t$.

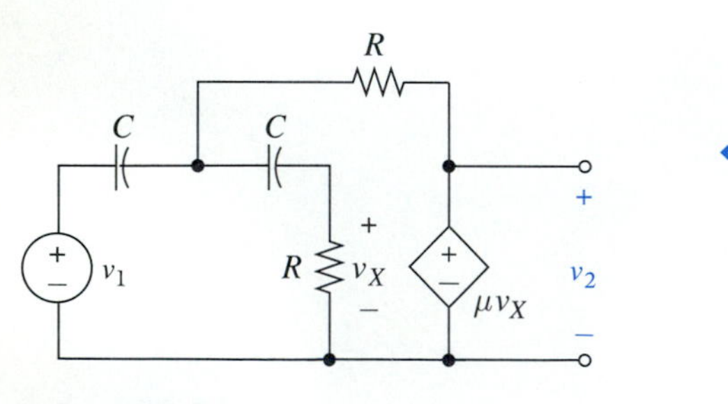

Figure P8-57

❖ **8-57** **(a)** **(A)** Derive the characteristic equation for the active RC circuit in Fig. P8-57.

(b) **(A)** For what range of values of μ is the circuit stable?

(c) **(D)** Select values of μ, R and C so that $\zeta = 0.5$ and $\omega_O = 1000$ rad/s.

(d) **(A)** Verify your design using SPICE or MICRO-CAP.

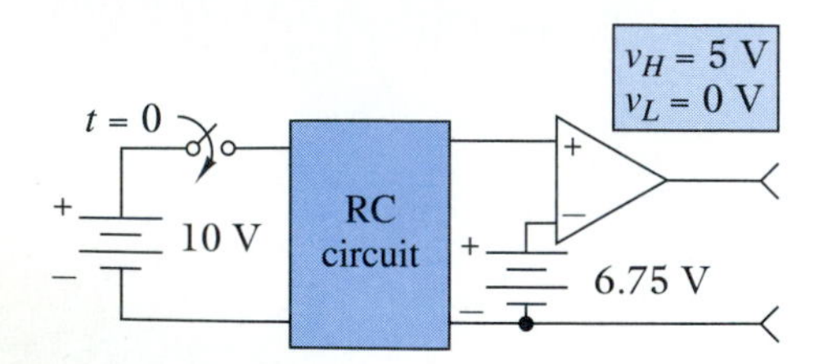

Figure P8-58

❖ **8-58** **(D)** The purpose of the RC circuit in Fig. P8-58 is to cause the comparator output to change from V_L to V_H at $t = 1.25$ s after the switch is closed. Design a suitable RC circuit using the 10-V supply and any of the standard values of resistance and capacitance defined in Appendix A.

8-59 **(A)** Figure P8-59 shows a sample-hold circuit at the input of an analog-to-digital converter. In the sample mode the converter commands the analog switch to close (ON) and the capacitor charges up to the value of the input signal. In the hold mode the converter commands the analog switch to open (OFF), the capacitor holds and feeds the input signal value to the converter via the OP AMP voltage follower. When conversion is completed the analog switch is turned ON again and the sample-hold cycle repeats.

(a) If $C = 10$ pF and series resistances of the analog switch are $R_{ON} = 100\ \Omega$ and $R_{OFF} = 100$ MΩ, what is the time constant in the sample mode and the time constant in the hold mode?

(b) The number of sample-hold cycles per second must be at least twice the highest frequency in the analog input signal. What is the minimum number of sample-hold cycles per second for an input $v_S(t) = 5 + 5 \sin 2\pi 1000t$?

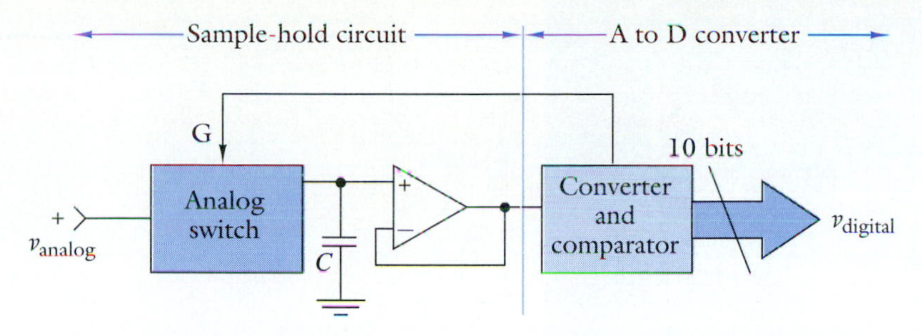

Figure P8-59

(c) Sampling at 10 times the minimum number of sample-hold cycles per second, what is the duration of the sample mode if the hold mode lasts 9 times as long as the sample mode?

(d) For the input in (b), will the capacitor voltage reach a steady-state condition during the sample mode?

(e) What fraction of the capacitor voltage will be lost during the hold mode?

8-60 **(A)** The circuit in Fig. P8-60 represents the ignition system of a vintage automobile. If the switch (the breaker points) has been closed for "a long time" and is suddenly opened at $t = 0$:

(a) Derive a second-order differential equation in $v_C(t)$ and determine the initial conditions.

(b) Show that the circuit damping ratio is very small.

(c) Assuming zero damping ratio, show that the voltage across the capacitor is approximately

$$v_C(t) = 2\sqrt{\frac{L}{C}}\sin(\sqrt{LC})t$$

(d) Show that the voltage across the capacitor is $-v_C(t)$.

(e) An arc occurs between the two electrodes (the spark plug) when the voltage between them exceeds 500 V. Will the spark plug fire?

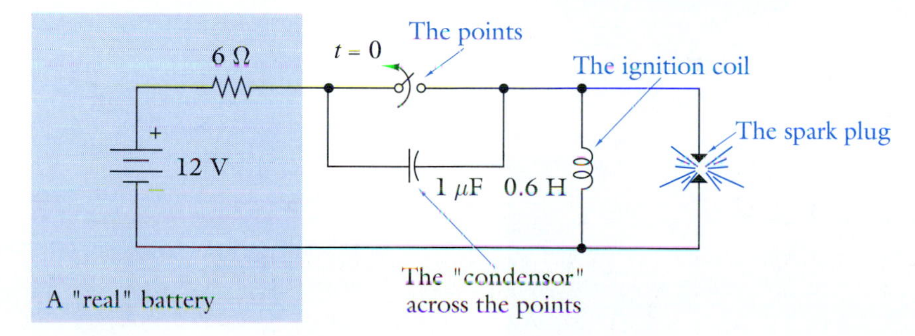

Figure P8-60

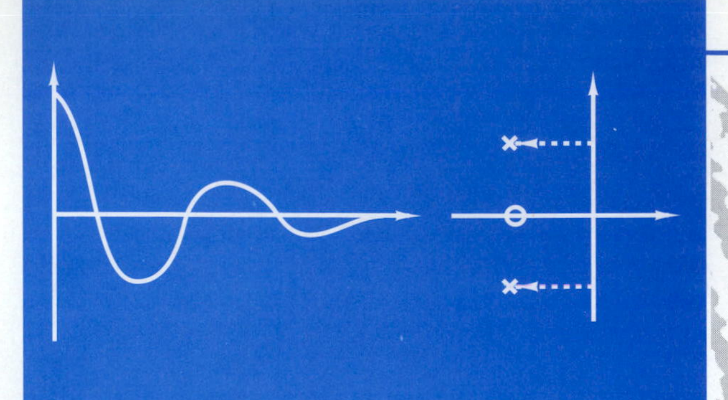

Chapter Nine

My method starts with a complex integral; I fear this sounds rather formidable; but it is really quite simple . . . I am afraid that no physical people will ever try to make out my method: but I am hoping that it may give them confidence to try your methods.

Thomas John Bromwich, 1915
British Mathematician

Laplace Transforms

Laplace transforms have their roots in the pioneering work of the eccentric British engineer Oliver Heaviside (1850–1925). His operational calculus was essentially a collection of intuitive rules that allowed him to formulate and solve a number of the important technical problems of his day. Heaviside was a practical man with no interest in mathematical elegance. His intuitive approach drew bitter criticism from the mathematicians of his day. However, mathematicians like Thomas John Bromwich and others eventually recognized the importance of Heaviside's methods and began to supply the necessary mathematical foundations. The quotation on the opposite page is taken from a 1915 letter from Bromwich to Heaviside in which he described what we now call the Laplace transformation. The transformation is named for Laplace because a complete mathematical development of Heaviside's methods was eventually found in the 1780 writings of the French mathematician Pierre Simon Laplace.

The Laplace transformation provides a new and important method of representing circuits and signals. The transform approach offers a viewpoint and terminology that pervades electrical engineering, particularly in linear circuit analysis and design. The first two sections of this chapter present the basic properties of the Laplace transformation and the concept of converting signals from the time domain to the frequency domain. The third section introduces frequency domain signal description via the pole-zero diagram. The fourth and fifth sections treat the inverse procedure for transforming signals from the frquency domain back into the time domain. In the sixth section, transform methods are used to solve differential equations that describe the response of linear circuits. The last three sections treat additional properties of the Laplace transformation, which provide important insights into the relationship between the time and frequency domains.

9-1 SIGNAL WAVEFORMS AND TRANSFORMS

A mathematical transformation employs rules to change the form of data without altering their meaning. An example of a transformation is the conversion of numerical data from decimal to binary form. In engineering circuit analysis, transformations are used to obtain alternative representations of circuits and signals. These alternate forms provide a different perspective that can be quite useful or even essential. Examples of the transform methods used in circuit analysis are Fourier transforms, fast Fourier transforms (FFTs), discrete Fourier transforms (DFTs), Z-transforms, and Laplace transforms. These methods all involve specific transformation rules, make certain analysis techniques more manageable, and provide a useful viewpoint for circuit design.

This chapter deals with the Laplace transformation. The discussion of the Laplace transformation follows the path shown in Fig. 9-1 by the solid blue arrow. The process begins with a linear circuit. We derive a differential equation describing the circuit response and then transform this equation into the frequency domain where it becomes an algebraic equation. Algebraic techniques are then used to solve the transformed equation

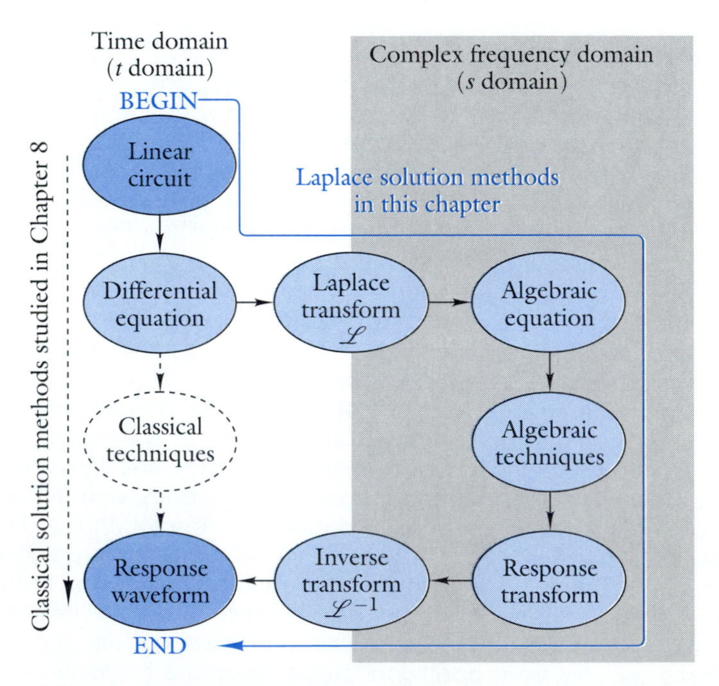

Figure 9-1 Flow diagram for solving dynamic circuits using Laplace transforms.

for the circuit response. The inverse Laplace transformation then changes the frequency domain response into the response waveform in the time domain. The dashed arrow in Fig. 9-1 shows there is another route to the time domain response using the classical technques in Chapter 8. The classical approach appears to be more direct, but the advantage of the Laplace transformation is that solving a differential equation becomes an algebraic process.

Symbolically, we represent the Laplace transformation as

$$\mathscr{L}\{f(t)\} = F(s) \tag{9-1}$$

This expression states that $F(s)$ is the Laplace transform of the waveform $f(t)$. The transformation operation involves two domains: (1) the time domain in which the signal is characterized by its *waveform* $f(t)$, and (2) the complex frequency domain in which the signal is represented by its *transform* $F(s)$. The symbol s stands for the *complex frequency variable*, which is the independent variable in the s-domain, just as time is the independent variable in the t domain. The symbol s is a complex variable written as $s = \sigma + j\omega$, where $\sigma = Re\{s\}$ is the real part and $\omega = Im\{s\}$ is the imaginary part.

A signal can be expressed as a waveform or a transform. Collectively, $f(t)$ and $F(s)$ are called a *transform pair*, where the pair involves two representations of the signal. To distinguish between the two forms, a lower-case letter denotes a waveform and an upper-case letter a transform. For electrical waveforms such as current $i(t)$ or voltage $v(t)$, the corresponding transforms are denoted $I(s)$ and $V(s)$. In this chapter we will use $f(t)$ and $F(s)$ to stand for signal waveforms and transforms in general.

The *Laplace transformation* is defined by the integral

$$F(s) = \int_{0-}^{\infty} f(t)e^{-st}dt \tag{9-2}$$

Since the definition involves an improper integral (the upper limit is infinite), we must discuss the conditions under which the integral exists (converges). The integral exists if the waveform $f(t)$ is piecewise continuous and of exponential order. *Piecewise continuous* means that $f(t)$ has a finite number of steplike discontinuities in any finite interval. *Exponential order* means that constants K and b exist such that $|f(t)| < Ke^{bt}$ for all $t > 0$. As a practical matter, the signals encountered in engineering applications meet these conditions.

Equation (9-2) uses a lower limit denoted $t = 0-$ to indicate a time just a whisker before $t = 0$. We use $t = 0-$ because in dealing with circuit responses $t = 0$ is defined by a discrete event, such as closing a switch. Such an event may cause a discontinuity in $f(t)$ at $t = 0$. In order to capture this discontinuity we set the lower limit at $t = 0-$, just prior to the event. Fortunately, in many situations there is no discontinuity, so we will not distinguish between $t = 0-$ and $t = 0$ unless it is crucial.

Equally fortunate is the fact that the number of different signals encountered in linear circuits is relatively small. The list includes the three basic waveforms from Chapter 6 (the step, exponential, and sinusoid), as well as composite waveforms such as the impulse, ramp, damped ramp, and damped sinusoid. Since the number of signals of interest is relatively small, we do not often evaluate the integral definition Laplace transforms in Eq. (9-2). Once a transform pair has been found it can be cataloged in a table for future reference and use. Tables 9-2 and 9-3 in this chapter are sufficient for our purposes.

EXAMPLE 9-1

Find the Laplace transform of the unit step function $f(t) = u(t)$.

SOLUTION:

Applying Eq. (9-2) yields

$$F(s) = \int_0^\infty u(t)e^{-st}\,dt$$

Since $u(t) = 1$ throughout the range of integration this integral becomes

$$F(s) = \int_0^\infty e^{-st}\,dt = -\left.\frac{e^{-st}}{s}\right|_0^\infty = -\left.\frac{e^{-(\sigma+j\omega)t}}{\sigma + j\omega}\right|_0^\infty$$

The last expression on the right side vanishes at the upper limit, since $e^{-\sigma t}$ goes to zero as t approaches infinity provided $\sigma > 0$. At the lower limit the expression reduces to $1/s$. We conclude that $\mathscr{L}\{u(t)\} = 1/s$.

EXAMPLE 9-2

Find the Laplace transform of the exponential waveform

$$f(t) = \left[e^{-\alpha t}\right]u(t)$$

SOLUTION:

Applying Eq. (9-2) yields

$$F(s) = \int_0^\infty e^{-\alpha t}e^{-st}\,dt = \int_0^\infty e^{-(s+\alpha)t}\,dt = \left.\frac{e^{-(s+\alpha)t}}{-(s+\alpha)}\right|_0^\infty$$

The last term on the right side vanishes at the upper limit, since $e^{-(s+\alpha)t}$ vanishes as t approaches infinity provided $\sigma > \alpha$. At the lower limit it reduces to $1/(s+\alpha)$. We conclude that $\mathscr{L}\{[e^{-\alpha t}]u(t)\} = 1/(s+\alpha)$.

EXAMPLE 9-3

Find the Laplace transform of the impulse function $f(t) = \delta(t)$.

SOLUTION:

Applying Eq. (9-2) yields

$$F(s) = \int_{0-}^{\infty} \delta(t)e^{-st}dt = \int_{0-}^{0+} \delta(t)e^{-st}dt = \int_{0-}^{0+} \delta(t)dt = 1$$

In this case the difference between $t = 0-$ and $t = 0$ is important, since the impulse is zero everywhere except at $t = 0$. To capture the impulse in the integration, we take a lower limit at $t = 0-$ and an upper limit at $t = 0+$. We conclude that $\mathscr{L}\{\delta(t)\} = 1$, since the term $e^{-st} = 1$ on this integration interval and the area under a unit impulse is unity.

Inverse Transformation

So far we have used the direct transformation to convert waveforms into transforms. But Fig. 9-1 points out the need to perform the inverse transformation to convert transforms into waveforms. Symbolically we represent the inverse process as

$$\mathscr{L}^{-1}\{F(s)\} = f(t) \tag{9-3}$$

This equation states that $f(t)$ is the inverse Laplace transform of $F(s)$. The *inverse Laplace transformation* is defined by the complex inversion integral

$$f(t) = \frac{1}{2\pi j}\int_{\alpha-j\infty}^{\alpha-j\infty} F(s)e^{st}ds \tag{9-4}$$

The Laplace transformation is an integral transformation, since the direct process in Eq. (9-2) and inverse process in Eq. (9-4) involve integrations. The Fourier transformation treated in Chapter 16 is another example of an integral transformation.

Happily, formal evaluation of the complex inversion integral is not necessary because of the uniqueness property of the Laplace transformation. A symbolical statement of the *uniqueness property* is

$$\text{IF } \mathscr{L}\{f(t)\} = F(s) \text{ THEN } \mathscr{L}^{-1}\{F(s)\} \,(=)\, u(t)f(t)$$

The mathematical justification for this statement is beyond the scope of our treatment. However, the notation (=) means equal almost everywhere. The only points where equality may not hold is at the discontinuities of $f(t)$.

The waveform produced by the inverse transformation is zero for $t < 0$; that is, $f(t)$ is causal. The transform pair $[f(t) \leftrightarrow F(s)]$ is unique if and only if $f(t)$ is causal. For this reason Laplace transform-related waveforms are written as $[f(t)]u(t)$ to make their causality visible. For example, in the next section we find the Laplace transform of the sinusoid waveform $\cos \beta t$. In the context of Laplace transforms this signal is not the eternal sinusoid but a causal waveform $f(t) = [\cos \beta t]u(t)$. It is important to remember that causality and Laplace transforms go hand in hand when interpreting the results of circuit analysis.

9-2 BASIC PROPERTIES AND PAIRS

The previous section gave the definition of the Laplace transformation and showed that the transforms of some basic signals can be found using the integral definition. In this section we develop the basic properties of the Laplace transformation and show how these properties can be used to obtain additional transform pairs.

The *linearity property* of the Laplace transformation states that

$$\mathscr{L}\{Af_1(t) + Bf_2(t)\} = AF_1(s) + BF_2(s) \tag{9-5}$$

where A and B are constants. This property is easily established for the direct Laplace transformation integral in Eq. (9-2):

$$\begin{aligned}\mathscr{L}\{Af_1(t) + Bf_2(t)\} &= \int_0^\infty [Af_1(t) + Bf_2(t)]e^{-st}dt \\ &= A\int_0^\infty f_1(t)e^{-st}dt + B\int_0^\infty f_2(t)e^{-st}dt \\ &= AF_1(s) + BF_2(s)\end{aligned}$$

The formal definition of the inverse transformation in Eq. (9-4) is also a linear integration. It follows that

$$\mathscr{L}^{-1}\{AF_1(s) + BF_2(s)\} = Af_1(t) + Bf_2(t) \tag{9-6}$$

An important consequence of linearity is that for any constant K

$$\mathscr{L}\{Kf(t)\} = KF(s) \qquad \text{and} \qquad \mathscr{L}^{-1}\{KF(s)\} = Kf(t) \tag{9-7}$$

The linearity property is an extremely important feature. Linearity means that Kirchhoff's laws are valid in both the time and complex frequency domain, a fact we will use in the next chapter. The next two examples show how this property can be used to obtain the transform of the exponential rise waveform and a sinusoidal waveform.

EXAMPLE 9-4

Find the Laplace transform of the exponential rise waveform

$$f(t) = A(1 - e^{-\alpha t})u(t)$$

SOLUTION:

This waveform is the difference between a step function and an exponential. We can use the linearity property of Laplace transforms to write

$$\mathscr{L}\{A(1 - e^{-\alpha t})u(t)\} = A\mathscr{L}\{u(t)\} - A\mathscr{L}\{e^{-\alpha t}u(t)\}$$

The transforms of the step and exponential functions were found in Examples 9-1 and 9-2. Using linearity we conclude that the transform of the exponential rise is

$$F(s) = \frac{A}{s} - \frac{A}{s+\alpha} = \frac{A\alpha}{s(s+\alpha)}$$

EXAMPLE 9-5

Find the Laplace transform of the sinusoid $f(t) = A[\sin(\beta t)]u(t)$.

SOLUTION:

Using Euler's relationship we can express the sinusoid as a sum of exponentials.

$$e^{+j\beta t} = \cos\beta t + j\sin\beta t$$

$$e^{-j\beta t} = \cos\beta t - j\sin\beta t$$

Subtracting the second equation from the first yields

$$f(t) = A\sin\beta t = \frac{A(e^{j\beta t} - e^{-j\beta t})}{2j} = \frac{A}{2j}e^{j\beta t} - \frac{A}{2j}e^{-j\beta t}$$

The transform pair $\mathscr{L}\{e^{-\alpha t}\} = 1/(s+\alpha)$ found in Example 9-2 is valid even if the exponent α is complex. Using this fact and the linearity property, we obtain the transform of the sinusoid as

$$\mathcal{L}\{A \sin \beta t\} = \frac{A}{2j}\mathcal{L}\{e^{j\beta t}\} - \frac{A}{2j}\mathcal{L}\{e^{-j\beta t}\}$$
$$= \frac{A}{2j}\left[\frac{1}{s - j\beta} - \frac{1}{s + j\beta}\right]$$
$$= \frac{A\beta}{s^2 + \beta^2}$$

Integration Property

In the time domain the *i-v* relationships for capacitors and inductors involve integration and differentiation. Since we will be working in the s-domain it is important to establish the s-domain equivalents of these mathematical operations. Applying the integral definition of the Laplace transformation to a time domain integration yields

$$\mathcal{L}\left[\int_0^t f(\tau)d\tau\right] = \int_0^\infty \left[\int_0^t f(\tau)d\tau\right] e^{-st}dt \qquad (9\text{-}8)$$

The right side of this expression can be integrated by parts using

$$y = \int_0^t f(\tau)d\tau \qquad \text{and} \qquad dx = e^{-st}dt$$

These definitions result in

$$dy = f(t) \qquad \text{and} \qquad x = \frac{-e^{-st}}{s}$$

Using these factors reduces the right side of Eq. (9-8) to

$$\mathcal{L}\left[\int_0^t f(\tau)d\tau\right] = \left[\frac{-e^{-st}}{s}\int_0^t f(\tau)d\tau\right]_0^\infty + \frac{1}{s}\int_0^\infty f(t)e^{-st}dt \qquad (9\text{-}9)$$

The first term on the right in Eq. (9-9) vanishes at the lower limit because the integral over a zero length interval is zero provided $f(t)$ is finite at $t = 0$. It vanishes at the upper limit because e^{-st} approaches zero as t goes to infinity for $\sigma > 0$. By the definition of the Laplace transformation the second term on the right is $F(s)/s$. We conclude that

$$\mathcal{L}\left[\int_0^t f(\tau)d\tau\right] = \frac{F(s)}{s} \qquad (9\text{-}10)$$

The *integration property* states that the integration of a waveform $f(t)$ in the time domain can be accomplished in the s domain by the algebraic process of dividing its transform $F(s)$ by s. The next example applies the integration property to obtain the transform of the ramp function.

EXAMPLE 9-6

Find the Laplace transform of the ramp function $r(t) = tu(t)$.

SOLUTION:

From our study of signals we know that the ramp waveform can be obtained from $u(t)$ by integration.

$$r(t) = \int_0^t u(\tau)d\tau$$

In Example 9-1 we found $\mathscr{L}\{u(t)\} = 1/s$. Using these facts and the integration property of Laplace transforms we obtain

$$\mathscr{L}\{r(t)\} = \mathscr{L}\left[\int_0^t u(\tau)d\tau\right] = \frac{1}{s}\mathscr{L}\{u(t)\} = \frac{1}{s^2}$$

Differentiation Property

The time domain differentiation operation transforms into the s domain as follows:

$$\mathscr{L}\left[\frac{df(t)}{dt}\right] = \int_{0-}^{\infty}\left[\frac{df(t)}{dt}\right]e^{-st}dt \qquad (9\text{-}11)$$

The right side of this equation can be integrated by parts using

$$y = e^{-st} \qquad \text{and} \qquad dx = \frac{df(t)}{dt}dt$$

These definitions result in

$$dy = -se^{-st}dt \qquad \text{and} \qquad x = f(t)$$

Inserting these factors reduces the right side of Eq. (9-11) to

$$\mathscr{L}\left[\frac{df(t)}{dt}\right] = \left[f(t)e^{-st}\right]_{0-}^{\infty} + s\int_{0-}^{\infty} f(t)e^{-st}dt \qquad (9\text{-}12)$$

For $\sigma > 0$ the first term on the right side of Eq. (9-12) is zero at the upper limit because e^{-st} approaches zero as t goes to infinity. At the lower limit it reduces to $-f(0-)$. By the definition of the Laplace transform the second term on the right side is $sF(s)$. We conclude

$$\mathscr{L}\left[\frac{df(t)}{dt}\right] = sF(s) - f(0-) \qquad (9\text{-}13)$$

The *differentiation property* states that differentiating a waveform $f(t)$ in the time domain is accomplished in the s domain by the algebraic process of multiplying the transform $F(s)$ by s and subtracting the constant $f(0-)$.

The s-domain equivalent of a second derivative is obtained by repeated application of Eq. (9-13). We first define a waveform $g(t)$ as

$$g(t) = \frac{df(t)}{dt} \qquad \text{hence} \qquad \frac{d^2 f(t)}{dt^2} = \frac{dg(t)}{dt}$$

Applying the differentiation rule to these two equations yields

$$G(s) = sF(s) - f(0-) \qquad \text{and} \qquad \mathscr{L}\left[\frac{d^2 f(t)}{dt^2}\right] = sG(s) - g(0-)$$

Substituting the first of these equations into the second results in

$$\mathscr{L}\left[\frac{d^2 f(t)}{dt^2}\right] = s^2 F(s) - sf(0-) - f'(0-)$$

where

$$f'(0-) = \left.\frac{df}{dt}\right|_{t=0-}$$

Repeated application of this procedure produces the nth derivative.

$$\mathscr{L}\left[\frac{d^n f(t)}{dt^n}\right] = s^n F(s) - s^{n-1} f(0-) - s^{n-2} f'(0-) \ldots - f^{(n)}(0-) \tag{9-14}$$

where $f^{(n)}(0-)$ is the nth derivative of $f(t)$ evaluated at $t = 0-$.

A hallmark feature of the Laplace transformation is the fact that time integration and differentiation change into algebraic operations in the s domain. This observation gives us our first hint as to why it is often easier to work with circuits and signals in the s domain. The next example shows how the differentiation rule can be used to obtain additional transform pairs.

EXAMPLE 9-7

Find the Laplace transform of $f(t) = [\cos \beta t]u(t)$.

SOLUTION:

We can express $\cos \beta t$ in terms of the derivative of $\sin \beta t$ as

$$\cos \beta t = \frac{1}{\beta} \frac{d}{dt} \sin \beta t$$

In Example 9-5 we found $\mathscr{L}\{\sin \beta t\} = \beta/(s^2 + \beta^2)$. Using these facts and the differentiation rule, we can find the Laplace transform of $\cos \beta t$ as follows:

$$\mathscr{L}\{\cos \beta t\} = \frac{1}{\beta}\mathscr{L}\left\{\frac{d}{dt}\sin \beta t\right\} = \frac{1}{\beta}\left[s\left(\frac{\beta}{s^2+\beta^2}\right) - \sin(0-)\right]$$
$$= \frac{s}{s^2+\beta^2}$$

In this section we derived the basic transform properties listed in Table 9-1. The Laplace transformation has other properties that are useful in signal-processing applications. We treat some of these properties in the last three sections of this chapter. However, the basic properties in Table 9-1 are used most frequently in circuit analysis and are sufficient for nearly all of the applications in this book.

Similarly, Table 9-2 lists a basic set of Laplace transform pairs that is sufficient for most of the applications in this book. Except for the damped ramp, damped sine, and damped cosine, all of these pairs were derived in the preceding two sections. The Laplace transforms of these damped waveforms are listed in Table 9-2, but are derived in Section 9-7.

The last example in this section shows how to use the properties and pairs in Tables 9-1 and 9-2 to obtain the transform of a waveform not listed in the tables.

TABLE 9-1 BASIC LAPLACE TRANSFORMATION PROPERTIES

Properties	Time Domain	Frequency Domain
Independent variable	t	s
Signal representation	$f(t)$	$F(s)$
Uniqueness	$\mathscr{L}^{-1}\{F(s)\}\ (=)\ [f(t)]u(t)$	$\mathscr{L}\{f(t)\} = F(s)$
Linearity	$Af_1(t) + Bf_2(t)$	$AF_1(s) + BF_2(s)$
Integration	$\int_0^t f(\tau)d\tau$	$\frac{F(s)}{s}$
Differentiation	$\frac{df(t)}{dt}$	$sF(s) - f(0-)$
	$\frac{d^2f(t)}{dt^2}$	$s^2F(s) - sf(0-) - f'(0-)$
	$\frac{d^3f(t)}{dt^3}$	$s^3F(s) - s^2f(0-) - sf'(0-) - f''(0-)$

TABLE 9-2

BASIC LAPLACE TRANSFORM PAIRS

Signal	Waveform $f(t)$	Transform F(s)
Impulse	$\delta(t)$	1
Step function	$u(t)$	$\frac{1}{s}$
Ramp	$tu(t)$	$\frac{1}{s^2}$
Exponential	$[e^{-\alpha t}]u(t)$	$\frac{1}{s+\alpha}$
Damped ramp	$[te^{-\alpha t}]u(t)$	$\frac{1}{(s+\alpha)^2}$
Sine	$[\sin \beta t]u(t)$	$\frac{\beta}{s^2+\beta^2}$
Cosine	$[\cos \beta t]u(t)$	$\frac{s}{s^2+\beta^2}$
Damped sine	$[e^{-\alpha t}\sin \beta t]u(t)$	$\frac{\beta}{(s+\alpha)^2+\beta^2}$
Damped cosine	$[e^{-\alpha t}\cos \beta t]u(t)$	$\frac{(s+\alpha)}{(s+\alpha)^2+\beta^2}$

EXAMPLE 9-8

Find the Laplace transform of the waveform

$$f(t) = 2u(t) - 5[e^{-2t}]u(t) + 3[\cos 2t]u(t) + 3[\sin 2t]u(t)$$

SOLUTION:

Using the linearity property we write the transform of $f(t)$ in the form

$$\mathscr{L}\{f(t)\} = 2\mathscr{L}\{u(t)\} - 5\mathscr{L}\{[e^{-2t}]u(t)\} + 3\mathscr{L}\{[\cos 2t]u(t)\} + 3\mathscr{L}\{[\sin 2t]u(t)\}$$

The transform of each term in this sum are listed in Table 9-2:

$$F(s) = \frac{2}{s} - \frac{5}{s+2} + \frac{3s}{s^2+4} + \frac{6}{s^2+4}$$

Normally a Laplace transform is written as a quotient of polynomials rather than as a sum of terms. Rationalizing the above sum yields

$$F(s) = \frac{16(s^2 + 1)}{s(s + 2)(s^2 + 4)}$$

EXERCISE 9-1

Find the Laplace transforms of the following waveforms.

(a) $f(t) = [e^{-2t}]u(t) + 4tu(t) - u(t)$

(b) $f(t) = [2 + 2\sin 2t - 2\cos 2t]u(t)$

ANSWERS:

(a) $F(s) = \dfrac{2(s + 4)}{s^2(s + 2)}$

(b) $F(s) = \dfrac{4(s + 2)}{s(s^2 + 4)}$

EXERCISE 9-2

Find the Laplace transforms of the following waveforms.

(a) $f(t) = [e^{-4t}]u(t) + 5\int_0^t \sin 4x \, dx$

(b) $f(t) = 5[e^{-40t}]u(t) + \dfrac{d[5te^{-40t}]u(t)}{dt}$

ANSWERS:

(a) $F(s) = \dfrac{s^3 + 36s + 80}{2(s + 4)(s^2 + 16)}$

(b) $F(s) = \dfrac{10s + 200}{(s + 40)^2}$

EXERCISE 9-3

Find the Laplace transforms of the following waveforms.

(a) $f(t) = A[\cos(\beta t - \phi)]u(t)$

(b) $f(t) = A[e^{-\alpha t}\cos(\beta t - \phi)]u(t)$

ANSWERS:

(a) $F(s) = A\cos\phi\left[\dfrac{s + \beta\tan\phi}{s^2 + \beta^2}\right]$

(b) $F(s) = A\cos\phi\left[\dfrac{s + \alpha + \beta\tan\phi}{(s + \alpha)^2 + \beta^2}\right]$

EXERCISE 9-4

Find the waveforms corresponding to the following Laplace transforms.

(a) $F(s) = \dfrac{10s}{s^2 + 25} - \dfrac{5}{s + 10} + 2$

(b) $F(s) = 2\dfrac{s + 20}{(s + 20)^2 + 20^2} + \dfrac{4}{s^2}$

ANSWERS:

(a) $f(t) = [10\cos 5t - 5e^{-10t}]u(t) + 2\delta(t)$

(b) $f(t) = [2e^{-20t}\cos 20t + 4t]u(t)$

9-3 POLE-ZERO DIAGRAMS

The transforms for signals in Table 9-2 are ratios of polynomials in the complex frequency variable s. Likewise, the transform found in Example 9-8 takes the form of a ratio of two polynomials in s. These observations are a general result for most signals, so transforms usually take the following form:

$$F(s) = \frac{b_m s^m + b_{m-1} s^{m-1} + \ldots + b_1 s + b_0}{a_n s^n + a_{n-1} s^{n-1} + \ldots + a_1 s + a_0} \tag{9-15}$$

If numerator and denominator polynomials are expressed in factored form, then $F(s)$ is written as

$$F(s) = K \frac{(s - z_1)(s - z_2) \ldots (s - z_m)}{(s - p_1)(s - p_2) \ldots (s - p_n)} \tag{9-16}$$

where the constant $K = b_m/a_n$ is called the *scale factor*

The roots of the numerator and denominator polynomials, together with the scale factor K, uniquely define a transform $F(s)$. The denominator roots are called *poles* because for $s = p_k$ $k = 1, 2, \ldots n$) the denominator vanishes and $F(s)$ becomes infinite. The roots of the numerator polynomial are called *zeros* because the transform $F(s)$ vanishes for $s = z_k$ ($k = 1, 2, \ldots m$). Collectively the poles and zeros are called *critical frequencies* because they are values of s at which $F(s)$ does dramatic things, like vanish, or blow up.

In the s domain we can specify a signal transform by listing the location of its critical frequencies together with the scale factor K. That is, in the frequency domain we describe signals in terms of poles and zeros. The description takes the form of a plot of the location of poles and zeros in the complex s plane. The pole locations in such plots are indicated by a small "×" and the zeros by a small "○". In the frequency domain the independent variable $s = \sigma + j\omega$ is complex and the poles or zeros can be complex as well. In the s plane we use a horizontal axis to plot the value of the real part of s and a vertical j axis to plot the imaginary part. The j axis is an important boundary in the s domain because it divides the s plane into two distinct half-planes. The real part of s is negative in the left-half plane and positive in the right-half plane. As we will soon see, the sign of the real part of a pole has a profound effect on the form of the corresponding waveform.

For example, Table 9-2 shows that $F(s) = 1/(s + 1)$ is the transform of the exponential waveform $f(t) = e^{-t}u(t)$. The exponential signal has a single pole at $s = -1$ and no finite zeros. We can portray this exponential in the s domain by the *pole-zero diagram* shown in Fig. 9-2(a). For this $F(s)$ the "×" for the

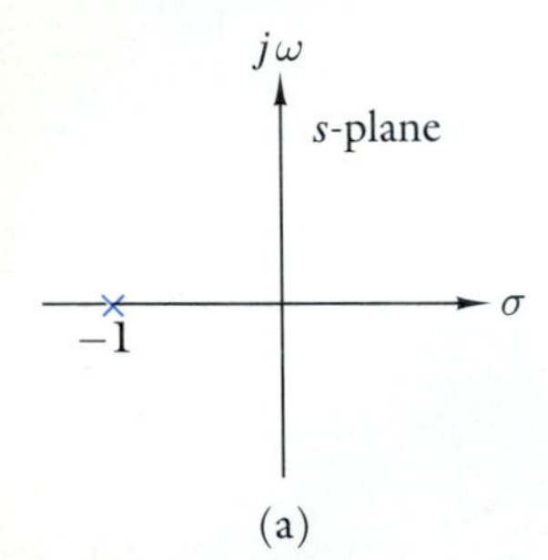

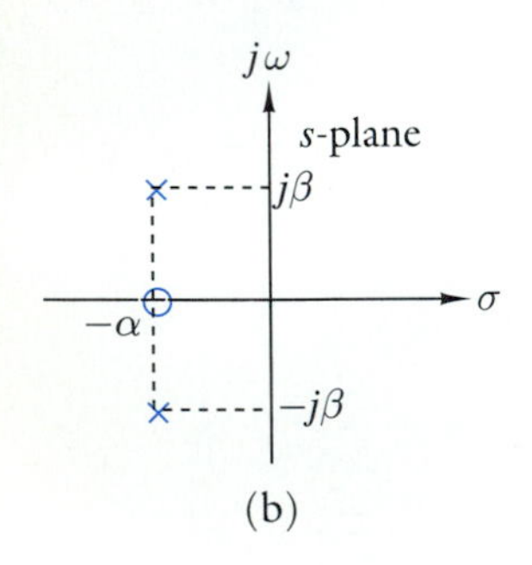

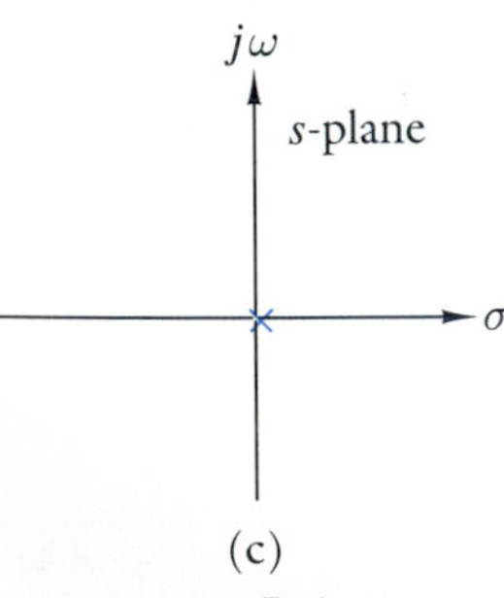

Figure 9-2 Pole-zero diagrams in the s plane.

pole is located at $-1 + j0$, which is on the negative real axis in the left-half plane.

The damped sinusoid $f(t) = [Ae^{-\alpha t} \cos \beta t]u(t)$ is an example of a signal with complex poles. From Table 9-2 the corresponding transform is

$$F(s) = \frac{A(s+\alpha)}{(s+\alpha)^2 + \beta^2}$$

The transform $F(s)$ has a finite zero on the real axis at $s = -\alpha$. The roots of the denominator polynomial are $s = -\alpha \pm j\beta$. The resulting pole-zero diagram is shown in Fig. 9-2(b). The poles of the damped cosine do not lie on either axis in the s plane because neither the real nor imaginary parts are zero.

Finally, the transform of a unit step function is $1/s$. The step function transform has no finite zeros and a pole at the origin ($s = 0 + j0$) in the s plane as shown in Fig. 9-2(c). The poles in all of the diagrams of Fig. 9-2 lie in the left-half plane or on the j-axis boundary.

The diagrams in Fig. 9-2 show the poles and zeros in the finite part of the s plane. Signal transforms may have poles or zeros at infinity as well. For example, the step function has a zero at infinity since $F(s) = 1/s$ approaches zero as $s \rightarrow \infty$. In general, a transform $F(s)$ given by Eq. (9-16) has a zero of order $n - m$ at infinity if $n > m$ and a pole of order $m - n$ at infinity if $n < m$. Thus, the number of zeros equals the number of poles if we include those at infinity.

These examples demonstrate that we can identify the signal represented in a pole-zero diagram without transforming it back into the time domain. In other words, the pole-zero diagram is the s domain portrayal of the signal, just as a plot of the waveform versus time depicts the signal in the t domain. The utility of a pole-zero diagram as a description of circuits and signals will become clearer as we develop additional s-domain analysis and design concepts.

EXAMPLE 9-9

Find the poles and zeros of the waveform

$$f(t) = \left[e^{-2t} + \cos 2t - \sin 2t\right] u(t)$$

SOLUTION:

Using the linearity property and the basic pairs in Table 9-2, we write the transform in the form

$$F(s) = \frac{1}{s+2} + \frac{s}{s^2+4} - \frac{2}{s^2+4}$$

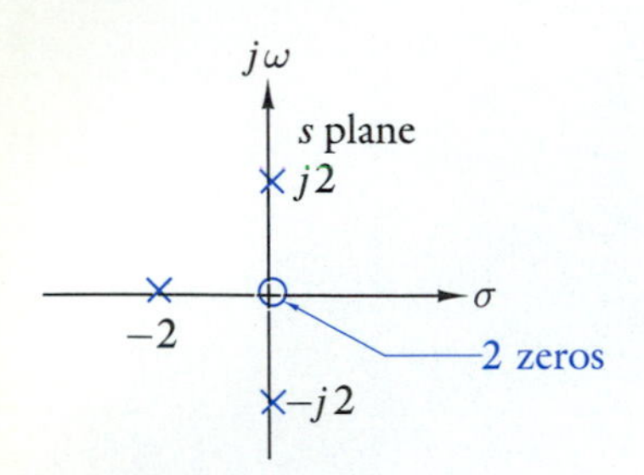

Figure 9-3

Rationalizing this expression yields $F(s)$:

$$F(s) = \frac{2s^2}{(s+2)(s^2+4)} = \frac{2s^2}{(s+2)(s+j2)(s-j2)}$$

This transform has three zeros and three poles. There are two zeros at $s = 0$ and one at $s = \infty$. There is a pole on the negative real axis at $s = -2 + j0$, and two poles on the imaginary axis at $s = \pm j2$. The resulting pole-zero diagram is shown in Fig. 9-3. Reviewing the analysis we can trace the poles to the components of $f(t)$. The pole on the real axis at $s = -2$ came from the exponential e^{-2t}, while the complex conjugate poles on the j axis came from the sinusoid $\cos 2t - \sin 2t$. The zeros, however, are not traceable to specific components. Their locations depend on all three components.

EXERCISE 9-5

Find the poles and zeros of the following waveforms.

(a) $f(t) = \left[-2e^{-t} - t + 2\right]u(t)$
(b) $f(t) = [4 - 3\cos\beta t]\,u(t)$
(c) $f(t) = \left[e^{-\alpha t}\cos\beta t + (\alpha/\beta)e^{-\alpha t}\sin\beta t\right]u(t)$

ANSWERS:

(a) Zeros: $s = 1,\ s = \infty$ (2). Poles: $s = 0$ (2), $s = -1$.
(b) Zeros: $s = \pm j2\beta,\ s = \infty$. Poles: $s = 0,\ s = \pm j\beta$.
(c) Zeros: $s = -2\alpha,\ s = \infty$. Poles: $s = -\alpha \pm j\beta$.

EXERCISE 9-6

A transform has poles at $s = -3 \pm j6$ and $s = -2$ and finite zeros at $s = 0$ and $s = -1$. Write $F(s)$ as a quotient of polynomials in s.

ANSWER:

$$F(s) = K\frac{s^2 + s}{s^3 + 8s^2 + 57s + 90}$$

9-4 INVERSE LAPLACE TRANSFORMS

The inverse transformation converts a transform $F(s)$ into the corresponding waveform $f(t)$. The uniqueness property of Laplace transforms makes it possible to go from a transform to a waveform when we can find $F(s)$ in a table of pairs. However, it does not take a very complicated circuit or signal before we exceed the capability of Table 9-2, or even the more extensive tables that are available. However, there is a general inverse transform method that decomposes $F(s)$ into terms that are listed in Table 9-2.

For linear circuits the transforms of interest are ratios of polynomials in s. In mathematics such functions are called *ratio-*

nal functions . To perform the inverse transformation we must find the waveform corresponding to rational functions of the form

$$F(s) = K\frac{(s - z_1)(s - z_2)\dots(s - z_m)}{(s - p_1)(s - p_2)\dots(s - p_n)} \tag{9-17}$$

where the K is the scale factor, $z_k (k = 1, 2, \dots m)$ are the zeros, and $p_k(k = 1, 2, \dots n)$ are the poles of $F(s)$.

If there are more finite poles than finite zeros ($n > m$), then $F(s)$ is called a *proper rational function* . If the denominator in Eq. (9-17) has no repeated roots ($p_i \neq p_j$ for $i \neq j$), then $F(s)$ is said to have *simple poles* . In this section we treat the problem of finding the inverse transform of proper rational functions with simple poles. The problem of improper rational functions and multiple poles is covered in the next section.

If a proper rational function has only simple poles, then it can be decomposed into a partial fraction expansion.

$$F(s) = \frac{k_1}{s - p_1} + \frac{k_2}{s - p_2} + \dots + \frac{k_n}{s - p_n} \tag{9-18}$$

In this case, $F(s)$ can be expressed as a linear combination of terms with one term for each of its n simple poles. The k's associated with each term are called *residues* .

Each term in the partial fraction decomposition has the form of the transform of an exponential signal. That is, we recognize that $\mathscr{L}^{1}\{k/(s+\alpha)\} = [ke^{-\alpha t}]u(t)$. We can now write the corresponding waveform using the linearity property

$$f(t) = \left[k_1 e^{p_1 t} + k_2 e^{p_2 t} + \dots + k_n e^{p_n t}\right] u(t) \tag{9-19}$$

In the time domain, the s-domain poles appear as the exponents in exponential waveforms and the residues at the poles become the waveform amplitudes.

Given the poles of $F(s)$, finding the inverse transform $f(t)$ reduces to finding the residues. To illustrate the procedure consider a case where $F(s)$ has three simple poles and one finite zero:

$$F(s) = K\frac{(s - z_1)}{(s - p_1)(s - p_2)(s - p_3)} = \frac{k_1}{s - p_1} + \frac{k_2}{s - p_2} + \frac{k_3}{s - p_3}$$

We find the residue k_1 by first multiplying this equation through by the factor $(s - p_1)$

$$(s-p_1)F(s) = K\frac{(s - z_1)}{(s - p_2)(s - p_3)} = k_1 + \frac{k_2(s - p_1)}{s - p_2} + \frac{k_3(s - p_1)}{s - p_3}$$

If we now set $s = p_1$ the last two terms on the right vanish leaving

$$k_1 = (s - p_1)F(s)|_{s=p_1} = K\frac{(s - z_1)}{(s - p_2)(s - p_3)}\bigg|_{s=p_1}$$

Using the same approach for k_2 yields

$$k_2 = (s - p_2)F(s)|_{s=p_2} = K\frac{(s - z_1)}{(s - p_1)(s - p_3)}\bigg|_{s=p_2}$$

The technique generalizes so that the residue at any simple pole p_i is

$$k_i = (s - p_i)F(s)\,|_{s=p_i} \qquad (9\text{-}20)$$

The process of determining the residue at any simple pole is sometimes called the *cover-up algorithm* because we temporarily remove (cover-up) the factor $(s - p_i)$ in $F(s)$, and then evaluate the remainder at $s = p_i$).

EXAMPLE 9-10

Find the waveform corresponding to the transform

$$F(s) = 2\frac{(s + 3)}{s(s + 1)(s + 2)}$$

SOLUTION:

$F(s)$ has simple poles at $s = 0$, $s = -1$, $s = -2$. Its partial fraction expansion is

$$F(s) = \frac{k_1}{s} + \frac{k_2}{s + 1} + \frac{k_3}{s + 2}$$

The cover-up algorithm yields the residues as

$$k_1 = sF(s)|_{s=0} = \frac{2(s + 3)}{(s + 1)(s + 2)}\bigg|_{s=0} = 3$$

$$k_2 = (s + 1)F(s)|_{s=-1} = \frac{2(s + 3)}{s(s + 2)}\bigg|_{s=-1} = -4$$

$$k_3 = (s + 2)F(s)|_{s=-2} = \frac{2(s + 3)}{s(s + 1)}\bigg|_{s=-2} = 1$$

The inverse transform $f(t)$ is

$$f(t) = \left[3 - 4e^{-t} + e^{-2t}\right]u(t)$$

EXERCISE 9-7

Find the waveforms corresponding to the transforms.

(a) $F(s) = \dfrac{4}{(s + 1)(s + 3)}$

(b) $F(s) = \dfrac{4(s + 2)}{(s + 1)(s + 3)}$

ANSWERS:

(a) $f(t) = [2e^{-t} - 2e^{-3t}]u(t)$

(b) $f(t) = [2e^{-t} + 2e^{-3t}]u(t)$ ■

■ EXERCISE 9-8

Find the waveforms corresponding to the following transforms.

(a) $F(s) = \dfrac{6(s+2)}{s(s+1)(s+4)}$

(b) $F(s) = \dfrac{4(s+1)}{s(s+1)(s+4)}$

ANSWERS:

(a) $f(t) = [3 - 2e^{-t} - e^{-4t}]u(t)$

(b) $f(t) = [1 - e^{-4t}]u(t)$ ■

Complex Poles

Special treatment is necessary when $F(s)$ has complex poles. In physical situations the function $F(s)$ is a ratio of polynomials with real coefficients. If $F(s)$ has a complex pole $p = -\alpha + j\beta$, then it must also have a pole $p^* = -\alpha - j\beta$; otherwise, the coefficients of the denominator polynomial would not be real. In other words, for physical signals the complex poles of $F(s)$ must occur in conjugate pairs. As a consequence, the partial fraction decomposition of $F(s)$ will contain two terms of the form

$$F(s) = \ldots + \frac{k}{s+\alpha-j\beta} + \frac{k^*}{s+\alpha+j\beta} + \ldots \tag{9-21}$$

The residues k and k^* at the conjugate poles are themselves conjugates because $F(s)$ is a rational function with real coefficients. These residues can be calculated using the cover-up algorithm and, in general, they turn out to be complex numbers. If the complex residues are written in polar form as

$$k = |k|e^{j\theta} \qquad \text{and} \qquad k^* = |k|e^{-j\theta}$$

then the waveform corresponding to the two terms in Eq. (9-21) are

$$f(t) = [\ldots + |k|e^{j\theta}e^{(-\alpha+j\beta)t} + |k|e^{-j\theta}e^{(-\alpha-j\beta)t} + \ldots]u(t)$$

This equation can be rearranged in the form

$$f(t) = [\ldots + 2|k|e^{-\alpha t}\left\{\frac{e^{+j(\beta t+\theta)} + e^{-j(\beta t+\theta)}}{2}\right\} + \ldots]u(t) \tag{9-22}$$

The expression inside the braces is of the form

$$\cos x = \left\{\frac{e^{+jx} + e^{-jx}}{2}\right\}$$

Consequently, we combine terms inside the braces as a cosine function with a phase angle:

$$f(t) = [\ldots + 2|k|e^{-\alpha t}\cos(\beta t + \theta) + \ldots]u(t) \tag{9-23}$$

In summary, if $F(s)$ has a complex pole, then in physical applications there must be an accompanying conjugate complex pole. The inverse transformation combines the two poles to produce a damped cosine waveform. We only need to compute the residue at one of these poles because the residues at conjugate poles must be conjugates. Normally, we calculate the residue for the pole at $s = -\alpha + j\beta$ because its angle equals the phase angle of the damped cosine. Note that the imaginary part of this pole is positive, which means that the pole lies in the upper half of the s plane.

The inverse transform of a proper rational function with simple poles can be found by the partial fraction expansion method. The residues (k) at the simple poles can be found using the cover-up algorithm. The resulting waveform is a sum of terms of the form $[ke^{-\alpha t}]u(t)$ for real poles and $[2|k|e^{-\alpha t}\cos(\beta t + \angle k)\}u(t)$ for a pair of complex conjugate poles. All of the data needed to construct the waveform are available in the partial fraction expansion of the transform.

EXAMPLE 9-11

Find the inverse transform of

$$F(s) = \frac{20(s+3)}{(s+1)(s^2+2s+5)}$$

SOLUTION:

$F(s)$ has a simple pole at $s = -1$ and a pair of conjugate complex poles located at the roots of the quadratic factor

$$(s^2 + 2s + 5) = (s + 1 - j2)(s + 1 + j2)$$

The partial fraction expansion of $F(s)$ is

$$F(s) = \frac{k_1}{s+1} + \frac{k_2}{s+1-j2} + \frac{k_2^*}{s+1+j2}$$

The residues at the poles are found from the cover-up algorithm.

$$k_1 = \left.\frac{20(s+3)}{s^2+2s+5)}\right|_{s=-1} = 10$$

$$k_2 = \left.\frac{20(s+3)}{(s+1)(s+1+j2)}\right|_{s=-1+j2}$$

$$= -5 - j5 = 5\sqrt{2}e^{+j5\pi/4}$$

We now have all of the data needed to construct the inverse transform.

$$f(t) = [10e^{-t} + 10\sqrt{2}e^{-t}\cos(2t + 5\pi/4)]u(t)$$

In this example we used k_2 to obtain the amplitude and phase angle of the damped cosine term. The residue k_2^* is not needed, but to illustrate a point we note that its value is

$$k_2^* = (-5 - j5)^* = -5 + j5 = 5\sqrt{2}e^{-j5\pi/4}$$

If k_2^* is used instead of k_2, we get the same amplitude for the damped sine but the wrong phase angle. *Caution:* Remember that Eq. (9-23) uses the residue at the complex pole with *a positive* . In this example this is the pole at $s = -1 + j2$, not the pole at $s = -1 - j2$.

Sums of Residues

The sum of the residues of a proper rational function are subject to certain conditions that are useful for checking the calculations in a partial fraction expansion. To derive these conditions we multiply Eqs. (9-17) and (9-18) by s and take the limit as $s \to \infty$. These operations yield

$$\lim_{s\to\infty} sF(s) = \lim_{s\to\infty} \frac{Ks^{m+1}}{s^n} = \lim_{s\to\infty}\left(\frac{k_1 s}{s+p_1} + \ldots + \frac{k_n s}{s+p_n}\right)$$

In the limit this equation reduces to

$$K\left[\lim_{s\to\infty}\frac{s^{m+1}}{s^n}\right] = k_1 + k_2 + \ldots + k_n$$

Since $F(s)$ is a proper rational function with $n > m$, the passing to the limit in this equation yields to following conditions:

$$\begin{aligned} &\text{IF } n > m+1 \text{ THEN } k_1 + k_2 + \ldots + k_n = 0 \\ &\text{ELSE} \qquad\qquad\qquad\qquad\qquad\qquad (9\text{-}24) \\ &\text{IF } n = m+1 \text{ THEN } k_1 + k_2 + \ldots + k_n = K \end{aligned}$$

For a proper rational function with simple poles the sum of residues is either zero or else equal to the transform scale factor K.

The conditions in Eq. (9-24) are used to check the residue calculations. For the examples in this section we obtain

Example 9-10: $n > m+1$, and $k_1 + k_2 + k_3 = 3 - 4 + 1 = 0$

Example 9-11: $n > m+1$ and $k_1 + k_2 + k_2^* =$

$$10 + (-5 - j5) + (-5 + j5) = 0$$

EXERCISE 9-9

Find the inverse transforms of the following rational functions

(a) $F(s) = \dfrac{16}{(s+2)(s^2+4)}$

(b) $F(s) = \dfrac{2(s+2)}{s(s^2+4)}$

ANSWERS:

(a) $f(t) = \left[2e^{-2t} + 2\sqrt{2}\cos(2t - 3\pi/4)\right]u(t)$

(b) $f(t) = \left[1 + \sqrt{2}\cos(2t - 3\pi/4)\right]u(t)$

EXERCISE 9-10

Find the inverse transforms of the following rational functions

(a) $F(s) = \dfrac{8}{s(s^2+4s+8)}$

(b) $F(s) = \dfrac{4s}{s^2+4s+8}$

ANSWERS:

(a) $f(t) = \left[1 + \sqrt{2}e^{-2t}\cos(2t + 3\pi/4)\right]u(t)$

(b) $f(t) = \left[4\sqrt{2}e^{-2t}\cos(2t + \pi/4)\right]u(t)$

9-5 SOME SPECIAL CASES

Most of the transforms encountered in physical applications are proper rational functions with simple poles. The inverse transforms of such functions can be handled by the partial fraction expansion method developed in the previous section. This section covers the problem of finding the inverse transform when $F(s)$ is an improper rational function or has multiple poles. These matters are treated as special cases because they rarely occur in real circuits. The reason is that they are unique situations that only occur for discrete values of circuit or signal parameters. However, some of these special cases are important, so we need to learn how to handle improper rational functions and multiple poles.

$F(s)$ is an *improper rational function* when the order of the numerator polynomial equals or exceeds the order of the denominator ($m \geq n$). For example, the transform

$$F(s) = \frac{s^3 + 6s^2 + 12s + 8}{s^2 + 4s + 3} \tag{9-25}$$

is improper because $m = 3$ and $n = 2$. This function can be reduced to a quotient and a remainder, which is a proper function by long division.

$$\begin{aligned} F(s) &= s + 2 + \frac{s+2}{s^2+4s+3} \\ &= \text{Quotient} + \text{Remainder} \end{aligned}$$

The remainder is a proper rational function, which can be expanded by partial fractions.

$$F(s) = s + 2 + \frac{1/2}{s+1} + \frac{1/2}{s+3}$$

All of the terms in this expansion are listed in Table 9-2 except the first term. The inverse transform of the first term is found using the transform of an impulse and the differentiation property. The Laplace transform of the derivative of an impulse is

$$\mathscr{L}\left[\frac{d\delta(t)}{dt}\right] = s\mathscr{L}[\delta(t)] - \delta(0-) = s$$

since $\mathscr{L}\{\delta(t)\} = 1$ and $\delta(0-) = 0$. By the uniqueness property of the Laplace transformation we have, $\mathscr{L}^{-1}\{s\} = d\delta(t)/dt$. The inverse transform of the improper rational function in Eq. (9-25) is

$$f(t) = \frac{d\delta(t)}{dt} + 2\delta(t) + \left[\frac{1}{2}e^{-t} + \frac{1}{2}e^{-3t}\right]u(t)$$

The method illustrated by this example generalizes in the following way. If $m = n$, long division produces a quotient K and a proper rational function remainder. The quotient K corresponds to a scaled impulse $K\delta(t)$, and partial fraction expansion identifies the waveform corresponding to the remainder. If $m > n$, then long division yields a quotient with terms like $s, s^2, \ldots s^{m-n}$ before a proper remainder function is obtained. These higher powers of s correspond to derivatives of the impulse. While these pathological waveforms are theoreticaly possible, they do not actually occur in real circuits.

However, improper rational functions can arise during mathematical manipulation of signal transforms. The problem is that partial fractions will appear to work when applied to an improper rational function. That is, the partial fraction method does not tell you that it gives the wrong answer when applied to an improper rational function. Therefore, it is essential to check $F(s)$ and reduce it by long division if it is improper, before performing a partial fraction expansion on the remainder.

Exercise 9-11

Find the inverse transforms of the following functions.

(a) $F(s) = \dfrac{s^2 + 4s + 5}{s^2 + 4s + 3}$

(b) $F(s) = \dfrac{s^2 - 4}{s^2 + 4}$

ANSWERS:

(a) $f(t) = \delta(t) + \left[e^{-t} - e^{-3t}\right]u(t)$

(b) $f(t) = \delta(t) - [4\sin(2t)]u(t)$ ■

EXERCISE 9-12

Find the inverse transforms of the following functions.

(a) $F(s) = \dfrac{2s^2 + 3s + 5}{s}$

(b) $F(s) = \dfrac{s^3 + 2s^2 + s + 3}{s + 2}$

ANSWERS:

(a) $f(t) = 2\dfrac{d\delta(t)}{dt} + 3\delta(t) + 5u(t)$

(b) $f(t) = \dfrac{d^2\delta(t)}{dt^2} + \delta(t) + \left[e^{-2t}\right]u(t)$ ■

Multiple Poles

Under certain special conditions transforms can have multiple poles. For example, the transform

$$F(s) = \frac{K(s - z_1)}{(s - p_1)(s - p_2)^2} \tag{9-26}$$

has a simple pole at $s = p_1$ and a pole of order 2 at $s = p_2$. Finding the inverse transform of this function requires special treatment because of the multiple pole. We first factor out one of the two multiple poles:

$$F(s) = \frac{1}{s - p_2}\left[\frac{K(s - z_1)}{(s - p_1)(s - p_2)}\right] \tag{9-27}$$

The quantity inside the bracket is a proper rational function with only simple poles and can be expanded by partial fractions using the method of the previous section.

$$F(s) = \frac{1}{s - p_2}\left[\frac{c_1}{s - p_1} + \frac{k_{22}}{s - p_2}\right]$$

We now multiply through by the pole factored out in the first step to get

$$F(s) = \frac{c_1}{(s - p_1)(s - p_2)} + \frac{k_{22}}{(s - p_2)^2}$$

The first term on the right is a proper rational function with only simple poles, so it too can be expanded by partial fractions as

$$F(s) = \frac{k_1}{s - p_1} + \frac{k_{21}}{s - p_2} + \frac{k_{22}}{(s - p_2)^2}$$

After two partial fraction expansions we have an expression in which every term is available in Table 9-2. The first two terms are simple poles that lead to exponential waveforms. The third term is of the form $k/(s+\alpha)^2$, which is the transform of a damped ramp waveform $[kte^{-\alpha t}]u(t)$. Therefore, the inverse transform of $F(s)$ in Eq. (9-26) is

$$f(t) = \left[k_1 e^{p_1 t} + k_{12} e^{p_2 t} + k_{22} t e^{p_2 t}\right] u(t) \qquad (9\text{-}28)$$

Caution: If $F(s)$ in Eq. (9-26) had another finite zero, then the term in the brackets in Eq. (9-27) would be an improper rational function. When this occurs, long division must be used to reduce the improper rational function before proceeding to the partial-fraction expansion in the next step.

As with simple poles, the s-domain location of multiple poles determines the exponents of the exponential waveforms. The residues at the poles are the amplitudes of the waveforms. The only difference here is that the double pole leads to two terms rather than a single waveform. The first term is an exponential, and the second term is a damped ramp.

The procedure outlined in this example can be applied to higher-order poles, although the process becomes quite tedious. For example, an nth order pole would require n partial fraction expansions, which is not an idea with irresistible appeal. But as a practical matter, multiplicities greater than 2 require unique circumstances which simply do not happen in real circuits. For practical purposes the contribution of multiple poles to inverse transforms can be covered using two pairs:

$$\mathscr{L}^{-1}\left[\frac{k}{(s+\alpha)^2}\right] = [kte^{-\alpha t}]u(t) \qquad (9\text{-}29)$$

for a real pole of multiplicity 2, and for a pair of complex poles of multiplicity 2:

$$\mathscr{L}^{-1}\left[\frac{k}{(s+\alpha-j\beta)^2} + \frac{k^*}{(s+\alpha+j\beta)^2}\right]$$
$$= \left[2|k|te^{-\alpha t}\cos(\beta t + \angle k)\right] u(t) \qquad (9\text{-}30)$$

EXAMPLE 9-12

Find the inverse transform of

$$F(s) = \frac{4(s+3)}{s(s+2)^2}$$

SOLUTION:

The given transform has a simple pole at $s = 0$ and a double pole at $s = -2$. Factoring out one of the multiple poles and expanding the remainder by partial fractions yields

$$F(s) = \frac{1}{s+2}\left[\frac{4(s+3)}{s(s+2)}\right] = \frac{1}{s+2}\left[\frac{6}{s} - \frac{2}{s+2}\right]$$

Multiplying through by the removed factor and expanding again by partial fractions produces

$$F(s) = \frac{6}{s(s+2)} - \frac{2}{(s+2)^2} = \frac{3}{s} - \frac{3}{s+2} - \frac{2}{(s+2)^2}$$

The last expansion on the right yields the inverse transform as

$$f(t) = \left[3 - 3e^{-2t} - 2te^{-2t}\right]u(t)$$

EXERCISE 9-13

Find the inverse transforms of the following functions.

(a) $F(s) = \dfrac{s}{(s+1)(s+2)^2}$

(b) $F(s) = \dfrac{16}{s^2(s+4)}$

ANSWERS:

(a) $f(t) = \left[-e^{-t} + e^{-2t} + 2te^{-2t}\right]u(t)$

(b) $f(t) = \left[4t - 1 + e^{-4t}\right]u(t)$

9-6 CIRCUIT RESPONSE USING LAPLACE TRANSFORMS

The payoff for learning about the Laplace transformation comes when we use it to find the response of dynamic circuits. The pattern for circuit analysis is shown by the solid line in Fig. 9-1. The basic analysis steps are as follows:

STEP 1: Develop the circuit differential equation in the time domain.

STEP 2: Transform this equation in the s domain and algebraically solve for the response transform.

STEP 3: Apply the inverse transformation to this transform to produce the response waveform.

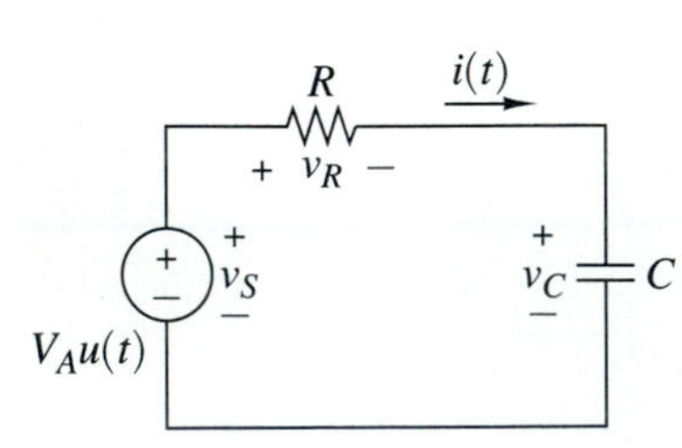

Figure 9-4 First-order RC circuit.

STEP 1: The first-order RC circuit in Fig. 9-4 will be used to illustrate these steps. The KVL equation around the loop and the element i-v relationship or element equations

$$\text{KVL:} \quad -v_S(t) + v_R(t) + v_C(t) = 0$$

$$\text{Source:} \quad v_S(t) = V_A u(t)$$

$$\text{Resistor:} \quad v_R(t) = i(t)R$$

$$\text{Capacitor:} \quad i(t) = C\frac{dv_C(t)}{dt}$$

Substituting the i-v relationships into the KVL equation and rearranging terms produces a first-order differential equation:

$$RC\frac{dv_C(t)}{dt} + v_C(t) = V_A u(t) \tag{9-31}$$

with an initial condition $v_C(0-) = V_O$.

STEP 2: The analysis objective is to determine the waveform $v_C(t)$, which satisfies this differential equation and the initial condition using Laplace transforms. We first apply the Laplace transformation to both sides of Eq. (9-31):

$$\mathscr{L}\left[RC\frac{dv_C(t)}{dt} + v_C(t)\right] = \mathscr{L}\left[V_A u(t)\right]$$

Using the linearity property leads to

$$RC\mathscr{L}\left[\frac{dv_C(t)}{dt}\right] + \mathscr{L}[v_C(t)] = V_A\mathscr{L}[u(t)]$$

Now, using the differentiation property and the transform of a unit step function produces

$$RC\left[sV_C(s) - V_O\right] + V_C(s) = V_A\frac{1}{s} \tag{9-32}$$

This result is an algebraic equation in $V_C(s)$, which is the transform of the response we seek. We rearrange Eq. (9-32) in the form

$$(s + 1/RC)V_C(s) = \frac{V_A/RC}{s} + V_O$$

and algebraically solve for $V_C(s)$:

$$V_C(s) = \frac{V_A/RC}{s(s + 1/RC)} + \frac{V_O}{s + 1/RC} \tag{9-33}$$

The function $V_C(s)$ is the transform of the waveform $v_C(t)$, which satisfies the differential equation and the initial condition. The initial condition appears explicitly in this equation as a result of applying the differentiation rule to obtain Eq. (9-32).

STEP 3: To obtain the waveform $v_C(t)$, we find the inverse transform of the right side of Eq. (9-33). The first term on the right is a proper rational function with two simple poles on the real axis in the s plane. The pole at the origin was introduced by the step function input. The pole at $s = -1/RC$ came from the circuit. The partial fraction expansion of the first term in Eq. (9-33) is

$$\frac{V_A/RC}{s(s + 1/RC)} = \frac{k_1}{s} + \frac{k_2}{s + 1/RC}$$

The residues k_1 and k_2 are found using the cover-up algorithm:

$$k_1 = \left.\frac{V_A/RC}{s+1/RC}\right|_{s=0} = V_A$$

and

$$k_2 = \left.\frac{V_A/RC}{s}\right|_{s=-1/RC} = -V_A$$

Using these residues we expand Eq. (9-33) by partial fractions as

$$V_C(s) = \frac{V_A}{s} - \frac{V_A}{s+1/RC} + \frac{V_O}{s+1/RC} \tag{9-34}$$

Each term in this expansion is recognizable: The first is a step function and next two are exponentials. Taking the inverse transform of Eq. (9-34) gives

$$\begin{aligned} v_C(t) &= \left[V_A - V_A e^{-t/RC} + V_O e^{-t/RC}\right] u(t) \\ &= \left[V_A - (V_O - V_A)\, e^{-t/RC}\right] u(t) \end{aligned} \tag{9-35}$$

The waveform $v_C(t)$ satisfies the differential equation in Eq. (9-31) and the initial condition $v_C(0-) = V_O$. The term $V_A u(t)$ is the forced response due to the step-function input and the term $\left[(V_O - V_A)\, e^{t/RC}\right] u(t)$ is the natural response. The complete response depends on three parameters: the input amplitude V_A, the circuit time constant RC, and the initial condition V_O.

These results are identical to those found using the classical methods in Chapter 8. The outcome is the same but the method is quite different. The Laplace transformation yields the complete response (forced and natural) by an algebraic process, which accounts for the initial conditions. The solid blue arrow in Fig. 9-1 shows the overall procedure. The reader should begin with Eq. (9-31) and relate each step leading to Eq. (9-35) to steps in Fig. 9-1.

EXAMPLE 9-13

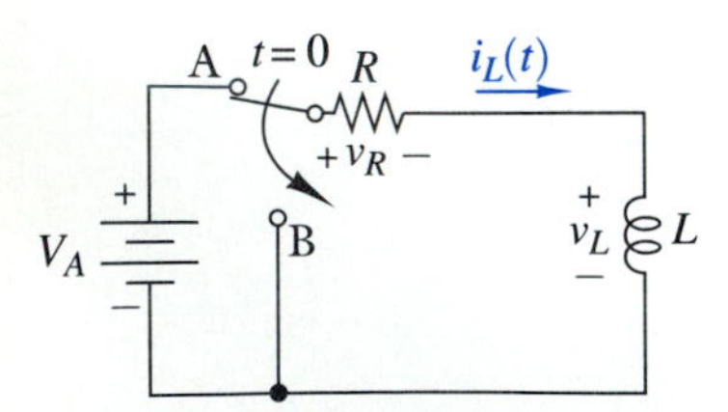

Figure 9-5

The switch in Fig. 9-5 has been in position A for a long time. At $t = 0$ it is moved to position B. Find $i_L(t)$ for $t \geq 0$.

SOLUTION:

STEP 1: The circuit differential equation is found by combining the KVL equation and element equations with the switch in position B.

$$\text{KVL:} \quad v_R(t) + v_L(t) = 0$$

$$\text{Resistor:} \quad v_R(t) = i_L(t)R$$

$$\text{Inductor:} \quad v_L(t) = L\frac{di_L(t)}{dt}$$

Substituting the element equations into the KVL equation yields

$$L\frac{di_L(t)}{dt} + Ri_L(t) = 0$$

Prior to $t = 0$ the circuit was in a dc steady-state condition with the switch in position A. Under dc conditions the inductor acts like a short circuit and the inductor current just prior to moving the switch is $i_L(0-) = I_O = V_A/R$.

STEP 2: Transforming the circuit differential equation into the s domain yields

$$L[sI_L(s) - I_O] + RI_L(s) = 0$$

To derive this result we used linearity and the differentiation property. Solving algebraically for $I_L(s)$

$$I_L(s) = \frac{I_O}{s + R/L}$$

STEP 3: The inverse transform of $I_L(s)$ is an exponential waveform:

$$i_L(t) = \left[I_O e^{-Rt/L}\right] u(t)$$

where $I_O = V_A/R$. Substituting $i_L(t)$ back into the differential equation yields

$$L\frac{di_L(t)}{dt} + Ri_L(t) = -RI_O e^{-Rt/L} + RI_O e^{-Rt/L} = 0$$

The waveform found using Laplace transforms does indeed satisfy the circuit differential equation and the initial condition.

EXAMPLE 9-14

The switch in Fig. 9-6 has been open for a long time. At $t = 0$ the switch is closed. Find $i(t)$ for $t \geq 0$.

SOLUTION:

The circuit in Fig. 9-6 is a second-order circuit. The circuit equation is found by combining the element equations and a KVL equation around the loop with the switch closed.

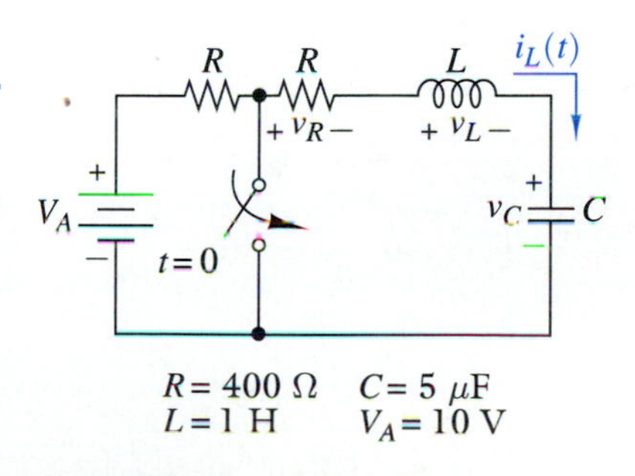

Figure 9-6

$$\text{KVL:} \quad v_R(t) + v_L(t) + v_C(t) = 0$$

$$\text{Resistor:} \quad v_R(t) = Ri(t)$$

$$\text{Inductor:} \quad v_L(t) = L\frac{di(t)}{dt}$$

$$\text{Capacitor:} \quad v_C(t) = \frac{1}{C}\int_0^t i(\tau)d\tau + v_C(0)$$

Substituting the element equations into the KVL equation yields

$$L\frac{di(t)}{dt} + Ri(t) + \frac{1}{C}\int_0^t i(\tau)d\tau + v_C(0) = 0$$

Transforming this second-order integrodifferential equation into the s domain yields

$$L[sI(s) - i_L(0)] + RI(s) + \frac{1}{C}\frac{I(s)}{s} + v_C(0)\frac{1}{s} = 0$$

To derive this result we used the linearity property, differentiation property, integration property, and transform of a step function. Solving for $I(s)$ results in

$$I(s) = \frac{si_L(0) - v_C(0)/L}{s^2 + \frac{R}{L}s + \frac{1}{LC}}$$

Prior to $t = 0$ the circuit was in a dc steady-state condition with the switch open. In dc steady state, the inductor acts like a short circuit and the capacitor like an open circuit, so the initial conditions are $i_L(0-) = 0$ and $v_C(0-) = V_A$. Inserting the initial conditions and the numerical values of the circuit parameters into the equation for $I(s)$ gives

$$I(s) = -\frac{10}{s^2 + 400s + 2 \times 10^5}$$

The denominator quadratic can be factored as $(s + 200)^2 + 400^2$ and $I(s)$ written in the following form:

$$I(s) = -\frac{10}{400}\left[\frac{400}{(s + 200)^2 + (400)^2}\right]$$

Comparing the quantity inside the bracket with the entries in the $F(s)$ column of Table 9-2, we find that $I(s)$ is a damped sine with $\alpha = 200$ and $\beta = 400$. By linearity, the quantity outside the bracket is the amplitude of the damped sine. The inverse transform is

$$i(t) = \left[-0.025e^{-200t}\sin 400t\right]u(t)$$

Substituting this result back into the circuit integrodifferential equation yields the following term by term tabulation:

$$L\frac{di(t)}{dt} = +5e^{-200t}\sin 400t - 10e^{-200t}\cos 400t$$

$$Ri(t) = -10e^{-200t}\sin 400t$$

$$\frac{1}{C}\int_0^t i(\tau)d\tau = +5e^{-200t}\sin 400t + 10e^{-200t}\cos 400t - 10$$

$$v_C(0) = +10$$

The sum of the right-hand sides of these equations is zero. This result shows that the waveform $i(t)$ found using Laplace transforms does indeed satisfy the circuit integrodifferential equation and the initial conditions.

EXERCISE 9-14

Find the transform $V(s)$ that satisfies the following differential equations and the initial conditions.

(a) $\frac{dv(t)}{dt} + 6v(t) = 4u(t) \quad v(0-) = -3$

(b) $4\frac{dv(t)}{dt} + 12v(t) = 16\cos 3t, \quad v(0-) = 2$

ANSWERS:

(a) $V(s) = \frac{4}{s(s+6)} - \frac{3}{s+6}$

(b) $V(s) = \frac{4s}{(s^2+9)(s+3)} + \frac{2}{s+3}$

EXERCISE 9-15

Find the $V(s)$ that satisfies the following equations.

(a) $\int_0^t v(\tau)d\tau + 10v(t) = 10u(t)$

(b) $\frac{d^2v(t)}{dt^2} + 4\frac{dv(t)}{dt} + 3v(t) = 5e^{-2t} \quad v'(0-) = 2 \quad v(0-) = -2$

ANSWERS:

(a) $V(s) = \frac{1}{s+0.1}$

(b) $V(s) = \frac{5}{(s+1)(s+2)(s+3)} - \frac{2}{s+1}$

Time-Varying Inputs

It is encouraging to find that the Laplace transformation yields results that agree with those obtained by classical methods. The transform method reduces solving circuit differential equations to an algebraic process that includes the initial conditions. However, before being overcome with euphoria we must remember that the Laplace transform method begins with the circuit differential equation and the initial conditions. It does not provide these quantities to us. The transform method simplifies the solution process, but it does not substitute for understanding how to formulate circuit differential equations.

The Laplace transform method particularly simplifies finding the circuit responses for time-varying driving forces. To illustrate, we return to the RC circuit in Fig. 9-4 and replace the step function input by a general input signal denoted $v_s(t)$. The right side of the circuit differential equation in Eq. (9-31) changes to accommodate the new input by taking the form

$$RC\frac{dv_C(t)}{dt} + v_C(t) = v_S(t) \qquad (9\text{-}36)$$

with an initial condition $v_C(0) = V_O$.

The only change here is that the driving force on the right side of the differential equation is a general time-varying waveform $v_S(t)$. The objective is to find the capacitor voltage $v_C(t)$ that satisfies the differential equation and the initial conditions. The classical methods of solving for the forced response depend on the form of $v_S(t)$. However, with the Laplace transform method we can proceed without actually specifying the form of the input signal.

We first transform Eq. (9-36) into the s domain:

$$RC\,[sV_C(s) - V_O] + V_C(s) = V_S(s)$$

The only assumption here is that the input waveform is Laplace transformable, a condition met by all causal signals of engineering interest. We now algebraically solve for the response $V_C(s)$:

$$V_C(s) = \frac{V_S(s)/RC}{s + 1/RC} + \frac{V_O}{s + 1/RC} \qquad (9\text{-}37)$$

The function $V_C(s)$ is the transform of the response of the RC circuit in Fig. 9-4 due to a general input signal $v_S(t)$. We can proceed this far without specifying the form of the input signal. In a sense, we have found the general solution in the s domain of the differential equation in Eq. (9-36) for any causal input signal.

All of the necessary ingredients are present in Eq. (9-37):

1. The transform $V_S(s)$ represents the applied input signal.
2. The pole at $s = -1/RC$ describes the circuit time constant.
3. The initial value $v_C(0-) = V_O$ summarizes all events prior to $t = 0$.

However, we must have a particular input in mind in order to solve for the waveform $v_C(t)$. The following examples illustrate the procedure for different input driving forces.

EXAMPLE 9-15

Find $v_C(t)$ in the RC circuit in Fig. 9-4 when the input is an exponential waveform $v_S(t) = [V_A e^{-\alpha t}]u(t)$.

SOLUTION:

The transform of the input is $V_S(s) = V_A/(s+\alpha)$. For the exponential input the response transform in Eq. (9-37) becomes

$$V_C(s) = \frac{V_A/RC}{(s+\alpha)(s+1/RC)} + \frac{V_O}{s+1/RC} \tag{9-38}$$

If $\alpha \neq 1/RC$, then the first term on the right is a proper rational function with two simple poles. The pole at $s = -\alpha$ came from the input and the pole at $s = -1/RC$ from the circuit. A partial fraction expansion of the first term has the form

$$\frac{V_A/RC}{(s+\alpha)(s+1/RC)} = \frac{k_1}{s+\alpha} + \frac{k_2}{s+1/RC}$$

The residues in this expansion are

$$k_1 = \left.\frac{V_A/RC}{s+1/RC}\right|_{s=-\alpha} = \frac{V_A}{1-\alpha RC}$$

$$k_2 = \left.\frac{V_A/RC}{s+\alpha}\right|_{s=-1/RC} = \frac{V_A}{\alpha RC - 1}$$

The expansion of the response transform $V_C(s)$ is

$$V_C(s) = \frac{V_A/(1-\alpha RC)}{s+\alpha} + \frac{V_A/(\alpha RC - 1)}{s+1/RC} + \frac{V_O}{s+1/RC}$$

The inverse transform of $V_C(s)$ is

$$v_C(t) = \left[\frac{V_A}{1-\alpha RC}e^{-\alpha t} + \frac{V_A}{\alpha RC - 1}e^{-t/RC} + V_O e^{-t/RC}\right]u(t)$$

The first term is the forced response and the last two terms are the natural response. The forced response is exponential because the input introduced a pole at $s = -\alpha$. The natural response also is an exponential, but its time constant depends on the circuit's pole at $s = -1/RC$.

If $\alpha = 1/RC$, then the response given above is no longer valid (k_1 and k_2 become infinity). To find the response for the singular condition we return to Eq. (9-38) and replace α by $1/RC$:

$$V_C(s) = \frac{V_A/RC}{(s+1/RC)^2} + \frac{V_O}{s+1/RC}$$

We now have a double pole at $s = -1/RC = -\alpha$. The double pole term is the transform of a damped ramp so the inverse transform is

$$v_C(t) = \left[V_A \frac{t}{RC} e^{-t/RC} + V_O e^{-t/RC}\right] u(t)$$

In this case we cannot separate the response into forced and natural components, since they have the same form. In the s domain the poles of the input and the circuit coincide.

EXAMPLE 9-16

Find $v_C(t)$ when the input to the RC circuit in Fig. 9-4 is a sinusoidal waveform $v_S(t) = [V_A \cos \beta t]\, u(t)$.

SOLUTION:

The transform of the input is $V_S(s) = V_A s/(s^2+\beta^2)$. For a cosine input the response transform in Eq. (9-37) becomes

$$V_C(s) = \frac{sV_A/RC}{(s^2+\beta^2)(s+1/RC)} + \frac{V_O}{s+1/RC}$$

The input sinusoid introduces a pair of poles located at $s = \pm j\beta$. The first term on the right is a proper rational function with three simple poles. The partial fraction expansion of the first term is

$$\frac{sV_A/RC}{(s-j\beta)(s+j\beta)(s+1/RC)} = \frac{k_1}{s-j\beta} + \frac{k_1^*}{s+j\beta} + \frac{k_2}{s+1/RC}$$

To find the response, we need to find the residues k_1 and k_2

$$k_1 = \left.\frac{sV_A/RC}{(s+j\beta)(s+1/RC)}\right|_{s=j\beta} = \frac{V_A/2}{1+j\beta RC} = |k_1|e^{j\theta}$$

where

$$|k_1| = \frac{V_A/2}{\sqrt{1+(\beta RC)^2}} \quad \text{and} \quad \theta = -\tan^{-1}(\beta RC)$$

The residue k_2 at the circuit pole is

$$k_2 = \left.\frac{sV_A/RC}{s^2+\beta^2}\right|_{s=-1/RC} = -\frac{V_A}{1+(\beta RC)^2}$$

We now perform the inverse transform to obtain the response waveform:

$$\begin{aligned} v_C(t) &= \left[2|k_1|\cos(\beta t+\theta) + k_2 e^{-t/RC} + V_O e^{-t/RC}\right]u(t) \\ &= \left[\frac{V_A}{\sqrt{1+(\beta RC)^2}}\cos(\beta t+\theta) - \frac{V_A}{1+(\beta RC)^2}e^{-t/RC}\right. \\ &\quad \left. + V_O e^{-t/RC}\right]u(t) \end{aligned}$$

The first term is the forced response and the remaining two are the natural response. The forced response is sinusoidal because the input introduced poles at $s = \pm j\beta$. The natural response is an exponential with a time constant determined by the location of the circuit's pole at $s = -1/RC$.

EXERCISE 9-16

Find $v_C(t)$ for the RC circuit in Fig. 9-4 when the input is:
(a) a ramp $v_S(t) = [V_A t/T]u(t)$
(b) a sinusoid $v_S(t) = [V_A \sin\beta t]u(t)$

ANSWERS:
(a) $v_C(t) = \left[V_A t/T - V_A RC/T + (V_O + V_A RC/T)\,e^{-tRC}\right]u(t)$
(b) $v_C(t) = \left[\frac{V_A}{\sqrt{1+(\beta RC)^2}}\cos(\beta t+\theta) + \left(\frac{\beta RCV_A}{1+(\beta RC)^2} + V_O\right)\right.$
$\left. e^{-t/RC}\right]u(t)$ where $\theta = -\pi/2 - \tan^{-1}(\beta RC)$. ■

9-7 TRANSLATION AND SCALING PROPERTIES

The four basic properties of the Laplace transformation are uniqueness, linearity, time integration, and time differentiation. These essential features are sufficient to cover most applications of Laplace transforms in circuit analysis and design. In the next three sections we introduce additional properties of the Laplace transformation that give further insight into the relationship between the time and frequency domains. We begin with properties that relate the origin and scale factors of the time and s domains.

Time Domain Translation Property

The *t-domain translation property* of the Laplace transformation is

IF $\mathscr{L}\{f(t)\} = F(s)$

$$\text{THEN for} \quad a > 0 \ \mathscr{L}\{f(t-a)u(t-a)\} = e^{-as}F(s) \qquad (9\text{-}39)$$

The theorem states that multiplying $F(s)$ by e^{-as} is equivalent to shifting $f(t)$ to the right in the t domain, that is, delaying $f(t)$ in time by an amount $a > 0$. Proof of this property follows from the definition of the Laplace transformation:

$$\begin{aligned} \mathscr{L}\{f(t-a)u(t-a)\} &= \int_{0-}^{\infty} f(t-a)u(t-a)e^{-st}dt \\ &= \int_{a}^{\infty} f(t-a)e^{-st}dt \end{aligned} \qquad (9\text{-}40)$$

In this equation we have used the fact that $u(t-a)$ is zero for $t < a$ and is unity for $t \geq a$. We now change the integration variable from t to $\tau = t - a$. This change of variable means $dt = d\tau$, $\tau = 0$ when $t = a$, and $\tau = \infty$ when $t = \infty$. We can write the last integral in Eq. (9-40) in the form

$$\begin{aligned} \mathscr{L}\{f(t-a)u(t-a)\} &= \int_{0}^{\infty} f(\tau)e^{-s\tau}e^{-as}d\tau \\ &= e^{-as}\int_{0}^{\infty} f(\tau)e^{-s\tau}d\tau = e^{-as}F(s) \end{aligned}$$

which is the result given in Eq. (9-39).

The rectangular pulse waveform in Fig. 9-7 provides an application of the time translation property. Our previous study of waveform synthesis showed that the rectangular pulse can be generated as the difference of a step functon and a delayed step function:

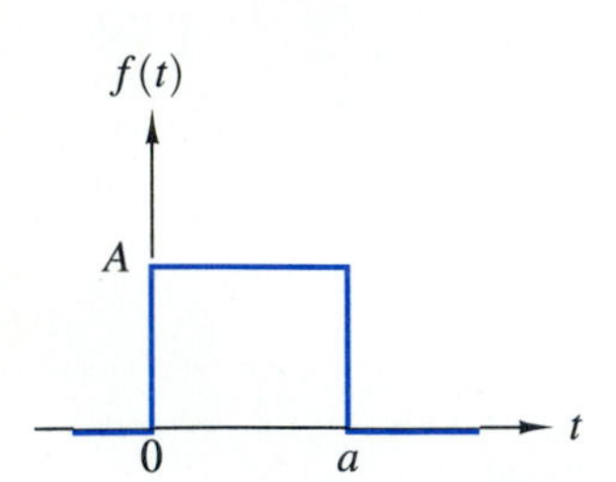

Figure 9-7 Rectangular pulse waveform.

$$f(t) = Au(t) - Au(t-a)$$

The t-domain translation property yields the transform of a rectangular pulse:

$$\begin{aligned} F(s) &= A\mathscr{L}\{u(t)\} - A\mathscr{L}\{u(t-a)\} \\ &= A\frac{1}{s} - Ae^{-as}\frac{1}{s} \\ &= \frac{A(1-e^{-as})}{s} \end{aligned}$$

The fact that the pulse transform is not a rational function does not cause any particular problem. Multiplying a rational function $F(s)$ by the irrational function e^{-as} simply tells us that the waveform $g(t) = \mathscr{L}^{-1}\{e^{-as}F(s)\}$ is a delayed version of $f(t) = \mathscr{L}^{-1}\{F(s)\}$.

The next example shows how to find circuit responses using delayed waveforms.

EXAMPLE 9-17

Find $v_C(t)$ in the RC circuit of Fig. 9-4 when the input is a rectangular pulse

$$v_S(t) = V_A u(t) - V_A u(t-a)$$

SOLUTION:

The input is a rectangular pulse that begins at $t = 0$ and ends at $t = a$. Using the t domain translation property the transform of the input is

$$V_S(s) = \frac{V_A(1 - e^{-as})}{s}$$

Using the rectangular pulse input the circuit response transform in Eq. (9-37) becomes

$$V_C(s) = V_A\left[\frac{1/RC}{s(s+1/RC)}\right] - V_A e^{-as}\left[\frac{1/RC}{s(s+1/RC)}\right] + \frac{V_O}{s+1/RC}$$

The rational functions inside the brackets can be expanded by partial fractions.

$$V_C(s) = V_A\left[\frac{1}{s} - \frac{1}{s+1/RC}\right]$$

$$-V_A e^{-as}\left[\frac{1}{s} - \frac{1}{s+1/RC}\right] + \frac{V_O}{s+1/RC}$$

The two rational functions have the same expansion whose inverse transform is

$$f(t) = \left[1 - e^{-t/RC}\right]u(t)$$

The factor e^{-as} multiplying the second expansion means that the corresponding waveform is delayed by a factor $t = a$. Using the t-domain translation property the response waveform for a rectangular pulse input is

$$v_C(t) = V_A\left[1 - e^{-t/RC}\right]u(t) + V_A\left[1 - e^{-(t-a)/RC}\right]u(t-a)$$
$$+ V_O e^{-t/RC}u(t)$$

The first factor is caused by the step function at $t = 0$ and the second by the step at $t = a$. The last term is the effect of the initial capacitor voltage at $t = 0$. Two plots of this response are shown in Fig. 9-8 for $V_O = 0$. For the short pulse $a = RC$ and the negative step function at $t = a$ occurs before the circuit has fully responded to the positive step function at $t = 0$. For the long pulse $a = 8RC$ and the step function at $t = a$ occurs after the circuit has reached a steady-state condition due to the step at $t = 0$.

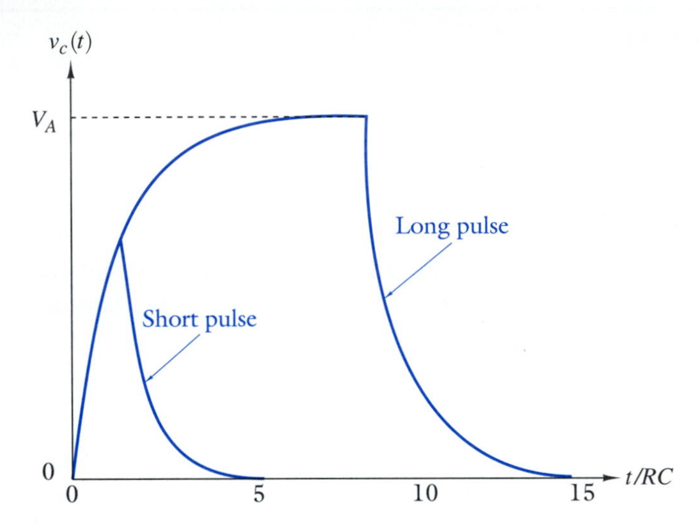

Figure 9-8 Effect of the *s*-domain shifting property.

EXERCISE 9-17

Find the Laplace transforms of the waveforms in Fig. 9-9.

ANSWERS:

(a) $F(s) = A\dfrac{\left(1 - e^{-\frac{sT}{2}}\right)^2}{s}$

(b) $F(s) = A\dfrac{\left(Ts - 1 + e^{-sT}\right)}{Ts^2}$ ■

EXERCISE 9-18

Find waveform corresponding to the following Laplace transforms

(a) $F(s) = e^{-2s}\dfrac{2(s+2)}{s^2 + 4s + 3}$

(b) $F(s) = \dfrac{4\pi}{T}\dfrac{(1 - e^{-Ts})}{s^2 + (2\pi/T)^2}$

ANSWERS:

(a) $f(t) = \left[e^{-(t-2)} + e^{-3(t-2)}\right]u(t-2)$

(b) $f(t) = 2\left[\sin\left(\dfrac{2\pi(t)}{T}\right)\right]u(t) - 2\left[\sin\left(\dfrac{2\pi(t-T)}{T}\right)\right]u(t-T)$ ■

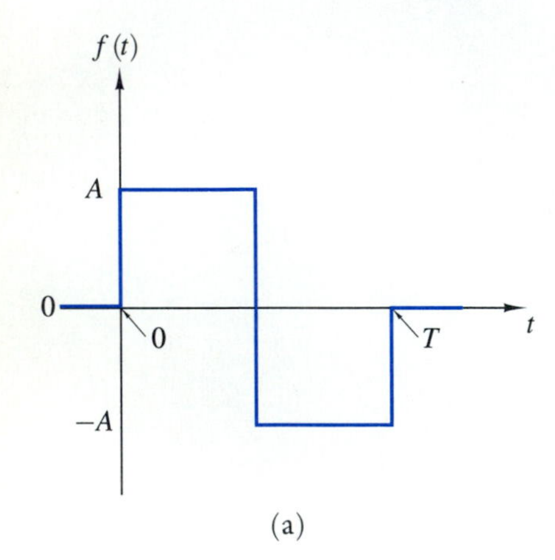

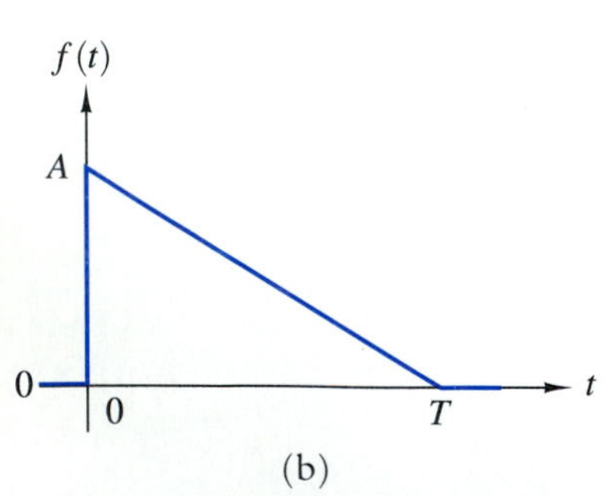

Figure 9-9

Frequency Domain Translation Property

The *s-domain translation property* of the Laplace transformation is

$$\text{IF } \mathscr{L}\{f(t)\} = F(s) \text{ THEN } \mathscr{L}\{e^{-\alpha t} f(t)\} = F(s+\alpha) \quad (9\text{-}41)$$

This theorem states that multiplying $f(t)$ by $e^{-\alpha t}$ is equivalent to replacing s by $s+\alpha$, that is, translating the origin in the s plane by an amount α. In engineering applications the parameter α is always a real number, but it can be either positive or negative so the origin in the s domain can be translated to the left or right. Proof of the theorem follows almost immediately from the definition of the Laplace transformation:

$$\mathscr{L}\{e^{-\alpha t} f(t)\} = \int_0^\infty e^{-\alpha t} f(t) e^{-st} dt$$
$$= \int_0^\infty f(t) e^{-(s+\alpha)} dt = F(s+\alpha)$$

The s-domain translation property can be used to derive transforms of damped waveforms from undamped prototypes. For instance, the Laplace transform of the ramp, cosine, and sine functions are

$$\mathscr{L}\{tu(t)\} = \frac{1}{s^2}$$
$$\mathscr{L}\{[\cos\beta t]u(t)\} = \frac{s}{s^2+\beta^2}$$
$$\mathscr{L}\{[\sin\beta t]u(t)\} = \frac{\beta}{s^2+\beta^2}$$

To obtain the damped ramp, damped cosine, and damped sine functions we multiply each waveform by $e^{-\alpha t}$. Using the s-domain translation property we replace s by $s+\alpha$ to obtain transforms of the damped waveforms:

$$\mathscr{L}\{te^{-\alpha t}u(t)\} = \frac{1}{(s+\alpha)^2}$$
$$\mathscr{L}\{[e^{-\alpha t}\cos\beta t]u(t)\} = \frac{s+\alpha}{(s+\alpha)^2+\beta^2}$$
$$\mathscr{L}\{[e^{-\alpha t}\sin\beta t]u(t)\} = \frac{\beta}{(s+\alpha)^2+\beta^2}$$

The s-domain translation property highlights the relationship between waveform damping and the location of poles in the s domain. Figure 9-10 shows several damped cosine waveforms and the corresponding pole-zero diagram for several values of α. For $\alpha = 0$, the poles lie on the j axis and the waveform is a sustained oscillaton that neither decays nor grows. For $\alpha > 0$ the

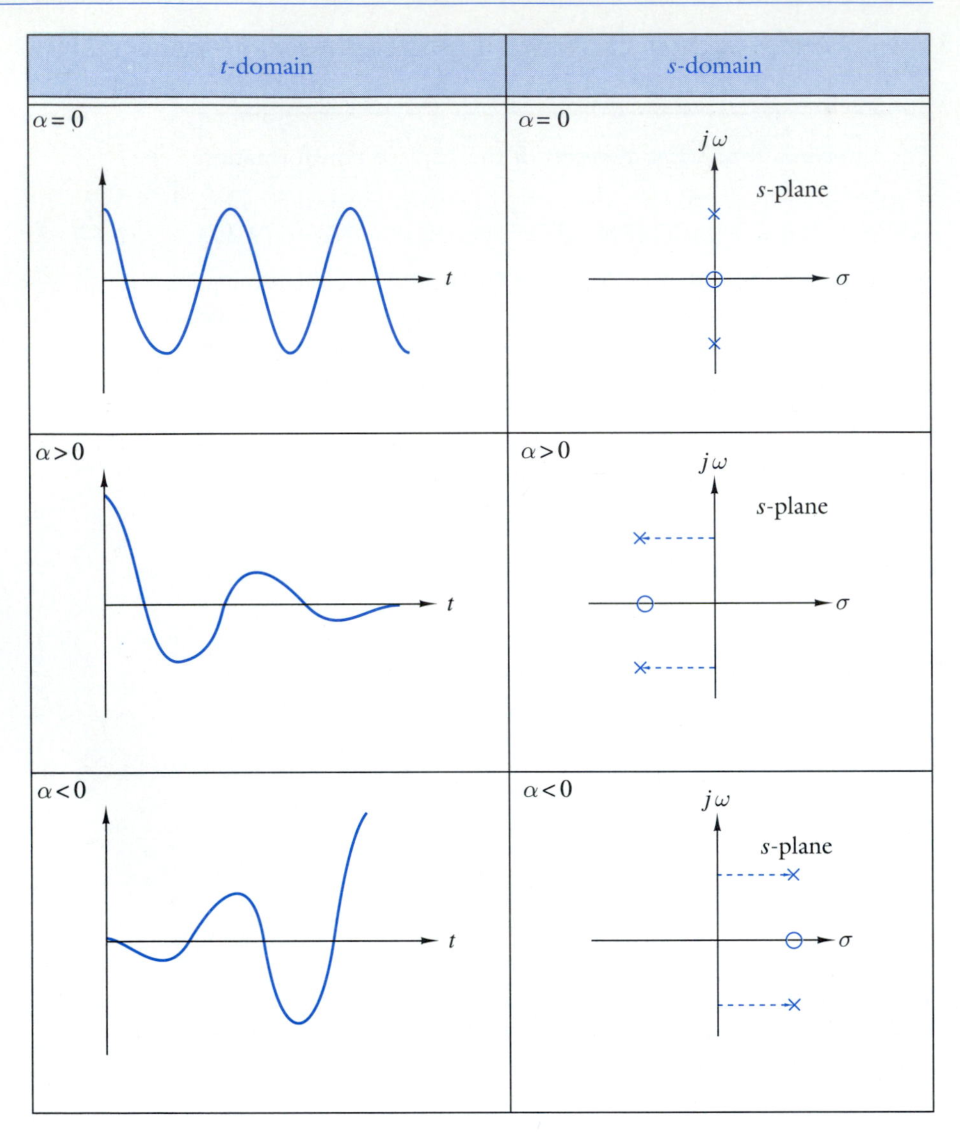

Figure 9-10

the poles shift horizontally into the left-half plane and the waveform decays exponentially to zero. For $\alpha < 0$ the poles shift horizontally into the right-half plane and the waveform grows exponentially without bound (blows up).

Replacing s by $s+\alpha$ shifts the critical frequencies of $F(s)$ to the left in the s plane if $\alpha > 0$, and to the right if $\alpha < 0$. Shifting poles to the left in the s plane increases waveform damping in the time domain, while shifting to the right decreases the damping. Whether s-domain translation causes the waveforms to decay or blow up depends on whether the poles end up in the left-half or right-half plane.

The j axis is an important boundary in the s plane, since it divides signals into two distinct classes. Signals with all of their poles in the left-half plane have waveforms that decay to zero as $t \to \infty$. Those with one or more poles in the right-half plane do not decay and become unbounded as $t \to \infty$. Half-plane location is revealed by the real part of the pole. If the real part is negative, then it is located in the left-half plane, and conversely.

EXERCISE 9-19

Consider the following transforms: (1) Classify them according to the half-plane location of their poles. (2) Classify them as to whether the corresponding waveforms are bounded or unbounded.

(a) $F(s) = \dfrac{s-1}{s^2+4s+3}$

(b) $F(s) = \dfrac{-2(s+7)}{s^2-2s+10}$

(c) $F(s) = \dfrac{s}{s^2+4}$

(d) $F(s) = \dfrac{s}{s^2-4}$

(e) $F(s) = \dfrac{s+2}{s^2+4s+4}$

ANSWERS:

(1) The transforms in (a) and (e) have all their poles in the left half of the s plane. The transform in (d) has one pole in the right half plane and one in the left half. The transform in (b) has both poles in the right-half plane. The transform in (c) has poles on the j axis.
(2) Waveforms corresponding to (a), (c), and (e) are bounded. Waveforms corresponding to (b) and (d) are unbounded. ■

Scaling

The *scaling property* is

$$\text{IF } \mathscr{L}\{f(t)\} = F(s) \text{ THEN for } a > 0 \;\; \mathscr{L}\{f(at)\} = \frac{1}{a}F\left(\frac{s}{a}\right) \tag{9-42}$$

The scaling property states that if t is replaced by at, then s is replaced by s/a and $F(s)$ is divided by a. Proof follows directly from the integral definition of the Laplace transformation:

$$\mathscr{L}\{f(at)\} = \int_0^\infty f(at)e^{-st}\,dt = \frac{1}{a}\int_0^\infty f(\tau)e^{-\tau s/a}\,d\tau = \frac{1}{a}F\left(\frac{s}{a}\right)$$

The second integral in this equation is obtained by changing the integration variable from t to $\tau = at$.

The scaling property is important in situations where a unity parameter prototype is adjusted to cover a large number of applications. For example, the transform of a cosine waveform with unity frequency is $\mathscr{L}\{\cos t\} = s/(s^2+1)$. The scaling property allows us to adjust this prototype to produce the transform at any frequency β as

$$\mathscr{L}\{\cos \beta t\} = \frac{1}{\beta}\frac{s/\beta}{(s/\beta)^2+1} = \frac{s}{s^2+\beta^2}$$

The scaling property highlights the ***reciprocal spreading*** relationship between the t domain and s domain. If the time scale is compressed ($a < 1$), then scale of the s plane expands, and vice versa. Figure 9-11 shows damped cosine waveforms and

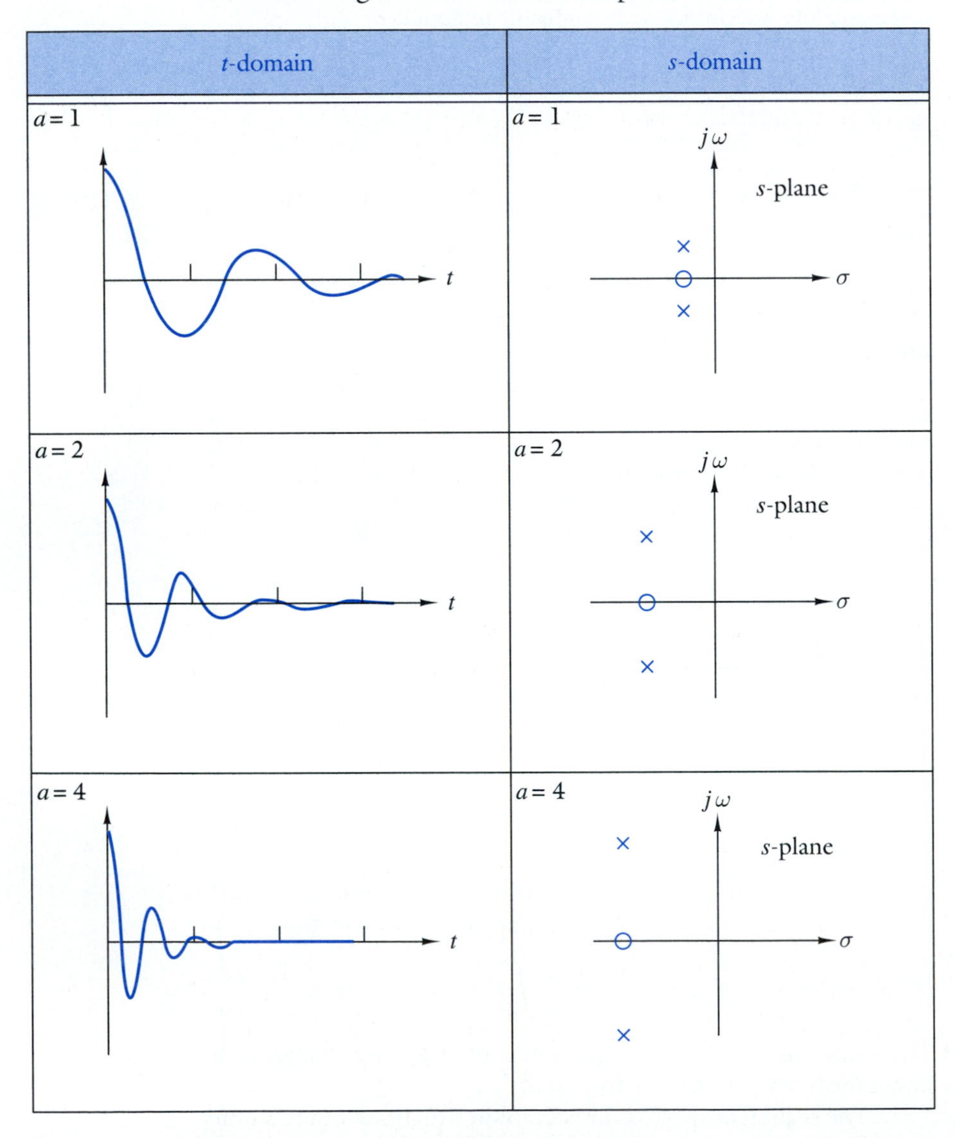

Figure 9-11 Effect of the scaling property.

pole-zero maps for different scale factors. As the figure shows, increasing a compresses the waveform in the t domain and expands the critical frequencies radially outward in the s domain. The reciprocal spreading concept is important to understanding the relationship of s-plane geometry to t-domain waveforms.

9-8 INITIAL-VALUE AND FINAL-VALUE PROPERTIES

The *initial-value* and *final-value properties* can be stated as follows.

$$\begin{aligned}\text{Initial Value: } \lim_{t\to 0+} f(t) &= \lim_{s\to\infty} sF(s)\\ \text{Final Value: } \lim_{t\to\infty} f(t) &= \lim_{s\to 0} sF(s)\end{aligned} \tag{9-43}$$

These properties display the relationship between the origin and infinity in the time and frequency domains. The value of $f(t)$ at $t = 0+$ in the time domain (initial value) is the same as the value of $sF(s)$ at infinity in the s plane. Conversely, the value of $f(t)$ as $t \to \infty$ (final value) is the same as the value of $sF(s)$ at the origin in the s plane.

Proof of both the initial-value and final-value properties starts with the differentiation property:

$$sF(s) - f(0-) = \int_{0-}^{\infty} \frac{df}{dt} e^{-st} dt \tag{9-44}$$

To establish the initial-value property we rewrite the integral on the right side of this equation and take the limit of both sides as $s \to \infty$.

$$\lim_{s\to\infty} [sF(s) - f(0-)] = \lim_{s\to\infty} \int_{0-}^{0+} \frac{df}{dt} e^{-st} dt + \lim_{s\to\infty} \int_{0+}^{\infty} \frac{df}{dt} e^{-st} dt \tag{9-45}$$

The first integral on the right side reduces to $f(0+) - f(0-)$, since e^{-st} is unity on the interval from $t = 0-$ to $t = 0+$. The second integral vanishes because e^{-st} goes to zero as $s \to \infty$. In addition, on the left side of Eq. (9-45) the $f(0-)$ is independent of s and can be taken outside the limiting process. Inserting all of these considerations reduces Eq. (9-45) to

$$\lim_{s\to\infty} sF(s) = \lim_{t\to 0+} f(t) \tag{9-46}$$

which completes the proof of the initial-value property.

Proof of the final-value theorem begins by taking the limit of both sides of Eq. (9-44) as $s \to 0$:

$$\lim_{s\to 0}[sF(s) - f(0-)] = \lim_{s\to 0}\int_{0-}^{\infty} \frac{df}{dt} e^{-st} dt \tag{9-47}$$

The integral on the right side of this equation reduces to $f(\infty) - f(0-)$ because e^{-st} becomes unity as $s \to 0$. Again, the $f(0-)$ on the left side is independent of s and can be taken outside of the limiting process. Inserting all of these considerations reduces Eq. (9-47) to

$$\lim_{s\to 0} sF(s) = \lim_{t\to\infty} f(t) \tag{9-48}$$

which completes the proof of the final-value property.

A damped cosine waveform provides an illustration of the application of these properties. The transform of the damped cosine is

$$\mathscr{L}\left\{\left[Ae^{-\alpha t}\cos\beta t\right]u(t)\right\} = \frac{A(s+\alpha)}{(s+\alpha)^2+\beta^2}$$

Applying the initial and final value limits we obtain

$$\text{Initial Value:}\ \lim_{t\to 0+} f(t) = \lim_{t\to 0+} Ae^{-\alpha t}\cos\beta t = A$$

$$\lim_{s\to\infty} sF(s) = \lim_{s\to\infty} \frac{sA(s+\alpha)}{(s+\alpha)^2+\beta^2} = A$$

$$\text{Final Value:}\ \lim_{t\to\infty} f(t) = \lim_{t\to\infty} Ae^{-\alpha t}\cos\beta t = 0$$

$$= \lim_{s\to 0} sF(s) = \lim_{s\to 0} \frac{sA(s+\alpha)}{(s+\alpha)^2+\beta^2} = 0$$

Note the agreement between the t-domain and s-domain limits in both cases.

There are restrictions on the initial- and final-value properties. The initial-value property is valid when $F(s)$ is a proper rational function, or, equivalently, when $f(t)$ does not have an impulse at $t = 0$. The final-value property is valid when the poles of $sF(s)$ are in the left-half plane or, equivalently, when $f(t)$ is a waveform that approaches a final value at $t \to \infty$. Note that the final value restrictions allows $F(s)$ to have a simple pole at the origin, since the limitation is on the poles of $sF(s)$.

The reader is cautioned that the initial- and final-value properties will appear to work when the restrictions listed above are not met. In other words, these properties will not tell you they are giving nonsense answers when the limitations on their validity are violated. Therefore, you must always check the restriction before applying either of these properties.

For example, applying the final-value property to a cosine waveform yields

$$\lim_{t\to\infty} \cos\beta t = \lim_{s\to 0} s\left[\frac{s}{s^2+\beta^2}\right] = 0$$

The final value property appears to say that $\cos\beta t$ approaches zero as $t \to \infty$. This conclusion is incorrect, since the waveform oscillates between ± 1. The problem is that the final-value property does not apply to cosine waveform because $sF(s)$ has poles on the j axis at $s = \pm j\beta$.

EXAMPLE 9-18

Find initial and final values of $v_C(t)$ in the RC circuit of Fig. 9-4 when the input is a step function $v_S(t) = V_A u(t)$.

SOLUTION:

The input transform is V_A/s and the circuit response transform in Eq. (9-37) becomes

$$V_C(s) = \frac{V_A/RC}{s(s+1/RC)} + \frac{V_O}{s+1/RC}$$

The initial-value property applies because $V_C(s)$ is a proper rational function. The initial value is

$$\begin{aligned}\lim_{t\to 0+} v_C(t) &= \lim_{s\to\infty} sV_C(s)\\ &= \lim_{s\to\infty}\left[\frac{V_A/RC}{s+1/RC}\right] + \lim_{s\to\infty}\left[\frac{sV_O}{s+1/RC}\right]\\ &= 0 + V_O = v_C(0)\end{aligned}$$

which is true by definition. The final-value property applies since the pole of $sF(s)$ is located at $s = -1/RC$, which is in the left-half plane. The final value is

$$\begin{aligned}\lim_{t\to\infty} v_C(t) &= \lim_{s\to 0} sV_C(s)\\ &= \lim_{s\to 0}\left[\frac{V_A/RC}{s+1/RC}\right] + \lim_{s\to 0}\left[\frac{sV_O}{s+1/RC}\right]\\ &= V_A\end{aligned}$$

A step function of amplitude V_A drives the circuit response from an initial value $v_C(0)$ to a final value V_A.

EXERCISE 9-20

Find the initial value at $t = 0+$ and final value of the waveforms corresponding to the transforms in Exercise 9-19.

ANSWERS:

(a) Initial Value = 1, Final Value = 0
(b) Initial Value = −2, Final Value does not exist
(c) Initial Value = 1, Final Value does not exist
(d) Initial Value = 1, Final Value does not exist
(e) Initial Value = 1, Final Value = 0 ■

■ EXERCISE 9-21

Find the initial and final values of the waveforms corresponding to the following transforms.

(a) $F(s) = 100\dfrac{s+3}{s(s+5)(s+20)}$

(b) $F(s) = e^{-10s}\dfrac{s(s+2)}{(s+1)(s+4)}$

ANSWERS:

(a) Initial Value = 0, Final Value = 3
(b) Initial Value = Final Value = 0 ■

◇ 9-9 CONVOLUTION PROPERTY

Because of the linearity property the inverse transform of the sum of two transforms is

$$\mathscr{L}^{-1}\{F_1(s) + F_2(s)\} = f_1(t) + f_2(t)$$

Linearity is the basis for the partial fraction expansion method of finding the inverse transform of a rational function. Another situation that occurs in practical problems is finding the inverse transform of the product of two transforms. The first thing to realize is that the product is not covered by the linearity property. That is,

$$\mathscr{L}^{-1}\{F_1(s)F_2(s)\} \neq f_1(t)f_2(t)$$

The inverse transform of the product of two transforms involves the *convolution property* of the Laplace transformation, which can be stated as

$$\begin{aligned}\mathscr{L}^{-1}\{F_1(s)F_2(s)\} &= \int_0^t f_1(\tau)f_2(t-\tau)d\tau \\ &= \int_0^t f_2(\tau)f_1(t-\tau)d\tau\end{aligned} \tag{9-49}$$

where τ is a dummy variable of integration. The two integrals in Eq. (9-49) are called the *convolution integrals*. The two convolution integrals are equivalent because one can be obtained from the other by interchanging $f_1(t)$ and $f_2(t)$. We say that the inverse transform of the product of two transforms involves *convolving* the corresponding waveforms. A special notation is used to indicate this operation

$$\mathscr{L}^{-1}\{F_1(s)F_2(s)\} = f_1(t) * f_2(t) = f_2(t) * f_1(t)$$

where the asterisk is a shorthand notation for the convolution integrals in Eq. (9-49).

Deriving the convolution property begins by writing the definition of $F_1(s)$ as

$$F_1(s) = \int_0^{\infty} f_1(\tau)e^{-s\tau}d\tau$$

Now multiplying this equation by $F_2(s)$ leads to

$$F_1(s)F_2(s) = \int_0^{\infty} f_1(\tau)\left[F_2(s)e^{-s\tau}\right]d\tau$$

The time translation property in Eq. (9-39) shows that the quantity inside the brackets on the right side of this equation is $\mathscr{L}\{[f_2(t-\tau)]u(t-\tau)\}$. Therefore, this equation can be written as

$$F_1(s)F_2(s) = \int_0^{\infty} f_1(\tau)\left[\int_0^{\infty} f_2(t-\tau)u(t-\tau)e^{-s\tau}dt\right]d\tau$$

By interchanging the order of integration we express this result in the form

$$F_1(s)F_2(s) = \int_0^{\infty}\left[\int_0^{t} f_1(\tau)f_2(t-\tau)dt\right]e^{-st}dt$$

The integration with respect to τ within the bracket only extends from 0 to t, since the integrand vanishes for $\tau > t$ because $u(t-\tau) = 0$ for $\tau > t$. By definition the outer integration in this equation is the Laplace transform of the quantity inside the bracket. That is,

$$F_1(s)F_2(s) = \mathscr{L}\left[\int_0^{t} f_1(\tau)f_2(t-\tau)dt\right]$$

Finally, evoking the uniqueness property of the Laplace transformation we have

$$\mathscr{L}^{-1}\{F_1(s)F_2(s)\} = \int_0^{t} f_1(\tau)f_2(t-\tau)d\tau$$

This establishes the first version of the convolution integral in Eq. (9-49). The second version is obtained by interchanging $F_1(s)$ and $F_2(s)$ in the derivation above.

We will encounter the convolution property again in Chapter 11 in connection with the impulse response of linear circuits. For the moment we will use the convolution property to derive Laplace transform pairs.

EXAMPLE 9-19

Use the convolution property to show that

$$\mathscr{L}^{-1}\left\{\frac{1}{(s+\alpha)^2}\right\} = te^{-\alpha t} \qquad t \geq 0$$

SOLUTION:

The function on the left is the product of $1/(s+\alpha)$ with itself. When $F_1(s) = F_2(s) = 1/(s+\alpha)$ we have $f_1(t) = f_2(t) = e^{-\alpha t}$. Using the convolution property we can write

$$\mathscr{L}^{-1}\left\{\frac{1}{(s+\alpha)(s+\alpha)}\right\} = \int_0^t e^{-\alpha\tau}e^{-\alpha(t-\tau)}d\tau$$

Evaluating the integral of the right side of this equation yields

$$\int_0^t e^{-\alpha\tau}e^{-\alpha t}e^{+\alpha\tau}d\tau = e^{-\alpha t}\int_0^t d\tau = te^{-\alpha t}$$

which is the result we set out to establish.

Table 9-3 lists the complex pole pairs developed in Section 9-4 plus the Laplace transform properties derived in the last three sections.

TABLE 9-3 ADDITIONAL LAPLACE TRANSFORM PROPERTIES AND PAIRS

Feature	Time Domain	Frequency Domain
Simple complex poles	$[2\lvert k\rvert e^{-\alpha t}\cos(\beta t + \angle k)]u(t)$	$\frac{k}{s+\alpha-j\beta} + \frac{k^*}{s+\alpha+j\beta}$
Double complex poles	$[2\lvert k\rvert te^{-\alpha t}\cos(\beta t + \angle k)]u(t)$	$\frac{k}{(s+\alpha-j\beta)^2} + \frac{k^*}{(s+\alpha+j\beta)^2}$
t translation	$[f(t-a)]u(t-a)$	$e^{-\alpha s}F(s)$
s translation	$e^{-\alpha t}f(t)$	$F(s+\alpha)$
Scaling	$f(at)$	$\frac{1}{a}F\left(\frac{s}{a}\right)$
Initial value	$\lim_{t\to 0+} f(t)$	$\lim_{s\to\infty} sF(s)$
Final value	$\lim_{t\to\infty} f(t)$	$\lim_{s\to 0} sF(s)$
Convolution	$\int_0^t f_1(\tau)f_2(t-\tau)d\tau$	$F_1(s)F_2(s)$

EXERCISE 9-22

Use the convolution property to find the inverse transforms of the following rational functions.

(a) $F(s) = \dfrac{1}{(s+\alpha)(s+\beta)}$

(b) $F(s) = \dfrac{1}{s^2(s+\alpha)}$

ANSWERS:

(a) $f(t) = \left[\dfrac{e^{-\alpha t} - e^{-\beta t}}{\beta - \alpha}\right] u(t)$

(b) $f(t) = \left[\dfrac{\alpha t - 1 + e^{-\alpha t}}{\alpha^2}\right] u(t)$

SUMMARY

- The Laplace transformation converts causal waveforms in the time domain to transforms in the s domain. The inverse transformation converts transforms into causal waveforms.
- The Laplace transforms of basic signals like the step function, exponential, and sinusoid are easily derived from the integral definition. Many other transform pairs can be derived using basic signal transforms and the uniqueness, linearity, time integration, and time differentiation properties of the Laplace transformation.
- Partial fraction expansions produce the inverse Laplace transforms of proper rational functions with simple poles. Simple real poles lead to exponential waveforms and simple complex poles to damped sinusoids.
- Partial fraction expansion of improper rational functions and functions with multiple poles requires special treatment.
- The response of a linear circuit is found using Laplace transforms by transforming the circuit differential equation into the s domain, algebraically solving for the response transform, and performing the inverse transformation to obtain the response waveform.
- The translation and scaling properties display the reciprocal spreading relationships between t-domain waveforms and s-domain transforms.
- The initial- and final-value properties determine the initial and final values of a waveform $f(t)$ from the value of $sF(s)$ at $s \to \infty$ and $s = 0$, respectively. The initial-value property applies if $F(s)$ is a proper rational function. The final-value property applies if all of the poles of $sF(s)$ are in the left-half plane.

- The convolution property relates the inverse transform of the product of two transforms to the time-domain convolution integral of the corresponding waveforms.

EN ROUTE OBJECTIVES AND ASSOCIATED PROBLEMS

ERO 9-1 Laplace Transform (Sects. 9-1, 9-2, 9-3)

Given a signal waveform, use the basic properties and pairs of the Laplace transformation to determine its Laplace transform and construct its pole-zero diagram.

9-1 (a) Find the Laplace transform of the waveforms below using the linearity property and transforms pairs in Table 9-2. (b) Express $F(s)$ as a rational function and construct its pole-zero diagram.
(a) $f(t) = A[1 - e^{-\alpha t}]u(t)$
(b) $f(t) = A[1 + 2\alpha t)e^{-\alpha t}]u(t)$
(c) $f(t) = A[e^{-\alpha t} - e^{-\beta t}]u(t)$
(d) $f(t) = A[\alpha e^{-\alpha t} - \beta e^{-\beta t}]u(t)$

9-2 Repeat Prob. 9-1 for each of the following waveforms.
(a) $f(t) = A[\sin(\beta t - \phi)]u(t)$
(b) $f(t) = A[\cos\beta t - \sin\beta t]u(t)$
(c) $f(t) = A[\cos\beta t - \cos\gamma t]u(t)$
(d) $f(t) = A[1 + 2\sin\beta t]u(t)$

9-3 Repeat Prob. 9-1 for each of the following waveforms.
(a) $f(t) = A[e^{-\alpha t}\sin(\beta t - \phi)]u(t)$
(b) $f(t) = A[e^{-\alpha t}(\cos\beta t - \sin\beta t)]u(t)$
(c) $f(t) = A[e^{-\alpha t}(\cos\beta t - \cos\gamma t)]u(t)$
(d) $f(t) = A[e^{-\alpha t}(1 + 2\sin\beta t)]u(t)$

9-4 The transforms of the waveforms below are given in Table 9-2. Find the transform $G(s) = \mathscr{L}\{g(t)\} = \mathscr{L}\{df(t)/dt\}$ by (a) using $F(s)$ and the time differentiation property and (b) differentiating $f(t)$ in the time domain and then transforming $g(t)$. (c) Express $F(s)$ and $G(s)$ as rational functions and construct their pole-zero diagrams. Are there any poles or zeros in $G(s)$ that are not in $F(s)$? Explain.
(a) $f(t) = A[e^{-\alpha t}]u(t)$
(b) $f(t) = A[\sin\beta t]u(t)$

9-5 Repeat Prob. 9-4 for the following waveforms.
(a) $f(t) = A[e^{-\alpha t}\sin\beta t]u(t)$
(b) $f(t) = A[\alpha t e^{-\alpha t}]u(t)$

9-6 The transforms of the waveforms below are given in Table 9-2. Find the transform $G(s) = \mathscr{L}\{g(t)\} = \mathscr{L}\{\int_0^t f(x)dx\}$ by (a) using $F(s)$ and the time integration property and (b) integrating $f(t)$ in the time domain and then transforming $g(t)$. (c) Express

$F(s)$ and $G(s)$ as rational functions and construct their pole-zero diagrams. Are there any poles or zeros in $G(s)$ that are not in $F(s)$? Explain.

(a) $f(t) = A[e^{-\alpha t}]u(t)$
(b) $f(t) = A[\sin \beta t]u(t)$

9-7 Repeat Prob. 9-6 for the following waveforms.

(a) $f(t) = A[e^{-\alpha t} \sin \beta t]u(t)$
(b) $f(t) = A[\alpha t e^{-\alpha t}]u(t)$

9-8 Find the Laplace transform of the waveforms listed below.

(a) $f(t) = A[\cos^2 \beta t]u(t)$
(b) $f(t) = A[(\sin \beta t)(\cos \beta t)]u(t)$

9-9 Find the Laplace transform of the waveforms listed below.

(a) $f(t) = A[\sinh \beta t]u(t)$
(b) $f(t) = A[(\beta t)(\sin \beta t)]u(t)$

ERO 9-2 Inverse Transforms (Sects. 9-4, 9-5)

Given a rational signal transform:

(a) Plot its pole-zero map,

(b) Determine the inverse transform and sketch the waveform

9-10 (a) Find $f(t)$ for the following transforms. (b) Plot the pole-zero map of $F(s)$ and sketch the waveform $f(t)$. Assume $a \neq b$.

(a) $F(s) = \dfrac{1}{(s+a)(s+b)}$

(b) $F(s) = \dfrac{s}{(s+a)(s+b)}$

(c) $F(s) = \dfrac{s^2}{(s+a)(s+b)}$

9-11 Repeat Prob. 9-10 for each of the following transforms.

(a) $F(s) = \dfrac{1}{(s+\alpha)^2 + \beta^2}$

(b) $F(s) = \dfrac{s}{(s+\alpha)^2 + \beta^2}$

(c) $F(s) = \dfrac{s^2}{(s+\alpha)^2 + \beta^2}$

9-12 Repeat Prob. 9-10 for each of the following transforms.

(a) $F(s) = \dfrac{1}{(s+\alpha)}$

(b) $F(s) = \dfrac{1}{s(s+\alpha)}$

(c) $F(s) = \dfrac{1}{s^2(s+\alpha)}$

9-13 Repeat Prob. 9-10 for each of the following transforms.

(a) $F(s) = \dfrac{4}{s^2 + 5s + 4}$

(b) $F(s) = \dfrac{4}{s^2 + 4s + 4}$

(c) $F(s) = \dfrac{4}{s^2 + 2s + 4}$

9-14 Verify that the following partial fraction expansions are correct. Find the unknown residue(s) and determine $f(t)$.

(a) $F(s) = \dfrac{6s^2 + 24s + 18}{(s+2)(s+4)(s+5)} = \dfrac{-1}{s+2} + \dfrac{K}{s+4} + \dfrac{16}{s+5}$

(b) $F(s) = \dfrac{s^2 + 5s + 6}{(s+2)(s+4)(s+5)} = \dfrac{K}{s+2} + \dfrac{-1}{s+4} + \dfrac{2}{s+5}$

(c) $F(s) = \dfrac{8s + 16}{s(s^2 + 4s + 8)} = \dfrac{2}{s} + \dfrac{K}{s+2-j2} + \dfrac{-1+j}{s+2+j2}$

(d) $F(s) = \dfrac{4}{(s+1)^2(s^2+1)} = \dfrac{A}{s+1} + \dfrac{2}{(s+1)^2} + \dfrac{-1}{s-j} + \dfrac{B}{s+j}$

(e) $F(s) = \dfrac{8(s^2+1)}{s(s^2+4)} = \dfrac{2}{s} + \dfrac{A}{s-j2} + \dfrac{B}{s+j2}$

9-15 Find $f(t)$ for each of the following transforms.

(a) $F(s) = \dfrac{(s+1)(s+3)}{s(s+2)(s+4)}$

(b) $F(s) = \dfrac{(s^2+4)(s^2+12)}{s(s^2+9)(s^2+16)}$

9-16 Find $f(t)$ for each of the following transforms.

(a) $F(s) = \dfrac{(s+2)}{s(s^2+2s+1)(s+3)}$

(b) $F(s) = \dfrac{(s+2)}{s^3 + 2s^2 + 2s + 1}$

9-17 Find $f(t)$ for each of the following transforms.

(a) $F(s) = \dfrac{(s+1)(s+4)}{s^2(s^2+2s+4)}$

(b) $F(s) = \dfrac{10(s+65)(s+90)}{s(s^2+100s+2500)}$

9-18 Find $f(t)$ for each of the following transforms.

(a) $F(s) = \dfrac{(s+10)(s+200)}{(s+20)(s+100)}$

(b) $F(s) = \dfrac{10(s+2)}{s^3+1}$

ERO 9-3 Circuit Response Using Laplace Transforms (Sect. 9-6)

Given a first- or second-order circuit:

(a) Determine the circuit differential equation and the initial conditions (if not given).

(b) Transform the differential equation into the s-domain and solve for the response transform.

(c) Use the inverse transformation to find the response waveform.

(d) Identify the forced and natural components of the response in its waveform and its transform.

9-19 Use the Laplace transformation to find the $y(t)$ that satisfies the following first-order differential equations.

(a) $\dfrac{dy}{dt} + 10y = 0, \quad y(0-) = 5$

(b) $10^{-4}\dfrac{dy}{dt} + y = 10u(t), \quad y(0-) = -10$

9-20 Use the Laplace transformation to find the $y(t)$ that satisfies the following first-order differential equations.

(a) $10^{-3}\dfrac{dy}{dt} + 10y = 10tu(t), \quad y(0-) = -2$

(b) $10^{-2}\dfrac{dy}{dt} + y = 5[\cos(100t)]u(t), \quad y(0-) = 0$

9-21 The switch in Fig. P9-21 has been closed for a "long time." At $t = 0$ the switch is opened. (a) Find the differential equation and initial condition for the inductor current $i_L(t)$. (b) Solve for $i_L(t)$ using the Laplace transformation. (c) Identify the forced and natural components in the response waveform and transform.

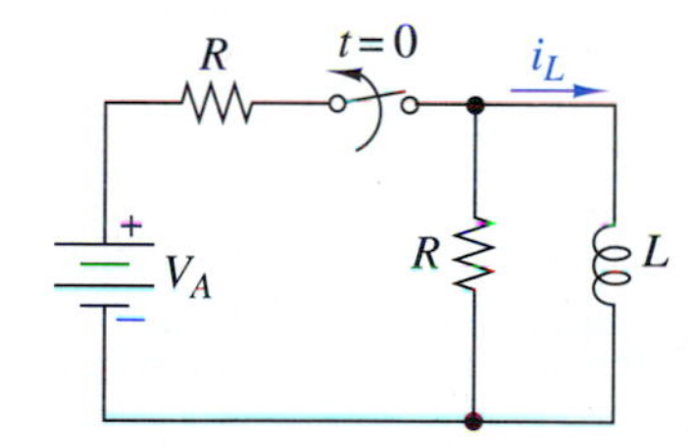

Figure P9-21

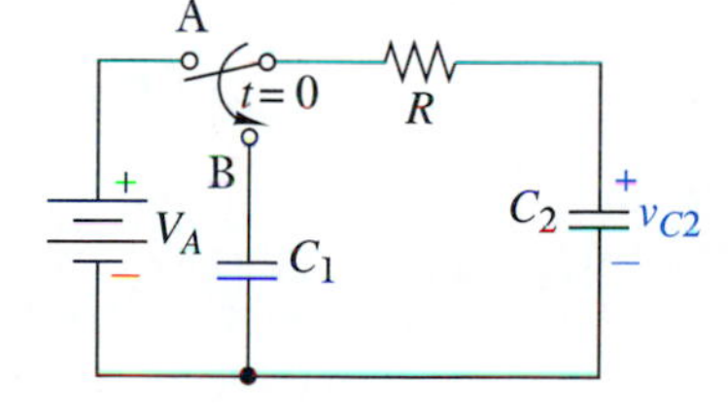

Figure P9-22

9-22 The switch in Fig. P9-22 has been in position A for a "long time." At $t = 0$ the switch is moved to position B. (a) Find the circuit differential equation and the initial condition. (b) Solve for $v_{C2}(t)$ using the Laplace transformation. (c) Identify the forced and natural components in the response waveform and transform.

9-23 The switch in Fig. P9-23 has been open for a "long time." At $t = 0$ the switch is closed. (a) Find the differential equation for the circuit and initial condition. (b) Find $v_O(t)$ using the Laplace transformation for $v_S(t) = 10\,u(t)$ and $v_S(t) = 10[e^{-1000t}]u(t)$.

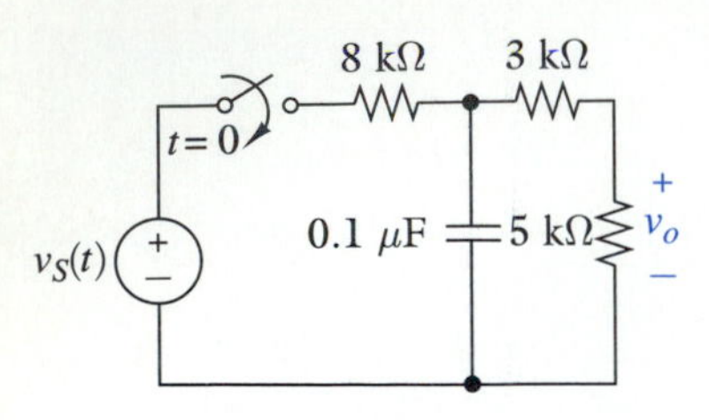

Figure P9-23

(c) Identify the forced and natural components in the response waveform and transform.

9-24 Repeat Prob. 9-23 for the input waveforms $v_S(t) = 10[\sin\ 1000t]u(t)$ and $v_S(t) = 10[e^{-2500t}]u(t)$.

9-25 Use the Laplace transformation to find the $y(t)$ that satisfies the following second-order differential equations.

(a) $\dfrac{d^2y}{dt^2} + 7\dfrac{dy}{dt} + 10y = 0, \quad y(0-) = 0, \quad y'(0-) = 15$

(b) $\dfrac{d^2y}{dt^2} + 4\dfrac{dy}{dt} + 4y = 0, \quad y(0-) = 0, \quad y'(0-) = 2$

9-26 Use the Laplace transformation to find the $y(t)$ that satisfies the following second-order differential equations.

(a) $\dfrac{d^2y}{dt^2} + 11\dfrac{dy}{dt} + 10y = 100u(t), \quad y(0-) = 0,\ y'(0-) = 0$

(b) $\dfrac{d^2y}{dt^2} + 11\dfrac{dy}{dt} + 10y = 100tu(t), \quad y(0-) = 0,\ y'(0-) = 0$

9-27 Use the Laplace transformation to find $y(t)$ that satisfies the following integrodifferential equations.

(a) $3\int_0^t y(x)dx + 15\dfrac{dy}{dt} + 12y = 0, \quad y(0-) = 0, \quad y'(0-) = 2$

(b) $2\int_0^t y(x)dx + 4\dfrac{dy}{dt} + 4y = 20u(t), \quad y(0-) = 0, \quad y'(0-) = 0$

9-28 Use the Laplace transformation to find the $y(t)$ that satisfies the following third-order equation with all initial conditions zero.

$$\frac{d^3y}{dt^3} + 5\frac{d^2y}{dt^2} + 6\frac{dy}{dt} = 2e^{-2t}u(t)$$

9-29 Use the Laplace transformation to find the $y(t)$ that satisfies the following third-order equation with all initial conditions zero.

$$\frac{d^3y}{dt^3} + 6\frac{d^2y}{dt^2} + 11\frac{dy}{dt} + 6y = 12u(t)$$

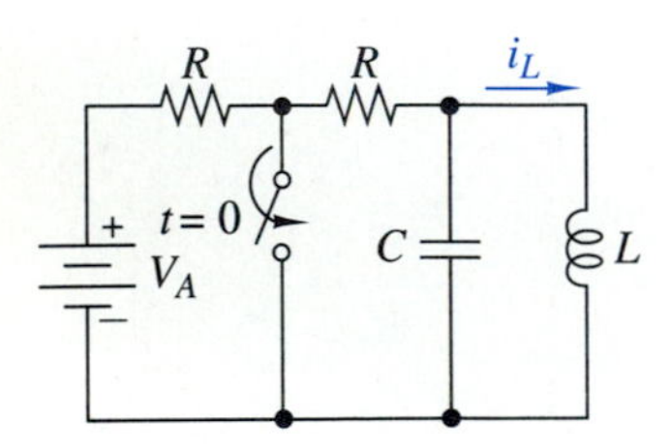

Figure P9-30

9-30 The switch in Fig. P9-30 has been open for a "long time" and is closed at $t = 0$. The circuit parameters are $R = 3$ kΩ, $L = 0.4$ H, $C = 0.1$ μF and $V_A = 10$ V: (a) Find the differential equation for the circuit and the initial conditions. (b) Use Laplace transforms to solve for the $i_L(t)$ for $t \geq 0$.

9-31 The switch in Fig. P9-30 has been closed for a "long time" and is opened at $t = 0$. The circuit parameters are $R = 3$ kΩ, $L = 0.4$ H, $C = 0.1$ μF and $V_A = 10$ V: (a) Find the differential equation for the circuit and the initial conditions. (b) Use Laplace transforms to solve for the $i_L(t)$ for $t \geq 0$.

ERO 9-4 Laplace Transformation Properties (Sects. 9-7, 9-8, 9-9)

(a) Use the translation properties to relate waveforms and transforms.

(b) Find the initial and final value of waveforms from their transforms using the initial- and final-value properties.

(c) Use the convolution property to find the waveform for a given transform.

9-32 Sketch the following waveforms and find their Laplace transforms using the time domain translation property.

(a) $f(t) = 10\,u(t) - 10\,u(t-1) + 10\,u(t-3) - 10\,u(t-4)$

(b) $f(t) = 10\,u(t) - 10\,u(t-1) - 10\,u(t-3) + 10\,u(t-4)$

9-33 Sketch the following waveforms and find their Laplace transforms using the time domain translation property.

(a) $f(t) = 10\,tu(t) - 10\,(t-1)u(t-1) - 10\,u(t-2)$

(b) $f(t) = 10\,tu(t) - 20\,(t-1)u(t-1) + 10\,(t-2)(t-2)$

9-34 **(a)** Write an expression for the waveform $f(t)$ in Fig. P9-34.

(b) Find the Laplace transform of the waveform found in part (a) using the time domain translation property.

(c) Verify your answer in part (b) by applying the integral definition of the Laplace transformation to $f(t)$ found in (a).

9-35 Repeat Problem 9-34 for the waveform in Fig. P9-35.

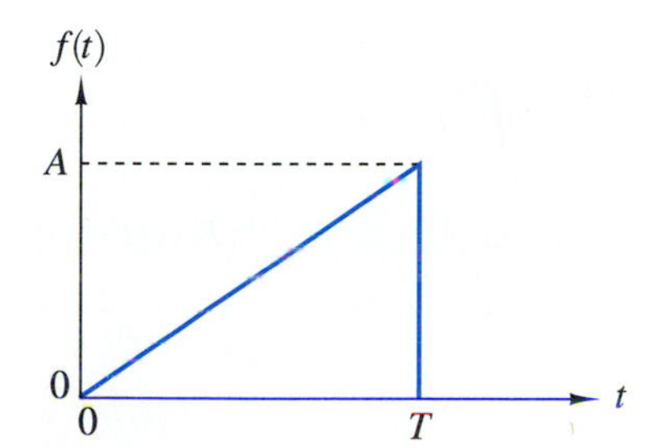

Figure P9-34

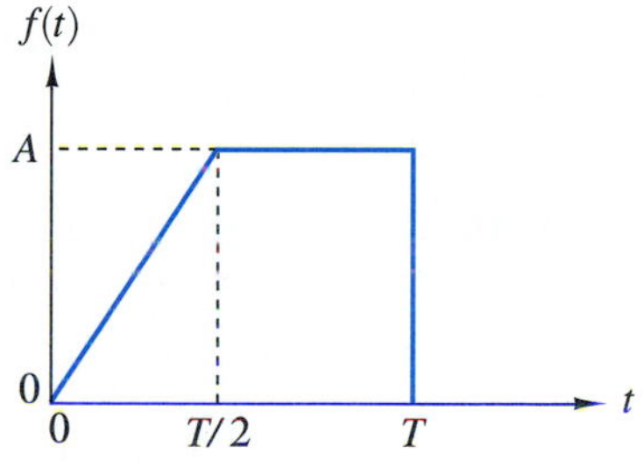

Figure P9-35

9-36 Repeat Problem 9-34 for the waveform in Fig. P9-36.

9-37 There is no initial energy storage in the circuit in Fig. P9-37. Find the current $i(t)$ when $v_S(t)$ is the waveform shown in Fig. P9-34.

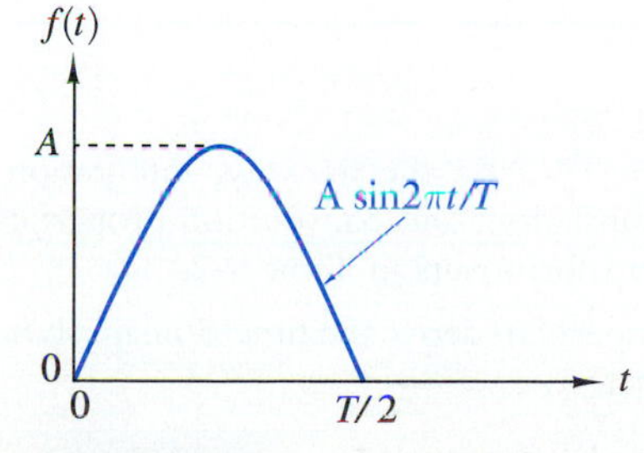

Figure P9-36

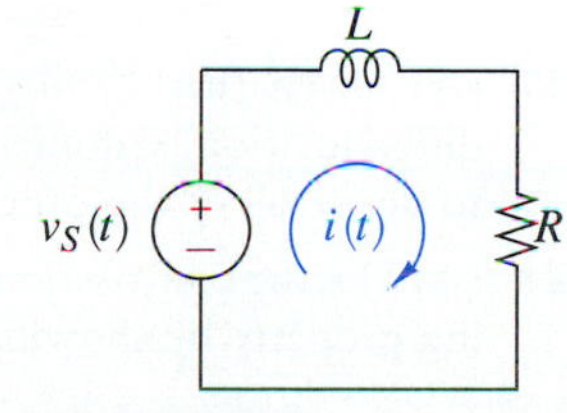

Figure P9-37

9-38 Find $f(t)$ for the following transforms.

(a) $F(s) = e^{-2s}\dfrac{10(s+2)}{s(s+2)}$

(b) $F(s) = \dfrac{1}{s+4} + e^{-5s}\dfrac{20}{s^2+4s+8}$

9-39 Given that $\mathscr{L}\{tu(t)\} = 1/s^2$, use the s-domain translation property to derive the Laplace transform of the damped ramp $f(t) = [te^{-\alpha t}]u(t)$.

9-40 Use the initial- and final-value properties to find the initial and final values of the waveform corresponding to the transforms in Prob. 9-15.

9-41 Use the initial- and final-value properties to find the initial and final values of the waveform corresponding to the transforms in Prob. 9-16.

9-42 Use the initial- and final-value properties to find the initial and final values of the waveform corresponding to the transforms in Prob. 9-17.

9-43 Use the initial- and final-value properties to find the initial and final values of the waveform corresponding to the transforms in Prob. 9-18.

9-44 Use the convolution property to show that

$$\mathscr{L}^{-1}\left\{\frac{F(s)}{s}\right\} = \int_0^t f(x)dx$$

9-45 Use the convolution property to find $f(t) = \mathscr{L}^{-1}\{F_1(s)F_2(s)\}$ for the following functions.

(a) $F_1(s) = \dfrac{1}{s+\alpha}$, $F_2(s) = \dfrac{1}{s-\alpha}$

(b) $F_1(s) = \dfrac{1}{s}$, $F_2(s) = \dfrac{\beta}{s^2+\beta^2}$

9-46 Use the convolution property to find $f(t) = \mathscr{L}^{-1}\{F_1(s)F_2(s)\}$ for the following functions.

(a) $F_1(s) = \dfrac{1}{(s+\alpha)^2}$, $F_2(s) = \dfrac{1}{s+\alpha}$

(b) $F_1(s) = F_2(s) = \dfrac{\beta}{s^2+\beta^2}$

CHAPTER INTEGRATING PROBLEMS

9-47 **(A)** Given that $\mathscr{L}\{u(t)\} = 1/s$, use the linearity, integration, differentiation, s-domain translation, and convolution properties to derive all of the other transform pairs in Table 9-2.

9-48 **(A)** Use the convolution property to derive the time-domain shifting property by showing that

$$\mathscr{L}^{-1}\left\{e^{-as}F(s)\right\} = [f(t-a)]u(t-a)$$

9-49 **(A)** A rational transform can be written as $F(s) = p(s)/q(s)$, where $p(s)$ and $q(s)$ are polynomials in s. If $F(s)$ has a simple pole at $s = \alpha$, show that the residue at the pole is $k_\alpha = p(s)/q'(s)\big|_{s=\alpha}$, where $q'(s)$ is the derivative of $q(s)$ with respect to s.

9-50 **(A)** Use Laplace transforms to solve for $y(t)$ in the equation

$$y(t) + \int_0^t y(\tau)e^{-\alpha(t-\tau)}d\tau = tu(t)$$

9-51 **(A)** The s-domain differentiation property of the Laplace transform states that

$$\text{IF } \mathscr{L}\{f(t)\} = F(s) \text{ THEN } \mathscr{L}\{tf(t)\} = -\frac{dF(s)}{ds}$$

Use this property to find the Laplace transforms of the following waveforms.

(a) $f(t) = [t^2]u(t)$
(b) $f(t) = [te^{-\alpha t}]u(t)$
(c) $f(t) = [t \sin \beta t]u(t)$

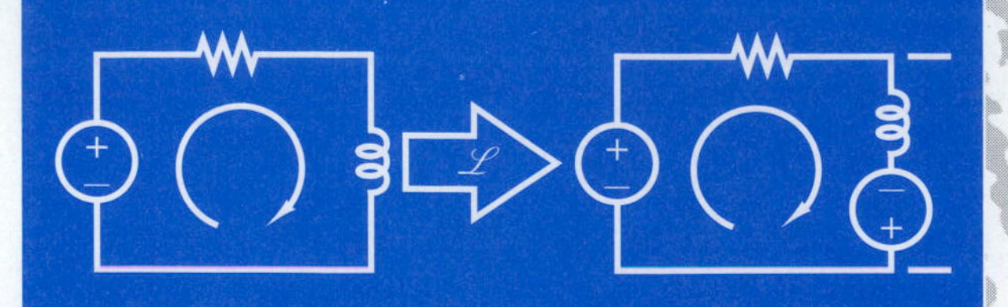

Chapter Ten

The resistance operator Z is a function of the electrical constants of the circuit components and of d/dt, the operator of time-differentiation, which will in the following be denoted by p simply.

Oliver Heaviside, 1887
British Engineer

s-Domain Circuit Analysis

The Laplace transform techniques used in this chapter have their roots in the works of Oliver Heaviside (1850–1925). The quotation on the opposite page was taken from his book *Electrical Papers* originally published in 1887. His resistance operator Z, which he later called impedance, is a unifying thread for much of electrical engineering. Heaviside does not usually receive the credit he deserves because his intuitive approach to mathematics was not accepted by most Victorian scientists of his day. Mathematical justification for his methods was eventually supplied by John Bromwich and others. However, no important errors were found in Heaviside's results.

In this chapter we transform a circuit directly from the time domain into the s domain without first developing the circuit differential equation. The transformed circuit can be analyzed using methods similar to those for resistance circuits because Kirchhoff's laws apply in the s domain, and element i-v characteristics become linear algebraic equations. The algebraic form of the i-v relationships leads to the concept of impedance as the s-domain generalization of resistance. s-domain circuit analysis closely parallels the resistance circuit analysis methods in Chapters 2 through 5 because the underlying element and connection constraints are similar.

◇ 10-1 Transformed Circuits

So far we have used the Laplace transformation to change waveforms into transforms and convert circuit differential equations into algebraic equations. These operations provide a useful introduction to the s domain. However, the real power of the Laplace transformation emerges when we transform the circuit itself and study its behavior directly in the s domain.

The solid blue arrow in Fig. 10-1 indicates the analysis path we will be following in this chapter. s-domain circuit analysis begins with a linear circuit in the time domain. We transform the circuit into the s domain, write the circuit equations directly in that domain, and then solve these algebraic equations for response transform. The inverse Laplace transformation then produces the response waveform. However, the s-domain approach is not just another way to derive response waveforms. This approach allows us work directly with the circuit model using analysis tools such as voltage and current division. By working directly with the circuit model, we gain insights into the interaction between circuits and signals that cannot be obtained using the classical approach indicated by the blue dotted path in Fig. 10-1.

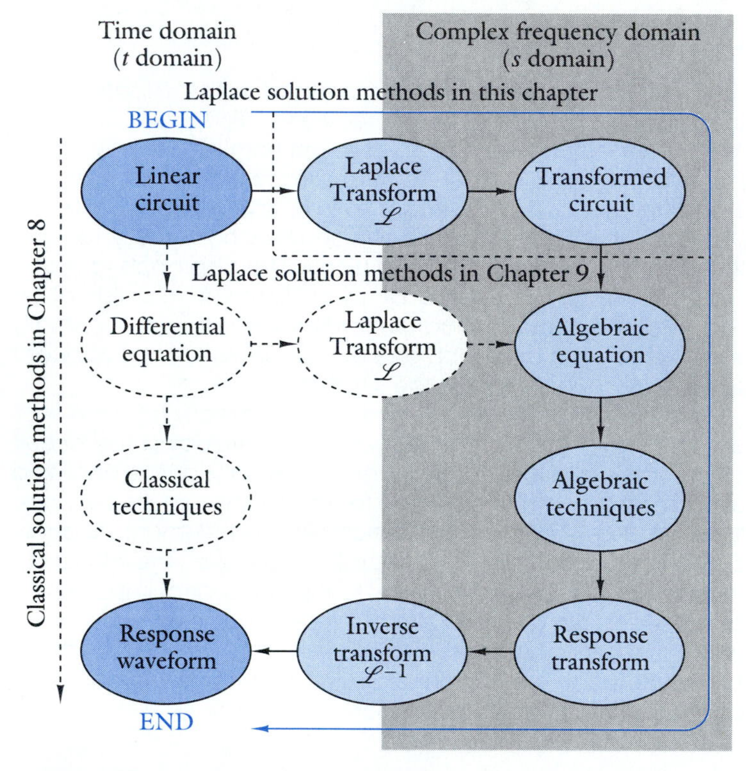

Figure 10-1 Flow diagram for s-domain circuit analysis.

How are we to transform a circuit? We have seen several times that circuit analysis is based on device and connection constraints. The connection constraints are derived from Kirchhoff's laws and the device constraints from the i-v relationships used to model the physical devices in the circuit. To transform circuits we must see how these two types of constraints are altered by the Laplace transformation.

Connection Constraints in the s Domain

A typical KCL connection constraint could be written as

$$i_1(t) + i_2(t) - i_3(t) + i_4(t) = 0 \tag{10-1}$$

This constraint requires that the algebraic sum of the current waveforms at a node be zero for all t. Using the linearity property, the Laplace transformation of this equation is

$$I_1(s) + I_2(s) - I_3(s) + I_4(s) = 0 \tag{10-2}$$

This form of the KCL connection constraint requires that the current transforms at a node sum to zero for all values of s. This idea generalizes to any number of currents at a node and any number of nodes. In addition, this idea applies to Kirchhoff's voltage law as well. The form of the connection constraints do not change because they are linear equations and the Laplace transformation is a linear operation. In summary, KCL and KVL apply to waveforms in the t domain and to transforms in the s domain.

Element Constraints in the s Domain

Turning now to the element constraints, we first deal with the independent signal sources shown in Fig. 10-2. The i-v relationships for these elements are

$$\begin{aligned} \text{Voltage Source:} \quad & v(t) = v_S(t) \\ & i(t) = \text{Depends on Circuit} \\ \text{Current Source:} \quad & i(t) = i_S(t) \\ & v(t) = \text{Depends on Circuit} \end{aligned} \tag{10-3}$$

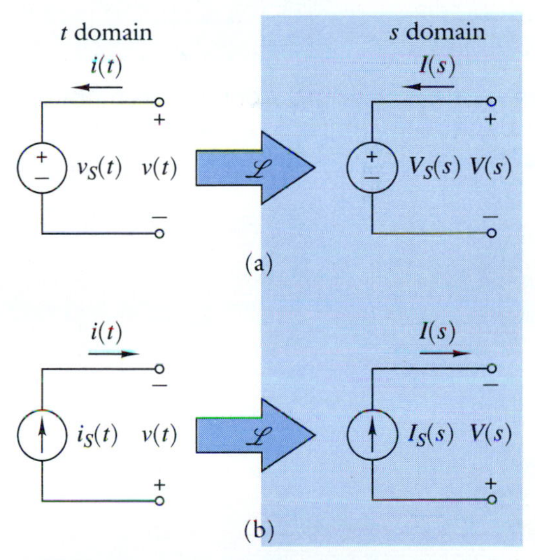

Figure 10-2 s-domain models of independent sources.

Independent sources are two-terminal elements. In the t domain they constrain the waveform of one signal variable and adjust the unconstrained variable to meet the demands of the external circuit. We think of an independent source as a generator of a specified voltage or current waveform. The Laplace transform of Eq. (10-3) is

$$\begin{aligned} \text{Voltage Source:} \quad & V(s) = V_S(s) \\ & I(s) = \text{Depends on Circuit} \\ \text{Current Source:} \quad & I(s) = I_S(s) \\ & I(s) = \text{Depends on Circuit} \end{aligned} \tag{10-4}$$

In the s domain independent sources function the same way as

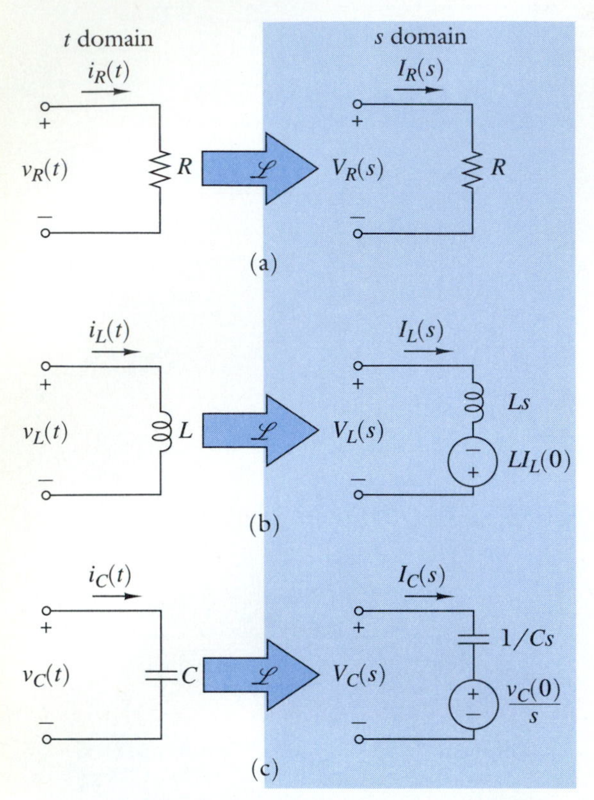

Figure 10-3 Time and *s*-domain element models using voltage sources for initial conditions.

in the t domain, except that we think of them as generating specified transforms rather than waveforms.

Next we consider the two-terminal passive circuit elements shown in Fig. 10-3. In the time domain their i-v relationships are

$$\begin{aligned}
&\text{Resistor:} && v_R(t) = Ri_R(t) \\
&\text{Inductor:} && v_L(t) = L\frac{di_L(t)}{dt} \\
&\text{Capacitor:} && v_C(t) = \frac{1}{C}\int_0^t i_C(\tau)d\tau + v_C(0)
\end{aligned} \tag{10-5}$$

These element constraints are transformed into the s domain by taking the Laplace transform of both sides of each equation using the linearity, differentiation, and integration properties:

$$\begin{aligned}
&\text{Resistor:} && V_R(s) = RI_R(s) \\
&\text{Inductor:} && V_L(s) = LsI_L(s) - Li_L(0) \\
&\text{Capacitor:} && V_C(s) = \frac{1}{Cs}I_C(S) + \frac{v_C(0)}{s}
\end{aligned} \tag{10-6}$$

As expected, the element relationships are algebraic equations in the s-domain. For the linear resistor the s domain version of Ohm's law says that the voltage transform $V_R(s)$ is proportional to the current transform $I_R(s)$. The element constraints for the inductor and capacitor also involve a proportionality between voltage and current, but include a term for the initial conditions as well.

The element constraints in Eq. (10-6) lead to the s-domain circuit models shown on the right side of Fig. 10-3. The t-domain parameters L and C are replaced by proportionality factors Ls and $1/Cs$ in the s domain. The initial conditions associated with the inductor and capacitor are modeled as voltage sources in series with these elements. The polarity of these sources are determined by the sign of the corresponding initial condition terms in Eq. (10-6). These initial condition voltage sources must be included when using these models to calculate the voltage transforms $V_L(s)$ or $V_C(s)$.

Impedance and Admittance

The concept of impedance is a basic feature of s-domain circuit analysis. For zero initial conditions, Eqs. (10-6) reduce to

$$\begin{aligned}
&\text{Resistor:} && V_R(s) = (R)I_R(s) \\
&\text{Inductor:} && V_L(s) = (Ls)I_L(s) \\
&\text{Capacitor:} && V_C(s) = \left(\frac{1}{Cs}\right)I_C(s)
\end{aligned} \tag{10-7}$$

In each case the voltage across the element is proportional to the current through it. The proportionality factor is called the element impedance, $Z(s)$. Stated formally,

Impedance is the ratio of a voltage transform to a current transform with all initial conditions set to zero.

It is important to remember that part of the definition of impedance is that the initial conditions are zero. The impedances of the three passive elements are

$$\begin{aligned}
&\text{Resistor:} \quad Z_R(s) = \frac{V_R(s)}{I_R(s)} = R \\
&\text{Inductor:} \quad Z_L(s) = \frac{V_L(s)}{I_L(s)} = Ls \qquad \text{with } i_L(0) = 0 \qquad (10\text{-}8) \\
&\text{Capacitor:} \quad Z_C(s) = \frac{V_C s}{I_C(s)} = \frac{1}{Cs} \qquad \text{with } v_C(0) = 0
\end{aligned}$$

Impedance is inherently an s-domain concept, since it is defined as the ratio of a voltage and current transform. Impedance is a generalization of the t-domain concept of resistance. The impedance of a resistor is its resistance R. The impedance of the inductor and capacitor depend on the inductance L and capacitance C, and the complex frequency variable s. A voltage transform has the units of V-s and current transform has units of A-s; therefore, the impedance has the units of ohms, since $(V\text{-}s)/(A\text{-}s) = V/A = \Omega$.

Algebraically solving Eq. (10-6) for the element currents in terms of the voltages produces alternative s-domain models:

$$\begin{aligned}
&\text{Resistor:} \quad I_R(s) = \frac{1}{R} V_R(s) \\
&\text{Inductor:} \quad I_L(s) = \frac{1}{Ls} V_L(s) + \frac{i_L(0)}{s} \qquad (10\text{-}9) \\
&\text{Capacitor:} \quad I_C(s) = CsV_C(s) - Cv_C(0)
\end{aligned}$$

In this form, the i-v relations lead to the s-domain models shown in the Fig. 10-4. The reference directions for the initial condition current sources are determined by the sign of the corresponding terms in Eq. (10-9). The initial condition sources are in parallel with what is called the element admittance.

Admittance $Y(s)$ is the s-domain generalization of the t-domain concept of conductance and can be thought of as the reciprocal of impedance.

$$Y(s) = \frac{1}{Z(s)} \qquad (10\text{-}10)$$

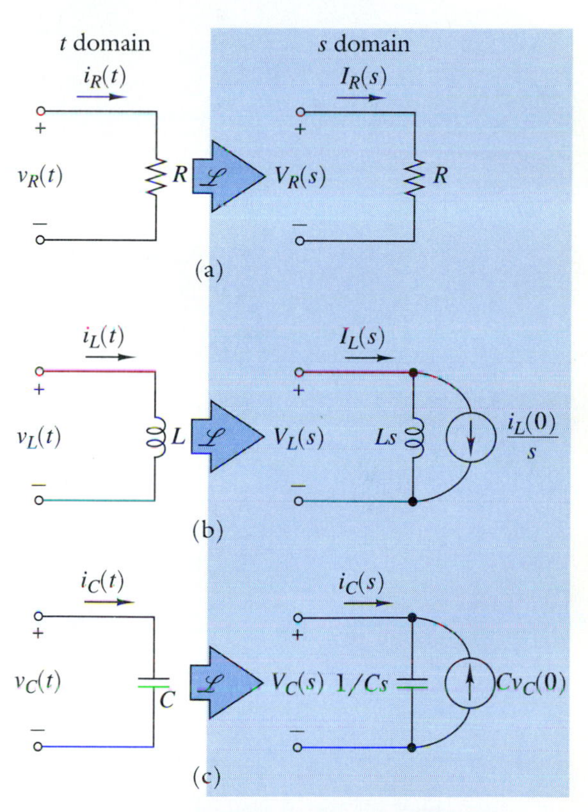

Figure 10-4 Time and s-domain element models using current sources for initial conditions.

Stated formally:

> **Admittance** is the ratio of a current transform to a voltage transform with all initial conditions set to zero.

Using this definition the admittances of the three passive elements are

$$\text{Resistor:}\quad Y_R(s) = \frac{1}{Z_R(s)} = \frac{I_R(s)}{V_R(s)} = \frac{1}{R} = G$$

$$\text{Inductor:}\quad Y_L(s) = \frac{1}{Z_L(s)} = \frac{I_L(s)}{V_L(s)} = \frac{1}{Ls} \qquad \text{with } i_L(0) = 0$$

$$\text{Capacitor:}\quad Y_C(s) = \frac{1}{Z_C(s)} = \frac{I_C(s)}{V_C(s)} = Cs \qquad \text{with } v_C(0) = 0 \tag{10-11}$$

Admittance $Y(s)$ is a ratio of current transform and a voltage transform so its units are siemens, since $(A\text{-}s)/(V\text{-}s) = A/V = S$.

In summary, to transform a circuit into the s domain we replace each element by an s-domain model. For independent sources and resistors the only change is that these elements now constrain transforms rather than waveforms. For inductors and capacitors we can use either the impedance model with a series initial condition voltage source (Fig. 10-3), or the admittance model with a parallel initial condition current source (Fig. 10-4). However, to avoid possible confusion we will always write the inductor impedance Ls and capacitor impedance $1/Cs$ beside the transformed element regardless of which initial condition source is used.

To analyze the transformed circuit we use the tools developed for resistance circuits in Chapters 2 through 5. These tools are applicable because KVL and KCL apply to transforms, and the s-domain element equations for inductors and capacitors are linear constraints similar to Ohm's law. These features make s-domain circuit analysis an algebraic process that is analogous to resistance circuit analysis.

EXAMPLE 10-1

(a) Transform the circuit in Fig. 10-5(a) into the s domain.

(b) Solve for the current transform $I(s)$.

(c) Find the waveform $i(t)$ by finding the inverse transform of $I(s)$.

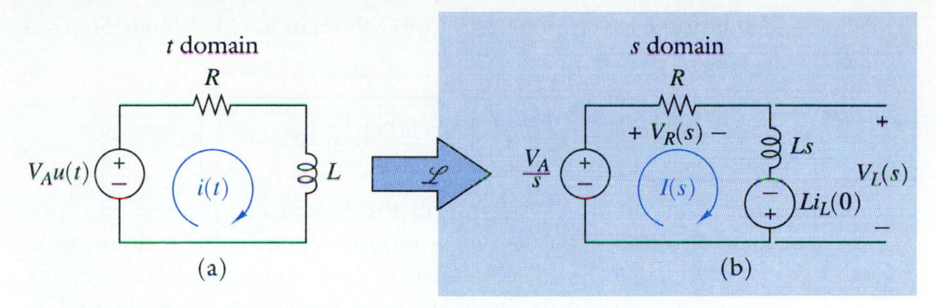

Figure 10-5

SOLUTION:

(a) Figure 10-5(b) shows the transformed circuit using a series voltage source $Li_L(0)$ to represent the inductor initial condition. The impedances of the two passive elements are R and Ls. The independent source voltage $V_Au(t)$ transforms as V_A/s.

(b) By KVL the sum of voltages around the loop is

$$-\frac{V_A}{s} + V_R(s) + V_L(s) = 0$$

Using the impedance models the s-domain element constraints are

$$\text{Resistor:} \quad V_R(s) = RI(s)$$

$$\text{Inductor:} \quad V_L(s) = LsI(s) - Li_L(0)$$

Substituting the element constraints into the KVL constraint and collecting terms yields

$$-\frac{V_A}{s} + (R + Ls)I(s) - Li_L(0) = 0$$

Solving for $I(s)$ produces

$$I(s) = \frac{V_A/L}{s(s + R/L)} + \frac{i_L(0)}{s + R/L}$$

$I(s)$ is the transform of the circuit response for a step function input. The response $I(s)$ is a rational function with simple poles at $s = 0$ and $s = -R/L$.

(c) The time-domain response is found by expanding $I(s)$ into partial fractions.

$$I(s) = \frac{V_A/R}{s} - \frac{V_A/R}{s + R/L} + \frac{i_L(0)}{s + R/L}$$

Taking the inverse transform of each term in this expansion gives

$$i(t) = \Bigg[\overbrace{\frac{V_A}{R}}^{\text{Forced}} - \overbrace{\frac{V_A}{R}e^{-Rt/L} + i_L(0)e^{-Rt/L}}^{\text{Natural}}\Bigg]u(t)$$

The step function input causes the forced response. The exponential terms in the natural response depend on the circuit time constant L/R. The step function and exponential components in $i(t)$ are related to the location of the two poles in $I(s)$. The pole at the origin came from the step function input and leads to the forced response. The pole at $s = -R/L$ came from the circuit and leads to the natural response.

EXAMPLE 10-2

Develop an s-domain model for the coupled coils in Fig. 10-6.

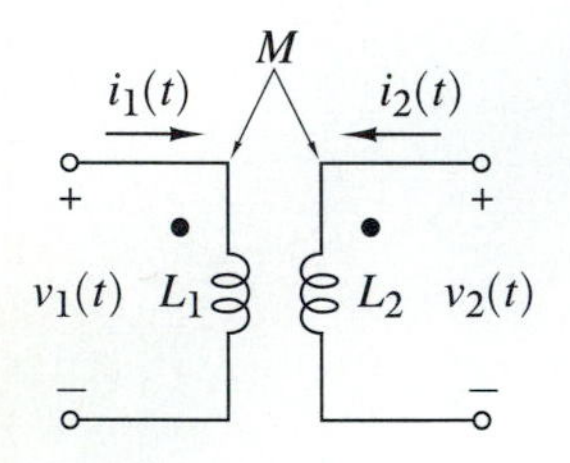

Figure 10-6

SOLUTION:

For the given reference marks the element i-v relationships are

$$v_1(t) = L_1\frac{di_1(t)}{dt} + M\frac{di_2(t)}{dt}$$

$$v_2(t) = M\frac{di_1(t)}{dt} + L_2\frac{di_2(t)}{dt}$$

These equations are transformed into the s domain using the linearity and differentiation properties of the Laplace transformation:

$$V_1(s) = L_1sI_1(s) - L_1i_1(0) + MsI_2(s) - Mi_2(0)$$

$$V_2(s) = MsI_1(s) - Mi_1(0) + L_2sI_2(s) - L_2i_2(0)$$

Collecting related terms rearranges these equations as follows:

$$V_1(s) = L_1sI_1(s) + MsI_2(S) - L_1i_1(0) - Mi_2(0)$$

$$V_2(s) = MsI_1(s) + L_2sI_2(s) - Mi_1(0) - L_2i_2(0)$$

The factors L_1s and L_2s are the impedance of the primary and secondary winding due to self inductance. The factor Ms is the mutual impedance caused by mutual inductance coupling. A mutual initial condition term for the secondary (primary) current appears in the equation for the primary (secondary) because mutual inductance couples the two windings. These equations lead to the s-domain model of coupled coils shown in Fig. 10-7(a).

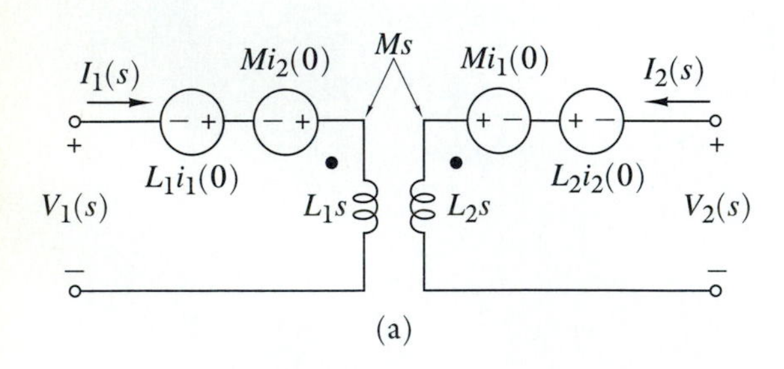

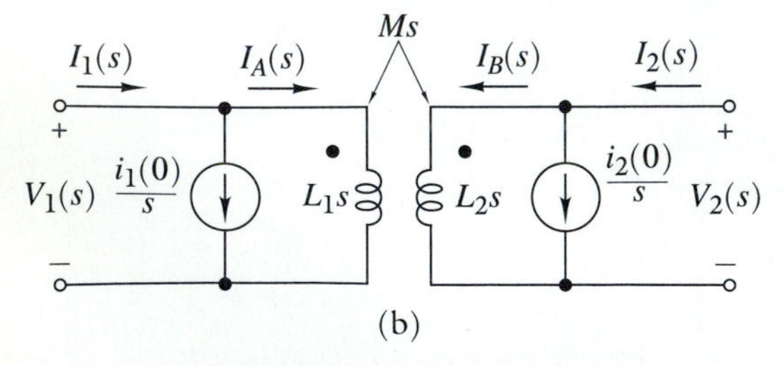

Figure 10-7

An alternative s-domain model is obtained by rearranging the element equations in the form

$$V_1(s) = L_1 s \underbrace{\left[I_1(s) - \frac{i_1(0)}{s} \right]}_{I_A(s)} + Ms \underbrace{\left[I_1(s) - \frac{i_2(0)}{s} \right]}_{I_B(s)}$$

$$V_2(s) = Ms \overbrace{\left[I_1(s) - \frac{i_1(0)}{s} \right]} + L_2 s \overbrace{\left[I_2(s) - \frac{i_2(0)}{s} \right]}$$

In this form we see that the currents $I_A(s)$ and $I_B(s)$ together with the impedances $L_1 s$, $L_2 s$, and Ms determine the terminal voltages $V_1(s)$ and $V_2(s)$. Inserting initial condition current sources as shown in Fig. 10-7(b) ensures that the terminal currents $I_1(s)$ and $I_2(s)$ are correct.

10-2 BASIC CIRCUIT ANALYSIS IN THE s DOMAIN

In this section we develop the s-domain versions series and parallel equivalence, and voltage and current division. These analysis techniques are the basic tools in s-domain circuit analysis, just as they were for resistance circuit analysis. These methods apply to circuits with elements connected in series or parallel and to ladder networks. General analysis methods using node-voltage or mesh-current equations are covered later in Sections 10-4 and 10-5.

Series Equivalence and Voltage Division

The concept of a series connection applies in the s domain because Kirchhoff's laws do not change under the Laplace transformation. In Fig. 10-8 the two elements with impedances $Z_1(s)$ and $Z_2(s)$ are connected in series so the same current $I(s)$ must exist in both elements. Using the element impedance we can express the element voltages in terms of this current:

$$V_1(s) = Z_1(s)I(s) \qquad V_2(s) = Z_2(s)I(s) \tag{10-12}$$

Applying KVL around the loop requires that $V(s) = V_1(s) + V_2(s)$. Therefore, $V(s)$ is related to $I(s)$ as

$$\begin{aligned} V(s) &= Z_1(s)I(s) + Z_2(s)I(s) \\ &= [Z_1(s) + Z_2(s)]\,I(s) \end{aligned} \tag{10-13}$$

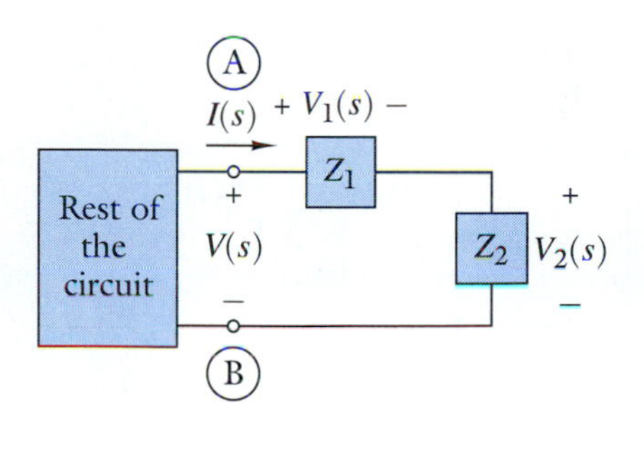

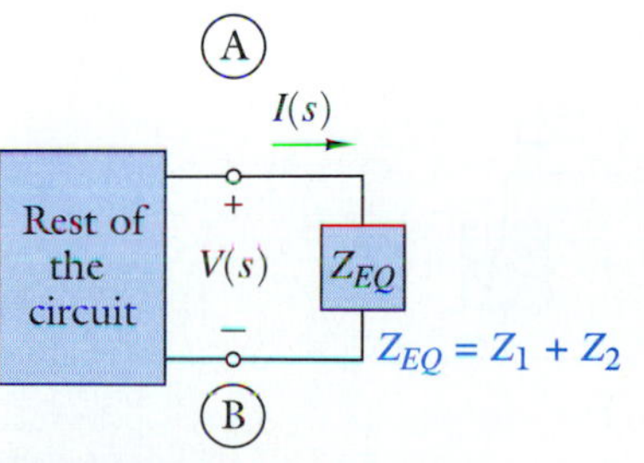

Figure 10-8 Series equivalence in the s domain.

This equation shows that the signals $V(s)$ and $I(s)$ in Fig. 10-8 do not change when $Z_1(s)$ and $Z_2(s)$ are replaced by an equivalent impedance $Z_{EQ}(s)$

$$Z_{EQ}(s) = Z_1(s) + Z_2(s) \tag{10-14}$$

This conclusion generalizes as:

> The **equivalent impedance** of two or more elements connected in series equals the sum of the individual element impedances.

Combining Eqs. (10-12), (10-13) and (10-14) we obtain the element voltages in the form

$$V_1(s) = \frac{Z_1(s)}{Z_{EQ}(s)} V(s) \qquad V_2(s) = \frac{Z_2(s)}{Z_{EQ}(s)} V(s) \tag{10-15}$$

These equations are examples of the voltage division principle in the s domain. The general statement of the voltage division principle is:

> Any element voltage in a series connection is equal to its impedance divided by the equivalent impedance of the connection times the voltage across the series circuit.

Parallel Equivalence and Current Division

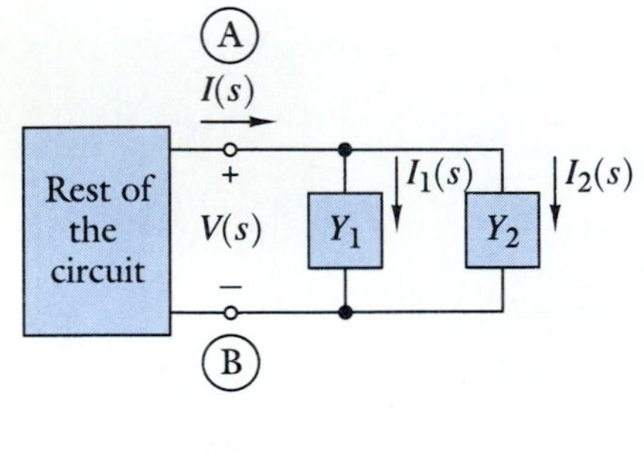

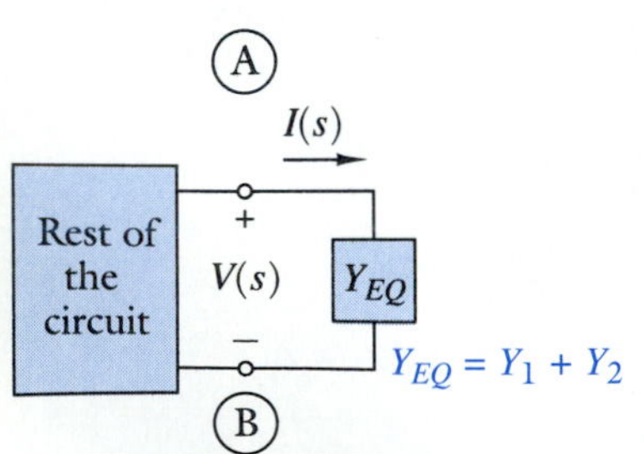

Figure 10-9 Parallel equivalence in the s domain.

In Fig. 10-9 the two elements with admittances $Y_1(s)$ and $Y_2(s)$ are connected in parallel so the same voltage $V(s)$ must appear across both. Using the element admittances we can express the element currents in terms of the voltage $V(s)$:

$$I_1(s) = Y_1(s)V(s) \qquad I_2(s) = Y_2(s)V(s) \tag{10-16}$$

Applying KCL at Node A requires that $I(s) = I_1(s) + I_2(s)$. Therefore, $I(s)$ is related to $V(s)$ as

$$\begin{aligned} I(s) &= Y_1(s)V(s) + Y_2(s)V(s) \\ &= [Y_1(s) + Y_2(s)]\, V(s) \end{aligned} \tag{10-17}$$

This equation shows that the signals $V(s)$ and $I(s)$ in Fig. 10-9 do not change when $Y_1(s)$ and $Y_2(s)$ are replaced by an equivalent admittance Y_{EQ}

$$Y_{EQ}(s) = Y_1(s) + Y_2(s) \tag{10-18}$$

This conclusion generalizes as:

> The **equivalent admittance** of two or more elements connected in parallel equals the sum of the individual element admittances.

Combining Eqs. (10-16), (10-17), and (10-18) we obtain the element currents in the form

$$I_1(s) = \frac{Y_1(s)}{Y_{EQ}(s)} I(s) \qquad I_2(s) = \frac{Y_2(s)}{Y_{EQ}(s)} I(s) \qquad (10\text{-}19)$$

These equations are examples of the current division principle in the s domain. The general statement of the current division principle is:

> Any element current in a parallel connection is equal to its admittance divided by the equivalent admittance of the connection times the current through the parallel circuit.

From these results we begin to see that s-domain circuit analysis involves basic concepts that closely parallel the analysis of resistance circuits in the t domain. Repeated application of equivalence and voltage/current division leads to an analysis approach called circuit reduction. The major difference here is that we use impedance and admittances rather than resistance and conductance, and the analysis yields voltage and current transforms rather than waveforms.

EXAMPLE 10-3

The inductor current and capacitor voltage in Fig. 10-10 are zero at $t = 0$.

(a) Transform the circuit into the s domain and find the equivalent impedance seen at terminals A and B.

(b) Use voltage division to solve for the output voltage transform $V_2(s)$.

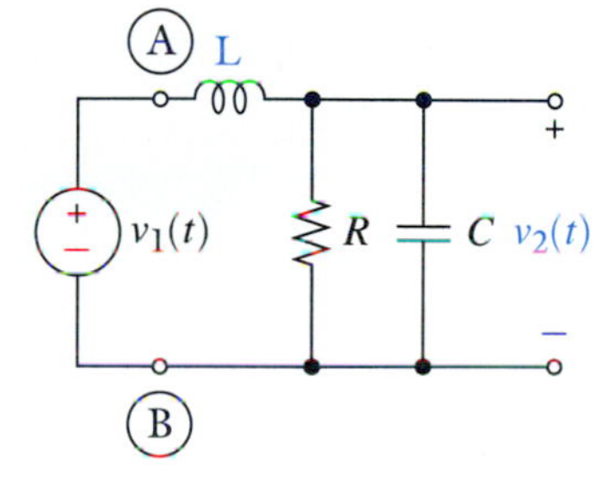

Figure 10-10

SOLUTION:

(a) Figure 10-11(a) shows the circuit in Fig. 10-10 transformed into the s domain. To find $Z_{EQ}(s)$ we replace the resistor and capacitor with an equivalent admittance:

$$Y_{EQ1}(s) = \frac{1}{Z_{EQ1}(s)} = \frac{1}{R} + Cs = \frac{RCs + 1}{R}$$

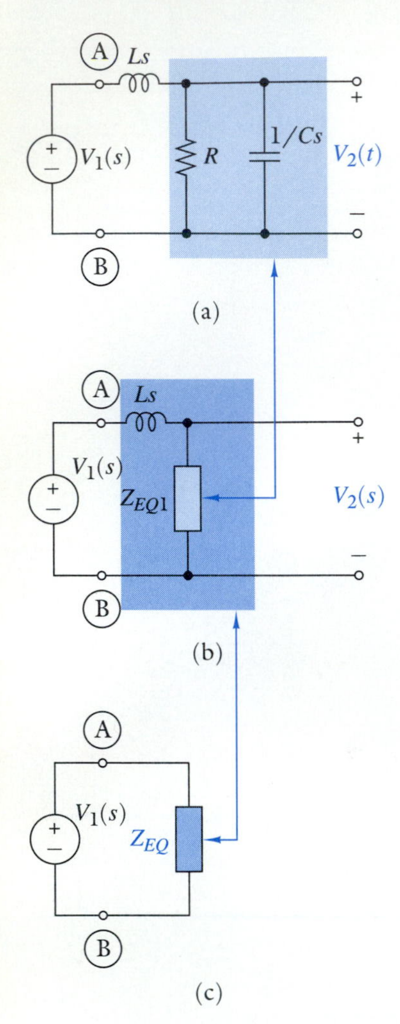

Figure 10-11

In Fig. 10-11(b) the equivalent impedance $Z_{EQ1}(s)$ is connected in series with the inductor. This series combination can be replaced by an equivalent impedance

$$Z_{EQ}(s) = Ls + Z_{EQ1}(s) = Ls + \frac{R}{RCs+1}$$

$$= \frac{RLCs^2 + Ls + R}{RCs+1}$$

as shown in Fig. 10-11(c). $Z_{EQ}(s)$ is the impedance seen between terminals A and B in Fig. 10-10.

(b) Using voltage division in Fig. 10-11(b) we find $V_2(s)$ as

$$V_2(s) = \left[\frac{Z_{EQ1}(s)}{Z_{EQ}(s)}\right] v_1(s) = \left[\frac{R}{RLCs^2 + Ls + R}\right] V_1(s)$$

Note that $Z_{EQ}(s)$ and $V_2(s)$ are rational functions of the complex frequency variable s.

EXERCISE 10-1

The inductor current and capacitor voltage in Fig. 10-12 are zero at $t = 0$.

(a) Transform the circuit into the s domain and find the equivalent admittance seen between terminals A and B.

(b) Solve for the output current transform $I_2(s)$ in terms of the input current $I_1(s)$.

ANSWERS:

(a) $Y_{EQ}(s) = \dfrac{LCs^2 + RCs + 1}{Ls(RCs+1)}$

(b) $I_2(s) = \left[\dfrac{LCs^2}{LCs^2 + RCs + 1}\right] I_1(s)$

EXERCISE 10-2

The inductor current and capacitor voltage in Fig. 10-13 are zero at $t = 0$.

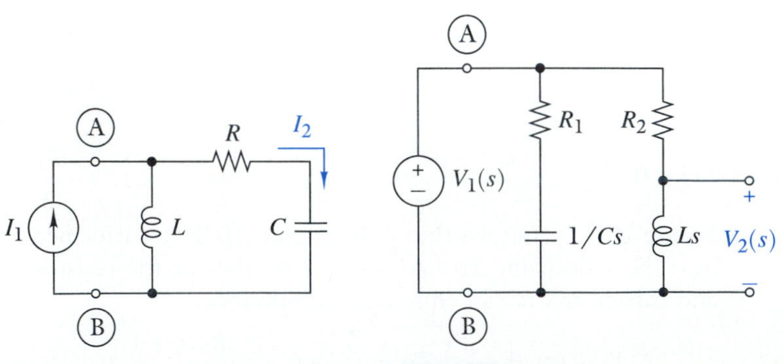

Figure 10-12 Figure 10-13

(a) Transform the circuit into the s domain and find the equivalent impedance seen between terminals A and B.

(b) Solve for the output voltage transform $V_2(s)$ in terms of the input voltage $V_1(s)$.

ANSWERS:

(a) $Z_{EQ}(s) = \dfrac{(R_1Cs + 1)(Ls + R_2)}{LCs^2 + (R_1 + R_2)Cs + 1}$

(b) $V_2(s) = \left[\dfrac{Ls}{Ls + R_2}\right] V_1(s)$ ■

10-3 CIRCUIT THEOREMS IN THE s DOMAIN

In this section we study the s-domain versions of circuit theorems on proportionality, superposition, and Thévenin/Norton equivalent circuits. These theorems define fundamental properties that provide conceptual tools for the analysis and design of linear circuits. With some modifications, all of the theorems studied in Chapter 3 apply to linear RLC circuits in the s domain.

Proportionality

For linear resistance circuits the *proportionality theorem* states that any output y is proportional to the input x:

$$y = Kx \tag{10-20}$$

The same concept applies to linear RLC circuits in the s domain except that the proportionality factor K is a rational function of s rather than a constant. For instance, in Example 10-3 we found the output voltage $V_2(s)$ to be

$$V_2(s) = \left[\frac{R}{RLCs^2 + Ls + R}\right] V_1(s)$$

where $V_1(s)$ is the transform of the input voltage. The quantity inside the bracket is a rational function that serves as the proportionality factor between the input and output transforms.

Rational functions that relate inputs and outputs in the s domain are called *network functions*. We will study network functions extensively beginning in Chapter 11. In this chapter we will utilize the proportionality theorem to develop the s-domain version of the unit-output method from Chapter 3. In the s domain the *unit-output method* assumes the output transform is unity, and determines the proportionality factor by calculating the input required to produce the unit output. Since $y = Kx$ the unit output method yields the proportionality factor as

$$K = \frac{\text{Output}}{\text{Input}} = \frac{1}{\text{Input for unit output}}$$

The following example illustrates the method in the s domain.

EXAMPLE 10-4

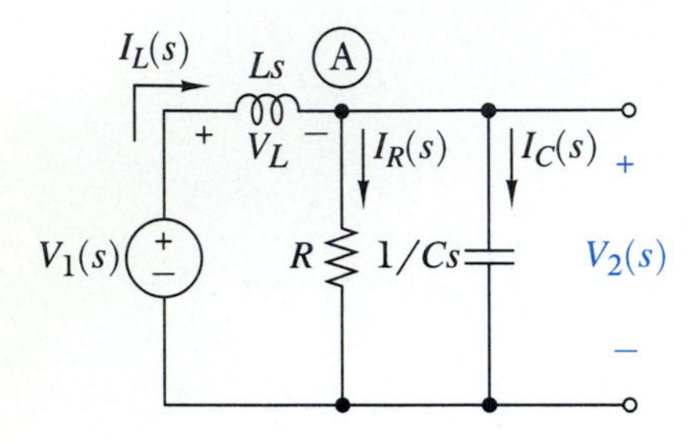

Figure 10-14

There is no initial energy stored in the circuit in Fig. 10-14. Transform the circuit into the s domain and use the unit output method to find the output voltage transform $V_2(s)$ in terms of the input $V_1(s)$.

SOLUTION:

The circuit in Fig. 10-14 was previously analyzed in Example 10-3 using voltage division. To apply the unit-output method we assume $V_2(s) = 1$ in Fig. 10-14. This assumption means that $V_R(s) = V_C(s) = 1$, since the capacitor and resistor are connected in parallel across the output. Using the resistor and capacitor impedances we calculate the element currents as

$$I_R(s) = \frac{V_R(s)}{Z_R} = \frac{1}{R} \qquad I_C(s) = \frac{V_C(s)}{Z_C} = \frac{1}{1/Cs} = Cs$$

Applying KCL to the current transforms at Node A we find the inductor current to be

$$I_L(s) = I_R(s) + I_C(s) = \frac{1}{R} + Cs = \frac{1 + RCs}{R}$$

The voltage across the inductor is

$$V_L(s) = Z_L(s) I_L(s) = Ls \left[\frac{RCs + 1}{R} \right] = \frac{RLCs^2 + Ls}{R}$$

Applying KVL around the input loop gives the input voltage.

$$V_1(s) = V_L(s) + 1 = \frac{RLCs^2 + Ls + R}{R}$$

This equation is the input voltage transform required to produce an output $V_2(s) = 1$. Applying the proportionality property yields the general input-output relationship.

$$V_2(s) = \left[\frac{1}{\text{Input for unit output}} \right] V_1(s)$$

$$= \left[\frac{R}{RLCs^2 + Ls + R} \right] V_1(s)$$

This result is the same as found in Example 10-3, using voltage division.

EXERCISE 10-3

Use the unit output method to express the output $V_2(s)$ of the circuit in Fig. 10-15 in terms of the input $I_1(s)$.

ANSWER:

$$V_2(s) = \left[\frac{R}{LCs^2 + RCs + 1}\right] I_1(s)$$

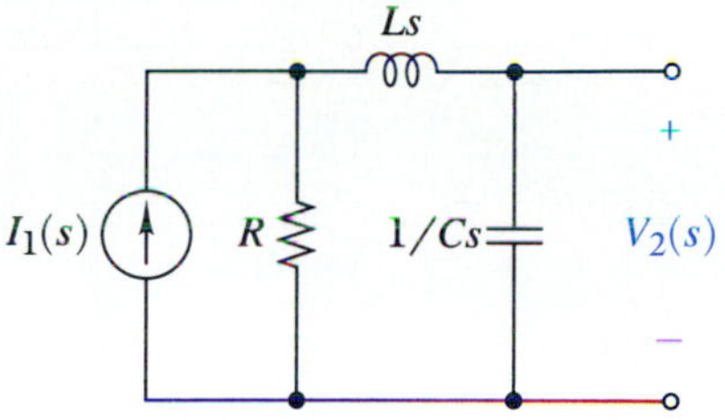

Ls

$I_1(s)$ R 1/Cs $V_2(s)$

Figure 10-15

Superposition

For linear resistance circuits the *superposition theorem* states that any output y of a linear circuit can be written as

$$y = K_1x_1 + K_2x_2 + K_3x_3 + \dots \tag{10-21}$$

where x_1, x_2, x_3, ... are circuit inputs and K_1, K_2, K_3, ... are weighting factors that depend on the circuit. The same concept applies to linear RLC circuits in the s domain except that the weighting factors are rational functions of s rather than constants.

Superposition is usually thought of as a way to find the response y by adding the component responses caused by each input acting alone. However, the principle applies to groups of inputs as well. For example, in the s-domain RLC circuits contain many different current and voltage sources. These sources can be lumped into one of two groups: (1) input voltage and current sources representing the external inputs for $t \geq 0$ or (2) initial condition voltage and current sources representing the energy stored at $t = 0$. The superposition principle states that the s-domain response can be found as the sum of two components:

1. The *zero-state response* caused by the input sources with the initial condition sources turned off
2. The *zero-input response* caused by the initial condition sources with the input sources turned off

Turning a source off means replacing voltage sources by short circuits $[V_S(s) = 0]$ and current sources by open circuits $[I_S(s) = 0]$.

The result is that voltage and current transform in a linear circuit can be found as the sum of two components of the form

$$V(s) = V_{zs}(s) + V_{zi}(s) \qquad I(s) = I_{zs}(s) + I_{zi}(s) \tag{10-22}$$

where the subscript zs stands for zero state and zi for zero input. An important corollary is that the time-domain response can also be partitioned into zero-state and zero-input components because the inverse Laplace transformation is a linear operation.

To illustrate we use superposition to analyzed the circuit previously treated in Example 10-1. The transformed circuit in Fig. 10-16 has two independent voltage sources: (1) an input voltage source and (2) a voltage source representing the initial inductor current. The resistor and inductor are in series so these two elements can be replaced by an impedance $Z_{EQ}(s) = Ls + R$.

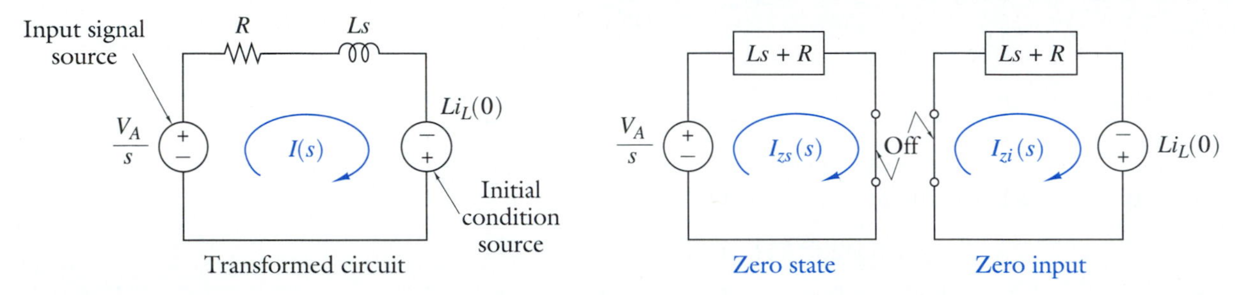

Figure 10-16 Using superposition to find the zero-state and zero-input responses.

First we turn off the initial condition source by replacing the voltage source replaced by a short circuit. Using the resulting zero-state circuit shown in Fig. 10-16 we obtain the zero-state response:

$$I_{zs}(s) = \frac{V_A/s}{Z_{EQ}(s)} = \frac{V_A/L}{s(s + R/L)} \tag{10-23}$$

The pole at $s = 0$ in $I_{zs}(s)$ comes from the input source and the pole at $s = -R/L$ comes from the RL circuit. Next, we turn off the input source and use the zero-input circuit shown in Fig. 10-16 to obtain the zero-input response

$$I_{zi}(s) = \frac{Li_L(0)}{Z_{EQ}(s)} = \frac{i_L(0)}{s + R/L} \tag{10-24}$$

The pole at $s = -R/L$ in $I_{zi}(s)$ comes from the circuit. The zero-input response does not have a pole at $s = 0$ because the step function input is turned off.

Superposition states that the total response is the sum of the zero-state component in Eq. (10-23) and the zero-input component in Eq. (10-24):

$$I(s) = I_{zs}(s) + I_{zi}(s) = \frac{V_A/L}{s(s + R/L)} + \frac{i_L(0)}{s + R/L} \tag{10-25}$$

The transform $I(s)$ in this equation is the same as found in Example 10-1. To derive the time-domain response we expand $I(s)$ by partial fractions

$$I(s) = \frac{V_A/R}{s} - \frac{V_A/R}{s + R/L} + \frac{i_L(0)}{s + R/L} \tag{10-26}$$

Performing the inverse transformation on each term yields

$$i(t) = \Big[\overbrace{\frac{V_A}{R}}^{\text{Forced}} \underbrace{- \frac{V_A}{R}e^{-Rt/L}}_{\text{Zero State}} \overbrace{}^{\text{Natural}} + \underbrace{i_L(0)e^{-Rt/L}}_{\text{Zero Input}}\Big]u(t) \tag{10-27}$$

Partitioning the waveform into zero-state and zero-input components using superposition produces the same result as Example 10-1. The zero-state component contains the forced response. The zero-state and zero-input components both contain an exponential term due to the pole at $s = -R/L$ because either type of source excites the circuit's natural response.

EXAMPLE 10-5

The switch in Fig. 10-17(a) has been open for a long time and is closed at $t = 0$.

(a) Transform the circuit into the s domain.

(b) Find the zero-state and zero-input components of V(s).

(c) Find $v(t)$ for $I_A = 1$ mA, $L = 2$ H, $R = 1.5$ kΩ, and $C = 1/6$ μF.

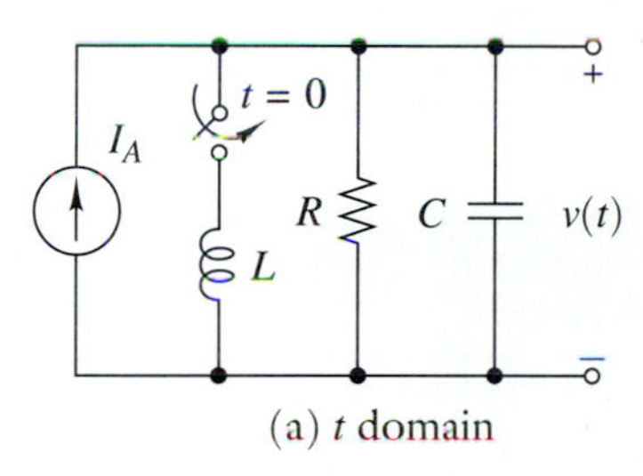

(a) t domain

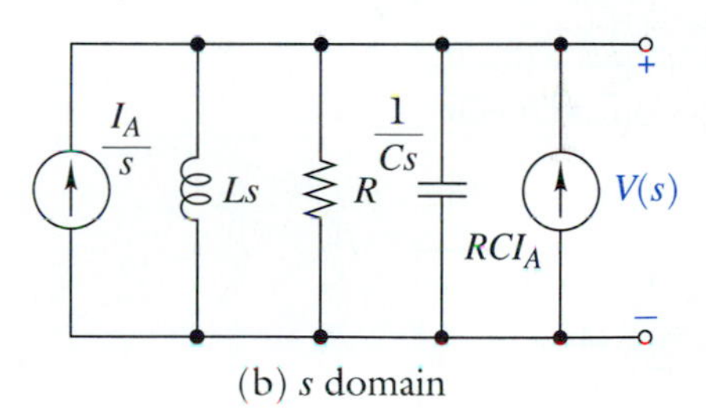

(b) s domain

Figure 10-17

SOLUTION:

(a) To transform the circuit into the s domain we must find the initial inductor current and capacitor voltage. For $t < 0$ the circuit is in a dc steady-state condition with the switch open. The inductor acts like a short circuit and the capacitor acts like an open circuit. By inspection the initial conditions at $t = 0-$ are $i_L(0) = 0$ and $v_C(0) = I_A R$. Figure 10-17(b) shows the s-domain circuit for these initial conditions. The current source version for the capacitor's initial condition is used here because the circuit elements are connected in parallel. The switch and constant current source combine to produce a step function $I_A u(t)$ whose transforms is I_A/s.

(b) The resistor, capacitor, and inductor can be replaced by an equivalent admittance

$$Y_{EQ} = \frac{1}{Z_{EQ}} = \frac{1}{Ls} + \frac{1}{R} + Cs$$

$$= \frac{RLCs^2 + Ls + R}{RLs}$$

The zero-state response is found with the capacitor initial condition source replaced by an open circuit and the step function input source on:

$$V_{zs}(s) = Z_{EQ}(s)\frac{I_A}{s} = \left[\frac{RLs}{RLCs^2 + Ls + R}\right]\frac{I_A}{s}$$

$$= \frac{I_A/C}{s^2 + \dfrac{s}{RC} + \dfrac{1}{LC}}$$

The zero-state response does not have a pole at $s = 0$ because the pole in the input is canceled by the zero at the origin in $Z_{EQ}(s)$. The zero-input response is found with the step function input source replaced by an open circuit and the capacitor's initial condition source on:

$$V_{zi}(s) = Z_{EQ}(s)CRI_A = \frac{RI_As}{s^2 + \dfrac{s}{RC} + \dfrac{1}{LC}}$$

(c) Inserting the given numerical values of the circuit parameters and expanding the zero-state and zero-input response transforms by partial fractions yields

$$V_{zs}(s) = \frac{6000}{(s+1000)(s+3000)} = \frac{3}{s+1000} + \frac{-3}{s+3000}$$

$$V_{zi}(s) = \frac{1.5s}{(s+1000)(s+3000)} = \frac{-0.75}{s+1000} + \frac{2.25}{s+3000}$$

Performing the inverse transform on each term in these expansions gives

$$v_{zs}(t) = [3e^{-1000t} - 3e^{-3000t}]u(t)$$

$$v_{zi}(t) = [-0.75e^{-1000t} + 2.25e^{-3000t}]u(t)$$

EXAMPLE 10-6

Use superposition to find the zero-state component of $I(s)$ in the s-domain circuit shown in Fig. 10-18(a).

SOLUTION:

Turning the voltage source off and replacing it by a short circuit produces the circuit in Fig. 10-18(b). In this circuit the resistor

(a) s-domain circuit

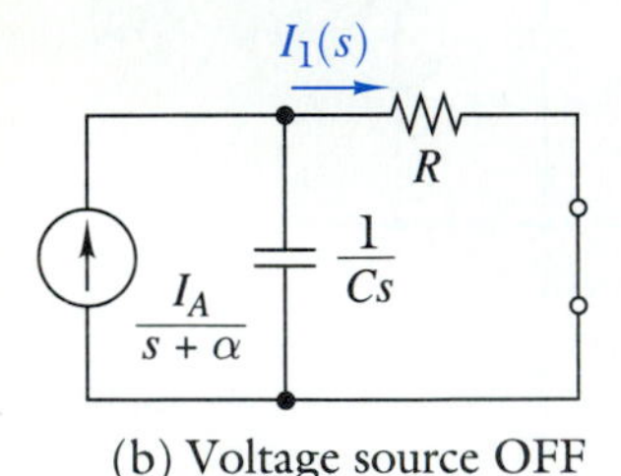

(b) Voltage source OFF

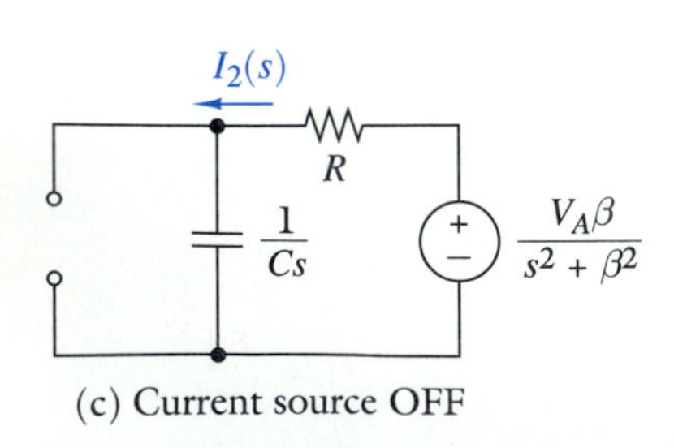

(c) Current source OFF

Figure 10-18

and capacitor are connected in parallel so current division yields $I_1(s)$ in the form

$$I_1(s) = \frac{Y_R}{Y_C + Y_R}\frac{I_A}{(s+\alpha)} = \frac{I_A}{(RCs+1)(s+\alpha)}$$

Turning the voltage source on and the current source off and replacing it by an open circuit produces the circuit in Fig. 10-18(c). In this case the resistor and capacitor are connected in series, so series equivalence gives the current $I_2(s)$:

$$I_2(s) = \frac{1}{Z_R + Z_C}\frac{V_A\beta}{s^2+\beta} = \frac{CsV_A\beta}{(RCs+1)(s^2+\beta^2)}$$

Using superposition the total response zero-state response is

$$\begin{aligned} I_{zs}(s) &= I_1(s) - I_2(s) \\ &= \frac{I_A}{(RCs+1)(s+\alpha)} - \frac{CsV_A\beta}{(RCs+1)(s^2+\beta^2)} \end{aligned}$$

There is a minus in this equation because $I_1(s)$ and $I_2(s)$ were assigned opposite reference directions in Figs. 10-18(b) and (c). The total zero-state response has four poles. The pole at $s = -1/RC$ came from the RC circuit. The pole at $s = -\alpha$ came from the current source and the two poles at $s = \pm j\beta$ came from the voltage source.

EXERCISE 10-4

The initial conditions for the circuit in Fig. 10-19 are $v_C(0-) = 0$ and $i_L(0-) = I_O$. Transform the circuit into the s domain and find the zero-state and zero-input components of $V(s)$.

ANSWER:

$$V_{zs}(s) = \left[\frac{1}{LCs^2 + RCs + 1}\right]\frac{V_A}{s}$$

$$V_{zi}(s) = \frac{LI_O}{LCs^2 + RCs + 1}$$

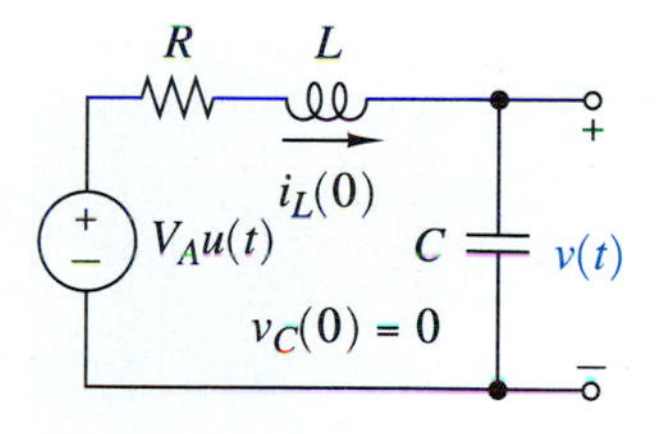

Figure 10-19

EXERCISE 10-5

The initial conditions for the circuit in Fig. 10-20 are $v_C(0-) = 0$ and the $i_L(0-) = I_O$. Transform the circuit into the s domain and find the zero-state and zero-input components of $I(s)$.

ANSWERS:

$$I_{zs}(s) = \left[\frac{LCs^2}{LCs^2 + RCs + 1}\right]\frac{I_A}{s}$$

$$I_{zi}(s) = -\left[\frac{LCs^2}{LCs^2 + RCs + 1}\right]\frac{I_O}{s}$$

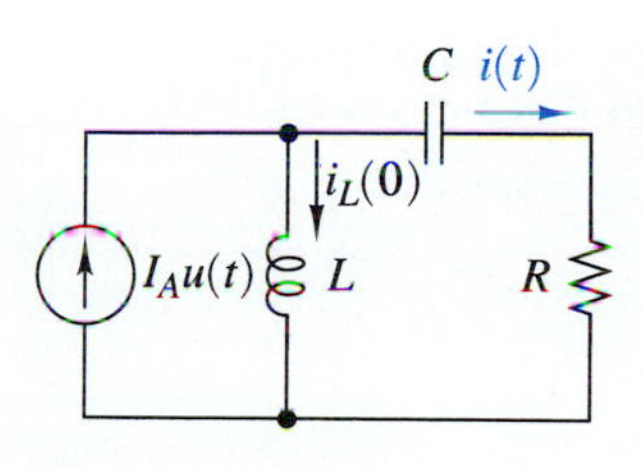

Figure 10-20

Thévenin and Norton Equivalent Circuits and Source Transformations

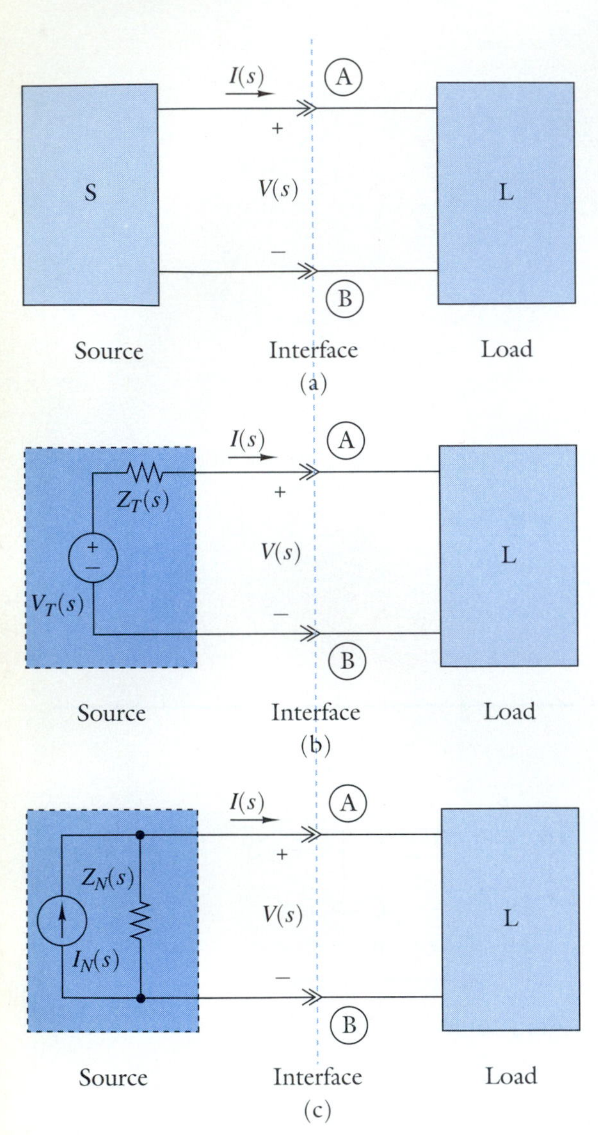

Figure 10-21 Thévenin and Norton equivalent circuits in the s domain.

Figure 10-21(a) shows a two-terminal interface in the s domain. When the source circuit S contains independent sources and linear elements, it can be replaced by the Thévenin equivalent circuit shown in Fig. 10-21(b) or the Norton equivalent circuit in Fig. 10-21(c). The methods of finding and using *Thévenin* or *Norton equivalent circuits* are the same in the s domain as in resistive circuits. The important difference here is that the interface signals are transforms rather than waveforms, the Thévenin source $V_T(s)$ and Norton source $I_N(s)$ are transforms, and the elements Z_T and Z_N are impedances.

The Thévenin and Norton circuits are obtained by finding the open-circuit voltage and short-circuit current at the interface. With an open-circuit load connected the interface voltages in Fig. 10-21 are

$$V_{OC}(s) = V_T(s) = I_N(s)Z_N(s) \tag{10-28}$$

With a short-circuit load connected the interface currents are

$$I_{SC}(s) = V_T(s)/Z_T(s) = I_N(s) \tag{10-29}$$

Taken together Eqs. (10-28) and (10-29) yield the conditions

$$V_T(s) = V_{OC}(s) \quad I_N(s) = I_{SC}(s) \quad Z_T(s) = Z_N(s) = \frac{V_{OC}(s)}{I_{SC}(s)} \tag{10-30}$$

Algebraically, the results in Eq. (10-30) are identical to the corresponding equations for resistance circuits. The important difference here is that the equations involve transforms and impedances rather than waveforms and resistances. In any case, the equations point out that the open-circuit voltage and short-circuit current are sufficient to define either the Thévenin or Norton equivalent circuit. The Thévenin and Norton equivalent circuits are related by an s-domain source transformation. The s-domain *source transformation* relations given in Eq. (10-30) transform a voltage source in series with an impedance into a current source in parallel with the same impedance, and vice versa.

We analyze the interface shown in Fig. 10-22 to illustrate finding the s-domain Thévenin equivalent circuit. The source circuit is given in the s domain with zero initial voltage on the capacitor. The open-circuit voltage at the interface is found by voltage division:

$$V_{OC}(s) = \frac{1/Cs}{R + 1/Cs}\frac{V_A}{s} = \frac{V_A}{s(RCs + 1)} \tag{10-31}$$

The pole in $V_{OC}(s)$ at $s = 0$ comes from the step function voltage source and the pole at $s = -1/RC$ from the RC circuit. Connecting a short circuit at the interface effectively removes the capacitor from the circuit. The short-circuit current is

$$I_{SC}(s) = \frac{1}{R}\frac{V_A}{s} \tag{10-32}$$

This current does not have a pole at $s = -1/RC$ because the capacitor has been shorted out. Taking the ratio of Eqs. (10-31) and (10-32) yields the Thévenin impedance:

$$Z_T(s) = \frac{V_{OC}(s)}{I_{SC}(s)} = \frac{R}{RCs + 1} \tag{10-33}$$

The Thévenin impedance can also be found by turning off the input source and finding the equivalent impedance looking back into the source circuit. Replacing the voltage source in Fig. 10-22 by a short circuit reduces the source circuit to a parallel combination of a resistor and capacitor. The equivalent look back impedance of the combination is

$$Z_T(s) = \frac{1}{Y_R + Y_C} = \frac{1}{\dfrac{1}{R} + Cs} = \frac{R}{RCs + 1} \tag{10-34}$$

which is the same result as Eq. (10-33).

This example illustrates that in the s domain Thévenin or Norton theorems involve basically the same concepts discussed in Chapter 3. We can determine the Thévenin or Norton equivalent by finding any two of the following quantities: (1) the open-circuit voltage, (2) the short-circuit current, or (3) the Thévenin impedance.

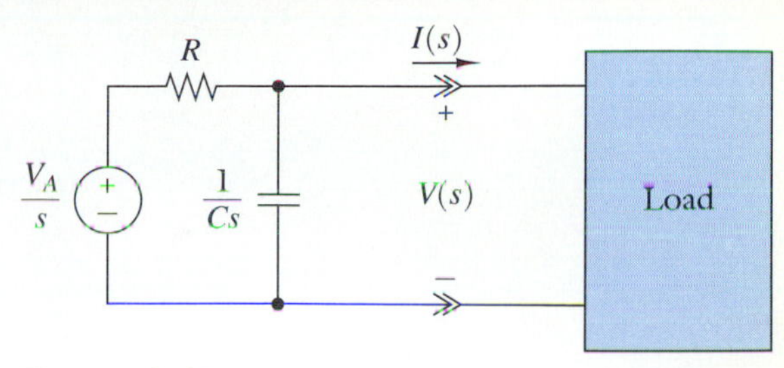

Figure 10-22 A source/load interface in the s domain.

EXAMPLE 10-7

Use the Thévenin equivalent of the source circuit in Fig. 10-22 to find the interface voltage $v(t)$ when the load is (a) a resistance R and (b) a capacitance C.

SOLUTION:

Figure 10-23 shows the s-domain circuit model of the source-load interface. The parameters of the Thévenin equivalent are given in Eqs. (10-31) and (10-33).

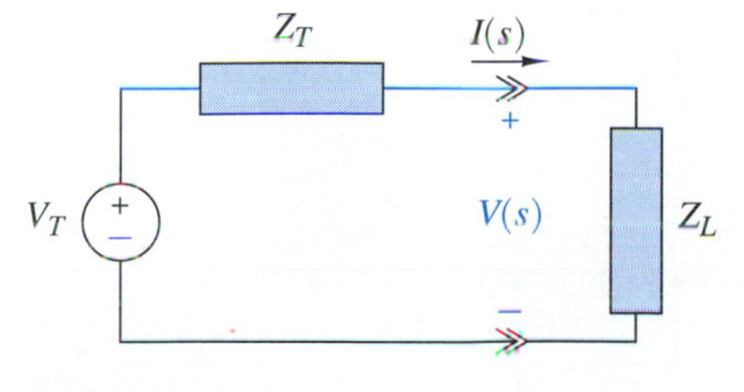

Figure 10-23

(a) For a resistance load $Z_L(s) = R$. The interface voltage is found by voltage division:

$$V(s) = \frac{Z_L(s)}{Z_T(s) + Z_L(s)} V_T(s)$$

$$= \left(\frac{R}{\dfrac{R}{RCs+1} + R} \right) \frac{V_A}{s(RCs+1)}$$

$$= \frac{V_A/RC}{s(s+2/RC)}$$

The partial fraction expansion of $V(s)$ is

$$V(s) = \frac{V_A/2}{s} - \frac{V_A/2}{s+2/RC}$$

Applying the inverse transform to each term yields the interface voltage waveform.

$$v(t) = \frac{V_A}{2}\left[1 - e^{-2t/RC}\right] u(t)$$

(b) For a capacitance load $Z_L = 1/Cs$ and the interface voltage transform is

$$V(s) = \frac{Z_L(s)}{Z_T(s) + Z_L(s)} V_T(s)$$

$$= \left(\frac{\dfrac{1}{Cs}}{\dfrac{R}{RCS+1} + \dfrac{1}{Cs}} \right) \frac{V_A}{s(RCs+1)}$$

$$= \frac{V_A/2RC}{s(s+1/2RC)}$$

The partial fraction expansion of $V(s)$ is

$$V(s) = \frac{V_A}{s} - \frac{V_A}{s+2/RC}$$

Applying the inverse transform to each term yields the interface voltage waveform for the capacitance load as

$$v(t) = V_A\left[1 - e^{-t/2RC}\right] u(t)$$

EXERCISE 10-6

Find the Norton equivalent of the s-domain circuits in Fig. 10-24.

ANSWERS:

(a) $I_N(s) = \dfrac{I_A}{(RCs+1)(s+\alpha)}$ $\quad Z_N(s) = \dfrac{RCs+1}{Cs}$

(b) $I_N(s) = \dfrac{RI_A}{(Ls+R)(s+\alpha)}$ $\quad Z_N(s) = \dfrac{Ls+R}{LCs^2+RCs+1}$

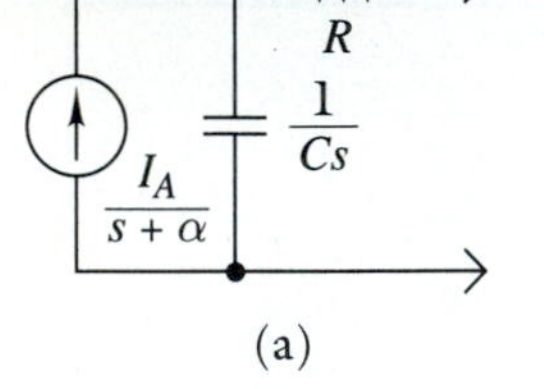

(a)

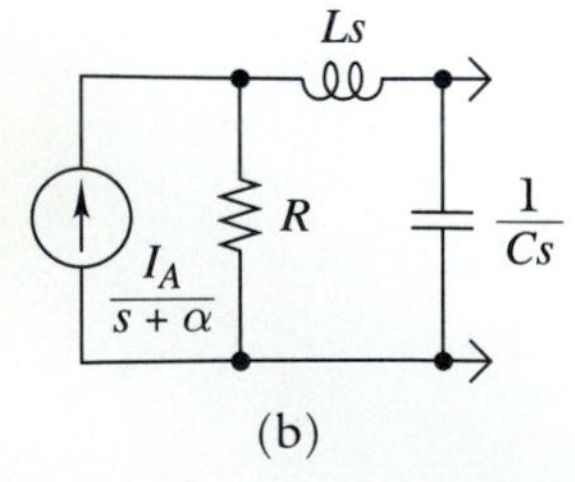

(b)

Figure 10-24

◇ 10-4 NODE-VOLTAGE ANALYSIS IN THE s DOMAIN

s-domain circuit analysis can be accomplished using the node-voltage or mesh-current methods developed for resistance circuits. These two general methods are more systematic and less intuitive than the approaches based on circuit reduction. The node-voltage and mesh-current methods both involve formulating and solving circuit equations in the s domain.

Formulating Node-Voltage Equations

For most circuits the node voltages are the appropriate set of circuit variables. These variables are defined by selecting a reference (or ground) node, and then using the voltages between the remaining nodes and the reference node as the unknowns. Because of the way that node voltages are defined, KVL allows the voltage across any two-terminal element to be expressed as the difference of the two node voltages at the terminal of the element. This fundamental property of node voltages applies in the s domain because the Laplace transformation does not change KVL.

In Chapter 5 we developed a four-step method of formulating the node-voltage equations for resistance circuits. The same steps can be used here except that a preliminary step must be added to transform the circuit into the s domain. The steps needed to develop s-domain node-voltage equations are as follows:

STEP 0: Transform the circuit into the s domain using current sources to represent capacitor and inductor initial conditions.

STEP 1: Select a reference node. Identify a node voltage at each of the nonreference nodes and a current with every element in the circuit.

STEP 2: Write KCL connection constraints in terms of the element currents at the nonreference nodes.

STEP 3: Use the element admittances and the fundamental property of node voltages to express the element currents in terms of the node voltages.

STEP 4: Substitute the device constraints from Step 3 into the KCL connection constraints from Step 2 and arrange the resulting equations in a standard form.

Steps 1 through 4 are identical to those used for resistance circuits, except that in the s-domain admittances are used to represent the device i-v characteristics in Step 3. Note that initial conditions are represented by current sources in Step 0 because node analysis emphasizes KCL and therefore favors current sources over voltage sources.

The following example illustrates using these steps to formulate node equations in the s domain.

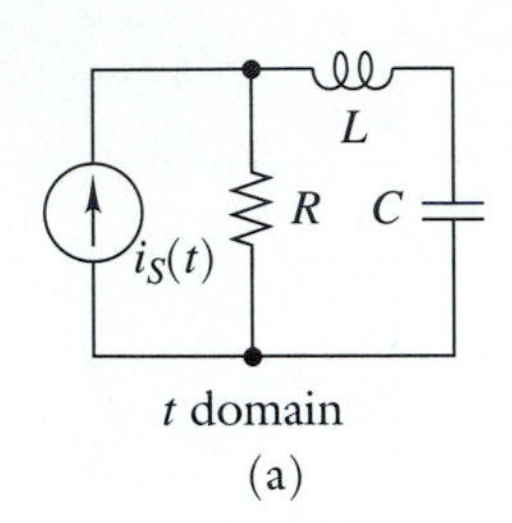

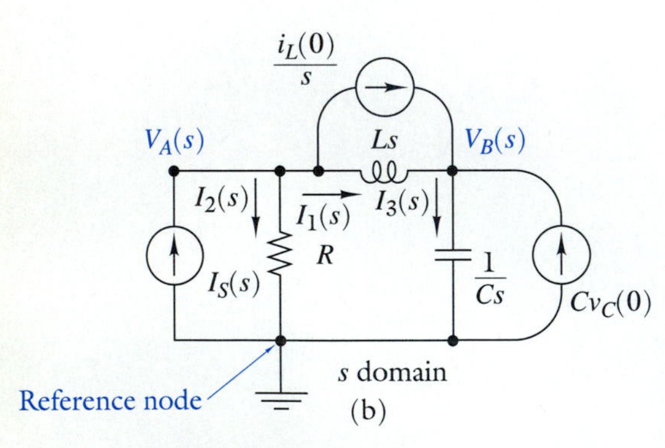

Figure 10-25

EXAMPLE 10-8

Formulate s-domain node-voltage equations for the circuit in Fig. 10-25(a).

SOLUTION:

Formulating node equations involves the following steps.

STEP 0: Figure 10-25(b) shows the circuit in the s domain with the element impedances and the initial conditions represented by current sources.

STEP 1: Figure 10-25(b) also identifies $N-1=2$ node voltages and a current with each element.

STEP 2: The KCL constraints at the two nonreference nodes are

$$\text{Node A:}\quad I_S(s) - \frac{i_L(0)}{s} - I_1(s) - I_2(s) = 0$$

$$\text{Node B:}\quad Cv_C(0) + \frac{i_L(0)}{s} + I_1(s) - I_3(s) = 0$$

STEP 3: The element equations are expressed in terms of the node voltages, using the element admittances and the fundamental property of node voltages:

$$I_1(s) = Y_L(s)[V_A(s) - V_B(s)] = \frac{1}{Ls}[V_A(s) - V_B(s)]$$

$$I_2(s) = Y_R(s)V_A(s) = GV_A(s)$$

$$I_3(s) = Y_C(s)V_B(s) = CsV_B(s)$$

where $G = 1/R$ is the conductance of the resistor.

STEP 4: Substituting the device constraints into the KCL connection constraints and collecting common terms yields the node-voltage equations:

$$\text{Node A:}\quad \left(G + \frac{1}{Ls}\right)V_A(s) - \left(\frac{1}{Ls}\right)V_B(s) = I_S(s) - \frac{i_L(0)}{s}$$

$$\text{Node B:}\quad -\left(\frac{1}{Ls}\right)V_A(s) + \left(\frac{1}{LS} + Cs\right)V_B(s) = Cv_C(0) + \frac{i_L(0)}{s}$$

The s-domain node-voltage equations are two linear algebraic equations in the two unknowns $V_A(s)$ and $V_B(s)$. When solved for the transforms $V_A(s)$ and $V_B(s)$, these equations provide enough information about the circuit response to determine any other voltage or current using only the device and connection constraints.

Formulating Node-Voltage Equations by Inspection

The final form of the node equations in Example 10-8 exhibits a symmetrical pattern seen before in the analysis of resistance circuits. The coefficients of $V_B(s)$ in the Node A equation and V_A in the Node B are the negative of the admittance connected between Nodes A and B. The coefficients of $V_A(s)$ in the Node A equation and $V_B(s)$ in the Node B equation are the sum of the admittances connected to the node. The terms on the right side of the equations are either input current sources or initial condition current sources.

This pattern always appears when node-voltage equations are written in a systematic way. As a result we can formulate node equations by inspection without going through the formality of writing the underlying device and connection equations.

Node-voltage equations focus on element admittances and KCL. In Fig. 10-26 the sum of currents leaving Node A through this admittance is

1. $V_A(s)$ times the sum of admittances connected to Node A
2. Minus V_B times the sum of admittances connected between Node A and Node B
3. Minus similar terms for other nodes connected to Node A via admittances

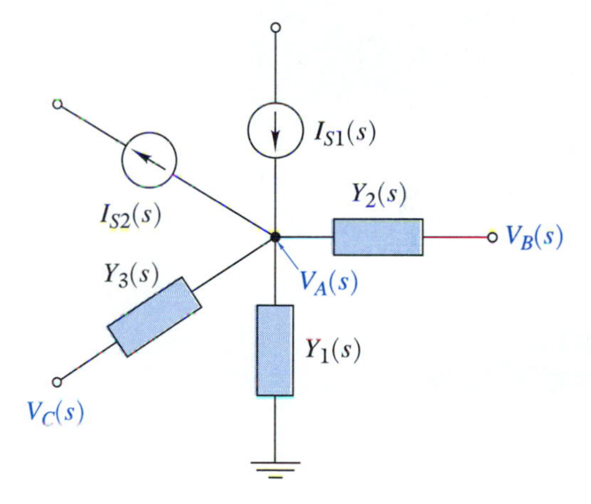

Figure 10-26 An example node.

The Node A equation is obtained by equating sum of currents leaving via admittances to the sum of currents directed into the node by current sources. For example, the s-domain equation for Node A in Fig. 10-26 is

$$(Y_1 + Y_2 + Y_3)\,V_A(s) - Y_2 V_B(s) - Y_3 V_C(s) = I_{S1}(s) - I_{S2}(s)$$

where Y_1 is the admittance connected from Node A to ground, Y_2 is the admittance connected between Nodes A and B, and Y_3 is the equivalent admittance between Nodes A and C. The currents $I_{S1}(s)$ and $I_{S2}(s)$ are the transforms of the input or initial condition sources connected to Node A.

The inspection method outlined above assumes that there are no voltage sources in the circuit. Initial condition sources are selected as current sources when node voltages are used. However, dependent or independent voltage sources can be treated using one of the following methods:

Method 1: Convert voltage sources into equivalent current sources using a source transformation.

Method 2: Select the reference node so that one terminal of the voltage source is connected to ground. The source voltage then determines the node voltage at the other source terminal.

Method 3: Create a supernode surrounding any voltage source that cannot be handled by Method 1 or 2.

Some circuits may require more than one of these methods.

EXAMPLE 10-9

Formulate node-voltage equations for the circuit in Fig. 10-27(a).

SOLUTION:

Figure 10-27(b) shows the circuit in the s domain using current sources for the initial conditions. This circuit has $N - 1 = 4$ nonreference nodes, so it appears that four node equations are needed. However, for the indicated reference node the two voltage sources determine the voltages at Nodes A and D. The only unknowns are the voltages at Nodes B and C. By inspection the node equations at Nodes B and C are

$$-G_1 V_A(s) + (C_1 s + C_2 s + G_1 + G_2)\, V_B(s)$$
$$- C_2 s V_C(s) - G_2 V_D(s) = C_1 v_{C_1}(0) + C_2 v_{C_2}(0)$$
$$-C_2 s V_B(s) + (C_2 s + G_3)\, V_C(s) = -C_2 v_{C_2}(0)$$

For the selected reference node $V_A(s) = V_1(s)$ and $V_D(s) = \mu V_X(s) = \mu V_C(s)$. Substituting these relations into the node equations yields

$$(C_1 s + C_2 s + G_1 + G_2)\, V_B(s) - (C_2 s + \mu G_2)\, V_C(s)$$
$$= G_1 V_1(s) + C_1 v_{C_1}(0) + C_2 v_{C_2}(0)$$
$$-C_2 s V_B(s) + (C_2 s + G_3)\, V_C(s)$$
$$= -C_2 v_{C_2}(0)$$

We now have two equations in the two unknown node voltages

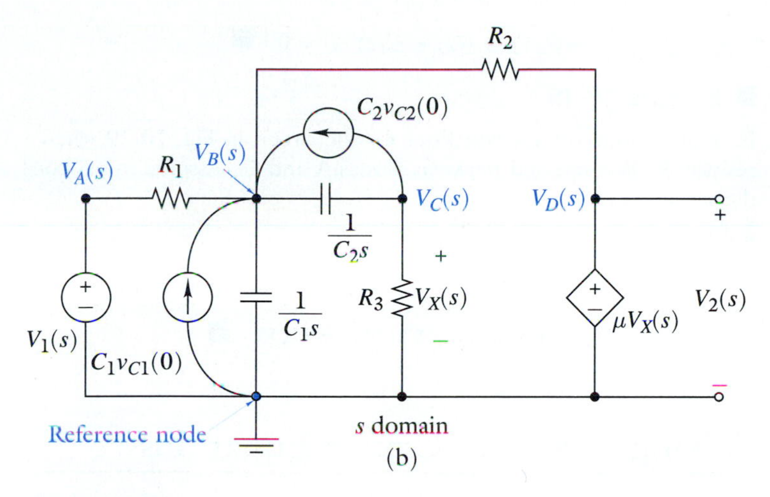

Figure 10-27

$V_B(s)$ and $V_C(s)$. The final node equations are not symmetrical because the dependent source is not a bilateral element like resistors, capacitors, and inductors.

EXERCISE 10-7

Formulate node-voltage equations for the circuit in Fig. 10-28. Assume the initial conditions are zero.

ANSWER:

$$\left(G_1 + G_2 + \frac{1}{Ls}\right) V_B - G_2 V_C = G_1 V_S$$

$$-G_2 V_B + (G_2 + Cs)\, V_C = Cs V_S$$

EXERCISE 10-8

Formulate node-voltage equations for the circuit in Fig. 10-28 when the initial conditions are not zero and a resistor R_3 is connected between Nodes A and C.

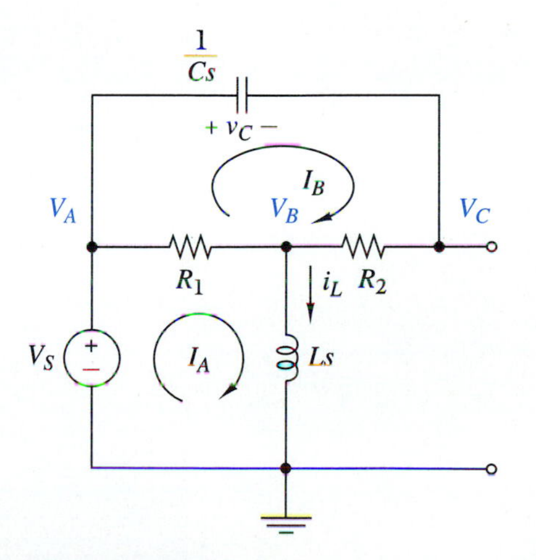

Figure 10-28

ANSWER:

$$\left(G_1 + G_2 + \frac{1}{Ls}\right) V_B - G_2 V_C = G_1 V_S - \frac{i_L(0)}{s}$$

$$-G_2 V_B + (G_2 + G_3 + Cs)\, V_C = (G_3 + +Cs)\, V_S - C v_C(0)$$ ■

■ EXERCISE 10-9

Formulate node-voltage equations for the circuit in Fig. 10-29. Assume the initial conditions are zero.

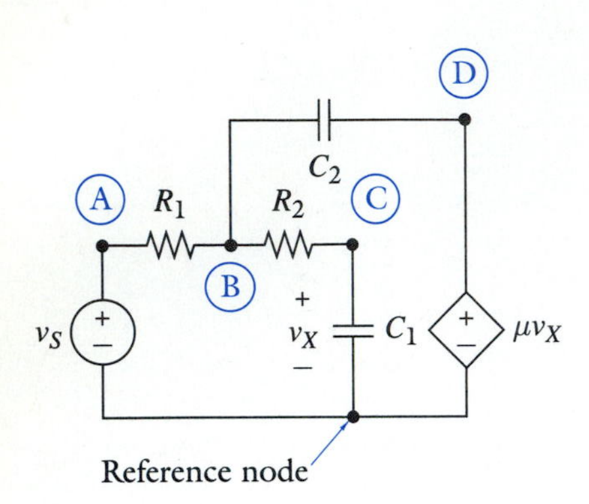

Figure 10-29

ANSWER:

$$(G_1 + G_2 + C_2 s)\, V_B - (G_2 + \mu C_2 s)\, V_C = G_1 V_S$$

$$-G_2 V_B + (G_2 + C_1 s)\, V_C = 0$$ ■

■ EXERCISE 10-10

Formulate node-voltage equations for the circuit in Fig. 10-29 when a resistor R_3 is connected between Nodes A and C. Assume initial conditions are zero.

ANSWER:

$$(G_1 + G_2 + C_2 s)\, V_B - (G_2 + \mu C_2 s)\, V_C = G_1 V_S$$

$$-G_2 V_B + (G_2 + G_3 + C_1 s)\, V_C = G_3 V_S$$ ■

Solving s-Domain Circuit Equations

Examples 10-8 and 10-9 show that formulating node-voltage equations leads to linear algebraic equations in the unknown node voltages. In principle, solving these node equations is straightforward using techniques such as Cramer's rule or Gaussian reduction. Usually Cramer's rule is the better approach when the element parameters are in symbolic form. Gaussian reduction offers some advantages for numerical problems. However, the algebra becomes somewhat tedious because the coefficients in the linear equations involve polynomials in the complex frequency variable.

We will illustrate the solution process using an example from earlier in this section. In Example 10-8 we formulated the following node voltage equations for the circuit in Fig. 10-25:

$$\left(G + \frac{1}{Ls}\right) V_A(s) - \frac{1}{Ls} V_B(s) = I_S(s) - \frac{i_L(0)}{s}$$

$$-\frac{1}{Ls} V_A(s) + \left(Cs + \frac{1}{Ls}\right) V_B(s) = \frac{i_L(0)}{s} + C v_C(0)$$

When using Cramer's rule it is convenient to first find the determinant of these equations:

$$\Delta(s) = \begin{vmatrix} G + 1/Ls & -1/Ls \\ -1/Ls & Cs + 1/LS \end{vmatrix}$$
$$= (G + 1/Ls)(Cs + 1/Ls) - (1/Ls)^2$$
$$= \frac{GLCs^2 + Cs + G}{Ls}$$

We call $\Delta(s)$ the *circuit determinant* because it only depends on the element parameters L, C, and $G = 1/R$. The determinant $\Delta(s)$ characterizes the circuit and does not depend on the input driving forces or initial conditions.

The node voltage $V_A(s)$ is found using Cramer's rule:

$$V_A(s) = \frac{\Delta_A(s)}{\Delta(s)} = \frac{\begin{vmatrix} I_S(s) - \dfrac{i_L(0)}{s} & -\dfrac{1}{Ls} \\ \dfrac{i_L(0)}{s} + Cv_C(0) & Cs + \dfrac{1}{Ls} \end{vmatrix}}{\Delta(s)} \tag{10-35}$$

$$= \underbrace{\frac{(LCs^2 + 1)I_S(s)}{GLCs^2 + Cs + G}}_{\text{Zero State}} + \underbrace{\frac{-LCsi_L(0) + Cv_C(0)}{GLCs^2 + Cs + G}}_{\text{Zero Input}}$$

Solving for the other node voltage $V_B(s)$ yields

$$V_B(s) = \frac{\Delta_B(s)}{\Delta(s)} = \frac{\begin{vmatrix} G + \dfrac{1}{Ls} & I_S(s) - \dfrac{i_L(0)}{s} \\ -\dfrac{1}{Ls} & \dfrac{i_L(0)}{s} + Cv_C(0) \end{vmatrix}}{\Delta(s)}$$

$$= \underbrace{\frac{I_S(s)}{GLCs^2 + Cs + G}}_{\text{Zero State}} + \underbrace{\frac{GLi_L(0) + (GLs + 1)Cv_C(0)}{GLCs^2 + Cs + G}}_{\text{Zero Input}} \tag{10-36}$$

Cramer's rule gives both the zero-input and zero-state components of the response transforms $V_A(s)$ and $V_B(s)$.

Note that Cramer's rule yields the node voltages as a ratio of determinants of the form

$$V_X(s) = \frac{\Delta_X(s)}{\Delta(s)} \tag{10-37}$$

The response transform is a rational function of s whose poles are either zeros of the circuit determinant or poles of the determinant $\Delta_X(s)$. That is, $V_X(s)$ in Eq. (10-37) has poles when $\Delta(s) = 0$ or $\Delta_X(s) \to \infty$. The partial fraction expansion of $V_X(s)$ will contain terms for each of these poles. We call the zeros of $\Delta(s)$ the *natural poles* because they depend only on the circuit elements and give rise to the natural response terms in the partial fraction expansion. We call the poles of $\Delta_X(s)$ the *forced poles* because they depend on the form of the input signal and give rise to the forced response terms in the partial fraction expansion.

The responses in Eqs. (10-35) and (10-36) were found without specifying the form of the input signal $i_S(t)$ or numerical values for the circuit parameters. These equations are the general solution for the response transform for any input, initial conditions, and values of R, L, and C. To obtain a specific response waveform we insert numerical values for the various parameters and use a partial fraction expansion to obtain response waveforms.

The following example illustrates the inverse transform step of the s-domain analysis process.

EXAMPLE 10-10

Find the zero-state component for $v_B(t)$ in Eq. (10-36) for $R = 1\ \text{k}\Omega$, $L = 0.5$ H, $C = 1\ \mu\text{F}$, and $i_S(t) = 10u(t)$ mA.

SOLUTION:

To find the required response we insert $I_S(s) = 10^{-2}/s$ and the numerical values of circuit parameters in the zero-state component in Eq. (10-36):

$$V_B(s) = \frac{10^{-2}/s}{5\times 10^{-10}s^2 + 10^{-6}s + 10^{-3}}$$

$$= \frac{2\times 10^7}{s\left(s^2 + 2\times 10^3 s + 2\times 10^6\right)}$$

$$= \frac{2\times 10^7}{s(s+1000-j1000)(s+1000+j1000)}$$

The response transform has three poles. A forced pole at $s = 0$ and two natural poles at $s = -1000 \pm j1000$. The forced pole comes from the step function input and the two natural poles are zeros of the circuit determinant. Expanding this rational function into partial fractions yields

$$V_B(s) = \frac{10}{s} + \frac{(10/\sqrt{2})e^{+j135^\circ}}{s+1000-j1000} + \frac{(10/\sqrt{2})e^{-j135^\circ}}{s+1000+j1000}$$

Taking the inverse transforms of each term from Tables 9-2 and 9-3 yields the required zero-state response waveform:

$$v_B(t) = 10u(t) + \left[10\sqrt{2}e^{-1000t}\cos(1000t + 135^\circ)\right]u(t)$$

The step function in $v_B(t)$ is the forced response caused by the forced pole at $s = 0$. The damped cosine is the natural response determined by the natural poles in $V_B(s)$.

❖ Design Example

EXAMPLE 10-11

The circuit in Fig. 10-30 is designed to produce a pair of complex poles. To simplify production the final design must have equal element values $R_1 = R_2 = R_3 = R$ and $C_1 = C_2 = C$. Select the values of R, C, and the gain μ so that the circuit has natural poles at $s = -1000 \pm j1000$ rad/s.

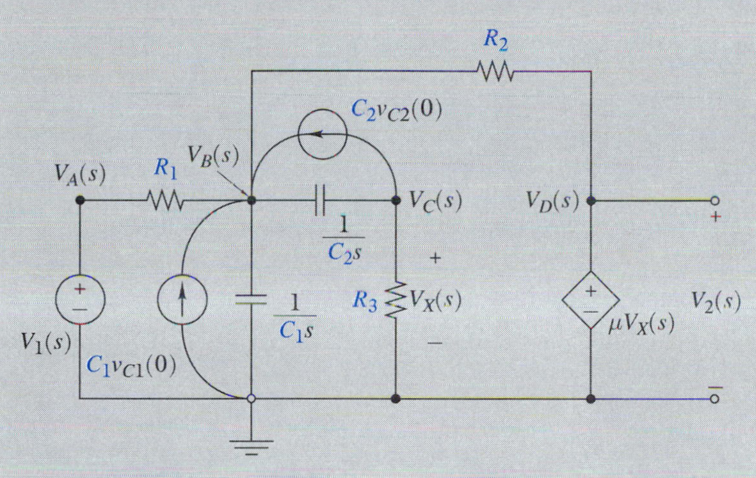

Figure 10-30

SOLUTION:

In Example 10-9 we formulated the following node-voltage equations for this circuit:

$$(C_1s + C_2s + G_1 + G_2)\,V_B(s) - (C_2s + \mu G_2)\,V_C(s) = G_1V_1(s) + C_1v_{C_1}(0) + C_2v_{C_2}(0)$$

$$-C_2sV_B(s) + (C_2s + G_3)\,V_C(s) = -C_2v_{C_2}(0)$$

The natural poles are roots of the circuit determinant:

$$\Delta(s) = \begin{vmatrix} C_1s + C_2s + G_1 + G_2 & -C_2s - \mu G_2 \\ -C_2s & C_2s + G_3 \end{vmatrix}$$
$$= C_1C_2s^2 + (C_1G_3 + C_2G_1 + C_2G_2 + C_2G_3 - \mu G_2C_2)s + G_1G_3 + G_2G_3$$

For equal resistances $R_1 = R_2 = R_3 = R$ and equal capacitances $C_1 = C_2 = C$ the circuit determinant reduces to

$$\frac{\Delta(s)}{C^2} = s^2 + \frac{4-\mu}{RC}s + \frac{2}{(RC)^2}$$

To produce the required natural poles the quadratic on the right side of this equation must be of the form

$$(s + 1000 - j1000)(s + 1000 + j1000) = s^2 + 2000s + 2 \times 10^6$$

Equating the coefficients in the polynomials yields the following design constraints:

$$\frac{4-\mu}{RC} = 2000 \quad \text{and} \quad \frac{2}{(RC)^2} = 2 \times 10^6$$

These constraints yield $RC = 10^{-3}$ and $\mu = 2$. Selecting $R = 10\ \text{k}\Omega$ makes $C = 100$ nF.

EXERCISE 10-11

(a) Solve for the determinant of the circuit in Fig. 10-28 using the node-equations in Exercise 10-7.
(b) Solve for the zero-state component of $V_B(s)$.

ANSWERS:

(a) $\Delta(s) = \dfrac{(G_1 + G_2)LCs^2 + (G_1G_2L + C)s + G_2}{Ls}$

(b) $V_B(s) = \left[\dfrac{(G_1 + G_2)LCs^2 + G_1G_2Ls}{(G_1 + G_2)LCs^2 + (G_1G_2L + C)s + G_2}\right] V_S(s)$ ■

EXERCISE 10-12

(a) Solve for the determinant of the circuit in Fig. 10-29 using the node equations in Exercise 10-9.
(b) Solve for the zero-state component of $V_D(s)$.

ANSWERS:

(a) $\Delta(s) = C_1C_2s^2 + (G_1C_1 + G_2C_1 + G_2C_2 - \mu G_2C_2)s + G_1G_2$
(b) $V_D(s) =$

$$\left[\frac{\mu G_1G_2}{C_1C_2s^2 + (G_1C_1 + G_2C_1 + G_2C_2 - \mu G_2C_2)s + G_1G_2}\right] V_S(s)$$ ■

Node Equations for OP AMP Circuits

Node-voltage equations are particularly suited to OP AMP circuits. In the time domain the ideal OP AMP i-v constraints are linear algebraic equations. Because of the linearity property, the Laplace transformation does not change the form of these constraints when they are transformed into the s domain.

$$\begin{array}{ll} \text{t domain} & \text{s domain} \\ v_P(t) = v_N(t) & V_P(s) = V_N(s) \\ i_N(t) = 0 & I_N(s) = 0 \\ i_P(t) = 0 & I_P(s) = 0 \end{array} \tag{10-38}$$

The only difference is that the s-domain constraints apply to transforms rather than waveforms. Since the element constraints have the same form we can formulate s-domain node equations for OP AMP circuits using the approach developed in Chapter 5 for resistance circuits. Basically that approach involves two steps:

STEP 1: Identify a node voltage at all OP AMP outputs but do not formulate node equations at these nodes.

STEP 2: Formulate node equations at the remaining nonreference nodes and use the OP AMP voltage constraint $[V_P(s) = V_N(s)]$ to reduce the number of unknowns.

The final node-voltage example involves an OP AMP circuit in the s domain.

EXAMPLE 10-12

(a) Formulate s-domain node-voltage equations for the OP AMP circuit in Fig. 10-31(a).

(b) Solve for the s-domain input-output relationship between the input $V_S(s)$ and the zero-state component of the output $V_O(s)$

(c) Solve for the zero-state output $v_O(t)$ when the input is a step function $v_S(t) = u(t)$.

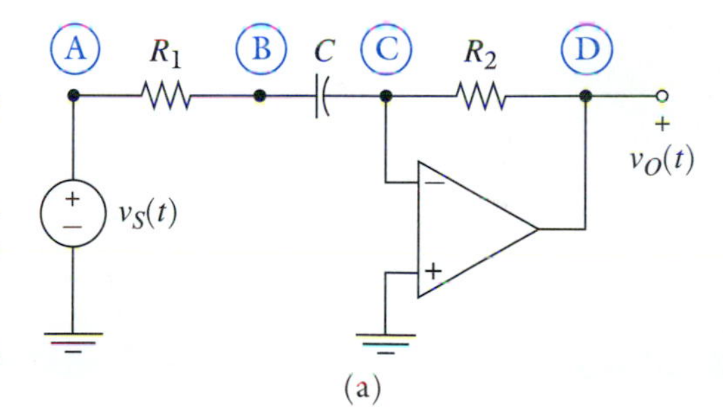

SOLUTION:

(a) Figure 10-31(b) shows the transformed circuit in the s domain. Node equations are not required at Nodes A and D because the selected reference node makes $V_A = V_S$ and V_D is the OP AMP output. By inspection the equations at Nodes B and C are

$$\text{Node B:} \quad -G_1V_A + (G_1 + Cs)V_B - CsV_c = Cv_C(0)$$

$$\text{Node C:} \quad -CsV_B + (G_2 + Cs)V_C - G_2V_D = -Cv_C(0)$$

where $G_1 = 1/R_1$ and $G_2 = 1/R_2$. The result above yields two node equations in four node voltages. However, as

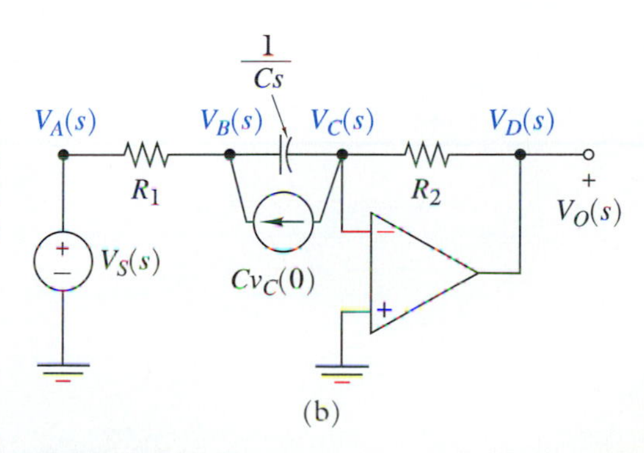

Figure 10-31

noted above $V_A(s) = V_S(s)$ and $V_D(s) = V_O(s)$. The OP AMP voltage constraint requires $V_C(s) = 0$, since the noninverting input is grounded. Inserting the conditions $V_A(s) = V_S(s)$ and $V_C(s) = 0$ reduces node-voltage equations to the following form

$$\begin{aligned} \text{Node B:} \quad & (G_1 + Cs)V_B = Cv_C(0) + G_1 V_S \\ \text{Node C:} \quad & -CsV_B - G_2 V_O = -Cv_C(0) \end{aligned}$$

We now have two equations in two unknowns $V_B(s)$ and $V_O(s)$.

(b) For the zero-state response the initial condition $v_C(0) = 0$. Solving the Node B equation above for $V_B(s)$ in terms of $V_S(s)$:

$$V_B(s) = \left[\frac{G_1}{G_1 + Cs}\right] V_S(s)$$

Substituting this result into the Node C equation and solving for $V_O(s)$ yields

$$\begin{aligned} V_O(s) &= -\left[\frac{G_1 Cs/G_2}{G_1 + Cs}\right] V_S(s) \\ &= -\left[\frac{R_2}{R_1}\frac{s}{s + 1/R_1C}\right] V_S(s) \end{aligned}$$

This equation is the s-domain input-output relationship for the circuit. The circuit is linear, so the output transform is proportional to the input transform. The proportionality factor (the function inside the bracket) is a function of the circuit parameters and the complex frequency variable. The factor has a natural pole at $s = -1/R_1C$ and a zero at $s = 0$.

(c) A step function input $V_S(s) = 1/s$ produces a forced pole at $s = 0$. However, the zero in the proportionality factor cancels the forced pole. The net result is that the zero-state output contains only a natural pole and has the form

$$v_O(t) = \mathscr{L}^{-1}\left\{-\frac{R_2}{R_1}\frac{1}{s + 1/R_1C}\right\} = -\left[\frac{R_2}{R_1}e^{-t/R_1C}\right]u(t)$$

◇ 10-5 MESH-CURRENT ANALYSIS IN THE s DOMAIN

The mesh-current method only works for circuits whose schematic can be drawn on a flat surface without crossovers. These planar circuits have special loops called meshes, which are closed paths that do not enclose any elements. The mesh-current variables are the loop currents assigned to each mesh in a planar circuit. Because of the way mesh currents are defined, KCL allows the current through any two-terminal element to be expressed as the difference of two adjacent mesh currents. This fundamental property of mesh currents applies in the s domain because the Laplace transformation does not change KCL.

In Chapter 5 we developed a four-step method of formulating the mesh-current equations for resistance circuits. The same steps can be used here except that a preliminary step must be added to transform the circuit into the s domain. The steps needed to develop s-domain mesh-current equations are as follows:

STEP 0: Transform the circuit into the s domain using voltage sources to represent capacitor and inductor initial conditions.

STEP 1: Identify a mesh current with every mesh and a voltage across every circuit element.

STEP 2: Use the element voltages to write KVL connection constraints around every mesh.

STEP 3: Use the element impedances and the fundamental property of mesh currents to express the element voltages in terms of the mesh currents.

STEP 4: Substitute the element equations from Step 3 into the KVL connection equation from Step 2 and arrange the resulting equations in standard form.

Steps 1 through 4 are identical to those used for resistance circuits in Chapter 5, except that s-domain impedances are used to represent the element i-v characteristics in Step 3. Note that voltage sources are used to represent the initial conditions in Step 0 because mesh analysis emphasizes KVL and therefore favors voltage sources of current sources.

EXAMPLE 10-13

(a) Formulate mesh-current equations for the circuit in Fig. 10-32(a).

(b) Solve for the zero-state component of $I_B(s)$.

(c) Find the zero-state component of $i_B(t)$ for $R_1 = 2$ kΩ, $R_2 = 3$ kΩ, $L_1 = L_2 = 1$ H, and an input voltage of $v_S(t) = 10u(t)$ V.

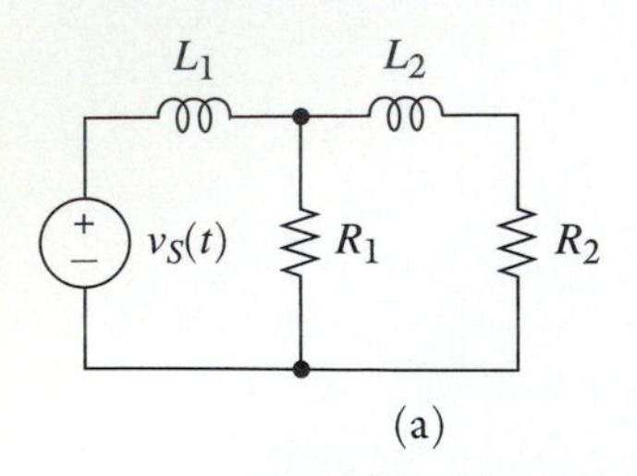

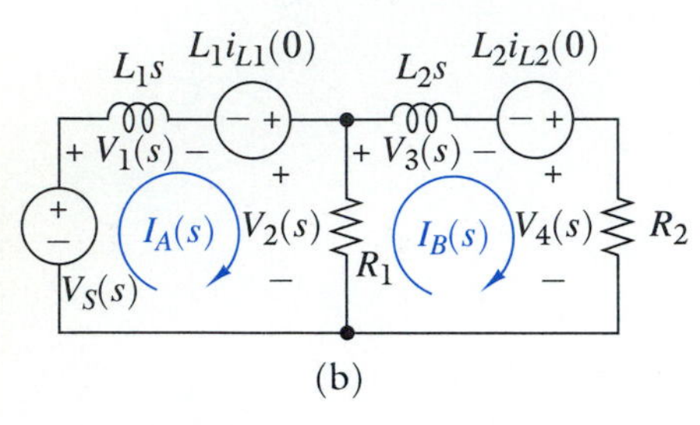

Figure 10-32

SOLUTION:

(a) Formulating the mesh-current equations involves the following steps.

STEP 0: Figure 10-32(b) shows the transformed circuit with voltage sources representing the initial conditions.

STEP 1: Figure 10-32(b) also identifies two mesh currents and element voltages.

STEP 2: The KVL constraints around the two meshes are

$$\text{Mesh A:} \quad -V_S(s) + V_1(s) - L_1 i_{L_1}(0) + V_2(s) = 0$$

$$\text{Mesh B:} \quad -V_2(s) + V_3(s) - L_2 i_{L_2}(0) + V_4(s) = 0$$

STEP 3: The element equations are written in terms of the mesh currents as

$$V_1(s) = L_1 s I_A(s)$$

$$V_2(s) = R_1[I_A(s) - I_B(s)]$$

$$V_3(s) = L_2 s I_B(s)$$

$$V_4(s) = R_2 I_B(s)$$

STEP 4: Substituting the element constraints into the KVL connection constraints and collecting common terms yields

$$\text{Mesh A:} \quad (L_1 s + R_1)I_A(s) - R_1 I_B(s) = V_S(s) + L_1 i_{L_1}(0)$$

$$\text{Mesh B:} \quad -R_1 I_A(s) + (L_2 s + R_1 + R_2)I_B(s) = L_2 i_{L_2}(0)$$

The s-domain circuit equations are two linear algebraic equations in the two unknown mesh-currents $I_A(s)$ and $I_B(s)$.

(b) To solve the mesh equations we first find the circuit determinant:

$$\Delta(s) = \begin{vmatrix} L_1 s + R_1 & -R_1 \\ -R_1 & L_2 s + R_1 + R_2 \end{vmatrix}$$

$$= L_1 L_2 s^2 + (R_1 L_2 + R_1 L_1 + R_2 L_1)s + R_1 R_2$$

To find the zero-state component $I_B(s)$ we let $i_{L_1}(0) = i_{L_2}(0) = 0$ and use Cramer's rule:

$$I_B = \frac{\begin{vmatrix} L_1 s + R_1 & V_S(s) \\ -R_1 & 0 \end{vmatrix}}{\Delta(s)}$$

$$= \frac{R_1 V_S(s)}{L_1 L_2 s^2 + (R_1 L_2 + R_1 L_1 + R_2 L_1)s + R_1 R_2}$$

(c) To find the time domain response we insert the input $V_S(s) = 10/s$ and the numerical parameters into the equation for $I_B(s)$ to produce the following rational function:

$$I_B(s) = \frac{(2000)(10/s)}{s^2 + 7 \times 10^3 s + 6 \times 10^6}$$

$$= \frac{2 \times 10^4}{s(s + 1000)(s + 6000)}$$

This function has three poles. The forced pole at $s = 0$ and two natural poles at $s = -1000$ and -6000 rad/s. Expanding this rational function by partial fractions yields

$$I_B(s) = \frac{1/3 \times 10^{-2}}{s} - \frac{0.4 \times 10^{-2}}{s + 1000} + \frac{2/3 \times 10^{-3}}{s + 6000}$$

The inverse transform of this expansion is the required zero-state response waveform:

$$i_B(s) = \left[\frac{10}{3} - 4e^{-1000t} + \frac{2}{3}e^{-6000t}\right] u(t) \text{ mA}$$

Writing Mesh-Current Equations by Inspection

The final form of the mesh equations in Example 10-13 exhibits a symmetrical pattern seen before in the analysis of resistance circuits. The coefficients of $I_B(s)$ in the Mesh A equation and I_A in the Mesh B equation are the negative of the impedance common to both meshes. The coefficients of $I_A(s)$ in the Mesh A equation and $I_B(s)$ in the Mesh B equation are the sum of the impedances in the mesh. The terms on the right side of the equations are independent sources or initial condition sources. This symmetrical pattern occurs when all mesh currents are defined in the same clockwise (or counterclockwise) sense.

The mesh analysis focuses on element impedances and KVL. The mesh equations can be written by inspection as the sum of voltages around a mesh. The sum of voltages across impedances in Mesh A in Fig. 10-33 is:

1. $I_A(s)$ times the sum of impedances in Mesh A
2. Minus $I_B(s)$ times the sum of impedances common to Mesh A and Mesh B
3. Minus similar terms for all other meshes adjacent to Mesh A

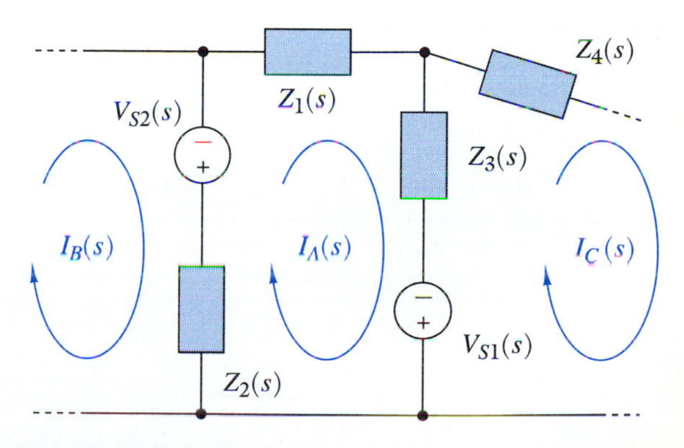

Figure 10-33 An example mesh.

The Mesh A equation is obtained by equating this sum to the sum of voltages produced by voltage sources in Mesh A. For example, the equation for Mesh A in Fig. 10-33 is

$$(Z_1 + Z_2 + Z_3)\, I_A(s) - Z_2 I_B(s) - Z_3 I_C(s) = V_{S1}(s) - V_{S2}(s)$$

where Z_1 is the impedance contained only in Mesh A, Z_2 is the impedance common to Meshes A and B, and Z_3 is common to Meshes A and C. The voltages $V_{S1}(s)$ and $V_{S2}(s)$ are the transforms of the input or initial condition voltage sources in Mesh A.

The method outlined above allows us to formulate mesh-current equations by inspection without writing the underlying connection and element equations. In describing the process, we assume that there are no current sources in the circuit. Initial condition sources are selected as voltage sources when mesh-current equations are used. However, if the circuit contains a dependent or independent current source, we use one of the following methods to deal with the current source:

Method 1: Convert current source into equivalent voltage source using a source transformation.

Method 2: Draw the circuit diagram so that only one mesh circulates through the current source. The one mesh current then is determined by the source current.

Method 3: Create a supermesh for any current source that cannot be handled by Methods 1 or 2.

Some circuits may require more than one of these methods.

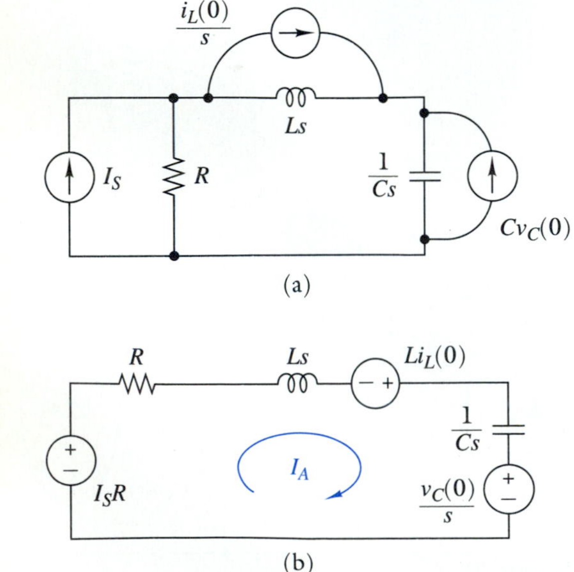

Figure 10-34

EXAMPLE 10-14

(a) Formulate mesh-current equations for the circuit in Fig. 10-34(a).

(b) Solve for the zero-input component of $i_A(t)$ for $i_L(0) = 0$, $v_C(0) = 10$ V, $L = 250$ mH, $C = 1\ \mu$F, and $R = 1$ kΩ.

SOLUTION:

(a) Figure 10-34(a) is the s-domain circuit used to develop node equations in Example 10-8. Each current source is connected in parallel with an impedance. We use source transformations to convert these current sources into the equivalent voltage sources shown in Fig. 10-34(b). Note that the circuit in Fig. 10-34(b) is a series RLC circuit of the type treated in Chapter 8. By inspection the equation for the single mesh in this circuit is

$$\left[R + Ls + \frac{1}{Cs}\right] I_A(s) = RI_S(s) + Li_L(0) - \frac{v_C(0)}{s}$$

The factor $R+Ls+1/Cs = (LCs^2+RCs+1)/Cs$ multiplying $I_A(s)$ in this equation is the circuit determinant. The zeros of the circuit determinant are natural poles of the circuit and are roots of the quadratic equation $LCs^2+RCs+1 = 0$.

(b) Solving the mesh equation for the zero-input component yields

$$I_A(s) = \frac{LCsi_L(0) - Cv_C(0)}{LCs^2 + RCs + 1}$$

Inserting the given numerical values produces

$$\begin{aligned} I_A(s) &= -\frac{10^{-5}}{0.25 \times 10^{-6}s^2 + 10^{-3}s + 1} \\ &= -\frac{40}{s^2 + 4 \times 10^3 s + 4 \times 10^6} \\ &= -\frac{40}{(s + 2000)^2} \end{aligned}$$

The zero-state response has two natural poles at $s = -2000$. The inverse transform of $I_B(s)$ is found in Table 9-2 to be a damped ramp waveform:

$$i_B(t) = -\left[40te^{-2000t}\right] u(t) \text{ A}$$

The damped ramp response indicates a critically damped second-order circuit. The minus sign means the direction of the actual current is opposite to the reference mark assigned to $I_B(s)$ in Fig. 10-34.

EXAMPLE 10-15

(a) Formulate mesh-current equations for the coupled coil circuit in Fig. 10-35(a).

(b) Find the input impedance seen by the voltage source $V_1(s)$ in Fig 10-35(b).

SOLUTION:

(a) The transformed circuit in Fig. 10-35(b) uses the *s*-domain model developed in Example 10-2. This circuit contains only voltage sources so we proceed directly to formulating the mesh equations. The total impedances in Meshes A and B are $R_1 + L_1 s$ and $R_2 + L_2 s$, respectively. The impedance common to Meshes A and B is the mutual inductance Ms.

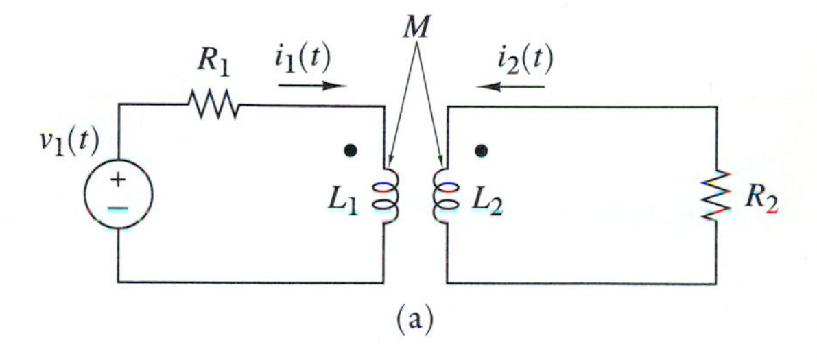

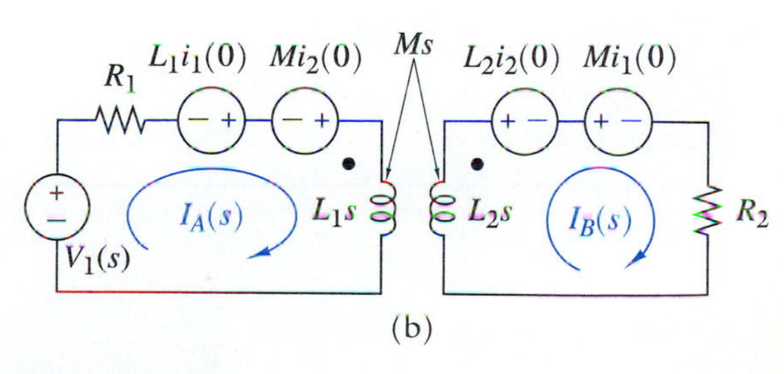

Figure 10-35

By inspection the mesh equations are

$$(L_1 s + R_1)\, I_A(s) - Ms I_B(s) = V_1(s) + L_1 i_1(0) + M i_2(0)$$

$$-Ms I_A(s) + (L_2 s + R_2)\, I_B(s) = -L_2 i_2(0) - M i_1(0)$$

The negative sign on the mutual inductance term follows the dot convention in Chapter 7. The negative sign is proper because reference direction for the mesh current $I_A(s)$ is directed toward a dot while reference direction for $I_B(s)$ is directed away from a dot. *Caution:* The signs of the initial condition terms depend on the reference directions originally assigned to i_1 and i_2 in Fig. 10-35(a).

(b) We first find the circuit determinant:

$$\Delta(s) = \begin{vmatrix} L_1 s + R_1 & Ms \\ -Ms & L_2 s + R_2 \end{vmatrix}$$

$$= (L_1 L_2 - M^2)s^2 + (L_1 R_2 + L_2 R_1)s + R_1 R_2$$

To find the input impedance all initial conditions must be zero. *Caution:* Zero initial conditions is part of the definition of impedance. To find the input impedance we use Cramer's rule to find the zero-state component of $I_A(s)$:

$$I_A(s) = \frac{\Delta_A(s)}{\Delta(s)} = \frac{\begin{vmatrix} V_1(s) & -Ms \\ 0 & L_2 s + R_2 \end{vmatrix}}{\Delta(s)}$$

$$= \frac{(L_2 s + R_2) V_1(s)}{(L_1 L_2 - M^2)s^2 + (L_1 R_2 + L_2 R_1)s + R_1 R_2}$$

The input impedance of the coupled coil circuit is

$$Z_{IN}(s) = \frac{V_1(s)}{I_A(s)}$$

$$= \frac{(L_1 L_2 - M^2)s^2 + (L_1 R_2 + L_2 R_1)s + R_1 R_2}{L_2 s + R_2}$$

In Chapter 7 we showed that perfect coupling occurs when the coupling coefficient $k^2 = M^2/L_1 L_2$ is unity. Perfect coupling means that $M^2 = L_1 L_2$ and the input impedance reduces to

$$Z_{IN}(s) = \frac{(L_1 R_2 + L_2 R_1)s + R_1 R_2}{L_2 s + R_2}$$

EXERCISE 10-13

(a) Formulate mesh-current equations for the circuit in Fig. 10-28. Assume that the initial conditions are zero.

(b) Find the circuit determinant.

(c) Solve for the zero-state component of $I_B(s)$.

ANSWERS:

(a) $(R_1 + Ls)\,I_A(s) - R_1 I_B(s) = V_S(s)$

$$-R_1 I_A(s) + \left(R_1 + R_2 + \frac{1}{Cs}\right) I_B(s) = 0$$

(b) $\Delta(s) = \dfrac{(R_1 + R_2)\,LCs^2 + (R_1 R_2 C + L)\,s + R_1}{Cs}$

(c) $I_B(s) = \dfrac{R_1 C s V_S(s)}{(R_1 + R_2)\,LCs^2 + (R_1 R_2 C + L)\,s + R_1}$ ■

■ EXERCISE 10-14

Formulate mesh-current equations for the circuit in Fig. 10-28 when a resistor R_3 is connected between Node C and ground. Assume that the initial conditions are zero.

ANSWER:

$$(R_1 + Ls)\,I_A(s) - R_1 I_B(s) - LsI_C(s) = V_S(s)$$

$$-R_1 I_A(s) + \left(R_1 + R_2 + \frac{1}{Cs}\right) I_B(s) - R_2 I_C(s) = 0$$

$$-LsI_A(s) - R_2 I_B(s) + (R_2 + R_3 + Ls)\,I_C(s) = 0$$ ■

10-6 SUMMARY OF s-DOMAIN CIRCUIT ANALYSIS

At this point we will review our progress and put s-domain circuit analysis into perspective. We have shown that linear circuits can be transformed from the time into the s domain. In this domain KCL and KVL apply to transforms and the element i-v characteristics become impedances with series or parallel initial condition sources. We use basic analysis methods such as equivalence and voltage division to solve for response transforms in simple circuits. For more complicated circuits we use systematic procedures such as the node-voltage or mesh-current method to solve for response transforms.

The response transform $Y(s)$* is a rational function whose partial fraction expansion leads directly to a response waveform of the form

$$y(t) = \sum k_j e^{p_j t}$$

where k_j is the residue of the pole in $Y(s)$ located at $s = -p_j$. The natural poles in $Y(s)$ are zeros of the s-domain circuit determinant and lead to the natural response terms in $y(t)$. The

*$Y(s)$ in this context is not admittance but the transform of the circuit output.

forced poles in $Y(s)$ depend on the input driving forces and lead to the force response components in $y(t)$.

In theory, we can perform s-domain analysis on circuits of any complexity. In practice, the algebraic burden of hand computations gets out of hand for circuits with more than three or four nodes or meshes. Of what practical use is an analysis method that becomes impractical at such a modest level of circuit complexity? Why not just appeal to computer-aided analysis tools like SPICE or MICRO-CAP in the first place?

Unquestionably, large-scale circuits are best handled by circuit analysis programs. Computer-aided analysis is probably the right tool for small-scale circuits when numerical values for all circuit parameters are given and a plot or numerical listing of the response waveform is the desired end product. Simply put, s-domain circuit analysis is not a computer algorithm for numerically generating response waveforms for a specific set of circuit parameters.

What s-domain analysis does that SPICE cannot do, is to generate a circuit response in symbolic form for all possible initial conditions and input signals. In effect, the analytical solution in symbolic form is a characterization of the circuit's signal-processing attributes. In particular, s-domain analysis leads to the concept of network functions that we study in Chapters 11 and 12 to analyze and design signal-processing circuits.

The purpose of circuit analysis is to gain insight into circuit behavior, not to grind out particular response waveforms. In this regard s-domain circuit analysis complements programs like SPICE or MICRO-CAP. It offers a way of characterizing circuits in very general terms. It provides insight into circuit behavior that allow us to use computer-aided analysis tools intelligently. In a circuit analysis book like this one, we may use s-domain analysis to generate specific numerical results. These hand computations are not the end product, but a means of gaining an understanding of circuit performance.

SUMMARY

- Kirchhoff's laws apply to waveforms in the time domain and transforms in the s domain.
- In the s domain the circuit models for the passive elements include initial condition sources and the element impedance or admittance. The impedances of the three passive elements are $Z_R = R$, $Z_L = Ls$, and $Z_C = 1/Cs$.
- Circuit analysis in the s domain uses techniques that closely parallel analysis methods for resistance circuits, because Kirchhoff's laws apply to voltage and current transforms

and the element *i*-*v* relationships are linear algebraic equations.

- Basic analysis techniques such as circuit reduction, Thévenin's and Norton's theorems, unit-output method, or superposition can be used to solve for response transforms in simple circuits. More complicated networks require a general approach such as the node-voltage or mesh-current methods.
- Response transforms are rational functions whose poles are zeros of the circuit determinant or poles of the transform of the input driving forces. The response waveform can be obtained from the partial fraction expansion of the response transform.
- The response transforms and waveforms in linear circuits can be separated into zero-state and zero-input components. The zero-state component is found by setting the initial capacitor voltages and inductor currents to zero. The zero-input component is found by setting all input driving forces to zero.
- The main purpose of *s*-domain circuit analysis is to gain insight into circuit performance without formally finding the time domain response. The circuit determinant contains considerable information about the circuit, since its zeros determine the form of the response.

EN ROUTE OBJECTIVES AND ASSOCIATED PROBLEMS

ERO 10-1 Equivalent Impedance (Sects. 10-1, 10-2)

Given a linear circuit: use series and parallel equivalence to find the equivalent impedance at a specified terminal pair.

10-1 Show that the two circuits in Fig. P10-1 have the same equivalent impedance between terminals 1 and 2. Express the impedance as a rational function and construct a pole-zero plot for this function.

10-2 A 2-kΩ resistor and a 2-H inductor are connected in series. The series combination is connected in parallel with a 0.5-μF capacitor. Find the equivalent admittance of the parallel combination as a rational function. Construct a pole-zero map for this function.

10-3 A 2-pF capacitor and a 0.5-H inductor are connected in series. The series combination is connected in parallel with a 5-kΩ

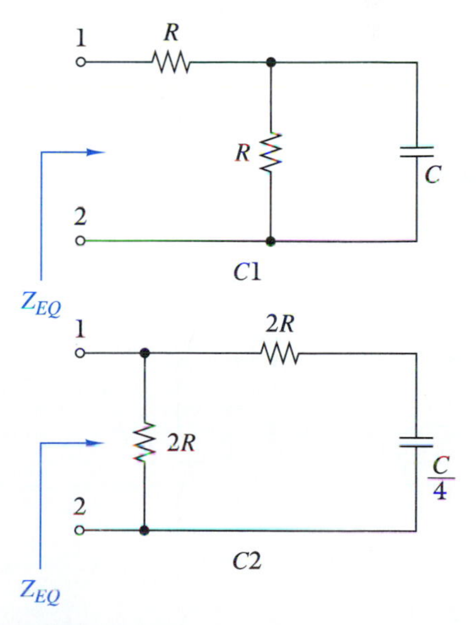

Figure P10-1

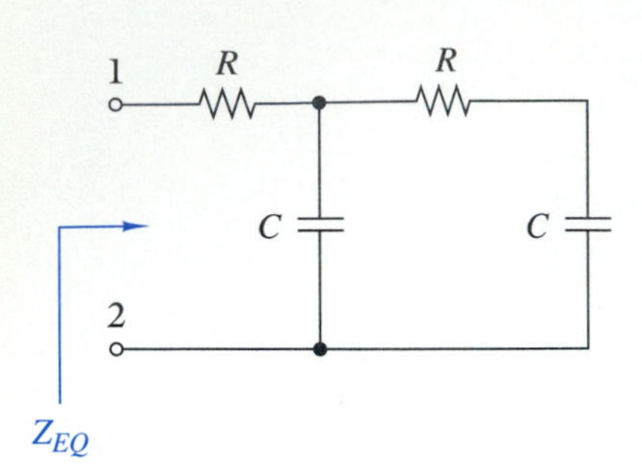

Figure P10-5

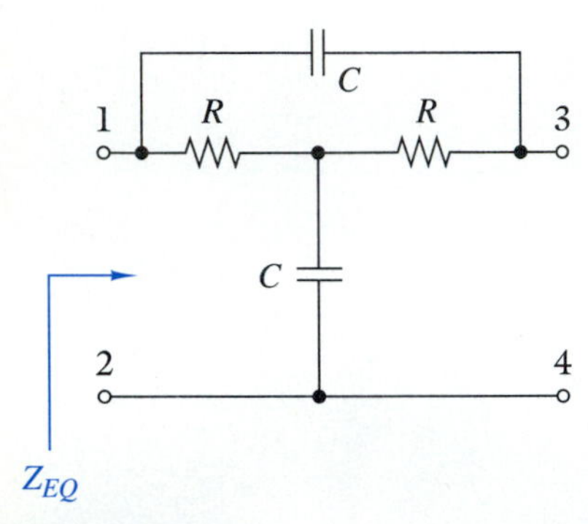

Figure P10-6

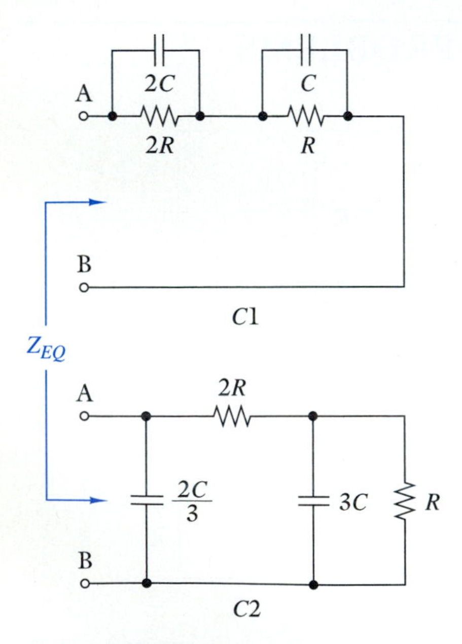

Figure P10-10

resistor. Find the equivalent impedance of the parallel combination as a rational function. Construct a pole-zero map for this function.

10-4 A 100-Ω resistor and a 10-mH inductor are connected in parallel. The parallel combination is connected in series with a 0.25-μF capacitor. Find the equivalent impedance of the parallel combination as a rational function. Construct a pole-zero map for this function.

10-5 Find the equivalent impedance between terminals 1 and 2 in the circuit of Fig. P10-5. Express the impedance as a rational function and construct its pole-zero map.

10-6 Find the equivalent impedance between terminals 1 and 2 in the circuit of Fig. P10-6. Express the impedance as a rational function and construct its pole-zero map.

10-7 Find the equivalent impedance between terminals 1 and 3 in the circuit of Fig. P10-6. Express the impedance as a rational function and construct its pole-zero map.

10-8 Repeat Prob. 10-6 with a short circuit connected between terminals 3 and 4.

10-9 Repeat Prob. 10-7 with a short circuit connected between terminals 3 and 4.

❖ **10-10** **(D)** The two circuits in Fig. P10-10 have the same equivalent impedance between terminals A and B.

(a) Show that this impedance is

$$Z_{EQ} = 3R\left[\frac{2RCs+1}{(4RCs+1)(RCs+1)}\right]$$

(b) Select values for R and C so that the poles of Z_{EQ} are located at $s = -1000$ rad/s, $s = -4000$ rad/s and $C \leq 1$ μF.

(c) Construct a pole-zero map of Z_{EQ} for the element values selected in (b).

❖ **10-11** **(D)** The two RC circuits in Fig. P10-11 have the same equivalent admittance between terminals A and B.

(a) Show that this admittance is

$$Y_{EQ} = Cs\left[\frac{3RCs+2}{(2RCs+1)(RCs+1)}\right]$$

(b) Select values for R and C so that the poles of Y_{EQ} are located at $s = -1000$ rad/s, $s = -4000$ rad/s and $C \leq 1$ μF.

(c) Construct a pole-zero map of Y_{EQ} for the element values selected in (b).

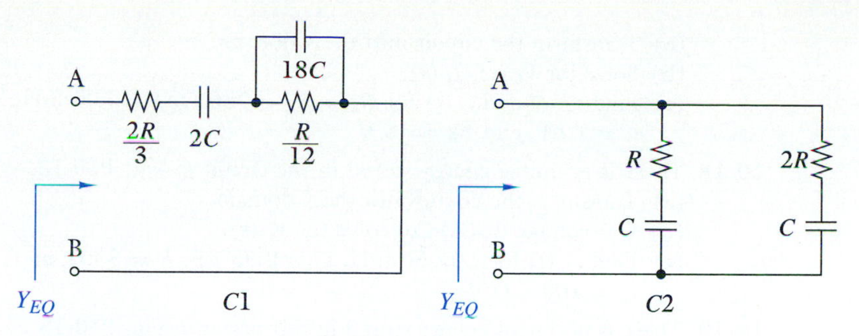

Figure P10-11

ERO 10-2 Basic Circuit Analysis Techniques (Sects. 10-2, 10-3)

Given a linear circuit:

(a) Determine the initial conditions (if not given).

(b) Transform the circuit into the s domain.

(c) Use circuit reduction, unit output, Thévenin/Norton equivalent circuits, or superposition to solve for zero-state and/or zero-input response transforms and waveforms.

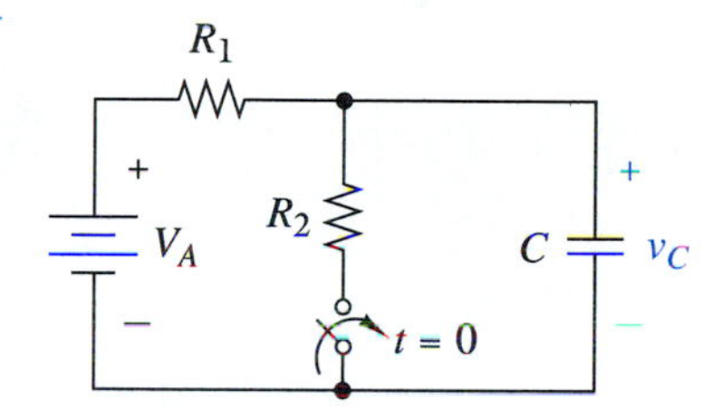

Figure P10-12

10-12 The switch in Fig. P10-12 has been open for a long time and is closed at $t = 0$.
(a) Transform the circuit into the s domain.
(b) Solve for $V_C(s)$ and $v_C(t)$.

10-13 The switch in Fig. P10-12 has been closed for a long time and is opened at $t = 0$.
(a) Transform the circuit into the s domain.
(b) Solve for $V_C(s)$ and $v_C(t)$.

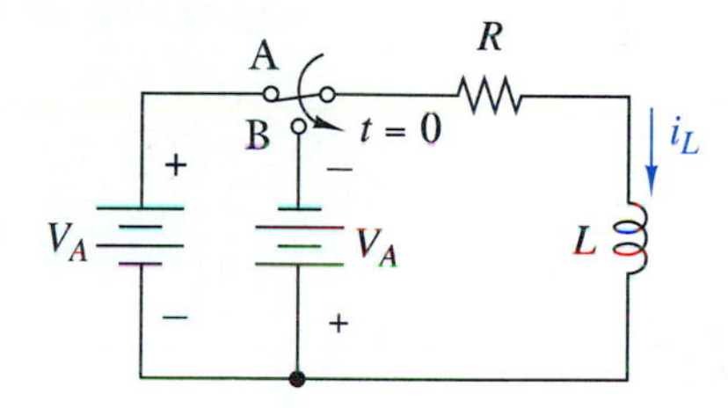

Figure P10-14

10-14 The switch in Fig. P10-14 has been in position A for a long time and is moved to position B at $t = 0$.
(a) Transform the circuit into the s domain.
(b) Solve for $I_L(s)$ and $i_L(t)$.

10-15 The switch in Fig. P10-14 has been in position B for a long time and is moved to position A at $t = 0$.
(a) Transform the circuit into the s domain.
(b) Solve for $I_L(s)$ and $i_L(t)$.

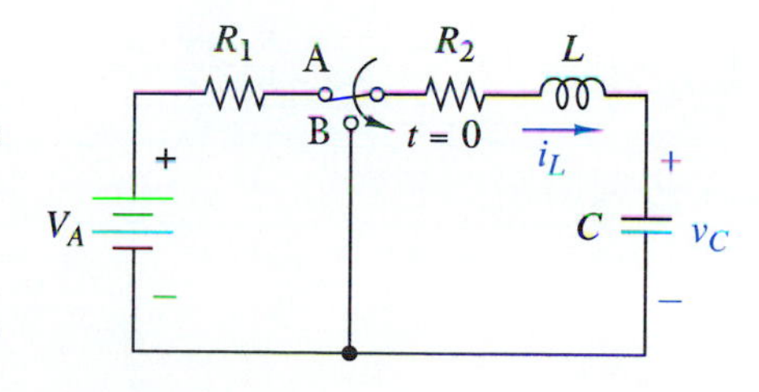

Figure P10-16

10-16 The switch in Fig. P10-16 has been in position A for a long time and is moved to position B at $t = 0$.
(a) Transform the circuit into the s domain.
(b) Solve for $V_C(s)$ and $I_L(s)$.
(c) Find $v_C(t)$ and $i_L(t)$ for $R_1 = R_2 = 500\ \Omega$, $L = 250$ mH, $C = 2\ \mu$F, and $V_A = 15$ V.

10-17 The switch in Fig. P10-16 has been in position B for a long time and is moved to position A at $t = 0$.

(a) Transform the circuit into the s domain.
(b) Solve for $V_C(s)$, $I_L(s)$.
(c) Find $v_C(t)$ and $i_L(t)$ for $R_1 = R_2 = 750\ \Omega$, $L = 500$ mH, $C = 1\ \mu$F, and $V_A = 15$ V.

10-18 There is no initial energy stored in the circuit in Fig. P10-18.
(a) Transform the circuit into the s domain.
(b) Use voltage division to solve for $V_C(s)$.
(c) Find $v_C(t)$ for $L = 50$ mH, $C = 0.05\ \mu$F, $R = 5$ kΩ, and $v_S = 100\ u(t)$ V.

10-19 There is no initial energy stored in the circuit in Fig. P10-18.
(a) Transform the circuit into the s domain.
(b) Use voltage division to solve for $V_R(s)$.
(c) Find $v_R(t)$ for $L = 50$ mH, $C = 0.05\ \mu$F, $R = 500\ \Omega$, and $v_S(t) = [10\ \cos\ 20000t]u(t)$ V.

10-20 The initial capacitor voltage in Fig. P10-20 is V_O.
(a) Transform the circuit into the s domain.
(b) Find the Thévenin equivalent at the indicated interface.
(c) Use the Thévenin equivalent to determine the zero-state and zero-input components of $I(s)$ and $i(t)$.

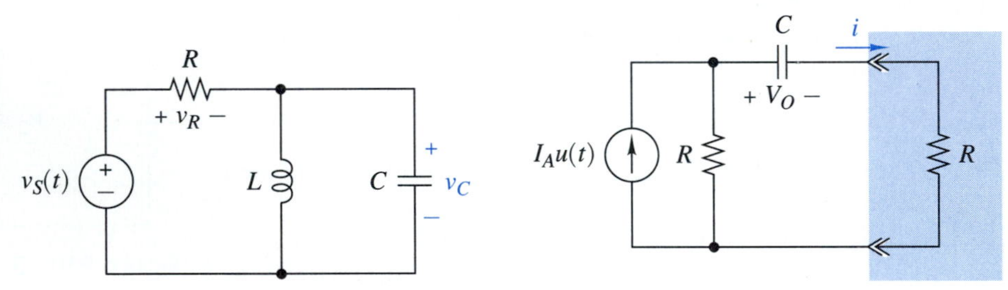

Figure P10-18

Figure P10-20

10-21 Repeat Prob. 10-20 when the current source delivers $I_A[\cos \beta t]u(t)$.

10-22 The initial inductor current in Fig. P10-22 is I_O.
(a) Transform the circuit into the s domain.
(b) Find the Thévenin equivalent at the indicated interface.
(c) Use the Thévenin equivalent to determine the zero-state and zero-input components of $I(s)$ and $i(t)$.

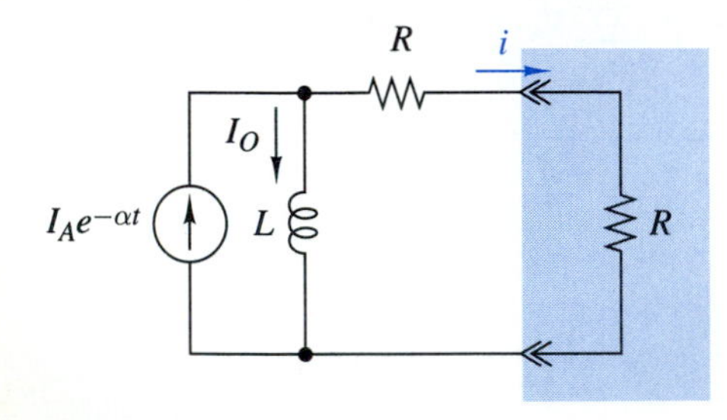

Figure P10-22

10-23 Repeat Prob. 10-22 when the current source delivers $I_A[\cos \beta t]u(t)$.

10-24 There is no initial energy stored in the circuit in Fig. P10-24.
(a) Transform the circuit into the s domain.
(b) Use superposition to solve for the zero-state component of $V(s)$.

10-25 The circuit in Fig. P10-25 has no initial stored energy.
(a) Transform the circuit into the s domain.
(b) Use superposition to solve for the zero-state component of $I(s)$.

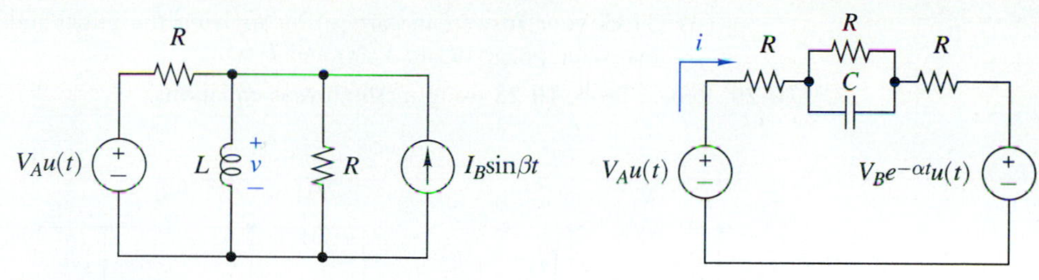

Figure P10-24

Figure P10-25

10-26 There is no initial energy stored in the circuit in Fig. P10-26. Use the unit-output method to find $V_2(s)$ in terms of the input voltage $V_1(s)$.

10-27 There is no initial energy stored in the circuit in Fig. P10-27. Use the unit-output method to find $V_2(s)$ in terms of the input voltage $V_1(s)$.

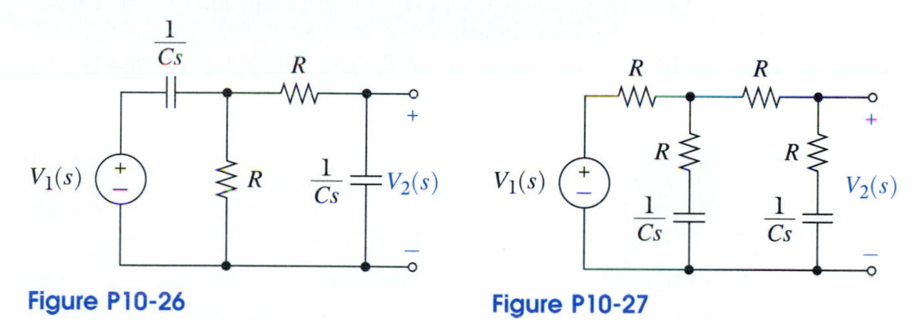

Figure P10-26

Figure P10-27

ERO 10-3 General Circuit Analysis (Sects. 10-4, 10-5, 10-6)

Given a linear circuit:

(a) Determine the initial conditions (if not given).

(b) Transform the circuit into the s domain and formulate node-voltage or mesh-current equations.

(c) Solve the circuit equations for the zero-state and/or zero-input response transforms and waveforms.

10-28 There is no initial energy stored in the circuit in Fig. P10-28.

(a) Transform the circuit into the s domain and formulate node-voltage equations.

(b) Solve the node-voltage equations for the zero-state and the zero-input components of $V_2(s)$ and $I_2(s)$ in terms of the input voltage $V_1(s)$.

(c) Find the zero-state component of $i_2(t)$ for $v_1(t) = 10u(t)$, $R_1 = R_2 = 10$ kΩ, $L = 10$ mH, and $C = 1$ μF.

(d) Check your answers in part (c) by applying the initial and final value properties to $V_2(s)$ and $I_2(s)$.

10-29 Repeat Prob. 10-28 using mesh-current equations.

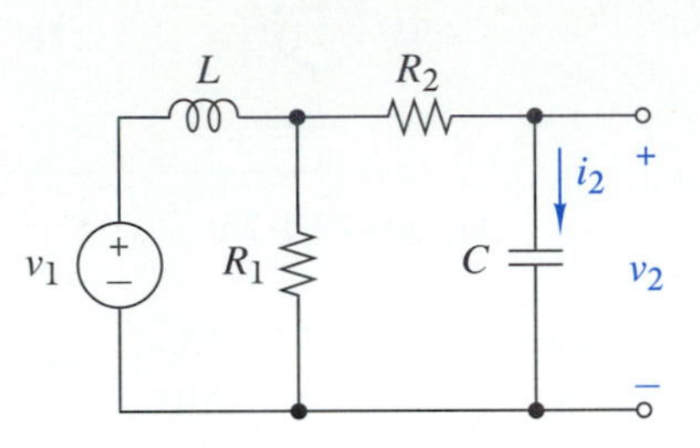

Figure P10-28

Figure P10-30

10-30 There is no initial energy stored in the circuit in Fig. P10-30.

(a) Transform the circuit into the s domain and formulate node-voltage equations.

(b) Solve the node-voltage equations for the zero-state and the zero-input components of $V_2(s)$ and $I_2(s)$ in terms of the input voltage $V_1(s)$.

(c) Find the zero-state component of $v_2(t)$ for $v_1(t) = 10u(t)$, $R_1 = R_2 = 10\ \text{k}\Omega$, and $C_1 = C_2 = 1\ \mu\text{F}$.

(d) Check your answers in part (c) by applying the initial and final value properties to $V_2(s)$ and $I_2(s)$.

10-31 Repeat Prob. 10-30 using mesh-current equations.

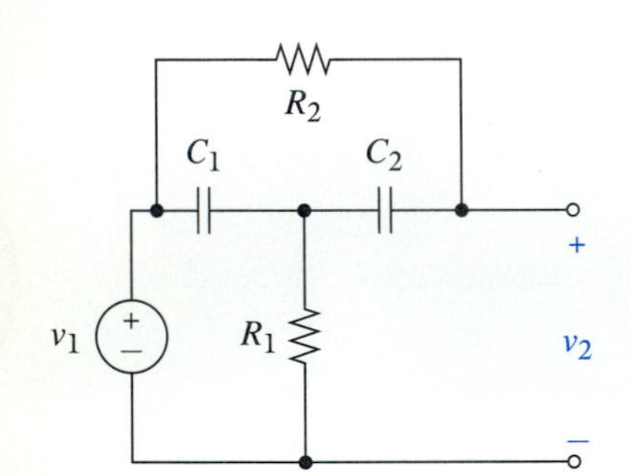

Figure P10-32

10-32 The circuit in Fig. P10-32 is called a bridged-T.

(a) Transform the circuit into the s domain and formulate either node-voltage or mesh-current circuit equations.

(b) Show that the zero-state component of $V_2(s)$ has a double zero at $s = -1/RC$ when $R_1 = R_2 = R$ and $C_1 = C_2 = C$.

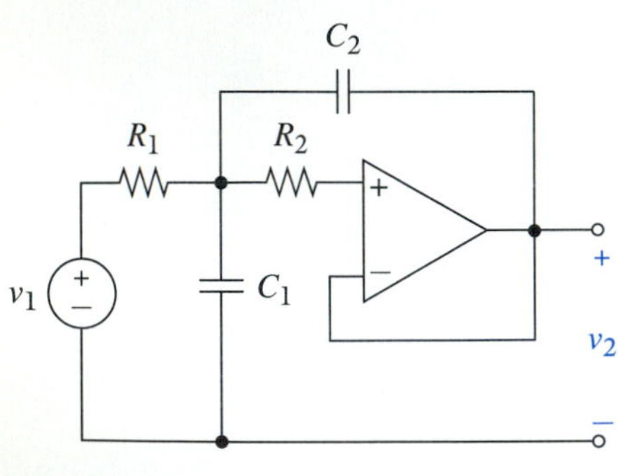

Figure P10-33

10-33 The circuit in Fig. P10-33 is an active RC filter using an OP AMP voltage follower.

(a) Transform the circuit into the s domain and formulate node-voltage equations.

(b) Solve the node-voltage equations for the zero-state component of the output voltage $V_2(s)$ in terms of input voltage $V_1(s)$.

(c) Show that $V_2(s)$ is independent of the value of C_2 and explain why.

10-34 The circuit in Fig. P10-34 is an active RC filter with two feedback paths around the OP AMP.

(a) Transform the circuit into the s domain and formulate node-voltage equations.

(b) Show that the circuit has a pair of complex conjugate poles when $R_1 = R_2 = R_3 = R$ and $C_1 = C_2 = C$.

10-35 **(D)** The circuit in Fig. P10-35 is an active RC highpass filter originally proposed by R. P. Sallen and E. K. Key in 1955.

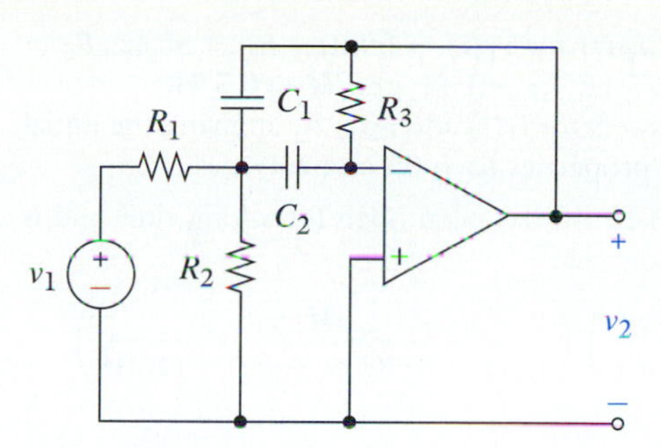

Figure P10-34

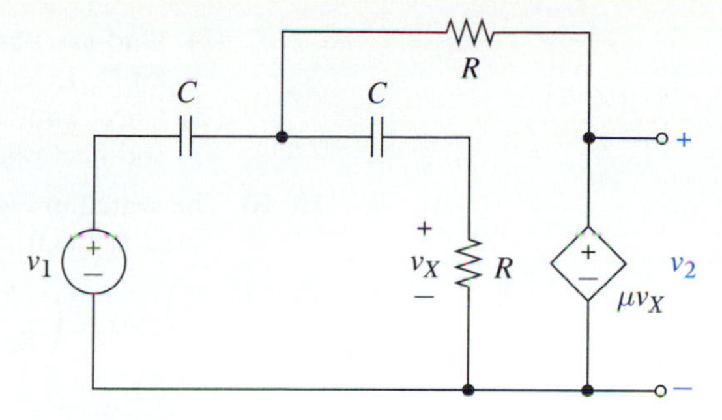

Figure P10-35

(a) Transform the circuit into the s domain and formulate either node-voltage or mesh-current circuit equations and find the circuit determinant.

(b) Select values of R, C, and μ so the circuit has natural poles at $s = \pm j1000$ rad/s.

❖ **10-36** **(D)** The circuit in Fig. P10-36 is an active RC bandpass filter proposed by J. J. Friend in 1970.

(a) Transform the circuit into the s domain and formulate either node-voltage or mesh-current circuit equations and find the circuit determinant.

(b) Select values or R_1, R_2, C_1, C_2, and μ so the circuit has two natural poles at $s = -1000$ rad/s.

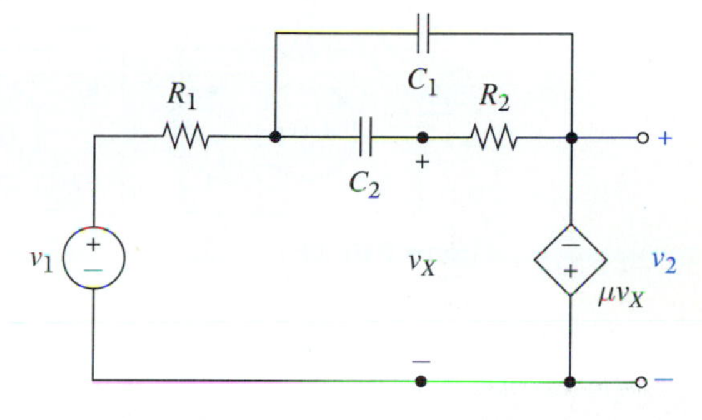

Figure P10-36

10-37 The circuit in Fig. P10-37 is a low-frequency model of a transistor in the common emitter configuration.

(a) Transform the circuit into the s domain.

(b) Use any method of analysis to solve for the zero-state component of the output voltage $V_2(s)$ in terms of the input $V_1(s)$.

(c) Find the zero-state component of $v_2(t)$ for $v_1(t) = 10u(t)$ mV, $R_1 = 10$ kΩ, $R_\pi = 1$ kΩ, $\beta = 100$, $R_O = 10$ kΩ, $C = 0.1$ μF, and $R_L = 100$ kΩ.

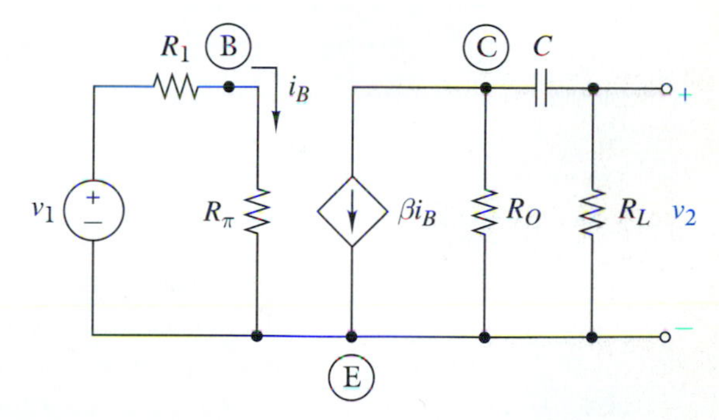

Figure P10-37

10-38 The circuit in Fig. P10-38 is a high-frequency model of a transistor in the common emitter configuration.

(a) Transform the circuit into the s domain and formulate node-voltage or mesh-current equations.

(b) Solve your circuit equations for the zero-state component of the output voltage $V_2(s)$ in terms of the input voltage $V_1(s)$.

(c) Find the zero-state component of $v_2(t)$ for $v_1(t) = 10u(t)$ mV, $R_1 = 10$ kΩ, $R_\pi = 1$ kΩ, $C_\mu = 2$ pF, $C_\pi = 20$ pF, $\beta = 50$, and $R_2 = 10$ kΩ.

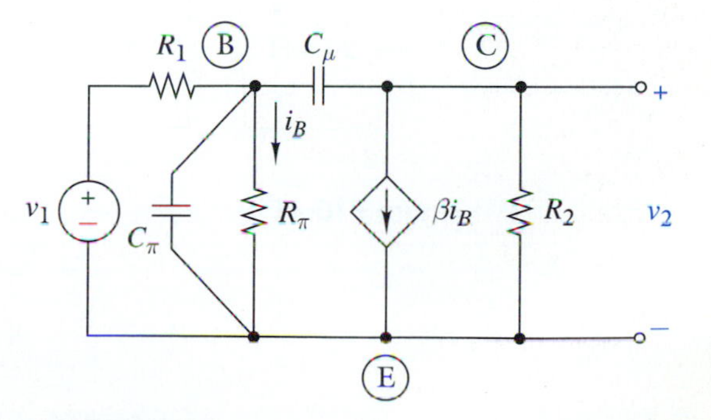

Figure P10-38

10-39 There is no initial energy stored in the circuit in Fig. P10-39.

(a) Transform the circuit into the s domain and formulate mesh-current equations.

(b) Solve the mesh-current equations for $I_A(s)$ and $I_B(s)$ in terms of the input voltage $V_1(s)$.

(c) Find $i_A(t)$ and $i_B(t)$ for $v_1(t) = 50u(t)$, $R_1 = 50\ \Omega$, $R_2 = 80\ \Omega$, $L_1 = 0.75$ H, $L_2 = 1$ H, and $M = 0.5$ H.

(d) Check your answers for $i_1(t)$ and $i_2(t)$ by applying the initial and final value properties to $I_1(s)$ and $I_2(s)$.

10-40 The switch in Fig. P10-40 has been open for a long time and is closed at $t = 0$. Show that

$$I_B(s) = \left(\frac{R_1 V_A}{R_0 + R_1}\right)\left(\frac{M}{(L_1 s + R_1)(L_2 s + R_2) - (Ms)^2}\right)$$

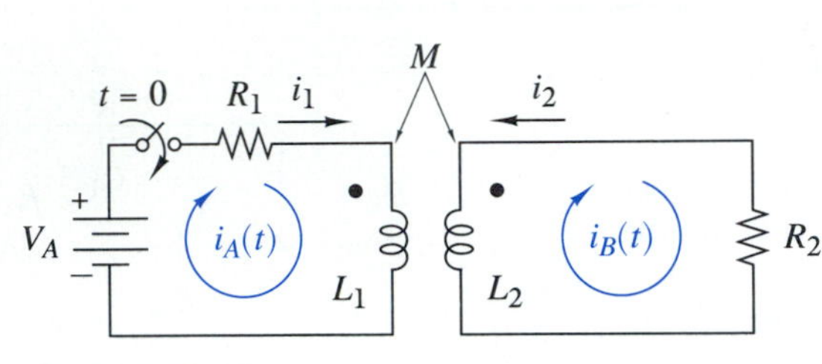

Figure P10-39

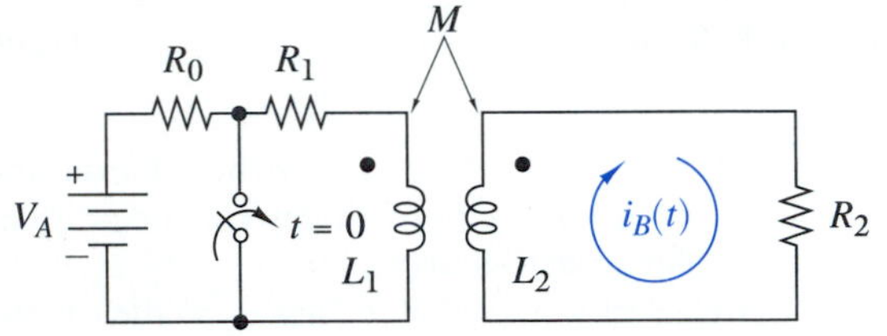

Figure P10-40

10-41 The equivalent impedance between terminals A and B in the circuit in Fig. P10-41 is

$$Z_{EQ} = \frac{0.02s(0.2s + 25000)}{s + 25000}$$

Find the value of the mutual inductance.

10-42 The switch in Fig. P10-42 has been open for a long time and is closed at $t = 0$.

(a) Transform the circuit into the s domain and formulate mesh-current circuit equations.

(b) Solve the mesh-current equations for $I_2(s)$ and $i_2(t)$.

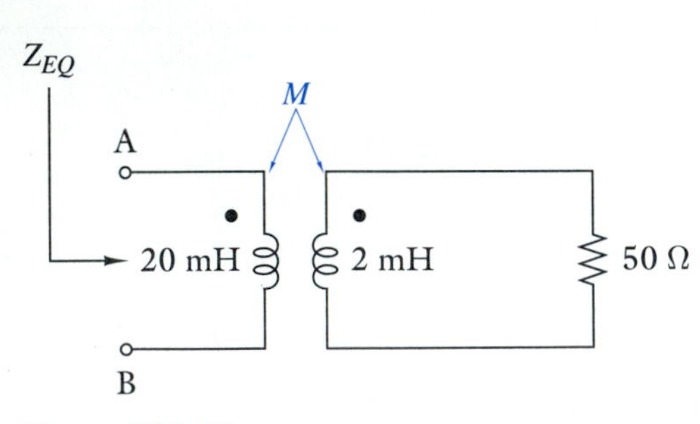

Figure P10-41

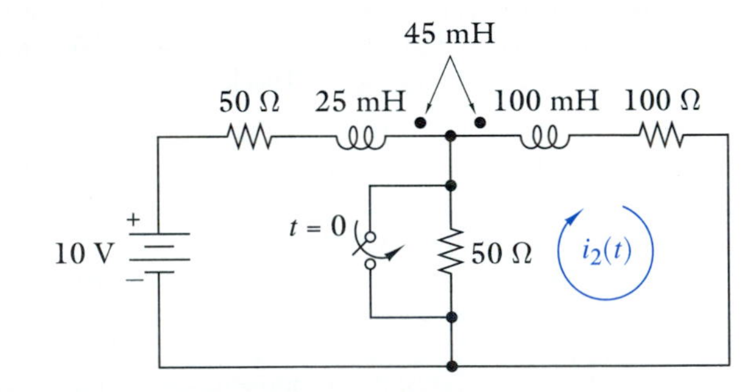

Figure P10-42

10-43 The switch in Fig. P10-42 has been closed for a long time and is opened at $t = 0$.

(a) Transform the circuit into the s domain and formulate mesh-current circuit equations.

(b) Solve the mesh-current equations for $I_2(s)$ and $i_2(t)$.

CHAPTER INTEGRATING PROBLEMS

10-44 **(A)** A black box containing a linear circuit has an on-off switch and a pair of external terminals. The voltage between the open-circuited external terminals is observed to be $10u(t)$ V when the switch is turned on. The short-circuit current is observed to be $[0.2e^{-1000t}]u(t)$ A when the switch is again turned on. Find the current the box would deliver to a 50-Ω resistive load.

10-45 **(A)** A current of $u(t)$ amperes is injected at a two-terminal interface and the interface voltage is observed to be $[10+20\ \sin 1000t]u(t)$ V. Find the interface current when a unit-step function voltage is applied across the interface.

10-46 **(A)**

(a) Find the Thévenin equivalent circuit seen by the load impedance in Fig. (10-46).

(b) Show that the Thévenin impedance is zero for perfect coupling between the two coils.

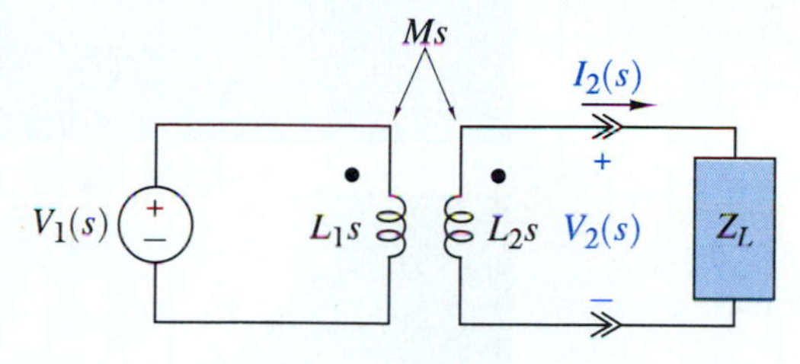

Figure P10-46

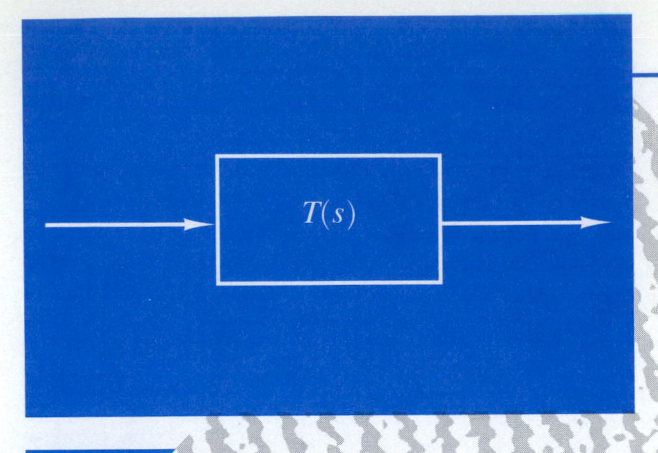

Chapter Eleven

"The driving-point impedance of a network is the ratio of an impressed electromotive force at a point in a branch of the network to the resulting current at the same point."

Ronald M. Foster, 1924
American Engineer

NETWORK FUNCTIONS

The concept of network functions emerged in the 1920s during the development of systematic methods of designing electric filters for long-distance telephone systems. The filter design effort eventually evolved into a formal realizability theory that came to be known as *network synthesis*. The objective of network synthesis is to design one or more networks that realize a given network function. The quotation on the opposite page is taken from one of the first papers to approach network design from a synthesis viewpoint.[1] Ronald Foster along with Sidney Darlington, Hendrik Bode, Wilhelm Cauer, and Otto Brune are generally considered the founders of the field of network synthesis. The classical synthesis methods they developed emphasized passive realizations using only resistors, capacitors, and inductors because the active elements of their era were costly and unreliable vacuum tubes. With the advent of semiconductor electronics, emphasis shifted to realizations using resistors, capacitors, and an active element such as the OP AMP.

In the last chapter we found that the transform of the output of a linear circuit is proportional to the transform of the input signal. In the s domain the proportionality factor is a rational function of s called a *network function*. The first section of this chapter shows how to define and calculate network functions for different types of circuits. The second section treats important properties of network functions including their impulse response, step response, and stability. The third section begins a study of the design of dynamic circuits by treating some basic methods of synthesizing circuits that realize a given network function. The fourth section deals with the relationship between network functions and the step response characteristics that define the performance of circuits. The last section shows how to use SPICE to calculate the step response of linear circuits.

[1] R. M. Foster, "A Reactance Theorem," Bell System Technical Journal, no. 3, pp 259–267, 1924.

11-1 INTRODUCTION

The proportionality property of linear circuits requires that the output be proportional to the input. In the s domain the proportionality factor is a rational function of s called a *network function*. More formally, a network function is defined in the s-domain as the ratio of the zero-state response (output) to the excitation (input):

$$\text{Network Function} = \frac{\text{Zero-State Response Transform}}{\text{Input Signal Transform}} \qquad (11\text{-}1)$$

Note carefully that this definition requires zero initial conditions and only one input.

The two major types of network functions are driving-point functions and transfer functions. A *driving-point function* relates the voltage and current at a given pair of terminals called a port. The driving-point impedance $Z(s)$ and admittance $Y(s)$ of the one-port circuit in Fig. 11-1 are defined as

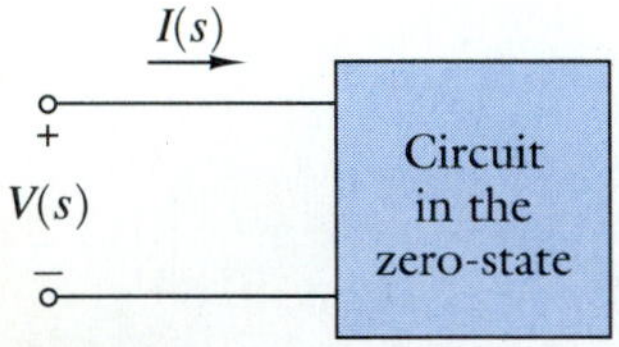

$I(s)$
$V(s)$
Circuit in the zero-state

Figure 11-1 One-port circuit for defining a driving-point function.

$$Z(s) = \frac{V(s)}{I(s)} = \frac{1}{Y(s)} \qquad (11\text{-}2)$$

Since $Z(s)$ and $Y(s)$ are reciprocals, we can calculate one of these functions and then obtain the other by inversion. A driving-point function determines the voltage response caused by an input current, or vice versa. In other words, a driving-point function is a network function whether upside down or right-side up.

The term *driving point* means that the circuit is driven at one port and the response is observed at the same port. The element impedances defined in Section 10-1 are elementary examples of driving-point impedances. The equivalent impedances found by combining elements in series and parallel are also driving-point impedances. Driving-point functions are the s-domain generalization of the concept of the input resistance. The terms *driving-point impedance*, *input impedance*, and *equivalent impedance* are synonymous.

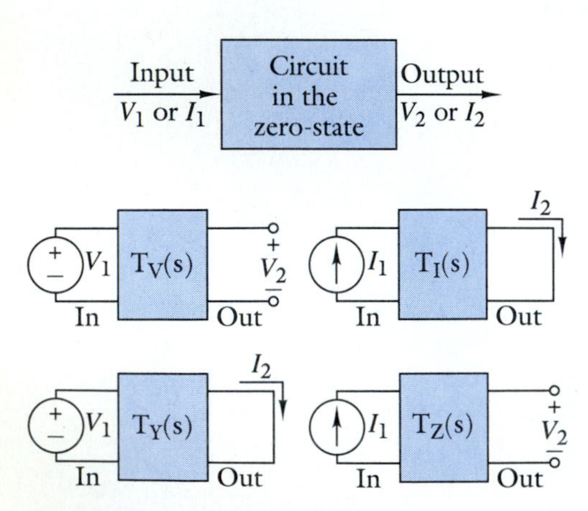

Figure 11-2 Two-port circuit configurations for defining transfer functions.

Transfer functions are of greater interest in signal-processing applications because they describe how a signal is modified by passing through a circuit. A *transfer function* relates an input and response (or output) at different ports in the circuit. Figure 11-2 shows all possible input-output configurations. Since the input and output signals can be either a current or a voltage, we can define four kinds of transfer functions:

$$T_V(s) = \text{Voltage Transfer Function} = \frac{V_2(s)}{V_1(s)}$$

$$T_I(s) = \text{Current Transfer Function} = \frac{I_2(s)}{I_1(s)} \qquad (11\text{-}3)$$

$$T_Y(s) = \text{Transfer Admittance} = \frac{I_2(s)}{V_1(s)}$$

$$T_Z(s) = \text{Transfer Impedance} = \frac{V_2(s)}{I_1(s)}$$

The functions $T_V(s)$ and $T_I(s)$ are dimensionless, since the input and output signals have the same units. The function $T_Z(s)$ has units of ohms and $T_Y(s)$ has unit of siemens.

Transfer functions always involve an input applied at one port and a response observed at a different port in the circuit. It is important to realize that a transfer function is only valid for a given input port and output port. They cannot be turned upside down like driving-point functions. For example, the voltage transfer function $T_V(s)$ relates the voltage $V_1(s)$ applied at the input port to the voltage response (V_2) observed at the output port in Fig. 11-2. The voltage transfer function for signal transmission in the opposite direction is not $1/T_V(s)$.

Calculating Network Functions

The rest of this section illustrates analysis techniques used to derive network functions for linear circuits. The application of network functions in circuit analysis and design begins in the next section and continues throughout the rest of the book. But first, we provide a short discussion of the methods used to determine the network functions of a given circuit.

The divider circuits in Fig. 11-3 occur so frequently that it is worth taking time to develop their transfer functions in general terms. Using s-domain voltage division in Fig. 11-3(a) we can write

$$V_2(s) = \left[\frac{Z_2(s)}{Z_1(s) + Z_2(s)}\right] V_1(s)$$

Therefore, the voltage transfer function of a voltage divider circuit is

$$T_V(s) = \frac{V_2(s)}{V_1(s)} = \frac{Z_2(s)}{Z_1(s) + Z_2(s)} \tag{11-4}$$

Using s-domain current division in Fig. 11-3(b) we have

$$I_2(s) = \left[\frac{Y_2(s)}{Y_1(s) + Y_2(s)}\right] I_1(s)$$

Therefore, the current transfer function of a current divider circuit is

$$T_I(s) = \frac{I_2(s)}{I_1(s)} = \frac{Y_2(s)}{Y_1(s) + Y_2(s)} \tag{11-5}$$

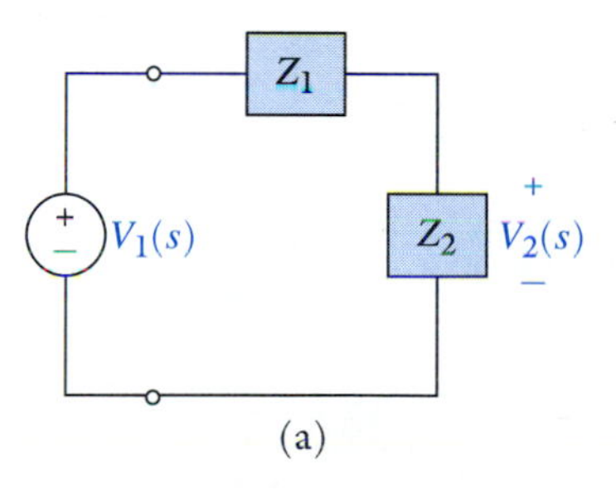

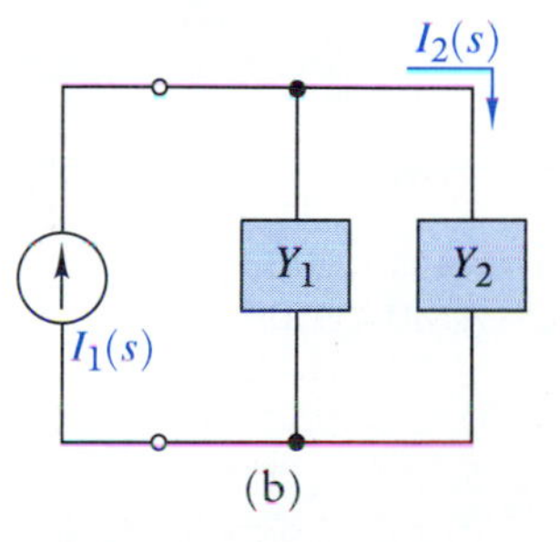

Figure 11-3 (a) Voltage divider. (b) Current divider.

By series equivalence the driving-point impedance at the input of the voltage divider is $Z_{EQ}(s) = Z_1(s) + Z_2(s)$. By parallel equivalence the driving-point admittance at the input of the current divider is $Y_{EQ}(s) = Y_1(s) + Y_2(s)$.

Two other very common circuits are the inverting and noninverting OP AMP configurations shown in Fig. 11-4. To determine the voltage transfer function of the inverting circuit in Fig. 11-4(a) we write a node-voltage equation at Node B:

$$-Y_1(s)V_A(s) + [Y_1(s) + Y_2(s)]\,V_B(s) - Y_2(s)V_C(s) = 0 \quad (11\text{-}6)$$

But the OP AMP device constraint requires that $V_B(s) = 0$, since the noninverting input is grounded. By definition the output voltage $V_2(s)$ equals node voltage $V_C(s)$ and the voltage source forces $V_A(s)$ to equal the input voltage $V_1(s)$. Combining all of these considerations reduces Eq. (11-6) to

$$-Y_1(s)V_1(s) - Y_2(s)V_2 = 0$$

Solving for the voltage transfer function yields

$$T_V(s) = \frac{V_2(s)}{V_1(s)} = -\frac{Y_1(s)}{Y_2(s)} = -\frac{Z_2(s)}{Z_1(s)} \quad (11\text{-}7)$$

From the study of OP AMP circuits in Chapter 4 the reader should recognize Eq. (11-7) as the s-domain generalization of the noninverting OP AMP circuit gain equation, $K = -R_2/R_1$. The driving-point impedance at the input to this circuit is impedance $Z_1(s)$, since $V_B = 0$.

For the noninverting circuit in Fig. 11-4(b) we write a node-voltage equation at Node B:

$$[Y_1(s) + Y_2(s)]\,V_B(s) - Y_2(s)V_C(s) = 0 \quad (11\text{-}8)$$

In the noninverting configuration the OP AMP device constraint requires that $V_B(s) = V_1(s)$. By definition the output voltage V_2 equals node voltage $V_C(s)$. Combining all of these considerations reduces Eq. (11-8) to

$$[Y_1(s) + Y_2(s)]\,V_1(s) - Y_2(s)V_2 = 0$$

Solving for the voltage transfer function yields

$$T_V(s) = \frac{V_2(s)}{V_1(s)} = \frac{Y_1(s) + Y_2(s)}{Y_2(s)} = \frac{Z_1(s) + Z_2(s)}{Z_1(s)} \quad (11\text{-}9)$$

Equation (11-9) is the s-domain version of the noninverting amplifier gain equation, $K = (R_1 + R_2)/R_1$. The transfer function of the noninverting configuration is the reciprocal of the transfer function of the voltage divider in the feedback path. The ideal OP AMP draws no current at its input terminals, and theoretically the driving-point impedance at the input to the noninverting circuit is infinite.

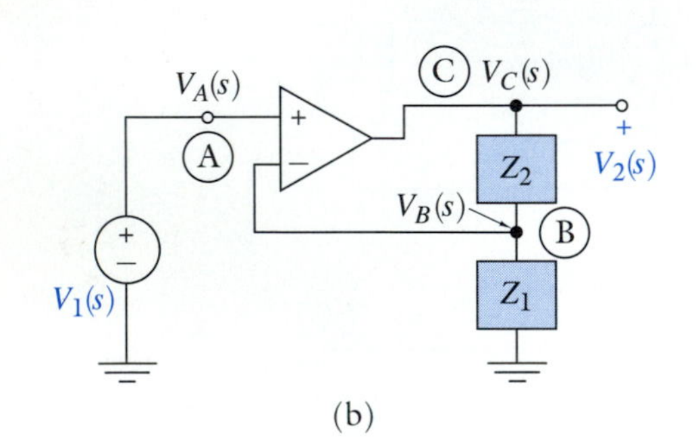

Figure 11-4 (a) Inverting OP AMP circuit. (b) Noninverting OP AMP circuit.

The transfer functions of divider circuits and the basic OP AMP configurations are useful analysis and design tools in many practical situations. However, a general method is needed to handle circuits of greater complexity than these basic building blocks. One general approach is to formulate either node-voltage or mesh-current equations with all initial conditions set to zero. These equations are then solved for network functions using Gaussian elimination or Cramer's rule. The algebra involved can be tedious at times, but the tedium is reduced because we only need the zero-state response for a single input source.

The following examples illustrate methods of calculating network functions.

EXAMPLE 11-1

(a) Find the transfer functions of the circuits in Fig. 11-5.

(b) Find the driving-point impedances seen by the input sources in these circuits.

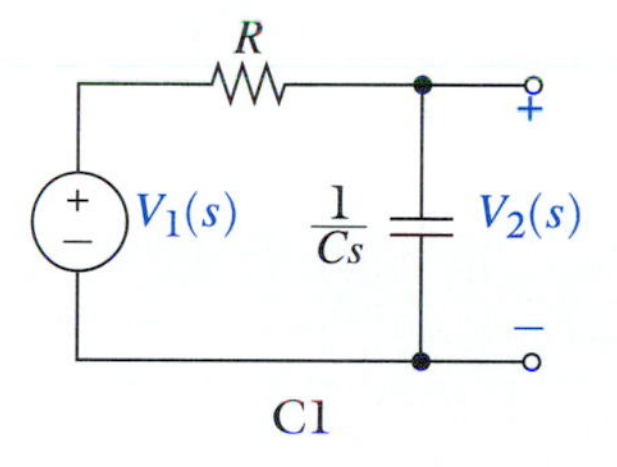

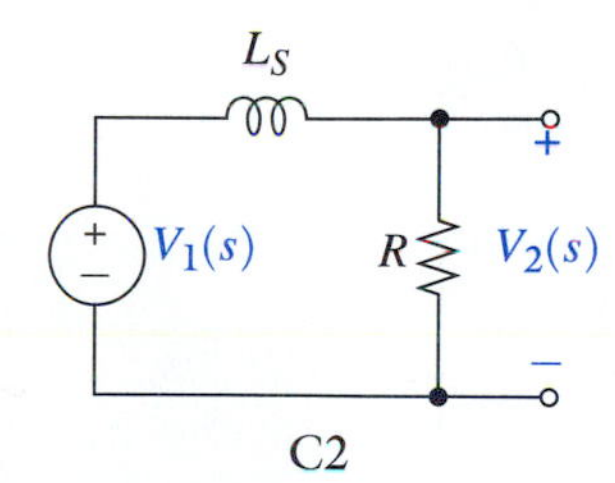

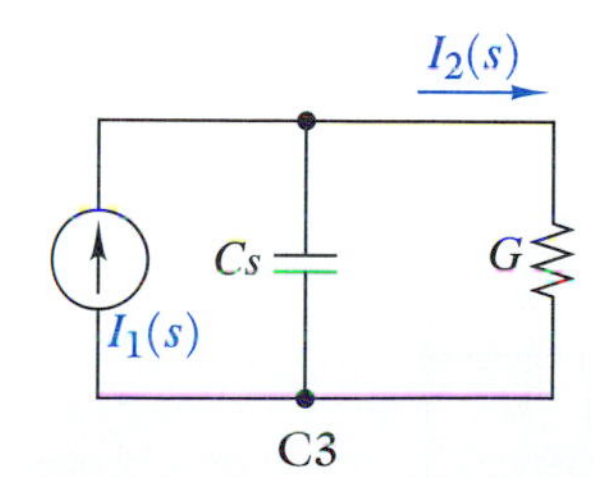

Figure 11-5

SOLUTION:

(a) These are all divider circuits so the required transfer functions can be obtained using Eq. (11-4) or (11-5).

For Circuit C1:

$$Z_1 = R,\ Z_2 = 1/Cs, \quad \text{and} \quad T_V(s) = 1/(RCs + 1).$$

For Circuit C2:

$$Z_1 = Ls,\ Z_2 = R, \quad \text{and} \quad T_V(s) = 1/(GLs + 1).$$

For Circuit C3:

$$Y_1 = Cs,\ Y_2 = G, \quad \text{and} \quad T_I(s) = 1/(RCs + 1).$$

These transfer functions are all of the form $1/(\tau s + 1)$, where τ is the circuit time constant. The general principle illustrated here is that two or more circuits can have the same transfer function. Put differently, a desired transfer function can be realized by several different circuits. This fact is important in design because circuits that produce the same transfer function may differ in other ways such as input impedance, power consumption, or cost.

(b) The driving-point impedances are found by series or parallel equivalence:

For circuit C1:

$$Z_1 = R, \; Z_2 = 1/Cs, \quad \text{and} \quad Z(s) = (RCs + 1)/Cs.$$

For circuit C2:

$$Z_1 = Ls, \; Z_2 = R, \quad \text{and} \quad Z(s) = Ls + R.$$

For circuit C3:

$$Y_1 = Cs, \; Y_2 = G, \text{ and } Z(s) = 1/(Cs+G) = R/(RCs+1).$$

The three circuits have different driving-point impedances even though they have the same transfer functions.

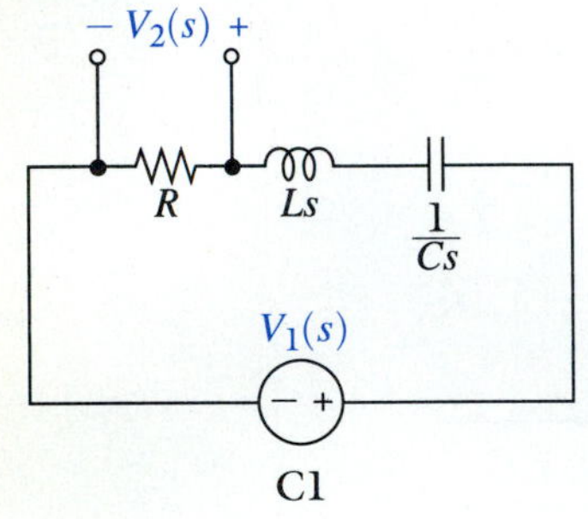

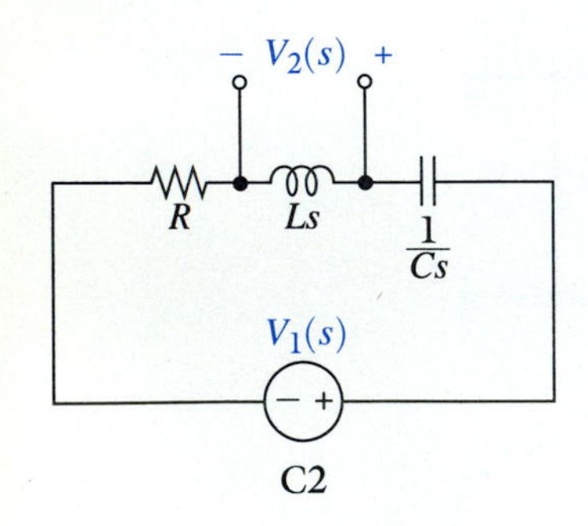

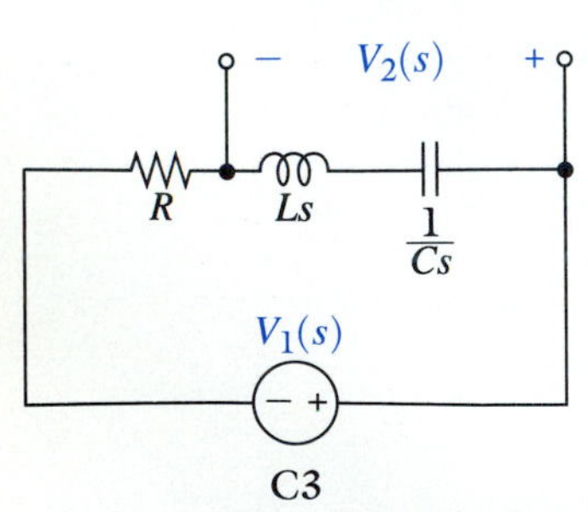

Figure 11-6

EXAMPLE 11-2

(a) Find the driving-point impedance seen by the voltage source in Fig. 11-6.

(b) Find the voltage transfer functions $T_V = V_2/V_1$ for each circuit.

SOLUTION:

These circuits are all voltage dividers in which the output is taken across different elements of a series RLC circuit.

(a) The driving-point impedance seen by the voltage source is the same in all three circuits:

$$Z_{EQ} = R + Ls + \frac{1}{Cs} = \frac{LCs^2 + RCs + 1}{Cs}$$

(b) For circuit C1 the output is taken across $Z_2(s) = R$, so by voltage division

$$T_V(s) = \frac{Z_2(s)}{Z_{EQ}(s)} = \frac{RCs}{LCs^2 + RCs + 1}$$

For circuit C2 the output is taken across $Z_2(s) = Ls$:

$$T_V(s) = \frac{Z_2(s)}{Z_{EQ}(s)} = \frac{LCs^2}{LCs^2 + RCs + 1}$$

Finally, for circuit C3 the output is taken across $Z_2(s) = Ls + 1/Cs$:

$$T_V(s) = \frac{Z_2}{Z_{EQ}} = \frac{LCs^2 + 1}{LCs^2 + RCs + 1}$$

The transfer functions for C1, C2, and C3 have the same poles but different zeros. The fact that a given circuit can have different transfer functions illustrates that a standard design can be adapted to meet several different performance requirements.

EXAMPLE 11-3

(a) Find the input impedance seen by the voltage source in Fig. 11-7.

(b) Find the voltage transfer function V_2/V_1 of the circuit.

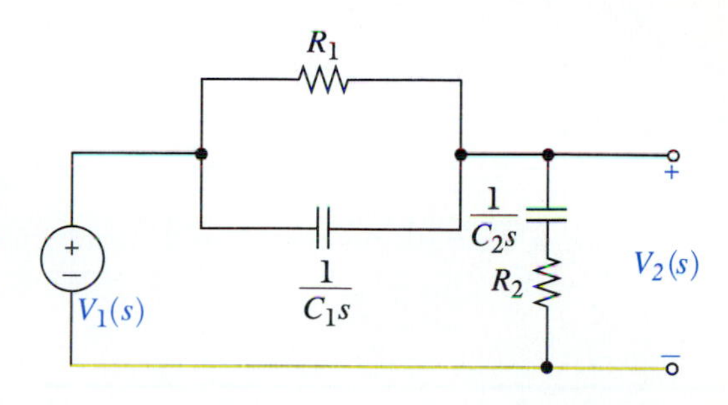

Figure 11-7

SOLUTION:

(a) The circuit is a voltage divider. We first calculate the equivalent impedances of the two legs of the divider. The two elements in parallel combine to produce the series leg impedance $Z_1(s)$:

$$Z_1(s) = \frac{1}{C_1 s + 1/R_1} = \frac{R_1}{R_1 C_1 s + 1}$$

The two elements in series combine to produce shunt leg impedance $Z_2(s)$:

$$Z_2(s) = R_2 + 1/C_2 s = \frac{R_2 C_2 s + 1}{C_2 s}$$

Using series equivalence the driving-point impedance seen at the input is

$$\begin{aligned} Z_{EQ}(s) &= Z_1(s) + Z_2(s) \\ &= \frac{R_1 C_1 R_2 C_2 s^2 + (R_1 C_1 + R_2 C_2 + R_1 C_2)s + 1}{C_2 s (R_1 C_1 s + 1)} \end{aligned}$$

(b) Now using voltage division the voltage transfer function is

$$\begin{aligned} T_V(s) &= \frac{Z_2(s)}{Z_{EQ}(s)} \\ &= \frac{(R_1 C_1 s + 1)(R_2 C_2 s + 1)}{R_1 C_1 R_2 C_2 s^2 + (R_1 C_1 + R_2 C_2 + R_1 C_2)s + 1} \end{aligned}$$

EXAMPLE 11-4

(a) Find the driving-point impedance seen by the voltage source in Fig. 11-8.

(b) Find the voltage transfer function V_2/V_1 of the circuit.

SOLUTION:

The circuit is an inverting OP AMP configuration of the form in Fig. 11-4(a). The input impedance and voltage transfer function

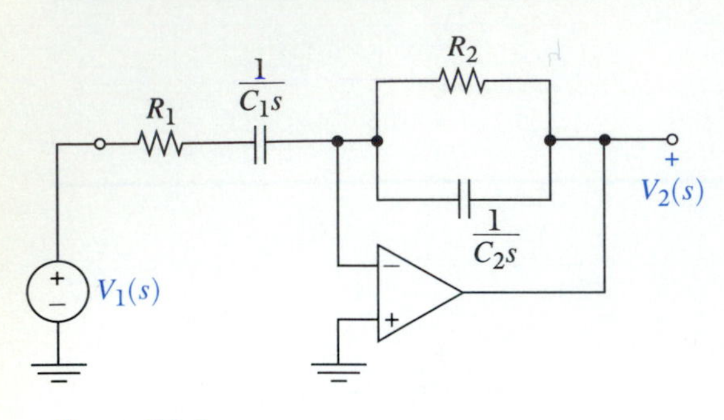

Figure 11-8

of this configuration are

$$Z_{IN}(s) = Z_1(s) \quad \text{and} \quad T_V(s) = -\frac{Z_2(s)}{Z_1(s)}$$

(a) The circuit input impedance is

$$Z_1(s) = R_1 + \frac{1}{C_1 s} = \frac{R_1 C_1 s + 1}{C_1 s}$$

(b) The feedback impedance is

$$Z_2(s) = \frac{1}{C_2 s + 1/R_2} = \frac{R_2}{R_2 C_2 s + 1}$$

and the voltage transfer function is

$$T_V(s) = -\frac{Z_2(s)}{Z_1(s)} = -\frac{R_2 C_1 s}{(R_1 C_1 s + 1)(R_2 C_2 s + 1)}$$

EXAMPLE 11-5

(a) Find the input admittance seen by the voltage source in Fig. 11-9.

(b) Find the transfer function $T_V = V_2/V_1$ for the circuit.

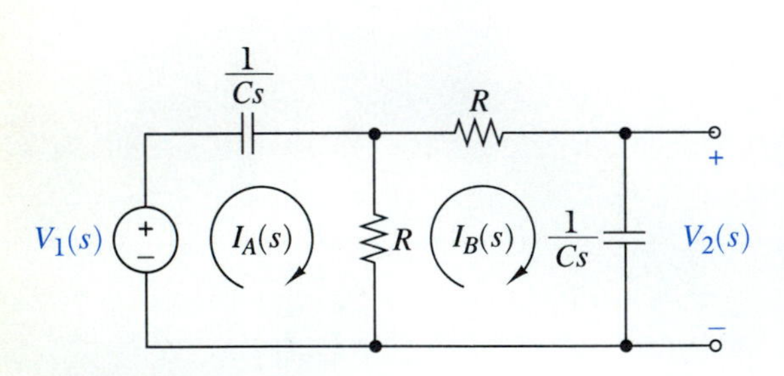

Figure 11-9

SOLUTION:

The circuit is not a divider, so we use mesh-current equations to illustrate the general approach to finding network functions. Using the inspection method discussed in Chapter 10, the mesh-current equations for this ladder circuit are

$$\left(R + \frac{1}{Cs}\right) I_A(s) - R I_B(s) = V_1(s)$$

$$-R I_A(s) + \left(2R + \frac{1}{Cs}\right) I_B(s) = 0$$

(a) Since $Y(s) = I_A(s)/V_1(s)$ we use Cramer's rule to solve for $I_A(s)$:

$$I_A(s) = \frac{\Delta_A}{\Delta} = \frac{\begin{vmatrix} V_1(s) & -R \\ 0 & 2R + 1/Cs \end{vmatrix}}{\begin{vmatrix} R + 1/Cs & -R \\ -R & 2R + 1/Cs \end{vmatrix}}$$

$$= \frac{Cs(2RCs + 1)}{(RCs)^2 + 3RCs + 1} V_1(s)$$

The input admittance of the circuit is

$$Y(s) = \frac{I_A(s)}{V_1(s)} = \frac{Cs(2RCs + 1)}{(RCs)^2 + 3RCs + 1}$$

(b) The mesh-current equations do not contain the output voltage directly. But, since $V_2 = I_B Z_C = I_B/Cs$, we can use Cramer's rule to solve for I_B:

$$I_B(s) = \frac{\Delta_B}{\Delta} = \frac{\begin{vmatrix} R + \dfrac{1}{Cs} & V_1(s) \\ -R & 0 \end{vmatrix}}{\begin{vmatrix} R + \dfrac{1}{Cs} & -R \\ -R & 2R + \dfrac{1}{Cs} \end{vmatrix}}$$

$$= \frac{RC^2s^2}{(RCs)^2 + 3RCs + 1} V_1(s)$$

The voltage transfer function of the circuit is

$$T_V(s) = \frac{V_2(s)}{V_1(s)} = \frac{I_B(s)/Cs}{V_1(s)} = \frac{RCs}{(RCs)^2 + 3RCs + 1}$$

EXAMPLE 11-6

Find the voltage transfer function V_2/V_1 of the circuit in Fig. 11-10.

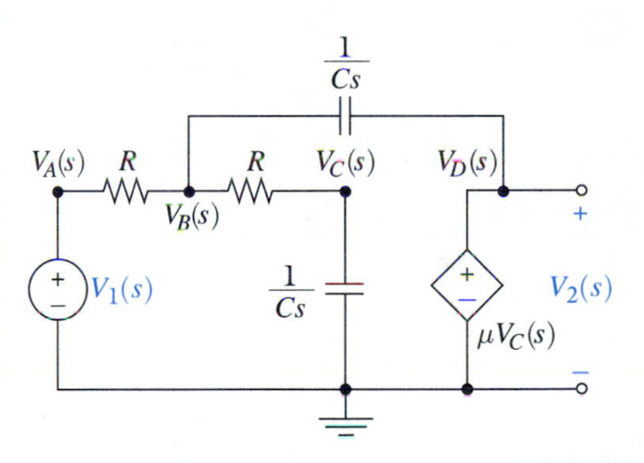

Figure 11-10

SOLUTION:

This an active RC circuit in which the voltage-controlled voltage source is a model of an active device such as a noninverting OP AMP amplifier. We use node-voltage equations in this case because the required output is a voltage. The circuit contains two voltage sources connected at a common node. Selecting this common node as the reference eliminates two node voltages as unknowns since $V_A(s) = V_1(s)$ and $V_D(s) = \mu V_C(s) = V_2(s)$. By inspection the equations at the two remaining nodes are

$$\text{Node B:} \quad -GV_1(s) + (2G + Cs)V_B(s) - GV_C(s) - Cs\,[\mu V_C(s)] = 0$$

$$\text{Node C:} \quad -GV_B(s) + (G + Cs)V_C(s) = 0$$

Multiplying both equations by $R = 1/G$ and rearranging terms produces

$$\text{Node B:} \quad (2 + RCs)V_B(s) - (1 + \mu RCs)V_C(s) = V_1(s)$$

$$\text{Node C:} \quad -V_B(s) + (1 + RCs)V_C(s) = 0$$

Using the Node C equation to eliminate $V_B(s)$ from the Node B equation leaves

$$(2 + RCs)(1 + RCs)V_C(s) - (1 + \mu RCs)V_C(s) = V_1(s)$$

Since the output $V_2(s) = \mu V_C(s)$, the required transfer function is

$$T_V(s) = \frac{V_2(s)}{V_1(s)} = \frac{\mu}{(RCs)^2 + (3 - \mu)RCs + 1}$$

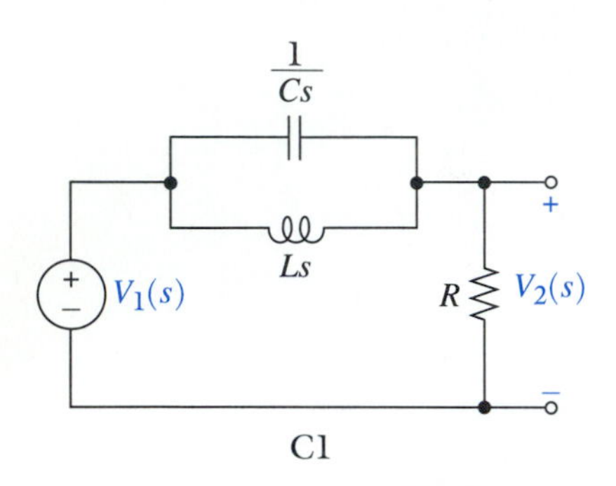

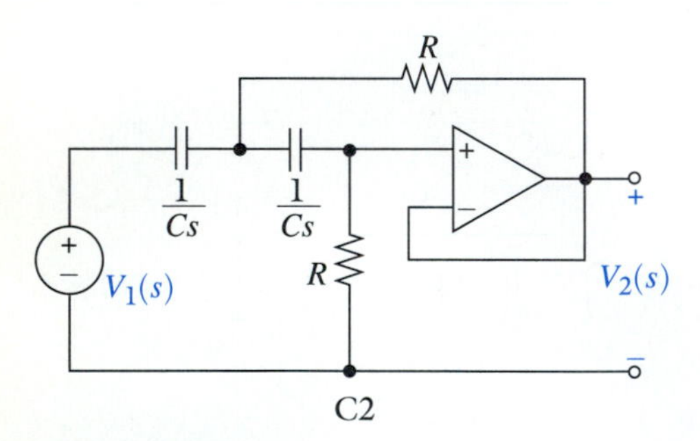

Figure 11-11

EXERCISE 11-1

Find the input admittances seen by the voltage sources in Fig. 11-11.

ANSWERS:

Circuit C1: $Y(s) = \dfrac{LCs^2 + 1}{RLCs^2 + Ls + R}$

Circuit C2: $Y(s) = \dfrac{Cs}{RCs + 1}$

EXERCISE 11-2

Find the voltage transfer functions V_2/V_1 of the circuits in Fig. 11-11.

ANSWERS:

Circuit C1: $T_V(s) = \dfrac{LCs^2 + 1}{LCs^2 + GLs + 1}$

Circuit C2: $T_V(s) = \dfrac{(RCs)^2}{(RCs + 1)^2}$

The Cascade Connection and the Chain Rule

Signal-processing circuits often involve a *cascade connection* in which the output voltage of one circuit serves as the input to the next stage. In some cases, the overall voltage transfer function of the cascade can be related to the transfer functions of the individual stages by a *chain rule*

$$T_V(s) = T_{V1}(s)T_{V2}(s)\ldots T_{Vk}(s) \tag{11-10}$$

where T_{V1}, T_{V2}, ... and T_{Vk} are the voltage transfer functions of the individual stages when operated separately. It is important to understand when the chain rule applies, since it greatly simplifies the analysis and design of cascade circuits.

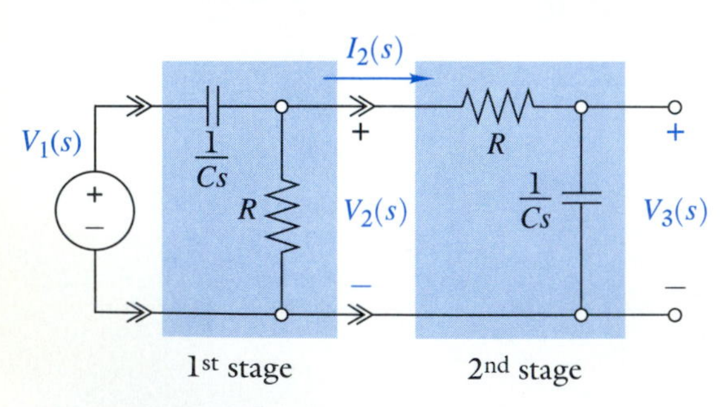

Figure 11-12 Two-port circuits connected in cascade.

Figure 11-12 shows two RC circuits or stages connected in cascade at an interface. When disconnected and operated separately the transfer functions of each stage are easily found using voltage division:

$$T_{V1}(s) = \frac{R}{R + 1/Cs} = \frac{RCs}{RCs + 1}$$
$$T_{V2}(s) = \frac{1/Cs}{R + 1/Cs} = \frac{1}{RCs + 1} \tag{11-11}$$

When connected in cascade the output of the first stage serves as the input to the second stage. If the chain rule applies we would obtain the overall transfer function as

$$T_V(s) = \frac{V_3(s)}{V_1(s)} = \left(\frac{V_2(s)}{V_1(s)}\right)\left(\frac{V_3(s)}{V_2(s)}\right) = (T_{V1}(s))\,(T_{V2}(s))$$
$$= \underbrace{\left(\frac{RCs}{RCs+1}\right)}_{1^{st}\ \text{stage}}\underbrace{\left(\frac{1}{RCs+1}\right)}_{2^{nd}\ \text{stage}} = \underbrace{\frac{RCs}{(RCs)^2 + 2RCs + 1}}_{\text{overall}} \tag{11-12}$$

However, in Example 11-5 the overall transfer function of this circuit was found to be

$$T_V(s) = \frac{RCs}{(RCs)^2 + 3RCs + 1} \tag{11-13}$$

which disagrees with the chain rule result in Eq. (11-12).

The reason for the discrepancy is that the second circuit "loads" the first circuit when they are connected in cascade. That is, the voltage-divider rule assumes that the current $I_2(s)$ in Fig. 11-12 is zero. The no-load condition $I_2(s) = 0$ is valid when the stages operate separately, but when connected together the current $I_2(s)$ is no longer zero. In other words, the chain rule does not apply here because loading caused by the second stage changes the transfer function of the first stage.

However, Fig. 11-13 shows how the loading problem can be overcome by inserting an OP AMP voltage follower between

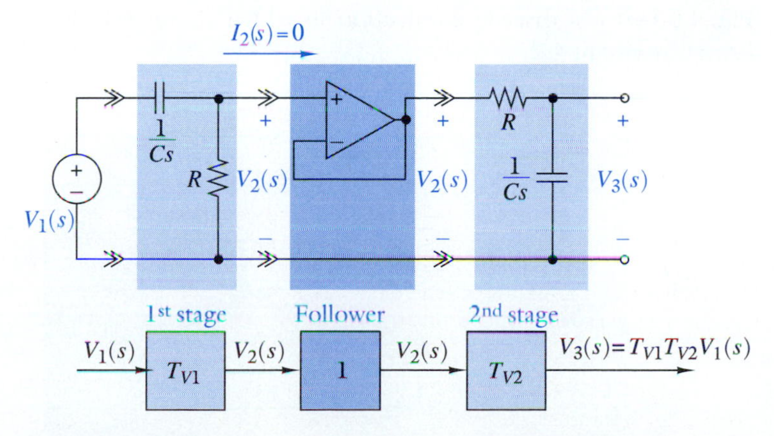

Figure 11-13 Cascade connection with OP AMP voltage follower isolation.

the RC circuit stages. With this modification the chain rule in Eq. (11-10) applies because the voltage follower isolates the two stages. The voltage follower isolates the two RC circuits because it has infinite input resistance and zero output resistance. Therefore, it does not draw any current from the first RC circuit ($I_2 = 0$) and applies $V_2(s)$ directly to the input of the second RC circuit.

The chain rule in Eq. (11-10) applies if connecting a stage does not change (load) the output of the preceding stage. Loading can be avoided by connecting an OP AMP voltage follower between stages as in Fig. 11-13. More importantly, loading does not occur if the output of the preceding stage is the output terminal of an OP AMP or controlled source. These elements act like ideal voltage sources whose outputs are unchanged by connecting the subsequent stage.

For example, the two circuits in Fig. 11-14 are connected in a cascade with circuit C1 appearing first in the cascade followed by circuit C2. The chain rule applies to this configuration because the output of circuit C1 is an OP AMP, which can handle the load presented by circuit C2. On the other hand, if the stages are interchanged so that circuit C2 follows circuit C1 in the cascade, then the chain rule does not apply because the input impedance of circuit C1 would then load the output of circuit C2.

EXAMPLE 11-7

Find the voltage transfer function of the cascade connection in Fig. 11-14 for a cascade connection in which circuit C1 is followed by circuit C2.

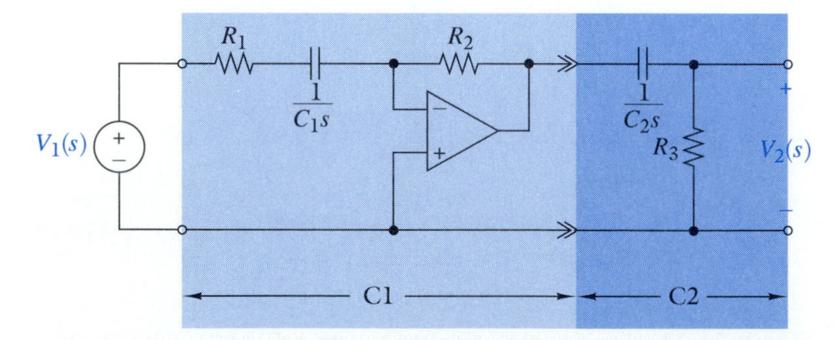

Figure 11-14 Cascade connection demonstrating the chain rule.

SOLUTION:

The chain rule applies to this configuration, since the output of circuit C1 is an OP AMP. Circuit C1 is an inverting OP AMP circuit and its transfer function is found from Eq. (11-7) as

$$T_{V1}(s) = -\frac{Z_2(s)}{Z_1(s)} = -\frac{R_2}{R_1 + 1/C_1 s} = -\frac{R_2 C_1 s}{R_1 C_1 s + 1}$$

Circuit C2 is a voltage divider whose voltage transfer function is given by the voltage-divider rule in Eq. (11-4) as

$$T_{V2}(s) = \frac{Z_2(s)}{Z_2(s) + Z_1(s)} = \frac{R_3}{R_3 + 1/C_2 s} = \frac{R_3 C_2 s}{R_3 C_2 s + 1}$$

Applying the chain rule in Eq. (11-10) yields the overall transfer function as

$$T_V(s) = T_{V1}(s) T_{V2}(s) = -\frac{R_2 C_1 R_3 C_2 s^2}{(R_1 C_1 s + 1)(R_2 C_2 s + 1)}$$

EXERCISE 11-3

Find the voltage transfer function when the output of circuit C2 in Fig. 11-11 serves as the input to circuit C1.

ANSWER:

$$T_V(s) = \frac{(RCs)^2(LCs^2 + 1)}{(RCs + 1)^2(LCs^2 + GLs + 1)}$$

EXERCISE 11-4

Find the voltage transfer function of the cascade connection in Fig. 11-14 for circuit C2 first followed by circuit C1.

ANSWER:

$$T_V(s) = -\frac{R_2 C_1 R_3 C_2 s^2}{R_1 C_1 R_3 C_2 s^2 + (R_1 C_1 + R_3 C_2 + R_3 C_1)s + 1}$$

11-2 PROPERTIES OF NETWORK FUNCTIONS

This section treats basic properties of network functions and their effect on circuit responses. By definition, a network function is the ratio of the zero-state response transform to an input-signal transform. To study the role of network functions in determining circuit responses we write the s-domain input-output relationship as

$$Y(s) = T(s)X(s) \tag{11-14}$$

where $T(s)$ is a circuit network function

$X(s)$ is the input signal transform

$Y(s)$ is the zero-state response or output

(note that Y(s) is a signal *not an admittance*)

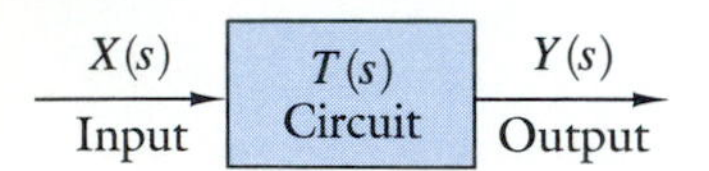

Figure 11-15 Block diagram for an s-domain input-output relationship.

Figure 11-15 shows block diagram representation of the s-domain input-output relationship in Eq. (11-14).

A *network function* is a rational function that can be written as a ratio of polynomials in s:

$$T(s) = \frac{r(s)}{q(s)} = \frac{b_m s^m + b_{m-1}s^{m-1} \ldots + b_1 s + b_0}{a_n s^n + a_{n-1}s^{n-1} \ldots + a_1 s + a_0} \quad (11\text{-}15)$$

The functions $r(s)$ and $q(s)$ are polynomials in s whose coefficients are a_k and b_k, respectively. The coefficients a_k and b_k must be real because they are sums, products, and quotients of real physical parameters like resistance, capacitance, and inductance.

When the polynomials $r(s)$ and $q(s)$ are expressed in factored form the network function is written as

$$T(s) = K\frac{(s - z_1)(s - z_2)\ldots(s - z_m)}{(s - p_1)(s - p_2)\ldots(s - p_n)} \quad (11\text{-}16)$$

where the numerator factors $(s - z_k)$ produce zeros at $s = z_k$, $k = 1, 2, \ldots\ m$, and the denominator factors $(s - p_j)$, produce poles at $s = p_j$, $j = 1, 2, \ldots\ n$. Collectively, the poles and zeros are called *critical frequencies*. The critical frequencies are either real or conjugate complex pairs, since the coefficients of the polynomials $r(s)$ and $q(s)$ are real. The real critical frequencies come from first-order factors of the form $(s + \alpha)$, where α is real. Conjugate complex critical frequencies come from second-order factors of the form

$$\begin{aligned}(s + \alpha + j\beta)(s + \alpha - j\beta) &= s^2 + 2\alpha s + \alpha^2 + \beta^2 \\ &= (s + \alpha)^2 + \beta^2 \quad (11\text{-}17)\end{aligned}$$

where α and β are real.

An alternative way to describe complex critical frequencies is to make use of the terms *undamped natural frequency* (ω_O) and *damping ratio* (ζ) introduced in Chapter 8. Using these terms we write a second-order factor as $s^2 + 2\zeta\omega_O s + \omega_O^2$. The roots of this expression are

$$s_{1,2} = \omega_O\left(-\zeta \pm \sqrt{\zeta^2 - 1}\right)$$

The quantity under the radical depends only on the damping ratio ζ. If $\zeta > 1$ the quantity is positive and the two roots are real and distinct, and the second-order factor becomes the product of two first-order terms. If $\zeta < 1$ the quantity is negative and the two roots are complex conjugates.

A pair of conjugate complex roots can be represented in terms of two parameters; either α and β, or ζ and ω_O. Using Eq. (11-17) we can relate the two sets of parameters, since

$$s^2 + 2\alpha s + \alpha^2 + \beta^2 = s^2 + 2\zeta\omega_O s + \omega_O^2$$

Equating the coefficients of like powers of s yields

$$\omega_O = \sqrt{\alpha^2 + \beta^2} \tag{11-18}$$

and

$$\zeta = \frac{\alpha}{\sqrt{\alpha^2 + \beta^2}} \tag{11-19}$$

These relationships are shown geometrically in Fig. 11-16. The parameters α and β are the rectangular coordinates of the roots. In a sense, the parameters ω_O and ζ are the polar coordinates of the root. The parameter ω_O is the radial distance from the origin to the root. Using the geometry in Fig. 11-16 shows that the damping ratio ζ determines the angle θ.

$$\theta = \cos^{-1}\zeta \tag{11-20}$$

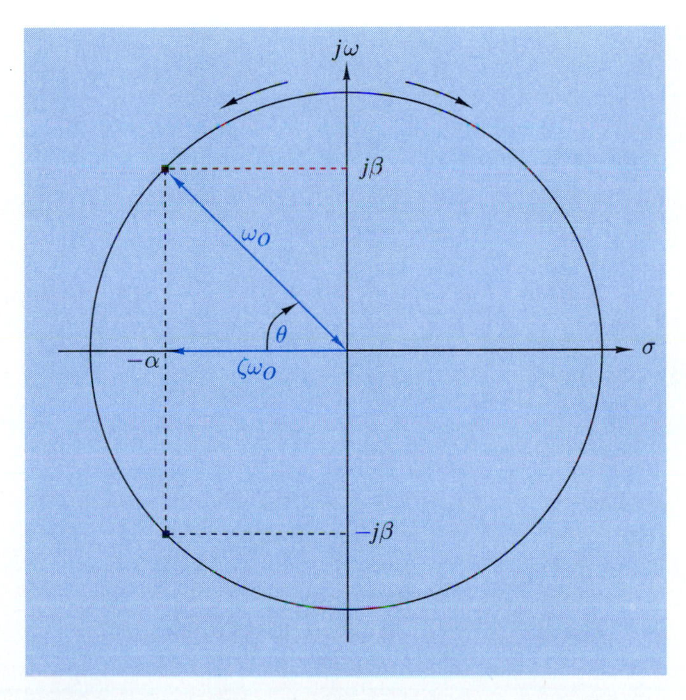

Figure 11-16 *s*-plane geometry relating α and β to ζ and ω_O.

Equation (11-14) points out that the critical frequencies of the response come from either the network function $T(s)$ or the input signal $X(s)$. The critical frequencies of $T(s)$ are *natural poles* and *zeros* because they are determined by circuit parameters and do not depend on the form of the input. The critical frequencies of $X(s)$ are *forced poles* and *zeros* because they are introduced by the input and do not depend on circuit parameters.

EXAMPLE 11-8

In Example 11-3 the transfer function of the circuit in Fig. 11-7 is found to be

$$T_V(s) = \frac{V_2(s)}{V_1(s)} = \frac{(R_1C_1s+1)(R_2C_2s+1)}{R_1C_1R_2C_2s^2 + (R_1C_1 + R_2C_2 + R_1C_2)s + 1}$$

(a) Find the output transform $V_2(s)$ when the input waveform is $v_1(t) = [\cos 2t]u(t)$ and the circuit parameters are related by $R_1C_1 = R_2C_2 = R_1C_2 = 1$.

(b) Locate and classify the critical frequencies of the response $V_2(s)$.

SOLUTION:

(a) The transform of the input waveform is $V_1(s) = s/(s^2+4)$. Using the numerical constraints produces an output transform of

$$V_2(s) = \frac{(s+1)^2}{(s^2+3s+1)}\frac{s}{(s^2+4)}$$

$$= \frac{(s+1)^2}{(s+0.382)(s+2.62)}\frac{s}{(s+j2)(s-j2)}$$

(b) The location and type of the critical frequencies in the response are:

Location	Type
$s=-1$	two real natural zeros
$s=0$	real forced zero
$s=-0.382$	real natural pole
$s=-2.62$	real natural pole
$s=+j2$	imaginary forced pole
$s=-j2$	imaginary forced pole

EXERCISE 11-5

(a) Find the transforms of the zero-state responses $V_2(s)$ of circuit C1 and circuit C2 in Fig. 11-11 when $v_1(t) = \cos t + \sin t$ and $LC = GL = RC = 1$.

(b) Locate and classify the critical frequencies of the responses $V_2(s)$.

ANSWERS:

(a)

Circuit C1

$$V_2(s) = \frac{s^2+1}{(s^2+s+1)}\frac{s+1}{(s^2+1)}$$

Circuit C2

$$V_2(s) = \frac{s^2}{(s+1)^2}\frac{s+1}{(s^2+1)}$$

(b)

Circuit C1

Location	Type
$s=-1$	real forced zero
$s=-0.5+j0.866$	complex natural pole
$s=-0.5-j0.866$	complex natural pole

The natural zeros at $s=\pm j$ cancel the forced poles at $s=\pm j$.

Circuit C2

Location	Type
$s=0$	two real natural zeros
$s=+j$	imaginary forced pole
$s=-j$	imaginary forced pole
$s=-1$	real natural pole

The forced zero at $s=-1$ cancels one of the natural poles at $s=-1$. ■

Impulse Response

An important special case occurs when the input signal is an impulse $x(t)=\delta(t)$. In this case $X(s)=\mathscr{L}\{\delta(t)\}=1$ and the input-output relationship Eq. (11-14) reduces to

$$Y(s) = T(s)\times 1 = T(s)$$

The *impulse response* transform equals the network function and we can treat $T(s)$ as if it were a signal transform. However, to avoid possible confusion between a network function (description of a circuit) and a transform (description of a signal), we denote the impulse response transform as $H(s)$ and the corresponding waveform as $h(t)$. That is,

Impulse Response

Transform	Waveform
$H(s)=T(s)$	$h(t)=\mathscr{L}^{-1}\{H(s)\}$

(11-21)

Since $X(s)=1$ the impulse input introduces no forced poles and the impulse response contains only natural poles. A good deal can be inferred about a circuit from the location of its natural poles. When a pole comes from a first-order factor $(s+\alpha)$, $h(t)$ contains an exponential component of the form

$$\left[ke^{-\alpha t}\right]u(t)$$

where k is the residue at the pole. When poles come from a second-order factor $(s+\alpha)^2+(\beta)^2$ then $H(s)$ has a pair of conjugate complex poles at $s=-\alpha\pm j\beta$. In Chapter 9 we found that a pair of conjugate poles corresponds to a damped cosine waveform of the form

$$\left[2|k|e^{-\alpha t}\cos(\beta t + \angle k)\right]u(t)$$

where k is the residue at the pole at $s = -\alpha + j\beta$.

The term stable suggests that the circuit response tends to return to some equilibrium condition. This simple idea can lead to complications, especially for nonlinear circuits. However, for our purposes we define stability in terms of the asymptotic behavior of the impulse response. A circuit is *stable* if its impulse response decays to zero as time approaches infinity.

The impulse response characterizes the stability of a circuit because it only contains natural poles contributed by the network function. Figure 11-17 shows the waveforms corresponding to complex and real poles in the s plane. Complex poles come in pairs that produce oscillatory waveforms, while real poles produce exponential waveforms. Complex or real poles on the j axis in the s plane produce sustained waveforms that neither increase or decay to zero. Poles in the right half of the s plane produce waveforms that increase without bound, while poles in the left half of the s plane give rise to waveforms that decay to zero.

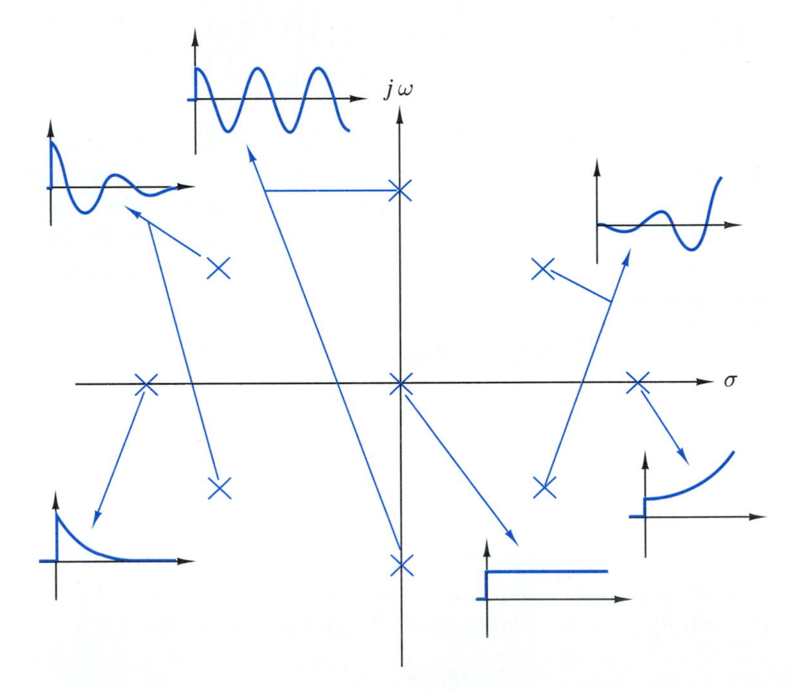

Figure 11-17 Location of natural poles and the corresponding waveforms.

Thus, a circuit is **stable** when *all* of its *natural* poles are located in the left half of the s plane. Note that stability requires all poles to be in the left-half plane. If one or more poles are in the right-half plane or on the j-axis, then the circuit is *unstable*, since the impulse response $h(t)$ does not decay to zero. A first-order factor of the form $(s + \alpha)$ leads to a pole at $s = -\alpha$ so $\alpha > 0$ is required for stability. A second-order factor of the form

$$(s + \alpha)^2 + \beta^2 = s^2 + 2\zeta\omega_O s + \omega_O^2$$

leads to a pair of conjugate poles at

$$s_{1,2} = -\alpha \pm j\beta = -\zeta\omega_O \pm j\omega_O\sqrt{1 - \zeta^2}$$

and again $\alpha > 0$, or equivalently $\zeta > 0$, is required for stability.

A positive damping ratio ($\zeta > 0$) is required for a circuit to be stable. The concepts of damping (energy absorption) and stability are closely related since the circuit losses are what cause the impulse response to decay to zero. Passive circuits containing only resistors, capacitors, and inductors are inherently stable because the resistors eventually dissipate the energy stored in the capacitors and inductors by an impulse input. However, dynamic circuits containing dependent sources or OP AMPs can be stable or unstable. The active elements in such circuits require dc power supplies, which may supply enough additional energy to overcome the inherent damping provided by resistors.

EXAMPLE 11-9

(a) Find the damping ratio and undamped natural frequency of the active RC circuit in Fig. 11-10.

(b) Determine the range of circuit parameters for which the circuit is stable.

SOLUTION:

(a) In Example 11-6 the transfer function of this circuit was found to be

$$T_V(s) = \frac{\mu}{(RCs)^2 + (3 - \mu)RCs + 1}$$

The natural poles of the circuit are roots of the polynomial

$$s^2 + \left(\frac{3 - \mu}{RC}\right)s + \frac{1}{(RC)^2} = 0$$

Using Eqs. (11-18) and (11-19) the damping ratio and natural frequency are

$$\omega_O = \frac{1}{RC} \qquad \text{and} \qquad \zeta = \frac{3 - \mu}{2}$$

(b) When $\mu < 3$ the damping ratio is positive, the poles are in the left half of the s plane, and the circuit is stable. For $\mu = 3$ the natural poles are located on the j-axis at $s = \pm j/RC$ and the impulse response is a sustained sinusoid. For $\mu > 3$ the poles are in the right-half plane and the impulse response is an exponentially growing sinusoid. The circuit is stable for $\mu < 3$.

EXERCISE 11-6

Find ζ and ω_O for circuits C1 and C2 in Fig. 11-11. Are the circuits stable or unstable?

ANSWERS:

Circuit C1: $\omega_O = \dfrac{1}{\sqrt{LC}}$ $\zeta = \dfrac{\sqrt{L/C}}{2R} > 0$ (stable)

Circuit C2: $\omega_O = \dfrac{1}{RC}$ $\zeta = 1 > 0$ (stable)

Step Response

The input for the step response is $x(t) = u(t)$. Under these conditions the input transform is $X(s) = \mathscr{L}\{u(t)\} = 1/s$. Using the s-domain input-output relationship in Eq. (11-14) the resulting *step response* is $T(s)/s$. The step response transform and waveform will be denoted by $G(s)$ and $g(t)$, respectively. That is

Step Response

Transform	Waveform
$G(s) = \dfrac{T(s)}{s}$	$g(t) = \mathscr{L}^{-1}\{G(s)\}$

(11-22)

The step response contains the circuit's natural poles contributed by the network function $T(s)$ and a forced pole at $s = 0$ introduced by the step function input.

We now show that there are relationships between the impulse and step responses. First, combining Eqs. (11-21) and (11-22) gives

$$G(s) = \frac{H(s)}{s}$$

The step response transform is the impulse response transform divided by s. Conversely, the impulse response transform can be obtained from the step response transform by multiplying by s.

The integration property of the Laplace transform tells us that division by s in the s domain corresponds to integration in the time domain. Therefore, we can relate the impulse and step response waveforms by integration in the time domain:

$$g(t) = \int_0^t h(\tau)d\tau \qquad (11\text{-}23)$$

Using the fundamental theorem of calculus the impulse response waveform is expressed in terms of the step response waveform

$$h(t)(=)\frac{dg(t)}{dt} \qquad (11\text{-}24)$$

where the symbol (=) means equal almost everywhere, a condition that excludes those points at which $g(t)$ has a discontinuity. In the time domain, the step response waveform is the integral of the impulse response waveform. Conversely, the impulse response waveform is (almost everywhere) the derivative of the step response waveform.

The key idea is that there are relationships between the network function $T(s)$ and the responses $H(s)$, $h(t)$, $G(s)$, and $g(t)$. If any one of these quantities is known, we can obtain any or all of the other four using relatively simple mathematical operations.

EXAMPLE 11-10

Find $h(t)$ and $g(t)$ of the circuit in Fig. 11-18.

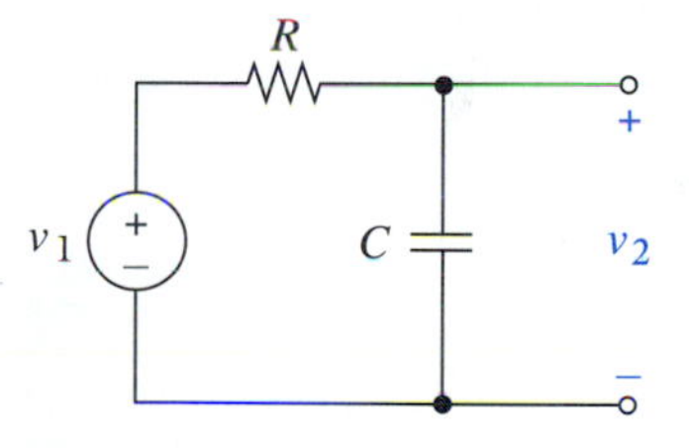

Figure 11-18

SOLUTION:

Using voltage division the transfer function is

$$T(s) = \frac{Z_2}{Z_1 + Z_2} = \frac{1/Cs}{R + 1/Cs} = \frac{1/RC}{s + 1/RC}$$

Since $H(s) = T(s)$ the impulse response waveform is

$$h(t) = \mathscr{L}^{-1}\left(\frac{1/RC}{s + 1/RC}\right) = \frac{1}{RC}e^{-t/RC} \qquad t \geq 0$$

The step response can be found by integrating $h(t)$.

$$g(t) = \int_0^t h(\tau)d\tau = \int_0^t \frac{1}{RC}e^{-\tau/RC}\,d\tau$$
$$= -e^{-\tau/RC}\,|_0^t = (1 - e^{-t/RC})u(t)$$

In the analysis above we first found the circuit transfer function, then using the fact that $H(s) = T(s)$, we found $h(t) = \mathscr{L}^{-1}H(s)$, and finally found $g(t)$ by integrating $h(t)$. An alternative approach is to express $G(s)$ in terms of $T(s)$ using

$$G(s) = \frac{H(s)}{s} = \frac{T(s)}{s} = \frac{1/RC}{s(s + 1/RC)} = \frac{1}{s} - \frac{1}{s + 1/RC}$$

and to calculate $g(t)$ from $G(s)$ using

$$g(t) = \mathscr{L}^{-1}\left(\frac{1}{s} - \frac{1}{s+1/RC}\right) = (1 - e^{-t/RC})u(t)$$

and then differentiating $g(t)$ to obtain $h(t)$.

$$\begin{aligned} h(t) &= \frac{dg(t)}{dt} = \frac{d}{dt}\left[\left(1 - e^{-t/RC}\right)u(t)\right] = \frac{d}{dt}\left[u(t) - e^{-t/RC}u(t)\right] \\ &= \delta(t) - \left(\frac{-1}{RC}\right)e^{-t/RC}u(t) - e^{-t/RC}\,\delta(t) \\ &= \delta(t) + \left(\frac{1}{RC}\right)e^{-t/RC}u(t) - (1)\delta(t) \\ &= \frac{1}{RC}e^{-t/RC}u(t) \end{aligned}$$

The point is that we can find $g(t)$ from $h(t)$ or $h(t)$ from $g(t)$. [Note that the sampling property was used to evaluate $e^{-t/RC}\delta(t)$]:

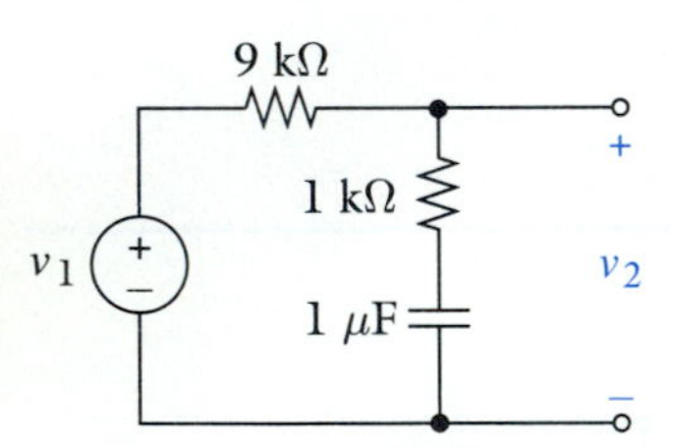

Figure 11-19

EXERCISE 11-7

Find $H(s)$, $h(t)$, $G(s)$, and $g(t)$ of the circuit in Fig. 11-19.

ANSWERS:

$$H(s) = \frac{1}{10}\frac{s+1000}{s+100} \qquad h(t) = \frac{\delta(t)}{10} + \left[90e^{-100t}\right]u(t)$$

$$G(s) = \frac{1}{10}\frac{s+1000}{s(s+100)} \qquad g(t) = \left[1 - 0.9e^{-1000t}\right]u(t)$$

EXERCISE 11-8

The step response of a linear circuit is $g(t) = 5[e^{-1000t}\sin(2000t)]u(t)$. Find the circuit transfer function $T(s)$.

ANSWER:

$$T(s) = \frac{10^4 s}{s^2 + 2000s + 5 \times 10^6}$$

Impulse Response and Convolution

In the s domain the circuit input-output relationship in Eq. (11-14) involves the product of transforms $T(s)X(s)$. In Chapter 9 we found that the inverse transform of such a product is a convolution integral. Using the fact that $h(t) = \mathscr{L}^{-1}\{T(s)\}$, we can write the time domain version of the input-output relationship in terms of the following convolution integrals:

$$\begin{aligned} y(t) = \mathscr{L}^{-1}\{T(s)X(s)\} &= \int_0^t h(\tau)x(t-\tau)d\tau \\ &= \int_0^t x(\tau)h(t-\tau)d\tau \end{aligned} \quad (11\text{-}25)$$

where τ is a dummy variable of integration. As we noted in Chapter 9, there are two equivalent forms of the convolution integral. Either form can be used because convolution is commutative, that is, $h(t)*x(t) = x(t)*h(t)$.

The *convolution integrals* in Eq. (11-25) provide t-domain relationships between the output waveform $y(t)$, the circuit impulse response $h(t)$, and the input waveform $x(t)$. Convolution applies to linear time-invariant circuits in the zero-state. When the lower limit is zero, as in Eq. (11-25), the input is assumed to be causal, i.e., $x(t) = 0$ for $t < 0$. When the upper limit is time it is assumed that $h(t)$ is also causal.

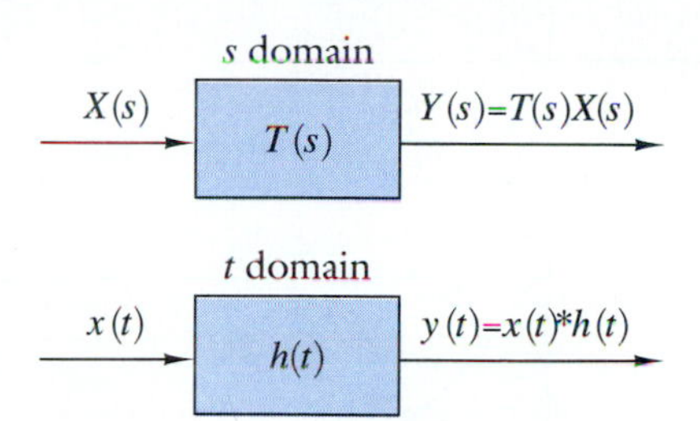

Figure 11-20 Block diagrams for s-domain and t-domain input-output relationships.

Figure 11-20 shows the parallelism between the input-output relationships in the s domain and t domain. This parallelism shows that we can determine the response $y(t)$ entirely in the t domain using the impulse response $h(t)$ and the input $x(t)$ in the convolution integral. However, it may be difficult to reduce the integral to closed form.

Convolution is a useful computational tool when $h(t)$ or $x(t)$ are obtained from experimental data that cannot be easily converted into the s domain. In discrete-time systems analog signals are characterized at equally spaced intervals of time. For such systems, there are efficient computer algorithms for calculating the discrete-time equivalent of the convolution integral. These applications are beyond the scope of this book and are treated in books aimed at subsequent courses in signal processing.[2]

Convolution usually is not the best computational tool for continuous-time circuits and signals. Given a linear circuit and an analytical expression for its input, it is generally easier to find the response in the s domain. With the circuit in the s domain the impulse response $H(s) = T(s)$ can be found using only algebraic methods. A good deal can then be inferred from the circuit's natural poles about the form of the impulse and step responses without formally performing the inverse transformation. Much of the viewpoint and terminology of electrical engineering uses the s-domain approach.

EXAMPLE 11-11

Use the convolution integral to find the output of the circuit in Fig. 11-18 when the input is the ramp $v_1(t) = tu(t)$.

[2]For example see, Alan V. Oppenheim and Alan S. Willsky, *Signals and Systems*, Prentice Hall, Englewood Cliffs, NJ, 1983.

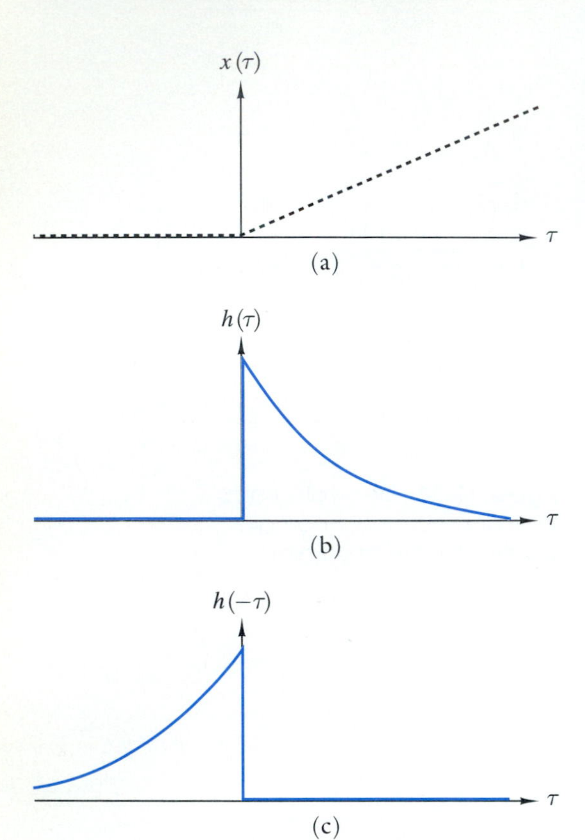

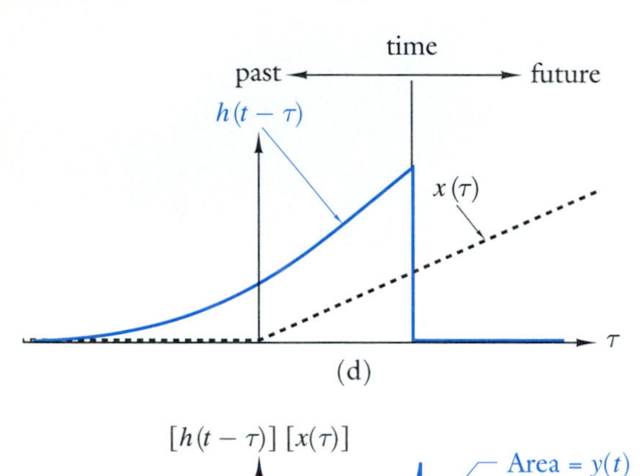

Figure 11-21 Graphical interpretation of convolution.

SOLUTION:

In Example 11-10 the impulse response of this circuit was found to be

$$h(t) = \left[\frac{e^{-t/RC}}{RC}\right]u(t)$$

For this $h(t)$ and input $v_1(t) = x(t) = tu(t)$ the second version of the convolution integral in Eq. (11-25) yields the output $y(t)$ in the form

$$\begin{aligned} y(t) &= \int_0^t \tau \frac{e^{-(t-\tau)/RC}}{RC} d\tau = \frac{e^{-t/RC}}{RC}\int_0^t \tau e^{\tau/RC} d\tau \\ &= \frac{e^{-t/RC}}{RC}\left[RCe^{\tau/RC}(\tau - RC)\right]_0^t \\ &= t - RC(1 - e^{-t/RC}) \qquad t \geq 0 \end{aligned}$$

Convolution supplies both the forced and natural components of the output.

EXERCISE 11-9

Use convolution to find the output of the circuit in Fig. 11-18 when the input is $v_1(t) = [V_0 e^{-\alpha t}]u(t)$. Assume $\alpha RC \neq 1$.

ANSWER:

$$v_2(t) = \frac{V_0}{1 - RC\alpha}\left[e^{-\alpha t} - e^{-tRC}\right]u(t)$$

Geometric Interpretation of Convolution

The waveforms in Example 11-11 can be used to illustrate the geometric interpretation of convolution. The process begins with the input and impulse response plotted against the integration variable τ as shown in Fig. 11-21(a) and (b). Forming $h(-\tau)$ reflects or folds the impulse response about the $\tau = 0$ axis as shown in Fig. 11-21(c). Forming $h(t - \tau)$ advances the reflected waveform by t seconds as shown in Fig. 11-21(d). The product $h(t - \tau)x(\tau)$ is shown in Fig. 11-21(e). The area under this product is the value of the output at time t. Increasing time t drags the function $h(t-\tau)$ further to the right yielding a different output $y(t)$ because the area under the product $h(t - \tau)x(\tau)$ changes.

Geometrically we visualize convolution as a process that reflects the impulse response across the origin and then drags it through the input $x(\tau)$ as t increases. The impulse response can be thought of as a weighting function. The output at any time t is a function of the input at that instant and all previous values of the input. The importance of the present and previous values of the input is determined by the impulse response.

For example, the impulse response in Fig. 11-21 is an exponential with a time constant of $T_C = RC$. Values of $x(\tau)$ for which $t - \tau > 5T_C$ will receive almost no weight in determining $y(t)$, because $h(t - \tau)$ is negligibly small in that range. The duration of the impulse response influences how much previous inputs are weighted in determining the present output. In other words, dynamic circuits have a memory time and tend to forget inputs that occurred in the distant past. The duration of the impulse response measures the circuit's memory time.

◇ 11-3 TRANSFER FUNCTION DESIGN

Finding the network function of a given circuit is an s-domain *analysis* problem. An s-domain *synthesis* problem involves finding a circuit that realizes a given network function. For linear circuits an analysis problem always has a unique solution. In contrast, a synthesis problem may have many solutions because different circuits can have the same network function. A transfer-function design problem involves synthesizing several circuits that realize a given function and evaluating the alternative designs, using criteria such as element spread, cost, and power consumption.

In this section we develop methods of synthesizing first- and second-order transfer functions using the circuits shown in Fig. 11-22. In an analysis problem we start with a diagram of one of these circuit configurations, transform the circuit into the s domain, determine the impedances $Z_1(s)$ and $Z_2(s)$, and derive the transfer function using the appropriate relationship in Fig. 11-22. The analysis process proceeds from a circuit diagram to a network function and always has a unique solution.

In a synthesis problem we reverse the analysis process. We now start with a transfer function, select one of the configurations in Fig. 11-22, assign values to the impedances $Z_1(s)$ and $Z_2(s)$, and then draw the circuit diagram as the final step. In some cases it may not be possible to realize the given transfer function using the selected circuit configuration. In other cases, restrictions on the transfer function may be needed to ensure that $Z_1(s)$ and $Z_2(s)$ are physically realizable. In any event, the synthesis process proceeds from a network function to a circuit diagram and may have many solutions or possibly no solutions.

Circuit	Transfer Function
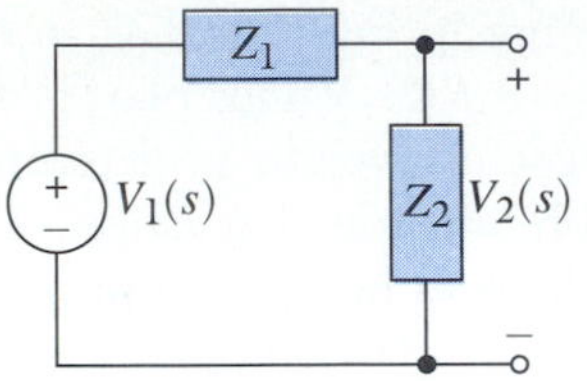 Voltage divider	$T_V(s) = \dfrac{Z_2(s)}{Z_1(s)+Z_2(s)}$
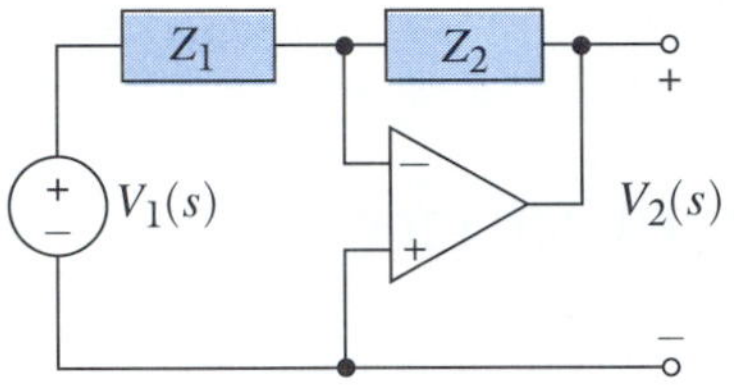 Inverting OP AMP	$T_V(s) = -\dfrac{Z_2(s)}{Z_1(s)}$
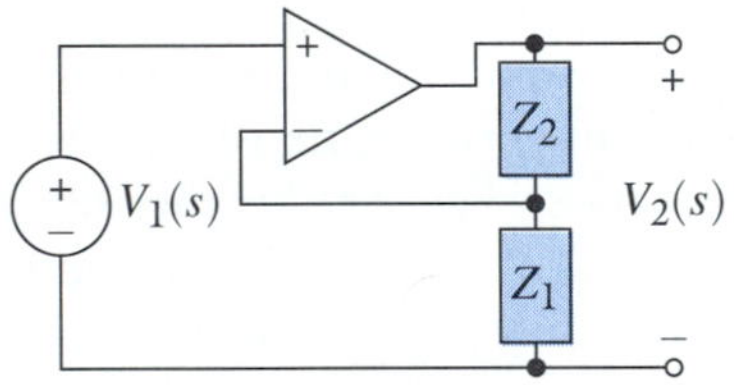 Noninverting OP AMP	$T_V(s) = \dfrac{Z_1(s)+Z_2(s)}{Z_1(s)}$

Figure 11-22 Voltage divider and OP AMP circuit building blocks.

First-Order Voltage-Divider Circuit Design

We begin our study of transfer-function design by developing a voltage-divider realization of a first-order transfer function of the form $K/(s+\alpha)$. The impedances $Z_1(s)$ and $Z_2(s)$ are related to the given transfer function using the voltage-divider relationship in Fig. 11-22:

$$T_V(s) = \frac{K}{s+\alpha} = \frac{Z_2(s)}{Z_1(s)+Z_2(s)} \tag{11-26}$$

To obtain a circuit realization we must assign part of the given

$T_V(s)$ to $Z_2(s)$ and the remainder to $Z_1(s)$. There are many possible realizations of $Z_1(s)$ and $Z_2(s)$ because there is no unique way to make this assignment. For example, simply equating the numerators and denominators in Eq. (11-26) yields

$$Z_2(s) = K \quad \text{and} \quad Z_1(s) = s + \alpha - Z_2(s) = s + \alpha - K \qquad (11\text{-}27)$$

Inspecting this result we see that $Z_2(s)$ is realizable as a resistance ($R_2 = K\ \Omega$) and $Z_1(s)$ as an inductance ($L_1 = 1$ H) in series with a resistance [$R_1 = (\alpha - K)\ \Omega$]. The resulting circuit diagram is shown in Fig. 11-23(a). For $K = \alpha$ the resistance R_1 can be replaced by a short circuit because its resistance is zero. A gain restriction $K \leq \alpha$ is necessary because a negative R_1 is not physically realizable as a single component.

An alternative synthesis approach involves factoring s out of the denominator of the given transfer function. In this case Eq. (11-26) is rewritten in the form

$$T_V(S) = \frac{K/s}{1 + \alpha/s} = \frac{Z_2(s)}{Z_1(s) + Z_2(s)} \qquad (11\text{-}28)$$

Equating numerators and denominators yields the branch impedances

$$Z_2(s) = \frac{K}{s} \quad \text{and} \quad Z_1(s) = 1 + \frac{\alpha}{s} - Z_2(s) = 1 + \frac{\alpha - K}{s} \qquad (11\text{-}29)$$

In this case we see that $Z_2(s)$ is realizable as a capacitance ($C_2 = 1/K$ F) and $Z_1(s)$ as a resistance ($R_1 = 1\ \Omega$) in series with a capacitance [$C_1 = 1/(\alpha - K)$ F]. The resulting circuit diagram is shown in Fig. 11-23(b). For $K = \alpha$ the capacitance C_1 can be replaced by a short circuit because its capacitance is infinite. A gain restriction $K \leq \alpha$ is required to keep C_1 from being negative.

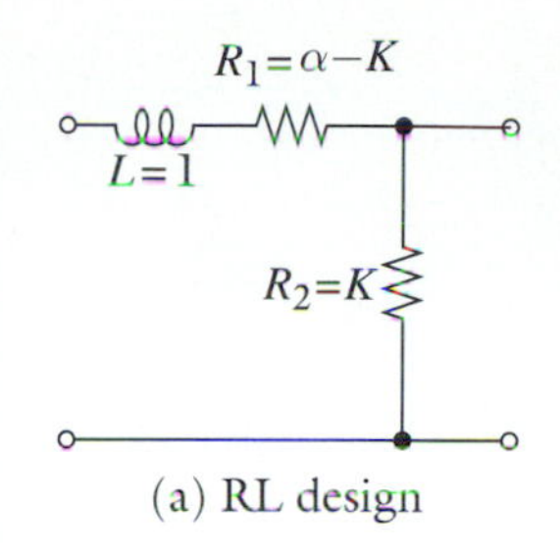

(a) RL design

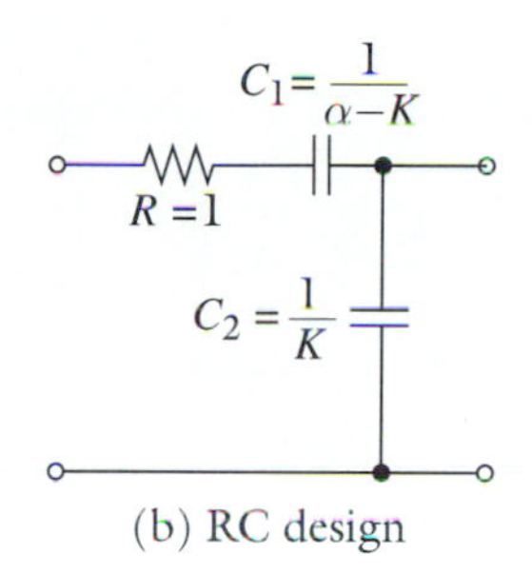

(b) RC design

Figure 11-23 Circuit realizations of $T(s) = K/(s + \alpha)$ for $K \leq \alpha$.

As a second design example consider a voltage-divider realization of the transfer function $Ks/(s + \alpha)$. We can find two voltage divider realizations by writing the given transfer function following two ways:

$$T(s) = \frac{Ks}{s + \alpha} = \frac{Z_2(s)}{Z_1(s) + Z_2(s)} \qquad (11\text{-}30a)$$

$$T(s) = \frac{K}{1 + \alpha/s} = \frac{Z_2(s)}{Z_1(s) + Z_2(s)} \qquad (11\text{-}30b)$$

Equation (11-30a) does not alter the form of the given transfer function while Eq. (11-30b) factors s out of the numerator and denominator. Equating the numerators and denominators in Eqs. (11-30a) and (11-30b) yields two possible impedance assignments:

Using Eq. (11-30a): $\quad Z_2 = Ks$

and $\quad Z_1 = s + \alpha - Z_2 = (1 - K)s + \alpha \qquad (11\text{-}31a)$

$$\text{Using Eq. (11-30b):} \qquad Z_2 = K$$

$$\text{and} \qquad Z_1 = 1 + \frac{\alpha}{s} - Z_2 = (1 - K) + \frac{\alpha}{s} \tag{11-31b}$$

The assignment in Eq. (11-31a) yields $Z_2(s)$ as an inductance ($L_2 = K$ H) and $Z_1(s)$ as an inductance [$L_1 = (1 - K)$ H] in series with a resistance ($R_1 = \alpha\ \Omega$). The assignment in Eq. (11-31b) yields $Z_2(s)$ as a resistance ($R_2 = K$) and $Z_1(s)$ as a resistance [$R_1 = (1 - K)\ \Omega$] in series with a capacitance ($C_1 = 1/\alpha$ F). The two realizations are shown in Fig. 11-24. Both realizations require $K \leq 1$ for the branch impedances to be realizable and both simplify when $K = 1$.

$R = \alpha$

$L_1 = 1 - K$

$L_2 = K$

(a) RL design

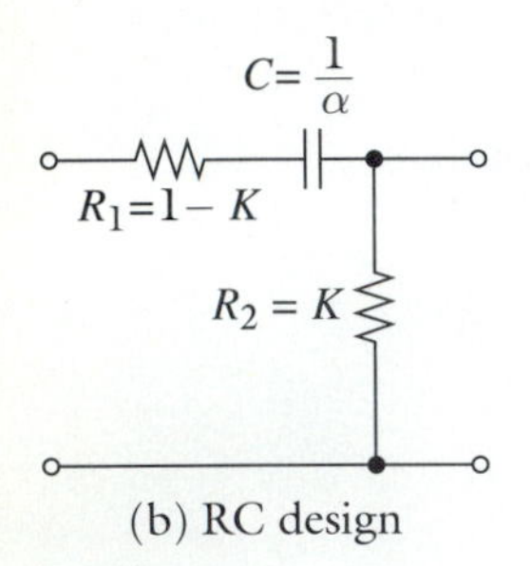

(b) RC design

Figure 11-24 Circuit realizations of $T(s) = Ks/(s + \alpha)$ for $K \leq 1$.

Voltage-divider and OP AMP Cascade Circuit Design

The examples in Fig. 11-23 and 11-24 illustrate an important feature of voltage-divider realizations. In general we can write a transfer function as a quotient of polynomials $T(s) = r(s)/q(s)$. A voltage-divider realization requires the impedances $Z_2(s) = r(s)$ and $Z_1(s) = q(s) - r(s)$ to be physically realizable. Physical realizability usually places a limitation on the gain K that can be achieved by a voltage-divider circuit. This gain limitation can be overcome by using an OP AMP circuit in cascade with a divider circuit.

For example, a voltage-divider realization of the transfer function in Eq. (11-26) requires $K \leq \alpha$. When $K > \alpha$, Eq. (11-27) shows that the circuit is not realizable, since $Z_2(s) = q(s) - r(s) = s + \alpha - K$ requires a negative resistance. However, the given transfer function can be written as a two-stage product

$$T_V(s) = \frac{K}{s + \alpha} = \underbrace{\left[\frac{K}{\alpha}\right]}_{\substack{1^{st} \\ \text{stage}}} \underbrace{\left[\frac{\alpha}{s + \alpha}\right]}_{\substack{2^{nd} \\ \text{stage}}} \tag{11-32}$$

When $K > \alpha$ the first stage is a positive gain greater than unity, which can be realized using the noninverting OP AMP configuration in Fig. 11-22. Using the transfer function from Fig. 11-22 for this circuit yields

$$\frac{Z_1(s) + Z_2(s)}{Z_1(s)} = \frac{K}{\alpha} \tag{11-33}$$

Arbitrarily choosing $Z_1(s) = R_1 = 1\ \Omega$, we solve for $Z_2(s)$ and find $R_2 = K/\alpha - 1\ \Omega$. The second stage in Eq. (11-32) is realizable as a voltage divider because $q(s) - r(s) = s$ is realizable as an inductance.

Alternatively, factoring s out of the second-stage transfer function yields a different RC divider realization:

$$\frac{\alpha/s}{1+\alpha/s} = \frac{Z_2(s)}{Z_1(s)+Z_2(s)} \tag{11-34}$$

Equating numerators and denominators yields $Z_2(s) = \alpha/s$ and $Z_1(s) = 1$. Figure 11-25 shows a cascade connection of a non-inverting first stage and the RC divider second stage. The chain rule applies to this circuit, since the first stage has an OP AMP output. The cascade circuit in Fig. 11-25 realizes the first-order transfer function $K/(s+\alpha)$ for $K > \alpha$, a gain requirement that cannot be met by the divider circuit alone.

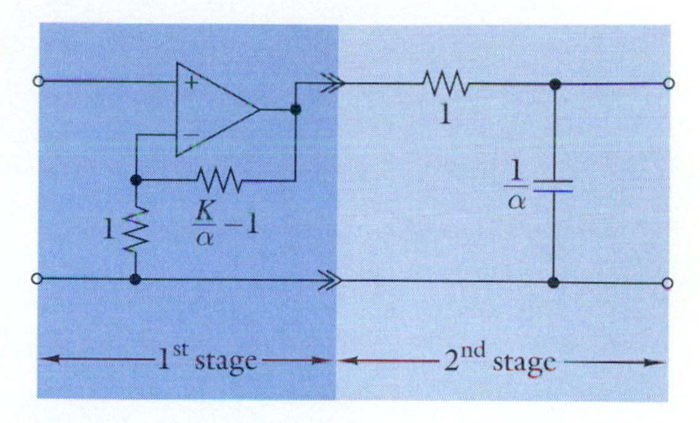

Figure 11-25 Circuit realization of $T(s) = K/(s+\alpha)$ for $K > \alpha$.

❖ Design Example

EXAMPLE 11-12

Design a circuit to realize the transfer function below using only resistors, capacitors and OP AMPs:

$$T_V(s) = \frac{3000s}{(s+1000)(s+4000)}$$

SOLUTION:

The given transfer function can be written as a three-stage product:

$$T_V(s) = \underbrace{\left[\frac{K_1}{s+1000}\right]}_{1^{st}\ \text{stage}} \underbrace{[K_2]}_{2^{nd}\ \text{stage}} \underbrace{\left[\frac{K_3 s}{s+4000}\right]}_{3^{rd}\ \text{stage}}$$

where the stage gains K_1, K_2, and K_3 have yet to be selected. Factoring s out of the denominator of the first-stage transfer function leads to an RC divider realization:

$$\frac{K_1/s}{1+1000/s} = \frac{Z_2(s)}{Z_1(s)+Z_2(s)}$$

Equating numerators and denominators yields

$$Z_2(s) = K_1/s \quad \text{and} \quad Z_1(s) = 1 + (K_1 - 1000)/s.$$

The first stage $Z_1(s)$ is simpler when we select $K_1 = 1000$. Factoring s out of the denominator of the third-stage transfer function leads to an RC divider realization:

$$\frac{K_3}{1+4000/s} = \frac{Z_2(s)}{Z_1(s)+Z_2(s)}$$

Equating numerators and denominators yields

$$Z_2(s) = K_3 \qquad \text{and} \qquad Z_1(s) = 1 - K_3 + 4000/s.$$

The third stage $Z_1(s)$ is simpler when we select $K_3 = 1$. The stage gains must meet the constraint $K_1K_2K_3 = 3000$, since the overall gain of the given transfer function is 3000. We have selected $K_1 = 1000$ and $K_3 = 1$, which means $K_2 = 3$. The second stage must have a positive gain greater than one and can be realized using a noninverting amplifier with $K_2 = (R_1 + R_2)/R_1 = 3$. Selecting $R_1 = 1\ \Omega$ requires that $R_2 = 2\ \Omega$.

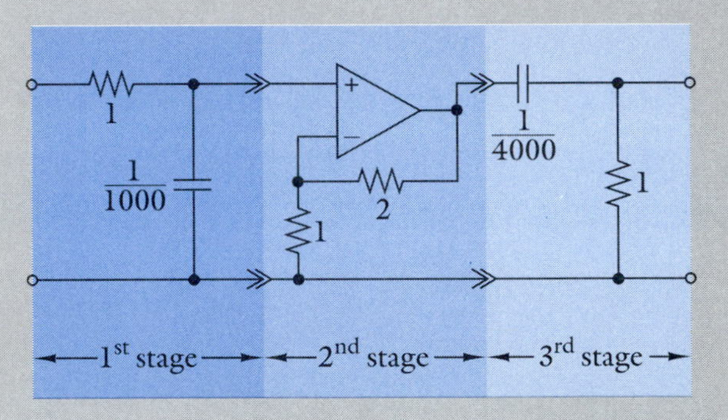

Figure 11-26

Figure 11-26 shows the three stages connected in cascade. The chain rule applies to this cascade connection because the OP AMP in the second stage isolates the RC voltage-divider circuits in the first and third stages. The circuit in Fig. 11-26 realizes the given transfer function but is not a realistic design because the values of resistance and capacitance are impractical. For this reason we call the circuit a *prototype design*. We will shortly discuss how to scale a prototype design to obtain a circuit with practical element values.

Inverting OP AMP Circuit Design

The inverting OP AMP circuit is more versatile than the voltage divider because it places fewer restrictions on the form of the transfer function. There are at least four different inverting OP AMP circuit designs for a general first-order transfer function of the form

$$T_V(s) = -K\frac{s+\gamma}{s+\alpha}$$

The first realization results from equating the transfer function $T_V(s)$ to the inverting OP AMP relationship in Fig. 11-22:

$$T_V(s) = -\frac{Ks + K\gamma}{s + \alpha} = -\frac{Z_2(s)}{Z_1(s)} \tag{11-35}$$

Equating numerators and denominators yields impedance $Z_2(s) = Ks + K\gamma$ as an inductance ($L_2 = K$ H) in series with a resistance ($R_2 = K\gamma\ \Omega$), and $Z_1(s) = s + \alpha$ as an inductance ($L_1 = 1$ H) in series with a resistance ($R_1 = \alpha\ \Omega$). These impedance identifications produce the RL circuit in Fig. 11-27(a).

A second inverting OP AMP realization is obtained by equating $Z_2(s)$ in Eq. (11-35) to the reciprocal of the denominator, and equating $Z_1(s)$ to the reciprocal of the numerator. This assignment yields the impedances $Z_2(s) = 1/(s+\alpha)$ as a capacitance ($C_2 = 1$ F) in parallel with a resistance ($R_2 = 1/\alpha\ \Omega$), and $Z_1(s) = 1/(Ks + K\gamma)$ as a capacitance ($C_1 = K$ F) in parallel with a resistance ($R_1 = 1/K\gamma\ \Omega$). These impedance identifications produce the RC circuit in Fig. 11-27(b).

Two more inverting realizations are obtained by factoring s out of the denominator of Eq. (11-35):

$$-\frac{K + K\gamma/s}{1 + \alpha/s} = -\frac{Z_2(s)}{Z_1(s)} \tag{11-36}$$

Equating numerators and denominators yields $Z_2(s) = K + K\gamma/s$ and $Z_1(s) = 1 + \alpha/s$. Both of these impedances are realizable as a resistance in series with a capacitance and lead to the RC circuit in Fig. 11-27(c). When we equate $Z_2(s)$ in Eq. (11-36) to the reciprocal of the denominator and $Z_1(s)$ to the reciprocal of the numerator we find

$$Z_2(s) = \frac{1}{1 + \alpha/s} \quad \text{and} \quad Z_1(s) = \frac{1}{K + K\gamma/s} \tag{11-37}$$

Both of these impedances are realizable as an inductor in parallel with a resistor and lead to the RL circuit in Fig. 11-27(d).

The four circuits in Fig. 11-27 show that there are several ways to partition a transfer function into realizable input and feedback impedances for an inverting OP AMP circuit. For a first-order circuit the only limitation is that the parameters K, γ, and α must be positive. These conditions are much less restrictive than the gain restrictions for the voltage-divider circuit.

Because it has fewer restrictions, it is often easier to realize transfer functions using the inverting OP AMP circuit. To use inverting circuits the given transfer function must require an inversion or be realized using an even number of inverting stages. In some cases the sign in front of the transfer function is immaterial and the required transfer function is specified as $\pm T_V(s)$.

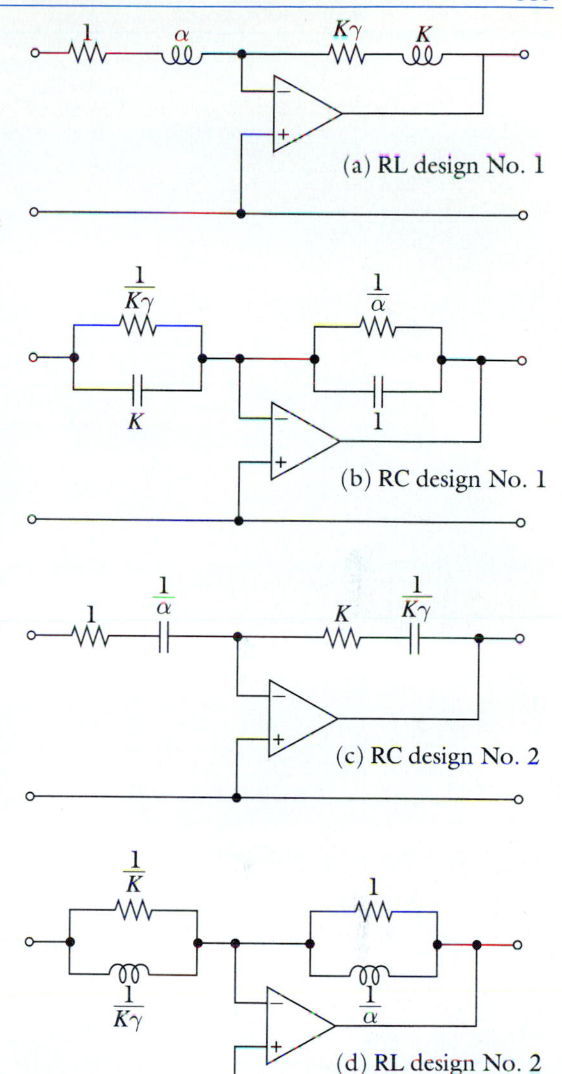

Figure 11-27 Inverting OP AMP circuit realizations of $T(s) = K(s + \gamma)/(s + \alpha)$.

Design Example

EXAMPLE 11-13

Design a circuit to realize the transfer function given in Example 11-12 using inverting OP AMP circuits.

SOLUTION:

The given transfer function can be expressed as the product of two inverting transfer functions.

$$T_V(s) = \frac{3000s}{(s+1000)(s+4000)} = \underbrace{\left[-\frac{K_1}{s+1000}\right]}_{1^{st}\text{ stage}}\underbrace{\left[-\frac{K_2 s}{s+4000}\right]}_{2^{nd}\text{ stage}}$$

where the stage gains K_1 and K_2 have yet to be selected. The first stage can be realized in an inverting OP AMP circuit, since

$$-\frac{K_1}{s+1000} = -\frac{K_1/1000}{1+s/1000} = -\frac{Z_2(s)}{Z_1(s)}$$

Equating the $Z_2(s)$ to the reciprocal of the denominator and $Z_1(s)$ to the reciprocal of the numerator yields

$$Z_2 = \frac{1}{1+s/1000} \quad \text{and} \quad Z_1 = 1000/K_1$$

The impedance $Z_2(s)$ is realizable as a capacitance ($C_2 = 1/1000$ F) in parallel with a resistance ($R_2 = 1\ \Omega$) and $Z_1(s)$ as a resistance ($R_1 = 1000/K_1\ \Omega$). We select $K_1 = 1000$, so that the two resistance in the first stage are equal. Since the overall gain requires $K_1K_2 = 3000$ this means that $K_2 = 3$. The second-stage transfer function can also be produced using an inverting OP AMP circuit:

$$-\frac{3s}{s+4000} = -\frac{3}{1+4000/s} = -\frac{Z_2(s)}{Z_1(s)}$$

Equating numerators and denominators yields $Z_2(s) = R_2 = 3$ and $Z_1(s) = R_1 + 1/C_1s = 1 + 4000/s$.

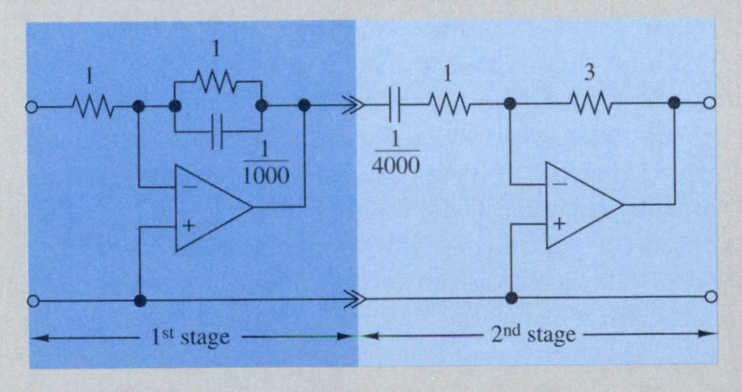

Figure 11-28

Figure 11-28 shows the cascade connection of the two RC OP AMP circuits that realize each stage. The overall transfer function is noninverting because the cascade uses an even number of inverting stages. The chain rule applies here, since the first stage has an OP AMP output. The circuit in Fig. 11-28 is a prototype design because the values of resistance and capacitance are impractical.

Magnitude Scaling

The circuits obtained in Examples 11-12 and 11-13 are called prototype designs because the element values are outside of practical ranges. The allowable ranges depend on the fabrication technology used to construct the circuits. For example, monolithic integrated circuit technology limits capacitances to a few hundred picofarads. An OP AMP circuit should have a feedback resistance greater than roughly 10 kΩ to keep the output current demand within the capabilities of general purpose OP AMP devices. Other technologies and applications place different constraints on element values.

There are no hard and fast rules here, but roughly speaking a circuit is probably realizable by some means if its passive element values fall in the following ranges:

Capacitors	1 pF to 100 μF
Inductors	10 μH to 100 mH
Resistors	10 Ω to 100 MΩ

The important idea here is that circuit designs like Fig. 11-28 are impractical because 1-Ω resistors and 1-mF capacitors are unrealistic.

It is possible to scale the magnitude of circuit impedances so that the element values fall into practical ranges. The key is to scale the element values in a way that does not change the transfer function of the circuit. Multiplying the numerator and denominator of the transfer function of a voltage-divider circuit by a *scale factor* k_m yields

$$T_V(s) = \frac{k_m}{k_m}\frac{Z_2(s)}{Z_1(s)+Z_2(s)} = \frac{k_m Z_2(s)}{k_m Z_1(s) + k_m Z_2(s)} \qquad (11\text{-}38)$$

Clearly this modification does not change the transfer function, but scales each impedance by a factor of k_m and changes each element values in the following way:

$$R_{after} = k_m R_{before} \quad L_{after} = k_m L_{before} \quad C_{after} = \frac{C_{before}}{k_m} \qquad (11\text{-}39)$$

Equation (11-39) was derived using the transfer function of a voltage-divider circuit. It is easy to show that we would reach the same conclusion if we had used the transfer functions of inverting or noninverting OP AMP circuits in Fig. 11-22.

In general, a circuit is magnitude scaled by multiplying all resistances, multiplying all inductances, and dividing all capacitances by a scale factor k_m. The scale factor must be positive, but can be greater than or less than one. Different scale factors can be used for each stage of a cascade design, but only one scale factor can be used for each stage. These scaling operations do not change the voltage transfer function realized by the circuit.

Our design strategy is to first create a prototype circuit whose element values may be unrealistically large or small. We then magnitude scale the prototype to obtain a design with practical element values. Sometimes there may be no scale factor that brings the prototype element values into a practical range. When this happens, we must find an alternative realization because an unscalable prototype is not a workable design candidate.

EXAMPLE 11-14

Magnitude scale the circuit in Fig. 11-28 so all resistances are at least 10 kΩ and all capacitances are less than 1 μF.

SOLUTION:

The resistance constraint requires $k_m R \geq 10^4$ Ω. The smallest resistance in the prototype circuit is 1 Ω; therefore the resistance condition requires $k_m \geq 10^4$. The capacitance constraint requires $C/k_m \leq 10^{-6}$ F. The largest capacitance in the prototype is 10^{-3} F; therefore, the capacitance condition requires $k_m \geq 10^3$. The resistance condition on k_m dominates the two constraints. Selecting $k_m = 10^4$ produces the scaled design in Fig. 11-29. This circuit realizes the same transfer function as the prototype in Fig. 11-28, but uses practical element values.

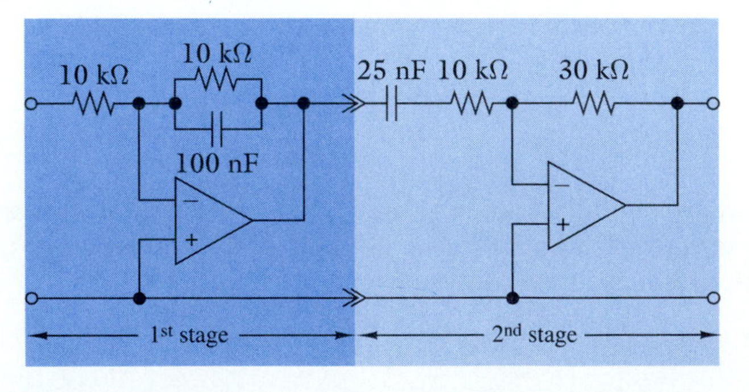

Figure 11-29

EXERCISE 11-10

Select a magnitude scale factor for each stage in Fig. 11-26 so that both capacitances are 10 nF and all resistances are greater than 10 kΩ.

ANSWER:
$k_m = 10^5$ for the first stage, $k_m = 10^4$ for the second stage, and $k_m = 10^5/4$ for the third stage.

Second-Order Circuit Design

The voltage divider and OP AMP circuits in Fig. 11-22 can also be used to realize second-order transfer functions. For example, the transfer function

$$T_V(s) = \frac{K}{s^2 + 2\zeta\omega_O s + \omega_O^2}$$

can be realized by factoring s out of the denominator and equating the result to the voltage-divider relationship in Fig. 11-22:

$$T_V(s) = \frac{K/s}{s + 2\zeta\omega_O + \omega_O^2/s} = \frac{Z_2(s)}{Z_1(s) + Z_2(s)} \quad (11\text{-}40)$$

Equating numerators and denominators yields

$$Z_2(s) = \frac{K}{s} \quad \text{and} \quad Z_1(s) = s + 2\zeta\omega_O + \frac{\omega_O^2 - K}{s} \quad (11\text{-}41)$$

The impedance $Z_2(s)$ is realizable as a capacitance ($C_2 = 1/K$ F) and $Z_1(s)$ as a series connection of an inductance ($L_1 = 1$ H), resistance ($R_1 = 2\zeta\omega_O$ Ω) and capacitance [$C_1 = 1/(\omega_O^2 - K)$ F]. The resulting voltage-divider circuit is shown in Fig. 11-30(a). The impedances in this circuit are physically realizable when $K \leq \omega_O^2$. Note that the resistance controls the damping ratio ζ because it is the element that dissipates energy in the circuit.

When $K > \omega_O^2$ we can partition the transfer function into a two-stage cascade of the form

$$T_V(s) = \underbrace{\left[\frac{K}{\omega_O^2}\right]}_{\substack{1^{st} \\ \text{stage}}} \underbrace{\left[\frac{\omega_O^2/s}{s + 2\zeta\omega_O + \omega_O^2/s}\right]}_{\substack{2^{nd} \\ \text{stage}}} \quad (11\text{-}42)$$

The first stage requires a positive gain greater than unity and can be realized using a noninverting OP AMP circuit. The second stage can be realized as a voltage divider with $Z_2(s) = \omega_O^2/s$ and $Z_1(s) = s + 2\zeta\omega_O$. The resulting cascade circuit is shown in Fig. 11-30(b).

An inverting OP AMP design of the second-order function also is possible provided the given transfer function requires inversion or its sign is not specified. Assuming $T_V = -K/(s^2 + 2\zeta\omega_O^2 s + \omega_O^2)$, we factor s out of the denominator and equate the

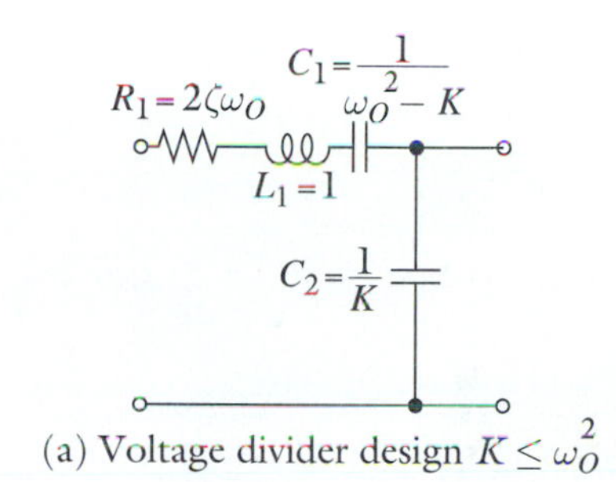

(a) Voltage divider design $K \leq \omega_O^2$

(b) Cascade design $K > \omega_O^2$

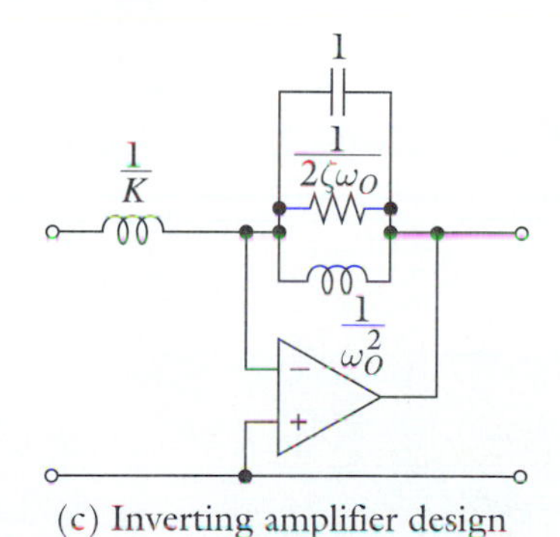

(c) Inverting amplifier design

Figure 11-30 Second-order circuit realizations of $T(s) = K/(s^2 + 2\zeta\omega_O s + \omega_O^2)$.

result to the inverting OP AMP transfer function in Fig. 11-22:

$$T_V(s) = -\frac{K/s}{s + 2\zeta\omega_O + \omega_O^2/s} = -\frac{Z_2(s)}{Z_1(s)} \qquad (11\text{-}43)$$

Equating $Z_2(s)$ to the reciprocal of the denominator and $Z_1(s)$ to the reciprocal of the numerator yields

$$Z_1(s) = \frac{s}{K} \qquad \text{and} \qquad Z_2(s) = \frac{1}{s + 2\zeta\omega_O + \omega_O^2/s} \qquad (11\text{-}44)$$

The impedance $Z_1(s)$ is realizable as an inductance ($L_1 = 1/K$ H) and the impedance $Z_2(s)$ as the parallel connection of a capacitance ($C_2 = 1$ F), resistance ($R_2 = 1/2\zeta\omega_O$ Ω), and inductance ($L_2 = 1/\omega_O^2$ H). The resulting inverting OP AMP realization is shown in Fig. 11-30(c).

The examples in Fig. 11-30 illustrate a general approach to designing prototype circuits for second-order transfer functions. The approach begins by factoring s out of the denominator. A voltage-divider realization is possible when the resulting denominator minus the numerator is a realizable impedance. When it is not realizable, either a cascade design with a noninverting gain stage or an inverting OP AMP realization can be used. Realizing a second-order transfer function requires resistors, inductors, and capacitors because factoring out an s leaves the denominator of the transfer function in the form

$$s + 2\zeta\omega_O + \omega_O^2/s = Ls + R + 1/Cs$$

The approach outlined above works when the second-order function is underdamped ($\zeta < 1$) or overdamped ($\zeta > 1$). However, when $T(s)$ is overdamped it could be factored into two first-order transfer functions and realized as a cascade connection. Conversely, a transfer function consisting of the product of two first-order functions can be realized as a single second-order stage.

❖ Design Example

EXAMPLE 11-15

Find a second-order realization of the transfer function given in Example 11-12.

SOLUTION:

The given transfer function can be written as

$$T_v(s) = \frac{3000s}{(s + 1000)(s + 4000)} = \frac{3000s}{s^2 + 5000s + 4 \times 10^6}$$

Factoring s out of the denominator and equating the result to the transfer function of a voltage divider gives

$$\frac{3000}{s + 5000 + 4 \times 10^6/s} = \frac{Z_2(s)}{Z_1(s) + Z_2(s)}$$

Equating the numerators and denominators yields

$$Z_2(s) = 3000 \qquad \text{and} \qquad Z_1(s) = s + 2000 + 4 \times 10^6/s$$

Both of these impedances are realizable, so a single-stage voltage-divider design is possible. The prototype impedance $Z_1(s)$ requires a 1-H inductor, which is a bit large. A more practical value is obtained using a scale factor of $k_m = 0.1$. The resulting scaled voltage divider circuit is shown in Fig. 11-31.

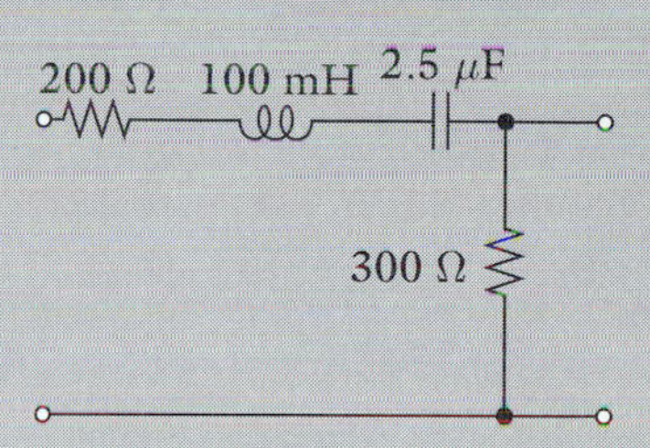

Figure 11-31

Design Summary and Evaluation

Examples 11-12, 11-13, and 11-15 show there are several ways to realize a given transfer function. Specifically, the transfer function in Example 11-12 was realized in the three ways summarized below.

Example	Figure	Description	Number of R	L	C	OP AMP
11-12	11-26	RC voltage-divider cascade	4	0	2	1
11-13	11-28	RC inverting cascade	4	0	2	2
11-5	11-31	RLC voltage divider	2	1	1	0

Selecting a final design from among these alternatives involves evaluation of each circuit using additional criteria. For example, inductors are heavy and lossy in low-frequency applications. These applications favor the first two circuits since they are "inductorless" realizations. However, the two RC circuits contain OP AMPs, which require a dc power supply. The RC voltage-divider cascade uses only one OP AMP, requiring less dc power than the RC inverting cascade, which uses two OP AMPs. On the other hand, the inverting circuit has an OP AMP output, so it could drive reasonable load impedances without changing the transfer function. Finally, in some applications the fact that the passive RLC voltage divider does not require a dc power supply could outweigh the disadvantage of the inductor.

A design problem involves more than simply finding a scalable prototype that realizes a given transfer function. In general, a design problem involves determining a transfer function that meets performance requirements such as the form of the step response or frequency response. In other words, transfer function design problems involve first designing a transfer function and then designing several circuits that realize the transfer function. To deal with all aspects of the design problem we must understand how performance characteristics like step response and frequency response are related to transfer functions. The next section provides this background for the step response. The relationship between network functions and frequency response is covered in Chapter 12.

Design Example

EXAMPLE 11-16

Given the step response $g(t) = [1 + 4e^{-500t}]u(t)$,

(a) Find the transfer function $T(s)$

(b) Design two RC OP AMP circuits that realize the $T(s)$ found in part (a).

(c) Compare the two designs found in (b) on the basis of element count, input impedance and output impedance.

SOLUTION:

(a) The transform of the step response is

$$G(s) = \mathscr{L}\{[1+4e^{-500t}]u(t)\} = \frac{1}{s} + \frac{4}{s+500} = \frac{5s+500}{s(s+500)}$$

and the required transfer function is found to be

$$T(s) = H(s) = sG(s) = \frac{5s+500}{s+500}$$

(b) For the first design we partition $T(s)$ as

$$T(s) = \underbrace{[5]}_{\substack{1^{st}\\ \text{stage}}} \times \underbrace{\left[\frac{s+100}{s+500}\right]}_{\substack{2^{nd}\\ \text{stage}}}$$

The first stage has a positive gain greater than one and can be realized by a noninverting OP AMP configuration with $R_2 = 4R_1$. The second stage can be realized as a voltage divider. Equating the second-stage transfer function to the voltage-divider transfer function expressed in terms of **admittances** gives

$$\frac{s+100}{s+500} = \frac{Y_1(s)}{Y_1(s)+Y_2(s)} = \frac{Z_2(s)}{Z_1(s)+Z_2(s)}$$

Equating numerators and denominators yields

$$Y_1(s) = s + 100 \quad \text{and} \quad Y_2(s) = s + 500 - Y_1(s) = 400$$

The admittance $Y_1(s)$ is realizable as a capacitance ($C_1 = 1$ F) in parallel with a resistor ($G_1 = 100$ S) and $Y_2(s)$ as a resistance ($G_2 = 400$ S). Using a scale factor of $k_m = 10^6$ produces circuit C1 in Fig. 11-32.

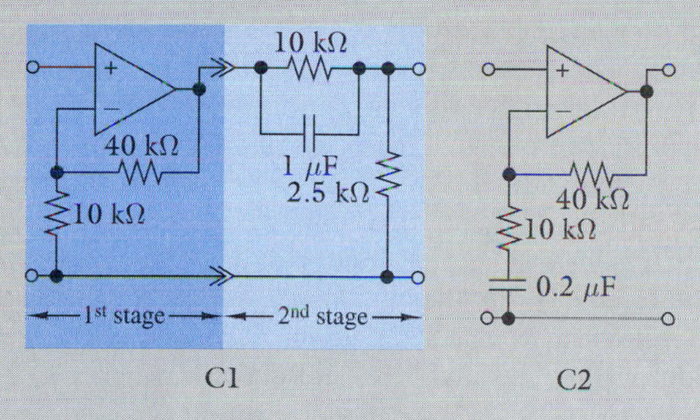

Figure 11-32

A second prototype is obtained by factoring s out of the denominator of $T(s)$ and equating the result to the transfer function of the noninverting OP AMP circuit in Fig. 11-22.

$$\frac{5+500/s}{1+500/s} = \frac{Z_1(s)+Z_2(s)}{Z_1(s)}$$

Equating numerators and denominators yields

$$Z_1(s) = 1 + \frac{500}{s} \quad \text{and} \quad Z_2(s) = 5 + \frac{500}{s} - Z_1(s) = 4$$

The impedance $Z_1(s)$ is realizable as a resistance ($R_1 = 1\ \Omega$) in series with a capacitance ($C_1 = 1/500$ F) and $Z_2(s)$ as a resistance ($R_2 = 4\ \Omega$). Using a scale factor of $k_m = 10^4$ produces circuit C2 in Fig. 11-32.

(c) Circuit C1 uses two more resistors than C2. Circuits C1 and C2 both have infinite input impedance since the input is connected to the noninverting terminal of an OP AMP. Circuit C2 has zero output resistance since it is connected to the output terminal of an OP AMP. Circuit C1 has finite output impedance and can only drive circuits with infinite input resistance.

11-4 TRANSFER FUNCTIONS AND STEP RESPONSE DESCRIPTORS

The step and sinusoidal responses provide standardized measures of the performance of circuits and systems. This section treats the relationship between transfer functions and important features of the step response transient. The relationship of the transfer function to sinusoidal response descriptors is treated in Chapter 12.

In Section 11-2 we found that a transfer function contains all of the data needed to construct the step response waveform $g(t)$. Usually we do not need a complete description of $g(t)$ but instead use partial descriptors of the waveform. The performance of devices, circuits, and systems are often specified in terms of these partial descriptors.

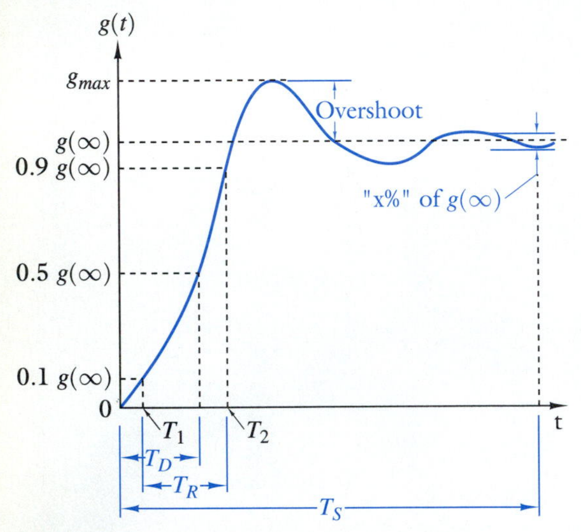

Figure 11-33 General form of a step response.

The step response of many circuits and systems takes on the stylized form shown in Fig. 11-33. The step response starts out at zero and is driven to a final value. The trajectory connecting the initial value to the final value may include overshooting the final value before the response finally settles down. Parameters commonly used to characterize the step response are the rise time, delay time, settling time, and percent overshoot.

The *rise time* (T_R) is defined as follows:

$$T_R = T_2 - T_1 \tag{11-45}$$

where T_1 is the time at which $g(t)$ reaches 10% of its final value and T_2 is the time at which $g(t)$ reaches 90% of its final value for the first time. The 10% and 90% values are the most common, although other values are sometimes used. Rise time is an important descriptor of the speed of response of a system. Step response rise times range from fractions of nanoseconds for high-speed digital circuits to several seconds for large electromechanical systems.

The step response delay time is often specified for systems in which the time between $t = 0$ and $t = T_1$ is large compared with the rise time. *Delay time* (T_D) is defined as the time interval between the application of the step input at $t = 0$ and the time at which $g(t)$ reaches 50% of its final value for the first time. In other words, it is the time required for the response to get halfway to its final value.

The settling time is a measure of the duration of the transient response, i.e., how long it takes for the response to settle down to its final value $g(\infty)$. *Settling time* (T_S) is defined as the time interval between $t = 0$ and the time at which the response remains within a specified percentage of its final value. This percentage is shown as "x% of $g(\infty)$ in Fig. 11-33. Typical values of this percentage are 2%, 5%, and 10%.

In many systems the step response displays a damped oscillation about its final value. The degree of damping can be controlled by specifying the amount by which the step response is allowed to overshoot its final value. A partial descriptor used for this purpose is called *percent overshoot* (PO), defined as

$$\text{PO} = \text{Percent Overshoot} = \frac{g_{MAX} - g(\infty)}{g(\infty)} \times 100\% \quad (11\text{-}46)$$

To be useful this definition requires that $g(0) = 0$ and $g(\infty) \neq 0$. Some step responses that do not meet these conditions are shown in Fig. 11-34.

The initial and final values of the step response waveform $g(t)$ can be determined directly from the transfer function $T(s)$. Since $G(s) = T(s)/s$, the initial and final value properties from Chapter 9 show that

$$\begin{aligned} \text{Initial Value: } \lim_{t\to 0} g(t) &= \lim_{s\to\infty} sG(s) \\ &= \lim_{s\to\infty} T(s) \\ \text{Final Value: } \lim_{t\to\infty} g(t) &= \lim_{s\to 0} sG(s) \\ &= \lim_{s\to 0} T(s) \end{aligned} \quad (11\text{-}47)$$

or

$$g(0) = T(\infty) \quad \text{and} \quad g(\infty) = T(0) \quad (11\text{-}48)$$

The values of $T(s)$ as $s \to \infty$ and at $s = 0$ give the values of $g(t)$ at $t = 0$ and as $t \to \infty$.

The step response parameters defined above (rise time, delay time, settling time, and percent overshoot) often adequately describe step responses for the purposes of analysis and design. The step response descriptors are relatively easy to measure in the laboratory or to calculate from numerical data generated by computer-aided circuit analysis. However, in most cases it is difficult to derive expressions for these descriptors for analytical step response waveforms. Those situations where analytical results are possible are especially important because they give insight into the relationships between the transfer function $T(s)$ and time domain step response descriptors.

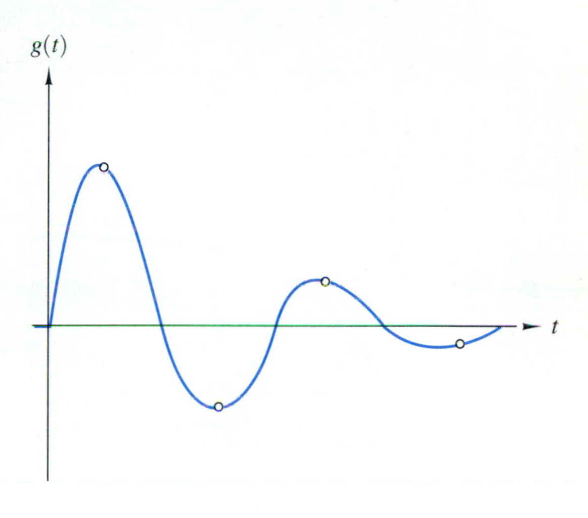

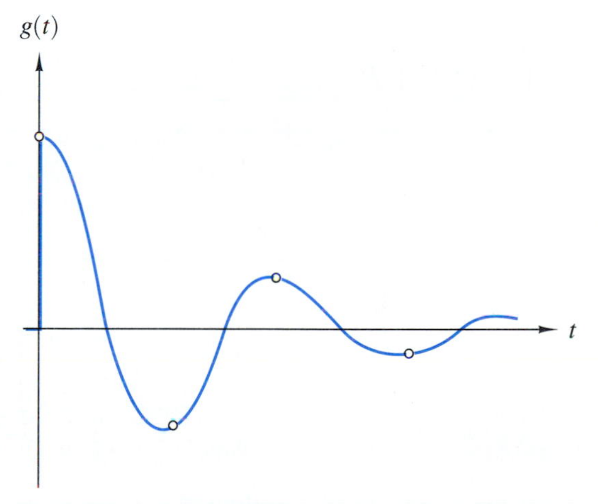

Figure 11-34 Step responses for which overshoot is not defined.

First-Order Circuits

A first-order transfer function has the form

$$T(s) = \frac{K}{s + \alpha} \quad (11\text{-}49)$$

The initial and final values of the step response are

$$\begin{aligned} \text{Initial Value: } g(0) &= T(\infty) = 0 \\ \text{Final Value: } g(\infty) &= T(0) = \frac{K}{\alpha} \end{aligned} \quad (11\text{-}50)$$

Since $h(t) = \mathscr{L}^{-1}\{T(s)\} = [Ke^{-\alpha t}]u(t)$, it follows that the step response is

$$g(t) = \int_0^t h(\tau)d\tau = \frac{K}{\alpha}\left[1 - e^{-\alpha t}\right]u(t) \tag{11-51}$$

Figure 11-35 shows how the first-order step response changes as the pole at $s = -\alpha$ moves to the left in the s plane. The rise time, delay time, and settling time all decrease as the pole moves to the left. The percent overshoot is always zero because $g(t)$ monotonically approaches its final value. Stated mathematically, $g_{MAX} = g(\infty)$, so Eq. (11-46) yields PO = 0%.

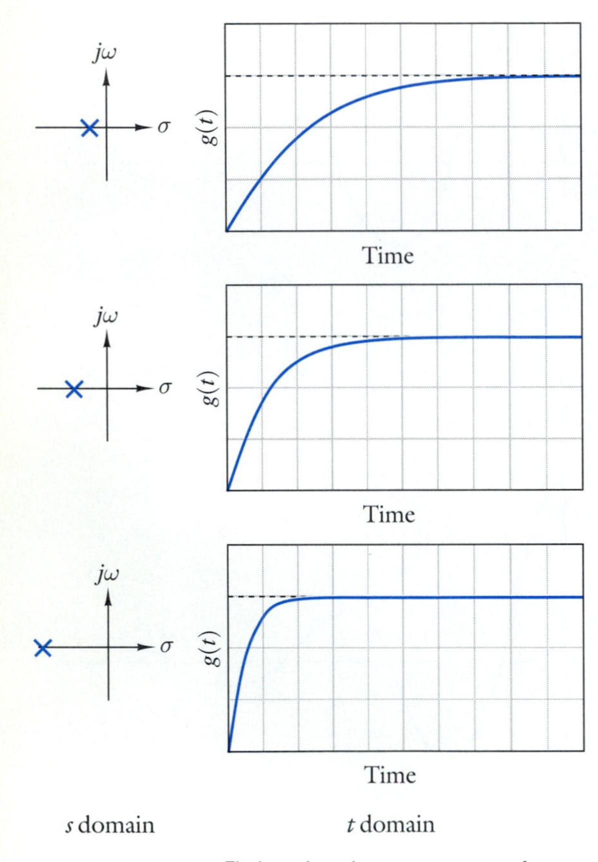

Figure 11-35 First-order step responses for different pole locations.

To develop an expression for the rise time of a first-order circuit we first find the time T_1 at which $g(t)$ reaches 10% of its final value:

$$g(T_1) = \frac{K}{\alpha}(1 - e^{-\alpha T_1}) = 0.1\frac{K}{\alpha}$$

This condition requires that

$$e^{-\alpha T_1} = 0.9 \qquad \text{or} \qquad T_1 = \frac{\ln 10}{\alpha} - \frac{\ln 9}{\alpha}$$

At time T_2 the step response $g(t)$ reaches 90% of its final value:

$$g(T_2) = \frac{K}{\alpha}(1 - e^{-\alpha T_2}) = 0.9\frac{K}{\alpha}$$

This condition requires that

$$e^{-\alpha T_2} = 0.1 \qquad \text{or} \qquad T_2 = \frac{\ln 10}{\alpha}$$

Combining the expressions for times T_1 and T_2 gives an expression for the 10/90 rise time:

$$T_R = T_2 - T_1 = \frac{\ln 9}{\alpha} \tag{11-52}$$

The delay time T_D is the time at which $g(t)$ reaches one-half of its final value:

$$g(T_D) = \frac{K}{\alpha}(1 - e^{-\alpha T_D}) = 0.5\frac{K}{\alpha}$$

This condition gives

$$e^{-\alpha T_D} = 0.5 \qquad \text{or} \qquad e^{\alpha T_D} = 2$$

and so the delay time of a first-order circuit is

$$T_D = \frac{\ln 2}{\alpha} \tag{11-53}$$

The settling time is the time at which $g(t)$ remains within a specified *settling error* (SE) of its final value. Mathematically the settling time requires that

$$\frac{g(T_S) - g(\infty)}{g(\infty)} = \frac{K/\alpha(1 - e^{-\alpha T_S}) - K/\alpha}{K/\alpha} = e^{-\alpha T_S} = SE$$

which requires

$$e^{\alpha T_S} = \frac{1}{SE}$$

The settling time for an $SE \times 100\%$ error criteria is

$$T_S = \frac{\ln(1/SE)}{\alpha} \tag{11-54}$$

where SE is the specified tolerance on the settling error. For example, settling within 1% error means $SE = 0.01$ so that $T_S \approx 4.61/\alpha$.

All response-time descriptors for the first-order case are inversely proportional to the distance α from $s = 0$ to the pole location. Moving the pole to the left in the s-plane increases the distance to the origin and decreases the response-time descriptors. The percent overshoot is zero since the first-order step response is an exponential rise, which monotonically approaches its final value.

EXAMPLE 11-17

For the circuit in Figure 11-36,

(a) Select C so that the rise time of the step response is less than 1 μs.

(b) Calculate the T_S (5% error) for the value of C selected in part (a).

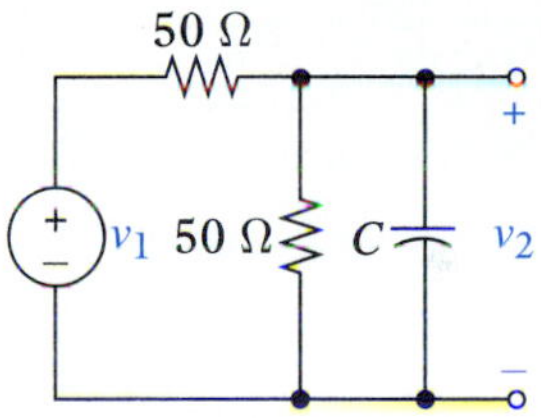

Figure 11-36

SOLUTION:

(a) By voltage division the transfer function of the circuit is

$$T(s) = \frac{Z_2(s)}{Z_1(s) + Z_2(s)} = \frac{\dfrac{1}{Cs + 1/50}}{50 + \dfrac{1}{Cs + 1/50}} = \frac{\dfrac{1}{50C}}{s + \dfrac{1}{25C}}$$

Comparing $T(s)$ with the first-order form $K/(s+\alpha)$ shows that $\alpha = 1/25C$. Using Eq. (11-52) gives the following requirement:

$$T_R = \frac{\ln 9}{\alpha} < 1\ \mu\text{s}$$

But since $\alpha = 1/25C$, this constraint requires

$$C < \frac{1}{25 \times 10^6 \times \ln 9} = 18.2\ \text{nF}$$

Selecting $C = 10$ nF meets the rise-time constraint.

(b) For the selected value of C, $\alpha = 1/25C = 4 \times 10^6$ rad/s. For a 5% settling error $SE = 0.05$ and Eq. (11-54) gives

$$T_S = \frac{-\ln SE}{\alpha} = \frac{2.996}{4 \times 10^{-6}} = 0.749\ \mu\text{s}$$

❖ Design Example

EXAMPLE 11-18

(a) Construct a first-order transfer function whose step response has $T_R \geq 1$ ms and $T_D \leq 0.5$ ms.

(b) Design an RC circuit that realizes the transfer function found in (a).

SOLUTION:

(a) The circuit must have a transfer function of the form $K/(s+\alpha)$. Using Eqs. (11-52) and (11-53) gives two design constraints in the following form:

$$T_R = \frac{\ln 9}{\alpha} \geq 10^{-3} \quad \text{and} \quad T_D = \frac{\ln 2}{\alpha} \leq 0.5 \times 10^{-3}$$

These constraints restrict α to the range

$$1386 = \frac{\ln 2}{0.5 \times 10^{-3}} \leq \alpha \leq \frac{\ln 9}{10^{-3}} = 2197$$

Any value of α in the range from 1386 rad/s to 2197 rad/s will meet the design constraints. For convenience we select $\alpha = 2000$ rad/s. Note that the design constraints do not specify the value of the gain K.

(b) To design a circuit we first try a voltage divider circuit:

$$T(s) = \frac{K}{s + 2000} = \frac{K/s}{1 + 2000/s} = \frac{Z_2(s)}{Z_1(s) + Z_2(s)}$$

Equating numerator and denominator gives the divider impedances:

$$Z_2 = \frac{K}{s} \quad \text{and} \quad Z_1 = 1 + \frac{2000 - K}{s}$$

Since K is not specified a simpler design results when $K = 2000$. The prototype and the scaled design ($k_m = 10^4$) are shown in Fig. 11-37.

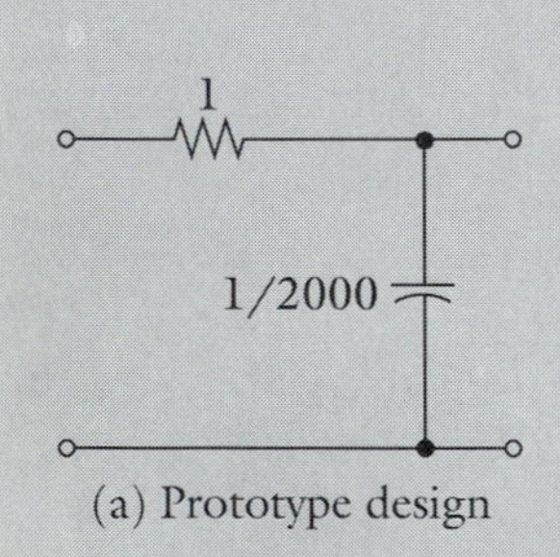

(a) Prototype design

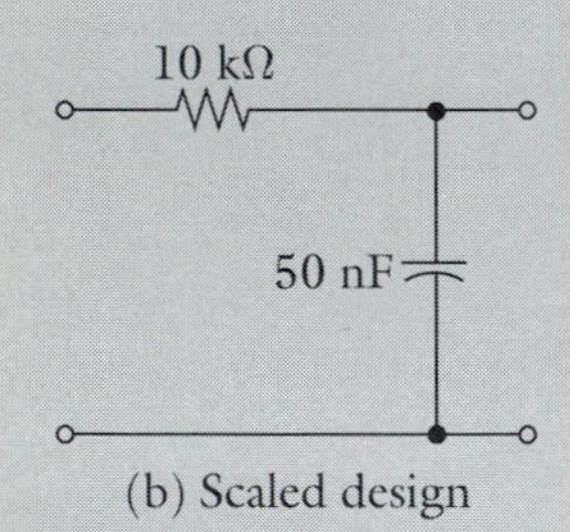

(b) Scaled design

Figure 11-37

DESIGN EXERCISE: EXERCISE 11-11

(a) Construct a first-order transfer function such that $g(\infty) = 5$ and T_S (5% error) is less than 0.2 ms.
(b) Design an RC circuit that realizes the transfer function found in (a).

ANSWERS:
(a) One possible solution is $T(s) = 10^5/(s + 2 \times 10^4)$.
(b) One possible design is shown in Fig. 11-38.

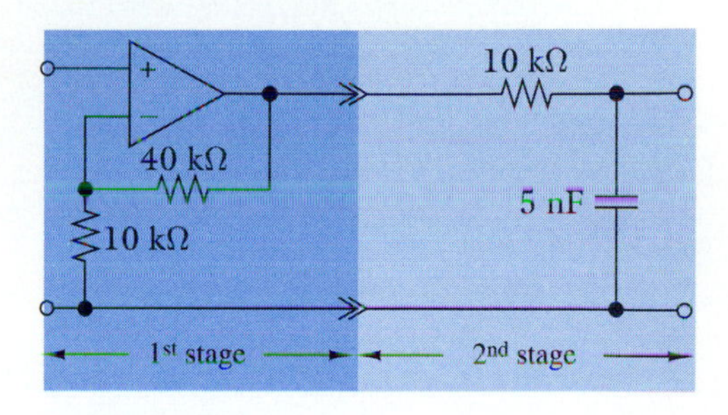

Figure 11-38

Second-Order Circuits

Analytical expressions for some of the step response descriptors can be derived for the underdamped second-order transfer function of the form

$$T(s) = \frac{K}{(s+\alpha)^2 + \beta^2} \tag{11-55}$$

The impulse response for this transfer function is

$$\begin{aligned} h(t) &= \mathscr{L}^{-1}\{T(s)\} = \frac{K}{\beta}\mathscr{L}^{-1}\left\{\frac{\beta}{(s+\alpha)^2+\beta^2}\right\} \\ &= \frac{K}{\beta}e^{-\alpha t}\sin\beta t, \qquad t \geq 0 \end{aligned} \tag{11-56}$$

and therefore the step response is

$$\begin{aligned} g(t) &= \int_0^t h(\tau)d\tau = \frac{K}{\beta}\left[\frac{-\alpha e^{-\alpha\tau}\sin\beta\tau - \beta e^{-\alpha\tau}\cos\beta\tau}{\alpha^2+\beta^2}\right]_0^t \\ &= \frac{K}{\alpha^2+\beta^2}\left[1 - e^{-\alpha t}\cos\beta t - \frac{\alpha}{\beta}e^{-\alpha t}\sin\beta t\right]u(t) \end{aligned} \tag{11-57}$$

Figure 11-39 shows how the underdamped second-order step response changes as the poles of $T(s)$ are moved to the left in the s plane.

The study of the s-domain shifting property in Chapter 9 points out that shifting poles to the left in the s domain increases the waveform damping in the t domain. The increase in damping causes the overshoot and settling time to decrease. In general, shifting poles to the left causes the step response to be less oscillatory and to settle faster. The effect of pole shifting on the rise time is more complicated because a highly oscillatory response can have a fast rise time but a very long settling time. For a second-order circuit there is a tradeoff between rise time and settling time.

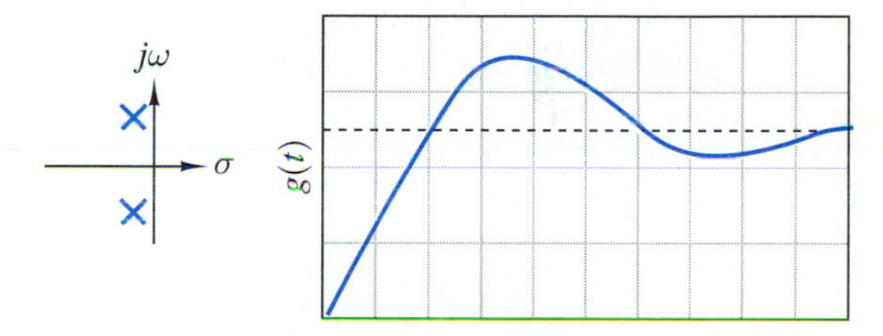

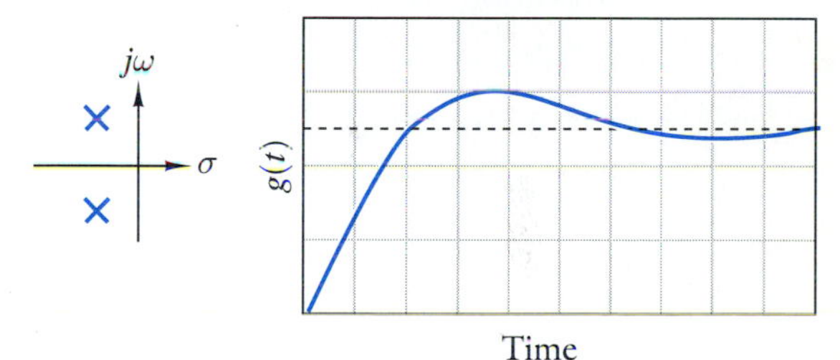

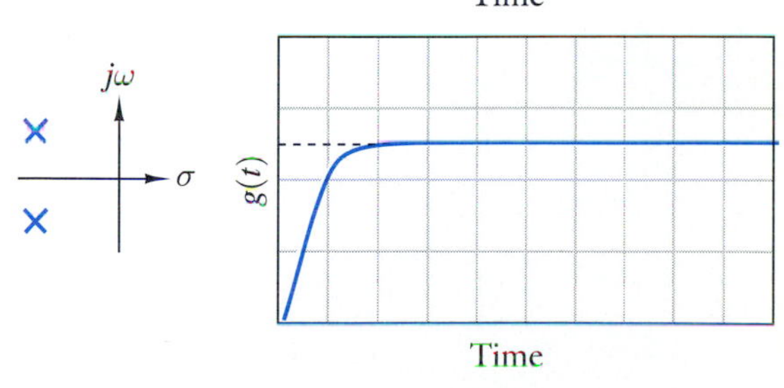

Figure 11-39 Second-order step responses for different pole locations.

To derive an expression for the percent overshoot we need to determine $g(0)$, $g(\infty)$, and g_{MAX}. The initial and final values of the step response are

$$\text{Initial Value: } g(0) = T(\infty) = 0$$
$$\text{Final Value: } g(\infty) = T(0) = \frac{K}{\alpha^2 + \beta^2} \tag{11-58}$$

The maximum and minimum values of $g(t)$ occurs when $h(t) = 0$ because the impulse response is the derivative of the step response. From Eq. (11-56) we see that $h(t) = 0$ when $\sin \beta t = 0$, which happens whenever $t = n\pi/\beta$, $n = 1, 2, 3, \ldots$. The maximum value of $g(t)$ occurs at the first of these extrema when $n = 1$. Substituting $t = \pi/\beta$ into Eq. (11-57) gives the maximum value of $g(t)$:

$$g_{MAX} = \frac{K}{\alpha^2 + \beta^2}\left(1 + e^{-\alpha\pi/\beta}\right) \tag{11-59}$$

Now substituting $g(\infty)$ from Eq. (11-58) and g_{MAX} from Eq. (11-59) into the definition of percent overshoot in Eq. (11-46) yields

$$\begin{aligned} \text{PO} &= \frac{g_{MAX} - g(\infty)}{g(\infty)} \times 100 \\ &= \left(\frac{\dfrac{K}{\alpha^2 + \beta^2}(1 + e^{-\alpha\pi/\beta}) - \dfrac{K}{\alpha^2 + \beta^2}}{\dfrac{K}{\alpha^2 + \beta^2}} \right) \times 100 \\ &= \left(e^{-\alpha\pi/\beta}\right) \times 100 \end{aligned} \tag{11-60}$$

The percent overshoot is a function of the ratio α/β. That is, for an underdamped second-order system the overshoot is a function of the ratio of the real part of the pole to the imaginary part. Shifting poles to the left reduces the overshoot because it increases α while holding β constant.

For an underdamped step response it is difficult to derive an exact expression for settling time (T_S). Settling time is determined when the following condition

$$f(t) = \left| \frac{g(t) - g(\infty)}{g(\infty)} \right| = e^{-\alpha t} \left| \cos \beta t + \frac{\alpha}{\beta} \sin \beta t \right| < SE \tag{11-61}$$

is satisfied for all $t > T_S$. It is difficult to find T_S from Eq. (11-61) because the function $f(t)$ is periodically zero. However, Eq. (11-61) points out the settling process is controlled by the exponential term $e^{-\alpha t}$. The envelope of the step response will settle within specified limits when

$$T_S = \frac{\ln(1/SE)}{\alpha} \tag{11-62}$$

where SE is the specified settling-time error as in the first-order response. The settling time expression in Eq. (11-62) has the same form as the first-order case in Eq. (11-54). The difference here is that in the present case α represents the real part of the pair of complex poles. In either case, T_S is inversely proportional to α. Moving poles to the left in the s plane increases damping and decreases the settling time.

❖ Design Example

EXAMPLE 11-19

(a) Construct a second-order transfer function whose step response has $T_S < 25\ \mu$s (5% error) and PO < 25%.

(b) Design a circuit which realizes the transfer function found in (a).

SOLUTION:

(a) The transfer function must be of the form

$$T(s) = \frac{K}{(s+\alpha)^2 + \beta^2}$$

The settling-time requirement places the constraint.

$$T_S = \frac{-\ln 0.05}{\alpha} \leq 25 \times 10^{-6} \qquad \text{or} \qquad \alpha \geq 1.2 \times 10^5$$

The overshoot requirement places the constraint.

$$\text{PO} = e^{-\alpha\pi/\beta} \times 100 \leq 25 \qquad \text{or} \qquad \beta \leq \frac{\alpha\pi}{\ln 4}$$

Selecting $\alpha = 2 \times 10^5$ rad/s meets the settling time constraint. The overshoot constraint then reduces to $\beta \leq 4.53 \times 10^5$ rad/s. With $\alpha = 2 \times 10^5$ rad/s selecting $\beta = 4 \times 10^5$ rad/s meets both constraints. The resulting $T(s)$ is

$$T(s) = \frac{K}{(s + 2 \times 10^5)^2 + (4 \times 10^5)^2}$$

$$= \frac{K}{s^2 + 4 \times 10^5 s + 2 \times 10^{11}}$$

(b) To design a circuit we factor s out of the denominator of $T(s)$ and equate the result to the voltage divider relationship

$$T(s) = \frac{K/s}{s + 4 \times 10^5 + 2 \times 10^{11}/s} = \frac{Z_2(s)}{Z_1(s) + Z_2(s)}$$

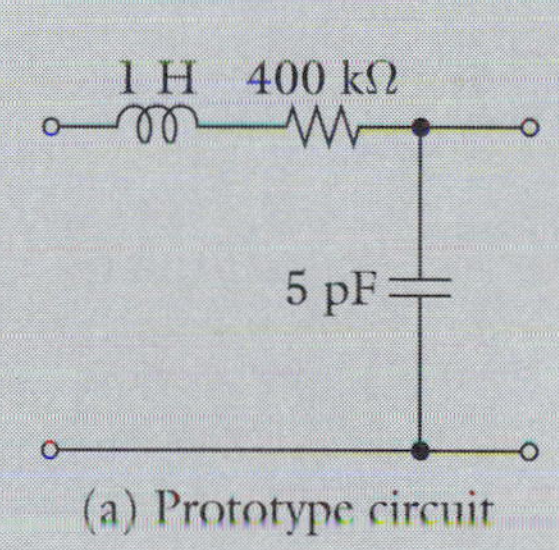

(a) Prototype circuit

(b) Scaled circuit

Figure 11-40

Equating numerators and denominators yields the branch impedances

$$Z_2 = K/s \quad \text{and} \quad Z_1 = s + 4 \times 10^5 + (2 \times 10^{11} - K)/s$$

Since K is not specified, the design is simplified by selecting $K = 2 \times 10^{11}$. The prototype circuit is shown in Fig. 11-40(a). A scale factor of $k_m = 10^{-3}$ produces the scaled design in Fig. 11-40(b).

EXERCISE 11-12

Find the percent overshoot and T_S (1% error) for the second-order systems defined by the following transfer functions.

(a) $T(s) = \dfrac{K}{s^2 + 200s + 2.6 \times 10^5}$

(b) $T(s) = \dfrac{K}{s^2 + 400s + 2 \times 10^5}$

(c) $T(s) = \dfrac{K}{s^2 + 800s + 2 \times 10^5}$

ANSWERS:

(a) PO = 53.3%, $T_S = 46.1$ ms.
(b) PO = 20.8%, $T_S = 23.0$ ms.
(c) PO = 0.2%, $T_S = 11.5$ ms.

DESIGN EXERCISE: EXERCISE 11-13

Design a second-order transfer function whose step response has $g(\infty) = 10$, $T_S > 10\ \mu s$ (1% error) and PO $< 10\%$.

ANSWER:

An example solution is

$$T(s) = \frac{3.2 \times 10^{12}}{s^2 + 8 \times 10^5 s + 3.2 \times 10^{11}}$$

11-5 STEP RESPONSE USING SPICE

The transient analysis features of computer-aided circuit analysis programs produce step responses when numerical values of the circuit elements are known. This capability is a useful computational tool for characterizing the pulse response of a given circuit or for comparing performance against design constraints. This section uses the SPICE syntax to illustrate the input data needed to describe a step response problem to an circuit analysis program.

In Chapter 8 we discussed the SPICE syntax for generating a transient analysis. The key element of the syntax is the transient analysis command state written as

```
.TRAN  TSTEP  TSTOP  [TSTART] [TMAX] [UIC]
```

This statement causes SPICE to perform a transient analysis beginning at $t = 0$ and ending to $t =$ TSTOP. The calculated responses designated in a .PRINT command are written to the output file every TSTEP seconds beginning at $t =$ TSTART and ending at $t =$ TSTOP. TMAX is the maximum internal step size allowed in the transient calculation. The UIC element directs SPICE to *U*se the *I*nitial *C*onditions specified in the inductor and capacitor element statements. The TSTEP and TSTOP parameters are mandatory, and the other three are optional.

TSTOP should be greater than the duration of the transient, which in the present context is the settling time of the step response. A reasonable starting place is TSTEP = TSTOP/100. This choice generates 101 equally spaced samples of $g(t)$ in the output file, which is adequate for graphical purposes unless the step response is heavily overdamped or underdamped.

The source statement for a unit voltage step function has the form

```
VS 1 0 DC 1
```

This statement generates a unit step function. Remember that SPICE always performs a dc analysis prior to starting the transient run. The dc source generating the step function could change the initial conditions on the capacitor and inductors unless the .TRAN command includes the UIC element and the initial conditions on capacitor and inductors are set to zero.

In the following examples we use SPICE generated step responses to verify circuit designs developed earlier in this chapter.

EXAMPLE 11-20

Use SPICE to verify the step response performance of the first-order RC circuit designed in Example 11-18.

SOLUTION:

The RC circuit in Fig. 11-41 was designed in Example 11-18 to have a rise time greater than 1 ms and a delay time less than 0.5 ms. The transfer function constructed to meet these constraints was

$$T(s) = \frac{2000}{s + 2000}$$

For this function the pole is located at $\alpha = 2000$. Using Eq. (11-54) we get the settling time for a 1% error criteria as

$$T(s) = \frac{\ln 0.01}{2000} = 2.30 \text{ ms}$$

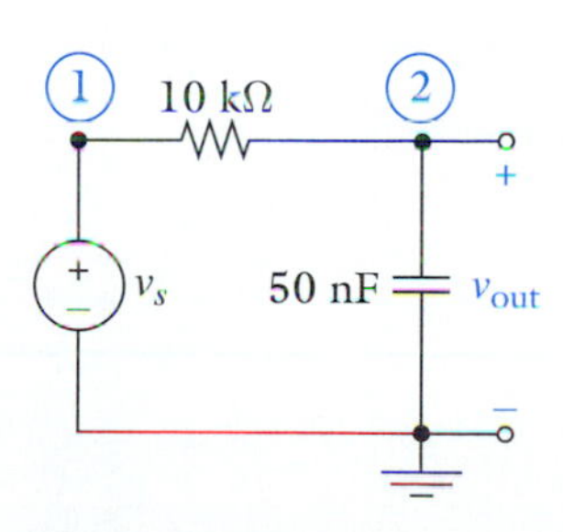

Figure 11-41

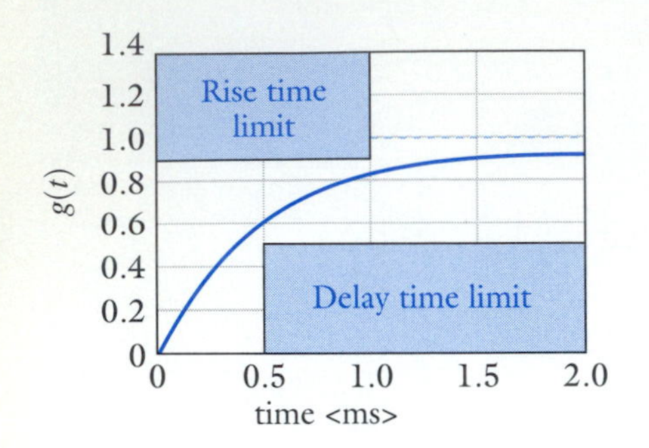

Figure 11-42

Reasonable control times for the analysis are TSTOP = 2.5 ms and TSTEP = TSTOP/100 = 25 μs. A SPICE input file for this analysis is shown below:

```
*EXAMPLE 11-20
VS 1   0   DC   1
R  1   2   10K
C  2   0   50N   IC=0
.TRAN   25U   2.5M   UIC
.PRINT  TRAN  V(2)
.END
```

This SPICE input file produces the step response in Fig. 11-42. Comparing the response with the specified constraints verifies that the circuit meets the design requirements.

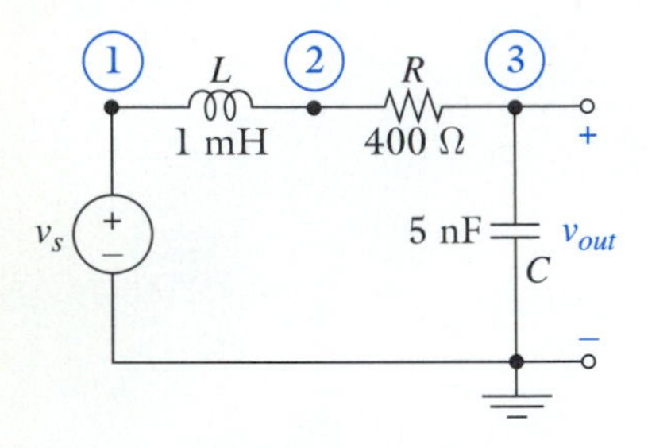

Figure 11-43

EXAMPLE 11-21

Use SPICE to verify the step response performance of the second-order RLC circuit designed in Example 11-19.

SOLUTION:

The RLC circuit in Fig. 11-43 was designed in Example 11-19 to have less than 25% overshoot and a 5% error settling time of less than 25 μs. Given the specified settling time, reasonable control times for the analysis are TSTOP = 30 μs and TSTEP = TSTOP/100 = 0.3 μs. An appropriate SPICE circuit file for this analysis is

```
*EXAMPLE 11-21
VS 1   0   DC   1
L  1   2   1M   IC=0
R  2   3   400
C  3   0   5N   IC=0
.TRAN   0.3U   30U   UIC
.PRINT  TRAN   V(3)
.END
```

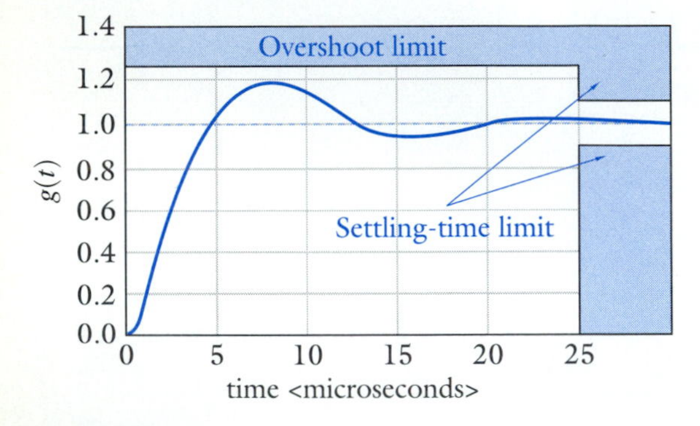

Figure 11-44

Figure 11-44 shows a plot of the step response generated by this circuit file. Comparing the response with the specified constraints verifies that the circuit meets the design requirements.

SUMMARY

- A network function is defined as the ratio of the zero-state response transform to the input transform. Network functions are either driving-point functions or transfer functions.

- Network functions are rational functions of s with real coefficients whose complex poles and zeros occur in conjugate pairs. A network function is defined by its denominator and numerator polynomials or by its poles and zeros together with the scale factor.

- Network functions for simple circuits like voltage and current dividers, and inverting and noninverting OP AMPs are easy to derive and often useful. Node-voltage or mesh-current methods can be used to find the network functions for more complicated circuits. The transfer function of a cascade connection obeys the chain rule when each circuit does not load its predecessor in the cascade.

- Forced poles and zeros are the critical frequencies introduced by the input signal. Critical frequencies introduced by the network function are called natural poles and zeros. The impulse response of a circuit contains only natural poles. A circuit is stable if all of its natural poles are in the left half of the s plane.

- The step response of a linear circuit can be found by integrating its impulse response waveform. The zero-state response of a linear circuit for any causal input $x(t)$ can be found in the time domain using its impulse response waveform and a convolution integral.

- First- and second-order transfer functions can be designed using voltage dividers, and inverting or noninverting OP AMP circuits. Higher-order transfer functions can be realized using a cascade connection of first- and second-order circuits. Prototype designs usually require magnitude scaling to obtain practical element values.

- The step responses of devices, circuits, and systems are often specified in terms of rise time, delay time, settling time, and overshoot. These descriptors are easily measured in the laboratory but may be difficult to calculate analytically.

- Circuit analysis programs like SPICE numerically generate step responses when the values of the circuit parameters are known. The user must have a rough idea of the duration and form of the transient in order to specify the control times in the transient analysis command.

EN ROUTE OBJECTIVES AND ASSOCIATED PROBLEMS

ERO 11-1 Network Functions (Sects. 11-1 and 11-2)

Given a linear circuit in the zero-state:

(a) Transform the circuit into the s domain and find specified network functions.

(b) Construct a pole-zero map and classify or adjust the critical frequencies in the output.

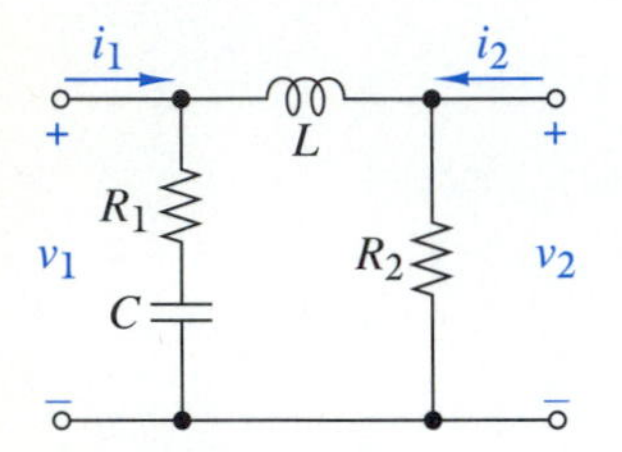

Figure P11-1

11-1 Connect a voltage source $v_1(t)$ at the input port and an open circuit at the output port of the circuit in Fig. P11-1.

(a) Transform the circuit into the s domain and find the driving-point admittance $I_1(s)/V_1(s)$ and the transfer function $V_2(s)/V_1(s)$.

(b) Construct a pole-zero map of $I_1(s)$ and $V_2(s)$ for $R_1 = R_2 = 1$ kΩ, $L = 1$ H, $C = 0.5$ μF, and $v_1(t) = 5u(t)$ V and classify the critical frequencies in these outputs.

11-2 Connect a current source $i_1(t)$ at the input port and a short circuit at the output port of the circuit in Fig. P11-1.

(a) Transform the circuit into the s domain and find the driving-point impedance $V_1(s)/I_1(s)$ and the transfer function $I_2(s)/I_1(s)$.

(b) Construct a pole-zero map of $V_1(s)$ and $I_2(s)$ for $R_1 = 2$ kΩ, $L = 1$ H, $C = 0.5$ μF, and $i_1(t) = [5e^{-1000t}]u(t)$ mA and classify the critical frequencies in these outputs.

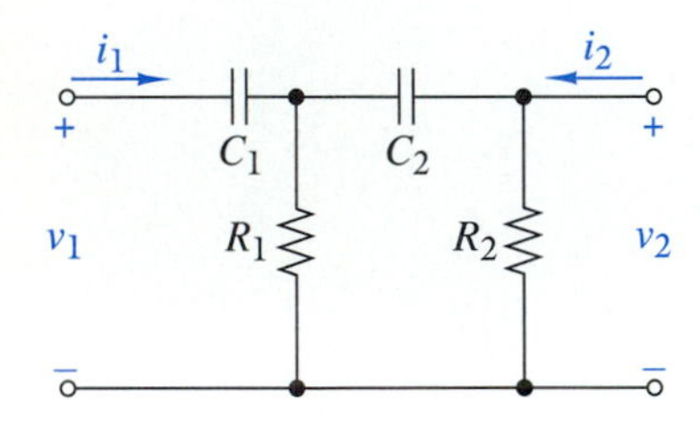

Figure P11-3

11-3 Connect a voltage source $v_1(t)$ at the input port and an open circuit at the output port of the circuit in Fig. P11-3.

(a) Transform the circuit into the s domain and find the driving-point admittance $I_1(s)/V_1(s)$ and the transfer function $V_2(s)/V_1(s)$.

(b) Construct a pole-zero map of $I_1(s)$ and $V_2(s)$ when $R_1 = R_2 = 10$ kΩ, $C_1 = C_2 = 1$ μF, and $v_1(t) = 10[\sin 100t]u(t)$ V. Classify the critical frequencies in the outputs.

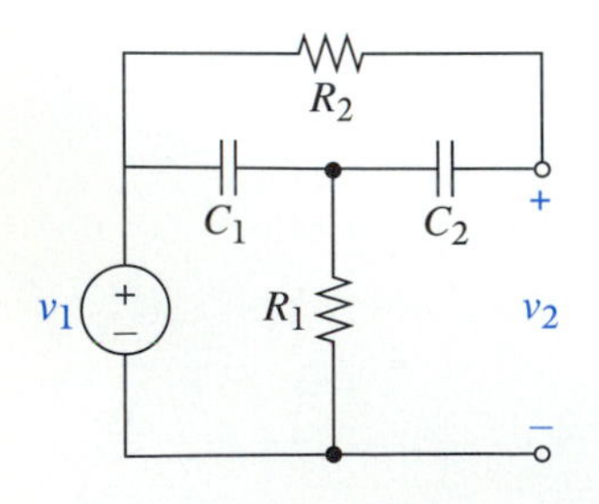

Figure P11-4

11-4 (a) Transform the circuit in Fig. P11-4 into the s domain and show that

$$T_V(s) = \frac{V_2(s)}{V_1(s)}$$

$$= \frac{R_1C_1R_2C_2s^2 + (R_1C_1 + R_1C_2)s + 1}{R_1C_1R_2C_2s^2 + (R_1C_1 + R_2C_2 + R_1C_2)s + 1}$$

(b) Construct a pole-zero map of $V_2(s)$ for $R_1 = R_2 = 10$ kΩ, $C_1 = C_2 = 1$ μF, and $v_1(t) = [e^{-100t}]u(t)$ V and classify the critical frequencies in the output.

❖ **11-5** (D)

(a) Transform the circuit in Fig. P11-5 into the s domain and solve for the driving-point impedance $V_1(s)/I_1(s)$ and the transfer function $V_2(s)/V_1(s)$.

(b) Select the element values so that the transfer function has a scale factor of -10, a pole at $s = -1000$ rad/s, and a zero at $s = -2000$ rad/s.

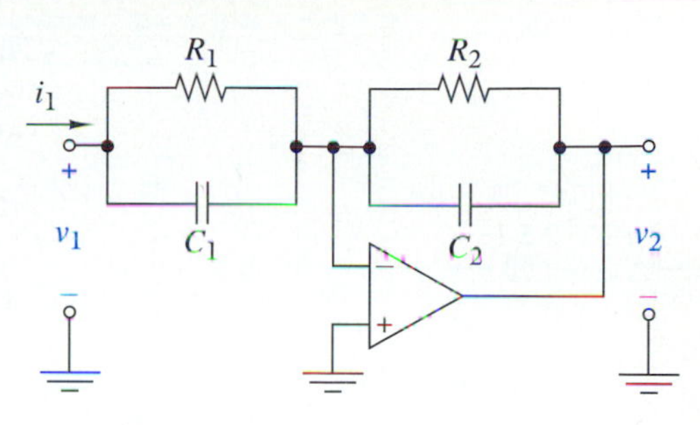

Figure P11-5

11-6 (D)

(a) Transform the circuit in Fig. P11-6 into the s domain and solve for the transfer function $V_2(s)/V_1(s)$.

(b) For what range of gain μ is the circuit stable?

(c) Select μ and the RC product so that the circuit has natural poles at $s = -1000 \pm j1000$ rad/s.

(d) Select μ and RC so that $\zeta = 0.707$ and $\omega_0 = 1000$ rad/s.

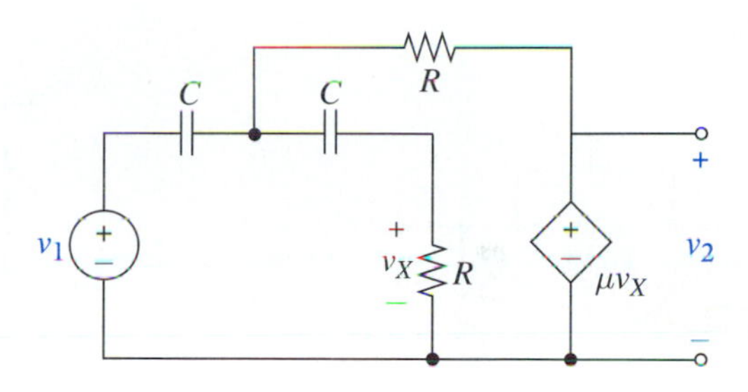

Figure P11-6

The following problems use different combinations of the branches B0 through B6 and circuits C1 through C4 shown in Fig. P11-7.

(a) For each problem connect the specified branches into the circuit indicated and transform the resulting circuit into the s domain.

(b) Solve for the network function relating the specified input and output.

(c) For the specified input construct a pole-zero map of the output and classify the critical frequencies.

Problem	Circuit	Branch A	Branch B	Input	Output
11-7	C1	B0	B2	$v_1(t) = u(t)$	$v_2(t)$
11-8	C2	B0	B1	$i_1(t) = u(t)$	$i_2(t)$
11-9	C2	B0	B3	$i_1(t) = u(t)$	$i_2(t)$
11-10	C3	B0	B1	$v_1(t) = u(t)$	$v_2(t)$
11-11	C4	B0	B1	$v_1(t) = e^{-250t}$	$v_2(t)$
11-12	C4	B2	B0	$v_1(t) = e^{-1000t}$	$v_2(t)$
11-13	C2	B2	B3	$i_1(t) = \cos 1000t$	$i_2(t)$
11-14	C1	B3	B2	$v_1(t) = u(t)$	$v_2(t)$
11-15	C1	B1	B4	$v_1(t) = u(t)$	$v_2(t)$
11-16	C2	B4	B1	$i_1(t) = u(t)$	$i_2(t)$
11-17	C1	B5	B0	$v_1(t) = u(t)$	$v_2(t)$
11-18	C1	B0	B5	$v_1(t) = \cos 1000t$	$v_2(t)$
11-19	C2	B0	B6	$i_1(t) = \cos 1000t$	$i_2(t)$
11-20	C1	B0	B6	$v_1(t) = u(t)$	$v_2(t)$
11-21	C4	B1	B3	$v_1(t) = u(t)$	$v_2(t)$
11-22	C4	B3	B1	$v_1(t) = u(t)$	$v_2(t)$

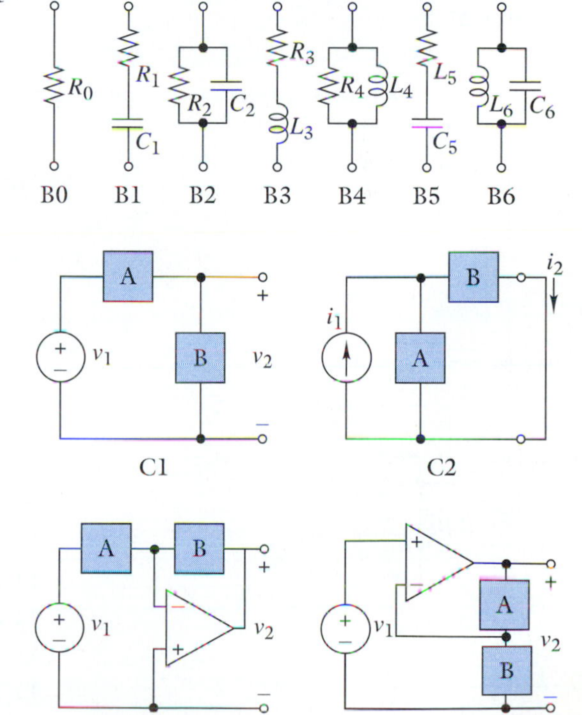

Figure P11-7

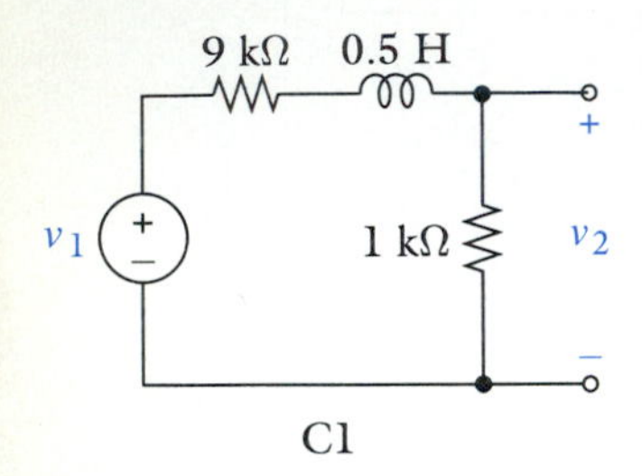

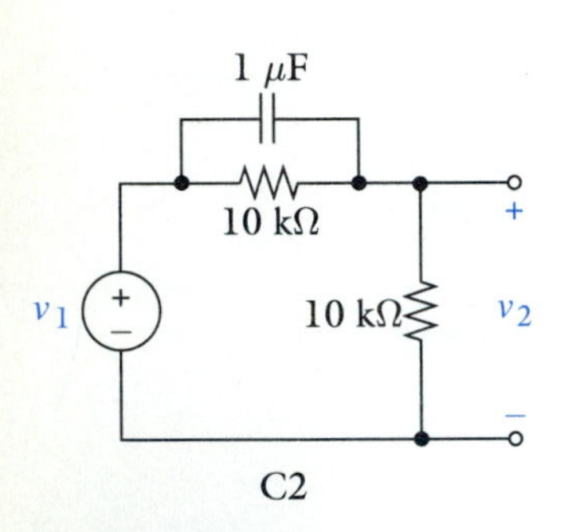

Figure P11-31

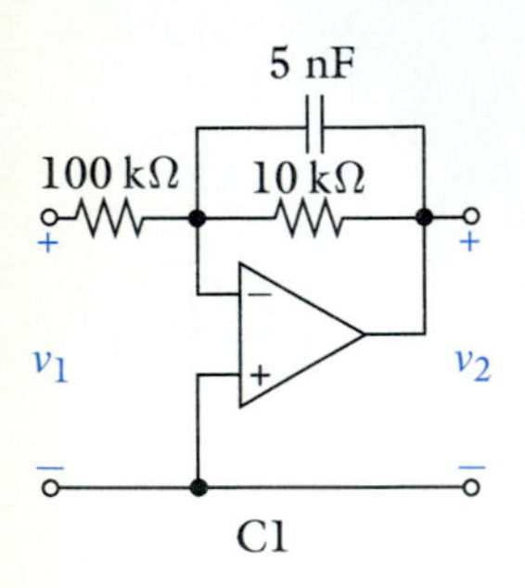

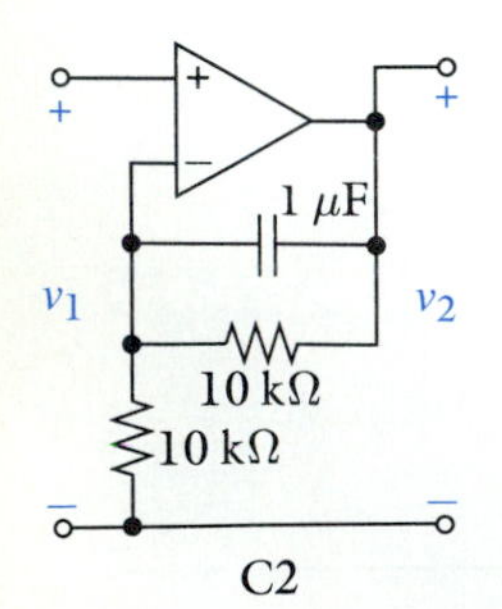

Figure P11-33

The following problems involve a cascade connection of up to four stages. The stages are constructed using branches B0 through B6 and circuits C1 through C4 in Fig. P11-7.

(a) Construct each stage by connecting the indicated branches into the circuit identified for that stage.

(b) Connect the stages in cascade and solve for the transfer function from the input voltage of the first stage to the output voltage of the last stage.

(c) Construct a pole-zero map of the transfer function and identify the stage that produced each critical frequency.

	1st Stage			2nd Stage			3rd Stage			4th Stage		
Problem	**Ckt**	**A**	**B**	**Ckt**	**A**	**B**	**Ckt**	**A**	**B**	**Ckt**	**A**	**B**
11-23	C3	B0	B2	C1	B0	B2						
11-24	C3	B3	B0	C1	B3	B0						
11-25	C1	B1	B0	C4	B0	B1						
11-26	C3	B0	B2	C1	B0	B2	C4	B0	B2			
11-27	C1	B2	B0	C4	B0	B1	C3	B1	B0			
11-28	C4	B0	B3	C1	B3	B0	C4	B0	B2			
11-29	C3	B1	B2	C1	B1	B0	C4	B1	B0	C1	B1	B0
11-30	C1	B0	B6	C4	B0	B0	C1	B5	B0	C3	B1	B0

ERO 11-2 Properties of Network Functions (Sect. 11-2)

(a) Find the impulse or step response waveform of a given first- or second-order linear circuit.

(b) Given one of the responses $H(s)$, $h(t)$, $G(s)$, or $g(t)$, find any of the other three.

(c) Given the impulse response of a linear circuit and an input waveform use the convolution integral to find the output waveform.

11-31 Find $T_V(s) = V_2/V_1$, $g(t)$, and $h(t)$ for circuit C1 in Fig. P11-31.

11-32 Find $T_V(s) = V_2/V_1$, $g(t)$, and $h(t)$ for circuit C2 in Fig. P11-31.

11-33 Find $T_V(s) = V_2/V_1$, $g(t)$, and $h(t)$ for circuit C1 in Fig. P11-33.

11-34 Find $T_V(s) = V_2/V_1$, $g(t)$, and $h(t)$ for circuit C2 in Fig. P11-33.

11-35 Find $T_V(s) = V_2/V_1$, $g(t)$, and $h(t)$ for the circuit in Fig. P11-35.

11-36 Find $T_V(s) = V_2/V_1$, $g(t)$, and $h(t)$ for the circuit in Fig. P11-36.

11-37 Find the transfer function and the impulse-response waveform corresponding to each of the following step responses.

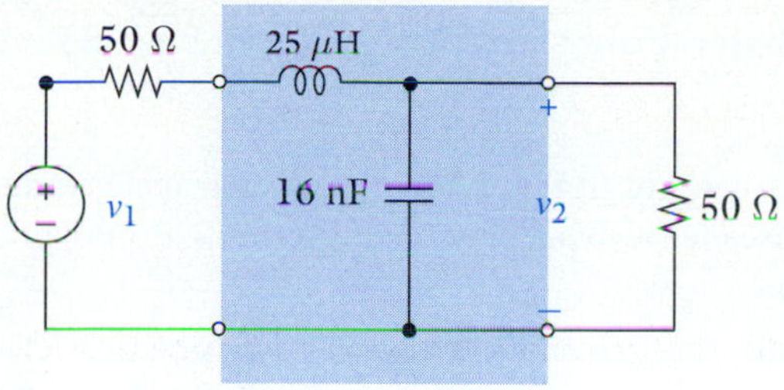

Figure P11-35

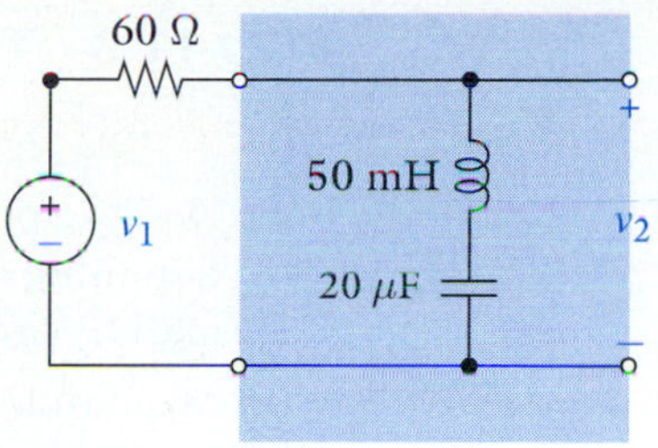

Figure P11-36

(a) $g(t) = [-e^{-2000t}]u(t)$
(b) $g(t) = [1 - e^{-2000t}]u(t)$
(c) $g(t) = [-1 + 0.8e^{-2000t}]u(t)$
(d) $g(t) = [-1 + 2e^{-2000t}]u(t)$

11-38 Find the transfer function and step response waveform corresponding to each of the following impulse responses.
(a) $h(t) = -1000[e^{-1000t}]u(t)$
(b) $h(t) = \delta(t) - 1000[e^{-1000t}]u(t)$
(c) $h(t) = \delta(t) - 500[e^{-1000t}]u(t)$
(d) $h(t) = \delta(t) - 2000[e^{-1000t}]u(t)$

11-39 Find the transfer function and the impulse-response waveform corresponding to each of the following step responses.
(a) $g(t) = 10[e^{-1000t} \sin 2000t]u(t)$
(b) $g(t) = [e^{-1000t} - e^{-2000t}]u(t)$

11-40 Find the transfer function and step response waveform corresponding to each of the following impulse responses.
(a) $h(t) = 10[e^{-1000t} \sin 2000t]u(t)$
(b) $h(t) = 10[e^{-1000t} \cos 2000t]u(t)$

11-41 The impulse response of a linear circuit is $h(t) = 10[u(t) - u(t - 10)]$. Use convolution to find the response due to an input $x(t) = u(t) - u(t - 5)$.

11-42 Repeat problem 11-41 for a linear circuit with $h(t) = \delta(t) + u(t - 10)$.

11-43 Repeat Prob. 11-41 for a linear circuit with $h(t) = u(t) + \delta(t - 10)$.

11-44 The impulse response of a linear circuit is $h(t) = [-1000e^{-1000t}]u(t)$. Use convolution to the find the response due to an input $x(t) = 10tu(t)$.

11-45 Repeat Prob. 11-44 for $x(t) = 10u(t) - 10u(t - 5 \times 10^{-3})$.

11-46 The impulse response of a linear circuit is $h(t) = \delta(t) - 1000[e^{-1000t}]u(t)$. Use convolution to find the response due to an input $x(t) = 10tu(t)$.

11-47 The impulse response of a linear circuit is $h(t) = 10[e^{-1000t} \sin 2000t]u(t)$. Use convolution to find the circuit response for an input $x(t) = 10tu(t)$.

11-48 Repeat Prob. 11-47 for $h(t) = [e^{-1000t} - e^{-2000t}]u(t)$.

ERO 11-3 Transfer Function Design (Sect. 11-3)

(a) Design a prototype circuit that realizes a given transfer function using a cascade of divider, inverting OP, or noninverting OP circuits.

(b) Magnitude scale the prototype circuit to get practical element values.

(c) Design a circuit that realizes a given impulse or step response waveform.

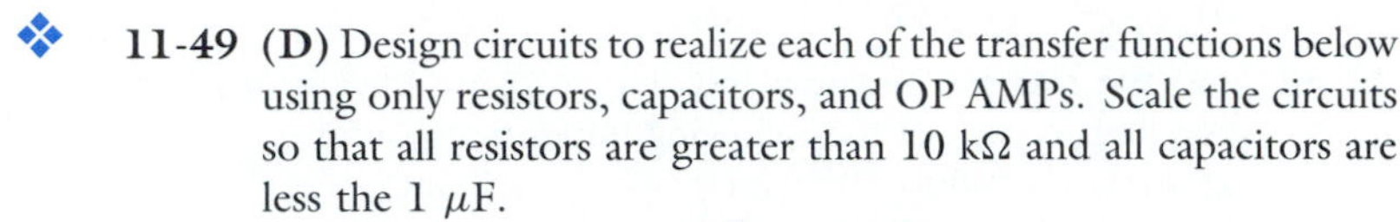

❖ **11-49** **(D)** Design circuits to realize each of the transfer functions below using only resistors, capacitors, and OP AMPs. Scale the circuits so that all resistors are greater than 10 kΩ and all capacitors are less the 1 μF.

(a) $T_V(s) = \pm \dfrac{10^5}{(s+100)(s+1000)}$

(b) $T_V(s) = \pm \dfrac{10^2(s+500)}{(s+100)(s+1000)}$

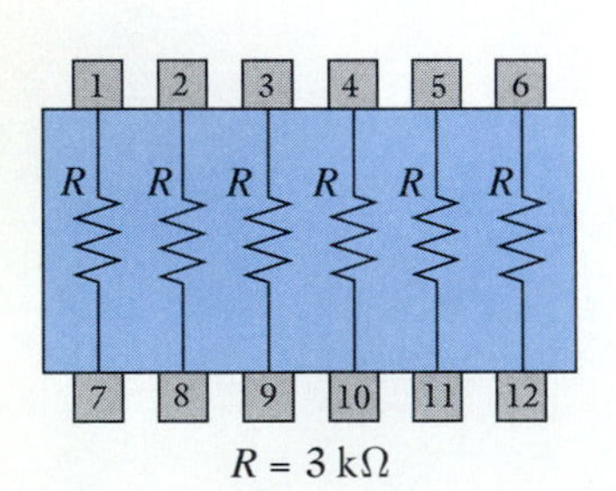

Figure P11-51

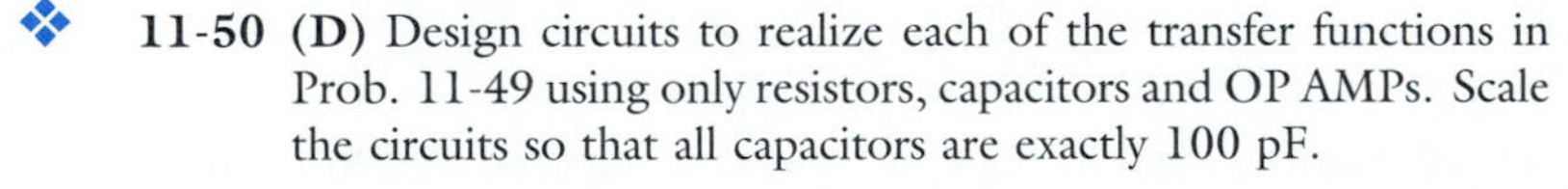

❖ **11-50** **(D)** Design circuits to realize each of the transfer functions in Prob. 11-49 using only resistors, capacitors and OP AMPs. Scale the circuits so that all capacitors are exactly 100 pF.

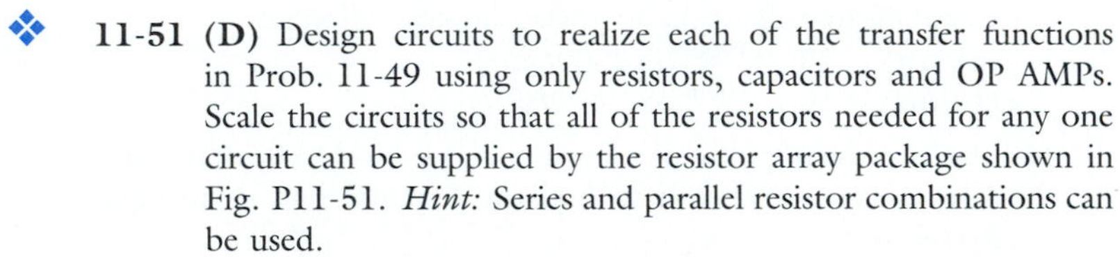

❖ **11-51** **(D)** Design circuits to realize each of the transfer functions in Prob. 11-49 using only resistors, capacitors and OP AMPs. Scale the circuits so that all of the resistors needed for any one circuit can be supplied by the resistor array package shown in Fig. P11-51. *Hint:* Series and parallel resistor combinations can be used.

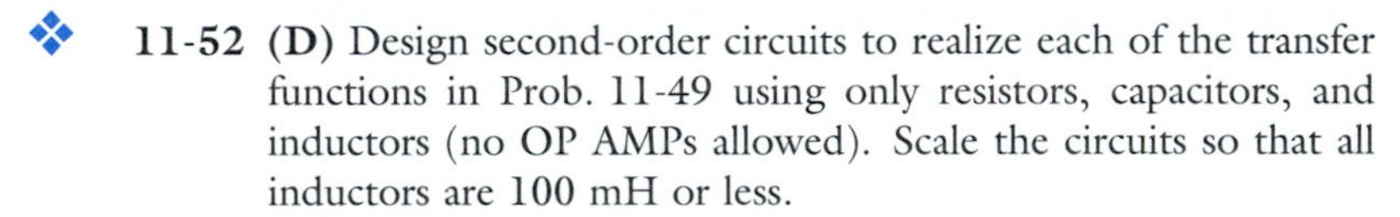

❖ **11-52** **(D)** Design second-order circuits to realize each of the transfer functions in Prob. 11-49 using only resistors, capacitors, and inductors (no OP AMPs allowed). Scale the circuits so that all inductors are 100 mH or less.

❖ **11-53** **(D)** Design circuits to realize each of the transfer functions listed below using only resistors, capacitors, and OP AMPs. Scale the circuits so that all resistors are greater than 10 kΩ and all capacitors are less the 1 μF.

(a) $T_V(s) = \pm \dfrac{(s+200)(s+500)}{(s+100)(s+1000)}$

(b) $T_V(s) = \pm \dfrac{s^2}{(s+100)(s+1000)}$

❖ **11-54** **(D)** Design circuits to realize each of the transfer functions in Prob. 11-53 using only resistors, capacitors and OP AMPs. Scale the circuits so that all capacitors are exactly 100 pF.

❖ **11-55** **(D)** Design circuits to realize each of the transfer functions in Prob. 11-53 using only resistors, capacitors, and OP AMPs. Scale the circuits so that all of the resistors needed for any one circuit can be supplied by the resistor array package shown in Fig. P11-51. *Hint:* Series and parallel resistor combinations can be used.

❖ **11-56** **(D)** Design second-order circuits to realize each of the transfer functions in Prob. 11-53 using only resistors, capacitors, and inductors (no OP AMPs allowed). Scale the circuits so that all inductors are 100 mH or less.

❖ **11-57** **(D)** Design circuits to realize each of the following transfer functions using only resistors, capacitors, and OP AMPs. Scale the circuits so that all resistors are exactly 10 kΩ.

(a) $T_V(s) = -\dfrac{100(s+400)}{s(s+200)}$

(b) $T_V(s) = -\dfrac{10^4}{s^2+200s+10^4}$

❖ **11-58** **(D)** Design circuits that produce each of the step responses in Prob. 11-37.

❖ **11-59** **(D)** Design circuits that produce each of the impulse responses in Prob. 11-38.

❖ **11-60** **(D)** Design circuits that produce each of the step responses in Prob. 11-39.

❖ **11-61** **(D)** Design circuits that produce each of the impulse responses in Prob. 11-40.

ERO 11-4 Transfer Functions and Step Response Descriptors (Sect. 11-4)

(a) Find the step response descriptors of a given first or second-order linear circuit.

(b) Find the step response descriptors of a given impulse response.

(c) Construct a transfer function whose step response has specified descriptors and design a circuit that realizes the transfer function.

11-62 Find the rise time, settling time (1% error), initial value, and final value of the step response of circuit C1 in Fig. P11-31. Sketch the step response waveform.

11-63 Find the rise time, settling time (1% error), initial value, and final value of the step response of circuit C1 in Fig. P11-33. Sketch the step response waveform.

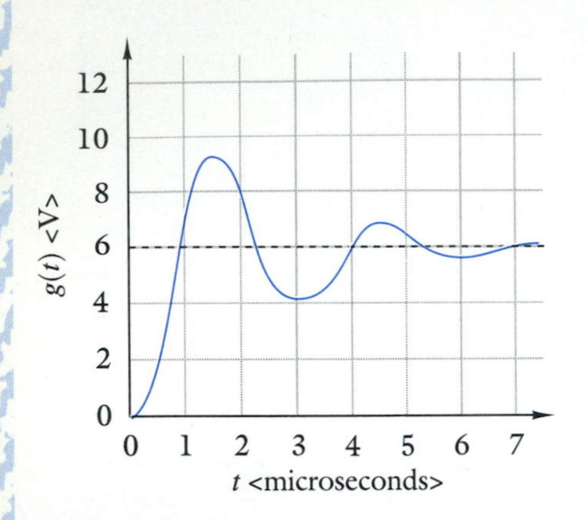

Figure P11-67

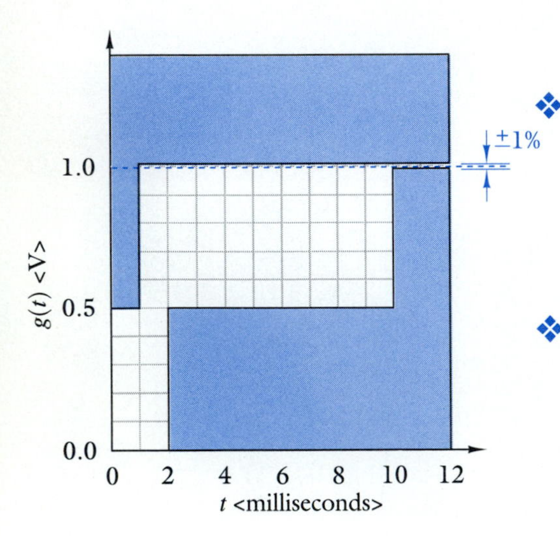

Figure P11-68

11-64 Find the settling time (5% error), percent overshoot, initial value, and final value of the step response of the circuit in Fig. P11-35. Sketch the step response waveform.

11-65 The impulse response of a first-order circuit is $h(t) = -2000[e^{-500t}]u(t)$. Find the rise time, settling time (5% error), initial value, and final value of the step response. Sketch the step response waveform.

11-66 The impulse response of a second-order circuit is $h(t) = 2000[e^{-1000t} \sin 1000t]u(t)$. Find the settling time (5% error), percent overshoot, initial value, and final value of the step response. Sketch the step response waveform.

❖ **11-67 (D)**
(a) Construct a transfer function corresponding to the second-order circuit step response shown in Fig. P11-67.
(b) Design a circuit to realize the $T(s)$ found in (a).

❖ **11-68 (D)**
(a) Construct a first-order transfer function so that its step responses lies entirely in the unshaded region in Fig. P11-68.
(b) Design an RC circuit to realize the transfer function found in (a).

❖ **11-69 (D)**
(a) Construct a first-order transfer function whose unit step response has the following properties: The rise time and settling time (5% error) are less than 1μs and the final value is 5 V.
(b) Design an RC circuit to realize the transfer function found in (a).

❖ **11-70 (D)** **(a)** Construct a second-order transfer function whose step response has less than 10% overshoot and a settling time (5% error) less than 1 μs.
(b) Design a circuit to realize the transfer function found in (a).

ERO 11-5 Step Response Using SPICE (Sect. 11-5)

Use the transient analysis option of SPICE to generate numerical values of the step response of a given linear circuit.

11-71 Use SPICE to generate the step response of circuit C2 in Fig. P11-31.

11-72 Use SPICE to generate the step response of circuit C1 in Fig. P11-33.

11-73 Use SPICE to generate the step response of circuit C2 in Fig. P11-33.

11-74 Use SPICE to generate the step response of the circuit in Fig. P11-35.

11-75 Use SPICE to generate the step response of the circuit in Fig. P11-36.

❖ **11-76** **(D)** Design a circuit to meet the requirements in Prob. 11-67 and use SPICE to verify the performance of the circuit.

❖ **11-77** **(D)** Design a circuit to meet the requirements in Prob. 11-68 and use SPICE to verify the performance of the circuit.

❖ **11-78** **(D)** Design a circuit to meet the requirements in Prob. 11-69 and use SPICE to verify the performance of the circuit.

❖ **11-79** **(D)** Design a circuit to meet the requirements in Prob. 11-70 and use SPICE to verify the performance of the circuit.

CHAPTER INTEGRATING PROBLEMS

11-80 **(A)** A circuit has a transfer function

$$T(s) = \frac{s}{(s+1)(s+2)}$$

(a) A certain input produced a zero-state output $y(t) = [-e^{-t} + \cos t + \sin t]u(t)$. What was the input waveform?

(b) A step function input produced $y(t) = [e^{-t} + e^{-2t}]u(t)$ as the output. Was the circuit in the zero state?

(c) An output $y(t) = [e^{-6t} + e^{-3t}]u(t)$ was observed. Was the input zero?

❖ **11-81** **(a)** **(A)** Show that the circuit in Fig. P11-81 has a voltage transfer function: $(Z_2 - Z_1)/(Z_2 + Z_1)$.

(b) **(D)** Use the result in (a) to design an RC circuit to realize: $T_V(s) = (s - \alpha)/(s + \alpha)$.

(c) **(A)** Explain why the transfer function in (a) cannot be realized using a voltage divider.

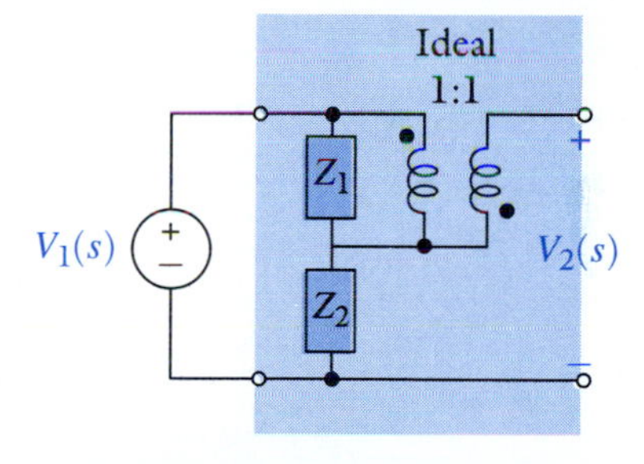

Figure P11-81

❖ **11-82** **(D)**

(a) Show that the transfer function $T_V(s) = (s + \gamma)/(s + \alpha)$ can be realized by a voltage divider if $\alpha > \gamma$, and by a noninverting OP-AMP circuit if $\alpha < \gamma$.

(b) Develop the RC circuit realizations (no inductors allowed) for $\alpha > \gamma$ and $\alpha < \gamma$.

(c) Discuss the advantages and disadvantages of the realization developed in (b).

(d) Applying the results in (a), (b), and (c) to generate a realization of the transfer function

$$T_V(s) = \frac{(s+100)(s+200)}{(s+50)(s+500)}$$

using no inductors and only one OP AMP.

11-83 **(A)** (a) In Chapter 7 we found the input-output relationship of the RC OP AMP integrator to be

$$v_{OUT}(t) = -\frac{1}{RC}\int_0^t v_{IN}(\tau)d\tau$$

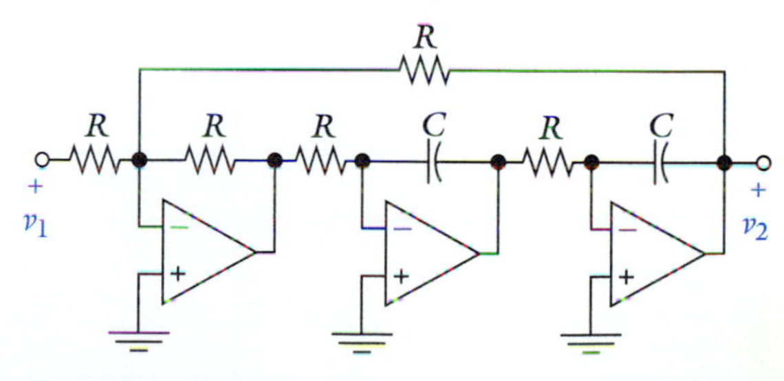

Figure P11-83

Find the voltage transfer function of the integrator. Is the circuit stable or unstable?

(b) Find the impulse and step responses of the OP AMP integrator. Is the impulse response consistent with your answers in part (a)?

(c) What is the voltage transfer function of a cascade connection of two OP AMP integrators? Does the chain rule apply in this case? Is the cascade circuit stable or unstable?

(d) Figure P11-83 shows an OP AMP summer and two OP AMP integrators connected in cascade with a feedback path from output to input. Find the voltage transfer function of this circuit. Is this circuit stable or unstable?

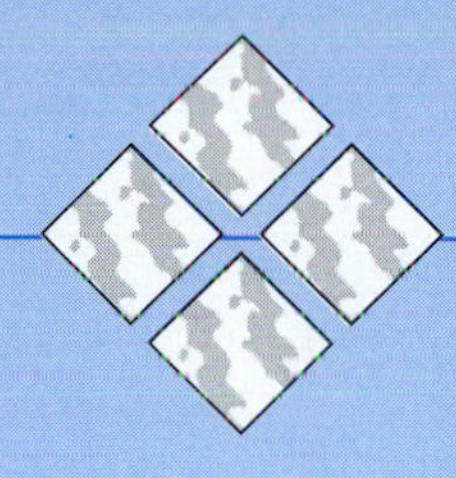

COURSE INTEGRATING PROBLEMS

PART II

Dynamic Circuits

Chapters 6 to 11

Course Objectives

Analysis **(A)**—Given a linear circuit with known input signals, select an analysis technique and determine output signals or input-output relationships.

Design **(D)**—Devise a linear circuit or modify an existing circuit to obtain a specified output signal or a specified input-output relationship.

Evaluation **(E)**—Given several circuits that perform the same signal-processing function, rank-order the circuits using criteria such as cost, parts count, or power dissipation.

I-10 The RC circuit in Fig. PI-10 is driven by the exponential voltage shown.

(a) **(A)** Determine $v_O(t)$ when there is no initial voltage on the capacitor when $C = 0.2\ \mu$F. Identify the forced response, the natural response, the zero-state response, and the zero-input response.

(b) **(D)** Determine the value of capacitance C required to cause the response to be of the form Kte^{-100t}.

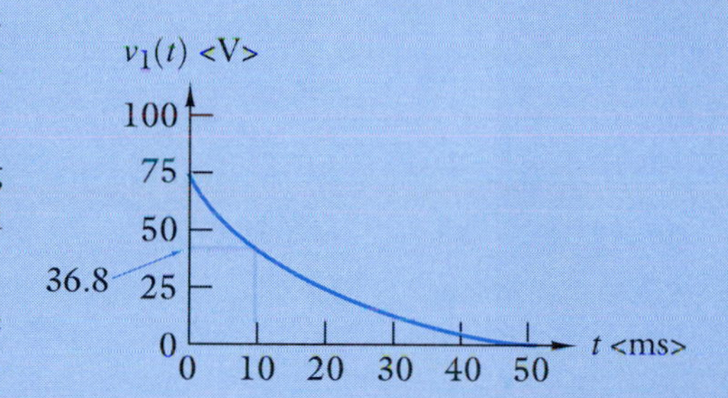

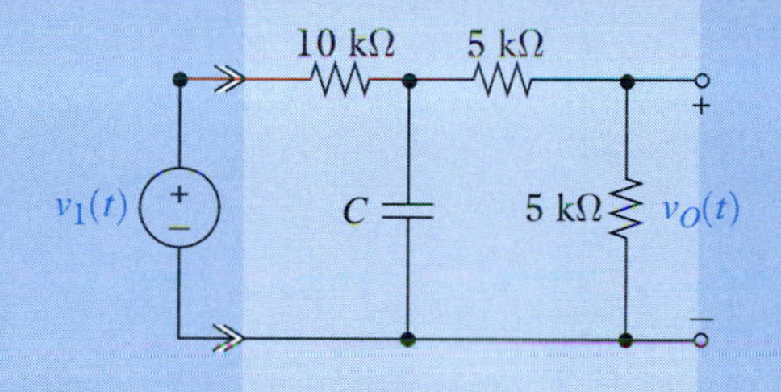

Figure PI-10

I-11 **(A)** The switch in Fig. PI-11 has been open for a long time and is closed at $t = 0$.

(a) Show that $i_L(t) = .5$ mA and $v_O(t) = -5$ V for all $t > 0$. That is, show that the circuit will remember the inductor current that existed at $t = 0$ when the switch was closed.

(b) Now replace the 100 mH inductor by the model shown in the figure. Again, let the switch be open for a long time and closed at $t = 0$. Determine $i_L(t)$ and $v_O(t)$ for $t > 0$. About how long does it take for the circuit to forget the conditions that existed at $t = 0$?

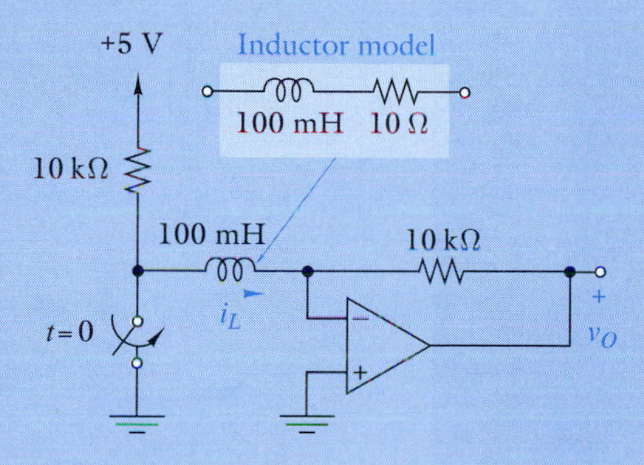

Figure PI-11

I-12 **(A)** The circuit in Fig. PI-12 is driven by an input $v_1(t) = 10\ u(t)$. For $v_C(0) = 5$ V and $i_L(0) = 0$ A.

(a) Determine the forced and natural responses of $v_O(t)$.

(b) Determine the initial conditions that would make the natural response to be zero.

(c) Determine input voltage that would make the forced response to be zero.

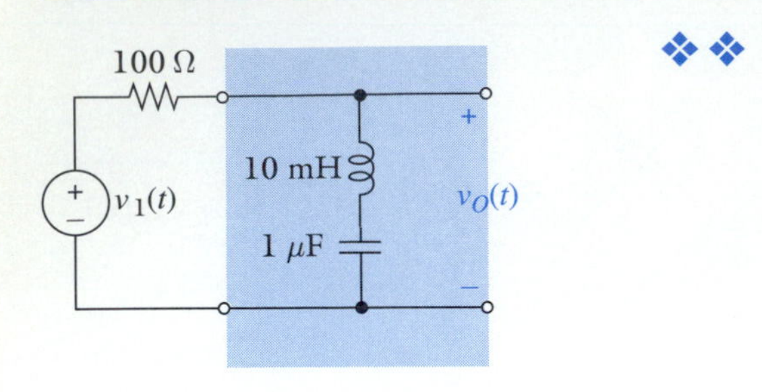

Figure PI-12

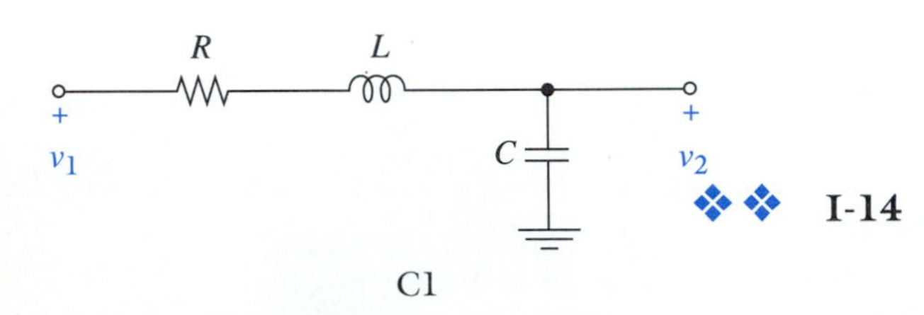

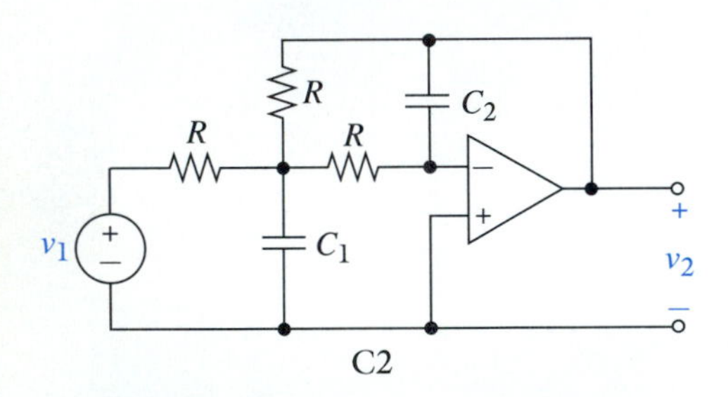

Figure PI-13

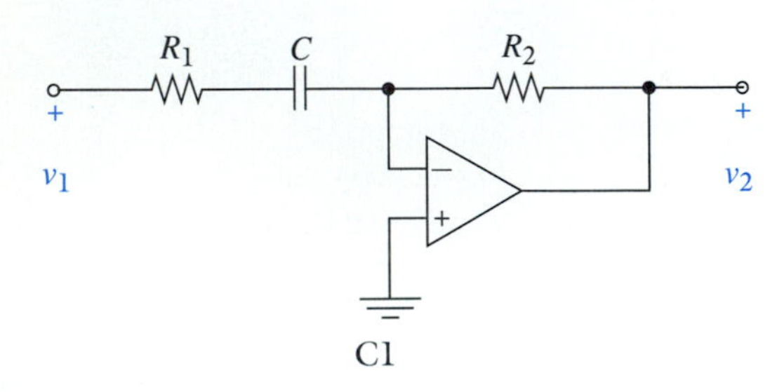

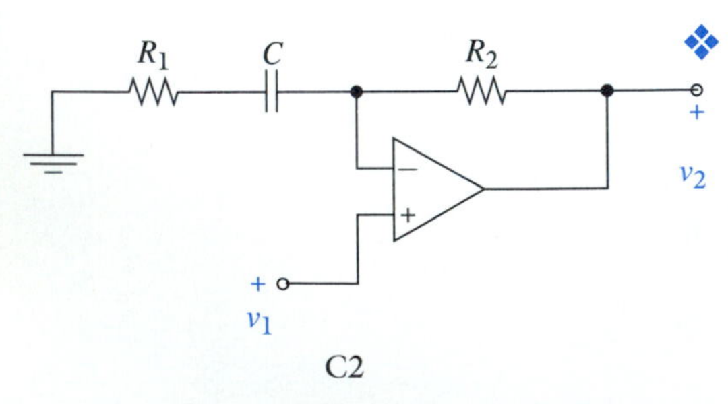

Figure PI-14

❖❖ **I-13** **(A, D, E)** Figure PI-13 shows an RLC and an RC OP AMP circuit.

(a) **(A)** Show that both circuits have a transfer function of the form

$$T_V(S) = \frac{V_2(s)}{V_1(s)} = \frac{\omega_O^2}{s^2 + 2\zeta\omega_O s + \omega_O^2}$$

(b) **(D)** Select the element values in both circuits so that $\zeta = 1$ and $\omega_0 = 100$ rad/s.

(c) **(D)** Repeat (b) for $\omega_0 = 10^8$.

(d) **(E)** The components available to build the two circuits have $L \le 1$ H, $C \le 1$ μF, and OP AMP bandwidths of 10 MHz. Which circuit should be used to meet the requirements in parts (b) and (c) above?

❖❖ **I-14** **(A, D, E)** This problem uses circuits C1 and C2 in Fig. PI-14.

(a) **(A)** Determine $T_V(s) = V_2(s)/V_1(s)$ for each circuit, construct a pole-zero diagram, and relate the pole-zero locations to the circuit parameters.

(b) **(D)** Use a cascade connection of one or more of these circuits to realize the following transfer functions. *Hint:* A circuit may be used more than once.

$$T_1(s) = \frac{s^2}{(s+10)(s+400)} \quad \text{and}$$

$$T_2(s) = \frac{s(s+200)}{(s+100)(2+400)}$$

(c) **(E)** Show that a cascade connection of these circuits *cannot* realize the following transfer function and explain why in terms of poles and zeros.

$$T_3(s) = \frac{s(s+500)}{(s+100)(s+400)}$$

(d) **(E)** Develop a design rule that restricts the locations of the poles and zeros of $T(s)$ to ensure that it can be obtained using a cascade connection of these circuits.

❖ **I-15** **(A, D)** The circuit in Fig. PI-15 is a two-input cascade connection of an inverting summer and two inverting integrators.

(a) **(A)** Find the s-domain relationship between the output voltage and the two inputs voltages.

(b) **(A)** Remove the v_2 input and make a feedback connection from the output to the former v_2 input, thereby forcing the condition $v_2 = v_O$. Use the relationship found in (a) to determine the transfer function $T(s) = V_O(s)/V_1(s)$.

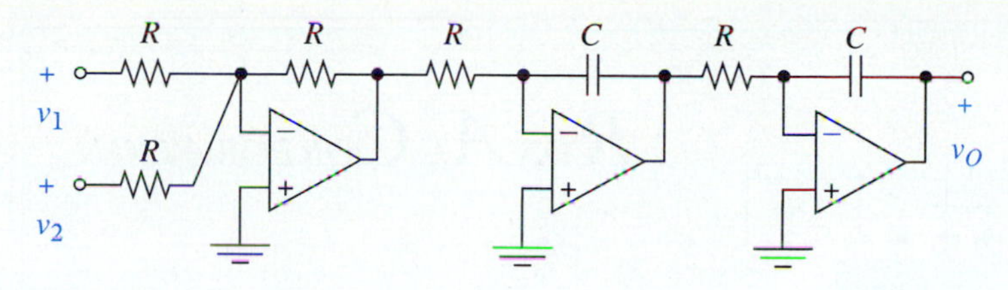

Figure PI-15

(c) **(A)** Construct a pole-zero diagram of $T(s)$ found in part (b) and show that the closed-loop circuit has poles on the j-axis in the s plane.

(d) **(D)** Select the values of R and C so the circuit oscillates with a period of 10 ms.

I-16 **(A, E)** The New Products Division of the RonAl Corporation (founded by two well-known authors) has announced the availability of an integrated circuit package called the Universal Single-Pole Transfer Function Synthesizer (USPTS). Figure PI-16 shows the first page of the data sheets on this new device. Unfortunately the second page, which describes how to interconnect the circuit pins to obtain each transfer function, has been lost in the mail.

(a) **(A)** Reconstruct the missing second page by showing how to interconnect the circuit pins to obtain each transfer function listed. All connections are short circuits from pin to pin, pin to ground, or pin to power. No external components other than the power supply are required. *Hint:* The output is always pin 7.

(b) **(E)** Explain why the transfer functions are independent of the load provided $RL > 2$ kΩ.

(c) **(E)** Explain why the RC products are controlled to ±1% whereas the OP AMP gain is only controlled to within a factor of 10.

I-17 **(A)** The RonAl Corporation has issued Application Note No. 1 on the USPTS integrated circuit described in Prob. PI-16 above. The note claims that interconnecting two circuits as shown in Fig. PI-17 produces a pair of complex poles with $\zeta = 1/\sqrt{2}$ and $\omega_O = \sqrt{2}/RC$.

(a) Verify or disprove the claim.

(b) The Application Note further claims that interchanging the connections at pins 2 and 3 on both circuits adds a zero at the origin in the s plane but does not change the poles. Verify or disprove this claim.

I-18 **(A)** Figure PI-18 shows an inverting amplifier and integrator with the same input and their outputs driving the inputs of a comparator. The comparator in this circuit

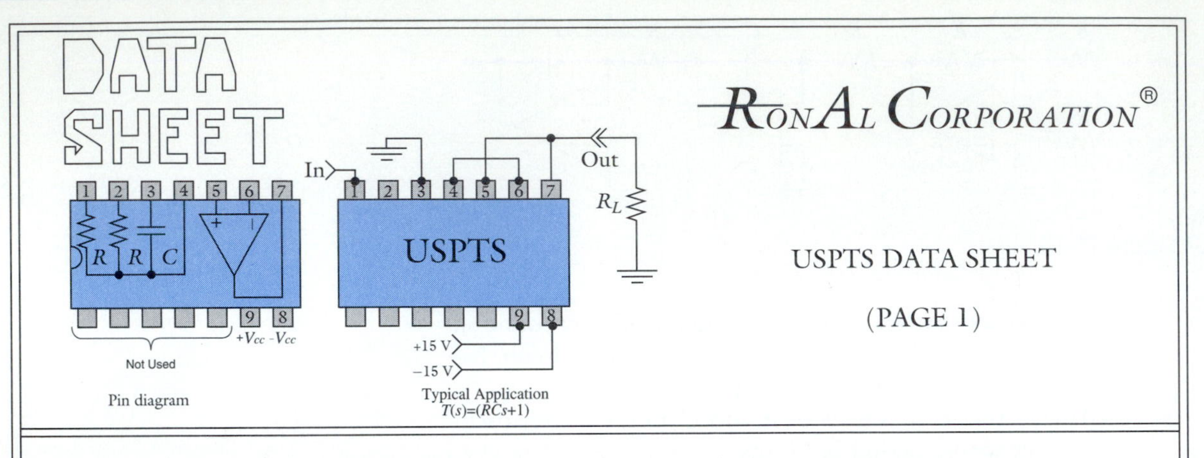

PERFORMANCE CHARACTERISTICS

1. OP AMP: For V_{CC} = 15 V output voltage swing = ±14 V output current (MAX) = 20 mA gain (MIN) = 50 V/mV gain (MAX) = 500 V/mV

2. For R_L > 2 kΩ the available transfer function are:

ALL POLE TYPE

$$\frac{1}{RCs+1},\quad \frac{2}{RCs+2},\quad \frac{1}{RCs+2},\quad -\frac{1}{RCs+1},\quad -\frac{1}{RCs},\quad -\frac{2}{RCs}$$

ALL ZERO TYPE

$$RCs+1,\quad \tfrac{1}{2}(RCs+2),\quad RCs+2,\quad -(RCs+1),\quad -RCs,\quad -\tfrac{1}{2}RCs$$

POLE-ZERO TYPE

$$\frac{RCs}{RCs+1},\quad \frac{RCs}{RCs+2},\quad \frac{RCs+1}{RCs+2},\quad \frac{RCs+1}{RCs},\quad \frac{RCs+2}{RCs},\quad \frac{RCs+2}{RCs+1}$$

3. To obtain different pole-zero locations order from standard models or order custom models in lots of 1000 or more.

Model	*RC(ms)*	*Tolerance*
USPTS - 0.1	0.1	±1%
USPTS - 0.3	0.3	±1%
USPTS - 1.0	1.0	±1%
USPTS - 3.0	3.0	±1%
USPTS - 10.0	10.0	±1%

Figure PI-16

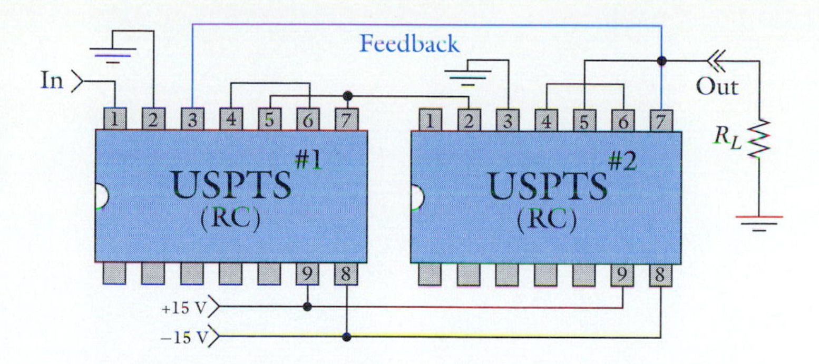

Figure PI-17

has an output $+V_{CC}$ when $v_P > v_N$ and an output $-V_{CC}$ when $v_P < v_N$. Throughout this problem assume that there is no initial voltage on the capacitor at $t = 0$.

(a) Show that when an input $v_1 = +V_{CC}$ is applied at $t = 0$ the comparator output is $+V_{CC}$ until $t = RC$ and is $-V_{CC}$ thereafter.

(b) Show that when an input $v_1 = -V_{CC}$ is applied at $t = 0$ the comparator output is $-V_{CC}$ until $t = RC$ and is $+V_{CC}$ thereafter.

(c) Now force the condition $v_1 = v_O$ by making a feedback connection from the comparator output to the input. Assume that the initial comparator output is $+V_{CC}$. Show that the comparator output is a square wave with a period of $4RC$ and the integrator output is a triangular wave with the same period.

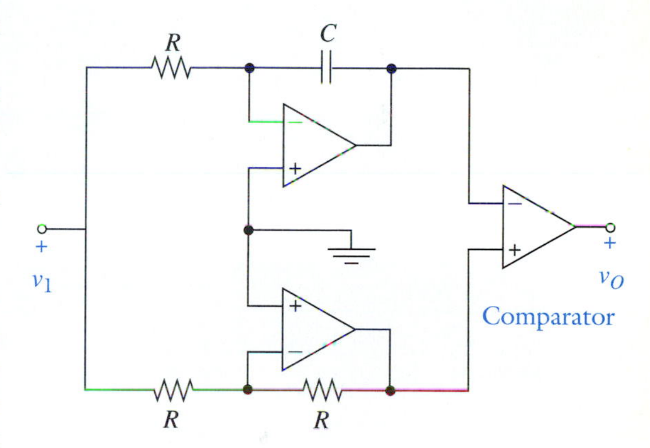

Figure PI-18

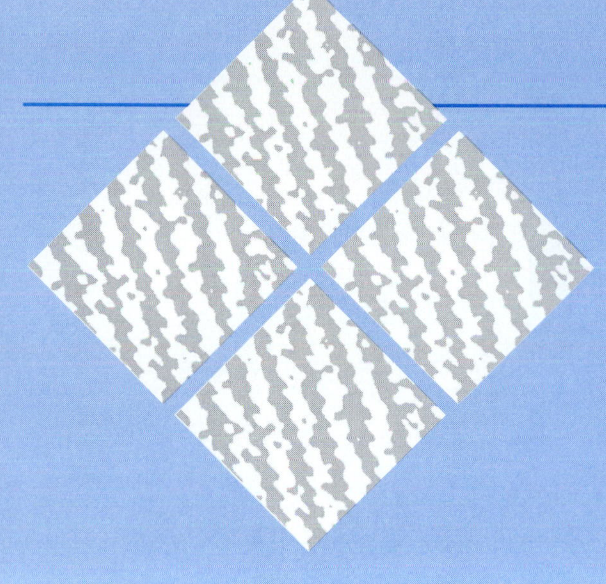

PART THREE

Applications

Course Objectives

- **Analysis (A)**—Given a linear circuit with known input signals, select an analysis technique and determine output signals or input-output relationships.
- **Design (D)**—Devise a linear circuit or modify an existing circuit to obtain a specified output signal or a specified input-output relationship.
- **Evaluation (E)**—Given several circuits that perform the same signal-processing function, rank-order the circuits using criteria such as cost, parts count, or power dissipation.

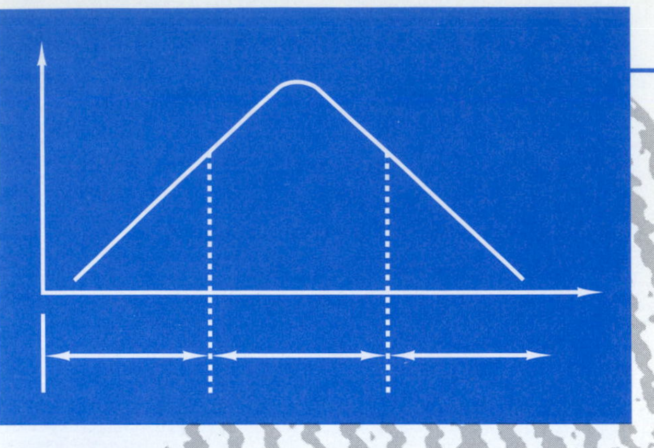

Chapter Twelve

"The advantage of the straight-line approximation is, of course, that it reduces the complete characteristics to a sum of elementary characteristics."

Hendrik W. Bode, 1945
American Engineer

FREQUENCY RESPONSE

In Chapter 11 we used network functions to study the transient response of linear circuits for impulse and step function inputs. In this chapter we turn our attention to the other standard test signal—the sinusoid. The sinusoidal steady-state response is the forced component of the zero-state response of a linear circuit for a sinusoidal input. In the steady state both the input and the output of a stable linear circuit are sinusoids of the same frequency. The relationships between the amplitudes and phase angles of the input and output sinusoids are frequency dependent. The frequency response of a circuit describes how these relationships change with the input frequency. The performance of devices and circuits are often specified in terms of frequency response descriptors.

The first section in this chapter shows that network functions contain all of the information needed to construct the sinusoidal steady-state response. The concept of frequency response and the parameters used to characterize the frequency response are discussed in the second section. The frequency response of first- and second-order circuits are discussed in the following two sections. The treatment of Bode diagrams in the fifth section is a generalization of the results obtained using basic first- and second-order building blocks. The final section introduces the frequency response analysis capability of SPICE.

Bode diagrams allow us to approximate the frequency response of a circuit using straight-line approximations. The name honors the American engineer Hendrik W. Bode (pronounced bow dee). Bode developed the straight-line approximations to simplify numerical calculations in the 1930s when the standard computational tool was a slide rule. The computational advantages of Bode diagrams are less important today with the advent of scientific calculators and personal computers. However, Bode plots are still useful because they describe frequency responses in a way that simplifies the analysis and design of circuits and systems. In fact, Bode plots are the industry-standard method of presenting frequency response information.

12-1 THE SINUSOIDAL STEADY STATE

In Chapter 6 we discussed two properties of the sinusoid that make it a very useful waveform for testing and describing linear circuits. The first is that the sum of two or more sinusoids of the same frequency is still a sinusoid. The second is that a sinusoid may be integrated or differentiated any number of times and the result remains a sinusoid. No other periodic waveform has these special properties. The result is that when a linear, time-invariant circuit is driven by a sinusoidal input, then all currents and voltages in the circuit contain a forced sinusoid of the same frequency as the input.

We use the familiar s-domain input-output relationship to mathematically describe the sinusoidal steady-state response:

$$Y(s) = T(s)X(s) \tag{12-1}$$

In the present case the input is a general sinusoidal waveform of the form

$$x(t) = X_A \cos(\omega t + \phi) \tag{12-2}$$

Expanding the right side of this expression by the trigonometric identity

$$\cos(x + y) = \cos x \cos y - \sin x \sin y$$

gives

$$x(t) = X_A(\cos \omega t \cos \phi - \sin \omega t \sin \phi) \tag{12-3}$$

The waveforms $\cos \omega t$ and $\sin \omega t$ in this expanded form are basic signals whose transforms are given in Table 9-2 as $\mathscr{L}\{\cos \omega t\} = s/(s^2 + \omega^2)$ and $\mathscr{L}\{\sin \omega t\} = \omega/(s^2 + \omega^2)$. Therefore, the input transform is

$$\begin{aligned} X(s) &= X_A \left[\frac{s}{s^2 + \omega^2} \cos \phi - \frac{\omega}{s^2 + \omega^2} \sin \phi \right] \\ &= X_A \left[\frac{s \cos \phi - \omega \sin \phi}{s^2 + \omega^2} \right] \end{aligned} \tag{12-4}$$

Equation (12-4) gives the Laplace transform of the general sinusoidal waveform in Eq. (12-2).

Using Eq. (12-1) we obtain the response transform for a general sinusoidal input:

$$Y(s) = X_A \left[\frac{s \cos \phi - \omega \sin \phi}{(s - j\omega)(s + j\omega)} \right] T(s) \tag{12-5}$$

The response transform contains forced poles at $s = \pm j\omega$ because the input is a sinusoid. As we saw in Chapter 10, a pair of conjugate poles on the j-axis produce a forced response that is sustained sinusoidal oscillation in the time domain. To describe the forced response we can expand Eq. (12-5) by partial fractions:

$$Y(s) = \underbrace{\frac{k}{s - j\omega} + \frac{k^*}{s + j\omega}}_{\text{forced poles}} + \underbrace{\frac{k_1}{s - p_1} + \frac{k_2}{s - p_2} + \cdots + \frac{k_N}{s - p_N}}_{\text{natural poles}}$$

where $p_1, p_2, \cdots p_N$ are the natural poles contained in the denominator of the transfer function $T(s)$. To obtain the response waveform we perform the inverse transformation:

$$y(t) = \underbrace{ke^{j\omega t} + k^* e^{-j\omega t}}_{\text{forced response}} + \underbrace{k_1 e^{p_1 t} + k_2 e^{p_2 t} + \cdots + k_N e^{p_N t}}_{\text{natural response}}$$

In the time domain the response contains exponential terms arising from the two forced poles and N exponential terms from the natural poles.

We now introduce a key assumption. If the circuit is stable, then all of the natural poles are in the left half of the s plane. Consequently, all of the exponential terms in the natural response decay to zero and only the steady-state component caused by the forced poles persists. In practical electrical circuits the steady-state condition is reached rather quickly, because the duration of the longest natural response term is usually no more than a few hundred milliseconds.

The persistent sinusoidal oscillation caused by the forced poles is called the *sinusoidal steady-state response*, or the *ac steady-state response*. In electrical engineering the term *steady state* means a condition reached after the natural response has died out. The steady-state concept only applies to stable circuits because the natural response of an unstable circuit does not decay to zero. The subscript (SS) is used to indicate the sinusoidal steady-state response. Since all of the natural components decay to zero, the steady-state response is

$$y_{SS}(t) = ke^{j\omega t} + k^* e^{-j\omega t} \tag{12-6}$$

To determine the amplitude and phase of the steady-state response we must find the residue k. Taking $Y(s)$ from Eq. (12-5) and using the cover-up method from Chapter 9 we determine k to be

$$k = (s - j\omega)\, X_A \left[\frac{s\cos\phi - \omega\sin\phi}{(s - j\omega)(s + j\omega)}\right] T(s)\bigg|_{s=j\omega}$$

$$= X_A \left[\frac{j\omega\cos\phi - \omega\sin\phi}{2j\omega}\right] T(j\omega)$$

$$= X_A \left[\frac{\cos\phi + j\sin\phi}{2}\right] T(j\omega)$$

Using Euler's relationship we can write $\cos\phi + j\sin\phi = e^{j\phi}$. Similarly, the complex quantity $T(j\omega)$ can be written in magnitude and angle form as $|T(j\omega)|e^{j\theta}$. Using these results the residue k becomes

$$k = X_A\left[\frac{1}{2}e^{j\phi}\right]|T(j\omega)|e^{j\theta}$$

$$= \frac{1}{2}X_A|T(j\omega)|e^{j(\phi+\theta)}$$

In Chapter 9 we showed that the time-domain waveform corresponding to a pair of conjugate poles is related to the residue by $2|k|\cos(\omega t + \angle k)$. Using the residue found above the steady-state response in Eq. (12-6) is

$$y_{SS}(t) = \underbrace{X_A|T(j\omega)|}_{\text{amplitude}}\cos(\omega t + \underbrace{\phi + \theta}_{\text{phase}}) \tag{12-7}$$

The steady-state response is a sinusoid with the **same frequency** as the input but with a **different amplitude** and **phase angle**. The relationships between the input and output sinusoids can be summarized in the following statements:

$$\text{output amplitude} = (\text{input amplitude}) \times (\text{magnitude of } T(j\omega))$$

$$\text{output phase} = (\text{input phase}) + (\text{angle of } T(j\omega))$$

Alternatively, we can express the transfer function in terms of the input and output sinusoids as

$$\text{magnitude of } T(j\omega) = \frac{\text{output amplitude}}{\text{input amplitude}}$$

$$\text{angle of } T(j\omega) = \text{output phase} - \text{input phase}$$

The next two examples use these results to calculate steady-state responses.

EXAMPLE 12-1

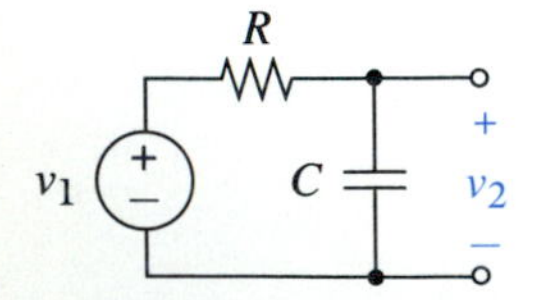

Figure 12-1

Find the sinusoidal steady-state output of the circuit in Fig. 12-1 for an input

$$v_1(t) = V_A\cos(\omega t + \phi).$$

SOLUTION:

The circuit transfer function is found by voltage division:

$$T(s) = \frac{1/Cs}{R + 1/Cs} = \frac{1}{RCs + 1}$$

The magnitude and angle of $T(j\omega) = 1/(1 + j\omega RC)$ are

$$|T(j\omega)| = \frac{1}{\sqrt{1 + (\omega RC)^2}}$$

$$\theta(\omega) = -\tan^{-1}(\omega RC)$$

From Eq. (12-7) the sinusoidal steady-state output is

$$v_{2SS}(t) = \frac{V_A}{\sqrt{1 + (\omega RC)^2}} \cos\left[\omega t + \phi - \tan^{-1}(\omega RC)\right]$$

Note that both the amplitude and phase angle of the steady-state response depend upon the frequency of the input sinusoid.

EXAMPLE 12-2

The step response of a linear circuit is

$$g(t) = 5\left[e^{-1000t} \sin\ 2000t\right]\ u(t).$$

(a) Find the sinusoidal steady-state response when

$$x(t) = 5\ \cos 1000t.$$

(b) Repeat part (a) when $x(t) = 5\ \cos\ 3000t$.

SOLUTION:

The transfer function corresponding to $g(t)$ is

$$\begin{aligned} T(s) &= sG(s) = s\mathscr{L}\{g(t)\} \\ &= s\left[5\frac{2000}{(s + 1000)^2 + (2000)^2}\right] \\ &= \frac{10^4 s}{s^2 + 2000s + 5 \times 10^6} \end{aligned}$$

(a) At $\omega = 1000$ rad/s the value of $T(j\omega)$ is

$$\begin{aligned} T(j1000) &= \frac{10^4(j1000)}{(j1000)^2 + 2000(j1000) + 5 \times 10^6} \\ &= \frac{j10^7}{(5 \times 10^6 - 10^6) + j2 \times 10^6} = \frac{j10}{4 + j2} \\ &= \frac{10e^{j90^\circ}}{\sqrt{20}e^{j26.6^\circ}} = 2.24e^{j63.4^\circ} \end{aligned}$$

and the steady-state response for $x(t) = 5\ \cos\ 1000t$ is

$$\begin{aligned} y_{SS}(t) &= 5 \times 2.24\ \cos(1000t + 0^\circ + 63.4^\circ) \\ &= 11.2\ \cos(1000t + 63.4^\circ) \end{aligned}$$

(b) At $\omega = 3000$ rad/s the value of $T(j\omega)$ is

$$T(j3000) = \frac{10^4(j3000)}{(j3000)^2 + 2000(j3000) + 5 \times 10^6}$$

$$= \frac{j3 \times 10^7}{5 \times 10^6 - 9 \times 10^6 + j6 \times 10^6}$$

$$= \frac{j30}{-4 + j6} = \frac{30e^{j90^\circ}}{\sqrt{52}e^{j123.7^\circ}} = 4.16e^{-j33.7^\circ}$$

and the steady-state response for $x(t) = 5 \cos 2000t$ is

$$y_{SS}(t) = 5 \times 4.16 \cos(2000t + 0^\circ - 33.7^\circ)$$

$$= 20.8 \cos(2000t - 33.7^\circ)$$

Note that the amplitude and phase angle of the steady-state response change, even though the amplitude and phase of the input do not change. Again we see that the steady-state amplitude and phase angle depend on the input frequency.

EXERCISE 12-1

The transfer function of a linear circuit is $T(s) = 5(s + 100)/(s + 500)$. Find the steady-state output for
(a) $x(t) = 3 \cos 100t$
(b) $x(t) = 2 \sin 500t$.

ANSWERS:
(a) $y_{SS}(t) = 4.16 \cos(100t + 33.7^\circ)$.
(b) $y_{SS}(t) = 7.21 \cos(500t - 56.3^\circ)$. ■

EXERCISE 12-2

The impulse response of a linear circuit is $h(t) = \delta(t) - 100[e^{-100t}]u(t)$. Find the steady-state output for
(a) $x(t) = 25 \cos 100t$
(b) $x(t) = 50 \sin 100t$.

ANSWERS:
(a) $y_{SS}(t) = 17.7 \cos(100t + 45^\circ)$.
(b) $y_{SS}(t) = 35.4 \cos(100t - 45^\circ)$. ■

12-2 FREQUENCY-RESPONSE DESCRIPTORS

The circuit transfer function influences the sinusoidal steady-state response through the *gain* function $|T(j\omega)|$ and *phase* function $\theta(\omega)$.

$$\text{output amplitude} = |T(j\omega| \times (\text{input amplitude})$$

$$\text{output phase} = \text{input phase} + \theta(\omega)$$

The gain and phase functions show how the circuit modifies the input amplitude and phase angle to produce the output sinusoid. The two functions describe the *frequency response* of the circuit, since they depend on the frequency of the input sinusoid. The signal-processing performance of devices, circuits, and systems are often specified in terms of their frequency response. The gain and phase functions can be expressed mathematically or graphically as frequency-response plots. Figure 12-2 shows examples of gain and phase responses versus frequency ω.

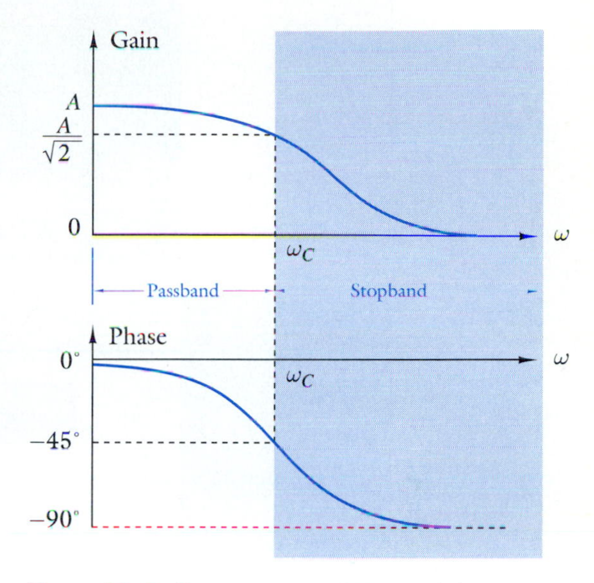

Figure 12-2 Frequency response plots.

The terminology used to describe the frequency response of circuits and systems is based on the form of the gain plot. For example, at high frequencies the gain in Fig. 12-2 falls off so that output signals in this frequency range are reduced in amplitude. The range of frequencies over which the output is significantly attenuated is called the *stopband*. At low frequencies the gain is essentially constant and there is relatively little attenuation. The frequency range over which there is little attenuation is called a *passband*. The frequency associated with the boundary between a passband and an adjacent stopband is called the *cutoff frequency* ($\omega_C = 2\pi f_C$). In general, the transition from the passband to the stopband is gradual so the precise location of the cutoff frequency is a matter of definition. The most widely used approach defines the cutoff frequency to be the frequency at which the gain has decreased by a factor of $1/\sqrt{2} = 0.707$ from its maximum value in the passband.

Again this definition is arbitrary, since there is no sharp boundary between a passband and an adjacent stopband. However, the definition is based on the fact that the power delivered to a resistance by a sinusoidal current or voltage waveform is proportional to the square of its amplitude. At a cutoff frequency the gain is reduced by a factor of $1/\sqrt{2}$ and the square of the output amplitude is reduced by a factor of one-half. For this reason the cutoff frequency is also called the *half-power frequency*.

Additional frequency-response descriptors are based on the four prototype gain characteristics shown in Fig. 12-3. A *low-pass* gain characteristic has a single passband extending from zero frequency (dc) to the cutoff frequency. A *high-pass* gain characteristic has a single passband extending from the cutoff frequency to infinite frequency. A *bandpass* gain has a single passband with two cutoff frequencies neither of which is zero or infinite. Finally, the *bandstop* gain has a single stopband with two cutoff frequencies neither of which is zero or infinite.

The *bandwidth* of a gain characteristic is defined as the frequency range spanned by its passband. For the bandpass case

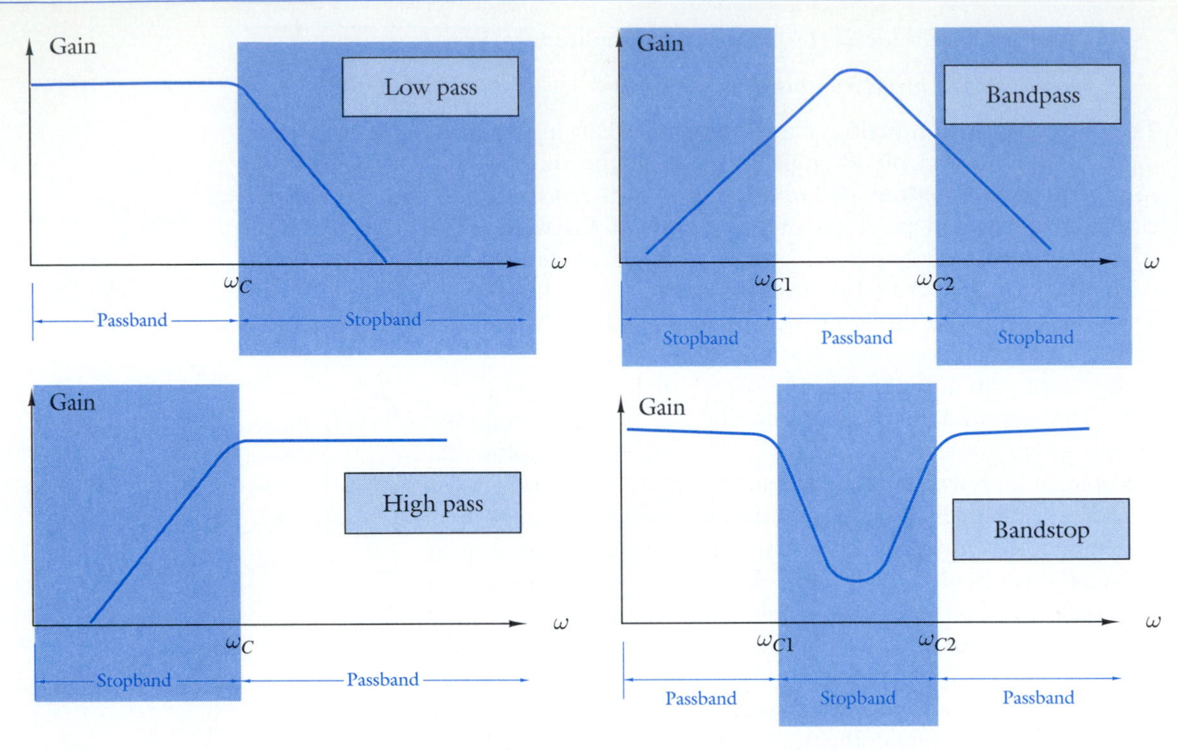

Figure 12-3 Prototype gain responses.

in Fig. 12-3 the bandwidth is the difference in the two cutoff frequencies.

$$\mathrm{BW} = \omega_{C2} - \omega_{C1} \tag{12-8}$$

This equation applies to the low-pass response with the lower cutoff frequency ω_{C1} set to zero. In other words, the bandwidth of a low-pass circuit is equal to its cutoff frequency ($BW = \omega_C$). The bandwidth of a high-pass characteristic is infinite since the upper cutoff frequency ω_{C1} is infinity. For the bandstop case Eq. (12-8) defines the bandwidth of the stopband rather than the passbands.

Frequency-response plots are almost always made using logarithmic scales for the frequency variable. The reason is that the frequency ranges of interest often span several orders of magnitude. A logarithmic frequency scale compresses the data range and highlights important features in the gain and phase responses. The use of a logarithmic frequency scale involves some special terminology. Any frequency range whose end points have a 2:1 ratio is called an *octave*. Any range whose endpoints have a 10:1 ratio is called a *decade*. For example, the frequency range from 10 Hz to 20 Hz is one octave, as is the range from 20 kHz to 40 kHz. The standard microwave L band used for radar and military communication spans one octave from 1 to 2 GHz, while the standard VHF (very high frequency) band used for FM radio and television broadcast spans one decade from 30 to 300 MHz.

In frequency-response plots the gain $|T(j\omega)|$ is usually expressed in *decibels* (dB), defined as

$$|T(j\omega)|_{dB} = 20\log_{10}|T(j\omega)| \tag{12-9}$$

To effectively use frequency-response plots the reader must have some familiarity with gains expressed in decibels. First note that the gain in dB can be either positive, negative or zero. A gain of zero dB means that $|T(j\omega)| = 1$; i.e, the input and output amplitudes are equal. Positive dB gains mean the output amplitude exceeds the input, since $|T(j\omega)| > 1$. A negative dB gains means the output amplitude is smaller than the input, since $|T(j\omega)| < 1$. A cutoff frequency occurs when the gain is reduced from its maximum passband value by a factor $1/\sqrt{2}$. Expressed in dB this is a gain reduction of

$$\begin{aligned} 20\log_{10}\left(\frac{1}{\sqrt{2}}|T|_{MAX}\right) &= 20\log_{10}|T_{MAX}| - 20\log_{10}\sqrt{2} \\ &= |T|_{MAX,\ dB} - 3\text{ dB} \end{aligned}$$

That is, cutoff occurs when the dB gain is reduced by about 3 dB. For this reason the cutoff is also called the *3-dB down frequency*.

The generalization of the above idea is that multiplying $|T(j\omega)|$ by a factor K means $20\log(K|T(j\omega)|) = |T(j\omega)|_{dB} + 20\log(K)$. That is, multiplying the gain by K changes the gain in dB by an additive factor K_{dB}. Some multiplicative factors whose dB equivalences are worth remembering are:

- A factor $K = 1$ changes $|T(j\omega)|_{dB}$ changes by 0 dB.
- A factor $K = \sqrt{2}$ $(1/\sqrt{2})$ changes $|T(j\omega)|_{dB}$ by about +3 dB (−3 dB).
- A factor $K = 2$ (1/2) changes $|T(j\omega)|_{dB}$ by about +6 dB (−6 dB).
- A factor $K = 10$ (1/10) changes $|T(j\omega)|_{dB}$ by +20 dB (−20 dB).
- A factor $K = 30$ (1/30) changes $|T(j\omega)|_{dB}$ by about +30 dB (−30 dB).
- A factor $K = 100$ (1/100) changes $|T(j\omega)|_{dB}$ by +40 dB (−40 dB).
- A factor $K = 1000$ (1/1000) changes $|T(j\omega)|_{dB}$ by +60 dB (−60 dB).

In summary, frequency-response plots use logarithmic frequency scales and linear scales for the gain (in dB) and phase (in degrees). Both plots are semilog graphs, although in effect the gain plot is log / log because of the logarithmic definition of $|T(j\omega)|_{dB}$. To create frequency-response plots we must calculate $|T(j\omega)|$ and $\theta(\omega)$ at a sufficient number of frequencies to adequately define the stop and passband characteristics. The most efficient way to generate accurate plots is to use one of the many available computer programs. Spread sheets, mathematical software packages, or scientific calculators can be used to calculate frequency-response curves for a given transfer function. Analysis processors such as SPICE and MICRO-CAP have ac analysis options that generate the frequency-response data for a circuit. We will treat the SPICE ac analysis option later in this chapter.

As always, some knowledge of the expected response is required to use these computational tools effectively. Using a computer to generate frequency-response plots requires some preliminary analysis or at least a rough sketch of the expected frequency response. To develop the required insight, we treat first- and second-order circuits and then use Bode plots to show how these building blocks combine to produce the frequency response of more complicated circuits.

Application Note

EXAMPLE 12-3

The use of the decibel as a measure of performance pervades the literature and folklore of electrical engineering. The decibel originally came from the definition of power ratios in *bels*.[1]

$$\text{number of bels} = \log_{10}\left(\frac{P_{OUT}}{P_{IN}}\right)$$

In practice, the decibel (dB) is a more convenient measure of power ratios since the bel is a rather large quantity. It follows that the number of dB is 10 times the number of bels:

$$\text{number of dB} = 10 \times (\text{number of bels}) = 10\log_{10}\frac{P_{OUT}}{P_{IN}}$$

When the input and output powers are delivered to equal input and output resistances R, the power ratio can be expressed in terms of voltages across the resistances:

$$\text{number of dB} = 10\log_{10}\frac{v_{OUT}^2/R}{v_{IN}^2/R} = 20\log_{10}\frac{v_{OUT}}{v_{IN}}$$

[1]The name of the unit honors Alexander Graham Bell (1847–1922), the inventor of the telephone.

or in terms of currents through the resistances:

$$\text{number of dB} = 10\log_{10}\frac{i_{OUT}^2 \times R}{i_{IN}^2 \times R} = 20\log_{10}\frac{i_{OUT}}{i_{IN}}$$

The definition of gain in dB in Eq. (12-9) is consistent with these results, since in the sinusoidal steady state the transfer function equals the ratio of output amplitude to input amplitude. The discussion above is not a derivation of Eq. (12-9) but simply a summary of its historical origin. In practice Eq. (12-9) is applied when the input and output are not measured across resistances of equal value.

When the chain rule applies to a cascade connection the overall transfer function is a product

$$T(j\omega) = T_1 \times T_2 \times \ldots T_N$$

where T_1, T_2, ... T_N are the transfer functions of the individual stages in the cascade. Expressed in dB the overall gain is

$$\begin{aligned}|T(j\omega)|_{dB} &= 20\log_{10}(|T_1| \times |T_2| \times \ldots |T_N|)\\ &= 20\log_{10}|T_1| + 20\log_{10}|T_2| + \ldots + 20\log_{10}|T_N|\\ &= |T_1|_{dB} + |T_2|_{dB} + \ldots + |T_N|_{dB}\end{aligned}$$

Because of the logarithmic definition, the overall gain (in dB) is the sum of the gains (in dB) of the individual stages in a cascade connection. The effect of altering a stage or adding an additional stage can be calculated by simply adding or subtracting the change in dB. Since summation is simpler than multiplication, the enduring popularity of the dB comes from its logarithmic definition and not its tenuous relationship to power ratios.

12-3 First-Order Circuit Frequency Response

First-Order Low-Pass Response

We begin the study of frequency response with the first-order low-pass transfer function:

$$T(s) = \frac{K}{s+\alpha} \qquad (12\text{-}10)$$

The constants K and α are real. The constant K can be positive or negative, but α must be positive so that the natural pole at

$s = -\alpha$ is in the left half of the s plane to ensure that the circuit is stable. Remember the concepts of sinusoidal steady state and frequency response do not apply to unstable circuits that have poles in the right half of the s plane or on the j axis.

To describe the frequency response of the low-pass transfer function, we replace s by $j\omega$ in Eq. (12-10).

$$T(j\omega) = \frac{K}{\alpha + j\omega} \tag{12-11}$$

and express the gain and phase functions as

$$\begin{aligned} |T(j\omega)| &= \frac{|K|}{\sqrt{\omega^2 + \alpha^2}} \\ \theta(\omega) &= \angle K - \tan^{-1}(\omega/\alpha) \end{aligned} \tag{12-12}$$

The gain function is a positive number. Since K is real the angle of $K(\angle K)$ is either $0°$ when $K > 0$ or $\pm 180°$ when $K < 0$. An example of a negative K occurs in an inverting OP AMP configuration where $T(s) = -Z_2(s)/Z_1(s)$.

Figure 12-4 shows the gain and phase functions versus normalized frequency ω/α. The maximum passband gain occurs at $\omega = 0$ where $|T(0)| = |K|/\alpha$. As frequency increases the gain gradually decreases until at $\omega = \alpha$

$$|T(j\alpha)| = \frac{|K|}{\sqrt{\alpha^2 + \alpha^2}} = \frac{|K|/\alpha}{\sqrt{2}} = \frac{|T(0)|}{\sqrt{2}} \tag{12-13}$$

That is, the cutoff frequency of the transfer function in Eq. (12-10) is $\omega_C = \alpha$. The graph of the gain function in Fig. 12-4(a) displays a low-pass characteristic with a passband between from zero frequency and $\omega = \alpha$.

The low- and high-frequency gain asymptotes shown in Fig. 12-4(a) are especially important. The low-frequency asymptote is the horizontal line and the high-frequency asymptote is the sloped line. At low frequencies ($\omega \ll \alpha$) the gain approaches $|T(j\omega)| \to |K|/\alpha$. At high frequencies ($\omega \gg \alpha$) the gain approaches $|T(j\omega)| \to |K|/\omega$. The intersection of the two asymptotes occurs when $|K|/\alpha = |K|/\omega$. The intersection forms a "corner" at $\omega = \alpha$, so the cutoff frequency is also called the *corner frequency*.

The high-frequency gain asymptote decreases by a factor of 10 (−20 dB) whenever the frequency increases by a factor of 10 (one decade). As a result the high-frequency asymptote has a slope of −20 dB per decade and the low-frequency asymptote has a slope of 0 db/decade. These two asymptotes provide a straight-line approximation to the gain response.

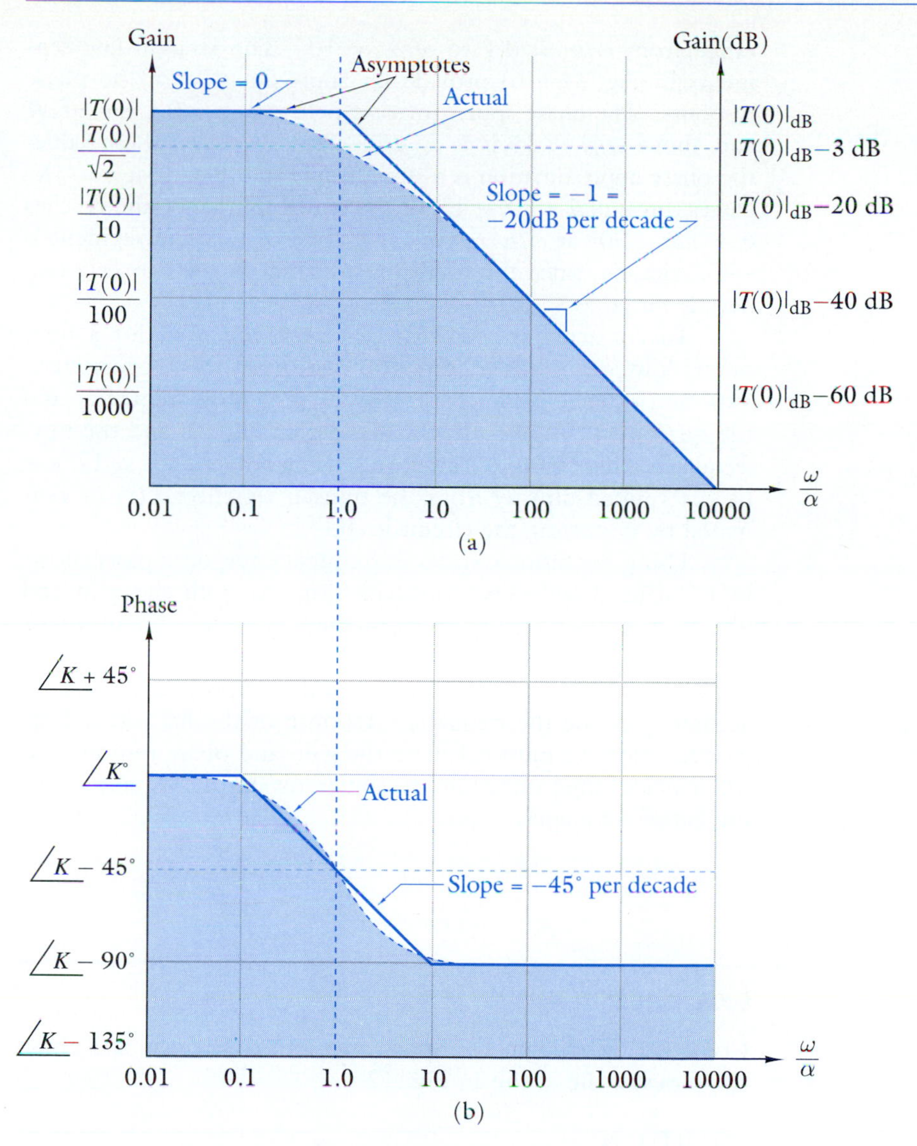

Figure 12-4 First-order low-pass frequency response plots.

The semilog plot of the phase shift of first-order low-pass transfer function is shown in Fig. 12-4(b). At $\omega = \alpha$ the phase angle in Eq. (12-12) is

$$\theta(\omega_C) = \angle K - \tan^{-1}\left(\frac{\alpha}{\alpha}\right)$$
$$= \angle K - 45^\circ$$

At low frequency ($\omega \ll \alpha$) the phase angle approaches $\angle K$ and at high frequencies ($\omega \gg \alpha$) the phase approaches $\angle K - 90^\circ$. Most of the total of -90° phase change occurs in the two decade range from $\omega/\alpha = 0.1$ to $\omega/\alpha = 10$. The straight-line segments in Fig. 12-4(b) provide an approximation of the phase response. The phase approximation below $\omega/\alpha = 0.1$ is $\theta = \angle K$ and above $\omega/\alpha = 10$ is $\theta = \angle K - 90°$. Between these values the phase approximation is a straight line that begins at $\theta = \angle K$, passes through $\theta = \angle K - 45°$ at the cutoff frequency and reaches $\theta = \angle K - 90°$ at $\omega/\alpha = 10$. The slope of this line segment is $-45°$/decade, since the total phase change is $-90°$ over a two decade range.

To construct the *straight-line approximations* for a first-order low-pass transfer function we need two parameters, $T(0)$ and α. The parameter α defines the cutoff frequency and the quantity $T(0)$ gives the passband gain $|T(\infty)|$ and the low-frequency phase $\angle T(0)$. The required quantities $T(0)$ and α can be determined directly from the transfer function $T(s)$ or estimated by inspecting the circuit itself.

Using logarithmic scales in frequency-response plots allows us to make straight-line approximations to both the gain and phase responses. These approximations provide a useful, quick estimate of a circuit frequency response. Often such graphical estimates are adequate for analysis and design purposes. To more accurately define the frequency response of the first-order low-pass function we must calculate the gain and phase responses at frequencies range from one decade below to one decade above the cutoff frequency.

EXAMPLE 12-4

Construct the straight-line approximations to the gain and phase responses of the circuit in Fig. 12-1.

SOLUTION:

Example 12-1 shows the voltage transfer function for the circuit to be

$$T(s) = \frac{1/RC}{s + 1/RC}$$

From Eq. (12-10) this transfer function has $\alpha = 1/RC$ and $T(0) = 1$. Therefore, $|T(0)|_{dB} = 0$ dB, $\omega_C = 1/RC$ and $\angle K = 0°$. Given these quantities we construct the straight-line approximations shown in Fig. 12-5. Note that the frequency scale in Fig. 12-5 is logarithmic and is normalized by multiplying ω by $RC = 1/\alpha$.

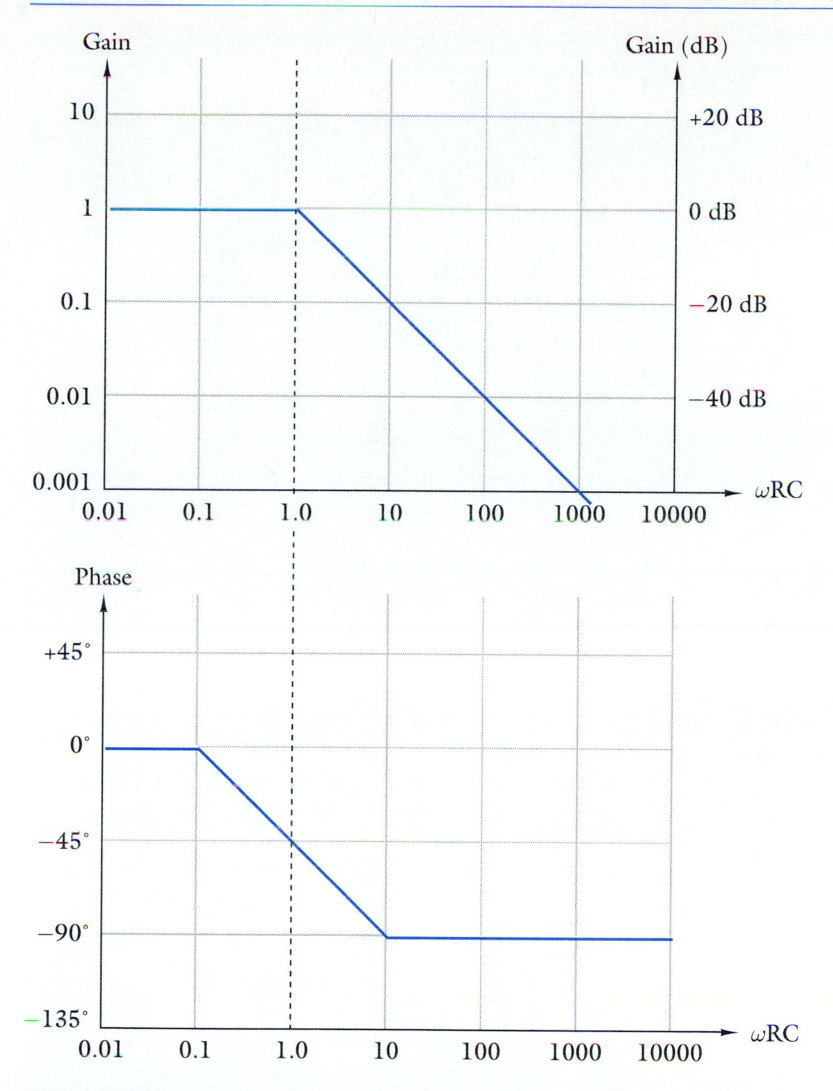

Figure 12-5

❖ Design Example

EXAMPLE 12-5

(a) Construct a first-order low-pass transfer function with a passband gain of 4 and a cutoff frequency of 20 kHz.

(b) Design an RC circuit to realize the transfer function found in (a).

SOLUTION:

(a) A first-order low-pass transfer function has the form given in Eq. (12-10).

$$T(s) = \frac{K}{s + \alpha}$$

The specified cutoff frequency is $f_C = 20$ kHz and therefore $\omega_C = \alpha = 2\pi(2 \times 10^4)$ rad/s. The specified passband gain is $|T(0)| = |K|/\alpha = 4$ and therefore $|K| = 4\alpha$. Since the scale factor K can be either positive or negative the required transfer function is

$$T(s) = \pm\frac{4(4\pi \times 10^4)}{s + 4\pi \times 10^4}$$

(b) A design method for the inverting OP AMP circuit is discussed in Section 11-3. Using that design method we factor $4\pi \times 10^4$ out of the numerator and denominator of the required transfer function and equate the result to the transfer function of the inverting circuit $(-Z_2/Z_1)$:

$$T(s) = -\frac{4}{\dfrac{s}{4\pi \times 10^4} + 1} = -\frac{Z_2(s)}{Z_1(s)}$$

Equating Z_1 to the reciprocal of the numerator and Z_2 to the reciprocal of the denominator in this expression yields the following branch impedance identifications:

$$Z_1(s) = \frac{1}{4} \quad \text{and} \quad Z_2(s) = \frac{1}{\dfrac{s}{4\pi \times 10^4} + 1}$$

The impedance $Z_1(s)$ is a resistance ($R_1 = 1/4\Omega$) and $Z_2(s)$ is a resistance ($R_2 = 1\ \Omega$) and a capacitance ($C_2 = 1/[4\pi \times 10^4]$F) in parallel. Fig. 12-6(a) shows a prototype circuit using these branch impedances. A magnitude scale factor of $k_m = 10^4$ produces the practical element values in the final design shown in Fig. 12-6(b).

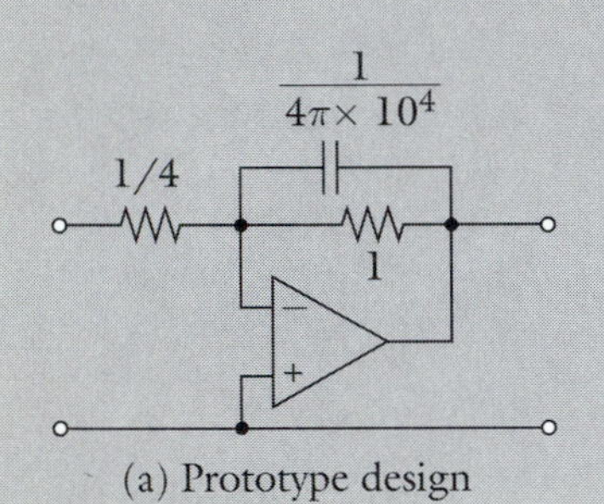

(a) Prototype design

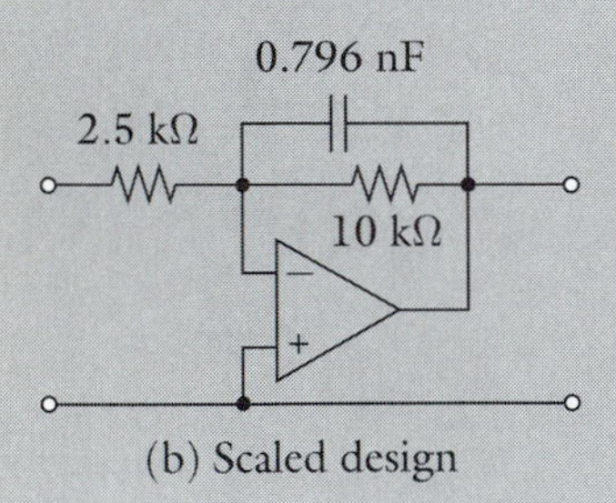

(b) Scaled design

Figure 12-6

Application Note

EXAMPLE 12-6

The ideal OP AMP model introduced in Chapter 4 assumes that the device has an infinite gain and an infinite bandwidth. A more realistic model of the device is shown in Fig. 12-7(a). The controlled source gain in Fig. 12-7(a) is a low-pass transfer function

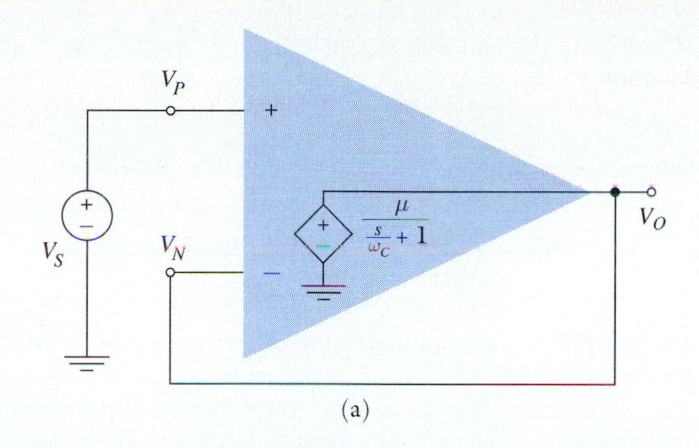

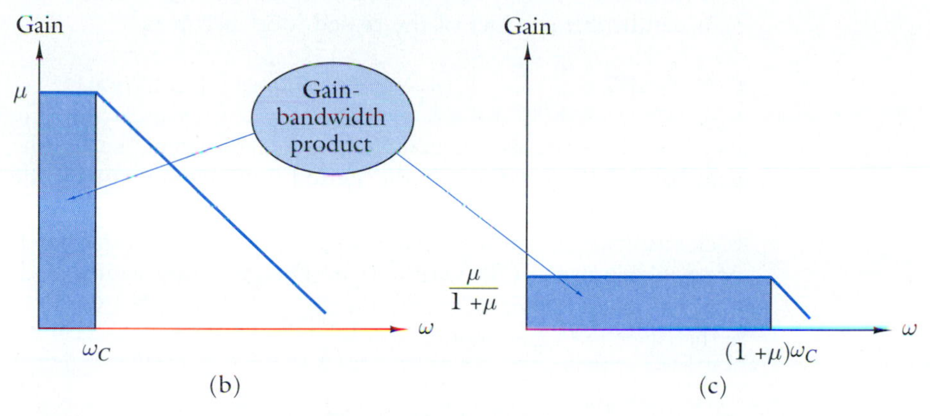

Figure 12-7

with a dc gain of μ and a cutoff frequency ω_C. The straight-line asymptotes of controlled source gain are shown in Fig. 12-7(b). The area under the gain response within the passband is called the *gain-bandwidth product* (G-BW).

With no feedback the OP AMP transfer function is the same as the controlled-source transfer function. The gain-bandwidth product of the open-loop transfer function is

$$\text{G-BW} = \mu\omega_C \text{ (open loop)}$$

The closed-loop transfer function of the circuit in Fig. 12-7(a) is found by writing the following device and connection equations.

$$\text{Device Equation:} \quad V_O = \frac{\mu}{s/\omega_C + 1}(V_P - V_N)$$

$$\text{Input Connection:} \quad V_P = V_S$$

$$\text{Feedback Connection:} \quad V_N = V_O$$

Substituting the connection equations into the OP AMP device equation yields

$$V_O = \frac{\mu}{s/\omega_C + 1}(V_S - V_O)$$

Solving for the closed-loop transfer function produces

$$T(s) = \frac{V_O(s)}{V_S(s)} = \frac{\mu}{\mu+1}\left[\frac{1}{\dfrac{s}{(\mu+1)\omega_C}+1}\right]$$

The straight-line asymptotes of the closed-loop transfer function are shown in Fig. 12-7(c).

The closed-loop circuit has a low-pass transfer function with a dc gain of $\mu/(\mu+1)$ and a cutoff frequency of $(\mu+1)\omega_C$. The gain-bandwidth product of the closed-loop circuit is

$$\text{G-BW} = \left[\frac{\mu}{\mu+1}\right][(\mu+1)\omega_C] = \mu\omega_C \ \text{(closed loop)}$$

which is the same as the open-loop case. In other words, the gain-bandwidth product of the closed-loop circuit equals the open-loop gain-bandwidth product of the OP AMP device. Feedback redistributes but does not change the gain-bandwidth product available in the OP AMP device. The gain-bandwidth product is a better descriptor of the limitations of an OP AMP than is the open-loop gain alone.

First-Order High-Pass Response

We next treat the first-order high-pass transfer function

$$T(s) = \frac{Ks}{s+\alpha} \tag{12-14}$$

The high-pass function differs from the low-pass case by the introduction of a zero at $s = 0$. Replacing s by $j\omega$ in $T(s)$ and solving for the gain and phase functions yields

$$|T(j\omega)| = \frac{|K|\omega}{\sqrt{\omega^2+\alpha^2}}$$
$$\theta(\omega) = \angle K + 90^\circ - \tan^{-1}(\omega/\alpha) \tag{12-15}$$

Fig. 12-8 shows the gain and phase functions versus normalized frequency ω/α. The maximum gain occurs at high frequency $(\omega \gg \alpha)$ where $|T(j\omega)| \to |K|$. At low frequency $(\omega \ll \alpha)$ the gain approaches $|K|\omega/\alpha$. At $\omega = \alpha$ the gain is

$$|T(j\alpha)| = \frac{|K|\alpha}{\sqrt{\alpha^2+\alpha^2}} = \frac{|K|}{\sqrt{2}} \tag{12-16}$$

which means the cutoff frequency is $\omega_C = \alpha$. The gain response

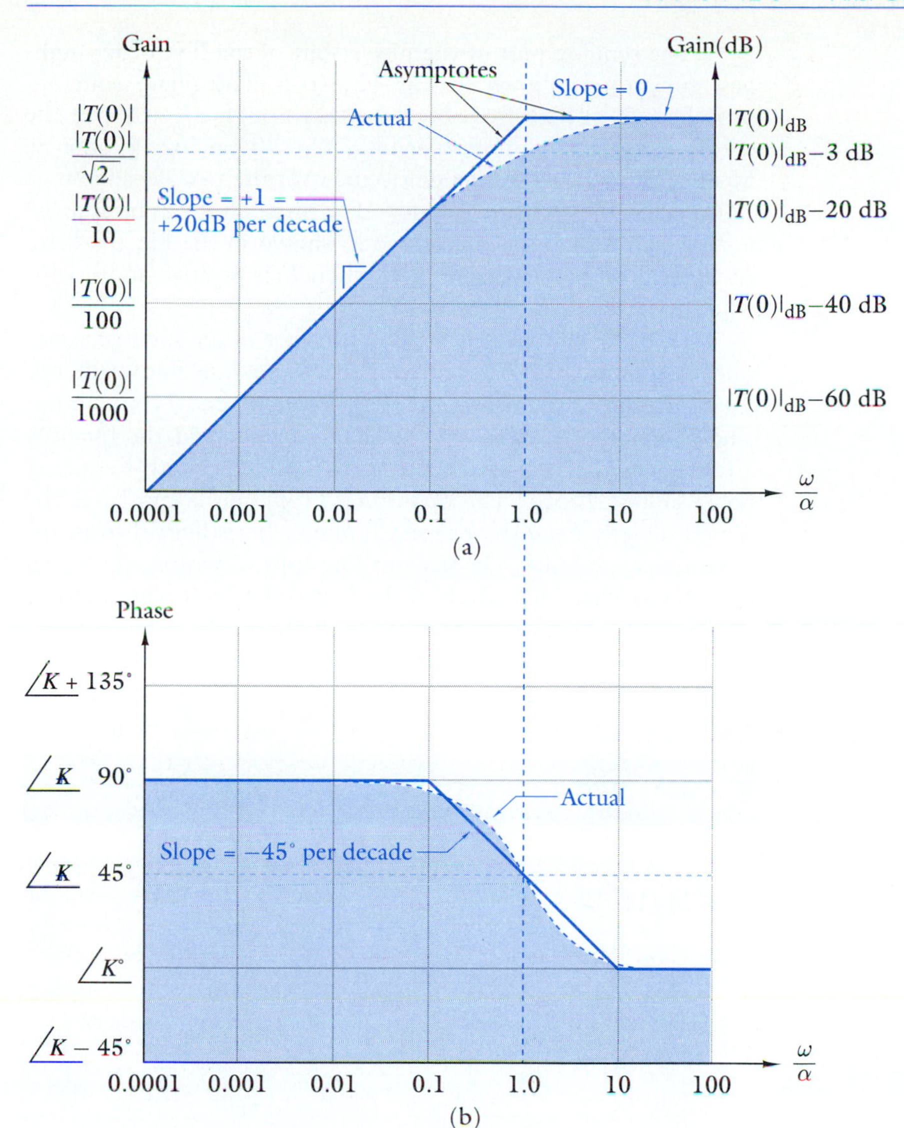

Figure 12-8 First-order high-pass frequency response plots.

plot in Fig. 12-8(a) displays a high-pass characteristic with a pass-band extending from $\omega = \alpha$ to infinity and a stop band between zero frequency and $\omega = \alpha$.

The low- and high-frequency gain asymptotes approximate the gain response in Fig. 12-8(a). The high-frequency asymptote ($\omega \gg \alpha$) is the horizontal line whose ordinate is $|K|$ (slope = 0 dB/decade). The low-frequency asymptote ($\omega \ll \alpha$) is a line of the form $|K|\omega/\alpha$ (slope = +20 dB/decade). The intersection of these two asymptotes occurs when $|K| = |K|\omega/\alpha$, which defines a corner frequency at $\omega = \alpha$.

The semilog plot of the phase shift of the first-order high-pass function is shown in Fig. 12-8(b). The phase shift approaches $\angle K$ at high frequency, passes through $\angle K + 45^\circ$ at the cutoff frequency, and approaches $\angle K + 90^\circ$ at low frequency. Most of the 90° phase change occurs over the two-decade range centered on the cutoff frequency. The phase shift can be approximated by the straight-line segments shown in the Fig. 12-8(b). As in the low-pass case, $\angle K$ is 0° when K is positive and ±180° when K is negative.

Like the low-pass function, the first-order high-pass frequency response can be approximated by straight-line segments. To construct these lines we need two parameters, $T(\infty)$, and α. The parameter α defines the cutoff frequency and the quantity $T(\infty)$ gives the passband gain $|T(\infty)|$ and the high-frequency phase angle $\angle T(\infty)$. The quantities $T(\infty)$ and α can be determined directly from the transfer function or estimated from the circuit in some cases. The straight-line approximations show that an accurate characterization of the first-order high-pass response requires that we calculate the gain and phase at frequencies on a two-decade range centered on the cutoff frequency.

❖ Design Example

EXAMPLE 12-7

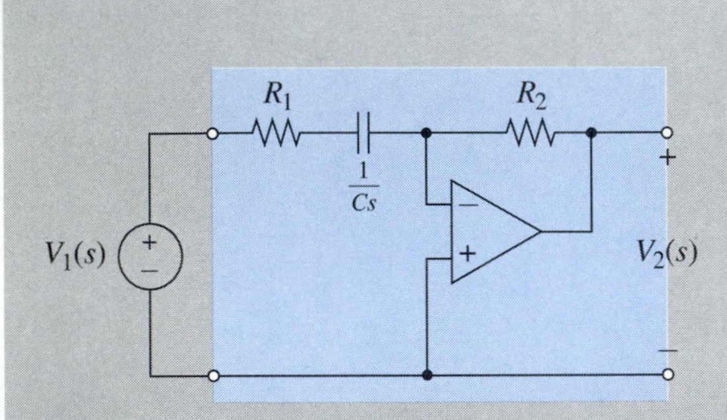

Figure 12-9

(a) Show that the transfer function $V_2(s)/V_1(s)$ of the circuit in Fig. 12-9 has a high-pass gain characteristic.

(b) Describe a method of selecting the circuit parameters to achieve a specified passband gain and cutoff frequency.

(c) Use the results in (b) to design a circuit with a passband gain of 2.5 and a cutoff frequency of 20 Hz.

SOLUTION:

(a) The branch impedances of the inverting OP AMP configuration in Fig. 12-9 are

$$Z_1(s) = R_1 + \frac{1}{Cs} = \frac{R_1Cs + 1}{Cs} \quad \text{and} \quad Z_2(s) = R_2$$

and the voltage transfer function is

$$T(s) = -\frac{Z_2 s}{Z_1(s)} = -\frac{R_2Cs}{R_1Cs + 1} = \frac{(-R_2/R_1)s}{s + 1/R_1C}$$

This results in a high-pass transfer function of the form $Ks/(s+\alpha)$ with $K = -R_2/R_1$ and $\alpha = \omega_C = 1/R_1C$.

(b) Given a specified gain and cutoff frequency, the element values can be selected as follows:

(1) Select a value of C

(2) Select $R_1 = 1/\omega_C C$

(3) Select $R_2 = |K|R_1$

There are lots of other design approaches, since the three circuit parameters are subject to only two design constraints.

(c) Applying the design approach from (b) to achieve a passband gain of 2.5 and cutoff of 20 Hz yields:

(1) Select $C = 1\ \mu\text{F}$

(2) $R_1 = 1/(2\pi \times 20 \times 10^{-6}) = 7.96\ \text{k}\Omega$

(3) $R_2 = 2.5 \times 7.96 = 19.9\ \text{k}\Omega$.

EXERCISE 12-3

For each circuit in Fig. 12-10 identify whether the gain response has low-pass or high-pass characteristics and find the passband gain and cutoff frequency.

ANSWERS:

(a) High pass, $|T(\infty)| = 1/3$, $\omega_C = 66.7$ rad/s.
(b) Low pass, $|T(0)| = 2/3$, $\omega_C = 300$ rad/s.
(c) Low pass, $|T(0)| = 1$, $\omega_C = 333$ krad/s.
(d) High pass, $|T(\infty)| = 1/3$, $\omega_C = 333$ krad/s.

EXERCISE 12-4

A first-order circuit has a step response $g(t) = 5[e^{-2000t}]u(t)$.
(a) Is the circuit frequency response low-pass or high-pass?
(b) Find the passband gain and cutoff frequency.

ANSWERS:

(a) High pass.
(b) $|T(\infty)| = 5$ and $\omega_C = 2000$ rad/s.

Bandpass Response Using First-Order Circuits

Figure 12-11 shows a cascade connection of first-order high-pass and low-pass circuits. When the second stage does not load the first, the overall transfer function can be found by the chain rule.

$$T(s) = T_1(s) \times T_2(s) = \underbrace{\left(\frac{K_1 s}{s+\alpha_1}\right)}_{\text{high pass}} \underbrace{\left(\frac{K_2}{s+\alpha_2}\right)}_{\text{low pass}} \tag{12-17}$$

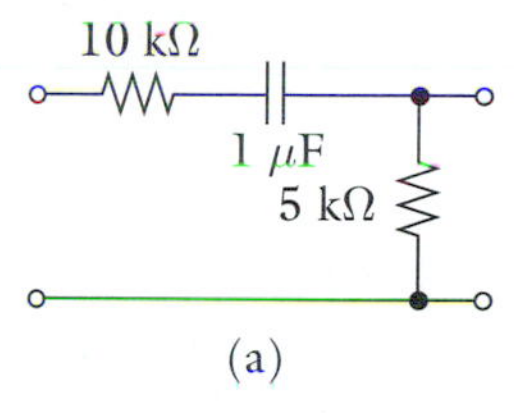

(a)

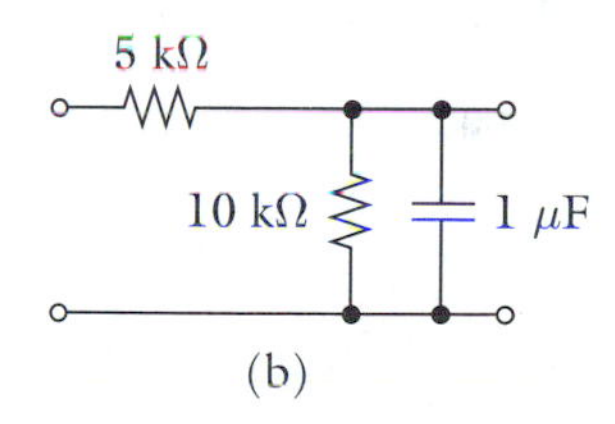

(b)

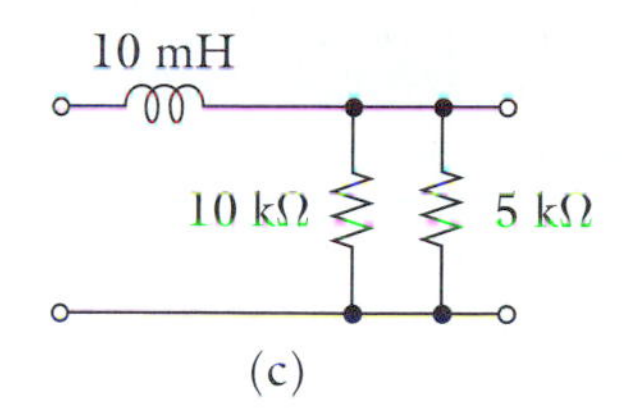

(c)

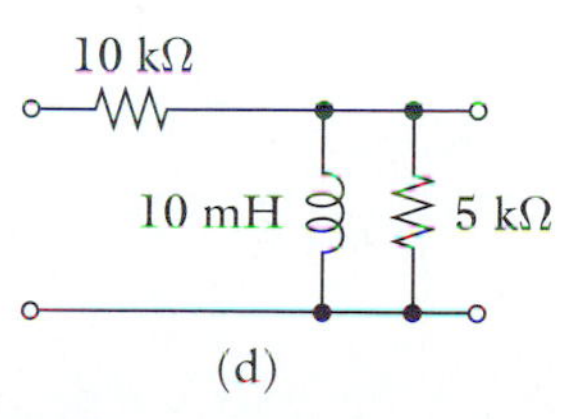

(d)

Figure 12-10

Replacing s by $j\omega$ in Eq. (12-17) and solving for the gain response yields

$$|T(j\omega)| = \underbrace{\left(\frac{|K_1|\omega}{\sqrt{\omega^2 + \alpha_1^2}}\right)}_{\text{high pass}} \underbrace{\left(\frac{|K_2|}{\sqrt{\omega^2 + \alpha_2^2}}\right)}_{\text{low pass}} \tag{12-18}$$

Note that the gain is zero at $\omega = 0$ and approaches zero as $\omega \to \infty$. This gain function appears to have stopbands at low and high frequencies with a passband in between.

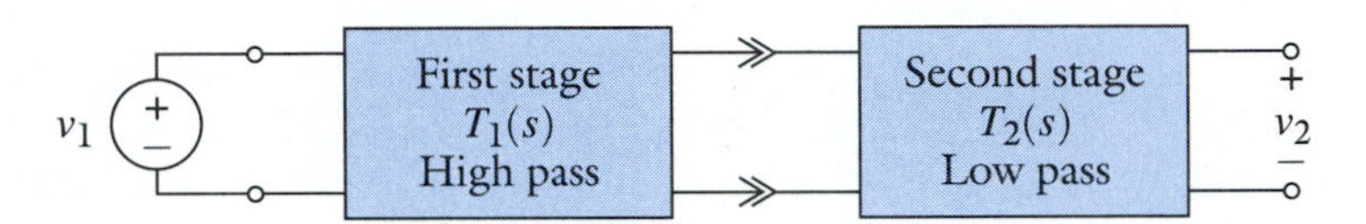

Figure 12-11 Bandpass circuit produced by a cascade connection of first-order high-pass and low-pass circuits.

When $\alpha_1 \ll \alpha_2$ the high-pass cutoff frequency is much lower than the low-pass cutoff frequency, and the overall transfer function has a bandpass characteristic. At low frequencies ($\omega \ll \alpha_1 \ll \alpha_2$) the gain approaches $|T(j\omega)| \to |K_1K_2|\omega/\alpha_1\alpha_2$. At mid frequencies ($\alpha_1 \ll \omega \ll \alpha_2$) the gain approaches $|T(j\omega)| \to |K_1K_2|/\alpha_2$. The low- and mid-frequency asymptotes intersect when $|K_1K_2|\omega/\alpha_1\alpha_2 = |K_1K_2|/\alpha_2$ at $\omega = \alpha_1$, that is, at the cutoff frequency of the high-pass stage. At high frequencies ($\alpha_1 \ll \alpha_2 \ll \omega$) the gain approaches $|T(j\omega)| \to |K_1K_2|/\omega$. The high- and mid-frequency asymptotes intersect when $|K_1K_2|/\omega = |K_1K_2|/\alpha_2$ at $\omega = \alpha_2$, that is, at the cutoff frequency of the low-pass stage. The plot of these asymptotic gains in Fig. 12-12 shows a passband between α_1 and α_2. Input sinusoids whose frequencies are outside of this range fall in one of the two stopbands.

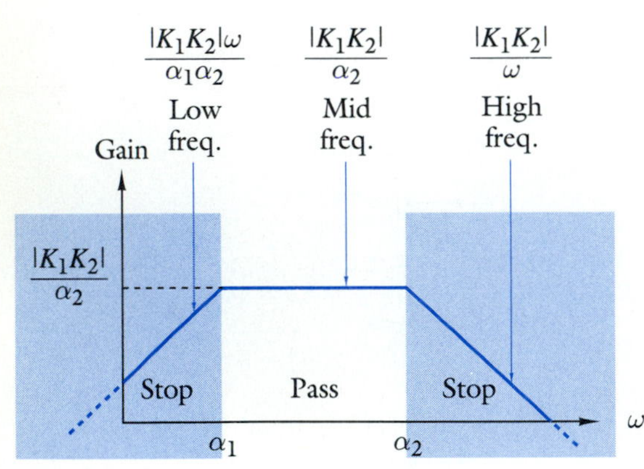

Figure 12-12 Asymptotic gain of a bandpass characteristic.

The asymptotic analysis of the transfer function in Eq. (12-17) illustrates using the responses of first-order transfer functions to estimate the frequency response of cascade connections. The asymptotic gain response in Fig. 12-12 is a reasonable approximation as long as two assumptions are valid: (1) The chain rule applies and (2) the high-pass cutoff frequency lies well below the low-pass cutoff ($\alpha_1 \ll \alpha_2$). The asymptotic analysis indicates that to accurately define the passband and stopband characteristics, we need to calculate gain and phase on a frequency range from a decade below α_1 to a decade above α_2. The range could be a very wide frequency range indeed, since the two cutoff frequencies may be separated by several decades.

❖ Design Example

EXAMPLE 12-8

(a) Construct a bandpass transfer function with a passband gain of 10 and cutoff frequencies at 20 Hz and 20 kHz.

(b) Design RC OP AMP circuits to realize the transfer function from (b).

SOLUTION:

(a) First-order low-pass and high-pass building blocks can be used in this case, since the lower cutoff frequency is three decades below the upper cutoff frequency. The required transfer function has the form

$$T(s) = \left(\frac{K_1 s}{s+\alpha_1}\right)\left(\frac{K_2}{s+\alpha_2}\right)$$

with the following constraints

lower cutoff frequency: $\alpha_1 = 2\pi(20) = 40\pi \text{ rad/s}$

upper cutoff frequency: $\alpha_2 = 2\pi(20 \times 10^3) = 4\pi \times 10^4 \text{ rad/s}$

midband gain: $\dfrac{|K_1 K_2|}{\alpha_2} = 10$

With numerical values inserted the required transfer function is

$$T(s) = \pm\frac{10(4\pi \times 10^4)s}{(s+40\pi)(a+4\pi \times 10^4)}$$

$T(s)$ has a plus or minus sign, since the sign of the passband gain is not specified.

(b) There are many ways to realize this transfer function. Two previously designed circuits can be used when the transfer function is partitioned as shown below.

$$T(s) = \underbrace{\left(-\frac{2.5s}{s+40\pi}\right)}_{\text{1st stage}}\underbrace{\left(-\frac{4(4\pi \times 10^4)}{s+4\pi \times 10^4}\right)}_{\text{2nd stage}}$$

The first stage is the high-pass function realized in Example 12-7. The second stage is the low-pass function realized in Example 12-5. The cascade connection of these two circuits is shown in Fig. 12-13. The chain rule applies here because the first stage output is an OP AMP.

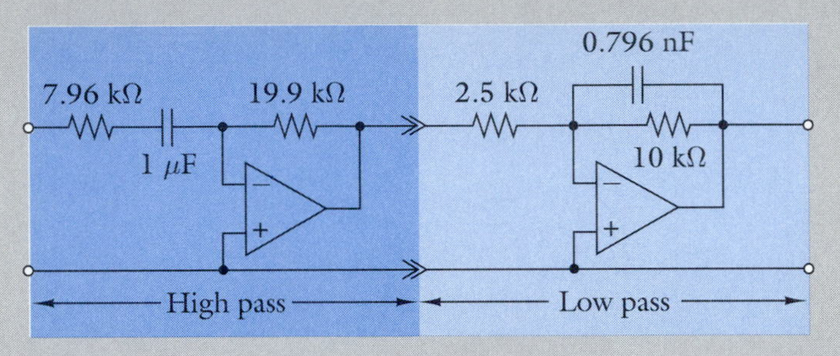

Figure 12-13

12-4 SECOND-ORDER CIRCUIT FREQUENCY RESPONSE

We begin our study with the second-order bandpass transfer function:

$$T(s) = \frac{Ks}{s^2 + 2\zeta\omega_O s + \omega_O^2} \tag{12-19}$$

The second-order polynomial in the denominator is expressed in terms of the damping ratio ζ and undamped natural frequency ω_O parameters defined in Chapter 10. These parameters play a key role in describing the frequency response of second-order circuits. The poles of $T(s)$ are complex conjugates when the damping ratio is less than one and these complex poles dramatically influence the frequency response.

To gain a qualitative understanding of the effect of complex poles we write Eq. (12-19) in factored form

$$T(s) = \frac{Ks}{(s - p_1)(s - p_2)} \tag{12-20}$$

where p_1 and p_2 are the poles of $T(s)$. The frequency-response gain function is

$$|T(j\omega)| = \frac{|K|\omega}{|j\omega - p_1|\,|j\omega - p_2|} = \frac{|K|\omega}{M_1 M_2} \tag{12-21}$$

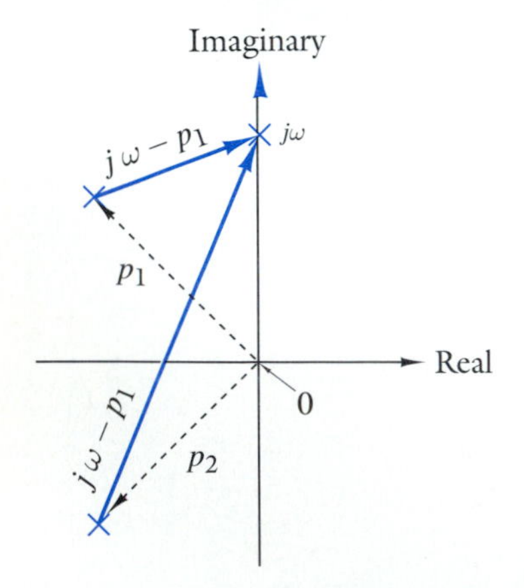

Figure 12-14 s-plane vectors defining $|T(j\omega)|$.

As shown in Fig. 12-14, the factors $(p_1 - j\omega)$ and $(p_2 - j\omega)$ can be interpreted as vectors from the natural poles at $s = p_1$ and

$s = p_2$ to the forced pole at $s = j\omega$. The length of these vectors are

$$M_1 = |j\omega - p_1|$$

$$M_2 = |j\omega - p_2|$$

The lengths M_1 and M_2 change when the frequency of the sinusoidal input changes. Changing the frequency moves the tips of vectors $p_1 - j\omega$ and $p_2 - j\omega$ up or down the j-axis while the tails remain fixed at the location of the natural poles. The length M_1 reaches a minimum when the forced pole at $s = j\omega$ is close to the natural pole at $s = p_1$. Since M_1 appears in the denominator of Eq. (12-21), this minimum tends to produce a maximum or peak in the gain response.

Figure 12-15 shows how different input frequencies change the lengths M_1 and M_2 to produce the gain response curve. At frequency ω_1 lengths M_1 and M_2 are approximately equal. At frequency ω_2 there is a peak in the gain response because length M_1 has decreased to a minimum and is very small compared with length M_2. When frequency increases to ω_3 the gain decreases since both lengths increase. The peak in the gain response is called a *resonance*. Resonance is caused by the fact that the forced pole at $s = j\omega$ and the natural pole at $s = p_1$ are close together, so the length M_1 is small.

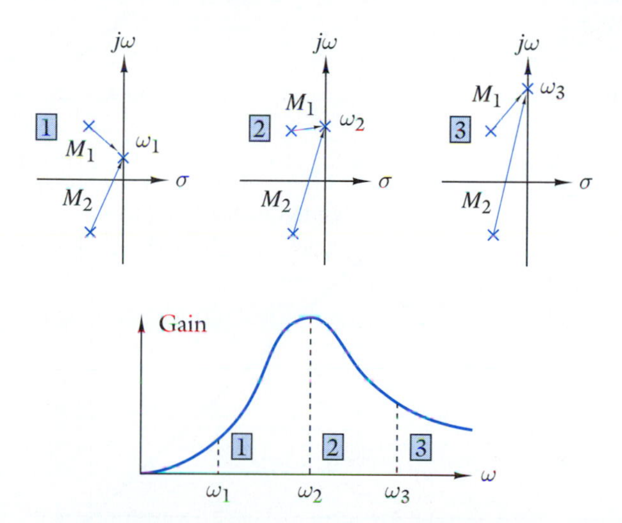

Figure 12-15 Gain response resonance from s-plane geometry.

The resonant peak decreases when p_1 and p_2 are shifted to the left in the s plane. Replacing s by $s + \alpha$ produces such a shift. The s-domain translation property

$$\mathscr{L}^{-1}\{F(s+\alpha)\} = e^{-\alpha t} f(t)$$

then shows that shifting the poles of $F(s)$ to the left increases the damping in the time domain. Qualitatively we expect resonant peaks in the frequency response to be less pronounced when circuit damping increases and more pronounced when it decreases.

Second-Order Bandpass Circuits

To develop a quantitative understanding of resonance we replace s by $j\omega$ in Eq. (12-19) to produce

$$T(j\omega) = \frac{Kj\omega}{-\omega^2 + 2\zeta\omega_O j\omega + \omega_O^2} \tag{12-22}$$

At low frequencies ($\omega \ll \omega_O$) the gain approaches $|T(j\omega)| \to |K|\omega/\omega_O^2$. At high frequencies ($\omega \gg \omega_O$) the gain approaches $|T(j\omega)| \to |K|/\omega$. The low-frequency asymptote is directly proportional to frequency (slope = +20 dB/decade), while the high-frequency asymptote is inversely proportional to frequency (slope = −20 dB/decade). As shown in Fig. 12-16, the two asymptotes intersect at $\omega = \omega_O$ and the gain at the intersection is $|T(j\omega)| = |K|/\omega_O$. Note that the frequency scale in Fig. 12-16 is logarithmic and normalized by dividing ω by the natural frequency ω_O.

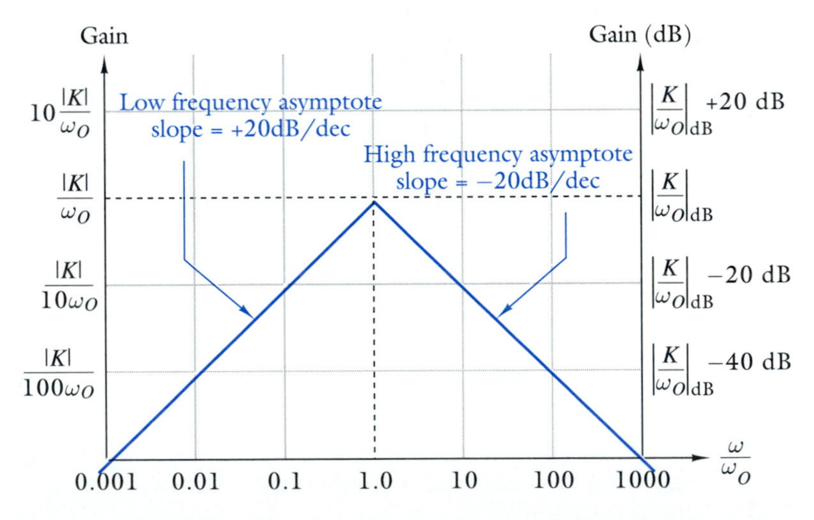

Figure 12-16 Second-order bandpass asymptotic gain response.

The two asymptotes display a bandpass frequency response centered on the natural frequency ω_O. For this reason, the ω_O of a bandpass circuit is also called the *center frequency*. To examine the gain response in more detail we must consider the effect of circuit damping. Factoring $\omega_O j\omega$ out of the denominator of Eq. (12-22) produces

$$T(j\omega) = \frac{K/\omega_O}{2\zeta + j\left(\dfrac{\omega}{\omega_O} - \dfrac{\omega_O}{\omega}\right)} \tag{12-23}$$

Expressing the transfer function in this form shows that only the imaginary part in the denominator varies with frequency. The maximum gain occurs when the imaginary part in the denominator of Eq. (12-23) vanishes at $\omega = \omega_O$ and the maximum gain is

$$|T(j\omega)|_{MAX} = \frac{|K|/\omega_O}{2\zeta} \tag{12-24}$$

Since the maximum gain is inversely proportional to the damping ratio, the resonant peak will clearly increase as circuit damping decreases, and vice versa. The quantity in the numerator of Eq. (12-24) is the gain at which the low- and high-frequency asymptotes intersect.

Figure 12-17 shows plots of the gain of the bandpass function for several values of the damping ratio ζ. For $\zeta < 0.5$ the gain curve lies above the asymptotes and has a narrow resonant peak. For $\zeta > 0.5$ the gain curve flattens out and lies below the asymptotes. When $\zeta = 0.5$ the gain curve falls close to the asymptotes.

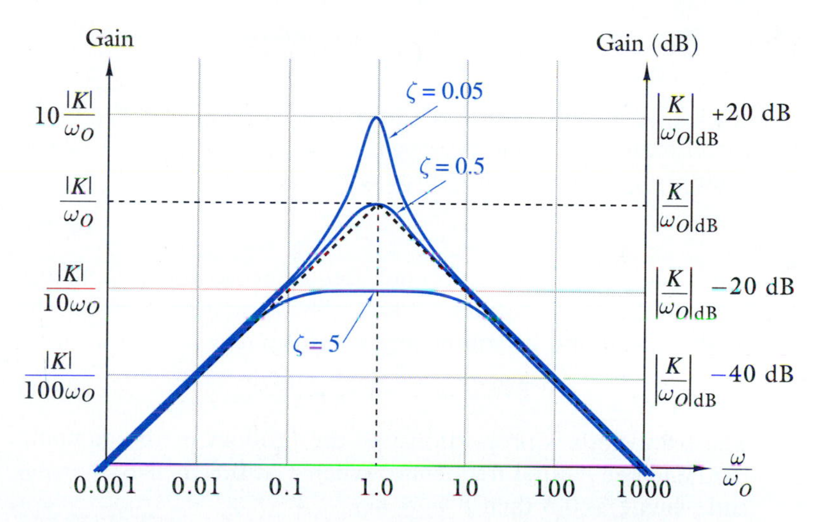

Figure 12-17 Second-order bandpass gain responses.

The gain response has a bandpass characteristic regardless of the value of the damping ratio. The passband is defined by a cutoff frequency on either side of the center frequency ω_O. To locate the two cutoff frequencies, we must find the values of ω

at which the gain is $|T(j\omega)|_{MAX}/\sqrt{2}$, where the maximum gain is given in Eq. (12-24). We note that when the imaginary part in the denominator of Eq. (12-23) equals $\pm 2\zeta$ the gain is

$$|T(j\omega)| = \frac{|K|/\omega_O}{|2\zeta \pm j2\zeta|} = \frac{\dfrac{|K|/\omega_O}{2\zeta}}{\sqrt{2}} = \frac{|T(j\omega)|_{MAX}}{\sqrt{2}} \tag{12-25}$$

The values of frequency ω that cause the imaginary part to be $\pm j2\zeta$ are the cutoff frequencies because at these points the gain is reduced by a factor of $1/\sqrt{2}$ from its maximum value at $\omega = \omega_O$.

To find the cutoff frequencies we set the imaginary part in the denominator in Eq. (12-23) equal to $\pm 2\zeta$:

$$\frac{\omega}{\omega_O} - \frac{\omega_O}{\omega} = \pm 2\zeta \tag{12-26}$$

which yields the quadratic equation

$$\omega^2 \mp 2\zeta\omega_O\omega - \omega_O^2 = 0 \tag{12-27}$$

whose roots are

$$\omega_{C1}, \omega_{C2} = \omega_O\left(\pm\zeta \pm \sqrt{1+\zeta^2}\right) \tag{12-28}$$

Only the two positive roots have physical significance. Since $[1+\zeta^2]^{\frac{1}{2}} > \zeta$ the two positive roots are

$$\begin{aligned} \omega_{C1} &= \omega_O\left(-\zeta + \sqrt{1+\zeta^2}\right) \\ \omega_{C2} &= \omega_O\left(+\zeta + \sqrt{1+\zeta^2}\right) \end{aligned} \tag{12-29}$$

Since $\zeta > 0$ these equations show that $\omega_{C1} < \omega_O$ is the lower cutoff frequency, while $\omega_{C2} > \omega_O$ is the upper cutoff frequency. Multiplying ω_{C1} times ω_{C2} produces

$$\omega_O^2 = \omega_{C1}\omega_{C2} \tag{12-30}$$

This results means that the center frequency ω_O is the geometric mean of the two cutoff frequencies. The ***bandwidth*** (BW) of the passband is found by subtracting ω_{C1} from ω_{C2}.

$$\text{BW} = \omega_{C2} - \omega_{C1} = 2\zeta\omega_O \tag{12-31}$$

The bandwidth is proportional to the product of the damping ratio and the natural frequency. When $\zeta < 0.5$ then BW $< \omega_O$ and when $\zeta > 0.5$ then BW $> \omega_O$.

We can think of $\zeta = 0.5$ as the boundary between two extreme cases. In the ***narrowband*** case ($\zeta \ll 0.5$) the complex natural poles are close to the j-axis and the gain response has a resonant peak that produces a bandwidth that is small compared to the circuit's natural frequency. In the ***wideband*** case ($\zeta \gg 0.5$) the gain response is relatively flat because the natural poles are further from the j-axis because the circuit has more damping. The narrow and wideband cases are both used in practice. The narrowband response is highly selective passing only a very restricted range of frequencies, often less than one octave. In contrast, the wideband response usually encompasses several decades within its passband.

The two narrow and broadband cases also show up in the phase response. From Eq. (12-23) the phase angle is

$$\theta(\omega) = \angle K - \tan^{-1}\left[\frac{\omega/\omega_O - \omega_O/\omega}{2\zeta}\right] \tag{12-32}$$

At low frequency ($\omega \ll \omega_O$) the phase angle approaches $\theta(\omega) \to \angle K + 90°$. At the center frequency ($\omega = \omega_O$) the phase reduces to $\theta(\omega_O) = \angle K$. Finally, at high frequencies ($\omega \gg \omega_O$) the phase approaches $\theta(\omega) \to \angle K - 90°$. Graphs of the phase response for several values of damping are shown in Fig. 12-18. The phase shift at the center frequency is 90° and the total phase change is $-180°$ regardless of the damping ratio. However, the transition from $+90°$ to $-90°$ is very abrupt for the narrowband case ($\zeta = 0.05$) and more gradual for the wideband case ($\zeta = 5$).

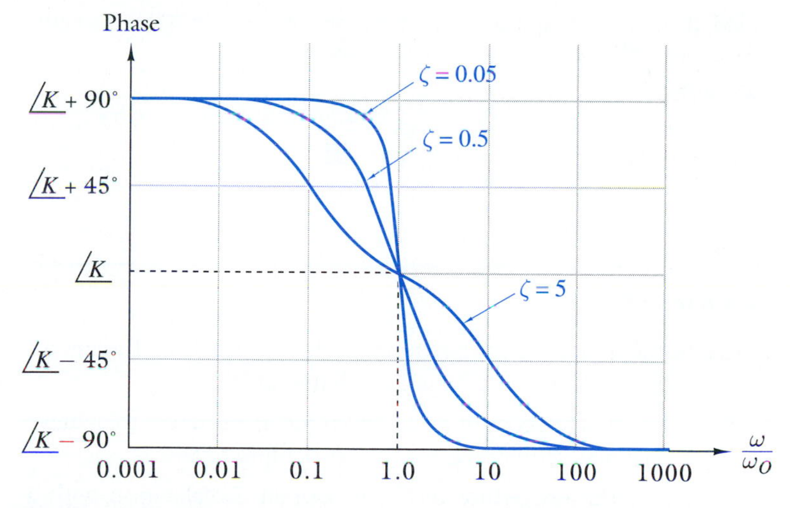

Figure 12-18 Second-order bandpass phase responses.

EXERCISE 12-5

Find the natural frequency ω_O, damping ratio ζ, cutoff frequencies, bandwidth BW, and maximum gain of the following bandpass transfer functions.

(a) $T(s) = \dfrac{500s}{s^2 + 200s + 10^6}$

(b) $T(s) = \dfrac{5 \times 10^{-3}s}{2.5 \times 10^{-5}s^2 + 5 \times 10^{-3}s + 1}$

ANSWERS:

	ω_O(rad/s)	ζ	ω_{C1}(rad/s)	ω_{C2}(rad/s)	BW(rad/s)	$\vert T(j\omega)\vert_{MAX}$
(a)	1000	0.1	905	1105	200	2.5
(b)	200	0.5	123.6	323.6	200	1

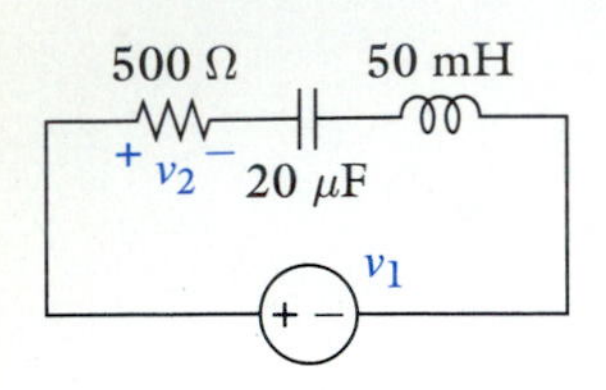

Figure 12-19

EXERCISE 12-6

(a) Find the transfer function $T(s) = V_2(s)/V_1(s)$ of the circuit in Fig. 12-19 when the output $V_2(s)$ is taken across the resistor.
(b) Find the natural frequency ω_O, damping ratio ζ, cutoff frequencies, bandwidth BW, and maximum gain of the transfer function.

ANSWERS:
(a) $T(s) = 10^4 s/(s^2 + 10^4 s + 10^6)$.
(b) $\omega_O = 1$ krad/s, $\zeta = 5$, $\omega_{C1} = 99.02$ rad/s, $\omega_{C2} = 10.099$ krad/s, BW $= 10$ krad/s, $|T(j\omega)|_{MAX} = 1$.

EXERCISE 12-7

The step response of a second-order bandpass circuit is

$$g(t) = 2\left[e^{-500t} \sin 2000t\right] u(t)$$

Find the natural frequency ω_O, damping ratio ζ, cutoff frequencies, bandwidth BW, and maximum gain of the circuit.

ANSWERS:
$\omega_O = 2062$ rad/s, $\zeta = 0.2425$, $\omega_{C1} = 1622$ rad/s, $\omega_{C2} = 2622$ rad/s, BW $= 1000$ rad/s, $|T(j\omega)|_{MAX} = 4$.

EXAMPLE 12-9

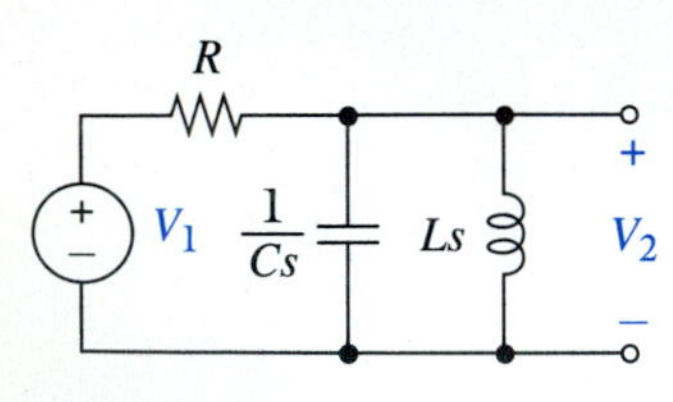

Figure 12-20

(a) Show that the transfer function $V_2(s)/V_1(s)$ of the circuit in Fig. 12-20 has a bandpass characteristic.
(b) Develop a procedure to select circuit parameters to achieve a specified center frequency and damping ratio.
(c) Apply the procedure in (b) to obtain a realization with a center frequency of 455 kHz and a bandwidth of 10 kHz.

SOLUTION:

(a) The given circuit is a voltage divider with branch impedances

$$Z_1 = R$$

$$Z_2 = \frac{1}{Cs + \dfrac{1}{Ls}} = \frac{Ls}{LCs^2 + 1}$$

By voltage division the transfer function is

$$T(s) = \frac{Z_2}{Z_1 + Z_2} = \frac{Ls}{RLCs^2 + Ls + R}$$

$$= \frac{s/RC}{s^2 + s/RC + 1/LC}$$

This transfer function is of the form $Ks/(s^2+2\zeta\omega_O s+\omega_O^2)$.

(b) Comparing the transfer function with the standard form yields the following constraints on the circuit parameters:

$$\omega_O^2 = \frac{1}{LC}$$

$$\zeta = \frac{\sqrt{L/C}}{2R}$$

One possible design approach is:

1. Select a value of C
2. Select $L = 1/\omega_O^2 C$
3. Select $R = \sqrt{L/C}/2\zeta$.

Many other design approaches are possible because there are three circuit parameters available and only two design constraints. In general the values of L and C determine ω_O and the resistance R is adjusted to produce the required damping ratio.

(c) The numerical values of the design parameters are

$$\omega_O = 2\pi \times 455 \times 10^3 = 2.86 \times 10^6$$

$$\zeta = \frac{BW}{2\omega_O} = \frac{2\pi \times 10 \times 10^3}{2(2\pi \times 455 \times 10^3)} = 0.0110$$

Applying the method in (b) we select $C = 10$ nF, (2) $L = 12.2\ \mu$H and (3) $R = 1.59$ kΩ. Since the damping ratio is very small, the natural poles are quite close to the $j\omega$-axis producing a very narrow resonance at 455 kHz.

Second-Order Low-Pass Circuits

The transfer function of a second-order low-pass prototype has the form

$$T(s) = \frac{K}{s^2 + 2\zeta\omega_O s + \omega_O^2} \qquad (12\text{-}33)$$

The frequency response is found by replacing s by $j\omega$:

$$T(j\omega) = \frac{K}{\omega_O^2 - \omega^2 + j2\zeta\omega_O\omega} \qquad (12\text{-}34)$$

At low frequencies ($\omega \ll \omega_O$) the gain approaches $|T(j\omega)| \to |K|/\omega_O^2 = |T(0)|$. At high frequencies ($\omega \gg \omega_O$) the gain approaches $|T(j\omega)| \to |K|/\omega^2$. The low-frequency asymptote is a constant equal to the zero frequency or dc gain. The high-frequency asymptote is inversely proportional to the square of the frequency so the gain decreases by a factor of 100 (−40 dB) when the frequency increases by a factor of 10 (one decade). The two asymptotes intersect to form a corner at $\omega = \omega_O$, as indicated in Fig. 12-21. The two asymptotes display a low-pass gain characteristic with a passband below ω_O and a stopband above ω_O. The passband gain is the dc gain $|T(0)|$ and the slope of the gain asymptote in the stopband is −40 dB/decade.

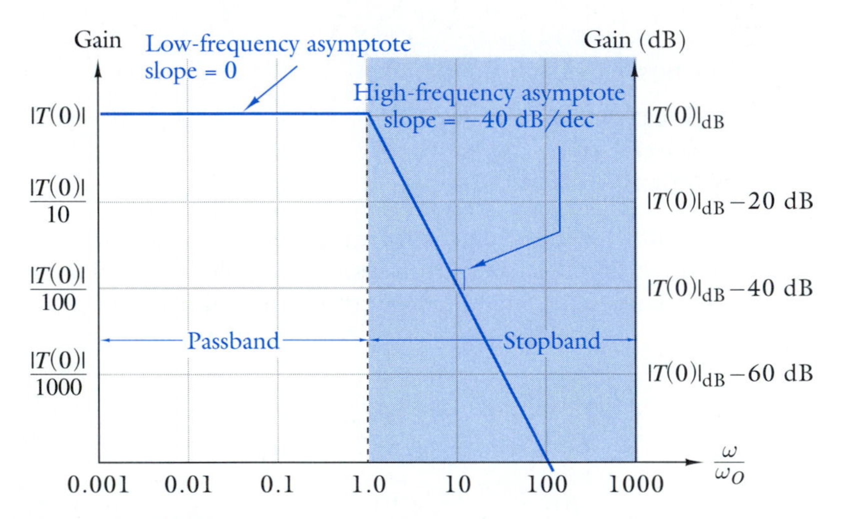

Figure 12-21 Second-order low-pass gain asymptotes.

The influence of the damping ratio on the gain response can be illustrated by evaluating the gain at $\omega = \omega_O$:

$$|T(j\omega_O)| = \frac{|K|/\omega_O^2}{2\zeta} = \frac{|T(0)|}{2\zeta} \tag{12-35}$$

This result and the plots in Fig. 12-22 show that the actual gain curve lies above the asymptotes when $\zeta < 0.5$, below the asymptotes when $\zeta > 0.5$, and close to the asymptotes when $\zeta = 0.5$. When $\zeta = 1/\sqrt{2}$ then the gain at $\omega = \omega_O$ is

$$|T(j\omega_O)| = \frac{|T(0)|}{\sqrt{2}} \quad \left(\zeta = 1/\sqrt{2}\right) \tag{12-36}$$

Since the passband gain is $|T(0)|$ this result means that the cutoff frequency is equal to ω_O when $\zeta = 1/\sqrt{2}$.

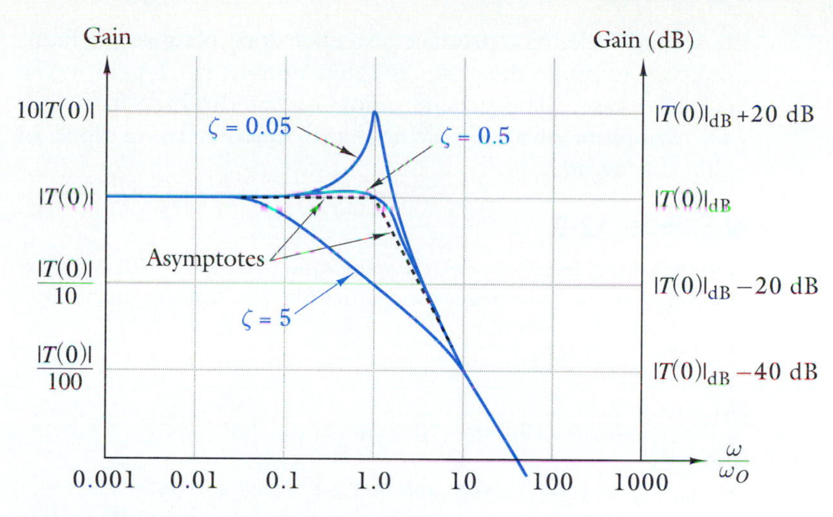

Figure 12-22 Second-order low-pass gain responses.

The distinctive feature of the low-pass case is a narrow resonant peak in the neighborhood of ω_O for lightly damped circuits with $\zeta \ll 0.5$. The maximum or peak gain response is

$$|T(j\omega)|_{MAX} = \frac{|T(0)|}{2\zeta\sqrt{1-\zeta^2}} \tag{12-37}$$

and occurs at

$$\omega_{MAX} = \omega_O\sqrt{1-2\zeta^2} \tag{12-38}$$

Note that the ω_{MAX} is always less than the natural frequency ω_O. Deriving Eqs. (12-37) and (12-38) is straightforward and is left as a problem for the student (see Prob. 12-36).

Figure 12-23 compares the response of the first-order low-pass prototype with the second-order gain response for $\zeta = 1/\sqrt{2}$. The important difference is that the gain slope in the stopband is −40 dB/decade for the second-order response and only −20 dB/decade for the first-order circuit. The two poles in

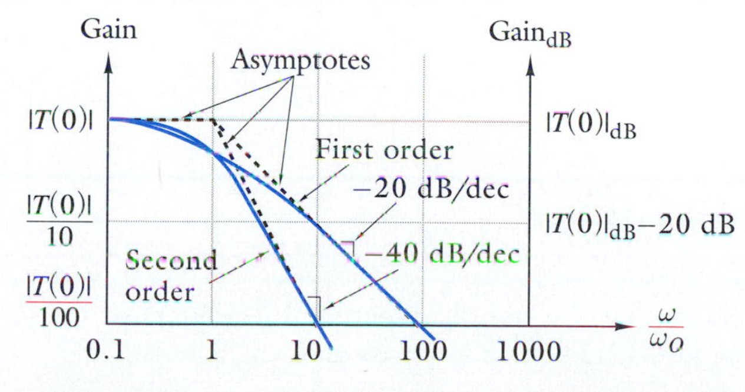

Figure 12-23 Comparison of first- and second-order gain responses.

the second-order case produce a steeper slope because the high-frequency response decreases as $1/\omega^2$, rather than $1/\omega$ as in the first-order case. The generalization is that the high-frequency gain asymptote of an n-pole low-pass function has a slope of $-20n$ dB/decade.

EXERCISE 12-8

Find the natural frequency ω_O, damping ratio ζ, dc gain $T(0)$, and maximum gain of the following second-order low-pass transfer functions.

(a) $T(s) = \dfrac{(4000)^2}{(s+3000)^2 + (4000)^2}$

(b) $T(s) = \dfrac{5}{0.25 \times 10^{-4}s^2 + 10^{-2}s + 1}$

ANSWERS:

(a) $\omega_O = 5$ krad/s, $\zeta = 0.6$, $T(0) = 0.64$, $T_{MAX} = 0.667$

(b) $\omega_O = 200$ krad/s, $\zeta = 1.00$, $T(0) = 5$, $T_{MAX} = T(0)$

EXERCISE 12-9

(a) Find the transfer function $T(s) = V_2(s)/V_1(s)$ of the circuit in Fig. 12-19 when the output $V_2(s)$ is taken across the capacitor.

(b) Find the natural frequency ω_O, damping ratio ζ, dc gain $|T(0)|$, and maximum gain of the transfer function.

ANSWERS:

(a) $T(s) = 10^6/(s^2 + 10^4 s + 10^6)$.

(b) $\omega_O = 10^3$ rad/s, $\zeta = 5$, $T(0) = 1$, $T_{MAX} = T(0)$.

EXERCISE 12-10

The impulse response of a second-order low-pass circuit is

$$h(t) = 2000\left[e^{-1000t} \sin\ 1000t\right]u(t)$$

Find the natural frequency ω_O, damping ratio ζ, dc gain $T(0)$, and maximum gain.

ANSWERS:

$\omega_O = 1414$ rad/s, $\zeta = 0.707$, $T(0) = T_{MAX} = 1$

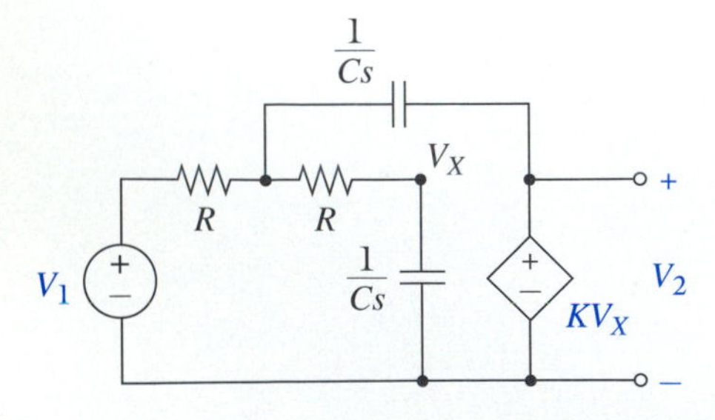

Figure 12-24

EXAMPLE 12-10

(a) Show that the transfer function $V_2(s)/V_1(s)$ of the circuit in Fig. 12-24 has a low-pass gain characteristic.

(b) Develop a procedure to select circuit parameters to achieve a specified natural frequency ω_O and damping ratio ζ.

SOLUTION:

(a) The transfer function of this circuit is found in Example 11-6 by writing two node-voltage equations and solving for $V_2(s)$:

$$T(s) = \frac{V_2(s)}{V_1(s)} = \frac{K}{(RCs)^2 + (3 - K)RCs + 1}$$

This is a second-order low-pass transfer function of the form $K/(s^2/\omega_O^2 + 2\zeta s/\omega_O + 1)$.

(b) Comparing the transfer function with the standard form yields the following design constraints on the circuit parameters:

$$\omega_O = \frac{1}{RC}$$

$$\zeta = \tfrac{1}{2}(3 - K)$$

One design approach is:

1. Select a value of C
2. Select $R = 1/(\omega_O C)$
3. Select the gain $K = 3 - 2\zeta$

In this approach R and C are determined by the specified ω_O and the gain K is then adjusted to produce the required damping ratio ζ. Many other approaches are possible, since there are three circuit parameters and two design constraints.

Second-Order High-Pass Circuits

To complete the study of second-order circuits we treat the high-pass case:

$$T(s) = \frac{Ks^2}{s^2 + 2\zeta\omega_O s + \omega_O^2} \tag{12-39}$$

The high-pass frequency response is a mirror image of the low-pass case reflected about $\omega = \omega_O$. At low frequencies ($\omega \ll \omega_O$) the gain approaches $|T(j\omega)| \to |K|\omega^2/\omega_O^2$. At high frequencies ($\omega \gg \omega_O$) the gain approaches $|T(j\omega)| \to |K| = |T(\infty)|$. The high-frequency gain is a constant $|T(\infty)|$ called the infinite-frequency gain. The low-frequency asymptote is inversely proportional to the square of the frequency (slope = +40 dB/decade). The two asymptotes intersect at the natural frequency ω_O as shown in Fig. 12-25. The two asymptotes display a high-pass gain response with a passband above ω_O and a stopband below ω_O. The passband gain is $|T(\infty)|$ and the slope of the gain asymptote in the stopband is +40 dB/decade.

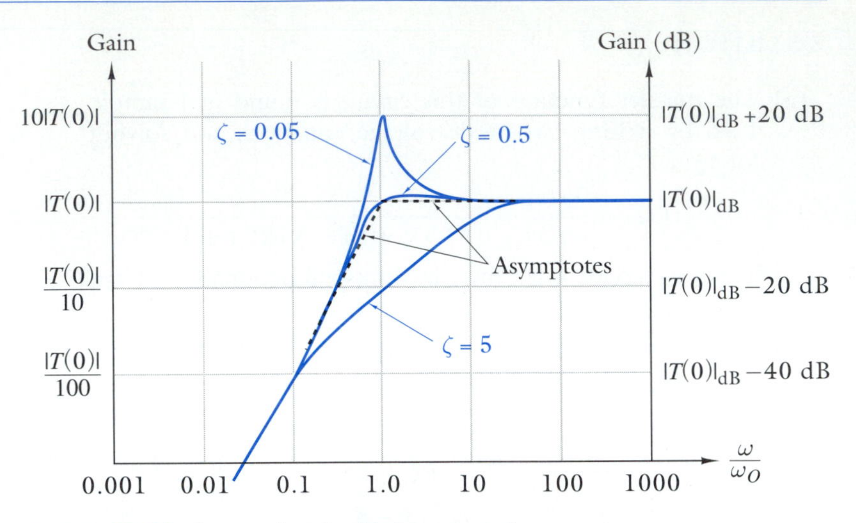

Figure 12-25 Second-order high-pass gain responses.

Figure 12-25 also shows the effect of the damping ratio on the actual gain curve. Evaluating the gain response at $\omega = \omega_O$ produces

$$|T(j\omega_O)| = \frac{|K|}{2\zeta} = \frac{|T(\infty)|}{2\zeta} \tag{12-40}$$

This result and the plots in Fig. 12-25 show that the gain response will be above the asymptotes for $\zeta < 0.5$, below for $\zeta > 0.5$, and close to the asymptotes for $\zeta = 0.5$. When $\zeta = 1/\sqrt{2}$ the gain in Eq. (12-40) is $|T(\infty)|/\sqrt{2}$. That is, ω_O is the cutoff frequency when $\zeta = 1/\sqrt{2}$.

Again, the distinctive feature of the second-order response is the narrow resonance in the neighborhood of ω_O for lightly damped circuits with $\zeta \ll 0.5$. For the second-order high-pass function the maximum gain is

$$|T(j\omega)|_{MAX} = \frac{|T(\infty)|}{2\zeta\sqrt{1-\zeta^2}} \tag{12-41}$$

and occurs at

$$\omega_{MAX} = \frac{\omega_O}{\sqrt{1-2\zeta^2}} \tag{12-42}$$

The frequency at which the gain maximum occurs is always higher than ω_O. Deriving Eqs. (12-41) and (12-42) is straightforward and is left as a problem for the student.

EXERCISE 12-11

(a) Find the transfer function $T(s) = V_2(s)/V_1(s)$ of the circuit in Fig. 12-19 when the output $V_2(s)$ is taken across the inductor.
(b) Find the natural frequency ω_O, damping ratio ζ, high-frequency gain $T(\infty)$, and maximum gain of the transfer function.

ANSWERS:

(a) $T(s) = s^2/(s^2 + 10^4 s + 10^6)$.

(b) $\omega_O = 10^3$ rad/s, $\zeta = 5$, $T(\infty) = 1$, $T_{MAX} = T(\infty)$. ■

■ EXERCISE 12-12

The step response of a second-order high-pass circuit is

$$g(t) = \left[e^{-1000t}\cos\ 2000t - 0.5e^{-1000t}\sin\ 2000t\right]u(t)$$

Find the natural frequency ω_O, damping ratio ζ, high-frequency gain $T(\infty)$, and maximum gain of the circuit.

ANSWERS:

$\omega_O = 2236$ rad/s, $\zeta = 0.447$, $T(\infty) = 1$, $T_{MAX} = 1.25$ ■

EXAMPLE 12-11

(a) Show that the transfer function $V_2(s)/V_1(s)$ of the circuit in Fig. 12-26 has a high-pass characteristic.

(b) Develop a procedure to select circuit parameters to achieve a specified natural frequency ω_O and damping ratio ζ.

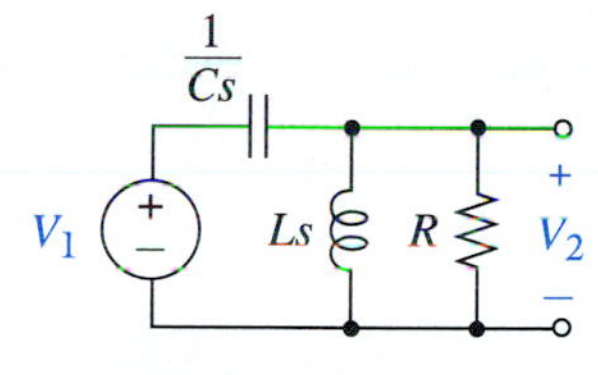

Figure 12-26

SOLUTION:

(a) The given circuit is a voltage divider whose branch impedances are

$$Z_1 = \frac{1}{Cs}$$

$$Z_2 = \frac{1}{\dfrac{1}{Ls} + \dfrac{1}{R}} = \frac{RLs}{Ls + R}$$

By voltage division the required transfer function is

$$T(s) = \frac{Z_2}{Z_1 + Z_2} = \frac{RLCs^2}{RLCs^2 + Ls + R} = \frac{s^2}{s^2 + s/RC + 1/LC}$$

This function is of the form $s^2/(s^2 + 2\zeta\omega_O s + \omega_O^2)$.

(b) Comparing the transfer function with the standard form yields the following constraints on the circuit parameters:

$$\omega_O^2 = \frac{1}{LC}$$

$$\zeta = \frac{\sqrt{L/C}}{2R}$$

One possible design procedure is:

1. Select a value of C
2. Select $L = 1/\omega_O^2 C$
3. Select $R = \sqrt{L/C}/2\zeta$

In this approach L and C are determined by the specified ω_O and the resistance R is adjusted to obtained the required damping ratio ζ.

12-5 BODE PLOTS

Bode plots are separate graphs of the gain $|T(j\omega)|_{dB}$ and phase $\theta(j\omega)$ versus log-frequency scales. Straight-line approximations to Bode plots can be quickly sketched from the poles and zeros of $T(s)$. In circuit design these straight-line plots serve as a shorthand notation for outlining design approaches or developing design requirements. The frequency-response performance of devices, circuits, and systems are often presented as Bode plots. The location of the poles and zeros of the transfer functions of these components can be estimated from the form of their Bode plots.

Bode Plots: Real Poles and Zeros

Bode plots are particularly useful when the poles and zeros are located on the real axis in the s-plane. As a starting place, consider the transfer function

$$T(s) = \frac{Ks(s+\alpha_1)}{(s+\alpha_2)(s+\alpha_3)} \tag{12-43}$$

where K, α_1, α_2, and α_3 are real. This function has zeros at $s = 0$ and $s = -\alpha_1$, and poles at $s = -\alpha_2$ and $s = -\alpha_3$. All of these critical frequencies lie on the real axis in the s-plane. To construct the Bode plots we express $T(j\omega)$ in a standard format by factoring out α_1, α_2, and α_3:

$$T(j\omega) = \left(\frac{K\alpha_1}{\alpha_2\alpha_3}\right)\frac{j\omega\,(1+j\omega/\alpha_1)}{(1+j\omega/\alpha_2)\,(1+j\omega/\alpha_3)} \tag{12-44}$$

Using the notation

$$\begin{aligned} \text{Magnitude} &= M = |1+j\omega/\alpha| = \sqrt{1+(\omega/\alpha)^2} \\ \text{Angle} &= \theta = \angle(1+j\omega/\alpha) = \tan^{-1}(\omega/\alpha) \\ \text{Scale Factor} &= K_O = \frac{K\alpha_1}{\alpha_2\alpha_3} \end{aligned} \tag{12-45}$$

we can write the transfer function in Eq. (12-44) in the form

$$\begin{aligned} T(j\omega) &= K_O\frac{\left(\omega e^{j90^\circ}\right)\left(M_1 e^{j\theta_1}\right)}{\left(M_2 e^{j\theta_2}\right)\left(M_3 e^{j\theta_3}\right)} \\ &= \frac{|K_O|\,\omega M_1}{M_2 M_3}e^{j(\angle K_O + 90^\circ + \theta_1 - \theta_2 - \theta_3)} \end{aligned} \tag{12-46}$$

The gain (in dB) and phase responses are

$$|T(j\omega)|_{dB} = \underbrace{20\log_{10}|K_O|}_{\text{scale factor}} + \underbrace{20\log_{10}\omega}_{\text{zero}} + \underbrace{20\log_{10}M_1}_{\text{zero}} - \underbrace{20\log_{10}M_2}_{\text{pole}} - \underbrace{20\log_{10}M_3}_{\text{pole}}$$
$$\theta(\omega) = \overbrace{\angle K_O} + \overbrace{90^\circ} + \overbrace{\theta_1} - \overbrace{\theta_2} - \overbrace{\theta_3} \tag{12-47}$$

The terms in Eq. (12-47) caused by zeros have positive signs and increase the gain and phase angle, while the pole terms have negative signs and decrease the gain and phase.

The summations in Eq. (12-47) illustrate a general principle. In a Bode plot the gain and phase responses are summations of terms produced by the following types of factors in $T(j\omega)$:

1. The scale factor K_O
2. A factor of the form $j\omega$ caused by a zero or a pole at the origin
3. Factors of the form $(1 + j\omega/\alpha)$ caused by zeros or poles at $s = -\alpha$

We can construct Bode plots by considering the contributions of three types of factors.

The Scale Factor. The gain and phase contributions of the scale factor are constants that are independent of frequency. The gain contribution $20\log_{10}|K_O|$ is a positive constant when $|K_O| > 1$ and a negative constant when $|K_O| < 1$. The phase contribution $\angle K_O$ is 0° when $K_O > 0$ and 180° when $K_O < 0$.

The Factor $j\omega$. A simple zero or pole at the origin contributes $\pm 20\ \log_{10}\omega$ to the gain and $\pm 90^\circ$ to the phase, where the plus sign applies to a zero and the minus to a pole. When $T(s)$ has a factor s^n in the numerator (denominator), it has a zero (pole) of order n at the origin. Multiple zeros or poles at $s = 0$ contribute $\pm 20n\ \log_{10}\omega$ to the gain and $\pm n90^\circ$ to the phase. Fig. 12-27 shows that the gain factors contributed by zeros and poles at the origin are straight lines that pass through 0 dB at $\omega = 1$ and have slopes of $\pm 20n$ dB/decade.

The Factor $1 + j\omega/\alpha$. The gain contributions of first-order zeros and poles are shown in Fig. 12-28. Like the first-order transfer functions studied earlier in this chapter, these factors produce straight-line gain asymptotes at low and high frequency. In a Bode plot the low-frequency ($\omega \ll \alpha$) asymptotes are horizontal lines at 0-dB gain. The high-frequency ($\omega \gg \alpha$) asymptotes are straight-lines of the form $\pm 20\ \log(\omega/\alpha)$, where the plus sign applies to a zero and the minus to a pole. The high-frequency gain asymptote has a slope of +20 dB/decade for a zero and −20 dB/decade for a pole. In either case, the low- and high-frequency asymptotes intersect at the **corner frequency** $\omega = \alpha$.

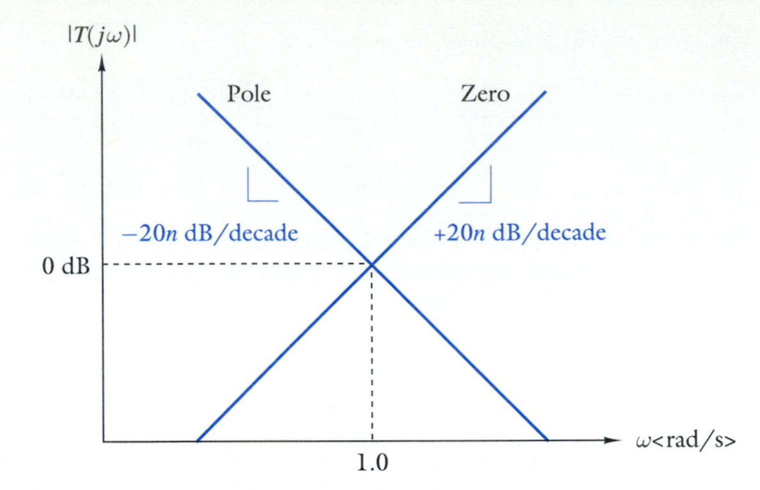

Figure 12-27 Gain responses of poles and zeros at $s = 0$.

Figure 12-29 shows the phase contributions of first-order zeros and poles. The straight-line approximations are similar to the gain asymptotes except that there are two slope changes. The first phase slope change occurs a decade below the corner frequency, and the second occurs a decade above the corner frequency. Since the phase changes by 90° over a two-decade range the straight-line approximations have slopes of ±45° per decade, where the plus sign applies to a zero and the minus to a pole.

To construct straight-line approximations to Bode plots it is convenient to start with the low-frequency gain and phase asymptotes. At low frequency ($\omega \ll \alpha$) first-order factors of the form $(1 + j\omega/\alpha)$ approach gains of 0 dB and phase angles of zero. Thus, at low frequency the only factors that influence the gain and phase are the scale factor K_O and poles or zeros at the origin (if any). Therefore, the low frequency asymptotes of the gain and phase are

$$|T(j\omega)|_{dB} \rightarrow 20\log_{10}|K_O| \pm 20n\log_{10}\omega \qquad (12\text{-}48a)$$

$$\theta(\omega) \rightarrow \angle K_O \pm n90^\circ \qquad (12\text{-}48b)$$

where n is the order of any zero or pole at the origin, and the plus (minus) sign applies to zeros (poles) at $s = 0$. In a Bode plot the low-frequency gain asymptote in Eq. (12-48a) is a straight line passing through a gain of $20\log_{10}(|K_O|)$ at $\omega = 1$ with a constant slope of $\pm 20n$ dB/decade. The low-frequency phase asymptote in Eq. (12-48b) is a horizontal line in a Bode plot.

To generate the straight-line gain plot we begin with the low-frequency gain and phase asymptotes. These low-frequency baselines account for the effect of the scale factor K_O and any poles or zeros at the origin. We account for the effect of the other

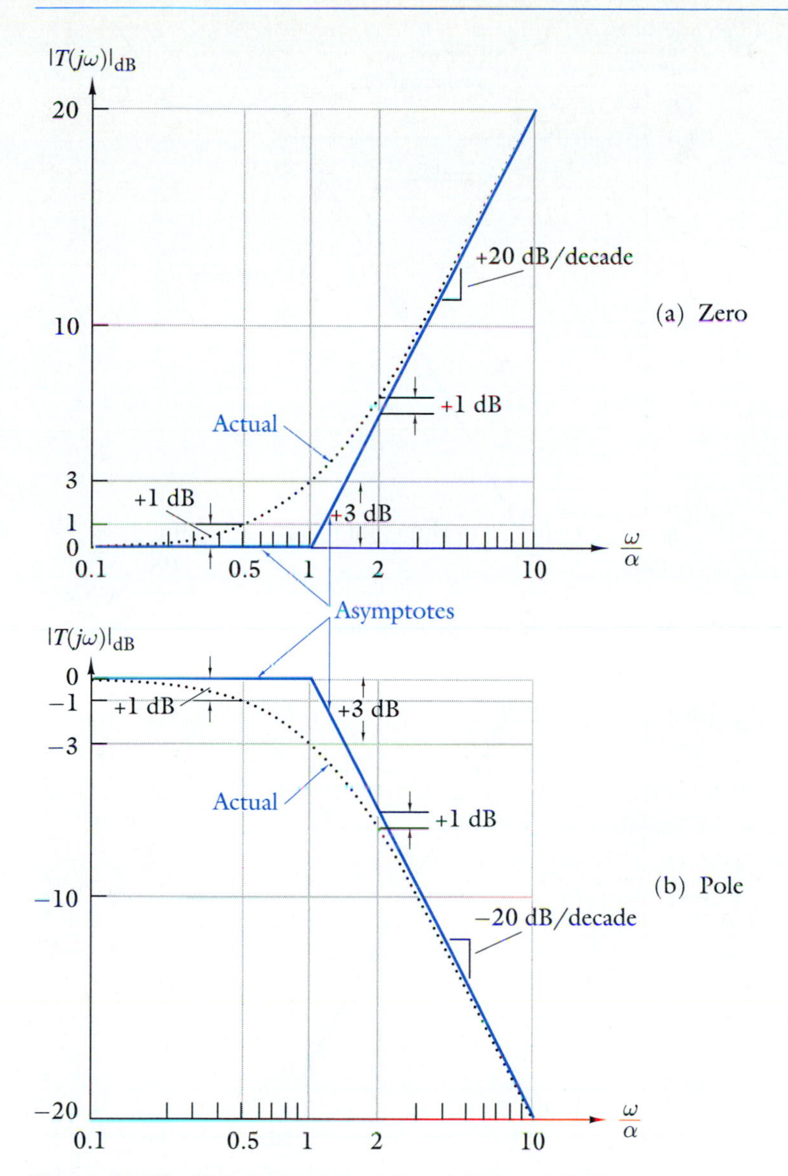

Figure 12-28 Gain responses of real poles and zeros.

poles and zeros by introducing the slope changes associated with their corner frequencies. For the gain plot these slope changes are ±20 dB/decade at each corner frequency. For the phase angle plot we change the slope by ±45°/decade one decade below and one decade above each corner frequency. Proceeding from low to high frequency and inserting the slope changes associated with each corner frequency in turn produces straight-line approximations to the gain and phase responses.

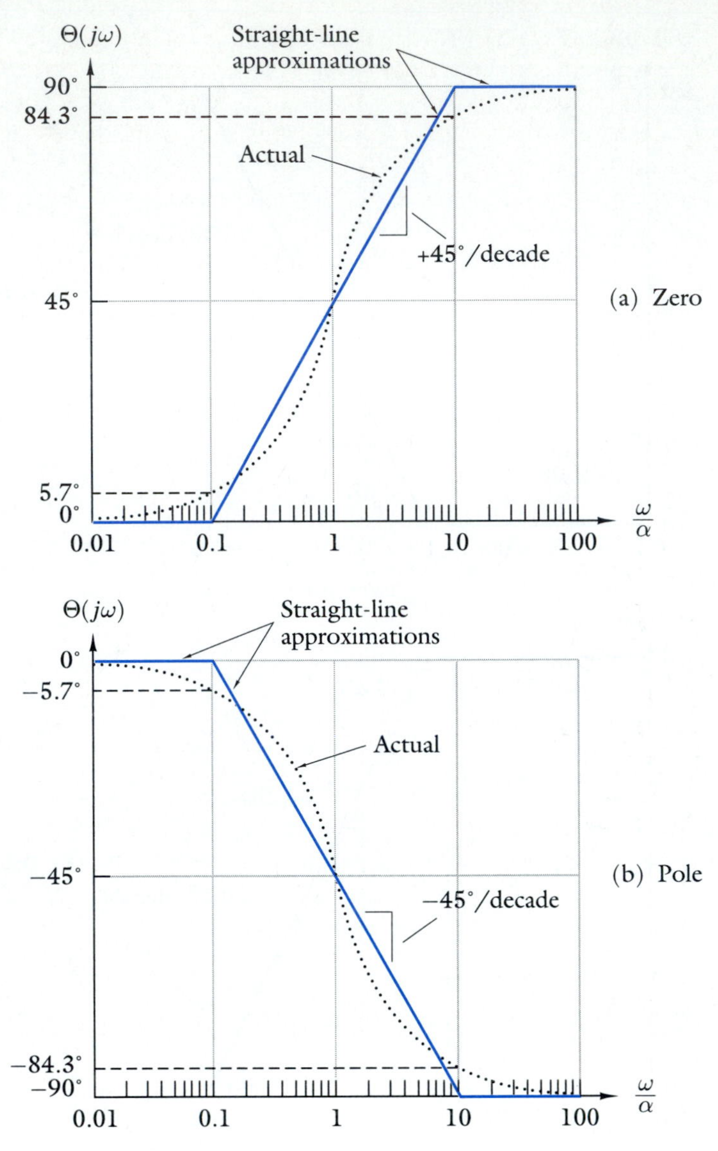

Figure 12-29 Phase response of real poles and zeros.

In many applications the straight-line gain and phase plots are all we need. When greater accuracy is required the straight-line gain plot can be refined by including gain corrections in the neighborhood of the corner frequency. Figure 12-28 shows the differences between the actual gains and the straight-line approximations. These gain corrections are ±3 dB at the corner frequency and ±1 dB an octave above or below the corner frequency, where the plus (minus) sign applies to zeros (poles). To estimate the actual gain we apply these corrections to the straight-line plot and sketch the actual gain. The graphical approach to gain correction is useful when the corner frequencies are separated by a decade or more so that the correction frequencies do not overlap. When the corner frequencies are not widely separated it is better to use the straight-line approximation as a first estimate and find the corrections at a few frequencies by direct calculation.

EXAMPLE 12-12

(a) Construct the Bode plot of the straight-line approximation of the gain of the following transfer function:

$$T(s) = \frac{12500(s + 10)}{(s + 50)(s + 500)}$$

(b) Find the gain corrections for each corner frequency and sketch the actual gain.

SOLUTION:

(a) Written in the standard form for a Bode plot the transfer function is

$$T(j\omega) = \frac{5(1 + j\omega/10)}{(1 + j\omega/50)(1 + j\omega/500)}$$

The scale factor is $K_O = 5$ and the corner frequencies are at $\omega_C = 10$, 50, and 500 rad/s. At low frequency ($\omega \ll 10$ rad/s)$T(j\omega) \to 5$ so the low frequency gain asymptote is

$$|T(j\omega)|_{dB} \to 20 \log_{10} 5 = 13.98 \approx 14 \text{ dB}$$

Since there are no poles or zeros at the origin the low-frequency gain asymptote is a horizontal line (slope = 0 dB/decade) at 14 dB.

Proceeding from one decade below the lowest corner frequency ($\omega = 1$ rad/s) to one decade above the highest corner frequency ($\omega = 5000$ rad/s) we encounter the following corner frequencies and slope changes.

FREQUENCY	CAUSED BY	SLOPE CHANGE	NET GAIN SLOPE
1 rad/s			0 db/decade
10 rad/s	zero at $s = -10$	+20 dB/decade	+20 dB/decade
50 rad/s	pole at $s = -50$	−20 dB/decade	0 dB/decade
500 rad/s	pole at $s = -500$	−20 dB/decade	−20 dB/decade
5000 rad/s			−20 dB/decade

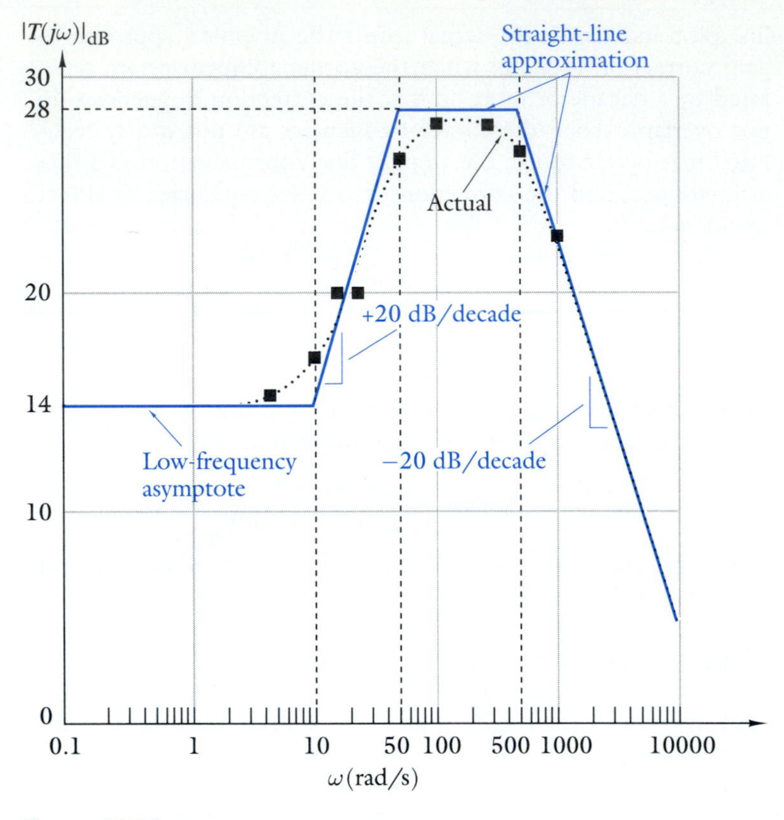

Figure 12-30

The solid lines in Fig. 12-30 show the resulting straight-line gain plot.

Below the first corner at $\omega = 10$ rad/s the straight line gain is the low-frequency asymptote at 14 dB. Between the first corner frequency at $\omega = 10$ rad/s and the second at $\omega = 50$ rad/s the straight-line gain is

$$|T(j\omega)|_{dB} = 14 + 20\log_{10}\left(\frac{\omega}{10}\right)\ \text{dB}$$

Using this equation, we calculate the straight-line gain at $\omega = 50$:

$$\left|T(j50)\right|_{dB} = 14 + 20\log_{10}\left(\frac{50}{10}\right) = 28\ \text{dB}$$

The gain plot continues (slope = 0 dB/decade) at a gain of 28 dB until the final corner frequency at $\omega = 500$ rad/s. Thereafter, the high-frequency gain asymptote is

$$|T(j\omega)|_{dB} = 28 - 20\log_{10}\left(\frac{\omega}{500}\right)\ \text{dB}$$

The high-frequency asymptote decreases to 0 dB when

$$|T(j\omega)|_{dB} = 28 - 20\log_{10}\left(\frac{\omega}{500}\right) = 0 \text{ dB}$$

which occurs at $\omega = 500 \times 10^{28/20} = 12.6 \times 10^3$ rad/s.

(b) The gain corrections are obtained by adding (zeros) or subtracting (poles) 3 dB at the corner frequency and 1 dB at an octave above and below the corner. The gain corrections are:

FREQUENCY	CAUSED BY	GAIN CORRECTION
5	zero at $s = -10$	+1 dB
10	zero at $s = -10$	+3 dB
20	zero at $s = -10$	+1 dB
25	pole at $s = -50$	−1 dB
50	pole at $s = -50$	−3 dB
100	pole at $s = -50$	−1 dB
250	pole at $s = -500$	−1 dB
500	pole at $s = -500$	−3 dB
1000	pole at $s = -500$	−1 dB

Figure 12-30 shows the values of the gain obtained by applying these corrections to the straight-line approximation. The dashed line in Fig. 12-30 shows a sketch of the actual gain response obtained using the corrected gains. The correction at $\omega = 20$ rad/s due to the zero at $s = -10$ and the correction at $\omega = 25$ rad/s due to the pole at $s = -50$ nearly overlap. The correction caused by the zero essentially cancels the pole correction, so the actual gain ends up close to the straight-line approximation.

EXERCISE 12-13

Sketch the Bode plot of the straight-line approximation to the gain response of the following transfer function:

$$T(s) = \frac{500(s + 50)}{(s + 20)(s + 500)}$$

(a) Find the straight-line gains at $\omega = 10$, 30, and 100 rad/s.
(b) Estimate the actual gain at the three corner frequencies.
(c) Find the frequency at which the high-frequency gain asymptote falls below −20 dB.

ANSWERS:
(a) 7.96 dB, 4.44 dB, 0 dB.
(b) 6.96 dB, +3 dB, −3 dB.
(c) 5000 rad/s. ■

EXAMPLE 12-13

Find the straight-line approximation to the phase response of the transfer function in Example 12-12.

SOLUTION:

In Example 12-12 the standard from of $T(j\omega)$ is shown to be

$$T(j\omega) = \frac{5(1 + j\omega/10)}{(1 + j\omega/50)(1 + j\omega/500)}$$

The scale factor is $K_O = 5$ and the corner frequencies are $\omega_C =$ 10, 50, and 500 rad/s. At low-frequency $T(j\omega) \rightarrow K_O = 5$, so the low-frequency phase asymptote is $\theta(\omega) \rightarrow \angle K_O = 0°$. Proceeding from one decade below the lowest corner frequency to one decade above the highest corner frequency we encounter the following corner frequencies and slope changes:

FREQUENCY	CAUSED BY	SLOPE CHANGE	NET SLOPE
1	zero at $s = -10$	+45°/decade	+45°/decade
5	pole at $s = -50$	−45°/decade	0°/decade
50	pole at $s = -500$	−45°/decade	−45°/decade
100	zero at $s = -10$	−45°/decade	−90°/decade
500	pole at $s = -50$	+45°/decade	−45°/decade
5000	pole at $s = -500$	+45°/decade	0°/decade

Figure 12-31 shows the straight-line approximation and the actual phase response.

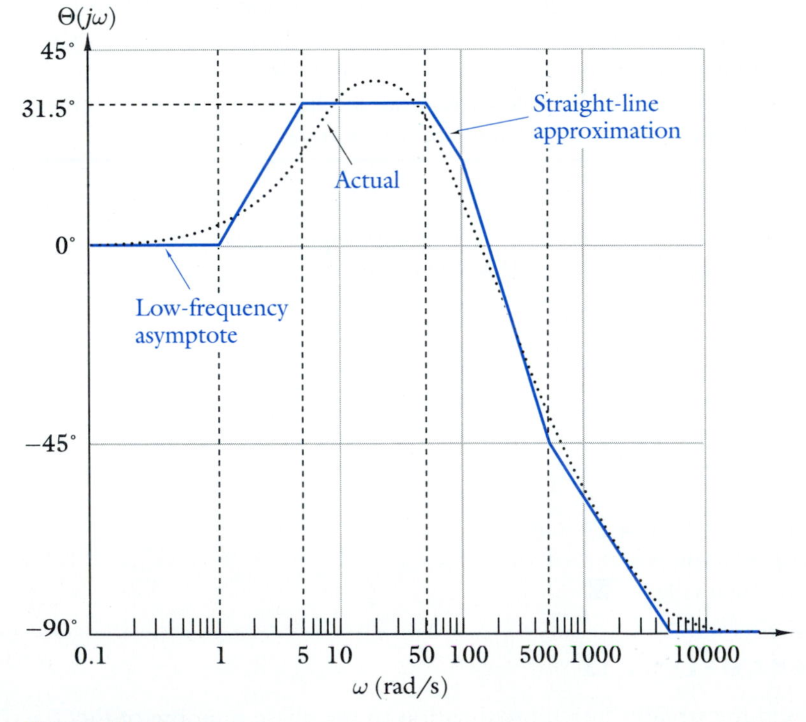

Figure 12-31

■ EXERCISE 12-14

Construct a Bode plot of the straight-line approximation to the phase response of the transfer function in Exercise 12-13. Use the plot to estimate the phase angles at $\omega = 1$, 15, 300, and 10^4 rad/s.

ANSWERS:

0°, −17.9°, −45°, −90° ■

EXAMPLE 12-14

Construct Bode plots of the straight-line approximations to the gains and phase response of the following transfer function:

$$T(s) = \frac{10s^2}{(s+40)(s+200)}$$

SOLUTION:

Writing $T(j\omega)$ in standard form produces

$$T(j\omega) = \frac{1}{800}\left[\frac{(j\omega)^2}{(1+j\omega/40)(1+j\omega/200)}\right]$$

In standard form $T(j\omega)$ has a scale factor of $K_O = 1/800$ and a zero of order 2 ($n = 2$) at the origin. At low-frequency $T(j\omega) \rightarrow (j\omega)^2/800$ so the low-frequency gain asymptote is

$$|T(j\omega)|_{dB} \rightarrow 20\log_{10}(\omega^2/800) = 20\log_{10}\omega^2 + 20\log_{10}(1/800)$$
$$= 40\log_{10}\omega - 58 \text{ dB}$$

In a Bode plot this asymptote is a straight line with a slope of +40 dB/decade. At $\omega = 10$ rad/s the asymptotic gain is

$$|T(j\omega)|_{dB} = 40\log_{10}(10) - 58 = -18 \text{ dB}$$

The starting place for our straight line gain plot is at $\omega = 10$ rad/s with a gain of −18 dB and a slope of +40 dB/decade. Proceeding to higher frequencies the gain response slope changes occur at the following frequencies:

FREQUENCY	CAUSED BY	SLOPE CHANGE	NET GAIN SLOPE
10			+40 dB/decade
40	pole at $s = -40$	−20 dB/decade	+20 dB/decade
200	pole at $s = -200$	−20 dB/decade	0 dB/decade
2000			0 dB/decade

Figure 12-32(a) shows the resulting straight-line gain plot.

At $\omega = 40$ the gain on low-frequency asymptote is

$$|T(j\omega)|_{dB} = 40\log_{10}(40) - 58 \text{ dB} = 6 \text{ dB}$$

Between $\omega = 40$ and 200 rad/s the straight-line gain is

$$|T(j\omega)|_{dB} = 6 + 20\log_{10}\left(\frac{\omega}{40}\right) \text{ dB}$$

Using this equation the gain at $\omega = 200$ rad/s is

$$|T(j\omega)|_{dB} = 6 + 20\log_{10}\left(\frac{200}{40}\right) = 20 \text{ dB}$$

For $\omega > 200$ rad/s the straight-line gain is a horizontal line (slope = 0 dB/decade) at +20 dB.

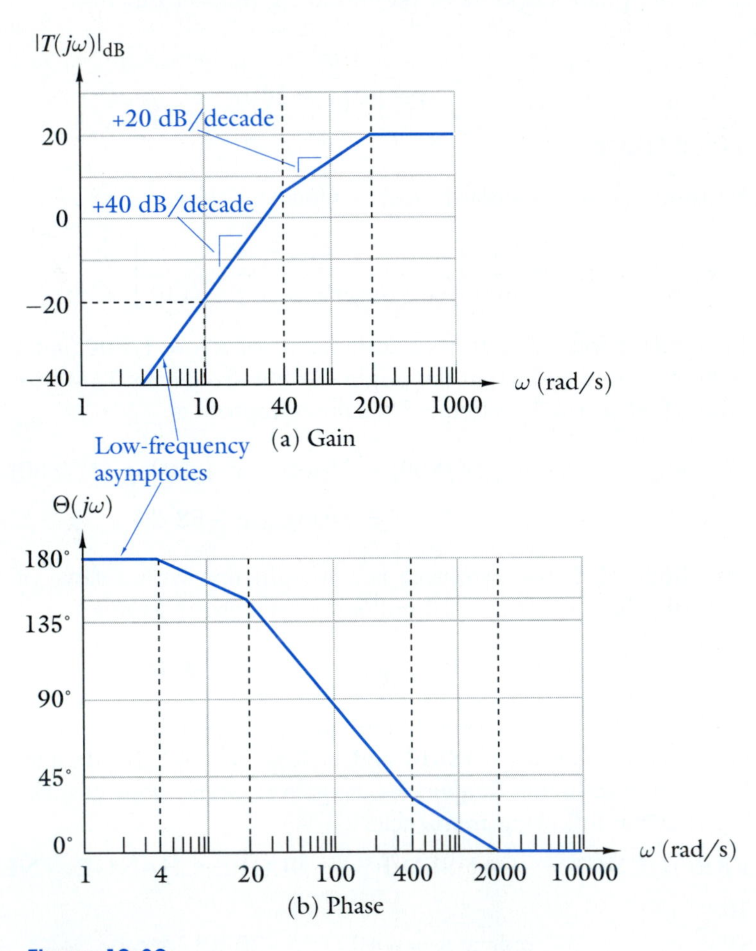

Figure 12-32

At low-frequency $T(j\omega) \rightarrow (j\omega)^2/800 = (\omega\angle 90°)^2/800$ so the low-frequency phase asymptote is $2(90°) = +180°$. Proceeding to higher frequency the phase response slope changes occur at the following frequencies:

FREQUENCY	CAUSED BY	SLOPE CHANGE	NET SLOPE
4	pole at $s = -40$	$-45°$/decade	$-45°$/decade
20	pole at $s = -200$	$-45°$/decade	$-90°$/decade
400	pole at $s = -40$	$+45°$/decade	$-45°$/decade
2000	pole at $s = -200$	$+45°$/decade	$0°$/decade

Figure 12-32(b) shows the Bode plot of the resulting straight-line approximation of the phase response.

EXAMPLE 12-15

(a) Find the transfer function $T(s)$ corresponding to the straight-line gain plot in Fig. 12-33.

(b) Find the time-domain step response corresponding to the gain plot.

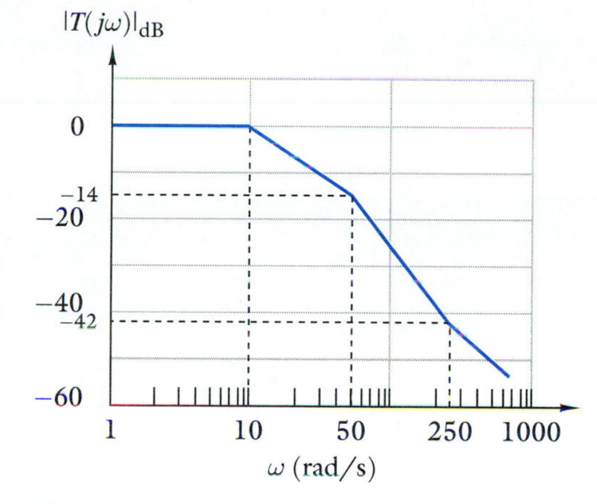

Figure 12-33

SOLUTION:

(a) The gain plot shows slope changes at $\omega = 10$, 50, and 250 rad/s. The slope changes at $\omega = 10$ and 50 rad/s are -20 dB/decade indicating that $T(s)$ has poles $s = -10$ and $s = -50$ rad/s. The slope change at $\omega = 250$ is $+20$ dB/decade indicating a zero at $s = -250$ rad/s. Therefore, the transfer function is of the form

$$T(s) = K\frac{(1 + s/250)}{(1 + s/10)(1 + s/50)} = \frac{2K(s + 250)}{(s + 10)(s + 50)}$$

where the scale factor K is yet to be determined. The gain plot in Fig. 12-33 has a low-frequency asymptote of 0 dB, so we conclude that $K = 1$.

(b) The transform of the step response is

$$G(s) = \frac{T(s)}{s} = \frac{2(s + 250)}{s(s + 10)(s + 50)}$$

Expanding $G(s)$ by partial fractions

$$G(s) = \frac{1}{s} - \frac{48/40}{s + 10} + \frac{4/20}{s + 50}$$

The inverse Laplace transformation produces the step response waveform.

$$g(t) = \left[1 - 1.2e^{-10t} + 0.2e^{-50t}\right]u(t)$$

EXERCISE 12-15

Given the following transfer function

$$T(s) = \frac{4000s}{(s+100)(s+2000)}$$

(a) Find the straight-line approximation to the gain at $\omega = 50$, 100, 500, 2000, 4000 rad/s.

(b) Estimate the actual gain at the frequencies in (a).

ANSWERS:

(a) 0 dB, 6 dB, 6 dB, 6 dB, 0 dB

(b) −1 dB, 3 dB, 6 dB, 3 dB, −1 dB

Bode Plots: Complex Poles and Zeros

The straight-line gain plots for transfer functions with complex poles and zeros are constructed in the same manner as the plots for real poles and zeros. However, complex critical frequencies produce resonant peaks in the gain response, which can cause the actual gain to depart significantly from the straight-line approximation. Nonetheless, straight-line plots are a good starting place because they provide a baseline for describing the frequency response of circuits with complex poles.

Complex poles and zeros occur in conjugate pairs and are introduced in $T(s)$ by quadratic factors of the form

$$s^2 + 2\zeta\omega_O s + \omega_O^2$$

where ζ and ω_O are the damping ratio and undamped natural frequency. The standard form of this quadratic factor for Bode plots is obtained by factoring ω_O^2 and replacing s by $j\omega$:

$$1 - (\omega/\omega_O)^2 + j2\zeta(\omega/\omega_O) \tag{12-49}$$

In a Bode plot summation the quadratic factor in Eq. (12-49) introduces gain and phase terms of the following form

$$|T(j\omega)|_{dB} = \pm 20\log_{10}\sqrt{\left[1-(\omega/\omega_O)^2\right]^2 + (2\zeta\omega/\omega_O)^2} \tag{12-50a}$$

$$\theta(\omega) = \pm\tan^{-1}\frac{2\zeta\omega/\omega_O}{1-(\omega/\omega_O)^2} \tag{12-50b}$$

where the plus sign applies to complex zeros and the minus sign to poles. The zeros enter the gain and phase summations with positive signs and tend to increase the gain and phase angle. Terms due to complex poles enter with negative signs and tend to decrease gain and phase.

Figure 12-34 shows the gain contribution of complex poles and zeros for several values of ζ. Like second-order transfer functions, complex critical frequencies have low- and high-frequency asymptotes that intersect at $\omega = \omega_O$. At low frequency ($\omega \ll \omega_O$) the asymptotes in a Bode plot are horizontal lines at 0 dB. At high

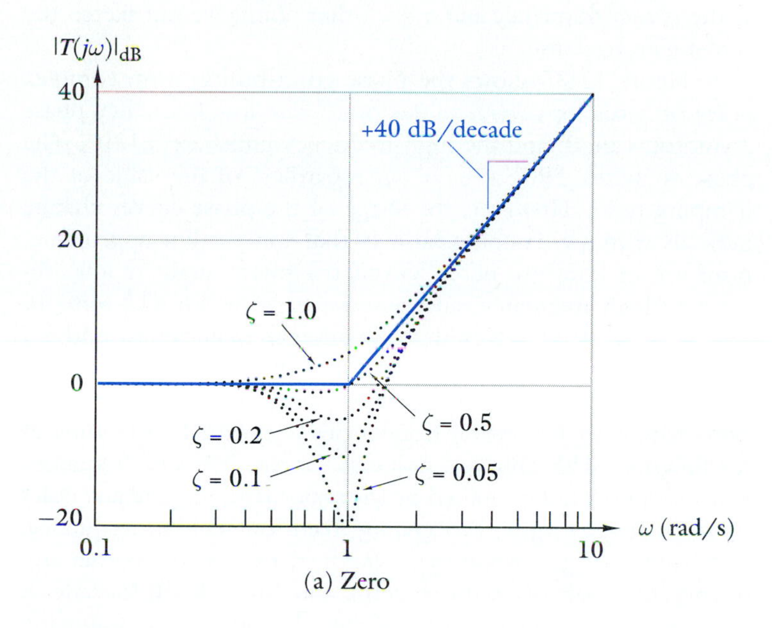

(a) Zero

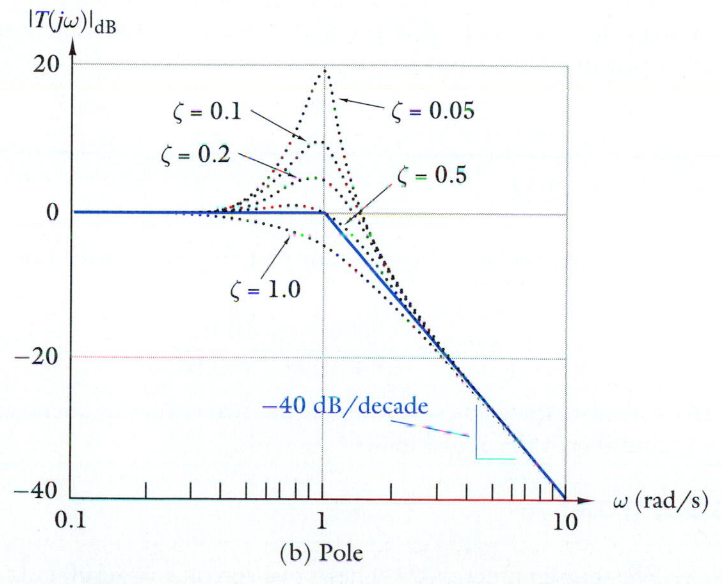

(b) Pole

Figure 12-34 Gain response of complex poles and zeros.

frequency ($\omega \gg \omega_O$) the gain asymptotes are $\pm 40 \log_{10}(\omega/\omega_O)$. In a Bode plot the high-frequency asymptote is a straight line with a slope of ±40 dB/decade, where the plus sign applies to zeros and the minus to poles. The gain corrections are functions of ζ, since the shape of the actual gain curve changes with the damping ratio. For example, at $\omega = \omega_O$ the gain correction is $\pm 20 \log(2\zeta)$. Equation (12-50a) can be used to calculate the gain corrections at other frequencies. Given the gain corrections at the corner frequency and a few other points we can sketch the actual gain response.

Figure 12-35 shows the phase contribution from complex poles or zeros for several values of ζ. The low-frequency phase asymptotes are 0° and the high-frequency limits are ±180°. The phase is always ±90° at $\omega = \omega_O$ regardless of the value of the damping ratio. However, the shape of the phase curves change radically with the damping ratio so that straight-line approximations are of little use here. Given the phase angle at low frequency, high frequency, and ω_O, we can use Eq. (12-50b) to calculate the phase angles at a few other frequencies around ω_O and then sketch the phase response curve.

Constructing the straight-line gain plot for transfer functions with complex critical frequencies is essentially the same as for functions with only real poles and zeros. The low-frequency gain asymptote is determined by the scale factor K_O and any poles or zeros at the origin. Proceeding from low to high frequency, we change the gain slope by +40 dB/decade at the corner frequency of a pair of complex zeros and by −40 dB/decade at the corner frequency of a pair of complex poles. The following examples illustrate Bode plot generation for a transfer function with a pair of complex poles.

EXAMPLE 12-16

(a) Construct the straight-line gain plot for the transfer function

$$T(s) = \frac{5000(s+100)}{s^2 + 100s + (500)^2}$$

(b) Calculate the gain corrections at the two corner frequencies and sketch the actual gain.

SOLUTION:

(a) The transfer function $T(s)$ has a real zero at $s = -100$ rad/s and a pair of complex poles with $\zeta = 0.1$ and $\omega_O =$

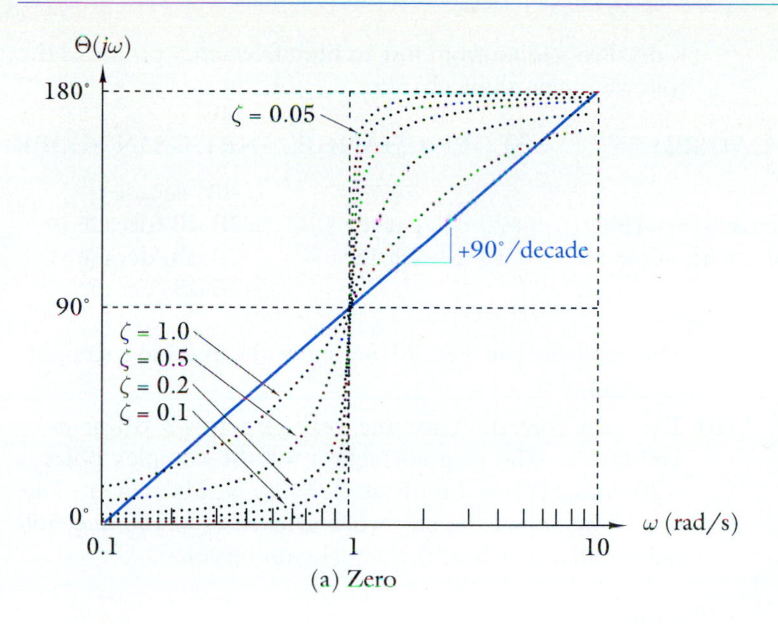

(a) Zero

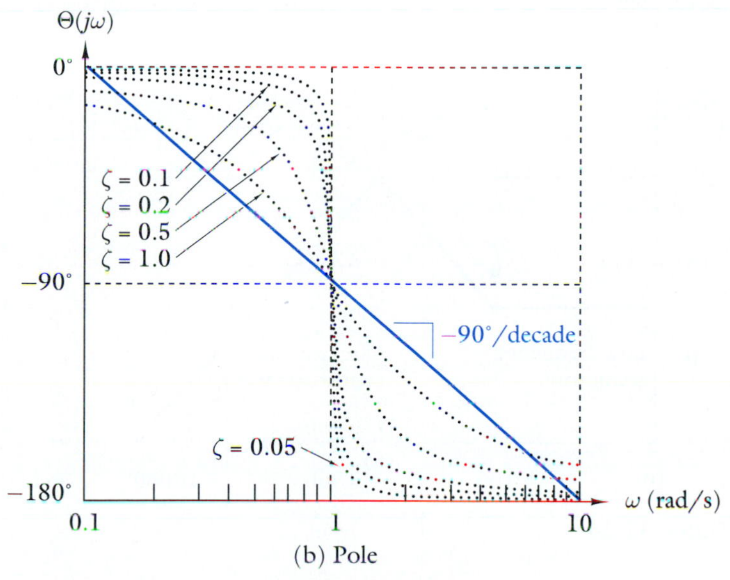

(b) Pole

Figure 12-35 Phase response of complex poles and zeros.

500 rad/s. Written in standard form $T(j\omega)$ is

$$T(j\omega) = 2\left(\frac{1 + j\omega/100}{1 - (\omega/500)^2 + j0.2\omega/500}\right)$$

The scale factor is $K_O = 2$ and there are corner frequencies at $\omega = 100$ rad/s due to the zero and $\omega = 500$ rad/s due to the pair of complex poles. At low frequency $T(j\omega) \rightarrow 2$, so the low-frequency gain asymptote is $20 \log_{10}(2) \approx$

6 dB. Proceeding from low to high frequency produces the following gain slope changes:

FREQUENCY	CAUSED BY	SLOPE CHANGE	NET GAIN SLOPE
10			0 dB/decade
100	zero at $s = -100$	+20 dB/decade	+20 dB/decade
500	poles with $\omega_O = 500$	−40 dB/decade	−20 dB/decade
5000			−20 dB/decade

The solid lines in Fig. 12-36 show the resulting straight-line gain plot.

(b) The gain correction for the real zero is +3 dB at $\omega = 100$ rad/s. The gain correction for the complex poles is $-20 \log_{10}(2\zeta) \approx 14$ dB at $\omega = \omega_O = 500$ rad/s. Figure 12-36 shows the corrected gains at $\omega = 100$ and 500 rad/s and a sketch of the actual gain response.

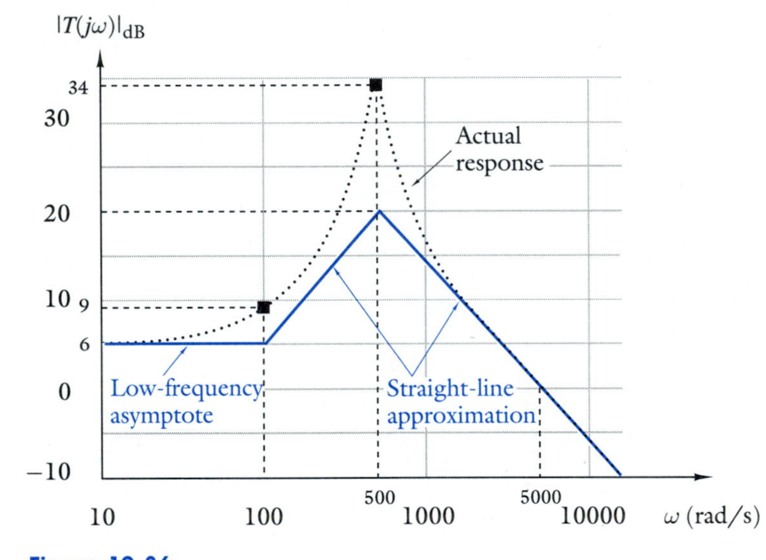

Figure 12-36

EXAMPLE 12-17

Sketch the phase response of the transfer function

$$T(s) = \frac{500(s + 100)}{s^2 + 100s + (500)^2}$$

SOLUTION:

The gain plot for this transfer function is found in Example 12-16. The function has a zero at $s = -100$ rad/s and a pair of complex poles with $\zeta = 0.1$ and $\omega_O = 500$ rad/s. Written in standard form $T(j\omega)$ is

$$T(j\omega) = 2\left(\frac{1 + j\omega/100}{1 - (\omega/500)^2 + j0.2\omega/500}\right)$$

This function has corner frequencies at $\omega = 100$ rad/s and $\omega = 500$ rad/s. At low frequency $T(j\omega) \rightarrow K_O = 2$, so the low-frequency phase asymptote is $\angle K_O = 0°$. The analytical expression for the phase angle of $T(j\omega)$ is

$$\theta(\omega) = +\tan^{-1}(\omega/100) - \tan^{-1}\left(\frac{0.2\omega/500}{1 - (\omega/500)^2}\right)$$

The frequency range of interest extends from 10 to 5000 rad/s. We need to calculate the phase at the end points of this range and at the two corner frequencies. The phase contribution of complex poles with a damping ratio of $\zeta = 0.1$ changes rapidly around $\omega_O = 500$ rad/s. Therefore, we also calculate the phase an octave above and below ω_O.

FREQUENCY	PHASE OF ZERO	PHASE OF COMPLEX POLES	TOTAL PHASE
10	+5.71°	−0.229°	+5.48°
100	+45.0°	−2.386°	+42.6°
250	+68.2°	−7.595°	+60.6°
500	+78.7°	−90.00°	−11.3°
1000	+84.3°	−172.4°	−88.1°
5000	+89.0°	−178.8°	−89.8°

Figure 12-37 shows a sketch of the phase response using these values.

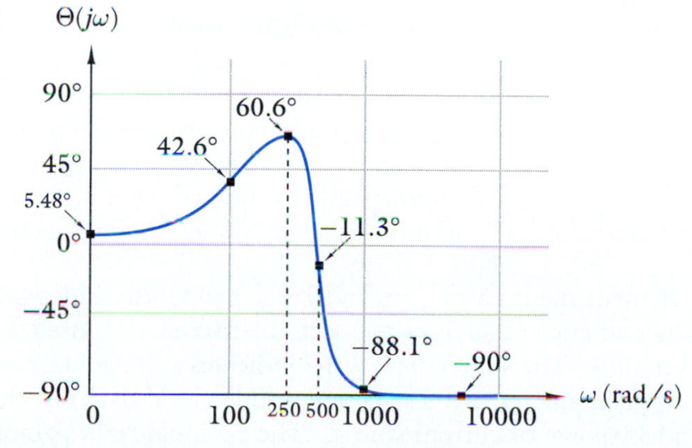

Figure 12-37

EXERCISE 12-16

Construct the gain corner plot for the transfer function

$$T(s) = \frac{10^5}{\left[s^2 + 100s + 10^4\right](s + 10)}$$

(a) Find the straight-line gain at $\omega = 1, 10, 100, 1000$ rad/s.
(b) Estimate the actual gain at the frequencies given in (a).

ANSWERS:
(a) 0 dB, 0 dB, −20 dB, −80 dB.
(b) 0 dB, −3 dB, −20 dB, −80 dB ■

EXERCISE 12-17

Calculate the phase response of the transfer function in Exercise 12-16 at $\omega = 1, 10, 100, 1000$ rad/s.

ANSWERS:
−6.28°, −50.8°, −174°, −264° ■

12-6 FREQUENCY RESPONSE USING SPICE

Computer analysis provides a useful computational aid for determining frequency response and comparing circuit performance with design objectives. Circuit analysis programs normally have a feature called *AC analysis*, which calculates circuit frequency responses. This section introduces the SPICE syntax for AC analysis to illustrate the input data needed to describe a frequency response problem to a circuit analysis program.

To begin with we need element descriptions for the sinusoidal sources that provide ac inputs. In SPICE element statements for independent ac voltage and current sources are

```
Vxx...xx <1st node> <2nd node> AC mag [phase]
Ixx...xx <1st node> <2nd node> AC mag [phase]
```

The element names Vxx. . .xx and Ixx. . .xx identify independent voltage and current sources connected between <1st node> and <2nd node>. The source type = AC indicates a sinusoidal source. The *mag* parameter defines the amplitude (in V or A) of the sinusoidal voltage or current source. The optional *phase* parameter defines the source phase angle (in degrees) and has a default value of 0° when no value is assigned.

Note that the source element statements do not specify the frequency of the sinusoidal input. The frequency assigned to all independent sources is defined in the .AC analysis command statement.

```
.AC <sweep> NFREQ FSTART FSTOP
```

This statement causes SPICE to perform sinusoidal steady-state analysis as the frequency of all independent ac sources is varied (swept) from FSTART to FSTOP (in hertz). There are three types of sweeps:

1. <sweep> = LIN is a linear sweep. The frequency is varied linearly from FSTART to FSTOP generating circuit responses at NFREQ equally spaced frequencies.

2. <sweep> = OCT is an octave sweep. The frequency is varied logarithmically from FSTART to FSTOP generating circuit responses at NFREQ frequencies per octave.

3. <sweep> = DEC is an decade sweep. The frequency is varied logarithmically from FSTART to FSTOP generating circuit responses at NFREQ frequencies per decade.

The integer NFREQ controls the number of frequencies at which the calculated responses are written to the output file. For a linear sweep the output file will contain responses at NFREQ frequencies. For a decade or octave sweep the number of output frequencies will be NFREQ times the number of decades or octaves between FSTART and FSTOP plus 1.

The user must have some idea of the frequency range of interest and how densely to sample the range. Otherwise, the response data in the output file may not adequately characterize the circuit frequency response. A rough sketch or other preliminary analysis is necessary in order to use the ac analysis capability of SPICE or any other program effectively.

The sinusoidal steady-state node voltages or device currents listed in the .PRINT command are written to the output file. The type of output data is controlled by adding one of the following suffixes to the V or I symbol:

M Amplitude of the sinusoid in V or A

DB Amplitude in decibels: $20 \times \log_{10}(\text{Amplitude})$

P Phase angle of the sinusoid in degrees

For example, the print command

```
.PRINT AC VM(1) VDB(4) IP(VIN)
```

causes SPICE to list the following responses in the output file:

1. Amplitude of node voltage V(1)
2. $20 \times \log_{10}$(Amplitude) of node voltage V(4)
3. The phase angle of the current through voltage source VIN.

Note that the analysis type (AC in this case) must be included in the print command.

In the following examples we use the SPICE ac analysis to calculate frequency responses of circuits designed earlier in this chapter.

EXAMPLE 12-18

Use SPICE to calculate the frequency response of the circuit shown in Fig. 12-38.

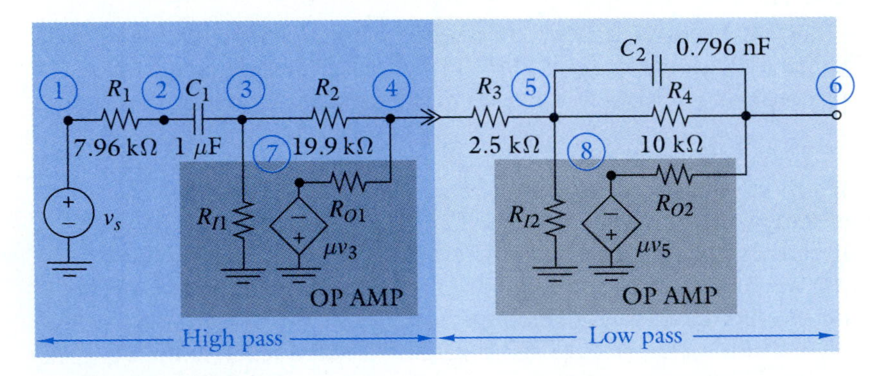

Figure 12-38

SOLUTION:

The RC OP AMP circuit in Fig. 12-38 was designed in Example 12-8 to have a bandpass characteristic with a passband gain of 10 and cutoff frequencies at 20 Hz and 20 kHz. The center frequency for this bandpass response is

$$f_O = \sqrt{f_{UPPER} f_{LOWER}} = \sqrt{20 \times 10^3 \times 20} = 632 \text{ Hz}$$

Therefore, the circuit damping ratio is

$$\zeta = \frac{BW}{2\omega_O} = \frac{f_{UPPER} - f_{LOWER}}{2f_O} = \frac{20 \times 10^3 - 20}{2 \times 632} = 15.8 \gg 0.5$$

Since the damping ratio is much greater than 0.5, the circuit has a very broadband response. To characterize the response we need to sweep the frequency from one decade below the lower-cutoff frequency to one decade above the upper cutoff frequency. The range from 2 Hz to 200 kHz spans five decades. Using a decade sweep and specifying NFREQ = 20 produces calculated responses in the output files at a total of $5 \times 20 + 1 = 101$ frequencies.

A two-column listing (to save space) of a SPICE input file for this analysis is

```
*EXAMPLE 12-18               RI1  3  0  1MEG
VS  1  0  AC  1  0           E1   0  7  3  0  1MEG
R1  1  2  7.96K              RO1  7  4  100
C1  2  3  1U                 RI2  5  0  1MEG
R2  3  4  19.9K              E2   0  8  5  0  1MEG
R3  4  5  2.5K               RO2  8  6  100
C2  5  6  0.796N             .AC  DEC  20  2  200K
R4  5  6  10K                .PRINT  AC  VM(6)  VP(6)
                             .END
```

The element statements in the second column model the two OP AMPs in Fig. 12-38. The AC command specifies a decade sweep from 2 Hz to 200 kHz with 20 points per decade. The input defined in the VS element statement has a 1-volt amplitude and a phase angle of zero. The print command specifies the outputs as the magnitude (M) and phase (P) of the voltage at Node 6. In effect, VM(6) and VP(6) are the gain and phase response of the circuit transfer function, since the input sinusoid has 1-V amplitude and 0° phase.

Figure 12-39 shows a plot of the gain and phase responses generated by this circuit file. The gain response has a bandpass shape with a passband gain of 10 and cutoff frequencies at about 20 Hz and 20 kHz. To check the cutoff frequencies we examine the numerical data in the output files. The relevant portion of SPICE generated output file is

```
   FREQ          VM(6)         VP(6)
    •              •             •
    •              •             •
    •              •             •
 1.589E+01     6.221E+00     5.149E+01
 1.783E+01     6.654E+00     4.823E+01
 2.000E+01     7.071E+00     4.493E+01
 2.244E+01     7.466E+00     4.164E+01
 2.518E+01     7.831E+00     3.838E+01
    •              •             •
    •              •             •
    •              •             •
 1.589E+04     7.829E+00    -3.840E+01
 1.783E+04     7.464E+00    -4.165E+01
```

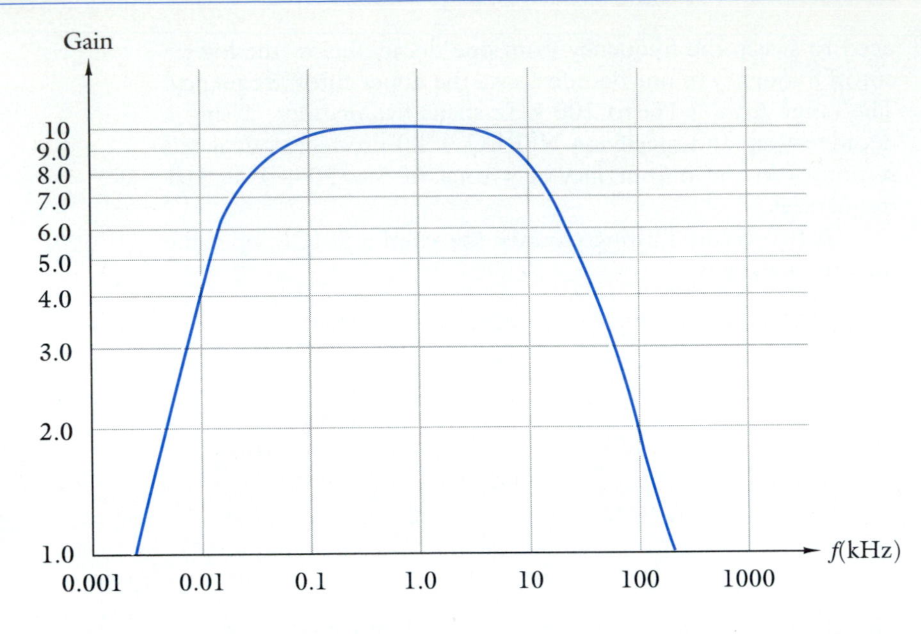

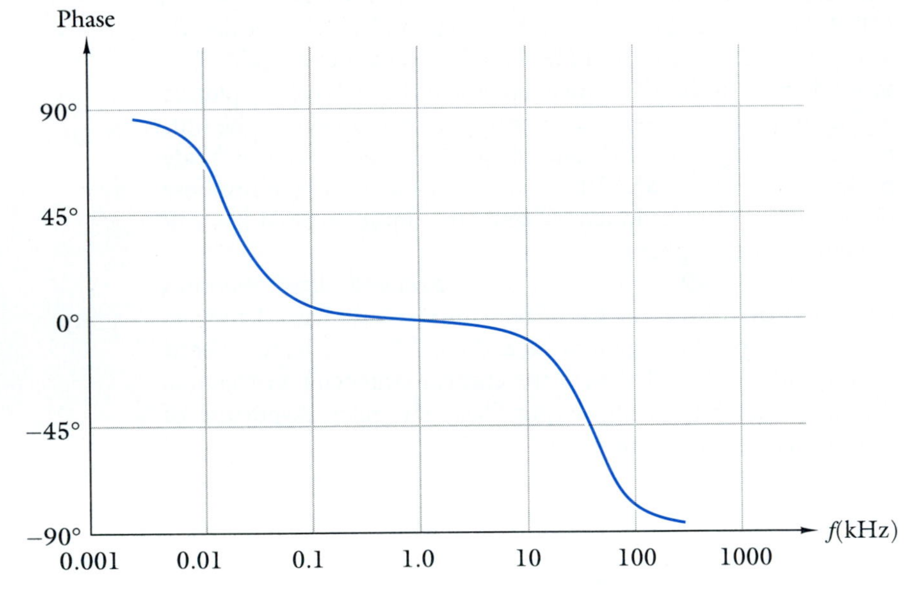

Figure 12-39

```
2.000E+04   7.070E+00   -4.495E+01
2.244E+04   6.652E+00   -4.825E+01
2.518E+04   6.219E+00   -5.150E+01
    •           •            •
    •           •            •
    •           •            •
```

The two cutoff frequencies occur when VM(6) $= 10/\sqrt{2} =$ 0.7071, since passband gain is 10. The SPICE output data confirm that the cutoff frequencies are very close to the 20-Hz and 20-kHz values specified in the design requirements.

EXAMPLE 12-19

Use SPICE to calculate the frequency response of the RLC circuit shown in Fig. 12-40.

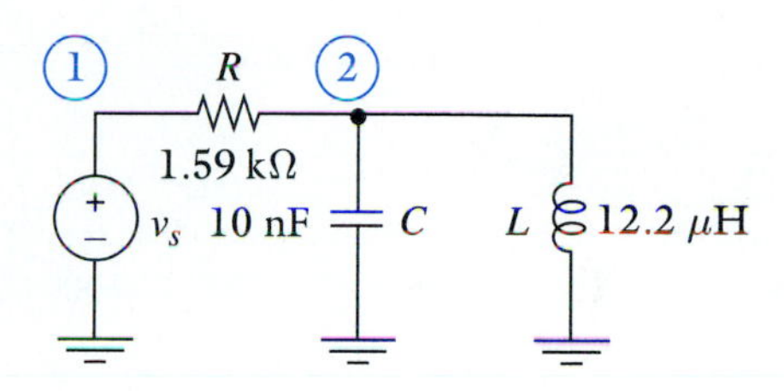

Figure 12-40

SOLUTION:

The circuit in Fig. 12-40 was designed in Example 12-9 to have a bandpass characteristic with a center frequency at 455 kHz and a bandwidth of 10 kHz. The damping ratio of this circuit is

$$\zeta = \frac{BW}{2\omega_O} = \frac{10 \times 10^3}{2 \times 455 \times 10^3} = 0.011 \ll 0.5$$

Since $\zeta \ll 0.5$ the circuit has a very narrowband response with a resonant peak at 455 kHz. To characterize this response we use a linear sweep from 50 kHz below to 50 kHz above the center frequency at 455 kHz. Specifying NFREQ = 101 causes SPICE to write calculated response data to the output file at 101 equally spaced frequencies.

A SPICE input file for ac analysis of the circuit is

```
*EXAMPLE 12-19
VS   1   0   AC   1   0
R    1   2   1.59K
C    2   0   10N
L    2   0   12.2U
.AC   LIN   101   405K   505K
.PRINT AC   VM(2)   VP(2)
.END
```

In effect, VM(6) and VP(6) are the gain and phase of the circuit transfer function, since the input is a 1-volt ac source with zero phase.

Figure 12-41 shows a plot of the frequency response generated by SPICE. The logarithmic frequency scale in the figure highlights the fact that the circuit frequency response has a very narrow resonant peak around 455 kHz. To check the circuit bandwidth we examine the numerical data in the output files.

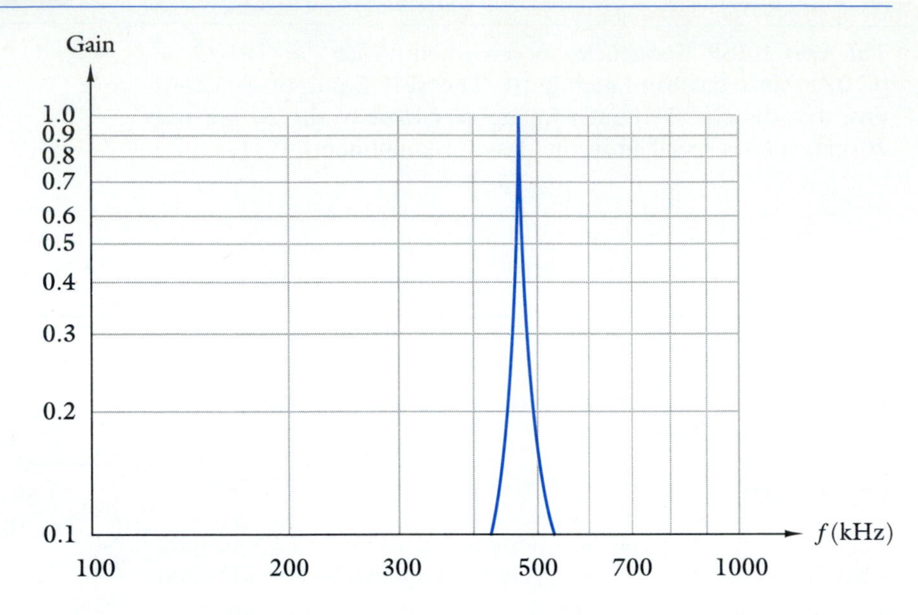

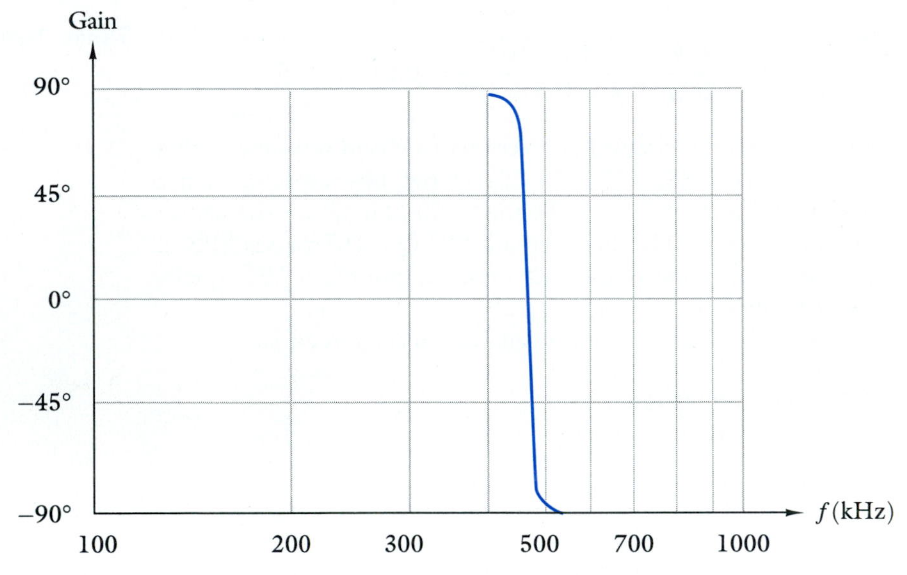

Figure 12-41

The relevant portion of SPICE-generated output file is

```
FREQ        VM(2)       VP(2)
  •           •           •
  •           •           •
  •           •           •
4.500E+05   6.601E-01   4.869E+01
4.510E+05   7.302E-01   4.310E+01
4.520E+05   8.061E-01   3.628E+01
```

```
4.530E+05   8.825E-01   2.805E+01
4.540E+05   9.490E-01   1.837E+01
4.550E+05   9.914E-01   7.510E+00
4.560E+05   9.977E-01  -3.893E+00
4.570E+05   9.660E-01  -1.498E+01
4.580E+05   9.062E-01  -2.501E+01
4.590E+05   8.327E-01  -3.363E+01
4.600E+05   7.570E-01  -4.080E+01
4.610E+05   6.859E-01  -4.669E+01
    •           •           •
    •           •           •
    •           •           •
```

The two cutoff frequencies occur when VM(2) $= 1/\sqrt{2} = 0.7071$, since the passband gain is 1.00. Interpolating between the data points given above yields cutoff frequencies at 450.7 kHz and 460.3 kHz for a bandwidth of 9.6 kHz. The bandwidth is only 4% less than the design requirement.

SUMMARY

- The sinusoidal steady-state response is the forced component of the zero-state response of a stable, linear circuit for a sinusoidal input. The circuit transfer function $T(s)$ contains the information required to derive the sinusoidal steady-state response.
- The frequency response of a circuit is the variation of the gain $|T(j\omega)|$ and phase $\angle T(j\omega)$ functions. The gain function is usually expressed in dB in frequency response plots. Logarithmic frequency scales are used on frequency response plots of the gain and phase functions.
- A passband is a range of frequencies over which the steady-state output is essentially constant and there is very little attenuation. A stop band is a range of frequencies over which the steady-state output is significantly attenuated. The cutoff frequency is the boundary between a passband and the adjacent stop band.
- Circuit frequency responses are classified as low pass, high pass, bandpass, and bandstop depending on the number and location of the stop and passbands. The performance of devices and circuits are often specified in terms of frequency response descriptors such as bandwidth, passband gain, and cutoff frequency.
- The low- and high-frequency gain asymptotes of first- and second-order circuits intersect at a corner frequency de-

termined by the location of the pole(*s*). The total phase change from low to high frequency is 90° for first-order circuits and 180° for second-order circuits. Second-order circuits with complex poles exhibit resonant peaks and narrow-band gain response when the damping ratio is less than 0.5.

- Bode plots are graphs of the gain (in dB) and phase angle (in degrees) versus log-frequency scales. Straight-line approximations to the gain and phase can be constructed using the corner frequencies defined by the poles and zeros of $T(s)$. The slopes of straight-line gain plots are integer multiples of ±20 dB/decade.
- Computer-aided circuit analysis programs such as SPICE or MICRO-CAP can generate frequency response data when circuit element values are given. The user must have a rough idea of the frequency range of interest in order to specify the frequency sampling parameters in the analysis command.

EN ROUTE OBJECTIVES AND ASSOCIATED PROBLEMS

ERO 12-1 The Sinusoidal Steady-State Response (Sect. 12-1)

Given a linear circuit or its transient response: Find the amplitude and phase of the sinusoidal steady-state response for a specified input sinusoid.

12-1 Find the amplitude and phase angle of the steady-state output of the circuit in Fig. P12-1 for (a) $v_1(t) = 5 \cos 10^4 t$ and (b) $v_1(t) = 10 \sin 10^3 t$.

12-2 Find the amplitude and phase angle of the steady-state output of the circuit in Fig. P12-2 for (a) $v_1(t) = 10 \sin 100t$ and (b) $v_1(t) = 3 \cos 200t$.

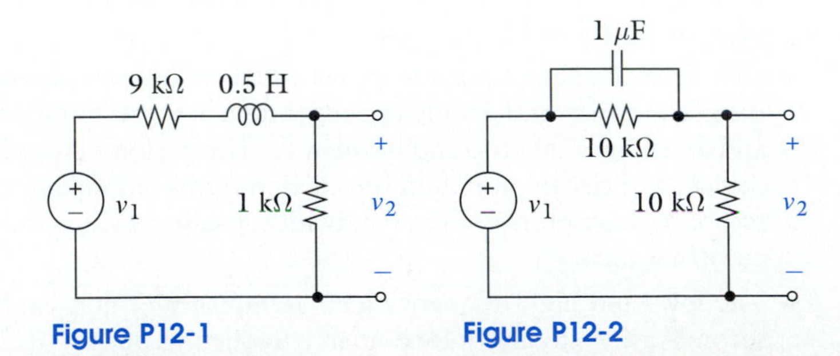

Figure P12-1

Figure P12-2

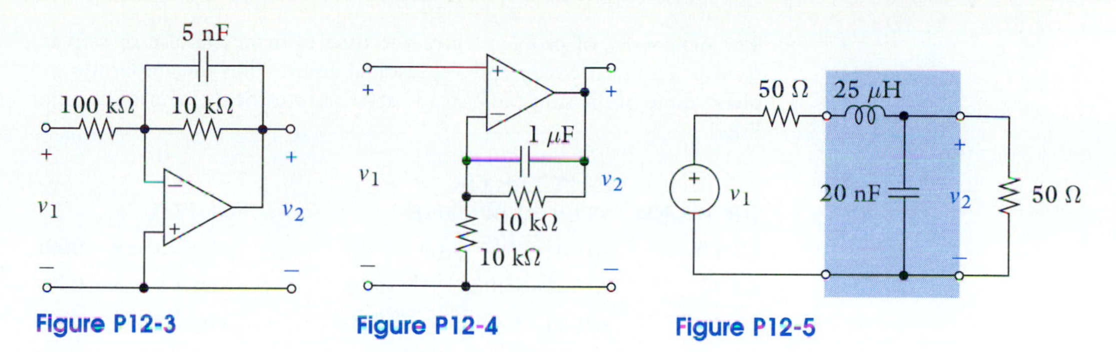

Figure P12-3 Figure P12-4 Figure P12-5

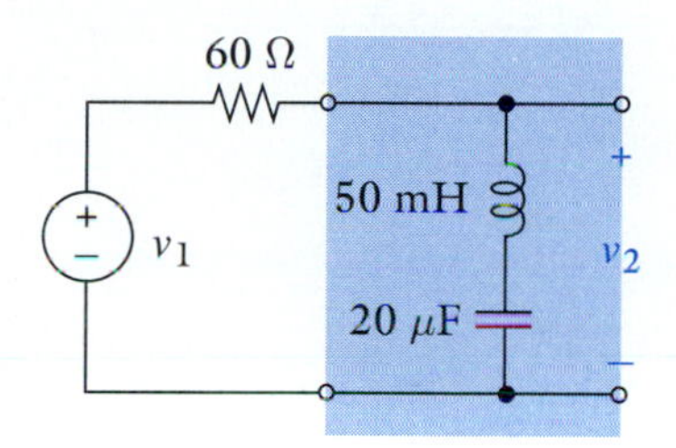

Figure P12-6

12-3 Repeat Prob. 12-1 for the circuit in Fig. P12-3.

12-4 Repeat Prob. 12-2 for the circuit in Fig. P12-4.

12-5 Find the amplitude and phase angle of the steady-state output of the circuit in Fig. P12-5 for (a) $v_1(t) = 10 \cos 10^6 t$ and (b) $v_1(t) = 5 \cos 2 \times 10^6 t$.

12-6 Find the amplitude and phase angle of the steady-state output of the circuit in Fig. P12-6 for (a) $v_1(t) = 15 \cos 10^3 t$ and (b) $v_1(t) = 5 \cos 2 \times 10^3 t$.

12-7 Find the amplitude and phase angle of the steady-state output of the circuit in Fig. P12-7 for $v_1(t) = 2 \cos 100t$. Repeat for $v_1(t) = 5 \sin 200t$.

12-8 Find the amplitude and phase angle of the steady-state output of the circuit in Fig. P12-8 for (a) $v_1(t) = 2 \cos 100t$ and (b) $v_1(t) = 2 \cos 1000t$.

12-9 Find the amplitude and phase angle of the steady-state output of the circuit in Fig. P12-9 for (a) $v_1(t) = 2 \cos 1000t$ and (b) $v_1(t) = 2 \cos 2000t$.

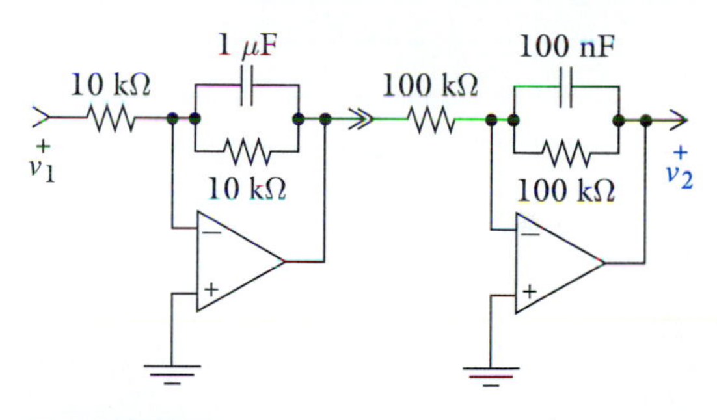

Figure P12-7

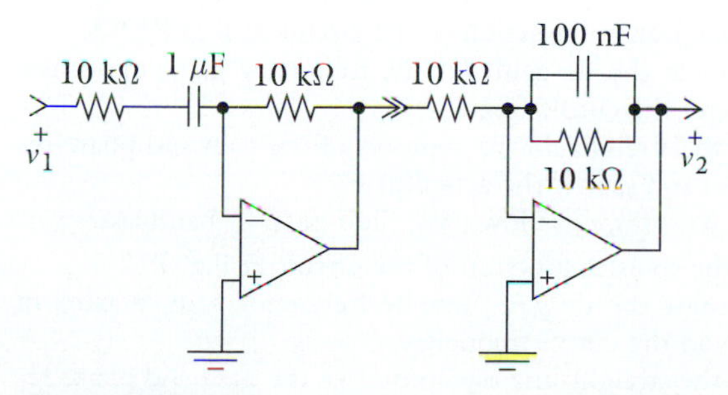

Figure P12-8

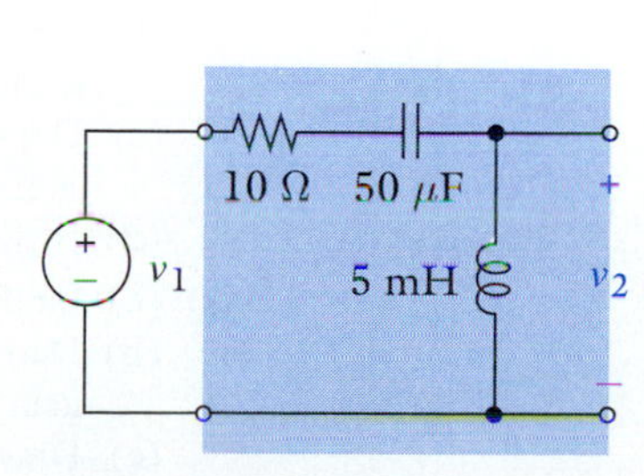

Figure P12-9

The next series of problems give the time domain impulse or step response of a linear circuit and a sinusoidal input. Find the amplitude and phase angle of the sinusoidal steady-state output for the sinusoidal input listed.

PROBLEM	IMPULSE OR STEP RESPONSE	INPUT
12-10	$g(t)=[e^{-1000t}]u(t)$	$x(t)=5\cos 1000t$
12-11	$h(t)=[-1000e^{-1000t}]u(t)$	$x(t)=5\cos 1000t$
12-12	$g(t)=[-1+0.8e^{-2000t}]u(t)$	$x(t)=4\cos 10^4 t$
12-13	$g(t)=[1+4\,e^{-2000t}]u(t)$	$x(t)=5\cos 2000t$
12-14	$h(t)=[100e^{-1000t}\sin 1000t]u(t)$	$x(t)=-3\sin 2000t$
12-15	$g(t)=[500e^{-1200t}\sin 1600t]u(t)$	$x(t)=4\cos 2000t$
12-16	$h(t)=5000[e^{-1000t}-e^{-5000t}]u(t)$	$x(t)=5\cos 1000t$
12-17	$g(t)=[e^{-100t}\cos 1000t]u(t)$	$x(t)=5\cos 1000t$

ERO 12-2 Basic Frequency Response (Sects. 12-2, 12-3, 12-4)

(a) Given a first- or second-order linear circuit or its transient response: Draw the straight-line asymptotes of the gain and phase responses, sketch the actual response, and determine frequency response descriptors.

(b) Given the straight-line asymptotes or descriptors of the gain or phase responses of a linear circuit: Find $T(s)$ and design a circuit that realizes $T(s)$.

12-18 (a) Find the transfer function of the circuit in Fig. P12-1.
(b) Determine the dc gain, infinite-frequency gain, maximum gain, and the cutoff frequency.
(c) Draw the straight-line asymptotes of the gain and phase responses and sketch the actual gain.
(d) Is the gain response low-pass, high-pass, or bandpass?

12-19 (a) Find the transfer function of the circuit in Fig. P12-3.
(b) Determine the dc gain, infinite-frequency gain, maximum gain, and the cutoff frequency.
(c) Draw the straight-line asymptotes of the gain and phase responses and sketch the actual gain.
(d) Is the gain response low-pass, high-pass, or bandpass?

12-20 (a) Find the transfer function of the circuit in Fig. P12-5.
(b) Determine the dc gain, infinite-frequency gain, maximum gain, and the cutoff frequency.
(c) Draw the straight-line asymptotes of the gain and phase responses and sketch the actual gain.

(d) Is the gain response low-pass, high-pass, or bandpass?

12-21 (a) Find the transfer function of the circuit in Fig. P12-9.
(b) Determine the dc gain, infinite-frequency gain, maximum gain, and the cutoff frequency.
(c) Draw the straight-line asymptotes of the gain and phase responses and sketch the actual gain.
(d) Is the gain response low-pass, high-pass, or bandpass?

12-22 Given the step response in Prob. 12-10:
(a) Find the circuit transfer function.
(b) Determine the dc gain, infinite-frequency gain, maximum gain, and the cutoff frequency.
(c) Draw the straight-line asymptotes of the gain and phase responses and sketch the actual gain.
(d) Is the gain response low-pass, high-pass, or bandpass?

12-23 Given the impulse response in Prob. 12-11:
(a) Find the circuit transfer function.
(b) Determine the dc gain, infinite-frequency gain, maximum gain, and the cutoff frequency.
(c) Draw the straight-line asymptotes of the gain and phase responses, and sketch the actual gain.
(d) Is the gain response low-pass, high-pass, or bandpass?

12-24 Given the impulse response in Prob. 12-14:
(a) Find the circuit transfer function.
(b) Determine the dc gain, infinite-frequency gain, maximum gain, and the cutoff frequency.
(c) Draw the straight-line asymptotes of the gain and phase responses, and sketch the actual gain.
(d) Is the gain response low-pass, high-pass, or bandpass?

12-25 Given the step response in Prob. 12-15:
(a) Find the circuit transfer function.
(b) Determine the dc gain, infinite-frequency gain, maximum gain, and the cutoff frequencies.
(c) Draw the straight-line asymptotes of the gain and phase responses and sketch the actual gain.
(d) Is the gain response low-pass, high-pass, or bandpass?

❖ **12-26 (D)**
(a) Construct a first-order high-pass transfer function that has a cutoff frequency of 2000 rad/s and a passband gain of 5.
(b) Design an RC circuit to realize the transfer

❖ **12-27 (D)**
(a) Construct a first-order low-pass transfer function that has a bandwidth less than 2000 rad/s, step function rise time less than 2 ms, and a dc gain of 10.
(b) Design an RC circuit to realize the transfer function in (a).

❖ **12-28 (D)**
(a) Construct a second-order bandpass transfer function with a center frequency and bandwidth of 1 MHz, and a maximum passband gain of 10.

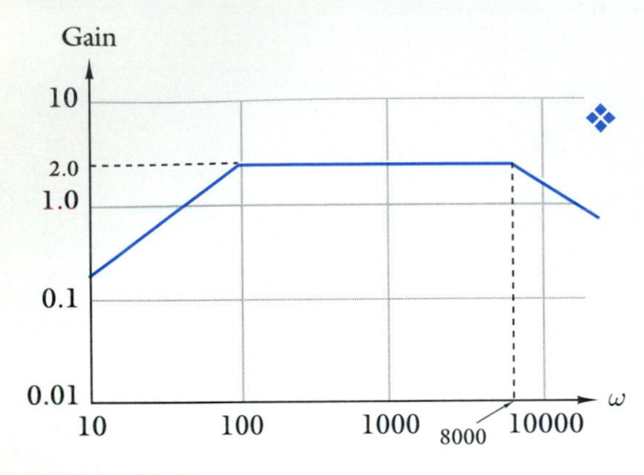

Figure P12-31

(b) Design a circuit that realizes the transfer function in (a). function in (a).

❖ **12-29 (D)**

(a) Construct a second-order low-pass transfer function with a cutoff frequency at 50 kHz, a dc gain of 10, and a maximum gain of 12.5.

(b) Design a circuit to realize the transfer function in (b).

12-30 A first-order low-pass circuit has a dc gain of 20 dB and cutoff frequency of 10 kHz. Find the circuit time-domain step response $g(t)$.

12-31 Figure P12-31 shows the straight lines approximation of the gain of a linear circuit. Find the circuit time-domain step response $g(t)$.

❖ **12-32 (D)**

(a) Construct a first or second-order transfer function whose gain response lies in the unshaded region in Fig. P12-32.

(b) Design a circuit that realizes the transfer function in (a).

12-33 **(a)** Show that the circuit in Fig. P12-33 has a second-order low-pass gain characteristic.

(b) Develop a procedure for selecting circuit parameters to achieve a specified corner frequency and damping ratio.

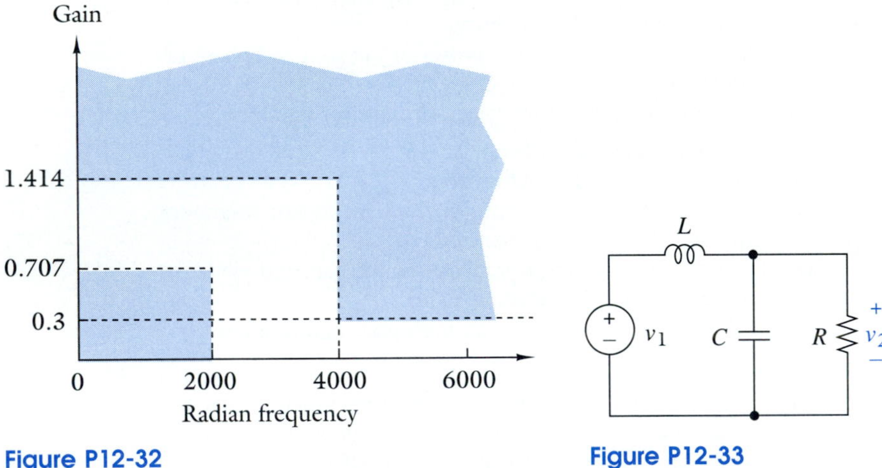

Figure P12-32

Figure P12-33

12-34 **(a)** Show that the circuit in Fig. P12-34 has a second-order bandpass gain characteristic.

(b) Develop a procedure for selecting circuit parameters to achieve a specified corner frequency and damping ratio.

12-35 **(a)** Show that the circuit in Fig. P12-35 has a second-order high-pass gain characteristic.

(b) Develop a procedure for selecting circuit parameters to achieve a specified corner frequency and damping ratio.

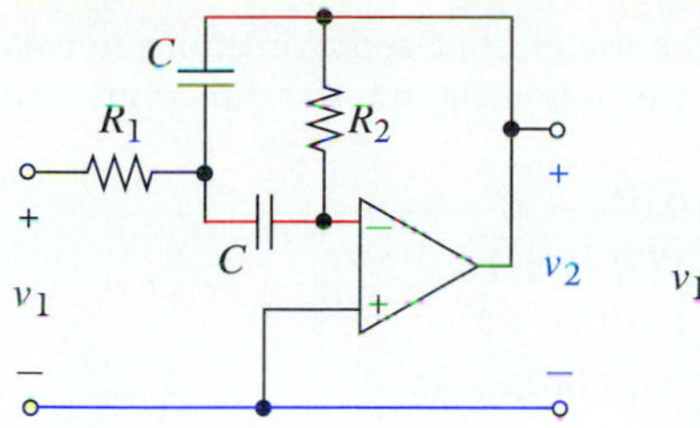

Figure P12-34

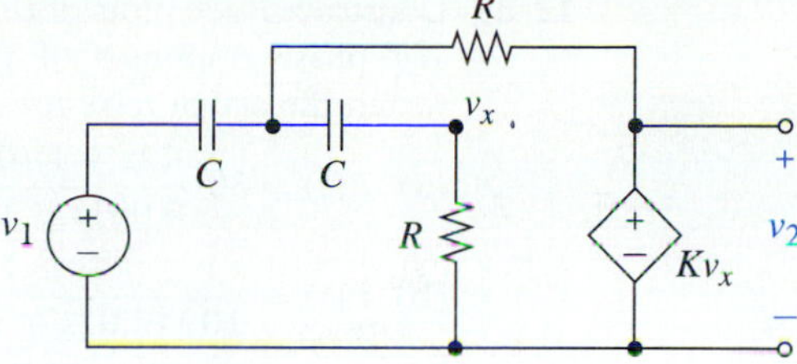

Figure P12-35

12-36 Given a second-order low-pass transfer function of the form

$$T(s) = \frac{K}{s^2 + 2\zeta\omega_O s + \omega_O^2}$$

(a) Show that the maximum gain is

$$|T(j\omega)|_{MAX} = \frac{|T(0)|}{2\zeta\sqrt{1-\zeta^2}}$$

(b) Show that the 3-dB cutoff frequency is

$$\omega_C = \omega_O\left[1 - 2\zeta^2 + \sqrt{(1-2\zeta^2)^2 + 1}\right]^{1/2}$$

ERO 12-3 Bode Plots (Sect. 12-5)

Given a linear circuit or its transfer function or transient response: Construct the Bode plots of the asymptotic gain or phase responses and sketch the actual response using gain corrections.

12-37 Construct Bode plots of the straight-line approximations to gain and phase responses of the following transfer functions and sketch the actual responses.

(a) $T(s) = \dfrac{(0.02s+1)(0.002s+1)}{(0.2s+1)(2s+1)}$

(b) $T(s) = \dfrac{(0.002s+1)(0.2s+1)}{(0.02s+1)(2s+1)}$

12-38 Construct Bode plots of the straight-line approximations to gain and phase responses of the following transfer functions and sketch the actual responses.

(a) $T(s) = 10\dfrac{(5s+1)(0.005s+1)}{(0.05s+1)(0.5s+1)}$

(b) $T(s) = 10\dfrac{(0.05s+1)(0.5s+1)}{(0.0005s+1)(5s+1)}$

12-39 Construct Bode plots of the straight-line approximations to gain and phase responses of the following transfer functions and sketch the actual responses.

(a) $T(s) = 30\dfrac{(2s+1)(0.02s+1)}{(0.001s+1)(0.4s+1)}$

(b) $T(s) = \dfrac{1}{100}\dfrac{(4s+1)((0.1s+1)}{(0.02s+1)(0.001s+1)}$

12-40 Construct Bode plots of the straight-line approximations to gain and phase responses of the following transfer functions and sketch the actual responses.

(a) $T(s) = 5\dfrac{(0.02s+1)}{(0.001s+1)(0.4s+1)}$

(b) $T(s) = \dfrac{1}{50}\dfrac{(0.004s+1)}{(0.02s+1)(0.1s+1)}$

12-41 Construct Bode plots of the straight-line approximations to gain and phase responses of the following transfer functions and sketch the actual responses.

(a) $T(s) = 4\dfrac{s}{(0.01s+1)(0.4s+1)}$

(b) $T(s) = \dfrac{1}{50}\dfrac{s^2}{(0.02s+1)(0.1s+1)}$

12-42 Construct Bode plots of the straight-line approximations to gain and phase responses of the following transfer functions and sketch the actual responses.

(a) $T(s) = 50\dfrac{s(0.02s+1)}{(0.001s+1)(0.4s+1)}$

(b) $T(s) = \dfrac{1}{400}\dfrac{s(0.1s+1)}{(0.02s+1)(0.001s+1)}$

12-43 Construct Bode plots of the straight-line approximations to gain and phase responses of the circuit in Fig. P12-2.

12-44 Construct Bode plots of the straight-line approximations to gain and phase responses of the circuit in Fig. P12-4.

12-45 Construct Bode plots of the straight-line approximations to gain and phase responses of the circuit in Fig. P12-7.

12-46 Construct Bode plots of the straight-line approximations to gain and phase responses of the circuit in Fig. P12-8.

12-47 Construct Bode plots of the straight-line approximations to gain and phase responses of the following transfer functions and sketch the actual responses.

(a) $T(s) = \dfrac{5}{(0.01s^2+0.1s+1)}$

(b) $T(s) = \dfrac{40}{(4s^2+0.1s+1)}$

12-48 Construct Bode plots of the straight-line approximations to gain and phase responses of the following transfer functions and sketch the actual responses.

(a) $T(s) = \dfrac{5(0.5s+1)}{(0.04s^2+0.04s+1)}$

(b) $T(s) = \dfrac{5s}{(0.25s^2+0.5s+1)}$

12-49 Construct Bode plots of the straight-line approximations to gain and phase responses of the circuit in Fig. P12-9 and sketch the actual response.

12-50 **(a)** Find the transfer function corresponding to the straight-line gain plot shown in Fig. P12-50.
(b) Determine the time-domain step response for the transfer function found in part (a).

12-51 **(a)** Find the transfer function corresponding to the straight-line gain plot shown in Fig. P12-51.
(b) Determine the time-domain step response for the transfer function found in part (a).

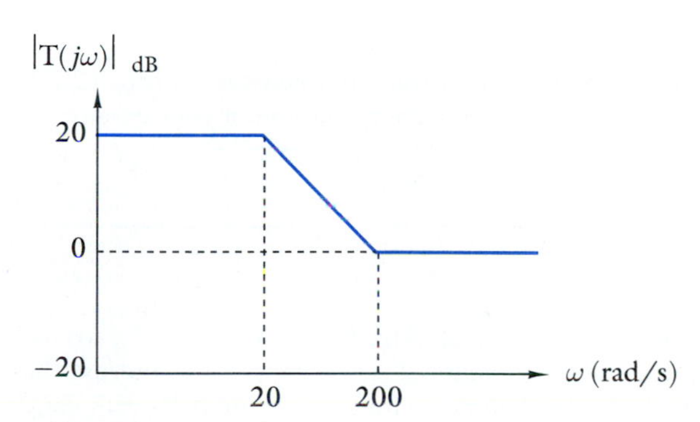

Figure P12-50

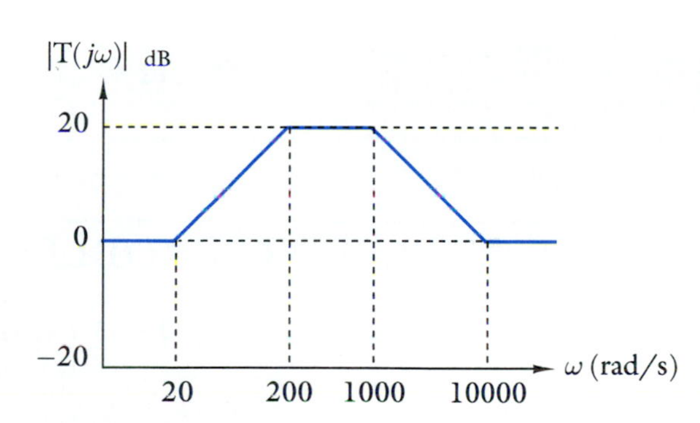

Figure P12-51

12-52 Construct Bode plots of the straight-line approximations to gain and phase responses of the circuit whose time-domain step response is given in Prob. 12-12.

12-53 Construct Bode plots of the straight-line approximations to gain and phase responses of the circuit whose time-domain step response is given in Prob. 12-13.

12-54 Construct Bode plots of the straight-line approximations to gain and phase responses of the circuit whose time-domain step response is given in Prob. 12-15.

12-55 Construct Bode plots of the straight-line approximations to gain and phase responses of the circuit whose time-domain step response is given in Prob. 12-17.

ERO 12-4 Frequency Response Using SPICE (Sect. 12-6)

Use the AC analysis option of SPICE to calculate the frequency response of linear circuits.

12-56 Use SPICE to generate the gain and phase Bode plots for the circuit Prob. 12-2.

12-57 Use SPICE to generate the gain and phase Bode plots for the circuit Prob. 12-4.

12-58 Use SPICE to generate the gain and phase Bode plots for the circuit Prob. 12-5.

12-59 Use SPICE to generate the gain and phase Bode plots for the circuit Prob. 12-7.

❖ **12-60** **(D)** Design a circuit to meet the requirements in Prob. 12-26. Use SPICE to verify the frequency response of your design.

❖ **12-61** **(D)** Design a circuit to meet the requirements in Prob. 12-28. Use SPICE to verify the frequency response of your design.

❖ **12-62** **(D)** Design a circuit to meet the requirements in Prob. 12-32. Use SPICE to verify the frequency response of your design.

CHAPTER INTEGRATING PROBLEMS

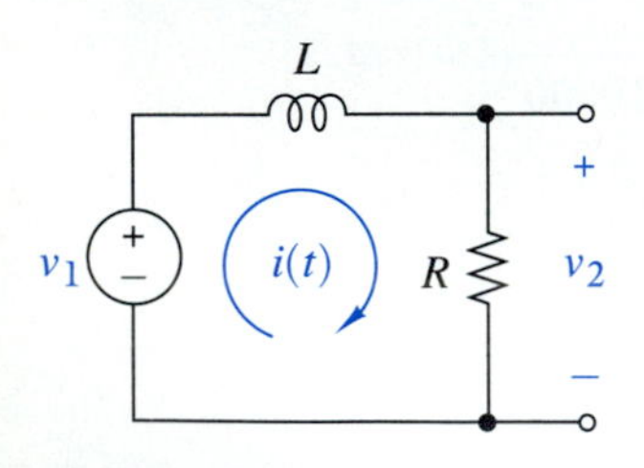

Figure P12-63

12-63 **(A)** The input to the RL circuit Fig. P12-63 is $v_1(t) = V_A \cos \omega t$.

(a) Find the transfer function $T(s) = V_2(s)/V_1(s)$ and use the method of this chapter to find the sinusoidal steady-state response of $i(t)$.

(b) Use the classical method of Chapter 8 to write a first-order differential equation in the current $i(t)$. Use the method of undetermined coefficients from Chapter 8 to find the forced component of $i(t)$.

(c) Show that the responses found in (a) and (b) are identical. Which do you think is easier to apply? Discuss.

12-64 **(A)** Figure P12-64 shows a transformer modeled as two coupled coils.

(a) Use the s-domain i-v relationships for the coupled coils as derived in Example 10-1 to find the transfer function $V_2(s)/V_1(s)$.

(b) Show that the transfer function found in (a) has a second-order bandpass characteristic.

(c) Using $L_1 = 0.25$ H, $L_2 = 1$ H, $M = 0.499$ H, $R_1 = 50\ \Omega$, and $R_L = 200\ \Omega$ calculate the center frequency, upper and lower cutoff frequencies, and the bandwidth.

(d) Construct a Bode plot of the gain response. Does the transformer have a wideband or narrowband bandpass frequency response?

(e) Verify your Bode plot in (d) using SPICE or MICRO-CAP.

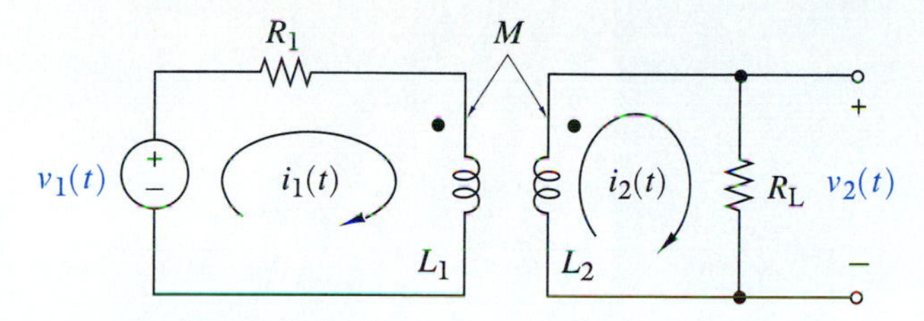

Figure P12-64

12-65 (A) Figure P12-65 shows a circuit which models an OP AMP with finite bandwidth.

(a) Show that the open-loop bandwidth of the OP AMP is $1/RC$.

(b) Show that connecting the output to the inverting (−) input produces a voltage follower amplifier with a bandwidth of $(\mu + 1)/RC$.

(c) Show that connecting the output to the noninverting (+) input produces an unstable circuit.

(d) Select the element values in the model to simulate an OP AMP with an open-loop bandwidth of 10 Hz.

(e) Calculate the bandwidth and the rise time of the step response of a voltage follower using the element values obtained in (d).

(f) Use SPICE or MICRO-CAP to check your answers in (e).

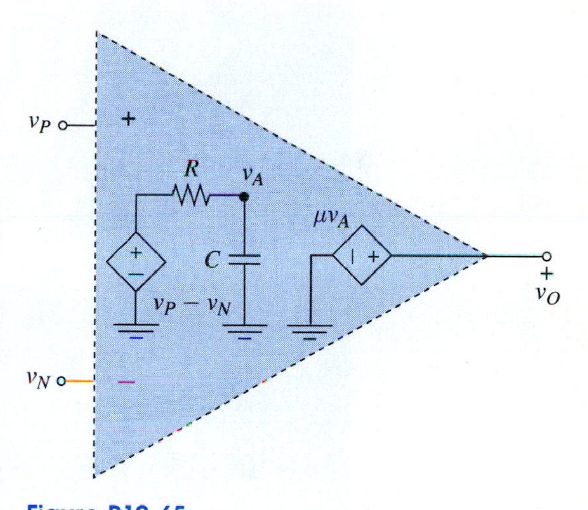

Figure P12-65

Chapter Thirteen

"The vector diagram of sine waves gives the best insight into the mutual relationships of alternating currents and emf's."

Charles P. Steinmetz, 1893
American Engineer

Sinusoidal Steady-State Response

In this chapter we introduce the vector representation of sinusoids as a tool for studying sinusoidal steady-state response at the device level. The vector model was first discussed in detail by Charles Steinmetz (1865–1923) at the International Electric Congress of 1893. Although there is some evidence of earlier use by Oliver Heaviside, Steinmetz is generally credited with popularizing the vector approach by demonstrating its many applications to alternating current devices and systems. By the turn of the century the concept was well established in engineering practice and education. In the 1950s the Steinmetz vector representation of sinusoids came to be called phasors in order to avoid possible confusion with the space vectors associated with electromagnetic fields.

The first section of this chapter defines and illustrates the phasor concept. The second section examines the form taken by the device and connection constraints when sinusoidal voltages and currents are represented by phasors. It turns out that these constraints have the same form previously encountered in Chapters 5 and 10. As a result, we can analyze circuits in the sinusoidal steady state using the same analysis tools as resistance circuits. The basic circuit analysis techniques such as series equivalence and voltage division are treated in the third section. The phasor domain versions of circuit theorems such as superposition and Thévenin's theorem are treated in the fourth section. General analysis methods using node voltages and mesh currents in the phasor domain are covered in the fifth section. The next two sections treat resonance and linear transformers, which are important applications of phasor circuit analysis. The last section discusses the SPICE syntax for phasor circuit analysis problems.

13-1 SINUSOIDS AND PHASORS

The phasor concept is the foundation for the analysis of linear circuits in the sinusoidal steady state. Simply put, a *phasor* is a complex number representing the amplitude and phase angle of a sinusoidal voltage or current. The connection between sinewaves and complex numbers is provided by Euler's relationship.

$$e^{j\theta} = \cos\,\theta + j\sin\,\theta \tag{13-1}$$

Equation (13-1) relates the sine and cosine functions to the complex exponential $e^{j\theta}$. To develop the phasor concept it is necessary to adopt the point of view that the cosine and sine functions can be written in the form

$$\cos\,\theta = Re\left\{e^{j\theta}\right\} \tag{13-2}$$

and

$$\sin\,\theta = Im\left\{e^{j\theta}\right\} \tag{13-3}$$

where *Re* stands for the "real part of" and *Im* for the "imaginary part of." Either Eq. (13-2) or (13-3) can be used to develop the phasor concept. The choice between the two involves deciding whether to describe the eternal sinewave using a sine or cosine function. In Chapter 6 we chose the cosine, so we will reference our phasors to the cosine function.

When Eq. (13-2) is applied to the general sinusoid defined in Chapter 6 we obtain

$$\begin{aligned} v(t) &= V_M\cos(\omega t + \phi) \\ &= V_M Re\left\{e^{j(\omega t+\phi)}\right\} = V_M Re\left\{e^{j\omega t}e^{j\phi}\right\} \\ &= Re\left\{(V_M e^{j\phi})e^{j\omega t}\right\} \end{aligned} \tag{13-4}$$

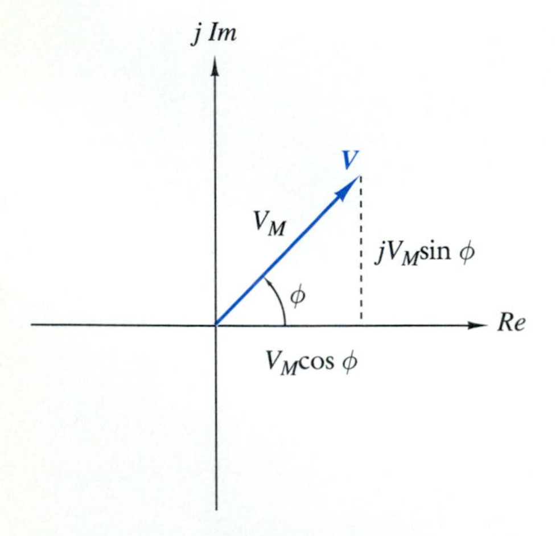

Figure 13-1 Phasor diagram.

In the last line of Eq. (13-4) we moved the amplitude V_M inside the real part operation because it is real and does not change the final result.

By definition, the quantity $V_M e^{j\phi}$ in the last line of Eq. (13-4) is the *phasor representation* of the sinusoid $v(t)$. The phasor $\boldsymbol{V}$ is written as

$$\boldsymbol{V} = V_M e^{j\phi} = V_M\cos\,\phi + jV_M\sin\,\phi \tag{13-5}$$

Note that $\boldsymbol{V}$ is a complex number determined by the amplitude and phase angle of the sinusoid. Figure 13-1 shows a graphical representation commonly called a *phasor diagram*.

The phasor is a complex number that can be written in either polar or rectangular form. An alternative way to write the polar form is to replace the exponential $e^{j\phi}$ by the shorthand notation $\angle\phi$. In subsequent discussions we will often express phasors as $\boldsymbol{V} = V_M\angle\phi$, which is equivalent to the polar form in Eq. (13-5).

Two features of the phasor concept need emphasis:

1. Phasors are written in bold face type such as $\mathbf{V}$ or $\mathbf{I}_A$ to distinguish them from other signal representations such as $v(t)$ and $V(s)$.
2. A phasor is determined by amplitude and phase angle and does not contain any information about the frequency of the sinusoid.

The first feature reminds us that signals and systems can be described in different ways. Although the phasor $\mathbf{V}$, waveform $v(t)$, and transform $V(s)$ are related concepts, they have quite different physical interpretations and our notation must clearly distinguish among them. The absence of frequency information in the phasors results from the fact that in the sinusoidal steady state all currents and voltages are sinusoids with the same frequency. Carrying frequency information in the phasor would be redundant, since it is the same for all phasors in any given steady-state circuit problem.

In summary, given a sinusoidal waveform $v(t) = V_M \cos(\omega t + \phi)$, the corresponding phasor representation is $\mathbf{V} = V_M e^{j\phi}$. Conversely, given the phasor $\mathbf{V} = V_M e^{j\phi}$, the corresponding sinusoid waveform is found by multiplying the phasor by $e^{j\omega t}$ and reversing the steps in Eq. (13-4) as follows:

$$\begin{aligned} v(t) &= Re\left\{(\mathbf{V}e^{j\omega t}\right\} \\ &= Re\left\{(V_M e^{j\phi})e^{j\omega t}\right\} = V_M Re\left\{e^{j(\omega t+\phi)}\right\} \\ &= V_M \cos(\omega t + \phi) \end{aligned} \tag{13-6}$$

The frequency ω in the complex exponential $\mathbf{V}e^{j\omega t}$ in Eq. (13-6) must be expressed or implied in a problem statement, since by definition it is not contained in the phasor. Figure 13-2 shows a geometric interpretation of the complex exponential $\mathbf{V}e^{j\omega t}$ as a vector in the complex plane of length V_M, which rotates counterclockwise with a constant angular velocity of ω. The real part operation projects the rotating vector onto the horizontal (real) axis and thereby generates $v(t) = V_M \cos(\omega t + \phi)$. The complex exponential is sometimes called a *rotating phasor*, and the phasor $\mathbf{V}$ is viewed as a snap shot of the situation at $t = 0$.

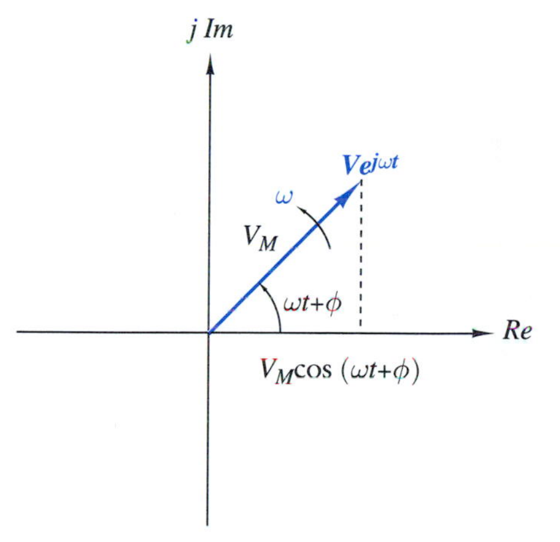

Figure 13-2 Complex exponential $\mathbf{V}e^{j\omega t}$.

Properties of Phasors

The utility of the phasor is based on its additive and network function properties. The *additive property* states that the phasor representing a sum of sinusoids of the same frequency is obtained by adding the phasor representations of the component sinusoids. To establish this property we write the expression

$$\begin{aligned} v(t) &= v_1(t) + v_2(t) + \ldots + v_N(t) \\ &= Re\left\{\mathbf{V}_1 e^{j\omega t}\right\} + Re\left\{\mathbf{V}_2 e^{j\omega t}\right\} + \ldots + Re\left\{\mathbf{V}_N e^{j\omega t}\right\} \end{aligned} \tag{13-7}$$

where $v_1(t)$, $v_2(t), \ldots$ and $v_N(t)$ are sinusoids of the same frequency whose phasor representations are $\mathbf{V}_1, \mathbf{V}_2, \ldots$ and $\mathbf{V}_N$. The real part operation is additive, so the sum of real parts equals the real part of the sum. Consequently, Eq. (13-7) can be written in the form

$$\begin{aligned} v(t) &= Re\left\{\mathbf{V}_1 e^{j\omega t} + \mathbf{V}_2 e^{j\omega t} + \ldots + \mathbf{V}_N e^{j\omega t}\right\} \\ &= Re\left\{(\mathbf{V}_1 + \mathbf{V}_2 + \ldots + \mathbf{V}_N) e^{j\omega t}\right\} \end{aligned} \tag{13-8}$$

Comparing the last line in Eq. (13-8) with the definition of a phasor we conclude that the phasor $\mathbf{V}$ representing $v(t)$ is

$$\mathbf{V} = \mathbf{V}_1 + \mathbf{V}_2 + \ldots + \mathbf{V}_N \tag{13-9}$$

The result in Eq. (13-9) applies only if the component sinusoids all have the same frequency so that $e^{j\omega t}$ can be factored out as shown in the last line in Eq. (13-8).

The *network function* property states that in the sinusoidal steady state the network function $T(j\omega)$ provides a relationship between input and output phasors. In Chapter 12 we established that a sinusoidal input of the form

$$v_1(t) = V_{M1}\cos(\omega t + \phi) = Re\left\{(V_{M1} e^{j\phi}) e^{j\omega t}\right\} = Re\left\{\mathbf{V}_1 e^{j\omega t}\right\}$$

produces a steady-state output of the form

$$v_{2SS}(t) = V_{M1}\,|T(j\omega)|\cos(\omega t + \phi + \theta) \tag{13-10}$$

where $T(j\omega) = |T(j\omega)|e^{j\theta}$ is the network function $T(s) = V_2(s)/V_1(s)$ evaluated with $s = j\omega$. Using Euler's relationship we can rewrite Eq. (13-10) in the form

$$\begin{aligned} v_{2SS}(t) &= Re\left\{V_{M1}\,|T(j\omega)|\,e^{j(\omega t + \phi + \theta)}\right\} \\ &= Re\left\{\left(V_{M1} e^{j\phi}\right)\left(|T(j\omega)|\,e^{j\theta}\right) e^{j\omega t}\right\} \\ &= Re\left\{[\mathbf{V}_1 T(j\omega)]\,e^{j\omega t}\right\} \end{aligned} \tag{13-11}$$

Comparing the last line in Eq. (13-11) with the definition of a phasor we conclude that the phasor $\mathbf{V}_2$ representing the output $v_{2SS}(t)$ is

$$\mathbf{V}_2 = \mathbf{V}_1 T(j\omega) \tag{13-12}$$

In the sinusoidal steady state the output phasor equals the input phasor times the network function $T(s)$ evaluated with $s = j\omega$, where ω is the frequency of the input sinusoid.

The next three examples illustrate these basic properties of phasors.

EXAMPLE 13-1

(a) Construct the phasors for the following waveforms

$$v_1(t) = 10\cos(1000t - 45°)$$
$$v_2(t) = 5\cos(1000t + 30°)$$

(b) Use the additive property of the phasors and the phasors found in (a) to find the waveform $v(t) = v_1(t) + v_2(t)$.

SOLUTION:

(a) The phasor representations of $v_1(t)$ and $v_2(t)$ are

$$\mathbf{V}_1 = 10e^{-j45°} = 10\cos(-45°) + j10\sin(-45°)$$
$$= 7.07 - j7.07$$
$$\mathbf{V}_2 = 5e^{+j30°} = 5\cos(30°) + j5\sin(30°)$$
$$= 4.33 + j2.5$$

(b) The two sinusoids have the same frequency so the additive property of phasors can be used to obtain their sum:

$$\mathbf{V} = \mathbf{V}_1 + \mathbf{V}_2 = 11.4 - j4.57 = 12.3e^{-j21.8°}$$

The waveform corresponding to this phasor sum is

$$v(t) = Re\left\{\left(12.3e^{-j21.8°}\right)e^{j1000t}\right\}$$
$$= 12.3\cos(1000t - 21.8°)$$

The phasor diagram in Fig. 13-3 shows that the process of summing sinusoids can be viewed geometrically in terms of phasors.

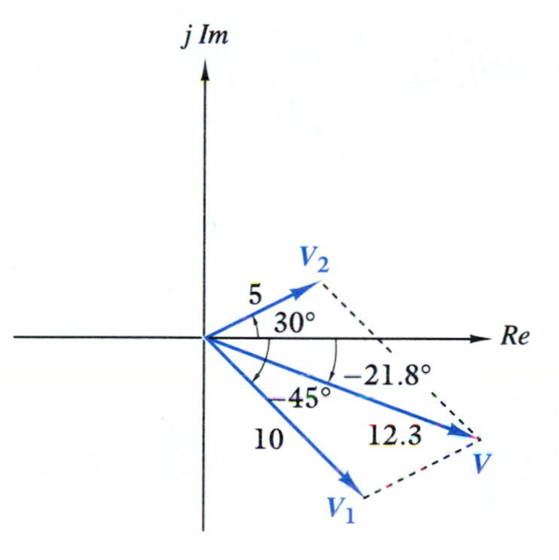

Figure 13-3

EXAMPLE 13-2

(a) Construct the phasors representing the following waveforms.

$$i_A(t) = 5\cos(377t + 50°)$$
$$i_B(t) = 5\cos(377t + 170°)$$
$$i_C(t) = 5\cos(377t - 70°)$$

(b) Use the additive property of phasors and the phasors found in (a) to find the sum of these waveforms.

SOLUTION:

(a) The phasor representation of the three current waveforms are

$$\boldsymbol{I_A} = 5e^{j50^\circ} = 5\cos(50^\circ) + j5\sin(50^\circ) = 3.21 + j3.83$$

$$\boldsymbol{I_B} = 5e^{j170^\circ} = 5\cos(170^\circ) + j5\sin(170^\circ)$$
$$= -4.92 + j0.87$$

$$\boldsymbol{I_C} = 5e^{-j70^\circ} = 5\cos(-70^\circ) + j5\sin(-70^\circ)$$
$$= 1.71 - j4.70$$

(b) The waveforms have the same frequency so the additive property of phasors applies. The phasor representing the sum of these currents is

$$\boldsymbol{I_A} + \boldsymbol{I_B} + \boldsymbol{I_C} = (3.21 - 4.92 + 1.71) + j(3.83 + 0.87 - 4.70) = 0 + j0$$

It is not obvious by examining the waveforms that these three currents add to zero. However, the phasor diagram in Fig. 13-4 makes this fact clear, since the sum of any two phasors is equal and opposite to the third. Phasors of this type occur in balanced three-phase power systems, which we will study in the next chapter. The balanced condition occurs when three equal-amplitude phasors are displaced in phase by exactly 120°.

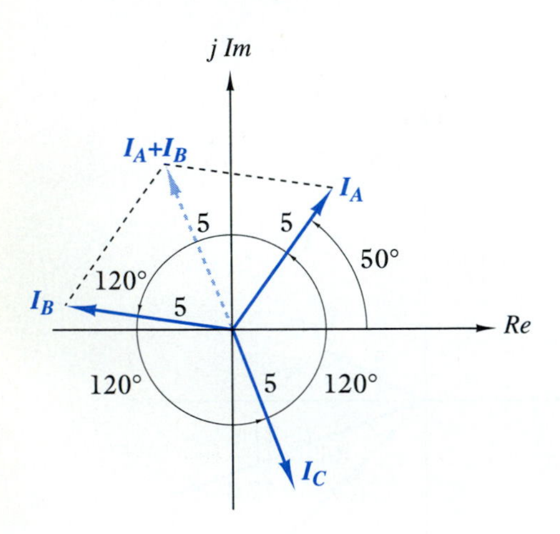

Figure 13-4

EXAMPLE 13-3

A linear circuit has a transfer function $V_2(s)/V_1(s) = s/(s + 100)$.

(a) Determine the input and output phasors for $v_1 = 10\cos(200t - 30^\circ)$.

(b) Determine the output waveform in the sinusoidal steady state.

SOLUTION:

(a) The phasor for the input sinusoid is $V_1 = 10\angle{-30^\circ}$. The value of the transfer function at the input frequency is

$$T(j200) = \frac{j200}{j200 + 100} = \frac{200\angle 90^\circ}{224\angle 63.4^\circ} = 0.894\angle 26.6^\circ$$

Using the network function property of phasors the output phasor is

$$\begin{aligned} V_2 &= V_1 T(j200) = \left(10e^{-j30^\circ}\right)\left(0.894e^{j26.6^\circ}\right) \\ &= 8.94e^{-j3.4^\circ} \end{aligned}$$

(b) The sinusoidal waveform corresponding to the phasor V_2 is

$$v_2(t) = Re\left\{\left(8.94e^{-j3.4^\circ}\right)e^{j200t}\right\} = 8.94\cos(200t - 3.4^\circ)$$

EXERCISE 13-1

Convert the following sinusoids to phasors in polar and rectangular form.

(a) $v(t) = 20\cos(150t - 60^\circ)$ V
(b) $v(t) = 10\cos(1000t + 180^\circ)$ V
(c) $i(t) = -4\cos 3t + 3\cos(3t - 90^\circ)$ A

ANSWERS:

(a) $V = 20\angle{-60^\circ} = 10 - j17.3$ V
(b) $V = 10\angle 180^\circ = -10 + j0$ V
(c) $I = 5\angle{-143^\circ} = -4 - j3$ A ■

EXERCISE 13-2

Convert the following phasors to sinusoidal waveforms.

(a) $V = 169\angle{-45^\circ}$ V at $f = 60$ Hz
(b) $V = 10\angle 90^\circ + 66 - j10$ V at $\omega = 10$ krad/s
(c) $I = 15 + j5 + 10\angle 180^\circ$ mA at $\omega = 1000$ rad/s

ANSWERS:

(a) $v(t) = 169\cos(377t - 45^\circ)$ V
(b) $v(t) = 66\cos 10^4 t$ V
(c) $i(t) = 7.07\cos(1000t + 45^\circ)$ mA ■

EXERCISE 13-3

Express the phasors for $V_M\cos\omega t$, $V_M\sin\omega t$, $-V_M\cos\omega t$, and $-V_M\sin\omega t$ in rectangular form. *Hint:* Use the identity $\sin\theta = \cos(\theta - 90^\circ)$.

ANSWERS:

$V_M + j0,\ 0 - jV_M,\ -V_M + j0,\ 0 + jV_M$. ■

13-2 PHASOR CIRCUIT ANALYSIS

Phasor circuit analysis involves solving for steady-state responses using phasors to represent the sinusoidal signals. How do we perform phasor circuit analysis? At several points in our study we have seen that circuit analysis of any type is based on two kinds of constraints:

1. Connection constraints (Kirchhoff's laws)
2. Device constraints (element equations)

To analyze phasor circuits we must see how these constraints are expressed in phasor form.

In the sinusoidal steady state the circuits' natural response has decayed to zero and all of the voltages and currents are sinusoids with the same frequency as the input. In the steady state the application of KVL around a loop would lead to an equation of the form

$$V_1 \cos(\omega t + \phi_1) + V_2 \cos(\omega t + \phi_2) - V_3 \cos(\omega t + \phi_3) = 0$$

These waveforms have the same frequency but have different amplitudes and phase angles. The additive property of phasors discussed in the preceding section shows that there is a one-to-one correspondence between waveform sums and phasor sums. Therefore, if the sum of the waveforms is zero then the phasors must also sum to zero:

$$\mathbf{V}_1 + \mathbf{V}_2 - \mathbf{V}_3 = 0$$

Clearly KVL applies to loops with any number of phasor voltages, and KCL applies to any number of phasor currents at a node. In other words, we can state *Kirchhoff's laws in phasor form* as follows.

KVL: The algebraic sum of phasor voltages around a loop is zero.

KCL: The algebraic sum of phasor currents at a node is zero.

Turning now to the device constraints, we note that in the *s* domain the element equations for the three passive elements are expressed in terms of their impedances:

$$\begin{aligned} \text{Resistor:} \quad & V_R(s) = Z_R(s)I_R(s) = RI_R(s) \\ \text{Inductor:} \quad & V_L(s) = Z_L(s)I_L(s) = LsI_L(s) \\ \text{Capacitor:} \quad & V_C(s) = Z_C(s)I_C(s) = \frac{1}{Cs}I_C(s) \end{aligned} \tag{13-13}$$

The network function property of phasors discussed in the preceding section points out that an s-domain transfer function $T(s)$ evaluated with $s = j\omega$ provides a relationship between input and output phasors. We can think of the element impedances in Eq. (13-13) as miniature transfer functions, with current through the element acting as the input and voltage across the element serving as the output. Applying the network function property of phasors to the element impedances yields the *element constraints in phasor form*:

$$\begin{aligned} \text{Resistor:} \quad & \mathbf{V}_R = R\mathbf{I}_R \\ \text{Inductor:} \quad & \mathbf{V}_L = j\omega L\mathbf{I}_L \\ \text{Capacitor:} \quad & \mathbf{V}_C = \frac{1}{j\omega C}\mathbf{I}_C \end{aligned} \tag{13-14}$$

In phasor form the passive element i-v constraints are simply the s-domain impedances with s replaced with $j\omega$, where ω is the frequency of the sinusoidal driving force that produces the steady-state condition.

Although Equations (13-13) and (13-14) are similar in form they apply in quite different situations. Equation (13-13) contains s-domain equations that relate voltage and current transforms. Equation (13-14) contains phasor domain equations that relate complex numbers describing the amplitude and phase angles of the sinusoidal steady-state responses. It is important to remember that the i-v relationships of these two-terminal elements assume the passive sign convention is used to assign reference marks for the voltages and currents.

To explore the phasor element equations in more detail, let us assume the current through a resistor is $i_R(t) = I_M \cos(\omega t + \phi)$. Using the impedance for a resistor from Eq. (13-14) we find that the phasor voltage across the resistor is

$$\mathbf{V}_R = R\mathbf{I}_R = RI_M e^{j\phi} \tag{13-15}$$

This equations shows that the phasor voltage and current have the same phase angle. Phasors are said to be *in phase* when they have the same phase angle and *out of phase* when their phase angles are different. Figure 13-5 shows the phasor diagram of the resistor current and voltage. When constructing a phasor diagram showing both voltage and current, we use two scale factors since the two phasors do not have the same dimensions.

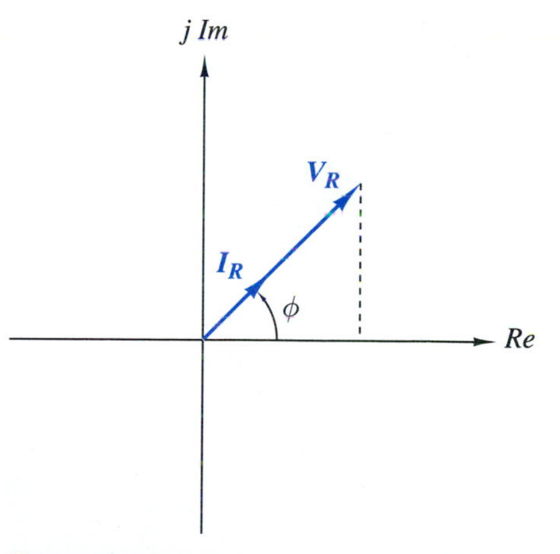

Figure 13-5 Phasor i-v characteristics of the resistor.

Next, let us assume that the current through an inductor is $i_L(t) = I_M \cos(\omega t + \phi)$. Using the inductor impedance from

Eq. (13-14) we find that the phasor voltage across the inductor is

$$\begin{aligned} \mathbf{V}_L &= j\omega L \mathbf{I}_L = \left(\omega L e^{j90^\circ}\right)\left(I_M e^{j\phi}\right) \\ &= \omega L I_M e^{j(\phi+90^\circ)} \end{aligned} \tag{13-16}$$

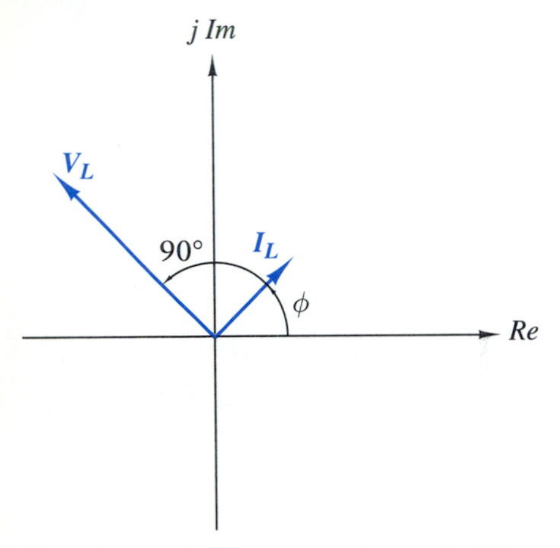

Figure 13-6 Phasor *i-v* characteristics of the inductor.

The resulting phasor diagram in Fig. 13-6 shows that the inductor voltage and current are 90° out of phase. The voltage phasor is advanced by 90° counterclockwise, which is in the direction of rotation of the complex exponential $e^{j\omega t}$. When the voltage phasor is advanced counterclockwise, that is, ahead of the rotating current phasor, we say the voltage phasor *leads* the current phasor by 90° or equivalently the current *lags* the voltage by 90°.

Finally, let us assume that the current through a capacitor is $i_C(t) = I_M \cos(\omega t + \phi)$. Using the capacitor impedance from Eq. (13-14) we find that the phasor voltage across the capacitor is

$$\begin{aligned} \mathbf{V}_C &= \frac{1}{j\omega C}\mathbf{I}_C = \left(\frac{1}{\omega C} e^{-j90^\circ}\right)\left(I_M e^{j\phi}\right) \\ &= \frac{I_M}{\omega C} e^{j(\phi-90^\circ)} \end{aligned} \tag{13-17}$$

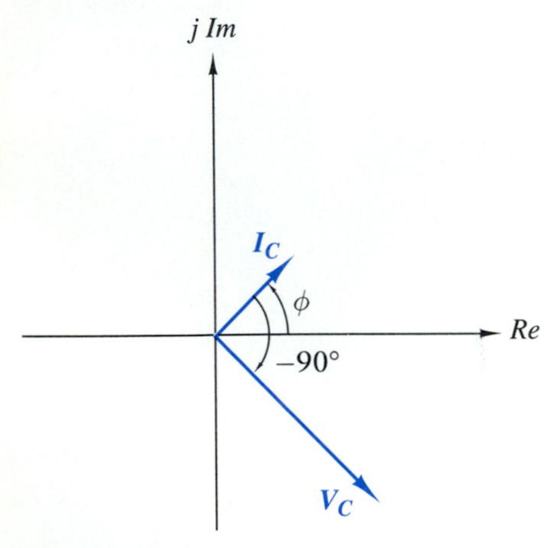

Figure 13-7 Phasor *i-v* characteristics of the capacitor.

The resulting phasor diagram in Fig. 13-7 shows that voltage and current are 90° out of phase. In this case the voltage phasor is retarded by 90° clockwise, which is in a direction opposite to rotation of the complex exponential $e^{j\omega t}$. When the voltage is retarded clockwise, that is behind the rotating current phasor, we say the voltage phasor *lags* the current phasor by 90° or equivalently the current *leads* the voltage by 90°.

To obtain the phasor impedances in Eq. (13-14) we use current as the input and the voltage as the output. However, the s-domain element equations in Eq. (13-13) relate a voltage and current at the same pair of terminals, so we could use voltage as the input and the current as the output. Using the network function property of phasors with voltage as the input and current as the output yields the following phasor i-v relationships:

$$\begin{aligned} &\text{Resistor:} && \mathbf{I}_R = \frac{1}{R}\mathbf{V}_R \\ &\text{Inductor:} && \mathbf{I}_L = \frac{1}{j\omega L}\mathbf{V}_L \\ &\text{Capacitor:} && \mathbf{I}_C = j\omega C \mathbf{V}_C \end{aligned} \tag{13-18}$$

The element constraints in Eq. (13-18) involve the s-domain admittances of the elements with s replaced with $j\omega$. In other words, in phasor circuit analysis an element constraint can be written in terms of an impedance Z or admittance Y:

$$\begin{aligned}
&\text{Resistor:} \quad && Z_R = R && \text{or} \quad Y_R = \frac{1}{R} \\
&\text{Inductor:} \quad && Z_L = j\omega L && \text{or} \quad Y_L = \frac{1}{j\omega L} \qquad (13\text{-}19) \\
&\text{Capacitor:} \quad && Z_C = \frac{1}{j\omega C} && \text{or} \quad Y_C = j\omega C
\end{aligned}$$

In general, $Y = 1/Z$ so we can use either impedance or admittance in any given problem. However, *to avoid possible confusion we will always show the element impedance in phasor circuit diagrams.*

Caution: Although impedances or admittances can be complex numbers they are **not** phasors. This important distinction is not a new concept. We have seen that transforms and network functions are both rational functions of s but have quite different physical meanings. A transform is a description of a signal, while a network function is a description of a circuit. Similarly, phasors and impedance are complex numbers, but phasors represent sinusoidal signals while impedances describe circuit elements.

EXERCISE 13-4

The circuit in Fig. 13-8 is operating in the sinusoidal steady state with $v_A(t) = 10 \cos \omega t$ V, $v_B(t) = 10 \sin \omega t$ V, $i_1 = \sqrt{2}\cos(\omega t + 135°)$ A, and $i_4 = \cos \omega t$ A. Use KCL and KVL to find the phasor voltage across and current through each element. Use the passive sign convention.

ANSWERS:

$\mathbf{V}_1 = 10\angle 0°$ V, $\mathbf{V}_2 = 10 + j10$ V, $\mathbf{V}_3 = \mathbf{V}_4 = 10\angle{-90°}$ V, $\mathbf{I}_1 = -1 + j$ A, $\mathbf{I}_2 = 1 - j$ A, $\mathbf{I}_3 = 1\angle{-90°}$ A, $\mathbf{I}_4 = 1 + j0$ A.

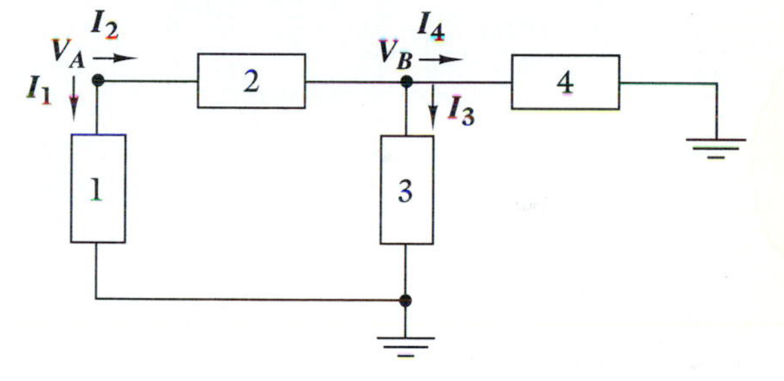

Figure 13-8

EXERCISE 13-5

The current through a 12-mH inductor is $i_L(t) = 20 \cos(10^6 t)$ mA. Determine:

(a) The impedance of the inductor
(b) The phasor voltage across the inductor
(c) The waveform of the voltage across the inductor

ANSWERS:

(a) $j12$ kΩ
(b) $240\angle 90°$ V
(c) $v_L = 240 \cos(10^6 t + 90°)$ V

EXERCISE 13-6

The current through a 20-pF capacitor is $i_C(t) = 0.3 \cos(10^6 t)$ mA. Determine:

(a) the impedance of the capacitor
(b) the phasor voltage across the capacitor
(c) the waveform of the voltage across the capacitor.

ANSWERS:
(a) $-j50$ kΩ
(b) $15\angle -90^\circ$ V
(c) $v_C = 15 \cos(10^6 t - 90^\circ)$ V ■

13-3 BASIC CIRCUIT ANALYSIS IN THE PHASOR DOMAIN

In the preceding section we developed the device and connection constraints in phasor form. The phasor constraints have the same format as the constraints for resistance circuits in Chapter 2 and s-domain circuits in Chapter 10. Therefore, in phasor circuit analysis we can use familiar tools such as series and parallel equivalence, voltage and current division, proportionality and superposition, Thévenin and Norton equivalent circuits, and node analysis or mesh analysis. In other words, we do not need new analysis techniques to handle phasor circuits. The major difference is that the circuit responses are complex numbers (phasors) and not waveforms or transforms.

We can think of phasor circuit analysis in terms of the flow diagram in Figure 13-9. Phasor circuit analysis begins in the time domain with a linear circuit operating in the sinusoidal steady state and involves three major steps:

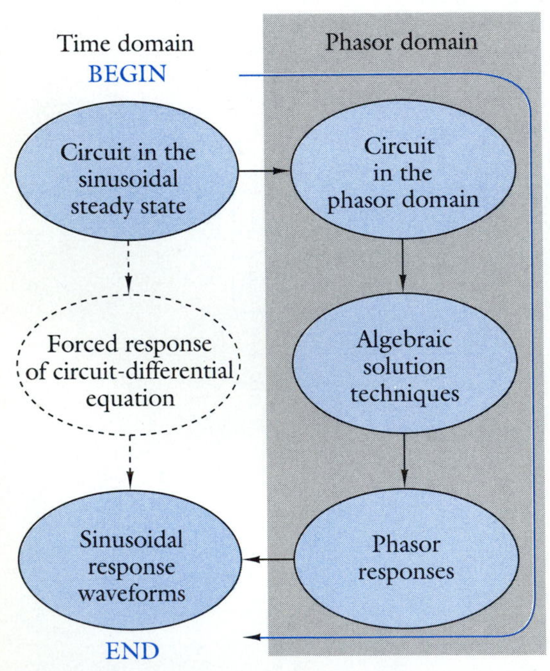

Figure 13-9 Flow diagram for phasor analysis.

STEP 1: The circuit is transformed into the phasor domain by replacing each input signal by a phasor source and each circuit element by an impedance

STEP 2: Algebraic circuit analysis techniques are used to solve for the unknown phasor responses.

STEP 3: The phasor responses are converted into time-domain sinewaves to obtain response waveforms.

The third step assumes that the required end product is a sinusoidal waveform. However, a phasor is just another representation of a sinusoid. With some experience we learn to think of the response as a phasor without converting it back into a time-domain waveform.

Figure 13-9 points out that there is another route to time-domain response using the classical method to find the forced response of the circuit differential equation. A quick review of these methods described in Section 8-4 will convince the reader that the phasor approach is far simpler. More importantly, phasor circuit analysis provides insight into steady-state responses that are essential to understanding much of the terminology and viewpoint of electrical engineering.

Series Equivalence and Voltage Division

We begin the study of phasor circuit analysis with two basic analysis tools—series equivalence and voltage division. In Fig. 13-10 the two-terminal elements are connected in series so the same phasor current $\mathbf{I}$ exists in impedances Z_1, Z_2, ... Z_N. Using KVL and the element constraints, the voltage across the series connection can be written as

$$\begin{aligned}\mathbf{V} &= Z_1\mathbf{I} + Z_2\mathbf{I} + \ldots + Z_N\mathbf{I} \\ &= (Z_1 + Z_2 + \ldots + Z_N)\,\mathbf{I}\end{aligned} \tag{13-20}$$

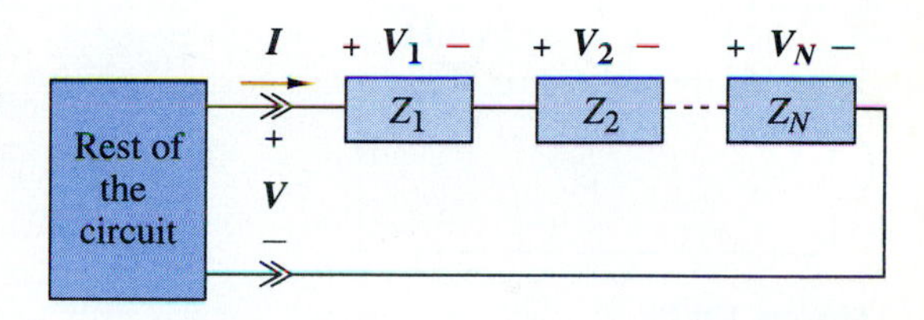

Figure 13-10 Series connection.

The same phasor responses $\mathbf{V}$ and $\mathbf{I}$ exist when the series connected elements are replaced by an equivalent impedance.

$$Z_{EQ} = \frac{\mathbf{V}}{\mathbf{I}} = Z_1 + Z_2 + \ldots + Z_N \tag{13-21}$$

In general, the equivalent impedance Z_{EQ} is a complex quantity of the form

$$Z_{EQ} = R + jX$$

where R is the real part and X is the imaginary part. The real part of Z is called *resistance* and the imaginary part (X not jX) is called *reactance*. Both resistance and reactance are expressed in ohms (Ω). For passive circuits resistance is always positive, while reactance X can be either positive or negative. A positive X is called an *inductive reactance* because the reactance of an inductor is ωL, which is always positive. A negative X is called a *capacitive reactance* because the reactance of a capacitor is $-1/\omega C$ which is always negative.

Combining Eqs. (13-20) and (13-21) we can write the phasor voltage across the kth element in the series connection as

$$\mathbf{V}_k = Z_k\mathbf{I}_k = \frac{Z_k}{Z_{EQ}}\mathbf{V} \tag{13-22}$$

Equation (13-22) is the phasor version of the voltage division principle. The phasor voltage across any element in a series connection equals the ratio of its impedance to the equivalent impedance of the connection, times the total phasor voltage across the connection.

The next example illustrates using these phasor circuit analysis tools.

EXAMPLE 13-4

The circuit in Fig. 13-11 is operating in the sinusoidal steady state with $v_S(t) = 35 \cos 1000t$ V.

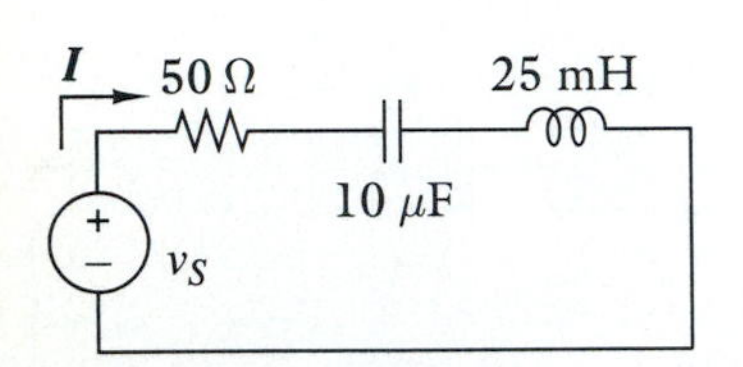

Figure 13-11

(a) Transform the circuit into the phasor domain.

(b) Solve for the phasor current $\mathbf{I}$.

(c) Solve for the phasor voltage across each element.

(d) Construct the waveforms corresponding to the phasors found in (b) and (c).

SOLUTION:

(a) The phasor representing the input source voltage is $\mathbf{V}_S = 35\angle 0°$. The impedances of the three passive elements are

$$Z_R = R = 50\ \Omega$$

$$Z_L = j\omega L = j1000 \times 25 \times 10^{-3} = j25\ \Omega$$

$$Z_C = \frac{1}{j\omega C} = \frac{1}{j1000 \times 10^{-5}} = -j100\ \Omega$$

Using these results we obtain the phasor-domain circuit in Fig. 13-12.

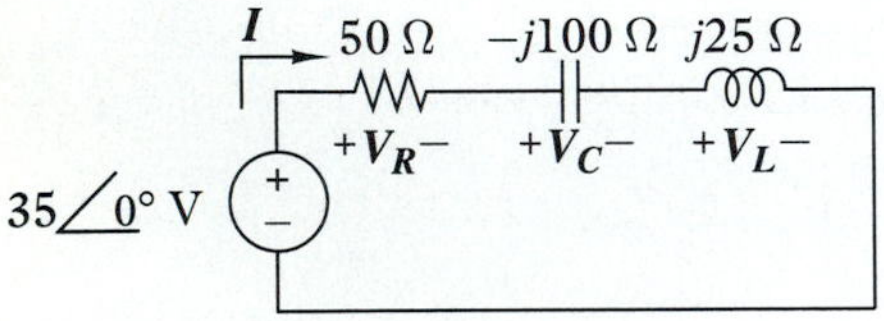

Figure 13-12

(b) The equivalent impedance of the series connection is

$$Z_{EQ} = 50 + j25 - j100 = 50 - j75 = 90.1\angle -56.3°\ \Omega$$

The current in the series circuit is

$$\mathbf{I} = \frac{\mathbf{V}_S}{Z_{EQ}} = \frac{35\angle 0°}{90.1\angle -56.3°} = 0.388\angle 56.3°\ \text{A}$$

(c) The current $\mathbf{I}$ exists in all three series elements so the voltage across each passive element is

$$\mathbf{V}_R = Z_R\mathbf{I} = 50 \times 0.388\angle 56.3° = 19.4\angle 56.3°\ \text{V}$$

$$\mathbf{V}_L = Z_L\mathbf{I} = j25 \times 0.388\angle 56.3° = 9.70\angle 146.3°\ \text{V}$$

$$\mathbf{V}_C = Z_C\mathbf{I} = -j100 \times 0.388\angle 56.3° = 38.8\angle -33.7°\ \text{V}$$

(d) The sinusoidal steady-state waveforms corresponding to the phasors in (b) and (c) are

$$i(t) = Re\left\{0.388e^{j56.3^\circ}e^{j1000t}\right\}$$
$$= 0.388\cos(1000t + 56.3^\circ)\ \text{A}$$
$$v_R(t) = Re\left\{19.4e^{j56.3^\circ}e^{j1000t}\right\}$$
$$= 19.4\cos(1000t + 56.3^\circ)\ \text{V}$$
$$v_L(t) = Re\left\{9.70e^{j146.3^\circ}e^{j1000t}\right\}$$
$$= 9.70\cos(1000t + 146.3^\circ)\ \text{V}$$
$$v_C(t) = Re\left\{38.8e^{-j33.7^\circ}e^{j1000t}\right\}$$
$$= 38.8\cos(1000t - 33.7^\circ)\ \text{V}$$

Parallel Equivalence and Current Division

These analysis tools are duals of series equivalence and voltage division. In Fig. 13-13 the two-terminal elements are connected in parallel so the same phasor voltage $\mathbf{V}$ appears across the admittances $Y_1, Y_2, \ldots Y_N$. Using KCL and the element constraints the current entering the parallel connection is

$$\begin{aligned}\mathbf{I} &= Y_1\mathbf{V} + Y_2\mathbf{V} + \ldots + Y_N\mathbf{V} \\ &= (Y_1 + Y_2 + \ldots + Y_N)\mathbf{V}\end{aligned} \tag{13-23}$$

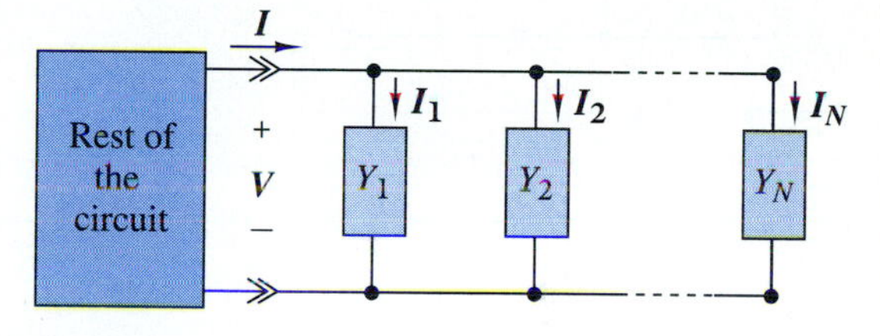

Figure 13-13 Parallel connection.

The same phasor responses $\mathbf{V}$ and $\mathbf{I}$ exists when the parallel connected elements are replaced by an equivalent admittance.

$$Y_{EQ} = \frac{\mathbf{I}}{\mathbf{V}} = Y_1 + Y_2 + \ldots + Y_N \tag{13-24}$$

In general, the equivalent admittance Y_{EQ} is a complex quantity of the form

$$Y_{EQ} = G + jB$$

The real part of Y or G is called *conductance* and the imaginary part (B **not** jB) is called *susceptance*, both of which are expressed in units of siemens (S). Given that

$$Y = G + jB = \frac{1}{Z} = \frac{1}{R + jX} = \frac{R}{R^2 + X^2} - j\frac{X}{R^2 + X^2}$$

it follows that resistance (R) and conductance (G) have the same sign, while reactance (X) and susceptance (B) have opposite signs in any circuit.

Combining Eqs. (13-23) and (13-24) we find that the phasor current through the kth element in the parallel connection is

$$\boldsymbol{I}_k = Y_k \boldsymbol{V}_k = \frac{Y_k}{Y_{EQ}} \boldsymbol{I} \qquad (13\text{-}25)$$

Equation (13-25) is the phasor version of the current division principle. The phasor current through any element in a parallel connection equals the ratio of its admittance to the equivalent admittance of the connection, times the total phasor current entering the connection.

The next example illustrates using these phasor circuit analysis tools.

EXAMPLE 13-5

The circuit in Fig. 13-14 is operating in the sinusoidal steady state with $i_S(t) = 50 \cos 2000t$ mA.

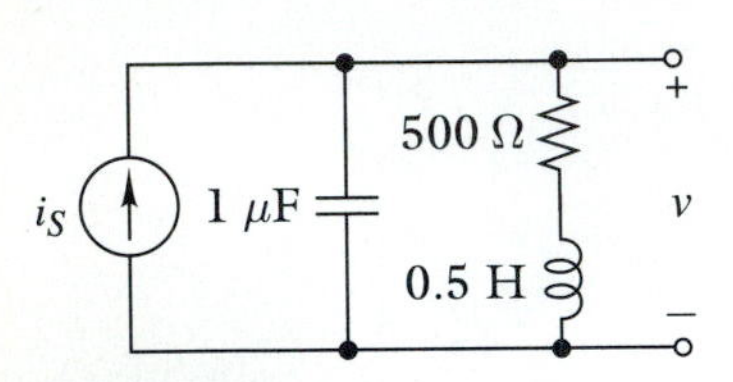

Figure 13-14

(a) Transform the circuit into the phasor domain.

(b) Solve for the phasor voltage $\boldsymbol{V}$.

(c) Solve for the phasor current through each element.

(d) Construct the waveforms corresponding to the phasors found in (b) and (c).

SOLUTION:

(a) The phasor representing the input source current is $\boldsymbol{I}_S = 0.05\angle 0°$ A. The impedances of the three passive elements are

$$Z_R = R = 500\ \Omega$$

$$Z_L = j\omega L = j2000 \times 0.5 = j1000\ \Omega$$

$$Z_C = \frac{1}{j\omega C} = \frac{1}{j2000 \times 10^{-6}} = -j500\ \Omega$$

Using these results we obtain the phasor-domain circuit in Fig. 13-15.

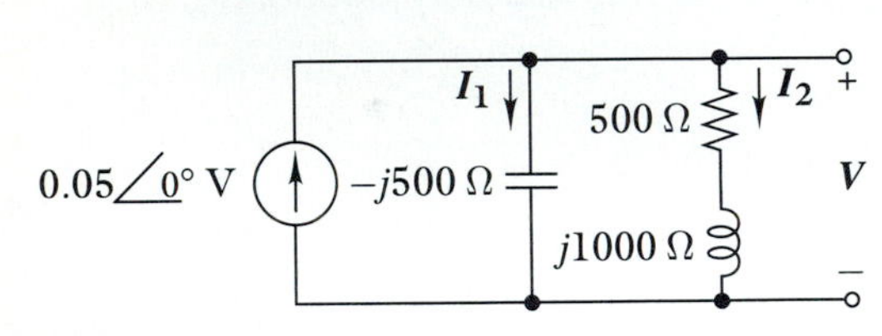

Figure 13-15

(b) The admittances of the two parallel branches are

$$Y_1 = \frac{1}{-j500} = j2 \times 10^{-3}\ \text{S}$$

$$Y_2 = \frac{1}{500 + j1000} = 4 \times 10^{-4} - j8 \times 10^{-4}\ \text{S}$$

The equivalent admittance of the parallel connection is

$$Y_{EQ} = Y_1 + Y_2 = 4 \times 10^{-4} + j12 \times 10^{-4}$$
$$= 12.6 \times 10^{-4}\angle 71.6°\ \text{S}$$

and the voltage across the parallel circuit is

$$V = \frac{I_S}{Y_{EQ}} = \frac{0.05\angle 0^\circ}{12.6 \times 10^{-4}\angle 71.6^\circ} = 39.7\angle -71.6^\circ \text{ V}$$

(c) The current through each parallel branch is

$$I_1 = Y_1 V = j2 \times 10^{-3} \times 39.7\angle -71.6^\circ = 79.4\angle 18.4^\circ \text{ mA}$$

$$I_2 = Y_2 V = \left(4 \times 10^{-4} - j8 \times 10^{-4}\right) \times 39.7\angle -71.6^\circ$$
$$= 35.5\angle -135^\circ \text{ mA}$$

(d) The sinusoidal steady-state waveforms corresponding to the phasors in (b) and (c) are

$$v(t) = Re\left\{39.7e^{-j71.6^\circ}e^{j2000t}\right\} = 39.7\cos(2000t - 71.6^\circ) \text{ V}$$

$$i_1(t) = Re\left\{79.4e^{j18.4^\circ}e^{j2000t}\right\} = 79.4\cos(2000t + 18.4^\circ) \text{ mA}$$

$$i_2(t) = Re\left\{33.9e^{-j135^\circ}e^{j2000t}\right\} = 33.9\cos(2000t - 135^\circ) \text{ mA}$$

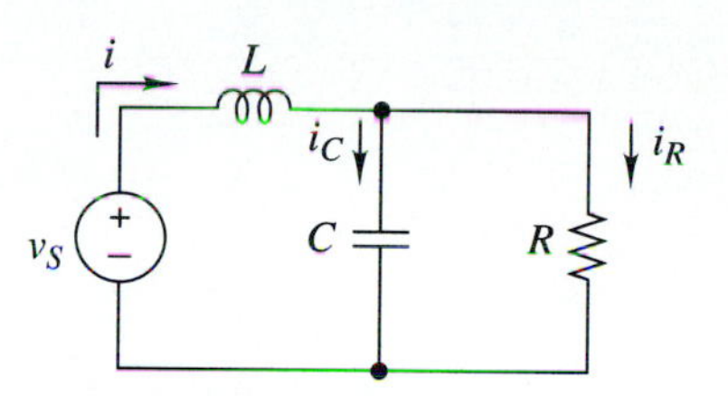

Figure 13-16

EXAMPLE 13-6

Find the steady-state currents $i(t)$, $i_C(t)$, and $i_R(t)$ in the circuit of Fig. 13-16 for $v_S = 100\cos 2000t$ V, $L = 250$ mH, $C = 0.5\ \mu$F, and $R = 3$ kΩ.

SOLUTION:

The phasor representation of the input voltage is $100\angle 0^\circ$. The impedances of the passive elements are

$$Z_L = j500\ \Omega \qquad Z_C = -j1000\ \Omega \qquad Z_R = 3000\ \Omega$$

Figure 13-17(a) shows the phasor-domain circuit.

To solve for the required phasor responses we reduce the circuit using a combination of series and parallel equivalence. Using parallel equivalence we find that the capacitor and resistor can be replaced by an equivalent impedance

$$Z_{EQ1} = \frac{1}{Y_{EQ1}} = \frac{1}{\dfrac{1}{-j1000} + \dfrac{1}{3000}}$$
$$= 300 - j900\ \Omega$$

The resulting circuit reduction is shown in Fig. 13-17(b). The equivalent impedance Z_{EQ1} is connected in series with the impedance $Z_L = j500$. This series combination can be replaced by

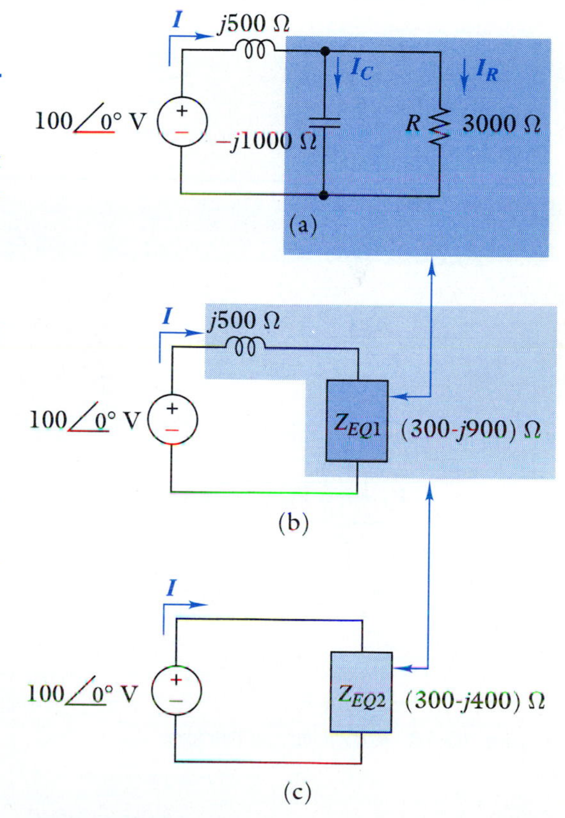

Figure 13-17

an equivalent impedance

$$Z_{EQ2} = j500 + Z_{EQ1} = 300 - j400\ \Omega$$

This step reduces the circuit to the equivalent input impedance shown in Figure 13-17(c). The phasor input current in Fig. 13-17(c) is

$$\boldsymbol{I} = \frac{100\angle 0^\circ}{Z_{EQ2}} = \frac{100\angle 0^\circ}{300 - j400} = 0.12 + j0.16 = 0.2\angle 53.1^\circ\ \text{A}$$

Given the phasor current $\boldsymbol{I}$, we use current division to find $\boldsymbol{I_C}$

$$\boldsymbol{I_C} = \frac{Y_C}{Y_C + Y_R}\boldsymbol{I} = \frac{\dfrac{1}{-j1000}}{\dfrac{1}{-j1000} + \dfrac{1}{3000}} 0.2\angle 53.1^\circ$$

$$= 0.06 + j0.18 = 0.19\angle 71.6^\circ\ \text{A}$$

By KCL $\boldsymbol{I} = \boldsymbol{I_C} + \boldsymbol{I_R}$, so the remaining unknown current is

$$\boldsymbol{I_R} = \boldsymbol{I} - \boldsymbol{I_C} = 0.06 - j0.02 = 0.0632\angle -18.4^\circ\ \text{A}$$

The waveforms corresponding to the phasor currents are

$$i(t) = Re\left\{\boldsymbol{I}e^{j2000t}\right\} = 0.2\cos(2000t + 53.1^\circ)\ \text{A}$$

$$i_C(t) = Re\left\{\boldsymbol{I_C}e^{j2000t}\right\} = 0.19\cos(2000t + 71.6^\circ)\ \text{A}$$

$$i_R(t) = Re\left\{\boldsymbol{I_R}e^{j2000t}\right\} = 0.0632\cos(2000t - 18.4^\circ)\ \text{A}$$

Application Note

EXAMPLE 13-7

The purpose of the impedance bridge in Fig. 13-18 is to measure the unknown impedance Z_X by adjusting known impedances Z_1, Z_2, and Z_3 until the detector voltage $\boldsymbol{V_{DET}}$ is zero. The circuit consists of a sinusoidal source $\boldsymbol{V_S}$ driving two voltage dividers connected in parallel. Using the voltage division principle we find that the detector voltage is

$$\boldsymbol{V_{DET}} = \boldsymbol{V_A} - \boldsymbol{V_B} = \frac{Z_2}{Z_1 + Z_2}\boldsymbol{V_S} - \frac{Z_X}{Z_3 + Z_X}\boldsymbol{V_S}$$

$$= \left[\frac{Z_2 Z_3 - Z_1 Z_X}{(Z_1 + Z_2)(Z_3 + Z_X)}\right]\boldsymbol{V_S}$$

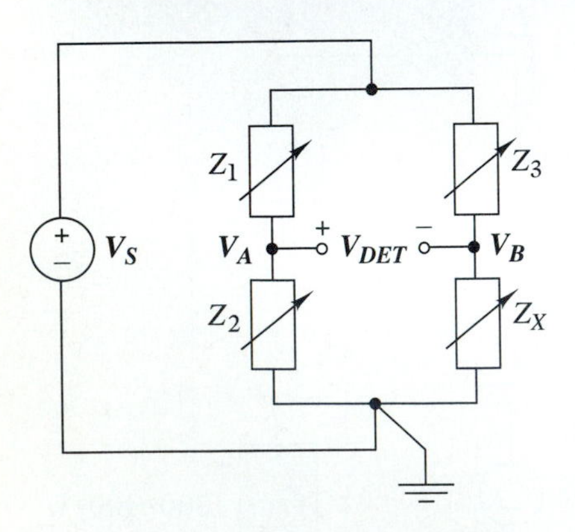

Figure 13-18 Impedance bridge.

This equation shows that the detector voltage will be zero when $Z_2 Z_3 = Z_1 Z_X$. When the branch impedances are adjusted so that the detector voltage is zero, the unknown impedance can be written in terms of the known impedances as follows:

$$Z_X = R_X + jX_X = \frac{Z_2 Z_3}{Z_1}$$

This equation is called the *balanced bridge* condition. Since the equality involves complex quantities, at least two of the known impedances must be adjustable to balance both the resistance R_X and the reactance X_X of the unknown impedance. In practice, bridges are designed assuming that the sign of the unknown reactance is known. Bridges that measure only positive reactance are called inductance bridges, while those that measure only negative reactance are called capacitance bridges.

The *Maxwell inductance bridge* in Fig. 13-19 is used to measure the resistance R_X and inductance L_X of an inductive device by alternately adjusting resistances R_1 and R_2 to balance the bridge circuit. The impedances of the legs of this bridge are

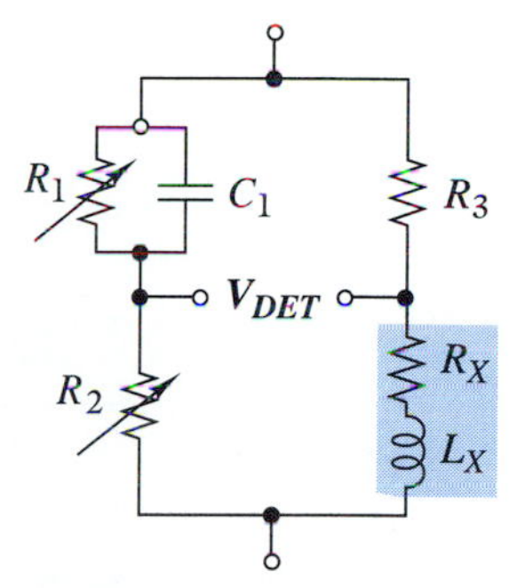

Figure 13-19 Maxwell bridge.

$$Z_1 = \frac{1}{j\omega C_1 + \dfrac{1}{R_1}}$$

$$Z_2 = R_2 \qquad Z_3 = R_3$$

For the Maxwell bridge the balance condition $Z_X = Z_2 Z_3 / Z_1$ yields

$$R_X + j\omega L_X = \frac{R_2 R_3}{R_1} + j\omega C_1 R_2 R_3$$

Equating the real and imaginary parts on each side of this equation yields the parameters of the unknown impedance in terms of the known impedances.

$$R_X = \frac{R_2 R_3}{R_1} \qquad \text{and} \qquad L_X = R_2 R_3 C_1$$

Note that adjusting R_1 only affects R_X. The Maxwell bridge measures inductance by balancing the positive reactance of an unknown inductive device with a calibrated fraction of negative reactance produced by the known capacitor C_1. If the reactance of the unknown device is actually capacitive (negative), then the Maxwell bridge cannot be balanced.

■ EXERCISE 13-7

A 1-μF capacitor and 1-kΩ resistor are connected in parallel and this parallel combination is connected in series with a 200-mH inductor.
(a) Find the equivalent impedance of the connection at $\omega = 1$ krad/s.
(b) Repeat (a) with $\omega = 4$ krad/s.

ANSWERS:
(a) $500 - j300\ \Omega$
(b) $58.8 + j565\ \Omega$. ■

EXERCISE 13-8

A voltage source $v_S = 15 \cos 2000t$ is applied across the circuit in Exercise 13-7.

(a) Find the steady-state current through the inductor.

(b) Find the steady-state voltage across the 1-kΩ resistor.

ANSWERS:

(a) $75 \cos 2000t$ mA

(b) $33.5 \cos(2000t - 63.4°)$ V

EXERCISE 13-9

Two branches with impedances $500 - j125$ Ω and $100 + j400$ Ω are connected in parallel.

(a) Find the equivalent impedance of the parallel combination.

(b) What element should be connected in series with the parallel combination to make the total reactance zero at $\omega = 10$ krad/s?

ANSWERS:

(a) $322\angle 37.3°$ Ω

(b) A capacitor with $C = 513$ nF

$Y \rightleftharpoons \Delta$ Transformations

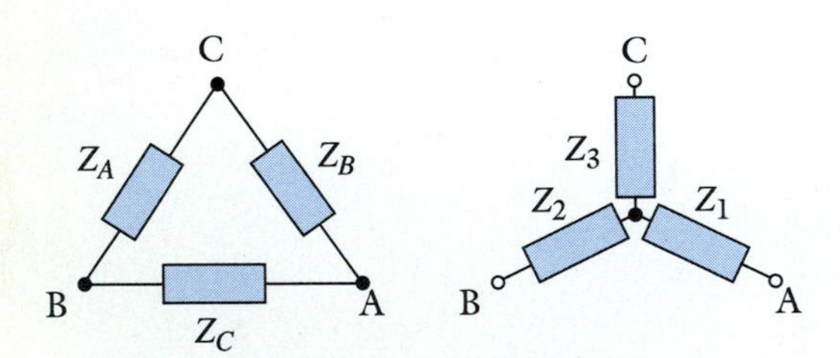

Figure 13-20 Δ- and Y-connected subcircuits.

In Chapter 2 we studied the equivalence of Δ and Y-connected resistors as a way to simplify resistance circuits with no series or parallel connected branches. The same basic concept applies to the Δ and Y connected impedances shown in Fig. 13-20. Replacing a Δ-connected subcircuit by an equivalent Y, or vice versa, changes circuit connections so that series and parallel equivalence can be used to reduce the circuit.

The equations for the Δ to Y transformation are

$$\begin{aligned} Z_1 &= \frac{Z_B Z_C}{Z_A + Z_B + Z_C} \\ Z_2 &= \frac{Z_C Z_A}{Z_A + Z_B + Z_C} \\ Z_3 &= \frac{Z_A Z_B}{Z_A + Z_B + Z_C} \end{aligned} \tag{13-26}$$

The equations for a Y to Δ transformation are

$$\begin{aligned} Z_A &= \frac{Z_1 Z_2 + Z_2 Z_3 + Z_1 Z_3}{Z_1} \\ Z_B &= \frac{Z_1 Z_2 + Z_2 Z_3 + Z_1 Z_3}{Z_2} \\ Z_C &= \frac{Z_1 Z_2 + Z_2 Z_3 + Z_1 Z_3}{Z_3} \end{aligned} \tag{13-27}$$

Equations (13-26) and (13-27) have the same form as Eqs. (2-26) and (2-27) except that here they involve impedances rather than resistances. Derivation of the impedance equations uses the same approach given for resistance circuits in Chapter 2. It is important to remember that the Y and Δ configurations are defined by electrical connections and not by the geometric shapes in the circuit diagram. That is, the circuit diagram does not have to show the subcircuits with geometric Y or Δ shapes for the transformation equations to apply.

Applying the transformation equations in phasor circuits involves manipulating complex numbers representing the impedances. In this regard the expressions in Eq. (13-26) are simpler than those in Eq. (13-27) because each equation involves only one multiplication rather than three. However, when the expressions in Eq. (13-27) are written in terms of admittances the Y to Δ transformation equations become

$$Y_A = \frac{Y_2 Y_3}{Y_1 + Y_2 + Y_3}$$
$$Y_B = \frac{Y_1 Y_3}{Y_1 + Y_2 + Y_3} \qquad (13\text{-}28)$$
$$Y_C = \frac{Y_1 Y_2}{Y_1 + Y_2 + Y_3}$$

These equations have the same mathematical form as those in Eq. (13-26) and are somewhat easier to apply.

A Y or Δ subcircuit is said to be *balanced* when $Z_1 = Z_2 = Z_3 = Z_Y$ or $Z_A = Z_B = Z_C = Z_\Delta$. Under balanced conditions the transformation equations reduce to $Z_Y = Z_\Delta/3$ and $Z_\Delta = 3Z_Y$. We make use of the balanced circuit transformation equations in the study of three-phase power systems in the next chapter.

EXAMPLE 13-8

Use a Δ-Y transformation to solve for the phasor current I_X in Figure 13-21.

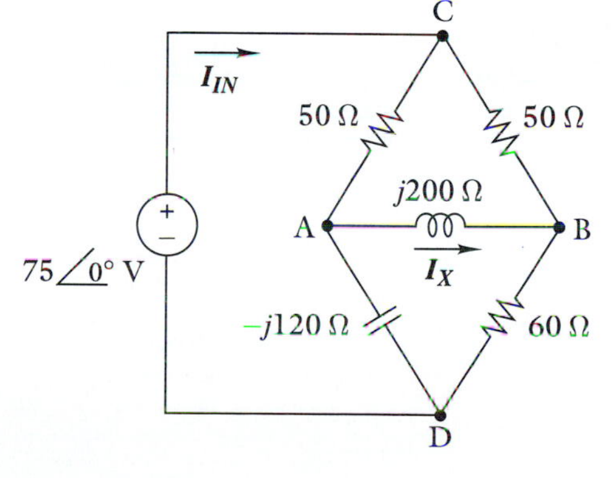

Figure 13-21

SOLUTION:

We cannot use basic reduction tools on the circuit in Fig. 13-21 because no elements are connected in series or parallel. However, if we replace either the upper delta (A, B, C) or lower delta (A, B, D) by an equivalent Y subcircuit, then we can apply series and parallel reduction methods. We choose to transform the upper delta because it has two equal resistors, which simplifies the transformation equations. The sum of the impedance in the upper delta is $100 + j200\ \Omega$. This sum is the denominator of the expressions in Eq. (13-26). Using the first equation in Eq. (13-26) we find the equivalent Y impedance connected at Node A to be

$$Z_1 = \frac{(50)(j200)}{100 + j200} = 40 + j20\ \Omega$$

Using the second equation in Eq. (13-26) we obtain the equivalent Y impedance connected at Node B:

$$Z_2 = \frac{(50)(j200)}{100 + j200} = 40 + j20\ \Omega$$

Using the third equation in Eq. (13-26) we obtain the equivalent Y impedance connected at Node C:

$$Z_3 = \frac{(50)(50)}{100 + j200} = 5 - j10\ \Omega$$

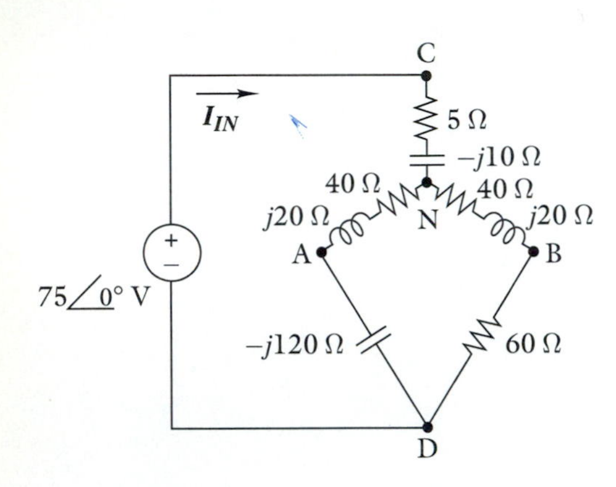

Figure 13-22

Figure 13-22 shows the revised circuit with the equivalent Y inserted in place of the upper delta in Fig. 13-21. Note that the transformation introduces a new node labeled N in Fig. 13-22.

The revised circuit in Fig. 13-22 can be reduced by series and parallel equivalence. The total impedance of the path NAD is $40 - j100\ \Omega$. The total impedance of the path NBD is $100 + j20\ \Omega$. These paths are connected in parallel so the equivalent impedance between Nodes N and D is

$$Z_{ND} = \frac{1}{\dfrac{1}{40 - j100} + \dfrac{1}{100 + j20}} = 60.6 - j31.1\ \Omega$$

The impedance Z_{ND} is connected in series with the remaining leg of the equivalent Y, so the equivalent impedance seen by the voltage source is

$$Z_{EQ} = 5 - j10 + Z_{ND} = 65.6 - j41.1\ \Omega$$

The input current shown in Fig. 13-22 is

$$\mathbf{I}_{IN} = \frac{75\angle 0^\circ}{Z_{EQ}} = 0.821 + j0.514\ \text{A}$$

Given the input current, we work backward through the equivalent circuit to find the required current $\mathbf{I}_X$. To calculate $\mathbf{I}_X$ we need the voltage $\mathbf{V}_{AB}$ between Nodes A and B. To find $\mathbf{V}_{AB}$ we first calculate the voltage $\mathbf{V}_{CN}$ as

$$\begin{aligned}\mathbf{V}_{CN} &= \mathbf{I}_{IN} Z_{CN} = (0.821 + j0.514)(5 - j10) \\ &= 9.24 - j5.64\ \text{V}\end{aligned}$$

By KVL $\mathbf{V}_{CN} + \mathbf{V}_{ND} = 75\angle 0^\circ$, therefore, we can write

$$\mathbf{V}_{ND} = 75\angle 0^\circ - \mathbf{V}_{CN} = 65.8 + j5.64\ \text{V}$$

The voltage $\mathbf{V}_{ND}$ appears across two parallel voltage dividers. Using voltage division, the voltages $\mathbf{V}_{AD}$ and $\mathbf{V}_{BD}$ are

$$\mathbf{V}_{AD} = \frac{-j120}{40 - j100}\mathbf{V}_{ND} = 70.4 - j21.4 \text{ V}$$

$$\mathbf{V}_{BD} = \frac{60}{100 + j20}\mathbf{V}_{ND} = 38.6 - j4.34 \text{ V}$$

Using KVL the voltage $\mathbf{V}_{AB}$ is

$$\mathbf{V}_{AB} = \mathbf{V}_{AD} - \mathbf{V}_{BD}$$
$$= 31.8 - j17.1 = 36.1\angle -28.3^\circ \text{ V}$$

Using the voltage $\mathbf{V}_{AB}$ the unknown current $\mathbf{I}_X$ is found to be

$$\mathbf{I}_X = \frac{\mathbf{V}_{AB}}{j200} = 0.18\angle -118^\circ \text{ A}$$

EXERCISE 13-10

Use a Y-Δ or Δ-Y transformation to find the equivalent impedance of the circuit in Fig. 13-23.

ANSWER:
$Z_{EQ} = 97.3 + j13.3\ \Omega$

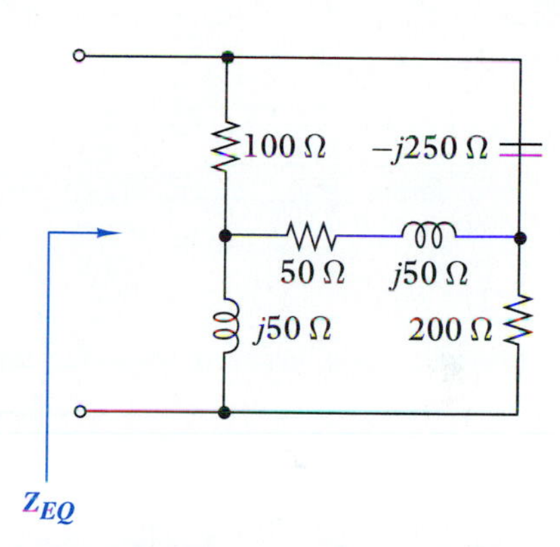

Figure 13-23

13-4 CIRCUIT THEOREMS IN THE PHASOR DOMAIN

In this section we treat basic properties of phasor circuits that parallel those the resistance circuit theorems developed in Chapter 3. Circuit linearity is the foundation for all of these properties. The proportionality and superposition properties are two fundamental consequences of linearity.

Proportionality

The *proportionality* property states that phasor output responses are proportional to the input phasor. Mathematically proportionality means that

$$\mathbf{Y} = K\mathbf{X}$$

where $\mathbf{X}$ is the input phasor, $\mathbf{Y}$ the output phasor, and K is the proportionality constant. In phasor circuit analysis the proportionality constant is generally a complex number.

The unit-output method discussed in Chapter 3 is based on the proportionality property. To apply the unit-output method in the phasor domain we assume the output is a unit phasor $\mathbf{Y} = 1\angle 0°$. By successive application of KCL, KVL and the element impedances, we solve for the input phasor required to produce the unit output. Because the circuit is linear the proportionality constant relating input and output is

$$K = \frac{\text{output}}{\text{input}} = \frac{1\angle 0°}{\text{Input Phasor for Unit Output}}$$

Once we have the constant K we can find the output for any input or the input required to produce any specified output.

The next example illustrates the unit-output method for phasor circuits.

EXAMPLE 13-9

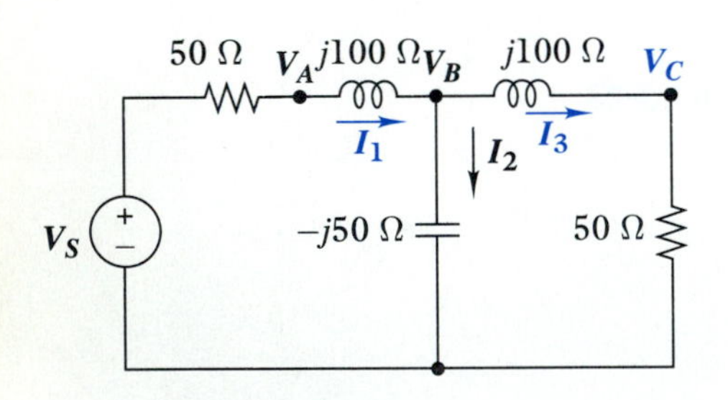

Figure 13-24

Use the unit-output method to find the input impedance, current $\mathbf{I}_1$, output voltage $\mathbf{V}_C$, and current $\mathbf{I}_3$ of the circuit in Fig. 13-24 for $\mathbf{V}_S = 10\angle 0°$.

SOLUTION:

The following steps implement the unit-output method for the circuit in Figure 13-24.

1. Assume a unit-output voltage $\mathbf{V}_C = 1 + j0$ V.
2. By Ohms's Law $\mathbf{I}_3 = \mathbf{V}_C/50 = 0.02 + j0$ A.
3. By KVL $\mathbf{V}_B = \mathbf{V}_C + (j100)\mathbf{I}_3 = 1 + j2$ V.
4. By Ohm's Law $\mathbf{I}_2 = \mathbf{V}_B/(-j50) = -0.04 + j0.02$ A.
5. By KCL $\mathbf{I}_1 = \mathbf{I}_2 + \mathbf{I}_3 = -0.02 + j0.02$ A.
6. By KVL $\mathbf{V}_S = (50 + j100)\mathbf{I}_1 + \mathbf{V}_B = -2 + j1$ V.

Given $\mathbf{V}_S$ and $\mathbf{I}_1$ the input impedance is

$$Z_{IN} = \frac{\mathbf{V}_S}{\mathbf{I}_1} = \frac{-2 + j1}{-0.02 + j0.02} = 75 + j25\ \Omega$$

The proportionality factor between the input $\mathbf{V}_S$ and output voltage $\mathbf{V}_C$ is

$$K = \frac{1}{\mathbf{V}_S} = \frac{1}{-2 + j} = -0.4 - j0.2$$

Given K and Z_{IN}, we can now calculate the required responses for an input $\mathbf{V}_S = 10\angle 0°$.

$$\mathbf{V}_C = K\mathbf{V}_S = -4 - j2 = 4.47\angle -153° \text{ V}$$

$$\mathbf{I}_1 = \frac{\mathbf{V}_S}{Z_{IN}} = 0.12 - j0.04 = 0.126\angle -18.4° \text{ A}$$

$$\mathbf{I}_3 = \frac{\mathbf{V}_C}{50} = -0.8 - j0.04 = 0.0894\angle -153° \text{ A}$$

Superposition

The superposition principle applies to phasor responses only if all of the independent sources driving the circuit have the **same frequency**. That is, when the input sources have the same frequency we can find the phasor response due to each source acting alone and obtain the total response by adding the individual phasors. If the sources have different frequencies, then each source must be treated in a separate steady-state analysis because the element impedances change with frequency. The phasor response for each source must be changed into waveforms and then superposition applied in the time domain. In other words, superposition applies in the time domain to any linear circuit. It also applies in the phasor domain when all independent sources have the same frequency. The following examples illustrate both cases.

EXAMPLE 13-10

Use superposition to find the steady-state voltage $v_R(t)$ in Fig. 13-25 for $R = 20\ \Omega$, $L_1 = 2$ mH, $L_2 = 6$ mH, $C = 20\ \mu$F, $v_{S1} = 100 \cos 5000t$ V, and $v_{S2} = 120 \cos(5000t + 30°)$.

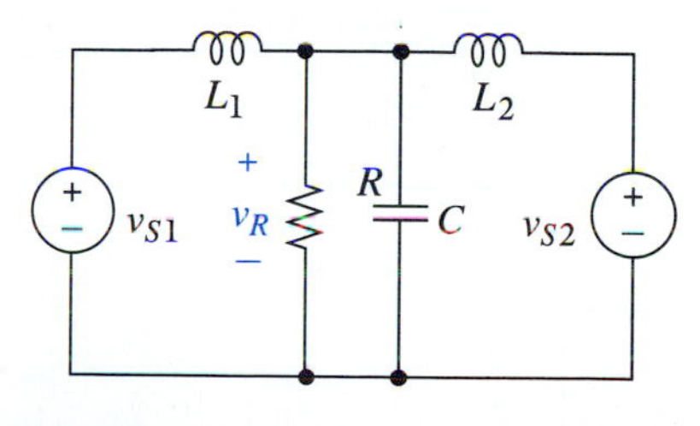

Figure 13-25

SOLUTION:

In this example the two sources operate at the same frequency. Fig. 13-26(a) shows the phasor-domain circuit with source No. 2 turned off and replaced by a short circuit. The three elements in parallel in Fig. 13-26(a) produce an equivalent impedance of

$$Z_{EQ2} = \frac{1}{\dfrac{1}{20} + \dfrac{1}{-j10} + \dfrac{1}{j30}} = 7.20 - j9.60\ \Omega$$

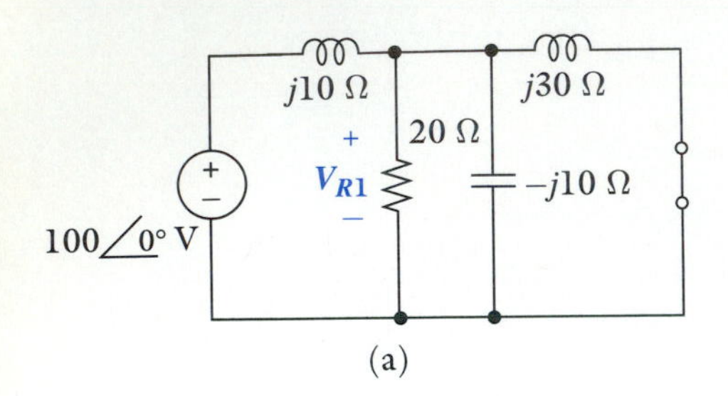

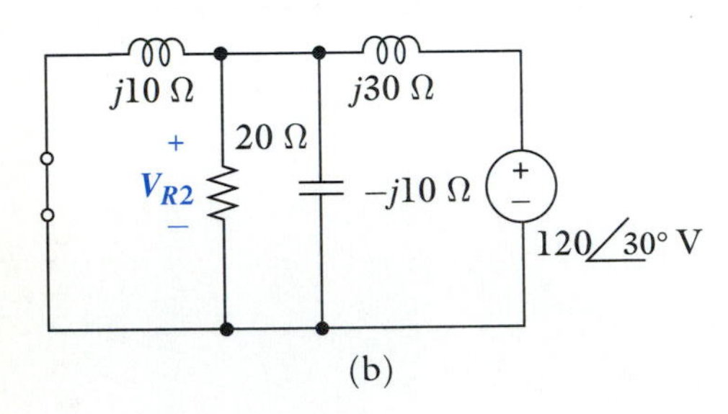

Figure 13-26

By voltage division the phasor response $\mathbf{V}_{R1}$ is

$$\mathbf{V}_{R1} = \frac{Z_{EQ1}}{j10 + Z_{EQ1}} 100\angle 0^\circ$$

$$= 92.3 - j138 = 166\angle{-56.3^\circ} \text{ V}$$

Figure 13-26(b) shows the phasor-domain circuit with source No. 1 turned off and source No. 2 on. The three elements in parallel in Fig. 13-26(b) produce an equivalent impedance of

$$Z_{EQ1} = \frac{1}{\dfrac{1}{20} + \dfrac{1}{-j10} + \dfrac{1}{j10}} = 20 - j0 \ \Omega$$

By voltage division the response $\mathbf{V}_{R2}$ is

$$\mathbf{V}_{R2} = \frac{Z_{EQ2}}{j30 + Z_{EQ2}} 120\angle 30^\circ$$

$$= 59.7 - j29.5 = 66.6\angle{-26.3^\circ} \text{ V}$$

Since the sources have the same frequency the total response can be found by adding the individual phasor responses $\mathbf{V}_{R1}$ and $\mathbf{V}_{R2}$:

$$\mathbf{V}_R = \mathbf{V}_{R1} + \mathbf{V}_{R2} = 152 - j167 = 226\angle{-47.8^\circ} \text{ V}$$

The waveform corresponding to the phasor sum is

$$v_R(t) = Re\left\{\mathbf{V}_R e^{j5000t}\right\} = 226\cos(5000t - 47.8^\circ) \text{ V}$$

The total response can also be obtained by adding the time-domain waveforms corresponding to the individual phasor responses $\mathbf{V}_{R1}$ and $\mathbf{V}_{R2}$.

$$v_R(t) = Re\left\{\mathbf{V}_{R1} e^{j5000t}\right\} + Re\left\{\mathbf{V}_{R2} e^{j5000t}\right\}$$

$$= 166\cos(5000t - 56.2^\circ) + 66.6\cos(5000t - 26.3^\circ) \text{ V}$$

The reader is encouraged to show that the two expressions for $v_R(t)$ are equivalent using the additive property of sinusoidal waveforms.

EXAMPLE 13-11

Use superposition to find the steady-state current $i(t)$ in Fig. 13-27 for $R = 10$ kΩ, $L = 200$ mH, $v_{S1} = 24\cos 20000t$ V, and $v_{S2} = 8\cos(60000t + 30^\circ)$.

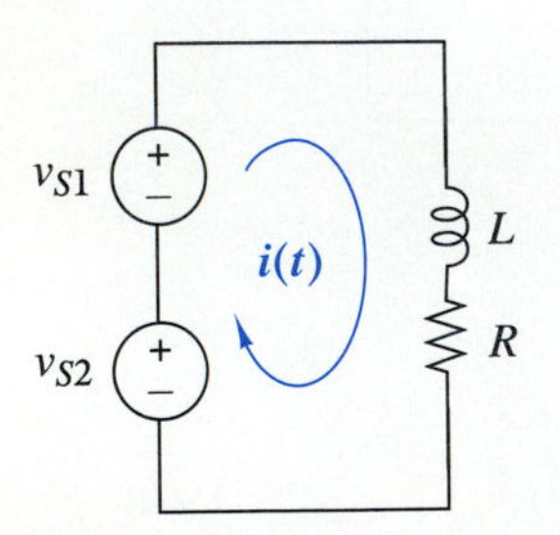

Figure 13-27

SOLUTION:

In this example the two sources operate at different frequencies. With source No. 2 off, the input phasor is $\mathbf{V}_{S1} = 24\angle 0^\circ$ V at a frequency of $\omega = 20$ krad/s. At this frequency the equivalent impedance of the inductor and resistor is

$$Z_{EQ1} = R + j\omega L = 10 + j4 \text{ k}\Omega$$

The phasor current due to source No. 1 is

$$\boldsymbol{I}_1 = \frac{\boldsymbol{V}_{S1}}{Z_{EQ1}} = \frac{24\angle 0^\circ}{10000 + j4000} = 2.23\angle{-21.8^\circ} \text{ mA}$$

With source No. 1 off and source No. 2 on, the input phasor $\boldsymbol{V}_{S2} = 8\angle 30^\circ$ V at a frequency of $\omega = 60$ krad/s. At this frequency the equivalent impedance of the inductor and resistor is

$$Z_{EQ1} = R + j\omega L = 10 + j12 \text{ k}\Omega$$

The phasor current due to source No. 2 is

$$\boldsymbol{I}_2 = \frac{\boldsymbol{V}_{S2}}{Z_{EQ2}} = \frac{8\angle 30^\circ}{10000 + j12000} = 0.512\angle{-20.2^\circ} \text{ mA}$$

The two input sources operate at different frequencies so the phasors responses $\boldsymbol{I}_1$ and $\boldsymbol{I}_2$ cannot be added to obtain the total response. However, the corresponding time-domain waveforms can be added to obtain the total response:

$$i(t) = Re\left\{\boldsymbol{I}_1 e^{j20000t}\right\} + Re\left\{\boldsymbol{I}_2 e^{j60000t}\right\}$$
$$= 2.23\cos(20000t - 21.8^\circ) + 0.512\cos(60000t - 20.2^\circ) \text{ mA}$$

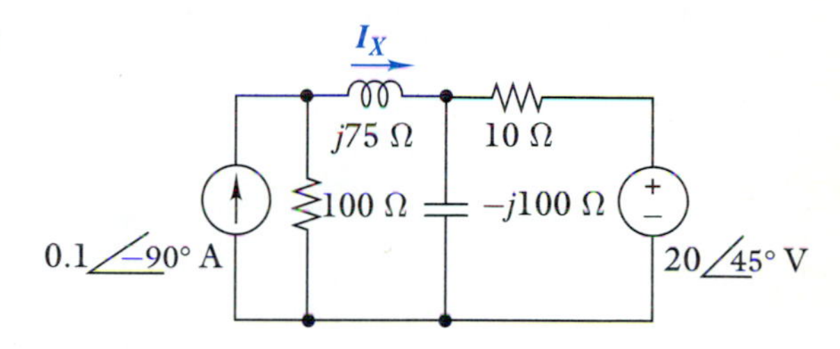

Figure 13-28

EXERCISE 13-11

The two sources in Fig. 13-28 have the same frequency. Use superposition to find the phasor current $\boldsymbol{I}_X$.

ANSWER:
$\boldsymbol{I}_X = 0.206\angle{-158^\circ}$ A

Thévenin and Norton Equivalent Circuits

In the phasor domain a two-terminal circuit containing linear elements and sources can be replaced by the Thévenin or Norton equivalent circuits shown in Figure 13-29. The general concept of Thévenin's and Norton's theorems and their restrictions are the same as in the resistive circuit studied in Chapter 3. The important difference here is that the signals $\boldsymbol{V}_T$, $\boldsymbol{I}_N$, $\boldsymbol{V}$, and $\boldsymbol{I}$ are phasors, and $Z_T = 1/Y_N$ and Z_L are complex numbers representing the source and load impedances.

Finding the Thévenin or Norton equivalent of a phasor circuit involves the same process as for resistance circuits, except

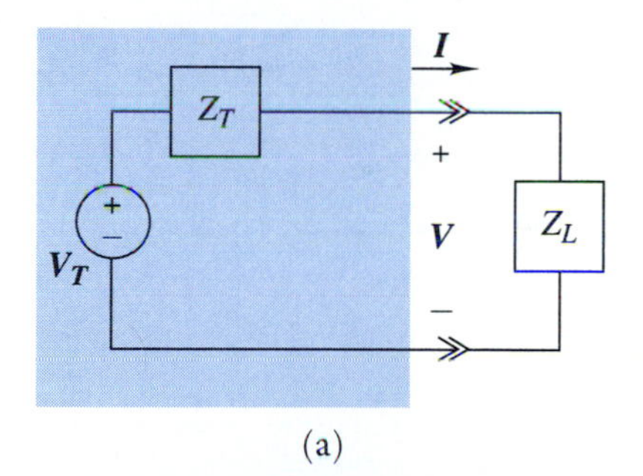

(a)

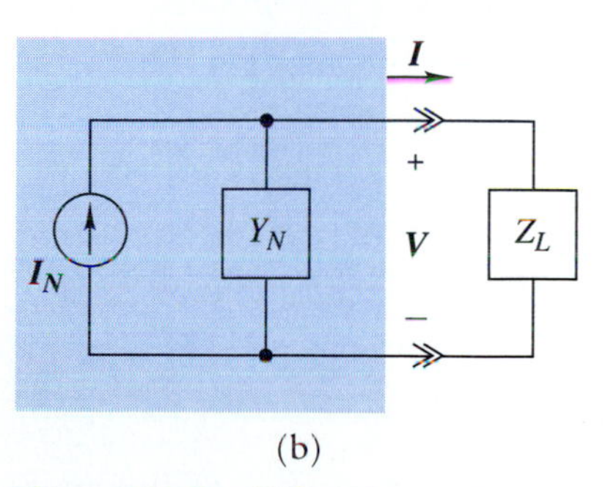

(b)

Figure 13-29 Thévenin and Norton equivalent circuits in the phasor domain.

that now we must manipulate complex numbers. The *Thévenin* and *Norton* circuits are equivalent to each other, so their circuit parameters are related as follows:

$$\begin{aligned} \boldsymbol{V}_{OC} &= \boldsymbol{V}_T = \boldsymbol{I}_N Z_T \\ \boldsymbol{I}_{SC} &= \frac{\boldsymbol{V}_T}{Z_T} = \boldsymbol{I}_N \\ Z_T &= \frac{1}{Y_N} = \frac{\boldsymbol{V}_{OC}}{\boldsymbol{I}_{SC}} \end{aligned} \qquad (13\text{-}29)$$

Algebraically, the results in Eq. (13-29) are identical to the corresponding equations for resistance circuits. These equations point out that we can determine a Thévenin or Norton equivalent by finding any two of the following quantities: (1) the open-circuit voltage $\boldsymbol{V}_{OC}$, (2) the short-circuit current $\boldsymbol{I}_{SC}$, or (3) the impedance Z_T looking back into the source circuit with all independent sources turned off.

The relationships in Eq. (13-29) define source transformations that allow us to convert a voltage source in series with an impedance into a current source in parallel with the same impedance, or vice versa. Phasor-domain source transformations simplify circuits and are useful in formulating general node-voltage or mesh-current equations discussed in the next section.

The next two examples illustrate applications of source transformation and Thévenin equivalent circuits.

EXAMPLE 13-12

Both sources in Fig. 13-30(a) operate at a frequency of $\omega = 5000$ rad/s. Find the steady-state voltage $v_R(t)$ using source transformations.

SOLUTION:

Example 13-10 solves this problem using superposition. In this example we use source transformations. We observe that the voltage sources in Fig. 13-30(a) are connected in series with an impedance and can be converted into the following equivalent current sources:

$$\boldsymbol{I}_{EQ1} = \frac{100\angle 0^\circ}{j10} = 0 - j10 \text{ A}$$

$$\boldsymbol{I}_{EQ2} = \frac{120\angle 30^\circ}{j30} = 2 - 3.46 \text{ A}$$

Fig. 13-30(b) shows the circuit after the two source transformations. The two current sources are connected in parallel and can

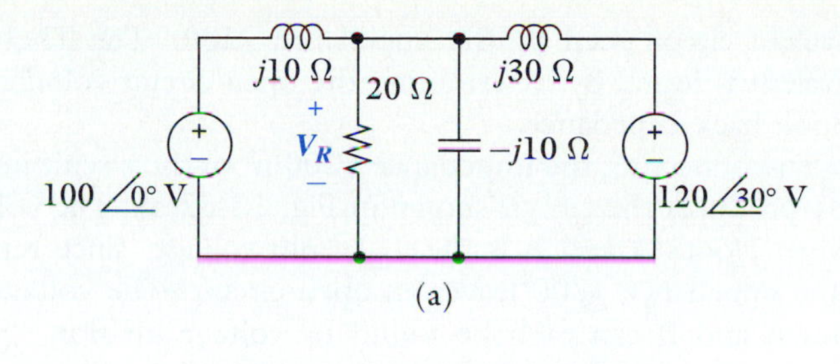

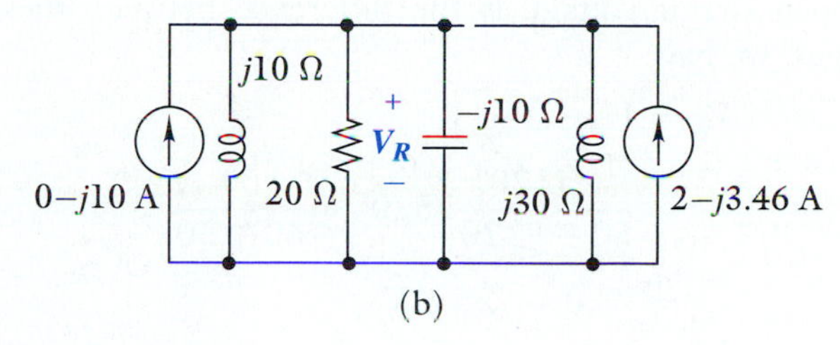

Figure 13-30

be replaced by a single equivalent current source

$$\mathbf{I}_{EQ} = \mathbf{I}_{EQ1} + \mathbf{I}_{EQ2} = 2 - j13.46 = 13.6\angle{-81.5^\circ} \text{ A}$$

The four passive elements are connected in parallel and can be replaced by an equivalent impedance

$$Z_{EQ} = \frac{1}{\dfrac{1}{20} + \dfrac{1}{-j10} + \dfrac{1}{j10} + \dfrac{1}{j30}} = 16.6\angle 33.7^\circ \ \Omega$$

The voltage across this equivalent impedance equals $\mathbf{V}_R$, since one of the parallel elements is the resistor R. Therefore, the unknown phasor voltage is

$$\mathbf{V}_R = \mathbf{I}_{EQ} Z_{EQ} = (13.6\angle{-81.5^\circ}) \times (16.6\angle 33.7^\circ) = 226\angle{-47.8^\circ} \text{ V}$$

The value of $\mathbf{V}_R$ found above is the same as the answer found in Example 13-10 using superposition. The corresponding waveform is

$$v_R(t) = Re\left\{\mathbf{V}_R e^{j5000t}\right\} = 226\cos(5000t - 47.8^\circ) \text{ V}$$

EXAMPLE 13-13

Use Thévenin's theorem to find the current $\mathbf{I}_X$ in the bridge circuit shown in Fig. 13-31.

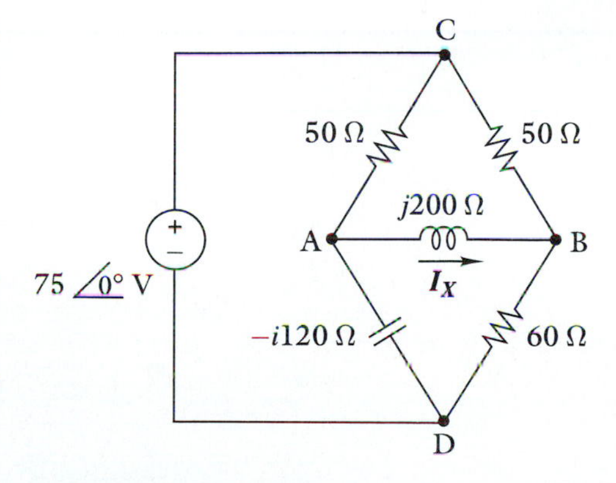

Figure 13-31

SOLUTION:

Example 13-8 solves this problem using a Δ to Y transformation. In this example we determine $\mathbf{I}_X$ by finding the Thévenin

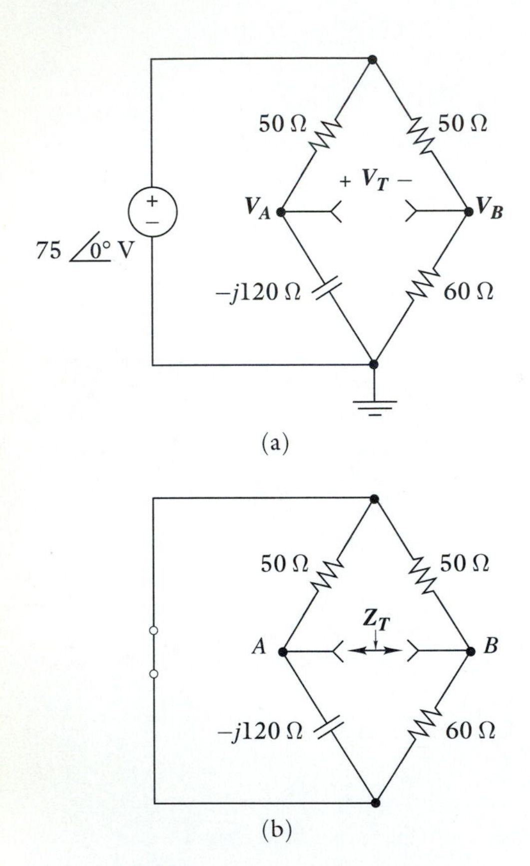

Figure 13-32

equivalent circuit seen by the impedance $j200$. The Thévenin equivalent is found by determining the open circuit voltage and the look-back impedance.

Disconnecting the impedance $j200$ from the circuit in Fig. 13-31 produces the circuit shown in Fig. 13-32(a). The voltage between Nodes A and B is the Thévenin voltage, since removing the impedance $j200$ leaves an open-circuit. The voltages at Nodes A and B can each be found by voltage division. Since the open circuit voltage is the difference between these node voltages, we have

$$\begin{aligned}\mathbf{V}_T &= \mathbf{V}_A - \mathbf{V}_B \\ &= \frac{-j120}{50 - j120} 75\angle 0^\circ - \frac{60}{60+50} 75\angle 0^\circ \\ &= 23.0 - j26.6 \text{ V}\end{aligned}$$

Turning off the voltage source in Fig. 13-32(a) and replacing it by a short circuit produces the situation shown in Fig. 13-32(b). The look-back impedance seen at the interface is a series connection of two pairs of elements connected in parallel. The equivalent impedance of the series/parallel combination is

$$Z_T = \frac{1}{\dfrac{1}{50} + \dfrac{1}{-j120}} + \frac{1}{\dfrac{1}{50} + \dfrac{1}{60}} = 69.9 - j17.8\ \Omega$$

Given the Thévenin equivalent circuit, we treat the impedance $j200$ as a load connected at the interface and calculate the resulting load current $\mathbf{I}_X$ as

$$\mathbf{I}_X = \frac{\mathbf{V}_T}{Z_T + j200} = \frac{23.0 - j26.6}{69.9 + j182} = 0.180\angle -118^\circ$$

The value of $\mathbf{I}_X$ found here is the same as the answer obtained in Example 11-8 using a Δ to Y transformation.

EXERCISE 13-12

(a) Find the Thévenin equivalent circuit seen by the inductor in Fig. 13-28.

(b) Use the Thévenin equivalent to calculate the current $\mathbf{I}_X$.

ANSWERS:

(a) $\mathbf{V}_T = -15.4 - j22.6$ V, $Z_T = 109.9 - j0.990\ \Omega$

(b) $\mathbf{I}_X = 0.206\angle -158^\circ$ A

EXERCISE 13-13

By inspection determine the Thévenin equivalent circuit seen by the capacitor in Fig. 13-24 for $\mathbf{V}_S = 10\angle 0^\circ$ V.

ANSWER:

$\mathbf{V}_T = 5\angle 0^\circ$ V, $Z_T = 25 + j50\ \Omega$

13-5 GENERAL CIRCUIT ANALYSIS IN THE PHASOR DOMAIN

The last three sections discuss basic analysis methods using equivalence, reduction, and circuit properties called theorems. These methods are valuable because we work directly with element impedances and thereby gain insight into steady-state circuit behavior. We also need general methods such as node and mesh analysis to deal with more complicated circuits than the basic methods can easily handle. General methods are founded on the concept of writing a complete set of device and connection equations, and then using node-voltage or mesh-current variables to reduce the number of equations that must be solved simultaneously.

When followed systematically, node or mesh methods lead to circuit equations with a well-defined pattern first observed in Chapter 5 and again in Chapter 10. This pattern allows us to develop node or mesh equations by inspection without going through the formality of writing the underlying device and connection equations.

Node-voltage equations focus on element admittances and KCL. These equations can be written by inspection as the sum of currents at a node. The sum of currents leaving Node A via admittances is

1. $\mathbf{V}_A$ times the sum of admittances connected to Node A
2. Minus $\mathbf{V}_B$ times the sum of admittances connected between Nodes A and B
3. Minus similar terms for other nodes connected to Node A by admittances

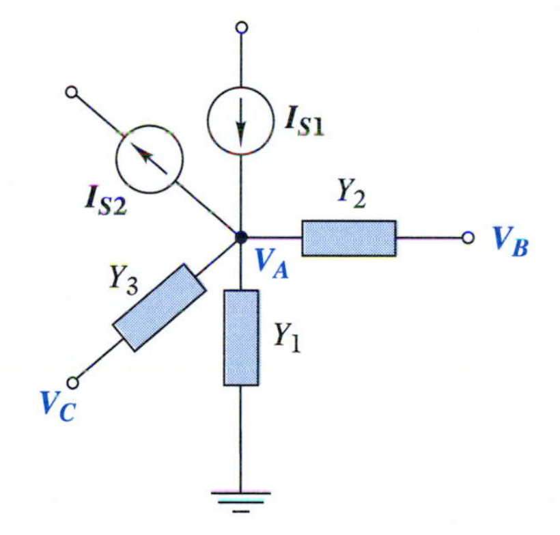

Figure 13-33 An example node.

The Node A equation is obtained by equating this sum to the sum of currents directed into the node by current sources.

For example, the equation for Node A in Fig. 13-33 is

$$\mathbf{V}_A(Y_1 + Y_2 + Y_3) - \mathbf{V}_B(Y_2) - \mathbf{V}_C(Y_3) = \mathbf{I}_{S1} - \mathbf{I}_{S2}$$

The phasors $\mathbf{V}_A$, $\mathbf{V}_B$, and $\mathbf{V}_C$ are node voltages. The admittance Y_1 is connected from Node A to ground, Y_2 is connected between Nodes A and B, and Y_3 is connected between Nodes A and C. Finally, $\mathbf{I}_{S1}$ and $\mathbf{I}_{S2}$ are the phasor current sources connected to Node A, with $\mathbf{I}_{S1}$ directed into and $\mathbf{I}_{S2}$ directed away from the node.

By duality, mesh-current equations focus on element impedances and KVL. These equations can be written by inspection as the sum of voltages around a mesh. The sum of voltages across impedances in Mesh A is

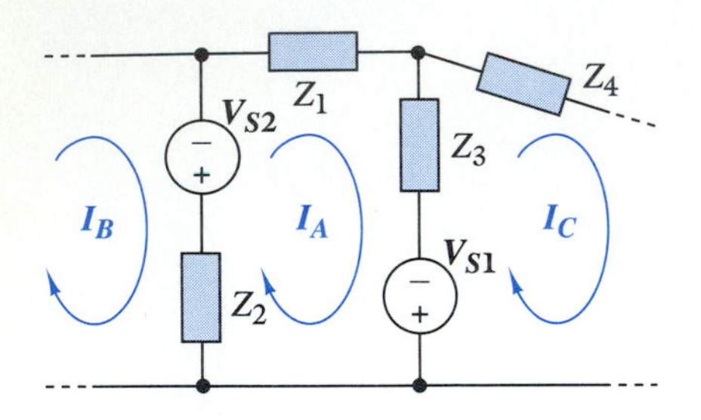

Figure 13-34 An example mesh.

1. $\boldsymbol{I}_A$ times the sum of impedances in Mesh A
2. Minus $\boldsymbol{I}_B$ times the sum of impedances common to Mesh A and Mesh B
3. Minus similar terms for other meshes adjacent to Mesh A

The Mesh A equation is obtained by equating this sum to the sum of the source voltages produced in the Mesh A.

For example, the equation for Mesh A in Fig. 13-34 is

$$\boldsymbol{I}_A\,(Z_1 + Z_2 + Z_3) - \boldsymbol{I}_B\,(Z_2) - \boldsymbol{I}_C\,(Z_3) = \boldsymbol{V}_{S1} - \boldsymbol{V}_{S2}$$

The phasors $\boldsymbol{I}_A$, $\boldsymbol{I}_B$, and $\boldsymbol{I}_C$ are mesh currents. The impedance Z_1 is contained in Mesh A only, Z_2 is in Meshes A and B, and Z_3 is in Meshes A and C. Finally, $\boldsymbol{V}_{S1}$ and $\boldsymbol{V}_{S2}$ are the phasor voltage sources in Mesh A.

The discussion above assumes that the circuit contains only current sources in the case of node analysis and voltage sources in mesh analysis. If there is a mixture of sources, we may be able to use the source transformations discussed in Section 13-4 to convert from voltage to current sources, or vice versa. A source transformation is possible only when there is an impedance connected in series with a voltage source or an admittance in parallel with a current source. When a source transformation is not possible, we use the phasor version of the modified node and mesh-analysis methods described in Chapter 5.

Formulating a set of equilibrium equations in phasor form is a straightforward process involving concepts that we have used before in Chapter 5 and Chapter 10. Solving these equations for phasor responses can be accomplished using Cramer's rule or Gaussian reduction, although these methods require manipulating linear equations with complex coefficients. In principle the solution process can be done by hand, but as a practical matter circuits with more than two or three nodes or meshes require computational aids. Modern hand-held scientific calculators can deal with complex arithmetic and solve simultaneous linear equations with complex coefficients. Circuit analysis programs such as SPICE or MICRO-CAP have ac analysis options that handle steady-state circuit analysis problems.

If computational aids are required why bother with general methods such as node and mesh analysis? Why not just use SPICE or MICRO-CAP? The answer is that hand analysis and computer-aided analysis are complementary rather than competitive. Computer-aided analysis excels at generating numerical responses when numerical values of circuit parameters are given. With hand analysis we can generate responses in symbolic form for all possible element values and input frequency. The hand-generated solutions for simple circuits give us the insight needed to intelligently use the numerical profusion generated by computer-aided analysis of large-scale circuits.

The following examples illustrate node and mesh analysis in the phasor domain.

EXAMPLE 13-14

Use node analysis to find the node voltages $\mathbf{V}_A$ and $\mathbf{V}_B$ in Fig. 13-35(a).

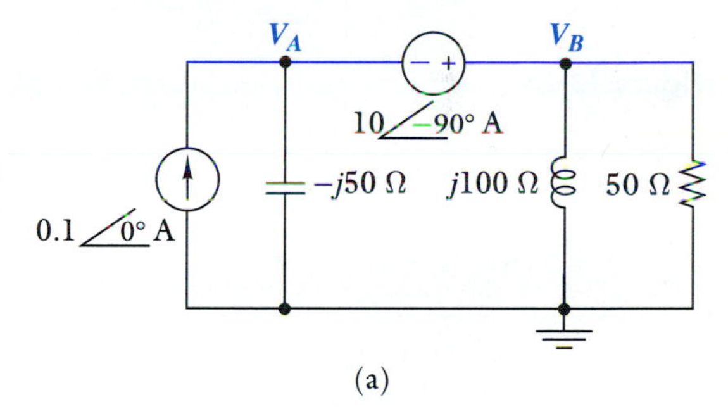

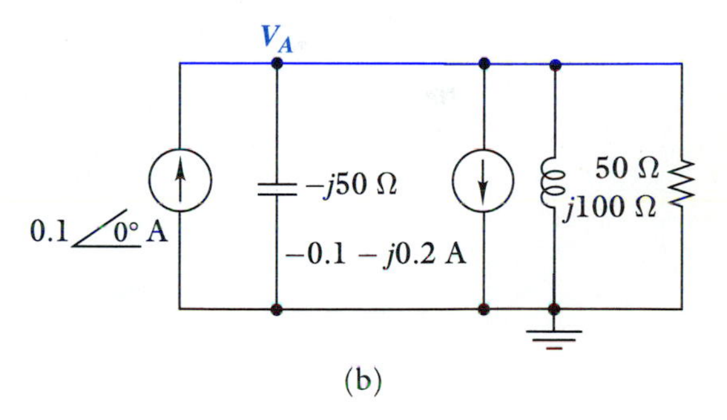

Figure 13-35

SOLUTION:

The voltage source in Fig. 13-35(a) is connected in series with an impedance consisting of a resistor and inductor connected in parallel. The equivalent impedance of this parallel combination is

$$Z_{EQ} = \frac{1}{\dfrac{1}{50} + \dfrac{1}{j100}} = 40 + j20\ \Omega$$

Applying a source transformation produces an equivalent current source of

$$\mathbf{I}_{EQ} = \frac{10\angle{-90^\circ}}{40 + j20} = -0.1 - j0.2\ \text{A}$$

Fig. 13-35(b) shows the circuit produced by the source transformation. Note that the transformation eliminates Node B. The node-voltage equation at the remaining nonreference node in Fig. 13-35(b) is

$$\left(\frac{1}{-j50} + \frac{1}{j100} + \frac{1}{50}\right)\mathbf{V}_A = 0.1\angle 0^\circ - (-0.1 - j0.2)$$

Solving for $\mathbf{V}_A$ yields

$$\mathbf{V}_A = \frac{0.2 + j0.2}{0.02 + j0.01} = 12 + j4 = 12.6\angle 18.4^\circ\ \text{V}$$

Referring to Fig. 13-35(a), we see that KVL requires $\mathbf{V}_B = \mathbf{V}_A + 10\angle{-90^\circ}$. Therefore, $\mathbf{V}_B$ is found to be

$$\mathbf{V}_B = (12 + j4) + 10\angle{-90^\circ} = 12 - j6 = 13.4\angle{-26.6^\circ}\ \text{V}$$

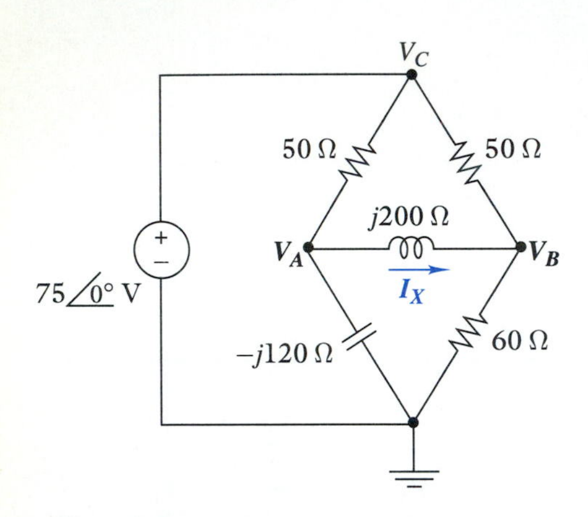

Figure 13-36

EXAMPLE 13-15

Use node analysis to find the current $\boldsymbol{I_X}$ in Fig. 13-36.

SOLUTION:

In this example we use node analysis on a problem solved in Example 13-8 using a Δ to Y transformation and solved again in Example 13-13 using a Thévenin equivalent circuit. The voltage source cannot be replaced by source transformation because it is not connected in series with an impedance. By inspection the node equations at Nodes A and B are

$$\text{Node A:}\ \left(\frac{1}{50}+\frac{1}{-j120}+\frac{1}{j200}\right)\boldsymbol{V_A}-\left(\frac{1}{j200}\right)\boldsymbol{V_B}-\left(\frac{1}{50}\right)\boldsymbol{V_C}=0$$

$$\text{Node B:}\ -\left(\frac{1}{j200}\right)\boldsymbol{V_A}+\left(\frac{1}{50}+\frac{1}{60}+\frac{1}{j200}\right)\boldsymbol{V_B}-\left(\frac{1}{50}\right)\boldsymbol{V_C}=0$$

We do not need a node equation at Node C because for the given reference node the voltage source forces the condition $\boldsymbol{V_C} = 75\angle 0°$. Substituting this constraint into the equations of Nodes A and B yields two equations in two unknowns:

$$\text{Node A:}\ \left(\frac{1}{50}+\frac{1}{-j120}+\frac{1}{j200}\right)\boldsymbol{V_A}-\left(\frac{1}{j200}\right)\boldsymbol{V_B}=\left(\frac{75\angle 0°}{50}\right)$$

$$\text{Node B:}\ -\left(\frac{1}{j200}\right)\boldsymbol{V_A}+\left(\frac{1}{50}+\frac{1}{60}+\frac{1}{j200}\right)\boldsymbol{V_B}=\left(\frac{75\angle 0°}{50}\right)$$

Solving these equations for $\boldsymbol{V_A}$ and $\boldsymbol{V_B}$ yields

$$\boldsymbol{V_A} = 70.4 - j21.4 \text{ V}$$

$$\boldsymbol{V_B} = 38.6 - j4.33 \text{ V}$$

Using the values for $\boldsymbol{V_A}$ and $\boldsymbol{V_B}$, the unknown current is found to be

$$\boldsymbol{I_X} = \frac{\boldsymbol{V_A}-\boldsymbol{V_B}}{j200} = \frac{31.8 - j17.1}{j200} = 0.180\angle -118° \text{ A}$$

This value of $\boldsymbol{I_X}$ is the same as the answer obtained in Example 13-8 and again in Example 13-13. The reader should review these three examples together to gain perspective on different approaches to phasor circuit analysis.

EXAMPLE 13-16

Use node-voltage analysis to determine the output voltage $\boldsymbol{V_2}$ in the circuit in Fig. 13-37(a).

SOLUTION:

Figure 13-37(b) shows the circuit after a source transformation. Using the node voltages and ground reference in Fig. 13-37 we can write the following node equations by inspection:

$$\text{Node A:} \quad (G_1 + G_2 + j\omega C_1)\,V_A - G_2 V_B = G_1 V_1$$

$$\text{Node B:} \quad -G_2 V_A + (G_2 + j\omega C_2)\,V_B = 0$$

The required output V_2 equals the node-voltage V_B. Using Cramer's rule we obtain

$$V_2 = V_B = \frac{\begin{vmatrix} G_1 + G_2 + j\omega C_1 & G_1 V_1 \\ -G_2 & 0 \end{vmatrix}}{\begin{vmatrix} G_1 + G_2 + j\omega C_1 & -G_2 \\ -G_2 & G_2 + j\omega C_2 \end{vmatrix}}$$

$$= \frac{G_1 G_2 V_1}{G_1 G_2 - \omega^2 C_1 C_2 + j\omega(G_1 C_2 + G_2 C_2 + G_2 C_1)}$$

We can describe the circuit frequency response since the response V_2 is in symbolic form. At $\omega = 0$ the output voltage equals the input voltage. As $\omega \to \infty$ the output voltage approaches $G_1 G_2 V_1/(-\omega^2 C_1 C_2)$. At high frequency the output amplitude decreases as $1/\omega^2$ and the phase shift approaches $-180°$. At $\omega^2 = G_1 G_2/C_1 C_2$ the real part of the denominator vanishes and the phase shift is $-90°$. The circuit appears to have a low-pass characteristic with a dc gain of unity and a cutoff in the neighborhood of the frequency at which the phase shift is $-90°$.

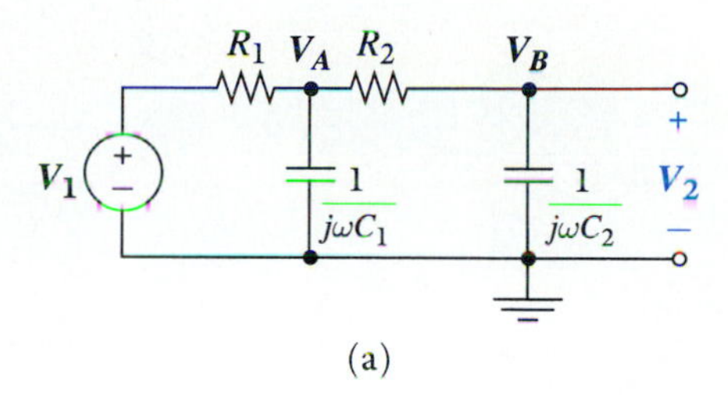

(a)

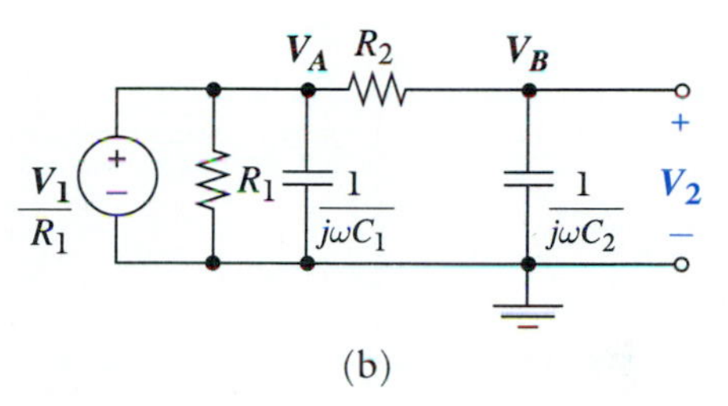

(b)

Figure 13-37

EXAMPLE 13-17

The circuit in Fig. 13-38 is an equivalent circuit of an ac induction motor. The current I_S is called the stator current, I_R the rotor current, and I_M the magnetizing current. Use the mesh-current method to solve for the branch currents I_S, I_R, and I_M.

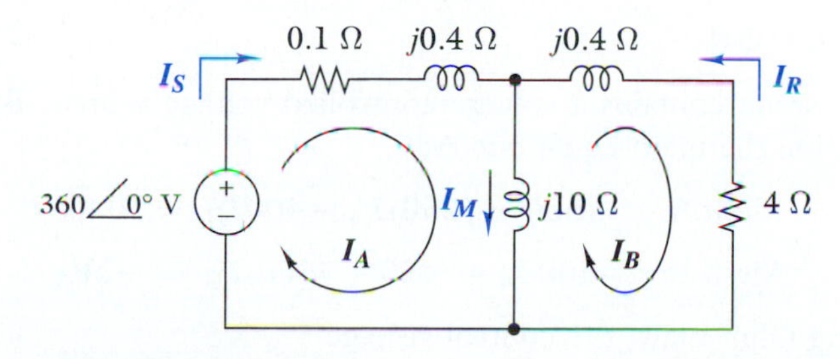

Figure 13-38

SOLUTION:

By inspection the mesh-current equations for the circuit in Fig. 13-38 are

$$(0.1 + j10.4)\boldsymbol{I_A} - (j10)\boldsymbol{I_B} = 360\angle 0^\circ$$

$$-(j10)\boldsymbol{I_A} + (4 + j10.4)\boldsymbol{I_B} = 0\angle 0^\circ$$

Solving these equations for $\boldsymbol{I_A}$ and $\boldsymbol{I_B}$ produces

$$\boldsymbol{I_A} = 79.0 - j48.2 \text{ A}$$

$$\boldsymbol{I_B} = 81.7 - j14.9 \text{ A}$$

The required stator, rotor and magnetizing currents are related to these mesh current as follows:

$$\boldsymbol{I_S} = \boldsymbol{I_A} = 92.5\angle 31.4^\circ \text{ A}$$

$$\boldsymbol{I_R} = -\boldsymbol{I_B} = -81.8 + j14.9 = 83.0\angle 170^\circ \text{ A}$$

$$\boldsymbol{I_M} = \boldsymbol{I_A} - \boldsymbol{I_B} = -2.68 - j33.3 = 33.4\angle{-94.6^\circ} \text{ A}$$

EXAMPLE 13-18

Use the mesh-current method to solve for output voltage $\boldsymbol{V_2}$ and input impedance Z_{IN} of the circuit in Fig. 13-39.

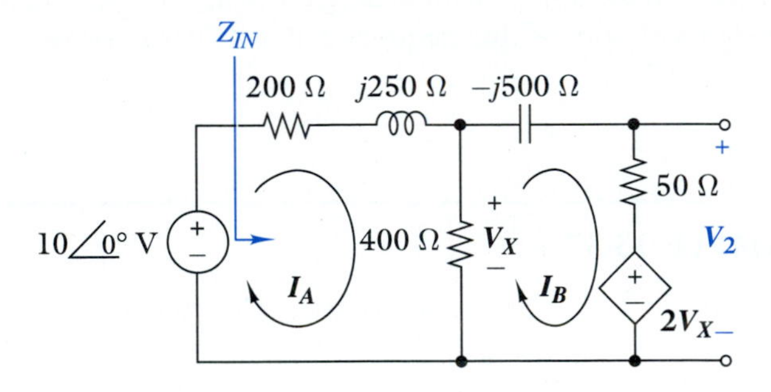

Figure 13-39

SOLUTION:

The circuit contains a voltage-controlled voltage source. By inspection the mesh equations are

$$\text{Mesh A:} \quad (600 + j250)\boldsymbol{I_A} - 400\boldsymbol{I_B} = 10\angle 0^\circ$$

$$\text{Mesh B:} \quad -400\boldsymbol{I_A} + (450 - j500)\boldsymbol{I_B} = -2\boldsymbol{V_X}$$

Using Ohm's law, the control voltage $\boldsymbol{V_X}$ is

$$\boldsymbol{V_X} = 400(\boldsymbol{I_A} - \boldsymbol{I_B})$$

Eliminating V_X from the mesh equations yields

$$\text{Mesh A:} \quad (600 + j250)I_A - 400I_B = 10\angle 0^\circ$$

$$\text{Mesh A:} \quad 400I_A + (-350 - j500)I_B = 0\angle 0^\circ$$

Solving for the two mesh currents produces

$$I_A = 10.8 - j11.1 \text{ mA}$$

$$I_B = -1.93 - j9.95 \text{ mA}$$

Using these values of the mesh currents the output voltage and input impedance are

$$\begin{aligned} V_2 &= 2V_X + 50I_B = 800\,(I_A - I_B) + 50I_B \\ &= 800I_A - 750I_B = 10.1 - j1.42 \\ &= 10.2\angle -8.00^\circ \text{ V} \end{aligned}$$

$$Z_{IN} = \frac{10\angle 0^\circ}{I_A} = \frac{10\angle 0^\circ}{0.0108 - j0.0111} = 450 + j463\ \Omega$$

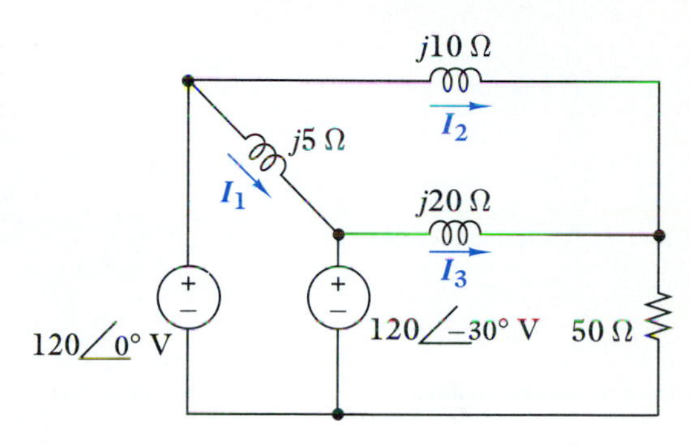

Figure 13-40

EXERCISE 13-14

Use the mesh-current or node-voltage method to find the branch currents I_1, I_2, and I_3 in Fig. 13-40.

ANSWERS:

$I_1 = 12.4\angle -15^\circ$ A, $I_2 = 3.61\angle -16.1^\circ$ A, $I_3 = 1.31\angle 166^\circ$

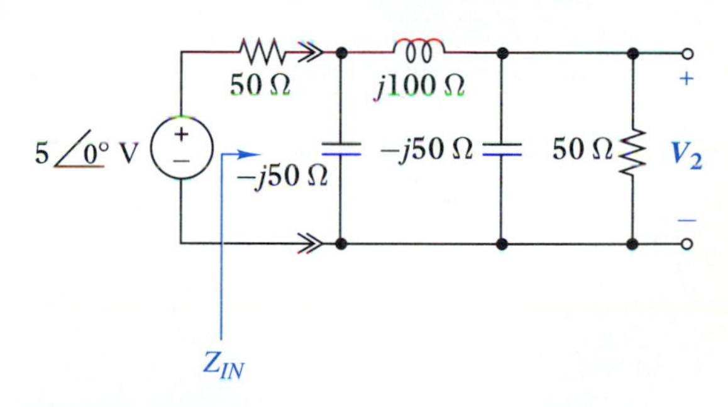

Figure 13-41

EXERCISE 13-15

Use the mesh-current or node-voltage method to find the output voltage V_2 and input impedance Z_{IN} in Fig. 13-41.

ANSWERS:

$V_2 = 1.77\angle -135^\circ$ V, $Z_{IN} = 50 - j100\ \Omega$

EXERCISE 13-16

Use the mesh-current or node-voltage method to find the current I_X in Fig. 13-42.

ANSWERS:

$I_X = 1.44\angle 171^\circ$ mA

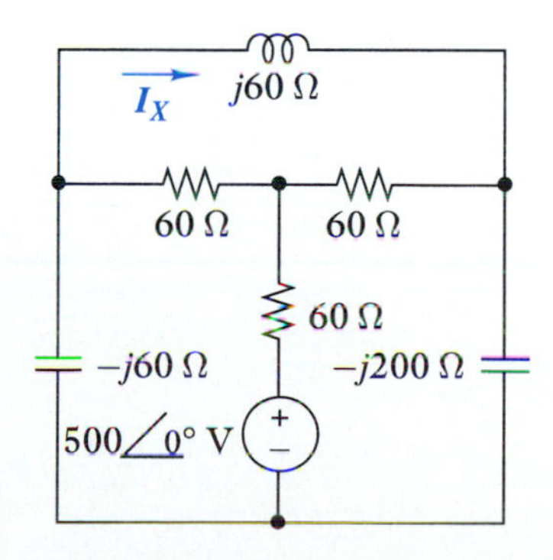

Figure 13-42

EXERCISE 13-17

Use the mesh-current or node-voltage method to find the current I_X in Fig. 13-28.

ANSWERS:

$I_X = 0.206\angle -158.2^\circ$ A

 13-6 RESONANT CIRCUITS

Element impedances play key roles in determining the behavior of circuits in the sinusoidal steady state. The impedances of the three passive elements are

$$\begin{aligned} \text{Resistor:} \quad & Z_R = R + j0 \\ \text{Inductor:} \quad & Z_L = 0 + j\omega L \\ \text{Capacitor:} \quad & Z_C = 0 - j\frac{1}{\omega C} \end{aligned} \tag{13-30}$$

The impedance of a resistor is purely real while the impedances of the other two are purely imaginary. When these elements are interconnected to produce a passive circuit the driving-point impedance seen at any pair of terminals can be written in rectangular form as

$$Z(j\omega) = R(\omega) + jX(\omega) \tag{13-31}$$

As discussed in Sect. 13-3, the real part of an impedance is called *resistance* and the imaginary part is called *reactance*. Both resistance and reactance have the dimensions of ohms (Ω).

Equation (13-31) shows that in general resistance and reactance vary with frequency. The variation of the rectangular components of impedance is what gives circuits their frequency selective characteristics. For passive circuits $R(\omega) \geq 0$, for reasons that will be made clear in the next chapter. On the other hand, the reactance of a passive circuit can be either positive or negative. Again as in Sect. 13-3, a positive reactance $X(\omega)$ is called an *inductive reactance* and a negative $X(\omega)$ is called a *capacitive reactance*.

In general, the reactance at a given pair of terminals can be inductive at some frequencies and capacitive at others. The frequency at which $X(\omega)$ changes from inductive to capacitive is given special attention. More specifically, when $X(\omega) = 0$ the driving-point impedance is purely resistive and the circuit is said to be in *resonance*. The frequency at which $X(\omega) = 0$ is called a *resonant frequency* and is denoted by ω_O.

EXAMPLE 13-19

(a) Derive expressions for the resistance and reactance of the driving-point impedance Z in Fig. 13-43.

(b) Derive an expression for the resonant frequency of the circuit.

SOLUTION:

(a) To find the driving-point impedance we first replace the parallel resistor and capacitor by an equivalent impedance

$$Z_{RC} = \frac{1}{Y_R + Y_C} = \frac{1}{\dfrac{1}{R} + j\omega C} = \frac{R}{1 + j\omega RC}$$

This expression for Z_{RC} can be put into rectangular form by multiplying and dividing by the conjugate of the denominator:

$$Z_{RC} = \frac{R}{1 + j\omega RC}\,\frac{1 - j\omega RC}{1 - j\omega RC}$$

$$= \frac{R}{1 + (\omega RC)^2} - j\frac{\omega R^2 C}{1 + (\omega RC)^2}$$

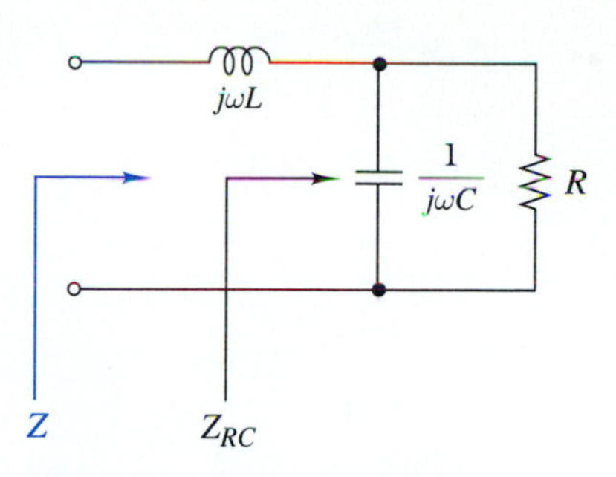

Figure 13-43

The equivalent impedance Z_{RC} is connected in series with the inductor. Therefore, the required driving-point impedance is

$$Z = Z_L + Z_{RC}$$

$$= \frac{R}{1 + (\omega RC)^2} + j\left[\omega L - \frac{\omega R^2 C}{1 + (\omega RC)^2}\right]$$

$$= R(\omega) + jX(\omega)$$

(b) Note that the real part of $Z(j\omega)$, the resistance $R(\omega)$, is positive for all ω. However, the reactance $X(\omega)$ can be positive or negative. The resonant frequency is found by setting $X(\omega_O)$ to zero:

$$X(\omega_O) = \omega_O L - \frac{\omega_O R^2 C}{1 + (\omega_O RC)^2} = 0$$

Solving for the resonant frequency gives

$$\omega_O = \sqrt{\frac{1}{LC} - \frac{1}{(RC)^2}}$$

Note the reactance $X(\omega)$ is inductive (positive) when $\omega > \omega_O$ and capacitive (negative) with $\omega < \omega_O$.

Exercise 13-18

A 600-Ω resistor and a 400-mH inductor are connected in parallel.

(a) Calculate the resistance and reactance of the parallel combination at $\omega = 2000$ rad/s.

(b) Calculate the capacitance that must be connected in series with the parallel combination to produce resonance at $\omega = 4000$ rad/s.

ANSWERS:

(a) $Z = 384 + j288\ \Omega$

(b) 1.27 μF.

EXERCISE 13-19

Two devices connected in parallel have impedances of $600 + j1000\ \Omega$ and $400 - j200\ \Omega$.

(a) Calculate the resistance and reactance of the parallel connection.

(b) Calculate the equivalent resistance and inductance or capacitance of the parallel combination if the frequency is $\omega = 3000$ rad/s.

ANSWERS:

(a) $Z_{EQ} = 405 - j43.9\ \Omega$

(b) $R_{EQ} = 405\ C_{EQ} = 7.59\ \mu\text{F}$.

Series Resonance

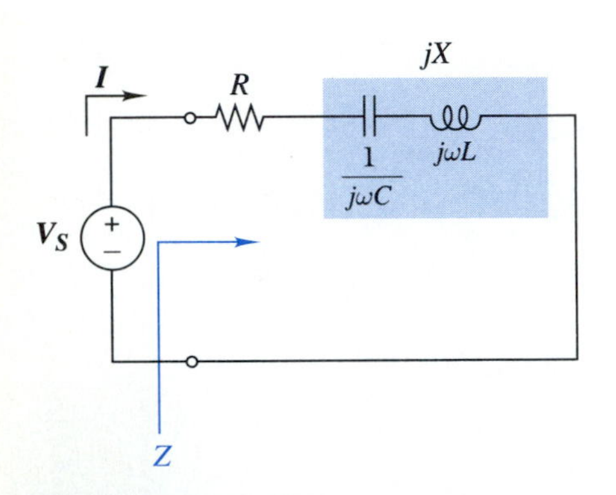

Figure 13-44 Series RLC circuit.

The series RLC circuit and parallel RLC circuit are standard forms traditionally used in electrical engineering to illustrate the response of second-order circuits. We first encountered these circuits in Chapter 8, where they served as vehicles in the study of the solution of second-order differential equations. In this chapter we use these circuits to study the resonance phenomena, starting with the series case in Fig. 13-44.

The driving-point impedance of the series RLC connection in Fig. 13-44 is

$$\begin{aligned} Z &= R + j\omega L + \frac{1}{j\omega C} \\ &= R + j\left[\omega L - \frac{1}{\omega C}\right] \\ &= R + jX(\omega) \end{aligned} \tag{13-32}$$

The resistive component of Z is R and the reactive component is

$$X(\omega) = \left[\omega L - \frac{1}{\omega C}\right] \tag{13-33}$$

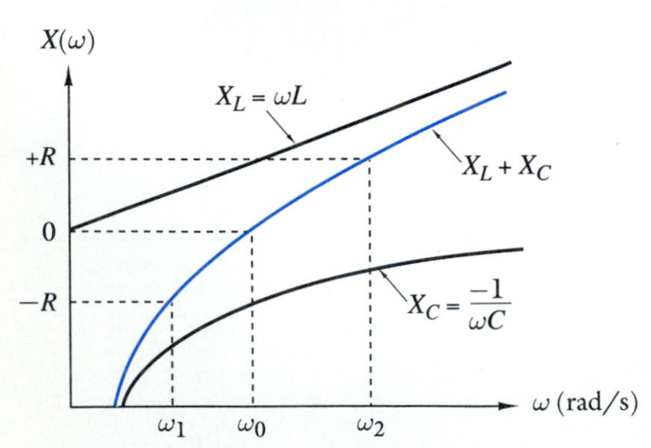

Figure 13-45 Reactance of the series RLC circuit.

Fig. 13-45 shows the variation $X(\omega)$ with frequency in graphical form. At low frequency the capacitor dominates so the reactance is negative. At high frequencies the inductor prevails and the reactance is positive. Between these two extremes there is a resonant frequency at which $X = 0$. Setting $X = 0$ in Eq. (13-33) yields a resonant frequency of

$$\omega_O = \frac{1}{\sqrt{LC}} \tag{13-34}$$

Figure 13-46 shows the variation of the magnitude of $Z(j\omega)$

$$|Z(j\omega)| = |R + jX| = \sqrt{R^2 + X^2}$$

with frequency. At the resonant frequency the reactance is zero and the driving-point impedance reduces to the resistance R. Above or below resonance the magnitude of the impedance increases because the reactance of the inductor and capacitor no longer cancel.

Several parameters are used to describe the resonance in more detail. First, the "quality" of the resonance is measured by a *quality factor* Q defined as the ratio of the inductive reactance at resonance to the resistance. Since $|X_L| = |X_C|$ at resonance, we find that Q is related to the circuit parameters as follows:

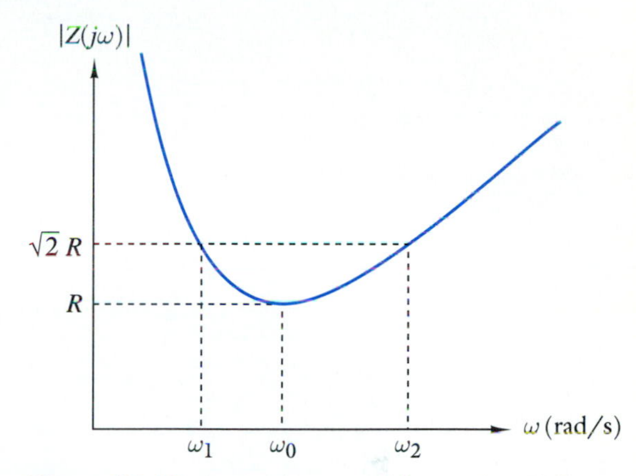

Figure 13-46 Impedance of the series RLC circuit.

$$Q = \frac{\omega_O L}{R} = \frac{1}{\omega_O RC} = \frac{\sqrt{L/C}}{R} \tag{13-35}$$

In Chapter 8 we show that the damping ratio of the series RLC circuit is related to circuit parameters by $2\zeta = R/\omega_O L$. Thus, the quality factor $Q = 1/2\zeta$ is inversely proportional to the damping ratio and can be a very large number when $\zeta \ll 1$ in circuits with little damping.

At resonance the net reactance is zero, so the current through the series circuit is $I = \mathbf{V_S}/R$, where $\mathbf{V_S}$ is the voltage source driving the series circuit in Fig. 13-44. At resonance the voltage across the inductor and capacitor are

$$\mathbf{V_L} = (j\omega_o L)\frac{\mathbf{V_S}}{R} = j\frac{\omega_O L}{R}\mathbf{V_S} = jQ\mathbf{V_S}$$

$$\mathbf{V_C} = \left(\frac{-j}{\omega_O C}\right)\frac{\mathbf{V_S}}{R} = -j\left(\frac{1}{\omega_O RC}\right)\mathbf{V_S} = -jQ\mathbf{V_S}$$

$\mathbf{V_L} = jQ\mathbf{V_S}$

$\mathbf{I}$ $\mathbf{V_R}$&$\mathbf{V_S}$

$\mathbf{V_C} = -jQ\mathbf{V_S}$

Figure 13-47 Phasor diagram of the series RLC circuit at resonance.

Fig. 13-47 shows the phasor diagram of the voltages in the series circuit at resonance. The fact that $\mathbf{V_L} + \mathbf{V_C} = 0$ at resonance does not mean the voltage across each reactive element is zero. In fact, the voltage across the inductor or capacitor can be much larger than the source voltage because Q can be a very large number.

The *resonant bandwidth* is defined as the frequency range over which the impedance does not change greatly from its value $Z = R + j0$ at resonance. By tradition the bandwidth is defined as the frequency range over which the magnitude of the reactance is less than resistance at resonance. The end points of the bandwidth occur when $Z = R \pm jR$, that is, when $|X| = R$. Fig. 13-45 shows that there are two such frequencies: (1) ω_1 below resonance where $X = -R$ and (2) ω_2 above resonance where $X = +R$. These two points are called the *resonant cutoff frequencies*

The cutoff frequencies can be expressed in terms of other parameters by equating the reactance in Eq. (13-33) to $\pm R$:

$$\left[\omega L - \frac{1}{\omega C}\right] = \pm R$$

which leads to the quadratic equation

$$\omega^2 LC \mp \omega RC - 1 = 0$$

At this point we express the coefficients in this quadratic in terms of the other descriptive parameters. To do so we note that $LC = 1/\omega_O^2$ and from Eq. (13-35) $RC = 1/Q\omega_O$, hence the quadratic equation can be rewritten as

$$\left(\frac{\omega}{\omega_O}\right)^2 \mp \frac{1}{Q}\left(\frac{\omega}{\omega_O}\right) - 1 = 0$$

Because of the $\mp$ sign this equation has four roots, but the only two that make physical sense are those that yield positive values.

$$\omega_1 = \omega_O \left[\sqrt{1 + \left(\frac{1}{2Q}\right)^2} - \frac{1}{2Q}\right]$$

$$\omega_2 = \omega_O \left[\sqrt{1 + \left(\frac{1}{2Q}\right)^2} + \frac{1}{2Q}\right] \tag{13-36}$$

Subtracting the two cutoff frequencies in Eq. (13-36) we find that the ***resonant bandwidth*** is

$$\text{BW} = \omega_2 - \omega_1 = \frac{\omega_O}{Q} \tag{13-37}$$

The bandwidth is inversely proportional to Q so that narrow bandwidths require a high value of Q. Multiplying the two cutoff frequencies in Eq. (13-36) yields

$$\omega_O = \sqrt{\omega_1 \omega_2} \tag{13-38}$$

which shows that the resonant frequency is the geometric mean of the cutoff frequencies.

Series resonance occurs when the reactance of the inductor and the reactance of the capacitor cancel. At the resonant frequency the current is limited only by the circuit resistance. As a result, the voltage across the inductor and across the capacitor can be much larger than the source voltage, even though the sum of these two voltages is zero. The series RLC circuit has five descriptive parameters (ω_O, Q, ω_1, ω_2, BW) and three circuit parameters (R, L, C). These parameters are interrelated by five equations: Eqs. (13-34), (13-35), (13-36), (13-37), and (13-38). Thus, only three of the eight parameters can be specified independently in an analysis or design problem involving series RLC resonance.

EXERCISE 13-20

Calculate the five descriptive parameters of the resonance in a series RLC circuit with $R = 20\ \Omega$, $L = 100$ mH, and $C = 2.5\ \mu$F.

ANSWERS:
$\omega_O = 2$ krad/s, $Q = 10$, $\omega_1 = 1902$ rad/s, $\omega_2 = 2102$ rad/s, and BW $= 200$ rad/s

EXERCISE 13-21

Calculate L, C, Q, ω_1, and ω_2 for a series RLC circuit with $R = 20\ \Omega$, $\omega_O = 100$ krad/s and BW $= 2$ krad/s.

ANSWERS:
$Q = 50$, $L = 10$ mH, $C = 10$ nF, $\omega_1 = 99$ krad/s, and $\omega_2 = 101$ krad/s

Parallel Resonance

The treatment of parallel resonance can be brief because the parallel RLC circuit is the dual of the series RLC circuit, which we treated above in some detail. The resonance in the parallel RLC circuit is described in terms of the element admittances. In general, the driving-point admittance seen at any pair of terminals can be written in the form

$$Y(j\omega) = G(\omega) + jB(\omega) \tag{13-39}$$

As first noted in Sect. 13-3, the real part of an admittance is called *conductance* and the imaginary part is called *susceptance*, both of which have the dimensions of seimens (S). The conductance and susceptance vary with frequency. For passive circuits the conductance is greater than 0 ($G(\omega) \geq 0$) while susceptance can be positive at some frequencies and negative at others. Analogous to when $X(\omega) = 0$, when $B(\omega) = 0$ the driving-point admittance is purely real and the circuit is in *resonance*. The frequency at which $B(\omega) = 0$ is a *resonant frequency*, which is denoted by ω_O.

The driving-point admittance of the parallel RLC circuit in Fig. 13-48 is

$$\begin{aligned} Y &= G + j\omega C + \frac{1}{j\omega L} \\ &= G + j\left[\omega C - \frac{1}{\omega L}\right] \\ &= G + jB(\omega) \end{aligned} \tag{13-40}$$

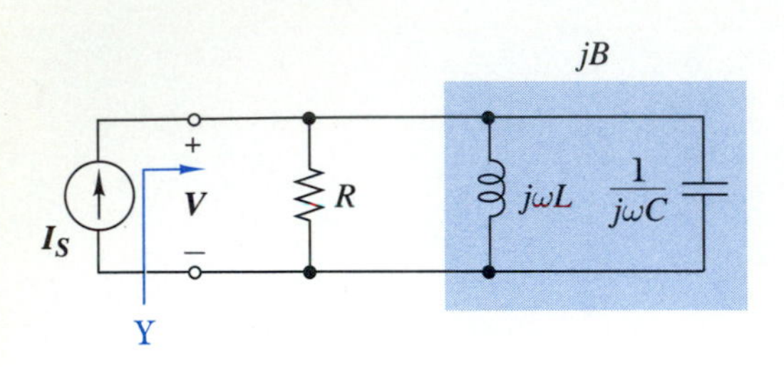

Figure 13-48 Parallel RLC circuit.

As with the series case, only the imaginary part of the driving-point function varies with frequency. At low frequency the inductor dominates and the susceptance is negative. Conversely at high frequency the capacitor dominates the admittance and the susceptance is positive. Between these extremes the susceptance is zero at the resonant frequency. Setting the susceptance B in Eq. (13-40) to zero yields a resonant frequency of

$$\omega_O = \frac{1}{\sqrt{LC}} \tag{13-41}$$

We observe that the resonant frequency for the parallel RLC circuit is the same as the resonant frequency in the series circuit. In fact, when we compare the admittance of the parallel circuit in Eq. (13-40) with the impedance of the series circuit in Eq. (13-32), we find that one driving-point function can be turned into the other by interchanging the circuit parameters as follows:

$$\begin{array}{ccc} \text{Series} & & \text{Parallel} \\ R & \Longleftrightarrow & G \\ L & \Longleftrightarrow & C \\ C & \Longleftrightarrow & L \end{array} \tag{13-42}$$

In other words, the series RLC and parallel RLC circuits are duals. We need not derive the equations for the parallel RLC circuit because they can be found by inserting the parameter interchanges in Eq. (13-42) into the Eqs. (13-34) through (13-38) for the series circuit. Using duality, the five descriptive parameters for resonance in the parallel RLC circuit are interrelated by the following equations:

Quality Factor: $$Q = \frac{\omega_O C}{G} = \frac{1}{\omega_O G L} = \frac{R}{\sqrt{L/C}} \tag{13-43}$$

Cutoff Frequencies: $$\omega_1 = \omega_O\left[\sqrt{1+\left(\frac{1}{2Q}\right)^2} - \frac{1}{2Q}\right]$$

$$\omega_2 = \omega_O\left[\sqrt{1+\left(\frac{1}{2Q}\right)^2} + \frac{1}{2Q}\right] \tag{13-44}$$

Bandwidth: $$\text{BW} = \omega_2 - \omega_1 = \frac{\omega_O}{Q} \tag{13-45}$$

Center Frequency: $$\omega_O = \sqrt{\omega_1 \omega_2} \tag{13-46}$$

These parameters describe parallel resonance in the same way as the corresponding series parameters. The quality factor is the ratio of susceptance to conductance at resonance and equals

$1/2\zeta$, where ζ is the damping ratio of the parallel RLC circuit. At resonance the current through the inductor and capacitor are

$$\mathbf{I}_L = -jQ\mathbf{I}_S$$

$$\mathbf{I}_C = jQ\mathbf{I}_S$$

where $\mathbf{I}_S$ is the current source driving the parallel circuit shown in Fig. 13-48. Since Q can be a very large number these results mean that the current in the loop formed by the inductor and capacitor can be much larger than the source current.

The resonance bandwidth is the range over which the magnitude of the susceptance is less than the conductance. The two cutoff frequencies and the resonant bandwidth are found by setting $B = \pm G$. Finally, Eq. (13-46) shows that the resonant frequency is the geometric mean of the cutoff frequencies. Note that the results in Eqs. (13-44), (13-45), and (13-46) are identical to those for the series circuit, because these equations only involve the descriptive parameters that are not changed by the duality transformation in Eq. (13-42).

EXERCISE 13-22

Calculate the five descriptive parameters of resonance in a parallel RLC circuit with $R = 10$ kΩ, $L = 50$ mH, and $C = 100$ nF.

ANSWERS:
$\omega_O = 14.1$ krad/s, $Q = 14.1$, $\omega_1 = 13.65$ krad/s, $\omega_2 = 14.65$ krad/s, and BW = 1 krad/s

EXERCISE 13-23

A parallel RLC circuit with $R = 10$ kΩ, $BW = 2$ krad/s, $\omega_O = 50$ krad/s is driven by a current source with $\mathbf{I}_S = 10\angle 0°\mu$A. Calculate the phasor current through each passive element at the resonant frequency.

ANSWERS:
$\mathbf{I}_R = 10\angle 0°\mu$A, $\mathbf{I}_L = 250\angle -90°\ \mu$A, $\mathbf{I}_C = 250\angle 90°\ \mu$A.

13-7 THE LINEAR TRANSFORMER

A linear transformer is a circuit element based on the mutual inductance coupling between two coils. In Chapter 7 we introduced the transformer using an ideal model that assumes perfect coupling between the coils. In this section we study the sinusoidal steady-state response of transformers using a model that does not assume perfect coupling.

Coupled coils form a two-port element whose i-v characteristics in the phasor domain are

$$\begin{aligned} \mathbf{V}_1 &= j\omega L_1\mathbf{I}_1 \pm j\omega M\mathbf{I}_2 \\ \mathbf{V}_2 &= \pm j\omega M\mathbf{I}_1 + j\omega L_1\mathbf{I}_2 \end{aligned} \qquad (13\text{-}47)$$

where:

1. V_1, I_1, and L_1 are the voltage, current and self-inductance of coil 1
2. V_2, I_2, and L_2 are the voltage, current, and self-inductance of coil 2
3. M is the mutual inductance of the pair of coils

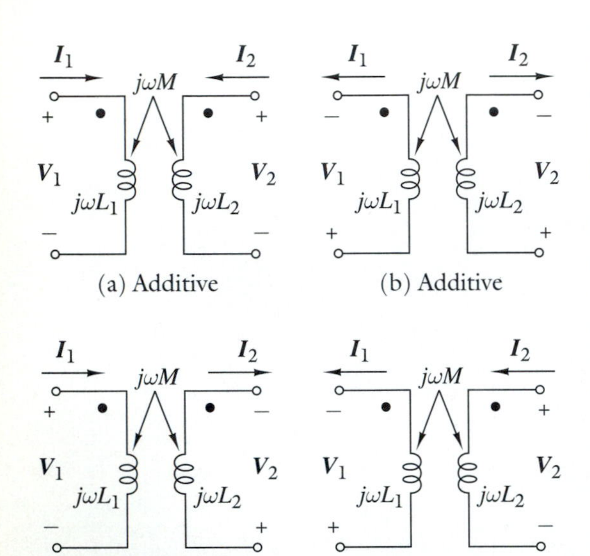

Figure 13-49 Voltage and current reference assignments for a pair of coupled coils.

These i-v relationships were first derived in the time domain in Chapter 7 and transformed into the s domain in Chapter 10. Replacing s by $j\omega$ in the s-domain equations from Chapter 10 (Example 10-2) leads directly to the phasor domain equations given in Eqs. (13-47) above.

The $\pm$ signs in Eq. (13-47) remind us that mutual inductance coupling can be either additive (the $+$ sign) or subtractive (the $-$ sign). The appropriate sign depends on the orientation of the reference marks assigned to the voltages and currents relative to dots indicating the spatial orientation of the two coils. Fig. 13-49 shows a given pair of coil dots and all possible reference mark assignments for the passive sign convention. The dot convention developed in Chapter 7 states that mutual coupling is additive when **both** current reference marks enter or leave a dotted terminal; otherwise, the coupling is subtractive. When applying this convention, it is critical to remember that the coil dots cannot be changed because they describe the physical orientation of the coils. The reference marks for voltage and current can be changed but must adhere to the passive sign convention for the dot convention stated above to apply.

The self-inductances relate voltage and current at the same port while the mutual inductance relates voltage at one port to the current at the other. The degree of coupling between coils is specified in terms of the coupling coefficient k:

$$k = \frac{M}{\sqrt{L_1 L_2}}$$

When deriving this relationship in Chapter 7 we found that energy considerations require that $0 \leq k \leq 1$. The ideal transformer model developed in Chapter 7 assumes that $k = 1$. In this chapter we use the coupled-coil model of the transformer and assume that the coupling coefficient is less than one.

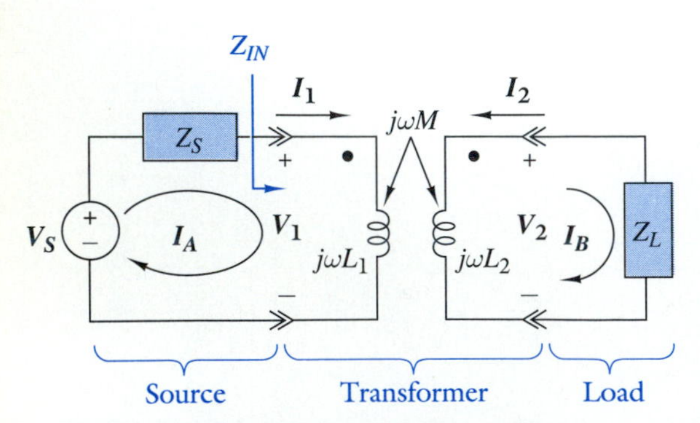

Figure 13-50 Phasor-domain model of transformer coupling.

Figure 13-50 shows a phasor-domain model for transformer coupling between a source and a load. The coil connected to the source is called the ***primary winding*** and the coil connected to the load is called the ***secondary winding***. We think of signal transfer passing from the primary to the secondary winding, although signals can pass in either direction.

Our immediate objective is to write circuit equations for the transformer using the mesh currents $\mathbf{I}_A$ and $\mathbf{I}_B$ in Fig. 13-50. Applying KVL around the primary circuit (Mesh A) and secondary circuit (Mesh B), we obtain the following equations:

$$\begin{aligned} \text{Mesh A:} \quad & Z_S \mathbf{I}_A + \mathbf{V}_1 = \mathbf{V}_S \\ \text{Mesh B:} \quad & -\mathbf{V}_2 + Z_L \mathbf{I}_B = 0 \end{aligned} \tag{13-48}$$

The reference directions for the coil currents in Fig. 13-50 are both directed in at dotted terminals, so the mutual inductance coupling is additive and the plus signs in Eq. (13-47) apply. By KCL, we see that the reference directions for the currents lead to the relations $\mathbf{I}_1 = \mathbf{I}_A$ and $\mathbf{I}_2 = -\mathbf{I}_B$. The i-v relationships of the coupled coils in terms of the mesh currents are

$$\begin{aligned} \mathbf{V}_1 &= +j\omega L_1 \mathbf{I}_A + j\omega M(-\mathbf{I}_B) \\ \mathbf{V}_2 &= +j\omega M \mathbf{I}_A + j\omega L_2(-\mathbf{I}_B) \end{aligned} \tag{13-49}$$

Substituting the coil voltages from Eq. (13-49) into the KVL constraints in Eq. (13-48) yields

$$\begin{aligned} \text{Mesh A:} \quad & (Z_S + j\omega L_1)\,\mathbf{I}_A - j\omega M \mathbf{I}_B = \mathbf{V}_S \\ \text{Mesh B:} \quad & -j\omega M \mathbf{I}_A + (Z_L + j\omega L_2)\,\mathbf{I}_B = 0 \end{aligned} \tag{13-50}$$

This set of mesh equations provides a complete description of the circuit response. Once we solve for the mesh currents we can calculate every voltage and current using Kirchhoff's laws and element equations.

The method used to develop Eq. (13-50) illustrates a general approach to the analysis of transformer circuits. The steps in the method are:

STEP 1: Write KVL equations around the primary and secondary circuits using assigned mesh currents, source voltages, and coil voltages.

STEP 2: Write the i-v characteristics of the coupled coils in terms of the mesh currents using the dot convention to determine whether the coupling is additive or subtractive.

STEP 3: Obtain mesh-current equations using the i-v relationships from Step (2) to eliminate the coil voltages from the KVL equations obtained in Step (1).

The next two examples illustrate this method of formulating mesh equations.

EXAMPLE 13-20

The source circuit in Fig. 13-50 has $Z_S = 600 + j800\ \Omega$ and $\boldsymbol{V_S} = 25\angle 0^\circ$ at $\omega = 1$ krad/s. The transformer has $L_1 = 450$ mH, $L_2 = 50$ mH, and $M = 100$ mH. The load is a 60-Ω resistor in series with a 5-μF capacitor. Find $\boldsymbol{I_A}$, $\boldsymbol{I_B}$, $\boldsymbol{V_1}$, $\boldsymbol{V_2}$, and the input impedance Z_{IN} seen by the source circuit.

SOLUTION:

The transformer and load impedances are

$$j\omega L_1 = j1000 \times 0.45 = j450\ \Omega$$

$$j\omega L_2 = j1000 \times 0.05 = j50\ \Omega$$

$$j\omega M = j1000 \times 0.1 = j100\ \Omega$$

$$Z_L = 60 - j\frac{1}{1000 \times 5 \times 10^{-6}} = 60 - j200\ \Omega$$

Using these impedances in Eqs. (13-50) yields the following mesh equations:

$$\text{Mesh A:}\quad (600 + j800 + j450)\boldsymbol{I_A} - j100\boldsymbol{I_B} = 25\angle 0^\circ$$

$$\text{Mesh B:}\quad -j100\boldsymbol{I_A} + (60 - j200 + j50)\boldsymbol{I_B} = 0\angle 0^\circ$$

Solving these equations for the mesh currents yields

$$\boldsymbol{I_A} = 7.425 \times 10^{-3} - j15.6 \times 10^{-3} = 17.3\angle{-64.5^\circ}\ \text{mA}$$

$$\boldsymbol{I_B} = -0.685 \times 10^{-3} + j10.7 \times 10^{-3} = 10.7\angle 93.7^\circ\ \text{mA}$$

The coil voltages are found to be

$$\boldsymbol{V_1} = \boldsymbol{V_S} - Z_S\boldsymbol{I_A} = 8.08 + j3.41 = 8.77\angle 22.9^\circ\ \text{V}$$

$$\boldsymbol{V_2} = Z_L\boldsymbol{I_B} = 2.10 + j0.779 = 2.24\angle 20.4^\circ\ \text{V}$$

The input impedance seen by the source circuit is

$$Z_{IN} = \frac{\boldsymbol{V_1}}{\boldsymbol{I_A}} = \frac{8.08 + j3.41}{(7.43 - j15.6) \times 10^{-3}} = 23.0 + j507.5\ \Omega$$

EXAMPLE 13-21

Find $\boldsymbol{I_A}$, $\boldsymbol{I_B}$, $\boldsymbol{V_1}$, $\boldsymbol{V_2}$, and the impedance Z_{IN} seen by the source circuit in Fig. 13-51.

SOLUTION:

The KVL equations around Meshes A and B are

$$\text{Mesh A:}\quad (50 - j75)\boldsymbol{I_A} - (-j75)\boldsymbol{I_B} + \boldsymbol{V_1} = 100\angle 0^\circ$$

$$\text{Mesh B:}\quad -(-j75)\boldsymbol{I_A} + (600 - j75)\boldsymbol{I_B} - \boldsymbol{V_2} = 0\angle 0^\circ$$

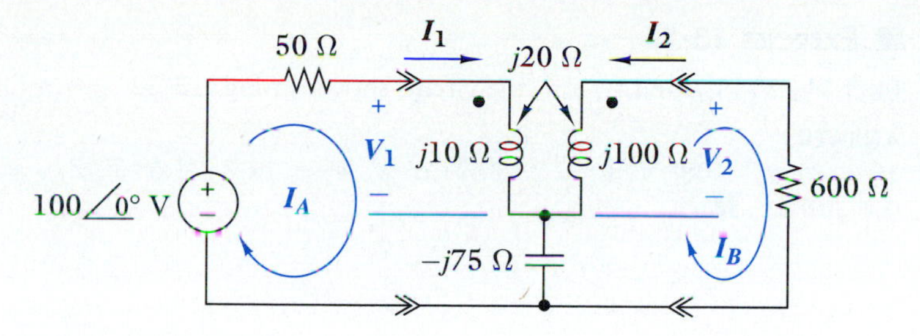

Figure 13-51

For the assigned reference directions the mutual coupling is additive and $I_1 = I_A$ and $I_2 = -I_B$. The coil i-v relations in terms of the mesh currents are

$$V_1 = j10I_A + j20(-I_B)$$

$$V_2 = j20I_A + j100(-I_B)$$

Using these equations to eliminate the coil voltages from the KVL equations yields the following mesh equations:

Mesh A: $(50 - j75 + j10)I_A + (j75 - j20)I_B = 100\angle 0^\circ$

Mesh B: $(j75 - j20)I_A + (600 - j75 + j100)I_B = 0\angle 0^\circ$

Solving these equations for the mesh currents produces

$$I_A = 0.756 + j0.896 = 1.17\angle 49.8^\circ \text{ A}$$

$$I_B = 0.0791 - j0.0726 = 0.107\angle -42.5^\circ \text{ A}$$

Given the mesh currents we find the coil voltages from the i-v relations:

$$V_1 = j10I_A - j20I_B = -10.4 + j5.98 = 12.0\angle 150^\circ \text{ V}$$

$$V_2 = j20I_A - j100I_B = -25.2 + j7.21 = 25.8\angle 164^\circ \text{ V}$$

Finally, the impedance seen by looking into the input interface is

$$Z_{IN} = \frac{V_S - 50I_A}{I_A} = 5.06 - j65.2 \ \Omega$$

EXERCISE 13-24

Repeat Example 13-20 when the dot on the secondary coil in Fig. 13-50 is moved to the bottom terminal and all other reference marks stay the same.

ANSWERS:

$I_A = 17.3\angle -64.5^\circ$ mA; $I_B = 10.7\angle -86.3^\circ$ mA

$V_1 = 8.77\angle 22.9^\circ$ V; $V_2 = 2.23\angle 159.6^\circ$ V; $Z_{IN} = 23.0 + j507\ \Omega$

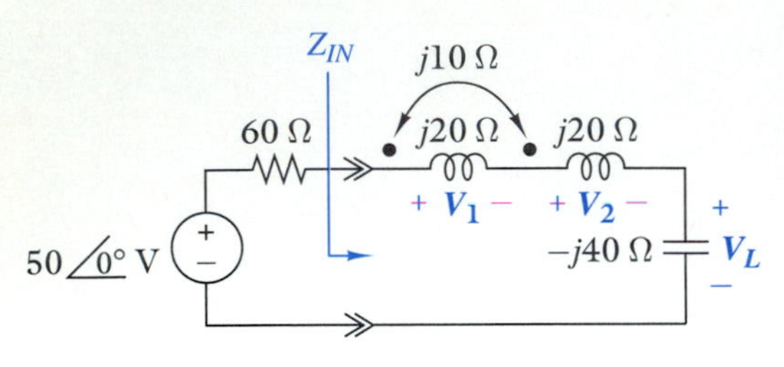

Figure 13-52

■ **EXERCISE 13-25**

Find V_L, V_1, V_2, and Z_{IN} in the circuit shown in Fig. 13-52.

ANSWERS:

$V_L = 31.6\angle{-108°}$ V, $V_1 = 23.7\angle 71.6°$ V, $V_2 = 23.7\angle 71.6°$ V, $Z_{IN} = 0 + j20\ \Omega$ ■

Reflected Impedance

The transformer in Fig. 13-50 serves as a coupling device that facilitates signal transfer between the source and the load. The effect of the transformer on source-to-load coupling can be seen by finding the input impedance Z_{IN} seen by the source. Without the transformer the input impedance is equal to the load impedance Z_L. What impedance does the source see when the transformer is inserted between the source and load?

To find $Z_{IN} = V_1/I_1$ with the transformer in place we first write the element equations for the transformer in Fig. 13-50:

$$\begin{aligned} V_1 &= +j\omega L_1 I_1 + j\omega M I_2 \\ V_2 &= +j\omega M I_1 + j\omega L_2 I_2 \end{aligned} \qquad (13\text{-}51)$$

The secondary coil voltage can be expressed in terms of the load impedance, since

$$V_2 = I_B Z_L = (-I_2) Z_L \qquad (13\text{-}52)$$

Using Eq. (13-52) to eliminate V_2 from the second i-v relation in Eq. (13-51), we find that I_2 is

$$I_2 = \frac{-j\omega M}{Z_L + j\omega L_2} I_1 \qquad (13\text{-}53)$$

Note that I_2 is zero when $\omega = 0$, which points out that a transformer does not pass dc signals. When I_2 in Eq. (13-53) is substituted into the first i-v relation in Eq. (13-51) we obtain

$$V_1 = \left[j\omega L_1 + j\omega M \frac{-j\omega M}{Z_L + j\omega L_2} \right] I_1 \qquad (13\text{-}54)$$

Therefore, the impedance seen by the source is

$$Z_{IN} = \frac{V_1}{I_1} = j\omega L_1 + \frac{(\omega M)^2}{j\omega L_2 + Z_L} \qquad (13\text{-}55)$$

The first term on the right in Eq. (13-55) is the impedance of the self-inductance of the primary winding. The self-inductance term is present in Z_{IN} even when there is no mutual coupling ($M = 0$). The second term on the right is called the *reflected impedance* because it represents the equivalent impedance inserted in the primary circuit as the result of the mutual coupling:

$$Z_R = \frac{(\omega M)^2}{j\omega L_2 + Z_L} \tag{13-56}$$

Although Eq. (13-56) was derived using additive coupling, the fact that the numerator is squared means that the result applies to subtractive coupling as well.

Figure 13-53 shows that the equivalent circuit for the primary winding is a series connection of the self-impedance of the primary and the reflected impedance due to mutual inductance coupling. When the load impedance is written as $Z_L = R_L + jX_L$ the reflected impedance becomes

$$\begin{aligned} Z_R &= \frac{(\omega M)^2}{R_L + j(X_L + \omega L_2)} \\ &= \left[\frac{(\omega M)^2}{R_L^2 + (X_L + \omega L_2)^2}\right][R_L - j(X_L + j\omega L_2)] \end{aligned} \tag{13-57}$$

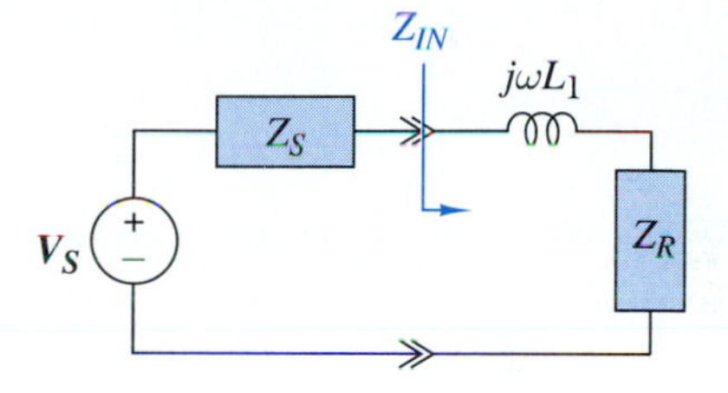

Figure 13-53

and the reflected resistance and reactance are

$$\begin{aligned} R_R &= \left[\frac{(\omega M)^2}{R_L^2 + (X_L + \omega L_2)^2}\right] R_L \\ X_R &= -\left[\frac{(\omega M)^2}{R_L^2 + (X_L + \omega L_2)^2}\right](X_L + \omega L_2) \end{aligned} \tag{13-58}$$

Both the reflected resistance R_R and reactance X_R are proportional to the positive scale factor $(\omega M)^2/[R_L^2 + (X_L + \omega L_2)^2]$. The reflected resistance is always positive, which reminds us that the transformer is a passive element. When $X_L + \omega L_2 = 0$ the self-impedance of the secondary circuit is in resonance and the reflected impedance reduces to a resistance $Z_R = R_R = (\omega M)^2/R_L$.

Because reflected reactance has a minus sign in Eq. (13-58) we see that X_R is negative when the secondary circuit reactance $X_L + \omega L_2$ is positive, and vice versa. In other words, the transformer reflects an impedance into the primary circuit which is a positive scale factor times the conjugate of the self-impedance of the secondary circuit.

EXERCISE 13-26

(a) Find the reflected impedance in the primary of the circuit in Example 13-20.

(b) At what frequency would this impedance be purely real?

ANSWERS:

(a) $Z_R = 22.99 + j57.47\ \Omega$

(b) 2 krad/s

Equivalent Circuits

In some analysis problems it may be advantageous to replace a pair of coupled coils by an equivalent circuit, which does not involve mutual inductance. Two circuits are equivalent if they have the same i-v characteristics at specified terminal pairs. Since coupled coils, involve two terminal pairs, equivalence requires that the i-v characteristics at both pairs be identical. There are several equivalent circuits for coupled coils, but we will only treat two of the simpler examples to illustrate the process.

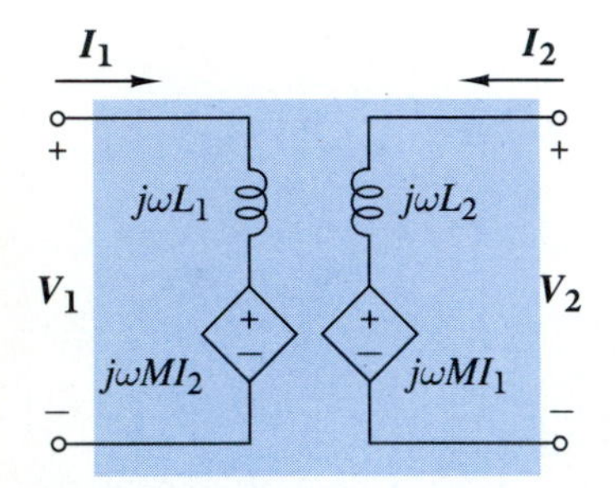

Figure 13-54 Controlled source equivalent circuit.

Figure 13-54 shows an equivalent circuit involving the self-inductances in series with current controlled voltages sources. Writing the voltage across each coil in terms of the current produces the i-v characteristics of the controlled source model:

$$\boldsymbol{V}_1 = j\omega L_1 \boldsymbol{I}_1 + j\omega M \boldsymbol{I}_2$$

$$\boldsymbol{V}_2 = j\omega L_2 \boldsymbol{I}_2 + j\omega M \boldsymbol{I}_1$$

These relations are identical to those in Eqs. (13-47) for two coils with additive coupling. To obtain a model for subtractive coupling we reverse the polarity of the two dependent sources.

Figure 13-55 shows a *T-equivalent circuit* involving three uncoupled inductances. Using the currents $\boldsymbol{I}_A$ and $\boldsymbol{I}_B$ we can write mesh equations for the T circuit by inspection as

$$\boldsymbol{V}_1 = [j\omega(L_1 \mp M) \pm j\omega M]\boldsymbol{I}_A - (\pm j\omega M)\boldsymbol{I}_B$$

$$-\boldsymbol{V}_2 = -(\pm j\omega M)\boldsymbol{I}_A + [j\omega(L_2 \mp M) \pm j\omega M]\boldsymbol{I}_B$$

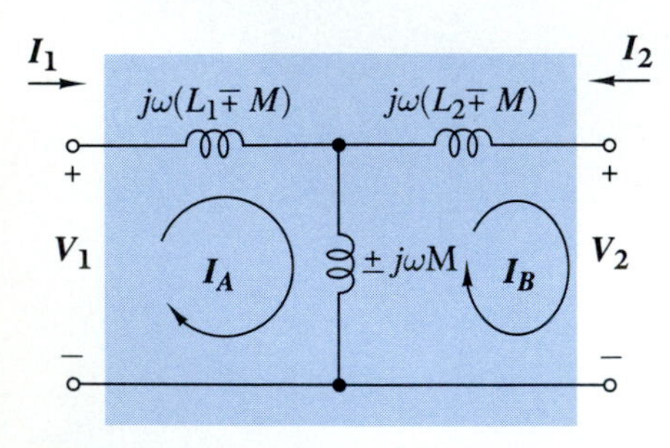

Figure 13-55 T-equivalent circuit.

But the coil currents are related to the mesh currents as $\boldsymbol{I}_1 = \boldsymbol{I}_A$ and $\boldsymbol{I}_2 = -\boldsymbol{I}_B$, so these mesh equations reduce to

$$\boldsymbol{V}_1 = j\omega L_1 \boldsymbol{I}_1 \pm j\omega M \boldsymbol{I}_2$$

$$\boldsymbol{V}_2 = \pm j\omega M \boldsymbol{I}_1 + j\omega L_2 \boldsymbol{I}_2$$

The equations are also identical to the coupled-coil relations in Eq. (13-47).

For additive coupling the T-equivalent circuit involves three uncoupled inductances whose values are $L_1 - M$, $L_2 - M$, and M. For subtractive coupling the three inductances are $L_1 + M$, $L_2 + M$, and $-M$. Thus, in general one of the inductances in the T-equivalent circuit is negative, which means that the circuit is not physically realizable. However, the negative inductance does not present a problem in analysis situations where negative inductance can be easily handled. For example, SPICE allows negative element values including negative inductances.

EXERCISE 13-27

Calculate the coupling coefficient and the T-equivalent circuit inductances of the coupled coils in Example 13-20.

ANSWERS:
0.667, 350 mH, −50 mH, 100 mH

13-8 AC CIRCUIT ANALYSIS USING SPICE

The SPICE ac circuit analysis option described in Chapter 12 can solve any of the numerical phasor circuit analysis problems presented in the chapter. In the SPICE syntax the ac analysis command has the form

```
.AC <sweep> NFREQ FSTART FSTOP
```

Phasor circuit analysis involves a single frequency f, which is specified in the .AC command statement by setting

```
NFREQ = 1 and FSTART = FSTOP = f (in hertz)
```

SPICE requires the frequency to be specified in hertz, whereas analysis problems usually specify the angular frequency $\omega = 2\pi f$ in rad/s. When only one value of frequency is involved the sweep control parameter in .AC analysis command has no effect and can be any one of the three types discussed in Chapter 12 (LIN, OCT, DEC).

The output file produced by an ac analysis run contains the phasors representing the sinusoidal steady-state responses of the circuit. The phasors can be expressed in either polar or rectangular form. The desired form is specified in a print command by adding one of the letters M, P, R, or I to the end of a variable name. The letter M stands for magnitude, P for phase, R for real part, and I for imaginary part. For example, the print statement

```
.PRINT AC VM(5),VP(5),IR(R2),II(R2)
```

produces an output file containing the magnitude VM(5) and phase VP(5) of the phasor voltage at Node No. 5, and the real part IR(R2) and the imaginary part II(R2) of the phasor current through resistor R_2. When no suffix letter is added to a variable name the default assignment is magnitude.

SPICE solves for voltage and current phasors. Circuit impedances must be calculated in the postanalysis phase using the phasor responses in the output file. If the purpose of a SPICE analysis is to determine the impedance at an interface, the user must specify the voltage and current at the interface as output phasors, so impedance can be calculated in the postanalysis phase. In addition, the reference directions for the voltage and current phasors must follow the passive sign convention for the ratio $\mathbf{V}/\mathbf{I}$ to be the impedance looking in at the interface.

The following examples illustrate SPICE ac circuit analysis.

EXAMPLE 13-22

Given the circuit in Fig. 13-56, use SPICE to find the driving-point impedance at the input interface and the amplitude of the voltage delivered to the output 50-Ω load resistor for $v_s = 10 \cos 10^5 t$ V.

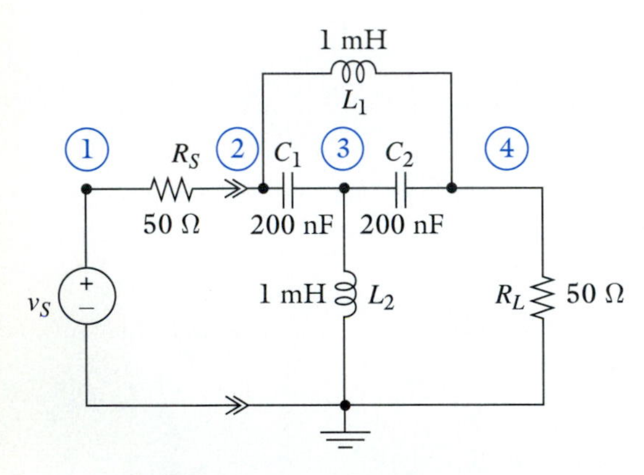

Figure 13-56

SOLUTION:

To calculate the required quantities we need the magnitude and phase of the current and voltage at the input interface, and the magnitude of the voltage at Node 4. Executing the following SPICE input file yields the required quantities:

```
EXAMPLE 13-22
VS   1  0  AC   10
RS   1  2  50
C1   2  3  200N
L1   2  4  1M
L2   3  0  1M
C2   3  4  200N
RL   4  0  50
.AC   LIN  1  15915  15915
.PRINT  AC VM(2)  VP(2)  IM(RS)  IP(RS)  VM(4)
.END
```

Note that the frequency $f = 15915 = 10^5/2\pi$ in the .AC command must be in hertz. The relevant entries in the SPICE output file are

```
FREQ         VM(20)       VP(20)       IM(RS)       IP(RS)       VM(4)
1.592E+04    9.618E+00    8.975E+00    3.163E-02    -7157E+01    1.581E-00
```

Using these output values we find that the input impedance is

$$Z_{IN} = \frac{9.619\angle 8.975^\circ}{0.03163\angle -71.57^\circ} = 49.96 + j300 \ \Omega$$

The magnitude of the voltage across the 50-Ω load is VM(4) = 1.581 V.

EXAMPLE 13-23

Given the circuit in Fig. 13-57, use SPICE to find the Thévenin equivalent circuit at the output interface for $v_s = 37 \cos 2500t$ V and $i_s = 0.75 \sin 2500t$ A.

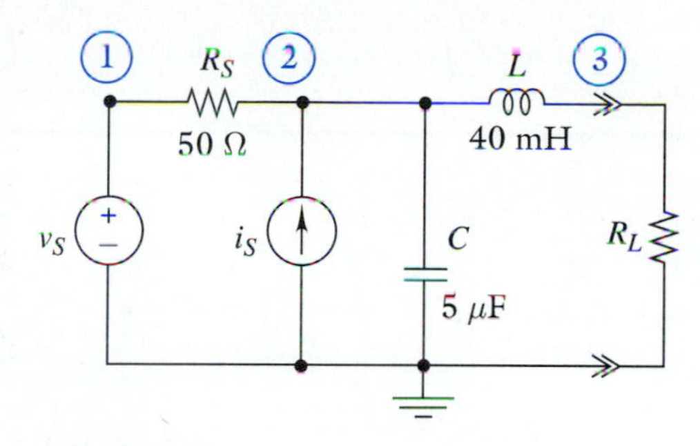

Figure 13-57

SOLUTION:

To find the Thévenin equivalent we need the magnitude and phase of the open-circuit voltage and the magnitude and phase the short-circuit current at the interface. Executing the following SPICE circuit file yields these quantites:

```
EXAMPLE 13-23
VS   1   0   AC   37   0
RS   1   2   50
C    2   0   5U
L    2   3   40M
IS   2   0   AC   0.75   90
RL   3   0   to be assigned
.AC   LIN   1   397.9   397.9
.PRINT   AC   to be assigned
.END
```

To obtain the open-circuit voltage we use the following assignments in the RL element statement and .PRINT statement:

```
RL   3   0   100MEG
.PRINT   AC   VM(3)   VP(3)
```

The RL element statement connects a 100-MΩ resistor across the interface, so the node voltage V(3) is essentially the open-circuit voltage. For this assignment the relevant data in the SPICE output file are

```
FREQ          VM(3)         VP(3)
3.979E+02     4.467E+01     1.337E+01
```

To obtain the short-circuit current we use the following assignments in the RL element statement and .PRINT statement:

```
RL   3   0   1U
.PRINT   AC   IM(RL)   IP(RL)
```

The RL element statements connects a 1-$\mu\Omega$ resistor across the interface so that the current through R_L is essentially the short-circuit current. For this assignment the relevant data in the SPICE output file are

```
FREQ          IM(RL)        IP(RL)
3.979E+02     5.227E-01     -5.174E+01
```

Using the values in the output files the parameters of the Thévenin equivalent circuit are found to be

$$V_T = 44.67\angle 13.37^\circ$$

$$Z_T = \frac{44.67\angle 13.37^\circ}{0.5227\angle -51.74^\circ} = 85.46\angle 65.11^\circ$$

EXAMPLE 13-24

The source v_{S1} in Fig. 13-58 is a 60-Hz sinusoid with a peak amplitude of 200 V. The source v_{S2} is a 180-Hz sinusoid whose peak amplitude is 10 V. The amplitude of the 180-Hz third harmonic is 5% of the 60-Hz fundamental. The purpose of the LC low-pass filter is to reduce the third harmonic content of the signal delivered to the load. Use SPICE to calculate the percent third harmonic in the load voltage.

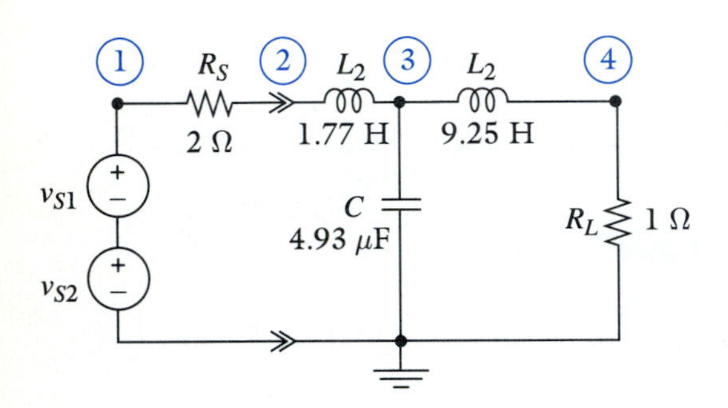

Figure 13-58

SOLUTION:

This problem involves two frequencies (60 Hz and 180 Hz), so we need two SPICE analysis runs. A SPICE circuit file for response to 60-Hz fundamental reads as follows:

```
EXAMPLE 13-24
VS  1  0  AC  200 0
RS  1  2  2
L1  2  3  1.77
C   3  0  4.93U
L2  3  4  9.25
Rl  4  0  1
.AC  LIN  1  60  60
.PRINT  AC  VM(4)
.END
```

The relevant portion of the SPICE output file for the 60-Hz input is

```
FREQ           VM(4)
6.000E+01      1.153E+02
```

To find the 180-Hz third harmonic response element statement for the source and the .AC analysis command are changed to

```
VS  1  0  AC  10  0
.AC  LIN  1  180  180
```

The relevant portion of the SPICE output file for the 180-Hz input is

```
FREQ           VM(4)
1.800E+02      9.587E-03
```

The percentage of third harmonic in the filter output is

$$\% \text{ 3rd harmonic} = \frac{0.009587}{115.3} 100\% = 0.00831\%$$

Thus the filter has virtually eliminated the 180-Hz component in the output.

SUMMARY

- A phasor is a complex number representing a sinusoidal waveform. The magnitude and angle of the phasor correspond to the amplitude and phase angle of the sinusoid. The phasor does not provide frequency information.
- Phasor analysis is an easy way to add two or more sinusoids of the same frequency. Phasors can be added only if they represent sinusoids with the same frequency. In the sinusoidal steady state, phasor currents, and voltages obey Kirchhoff's laws.
- In the sinusoidal steady state, the element i-v relationships are written in terms of an impedance or admittance. The sinusoidal steady-state response of a linear circuit can be found in a manner analogous to resistance circuits using the element impedances (or admittances) and Kirchhoff's laws.
- Phasor circuit analysis techniques include series equivalence, parallel equivalence, Y-Δ transformations, circuit reduction, Thévenin's and Norton's theorems, unit output method, superposition, node-voltage analysis, and mesh-current analysis.
- In the sinusoidal steady state, the driving-point impedance at a pair of terminals is $Z(j\omega) = R(\omega) + jX(\omega)$ where $R(\omega)$ is called resistance and $X(\omega)$ is called reactance. The driving-point admittance is $Y(j\omega) = G(\omega) + jB(\omega)$ where $G(\omega)$ is called conductance and $B(\omega)$ susceptance.
- A frequency at which a driving-point impedance or admittance is purely real is called a resonant frequency. The resonant frequency, bandwidth, cutoff frequencies, and quality factor are parameters that describe the resonance in a series or parallel RLC circuit. At the resonant frequency the current in (voltage across) a series (parallel) RLC circuit is determined by the driving source and the circuit resistance.
- A linear transformer is a two-port circuit element based upon the mutual inductance coupling between two coils. Linear transformers can be analyzed using a mesh-current approach. The impedance reflected into the primary of a

linear transformer is proportional to the conjugate of the self-impedance of the secondary circuit.

EN ROUTE OBJECTIVES AND ASSOCIATED PROBLEMS

ERO 13-1 Sinusoids and Phasors (Sect. 13-1)

(a) Given two or more sinusoids of the same frequency, convert the waveforms or their sum into phasor form.

(b) Given two or more phasors representing sinusoids of the same frequency; convert the phasors or their sum into waveforms.

13-1 Transform the following waveforms into phasor form. Use the phasors to find $v_1(t) + v_2(t)$. Construct a phasor diagram for these voltages and their sum.
(a) $v_1(t) = 100\cos(\omega t - 90°)$
(b) $v_2(t) = 50\cos\omega t + 200\sin\omega t$

13-2 Transform the following waveforms into phasor form. Use these phasors to find $i_1(t) + i_2(t)$. Construct the phasor diagram for these currents and their sum.
(a) $i_1(t) = 3\cos\omega t + 4\sin\omega t$
(b) $i_2(t) = 5\cos(\omega t - 223°)$

13-3 Convert the following phasors into sinusoidal waveforms at a frequency of 100 rad/s.
(a) $\mathbf{V}_1 = 10e^{j30°}$
(b) $\mathbf{V}_2 = 60e^{-j270°}$
(c) $\mathbf{I}_1 = 5e^{j180°}$
(d) $\mathbf{I}_2 = \dfrac{1}{100}e^{-j70°}$

13-4 Use the phasors in Prob. 13-3 to find the waveforms $v_1(t)+v_2(t)$ and $i_1(t) + i_2(t)$. Construct phasor diagrams for these sums.

13-5 Convert the following phasors into sinusoidal waveforms at a frequency of 200 rad/s.
(a) $\mathbf{V}_1 = \dfrac{10 + j10}{2 - j3}$
(b) $\mathbf{V}_2 = (3 - j8)5e^{-j60°}$
(c) $\mathbf{I}_1 = \dfrac{10}{1 + j3}$
(d) $\mathbf{I}_2 = \dfrac{1 + j3}{1 - j3}$

13-6 Given the waveforms

$$v_1(t) = 50\cos(\omega t - 45°) \quad \text{and} \quad v_2(t) = 25\sin\omega t$$

use phasors to find the waveform $v_3(t)$ such that $v_1 + v_2 + v_3 = 0$.

13-7 Given the phasor diagram in Figure P13-7, write time-domain expressions for the waveforms $v_1(t)$, $v_2(t)$, and $v_3(t) = v_1(t) + v_2(t)$.

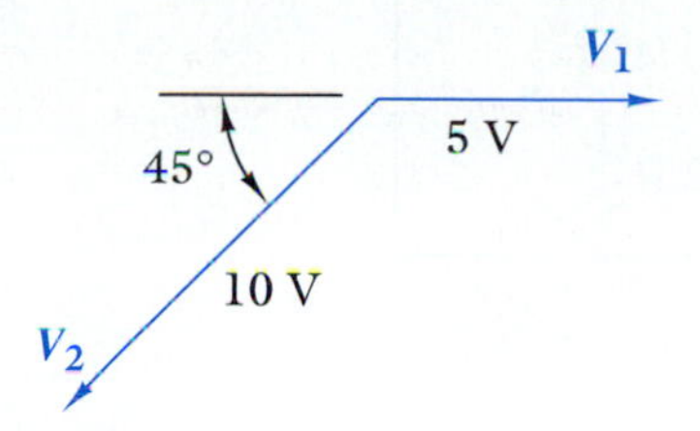

Figure P13-7

13-8 The phasor $V_1 = 2 + j6$ is rotated clockwise by 60°. Express the resulting phasor in rectangular form.

13-9 Given a phasor $V_1 = -3 + j4$, use phasor methods to find a voltage $v_2(t)$ that leads $v_1(t)$ by 90° and has an amplitude of 10 V.

ERO 13-2 Basic Phasor Circuit Analysis (Sects. 13-2, 13-3, 13-4)

Given a linear circuit operating in the sinusoidal steady state: Find driving-point impedances and phasor responses using basic analysis methods such as series and parallel equivalence, voltage and current division, circuit reduction, Thévenin's or Norton's theorem, and proportionality or superposition.

13-10 Express the impedance of the following circuits in rectangular and polar form.

(a) A 50-Ω resistor in series with a 20-mH inductor at $\omega = 2000$ rad/s

(b) A 50-Ω resistor in parallel with a 200-nF capacitor at $\omega = 200{,}000$ rad/s

(c) A 100-mH inductor in series with a 1-μF capacitor at $\omega = 2000$ rad/s

(d) Repeat (c) at $\omega = 4000$ rad/s.

13-11 A voltage $v(t) = 10 \cos 2500t$ V is applied across a series connection of a 100-Ω resistor and 40-mH inductor. Find the steady-state current $i(t)$ through the series connection. Draw a phasor diagram showing $\boldsymbol{V}$ and $\boldsymbol{I}$.

13-12 A voltage $v(t) = 50 \cos(1000t - 45°)$ V is applied across a parallel connection of a 1-kΩ resistor and a 200-nF capacitor. Find the steady-state current $i_C(t)$ through the capacitor and the steady-state current $i_R(t)$ through the resistor. Draw a phasor diagram showing $\boldsymbol{V}$, $\boldsymbol{I}_C$, and $\boldsymbol{I}_R$.

13-13 A current source delivering $i(t) = 300 \cos 377t$ mA is connected across a parallel combination of a 100-kΩ resistor and a 50-nF capacitor. Find the steady-state current $i_R(t)$ through the resistor and the steady-state current $i_C(t)$ through the capacitor. Draw a phasor diagram showing $\boldsymbol{I}$, $\boldsymbol{I}_C$, and $\boldsymbol{I}_R$.

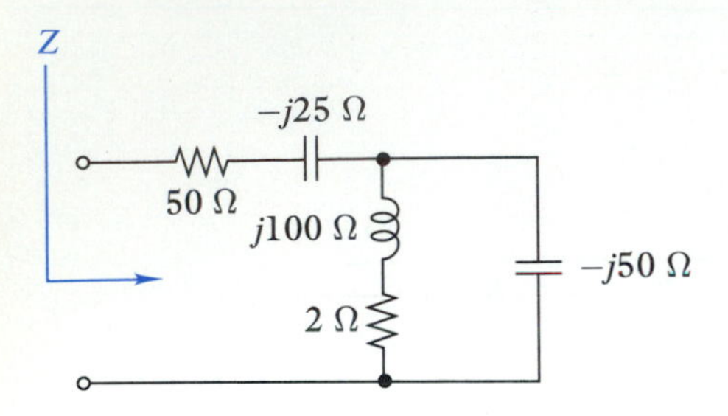

Figure P13-14

13-14 Find the driving-point impedance Z in Fig. P13-14. Express the result in both polar and rectangular form.

13-15 Find the driving-point impedance Z in Fig. P13-15. Express the result in both polar and rectangular form.

13-16 The circuit in Fig. P13-16 is operating in the sinusoidal steady state with $\omega = 377$ rad/s. Find the driving-point impedance Z. Express the result in polar and rectangular form. Repeat for $\omega = 6 \times 10^5$ rad/s.

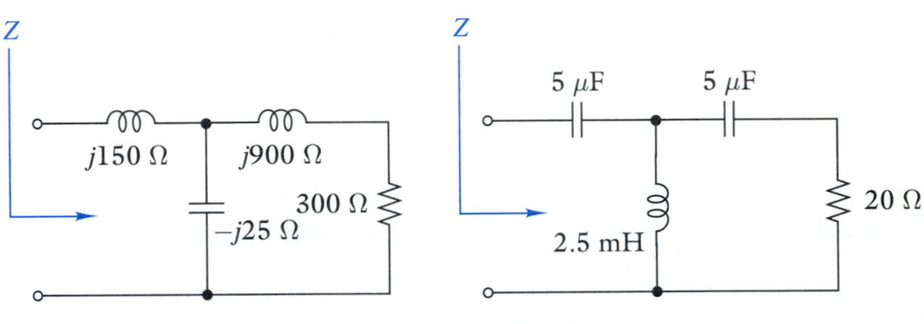

Figure P13-15

Figure P13-16

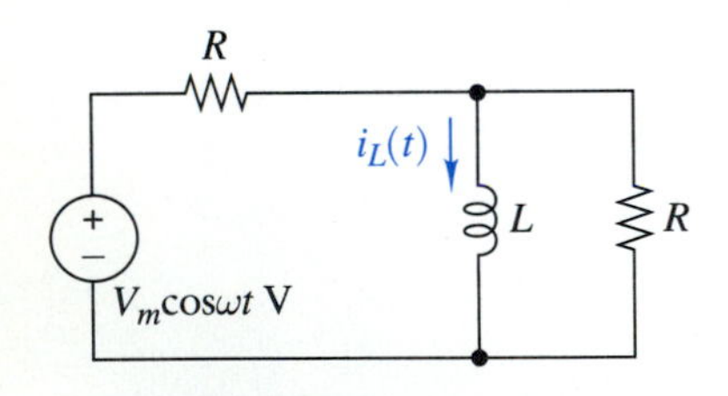

Figure P13-18

13-17 The voltage applied at the input to a linear circuit is $v(t) = 200\ \cos(1000t + 45°)$ V. In the sinusoidal steady state the input current is observed to be $i(t) = 20\ \sin\ 1000t$ mA. (a) Find the driving-point impedance at the input. (*b*) Find the steady-state current $i(t)$ for $v(t) = 150\ \sin(1000t - 70°)$.

13-18 The circuit in Fig. P13-18 is operating in the sinusoidal steady state. Use circuit reduction to derive an expression for the steady-state response $i_L(t)$.

13-19 The circuit in Fig. P13-19 is operating in the sinusoidal steady state. Use circuit reduction to derive an expression for the steady-state response $v_R(t)$.

13-20 The circuit in Fig. P13-20 is operating in the sinusoidal steady state. Use circuit reduction to find the input impedance seen by the voltage source and the steady-state response $v_X(t)$.

13-21 The circuit in Fig. P13-21 is operating in the sinusoidal steady state. Use circuit reduction to find the input impedance seen by the current source and steady-state response $v_X(t)$.

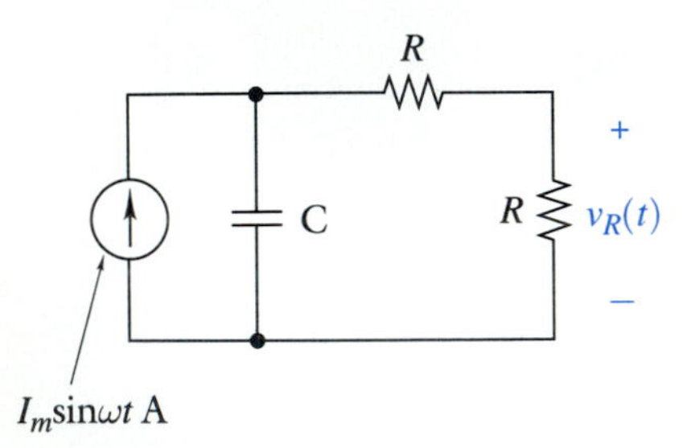

Figure P13-19

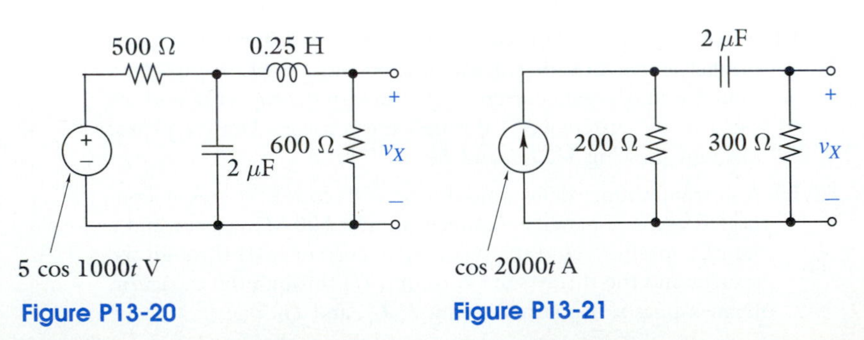

Figure P13-20

Figure P13-21

13-22 The circuit in Fig. P13-22 is operating in the sinusoidal steady state. Use circuit reduction to find the input impedance seen by the voltage source and the steady-state phasor response $\mathbf{V}_X$.

13-23 The circuit in Fig. P13-23 is operating in the sinusoidal steady state. Use superposition to find the phasor response $\mathbf{V}_X$.

13-24 The circuit in Fig. P13-24 is operating in the sinusoidal steady state. Use superposition to find the phasor response $\mathbf{I}_X$.

13-25 The circuit in Figure P13-25 is operating in the sinusoidal steady state. Use superposition to find the response waveforms $i_X(t)$ and $v_X(t)$. Note: *The sources do not have the same frequency.*

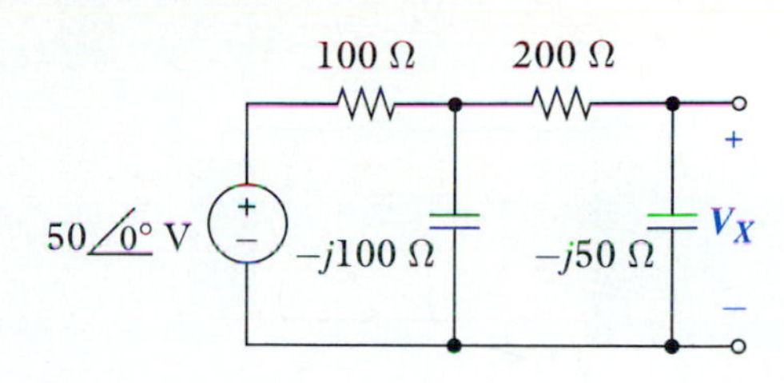

Figure P13-22

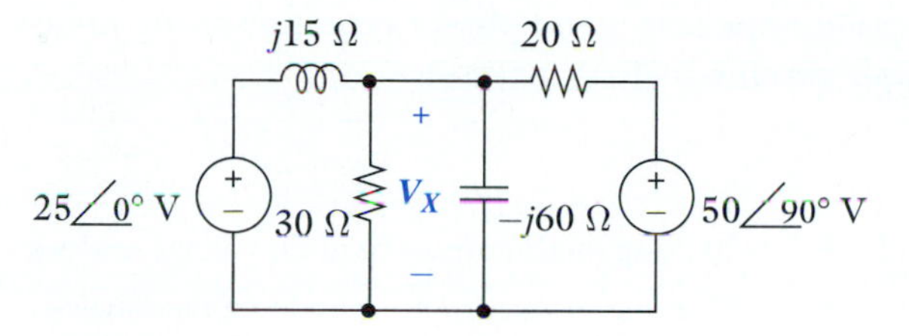

Figure P13-23

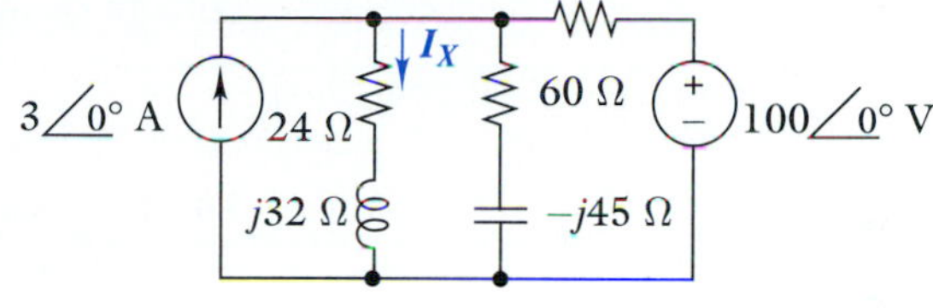

Figure P13-24

13-26 Repeat Prob. 13-20 using the unit output method.

13-27 Repeat Prob. 13-22 using the unit output method.

13-28 Repeat Prob. 13-23 using source transformations.

13-29 Repeat Prob. 13-24 using source transformations.

13-30 Repeat Prob. 13-23 using Thévenin's theorem.

13-31 Repeat Prob. 13-24 using Thévenin's theorem.

13-32 Find the phasor Thévenin equivalent of the source circuit to the left of the interface in Fig. P13-32. Then use the equivalent circuit to find the voltage $v(t)$ and current $i(t)$ delivered to the load.

13-33 Find the phasor Thévenin equivalent of the source circuit to the left of the interface in Fig. P13-33. Then use the equivalent circuit to find the phasor voltage $\mathbf{V}$ and current $\mathbf{I}$ delivered to the load.

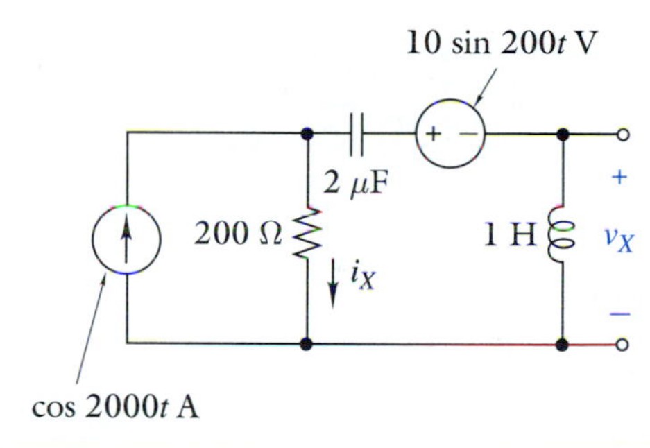

Figure P13-25

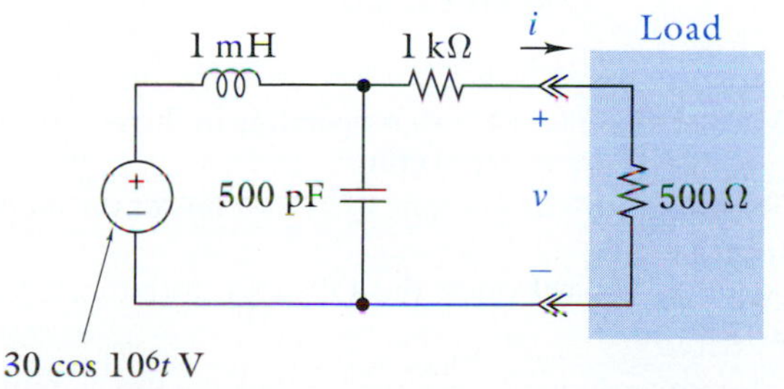

Figure P13-32

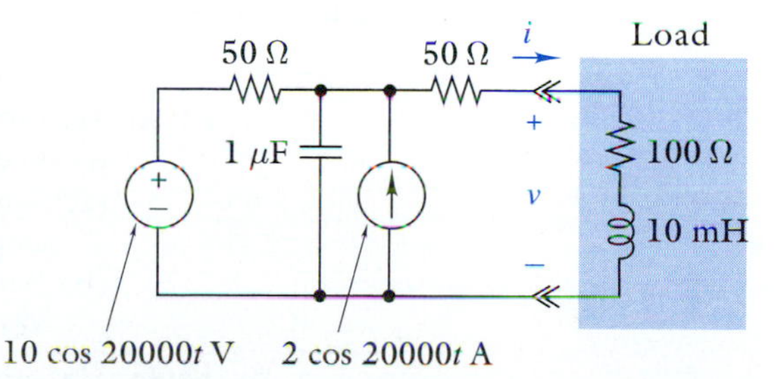

Figure P13-33

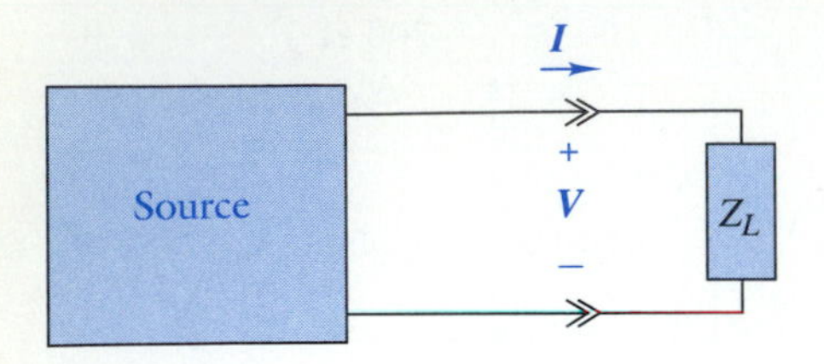

Figure P13-34

13-34 The circuit in Fig. P13-34 is operating in the sinusoidal steady state. When $Z_L = 0$ the phasor current at the interface is $\mathbf{I} = 1.22 - j0.976$ A. When $Z_L = -j40\ \Omega$ the phasor interface current is $\mathbf{I} = 2 + j0$ A. Find the Thévenin equivalent of the source circuit.

ERO 13-3 General Circuit Analysis in the Phasor Domain (Sect. 13-5)

Given a linear circuit operating in the sinusoidal steady state: Find driving-point impedances and phasor responses using node-voltage or mesh-current analysis methods.

13-35 Repeat Prob. 13-20 using mesh-current or node-voltage analysis.

13-36 Repeat Prob. 13-22 using mesh-current or node-voltage analysis.

13-37 Repeat Prob. 13-24 using mesh-current or node-voltage analysis.

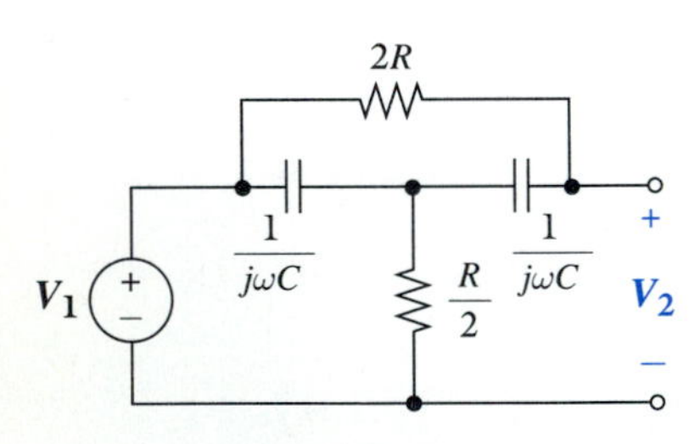

Figure P13-38

13-38 The circuit in Fig. P13-38 is operating in the sinusoidal steady state.

(a) Use node-voltage equations to solve for the phasor output voltage $\mathbf{V}_2$.

(b) Evaluate the output at $\omega = 0$, $\omega \to \infty$ and $\omega = 1/RC$. Comment on the frequency response characteristics of the circuit.

13-39 Repeat Prob. 13-38 using the OP AMP circuit in Fig. P13-39.

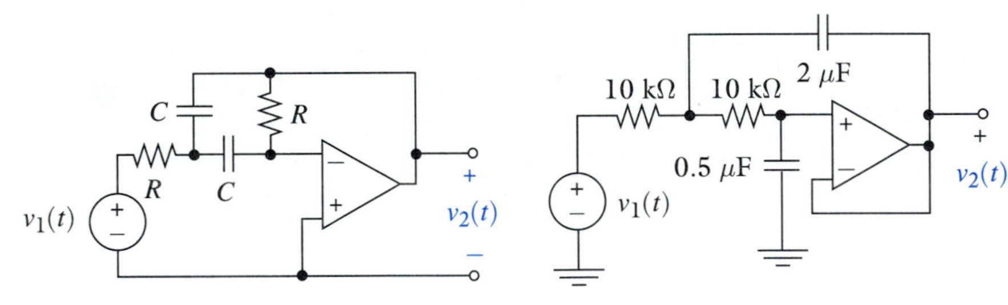

Figure P13-39

Figure P13-40

13-40 The OP AMP circuit in Fig. P13-40 is operating in the sinusoidal steady state with $v_1(t) = V_A \cos 1000t$.

(a) Solve for the steady-state output $v_2(t)$ in terms of the input amplitude V_A.

(b) What value of V_A will cause the OP AMP to saturate for $V_{CC} = 15$ V?

13-41 Use mesh-current analysis to solve for the steady-state currents $\mathbf{I}_A$ and $\mathbf{I}_B$ in Figure P13-41.

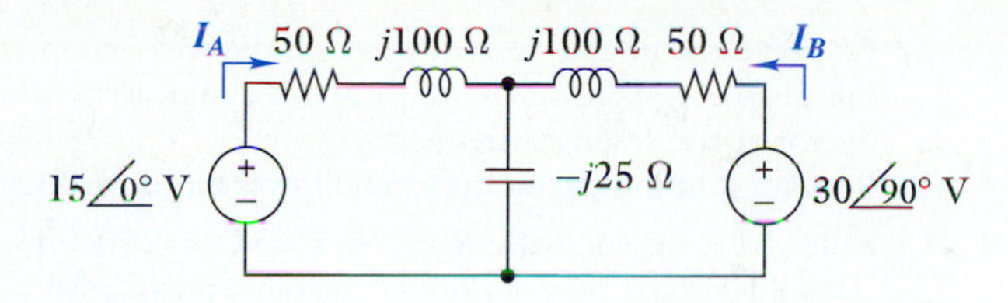

Figure P13-41

ERO 13-4 Resonant Circuits (Sect. 13-6)

(a) (Analysis) Determine the descriptive or circuit parameters of a parallel or series RLC circuit.

(b) (Design) Select the parameters of a series or parallel RLC circuit to achieve specified resonant characteristics.

13-42 The circuit in Fig. P13-42 is operating in the sinusoidal steady state with $\omega = 5000$ rad/s.

(a) Find the value of capacitance C that will cause the driving-point impedance Z to be purely resistive.

(b) Find the driving-point resistance for this value of C.

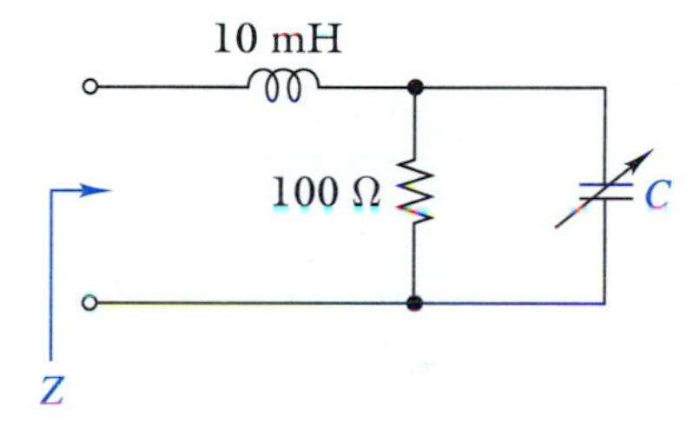

Figure P13-42

13-43 The circuit in Fig. P13-43 is operating in the sinusoidal steady state.

(a) Find the frequency at which the driving-point impedance is purely resistive.

(b) Find the driving-point resistance at the resonant frequency.

13-44 A series RLC circuit with $L = 0.5$ mH and $C = 200$ nF has a driving-point impedance of $10 + j10\ \Omega$ at its upper cutoff frequency. Calculate the resonant frequency ω_O, the bandwidth, the quality factor Q, and the upper and lower cutoff frequencies.

13-45 A series RLC circuit is designed to have a bandwidth of 8 Mrad/s and an impedance of 24 Ω at its resonant frequency of 50 Mrad/s. Determine the values L, C, Q, and the upper and lower cutoff frequencies.

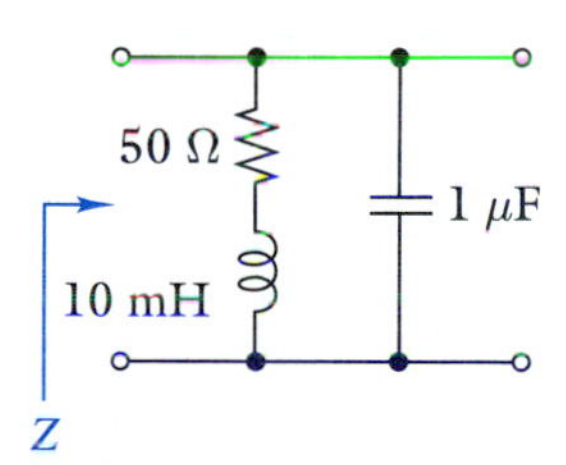

Figure P13-43

13-46 A 20-mH inductor with an internal series resistance of 20 Ω is connected in series with a capacitor and a voltage source with a Thévenin resistance of 50 Ω.

(a) Determine the value of capacitance C required to produce resonance at 7 krad/s.

(b) Calculate the Q and bandwidth of the circuit.

13-47 A series RLC circuit with $R = 100\ \Omega$, $L = 25$ mH, and $C = 100$ nF is driven by a sinusoidal voltage source with a peak amplitude of 10 V.

(a) Calculate the phasor current through the series connection, the phasor voltage across the capacitor and across the inductor at the resonant frequency.
(b) Repeat (a) at the circuit upper and lower cutoff frequencies.

13-48 A parallel RLC circuit with $R = 15$ kΩ, $L = 250$ mH, and $C = 10$ nF is driven by a sinusoidal current source with a peak amplitude of 10 mA.
(a) Calculate the phasor voltage across the parallel connection and the phasor current through the capacitor and through the inductor at the resonant frequency.
(b) Repeat (a) at the circuit upper and lower cutoff frequencies.

13-49 A parallel RLC circuit has a bandwidth of 5 krad/s and a driving-point impedance of 8 kΩ at its resonant frequency of 100 krad/s. Determine the values L, C, Q, and the upper and lower cutoff frequencies.

13-50 A parallel RLC circuit has a driving-point resistance of 4 kΩ at its resonant frequency of 455 kHz. Calculate the values of L and C required to produce a bandwidth of 10 kHz.

❖ **13-51** **(D)** An L and C are needed to produce series RLC resonance at $f_O = 100$ kHz. The series connected L and C are to be driven by a sinusoidal source with a Thévenin resistance of 50 Ω. The following standard capacitors are available in the stock room: 1 μF, 680 nF, 470 nF, 330 nF, 220 nF, and 150 nF. The inductor will be custom-designed to match the capacitor used. Select the capacitor which maximizes the circuit bandwidth.

❖ **13-52** **(D)** A series RLC circuit is to be used as a notch filter to eliminate a bothersome 60-Hz hum in an audio channel. The signal source has a Thévenin resistance of 600 Ω. Select values of L and C so the upper cutoff frequency is below 200 Hz.

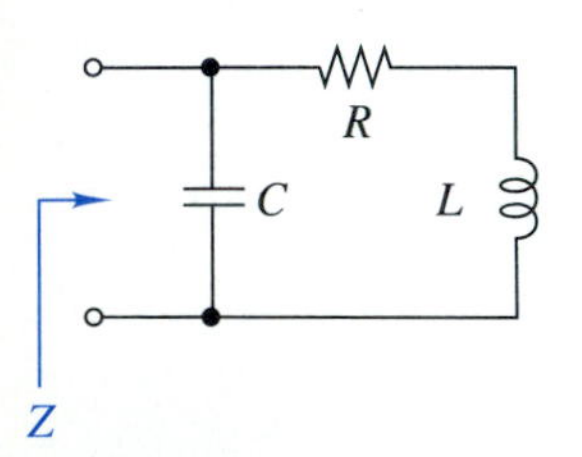

Figure P13-53

13-53 Express the driving-point impedance Z of the circuit in Fig. P13-53 in the form $R(\omega) + jX(\omega)$.
(a) Evaluate $R(0)$, $X(0)$, $R(\infty)$ and $X(\infty)$.
(b) Compare the values found in (a) with those obtained directly from the circuit using the values Z_L and Z_C at dc and infinite frequency.
(c) Derive an expression for the resonant frequency ω_O in terms of circuit parameters.

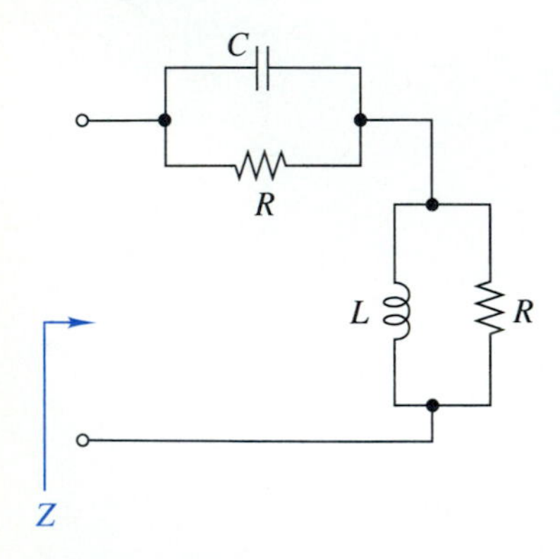

Figure P13-54

13-54 Express the driving-point impedance of the circuit in Fig. P13-54 in the form $R(\omega) + jX(\omega)$. Show that the circuit resonates at all frequencies when $L = R^2C$.

ERO 13-5 The Linear Transformer (Sect. 13-7)

Given a linear circuit with two coupled coils and operating in the sinusoidal steady state:

(a) Find driving-point impedances and phasor responses using mesh-current analysis methods

(b) Find transformer parameters that produce prescribed responses or input impedances.

13-55 The input voltage to the transformer in Fig. P13-55 is a sinusoid $v_1(t) = 10 \cos 2000t$. Use mesh-current analysis to find (a) the steady-state output voltage $v_2(t)$ and (b) the driving-point impedance seen by the input source.

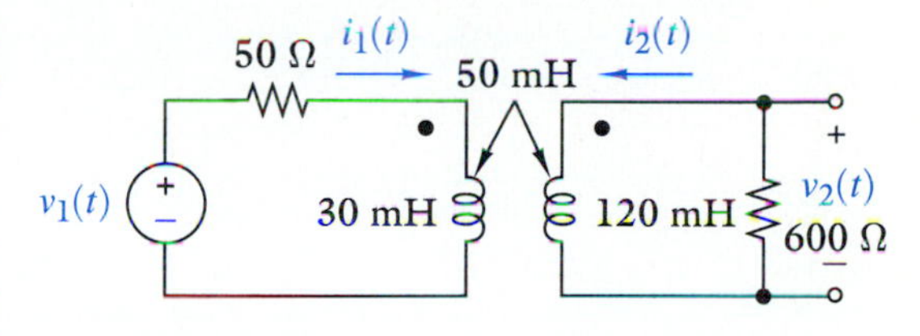

Figure P13-55

13-56 A sinusoidal source with an internal resistance of 80 Ω is connected to the primary of a linear transformer with $L_1 = 20$ mH, $L_2 = 45$ mH, and $k = 0.75$. The load connected to the secondary of the transformer is a 125-Ω resistor in series with 1-μ F capacitor. Assume additive coupling.

(a) Find $\boldsymbol{I}_1$, $\boldsymbol{I}_2$, $\boldsymbol{V}_1$, and $\boldsymbol{V}_2$ when the Thévenin voltage of the source is a sinusoid with a peak amplitude of 90 V and a frequency of 4 krad/s.

(b) Find the reflected impedance in the primary circuit.

(c) Find the frequency at which the reflected impedance would be purely resistive.

13-57 (a) Determine the input impedance seen by the source and the phasor voltage delivered to the load in the circuit shown in Fig. P13-57.

(b) Calculate the coupling coefficient of the coils.

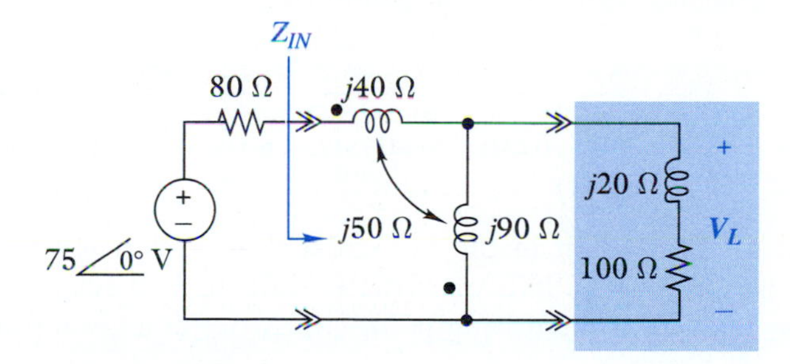

Figure P13-57

13-58 The voltage source in Fig. P13-58 generates $v_S(t) = 35 \cos 5000t$ V. The resistors in series with the coils represent parasitic winding resistances.

(a) Find the input impedance seen by the source.

(b) Find the phasor voltage delivered at the input to the transformer.

(c) Find the phasor voltage delivered to the capacitor load.

13-59 Find the voltage delivered to the 40-Ω resistor in Fig. P13-59 for $v_S = 65 \cos 250t$ V.

13-60 Repeat Prob. 13-59 with the coupling coefficient of the coils reduced to $k = 0.2$.

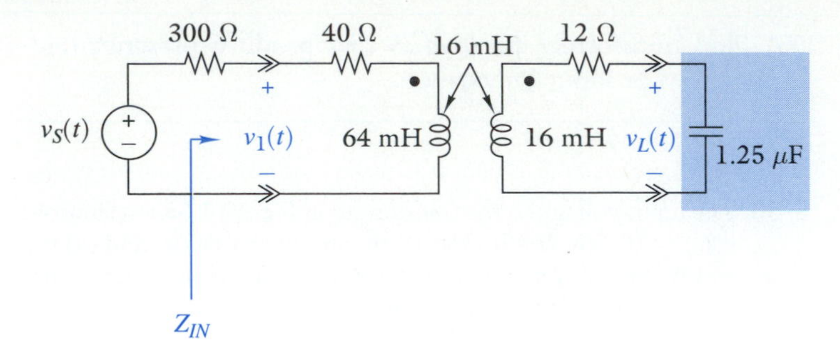

Figure P13-58

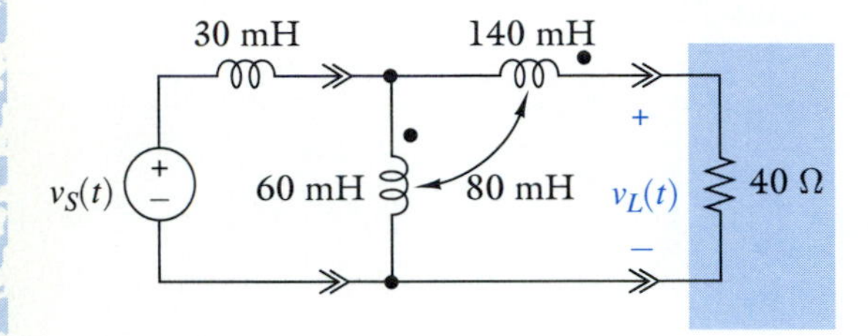

Figure P13-59

13-61 The self-inductances of a transformer are $L_1 = 80$ mH and $L_2 = 500$ mH. The load connected to the secondary is a 300-Ω resistor. The impedance seen looking into the primary winding is $96 - j64$ Ω when the primary is driven by a sinusoid source with $\omega = 800$ rad/s.

(a) What is the coupling coefficient of the transformer?

(b) What is the reflected impedance in the primary winding?

13-62 The inductance looking into the primary winding of a transformer with the secondary open circuited is 50 mH. The inductance looking into the primary winding with the secondary short circuited is 10 mH. The inductance looking into the secondary winding with the primary open circuited is 72 mH. What is the coupling coefficient of the transformer?

ERO 13-6 AC Circuit Analysis Using SPICE (Sect. 13-8)

Given a linear circuit in the sinusoidal steady state, use the ac analysis option in SPICE or MICRO-CAP to find selected response phasors or the driving-point impedance at a given pair of terminals.

13-63 Solve Prob. 13-20 using SPICE or MICRO-CAP.

13-64 Solve Prob. 13-21 using SPICE or MICRO-CAP.

13-65 Solve Prob. 13-32 using SPICE or MICRO-CAP.

13-66 Solve Prob. 13-33 using SPICE or MICRO-CAP.

13-67 Solve Prob. 13-40 using SPICE or MICRO-CAP.

13-68 Solve Prob. 13-57 using SPICE or MICRO-CAP.

13-69 Solve Prob. 13-58 using SPICE or MICRO-CAP.

13-70 Solve Prob. 13-59 using SPICE or MICRO-CAP.

13-71 Solve Prob. 13-60 using SPICE or MICRO-CAP.

CHAPTER INTEGRATING PROBLEMS

13-72 **(A)** An ac voltmeter measures only the amplitude of a sinusoid and not its phase angle. The magnitude and phase can be inferred by making several measurements and using KVL. For example, Fig. P13-72 shows a relay coil of unknown resistance and inductance. The following ac voltmeter readings are taken with the circuit operating in the sinusoidal steady state at $f = 1$ kHz: $|\mathbf{V}_S| = 10$ V, $|\mathbf{V}_1| = 4$ V, and $|\mathbf{V}_2| = 8$ V.

(a) Use these voltage magnitude measurements to solve for R and L.

(b) Determine the phasor voltage across each element and construct a phasor diagram.

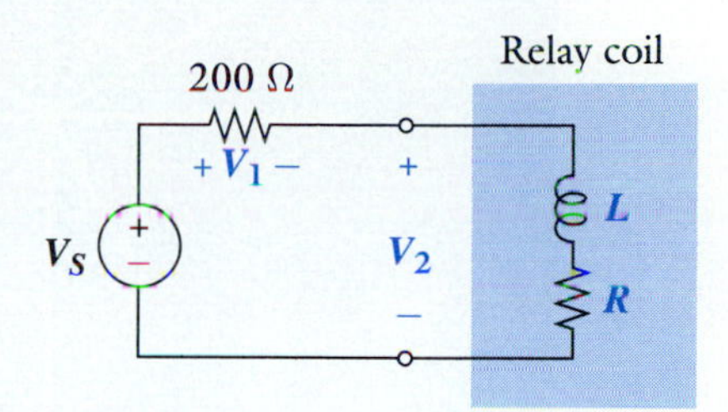

Figure P13-72

13-73 **(A)** The purpose of this problem is to demonstrate that the steady-state response obtained using phasor circuit analysis is the forced component of the solution of the circuit differential equation.

(a) Use phasor circuit analysis from this chapter to solve for the sinusoidal steady-state response $v(t)$ in Fig. P13-73.

(b) Formulate the circuit differential equation in the capacitor voltage $v(t)$.

(c) Show that the $v(t)$ from (a) satisfies the differential equation in (b).

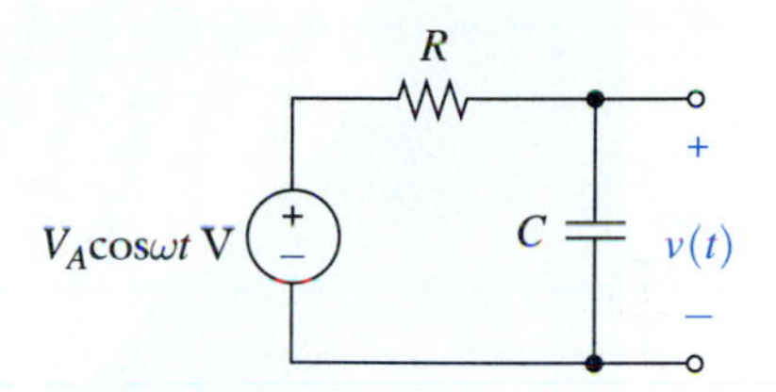

Figure P13-73

13-74 **(A)** The circuit in Fig. P13-74 includes two uncoupled inductances determined by the coupling coefficient k and the self-inductance L_1. The turns ratio of the ideal transformer is $n = L_2/M$. Show that this circuit is equivalent to two coupled coils with additive coupling.

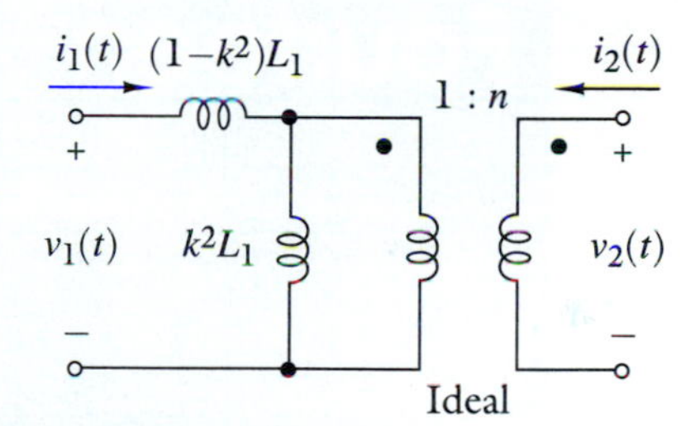

Figure P13-74

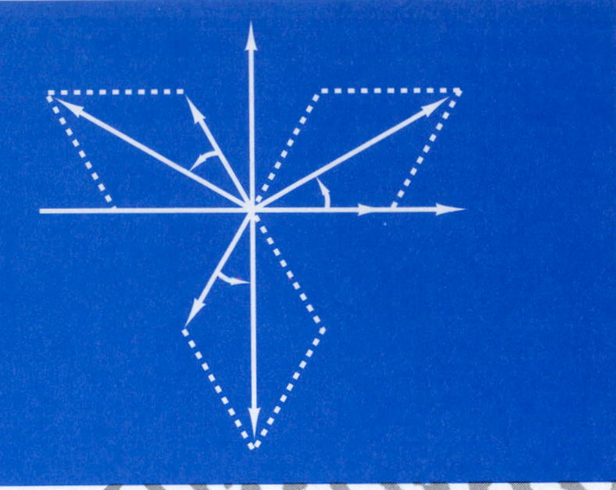

Chapter Fourteen

"George Westinghouse was in my opinion, the only man on the globe who could take my alternating current (power) system under the circumstances then existing and win the battle against prejudice and money power."

Nikola Tesla, 1932
American Engineer

Power in the Sinusoidal Steady State

In the last two decades of the 19th century there were two competing electrical power systems in this country. The direct current approach was developed by Thomas Edison, who installed the first of his dc power systems at the famous Pearl Street Station in New York City in 1882. By 1884 there were more than 20 such power stations operating in the United States. The competing alternating current technology was hampered by the lack of practical motors for the single-phase systems initially produced. The three-phase ac induction motor that met this need was the product of the mind of the brilliant Serbian immigrant Nikola Tesla (1856–1943). Tesla was educated in Europe and in 1884 emigrated to the United States, where he initially worked for Edison. In 1887 Tesla founded his own company to develop his inventions, which eventually led to 40 patents on three-phase equipment and controls. The importance of Tesla's work was recognized by George Westinghouse, a hard-driving electrical pioneer who purchased the rights to Tesla's patents. In the early 1890s the competition between the older dc system championed by Edison and the newer ac technology sponsored by Westinghouse developed into a heated controversy called the "war of the currents." The showdown came over the equipment to be installed in a large electric power station at Niagara Falls, New York. The choice of an ac system for Niagara Falls established a trend that led to large interconnected ac systems that form the backbone of modern industry today.

In this chapter we study the flow of electrical power in the sinusoidal steady state. The first section shows that the instantaneous power at an interface can be divided into two components called real and reactive power. The unidirectional real component produces a net transfer of energy from the source to the load. The oscillatory reactive power represents an interchange between the source and load with no net transfer of energy. The concept of complex power developed in the second section combines these two components into a single complex entity relating power flow to the interface voltage and current phasors. The basic tools for using complex power to analyze power flow in ac circuits are developed in the third section. These tools are then applied to the maximum power transfer and load-flow problems in the fourth and fifth sections. The next two sections discuss the three-phase circuits that transfer and utilize the large blocks of electrical power required in a modern industrial society. The chapter concludes by describing how to use the ac analysis capability of SPICE to solve power analysis problems.

14-1 AVERAGE AND REACTIVE POWER

We begin our study of electric power circuits with the two-terminal interface in Fig. 14-1. In power applications we normally think of one circuit as the source and the other as the load. Our objective is to describe the flow of power across the interface when the circuit is operating in the sinusoidal steady state. To this end, we write the interface voltage and current in the time domain as sinusoids of the form

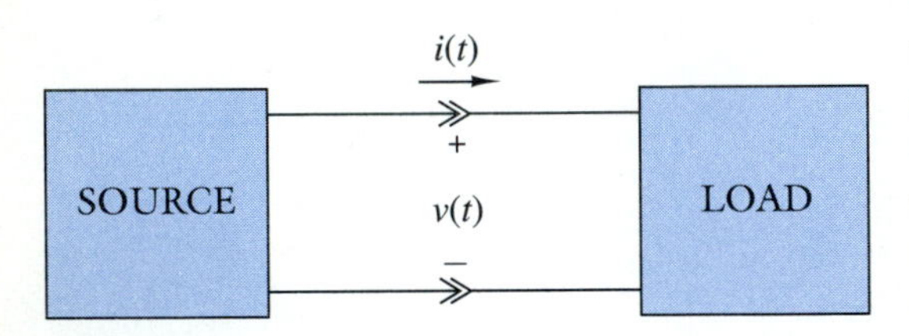

Figure 14-1 A two-terminal interface.

$$\begin{aligned} v(t) &= V_A \cos(\omega t + \theta) \\ i(t) &= I_A \cos \omega t \end{aligned} \tag{14-1}$$

In Eq. (14-1) V_A and I_A are real, positive numbers representing the peak amplitudes of the voltage and current respectively.

The forms of $v(t)$ and $i(t)$ in Eq. (14-1) are completely general. We have selected the $t = 0$ reference at the positive maximum of the current $i(t)$ and assigned a phase angle to $v(t)$ to account for the fact that the voltage maximum may not occur at the same time. In the phasor domain the angle $\theta = \phi_v - \phi_I$ is the angle between the phasors $\boldsymbol{V} = V_A \angle \phi_v$ and $\boldsymbol{I} = I_A \angle \phi_I$. In effect choosing $t = 0$ at the current maximum shifts the phase reference by an amount $-\phi_I$ so that the voltage and current phasors become $\boldsymbol{V} = V_A \angle \theta$ and $\boldsymbol{I} = I_A \angle 0$.

A method of relating power to phasor voltage and current will be presented in the next section, but at the moment we write the *instantaneous power* in the time domain:

$$\begin{aligned} p(t) &= v(t) \times i(t) \\ &= V_A I_A \cos(\omega t + \theta) \cos \omega t \end{aligned} \tag{14-2}$$

This expression for instantaneous power contains dc and ac components. To separate the components we first use the identity

$\cos(x + y) = \cos x \cos y - \sin x \sin y$ to write $p(t)$ in the form

$$\begin{aligned} p(t) &= V_A I_A[\cos \omega t \cos \theta - \sin \omega t \sin \theta] \cos \omega t \\ &= [V_A I_A \cos \theta] \cos^2 \omega t - [V_A I_A \sin \theta] \cos \omega t \sin \omega t \end{aligned} \tag{14-3}$$

Using the identities $\cos^2 x = 1/2(1 + \cos 2x)$ and $\cos x \sin x = 1/2 \sin 2x$, we write $p(t)$ in the form

$$p(t) = \underbrace{\left[\frac{V_A I_A}{2} \cos \theta\right]}_{\text{dc component}} + \underbrace{\left[\frac{V_A I_A}{2} \cos \theta\right] \cos 2\omega t - \left[\frac{V_A I_A}{2} \sin \theta\right] \sin 2\omega t}_{\text{ac component}} \tag{14-4}$$

Written this way, we see that the instantaneous power is the sum of a dc component and a double-frequency ac component. That is, the instantaneous power is the sum of a constant plus a sinusoid whose frequency is 2ω, which is twice the angular frequency of the voltage and current in Eq. (14-1).

Note that instantaneous power in Eq. (14-4) is periodic. In Chapter 6 we defined the average value of such a periodic waveform as

$$P = \frac{1}{T} \int_0^T p(t) dt$$

where $T = 2\pi/2\omega$ is the period of $p(t)$. In Chapter 6 we also showed that the average value of a sinusoid is zero, since the area under the waveform during a positive half-cycle is canceled by the area under a subsequent negative half-cycle. Therefore, the ***average value*** P of $p(t)$ is equal to the constant or dc term in Eq. (14-4):

$$P = \frac{V_A I_A}{2} \cos \theta \tag{14-5}$$

The amplitude of the $\sin 2\omega t$ in Eq. (14-4) has a form much like the average power in Eq. (14-5), except it involves $\sin \theta$ rather than $\cos \theta$. This amplitude factor is called the ***reactive power*** Q of $p(t)$ where Q is defined as[1]

$$Q = \frac{V_A I_A}{2} \sin \theta \tag{14-6}$$

Substituting Eqs. (14-5) and (14-6) into Eq. (14-4) yields the instantaneous power in terms of the average power and

[1]Note that the symbol Q has a different meaning in this chapter from its meaning in Chapter 13.

reactive power:

$$p(t) = \underbrace{P(1+\cos 2\omega t)}_{\text{unipolar}} - \underbrace{Q\sin 2\omega t}_{\text{bipolar}} \tag{14-7}$$

The first term in Eq. (14-7) is unipolar because the factor $1 + \cos 2\omega t$ never changes sign. As a result, the first term is either positive or always negative depending on the sign of P. The second term is bipolar because the factor $\sin 2\omega t$ alternates signs.

The energy transferred across the interface during one cycle $T = 2\pi/2\omega$ of $p(t)$ is

$$\begin{aligned} W &= \int_0^T p(t)dt \\ &= \underbrace{P\int_0^T (1+\cos 2\omega t)dt}_{\text{net energy}} - \underbrace{Q\int_0^T \sin 2\omega t dt}_{\text{no net energy}} \\ &= \qquad P\times T \qquad - \qquad 0 \end{aligned} \tag{14-8}$$

Only the unipolar term in Eq. (14-7) provides any net energy transfer and that energy is proportional to the average power P. With the passive sign convention the energy flows from source to load when $W > 0$. Equation (14-8) shows that the net energy will be positive if the average power $P > 0$. Equation (14-5) points out that the average power P is positive when $\cos\theta > 0$, which in turn means $|\theta| < 90°$. We conclude that:

> The net flow in Fig. 14-1 is from source to load when the angle between the interface voltage and current is bounded by $-90° < \theta < 90°$; otherwise, the net energy flow is from load to source.

The bipolar term in Eq. (14-7) is a power oscillation, which transfers no net energy across the interface. In the sinusoidal steady state the load in Fig. 14-1 borrows energy from the source circuit during part of a cycle and temporarily stores it in the load inductance or capacitance. In another part of the cycle the borrowed energy is returned to the source unscathed. The amplitude of the power oscillation is called reactive power because it invovles periodic energy storage and retrieval from the reactive elements of the load. The reactive power can be either positive or negative depending on the sign of $\sin\theta$. However, the sign of Q tells us nothing about the net energy transfer, which is controlled by the sign of P.

We are obviously interested in average power, since this component carries net energy from source to load. For most power system customers the basic cost of electrical service is proportional to the net energy delivered to the load. Large industrial users may also pay a service charge for their reactive power as well. This may seem unfair, since reactive power transfers no net energy. However, the electric energy borrowed and returned by the load is generated within a power system that has losses. From a power company's viewpoint the reactive power is not free because there are losses in the system connecting the generators in the power plant to source/load interface at which the lossless interchange of energy occurs.

In ac power circuit analysis, it is necessary to keep track of both the average power and reactive power. These two components of power have the same dimensions, but because they represent quite different effects they traditionally are given different units. The average power is expressed in watts (W), while reactive power is expressed in VARs, which is an acronym for "volt-amperes reactive."

EXERCISE 14-1

Using the reference marks in Fig. 14-1, calculate the average and reactive power for the following voltages and currents. State whether the load is absorbing or delivering net energy.

(a) $v(t) = 168\cos(377t + 45°)$ V, $i(t) = 0.88\cos 377t$ A
(b) $v(t) = 285\cos(2500t - 68°)$ V, $i(t) = 0.66\cos 2500t$ A
(c) $v(t) = 168\cos(377t + 45°)$ V, $i(t) = 0.88\cos(377t - 60°)$ A
(d) $v(t) = 285\cos(2500t - 68°)$ V, $i(t) = 0.66\sin 2500t$ A

ANSWERS:

(a) $P = +52.3$ W, $Q = +52.3$ VAR, absorbing
(b) $P = +35.2$ W, $Q = -87.2$ VAR, absorbing
(c) $P = -19.1$ W, $Q = +71.4$ VAR, delivering
(d) $P = +87.2$ W, $Q = +35.2$ VAR, absorbing

14-2 COMPLEX POWER

Relating average and reactive power to phasor quantities is important because ac circuit analysis is conveniently carried out using phasors. In the previous chapter the magnitude of a phasor represented the peak amplitude of a sinusoid. However, in power circuit analysis it is convenient to express phasor magnitudes in *rms* values. In this chapter phasor voltages and currents are expressed as

$$\mathbf{V} = V_{rms}e^{j\phi_V} \qquad \mathbf{I} = I_{rms}e^{j\phi_I} \tag{14-9}$$

where the phasors magnitudes represent the *rms* amplitude of the corresponding sinusoid.

Equations (14-5) and (14-6) express average and reactive power in terms of peak amplitudes V_A and I_A. In Chapter 6 we showed that the peak and *rms* values of a sinusoid are related by

$V_{rms} = V_A/\sqrt{2}$. The expression for average power easily can be converted to *rms* amplitudes, since we can write Eq. (14-5) as

$$P = \frac{V_A I_A}{2}\cos\theta = \frac{V_A}{\sqrt{2}}\frac{I_A}{\sqrt{2}}\cos\theta \\ = V_{rms}I_{rms}\cos\theta \tag{14-10}$$

where $\theta = \phi_V - \phi_I$ is the angle between the voltage and current phasors. By similar reasoning Eq. (14-6) becomes

$$Q = V_{rms}I_{rms}\sin\theta \tag{14-11}$$

Using *rms* phasors we define the *complex power* (S) at a two-terminal interface as follows:

$$S = \boldsymbol{V}\boldsymbol{I}^* \tag{14-12}$$

That is, the complex power at an interface is the product of the voltage phasor times the conjugate of the current phasor. Substituting Eq. (14-9) into this definition yields

$$S = \boldsymbol{V}\boldsymbol{I}^* = V_{rms}e^{j\phi_V}I_{rms}e^{-j\phi_I} \\ = [V_{rms}I_{rms}]\,e^{j(\phi_V-\phi_I)} \tag{14-13}$$

Using Euler's relationship and the fact that the angle is $\theta = \phi_V - \phi_I$ we can write complex power as

$$S = [V_{rms}I_{rms}]\,e^{j\theta} \\ = [V_{rms}I_{rms}]\cos\theta + j\,[V_{rms}I_{rms}]\sin\theta \\ = P + jQ \tag{14-14}$$

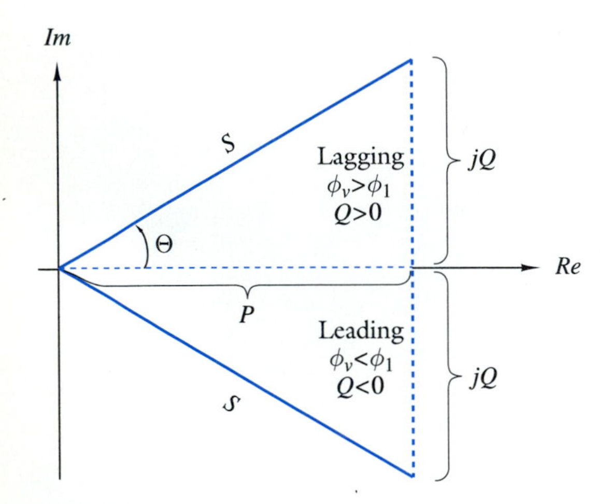

Figure 14-2 Power triangles.

The real part of the complex power S is the average power and the imaginary part is the reactive power. Although S is a complex number, it is not a phasor. However, it is a convenient variable for keeping track of the two component of powers when the voltage and currents are expressed as phasors.

The *power triangles* in Fig. 14-2 provide a convenient way to remember complex power relationships and terminology. We confine our study to cases in which net energy is transferred from source to load. In such cases $P > 0$ and the power triangles fall in the first or fourth quadrant as indicated in Fig. 14-2.

The magnitude $|S| = V_{rms}I_{rms}$ is called *apparent power* and is expressed using the unit volt-ampere (VA). The ratio of the average power to the apparent power is called the *power factor pf*. Using Eq. (14-10) we see that the power factor is

$$pf = \frac{P}{|S|} = \frac{V_{rms}I_{rms}\cos\theta}{V_{rms}I_{rms}} = \cos\theta \tag{14-15}$$

Since $pf = \cos\,\theta$ the angle θ is called the *power factor angle*.

When the power factor is unity the phasors $\boldsymbol{V}$ and $\boldsymbol{I}$ are in phase ($\theta = 0°$) and the reactive power is zero, since $\sin\,\theta = 0$.

When the power factor is less than unity, the reactive power is not zero and its sign is indicated by the modifiers lagging or leading. The term *lagging power factor* means the current phasor lags the voltage phasor so that $\theta = \phi_V - \phi_I > 0$. For a lagging power factor S falls in the first quadrant in Fig. 14-2 and the reactive power is positive, since $\sin\theta > 0$. The term *leading power factor* means the current phasor leads the voltage phasor so that $\theta = \phi_V - \phi_I < 0$. In this case S falls in the fourth quadrant in Fig. 14-2 and the reactive power is negative, since $\sin\theta < 0$. Most industrial and residential loads have lagging power factors.

The apparent power or VA rating of an electrical power device is an important design parameter. The wiring must be large enough to carry the required current and insulated well enough to withstand the rated voltage. However, only the average power is potentially available as useful output, since the reactive power represents a lossless interchange between the source and device. Because reactive power increases the apparent power rating without increasing the available output, it is desirable for electrical devices to operate as close to unity power factor (zero reactive power) as possible.

EXERCISE 14-2

Determine the average power, reactive power, and apparent power for the following voltage and current phasors. State whether the power factor is lagging or leading.

(a) $\mathbf{V} = 208\angle{-90^\circ}$ V(rms), $\mathbf{I} = 1.75\angle{-75^\circ}$ A(rms).
(b) $\mathbf{V} = 277\angle{+90^\circ}$ V(rms), $\mathbf{I} = 11.3\angle{0^\circ}$ A(rms).
(c) $\mathbf{V} = 120\angle{-30^\circ}$ V(rms), $\mathbf{I} = 0.30\angle{^-90^\circ}$ A(rms).
(d) $\mathbf{V} = 480\angle{+75^\circ}$ V(rms), $\mathbf{I} = 8.75\angle{+105^\circ}$ A(rms).

ANSWERS:

(a) $P = 352$ W, $Q = -94.2$ VAR, $|S| = 364$ VA, leading
(b) $P = 0$ W, $Q = +3.13$ kVAR, $|S| = 3.13$ kVA, lagging
(c) $P = 18$ W, $Q = +31.2$ VAR, $|S| = 36$ VA, lagging
(d) $P = 3.64$ kW, $Q = -2.1$ kVAR, $|S| = 4.20$ kVA, leading

Complex Power and Load Impedance

In many cases power circuit loads are described in terms of their power ratings at a specified voltage or current level. In order to find voltages and current elsewhere in the circuit it is necessary to know the load impedance. For this reason we need to relate complex power and load impedance.

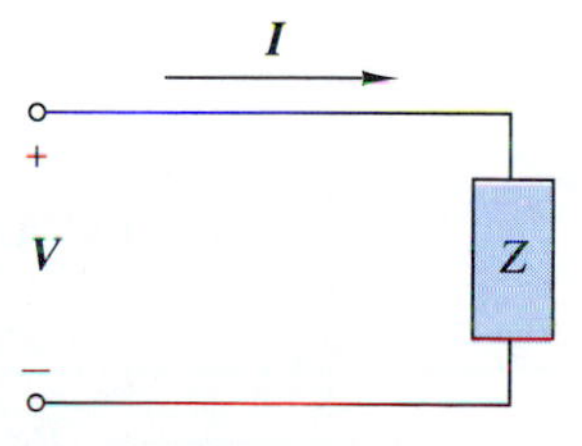

Figure 14-3 Two-terminal impedance.

Figure 14-3 shows the general case for a two-terminal load. For the assigned reference directions the load produces the element constraint $\mathbf{V} = Z\mathbf{I}$. Using this constraint in Eq. (14-12) we write the complex power of the load as

$$S = \mathbf{V}\mathbf{I}^* = Z\mathbf{I}\mathbf{I}^* = Z|\mathbf{I}|^2$$
$$= (R + jX)I_{rms}^2$$

where R and X are the resistance and reactance of the load respectively. Since $S = P + jQ$ we conclude that

$$R = \frac{P}{I_{rms}^2} \quad \text{and} \quad X = \frac{Q}{I_{rms}^2} \tag{14-16}$$

The load resistance and reactance are proportional to the average and reactive power of the load, respectively.

The first condition in Eq. (14-16) demonstrates that resistance cannot be negative, since P cannot be negative for a passive circuit. That is, a passive circuit cannot produce average power in the sinusoidal steady state; otherwise, perpetual motion would be possible and the energy crisis would be a small footnote in the great sweep of human history. The second condition in Eq. (14-16) points out that when the reactive power is positive the load is inductive, since $X_L = \omega L$ is positive. Conversely, when the reactive power is negative the load is capacitive since $X_C = -1/\omega C$ is negative. The terms *inductive load*, *lagging power factor*, and *positive reactive power* are synonymous as are the terms *capacitive load*, *leading power factor*, and *negative reactive power*.

EXAMPLE 14-1

At 440 V (rms) a two-terminal load draws 3 kVA of apparent power at a lagging power factor of 0.9. Find

(a) Irms,
(b) P,
(c) Q
(d) the load impedance. Draw the power triangle for the load.

SOLUTION:

(a) $I_{rms} = |S|/V_{rms} = 3000/440 = 6.82$ A (rms)
(b) $P_{AVE} = V_{rms}I_{rms} \cos \theta = 3000 \times 0.9 = 2.7$ kW
(c) For $\cos \theta = 0.9$ lagging, $\sin \theta = 0.436$ and $Q = V_{rms}I_{rms} \sin \theta = 1.31$ kVAR
(d) $Z = (P + j\,Q)/(I_{rms})^2 = (2700 + j1310)/46.5 = 58.0 + j28.2\ \Omega$

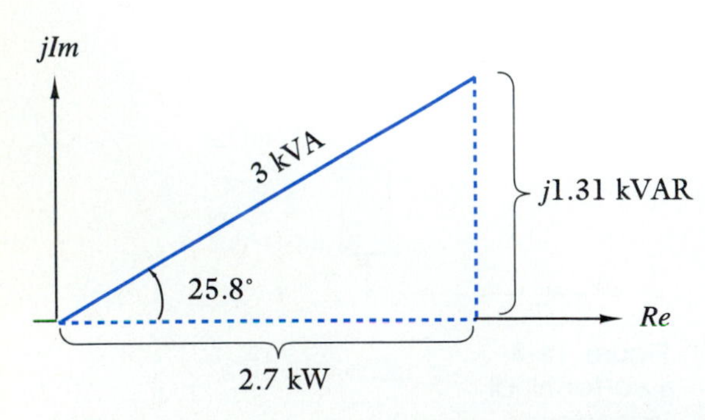

Figure 14-4

Figure 14-4 shows the power triangle for this load.

EXERCISE 14-3

Find the impedance of a two-terminal load under the following conditions.

(a) $V = 120\angle 30°$ V (rms) and $I = 20\angle 75°$ A (rms)
(b) $|S| = 3.3$ kVA, $Q = -1.8$ kVAR, and $I_{rms} = 7.5$ A
(c) $P_{AVE} = 3$ kW, $Q = 4$ kVAR, and $V_{rms} = 880$ V
(d) $V_{rms} = 208$ V, $I_{rms} = 17.8$ A, and $P_{AVE} = 3$ kW

ANSWERS:

(a) $Z = 4.24 - j4.24\ \Omega$
(b) $Z = 49.2 - j32\ \Omega$
(c) $Z = 92.9 + j124\ \Omega$
(d) $Z = 9.47 \pm j6.85\ \Omega$ ■

14-3 AC POWER ANALYSIS

The nature of ac power analysis can be modeled in terms of the phasor voltage, phasor current, and the complex power at the source/load interface in Fig. 14-1. Three distinctly different types of problems are treated using this model. In the *direct analysis* problem the source and load circuit are given and we are required to find the steady-state responses at one or more interfaces. This type of problem is essentially the same as the phasor circuit analysis problems discussed in Chapter 13, except that we calculate complex powers as well as phasor responses. In the *maximum power transfer* problem the source circuit is given and we are required to adjust the load so that maximum average power is transferred across the interface. This type of problem arises in communication systems, where the objective is to design the load circuit to extract the maximum available power from power-limited sources such as antennas or radio frequency transmitters. In the *load-flow* problem we are required to adjust the source so that a prescribed complex power is delivered to the load at a specified interface voltage magnitude. This type of problem arises in electrical power systems where the objective is to supply changing energy demands at a fixed voltage level.

The following examples are direct analysis problems that illustrate the computational tools needed to deal with the maximum power transfer and load-flow problems discussed in the next two sections. One of the useful tools is the principle of the *conservation of complex power*, which can be stated as follows.

In a linear circuit operating in the sinusoidal steady state, the sum of the complex powers produced by each independent source is equal to the sum of the complex power absorbed by all other two-terminal elements in the circuit.

To apply this principle it is important to distinguish between the complex power "produced by" and "absorbed by" a two-terminal element. When the reference marks for the element voltage $\boldsymbol{V}$ and current $\boldsymbol{I}$ are assigned in accordance with the passive sign convention, the complex power absorbed by the element is $S = \boldsymbol{V}\boldsymbol{I}^*$ and the power produced by an element is $S = -\boldsymbol{V}\boldsymbol{I}^*$.

EXAMPLE 14-2

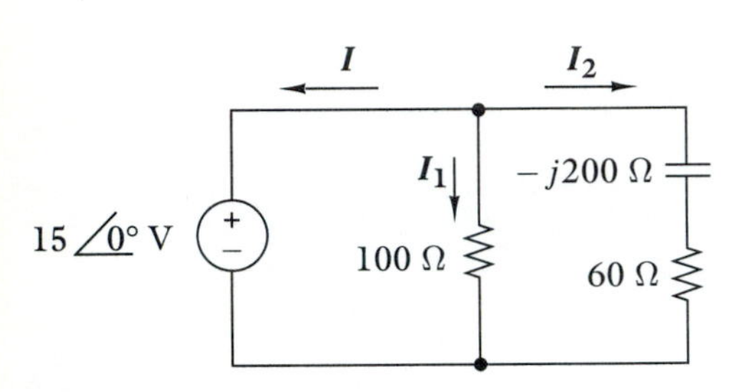

Figure 14-5

(a) Calculate the complex power absorbed by each parallel branch in Fig. 14-5.

(b) Calculate the complex power produced by the source and the power factor of the load seen by the source.

SOLUTION:

(a) The voltage across each branch is $15\angle 0°$ and the branch impedances are $Z_1 = 100$ and $Z_2 = 60 - j200$. Therefore, the branch currents are

$$\boldsymbol{I}_1 = \frac{15\angle 0°}{100} = 0.15\angle 0° \text{ A}$$

$$\boldsymbol{I}_2 = \frac{15\angle 0°}{60 - j200} = 0.0718\angle 73.3° \text{ A}$$

The complex power absorbed by each branch is

$$S_1 = (15\angle 0°)\boldsymbol{I}_1^* = (15\angle 0°)(0.15\angle{-0°}) = 2.25\angle 0° \text{ VA}$$

$$S_2 = (15\angle 0°)\boldsymbol{I}_2^* = (15\angle 0°)(0.0718\angle{-73.3°})$$
$$= 1.08\angle{-73.3°} \text{ VA}$$

(b) Using KCL the source current $\boldsymbol{I}$ is

$$\boldsymbol{I} = -(\boldsymbol{I}_1 + \boldsymbol{I}_2) = -(0.15\angle 0° + 0.0718\angle 73.3°)$$
$$= -0.171 - j0.0688 = 0.184\angle{-158.1°} \text{ A}$$

The source current $\boldsymbol{I}$ and source voltage $15\angle 0°$ conform to the passive sign convention so the complex power produced by the source is

$$S = -(15\angle 0°)\boldsymbol{I}^* = -(15\angle 0°)(0.184\angle 158.1°)$$
$$= 2.76\angle{-21.9°} \text{ VA}$$

The power factor is $\cos(-21.9°) = 0.928$ leading. Alternatively, we can use the conservation of complex power to obtain the complex power produced by the source as the sum of the complex powers delivered to the passive elements:

$$\begin{aligned}S_1 + S_2 &= 2.25\angle 0^\circ + 1.08\angle -73.3^\circ \\ &= 2.25 + j0 + 0.310 - j1.03 \\ &= 2.56 - j1.03 = 2.76\angle -21.9^\circ \text{ VA}\end{aligned}$$

This result is the same as the answer obtained above using $S = -\boldsymbol{VI}^*$.

EXAMPLE 14-3

Find the complex power produced by each source in Fig. 14-6.

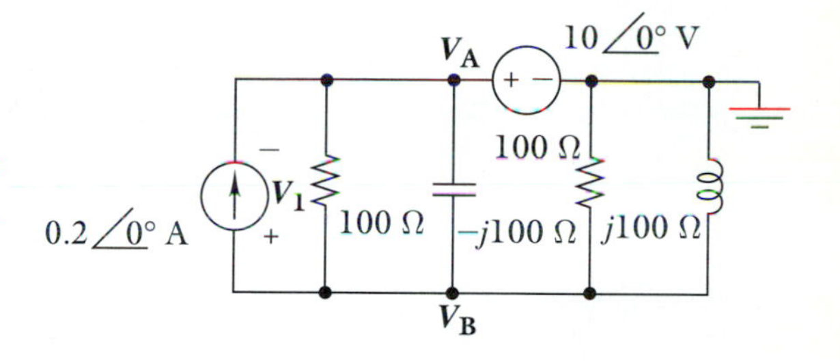

Figure 14-6

SOLUTION:

The Node B voltage is the only unknown node voltage in the circuit, since the voltage source forces the condition $\boldsymbol{V}_A = 10\angle 0^\circ$. By inspection the node equation at Node B is

$$\left[\frac{1}{100} + \frac{1}{100} + \frac{1}{-j100} + \frac{1}{j100}\right]\boldsymbol{V}_B - \left[\frac{1}{100} + \frac{1}{-j100}\right]\boldsymbol{V}_A = -0.2\angle 0^\circ$$

Since $\boldsymbol{V}_A = 10 + j0$, this node equation reduces to

$$\left[\frac{1}{50} + j0\right]\boldsymbol{V}_B = -0.1 + j0.1$$

which yields $\boldsymbol{V}_B = -5 + j5$ V. Using KCL at the reference node, the current through the voltage source is

$$\boldsymbol{I}_V = -\frac{\boldsymbol{V}_B}{100} - \frac{\boldsymbol{V}_B}{j100} = \frac{5 - j5}{100} - j\frac{5 - j5}{100} = 0 - j0.1 \text{ A}$$

Given the voltage across and the current through the voltage source the complex power produced by the source is

$$S_V = (10\angle 0^\circ)(-\boldsymbol{I}_V^*) = 10(0 - j0.1) = 0 - j1 \text{ VA}$$

The voltage across the current source is

$$\boldsymbol{V}_I = \boldsymbol{V}_B - \boldsymbol{V}_A = -15 + j5 \text{ V}$$

Given the voltage across and the current through the current source the complex power produced by the source is

$$S_I = (\boldsymbol{V}_I)(-0.2\angle 0^\circ)^* = 3 - j1 \text{ VA}$$

EXAMPLE 14-4

Find the complex power produced by the voltage source and absorbed by the 100-Ω resistor in the transformer circuit in Fig. 14-7.

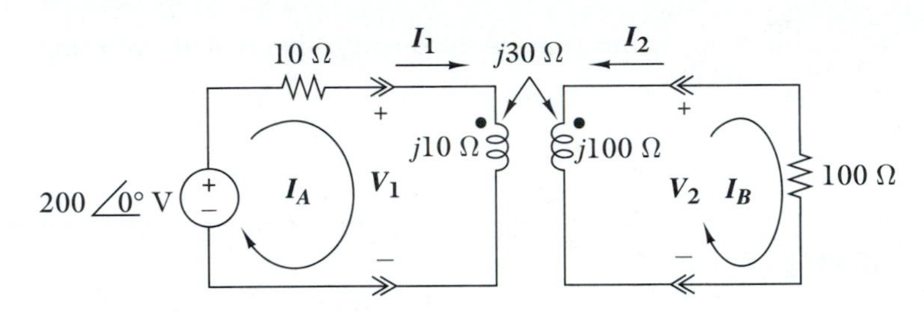

Figure 14-7

SOLUTION:

Using the coil voltages and mesh currents in Fig. 14-7 we can write KVL equations around the primary and secondary circuits.

$$10\boldsymbol{I}_A + \boldsymbol{V}_1 = 200\angle 0^\circ$$

$$100\boldsymbol{I}_B - \boldsymbol{V}_2 = 0$$

The coil currents are related to the mesh currents as $\boldsymbol{I}_1 = \boldsymbol{I}_A$ and $\boldsymbol{I}_2 = -\boldsymbol{I}_B$. Both coil currents are directed inward at the coil dots, so the mutual inductive coupling is additive and the i-v characteristics of the coupled coils in terms of the mesh currents are

$$\boldsymbol{V}_1 = j10\boldsymbol{I}_A + j30(-\boldsymbol{I}_B)$$

$$\boldsymbol{V}_2 = j30\boldsymbol{I}_A + j100(-\boldsymbol{I}_B)$$

Substituting the coil i-v characteristics into the KVL equation produces the two mesh current equations for the transformer:

$$(10 + j10)\boldsymbol{I}_A - j30\boldsymbol{I}_B = 200$$

$$-j30\boldsymbol{I}_A + (100 + j100)\boldsymbol{I}_B = 0$$

Solving for the two mesh current produces

$$\boldsymbol{I}_A = 12.1 - j4.57 \text{ A}$$

$$\boldsymbol{I}_B = 2.49 + j1.12 \text{ A}$$

Using these mesh currents, the input power produced by the source is

$$S_{IN} = (200\angle 0^\circ)\boldsymbol{I}_A^* = 2412 - j914 \text{ VA}$$

and the output power absorbed by the 100-Ω resistor is

$$S_{OUT} = (100\boldsymbol{I}_B)\boldsymbol{I}_B^* = 748 + j0 \text{ VA}$$

■ **EXERCISE 14-4**

Calculate the complex power delivered by each source in Fig. 14-8.

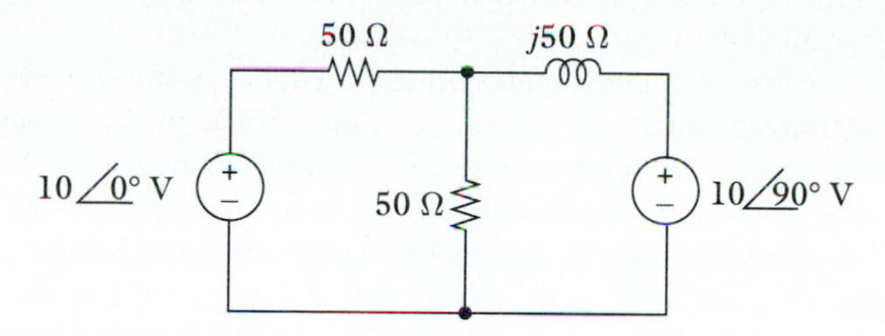

Figure 14-8

ANSWERS:

$S_1 = 0.4 + j0.8$ VA, $S_2 = 1.6 + j1.2$ VA ■

14-4 MAXIMUM POWER TRANSFER

In communication systems it often is desirable to transfer as much power as possible across an interface even though the resulting efficiency may be quite low. To address the maximum power transfer problem we use Thévenin's theorem to model the source/load interface as shown in Fig. 14-9. The source circuit is represented by a Thévenin equivalent circuit with source voltage $\mathbf{V}_T$ and source impedance $Z_T = R_T + jX_T$. The load circuit is represented by its driving-point impedance $Z_L = R_L + jX_L$. In the maximum power transfer problem the source parameters $\mathbf{V}_T$, R_T, and X_T are given, and the objective is to adjust the load impedance R_L and X_L so that P is maximized.

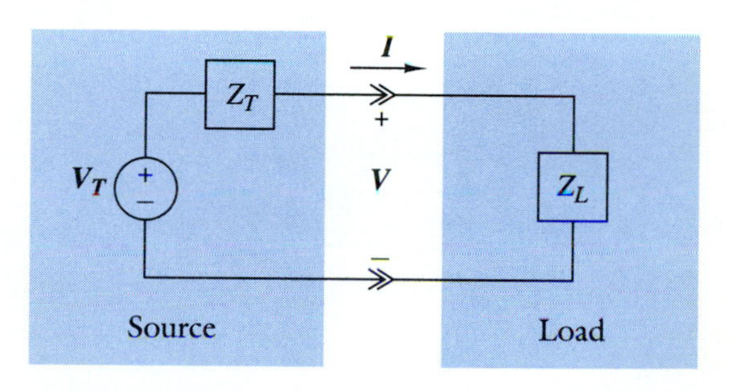

Figure 14-9 A source-load interface.

To determine maximum power transfer, we first express the average power in terms of the phasor current and load resistance:

$$P = |\mathbf{I}|^2 R_L$$

Then using series equivalence we express the magnitude of the interface current as

$$|\mathbf{I}| = \left|\frac{\mathbf{V}_T}{Z_T + Z_L}\right| = \frac{|\mathbf{V}_T|}{|(R_T + R_L) + j(X_T + X_L)|}$$

$$= \frac{|\mathbf{V}_T|}{\sqrt{(R_T + R_L)^2 + (X_T + X_L)^2}}$$

Combining the last two equations, we find the average power delivered across the interface in Fig. 14-9 is

$$P = \frac{|\mathbf{V}_T|^2 R_L}{(R_T + R_L)^2 + (X_T + X_L)^2} \qquad (14\text{-}17)$$

In the maximum power transfer problem the source is fixed and the load is adjustable. Hence, in Eq. (14-17) the quantities $|\mathbf{V}_T|$, R_T, and X_T are fixed. Our problem is to select R_L and X_L so as to maximize P.

Clearly the denominator in Eq. (14-17) is minimized and P maximized when $X_L = -X_T$. This choice of X_L always is possible because a reactance can be positive or negative. When the source Thévenin equivalent has an inductive reactance ($X_T > 0$), we select the load to have a capacitive reactance of the same magnitude, and vice versa. Note that this step reduces the net reactance of the series connection in Fig. 14-9 to zero and creates a resonant condition in which the net impedance seen by the Thévenin voltage source is purely resistive.

When the source and load reactances are canceled out, the expression for average power in Eq. (14-17) reduces to

$$P = \frac{|\mathbf{V}_T|^2 R_L}{(R_T + R_L)^2} \tag{14-18}$$

This equation has the same form encountered in Chapter 3 in dealing with maximum power transfer in resistance circuits. From the derivation in Section 3-4 we know P is maximized when we select $R_L = R_T$. In summary, to obtain maximum power transfer in the sinusoidal steady state, we select the load resistance and reactance so that

$$R_L = R_T \quad \text{and} \quad X_L = -X_T \tag{14-19}$$

These conditions can be compactly expressed in the following way:

$$Z_L = Z_T^* \tag{14-20}$$

The condition for maximum power transfer is called a ***conjugate match***, since the load impedance is the conjugate of the source impedance. When the conjugate-match conditions are inserted into Eq. (14-17), we find that the maximum average power available from the source circuit is

$$P_{MAX} = \frac{|\mathbf{V}_T|^2}{4R_T} \tag{14-21}$$

where $\mathbf{V}_T$ is the rms amplitude of the Thévenin equivalent source voltage.

It is important to remember the problem constraints to which conjugate matching applies. In the maximum power transfer problem the source is fixed and the load is adjusted. These conditions arise frequently in power-limited communication systems. However, conjugate matching does not apply to electrical power systems because the problem constraints are different, as discussed in the next section.

EXAMPLE 14-5

(a) Calculate the average power delivered to the load in the circuit shown in Fig. 14-10 for $v_S(t) = 5 \cos 10^6 t$, $R = 200\ \Omega$, $R_L = 200\ \Omega$, and $C = 10$ nF.

(b) Calculate the maximum average power available at the interface and specify the load required to draw the maximum power.

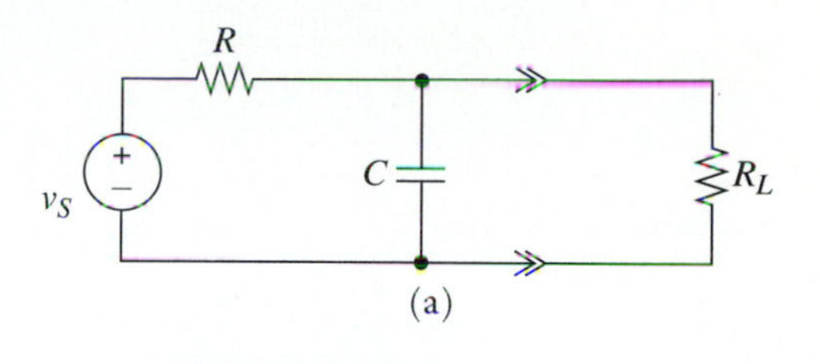

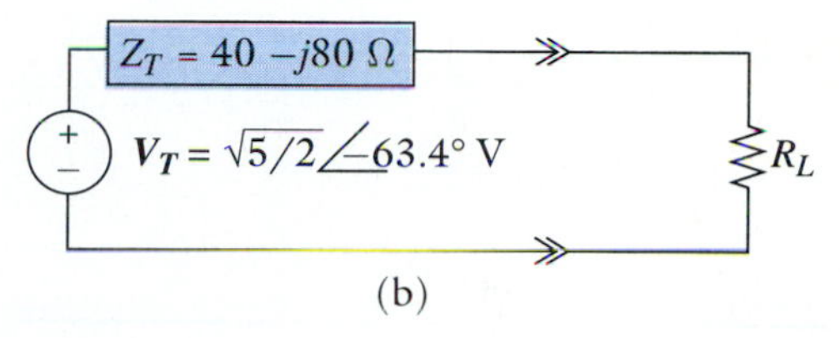

Figure 14-10

SOLUTION:

The input sinusoid $5 \cos 10^6 t$ has a peak amplitude of 5 V. In this chapter we use *rms* amplitudes so the phasor representing the input sinusoid is $\mathbf{V}_S = (5\angle 0°)/\sqrt{2}$ V (rms).

(a) To find the power delivered to the 200-Ω load resistor we use a Thévenin equivalent circuit. By voltage division the open-circuit voltage at the interface is

$$\mathbf{V}_T = \frac{Z_C}{Z_R + Z_C}\mathbf{V}_S = \frac{-j100}{200 - j100}\frac{5\angle 0°}{\sqrt{2}}$$

$$= \frac{1 - j2}{\sqrt{2}} = \sqrt{\frac{5}{2}}\angle -63.4° \text{ V(rms)}$$

By inspection the short-circuit current at the interface is

$$\mathbf{I}_N = \frac{(5\angle 0°)/\sqrt{2}}{200} = \frac{0.025 + j0}{\sqrt{2}} \text{ A(rms)}$$

Given $\mathbf{V}_T$ and $\mathbf{I}_N$ we calculate the Thévenin source impedance:

$$Z_T = \frac{\mathbf{V}_T}{\mathbf{I}_N} = \frac{1 - j2}{0.025} = 40 - j80\ \Omega$$

Using the Thévenin equivalent shown in Fig. 14-10(b), we find that the current through the 200-Ω resistor is

$$\mathbf{I} = \frac{\mathbf{V}_T}{Z_T + Z_L} = \frac{(\sqrt{5}\angle -63.4°)/\sqrt{2}}{40 - j80 + 200}$$

$$= \frac{8.84\angle -45°}{\sqrt{2}} \text{ mA(rms)}$$

and the average power delivered to the load resistor is

$$P_L = |\mathbf{I}|^2 R_L = \left(\frac{8.84 \times 10^{-3}}{\sqrt{2}}\right)^2 200 = 7.81 \text{ mW}$$

(b) Using Eq. (14-21) the maximum average power available at the interface is

$$P_{MAX} = \frac{1}{4}\frac{|V_T|^2}{R_T} = \frac{(\sqrt{5}/2)^2}{(4)(40)} = 15.6 \text{ mW}$$

The 200-Ω load resistor in part (a) draws about half of the maximum available power. To extract maximum power the load impedance must be

$$Z_L = Z_T^* = 40 + j80 \ \Omega$$

This impedance can be obtained using a 40-Ω resistor in series with a reactance of +80 Ω. The required reactance is inductive (positive) and can be produced by an inductance of

$$L = \frac{|X_T|}{\omega} = \frac{80}{10^6} = 80 \ \mu\text{H}$$

EXAMPLE 14-6

(a) Find the maximum average power available at the output interface of the transformer circuit in Fig. 14-11.

(b) Compare the result in part (a) with the average power delivered for $Z_L = 400\ \Omega$.

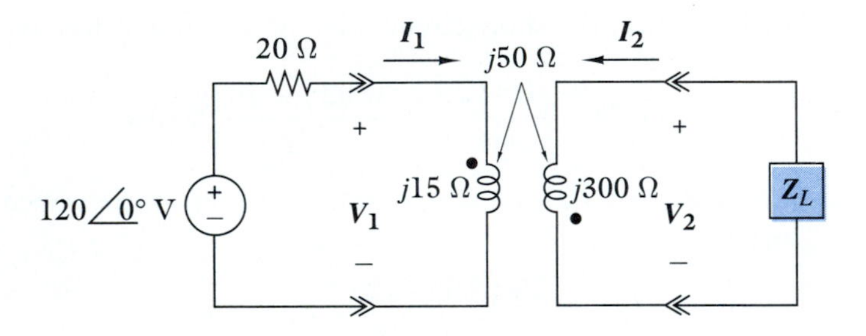

Figure 14-11

SOLUTION:

(a) To obtain the Thévenin equivalent seen by the load in Fig. 14-11 we need the open-circuit voltage and short-circuit current at the output interface. Given the reference directions in Fig. 14-11 the coupling is subtractive and the coupled coil element equations are

$$\mathbf{V}_1 = j15\mathbf{I}_1 - j50\mathbf{I}_2$$

$$\mathbf{V}_2 = -j50\mathbf{I}_1 + j300\mathbf{I}_2$$

The KVL equation around the primary circuit is

$$120\angle 0^\circ = 20\mathbf{I}_1 + \mathbf{V}_1$$

Given these KVL and element equations we can now determine the open-circuit voltage and short-circuit current.

With an open circuit at the output, $\mathbf{I}_2 = 0$ and the first element equation yields $\mathbf{V}_1 = j15\mathbf{I}_1$. When this result is substituted into the KVL equation we can solve for $\mathbf{I}_1$:

$$\mathbf{I}_1 = \frac{120\angle 0^\circ}{20 + j15} = 4.80\angle{-36.9^\circ}$$

Inserting this value into the second element equation yields the open-circuit voltage:

$$\mathbf{V}_T = \mathbf{V}_2 = -j50\mathbf{I}_1 = 240\angle{-126.9^\circ}\ \text{V(rms)}$$

With a short circuit at the output $\mathbf{V}_2 = 0$ and the second element equation yields $\mathbf{I}_1 = 6\mathbf{I}_2$. When this relationship for $\mathbf{I}_1$ is substituted into the first element equation we obtain

$$\mathbf{V}_1 = j15(6\mathbf{I}_2) - j50\mathbf{I}_2 = j40\mathbf{I}_2$$

Inserting $\mathbf{V}_1 = j40\mathbf{I}_2$ and $\mathbf{I}_1 = 6\mathbf{I}_2$ into the KVL equation allows us to solve for the short-circuit current:

$$\mathbf{I}_N = -\mathbf{I}_2 = -\frac{120\angle 0^\circ}{120 + j40} = -0.9 + j0.3$$

Given the open-circuit voltage and short-circuit current, the Thévenin impedance is

$$Z_T = \frac{\mathbf{V}_T}{\mathbf{I}_N} = \frac{240\angle{-126.9^\circ}}{-0.9 + j0.3} = 80 + j240\ \Omega$$

Finally, the Thévenin voltage and impedance allow us to calculate the maximum average power available at the load in Fig. 14-11:

$$P_{MAX} = \frac{|\mathbf{V}_T|^2}{4R_T} = \frac{(240)^2}{4 \times 80} = 180\ \text{W}$$

(b) For $Z_L = 400\ \Omega$ the current through the load is

$$\mathbf{I}_L = \frac{\mathbf{V}_T}{400 + Z_T} = \frac{240\angle{-126.9^\circ}}{480 + j240} = 0.447\angle{-153.4^\circ}\ \text{A(rms)}$$

The average power delivered to a 400-Ω load resistance is

$$P_L = |\mathbf{I}_L|^2 R_L = (0.447)^2 \times 400 = 79.9\ \text{W}$$

EXERCISE 14-5

Calculate the maximum average power available at the interface in Fig. 14-12.

ANSWER:
125 mW

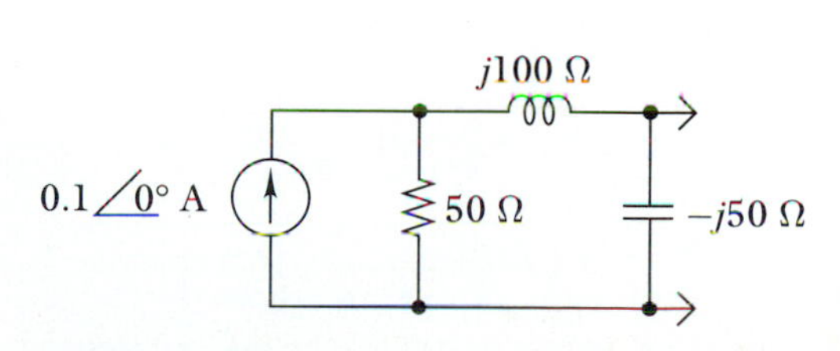

Figure 14-12

 14-5 LOAD-FLOW ANALYSIS

The analysis of ac electrical power systems is one of the major applications of phasor circuit analysis. Although the loads on power systems change during the day, these variations are extremely slow compared with the period of the 50/60 Hz sinusoid involved.[2] Consequently, electrical power system analysis can be carried out using steady-state concepts and phasors. In fact, it was the study of the steady-state performance of ac power equipment that lead Steinmetz to advocate phasors in the first place.

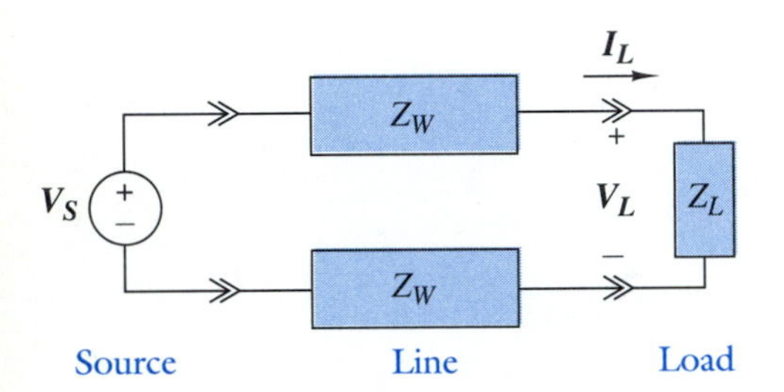

Figure 14-13 A simple electrical power system.

In this section we treat ac power analysis using the simplest possible model of an electrical power system. The model in Fig. 14-13 is a series circuit with an ac source connected to a load via power lines whose wire impedances are Z_W. In a *load-flow problem* we are asked to find the source voltage required to deliver a specified power and voltage to the load. The analysis approach is similar to the unit output method. That is, we begin with conditions at the load and work backward through the circuit to establish the required source voltage. The load-flow problem is different from the maximum power transfer problem studied in the previous section. In the maximum power transfer problem the source is fixed and the load is adjusted to achieve a conjugate match. Conjugate matching does not apply to power systems because the load is fixed and the source is adjusted to meet customer demands.

Large industrial customers are charged for their reactive power, so in some cases it is desirable to reduce the load reactance. Since power system loads almost always are inductive, the net reactive power of the load can be reduced by adding a capacitor in parallel with the load. The amount of the negative reactive power taken by the capacitor is selected to cancel most or all of the positive reactance power drawn by the inductive load. Physically, this means that the oscillatory interchange of energy represented by reactive power takes place between the capacitor and inductance in the load, rather than between the load inductance and the lossy power system.

Adding parallel capacitance is called *power factor correction*, since the net power factor of the composite load is made as close to unity as possible. If the power factor is increased to unity, then the net reactance is zero and the load is in resonance. Power factor correction reduces the reactive power but does not change the average power delivered to the load.

[2]In the United States commercial ac power systems operate at 60 Hz. In most of the rest of the world the standard operating frequency is 50 Hz.

The following examples illustrate ac power analysis problems including load-flow and power-factor correction.

EXAMPLE 14-7

In this problem the two parallel connected impedances in Fig. 14-14 are the load in the power system model in Fig. 14-13. With $V_L = 480$ V (rms), load Z_1 draws an average power of 10 kW at a lagging power factor of 0.8 and load Z_2 draws 12 kW at a lagging power factor of 0.75. The line impedances in Fig. 14-13 are $Z_W = 0.35 + j1.5\ \Omega$. Find the source complex power S_S and voltage $\mathbf{V}_S$.

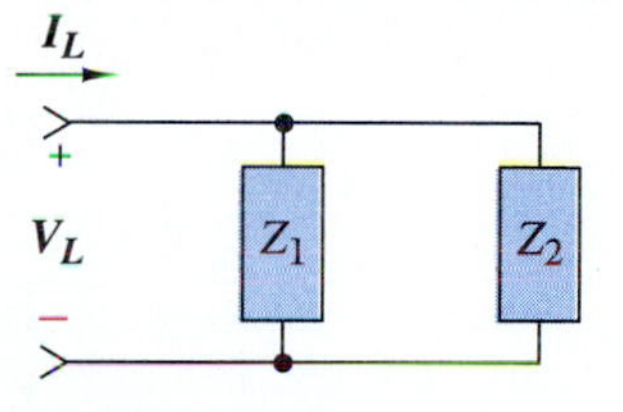

Figure 14-14

SOLUTION:

The complex powers delivered to each load are

$$S_{L1} = 10 + j10\tan(\cos^{-1} 0.8) = 10 + j7.5 \text{ kVA}$$

$$S_{L2} = 12 + j12\tan(\cos^{-1} 0.75) = 12 + j10.6 \text{ kVA}$$

The total complex power delivered to the composite load is

$$S_L = S_{L1} + S_{L2} = 22 + j18.1 = 28.5\angle 39.4^\circ \text{ kVA}$$

Using the load voltage as the phase reference we find the load current.

$$\mathbf{I}_L^* = \frac{S_L}{\mathbf{V}_L} = \frac{28500\angle 39.4^\circ}{480\angle 0^\circ} = 59.4\angle 39.4^\circ \text{ A (rms)}$$

or $\mathbf{I}_L = 59.4\angle -39.4^\circ$ A (rms). We need to find the complex power delivered to the line in order to obtain the required source power:

$$S_W = 2|\mathbf{I}_L|^2(R_W + jX_W) = 2 \times (59.4)^2(0.35 + j1.5)$$
$$= 2.47 + j10.6 \text{ kVA}$$

Using the conservation of complex power, we find the source must produce

$$S_S = S_L + S_W = 24.5 + j28.7 = 37.7\angle 49.5^\circ \text{ kVA}$$

For the series model in Fig. 14-13 the source and load currents are equal so the source power is $S_S = \mathbf{V}_S(\mathbf{I}_L)^*$. Given the source power and load current we find that the required source voltage is

$$\mathbf{V}_S = \frac{S_S}{\mathbf{I}_L^*} = \frac{37700\angle 49.5^\circ}{59.4\angle 39.4^\circ} = 635\angle 10.1^\circ \text{ V (rms)}$$

EXAMPLE 14-8

Using the loads and line impedances defined in Example 14-7:

(a) Find the parallel capacitance needed to correct the load power factor to unity.

(b) Find the required source outputs S_S and $\mathbf{V}_S$ with the capacitor connected. Assume the power system frequency is 60 Hz.

SOLUTION:

(a) To determine the capacitance it is necessary to relate the reactive power of a capacitor to the load voltage. Since the capacitor is in parallel with the load, the current through it is $\mathbf{I}_C = j\omega C\mathbf{V}_L$. Therefore, the reactive power of the capacitor can be written as

$$Q_C = |\mathbf{I}_C|^2 X_C = |j\omega C\mathbf{V}_L|^2 \left(\frac{-1}{\omega C}\right) = |\mathbf{V}_L|^2(-\omega C)$$

Note that Q_C is negative. In Example 14-7 the complex power of the two parallel loads is found to be

$$S_L = P_L + jQ_L = 22 + j18.1 \text{ kVA}$$

To correct the power factor to unity the capacitive reactive power must cancel the inductive reactive power of the load, that is, $Q_C = -Q_L$. Given the given load voltage, inductive reactive power of the load, and system frequency, the required capacitance is

$$C = \frac{Q_L}{V_L^2\omega} = \frac{18100}{(480)^2(2\pi 60)} = 208 \ \mu\text{F}$$

(b) With the power factor correction capacitance connected in parallel with the load, the complex power delivered to the composite load is $S_L = 22 + j0$ kVA. Using the load voltage for the phase reference we find that the load current is

$$\mathbf{I}_L^* = \frac{S_L}{\mathbf{V}_L} = \frac{22000 + j0}{489\angle 0^\circ} = 45.8\angle 0^\circ \ \text{ A (rms)}$$

The power absorbed by the line is

$$S_W = |\mathbf{I}_L|^2 2(R_W + jX_W) = (45.8)^2 \times 2(0.35 + j1.5)$$
$$= 1.47 + j6.3 \text{ kVA}$$

Thus, with power factor correction the source must produce

$$S_S = S_L + S_W = 23.5 + j6.3 = 24.3\angle 15^\circ \text{ kVA}$$

and the required source voltage is

$$\mathbf{V}_S = \frac{S_S}{\mathbf{I}_L^*} = \frac{24300\angle 15^\circ}{45.8\angle 0^\circ} = 531\angle 15^\circ \text{ V (rms)}$$

In Example 14-7 we found that the source outputs without power factor correcton are $S_S = 25.5 + j28.7$ kVA and $\mathbf{V}_S = 635\angle 10.1^\circ$ V(rms). With power factor delivering the same average power to the load requires lower outputs from the source. Thus, reactive power is a burden to a power system even though it represents a lossless interchange of energy at the terminals of the load.

EXERCISE 14-6

Find the source voltage and apparent power required to deliver 2400 V (rms) to a load that draws 25 kVA at a 0.85 lagging power factor from a line with a total line impedance of $2Z_W = 4 + j20\ \Omega$.

ANSWERS:
2.55 kV (rms) and 26.6 kVA at a lagging power factor of 0.82

Application Note

EXAMPLE 14-9

The electrical power for most residential customers in the United States is supplied by the 60-hertz, 110/220 V(rms) single-phase, three-wire system modeled in Fig. 14-15. The term *single phase* means that the phasors representing the two source voltages are in phase. The three lines connecting the sources and loads are labeled A, B, and N (for neutral). The impedances Z_W and Z_N

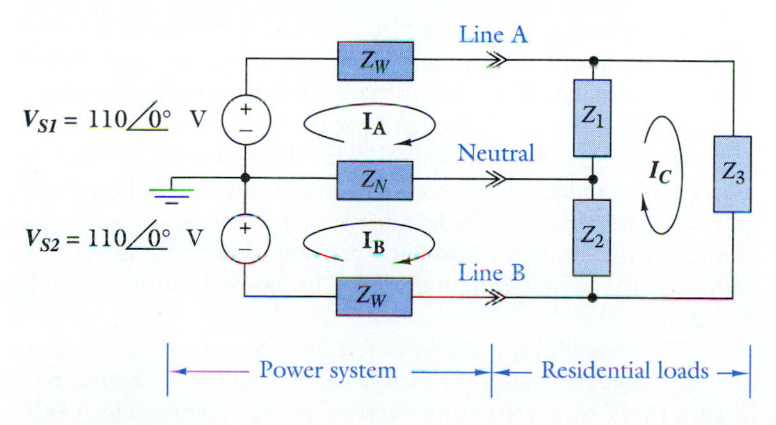

Figure 14-15

are small compared with the load impedances so the load voltages differ from the source voltages by only a few percent. The impedances Z_1 and Z_2 connected from Lines A or B to Neutral represent small appliance and lighting loads, which require 110 V(rms) service. The impedance Z_3 connected between Lines A and B are heavier loads such as water heaters, stoves, air conditioners, or clothes dryers, which require 220 V(rms) service.

When the two source voltages are exactly equal and $Z_1 = Z_2$, the system is said to be balanced. Under balanced conditons the current in the neutral wire is zero. To show why the neutral current is zero we write two mesh-current equations:

$$\text{Mesh A:} \quad (Z_W + Z_1 + Z_N)\boldsymbol{I_A} - Z_N\boldsymbol{I_B} - Z_1\boldsymbol{I_C} = \boldsymbol{V_{S1}}$$

$$\text{Mesh B:} \quad -Z_N\boldsymbol{I_A} + (Z_W + Z_2 + Z_N)\boldsymbol{I_B} - Z_2\boldsymbol{I_C} = \boldsymbol{V_{S2}}$$

For balanced conditions $\boldsymbol{V_{S1}} = \boldsymbol{V_{S2}}$ and $Z_1 = Z_2 = Z_L$. Subtracting the Mesh B equation from the Mesh A equation yields the condition

$$(Z_W + Z_L + 2Z_N)(\boldsymbol{I_A} - \boldsymbol{I_B}) = 0$$

This condition requires $\boldsymbol{I_A} - \boldsymbol{I_B} = 0$, since the impedance sum cannot be zero for all loads. Therefore, the net current in the neutral line is zero and theoretically the neutral wire can be disconnected. In practice the balance is never perfect and the neutral is included for safety reasons. But even so, the current in the neutral is usually less than the line currents, so losses in the feeder lines are reduced.

14-6 THREE-PHASE CIRCUITS

The three-phase system in Fig. 14-16 is the predominant method of generating and distributing ac electrical power. The system uses four lines (A, B, C, N) to transmit power from the source to the loads. The symbols stand for the three phases A, B, and C, and a neutral line labeled N. The three-phase generator in Fig. 14-16 is modeled as three independent sources, although the physical hardware is a single unit with three separate windings. Similarly, the loads are modeled as three separate impedances, although the actual equipment may be housed within a single container.

The terminology Y-connected and Δ-connected refers to the two ways the source and loads can be electrically connected. Figure 14-17 shows the same electrical arrangement as Fig. 14-16 with the elements rearranged to show the "Y" and "Δ" nature

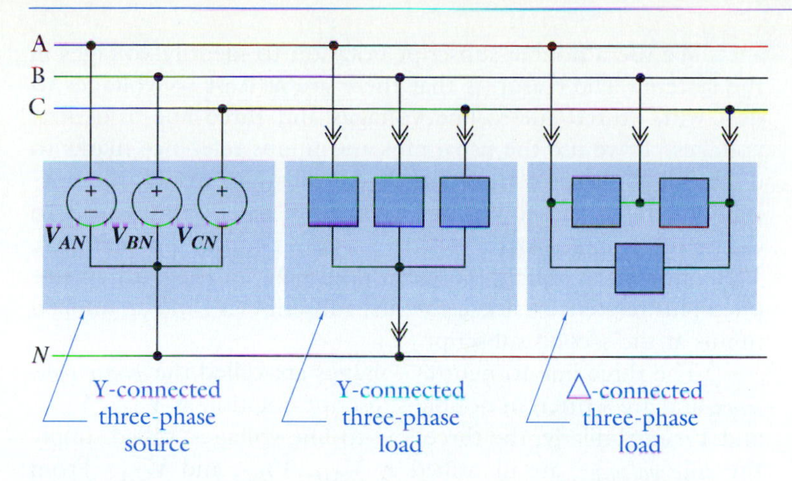

Figure 14-16 A three-phase ac electrical power system.

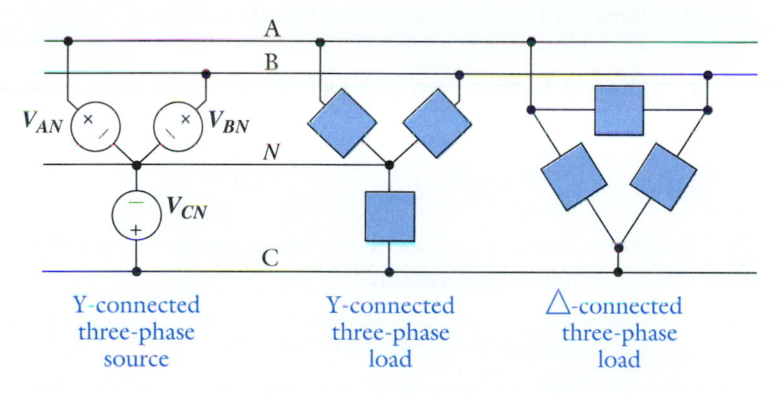

Figure 14-17 A three-phase power system with the loads rearranged.

of the connections (the "Δ" is upside down in the figure). The circuit diagrams in the two figures are electrically equivalent, but we will use the form in Fig. 14-16 because it highlights the purpose of the system. The student need only remember that in a Y connection the three elements are connected from line to neutral, while in the Δ connection they are connected from line to line. In most systems the source is Y connected while the loads can be either Y or Δ, although the latter is more common.

Three-phase sources usually are Y-connected because the Δ-connection involves a loop of voltage sources. Large currents may circulate in this loop if the three voltages do not exactly sum to zero. In analysis situations, Δ-connected sources are awkward because it is impossible to uniquely determine the current in each source.

We use a double-subscript notation to identify voltages in the system. The reason is that there are at least six voltages to deal with: three line-to-line voltages and three line-to-neutral voltages. If we use the usual plus and minus reference marks to define all of these voltages, our circuit diagram would be hopelessly cluttered and confusing. Hence, we use two subscripts to define the points across which a voltage is defined. For example, $\mathbf{V}_{XY}$ means the voltage between points X and Y, with an implied plus reference mark at the first subscript (X) and an implied minus at the second subscript (Y).

The three line-to-neutral voltages are called the *phase voltages*, and are written in double subscript notation as $\mathbf{V}_{AN}$, $\mathbf{V}_{BN}$, and $\mathbf{V}_{CN}$. Similarly, the three line-to-line voltages, called simply the *line voltages*, are identified as $\mathbf{V}_{AB}$, $\mathbf{V}_{BC}$, and $\mathbf{V}_{CA}$. From the definition of the double-subscript notation it follows that $\mathbf{V}_{XY} = -\mathbf{V}_{YX}$. Using this result and KVL we derive the relationships between the line voltages and phase voltages:

$$\begin{aligned} \mathbf{V}_{AB} &= \mathbf{V}_{AN} + \mathbf{V}_{NB} = \mathbf{V}_{AN} - \mathbf{V}_{BN} \\ \mathbf{V}_{BC} &= \mathbf{V}_{BN} + \mathbf{V}_{NC} = \mathbf{V}_{BN} - \mathbf{V}_{CN} \\ \mathbf{V}_{CA} &= \mathbf{V}_{CN} + \mathbf{V}_{NA} = \mathbf{V}_{CN} - \mathbf{V}_{AN} \end{aligned} \tag{14-22}$$

$\mathbf{V}_{CN} = V_P \angle 240°$ V $j\,Im$

$\mathbf{V}_{AN} = V_P \angle 0°$ V Re

$\mathbf{V}_{BN} = V_P \angle 120°$ V

(a)

A balanced three-phase source produces phase voltages that obey the following two constraints:

$$|\mathbf{V}_{AN}| = |\mathbf{V}_{BN}| = |\mathbf{V}_{CN}| = V_P$$

$$\mathbf{V}_{AN} + \mathbf{V}_{BN} + \mathbf{V}_{CN} = 0 + j0$$

That is, the phase voltages have equal amplitudes (V_P) and sum to zero. There are two ways to satisfy these constraints:

POSITIVE PHASE SEQUENCE	NEGATIVE PHASE SEQUENCE
$\mathbf{V}_{AN} = V_P\angle 0°$	$\mathbf{V}_{AN} = V_P\angle 0°$
$\mathbf{V}_{BN} = V_P\angle -120°$	$\mathbf{V}_{BN} = V_P\angle -240°$
$\mathbf{V}_{CN} = V_P\angle -240°$	$\mathbf{V}_{CN} = V_P\angle -120°$

(14-23)

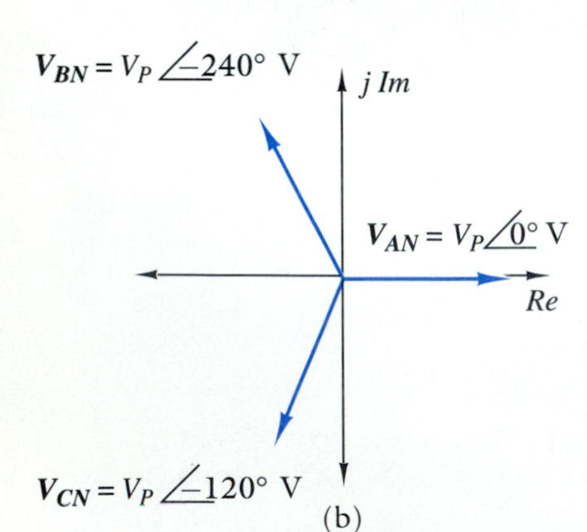

Figure 14-18 Two possible phase sequences: (a) Positive. (b) Negative.

Figure 14-18 shows the phasor diagrams for the positive and negative phase sequences. It is apparent that both sequences involve three equal-length phasors that are separated by an angle of 120°. As a result, the sum of any two phasors cancels the third. In the positive sequence the Phase B voltage lags the Phase A voltage by 120°. In the negative sequence Phase B lags by 240°. It also is apparent that we can convert one phase sequence into the other by simply interchanging the labels on Lines B and C.

From a circuit analysis viewpoint there is no conceptual difference between the two sequences. Consequently, in analysis problems we will use the positive phase sequence unless otherwise stated.

However, the reader is cautioned that "no conceptual difference" does not mean that phase sequence is unimportant. It turns out that three-phase motors run in one direction when the positive sequence is applied, and in the opposite direction for the negative sequence. This could be a matter of some importance if the motor is driving a conveyor belt at a sewage treatment facility. In practice, it is essential that there be no confusion about which is line A, B, and C, and whether the source phase sequence is positive or negative.

A simple relationship between the line and phase voltages is obtained by substituting the positive phase sequence voltages from Eq. (14-23) into the phasor sums in Eq. (14-22). For the first sum

$$\begin{aligned}\mathbf{V}_{AB} &= \mathbf{V}_{AN} - \mathbf{V}_{BN}\\ &= V_P\angle 0^\circ - V_P\angle{-120^\circ}\\ &= V_P(1+j0) - V_P\left(-\frac{1}{2} - j\frac{\sqrt{3}}{2}\right)\\ &= V_P\left(\frac{3}{2} + j\frac{\sqrt{3}}{2}\right)\\ &= \sqrt{3}V_P\angle 30^\circ\end{aligned} \tag{14-24}$$

Using the other sums we find the other two positive sequence line voltages as

$$\begin{aligned}\mathbf{V}_{BC} &= \sqrt{3}V_P\angle{-90^\circ}\\ \mathbf{V}_{CA} &= \sqrt{3}V_P\angle{-210^\circ}\end{aligned} \tag{14-25}$$

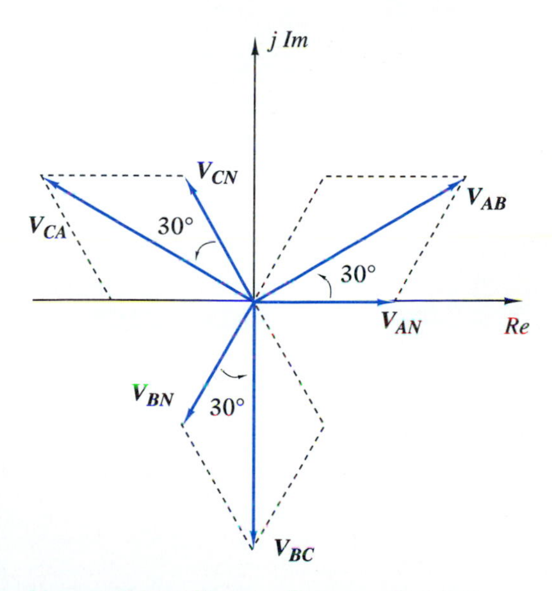

Figure 14-19 Phasor diagram showing phase and line voltages for the positive phase sequence.

Figure 14-19 shows the phasor diagram of these results. The line voltage phasors have the same amplitude and are displaced from each other by 120°. Hence, they obey equal-amplitude and zero-sum constraints like the phase voltages.

If we denote the amplitude of the line voltages as V_L, then

$$V_L = \sqrt{3}V_P \tag{14-26}$$

In a balanced three-phase system the line voltage amplitude is $\sqrt{3}$ times the phase voltage amplitude. This ratio appears in equipment descriptions such as 277/480 V three phase, where 277 is the phase voltage and 480 the line voltage.

It is necessary to choose one of the phasors as the zero-phase reference when defining three-phase voltages and currents. Usually the reference is the Line A phase voltage (i.e., $\mathbf{V}_{AN} =$

$V_P\angle 0°$) as illustrated in Figs. 14-18 and 14-19. Unless otherwise stated, $\mathbf{V}_{AN}$ will be used as the phase reference in this chapter.

EXERCISE 14-7

A balanced Y-connected three-phase source produces positive sequence phase voltages with 2400-V (rms) amplitudes. Write expressions for the phase and line voltage phasors.

ANSWERS:

$\mathbf{V}_{AN} = 2400\angle 0°$ $\mathbf{V}_{BN} = 2400\angle -120°$ $\mathbf{V}_{CN} = 2400\angle -240°$

$\mathbf{V}_{AB} = 4160\angle +30°$ $\mathbf{V}_{BC} = 4160\angle -90°$ $\mathbf{V}_{CA} = 4160\angle -210°$

EXERCISE 14-8

Given that $\mathbf{V}_{BC} = 480\angle +135°$ in a balanced, positive sequence, three-phase system, write expressions for the three phase-voltage phasors.

ANSWERS:

$\mathbf{V}_{AN} = 277\angle -135°$ $\mathbf{V}_{BN} = 277\angle +105°$ $\mathbf{V}_{CN} = 277\angle -15°$

14-7 THREE-PHASE AC POWER ANALYSIS

This section treats the analysis of balanced three-phase circuits. We first treat the direct analysis problem beginning with the Y-connected source and load shown in Fig. 14-20. In a direct analysis problem we are given the source phase voltages $\mathbf{V}_{AN}$, $\mathbf{V}_{BN}$, and $\mathbf{V}_{CN}$ and the load impedances Z. Our objective is to determine the three line currents $\mathbf{I}_A$, $\mathbf{I}_B$, and $\mathbf{I}_C$ and the total complex power delivered to the load.

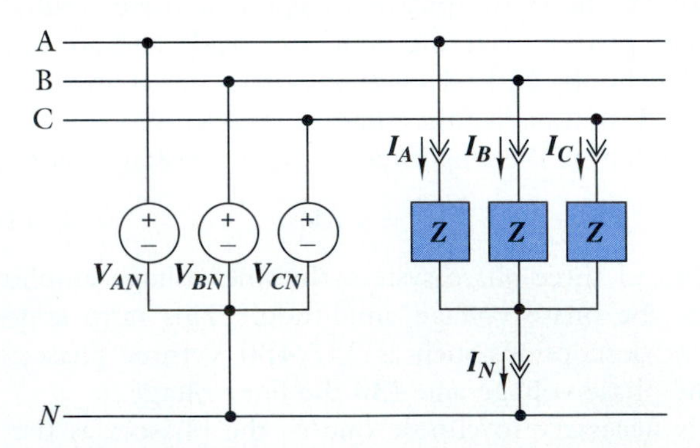

Figure 14-20 A balanced three-phase power system with Y-connected source and load.

Y-Connected Source and Y-Connected Load

The load in Fig. 14-20 is balanced because the phase impedances in the legs of the Y are equal. With the neutral point at the load connected the voltage across each phase impedance is a phase voltage. Using $\mathbf{V}_{AN}$ as the phase reference we find that the line currents are

$$\begin{aligned}\mathbf{I}_A &= \frac{\mathbf{V}_{AN}}{Z} = \frac{V_P\angle 0^\circ}{|Z|\angle\theta} = \frac{V_P}{|Z|}\angle -\theta \\ \mathbf{I}_B &= \frac{\mathbf{V}_{BN}}{Z} = \frac{V_P\angle -120^\circ}{|Z|\angle\theta} = \frac{V_P}{|Z|}\angle(-120^\circ - \theta) \\ \mathbf{I}_C &= \frac{\mathbf{V}_{CN}}{Z} = \frac{V_P\angle -240^\circ}{|Z|\angle\theta} = \frac{V_P}{|Z|}\angle(-240^\circ - \theta)\end{aligned} \tag{14-27}$$

Figure 14-21 shows the phasor diagram of the line currents and phase voltages.

The line current phasors in Eq. (14-27) and Fig. 14-21 have the same amplitude I_L, where

$$I_L = \frac{V_P}{|Z|} \quad \text{(Y-connected load)} \tag{14-28}$$

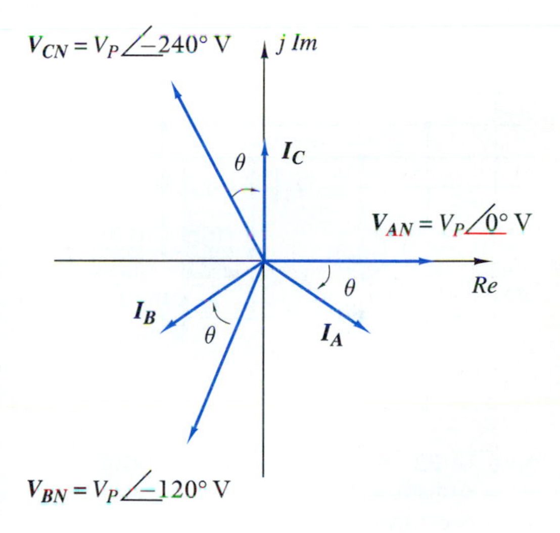

Figure 14-21 Line currents and phase voltages in a balanced three-phase power system.

The line currents have equal amplitudes and are symmetrically disposed at 120° intervals so they obey the zero-sum condition $\mathbf{I}_A + \mathbf{I}_B + \mathbf{I}_C = 0$. Applying KCL at the neutral point of the load in Fig. 14-20, we find that $\mathbf{I}_N = \mathbf{I}_A + \mathbf{I}_B + \mathbf{I}_C = 0$.

Thus, in a balanced *Y-Y* circuit there is no current in the neutral line. The neutral connection could be replaced by any impedance whatsoever, including infinity, without affecting the power delivered to the load. In other words, the neutral wire can be disconnected without changing the circuit response. Real systems may or may not have a neutral wire, but in working problems it usually is helpful to draw the neutral line, since it serves as the reference point for the phase voltages.

The total complex power delivered to the load is

$$\begin{aligned}S_L &= \mathbf{V}_{AN}\mathbf{I}_A^* + \mathbf{V}_{BN}\mathbf{I}_B^* + \mathbf{V}_{CN}\mathbf{I}_C^* \\ &= (V_P\angle 0^\circ)(I_L\angle\theta) + (V_P\angle -120^\circ)(I_L\angle 120^\circ + \theta) \\ &\quad + (V_P\angle -240^\circ)(I_L\angle 240^\circ + \theta) \\ &= 3V_P I_L\angle\theta\end{aligned} \tag{14-29}$$

Since $V_P = V_L/\sqrt{3}$, the expression for complex power can also be written using the line voltage:

$$S_L = \sqrt{3}V_L I_L\angle\theta \tag{14-30}$$

In either Eq. (14-29) or (14-30) the power factor angle θ is the angle of the per phase impedance of the Y-connected load.

EXERCISE 14-9

A Y-connected load with $Z = 10 + j4\ \Omega$/phase is driven by a balanced, positive sequence three-phase generator with $V_L = 4.16$ kV (rms). Using $\mathbf{V}_{AN}$ as the phase reference:

(a) Find the line currents.

(b) Find the average and reactive power delivered to the load.

ANSWERS:

(a) $\mathbf{I}_A = 223\angle{-21.8^\circ}$ A $\quad \mathbf{I}_B = 223\angle{-141.8^\circ}$ A
$\mathbf{I}_C = 223\angle{-261.8^\circ}$ A

(b) $P_L = 1.49$ MW $Q_L = 0.597$ MVAR.

Y-Connected Source and Δ-Connected Load

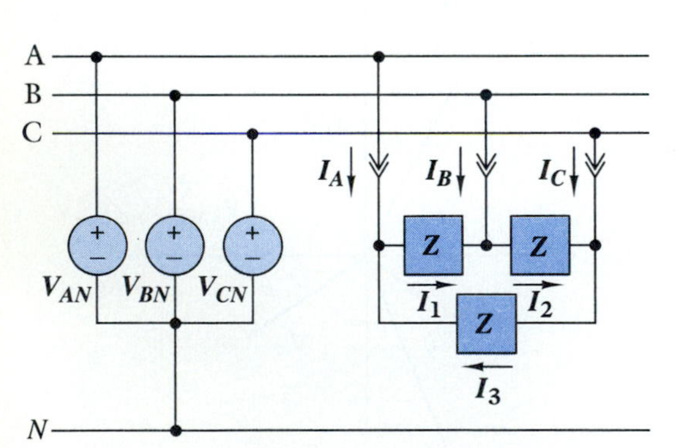

Figure 14-22 A balanced three-phase power system with a Y-connected source and a Δ-connected load.

We now turn to the balanced Δ-connected load shown in Fig. 14-22. When the objective is to determine the line currents and total complex power, we can convert the load into an equivalent wye using a Δ-to-Y transformation. As discussed in Section 13-3, the balanced Δ-connected load in Fig. 14-22 can be replaced by an the equivalent Y-connected load with leg impedances

$$Z_Y = \frac{Z}{3} \tag{14-31}$$

Using the Δ-to-Y transformation, we reduce the problem to a circuit in which the source and load are Y-connected. The resulting Y-Y configuration can then be analyzed to determine line currents and total power as discussed above.

However, when we need to know the current or power delivered to each leg of the delta load we must determine the phase currents $\mathbf{I}_1$, $\mathbf{I}_2$, and $\mathbf{I}_3$ shown in Fig. 14-22. The phase currents can be expressed in terms of the phase impedances and the line voltage. Assuming the positive phase sequence and using V_{AN} as the phase reference these expressions are

$$\begin{aligned}
\mathbf{I}_1 &= \frac{\mathbf{V}_{AB}}{Z} = \frac{V_L\angle 30^\circ}{|Z|\angle\theta} = \frac{V_L}{|Z|}\angle(30^\circ - \theta)\\
\mathbf{I}_2 &= \frac{\mathbf{V}_{BC}}{Z} = \frac{V_L\angle{-90^\circ}}{|Z|\angle\theta} = \frac{V_L}{|Z|}\angle(-90^\circ - \theta)\\
\mathbf{I}_3 &= \frac{\mathbf{V}_{CA}}{Z} = \frac{V_L\angle{-210^\circ}}{|Z|\angle\theta} = \frac{V_L}{|Z|}\angle(-210^\circ - \theta)
\end{aligned} \tag{14-32}$$

The phase currents have the same amplitude I_P defined as

$$I_P = \frac{V_L}{|Z|} \quad (\Delta\text{-connected load}) \tag{14-33}$$

Although of no immediate physical importance, note that the phase currents sum to zero because they have equal amplitudes and are symmetrically disposed at 120° intervals.

Using the results in Eq. (14-32), the complex power delivered to each leg of the delta load is

$$\begin{aligned} S_1 &= \mathbf{V}_{AB} \times \mathbf{I}_1^* = (V_L\angle 30°)(I_P\angle\theta - 30°) = V_L I_P\angle\theta \\ S_2 &= \mathbf{V}_{BC} \times \mathbf{I}_2^* = (V_L\angle -90°)(I_P\angle\theta + 90°) = V_L I_P\angle\theta \\ S_3 &= \mathbf{V}_{CA} \times \mathbf{I}_3^* = (V_L\angle -210°)(I_P\angle\theta + 210°) = V_L I_P\angle\theta \end{aligned} \tag{14-34}$$

Clearly the total complex power delivered to the Δ-connected load is

$$S_L = S_1 + S_2 + S_3 = 3V_L I_P\angle\theta \tag{14-35}$$

where the power factor angle θ is the angle of the per phase impedance of the Δ-connected load.

The result in Eq. (14-35) can be put in the same form as Eq. (14-30) by replacing the phase current I_P by the line current I_L. As noted above, the line currents for a balanced Δ-connected load can be calculated using a Δ-to-Y transformation. In the transformed circuit the line current amplitude is $I_L = V_P/|Z_Y|$, where Z_Y is the impedance in each leg of the equivalent Y-connected load. But in a balanced system $V_P = V_L/\sqrt{3}$, and according to Eq. (14-31) $Z_Y = Z/3$. Therefore, the amplitudes of the line and phase currents in a Δ-connected load are related as follows:

$$I_L = \frac{V_L/\sqrt{3}}{|Z/3|} = \sqrt{3}\frac{V_L}{|Z|} = \sqrt{3}I_P \tag{14-36}$$

When $I_P = I_L/\sqrt{3}$ is substituted into Eq. (14-35) we obtain

$$S_L = \sqrt{3}V_L I_L\angle\theta \tag{14-37}$$

Equations (14-30) and (14-37) are identical, which means that the relationship applies to balanced three-phase loads whether Y- or Δ-connected. In either case, the power factor angle θ is the angle of the per phase impedance of the load because the transformation $Z_Y = Z/3$ does not alter the phase angle of the phase impedance.

EXAMPLE 14-10

A Δ-connected load with $Z = 40 + j30\ \Omega$/phase is driven by a balanced, positive sequence three-phase generator with $V_L =$ 2400 V (rms). Using $\mathbf{V}_{AN}$ as the phase reference:

(a) Find the phase currents.

(b) Find the line currents.

(c) Find the average and reactive power delivered to the load.

SOLUTION:

(a) The first phase current is

$$\mathbf{I}_1 = \frac{\mathbf{V}_{AB}}{Z} = \frac{2400\angle 30^\circ}{40 + j30} = 48.0\angle - 6.87^\circ \text{ A (rms)}$$

Since the circuit is balanced the other phase currents all have amplitudes of $I_P = 48.0$ A and are displaced at 120° intervals:

$$\mathbf{I}_2 = 48\angle{-126.87^\circ} \text{ A (rms)}$$

$$\mathbf{I}_3 = 48\angle{-246.87^\circ} \text{ A (rms)}$$

(b) Using $V_P = 2400/\sqrt{3}$ and Δ-Y transformation with $Z_Y = (40 + j30)/3$ we find the phase A line current is

$$\mathbf{I}_A = \frac{\mathbf{V}_{AN}}{Z_Y} = \frac{2400/\sqrt{3}}{(40 + j30)/3} = 83.1\angle{-36.9^\circ} \text{ A (rms)}$$

Since the circuit is balanced the other line currents have amplitudes of $I_L = 83.1$ A and are displaced at 120° intervals.

$$\mathbf{I}_B = 83.1\angle{-156.9^\circ} \text{ A (rms)}$$

$$\mathbf{I}_C = 83.1\angle{-276.9^\circ} \text{ A (rms)}$$

(c) The complex power delivered to the load is

$$S_L = \sqrt{3}V_L I_L \angle\theta = \sqrt{3} \times 2400 \times 83.1\angle 36.9^\circ = 345\angle 36.9^\circ \text{ kVA}$$

Therefore, $P_L = 276$ kW and $Q_L = 207$ kVAR.

EXERCISE 14-10

Exercise 14-9 involved a balanced Y-connected load with $Z = 1 - +j4\ \Omega$/phase and a line voltage of $V_L = 4.16$ kV (rms). In this exercise these same parameters apply to a Δ-connected load. Using $\mathbf{V}_{AN}$ as the phase reference:

(a) Find the phase currents.

(b) Find the line currents.

(c) Find the average and reactive power delivered to the load.

ANSWERS:

(a) $I_1 = 386\angle+8.20^\circ$ A $I_2 = 386\angle-111.8^\circ$ A
$I_3 = 386\angle-231.8^\circ$ A

(b) $I_A = 669\angle-21.8^\circ$ A $I_B = 669\angle-141.8^\circ$ A
$I_C = 669\angle-261.8^\circ$ A

(c) $P_L = 4.47$ MW $Q_L = 1.79$ MVAR ■

Three-Phase Load Flow

The direct analysis method determines the currents and power delivered to the load when the load impedances and source voltages are given. In a load-flow problem, we are required to find the source outputs required to deliver a prescribed complex power at a specified voltage level.

A simplified model of the three-phase circuit for this problem is shown in Fig. 14-23. The impedance Z_W represents the wire impedances of the power lines connecting the source and load. The load configuration (Y or Δ) or per phase impedances are not required because the line currents can be found from Eq. (14-37) using the specified load power and voltage. Given the line currents we next find the line losses (S_W), and then the complex power required from the source using the conservation of complex power $S_S = S_L + S_W$. Finally, with the line currents and the source complex power known, we again use Eq. (14-37), this time to find the required source voltages.

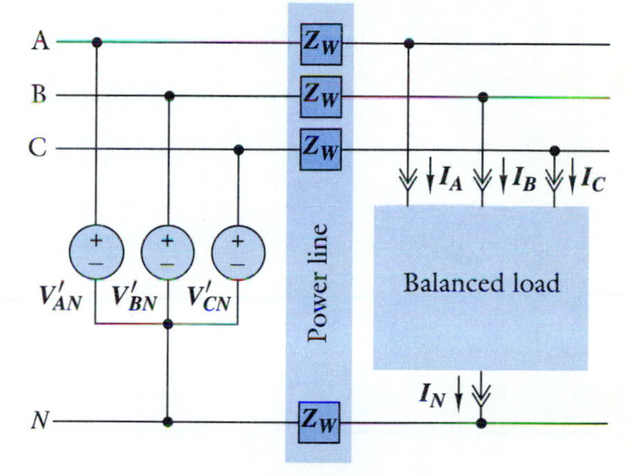

Figure 14-23 A balanced three-phase power system with line impedances.

The approach outlined above is basically the same as the method used to handle single-phase two-wire load-flow problems in Section 14-5. The only complication here is that we must account for the power in all three phases as we work our way from the load to the source. Also, we must distinguish between the line and phase voltages at the load and those at the source. We use a prime to indicate source voltages as indicated in Fig. 14-23. The next example is a three-phase load-flow problem.

EXAMPLE 14-11

At a line voltage of 4800 V (rms) a balanced three-phase load draws 100 kW at 0.8 lagging power factor from a balanced three-phase source via power lines each with $Z_W = 2 + j20\ \Omega$. Find the complex power produced by the source and the source line voltage.

SOLUTION:

The complex power delivered to the load is

$$S_L = 100 + j100\tan(\cos^{-1} 0.8) = 100 + j75 = 125\angle 36.9^\circ \text{ kVA}$$

We now use Eq. (14-37) to calculate the line current for the given line voltage at the load:

$$I_L = \frac{|S_L|}{\sqrt{3}V_L} = \frac{125000}{\sqrt{3} \times 4800} = 15.04 \text{ A (rms)}$$

The total complex power delivered to all three power lines is

$$S_W = 3I_L^2(R_W + jX_W) = 3(15.04)^2(2 + j20)$$
$$= 1356 + j13563 \text{ VA}$$

The above computation includes only three lines because the circuit is balanced and there is no current in the neutral line. The complex power required from the source is found using conservation of complex power as

$$S_S = S_L + S_W = 101 + j88.6 = 134.6\angle 41.15^\circ \text{ kVA}$$

We now use Eq. (14-37) again to find the required source line voltage:

$$V_L' \frac{|S_S|}{\sqrt{3}I_L} = \frac{134600}{\sqrt{3} \times 15.04} = 5168 \text{ V (rms)}$$

EXERCISE 14-11

When the phase voltage is 277 V (rms), a balanced three-phase load draws 8 kVA at lagging power factor of 0.9. The load is fed by a balanced three-phase source via lines with $Z_W = 0.3 + j5\ \Omega$. Find

(a) the line current

(b) the line voltage required at the source

ANSWERS:

(a) 9.63 A

(b) 526 V

Application Note

EXAMPLE 14-12

Electrical power is generated and transmitted in three-phase form even though most residential and small commercial customers are single-phase loads. The primary reason is that three-phase devices and transmission lines (see Problem 14-58) are smaller and more efficient than the corresponding single-phase equipment handling the same power. Part of the advantage enjoyed by three phase is that the instantaneous power in a balanced three-phase circuit or device is constant.

To show that $p_T(t)$ is constant we write the instantaneous power in each phase of the balanced three-phase circuit:

$$p_A(t) = v_{AN}(t) \times i_A(t)$$
$$= \left[\sqrt{2}V_P \cos(\omega t)\right] \times \left[\sqrt{2}I_L \cos(\omega t - \theta)\right]$$
$$p_B(t) = v_{BN}(t) \times i_B(t)$$
$$= \left[\sqrt{2}V_P \cos(\omega t - 120°)\right] \times \left[\sqrt{2}I_L \cos(\omega t - 120° - \theta)\right]$$
$$p_C(t) = v_{CN}(t) \times i_C(t)$$
$$= \left[\sqrt{2}V_P \cos(\omega t - 240°)\right] \times \left[\sqrt{2}I_L \cos(\omega t - 240° - \theta)\right]$$

where $\sqrt{2}V_P$ is the peak amplitude of each line-to-neutral voltage and $\sqrt{2}I_L$ is the peak amplitude of each line current. Using the trigonometric identity

$$[\cos x] \times [\cos y] = \frac{1}{2}\cos(x - y) + \frac{1}{2}\cos(x + y)$$

the phase powers can be put into the form

$$p_A(t) = V_P I_L \cos\theta + V_P I_L \cos(2\omega t - \theta)$$
$$p_B(t) = V_P I_L \cos\theta + V_P I_L \cos(2\omega t - 240° - \theta)$$
$$p_C(t) = V_P I_L \cos\theta + V_P I_L \cos(2\omega t - 480° - \theta)$$

Each phase power is a constant or dc term $V_P I_L \cos\,\theta$ plus a double-frequency ac term. The double-frequency sinusoids all have the same amplitude $V_P I_L$ and are symmetrically disposed at 120° intervals because −480° is the same as −120°. Therefore, the double-frequency sinusoids sum to zero and the total instantaneous power is

$$p_T(t) = p_A(t) + p_B(t) + p_C(t) = 3V_P I_L \cos\theta$$

The fact that the total instantaneous power is constant means that three-phase generators and motors involve constant mechanical torque. As a result, there is smoother operation with less vibration at the electromechanical interfaces of the system.

14-8 AC Power Analysis Using SPICE

As previously discussed in Section 13-8, SPICE can be used to solve sinusoidal steady-state circuit analysis problems. SPICE is not particularly well suited to ac power analysis because it only solves for phasor voltages and currents. As a result, average and reactive power calculations must be carried out in the postanalysis phase. In addition, SPICE requires that the magnitude and phase of all independent sources be specified in advance, and all loads have to be expressed in the form of resistance, inductance, or capacitance. In other words, SPICE only solves direct analysis-type problems and not the more interesting load-flow problems.

To solve a load-flow problem using SPICE we must recast it as a direct analysis problem. To do this we create circuit models for loads specified in terms of the complex power and voltage. Either the series or parallel models shown in Fig. 14-24 are used as equivalent circuits. In either case, the complex power delivered to the load is $S = \mathbf{V}\mathbf{I}^*$. For the series model we use the constraint $\mathbf{V} = Z_S\mathbf{I}$ to write the complex power in the form

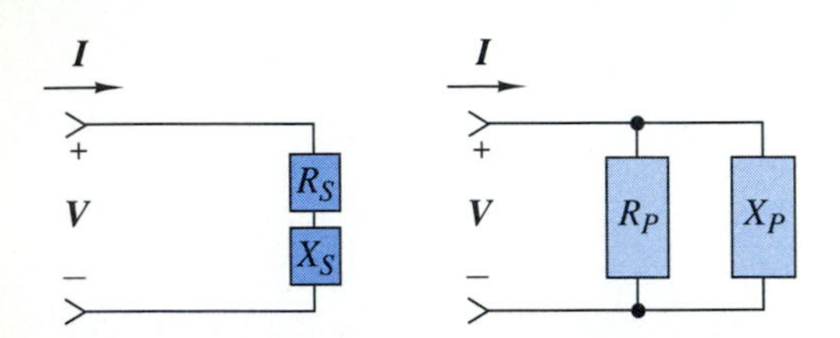

Figure 14-24 Series and parallel circuits for modeling electrical loads specified in terms of complex power.

$$S = Z_S\mathbf{I}\mathbf{I}^*$$

$$P + jQ = (R_S + jX_S)|\mathbf{I}|^2$$

Since $|\mathbf{I}|^2 = |S|^2/|\mathbf{V}|^2$ this equation can be written in the form

$$P + jQ = (R_S + jX_S)\frac{|S|^2}{|\mathbf{V}|^2}$$

Equating real and imaginary parts yields the series circuit parameters in terms of complex power and voltage magnitude.

$$R_S = P\frac{|\mathbf{V}|^2}{|S|^2} \quad \text{and} \quad X_S = Q\frac{|\mathbf{V}|^2}{|S|^2} \quad \text{(Series Circuit)} \qquad (14\text{-}38)$$

For the parallel circuit in Fig. 14-24 we use the constraint $\mathbf{I} = Y_P\mathbf{V}$ to write the complex power in the form

$$S = \mathbf{V}(Y_P\mathbf{V})^* = |\mathbf{V}|^2\left(\frac{1}{R_P} + \frac{1}{jX_P}\right)^*$$

$$P + jQ = |\mathbf{V}|^2\left(\frac{1}{R_P} + \frac{j}{X_P}\right)$$

Equating real and imaginary parts yields the parallel circuit parameters in terms of complex power and voltage magnitude.

$$R_P = \frac{|\mathbf{V}|^2}{P} \quad \text{and} \quad X_P = \frac{|\mathbf{V}|^2}{Q} \quad \text{(Parallel Circuit)} \qquad (14\text{-}39)$$

Equations (14-38) and (14-39) and the system frequency relate the element values of a series or parallel equivalent circuit to the complex power delivered to a two-terminal load. These equivalent circuits provide element values used in SPICE to support load-flow analysis.

Exercise 14-12

At 480 V (rms) a two-terminal load draws 10 kVA at lagging power factor of 0.85 from a 60-Hz line. Find the element values for the series and parallel equivalent circuits.

ANSWERS:
Series: A 19.6-Ω resistor in series with a 32.2-mH inductor
Parallel: A 27.1-Ω resistor in parallel with a 116-mH inductor ■

The direct analysis capability of SPICE can be used to solve load-flow problems using the proportionality property of linear circuits. We create a SPICE circuit model for the problem by relating complex power and voltage to element values as discussed above. We set the amplitudes of the independent sources to unity and use SPICE to solve for the voltage delivered to the load. In the postanalysis phase we find the scale factor required to raise the delivered voltage to the level specified in the load-flow problem. This scale factor can then be applied to all voltages and currents, including those provided by the sources. The scaled voltages and currents allow us to calculate the complex power that the source must deliver in the load-flow problem. The next example illustrates using SPICE to solve a load-flow problem.

Example 14-13

A balanced three-phase load draws $P = 120$ kW and $Q = 90$ kVAR at a phase voltage of $V_P = 2400$ V (rms). The load is connected to a balanced three-phase, 60-Hz source via four identical lines whose wire impedances are $Z_W = 4 + j20\ \Omega$. Use SPICE to find the line currents, phase voltages, and complex power that must be produced by the source.

SOLUTION:

To use SPICE we must create a circuit model for the load. Using the parallel model in Fig. 14-24, Eqs. (14-39) yields the per phase resistance and reactance of a Y-connected load.

$$R_P \frac{(2400)^2}{120000/3} = 144\ \Omega/\text{phase} \quad X_P = \frac{(2400)^2}{90000/3} = 192\ \Omega/\text{phase}$$

Each phase of the Y-connected load is simulated as a resistance $R_P = 144\ \Omega$ in parallel with an inductance $L_P = 192/(2\pi 60) = 0.509$ H. Likewise, the Phase A, B, C, and neutral lines are simulated using a resistance $R_W = 4\ \Omega$ in series with an inductance of $L_W = 20/(2\pi 60) = 53.0$ mH.

Figure 14-25 shows a circuit model for a SPICE solution of the direct analysis problem. The Y-connected source has bal-

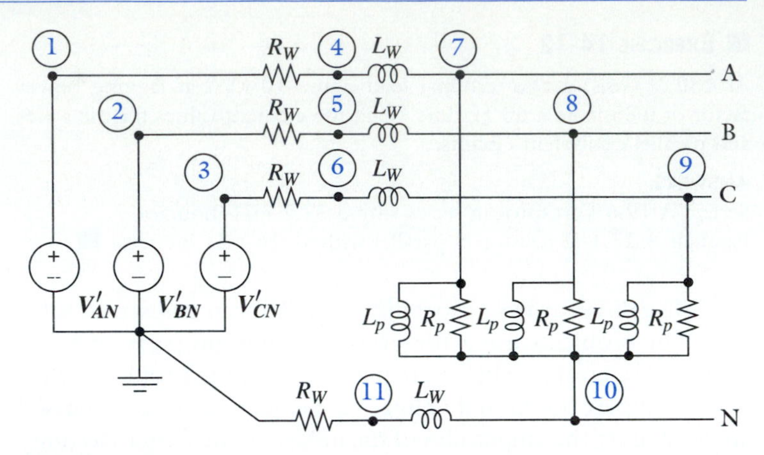

Figure 14-25

anced phase voltages with unit amplitudes:

$$V'_{AN} = 1\angle 0°, V'_{BN} = 1\angle -120°, \text{ and } V'_{CN} = 1\angle -240°.$$

The circuit in Fig. 14-25 includes a return line connecting the load and source neutral points. However, since both the source and load are balanced, we expect the current in the neutral line to be zero.

A SPICE circuit file for the network in Fig. 14-25 is listed below:

```
*EXAMPLE 14-13
VAN  1  0  AC  1 0
VBN  2  0  AC  1 -120
VCN  3  0  AC  1 -240
RW1  1  4  4
RW2  2  5  4
RW3  3  6  4
LW1  4  7  53M
LW2  5  8  53M
LW3  6  9  53M
RP1  7 10  144
RP2  8 10  144
RP3  9 10  144
LP1  7 10  509M
LP2  8 10  509M
LP3  9 10  509M
RW4 11  0  4
LW4 10 11  53M
.AC  LIN 1 60 60
.PRINT AC VM(7) IM(RW1) IP(RW1) IM(RW4)
.END
```

The important results listed in the SPICE output file are

```
FREQ VM(7)      IM(RW1)   IP(RW1)    IM(RW4)
60   8.787E-01 7.629E-03 -4.283E-01 5.301E-18
```

The SPICE output IM(RW4) = 5.3×10^{-18} confirms that the neutral current is zero within the simulation accuracy. The SPICE output VM(7) is the amplitude of the line-to-neutral load voltage on Phase A for source voltage amplitudes of 1 V. The load-flow problem requires the phase voltage at the load be 2400 V line to neutral. The scale factor needed to bring all voltages and currents up to the load-flow problem level is

$$K = \frac{2400}{0.8787} = 2731$$

The line-to-neutral source voltages required to produce $V_P =$ 2400 V at the load are obtained using the scale factor.

$$V'_{AN} = 2731\angle 0°,\ V'_{BN} = 2731\angle -120°, \text{ and}$$
$$V'_{CN} = 2731\angle -240° \text{ V (rms)}$$

The SPICE outputs IM(RW1) and IP(RW1) are the magnitude and phase angle (in degrees) of the line current in Phase A when the source voltages are 1 V. When the voltages are scaled, the Phase A line current becomes

$$\mathbf{I}_A = (2731)(0.007629)\angle -42.83° = 20.83°\angle -42.83° \text{ A(rms)}$$

Because the system is balanced the other line currents are

$$\mathbf{I}_B = 20.83\angle -42.83° - 120° = 20.83\angle -162.8° \text{ A (rms)}$$

$$\mathbf{I}_C = 20.83\angle -42.83° - 240° = 20.83\angle -282.8° \text{ A (rms)}$$

Given the Phase A line current and source Phase A line-to-neutral voltage, we find that the complex power the source must produce is

$$S_S = 3(2731\angle 0°)(20.83\angle +42.83°) = 171\angle 42.83° \text{ kVA}$$

The example above illustrates using SPICE to solve a three-phase load-flow problem. The method involves a fair amount of preprocessing to develop a circuit model and postanalysis processing to scale the SPICE results to the load-flow levels. The hand-computation effort required to use SPICE is not significantly different from the effort required to solve the whole problem by hand analysis in the first place. In other words, for simple single-phase or balanced three-phase circuits SPICE does not offer any particular advantage over hand computations.

SPICE offers some advantage when the circuit is unbalanced. The most severe unbalanced situation is a fault condition in which a short circuit occurs either from line to line or line to neutral. Designing protection measures for a power system involves calculating the line currents under *normal (balanced)* and *fault (unbalanced)* conditions. The final example illustrates fault current calculations using SPICE.

EXAMPLE 14-14

A balanced three-phase load draws $P = 120$ kW and $Q = 90$ kVAR at a phase voltage of $V_P = 2400$ V (rms). The load is connected to a balanced three-phase 60-Hz source via four identical lines whose wire impedances are $Z_W = 4 + j20\ \Omega$. Use SPICE to find the line currents when a line-to-neutral fault occurs on Phase A at the terminals of the load.

SOLUTION:

To calculate fault currents we must first solve a load-flow problem to find the source voltages under normal conditions prior to the fault. The load conditions given above are the same as in Example 14-13, so the circuit model in Fig. 14-25 applies here as well. The load-flow analysis in Example 14-13 shows that the required phase voltages at the source are

$$\mathbf{V}'_{AN} = 2731\angle 0^\circ, \mathbf{V}'_{BN} = 2731\angle -120^\circ, \text{ and}$$

$$\mathbf{V}'_{CN} = 2731\angle -240^\circ \text{ V (rms)}$$

Under normal conditions these source voltages deliver $S_L = 120 + j90$ kVA to the load at phase voltage of 2400 V (rms).

A line-to-neutral fault on Phase A is a zero-resistance path between Nodes 7 and 10 in Fig. 14-25. The fault can be simulated by setting the load resistance RP1 to 1 milliohm. Listed below is a SPICE circuit file for the circuit in Fig. 14-25 with the simulated fault and source voltages at the prefault load-flow levels:

```
*EXAMPLE 14-14
VAN   1  0   AC   2731 0
VBN   2  0   AC   2731 -120
VCN   3  0   AC   2731 -240
RW1   1  4   4
RW2   2  5   4
RW3   3  6   4
LW1   4  7   53M
LW2   5  8   53M
LW3   6  9   53M
RP1   7 10   1M   *SIMULATES LINE TO NEUTRAL FAULT
RP2   8 10   144
RP3   9 10   144
```

```
LP1  7 10  509M
LP2  8 10  509M
LP3  9 10  509M
RW4 11  0  4
LW4 10 11  53M
.AC  LIN 1 60 60
.PRINT AC IM(RW1) IP(RW1) IM(RW2) IP(RW2) IM(RW3) IP(RW3)
.END
```

The important results listed in the SPICE output file are

```
FREQIM(RW1)   IP(RW1)     IM(RW2)    IP(RW2)    IM(RW3)    IP(RW3)
60 8.333E+01 -7.209E+01 2.474E+01 1.792E+02 2.682E+01 9.043E+01
```

The SPICE output gives the following line currents under the fault condition.

$$\mathbf{I}_A = 83.33\angle{-72.09^\circ}$$

$$\mathbf{I}_B = 24.74\angle{179.2^\circ}$$

$$\mathbf{I}_C = 26.82\angle{90.43^\circ}\ \text{A (rms)}$$

In Example 14-13 the line currents under normal operation were found to be

$$\mathbf{I}_A = 20.83\angle{-42.83^\circ}$$

$$\mathbf{I}_B = 20.83\angle{-162.8^\circ}$$

$$\mathbf{I}_C = 20.83\angle{-282.8^\circ}\ \text{A (rms)}$$

Under the fault condition the line currents have unequal amplitudes and phase angles that are not 120° apart. The fault produces an unbalanced circuit in which the line currents do not sum to zero. By KCL the return current in the neutral line is

$$\mathbf{I}_N = \mathbf{I}_A + \mathbf{I}_B + \mathbf{I}_C = 52.14\angle{-89.25^\circ}\ \text{A (rms)}$$

SUMMARY

- In the sinusoidal steady state the instantaneous power at a two-terminal interface contains average and reactive power components. The average power represents a unidirectional transfer of energy from source to load. The reactive power represents a lossless interchange of energy between the source and load.

- In ac power analysis problems amplitudes of voltage and current phasors are expressed in rms values. Complex power is defined as the product of the voltage phasor and the conjugate of the current phasor. The real part of the complex power is the average power in watts (W), the imaginary part is the reactive power in volt-amperes reactive (VAR), and the magnitude is the apparent power in volt-amperes (VA).
- For inductive loads, the current phasor lags the voltage phasor and the reactive power is positive. For capacitive loads, the current phasor leads the voltage phasor and the reactive power is negative. The maximum average power is delivered by a fixed source to an adjustable load when the source and load impedances are conjugates.
- In a direct analysis problem the load impedance, line impedances, and source voltage are given, and the load voltage, current, and power are the unknowns. In a load-flow problem the line impedances and the load power and voltage are given, and the required source voltage, current, and power are the unknowns.
- Three-phase systems transmit power from source to load on four lines labeled A, B, C, and N. The three line-to-neutral voltages $\boldsymbol{V}_{AN}$, $\boldsymbol{V}_{BN}$, and $\boldsymbol{V}_{CN}$ are called the phase voltages. The three line-to-line voltages $\boldsymbol{V}_{AB}$, $\boldsymbol{V}_{BC}$, and $\boldsymbol{V}_{CA}$ are called line voltages.
- In a balanced three-phase system: (1) the neutral wire carries no current, (2) the line-voltage amplitude V_L is related to the phase-voltage amplitude V_P by $V_L = \sqrt{3}V_P$, (3) the three line currents $\boldsymbol{I}_A$, $\boldsymbol{I}_B$, and $\boldsymbol{I}_C$ have the same amplitude I_L, and (4) the total complex power delivered to a Y- or a Δ-connected load is $\sqrt{3}V_L I_L \angle\theta$, where θ is the angle of the phase impedance.
- SPICE solves only direct analysis ac power problems. The average and reactive power are calculated in postanalysis processing. To solve load-flow problems using SPICE, circuit models are found for loads and proportionality is used to scale the direct analysis results.

EN ROUTE OBJECTIVES AND ASSOCIATED PROBLEMS

ERO 14-1 Complex Power (Sects. 14-1, 14-2)

Given a linear circuit operating in the sinusoidal steady-state:

(a) Find the phasor voltage, current, and complex power delivered at a specified interface

(b) Find the phasor voltages or currents required to deliver a specified complex power at an interface

(c) Find the impedance required to absorb a specified complex power at a given voltage level.

14-1 Calculate average power, reactive power, and power factor for the circuit in Fig. P14-1 using voltages and currents listed below. State whether the circuit is absorbing or delivering net energy.

(a) $v(t) = 170 \cos(\omega t + 45°)$ V, $i(t) = 12 \cos(\omega t - 30°)$ A
(b) $v(t) = 283 \cos(\omega t + 90°)$ V, $i(t) = 0.5 \sin(\omega t - 30°)$ A
(c) $v(t) = 135 \cos(\omega t)$ V, $i(t) = 0.5 \sin(\omega t + 30°)$ A
(d) $v(t) = 37 \sin(\omega t)$ V, $i(t) = 0.115 \cos(\omega t + 130°)$ A

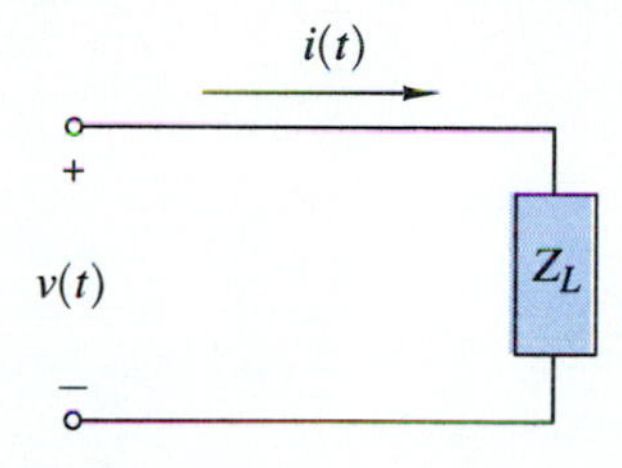

Figure P14-1

14-2 Repeat Prob. 14-1 for the conditions given below.

(a) $\mathbf{V} = 22.5\angle 0°$ V (rms), $\mathbf{I} = 250\angle -105°$ mA (rms)
(b) $\mathbf{V} = 120\angle 35°$ V (rms), $\mathbf{I} = 12.5\angle 115°$ A (rms)
(c) $\mathbf{V} = 2.4\angle 0°$ kV (rms), $Z_L = 250\angle -105°\ \Omega$
(d) $Z_L = 225\angle 0°\ \Omega$, $\mathbf{I} = 120\angle 125°$ mA (rms)

14-3 Find the power factor of the circuit in Fig. P14-1 under the following conditions. State whether the power factor is lagging or leading.

(a) $S = 1000 + j750$ VA
(b) $S = -750 + j1000$ VA
(c) $|\mathbf{V}| = 440$ V (rms), $|Z_L| = 30\ \Omega$, $P_{AVE} = 3$ kW
(d) $|S| = 10$ kVA, $Q = -8$ kVA, $\cos\theta > 0$

14-4 The instantaneous power absorbed by the load in Fig. P14-1 is $p(t) = 1200 + 1200 \cos 2\omega t + 750 \sin 2\omega t$ VA.

(a) Find P_{AVE}, Q, and the power factor.
(b) Find the impedance of the load if the voltage level is 622 V (peak).

14-5 At 2400 V (rms) the load in Fig. P14-1 absorbs an apparent power of 10 kVA at a power factor of 0.8 lagging.

(a) Find the average power, reactive power, and the rms current delivered to the load.
(b) Find the load driving-point impedance.

14-6 The load in Fig. P14-1 absorbs 33 kW of average power and draws 16.1 A (rms) from a 2400-V (rms) line.

(a) Find the apparent power delivered to the load and the power factor.
(b) Find the load driving-point impedance.

14-7 The load in Fig. P14-1 draws 12 A (rms) and absorbs 4.2 kVARS from a 440 V (rms) 60-Hz source. Find the load power factor and impedance.

14-8 An air-conditioning unit is rated at 220 V (rms) and 22 A (rms). What is the unit's driving-point impedance if its power factor is 0.9 lagging? Repeat for 0.8 lagging.

ERO 14-2 AC Power Analysis (Sect. 14-3)

Given a linear circuit operating in the sinusoidal steady state:

(a) Find the voltage, current, and complex power associated with any element.

(b) Find the load impedance required to draw a given complex power from a source.

14-9 A load consisting of a 50-ohm resistor in series with a 100-mH inductor is connected across a 60-Hz voltage source that delivers 240 V (rms). Find the phasor voltage, current, and complex power delivered to the load.

14-10 A load consisting of a 100-Ω resistor in parallel with a 40-μF capacitor is connected across a 400-Hz voltage source that delivers 110 V (rms). Find the phasor voltage, current, and complex power delivered to the load.

14-11 A load consisting of a resistor and capacitor connected in parallel absorbs a complex power $S = 10 - j126$ VA when connected to a 440-V (rms) 60-Hz line. Find the values of R and C.

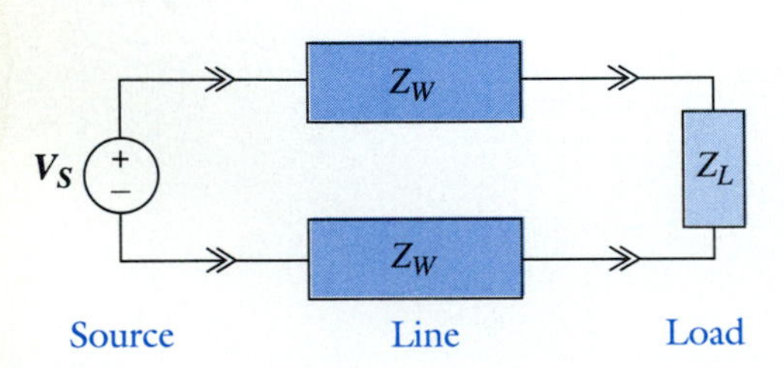

Figure P14-12

14-12 The load in Fig. P14-12 consists of a 400-Ω resistor in parallel with an inductor whose reactance is 800 Ω. The voltage source output is 440 V (rms) at 60 Hz, and the wire impedance of the line is $Z_W = 1 + j10$ Ω.
(a) Find the rms current in the line.
(b) Find the complex power absorbed by the load and line.
(c) Calculate the *transmission efficiency* (η) defined as $\eta = (P_{LOAD}/P_{SOURCE}) \times 100$.

14-13 The load in Fig. P14-12 draws 2.5 kVA at a power factor of 0.9 lagging. The 2400-V (rms) source is delivering 2.65 kVA at a power factor of 0.88 lagging. Find the line current, load impedance, and line impedance.

14-14 The three load impedances in Fig. P14-14 are $Z_1 = 25 + j6$ Ω, $Z_2 = 16 + j4$ Ω, and $Z_3 = 10 + j0$ Ω.
(a) Find the current in Lines A, B, and N.
(b) Find the complex power produced by each source.

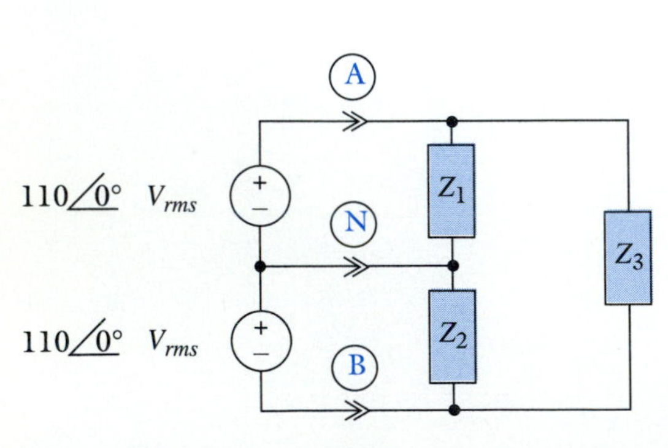

Figure P14-14

14-15 The three loads in Fig. P14-14 are absorbing the following complex powers:

$$S_1 = 1250 + j250 \text{ VA},$$

$$S_2 = 800 + j400 \text{ VA},$$

$$S_3 = 2000 + j0 \text{ VA}.$$

(a) Find the current in Lines A, B and N.
(b) Find the complex power produced by each source.

14-16 The voltage across three loads connected in parallel is 440 V (rms) at 60 Hz. The first load draws an apparent power of 10 kVA at 0.9 lagging power factor. The second load absorbs 8 kW of average power and 6 kVAR of reactive power. The third load absorbs 15 kW of average power and has a power factor of unity.
(a) Find the power factor of the three loads in parallel.
(b) Find the equivalent impedance of the three parallel loads.

ERO 14-3 Maximum Power Transfer (Sect. 14-4)

Given a linear circuit operating in the sinusoidal steady state:
(a) Find the complex power delivered to a load impedance.
(b) Find the maximum average power available at the load and determine the load impedance required to draw the maximum power.

14-17 (a) Find the complex power delivered to the 500-Ω load in Fig. P14-17.
(b) Find the maximum available average power at the interface shown in Fig. P14-17.
(c) Specify the load required to extract the maximum average power.

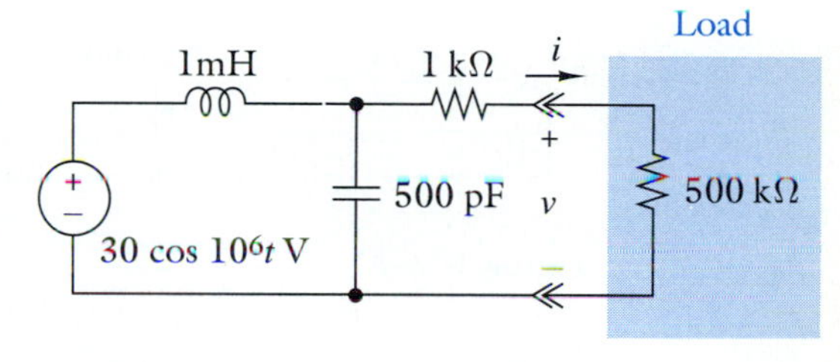

Figure P14-17

14-18 (a) Find the complex power delivered to the load impedance in Fig. P14-18.
(b) Find the maximum available average power at the interface shown in Fig. P14-18.
(c) Specify the load required to extract the maximum average power.

14-19 (a) Find the maximum average power available at the interface in Fig. P14-19.
(b) Specify the values of R and C that will extract the maximum power from the source circuit.

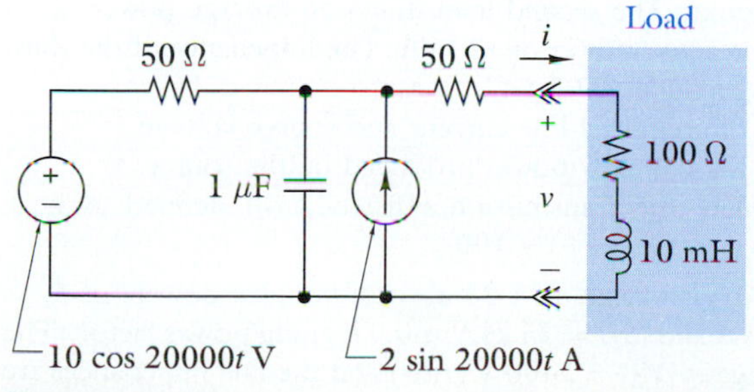

Figure P14-18

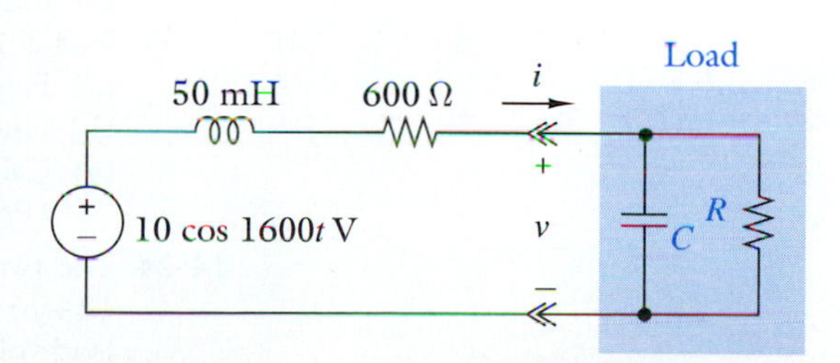

Figure P14-19

14-20 The steady-state open-circuit voltage at the output interface of a source circuit is $10\angle 0°$ V. When a 50-Ω resistor is connected across the interface the steady-state output voltage is $4.68 \angle -20.56°$ V. Find the maximum average power available at the interface and the resistance and reactance required to extract the maximum power.

14-21 (a) Find the complex power delivered to the load impedance in Fig. P14-21.
(b) Find the maximum available average power at the interface shown in Fig. P14-21.
(c) Specify the load required to extract the maximum average power.

14-22 (a) Find the maximum available average power at the interface in the secondary circuit of the transformer shown in Fig. P14-22.
(b) Specify the load required to extract the maximum average power.

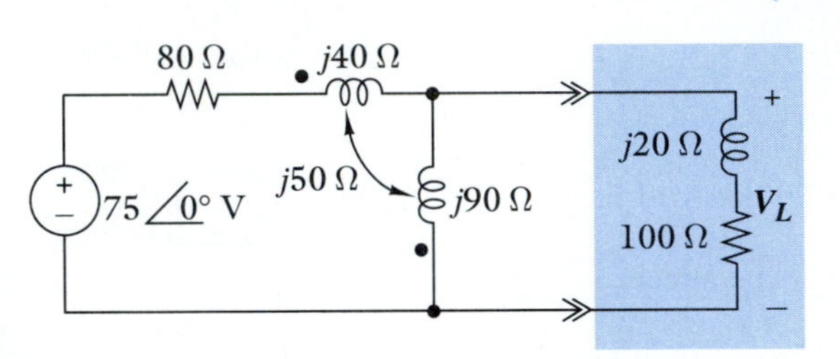

Figure P14-21

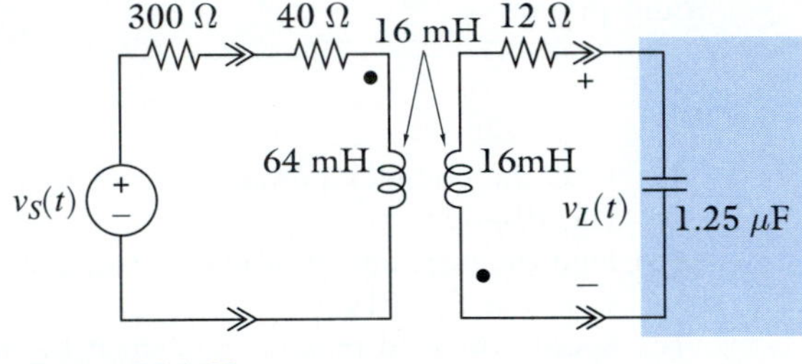

Figure P14-22

ERO 14-4 Load-Flow Analysis (Sect. 14-5)

Given a linear circuit operating in the sinusoidal steady-state: Find the source outputs required to deliver a specified complex power and voltage magnitude at a given interface.

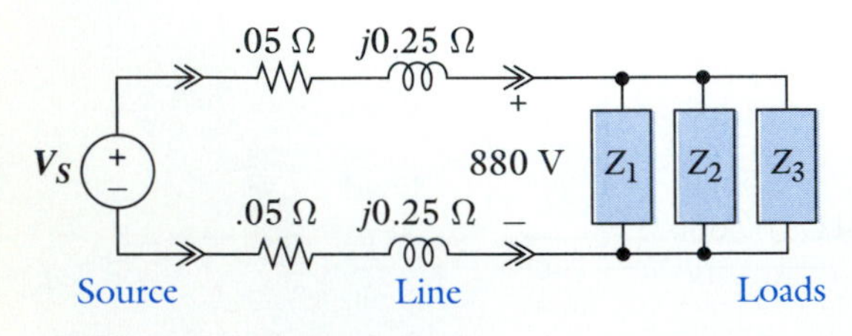

Figure P14-23

14-23 The voltage across the three loads in Fig. P14-23 is 880 V (rms). The first load draws an apparent power of 20 kVA at 0.8 power factor lagging. The second load draws an average power of 16 kW and a reactive power of 12 kVA. The impedance of the third load is $Z_3 = 30 + j22.5$ Ω.
(a) Find the required line current and source voltage.
(b) Find the complex power produced by the source.
(c) Calculate the transmission efficiency (η) defined as $\eta = (P_{LOAD}/P_{SOURCE}) \times 100$.

14-24 The two loads in Fig. P14-24 absorb complex powers of $S_1 = 12 + j6$ kVA and $|S_2| = 25$ kVA at 0.9 lagging power factor. The load voltage is $|V_L| = 2400$ V (rms) and the line impedances are $Z_W = 0.5 + j2$ Ω.

(a) Find the required line current and source voltage.
(b) Find the complex power produced by the source.
(c) Calculate the transmission efficiency (η) defined as $\eta = (P_{LOAD}/P_{SOURCE}) \times 100$.

14-25 The two loads in Fig. P14-24 absorb complex powers of $|S_1| = 16$ kVA at 0.75 lagging power factor and $|S_2| = 25$ kVA at unity power factor. The load voltage is $|\mathbf{V}_L| = 277$ V (rms) and the line impedances are $Z_W = 0.1 + j0.5\ \Omega$.
(a) Find the required line current and source voltage.
(b) Find the complex power absorbed by the line and the source power factor.

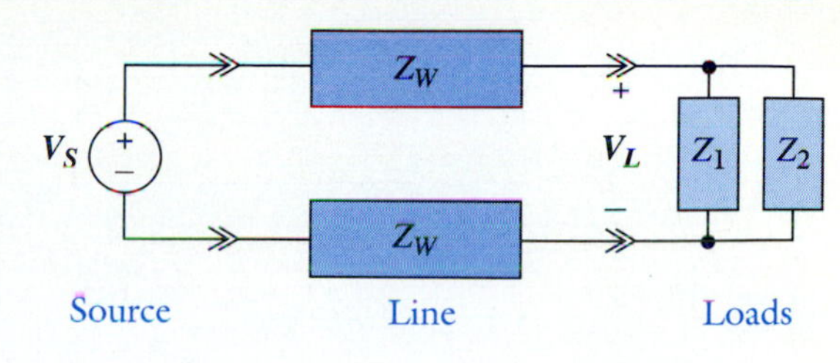

Figure P14-24

14-26 A load is rated at 10 kW at 0.75 lagging power factor when the voltage across the load is 440 V (rms). The load is supplied by a line with a source voltage of 480 V (rms) and line impedances of $0.3 + j2\ \Omega$. Find the actual current, voltage, and complex power delivered to the load.

14-27 The three loads in Fig. P14-27 draw $S_1 = 4 + j1.2$ kVA, $S_2 = 4 + j1.2$ kVA, and $S_3 = 6.5 + j0$ kVA.
(a) Find the current in Lines A, B, and N.
(b) Find the required source voltages.
(c) Find the complex power produced by each source.

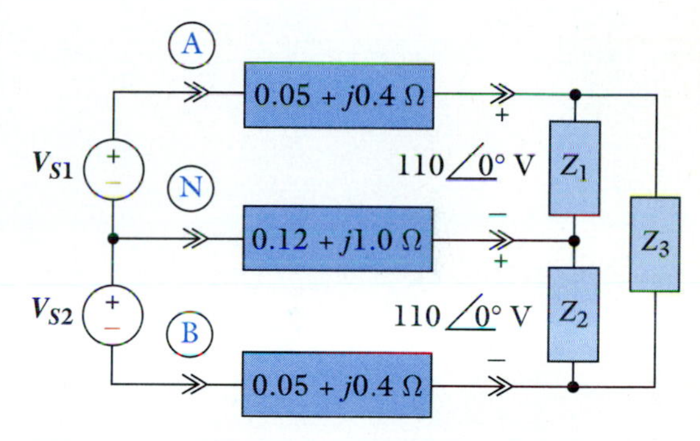

Figure P14-27

14-28 The load in Fig. P14-28 operates at 440 V (rms), 60 Hz and draws 33 kVA at 0.7 power factor lagging.
(a) Calculate the value of C required to produce an overall power factor of 0.95.
(b) Calculate the transmission efficiency with and without the capacitor found in (a).

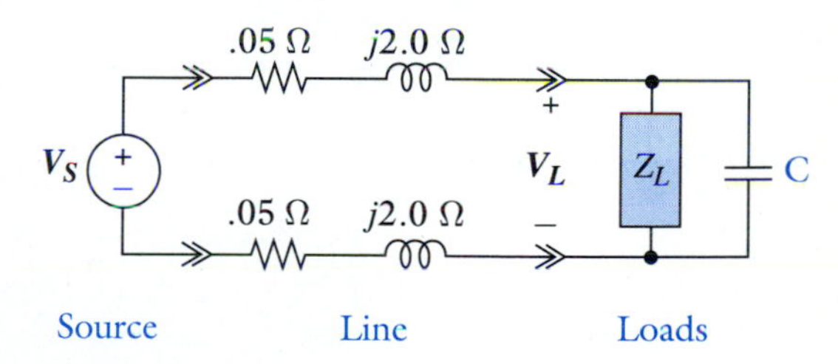

Figure P14-28

14-29 The load in Fig. P14-29 operates at 60 Hz. With $\mathbf{V}_L = 12\angle 0°$ kV (rms) the load draws 1.2 MVA at 0.9 power factor lagging. The first source voltage is $V_{S1} = 12.6 + j1.2$ kV (rms). Find the complex power supplied by each of the sources.

14-30 The load in Fig. P14-30 operates at $|\mathbf{V}_L| = 220$ V (rms) and draws 40 kVA at 0.8 power factor lagging. The turns ratio of the ideal transformer is 10:1 and the line impedances are $Z_W = 0.4 + j2\ \Omega$.
(a) Find the line current on the primary side of the transformer.
(b) Find the source voltage and the complex power produced by the source.
(c) Calculate the transmission efficiency (η) defined as $\eta = (P_{LOAD}/P_{SOURCE}) \times 100$.

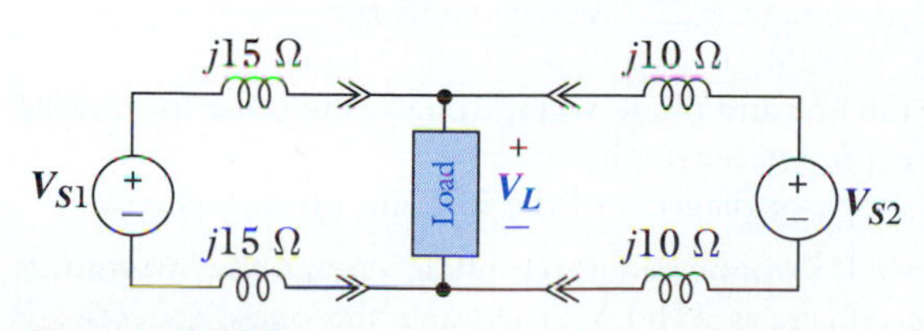

Figure P14-29

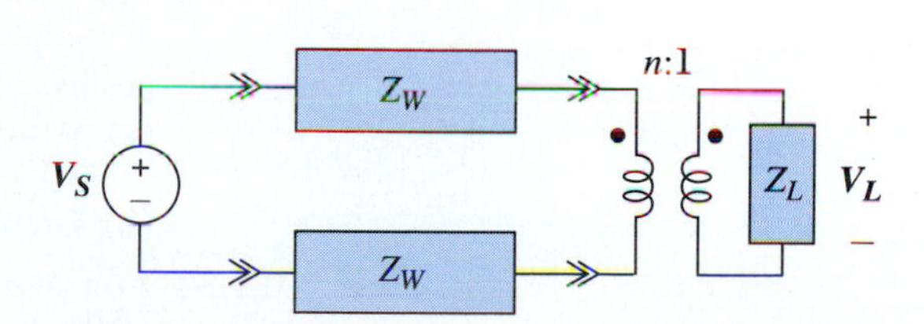

Figure P14-30

14-31 Repeat Prob. 14-30 with a turns ratio of 60:1.

14-32 The transformer in Fig. P14-32 is ideal. With $|\mathbf{V}_L| = 220$ V (rms) the load draws 10 kVA at a power factor of 0.8 lagging. Find the current, voltage, and complex power produced by the source.

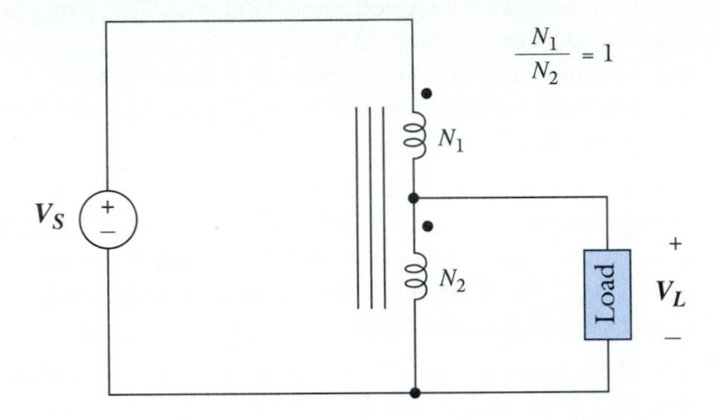

Figure P14-32

ERO 14-5 Three-Phase Power (Sects. 14-6, 14-7)

Given a balanced three-phase circuit operating in the sinusoidal steady state:

(a) Find the line (phase) voltages or currents given the magnitude of the phase (line) voltage or current and the phase sequence.

(b) Find the line currents and total complex power given the line or phase voltage and load impedances.

(c) Find the line currents and phase impedance given the total complex power, and the line or phase voltage.

(d) Find the source outputs required to deliver a specified complex power and voltage magnitude at a given interface.

14-33 In a balanced Y-connected three-phase circuit the magnitude of the phase voltage is 277 V (rms) and the phase sequence is positive.

(a) Write the line and phase-voltage phasors in polar form using $\mathbf{V}_{AN}$ as the reference phasor.

(b) Draw a phasor diagram of the line and phase voltages.

14-34 In a balanced Y-connected three-phase circuit the magnitude of the line voltage is 4160 V (rms) and the phase sequence is positive.

(a) Write the line and phase voltage phasors in polar form using $\mathbf{V}_{AB}$ as the reference phasor.
(b) Draw a phasor diagram of the line and phase voltages.

14-35 In a balanced Δ-connected three-phase circuit the magnitude of the phase current is 57.7 A (rms) and the phase sequence is negative.
(a) Write the line and phase current phasors in polar form using $\mathbf{I}_A$ as the reference phasor.
(b) Draw a phasor diagram of the line and phase currents.

14-36 A balanced Y-connected three-phase load with a per-phase impedance of $10 + j5\ \Omega$ operates with a line voltage magnitude of 208 V (rms) using a positive phase sequence. Using $\mathbf{V}_{AN}$ as the reference phasor:
(a) Find the line current phasors in rectangular form.
(b) Calculate the total complex power delivered to the load.
(c) Draw a phasor diagram showing the line currents and voltages.

14-37 A balanced Y-connected three-phase load with a per-phase impedance of $5\text{-}j2\ \Omega$ operates with a line voltage magnitude of 480 V (rms) using a negative phase sequence. Using $\mathbf{V}_{AN}$ as the reference:
(a) Find the line current phasors in rectangular form.
(b) Calculate the total complex power delivered to the load.
(c) Draw a phasor diagram showing the line currents and voltages.

14-38 A balanced Δ-connected three-phase load with a per-phase impedance of $100\angle 30^\circ\ \Omega$ operates with a line voltage magnitude of 4.8 kV (rms) using a positive phase sequence. Using $\mathbf{V}_{AB}$ as the reference:
(a) Find the line current phasors in rectangular form.
(b) Calculate the total complex power delivered to the load.
(c) Draw a phasor diagram showing the line currents and voltages.

14-39 A balanced Δ-connected three-phase load with a per-phase impedance of $8 + j6\ \Omega$ is connected in parallel with a balanced Y-connected three-phase load with a per-phase impedance of $15\angle 30^\circ\ \Omega$. The phase voltage is $V_P = 277$ V (rms). Find the magnitude of the line currents and the power factor of the combined loads.

14-40 In a balanced Y-connected three-phase load the phase A line current is $\mathbf{I}_A = 100\angle -30^\circ$ A (rms) with a line voltage of $\mathbf{V}_{AB} = 480\angle 30^\circ$ V(rms). Find the phase impedance of the load assuming a positive phase sequence.

14-41 A balanced Y-connected three-phase load absorbs 50 kVA at a power factor of 0.9 lagging when the line voltage magnitude is

480 V (rms).

(a) Find the magnitude of the line current.

(b) Calculate the resistance and reactance of the per-phase impedance.

14-42 A balanced Y-connected three-phase load absorbs 20 kVA when the line current magnitude is 83 A (rms) and the line voltage magnitude is 480 V (rms). Find the resistance and reactance of the per-phase impedance assuming a lagging power factor.

14-43 A balanced Δ-connected three-phase load absorbs 37 kVA at a power factor of 0.8 lagging when the line voltage magnitude is 2400 V (rms).

(a) Find the magnitude of the line current.

(b) Calculate the resistance and reactance of the per-phase impedance.

14-44 Two three-phase loads are connected in parallel. The first is a balanced Y-connected circuit absorbing 28 kVA at a power factor of 0.8 lagging. The second is a balanced Δ-connected load with a per-phase impedance of $40 + j0\ \Omega$. The magnitude of the line voltage at the loads is 480 V (rms). Find the magnitude of the total line current and the total complex power delivered to the loads.

14-45 The wire impedances of a three-phase line are $Z_W = 0.6 + j4\ \Omega$ per phase. The line feeds a balanced load that absorbs a total complex power of $S_T = 16 + j12$ kVA. At the load end of the line the magnitude of the line voltage is 440 V (rms). Find the magnitude of the line voltage at the source end of the line and the source power factor.

14-46 A three-phase line with a line impedances of $Z_W = 0 + j8\ \Omega$ per phase feeds a balanced Y-connected load which absorbs 800 kVA at 0.8 power factor lagging. The magnitude of the line voltage at the load end of the line is 3.8 kV. Find the required line currents and voltages at the source end of the line and the source power factor.

14-47 Figure P14-47 shows a model of a Y-connected generator with an internal impedance of $j2\Omega$/phase. The generator supplies a balanced Y-connected three-phase load via power lines with wire impedances of $1 + j5\ \Omega$/phase.

(a) Find the line currents $\mathbf{I}_A$, $\mathbf{I}_B$, and $\mathbf{I}_C$.

(b) Find the magnitude of the phase voltage at the terminals of the load.

(c) Find the total complex power produced by the generator and absorbed by the load.

(d) Find the transmission efficiency (η) defined as $\eta = (P_{LOAD}/P_{SOURCE}) \times 100$.

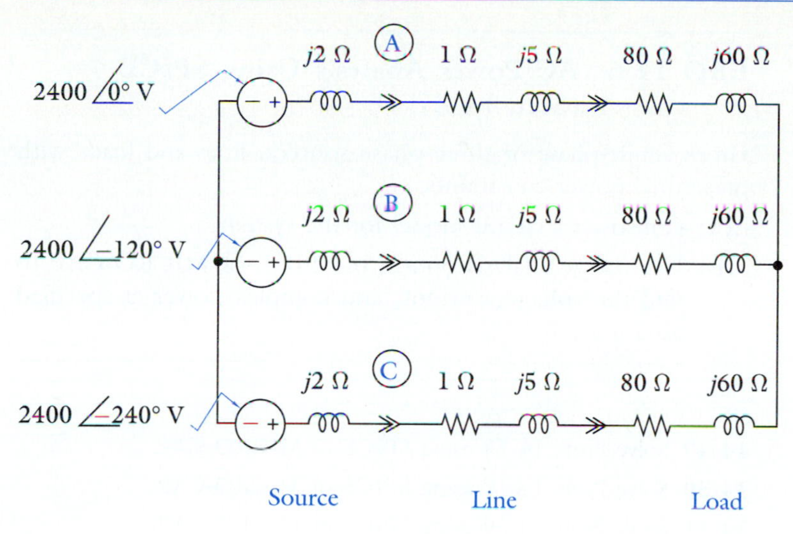

Figure P14-47

14-48 The three-phase load in Fig. P14-48 draws an average power of 100 MW at a power factor of 0.8 lagging. At the terminals of the load the Phase A voltage is $\mathbf{V}_{AN} = 90\angle 0°$ kV (rms). At the terminals of source No. 2 the Phase A voltage is $\mathbf{V}'_{AN} = 99.9\angle 10°$ kV (rms). Find the complex power produced by each source.

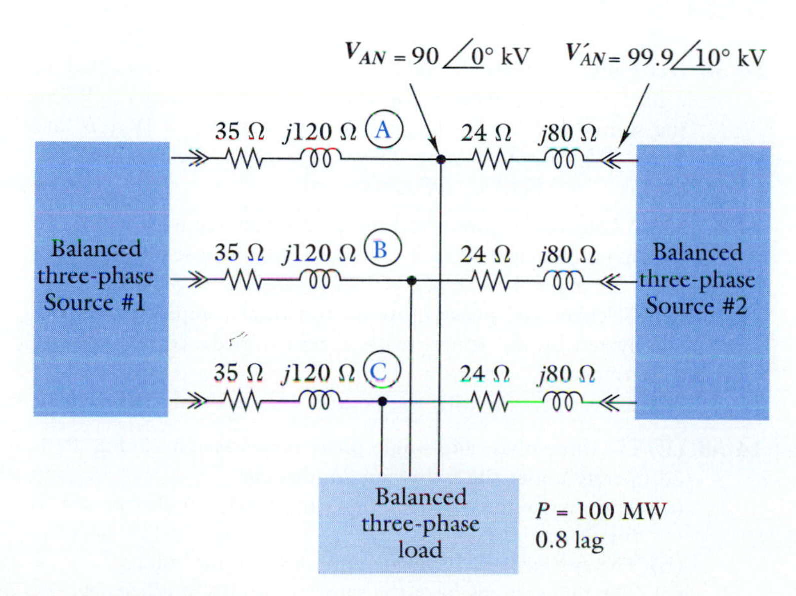

Figure P14-48

ERO 14-6 AC Power Analysis Using SPICE (Sect. 14-8)

Given single-phase or three-phase sources, lines and loads with prescribed power constraints:

(a) Construct a circuit model for the system.

(b) Use the ac analysis option of SPICE or MICRO-CAP to find the voltage, current, and complex power at specified interfaces.

14-49 Solve Prob. 14-15 using SPICE or MICRO-CAP.

14-50 Solve Prob. 14-17 using SPICE or MICRO-CAP.

14-51 Solve Prob. 14-39 using SPICE or MICRO-CAP.

14-52 Solve Prob. 14-44 using SPICE or MICRO-CAP.

14-53 Solve Prob. 14-45 using SPICE or MICRO-CAP.

14-54 Solve Prob. 14-46 using SPICE or MICRO-CAP. Then connect a short-circuit between Lines A and B at the load terminals and use SPICE or MICRO-CAP to solve for the line currents.

14-55 Solve Prob. 14-47 using SPICE or MICRO-CAP. Then connect a short-circuit between Lines A and Neutral at the load terminals and use SPICE or MICRO-CAP to solve for the line currents.

CHAPTER INTEGRATING PROBLEMS

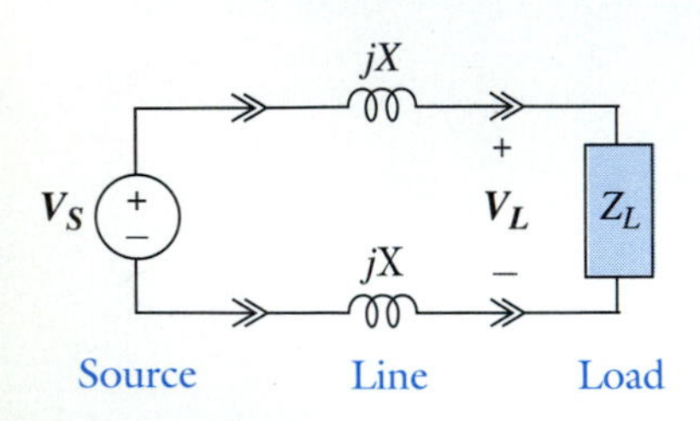

Figure P14-56

14-56 **(A)** Figure P14-56 shows a single-phase source connected to a load via a line whose impedance is purely reactive. When the source and load voltages are written as $\mathbf{V}_S = |\mathbf{V}_S|\angle\alpha$ and $\mathbf{V}_L = |\mathbf{V}_L|\angle\beta$, show that the average power delivered to the load is $P = (|\mathbf{V}_L||\mathbf{V}_S|/2X)\sin(\beta - \alpha)$.

14-57 **(A)** A balanced Y-connected three-phase source with $V_P = 120$ V (rms) is connected to a Y-connected load whose phase impedances are $Z_A = 100\ \Omega$, $Z_B = 100\ \Omega$, and $Z_C = j100\ \Omega$.

(a) Calculate the phase currents and total complex power delivered by the source with a zero impedance neutral wire connected.

(b) Repeat part (a) with the neutral wire disconnected.

14-58 **(E)** The three-phase and single-phase power systems in Fig. P14-58 operate under the following conditions:

(a) The two systems deliver the same total complex power to the load.

(b) The two systems have the same line-to-line voltage.

(c) The two systems have the same transmission efficiency.

(d) The distance from source to load is the same.

(e) The resistance of a line is proportional to its length divided by the cross-sectional area of the wire.

Show that the three-phase system requires 25% less copper in the transmission lines than the single-phase system.

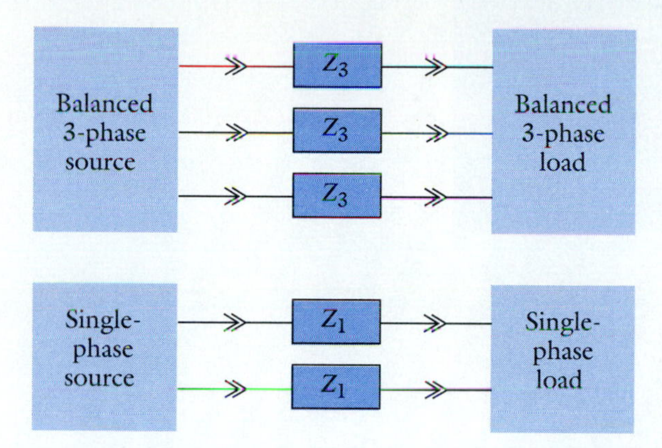

Figure P14-58

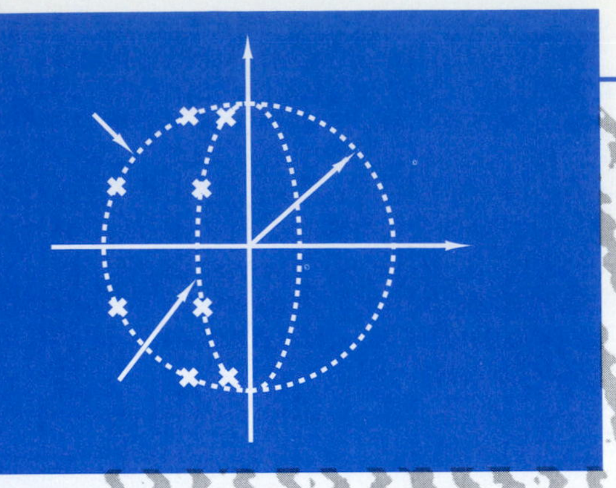

Chapter Fifteen

"In its usual form the electric wave-filter transmits currents of all frequencies lying within one or more specified ranges, and excludes currents of all other frequencies."

George A. Campbell, 1922
American Engineer

FILTER DESIGN

The electric filter was independently invented during World War I by George Campbell in the United States and by K. W. Wagner in Germany. Electric filters and vacuum tube amplifiers were the key technological developments that triggered the growth of telephone and radio communication systems in the 1920s and 1930s. The emergence of semiconductor electronics in the 1960s, especially the integrated circuit OP AMP, facilitated combining the functions of filtering and amplification into what are now called active filters.

This chapter is concerned with the design of active filters using integrated circuit OP AMPs, but the design methodology can serve other electronic applications where a modular approach to synthesis can be applied. The signal-processing functions performed by filters are outlined in the first section. The second section introduces filter design techniques using a cascade connection of first-order circuits. The third and fourth sections extend the cascade design to second-order circuits and the realization of higher-order low-pass filters. The fifth section shows how the low-pass filter serves as a prototype for designing high-pass, bandpass, and bandstop filters. The final section describes using SPICE to simulate and evaluate active RC filter circuits.

15-1 FREQUENCY-DOMAIN SIGNAL PROCESSING

Any signal can be described by a spectrum giving the amplitude and phase angle of its sinusoidal components as a function of frequency. In the frequency domain, signal characterization can be thought of as a large number of voltage sources in series, each producing a sinusoid with a specified frequency, amplitude, and phase angle. In frequency-domain signal processing the output spectrum is obtained by selectively altering the amplitudes and phase angles of the frequencies contained in the input spectrum. The idea that signals can be described and processed in terms of their frequency components is one of the fundamental concepts in electrical engineering.

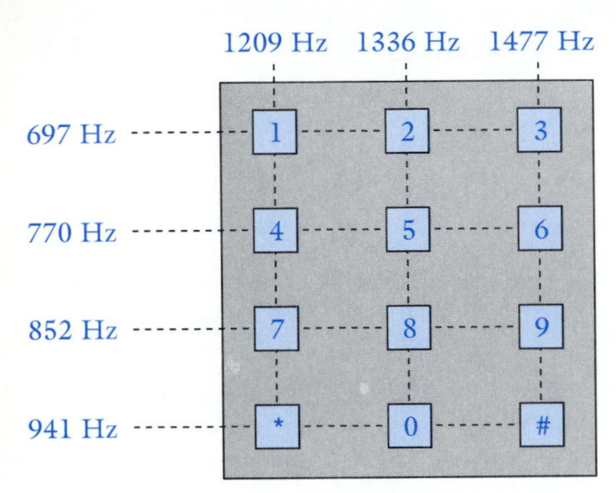

Figure 15-1 Touch-tone dialing frequencies.

For example, the touch-tone telephone shown in Fig. 15-1 transmits two frequencies when the button for a digit is pressed. One frequency comes from the low group (697, 770, 852, and 941 Hz) and the other from the high group (1209, 1336, and 1477 Hz). When the No. 3 button is pressed the instrument transmits a sum of sinusoids at 697 Hz and 1477 Hz. When the No. 5 button is pressed the dial signal is a combination of 770 Hz and 1336 Hz. The details of the time-domain waveform produced are not important. The important point is that the touch-tone telephone produces signals whose frequency spectrum lies in the range from 697 Hz to 1477 Hz. The telephone-switching equipment in the central office decodes a dial signal by detecting the frequencies it contains. Thus, the touch-tone signal can be described and processed based on its frequency content.

The touch-tone signals contain well-defined discrete frequencies. Other types of signals have continuous spectra that concentrate in certain frequency ranges. For example, most of the spectral content of human speech concentrates in the range from about 100 Hz to roughly 5 kHz. This range is much greater than is required for intelligibility. Signal filters in the telephone system limit voice signals to the range from about 300 Hz to 3300 Hz. The human voice can produce and the ear can detect signal frequencies well outside this range. However, most listeners can recognize speakers and understand conversations when the voice spectrum is limited to the range from 300 Hz to 3300 Hz. The important idea is that the telephone voice signal can be described and processed in terms of its frequency content.

A *filter* is a signal processor that transmits or even amplifies signals in a band of frequencies and rejects or attenuates signals in other bands. Figure 15-2 shows how an input spectrum containing three frequencies is modified by the four prototype filter responses described in Chapter 12. The low-pass filter passes frequencies below its cutoff frequency ω_C and attenuates the high-frequency components in the stopband above the cutoff frequency. The high-pass filter passes frequencies above its cutoff frequency and attenuates the low-frequency components below cutoff. The bandpass filter passes the intermediate frequency and attenuates the low- and high-frequency components on either side. Finally, the bandstop filter is the inverse

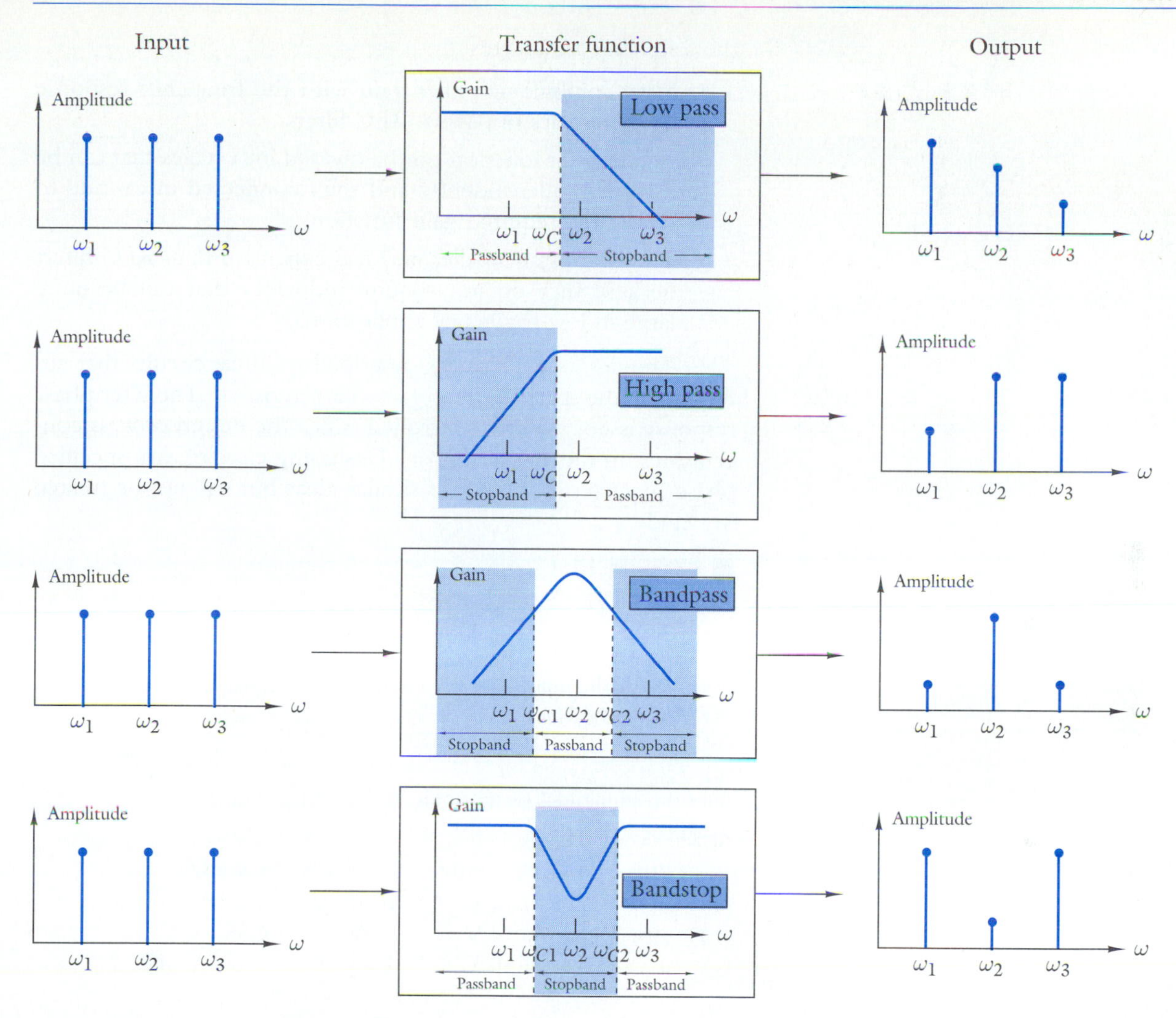

Figure 15-2 Four prototype filter characteristics.

of the bandpass case, attenuating the intermediate frequency and passing the low- and high-frequency components.

This chapter introduces methods of designing filter circuits with prescribed frequency-domain signal processing characteristics. A filter design problem is defined by specifying a frequency response gain function $|T(j\omega)|$. The objective is to design one or more circuits that produce the desired gain function. In general, a given gain response can be realized by many different circuits, including passive circuits containing only resistors, capacitors, and inductors. We limit our study to active filter realization involving resistors, capacitors, and OP AMPs. These *active RC filters* offer the following advantages:

1. They combine amplifier gain with the frequency response characteristics of passive RLC filters.
2. The transfer function can be divided into stages that can be designed independently and then connected in cascade to realize the required gain function.
3. They are often smaller and less expensive than RLC filters because they do not require inductors that can be quite large in low-frequency applications.

The design approaches described produce circuits that approximate the specified filter gain characteristics. The filter phase response is not directly controlled, since the design process concentrates on the gain response. Designing circuits with specified phase characteristics involves similar ideas but will not be treated in detail.

EXERCISE 15-1

The waveform

$$v_1(t) = 10\cos 200t + 5\cos 800t + 2\cos 2000t - 5\cos 4000t \text{ V}$$

is applied at the input of linear circuits whose transfer functions are
(a) $T_1(s) = 1/(2.5 \times 10^{-3}s + 1)$
(b) $T_2(s) = s/(s + 1000)$
(c) $T_1(s) \times T_2(s)$
Find the amplitudes of sinusoids in the output signals.

ANSWERS:

	200 rad/s	800 rad/s	2 krad/s	4 krad/s
(a)	8.94 V	2.24 V	0.392 V	0.498 V
(b)	1.96 V	3.12 V	1.79 V	4.85 V
(c)	1.75 V	1.40 V	0.351 V	0.483 V

15-2 CASCADE DESIGN WITH FIRST-ORDER CIRCUITS

In this section we begin the study of filter design problems in which characteristics of the gain response $|T(j\omega)|$ are specified. The end product we seek is a circuit whose transfer function has the specified gain response and whose element values are in practical ranges. The design process begins by constructing a transfer function whose gain response meets the specification. Next we partition the transfer function into a product of the form

$$T(s) = T_1(s) \times T_2(s) \times T_3(s) \times \ldots T_n(s) \qquad (15\text{-}1)$$

where each $T_k(s)$ in this equation is a first-order transfer function with one pole and no more than one finite zero. In Chapter 12 we learned how to design prototype first-order circuits using

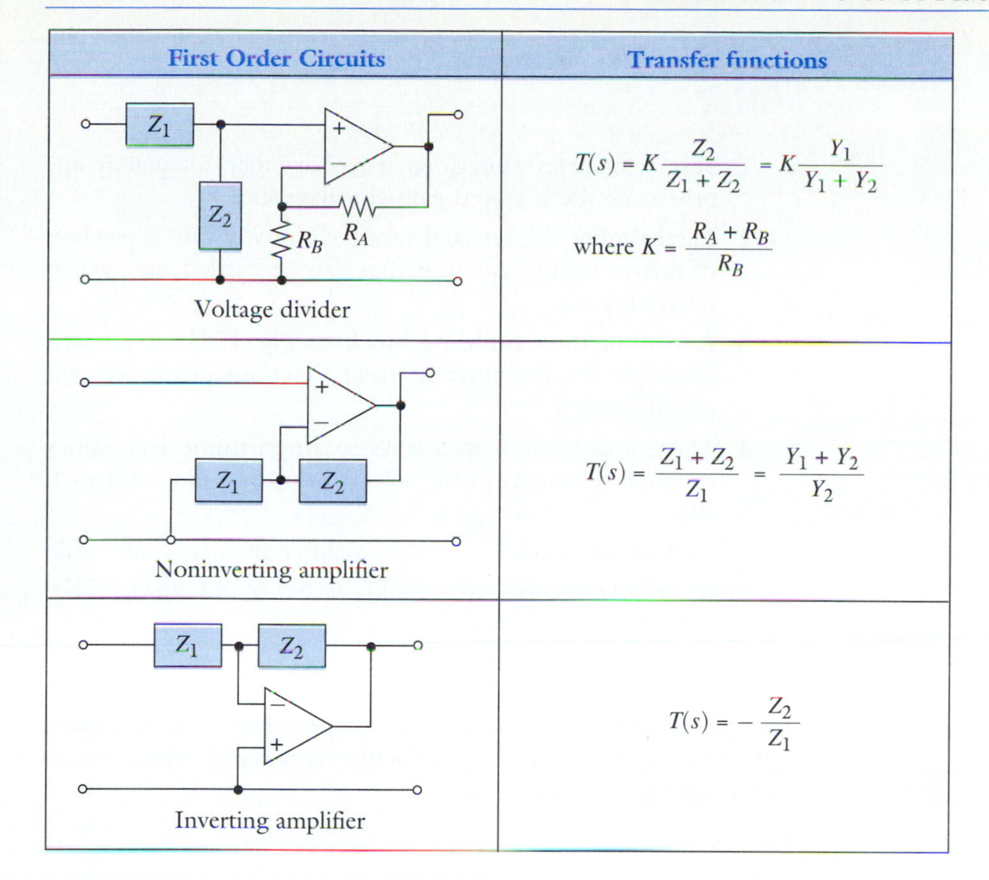

First Order Circuits
Transfer functions
Z_1
Z_2
R_B
R_A
Voltage divider
$T(s) = K\frac{Z_2}{Z_1 + Z_2} = K\frac{Y_1}{Y_1 + Y_2}$
where $K = \frac{R_A + R_B}{R_B}$
Noninverting amplifier
$T(s) = \frac{Z_1 + Z_2}{Z_1} = \frac{Y_1 + Y_2}{Y_2}$
Inverting amplifier
$T(s) = -\frac{Z_2}{Z_1}$

Figure 15-3 First-order circuit building blocks.

the voltage divider, noninverting amplifier, and the inverting amplifier building blocks in Fig. 15-3. The circuit element values in the prototype usually are unrealistically large or small. Applying a magnitude scale factor to each stage brings the element values into practical ranges.

The required transfer function $T(s)$ is realized by connecting the scaled circuit for each stage in cascade as shown in Fig. 15-4. The cascade connection produces the required $T(s)$ because the building blocks in Fig. 15-3 have low-impedance OP AMP outputs. As a result, interstage loading does not change the stage transfer functions and the chain rule in Eq. (15-1) applies. The noninverting and inverting amplifier circuits in Fig. 15-3 inherently have low-impedance outputs. Adding a noninverting amplifier to the output of the basic voltage-divider circuit in Fig. 15-3 produces a circuit that can be connected in cascade without changing the stage transfer function.

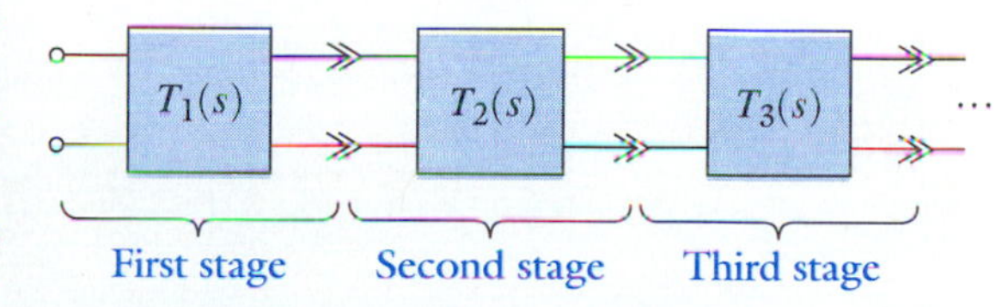

Figure 15-4 A cascade connection.

The end result is a cascade circuit whose gain response has the attributes required by the specification. The procedure outlined above does not produce a unique solution for the following reasons:

1. Several **transfer functions** may exist that adequately approximate the required gain characteristics.
2. **Partitioning** the selected transfer function into a product of first-order transfer functions can be carried out several different ways.
3. Any of the three **building blocks** in Fig. 15-3 can be used to realize the first-order transfer functions in the selected partitioning.
4. There usually are several ways to assign **numerical values** to circuit elements in the selected building block for each stage.
5. Each stage normally requires a different magnitude **scale factor** to bring the assigned numerical values into practical ranges.

In summary, since there are many choices there are no unique answers to the circuit design problem. The following examples illustrate the design of active RC circuits, which realize prescribed gain characteristics.

Design Example

EXAMPLE 15-1

(a) Construct a transfer function $T(s)$ with the asymptotic gain response shown in Fig. 15-5.

(b) Design an active RC circuit to realize the $T(s)$ found in (a).

SOLUTION:

(a) The asymptotic gain response shows slope changes at $\omega = 20$ rad/s, 100 rad/s, and 1000 rad/s. Since the net slope changes at $\omega = 20$ rad/s and 1000 rad/s are both -20 dB/decade, the denominator of $T(s)$ must have poles at $s = -20$ rad/s and $s = -1000$ rad/s. Since the net slope change at $\omega =$ 100 rad/s is $+20$ dB/decade, the numerator of $T(s)$ must have a zero at $s = -100$ rad/s. These critical frequencies

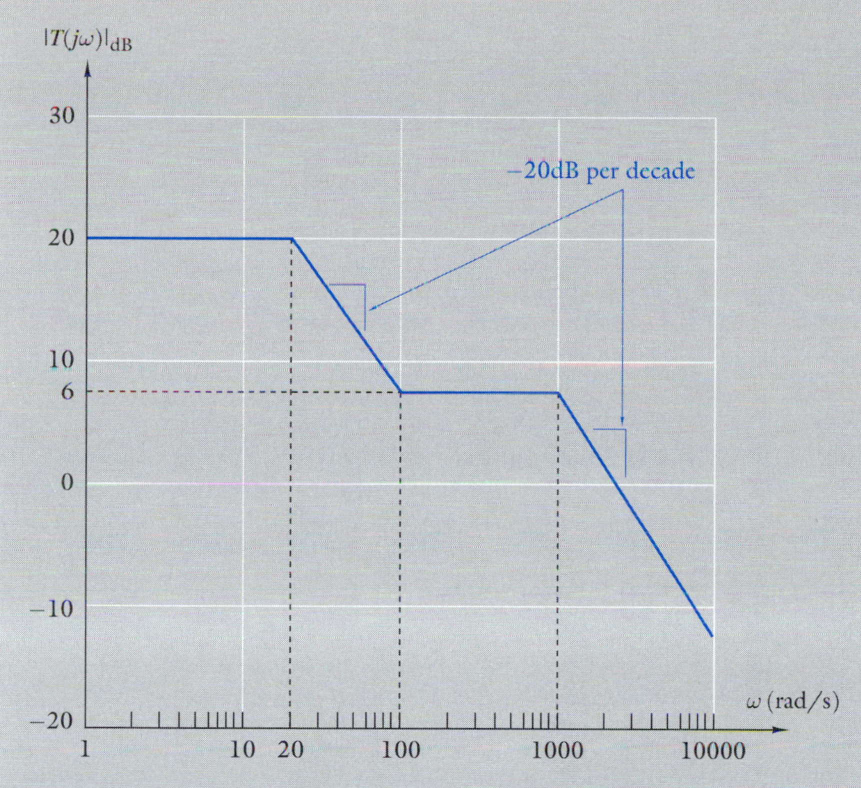

Figure 15-5

account for all of the slope changes so the required transfer function has the form

$$T(s) = K\frac{s/100 + 1}{(s/20 + 1)(s/1000 + 1)}$$

where the gain K is yet to be determined. The dc gain of the proposed transfer function is $T(0) = K$ and the required dc gain in Fig. 15-5 is +20 dB. We conclude that $20 \log_{10} K = 20$ dB or $K = 10$.

(b) To design filter circuits we partition $T(s)$ into the following product of first-order factors:

$$T(s) = T_1(s)T_2(s) = \left(\frac{s/100 + 1}{s/20 + 1}\right)\left(\frac{10}{s/1000 + 1}\right)$$

This partitioning is not unique because the overall gain $K = K_1 \times K_2 = 10$ can be distributed between the two stages in any number of ways (for example, $K_1 = 2$ and $K_2 = 5$). We could move the zero from the first stage to the second stage or reverse the order in which the stages are listed. Using the partitioning given above as a starting place, we can develop two alternative designs as follows.

First Design: Selecting the inverting amplifier building block in Fig. 15-3 for both stages leads to the design sequence shown in Fig. 15-6. The minus signs on the stage transfer functions is not a problem, since there are two stages and the two phase inversions cancel. That is,

$$T(s) = [-T_1(s)][-T_2(s)] = T_1(s)T_2(s)$$

The transfer function of the inverting amplifier is $-Z_2(s)/Z_1(s)$ so the selected building block places constraints on Z_1 and Z_2 as indicated in the second row in Fig. 15-6. Equating $Z_1(s)$ to the reciprocal of the transfer function numerator and Z_2 to the reciprocal of the denominator yields impedances of the form $1/(Cs + G)$. An impedance of this form is a resistor in parallel with a capacitor as shown in the third and fourth rows in Fig. 15-6. The element values of the prototype designs shown in the fifth row are not within practical ranges. To reduce power dissipation we make all resistors 10 kΩ or greater by using a magnitude scale factor of $k_m = 10^4$ in the first stage and $k_m = 10^5$ in the second stage. Applying these scale factors to the prototype designs produces the final designs shown in the last row of Fig. 15-6.

Second Design: This design uses the same partitioning of $T(s)$ as in the first design. Using the voltage-divider building block in Fig. 15-3 for each stage leads to the design sequence shown in Fig. 15-7. Factoring s out of the numerator and denominator of each stage transfer function produces the design constraints shown in the second row of Fig. 15-7. Equating Z_2 to the numerator of the stage transfer function and Z_1+Z_2 to the denominator yields RC circuits shown in the fourth and fifth rows of Fig. 15-7. The required gain factor for each stage is shown in the third row. The first-stage gain is $K = 1$, so the first-stage OP AMP is a voltage follower as indicated in the sixth row of Fig. 15-7. The second stage requires a gain of $K = 10$, which is achieved by selecting its OP AMP feedback resistors so that $R_A = (K - 1)R_B = 9R_B$. Selecting $R_B = 1/1000$ yields $R_A = 9/1000$ and leads to the second-stage prototype shown in the sixth row. To make all resistors 10 kΩ or greater, we use $k_m = 10^6$ in the first stage and $k_m = 10^7$ in the second stage. Applying these magnitude scale factors to the prototypes in the sixth row produces the final designs in the last row of Fig. 15-7.

Item	First Stage	Second Stage
Transfer function	$T_1(s) = -\dfrac{(s/100)+1}{(s/20)+1}$	$T_2(s) = -\dfrac{10}{(s/1000)+1}$
Design constraints	Inverting Amplifier $\dfrac{Z_2}{Z_1} = \dfrac{(s/100)+1}{(s/20)+1}$	Inverting Amplifier $\dfrac{Z_2}{Z_1} = \dfrac{10}{(s/1000)+1}$
Z_1	$\dfrac{1}{(s/100)+1}$ 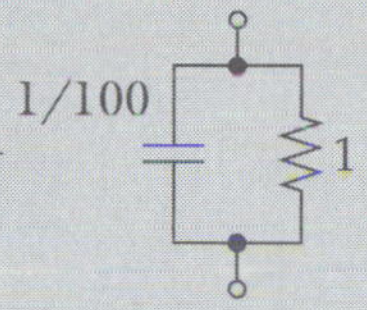 1/100 1	$\dfrac{1}{10}$ 1/10 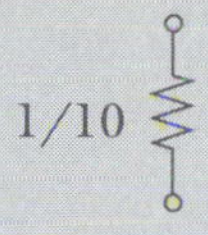
Z_2	$\dfrac{1}{(s/20)+1}$ 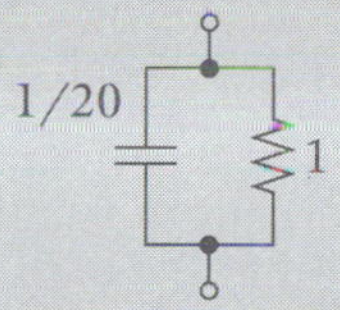 1/20 1	$\dfrac{1}{(s/1000)+1}$ 1/1000 1 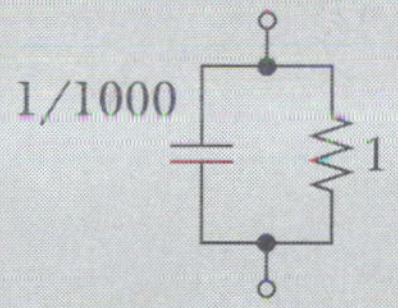
Prototype designs	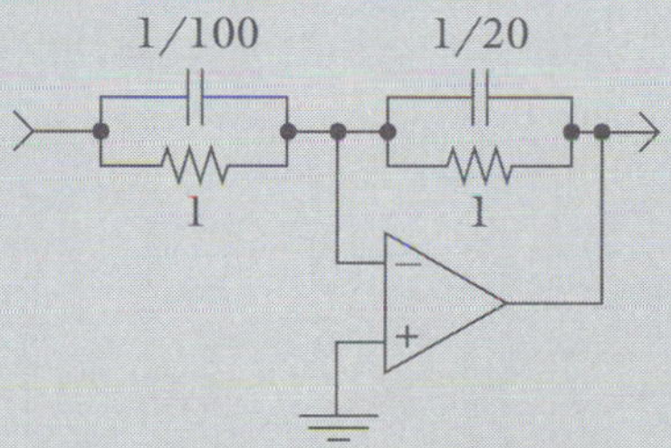 1/100 1 1/20 1	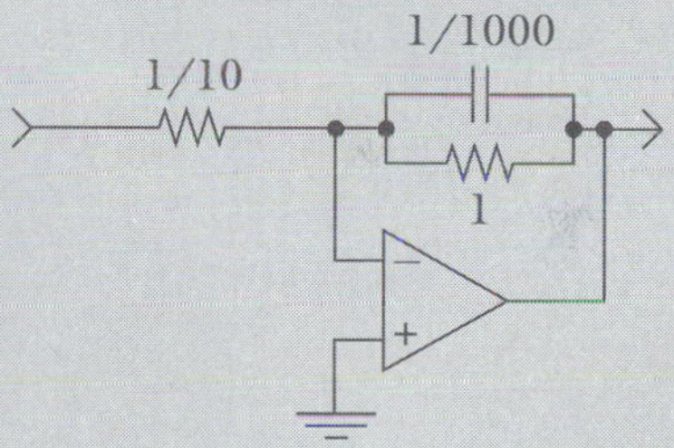 1/10 1/1000 1
Final designs	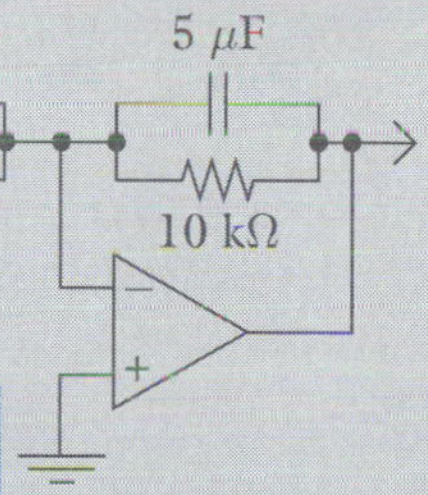 1 μF 10 kΩ 5 μF 10 kΩ $k_m = 10^4$	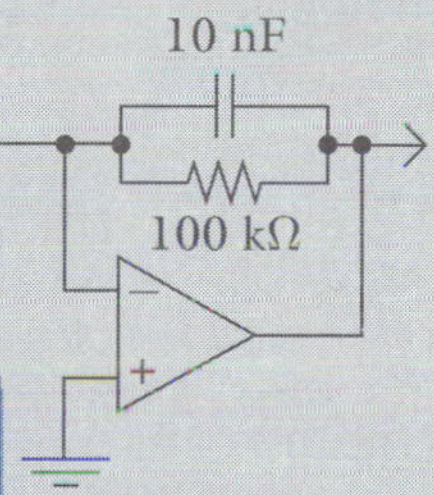 10 kΩ 10 nF 100 kΩ $k_m = 10^5$

Figure 15-6 First design sequence for Example 15-1.

Item	First Stage	Second Stage
Transfer function	$T_1(s) = -\dfrac{(s/100)+1}{(s/20)+1}$	$T_2(s) = \dfrac{10}{(s/100)+1}$
Design constraints	Voltage Divider $K\dfrac{Z_2}{Z_1+Z_2} = \dfrac{1/100 + 1/s}{(1/20)+1/s}$	Voltage Divider $K\dfrac{Z_2}{Z_1+Z_2} = 10\dfrac{1/s}{(1/1000)+1/s}$
K	1	10
Z_1	$\dfrac{1}{25}$ 1/25 	$\dfrac{1}{1000}$ 1/1000 
Z_2	$\dfrac{1}{s} + \dfrac{1}{100}$ 1/100, 1 	$\dfrac{1}{s}$ 1 
Prototype designs	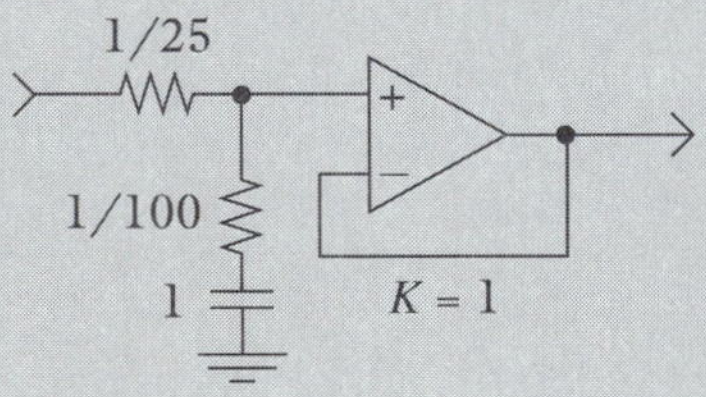 1/25, 1/100, 1, $K = 1$	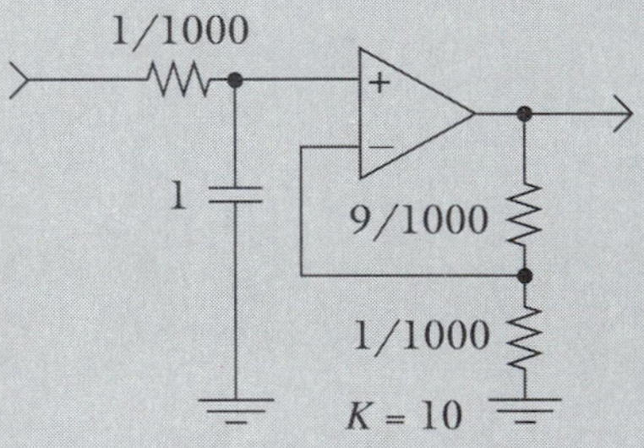 1/1000, 1, 9/1000, 1/1000, $K = 10$
Final designs	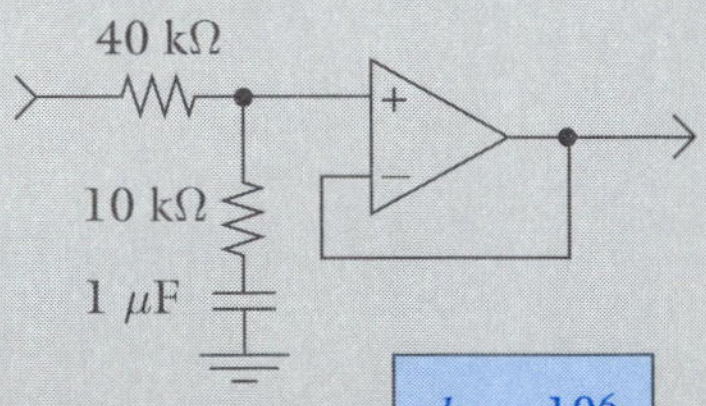 40 kΩ, 10 kΩ, 1 μF, $k_m = 10^6$	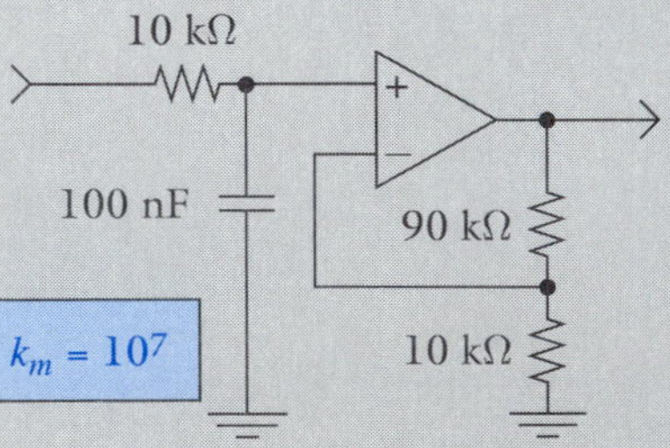 10 kΩ, 100 nF, 90 kΩ, 10 kΩ, $k_m = 10^7$

Figure 15-7 Second design sequence for Example 15-1.

❖ Design Example

EXAMPLE 15-2

(a) Construct a transfer function $T(s)$ with the asymptotic gain response shown in Fig. 15-8.

(b) Design an active RC circuit to realize the $T(s)$ found in (a).

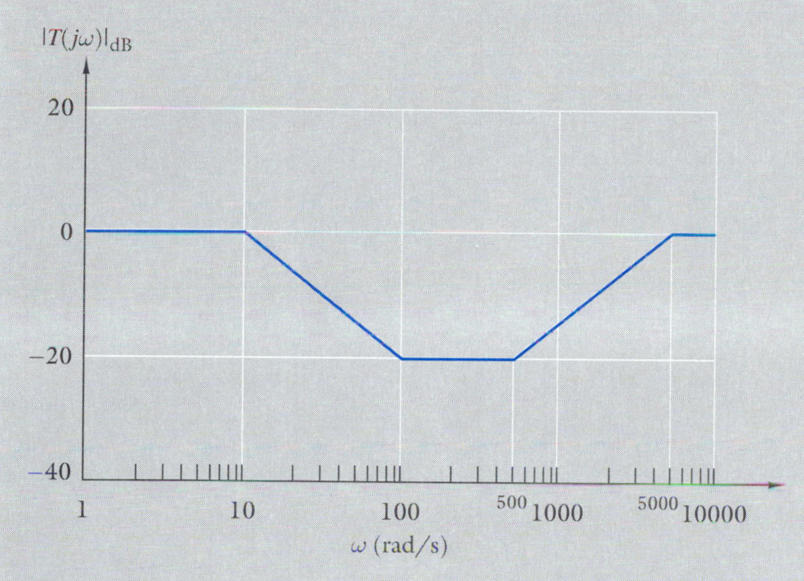

Figure 15-8

SOLUTION:

(a) The straight-line gain response shows slope changes at ω = 10 rad/s, 100 rad/s, 500 rad/s, and 5000 rad/s. The net slope changes at $\omega = 10$ rad/s and 5000 rad/s are -20 dB/decade. The net slope changes at $\omega = 100$ rad/s and 500 rad/s are $+20$ dB/decade. Therefore, $T(s)$ must have poles at $s = -10$ rad/s and $s = -5000$ rad/s and zeros at $s = -100$ rad/s and $s = -500$ rad/s. The required transfer function has the form

$$T(s) = K\frac{(s/100+1)(s/500+1)}{(s/10+1)(s/5000+1)}$$

where the gain K is yet to be determined. The gain plot in Fig. 15-8 indicates that the required dc gain is 0 dB. The dc gain of the proposed transfer function is $20\log_{10} K$, so we conclude that $K = 1$.

Item	First Stage	Second Stage
Transfer function	$T_1(s) = \dfrac{(s/100)+1}{(s/10)+1}$	$T_2(s) = \dfrac{(s/500)+1}{(s/5000)+1}$
Design constraints	Voltage Divider $K\dfrac{Z_2}{Z_1+Z_2} = \dfrac{1/100+1/s}{(1/10)+1/s}$	Noninverting Amplifier $\dfrac{Z_1+Z_2}{Z_1} = \dfrac{1/500+1/s}{(1/5000)+1/s}$
K	1	Not Applicable
Z_1	$\dfrac{9}{100}$ 9/100 	$\dfrac{1}{s} + \dfrac{1}{5000}$ 1/5000 1 
Z_2	$\dfrac{1}{s} + \dfrac{1}{1000}$ 1/100 1 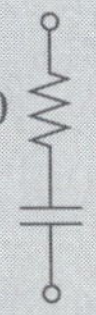	$\dfrac{9}{5000}$ 9/5000 
Prototype designs	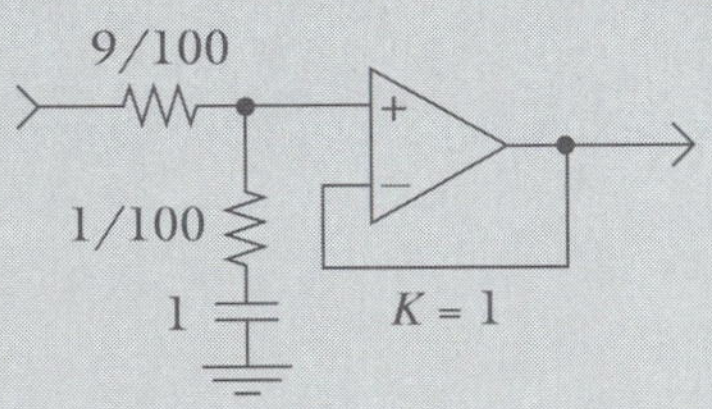 9/100 1/100 1 $K = 1$	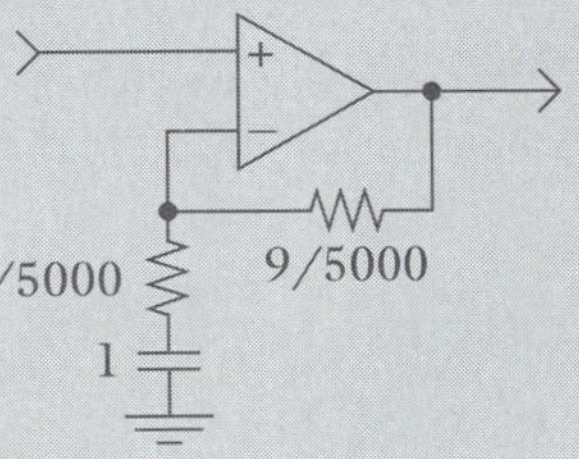 1/5000 1 9/5000
Final designs	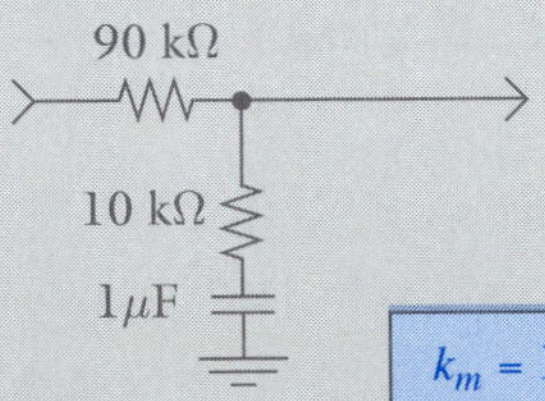 90 kΩ 10 kΩ 1μF $k_m = 10^6$	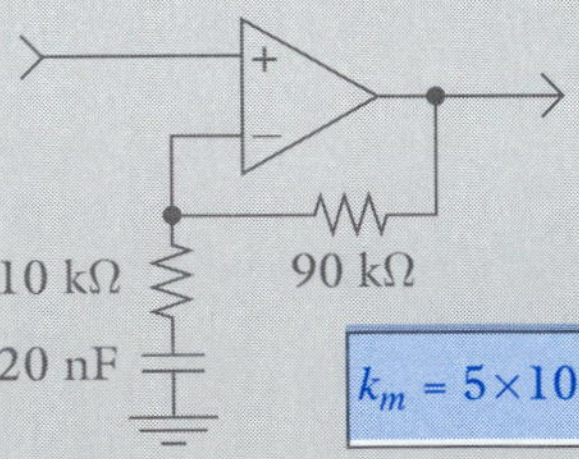 10 kΩ 20 nF 90 kΩ $k_m = 5\times10^7$

Figure 15-9 Design sequence for Example 15-2.

(b) To begin the circuit design we partition $T(s)$ into the following product of first-order transfer functions:

$$T(s) = T_1(s)T_2(s) = \left(\frac{s/100+1}{s/10+1}\right)\left(\frac{s/500+1}{s/5000+1}\right)$$

This partitioning of $T(s)$ is not unique, since we could, for example, interchange the zeros or reverse the order in which the stages are listed.

The design sequence shown in Fig. 15-9 uses a voltage-divider circuit for the first stage and a noninverting amplifier for the second stage. Factoring an s out of the numerator and denominator of the stage transfer functions produces the design constraints in the second row of Fig. 15-9. Equating numerators and denominators of the design constraint equations leads to the RC circuits shown in the fourth and fifth rows of Fig. 15-9. The gain of the voltage-divider stage is unity ($K = 1$), so the OP AMP in the first stage is a voltage follower as indicated in the prototype design. However, the output of the first stage drives the noninverting input of the second-stage OP AMP. Since OP AMP inputs draw negligible current no interstage loading occurs when the voltage follower in the first stage is eliminated. The final design in the last row of Fig. 15-9 eliminates the first-stage OP AMP and uses magnitude scale factors of $k_m = 10^6$ and $k_m = 5 \times 10^7$ to make all resistors greater than 10 kΩ.

❖ DESIGN EXERCISE: EXERCISE 15-2

(a) Construct a transfer function $T(s)$ with the asymptotic gain response shown in Fig. 15-10.

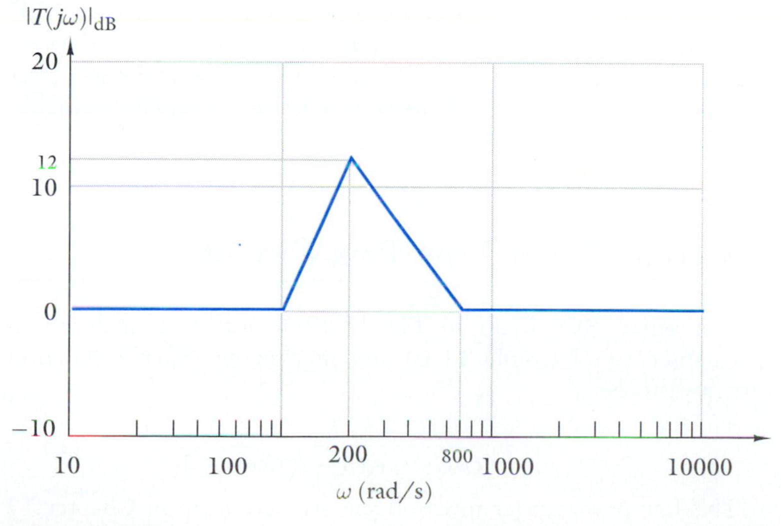

Figure 15-10

(b) Design an active RC circuit to realize the $T(s)$ found in (a).

ANSWERS:

(a) $T(s) = \dfrac{(s/100 + 1)^2(s/800 + 1)}{(s/200 + 1)^3}$

(b) There is no unique answer. One possible prototype design is shown in Fig. 15-11. ■

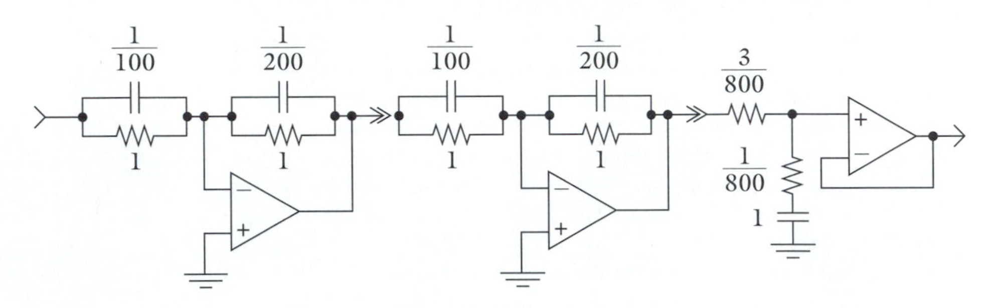

Figure 15-11

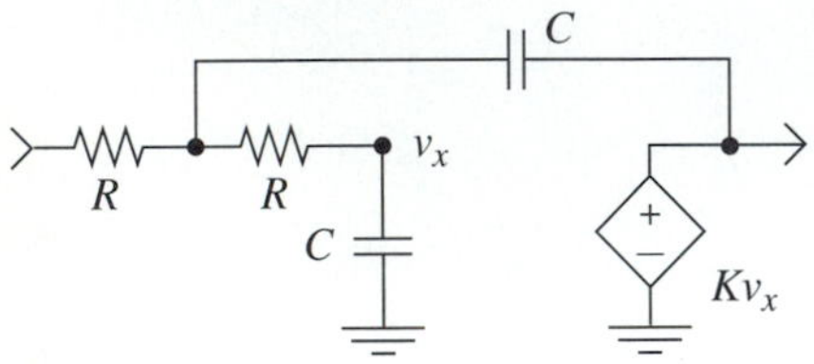

◇ 15-3 CASCADE DESIGN WITH SECOND-ORDER CIRCUITS

First-order circuit design methods can only realize transfer functions with poles on the negative real axis in the s plane. Second-order transfer functions with complex poles can produce filter characteristics that cannot be achieved using the same number of real poles. Complex poles can be produced by passive RLC circuits. However, our discussion of filter design concentrates on active RC realizations. To use complex poles in active RC filter design we need to identify some second-order circuits comparable to the first-order circuits in Fig. 15-3.

(a) Dependent source circuit

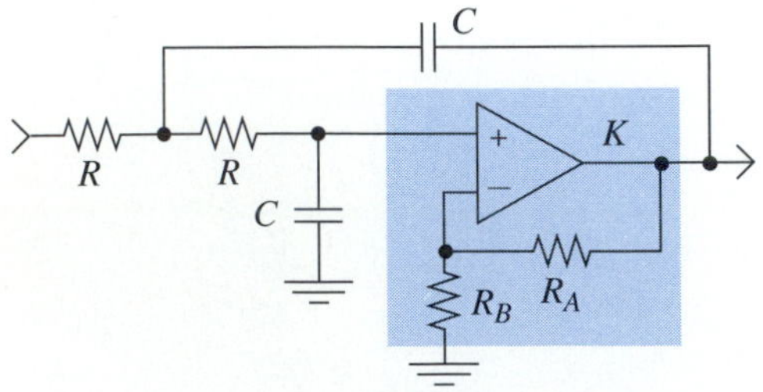

(b) OP AMP realization

Figure 15-12 A second-order low-pass circuit.

Second-Order Low-Pass Circuits

The active RC circuit in Fig. 15-12(a) was first analyzed in Chapter 11 (Example 11-6) and its voltage transfer function shown to be

$$T(s) = \frac{K}{(RCs)^2 + (3 - K)RCs + 1} \tag{15-2}$$

This low-pass transfer function was studied again in Chapter 12 (Example 12-10), where we developed a procedure for selecting circuit parameters to produce a specified natural frequency ω_O and damping ratio ζ. To utilize this circuit as a building block in practical filter designs we simulate the dependent source in Fig. 15-12(a) using the noninverting OP AMP circuit in Fig. 15-12(b). The OP AMP circuit is equivalent to the dependent source because it draws negligible input current and has very low output resistance. The OP AMP circuit duplicates the controlled source gain K when the feedback resistances are selected so that $R_A = (K-1)R_B$. Note that this means that K must be greater than or equal to one to ensure that R_A is physically realizable.

The denominator of a second-order transfer function is written in the form

$$(s/\omega_O)^2 + 2\zeta(s/\omega_O) + 1 \tag{15-3}$$

Comparing the denominator of Eq. (15-2) with Eq. (15-3) produces the following design constraints on the circuit parameters:

$$RC = \frac{1}{\omega_O} \quad \text{and} \quad K = 3 - 2\zeta \tag{15-4}$$

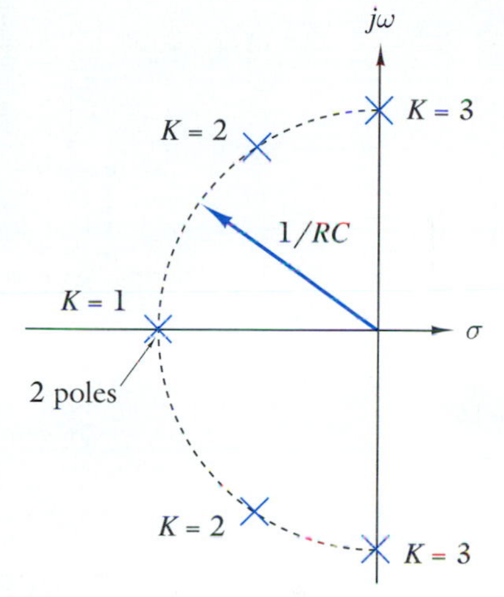

Figure 15-13 Complex pole locations.

Figure 15-13 shows how the circuit parameters influence the location of the complex poles produced by this circuit. The RC product controls natural frequency ω_O, which is the radial distance from the origin in the s plane to the complex poles. The gain K controls the damping ratio that defines the position of the poles on a circle of radius $1/RC$.

When $K = 1$ the two poles lie at the same location on the negative real axis and the damping ratio is $\zeta = 1$ (what we referred to as Case B back in Chapter 8). When $K = 2$ the poles are complex and lie on the s plane to the left of the j axis and the damping ratio is $\zeta = 0.5$ (Case C in Chapter 8). When $K = 3$ the poles lie on the j axis and the damping ratio is $\zeta = 0$. When $K > 3$ the poles are in the right half-plane and the circuit is of no use in filter design because it is unstable. Thus, $3 > K > 1$ is the gain range of interest for producing filters with complex poles using the building block in Fig. 15-12.

It is important to observe that the dc gain of the transfer function in Eq. (15-2) is $T(0) = K$. Since the gain parameter K is adjusted to control ζ we cannot independently control both the passband gain and the damping ratio. In some cases the filter requirements may not specify the passband gain because what really matters in a filter is the relative values of the passband and stopband gains. When the passband gain is specified, we may have to add another stage to the cascade to provide the additional gain or attenuation required.

Design Example

EXAMPLE 15-3

Use the circuit in Fig. 15-12 to design a second-order low-pass filter with $\omega_O = 1000$ rad/s and $\zeta = 0.5$.

SOLUTION:

The design constraints in Eq. (15-4) require that $RC = 10^{-3}$ s and $K = 2$. Selecting $C = 100$ nF requires that $R = 10$ kΩ. The gain requirement is achieved by selecting $R_B = 10$ kΩ and $R_A = (K-1)R_B = 10$ kΩ. Fig. 15-14 shows the resulting design. Note that all four resistances are 10 kΩ and both capacitances are 100 nF. The equal element values in this design simplify the production and maintenance of this circuit. The transfer function realized by the design is

$$T(s) = \frac{2}{(s/1000)^2 + (s/1000) + 1}$$

The filter specification did not specify the filter dc gain. The dc gain achieved by this design $[T(0) = K = 2]$ is determined by the damping ratio requirement of the specification.

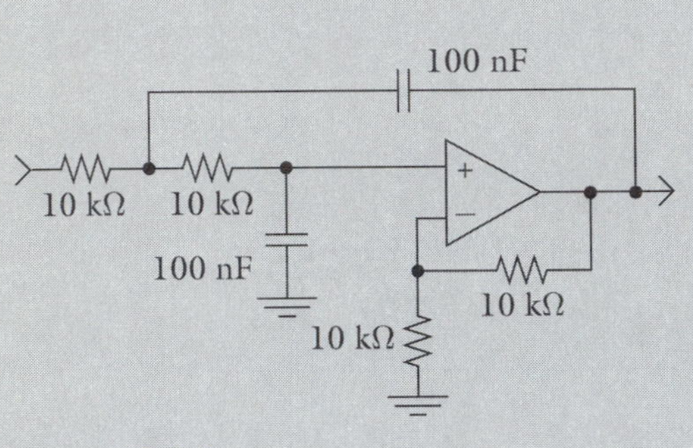

Figure 15-14

Design Example

EXAMPLE 15-4

(a) Construct a low-pass transfer function $T(s)$ that has the asymptotic gain response in Fig. 15-15 and has a cutoff frequency at $\omega_C = 1000$ rad/s.

(b) Design an active RC circuit to realize the transfer function found in (a).

SOLUTION:

(a) The slope of asymptotic gain at high frequency is −60 dB/decade. The required transfer function must have three poles, since each pole contributes −20 dB/decade to the high-frequency slope. One approach is to use one real pole and a pair of complex poles all with corner frequencies at $\omega =$

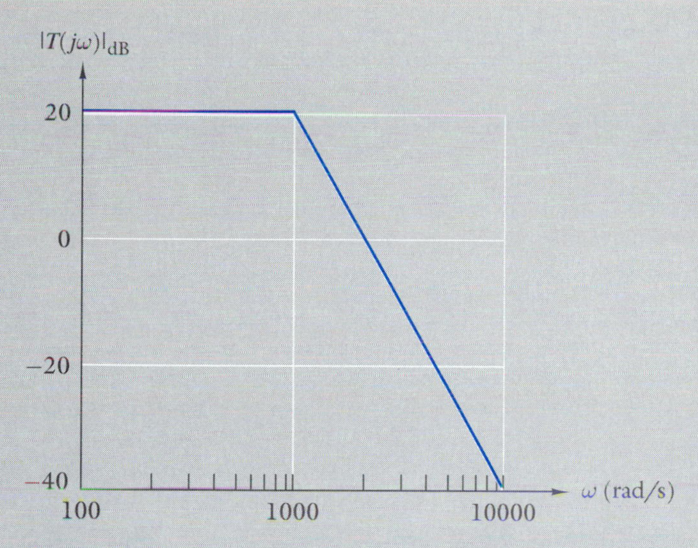

Figure 15-15

1000 rad/s. Such a transfer function has the form

$$T(s) = \frac{K_O}{(s/1000+1)[(s/1000)^2 + 2\zeta(s/1000)+1]}$$

where the constants K_O and ζ are yet to be determined. The dc gain of this transfer function is $T(0) = K_O$. The dc gain specified in Fig. 15-15 is 20 dB and we conclude that $K_O = 10$.

The specification requires that the cutoff frequency occur at $\omega = 1000$ rad/s which means that $|T(j1000)| = |K_O|/\sqrt{2}$. At $s = j1000$ the magnitude of the proposed transfer function is

$$|T(j1000)]|$$

$$= \frac{|K_0|}{|j1000/1000+1|\;\left|(j1000/1000)^2 + 2\zeta(j1000/1000)+1\right|}$$

$$= \frac{|K_0|}{|1+j||j2\zeta|}$$

$$= \frac{|K_0|}{(\sqrt{2})(2\zeta)}$$

To have the cutoff frequency at $\omega = 1000$ rad/s requires that the damping ratio of the second-order factor be $\zeta = 0.5$.

Thus, a transfer function that meets the low-pass filter specification is

$$T(s) = \frac{10}{(s/1000+1)\left[(s/1000)^2 + (s/1000) + 1\right]}$$

(b) The required transfer function can be partitioned in the following way:

$$\begin{aligned} T(s) &= T_1(s)T_2(s) \\ &= \left(\frac{5}{(s/1000)+1}\right)\left(\frac{2}{(s/1000)^2 + (s/1000) + 1}\right) \end{aligned}$$

The second-order low-pass transfer function was realized in Example 15-3 by the circuit in Fig. 15-14. We can include this circuit directly as the second-stage in our cascade design. The first-order transfer function can be realized using the methods discussed in the previous section. Using a voltage divider building block from Fig. 15-3 yields the following design constraint:

$$T_1(s) = K\frac{Z_2}{Z_1 + Z_2} = 5\left(\frac{1/s}{1/1000 + 1/s}\right)$$

This constraint requires $K = 5$, $Z_2 = 1/s$, $Z_1 + Z_2 = 1/1000 + 1/s$, hence $Z_1 = 1/1000$. Using a magnitude scale factor of $k_m = 10^7$ yields Z_1 as a 10-kΩ resistor and Z_2 as a 100-nF capacitor. The stage gain of $K = 5$ can be achieved by selecting $R_B = 10$ kΩ and $R_A = (K-1)R_B = 40$ kΩ. The resulting first-stage design, together with the previously designed second stage, are shown in Fig. 15-16. Note that we have maintained equal element values throughout the design except for one 40-kΩ resistor in the feedback path of the first-stage OP AMP.

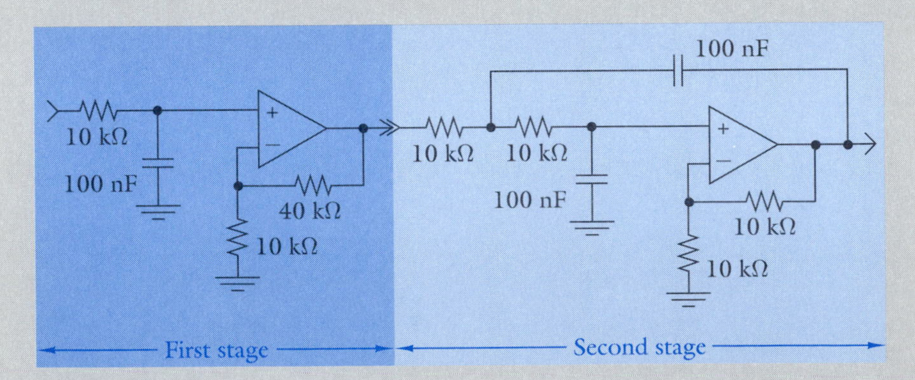

Figure 15-16

Second-Order High-Pass Circuits

The active RC circuit in Fig. 15-17(a) has a second-order high-pass voltage transfer function of the form

$$T(s) = \frac{K(RCs)^2}{(RCs)^2 + (3 - K)RCs + 1} \tag{15-5}$$

This high-pass circuit is obtained from the low-pass circuit in Fig. 15-14(a) by interchanging the locations of the resistors and capacitors.[1] Derivation of the high-pass transfer function in Eq. (15-5) is left as a exercise for the reader (see Problem 12-35).

To utilize the second-order high-pass circuit as a building block in practical filter designs we simulate the dependent source in Fig. 15-17(a) using the noninverting OP AMP circuit in Fig. 15-17(b). The OP AMP circuit duplicates the controlled source gain K when the feedback resistances are selected so that $R_A = (K - 1)R_B$. Again note that this requires that $K \geq 1$ for the feedback resistor R_A to be physically realizable.

The standard form for the denominator of a second-order transfer function is

$$(s/\omega_O)^2 + 2\zeta(s/\omega_O) + 1 \tag{15-6}$$

Comparing the denominator of Eq. (15-5) with Eq. (15-6) yields the following design constraints:

$$RC = \frac{1}{\omega_O} \quad \text{and} \quad K = 3 - 2\zeta \tag{15-7}$$

These two constraints are the same as found for the low-pass case in Eq. (15-4). The RC product controls the natural frequency ω_O, while the gain K controls the damping ratio ζ. The position of the poles for different values of K are the same as for the low-pass case in Fig. 15-13. The gain range $3 > K > 1$ is used in filter design to produce complex poles in the left half of the s plane.

The high-frequency passband gain of the transfer function in Eq. (15-5) is $T(\infty) = K$. Since the parameter K is adjusted to control the damping ratio ζ, we cannot independently control both the passband gain and the damping ratio. In some cases the filter design requirements may not specify the passband gain. When the passband gain is specified, we add another stage

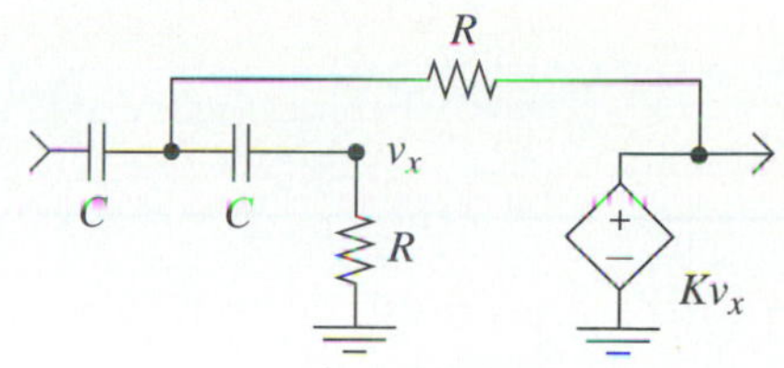

(a) Dependent source circuit

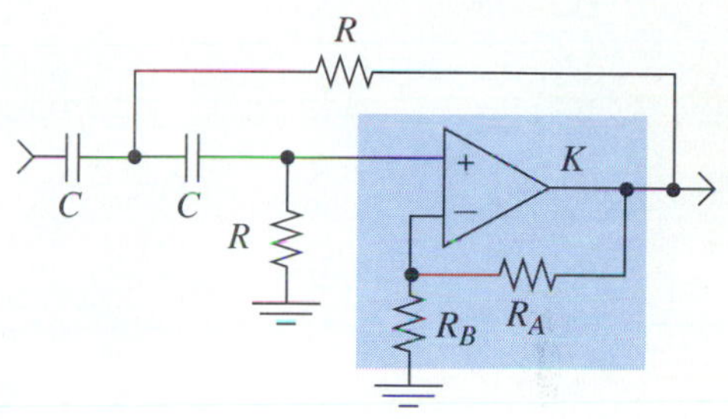

(b) OP AMP realization

Figure 15-17 A second-order high-pass circuit.

[1]Both circuits are members of a family of circuits originally proposed by R. P. Sallen and E. L. Key in "A Practical Method of Designing RC Active Filters," *IRE Transactions on Circuit Theory*, Vol. CT-2, pp. 74–85, 1955. In 1955, the controlled source was simulated using vacuum tubes.

to provide the additional gain or attentuation required. The next example illustrates the problem of achieving both a specified damping ratio and a passband gain.

Design Example

EXAMPLE 15-5

Design a second-order high-pass circuit with $\omega_C = \omega_O = 20$ rad/s and a passband gain of 0 dB.

SOLUTION:

A natural frequency at $\omega_O = 20$ rad/s requires a high-pass transfer function of the form

$$T(s) = \frac{K_\infty (s/20)^2}{(s/20)^2 + 2\zeta(s/20) + 1}$$

where the constants K_∞ and ζ are yet to be determined. The passband gain of this transfer function is $T(\infty) = K_\infty$. The specification requires a passband gain of 0 dB so $K_\infty = 1$. At the cutoff frequency the gain must be $|K_\infty|/\sqrt{2}$. At the natural frequency the gain of the proposed transfer function is

$$|T(j20)| = \frac{|K_\infty|\,|j20/20)|^2}{\left|(j20/20)^2 + 2\zeta(j20/20) + 1\right|} = \frac{|K_\infty|}{2\zeta}$$

To have a cutoff frequency at $\omega = 20$ rad/s requires that $2\zeta = \sqrt{2}$. A transfer function that meets a high-pass filter specification is

$$T(s) = \frac{(s/20)^2}{(s/20)^2 + \sqrt{2}(s/20) + 1}$$

The high-pass circuit in Fig. 15-17 realizes the poles and zeros of this function when the constraints in Eq. (15-7) are satisfied, namely, $RC = 1/20$ s and $K = 3 - 2\zeta = 1.586$. Selecting $C = 1\ \mu$F requires $R = 50$ kΩ. The gain constraint is met by selecting $R_B = 50$ kΩ and $R_A = (K - 1)R_B = 29.3$ kΩ.

However, the passband gain of the resulting second-order circuit is $T(\infty) = K = 1.586$ and not unity (0 dB) as required by the specification. To overcome this problem we partition the required transfer function as follows:

$$T(s) = \left(\frac{1}{1.586}\right)\left(\frac{1.586(s/20)^2}{(s/20)^2 + \sqrt{2}(s/20) + 1}\right)$$

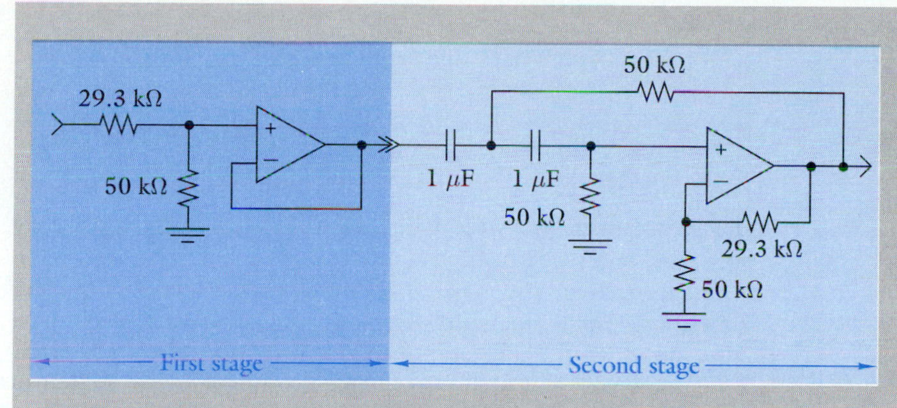

Figure 15-18

The second stage can be realized using the second-order high-pass circuit designed above with a passband gain of 1.586. The first stage is a voltage divider that provides an attentuation of 1/1.586 to bring the overall passband gain down to the 0 dB level required by the specification. Figure 15-18 shows this two-stage cascade design.

DESIGN EXERCISE: EXERCISE 15-3

Design an active RC circuit that has the same transfer function $T(s) = V_2/V_1$ as the RLC circuit in Fig. 15-19.

ANSWER:
A possible solution is shown in Fig. 15-20. ■

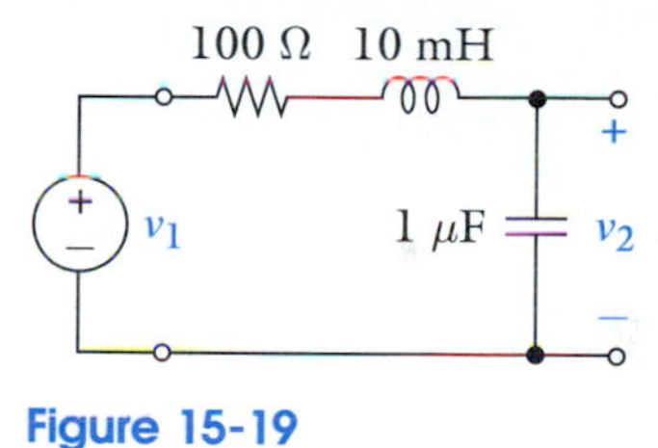

Figure 15-19

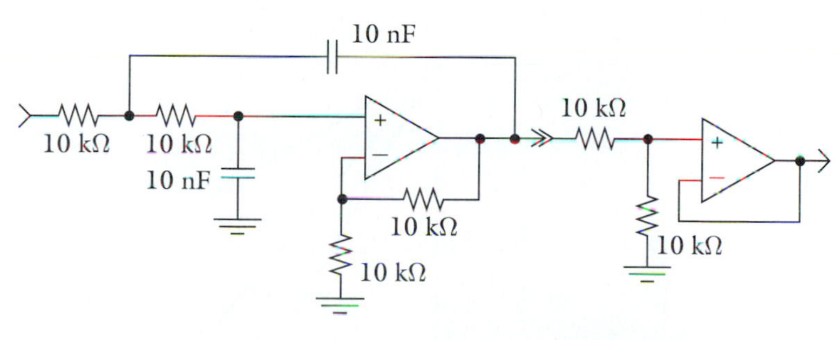

Figure 15-20

DESIGN EXERCISE: EXERCISE 15-4

Design a second-order active RC high-pass circuit with $\omega_0 = 500$ rad/s and $\zeta = 1$ using only 10-kΩ resistors.

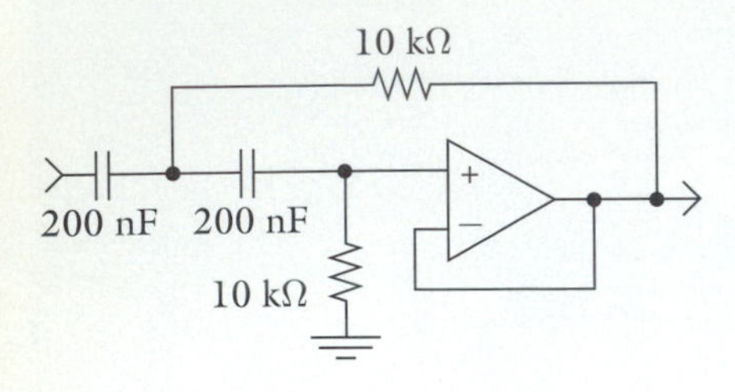

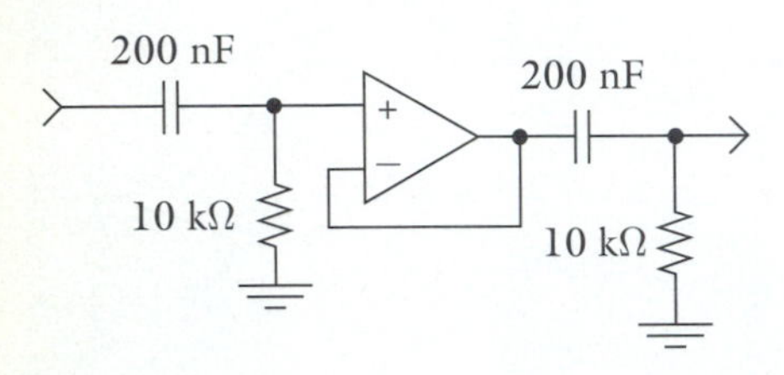

Figure 15-21

ANSWER:
Two possible solutions are shown in Fig. 15-21. ■

◇ 15-4 LOW-PASS FILTER DESIGN

There are many ways to design active RC circuits to produce low-pass, high-pass, bandpass, and bandstop filter characteristics. The approach we use is based on developing a low-pass transfer function. The low-pass prototype can be realized as a low-pass filter or converted into an equivalent high-pass filter. By combining low-pass and high-pass filters we can realize bandpass and bandstop filters. In other words, the low-pass filter design methods in this section are the foundation for designing other types of filters.

Figure 15-22 shows the asymptotic gain characteristics of ideal and real low-pass filters. The gain of the ideal filter is unity (0 dB) throughout the passband and zero ($-\infty$ dB) in the stopband. The asymptotic gain responses of real low-pass filters show that we can only approximate the ideal response. As the number of poles n increases, the approximation improves, since the asymptotic slope in the stopband is $-20n$ dB/decade. On the other hand, adding poles requires additional stages in a cascade realization, so there is a tradeoff between filter complexity and cost, and how closely the filter gain approximates the ideal response.

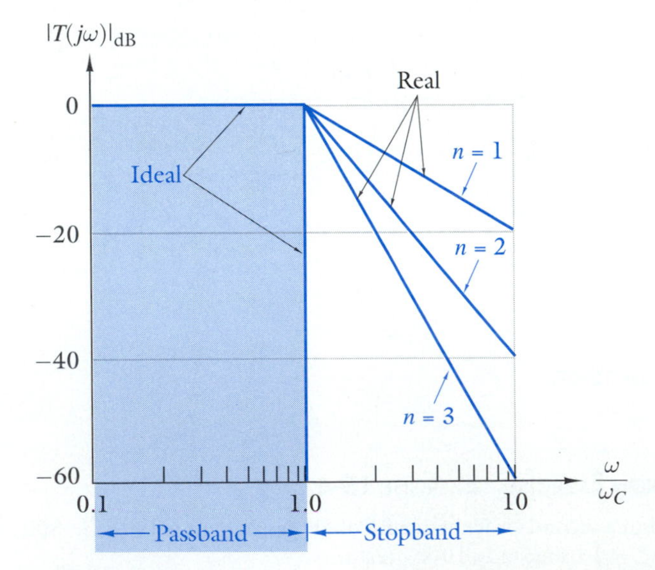

Figure 15-22 Ideal and real low-pass filter asymptotic responses.

Figure 15-23 shows how low-pass filter requirements are often specified. To meet the specification the gain response must lie within the unshaded region in the figure, as illustrated by the two example responses shown in Fig. 15-23. The parameter T_{MAX} is the *passband gain*. In the passband the gain must be within 3 dB of T_{MAX} and must equal $T_{MAX}/\sqrt{2}$ at the *cutoff frequency* ω_C. In the stopband the gain must decrease and remain below a gain of T_{MIN} for all $\omega \geq \omega_{MIN}$. A low-pass filter design requirement is defined by specifying values for these four parameters. The parameters T_{MAX} and ω_C define the passband response, and the parameters T_{MIN} and ω_{MIN} specify how rapidly the stopband response must decrease or *roll-off*

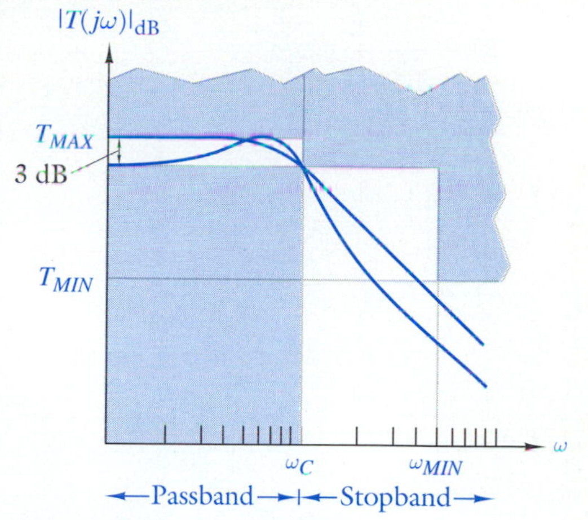

Figure 15-23 Low-pass filter specifications.

To design a low-pass filter we first construct a transfer function whose gain $|T(j\omega)|$ approximates the ideal filter response within the tolerances allowed by the four parameters T_{MAX}, T_{MIN}, ω_C, and ω_{MIN}. Low-pass transfer functions have the form

$$T(s) = \frac{K_O}{q(s)} \tag{15-8}$$

where $q(s)$ is a polynomial defining the poles of $T(s)$. To generate $T(s)$ we must select the gain factor K_O and the polynomial $q(s)$.

First-Order Cascade Responses

A simple way to design low-pass filters is to connect first-order low-pass filters in cascade. The denominator polynomial of a first-order low-pass function is $(s/\alpha + 1)$. Assuming that the chain rule applies, the denominator polynomial of n first-order low-pass function in cascade is $(s/\alpha + 1)^n$. The transfer function of an nth-order cascade of first-order circuits is

$$T_n(s) = \frac{K_O}{(s/\alpha + 1)^n} \tag{15-9}$$

In a design problem we must select K_O, α, and n such that the gain response meets the low-pass filter design requirements defined by T_{MAX}, ω_C, T_{MIN}, and ω_{MIN}.

The transfer function in Eq. (15-9) produces a gain response of

$$|T_n(j\omega)| = \frac{|K_O|}{\left[\sqrt{1 + (\omega/\alpha)^2}\right]^n} \tag{15-10}$$

The maximum gain occurs at dc where gain is $|T(0)| = |K_O|$. To meet the passband gain requirement we select $|K_O| = T_{MAX}$. By definition, the gain at the cutoff frequency is

$$|T_n(j\omega_C)| = \frac{|K_O|}{\left[\sqrt{1 + (\omega_C/\alpha)^2}\right]^n} = \frac{T_{MAX}}{\sqrt{2}} \tag{15-11}$$

Since $|K_O| = T_{MAX}$ we can equate the denominators on the right side of this expression and solve for α:

$$\alpha = \frac{\omega_C}{\sqrt{2^{1/n} - 1}} \tag{15-12}$$

The corner frequency of each first-order factor depends on the overall filter cutoff frequency and the number of poles n.

Figure 15-24 shows normalized gain responses of first-order low-pass cascades for $n = 1$ to $n = 8$. All of these responses fall within the passband tolerance and are 3 dB down at the cutoff frequency. However, the amount of attenuation provided in the stopband increases as the number of poles n increases. The graphs in Fig. 15-24 can be used to determine the number of poles required to meet stopband design requirements. For example, given a stopband requirement of $T_{MIN} = -40$ dB at $\omega_{MIN} = 10\omega_C$, we see in Fig. 15-24 that the response for $n = 2$ is about -32 dB, one decade above the cutoff frequency while the gain for $n = 3$ is about -42 dB. Thus, $n = 3$ is the smallest number of poles in a low-pass cascade that meets these stopband requirements.

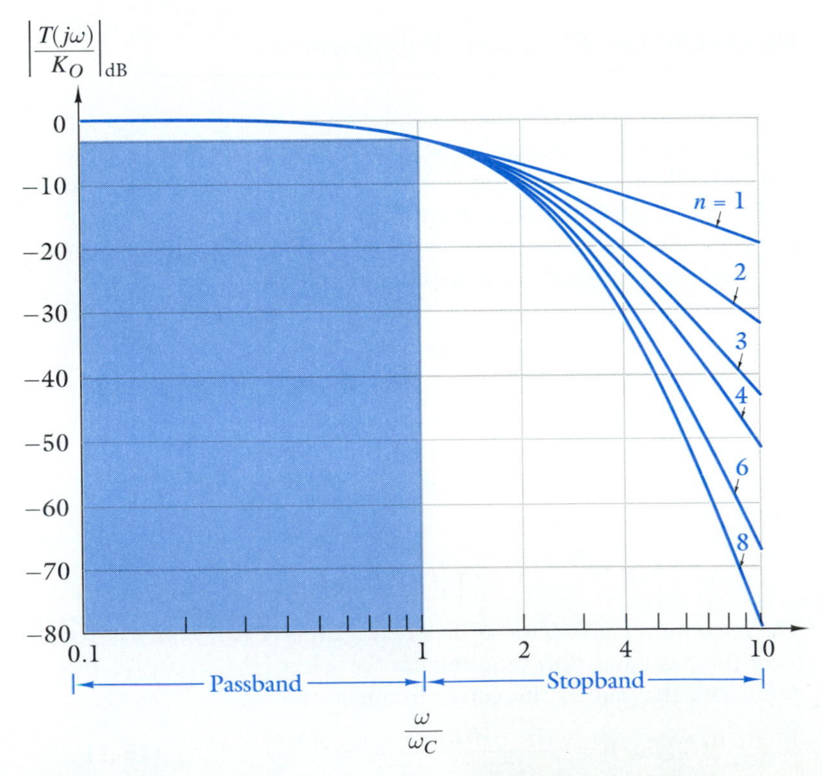

Figure 15-24 First-order low-pass cascade filter responses.

In summary, to construct the required $T_n(s)$ in Eq. (15-9), we first select $|K_O| = T_{MAX}$. We use Fig. 15-24 to determine the lowest value of n that meets the stopband requirements and then calculate α using Eq. (15-12). To realize $T_n(s)$ we partition it into a product of first factors that can be realized using the first-order building blocks in Fig. 15-3. Example 15-6 illustrates the design procedure for a cascade of first-order low-pass filters.

❖ Design Example

EXAMPLE 15-6

(a) Construct the transfer function of a first-order low-pass cascade with $T_{MAX} = 0$ dB, $\omega_C = 200$ rad/s, $T_{MIN} = -30$ dB, and $\omega_{MIN} = 800$ rad/s.

(b) Design a cascade of active RC circuits that realizes the transfer function developed in (a).

SOLUTION:

(a) The passband gain condition requires that $K_O = 1$. The stopband requirements dictate that the gain must be -30 dB at $\omega/\omega_C = 800/200 = 4$. Using Fig. 15-24 we find that $n = 8$ is the smallest number of poles that meets this stopband requirement. Using Eq. (15-12) we calculate corner frequency for each of the eight first-order factors:

$$\alpha = \frac{200}{\sqrt{2^{1/8} - 1}} = 665 \qquad \text{rad/s}$$

The required first-order cascade transfer function is

$$T(s) = \frac{1}{(s/665 + 1)^8}$$

(b) The transfer function developed in (a) can be partitioned into a product of eight identical first-order functions of the form $1/(s/665 + 1)$. Using the voltage-divider building block in Fig. 15-3 produces the following design constraints:

$$K\frac{Z_2}{Z_1 + Z_2} = \frac{1/s}{1/665 + 1/s}$$

These conditions yield $K = 1$, $Z_2 = 1/s$, and $Z_1 = 1/665$. A magnitude scale factor of $k_m = 10^7$ produces the first-order low-pass circuit in Fig. 15-25. Since $K = 1$ the OP AMP is a voltage follower. A cascade connection of eight such circuits produces a low-pass filter which meets the design requirements.

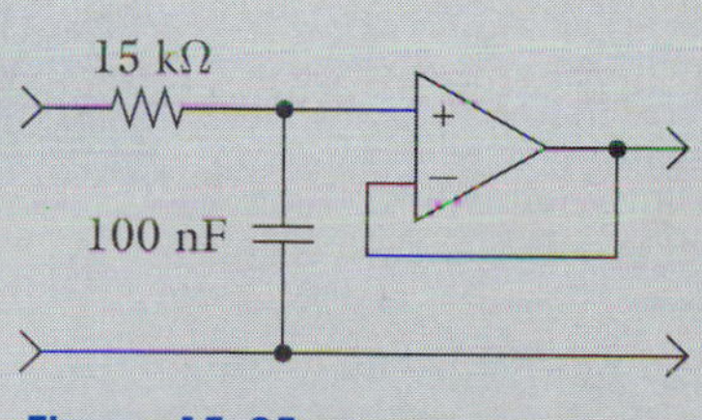

Figure 15-25

Butterworth Low-Pass Responses

The Butterworth low-pass filters are derived from a gain response of the form

$$|T_n(j\omega)| = \frac{|K_O|}{\sqrt{1 + (\omega/\omega_C)^{2n}}} \tag{15-13}$$

where ω_C is the cutoff frequency and n is the number of poles in $T_n(s)$. The maximum passband gain occurs at dc where the gain is $|T(0)| = |K_O|$. At the cutoff frequency the gain is $|T_n(j\omega_C)| = |K_O|\sqrt{2}$ for all n. When $K_0 = T_{MAX}$ the form of the Butterworth gain response in Eq. (15-13) ensures that the passband requirements are satisfied for any value of n.

Figure 15-26 compares normalized first-order cascade and Butterworth gain responses for $n = 4$. Both responses meet the passband requirements and have high-frequency asymptotic slopes of $-20n = -80$ dB/decade. However, the Butterworth response provides greater attenuation in the stopband because it approaches the high-frequency asymptote at lower frequencies. Although first-order cascade filters are easy to design, they are

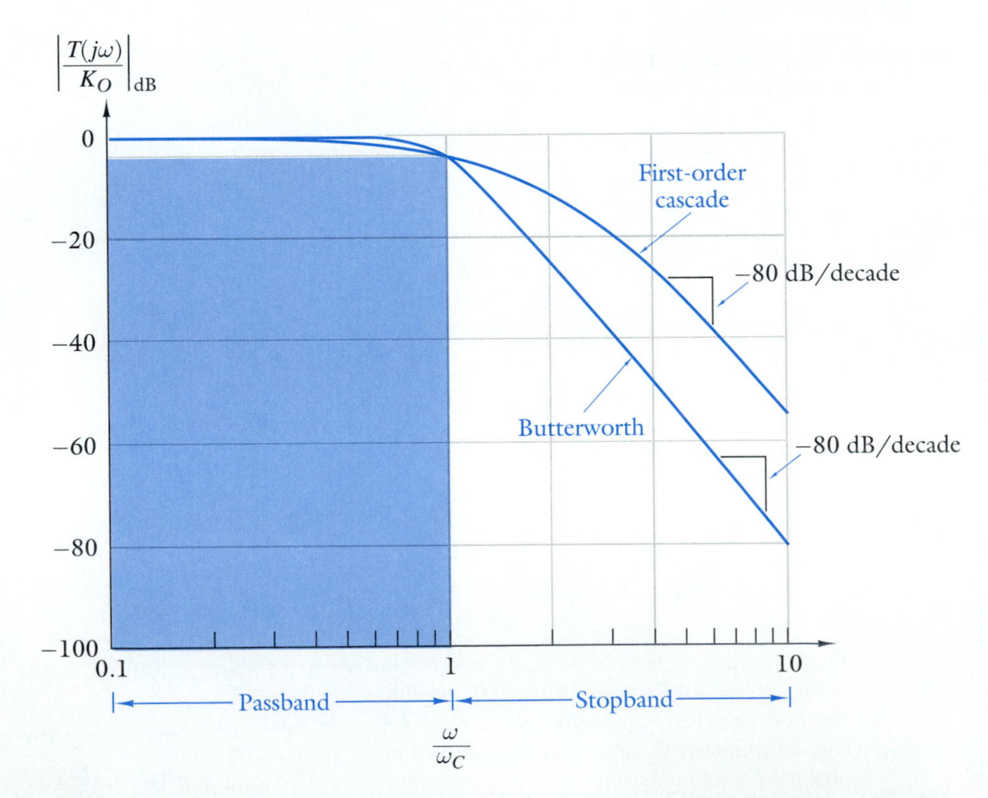

Figure 15-26 First-order cascade and Butterworth low-pass filter responses for $n = 4$.

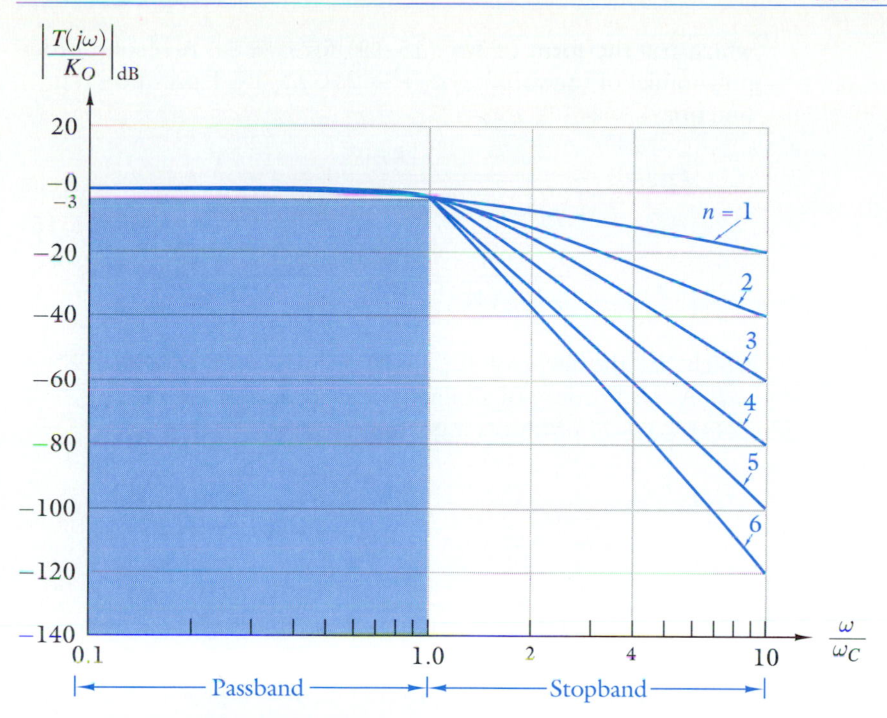

Figure 15-27 Butterworth low-pass filter responses.

not used very often because their stopband roll off is very poor compared with filters with complex poles such as the Butterworth response.

Figure 15-27 shows normalized plots of the Butterworth gain response in Eq. (15-13) for $n = 1$ to $n = 6$. These graphs can be used to determine the number of poles required to meet specified stopband requirements. For example, suppose we are given a stopband requirement of $T_{MIN} = -30$ dB at $\omega_{MIN} = 4\omega_C$. Figure 15-27 indicates that the response for $n = 2$ is about -24 dB at $\omega/\omega_C = 4$ while the gain for $n = 3$ is about -36 dB. Thus, $n = 3$ is the lowest order Butterworth response that meets the condition $T_{MIN} = -30$ dB at $\omega_{MIN} = 4\omega_C$. In Example 15-6 we found that $n = 8$ is the minimum number of poles required to meet the same stopband requirement using a cascade of first-order functions. The Butterworth realization is much simpler and usually less expensive than a cascade design, since the number of stages is closely related to the number of poles.

To construct a Butterworth transfer function $T_n(s)$ we need to determine the denominator polynomial $q_n(s)$ required to produce the gain response in Eq. (15-13). A first-order polynomial of the form $s/\omega_C + 1$ produces the gain function

$$|T_1(j\omega)| = \frac{|K_O|}{\sqrt{1 + (\omega/\omega_C)^2}} \tag{15-14}$$

which has the form of Eq. (15-13) for $n = 1$. A second-order polynomial of the form $(s/\omega_C)^2 + 2\zeta(s/\omega_C) + 1$ produces a gain function

$$|T_2(j\omega)| = \frac{|K_O|}{\sqrt{\left[1 - (\omega/\omega_C)^2\right]^2 + (2\zeta\omega/\omega_C)^2}} = \frac{|K_O|}{\sqrt{1 + (4\zeta^2 - 2)(\omega/\omega_C)^2 + (\omega/\omega_C)^4}} \tag{15-15}$$

which has the form of Eq. (15-13) for $n = 2$ provided $2\zeta = \sqrt{2}$. A third-order polynomial of the form $(s/\omega_C + 1)[(s/\omega_C)^2 + 2\zeta(s/\omega_C) + 1]$ produces a gain function

$$|T_3(j\omega)| = \frac{|K_O|}{\sqrt{1 + (\omega/\omega_C)^2}\sqrt{\left[1 - (\omega/\omega_C)^2\right]^2 + (2\zeta\omega/\omega_C)^2}} = \frac{|K_O|}{\sqrt{1 + (4\zeta^2 - 1)(\omega/\omega_C)^2 + (4\zeta^2 - 1)(\omega/\omega_C)^4 + (\omega/\omega_C)^6}} \tag{15-16}$$

which has the form of Eq. (15-13) for $n = 3$ provided $2\zeta = 1$.

The above analysis shows that the denominator polynomials

$$q_1(s) = s/\omega_C + 1$$

$$q_2(s) = (s/\omega_C)^2 + \sqrt{2}(s/\omega_C) + 1$$

$$q_3(s) = (s/\omega_C + 1)[(s/\omega_C)^2 + (s/\omega_C) + 1]$$

produce the Butterworth gain response in Eq. (15-13) for $n = 1$, 2, and 3. These polynomials are the first three of a set of polynomials that produce the Butterworth gain responses. Table 15-1 lists the normalized ($\omega_C = 1$) version of these polynomials through $n = 6$.

An nth-order Butterworth low-pass transfer function with a passband gain of $|K_O|$ and a cutoff frequency of ω_C is written as

$$T_n(s) = \frac{K_O}{q_n(s/\omega_C)} \tag{15-17}$$

where $q_n(s)$ is the nth-order normalized polynomial in Table 15-1. When $K_O = T_{MAX}$ this transfer function meets the passband and cutoff frequency conditions because of the way the normalized polynomials are constructed. The order of the polynomial, hence the number of poles, is determined from the stopband requirement using Fig. 15-27.

TABLE 15-1
NORMALIZED POLYNOMIALS DEFINING THE BUTTERWORTH POLES

Order	Normalized Denominator Polynomials
1	$(s+1)$
2	$(s^2+1.414s+1)$
3	$(s+1)(s^2+s+1)$
4	$(s^2+0.7654s+1)(s^2+1.848s+1)$
5	$(s+1)(s^2+0.6180s+1)(s^2+1.618s+1)$
6	$(s^2+0.5176s+1)(s^2+1.414s+1)(s^2+1.932s+1)$

Once we have constructed the required $T_n(s)$ we partition it into a product of first- and second-order factors and realize each factor using the building blocks developed in the previous sections. The next example illustrates the design procedure for a Butterworth low-pass filter.

Design Example

EXAMPLE 15-7

(a) Construct a Butterworth low-pass transfer function with $T_{MAX} =$ 20 dB, $\omega_C = 1000$ rad/s, and $T_{MIN} = -20$ dB at $\omega_{MIN} =$ 4000 rad/s.

(b) Design a cascade of active RC circuits that realizes the transfer function found in part (a).

SOLUTION:

(a) The passband gain condition means $K_O = 10$ in Eq. (15-17). The stopband requirement specifies that the gain must be 40 dB less than the passband gain at $\omega/\omega_C = 4000/1000 = 4$. Using Fig. 15-27 we see that $n = 4$ is the lowest-order polynomial that meets the stopband requirements. Using the fourth-order polynomial from Table 15-1 we obtain the required

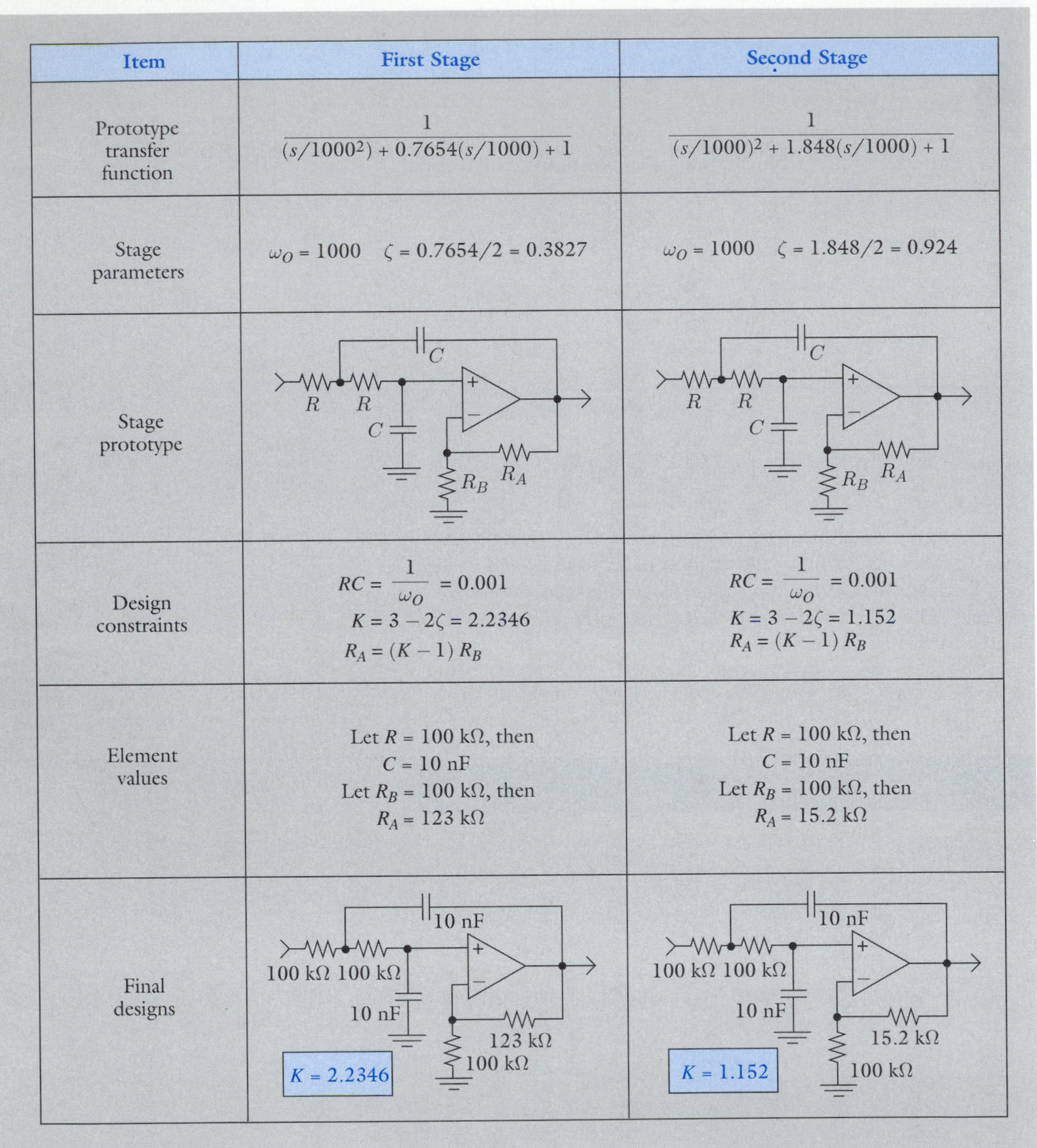

Item	First Stage	Second Stage
Prototype transfer function	$\dfrac{1}{(s/1000^2) + 0.7654(s/1000) + 1}$	$\dfrac{1}{(s/1000)^2 + 1.848(s/1000) + 1}$
Stage parameters	$\omega_O = 1000 \quad \zeta = 0.7654/2 = 0.3827$	$\omega_O = 1000 \quad \zeta = 1.848/2 = 0.924$
Stage prototype	R, R, C, C, R_B, R_A, +, –	R, R, C, C, R_B, R_A, +, –
Design constraints	$RC = \dfrac{1}{\omega_O} = 0.001$ $K = 3 - 2\zeta = 2.2346$ $R_A = (K-1)\,R_B$	$RC = \dfrac{1}{\omega_O} = 0.001$ $K = 3 - 2\zeta = 1.152$ $R_A = (K-1)\,R_B$
Element values	Let R = 100 kΩ, then C = 10 nF Let R_B = 100 kΩ, then R_A = 123 kΩ	Let R = 100 kΩ, then C = 10 nF Let R_B = 100 kΩ, then R_A = 15.2 kΩ
Final designs	10 nF, 100 kΩ, 100 kΩ, 10 nF, 123 kΩ, 100 kΩ, +, – $K = 2.2346$	10 nF, 100 kΩ, 100 kΩ, 10 nF, 15.2 kΩ, 100 kΩ, +, – $K = 1.152$

Figure 15-28 Design sequence for Example 15-7.

Butterworth low-pass transfer function:

$$T(s) = \frac{K_O}{q_4(s/1000)}$$

$$= \frac{10}{\left[\left(\frac{s}{1000}\right)^2 + 0.7654\left(\frac{s}{1000}\right) + 1\right]\left[\left(\frac{s}{1000}\right)^2 + 1.848\left(\frac{s}{1000}\right) + 1\right]}$$

(b) The transfer function developed in (a) can be partitioned as follows

$$\frac{T(s)}{10} = \left[\frac{1}{\left(\frac{s}{1000}\right)^2 + 0.7654\left(\frac{s}{1000}\right) + 1}\right]\left[\frac{1}{\left(\frac{s}{1000}\right)^2 + 1.848\left(\frac{s}{1000}\right) + 1}\right]$$

Figure 15-28 shows a design sequence for each of the second-order transfer functions in this partitioning. The first and second rows in the figure give the transfer function for each stage and the stage parameters. The active RC building blocks in the third row of the figure lead to the design constraints in the fourth row. The element values selected in the fifth row produce the final design shown in the last row of Fig. 15-28.

The two active RC circuit building blocks used in this design do not allow us to control the passband gain because the gain K controls the stage damping ratio. The final designs in Fig. 15-28 produce a passband gain of $K_O = 2.2346 \times 1.152 = 2.57$. Since this gain falls short of the $K_O = 10$ requirement, we add a third stage with a gain of $10/2.57 = 3.89$ to make up the difference. Figure 15-29 shows the final three-stage cascade design, which meets all of the low-pass filter requirements.

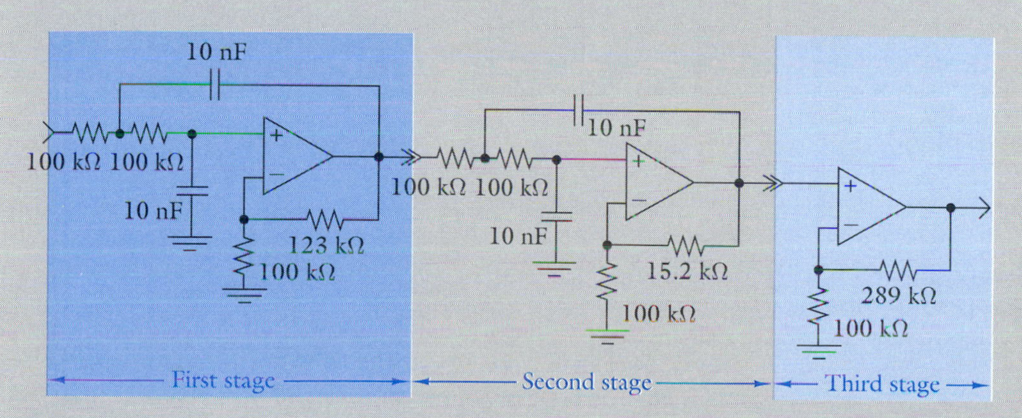

Figure 15-29

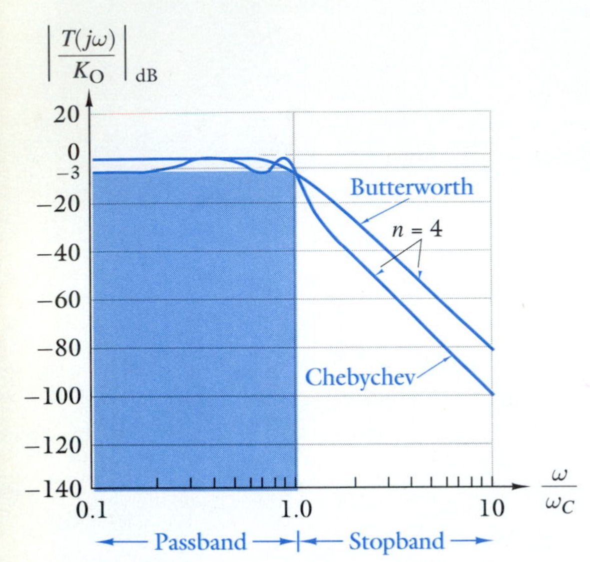

Figure 15-30 Butterworth and Chebychev low-pass filter responses for $n = 4$.

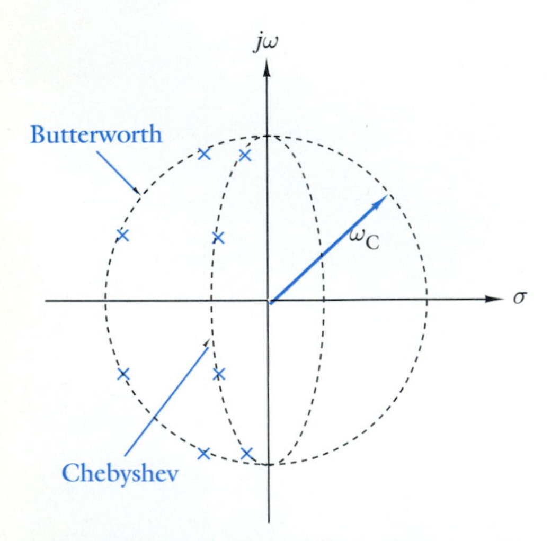

Figure 15-31 Butterworth and Chebychev low-pass filter pole locations for $n = 4$.

Chebychev Low-Pass Responses

The Chebychev response provides an alternative way to approximate ideal low-pass filter characteristics. Figure 15-30 compares the Butterworth and Chebychev gain responses for $n = 4$. Both responses have asymptotic slopes of -80 dB/decade in the stopband. The Butterworth response is very flat in the passband resulting in a more gradual transition to its high-frequency asymptote. In contrast, the Chebychev response has a series of peaks and valleys in the passband resulting in a more abrupt transition to the stopband asymptote. The net result is that the Chebychev response offers greater attenuation in the stopband than a Butterworth response of the same order.

The Butterworth response is called ***maximally flat*** because it emphasizes smoothness of the passband response. The Chebychev response is called ***equal ripple*** because the passband gain oscillations have the same amplitudes. Figure 15-31 shows the Butterworth and Chebychev pole locations for $n = 4$. The Butterworth poles are evenly distributed on a circle of radius ω_C. The Chebychev poles are located on a ellipse, which pushes them closer to the j axis. The Chebychev poles have lower damping ratios and the resulting resonant peaks produce the equal ripple response in the passband. The radial distance to any pair of complex poles is their natural frequency ω_O. The stages in a cascade realizing Butterworth poles are said to be ***synchronously tuned*** because they all have the same natural frequency $\omega_O = \omega_C$. The stages producing pairs of Chebychev poles are called ***stagger tuned*** because they all have different natural frequencies that are not equal to the overall cutoff frequency of the cascade.

Table 15-2 lists the normalized ($\omega_C = 1$) polynomials through $n = 6$ defining the pole locations for the Chebychev response. The transfer function of an nth-order Chebychev low-pass filter with a passband gain of K_O and a cutoff frequency of ω_C is written as follows:

$$T_n(s) = \frac{K_O}{q_n(s/\omega_C)} \quad n \text{ odd}$$
$$= \frac{K_O/\sqrt{2}}{q_n(s/\omega_C)} \quad n \text{ even} \qquad (15\text{-}18)$$

where $q_n(s)$ is the nth-order normalized polynomial in Table 15-2. When $K_O = T_{MAX}$, the $1/\sqrt{2}$ gain adjustment in Eq. (15-18) for even order polynomials ensures that the maximum gain in the passband is T_{MAX}. As with the Butterworth response, the order of the denominator polynomial is determined by the stopband requirements.

Figure 15-32 shows the normalized gain responses of Chebychev low-pass transfer functions through $n = 6$. Note the equal-ripple performance in the passband and the increased stopband attenuation compared with the Butterworth maximally flat responses in Fig. 15-27. The graphs in Fig. 15-32 can be used to determine the Chebychev order needed to meet stopband conditions. For example, $T_{MIN} = -20$ dB at one octave above cutoff requires $n = 3$ for Chebychev responses. The same stopband conditions requires $n = 4$ using Butterworth responses. The Chebychev approach may yield a lower-order transfer function requiring fewer stages than the Butterworth filter to meet the same stopband design specifications.

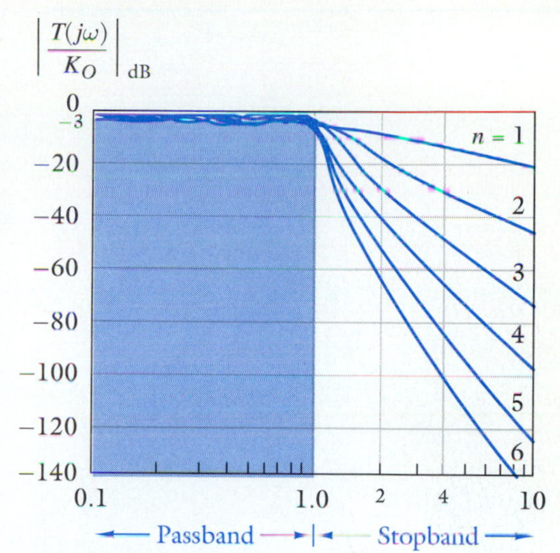

Figure 15-32 Chebychev low-pass filter responses.

TABLE 15-2 NORMALIZED POLYNOMIALS DEFINING THE CHEBYCHEV POLES

Order	Normalized Denominator Polynomials
1	$(s + 1)$
2	$[(s/0.8409)^2 + 0.7654(s/0.8409) + 1]$
3	$[(s/0.2980) + 1][(s/0.9159)^2 + 0.3254(s/0.9159) + 1]$
4	$[(s/0.9502)^2 + 0.1789(s/0.9502) + 1][(s/0.4425)^2 + 0.9276(s/0.4425) + 1]$
5	$[(s/0.1772) + 1][(s/0.9674)^2 + 0.01132(s/0.9674) + 1]$ $[(s/0.6139)^2 + 0.4670(s/0.6139) + 1]$
6	$[(s/0.9771)^2 + 0.0781(s/0.9771) + 1][(s/0.7223)^2 + 0.2886(s/0.7223) + 1]$ $[(s/0.2978)^2 + 0.9562(s/0.2978) + 1]$

Design Example

EXAMPLE 15-8

(a) Construct a Chebychev low-pass transfer function with a passband requirement of $T_{MAX} = 20$ dB and $\omega_C = 10$ rad/s, and a stopband requirement of $T_{MIN} = -20$ dB at $\omega_{MIN} = 40$ rad/s.

(b) Design a cascade of active RC circuits that realizes the transfer function found in part (a).

SOLUTION:

(a) The filter specification is the same as in Example 15-7 except that the cutoff frequency is two decades lower. The passband gain condition requires $K_O = 10$. The stopband requirement specifies that the gain must be 40 dB less than the passband gain at $\omega/\omega_C = 40/10 = 4$. The Chebychev responses in Fig. 15-32 show that $n = 3$ meets the stopband requirement. Note that the same stopband conditions in Example 15-7 required $n = 4$ using the Butterworth approach. Since the Chebychev order is odd, no gain correction is required. Using the third-order polynomial from Table 15-2, we construct the required third-order Chebychev low-pass transfer function:

$$T(s) = \frac{K_O}{q_3(s/10)}$$

$$= \frac{10}{\left[\left(\dfrac{s}{0.298 \times 10}\right) + 1\right]\left[\left(\dfrac{s}{0.9159 \times 10}\right)^2 + 0.3254\left(\dfrac{s}{0.9159 \times 10}\right) + 1\right]}$$

(b) The Chebychev transfer function developed in (a) can be partitioned in the following way:

$$\frac{T(s)}{10} = \left[\frac{1}{\left(\dfrac{s}{0.298 \times 10}\right) + 1}\right]\left[\frac{1}{\left(\dfrac{s}{0.9159 \times 10}\right)^2 + 0.3254\left(\dfrac{s}{0.9159 \times 10}\right) + 1}\right]$$

Realizing this partition requires a cascade connection of a first- and second-order building block. The design sequence in Fig. 15-33 begins with the required transfer functions in the first row. The next three rows show

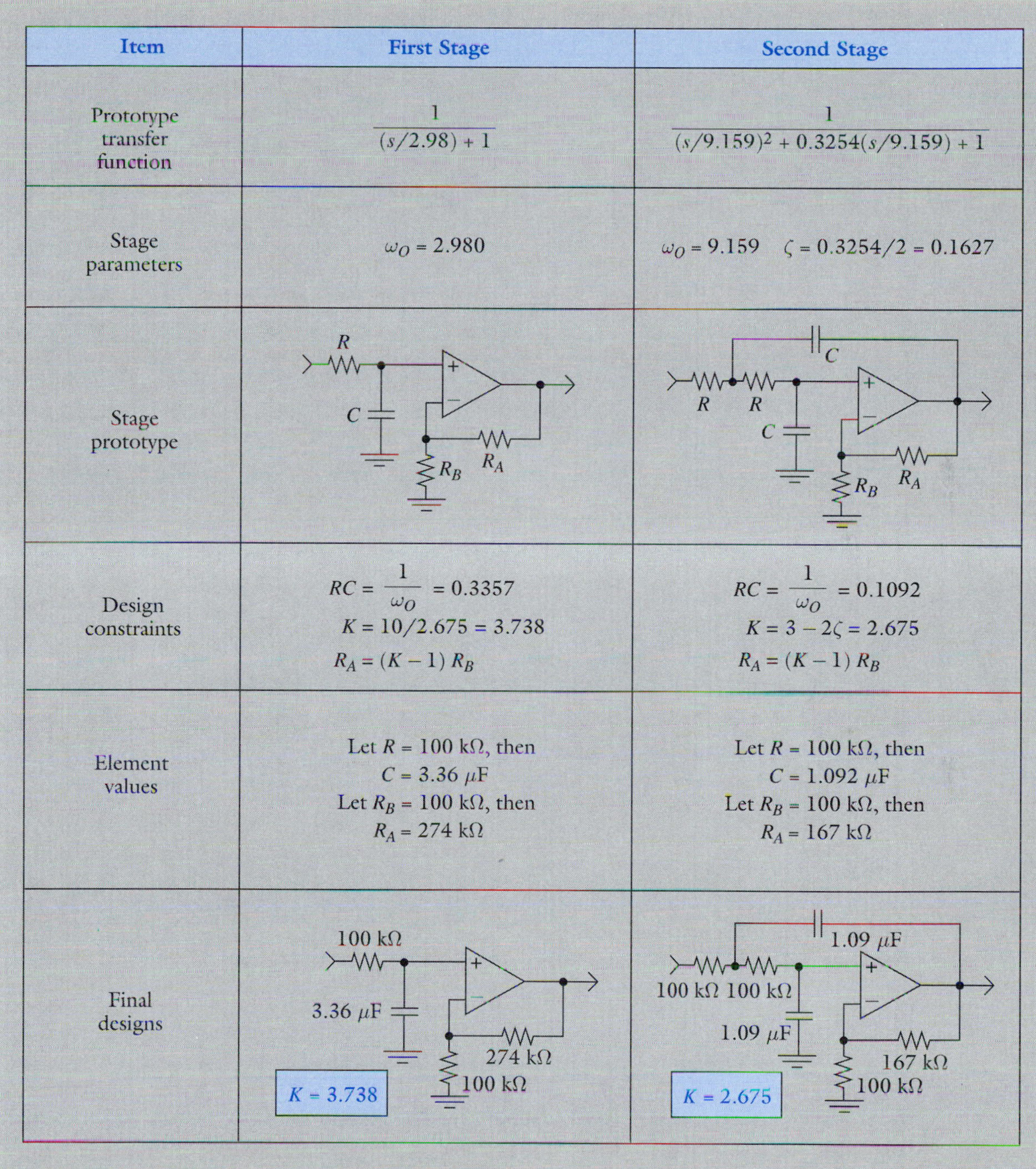

Item	First Stage	Second Stage
Prototype transfer function	$\dfrac{1}{(s/2.98)+1}$	$\dfrac{1}{(s/9.159)^2+0.3254(s/9.159)+1}$
Stage parameters	$\omega_O = 2.980$	$\omega_O = 9.159 \quad \zeta = 0.3254/2 = 0.1627$
Stage prototype	R, C, R_A, R_B	R, R, C, C, R_A, R_B
Design constraints	$RC = \dfrac{1}{\omega_O} = 0.3357$ $K = 10/2.675 = 3.738$ $R_A = (K-1)\,R_B$	$RC = \dfrac{1}{\omega_O} = 0.1092$ $K = 3 - 2\zeta = 2.675$ $R_A = (K-1)\,R_B$
Element values	Let R = 100 kΩ, then C = 3.36 μF Let R_B = 100 kΩ, then R_A = 274 kΩ	Let R = 100 kΩ, then C = 1.092 μF Let R_B = 100 kΩ, then R_A = 167 kΩ
Final designs	100 kΩ, 3.36 μF, 274 kΩ, 100 kΩ $K = 3.738$	1.09 μF, 100 kΩ 100 kΩ, 1.09 μF, 167 kΩ, 100 kΩ $K = 2.675$

Figure 15-33 Design sequence for Example 15-8.

the stage parameters, stage prototype, and design constraints for the selected building blocks. The constraints are then used to select element values that produce the final stage designs shown in the last row of Fig. 15-33. Note that the second-stage gain is determined by the stage damping ratio. However, the first stage is a voltage divider whose gain can be adjusted without changing its pole location. Once the second-stage gain is known the first-stage gain is selected so that an overall passband gain of $K_O = 10$ is achieved. A cascade connection of the final designs in Fig. 15-33 meets all design requirements without introducing the additional gain stage as was required in Example 15-7.

EXERCISE 15-5

What is the minimum order of the first-order cascade, Butterworth and Chebychev low-pass transfer functions that meet the following stopband conditions:

(a) $T_{MIN} = -20$ dB at $4\omega_C$
(b) $T_{MIN} = -80$ dB at $6\omega_C$
(c) $T_{MIN} = -80$ dB at $10\omega_C$
(d) $T_{MIN} = -120$ dB at $10\omega_C$

ANSWERS:

	Cascade	Butterworth	Chebychev
(a)	$n = 3$	$n = 2$	$n = 2$
(b)	$n > 8$	$n = 5$	$n = 4$
(c)	$n = 8$	$n = 4$	$n = 4$
(d)	$n \gg 8$	$n = 6$	$n = 5$

15-5 HIGH-PASS, BANDPASS, AND BANDSTOP FILTER DESIGN

High-Pass Filter Design

High-pass transfer functions can be derived from low-pass functions using a transformation of the complex frequency variable s. If a low-pass transfer function $T_{LP}(s)$ has a passband gain of T_{MAX} and cutoff frequency of ω_C, then the transfer function $T_{HP}(s)$ defined as

$$T_{HP}(s) = T_{LP}\left(\frac{\omega_C^2}{s}\right) \tag{15-19}$$

has high-pass characteristics with the same cutoff frequency and passband gain. In other words, replacing every s in $T_{LP}(s)$ by ω_C^2/s produces a high-pass transfer but does not change the parameters defining passband performance.

The first-order low-pass transfer function provides a simple example of the transformation:

$$T_{LP}(s) = \frac{T_{MAX}}{\dfrac{s}{\omega_C} + 1}$$

Replacing s by ω_C^2/s in this expression yields a first-order high-pass transfer function

$$T_{HP}(s) = \frac{T_{MAX}}{\dfrac{\omega_C^2/s}{\omega_C} + 1} = \frac{T_{MAX}s}{s + \omega_C}$$

Note that the cutoff frequency and passband gain of the high-pass transfer function are the same as those of the low-pass transfer function from which it was derived.

Figure 15-34 shows the high-pass responses derived from the fourth-order Butterworth and Chebychev low-pass responses in Fig. 15-30. From these plots we see that the transformation interchanges the pass and stopbands. The cutoff frequency is unchanged because it lies at the boundary between the two bands. Moreover, the same maximally flat or equal-ripple features appear in the passband of the high-pass response. Finally, the high-pass stopband gain is a mirror image about $\omega/\omega_C = 1$ of the low-pass stopband response. For example, if a low-pass function has -40 dB gain one octave above cutoff then the corresponding high-pass function will have -40 dB gain one octave below cutoff.

A high-pass filter design problem is defined by specifying four parameters: T_{MAX}, ω_C, T_{MIN}, and ω_{MIN}. To obtain a high-pass transfer function that meets these requirements, we first construct a low-pass prototype $T_{LP}(s)$. The passband parameters for the low-pass prototype are the same as T_{MAX} and ω_C for the high-pass. The stopband requirement for the low-pass prototype is T_{MIN} at ω_C^2/ω_{MIN}. Given these design requirements, we construct a low-pass prototype $T_{LP}(s)$ using the methods described in Section 15-4. Replacing s by ω_C^2/s transforms the low-pass prototype into a high-pass transfer function $T_{HP}(s)$. The high-pass function is then partitioned into first- and second-order factors, and realized using high-pass building blocks discussed in previous sections.

The next two examples illustrate the method of designing high-pass filters using a low-pass prototype.

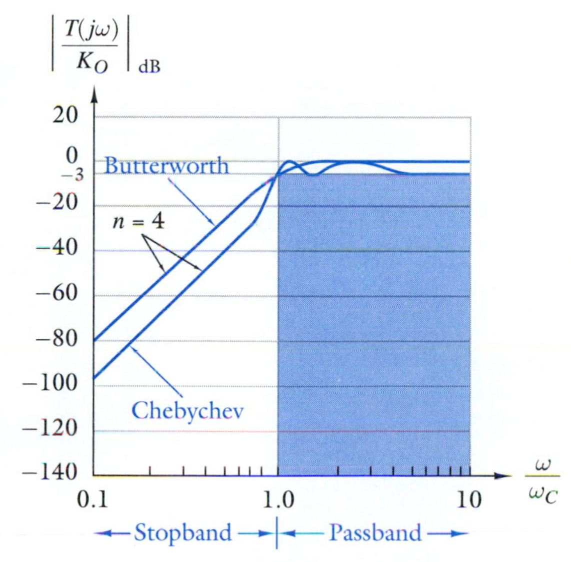

Figure 15-34 Butterworth and Chebychev high-pass filter responses for $n = 4$.

❖ Design Example

EXAMPLE 15-9

(a) Construct a Butterworth high-pass transfer function with $T_{MAX} = 20$ dB, $\omega_C = 10$ rad/s, and a stopband requirement of $T_{MIN} = -20$ dB at $\omega_{MIN} = 2.5$ rad/s.

(b) Design a cascade of active RC circuits that realizes the transfer function found in part (a).

SOLUTION:

(a) The passband requirements for the low-pass prototype are: $K_O = 10$ and $\omega_C = 10$ rad/s. The stopband requirement for the low-pass prototype is $T_{MIN} = -20$ dB at $\omega_C^2/\omega_{MIN} = 40 = 4\omega_C$. Using Fig. 15-27 we find that $n = 4$ is the lowest-order Butterworth response that meets the low-pass design requirements. Using the fourth-order polynomial from Table 15-1, the required low-pass prototype is

$$T_{LP}(s) = \frac{K_O}{q_4(s/10)}$$

$$= \frac{10}{\left[\left(\frac{s}{10}\right)^2 + 0.7654\left(\frac{s}{10}\right) + 1\right]\left[\left(\frac{s}{10}\right)^2 + 1.848\left(\frac{s}{10}\right) + 1\right]}$$

The high-pass transfer function is obtained by replacing s with $\omega_C^2/s = 100/s$.

$$T_{HP}(s) = \frac{10}{\left[\left(\frac{100/s}{10}\right)^2 + 0.7654\left(\frac{100/s}{10}\right) + 1\right]\left[\left(\frac{100/s}{10}\right)^2 + 1.848\left(\frac{100/s}{10}\right) + 1\right]}$$

$$= 10\left[\frac{(s/10)^2}{(s/10)^2 + 0.7654(s/10) + 1}\right]\left[\frac{(s/10)^2}{(s/10)^2 + 1.848(s/10) + 1}\right]$$

$T_{HP}(s)$ is a fourth-order high-pass transfer function that meets the filter design requirements.

(b) Figure 15-35 shows a design sequence for realizing $T_{HP}(s)$ using the second-order high-pass active RC building block. The stage transfer functions given in the first row yield the stage parameters in the second row. The stage parameters together with the stage prototype yield the design constraints in the fourth row. Selecting element values within these constraints produces the final stage designs shown in the last row of Fig. 15-35.

In the final design the passband gain is $|T_{HP}(\infty)| = 2.235 \times 1.152 = 2.57$, which is less than the gain 10 requirement

(T_{MAX} = 20 dB). To make up the desired gain we add a third stage with a gain of $10/2.57 = 3.89$. Figure 15-36 shows the final three-stage cascade design including the required gain stage.

Item	First Stage	Second Stage
Prototype transfer function	$\dfrac{(s/10)^2}{(s/10^2) + 0.7654(s/10) + 1}$	$\dfrac{(s/10)^2}{(s/10)^2 + 1.848(s/10) + 1}$
Stage parameters	$\omega_O = 10 \quad \zeta = 0.7654/2 = 0.3827$	$\omega_O = 10 \quad \zeta = 1848/2 = 0.924$
Stage prototype	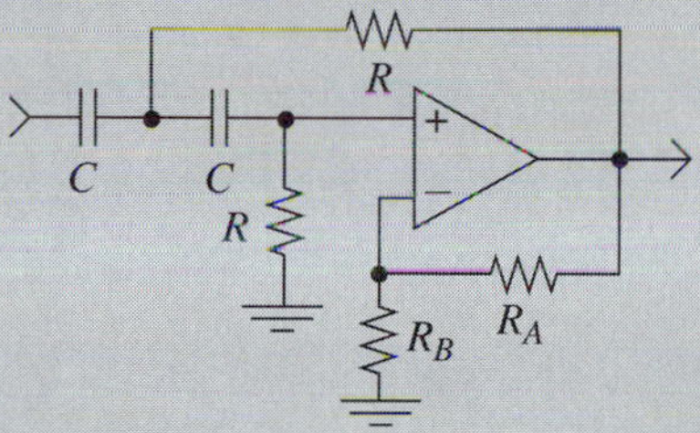 R, C, C, R, R_B, R_A	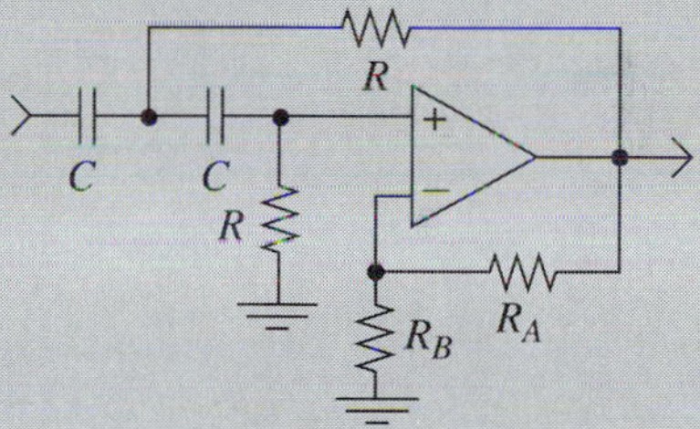 R, C, C, R, R_B, R_A
Design constraints	$RC = \dfrac{1}{\omega_O} = 0.1$ $K = 3 - 2\zeta = 2.2346$ $R_A = (K-1)\,R_B$	$RC = \dfrac{1}{\omega_O} = 0.1$ $K = 3 - 2\zeta = 1.152$ $R_A = (K-1)\,R_B$
Element values	Let R = 100 kΩ, then C = 1 μF Let R_B = 100 kΩ, then R_A = 123 kΩ	Let R = 100 kΩ, then C = 1 μF Let R_B = 100 kΩ, then R_A = 15.2 kΩ
Final designs	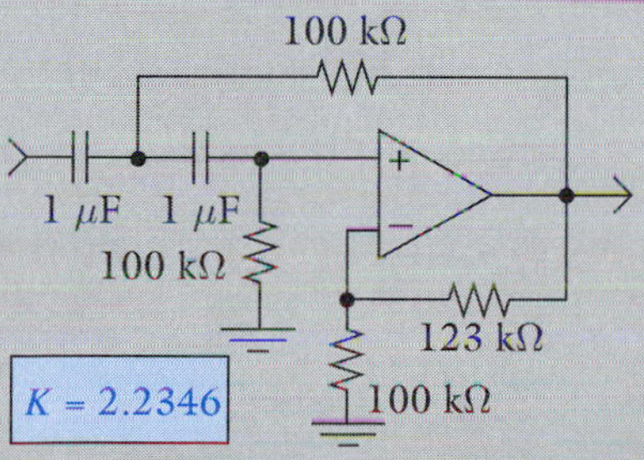 100 kΩ, 1 μF, 1 μF, 100 kΩ, 123 kΩ, 100 kΩ $K = 2.2346$	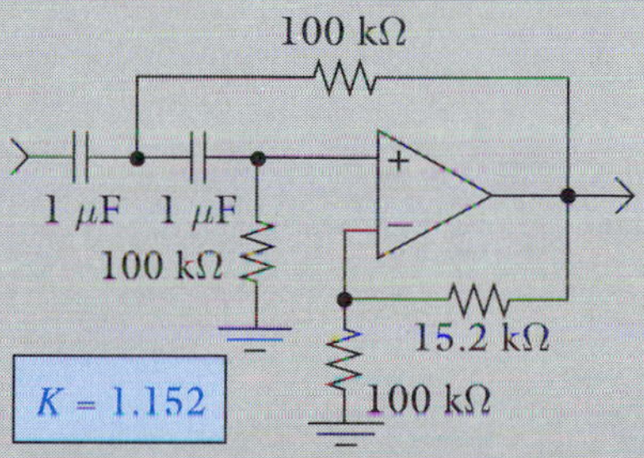 100 kΩ, 1 μF, 1 μF, 100 kΩ, 15.2 kΩ, 100 kΩ $K = 1.152$

Figure 15-35 Design sequence for Example 15-9.

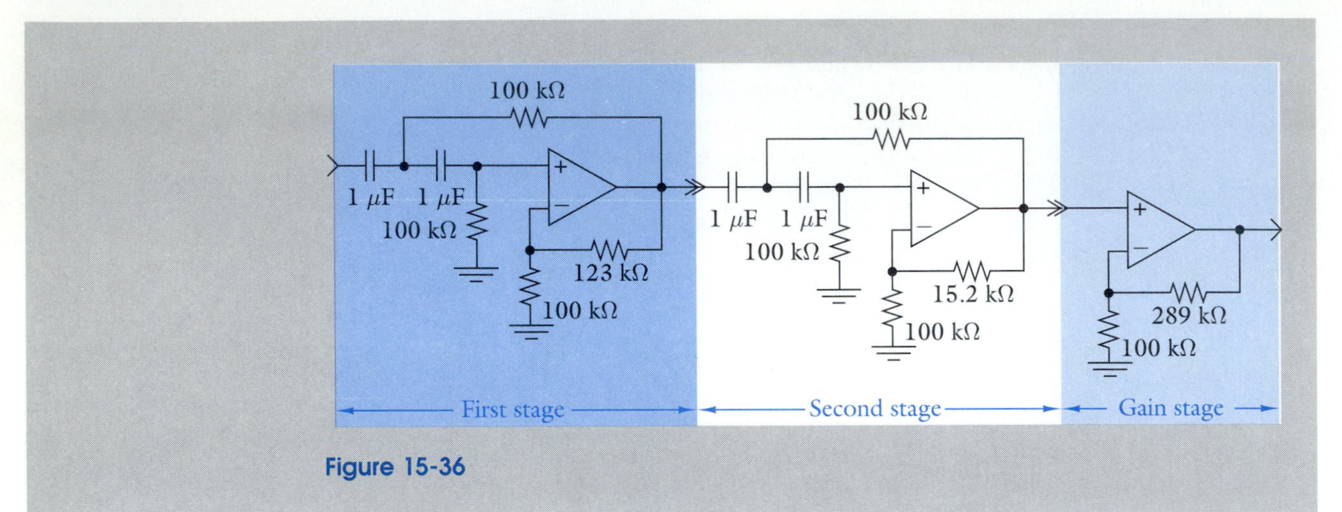

Figure 15-36

Design Example

EXAMPLE 15-10

(a) Construct a Chebychev high-pass transfer function with $T_{MAX} =$ 20 dB, $\omega_C = 1$ krad/s, and a stopband requirement of $T_{MIN} =$ -20 dB at $\omega_{MIN} = 250$ rad/s.

(b) Design a cascade of active RC circuits that realizes the transfer function found in part (a).

SOLUTION:

(a) The design sequence in Fig. 15-37 is based on a third-order high-pass function derived as follows. The passband requirements for the low-pass prototype are: $K_O = 10$ and $\omega_C =$ 1000 rad/s. The stopband requirement for the low-pass prototype is $T_{MIN} = -20$ dB at $\omega_C^2/\omega_{MIN} = 4000 = 4\omega_C$. Using Fig. 15-32 we find that $n = 3$ is the lowest-order Chebychev response that meets the low-pass prototype requirements. Using the third-order polynomial from Table 15-2 yields the required low-pass prototype:

$$T_{LP}(s) = \frac{K_O}{q_3(s/1000)}$$

$$= \frac{10}{\left[\left(\dfrac{s}{0.298 \times 1000}\right) + 1\right]\left[\left(\dfrac{s}{0.9159 \times 1000}\right)^2 + 0.3254\left(\dfrac{s}{0.9159 \times 1000}\right) + 1\right]}$$

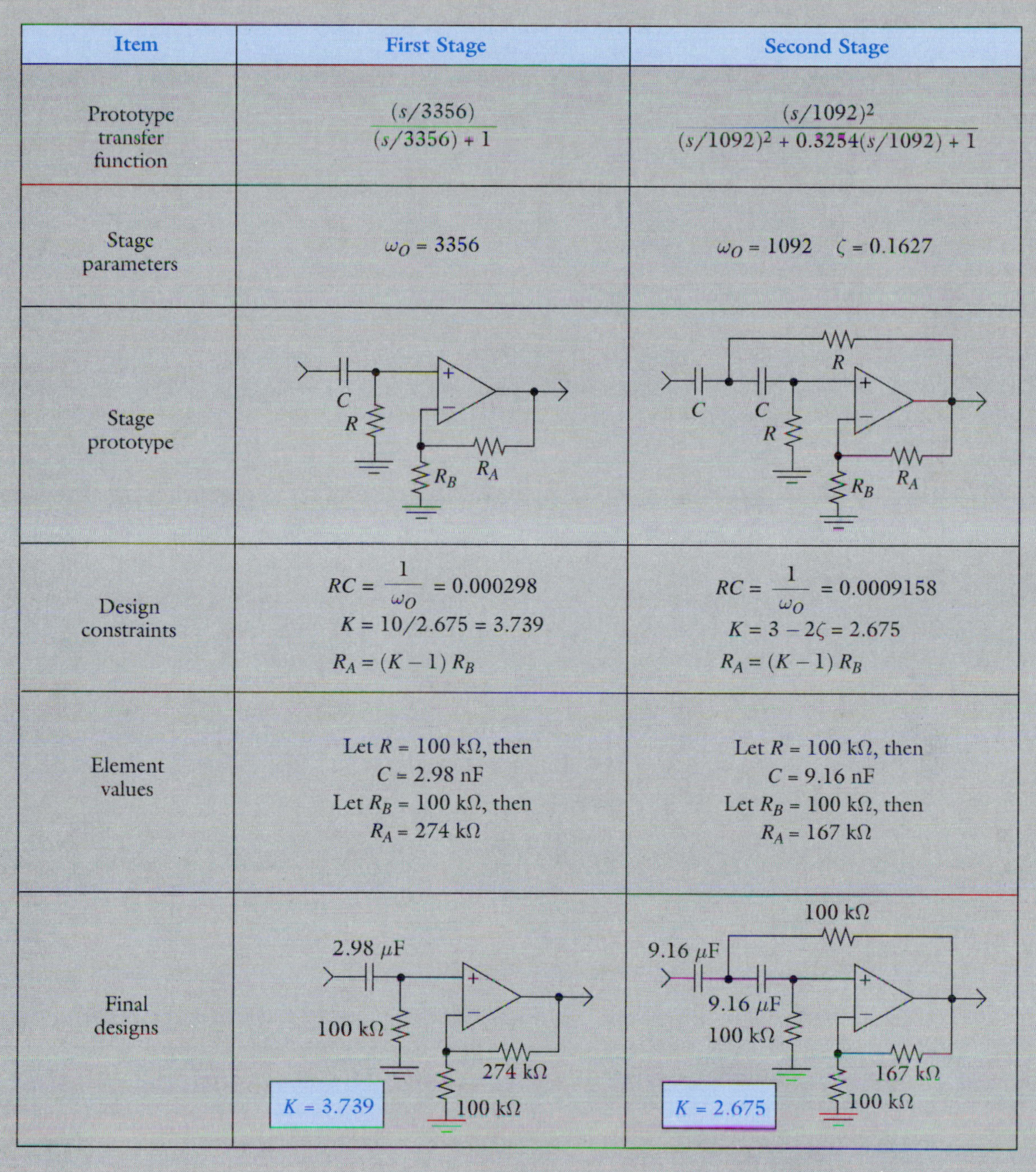

Item	First Stage	Second Stage
Prototype transfer function	$\dfrac{(s/3356)}{(s/3356)+1}$	$\dfrac{(s/1092)^2}{(s/1092)^2+0.3254(s/1092)+1}$
Stage parameters	$\omega_O = 3356$	$\omega_O = 1092 \quad \zeta = 0.1627$
Stage prototype	C, R, R_B, R_A	R, C, C, R, R_B, R_A
Design constraints	$RC = \dfrac{1}{\omega_O} = 0.000298$ $K = 10/2.675 = 3.739$ $R_A = (K-1)\,R_B$	$RC = \dfrac{1}{\omega_O} = 0.0009158$ $K = 3 - 2\zeta = 2.675$ $R_A = (K-1)\,R_B$
Element values	Let R = 100 kΩ, then C = 2.98 nF Let R_B = 100 kΩ, then R_A = 274 kΩ	Let R = 100 kΩ, then C = 9.16 nF Let R_B = 100 kΩ, then R_A = 167 kΩ
Final designs	2.98 μF, 100 kΩ, 274 kΩ, 100 kΩ $K = 3.739$	100 kΩ, 9.16 μF, 9.16 μF, 100 kΩ, 167 kΩ, 100 kΩ $K = 2.675$

Figure 15-37 Design sequence for Example 15-10.

The required high-pass function is obtained by replacing s with $(1000)^2/s$:

$$T_{HP}(s) = \frac{10}{\left[\left(\frac{1000/0.298}{s}\right)+1\right]\left[\left(\frac{1000/.9159}{s}\right)^2+0.3254\left(\frac{1000/0.9159}{s}\right)+1\right]}$$

$$= 10\left[\frac{s/3356}{(s/3356)+1}\right]\left[\frac{(s/1092)^2}{(s/1092)^2+0.3254(s/1092)+1}\right]$$

$T_{HP}(s)$ is a third-order high-pass transfer function that meets the design requirements.

(b) Figure 15-37 shows the design sequence realizing $T_{HP}(s)$ using first- and second-order active RC building blocks. The stage parameters and stage prototypes lead to the design constraints shown in the fourth row. Selecting element values within these constraints yields the final designs shown in the last row of the figure. The first-stage is a voltage divider building block whose gain is adjusted to achieve an overall passband gain of $3.738 \times 2.675 = 10.0$ as required by the specification. A cascade connection of the final stage designs in Fig. 15-37 meets all design requirements without using a gain stage as was required in Example 15-9.

EXERCISE 15-6

Construct Butterworth and Chebychev high-pass transfer functions that meet the following requirements: passband gain of 0 dB, cutoff frequency of 50 rad/s, and $T_{MIN} = -40$ dB at $\omega = 5$ rad/s.

ANSWER:

$$T_B(s) = \frac{s^2}{s^2+70.7s+2500} \qquad T_C(s) = \frac{s^2/\sqrt{2}}{s^2+45.5s+3535}$$

Bandpass and Bandstop Filter Design

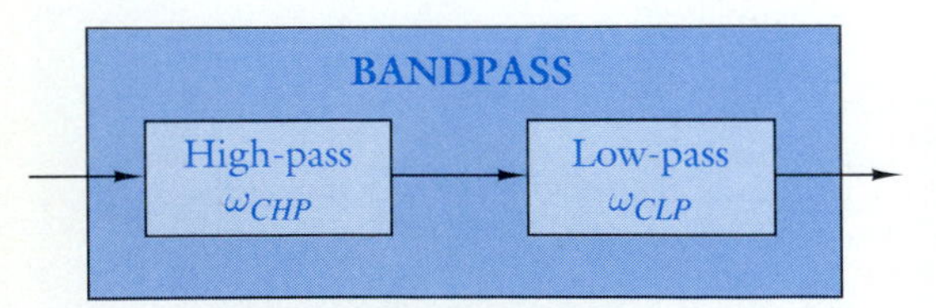

Figure 15-38 Bandpass filter realization using a cascade connection of high-pass and low-pass filters.

Figure 15-38 shows bandpass filter produced by connecting a high- and low-pass filter in cascade. If the cutoff frequency of the low-pass filter (ω_{CLP}) is higher than the cutoff frequency of the high-pass filter (ω_{CHP}), then the interval between the two frequencies is a passband separating two stopbands. In other words, the cascade connection in Fig. 15-38 produces a bandpass filter with two cutoff frequencies that are approximately $\omega_{C1} \approx \omega_{CHP}$ and $\omega_{C2} \approx \omega_{CLP}$.

In Example 15-7 we designed a fourth-order Butterworth low-pass filter with $\omega_{CLP} = 1000$ rad/s. In Example 15-9 we designed a fourth-order Butterworth high-pass filter with $\omega_{CHP} = 10$ rad/s. Connecting these two realizations in cascade produces the eighth-order bandpass gain characteristic in Fig. 15-39. Because the low-pass cutoff frequency is more than a decade above the high-pass cutoff, the bandpass filter cutoff frequencies and bandwidth are

$$\omega_{C1} = \omega_{CHP} = 10 \text{ rad/s}$$

$$\omega_{C2} = \omega_{CLP} = 1000 \text{ rad/s}$$

$$\text{BW} = \omega_{C2} - \omega_{C1} = 990 \text{ rad/s}$$

The low-pass filter provides the high-frequency stopband and the high-pass filter the low-frequency stopband. Frequencies between the two cutoffs fall in the passband of both filters and are transmitted through the cascade connection to produce the passband of the resulting bandpass filter.

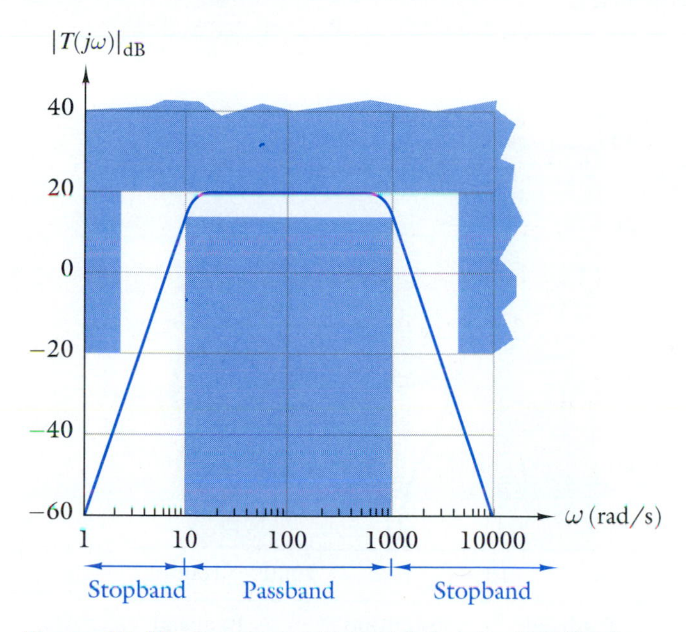

Figure 15-39 Bandpass filter obtained using the low-pass filter in Example 15-7 and high-pass filter in Example 15-9.

Figure 15-40 shows the dual situation in which a high-pass and a low-pass filter are connected in parallel. If $\omega_{CLP} \ll \omega_{CHP}$, then the region between the two cutoff frequencies is a stopband separating two passbands. That is, the parallel connection in Fig. 15-40 produces a bandstop filter whose cutoff frequencies are approximately $\omega_{C1} \approx \omega_{CLP}$ and $\omega_{C2} \approx \omega_{CHP}$.

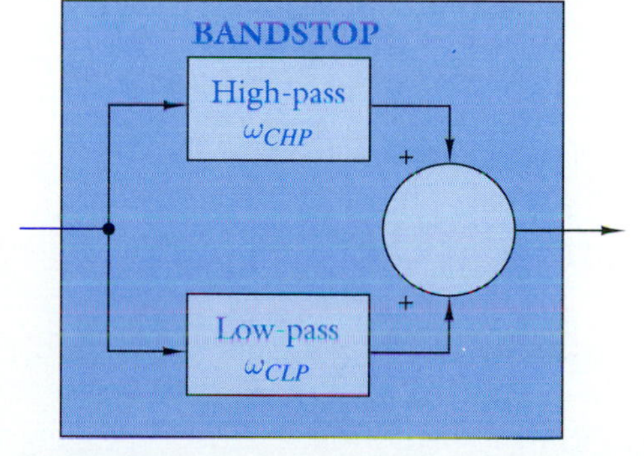

Figure 15-40 Bandstop filter realization using a parallel connection of high-pass and low-pass filters.

In Example 15-8 we designed a third-order Chebychev low-pass filter with $\omega_{CLP} = 10$ rad/s. In Example 15-10 we designed a third-order Chebychev high-pass filter with $\omega_{CHP} = 1000$ rad/s. Connecting these two realizations in parallel produces the bandstop gain response in Fig. 15-41. Because the cutoff frequencies and bandwidth of the notch in the bandstop response are

$$\omega_{C1} = \omega_{CLP} = 10 \text{ rad/s}$$

$$\omega_{C2} = \omega_{CHP} = 1000 \text{ rad/s}$$

$$\text{BW} = \omega_{C2} - \omega_{C1} = 990 \text{ rad/s}$$

The low-pass filter provides the low-frequency passband via the lower path in the Fig. 15-40 and the high-pass filter the high-frequency passband via the upper path. Frequencies between these passbands fall in the stopbands of both filters and are suppressed because neither path in Fig. 15-40 passes signals at these frequencies.

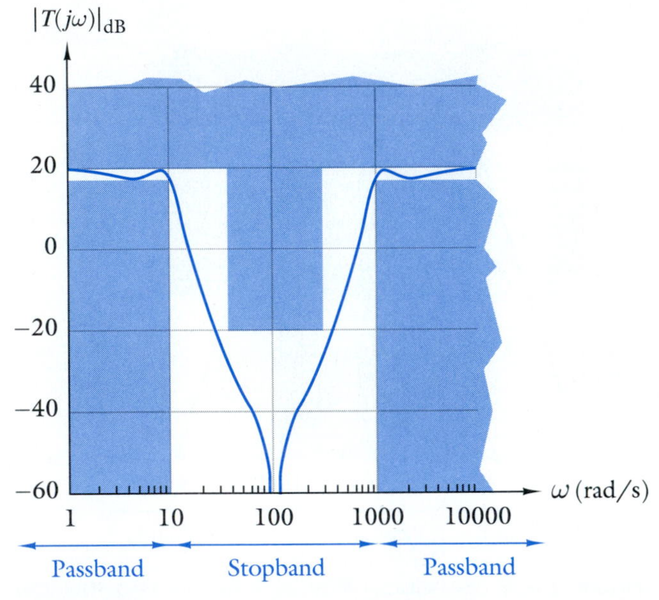

Figure 15-41 Bandstop filter obtained using the low-pass filter in Example 15-8 and the high-pass filter in Example 15-10.

Bandpass and bandstop filters can be designed using a cascade or parallel connection of low-pass and high-pass filters. When the two cutoff frequencies are widely separated the problem reduces to designing separate low-pass and high-pass filters and then interconnecting them in cascade or parallel to obtain the required overall response. Active RC circuits can be connected in cascade without interstage loading. The bandstop case requires a summation, which can be implemented using an OP AMP summer.

EXERCISE 15-7

Construct Butterworth low-pass and high-pass transfer functions whose cascade connection produces a bandpass function with cutoff frequencies at 20 rad/s and 500 rad/s, a passband gain of 0 dB, and a stopband gain less than −20 dB at 5 rad/s and 2000 rad/s.

ANSWER:

$$T(s) = \left[\frac{500^2}{s^2 + 707s + 500^2}\right]\left[\frac{s^2}{s^2 + 28.3s + 400}\right]$$

EXERCISE 15-8

Develop Butterworth low-pass and high-pass transfer functions whose parallel connection produces a bandstop filter with cutoff frequencies at 2 rad/s and 800 rad/s, passband gains of 20 dB, and stopband gains less than −50 dB at 20 rad/s and 80 rad/s.

ANSWER:

$$T_{LP}(s) = \frac{80}{(s^2 + 2s + 4)(s + 2)}$$

$$T_{HP}(s) = \frac{10s^3}{(s^2 + 800s + 800^2)(s + 800)}$$

15-6 SIMULATING ACTIVE RC CIRCUITS USING SPICE

The active RC circuit design techniques described in this chapter are based on an ideal OP AMP model. Using an ideal model greatly simplifies the development of analysis and design procedures. However, real devices depart from the ideal model in several ways. The finite gain and bandwidth of real devices can influence the response of an active RC filter. Computer-aided circuit analysis programs like SPICE are particularly useful for evaluating the effect of OP AMP gain and bandwidth on filter performance.

The circuit in Fig. 15-42 is an extension of the OP AMP model introduced in Chapter 5 (Example 5-14). The resistance R_I and R_O simulate the input and output resistance of the device. The OP AMP frequency response is simulated by the low-pass RC circuit between the two dependent sources. The two dependent sources and the RC circuit combine to produce an s-domain input-output relationship of the form

$$V_O = \frac{\mu}{RCs + 1}(V_P - V_N)$$

The factor $\mu/(RCs+1)$ is the open-loop gain of the model. The open-loop gain is a first-order low-pass transfer function with a pole at $s = -1/RC$. The dc open-loop gain is $T(0) = \mu$. The bandwidth of the gain function is $BW = 2\pi f_C = 1/RC$, where f_C is the cutoff frequency in hertz. Nominal values for these model parameters for the $\mu A741$ general purpose OP AMP (an industry standard) are: $R_I = 2$ MΩ, $R_O = 75$ Ω, $\mu = 200$ V/mV, and $f_C = 10$ Hz.

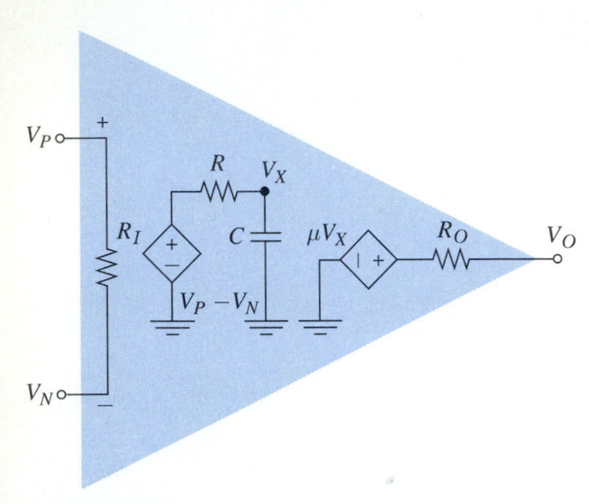

Figure 15-42 Circuit model simulating OP AMP frequency response.

Figure 15-42 is not a circuit diagram of an OP-AMP device. It is a circuit model that approximates the terminal behavior of devices operating in the linear mode. The circuit is called a *behavioral model*, since it simulates a real device in terms of externally observed behavior rather than internal construction. The behavioral model is particularly useful because it is easily implemented using SPICE.

The fact that the μA741 has an open-loop cutoff frequency of only 10 Hz may seem to indicate that the device can only be used in very low-frequency applications. However, linear applications always have feedback that greatly increases the bandwidth of the closed-loop circuit. The next example uses SPICE and the model in Fig. 15-42 to demonstrate gain-bandwidth effects in OP-AMP circuits.

EXAMPLE 15-11

Use SPICE to calculate the bandwidth of a noninverting amplifier circuit for closed-loop gains of 1, 10, and 100. Use the behavioral model in Fig. 15-42 with the μA741 parameter values.

SOLUTION:

Figure 15-43 shows the circuit diagram of a noninverting amplifier using the behavioral OP AMP model in Fig. 15-42. Nodes 1 and 2 are the noninverting and inverting inputs, respectively. Node 6 is the output terminal. The input source V_S drives the noninverting input and the resistors R_1 and R_2 provide the external feedback path from the output terminal to the inverting input.

An ideal OP AMP in the circuit in Fig. 15-43 produces a closed-loop gain of

$$K_O = \frac{R_1 + R_2}{R_1}$$

The constraint $R_2 = (K_O - 1)R_1$ determines the values of R_1 and R_2 needed to obtain the three specified closed-loop gains. By selecting $R_1 = 10$ kΩ we derive the following: (1) $K_O = 1$

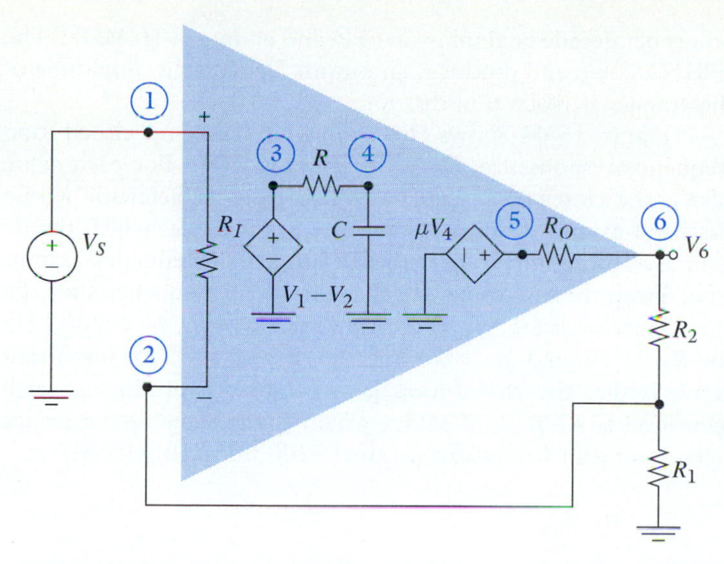

Figure 15-43

requires $R_2 = 0$, (2) $K_O = 10$ requires $R_2 = 90$ kΩ, and (3) $K_O = 100$ requires $R_2 = 990$ kΩ. The nominal open-loop cutoff frequency of the μA741 is 10 Hz. The cutoff frequency of the behavioral model is determined by an RC circuit with $f_C = 1/(2\pi RC)$. By selecting $R = 10$ kΩ we calculate the model capacitance to be $C = 1/(2\pi R f_C) = 1.5915$ μF. A SPICE input file for the μA741 in a noninverting circuit configuration is

```
*EXAMPLE 15-11
VS  1  0  AC  1  0
R1  2  0  10K
R2  6  2  1U (also 90K, 990K)
*uA741 BEHAVIORAL MODEL
     RI  1  2  2 MEG
     E1  3  0  1  2  1
     E2  5  0  4  0  200K
     R   3  4  10K
     C   4  0  1.5915U
     RO  5  6  75
.AC  DEC  20  1  10MEG
.PRINT  AC  VDB(6)
.END
```

The amplitude of the voltage at Node 6 represents the closed-loop gain, since the input source VS has an amplitude of one. The three required closed-loop gains are found using the different values of R2 shown. Note that R2 = 1 $\mu\Omega$ is used to obtain $K_O = 1$ because SPICE does not allow an element value of zero. The .AC command produces the circuit frequency response at 20 points per decade beginning at 1 Hz and ending at 10 MHz. The .PRINT command produces an output file with the amplitude of the voltage at Node 6 in dB.

Figure 15-44 shows the open-loop gain and closed-loop frequency responses at $K_O = 1$, 10, and 100. For each value of K_O the closed-loop gain has a low-pass characteristic whose passband extends from dc out to the frequency at which the dc gain asymptote intersects the open-loop high-frequency asymptote. From these plots we see that the cutoff frequencies for the three gain levels are: $f_C = 2$ MHz for $K_O = 1$, $f_C = 200$ kHz for $K_O = 10$, and $f_C = 20$ kHz for $K_O = 100$. An important result is that the closed-loop gain-bandwidth product at each gain level is $K_O f_C = 2$ MHz, which is the same as the device open-loop gain-bandwidth product $(200{,}000)(10) = 2$ MHz.

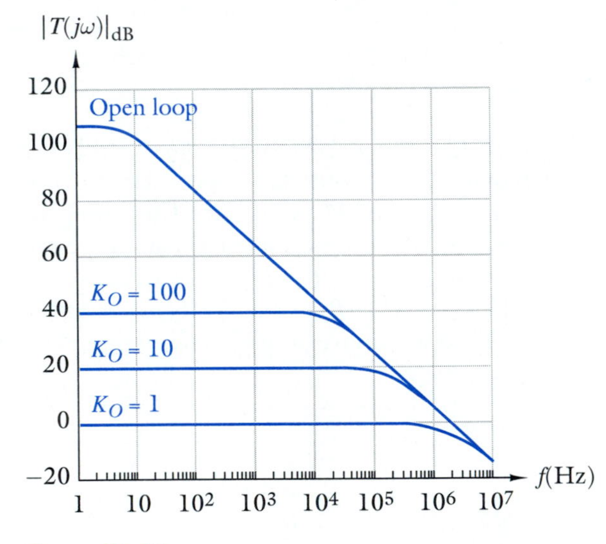

Figure 15-44

Example 15-11 illustrates that the device gain-bandwidth product limits the high-frequency performance of a feedback amplifier circuit. The gain-bandwidth limitation will affect the high-frequency performance of active RC filters when the gain-bandwidth required by the circuit exceeds the OP AMP's capability. For example, Fig. 15-16 shows a low-pass filter designed in Example 15-4. The filter cutoff frequency is 1000 rad/s (159 Hz) and the closed-loop gains of the two OP AMPs are 5 and 2. In this example the largest gain-bandwidth required by the circuit is $5 \times 1590 \approx 8$ kHz. The μA741 device has a nominal gain-bandwidth product of 2 MHz. In this case, the actual performance of the filter will closely follow the ideal OP

AMP model predictions because the required gain-width product is well within the capability of a general purpose device.

The gain of a low-pass or bandpass filter approaches zero at high frequency, so these filters require finite gain-bandwidth products. Theoretically, high-pass and bandstop filters require infinite gain-bandwidth products because their passbands extend to infinite frequency. Since no active device has infinite gain-bandwidth, the OP AMP device limits the high-frequency response of active RC realizations of high-pass and bandstop filters. The final example illustrates this point using a bandstop filter.

EXAMPLE 15-12

Use SPICE to evaluate the frequency response of the circuit designed in Example 15-2 using (a) an ideal OP AMP model and (b) the behavioral model with μA741 parameter values.

SOLUTION:

The circuit designed in Example 15-2 realized the transfer function

$$T(s) = \frac{(s/100+1)(s/500+1)}{(s/10+1)(s/5000+1)}$$

This transfer function produces a stopband in the range between 10 and 5000 rad/s as shown by the gain plot in Fig. 15-8. Figure 15-45(a) shows the active RC circuit designed in Example 15-2. Figures 15-45(b) and (c) show the ideal and behavioral models of the OP AMP.

A SPICE input file for the ideal OP AMP mode is

```
*EXAMPLE 15-12--IDEAL MODEL
VS   1   0   AC   1   0
R1   1   2   90K
R2   2   3   10K
C2   3   0   1U
R3   4   5   10K
C3   5   0   20N
R4   6   4   90K
E1   6   0   2   4   100MEG
.AC   DEC   20   1   10MEG
.PRINT   AC   VDB(6)
.END
```

A SPICE input file for the ideal OP AMP mode is

```
*EXAMPLE 15-12--BEHAVIORAL MODEL
VS   1   0   AC   1   0
R1   1   2   90K
```

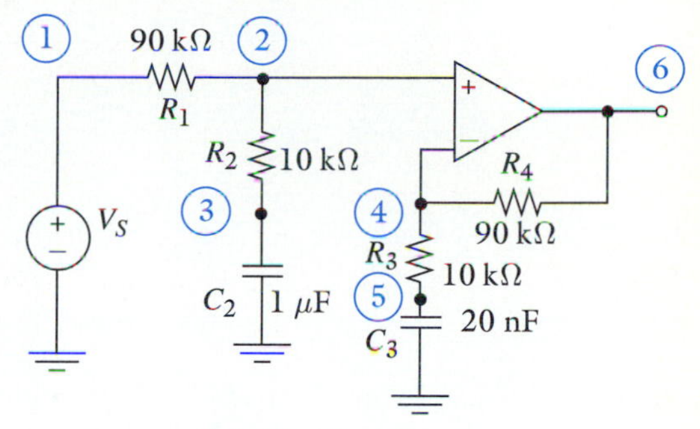

(a) Circuit realization

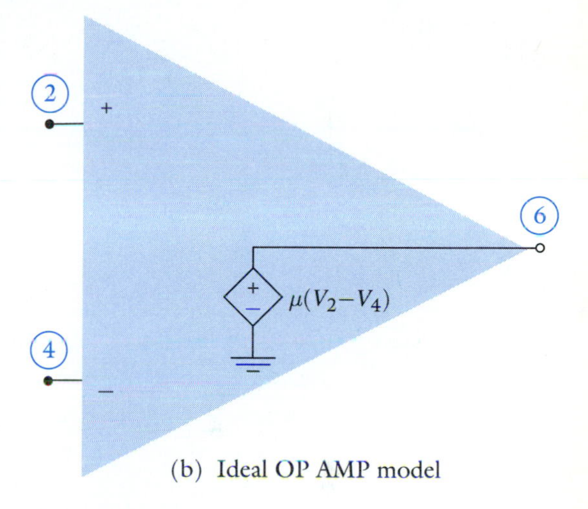

(b) Ideal OP AMP model

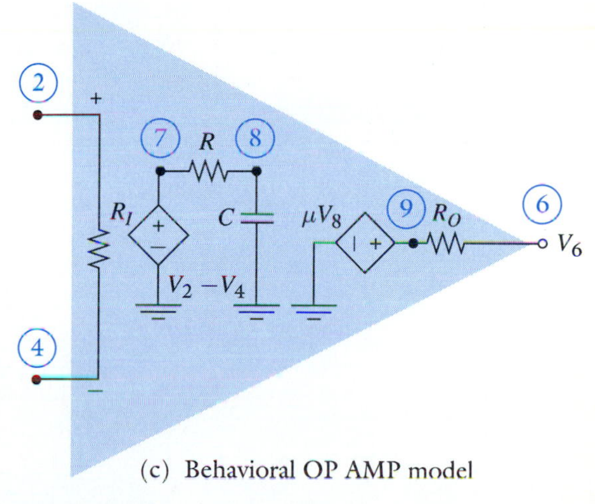

(c) Behavioral OP AMP model

Figure 15-45

```
R2  2  3  10K
C2  3  0  1U
R3  4  5  10K
C3  5  0  20N
R4  6  4  90K
RI  2  4  2K
E1  7  0  2  4  1
R 7 8 10K
C 8 0 1.5915U
E2  9  0  8  0  200K
RO  9  6  75
.AC  DEC  20  1  10MEG
.PRINT  AC  VDB(6)
.END
```

Figure 15-46 shows the SPICE generated frequency response plots for the two OP AMP models. Both models produce the required stopband in the range from 10 to 5000 rad/s. With the ideal model, the upper passband extends indefinitely to infinite frequency. With the μA741 behavioral model, the upper passband gain starts to roll off around 1 Mrad/s because the closed-loop gain of the OP AMP begins to decrease. The filter high-frequency response depends upon the OP AMP gain-bandwidth product.

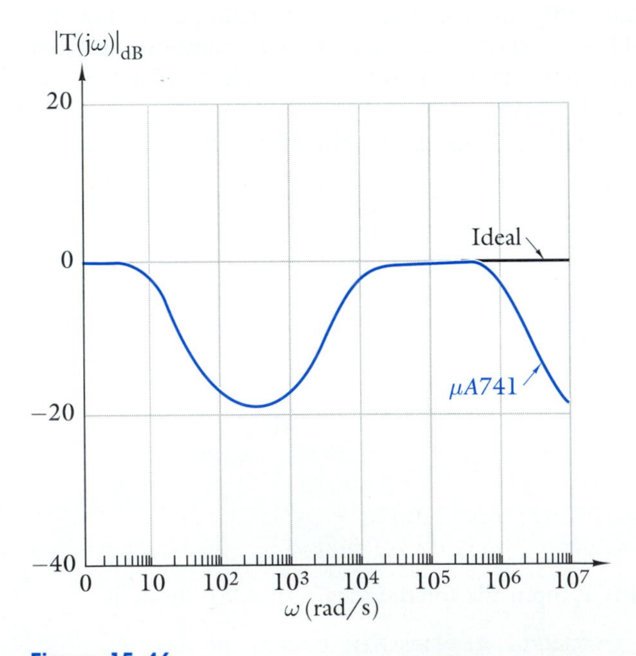

Figure 15-46

SUMMARY

- In the frequency domain, signals can be described and processed in terms of their frequency content.

- A filter design problem is defined by specifying attributes of the gain response such as an asymptotic gain plot, cutoff frequency, passband gains, and stopband attenuation. The first step in the design process is to construct a transfer function $T(s)$ whose gain response meets the specification requirements.

- In the cascade design approach, the required transfer function is partitioned into a product of first- and second-order transfer factors, which can be independently realized using basic active RC building blocks.

- Transfer functions with real poles and zeros can be realized using the voltage divider, noninverting amplifier, or inverting amplifier building blocks. Transfer functions with complex poles can be realized using second-order active RC circuits.

- Transfer functions meeting low-pass filter specifications can be constructed using first-order cascade, Butterworth or Chebychev poles. First-order cascade filters are easy to design but have poor stopband performance. Butterworth responses produce maximally flat passband responses and more stopband attenuation than a first-order cascade with the same number of poles. The Chebychev responses produce equal ripple passband responses and more stopband attenuation than the Butterworth response with the same number of poles.

- A high-pass transfer function can be constructed from a low-pass prototype by replacing s with ω_C^2/s. Bandpass (bandstop) filters can be constructed using a cascade (parallel) connection of a low-pass and a high-pass filter.

- Computer circuit analysis programs such as SPICE can be used to evaluate the effect of OP AMP parameters on active filter performance. The OP AMP gain-bandwidth product determines the limitations on the high-frequency performance of an active RC filter. The finite gain-bandwidth products of real devices can be simulated in SPICE using an RC circuit behavioral model.

EN ROUTE OBJECTIVES AND ASSOCIATED PROBLEMS

ERO 15-1 Cascade Design with First-Order Circuits (Sect. 15-2)

Given the straight-line gain or phase characteristics of a linear circuit:

(a) Construct a transfer function $T(s)$ that has the required characteristics.

(b) Design a cascade connection of first-order active RC circuits that realizes $T(s)$.

❖ **15-1** **(D)** The straight-line gain plot in Fig. P15-1 shows a two-pole roll-off low-pass characteristic.

(a) Construct a transfer function that has this gain response using only real poles and zeros.

(b) Design a cascade of first-order active RC circuits to realize the $T(s)$ found in (a). Scale the circuit so the element values are in a practical range.

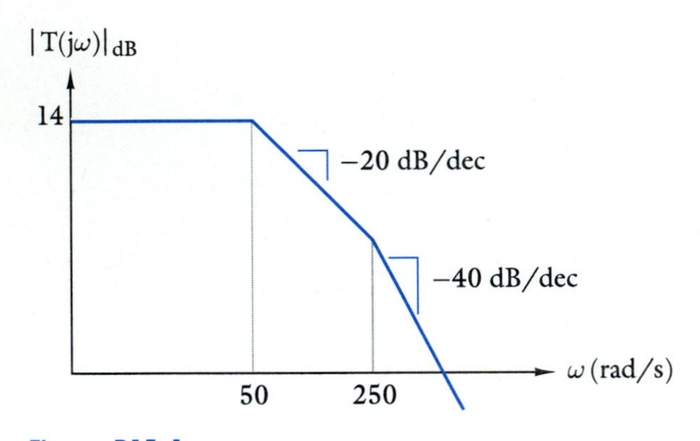

Figure P15-1

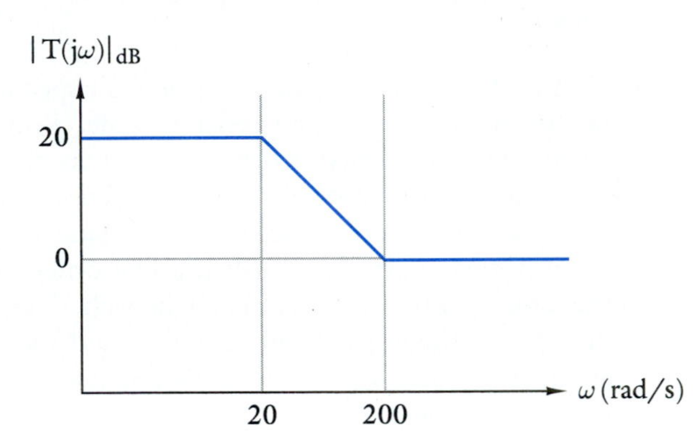

Figure P15-2

❖ **15-2** **(D)** The straight-line gain plot in Fig. P15-2 emphasizes the frequencies below 20 rad/s and deemphasizes the frequencies above 200 rad/s.

(a) Construct a transfer function $T(s)$ that has this gain response using only real poles and zeros.

(b) Design a single-stage first-order active RC circuit to realize the transfer function found in (a). Scale the circuit so the element values are in a practical range.

❖ **15-3** **(D)** The straight-line phase response in Fig. P15-3 provides a leading phase angle over the range from 10 rad/s to 10^5 rad/s.

(a) Construct a transfer function $T(s)$ that has this phase response using only real poles and zeros.

(b) Design a single-stage first-order active RC circuit to realize the $T(s)$ found in (a). Scale the circuit so the element values are in a practical range.

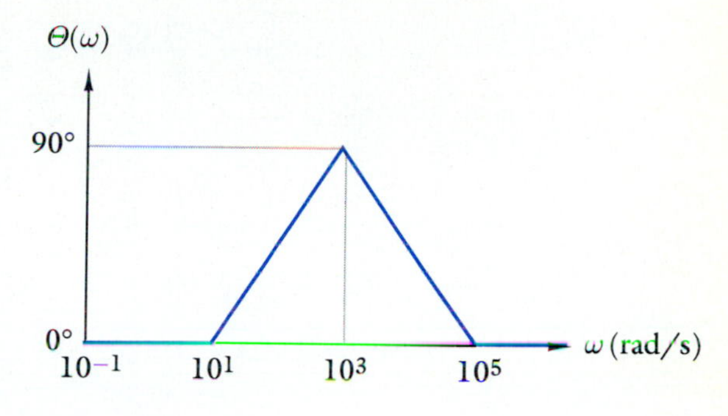

Figure P15-3

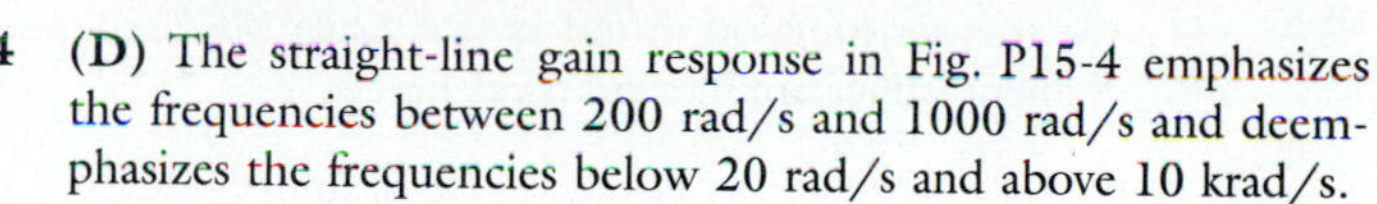

15-4 **(D)** The straight-line gain response in Fig. P15-4 emphasizes the frequencies between 200 rad/s and 1000 rad/s and deemphasizes the frequencies below 20 rad/s and above 10 krad/s.

(a) Construct a transfer function $T(s)$ that has this gain response using only real poles and zeros.

(b) Design a cascade of first-order active RC circuits to realize the $T(s)$ found in (a). Scale the circuits so the element values are in a practical range.

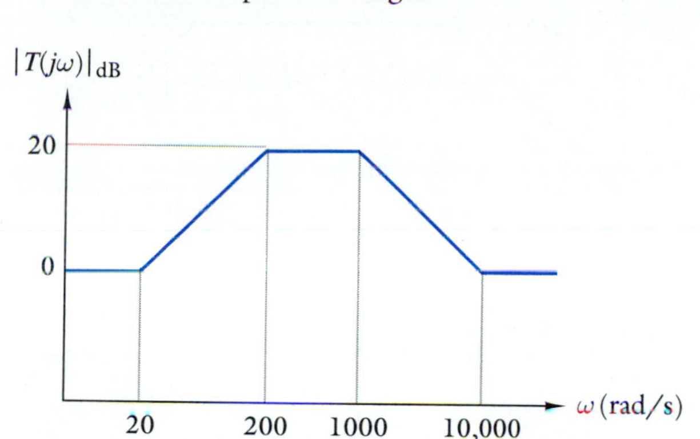

Figure P15-4

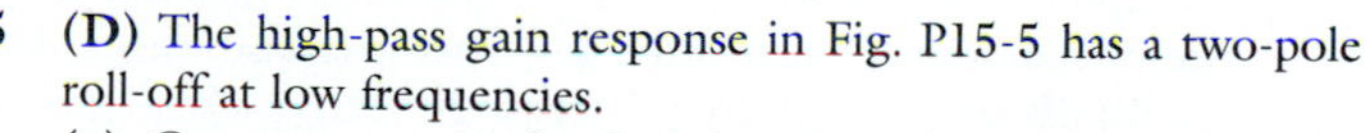

15-5 **(D)** The high-pass gain response in Fig. P15-5 has a two-pole roll-off at low frequencies.

(a) Construct a transfer function $T(s)$ that has this gain response.

(b) Design a cascade of first-order active RC circuits to realize the $T(s)$ found in (a). Scale the circuits so the element values are in a practical range.

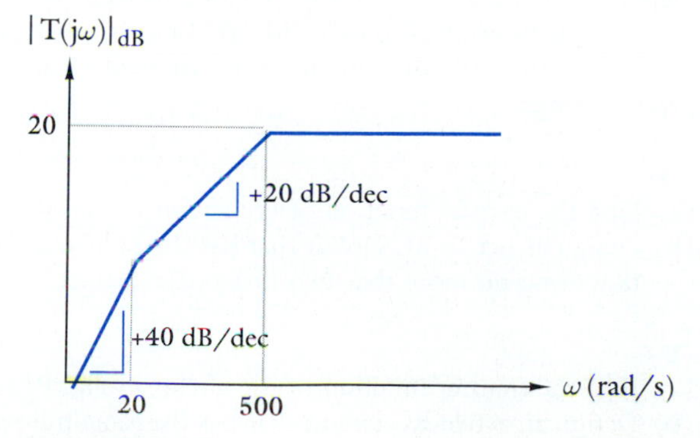

Figure P15-5

❖ **15-6** **(D)** The gain response in Fig. P15-6 has a passband between 40 rad/s and 2000 rad/s.

(a) Construct a transfer function $T(s)$ that has this gain response.

(b) Design a cascade of first-order active RC circuits to realize the transfer function found in (a). Scale the circuits so the element values are in a practical range.

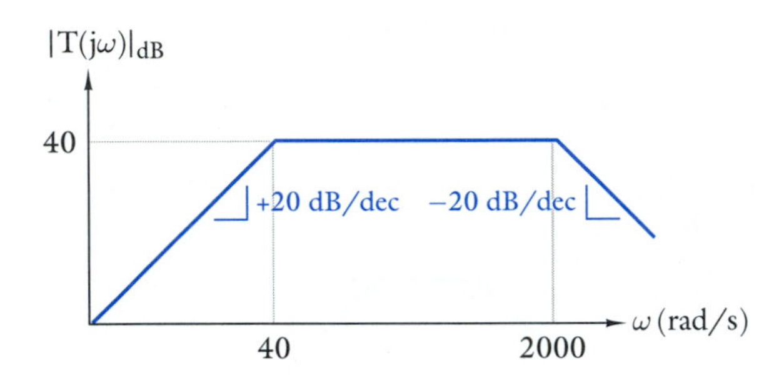

Figure P15-6

❖ **15-7** **(D)** A first-order low-pass amplifier has a dc gain of 40 dB and a cutoff frequency of 2 krad/s. Design an active RC circuit to connect in cascade with the amplifier so that the resulting combination has a first-order low-pass response with a dc gain of 60 dB and a cutoff frequency of 20 krad/s. Scale the circuit so the element values are in a practical range.

❖ **15-8** **(D)** A first-order low-pass amplifier has a dc gain of 10 dB and a cutoff frequency of 1 krad/s. Design an active RC circuit to connect in cascade with the amplifier so that the resulting combination has a second-order bandpass characteristics with cutoff frequencies at 10 rad/s and 1000 rad/s and a passband gain of 20 dB. Scale the circuit so the element values are in a practical range.

❖ **15-9** **(D)**

(a) Find the transfer function of the circuit in Fig. P12-1.

(b) Design an active RC circuit that has the same transfer function using no more that two first-order circuits.

❖ **15-10** **(D)**

(a) Find the transfer function of the circuit in Fig. P12-2.

(b) Design an active RC circuit that has the same transfer function using no more that two first-order circuits.

ERO 15-2 Cascade Design with Second-Order Circuits (Sect. 15-3)

(a) Design an active RC circuit to realize a specified second-order low-pass or high-pass transfer function with complex poles.

(b) Construct a transfer function $T(s)$ with prescribe gain characteristics using real and complex poles and design an active RC circuit to realize $T(s)$.

Design second-order active RC circuits to meet the following requirements.

PROBLEM	TYPE	ω_O(rad/s)	ζ	CONSTRAINTS
❖ **15-11 (D)**	Low-pass	10^3	0.2	Use 20-nF capacitors
❖ **15-12 (D)**	High-pass	2000	.5	Gain of 10 dB at 2000 rad/s
❖ **15-13 (D)**	High-pass	500	?	Cutoff frequency at 500 rad/s
❖ **15-14 (D)**	Low-pass	10^3	0.05	20 dB gain at corner frequency
❖ **15-15 (D)**	High-pass	20	0.75	Use 20-kΩ resistors
❖ **15-16 (D)**	Low-pass	50	0.707	dc gain of 40 dB
❖ **15-17 (D)**	High-pass	500	0.5	Gain of −40 dB at 5 rad/s
❖ **15-18 (D)**	Low-pass	30	0.6	Gain of −30 dB at 3000 rad/s

❖ **15-19 (D)** Design an active RC circuit that realizes the same transfer function as the RLC circuit in Fig. P12-5.

❖ **15-20 (D)** Design an active RC circuit that realizes the same transfer function as the RLC circuit in Fig. P12-9.

❖ **15-21 (D)** Two second-order low-pass circuits with natural frequencies of 1 krad/s are connected in cascade.

(a) Select the damping ratio of each circuit so that the cutoff frequency of the cascade connection is at 1 krad/s.

(b) Design active RC circuits to realize each of the circuits.

ERO 15-3 Low-Pass and High-Pass Filter Design (Sects. 15-4 and 15-5)

Given a low-pass or high-pass filter specification:

(a) Construct a transfer function $T(s)$ using first-order cascade, Butterworth or Chebychev poles that meets the specification.

(b) Design a cascade connection of first- and second-order circuits that realizes $T(s)$.

For the next series of problems design an active RC filter with a passband gain of 0 dB to meet the given requirements. All frequencies are in rad/s.

PROBLEM	TYPE	ω_C	ω_{MIN}	T_{MIN}	CONSTRAINTS
❖ **15-22** (D)	Low-pass	100	1000	−40 dB	Use 1st-order cascade poles
❖ **15-23** (D)	Low-pass	2000	20,000	−80 dB	Use Butterworth poles
❖ **15-24** (D)	Low-pass	50	500	−40 dB	Use 20-nF capacitors
❖ **15-25** (D)	Low-pass	25	75	−20 dB	Use 10-kΩ resistors
❖ **15-26** (D)	Low-pass	200	600	−20 dB	Use Chebychev poles
❖ **15-27** (D)	High-pass	4000	400	−60 dB	Use Butterworth poles
❖ **15-28** (D)	High-pass	500	250	−40 dB	Use Chebychev poles
❖ **15-29** (D)	High-pass	90	30	−20 dB	Use 10-kΩ resistors
❖ **15-30** (D)	High-pass	100	50	−30 dB	Use 200-nF capacitors

❖ **15-31** **(D)** A low-pass filter with a cutoff frequency of 2 krad/sec and a passband gain of 20 dB is needed to correct a noise problem in an existing circuit design. The existing design includes a quad OP AMP package two of which are not being used. Design an active RC filter using no more than two OP AMPs that provides as much attenuation as possible at 4000 rad/s.

❖ **15-32** **(D)** Design a low-pass filter with 0 dB passband gain, a cutoff frequency of 3.2 kHz, and stopband gains less than −20 dB at 6.4 kHz and −40 dB at 16 kHz. Calculate the stopband gains realized by your design at 3.2 kHz, 6.4 kHz, and 16 kHz.

❖❖ **15-33** **(D) (E)** A Butterworth low-pass filter is required to attenuate a sinusoidal tone at 2 kHz by at least 40 dB and to pass a tone at 500 Hz with no more than 3-dB attenuation?

(a) Design an active RC filter to meet the requirement.

(b) What are the advantages of using a Chebychev filter in this application?

ERO 15-4 Bandpass and Bandstop Filter Design (Sect. 15-5)

Given a bandpass or bandstop filter specification:

(a) Construct a transfer function $T(s)$ to meet the specification using low-pass and high-pass functions.

(b) Design a cascade or parallel connection of first and second-order circuits that realizes $T(s)$.

For the next series of problems design an active RC filter with a passband gain(s) of 0 dB to meet the given requirements. All frequencies are in rad/s.

PROBLEM	TYPE	ω_C	ω_{MIN}	T_{MIN}
❖ 15-34 (D)	Bandpass	100, 1000	25, 4000	−40 dB
❖ 15-35 (D)	Bandpass	2000, 30,000	1000, 60,000	−20 dB
❖ 15-36 (D)	Bandstop	100, 3000	400, 750	−30 dB
❖ 15-37 (D)	Bandstop	200, 20,000	2000	−40 dB

❖ **15-38** **(D)** Design a bandpass filter with cutoff frequencies at 100 Hz and 20 kHz. The passband gain must be at least 0 dB and the stopband gain must be at least 20 dB below the passband gain at 50 Hz and 40 kHz.

❖ **15-39** **(D)** Design a Butterworth bandpass filter that:
(a) Passes sinusoidal tones at 1175 Hz and 2275 Hz with no more than 3 dB attenuation.
(b) Attenuates tones at 60 Hz and 25 kHz by at least 40 dB.

❖ **15-40** **(D)** Design a bandstop filter that has passbands below 100 rad/s and above 2000 rad/s and has a stopband gain at least 20 dB below the passband gains at 500 rad/s.

ERO 15-5 Simulating Active RC Circuits Using SPICE (Sect. 15-6)

(a) Use SPICE or MICRO-CAP to verify an active RC circuit filter design using an ideal OP AMP model.

(b) Use SPICE or MICRO-CAP and the RC circuit behavioral model of the OP AMP to evaluate the effect of finite gain-bandwidth product on active RC circuit filter performance.

15-41 Use SPICE or MICRO-CAP to verify your active RC circuit design for Prob. 15-1 using an ideal OP AMP model.

15-42 Use SPICE or MICRO-CAP to verify your active RC circuit design for Prob. 15-2 using an ideal OP AMP model.

15-43 Use SPICE or MICRO-CAP to verify your active RC circuit design for Prob. 15-5 using an ideal OP AMP model.

15-44 Use SPICE or MICRO-CAP to verify your active RC circuit design for Prob. 15-6 using an ideal OP AMP model.

15-45 Use SPICE or MICRO-CAP to verify your active RC circuit design for Prob. 15-22 using an ideal OP AMP model.

15-46 Use SPICE or MICRO-CAP to verify your active RC circuit design for Prob. 15-23 using an ideal OP AMP model.

15-47 Use SPICE or MICRO-CAP to verify your active RC circuit design for Prob. 15-27 using an ideal OP AMP model.

15-48 Use SPICE or MICRO-CAP to verify your active RC circuit design for Prob. 15-35 using an ideal OP AMP model.

15-49 Use SPICE or MICRO-CAP to verify your active RC circuit design for Prob. 15-37 using an ideal OP AMP model.

❖❖ **15-50** **(E)** Use SPICE or MICRO-CAP to evaluate the effect of finite gain-bandwidth product on your active RC circuit design for Prob. 15-4 using an the RC circuit behavioral model of the μA741.

❖❖ **15-51** **(E)** Use SPICE or MICRO-CAP to evaluate the effect of finite gain-bandwidth product on your active RC circuit design for Prob. 15-6 using the RC circuit behavioral model of the μA741.

❖❖ **15-52** **(E)** Use SPICE or MICRO-CAP to evaluate the effect of finite gain-bandwidth product on your active RC circuit design for Prob. 15-23 using the RC circuit behavioral model of the μA741.

❖❖ **15-53** **(E)** Use SPICE or MICRO-CAP to evaluate the effect of finite gain-bandwidth product on your active RC circuit design for Prob. 15-27 using the RC circuit behavioral model of the μA741.

CHAPTER INTEGRATING PROBLEMS

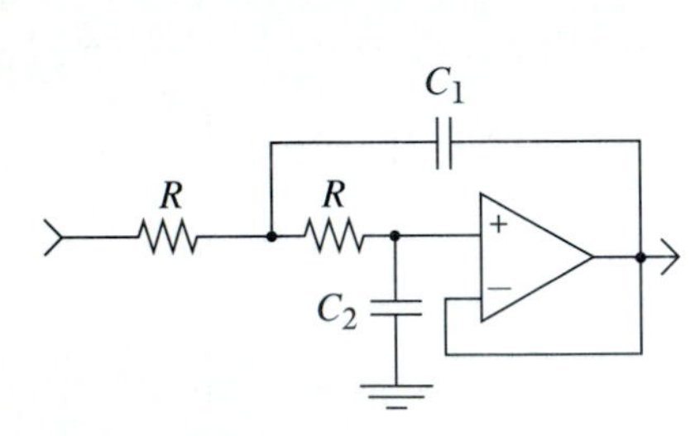

Figure P15-54

❖❖ **15-54** **(A, E)**

(a) Show that the circuit in Fig. P15-54 can realize a second-order low-pass transfer function with complex poles.

(b) Develop an algorithm for selecting circuit parameters to realize specified ω_O and ζ.

(c) Evaluate the advantages and disadvantages of this circuit versus the second-order building block discussed in Section 15-3. Discuss.

❖ **15-55** **(D)** Design a second-order Butterworth low-pass filter whose step response has a 1% settling time less than 10 ms and a cutoff frequency less than 200 Hz.

❖❖ **15-56** **(A, E)** A cascade connection of first-order transfer functions of the form $T(s) = 1/(s/\omega_O + 1)$ is called a synchronously tuned low-pass filter.

(a) Show that the cutoff frequency of a nth order synchronously tuned filter is

$$\omega_C = \omega_O\sqrt{2^{1/n} - 1}$$

(b) Evaluate the stopband attenuation one octave above the cutoff frequency for $n = 1$, 2, and 3.

(c) Compare the attenuation in (b) with the attenuation of Chebychev filters with the same order.

15-57 **(A)** The gain response of an nth-order Butterworth low-pass filter has the form

$$|T(j\omega)| = \frac{K_O}{\sqrt{1 + (\omega/\omega_C)^{2n}}}$$

where K_O is the passband gain and ω_C is the cutoff frequency. Derive an expression for the number of poles n required to meet a stopband requirement of $T_{MIN} = K_O/A$ at $\omega = \omega_{MIN}$.

15-58 **(A)** For $\omega > \omega_C$ the gain response of an nth-order Chebychev low-pass filter has the form

$$|T(j\omega)| = \frac{K_O}{\sqrt{1 + \{\cosh[n\cosh^{-1}(\omega/\omega_C)]\}^2}}$$

where K_O is the passband gain and ω_C is the cutoff frequency. Derive an expression for the number of poles n required to meet a stopband requirement of $T_{MIN} = K_O/A$ at $\omega = \omega_{MIN}$.

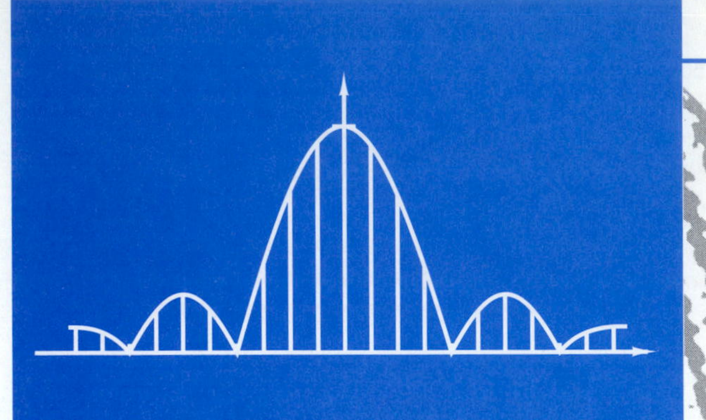

Chapter Sixteen

"The series formed of sines or cosines of multiple arcs are therefore adapted to represent between definite limits all possible functions, and the ordinates of lines or surfaces whose form is discontinuous."

Jean Baptiste Joseph Fourier, 1822
French Mathematician/Physicist

FOURIER TRANSFORMS

The mathematical techniques in this chapter have their origin in the works of the French mathematician Joseph Fourier (1768–1830). In his study of heat transfer Fourier found that discontinuous functions could be represented by an infinite series of harmonic sinusoids, or sines of multiple arcs as he called them. Initially his theory drew heavy criticism from the mathematicians of his day but was eventually accepted and is now firmly entrenched in all branches of science and engineering. Application to electrical systems began in 1855, when Sir William Thomson (Lord Kelvin) used Fourier techniques to analyze a proposed transatlantic telegraph system. His analysis showed the system to be feasible, and the Atlantic Telegraph Company began to lay a submarine cable on the ocean floor in 1857.

The study of the sinusoidal steady-state response and filter design in Chapters 12, 13, and 15 emphasized circuit frequency response. The Fourier transform methods in this chapter expands our view of frequency response to include signals as well as circuits. Why introduce another transform when we already have Laplace transforms? There are two major reasons: (1) Communication system applications use signal and circuit models for which Fourier transforms exist and Laplace transforms do not. (2) Digital signal processing is most often implemented using Fourier transforms rather than Laplace transforms. This chapter serves as a bridge between the Laplace transform-based circuit analysis and design topics in preceding chapters and the Fourier transform-based signal-processing topics in subsequent courses.

The trigonometric and exponential Fourier series treated in the first two sections of this chapter provide the foundation for the Fourier transform. In the third section a limiting process is shown to lead to definitions of the direct and inverse Fourier transformations. The fourth section treats the relationship between the Fourier transforms and the Laplace transforms. Transform pairs and important mathematical properties of the Fourier transformation are derived in the next two sections. The final two sections present applications of Fourier transforms.

16-1 THE FOURIER SERIES

The Fourier series is a stepping stone to Fourier transforms and signal spectra. The series expresses a periodic waveform as an infinite series of harmonically related sinusoids. More specifically, if $f(t)$ is periodic with period T_0 and is reasonably well behaved, then $f(t)$ can be expressed as a *Fourier* series of the form

$$f(t) = a_0 + a_1 \cos 2\pi f_0 t + a_2 \cos 2\pi 2 f_0 t + \ldots + a_n \cos 2\pi n f_0 t + \ldots + b_1 \sin 2\pi f_0 t + b_2 \sin 2\pi f_0 t + \ldots + b_n \sin 2\pi n f_0 t + \ldots \quad (16\text{-}1)$$

or more compactly,

$$f(t) = \underbrace{a_0}_{\text{dc}} + \underbrace{\sum_{n=1}^{\infty} [a_n \cos(2\pi n f_0 t) + b_n \sin(2\pi n f_0 t)]}_{\text{ac}} \quad (16\text{-}2)$$

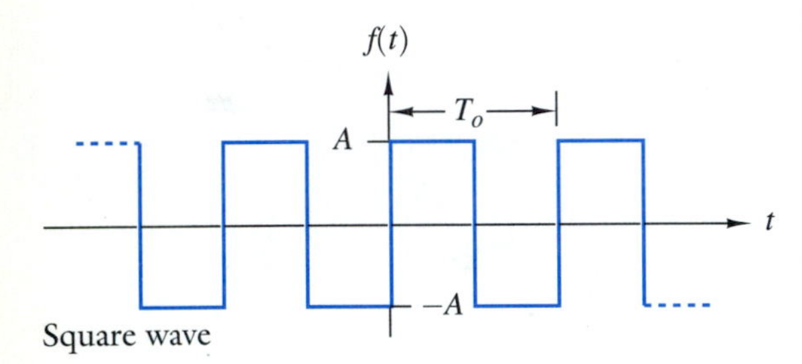

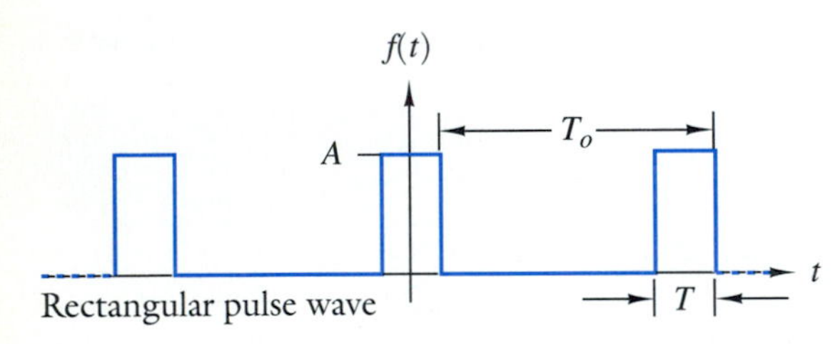

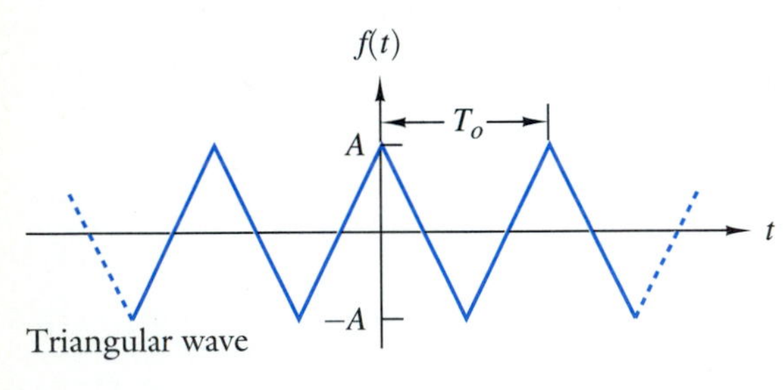

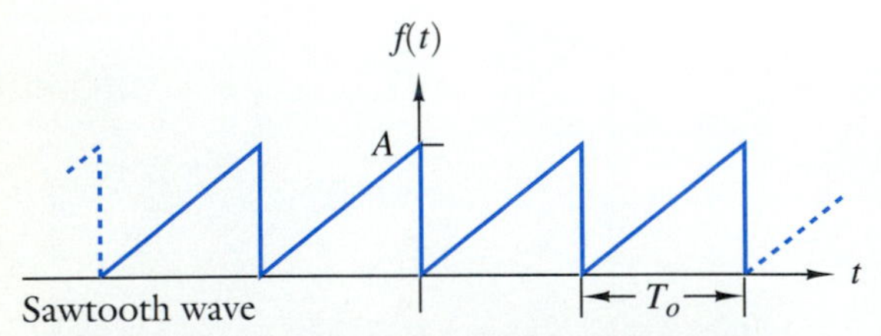

Figure 16-1 Some examples of periodic waveforms.

The coefficient a_0 is the dc component or average value of $f(t)$. The constants a_n and b_n ($n = 1, 2, 3, \ldots$) are the *Fourier coefficients* or amplitudes of the sinusoids in the ac component. The first frequency in the ac component is determined by the period of the waveform, i.e., $f_0 = 1/T_0$. For this reason f_0 is called the *fundamental frequency*. The remaining frequencies are integer multiples of f_0 called the second harmonic ($2f_0$), third harmonic ($3f_0$), and, in general, the nth harmonic (nf_0).

The Fourier series in Eq. (16-1) exists as long as $f(t)$ is reasonably well behaved. Basically, well behaved means $f(t)$ is single-valued, the integral of $|f(t)|$ over a period is finite and $f(t)$ has a finite number of discontinuities in any one period. These requirements, called the *Dirichlet conditions*, are sufficient to show that the Fourier series exists. The Fourier series converges for any periodic waveform that meets the Dirichlet conditions. However, there are waveforms that do not meet the Dirichlet conditions that have convergent Fourier series. That is, the Dirichlet conditions are sufficient; they are not necessary and sufficient. This limitation does not present a serious problem because the Dirichlet conditions are satisfied by the waveforms generated in physical systems. Some examples of periodic waveforms meeting these standards are shown in Fig. 16-1.

Before discussing how to determine the coefficients of the Fourier series, it is important to have an overview of why the series is important in electrical engineering. The Fourier series resolves a periodic waveform into dc and ac components. A periodic signal can be thought of as being generated by dc and ac signal sources as indicated by the conceptual block diagram in Fig. 16-2. Since linear circuits obey superposition, we can think of the steady-state response to a periodic input as the sum of the steady-state responses to each of these sources. Since we know how to find the steady-state response to dc inputs (Chapters 2 through 5) and sinusoidal (ac) inputs (Chapters 12, 13, and 14), at least in principle we can calculate the steady-state response to a periodic input by superposition.

Although Fig. 16-2 presents a useful concept, we do not actually generate square waves in the laboratory by summing infinitely many sinusoids. Fourier analysis is not a laboratory technique, but a point of view that pervades analog signal progressing. We think of periodic signals as having a spectrum containing components at the discrete frequencies nf_0 ($n = 0, 1, 2, 3, \ldots$). In theory the spectrum contains infinitely many frequencies, except that the harmonic amplitudes eventually decrease at high frequency, so the higher-order harmonics ultimately become negligibly small.

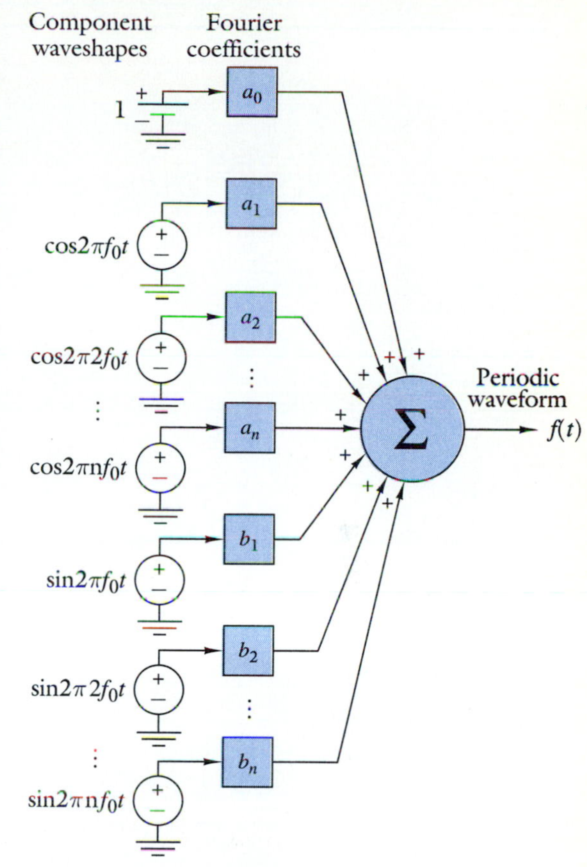

Figure 16-2 Conceptual block diagram for generating a periodic waveform by summing component waveshapes.

Determining Fourier Coefficients

The Fourier coefficients for any periodic waveform $f(t)$ satisfying the Dirichlet conditions can be obtained from the equations

$$a_0 = \frac{1}{T_0} \int_{-T_0/2}^{+T_0/2} f(t)\, dt$$

$$a_n = \frac{2}{T_0} \int_{-T_0/2}^{+T_0/2} f(t) \cos(2\pi n t / T_0)\, dt \qquad (16\text{-}3)$$

$$b_n = \frac{2}{T_0} \int_{-T_0/2}^{+T_0/2} f(t) \sin(2\pi n t / T_0)\, dt$$

The integration limits in these equations extend from $-T_0/2$ to $+T_0/2$. However, the limits can span any convenient interval as long as it is exactly one period. For example, the limits could be from 0 to T_0 or T_0 to $2T_0$. We will show where Eq. (16-3) comes from in a moment, but first we use the equations to obtain the Fourier coefficients of a square wave.

EXAMPLE 16-1

Find the Fourier coefficients for the square wave in Fig. 16-1.

SOLUTION:

An expression for a square wave on the interval $0 < t <$ to T_0 is

$$f(t) = \begin{cases} A & 0 < t < T_0/2 \\ -A & T_0/2 < t < T_0 \end{cases}$$

Because of the time interval covered by this expression we use 0 and T_0 as the limits in Eqs. (16-3). We use the first expression in Eq. (16-3) to find a_0 as follows:

$$\begin{aligned} a_0 &= \frac{1}{T_0}\int_0^{T_0/2} A\,dt + \frac{1}{T_0}\int_{T_0/2}^{T_0} (-A)dt \\ &= \frac{A}{T_0}\left[\frac{T_0}{2} - 0 - T_0 + \frac{T_0}{2}\right] = 0 \end{aligned}$$

The result $a_0 = 0$ means that the average or dc value of the square wave is zero, which is easy to see because the area under a positive half-cycle cancels the area under a negative half-cycle. We use the second expression in Eq. (16-3) to find a_n as follows:

$$\begin{aligned} a_n &= \frac{2}{T_0}\int_0^{T_0/2} A\cos(2\pi nt/T_0)dt + \frac{2}{T_0}\int_{T_0/2}^{T_0} (-A)\cos(2\pi nt/T_0)dt \\ &= \frac{2A}{T_0}\left[\frac{\sin(2\pi nt/T_0)}{2\pi n/T_0}\right]_0^{T_0/2} - \frac{2A}{T_0}\left[\frac{\sin(2\pi nt/T_0)}{2\pi n/T_0}\right]_{T_0/2}^{T_0} \\ &= \frac{A}{n\pi}[\sin(n\pi) - \sin(0) - \sin(2n\pi) + \sin(n\pi)] = 0 \end{aligned}$$

There are no cosine terms in the series, since $a_n = 0$ for all n. This result makes sense because the square wave in Fig. 16-1 resembles a sine more than a cosine. The b_n coefficients for the sine terms are found using the third expression in Eq. (16-3):

$$\begin{aligned} b_n &= \frac{2}{T_0}\int_0^{T_0/2} A\sin(2\pi nt/T_0)dt + \frac{2}{T_0}\int_{T_0/2}^{T_0} (-A)\sin(2\pi nt/T_0)dt \\ &= \frac{2A}{T_0}\left[-\frac{\cos(2\pi nt/T_0)}{2\pi n/T_0}\right]_0^{T_0/2} - \frac{2A}{T_0}\left[-\frac{\cos(2\pi nt/T_0)}{2\pi n/T_0}\right]_{T_0/2}^{T_0} \\ &= \frac{A}{n\pi}[-\cos(n\pi) + \cos(0) + \cos(2n\pi) - \cos(n\pi)] \\ &= \frac{2A}{n\pi}[1 - \cos(n\pi)] \end{aligned}$$

Given the coefficients a_n and b_n found above, the Fourier series representation of the square wave is

$$f(t) = \sum_{n=1}^{\infty} \frac{2A}{n\pi}[1 - \cos(n\pi)][\sin(2\pi n f_0 t)]$$

The term $[1 - \cos(n\pi)] = 2$ if n is odd, and zero if n is even. In other words, only the odd harmonic sine terms are present in the series. To illustrate this result we write the first three nonzero terms in the series:

$$f(t) = \frac{4A}{\pi}\left[\sin 2\pi\, f_0 t + \frac{1}{3}\sin 2\pi\, 3 f_0 t + \frac{1}{5}\sin 2\pi\, 5 f_0 t + \ldots\right]$$

Figure 16-3 shows how the harmonic sinusoids in this expression combine to produce a square wave. The sine wave in

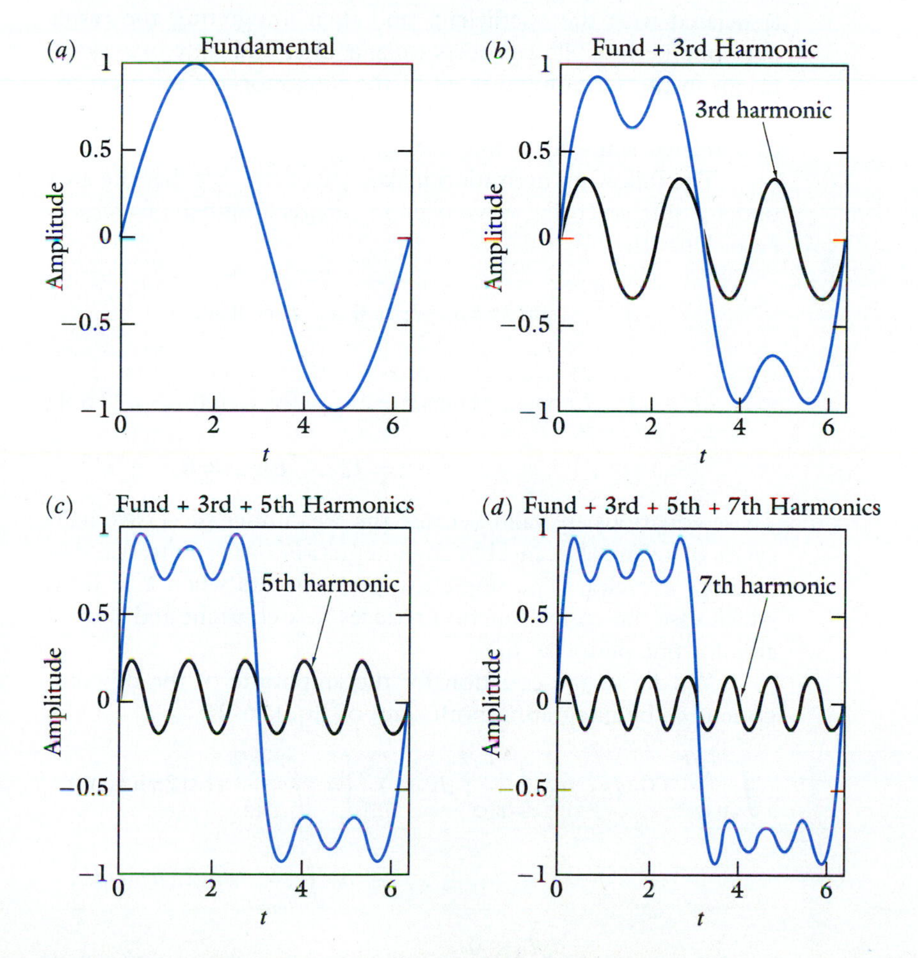

Figure 16-3 Using harmonics to generate a square wave.

Fig. 16-3(a) represents the ac component at the fundamental frequency. Adding the third harmonic produces the result in Fig. 16-3(b). The result in (b) plus the fifth harmonic produces (c), and so on. After just three terms the sum in Fig. 16-3(c) begins to look like a square wave. As we add more and more terms, the sum is closer and closer to a square wave.

Deriving Equations for a_n and b_n

The square wave example shows how the coefficients in the Fourier series are obtained using Eq. (16-3). We now turn to derivation of these equations. Any one Fourier coefficient can be found by multiplying both sides of Eq. (16-2) by the waveshape associated with the coefficient, and then integrating the result over one period. This process isolates one, and only one, coefficient in the series because all of the integrations vanish except one. Mathematicians call this special property of the harmonically related sinusoids *orthogonality*.

The following derivation makes use of the fact that the area under a sine or cosine wave over an integer number of cycles is zero. That is,

$$\int_{-T_0/2}^{+T_0/2} \sin(2\pi k f_0 t)dt = 0 \qquad \text{for all } k$$

$$\int_{-T_0/2}^{+T_0/2} \cos(2\pi k f_0 t)dt = 0 \qquad \text{for } k \neq 0 \tag{16-4}$$

$$= T_0 \qquad \text{for } k = 0$$

These equations are valid because the area under successive half-cycles of a sinusoid cancel in any integration over an interval that includes k cycles. The single exception occurs when $k = 0$, in which case the cosine function reduces to a constant and the net area for one period is T_0.

We derive the equation for the amplitude of the dc component a_0 by integrating both sides of Eq. (16-2):

$$\int_{-T_0/2}^{+T_0/2} f(t)dt = a_0 \int_{-T_0/2}^{+T_0/2} dt + \sum_{n=1}^{\infty} \left[a_n \int_{-T_0/2}^{+T_0/2} \cos(2\pi n f_0 t)dt + b_n \int_{-T_0/2}^{+T_0/2} \sin(2\pi n f_0 t)dt \right]$$

$$= a_0 T_0 + 0 \tag{16-5}$$

The integral of the dc component reduces to $a_0 T_0$. The integrals of the ac components vanish because of the properties in Eq. (16-4). Consequently, the right side of Eq. (16-5) reduces to $a_0 T_0$. Solving for a_0 yields the first relationship in Eq. (16-3).

To derive the expression for a_n we multiply Eq. (16-2) by $\cos(2\pi m f_0 t)$ and integrate over $-T_0/2$ to $+T_0/2$:

$$\int_{-T_0/2}^{+T_0/2} f(t)\cos(2\pi m f_0 t)dt = a_0 \int_{-T_0/2}^{+T_0/2} \cos(2\pi m f_0 t)dt$$
$$+\sum_{n=1}^{\infty}\left[a_n \int_{-T_0/2}^{+T_0/2} \cos(2\pi m f_0 t)\cos(2\pi n f_0 t)dt \right. \tag{16-6}$$
$$\left. +b_n \int_{-T_0/2}^{+T_0/2} \cos(2\pi m f_0 t)\sin(2\pi n f_0 t)dt\right]$$

All of the integrals on the right side of this equation are zero except one. To show this we use identities

$$\cos(x)\cos(y) = \frac{1}{2}\cos(x-y) + \frac{1}{2}\cos(x+y)$$
$$\cos(x)\sin(y) = \frac{1}{2}\sin(x-y) + \frac{1}{2}\sin(x+y)$$

to change Eq. (16-6) into the following form:

$$\int_{-T_0/2}^{+T_0/2} f(t)\cos(2\pi m f_0 t)dt = a_0 \int_{-T_0/2}^{+T_0/2} \cos(2\pi m f_0 t)dt$$
$$+\sum_{n=1}^{\infty}\left\{\frac{a_n}{2}\int_{-T_0/2}^{+T_0/2} \cos[2\pi(m-n)f_0 t]dt\right.$$
$$\left. +\frac{a_n}{2}\int_{-T_0/2}^{+T_0/2} \cos[2\pi(m+n)f_0 t]dt\right\} \tag{16-7}$$
$$+\sum_{n=1}^{\infty}\left\{\frac{b_n}{2}\int_{-T_0/2}^{+T_0/2} \sin[2\pi(m-n)f_0 t]dt\right.$$
$$\left. +\frac{b_n}{2}\int_{-T_0/2}^{+T_0/2} \sin[2\pi(m+n)f_0 t]dt\right\}$$

All of the integrals are now of the same form as Eq. (16-4). Using Eq. (16-4) we see all of the integrals on the right side of Eq. (16-7) vanish, except for one cosine integral when $m = n$. This one survivor corresponds to the $k = 0$ case in Eq. (16-4). In other words, the right side of Eq. (16-7) reduces to $a_n T_0/2$. When we solve the right side of Eq. (16-7) for a_n we obtain the second expression in Eq. (16-3).

To obtain the expression for b_n we multiply Eq. (16-2) by $\sin(2\pi m f_0 t)$ and integrate over $t = -T_0/2$ to $+T_0/2$. The derivation steps then parallel the approach used to find a_n. The end result is that the dc component integral vanishes and the ac component integrals reduce to $b_n T_0/2$, which yields the expression for b_n in Eq. (16-3). The details of the development for b_n are left as a problem for the reader.

The derivation of Eq. (16-3) focuses on the problem of finding the Fourier coefficients of a given periodic waveform. Some experience and practice are necessary to fully understand the implications of this procedure. On the other hand, it is not necessary to go through these mechanics for every newly encountered periodic waveform because tables of Fourier series expansions are available. For our purposes the listing in Fig. 16-4 will suffice. For each waveform defined graphically, the figure lists

Waveform	Fourier Coefficients	Waveform	Fourier Coefficients
Constant (dc) $f(t)$, A, t	$a_0 = A$ $a_n = 0$ all n $b_n = 0$ all n	Sawtooth wave $f(t)$, A, t, T_o	$a_0 = A/2$ $a_n = 0$ all n $b_n = -\frac{A}{n\pi}$ all n
Cosine wave $f(t)$, T_o, A, t	$a_0 = 0$ $a_1 = A$ $a_n = 0$ $n \neq 1$ $b_n = 0$ all n	Triangular wave $f(t)$, A, t, T_o	$a_0 = 0$ $a_n = \frac{8A}{(n\pi)^2}$ n odd $a_n = 0$ n even $b_n = 0$ all n
Sine wave $f(t)$, A, t, T_o	$a_0 = 0$ $a_n = 0$ all n $b_1 = A$ $b_n = 0$ $n \neq 1$	Half-wave sine $f(t)$, A, t, T_o	$a_0 = A/\pi$ $a_n = \frac{2A/\pi}{1 - n^2}$ n even $a_n = 0$ n odd $b_1 = \frac{A}{2}$, $b_n = 0$ $n \neq 1$

Figure 16-4 Fourier coefficients for some periodic waveforms.

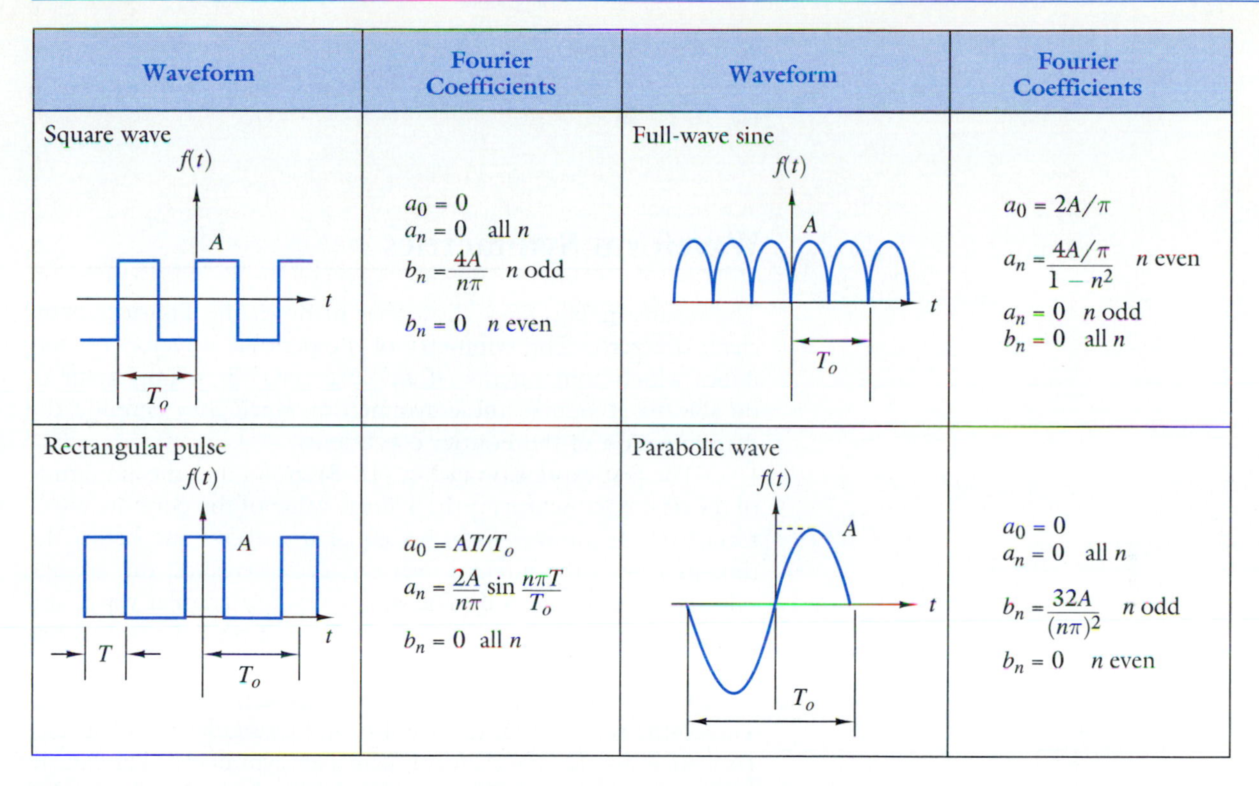

Waveform
Fourier Coefficients
Waveform
Fourier Coefficients
Square wave
f(t)
A
t
T_o
$a_0 = 0$
$a_n = 0$ all n
$b_n = \frac{4A}{n\pi}$ n odd
$b_n = 0$ n even
Full-wave sine
f(t)
A
t
T_o
$a_0 = 2A/\pi$
$a_n = \frac{4A/\pi}{1-n^2}$ n even
$a_n = 0$ n odd
$b_n = 0$ all n
Rectangular pulse
f(t)
A
t
T
T_o
$a_0 = AT/T_o$
$a_n = \frac{2A}{n\pi} \sin \frac{n\pi T}{T_o}$
$b_n = 0$ all n
Parabolic wave
f(t)
A
t
T_o
$a_0 = 0$
$a_n = 0$ all n
$b_n = \frac{32A}{(n\pi)^2}$ n odd
$b_n = 0$ n even

Figure 16-4 (concluded) Fourier coefficients for some periodic waveforms.

the expressions for a_0, a_n, and b_n as well as restrictions on the integer n.

In theory a Fourier series includes infinitely many harmonics, although the harmonics tend to decrease in amplitude as the frequency increases. For example, the results in Fig. 16-4 show that the amplitudes of the square wave decrease as $1/n$, the triangular wave as $1/n^2$, and the parabolic wave as $1/n^3$. The $1/n^3$ dependence means that the amplitude of the fifth harmonic in a parabolic wave is less than 1% of the amplitude of the fundamental. In practical signals the harmonic amplitudes decrease at high frequency so that at some point the higher-order components become negligibly small. In other words, the spectrum of a periodic waveform can be truncated after a finite number of terms, and we can restrict the bandwidth of the signal while retaining an adequate representation of the original signal.

EXERCISE 16-1

The triangular wave in Fig. 16-4 has a peak amplitude of $A = 10$ and $T_0 = 2$ ms. Calculate the Fourier coefficients of the first nine harmonics.

ANSWERS:

$a_1 = 8.11$, $a_2 = 0$, $a_3 = 0.901$, $a_4 = 0$, $a_5 = 0.324$, $a_6 = 0$, $a_7 = 0.165$, $a_8 = 0$, $a_9 = 0.100$, $b_n = 0$ for all n ■

EXERCISE 16-2

Use Eq. (16-3) to verify Fourier coefficients of the sawtooth wave in Fig. 16-4.

Waveform Symmetries

The results in Fig. 16-4 show that many of the Fourier coefficients are zero. The symmetry of the periodic waveform determines which components (if any) are zero. It is important to be able to recognize these symmetries, since they simplify the determination of the Fourier coefficients.

The first expression in Eq. (16-3) shows that the amplitude of the dc component a_0 is the average value of the periodic waveform $f(t)$. If the waveform has equal area above and below the time axis, then the integral over one cycle vanishes, the average value is zero, and $a_0 = 0$. The square wave, triangular wave, and parabolic wave in Fig. 16-4 are examples of periodic waveforms with zero average value.

A waveform is said to have *even symmetry* if $f(-t) = f(t)$. The cosine wave, rectangular pulse, and triangular wave[1] in Fig. 16-4 are examples of waveforms with even symmetry. The Fourier series of an even waveform is made up entirely of cosine terms; that is, all of the b_n coefficients are zero. To show this we write the Fourier series for $f(t)$ in the form

$$f(t) = a_0 + \sum_{n=1}^{\infty}[a_n \cos(2\pi n f_0 t) + b_n \sin(2\pi n f_0 t)] \qquad (16\text{-}8)$$

Given the Fourier series for $f(t)$, we use the identities $\cos(-x) = \cos(x)$ and $\sin(-x) = -\sin(x)$ to write the Fourier series for $f(-t)$ as follows:

$$f(-t) = a_0 + \sum_{n=1}^{\infty}[a_n \cos(2\pi n f_0 t) - b_n \sin(2\pi n f_o t)] \qquad (16\text{-}9)$$

When $f(t)$ has even symmetry, the right sides of Eq. (16-8) and (16-9) must be equal, since $f(t) = f(-t)$. Comparing the Fourier coefficients term by term, we find that equality requires $b_n = -b_n$. The only way this can happen is for $b_n = 0$ for all n.

[1]The rectangular pulse and the triangular wave are even functions because of the choice of the time origin. If the origin is selected differently, it is possible to create rectangular pulse and triangular wave signals that are odd functions. Likewise, the square wave and parabolic wave can be made even functions by an appropriate choice of time origin.

A waveform is said to have *odd symmetry* if $-f(-t) = f(t)$. The sine wave, square wave, and parabolic wave in Fig. 16-4 are examples of waveforms with this type symmetry. The Fourier series of odd waveforms are made up entirely of sine terms; that is, all of the a_n coefficients are zero. Given the Fourier series for $f(t)$ in Eq. (16-8), we use the identities $\cos(-x) = \cos(x)$ and $\sin(-x) = -\sin(x)$ to write the Fourier series for $-f(-t)$ in the form

$$-f(-t) = -a_0 + \sum_{n=1}^{\infty}[-a_n \cos(2\pi n f_0 t) + b_n \sin(2\pi n f_0 t)] \tag{16-10}$$

When $f(t)$ has odd symmetry the right sides of Eq. (16-8) and (16-10) must be equal, since $f(t) = -f(-t)$. Comparing the Fourier coefficients term by term we find that equality requires $a_0 = -a_0$ and $a_n = -a_n$. The only way this can happen is for $a_n = 0$ for all n, including $n = 0$.

A waveform is said to have *half-wave symmetry* if $-f(t - T_0/2) = f(t)$. This definition means that the waveform is unchanged if we invert it and then time-shift by half a cycle $(T_0/2)$. The sine wave, cosine wave, square wave, triangular wave, and parabolic wave in Fig. 16-4 are examples of waveforms with this symmetry. The amplitudes of all even harmonics are zero for waveforms with half-wave symmetry. Given the Fourier series for $f(t)$ in Eq. (16-8), we make use of the identities $\cos(x - n\pi) = (-1)^n \cos(x)$ and $\sin(x - n\pi) = (-1)^n \sin(x)$ to write the Fourier series of $-f(t - T_0/2)$ in the form

$$-f(t - T_0/2) = -a_0 + \sum_{n=1}^{\infty}[-(-1)^n a_n \cos(2\pi n f_0 t) - (-1)^n b_n \sin(2\pi n f_0 t)] \tag{16-11}$$

When $f(t)$ has half-wave symmetry the right sides of Eqs. (16-8) and (16-11) must be equal. Comparing the coefficients term by term we find that equality requires $a_0 = -a_0$, $a_n = -(-1)^n a_n$ and $b_n = -(-1)^n b_n$. The only way this can happen is for $a_0 = 0$ and for $a_n = b_n = 0$ when n is even. In other words, the only nonzero Fourier coefficients occur when n is odd.

A waveform may have more than one symmetry. For example, the triangular wave in Fig. 16-4 has even and half-wave symmetries, while the square wave has odd and half-wave symmetries. The sawtooth wave in Fig. 16-4 is an example where an underlying odd symmetry is masked by a dc component. A symmetry that is not apparent unless the dc component is removed is called a *hidden symmetry*.

EXERCISE 16-3

Identify the symmetries in the following waveforms.

(a) $f(t) = 5\cos 1000t + 3\cos 2000t - 0.5\cos 3000t + 0.25\cos 4000t$

(b) $f(t) = 20\cos 200t - 15\sin 600t + 1.5\cos 1000t - 0.75\sin 1400t$

(c) $f(t) = 50\sin 150t + 33\sin 450t - 5.5\sin 750t + 2.5\sin 1050t$

(d) $f(t) = 8\cos 2000t - 4\sin 6000t + 5\cos 8000t + 3\cos 10000t$

ANSWERS:

(a) Even, no sine terms present.

(b) Half-wave, no even harmonics are present.

(c) Odd and half-wave, no cosine terms and no even harmonics are present.

(d) None, sine and cosine terms, and both even and odd harmonics are present. ■

EXERCISE 16-4

An expression for the first cycle of the parabolic wave in Fig. 16-4 is

$$f(t) = \begin{cases} 32A(t/T_0)(1 - 2t/T_0) & 0 \leq t \leq T_0/2 \\ -32A(1 - t/T_0)(1 - 2t/T_0) & T_0/2 \leq t \leq T_0 \end{cases}$$

Show that it has zero average value and odd, half-wave symmetry. ■

16-2 THE EXPONENTIAL FOURIER SERIES

When the Fourier series of $f(t)$ is written using angular frequency $\omega_0 = 2\pi f_0$ it takes the form

$$f(t) = a_0 + \sum_{n=1}^{\infty} a_n \cos n\omega_0 t + b_n \sin n\omega_0 t \tag{16-12}$$

The coefficients a_n and b_n are the amplitudes of the cosine and sine components of the nth harmonic. The distribution of the amplitudes of the components of a signal as a function of frequency is called a *spectrum*. The trigonometric form of the Fourier series in Eq. (16-12) yields the spectrum of $f(t)$ in terms of two parameters a_n and b_n. The advantage of the exponential form of the series is that it compactly describes the spectrum in terms of a single parameter c_n.

To develop the exponential form of the Fourier series we begin with Euler's exponential representation of the cosine and sine functions:

$$\begin{aligned} \cos n\omega_0 t &= \frac{e^{jn\omega_0 t} + e^{-jn\omega_0 t}}{2} \\ \sin n\omega_0 t &= \frac{e^{jn\omega_0 t} - e^{-jn\omega_0 t}}{j2} \end{aligned} \tag{16-13}$$

Substituting these expressions into Eq. (16-12) and grouping the factors involving like exponential terms produce

$$f(t) = a_0 + \sum_{n=1}^{\infty}\left(\frac{a_n - jb_n}{2}\right)e^{jn\omega_0 t} + \sum_{n=1}^{\infty}\left(\frac{a_n + jb_n}{2}\right)e^{-jn\omega_0 t} \tag{16-14}$$

A complex parameter c_n is now defined in terms of a_n and b_n as follows:

$$c_n = \frac{a_n - jb_n}{2} \tag{16-15}$$

The parameters a_n and b_n are defined in Eq. (16-3) in terms of $f(t)$ and n. When n is replaced by $-n$ in these equations we see that $a_{-n} = a_n$ and $b_{-n} = -b_n$. From the definition of c_n in Eq. (16-15) it follows that

$$c_n^* = c_{-n} = \frac{a_n + jb_n}{2} \tag{16-16}$$

If we also define $c_0 = a_0$, the series in Eq. (16-14) has the form

$$f(t) = \sum_{n=0}^{\infty} c_n e^{jn\omega_0 t} + \sum_{n=1}^{\infty} c_{-n} e^{-jn\omega_0 t}$$

The first summation begins at $n = 0$ to include the dc component c_0. The second summation runs from $n = 1$ to ∞. However, each term in the second summation is unchanged if c_n replaces c_{-n} and the summation run from $n = -1$ to $n = -\infty$. In which case the expression for $f(t)$ takes the form

$$f(t) = \sum_{n=0}^{\infty} c_n e^{jn\omega_0 t} + \sum_{n=-1}^{-\infty} c_n e^{jn\omega_0 t}$$

which can be compactly written as follows:

$$f(t) = \sum_{n=-\infty}^{\infty} c_n e^{jn\omega_0 t}$$

In this form the summation extends over both positive ($n > 0$) and negative ($n < 0$) frequencies. The summation includes the index $n = 0$, which produces the amplitude of the zero frequency or dc component $c_0 = a_0$. Finally, we can relate c_n to $f(t)$ by substituting Eq. (16-3) and (16-13) into the definition of c_n in

Eq. (16-15) to obtain

$$c_n = \frac{1}{2}\left[\frac{2}{T_0}\int_{-T_0/2}^{T_0/2} f(t)\frac{e^{jn\omega_0 t} + e^{-jn\omega_0 t}}{2}dt - j\frac{2}{T_0}\int_{-T_0/2}^{T_0/2} f(t)\frac{e^{jn\omega_0 t} - e^{-jn\omega_0 t}}{j2}dt\right]$$

$$= \frac{1}{T_0}\int_{-T_0/2}^{T_0/2} f(t)e^{-jn\omega_0 t}dt$$

This completes the derivation of the exponential Fourier series. Summarizing the main results, the exponential form of the Fourier series is

$$f(t) = \sum_{n=-\infty}^{\infty} c_n e^{jn\omega_0 t} \tag{16-17}$$

where the complex Fourier coefficients c_n are obtained from a periodic waveform $f(t)$ by the equation

$$c_n = \frac{1}{T_0}\int_{-T_0/2}^{T_0/2} f(t)e^{-jn\omega_0 t}dt \tag{16-18}$$

Writing the Fourier series in exponential form highlights the basic elements of the Fourier transformation. We think of Eq. (16-18) as the direct transformation, which converts a periodic waveform $f(t)$ into a spectrum c_n. Conversely, we think of Eq. (16-17) as the inverse transformation, which converts the spectrum c_n into the periodic waveform $f(t)$.

The exponential form yields a *two-sided spectrum*, since both positive and negative frequencies are involved. The spectrum is called a discrete or line spectrum because it contains components only at the discrete harmonics $n\omega_0$ ($n = 0, \pm 1, \pm 2, \pm 3 \ldots$). The magnitude and angle of c_n are related to the Fourier coefficients a_n and b_n by

$$|c_n| = \frac{\sqrt{a_n^2 + b_n^2}}{2} \quad \text{and} \quad \angle c_n = \tan^{-1}(-b_n/a_n) \tag{16-19}$$

These equations show that $2|c_n|$ is the amplitude and $\angle c_n$ is the phase angle of the nth harmonic sinusoid in the Fourier series. Given the coefficients a_n and b_n the relations in Eq. (16-19) or the definition in Eq. (16-15) are used to determine c_n. Given a periodic waveform $f(t)$ we obtain c_n directly using Eq. (16-18). Both methods are illustrated in the following example.

EXAMPLE 16-2

(a) Use Eq. (16-18) to find the coefficient c_n for the periodic pulse train in Fig. 16-5.

(b) Repeat (a) using Eqs. (16-15) and the expressions for a_n and b_n from Fig. 16-4.

(c) Sketch the amplitude spectrum $|c_n|$ of the signal.

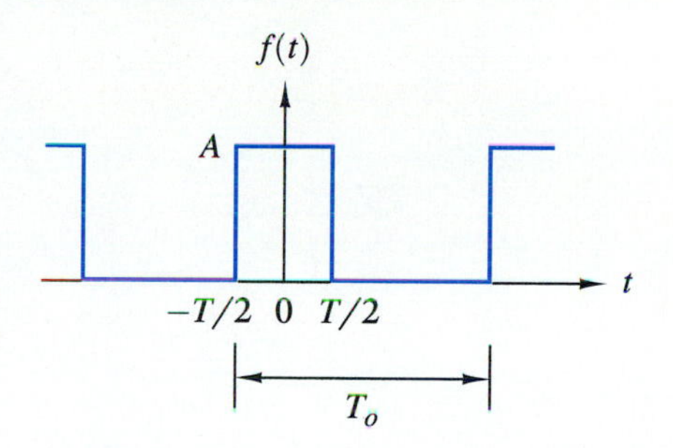

Figure 16-5

SOLUTION:

(a) The cycle centered at the origin ($t = 0$) has zero amplitude except on the range $-T/2 \le t \le +T/2$. The one cycle integration in Eq. (16-18) takes the form

$$c_n = \frac{1}{T_0}\int_{-T/2}^{T/2} Ae^{-jn\omega_0 t}\,dt = \left(\frac{A}{T_0}\right)\frac{e^{-jn\omega_0 t}}{-jn\omega_0}\Bigg|_{-T/2}^{T/2}$$

$$= \frac{2A}{n\omega_0 T_0}\left[\frac{e^{jn\omega_0 T/2} - e^{-jn\omega_0 T/2}}{j2}\right]$$

$$= \frac{2A}{T_0}\frac{\sin(n\omega_0 T/2)}{n\omega_0}$$

The pulse train in Fig. 16-5 has even symmetry, so $b_n = 0$. As a result c_n is real, since $c_n = (a_n - j\,b_n)/2$.

It is convenient to write c_n in the form of the function $(\sin x)/x$. Multiplying and dividing the right side of the expression for c_n by T and rearranging terms yields

$$c_n = \frac{AT}{T_0}\left[\frac{\sin(n\omega_0 T/2)}{(n\omega_0 T/2)}\right]$$

(b) Figure 16-4 gives the a_n and b_n coefficients for the pulse train as follows:

$$a_n = \frac{2A}{n\pi}\sin(n\pi T/T_0) \qquad \text{and} \qquad b_n = 0$$

Because $b_n = 0$ the definition in Eq. (16-15) yields $c_n = a_n/2$, or

$$c_n = \frac{A}{n\pi}\sin(n\pi T/T_0)$$

This expression can be put in the $(\sin x)/x$ form by rearranging terms:

$$n\pi T/T_0 = n\frac{2\pi}{T_0}T/2 = n\omega_0 T/2$$

and

$$\frac{A}{n\pi}\frac{T/T_0}{T/T_0} = \frac{AT}{T_0}\frac{1}{n\frac{2\pi}{T_0}T/2} = \frac{AT}{T_0}\frac{1}{n\omega_0 T/2}$$

producing

$$c_n = \frac{AT}{T_0}\left[\frac{\sin(n\omega_0 T/2)}{(n\omega_0 T/2)}\right]$$

which is the same result found in (a).

(c) Figure 16-6 shows a plot of $|c_n|$ versus angular frequency ω for $T = T_0/4$. The spacing between lines in the spectrum is the fundamental frequency ω_0. The envelope of the amplitude spectrum is defined by a function of the form $K\,|(\sin x)/x|$, where $x = \omega T/2$. Evaluating $f(x) = (\sin x)/x$ at $x = 0$ yields the indeterminant form 0/0. Applying l'Hopital's rule shows that $f(x) = 1$ at $x = 0$. The maximum height of the spectral envelope occurs at $\omega = 0$ and is $K = AT/T_0$. The null values in the envelope of $|c_n|$ occur when $\sin x = 0$. These nulls occur when $x = m\pi\,(m = \pm 1,\ \pm 2,\ \ldots)$ or at frequencies $\omega = 2\pi m/T$.

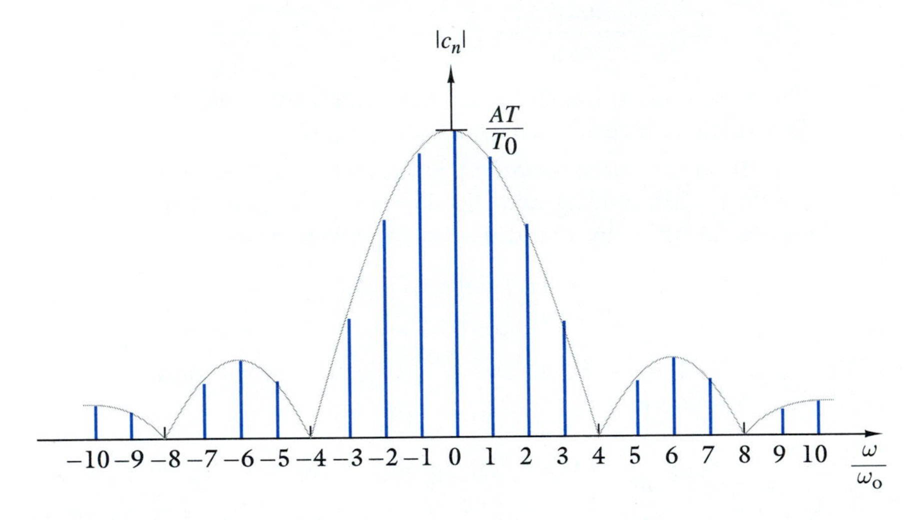

Figure 16-6

■ **EXERCISE 16-5**

Find c_n for the square wave in Fig. 16-1.

ANSWER:
$c_n = -jA(1 - \cos n\pi)/(n\pi)$. ■

16-3 FOURIER SERIES TO FOURIER TRANSFORMS

The Fourier series applies to periodic waveforms and expresses spectral content in terms of a line spectrum at discrete harmonic frequencies. Waveforms such as the exponential or step function are aperiodic and cannot be represented by a Fourier series. To treat aperiodic signals we need Fourier transforms and the concept of a continuous spectrum.

An aperiodic waveform can be thought of as a limiting case of a periodic signal as the period becomes infinite. For instance, in Example 16-2 we found the line spectrum of a periodic pulse train to be

$$c_n = \frac{AT}{T_0}\left[\frac{\sin(n\omega_0 T/2)}{n\omega_0 T/2}\right]$$

Figure 16-7 shows what happens to the pulse train and its amplitude spectrum as the period increases. As T_0 increases the envelope of the spectrum retains its $K\,|(\sin\,x)/x|$ form, but the height of the envelope ($K = AT/T_0$) and the spacing between spectral lines ($\omega_0 = 2\pi/T_0$) decrease. As T_0 approaches infinity

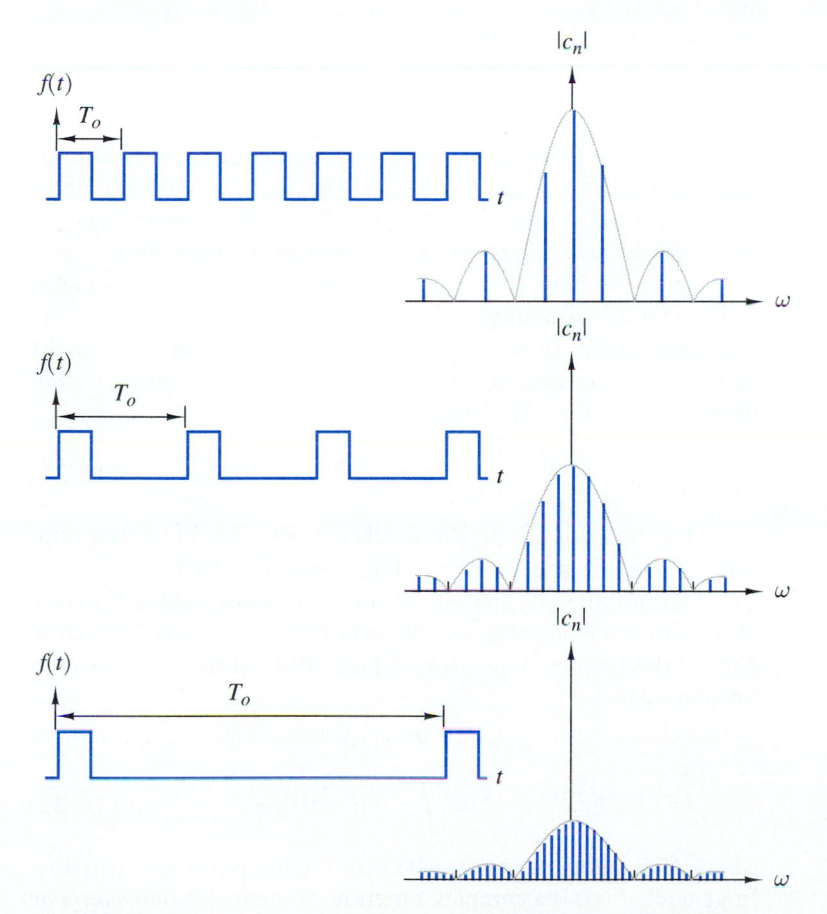

Figure 16-7 Changes in the rectangular pulse train waveform and amplitude spectrum as the period T_0 increases.

the envelope height becomes vanishingly small and the spectral lines become so densely packed that they merge into a continuous spectrum.

The Fourier transformation can be thought of as the result of applying the limiting condition $T_0 \to \infty$ to the exponential Fourier series. To develop this concept we begin with the transformation equations for the exponential Fourier series. Inserting Eq. (16-18) into (16-17) we obtain $f(t)$ in the form

$$f(t) = \sum_{n=-\infty}^{\infty} \left[\frac{\omega_0}{2\pi} \int_{-T_0/2}^{T_0/2} f(t) e^{-jn\omega_0 t} dt \right] e^{jn\omega_0 t}$$

$$= \frac{1}{2\pi} \sum_{n=-\infty}^{\infty} \left[\int_{-T_0/2}^{T_0/2} f(t) e^{-jn\omega_0 t} dt \right] \omega_0 e^{jn\omega_0 t} \quad (16\text{-}20)$$

The quantity inside the brackets is the line spectrum c_n, with $1/T_0$ written as $\omega_0/(2\pi)$. As $T_0 \to \infty$ the spacing between the lines approaches zero. For example, the distance between the kth and $(k+1)$th harmonic is

$$\Delta\omega = (k+1)\omega_0 - k\omega_0 = \omega_0 = \frac{2\pi}{T_0}$$

As $T_0 \to \infty$ the distance between harmonics becomes vanishingly small so that $\Delta\omega = \omega_0 \to d\omega$ and the frequency of any given harmonic $n\omega_0$ becomes indistinguishable from its neighbors. In other words, $n\omega_0$ merges into a continuous variable $n\omega_0 \to \omega$.

In addition, as $T_0 \to \infty$ the summation outside the bracket in Eq. (16-20) is adding up spectral components $c_n d\omega$ that are indistinguishably close together, so the summation sign can be replaced by a continuous integration. With these limiting conditions inserted, Eq. (16-20) becomes

$$f(t) = \frac{1}{2\pi} \int_{-\infty}^{\infty} \left[\int_{-\infty}^{\infty} f(t) e^{-j\omega t} dt \right] e^{+j\omega t} d\omega \quad (16\text{-}21)$$

The function inside the bracket in Eq. (16-21) is the remnant of the line spectrum c_n. The spectral remnant is denoted $F(\omega)$ because the integration within the bracket yields a function of a continuous variable ω. By definition the spectral function $F(\omega)$ is the Fourier transform of $f(t)$. That is, the *direct Fourier transformation* is

$$F(\omega) = \mathscr{F}\{f(t)\}$$

$$= \int_{-\infty}^{\infty} f(t) e^{-j\omega t} dt \quad (16\text{-}22)$$

where $\mathscr{F}$ is an operator that performs the Fourier transformation. In general, $F(\omega)$ is a complex function of the real variable ω. The magnitude $|F(\omega)|$ is called the *amplitude spectrum* and the angle $\angle F(\omega)$ the *phase spectrum*.

In Chapter 9 we found that the Laplace transformation will only handle causal waveforms, that is, waveforms that are zero for all $t \leq T$, where T is normally zero. However, the Fourier transformation can handle both causal and noncausal waveforms because the integration in Eq. (16-22) extends from $t = -\infty$ to $t = \infty$.

Given the notation $\mathscr{F}\{f(t)\} = F(\omega)$, we see that Eq. (16-21) displays the *inverse Fourier transformation*:

$$f(t) = \mathscr{F}^{-1}\{F(\omega)\} = \frac{1}{2\pi}\int_{-\infty}^{\infty} F(\omega)e^{j\omega t}\,d\omega \tag{16-23}$$

where $\mathscr{F}^{-1}$ is an operator that performs the inverse Fourier transformation. The integration in the inverse operation extends from $\omega = -\infty$ to $\omega = \infty$, which shows that the Fourier transformation yields a two-sided spectrum that spans both positive and negative frequencies.

The waveform $f(t)$ and transform $F(\omega)$ comprise a *Fourier transform pair*, since one can be derived from the other by the direct or inverse transformation. The transform $F(\omega)$ is the continuous spectrum representation of an aperiodic waveform $f(t)$ that can be causal or noncausal. For a voltage (current) waveform the transform has the dimensions of volt-seconds (ampere-seconds). For a voltage (current) signal the quantity $F(\omega)d\omega$ has the dimensions of volts (amperes) and represents the infinitesimal amplitude of the spectral frequencies in the range $d\omega$.

It should be clearly understood that the discussion leading to Eqs. (16-22) and (16-23) is **not** a derivation of the Fourier transformation. Rather, it is a plausibility argument suggesting that the definitions in these equations are reasonable. In physical science it is often more important to understand why something might be worth proving than it is to construct a formal proof.

The fact that the direct and inverse transformations involve improper integrals with infinite limits should alert the reader to the question of convergence and existence of $F(\omega)$. One set of sufficient conditions, called the *Dirichlet conditions*, can be stated as follows. For the Fourier transform to exist the waveform $f(t)$ must meet the following conditions:

1. Have a finite number of discontinuities
2. Have a finite number of maxima and minima in any finite interval
3. Be absolutely integrable, i.e.,

$$\int_{-\infty}^{\infty} |f(t)|\,dt < \infty \tag{16-24}$$

A number of useful waveforms meet the Dirichlet conditions, for example, the rectangular pulse $[u(t) - u(t - T)](|T| < \infty)$ and the causal exponential $u(t)e^{-\alpha t}(\alpha > 0)$. However, signals such as the step function $u(t)$ and the eternal sine wave $\cos \omega t$ do not. Since step functions and sinusoids are important in circuit analysis we cannot limit ourselves to the Dirichlet conditions. Bear in mind that the Dirichlet conditions are sufficient to establish the existence of $F(\omega)$. They are not necessary and sufficient. Happily, there are ways of extending the Fourier transformation to signals that do not meet the Dirichlet conditions.

Finally, whenever $F(\omega)$ exists it is unique. That is, there is a unique, one-to-one correspondence between a waveform $f(t)$ and its Fourier transform $F(\omega)$. Uniqueness means that once a transform pair has been found it can be used to go from waveform to transform or transform to waveform. Once we have developed a basic set of transforms we will summarize our results in a table of Fourier transform pairs similar to the table of Laplace transform pairs in Chapter 9.

EXAMPLE 16-3

Find the Fourier transform of the rectangular pulse in Fig. 16-8 and sketch its amplitude spectrum.

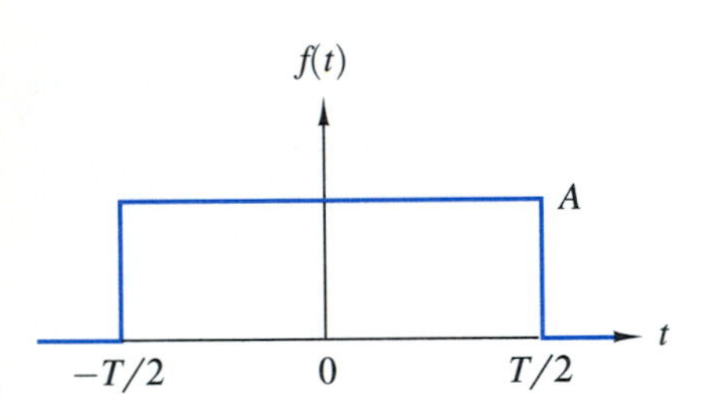

Figure 16-8

SOLUTION:

Since the waveform in Fig. 16-8 is zero everywhere except in the range from $t = -T/2$ to $t = +T/2$, the integration in Eq. (16-22) can be written in the form

$$F(\omega) = \int_{-T/2}^{T/2} Ae^{-j\omega t}\,dt = -\frac{A}{j\omega}e^{-j\omega t}\Big|_{-T/2}^{T/2}$$

$$= \frac{A}{\omega/2}\left[\frac{e^{j\omega T/2} - e^{-j\omega T/2}}{j2}\right]$$

$$= AT\frac{\sin(\omega T/2)}{\omega T/2}$$

The final result has been put in the form $(\sin x)/x$ because this function plays an important role in Fourier analysis. Figure 16-9 shows a plot of the amplitude spectrum $|F(\omega)|$ versus ω. Note that $|F(\omega)|$ is a two-sided spectrum containing both positive and negative frequencies. The nulls in the spectral plot occur when $\sin(\omega T/2) = 0$ at the frequencies $\omega = 2k\pi/T$ $(k = \pm 1, \pm 2, \ldots)$. The maximum value of $|F(\omega)|$ occurs at $\omega = 0$. At $x = 0$ the value of $\sin(x)/x = 1$ and so the maximum value of $|F(\omega)|$ is $|A|T$.

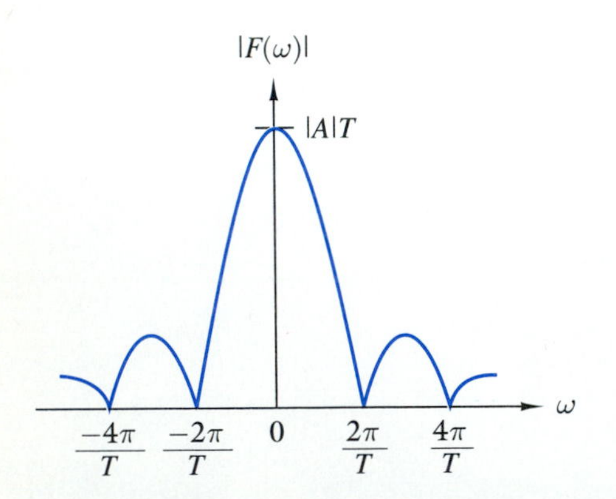

Figure 16-9

EXAMPLE 16-4

Find the Fourier transform of the exponential $f(t) = \{Ae^{-\alpha t}\}u(t)$ and plot its amplitude and phase spectra. Assume $\alpha > 0$.

SOLUTION:

For this waveform the integration in Eq. (16-22) extends from $t = 0$ to $t = +\infty$.

$$F(\omega) = \int_0^\infty Ae^{-\alpha t}e^{-j\omega t}dt = A\int_0^\infty e^{-(\alpha+j\omega)t}dt$$

$$= -A\frac{e^{-(\alpha+j\omega)t}}{\alpha+j\omega}\Bigg|_0^\infty$$

For $\alpha > 0$ the integral vanishes at the upper limit and $F(\omega)$ becomes

$$F(\omega) = \frac{A}{\alpha + j\omega} \qquad (\alpha > 0)$$

The amplitude and phase spectra of $F(\omega)$ are

$$|F(\omega)| = \frac{|A|}{\sqrt{\alpha^2+\omega^2}} \qquad \textit{(Amplitude)}$$

and

$$\phi(\omega) = \angle F(\omega) = -\tan^{-1}(\omega/\alpha) \qquad \textit{(Phase)}$$

The plots in Fig. 16-10 show the two-sided amplitude and phase spectra. The Fourier transform does not exist for $\alpha < 0$ because the exponential waveform becomes unbounded and the integral in Eq. (16-22) does not converge.

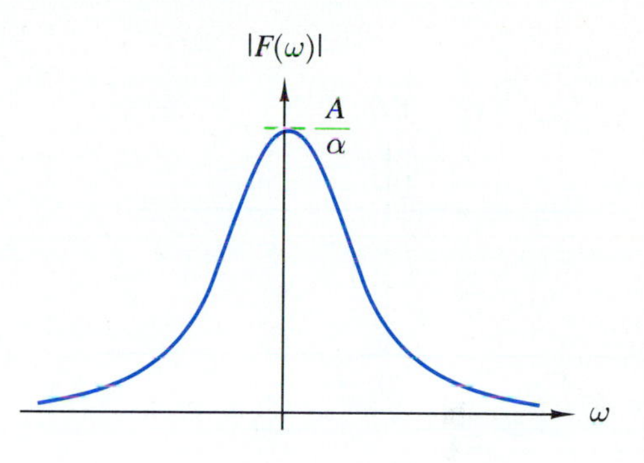

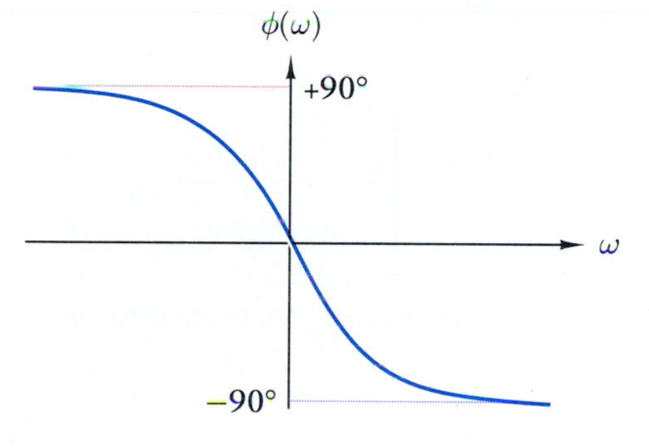

Figure 16-10

■ **EXERCISE 16-6**

Find the Fourier transform of the waveform shown in Fig. 16-11.

ANSWER:

$$F(\omega) = -jAT\frac{1-\cos(\omega T/2)}{\omega T/2}$$ ■

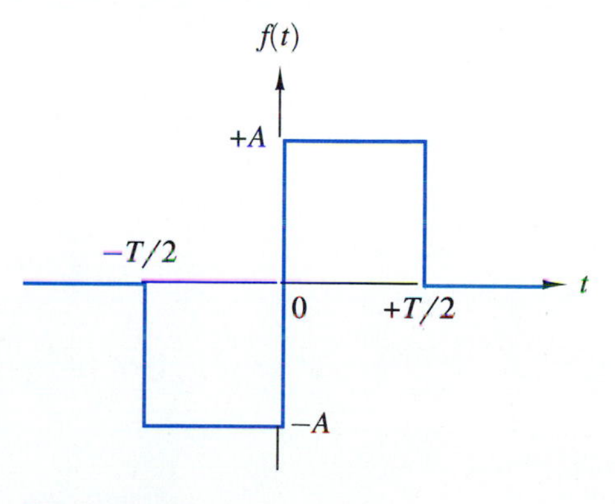

Figure 16-11

Inverse Fourier Transformation

In contrast to the Laplace transformation, the formal inversion integral for the Fourier transformation is often used because it is relatively easy to apply. As an example, consider the transform

$$F(\omega) = 2\pi\delta(\omega - \beta) \qquad (16\text{-}25)$$

The transform is a frequency-domain impulse located at $\omega = \beta$ with an area of 2π. The time-domain waveform $f(t)$ corresponding to this transform is found by applying the inversion integral in Eq. (16-23). Because $F(\omega)$ is zero except at $\omega = \beta$ the integration limits need only span the range from $\omega = \beta-$ to $\omega = \beta+$, just slightly above and below the frequency $\omega = \beta$:

$$f(t) = \frac{1}{2\pi}\int_{\beta-}^{\beta+} 2\pi\delta(\omega-\beta)e^{j\omega t}d\omega$$

$$= e^{j\beta t}\int_{\beta-}^{\beta+} \delta(\omega-\beta)d\omega = e^{j\beta t} \qquad (16\text{-}26)$$

By direct application of the inversion integral we find that the inverse transform of $2\pi\delta(\omega-\beta)$ is the complex exponential $e^{j\beta t}$. Because the Fourier transform is unique, we also know that $\mathscr{F}\{e^{j\beta t}\} = 2\pi\delta(\omega-\beta)$.

In particular, if $\beta = 0$, then $f(t) = e^{j0} = 1$ for all time t and $F(\omega) = 2\pi\delta(\omega)$. In other words, the Fourier transform of a dc waveform with amplitude $A = 1$ is an impulse of area 2π at $\omega = 0$. Figure 16-12 shows the waveform and Fourier transform of the unity amplitude dc signal. The dc waveform is a noncausal signal because it extends indefinitely in both the positive and negative time directions. Such a waveform does not have a Laplace transform because it is noncausal. The dc waveform does not meet the Dirichlet conditions because it is not absolutely integrable as defined in Eq. (16-24). But its Fourier transform does exist as an impulse at $\omega = 0$.

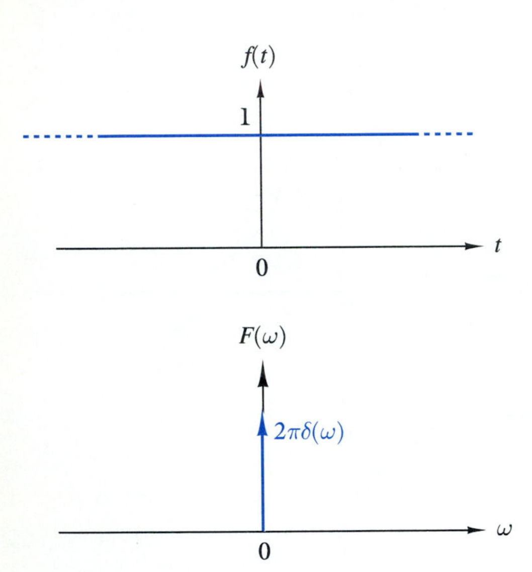

Figure 16-12 Waveform and transform of a unit dc signal.

EXAMPLE 16-5

Use the inversion integral to find the waveform corresponding to the rectangular transform

$$F(\omega) = \frac{\pi A}{\beta}[u(\omega+\beta) - u(\omega-\beta)]$$

SOLUTION:

Since $F(\omega)$ is zero everywhere except in the range from $\omega = -\beta$ to $\omega = +\beta$ the inversion integral in Eq. (16-23) takes the form

$$f(t) = \frac{1}{2\pi}\int_{-\beta}^{\beta} \pi\frac{A}{\beta}e^{j\omega t}d\omega = \frac{A}{2\beta}\frac{e^{j\omega t}}{jt}\bigg|_{-\beta}^{\beta}$$

$$= \frac{A}{\beta t}\frac{e^{j\beta t} - e^{-j\beta t}}{j2} = A\frac{\sin(\beta t)}{\beta t}$$

Figure 16-13 shows the given amplitude spectrum and the resulting waveform found from the inverse transformation. The waveform $f(t) = A\sin(\beta t)/(\beta t)$ is another example of a non-causal signal which does not have a Laplace transform. However, the Fourier transform of $f(t)$ exists and has a rectangular amplitude spectrum centered around $\omega = 0$ and extending from $\omega = -\beta$ to $\omega = \beta$.

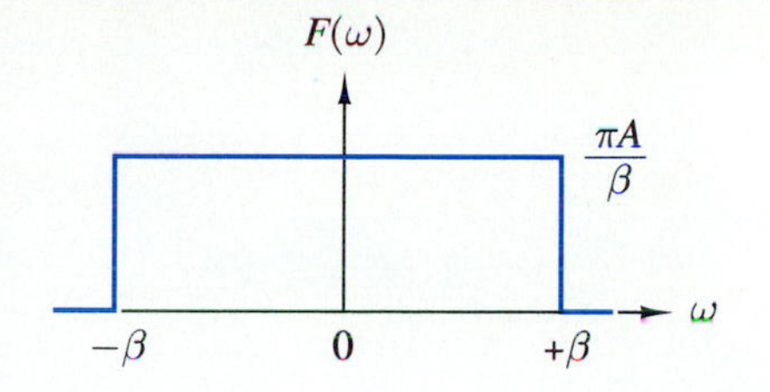

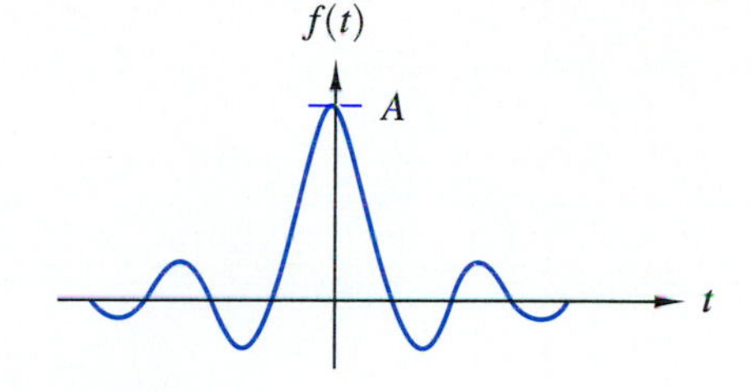

Figure 16-13

■ EXERCISE 16-7

Use the inversion integral to find the waveform corresponding to the transform $F(\omega) = \pi\alpha A e^{-\alpha|\omega|}$. Assume that $\alpha > 0$.

ANSWER:

$$f(t) = \frac{A}{1 + (t/\alpha)^2}$$ ■

16-4 LAPLACE TRANSFORMS TO FOURIER TRANSFORMS

For many useful signals there is a simple relationship between the Laplace and Fourier transforms. In Chapter 9 the integral definition of the direct Laplace transformation is given as

$$\mathscr{L}\{f(t)\} = \int_0^\infty f(t)e^{-st}dt = F(s) \qquad (16\text{-}27)$$

where $s = \sigma + j\omega$ is the complex frequency variable. The lower limit of this integration is $t = 0$, which reminds us that $f(t)$ must be causal. If $f(t)$ is absolutely integrable in the sense defined in Eq. (16-24), then the integration in Eq. (16-27) converges when $\sigma = 0$. When $\sigma = 0$ Eq. (16-27) becomes

$$\mathscr{L}\{f(t)\} = \int_0^\infty f(t)e^{-j\omega t}dt = F(s)\rceil_{\sigma=0} \qquad (16\text{-}28)$$

On the other hand, when $f(t)$ is causal and absolutely integrable its Fourier transform exists and is found from Eq. (16-22):

$$\mathscr{F}\{f(t)\} = \int_0^\infty f(t)e^{-j\omega t}dt = F(\omega) \qquad (16\text{-}29)$$

Comparing Eqs. (16-28) and (16-29) we conclude that

$$F(\omega) = F(s)\rceil_{\sigma=0} \qquad (16\text{-}30)$$

provided that $f(t)$ is causal and absolutely integrable.

For $f(t)$ to be absolutely integrable it must have finite duration or decay to zero rapidly enough so that the integral of $|f(t)|$ from $t = 0$ to $t = \infty$ converges. A sufficient condition for $f(t)$ to decay to zero is that all of the poles of $F(s)$ lie in the left-half plane. For example, the Laplace transform of the causal exponential $Ae^{-\alpha t}u(t)$ is $A/(s+\alpha)$. The transform has a pole at $s = -\alpha$, which lies in the left half of the s plane provided $\alpha > 0$. When $\alpha > 0$ the Fourier transform of the causal exponential is found from its Laplace transform to be

$$F(\omega) = F(s)]_{\sigma=0} = \frac{A}{j\omega + \alpha}$$

The $F(\omega)$ obtained above using Laplace transforms agrees with the conclusion reached in Example 16-4 using the integral definition of the Fourier transformation.

We can use the conditions in Eq. (16-30) to find $F(\omega)$ from $F(s)$ provided all poles of $F(s)$ lie in the left half of the s plane. The poles cannot lie in the right half plane or on the j-axis boundary. For example, the Laplace transform approach cannot be used to find the Fourier transform of a step function because $F(s) = 1/s$ has a pole on the j axis at $s = 0$. Likewise, the Laplace transform method will not work for the causal sine wave $f(t) = u(t)[\sin \beta t]$, since $F(s) = 1/(s^2 + \beta^2)$ has poles on the j axis at $s = \pm j\beta$.

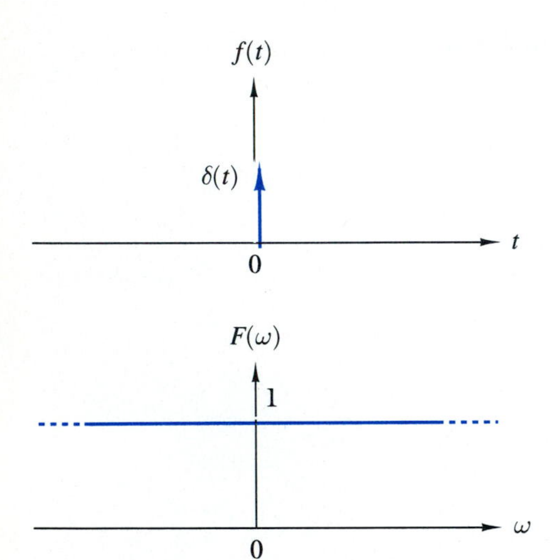

Figure 16-14 Waveform and transform of a unit impulse signal.

The Laplace transform method can be used to find the Fourier transform of the unit impulse because $\delta(t)$ is causal and absolutely integrable. Applying Eq. (16-30) we obtain

$$\mathscr{F}[\delta(t)] = \mathscr{L}[\delta(t)]]_{\sigma=0} = 1 \qquad (16\text{-}31)$$

The time-domain impulse has a spectrum $F(\omega)$ that is constant at all frequencies as shown in Fig. 16-14. Note the symmetry between plots in Fig. 16-14 and those in Fig. 16-12. A constant time-domain waveform leads to the impulsive frequency spectrum in Fig. 16-12. Conversely, an impulsive time-domain waveform leads to the constant frequency-domain spectrum in Fig. 16-14.

Laplace transform concepts can be used to find inverse Fourier transforms as well. Given a Fourier transform $F(\omega)$, we form the Laplace transform $F(s)$ by replacing $j\omega$ by s, or equivalently, replacing ω by $-js$. If the poles of $F(s)$ all lie in the left half of the s plane, then by the uniqueness property of Fourier and Laplace transforms we have

$$f(t) = \mathscr{F}^{-1}\{F(\omega)\} = \mathscr{L}^{-1}\{F(s)\} \qquad (t > 0) \qquad (16\text{-}32)$$

The inverse transform of $F(s)$ in this equation can be obtained using partial fraction expansion if need be. However, keep in mind that the Laplace transform method requires: (1) $f(t)$ to be causal and absolutely integrable, or (2) all of the poles of $F(s)$ to be in the left half of the s plane.

EXERCISE 16-8

Use Laplace transforms to find the Fourier transforms of the following causal waveforms. Assume $\alpha > 0$.

(a) $f(t) = A[e^{-\alpha t} \sin \beta t]u(t)$

(b) $f(t) = A[\alpha t e^{-\alpha t}]u(t)$

ANSWERS:

(a) $F(\omega) = \dfrac{A\beta}{\beta^2 + \alpha^2 - \omega^2 + j2\alpha\omega}$

(b) $F(\omega) = \dfrac{A\alpha}{\alpha^2 - \omega^2 + j2\alpha\omega}$ ■

EXAMPLE 16-6

Use Laplace transforms to find the inverse Fourier transform of

$$F(\omega) = \frac{10}{(j\omega + 2)(j\omega + 4)}$$

SOLUTION:

Replacing $j\omega$ by s yields $F(s)$:

$$F(s) = \frac{10}{(s + 2)(s + 4)}$$

$F(s)$ is a rational function of s with poles at $s = -2$ and $s = -4$. Both poles are in the left-half plane, so the inverse transform of $F(s)$ is the inverse transform of $F(\omega)$. $F(s)$ can be expanded by partial fractions because it is a proper rational function:

$$F(s) = \frac{k_1}{s + 2} + \frac{k_2}{s + 4}$$

The poles are simple so the residues k_1 and k_2 are

$$k_1 = (s + 2)F(s)]_{s=-2} = 5$$

and

$$k_2 = (s + 4)F(s)]_{s=-4} = -5$$

The inverse transform of $F(\omega)$ is

$$\begin{aligned} f(t) = \mathscr{L}^{-1}\{F(s)\} &= \mathscr{L}^{-1}\left\{\frac{5}{s + 2}\right\} + \mathscr{L}^{-1}\left\{\frac{-5}{s + 4}\right\} \\ &= \left[5e^{-2t} - 5e^{-4t}\right]u(t) \end{aligned}$$

EXERCISE 16-9

Use the Laplace transform method to find the causal waveform corresponding to the following Fourier transform:

$$F(\omega) = \frac{5(j\omega - 3)}{(j\omega + 1)(j\omega + 5)}$$

ANSWER:
$f(t) = \left[-5e^{-t} + 10e^{-5t}\right]u(t)$

EXERCISE 16-10

Explain why Laplace transforms cannot be used to find the Fourier transforms of the following waveforms. Assume $\alpha > 0$.

(a) $f(t) = A\alpha t u(t)$
(b) $f(t) = Ae^{-\alpha|t|}$
(c) $f(t) = A \sin \alpha t$

ANSWERS:
(a) $F(s)$ has a double pole at $s = 0$
(b) $f(t)$ is not causal
(c) $f(t)$ is not absolutely integrable

EXERCISE 16-11

Explain why Laplace transforms cannot be used to find the waveforms corresponding to the following Fourier transforms. Assume $\alpha > 0$.

(a) $F(\omega) = \dfrac{2\alpha}{\alpha^2 + \omega^2}$
(b) $F(\omega) = u(\omega + \alpha) - u(\omega - \alpha)$

ANSWERS:
(a) $F(s)$ has a pole at $s = +\alpha$
(b) $f(t)$ is not causal

16-5 BASIC TRANSFORM PROPERTIES AND PAIRS

This section introduces the important basic mathematical properties of the Fourier transformation. The basic properties are then applied to derive additional Fourier transform pairs.

Linearity

Like Laplace transforms, the foremost property of Fourier transforms is *linearity*. Stated formally

$$\mathscr{F}\{Af_1(t) + Bf_2(t)\} = AF_1(\omega) + BF_2(\omega) \qquad (16\text{-}32)$$

where A and B are constants. Proof of this property follows directly from the integral definition of the direct Fourier transformation in Eq. (16-22) using the same logic applied in Section 9-2 to derive the linearity property of the Laplace transforms. The following example illustrates an application of linearity.

EXAMPLE 16-7

Find the Fourier transforms of $\cos \beta t$ and $\sin \beta t$.

SOLUTION:

These waveforms are noncausal, eternal sinusoids, so we cannot use the Laplace transform method. In Section 16-3 we found that

$$\mathscr{F}\{e^{j\beta t}\} = 2\pi\delta(\omega - \beta)$$

Using Euler's relationship for $\cos \beta t$ and the linearity property of the Fourier transform we can write

$$\begin{aligned}\mathscr{F}\{\cos \beta t\} &= \mathscr{F}\left\{\frac{e^{j\beta t} + e^{-j\beta t}}{2}\right\} = \frac{1}{2}\mathscr{F}\{e^{j\beta t}\} + \frac{1}{2}\mathscr{F}\{e^{-j\beta t}\} \\ &= \pi\left[\delta(\omega - \beta) + \delta(\omega + \beta)\right]\end{aligned}$$

Similarly for $\sin \beta t$ we obtain

$$\begin{aligned}\mathscr{F}\{\sin \beta t\} &= \mathscr{F}\left\{\frac{e^{j\beta t} - e^{-j\beta t}}{j2}\right\} = \frac{1}{j2}\mathscr{F}\left\{e^{j\beta t}\right\} - \frac{1}{j2}\mathscr{F}\left\{e^{-j\beta t}\right\} \\ &= -j\pi\left[\delta(\omega - \beta) - \delta(\omega + \beta)\right]\end{aligned}$$

In Fig. 16-12 we found that the Fourier transform of a dc waveform contains an impulse at zero frequency ($\omega = 0$). Here we see that the transforms of ac waveforms have impulses located at $\omega = \pm\beta$, where β is the frequency of the sinusoids.

Time Differentiation and Integration

The *time differentiation* property of Fourier transforms states that

$$\mathscr{F}\left\{\frac{df(t)}{dt}\right\} = j\omega F(\omega) \qquad (16\text{-}34)$$

Derivation of this property of Fourier transforms begins with the inverse transformation integral:

$$f(t) = \frac{1}{2\pi}\int_{-\infty}^{\infty} F(\omega)e^{j\omega t}d\omega \qquad (16\text{-}35)$$

We first differentiate both sides of this equation with respect to time:

$$\frac{df(t)}{dt} = g(t) = \frac{d}{dt}\left[\frac{1}{2\pi}\int_{-\infty}^{\infty} F(\omega)e^{j\omega t}d\omega\right]$$

Assuming the order of integration and differentiation on the right side of this equation can be interchanged, we obtain

$$\frac{df(t)}{dt} = g(t) = \frac{1}{2\pi}\int_{-\infty}^{\infty}\left[\frac{d}{dt}F(\omega)e^{j\omega t}d\omega\right]$$

$$= \frac{1}{2\pi}\int_{-\infty}^{\infty} j\omega F(\omega)e^{j\omega t}d\omega$$

The right side of the last line in this equation implies $\mathscr{F}[g(t)] = j\omega F(\omega)$. Since $g(t) = df/dt$ we conclude that

$$\mathscr{F}\left[\frac{df(t)}{dt}\right] = j\omega F(\omega) \tag{16-36}$$

Differentiating $f(t)$ in the time domain corresponds to multiplying $F(\omega)$ by $j\omega$ in the frequency domain.

Given the differentiation property, it is reasonable to expect that integrating $f(t)$ in the time domain corresponds to dividing $F(\omega)$ by $j\omega$ in the frequency domain. This expectation is correct, except that integrating a waveform may produce a constant or dc component. As we have seen, the Fourier transform of a dc component is an impulse at $\omega = 0$ in the frequency domain. As a result, the statement of the *integration property* is

$$\mathscr{F}\left[\int_{-\infty}^{t} f(x)dx\right] = \frac{F(\omega)}{j\omega} + \pi F(0)\delta(\omega) \tag{16-37}$$

Integrating $f(t)$ in the time domain leads to division of $F(\omega)$ by $j\omega$ plus an additive term $\pi F(0)\delta(\omega)$ to account for the possibility of a dc component in the integral of $f(t)$. The factor $F(0)$ is related to $f(t)$ via the integral defining the direct Fourier transformation. Replacing ω by 0 reduces Eq. (16-22) to

$$F(0) = \int_{-\infty}^{\infty} f(t)dt \tag{16-38}$$

The zero frequency or dc component $F(0)$ is zero when the integral of $f(t)$ over all time is zero. In such a case, the second term on the right-hand side in Eq. (16-37) vanishes, and time integration corresponds to division by $j\omega$ in the frequency domain.

EXAMPLE 16-8

Use the integration property to find the Fourier transform of the step function $u(t)$.

SOLUTION:

We cannot use the Laplace transform method here because the step function is not absolutely integrable. The Fourier transform of a time-domain impulse was previously found to be $\mathscr{F}[\delta(t)] = 1$. Since the step function is the integral of an impulse the integration property yields the Fourier transform of the unit step function:

$$\mathscr{F}\{u(t)\} = \mathscr{F}\left\{\int_{-\infty}^{t} \delta(x)dx\right\} = \frac{1}{j\omega} + \pi\delta(\omega)$$

The Fourier and Laplace transform of the step function have the same form when s is replaced by $j\omega$, except for the frequency-domain impulse $\pi\delta(\omega)$ required to account for the dc component in $u(t)$.

Reversal

A waveform $f(t)$ is said to be reversed when t is replaced by $-t$. The waveform $f(-t)$ is reversed because everything happens in reverse order with events that previously happened for $t > 0$ now occurring for $t < 0$, and vice versa. If the Fourier transform of $f(t)$ is $F(\omega)$, what can be said about the transform of $f(-t)$? The *reversal property* of the Fourier transformation is formally stated as follows:

$$\text{If } \mathscr{F}[f(t)] = F(\omega) \text{ then } \mathscr{F}[f(-t)] = F(-\omega) \qquad (16\text{-}39)$$

Simply stated, reversing $f(t)$ reverses $F(\omega)$.

Deriving this property begins by noting that by definition the Fourier transform of $f(t)$ is

$$F(\omega) = \int_{-\infty}^{\infty} f(t)e^{-j\omega t}dt \qquad (16\text{-}40)$$

and the Fourier transform of the reversed waveform is

$$G(\omega) = \int_{-\infty}^{\infty} f(-t)e^{-j\omega t}dt \qquad (16\text{-}41)$$

Changing the dummy variable of integration in Eq. (16-41) to $\tau = -t$ produces

$$\begin{aligned} G(\omega) &= -\int_{\infty}^{-\infty} f(\tau)e^{j\omega\tau}d\tau \\ &= \int_{-\infty}^{\infty} f(\tau)e^{-j(-\omega)\tau}d\tau \end{aligned} \qquad (16\text{-}42)$$

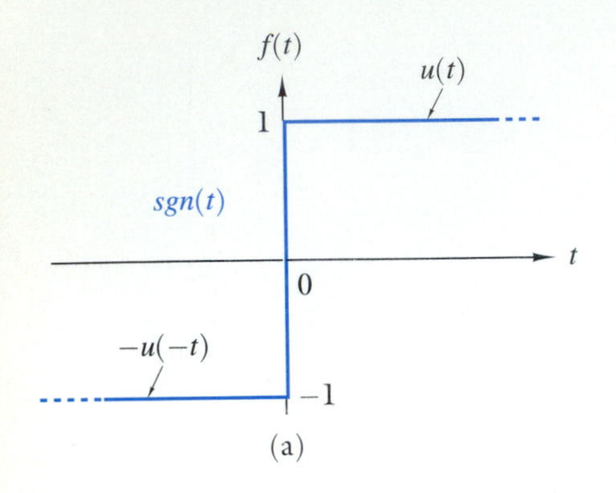

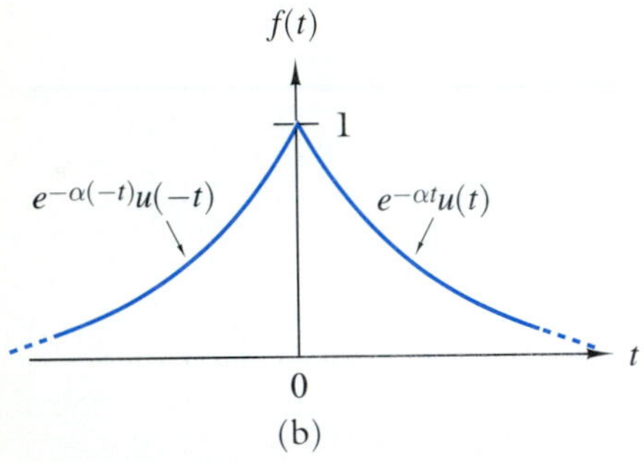

Figure 16-15 Waveforms of the signum and double-sided exponential signals.

Comparing the last line in Eq. (16-42) with the definition in Eq. (16-40) we conclude that $\mathscr{F}\{f(-t)\} = G(\omega) = F(-\omega)$.

The reversal property can be used to derive the Fourier transforms of waveforms of the type shown in Fig. 16-15. The mathematical expressions for these two waveforms are

$$sgn(t) = u(t) - u(-t)$$
$$e^{-\alpha|t|} = u(t)e^{-\alpha t} + u(-t)e^{-\alpha(-t)} \qquad (16\text{-}43)$$

The first waveform is called the *signum function*. By definition $sgn(t)$ is $+1$ for $t > 0$ and -1 for $t < 0$ and zero for $t = 0$. As indicated in Eq. (16-43), $sgn(t)$ can be constructed by adding a step $u(t)$ to a reversed step $-u(-t)$. The second waveform is called a *double-sided exponential*. This waveform is unity at $t = 0$ and exponentially decays (for $\alpha > 0$) to zero in both directions along the time axis. As shown in Eq. (16-43), the double-sided exponential is the sum of a causal exponential and a reversed causal exponential.

The signum and double-sided exponential are examples of noncausal waveforms that can be written as $f(t) = g(t) \pm h(-t)$, where $g(t)$ and $h(t)$ are causal. If the Fourier transforms of $g(t)$ and $h(t)$ are known, then using the reversal and linearity properties shows that $F(\omega) = G(\omega) \pm H(-\omega)$. If $g(t)$ and $h(t)$ are also absolutely integrable, then we can use Laplace transforms to find $G(\omega)$ and $H(\omega)$.

The next example shows how to use the reversal property to obtain the Fourier transforms of noncausal waveforms of the form $f(t) \pm h(-t)$.

EXAMPLE 16-9

Use the linearity and reversal properties to find the Fourier transform of

(a) $sgn(t)$

(b) $e^{-\alpha|t|}$

SOLUTION:

(a) The signum function is of the form $f(t) = g(t) - g(-t)$, where $g(t) = u(t)$. In Example 16-8 the transform of a unit step function is shown to be $\mathscr{F}\{u(t)\} = 1/j\omega + \pi\delta(\omega)$. Using the linearity and reversal properties yields the Fourier transform of $sgn(t)$:

$$\mathscr{F}\{sgn(t)\} = G(\omega) - G(-\omega)$$
$$= \left[\frac{1}{j\omega} + \pi\delta(\omega)\right] - \left[\frac{1}{j(-\omega)} + \pi\delta(-\omega)\right]$$
$$= \frac{2}{j\omega}$$

In the last line of this equation we have used the fact that $\pi\delta(\omega) - \pi\delta(-\omega) = 0$, since both $\delta(\omega)$ and $\delta(-\omega)$ are impulses at $\omega = 0$.

(b) For the double-sided exponential $f(t) = g(t) + g(-t)$, where $g(t) = [e^{-\alpha t}]u(t)$. The Fourier transform of $g(t)$ is $G(\omega) = 1/(\alpha + j\omega)$. Using the linearity and reversal properties yields the Fourier transform of the double-sided exponential:

$$\mathscr{F}\left\{e^{-\alpha|t|}\right\} = G(\omega) + G(-\omega)$$
$$= \left[\frac{1}{\alpha + j\omega}\right] + \left[\frac{1}{\alpha + j(-\omega)}\right]$$
$$= \frac{2\alpha}{\alpha^2 + \omega^2}$$

EXERCISE 16-12

Use linearity and reversal properties to find the Fourier transform of

$$f(t) = [e^{-\alpha t}]u(t) + u(-t)$$

Assume $\alpha > 0$.

ANSWER:

$$F(\omega) = \pi\delta(\omega) - \frac{\alpha}{j\omega(j\omega + \alpha)}$$ ■

16-6 MORE FOURIER TRANSFORM PROPERTIES

Uniqueness, linearity, time differentiation, time integration, and reversal are the basic properties of the Fourier transforms that are most commonly used in Fourier analysis problems. This section discusses additional properties that are useful.

Even and Odd Parts of $F(\omega)$

In general a Fourier transform $F(\omega)$ is a complex function that can be written in rectangular and polar form:

$$F(\omega) = A(\omega) + jB(\omega) = |F(\omega)|e^{j\theta(\omega)}$$

The functions $A(\omega)$ and $B(\omega)$ are the real and imaginary parts of $F(\omega)$, while $|F(\omega)|$ and $\theta(\omega)$ are the magnitude and angle of $F(\omega)$. The next example shows how to calculate these functions for a given transform.

EXAMPLE 16-10

Find the real part, imaginary parts, magnitude, and angle of the Fourier transform of the causal exponential $f(t) = [e^{-\alpha t}]u(t)$.

SOLUTION:

Example 16-4 shows that $F(\omega) = 1/(\alpha + j\omega)$. Multiplying the numerator and denominator of $F(\omega)$ by the conjugate of the denominator yields

$$F(\omega) = \frac{1}{\alpha + j\omega}\frac{\alpha - j\omega}{\alpha - j\omega} = \frac{\alpha - j\omega}{\alpha^2 + \omega^2} = \frac{\alpha}{\alpha^2 + \omega^2} + \frac{-j\omega}{\alpha^2 + \omega^2}$$

The real and imaginary parts are

$$A(\omega) = \frac{\alpha}{\alpha^2 + \omega^2} \quad \text{and} \quad B(\omega) = \frac{-\omega}{\alpha^2 + \omega^2}$$

The magnitude and angle of $F(\omega)$ are

$$|F(\omega)| = \frac{1}{\sqrt{\alpha^2 + \omega^2}} \quad \text{and} \quad \theta(\omega) = \tan^{-1}(-\omega/\alpha)$$

Note that the real part and magnitude are even functions of ω, while the imaginary part and angle are odd functions.

The conclusion found in the above example regarding the even and odd parts of $F(\omega)$ can be generalized as follows. Using Euler's equation $e^{-j\omega t} = \cos\omega t - j\sin\omega t$ in the integral defining the Fourier transformation yields

$$F(\omega) = \int_{-\infty}^{\infty} f(t)\cos\omega t\, dt - j\int_{-\infty}^{\infty} f(t)\sin\omega t\, dt$$

From which we conclude that

$$A(\omega) = \int_{-\infty}^{\infty} f(t)\cos\omega t\, dt \tag{16-44}$$

$$B(\omega) = -\int_{-\infty}^{\infty} f(t)\sin\omega t\, dt \tag{16-45}$$

and

$$|F(\omega)| = \sqrt{A^2(\omega) + B^2(\omega)} \tag{16-46}$$

$$\theta(\omega) = \tan^{-1}\left(\frac{B(\omega)}{A(\omega)}\right) \tag{16-47}$$

When $f(t)$ is a real-valued function, we can make the following observations:

1. The real part of $F(\omega)$ is an even function since replacing ω by $-\omega$ in Eq. (16-44) shows that $A(-\omega) = A(\omega)$.
2. The imaginary part of $F(\omega)$ is an odd function since replacing ω by $-\omega$ in Eq. (16-45) shows that $B(-\omega) = -B(\omega)$.
3. The magnitude of $F(\omega)$ is an even function, since replacing ω by $-\omega$ in Eq. (16-46) shows that $|F(-\omega)| = |F(\omega)|$ because $A^2(\omega)$ and $B^2(\omega)$ are even functions.
4. The angle of $F(\omega)$ is an odd function, since replacing ω by $-\omega$ in Eq. (16-47) shows that $\theta(-\omega) = -\theta(\omega)$ because $B(\omega)$ is an odd function and $A(\omega)$ is an even function.
5. As a result of (1) and (2), the reversal of $F(\omega)$ is its complex conjugate since $F(-\omega) = A(\omega) - jB(\omega) = F^*(\omega)$.
6. As a result of (5), the product of $F(\omega)$ and its reversal is its squared magnitude, since $F(\omega)F(-\omega) = A^2(\omega) + B^2(\omega) = |F(\omega)|^2$.
7. When $f(t)$ is an even function, then $B(\omega) = 0$ since the integrand $f(t)\sin\omega t$ in Eq. (16-45) is odd. That is, when $f(t)$ is even $F(\omega)$ is real.
8. When $f(t)$ is an odd function, then $A(\omega) = 0$ since the integrand $f(t)\cos\omega t$ in Eq. (16-44) is odd. That is, when $f(t)$ is odd $F(\omega)$ is imaginary.

For example, the double-sided exponential waveform in Fig. 16-15 is even and its Fourier transform $F(\omega) = 2\alpha/(\alpha^2 + \omega^2)$ is real. Conversely, the signum waveform in Fig. 16-15 is odd and its transform $F(\omega) = 2/j\omega$ is imaginary.

EXERCISE 16-13

(a) Find the real part, imaginary part, magnitude, and angle of the Fourier transforms in Exercise 16-6.

(b) Repeat for Exercise 16-7. Indicate whether the waveforms are even or odd or neither.

ANSWERS:

(a) $A(\omega) = 0,\ B(\omega) = -2A[1 - \cos(\omega T/2)]/\omega$
$|F(\omega)| = |B(\omega)|,\ \theta(\omega) = -90^\circ,\ f(t)$ is odd

(b) $A(\omega) = \pi\alpha A e^{-\alpha|\omega|},\ B(\omega) = 0$
$|F(\omega)| = A(\omega),\ \theta(\omega) = 0^\circ,\ f(t)$ is even

Symmetry

The general statement of *symmetry property* is

$$\text{If } \mathscr{F}\{f(t)\} = F(\omega) \text{ then } \mathscr{F}\{F(t)\} = 2\pi f(-\omega) \qquad (16\text{-}48)$$

If $\mathscr{F}\{f(t)\} = F(\omega)$, then the inverse Fourier transform can be written in the form

$$2\pi f(t) = \int_{-\infty}^{\infty} F(\omega)e^{j\omega t}\,d\omega$$

Replacing t by $-t$ yields

$$2\pi f(-t) = \int_{-\infty}^{\infty} F(\omega)e^{-j\omega t}\,d\omega$$

Now interchanging t and ω produces

$$2\pi f(-\omega) = \int_{-\infty}^{\infty} F(t)e^{-j\omega t}\,dt$$

This equation is the conclusion in the statement in Eq. (16-48).

The symmetry property states that waveforms and Fourier transforms can be interchanged within a factor of 2π. The origin of the symmetry property traces back to the similarity of the direct transformation in Eq. (16-22) and the inverse transformation in Eq. (16-23). Except for the factor of 2π, these two equations are similar in form. The symmetry shows up in transform pairs. For example, Fig. 16-12 shows that a constant waveform produces an impulsive transform, while Fig. 16-14 shows that an impulsive waveform produces a constant Fourier transform.

EXERCISE 16-14

Given that $\mathscr{F}\{e^{-\alpha|t|}\} = 2\alpha/(\alpha^2 + \omega^2)$, use the symmetry and linearity properties to find the Fourier transform of $f(t) = 1/\left[1 + (t/\alpha)^2\right]$.

ANSWER:

$F(\omega) = \pi\alpha e^{-\alpha|\omega|}$

Time and Frequency Shifting

Two important Fourier transform properties that are easily derived from the defining integrals are

$$\text{If } \mathscr{F}\{f(t)\} = F(\omega) \text{ then } \mathscr{F}\{f(t-T)\} = e^{-j\omega T}F(\omega)$$

$$\text{If } \mathscr{F}^{-1}\{F(\omega)\} = f(t) \text{ then } \mathscr{F}^{-1}\{F(\omega-\beta)\} = e^{j\beta t}f(t) \qquad (16\text{-}49)$$

The first statement says that shifting the time-domain origin by T results in multiplying the transform by $e^{-j\omega T}$. The *time-domain shifting property* means that delaying a waveform ($T > 0$) does not change the magnitude of its spectrum since $|e^{-j\omega T}| = 1$. In other words, time delay affects only the phase and not the spectral magnitude of a signal.

The second statement in Eq. (16-49) says that shifting the frequency-domain origin by β corresponds to multiplying the waveform by $e^{j\beta t}$. The *frequency-domain shifting property* is used to develop a signal-processing function called modulation, which shifts the spectral content of a signal from one frequency range to another.

Scaling

The scaling property of the Laplace transformation was discussed in Section 9-7. The integral definition of the Fourier transform leads to a similar reciprocal relationship between changing the scale of the time axis and the resulting scale change of the frequency axis. The *scaling property* of the Fourier transformation is

$$\text{If } \mathscr{F}[f(t)] = F(\omega) \text{ then } \mathscr{F}[f(at)] = \frac{1}{|a|}F(\omega/a) \qquad (16\text{-}50)$$

where a is a real constant. This statement is sometimes called the reciprocal spreading property because it means that compressing the time scale ($a < 1$) expands the frequency scale, and vice versa. Reciprocal spreading means that shortening (lengthening) the time-domain duration of a waveform increases (decreases) the frequency-domain bandwidth required to contain its spectrum.

Summary Tables

Table 16-1 lists a basic set of Fourier transform pairs. Additional pairs are easy to derive using Laplace transform methods or the transform properties summarized in Table 16-2.

16-7 CIRCUIT ANALYSIS USING FOURIER TRANSFORMS

Chapter 10 shows that circuit analysis is simplified by transforming a circuit into the s domain using the Laplace transformation. It should come as no surprise that we can transform circuits using the Fourier transformation. Circuit analysis using Fourier methods closely parallels the Laplace method, since both involve linear, integral transformations.

To see how the Fourier transformation applies to circuit analysis, we must examine how the transformation affects the

TABLE 16-1

Basic Fourier Transform Pairs

Signal	Waveform $f(t)$	Transform $F(\omega)$
Impulse	$\delta(t)$	1
Constant (dc)	1	$2\pi\delta(\omega)$
Step function	$u(t)$	$\frac{1}{j\omega}+\pi\delta(\omega)$
Signum	$sgn(t)$	$\frac{2}{j\omega}$
Causal exponential	$\left[e^{-\alpha t}\right]u(t)$	$\frac{1}{\alpha+j\omega}$
Two-sided exponential	$e^{-\alpha\|t\|}$	$\frac{2\alpha}{\alpha^2+\omega^2}$
Complex exponential	$e^{j\beta t}$	$2\pi\delta(\omega-\beta)$
Cosine (ac)	$\cos\beta t$	$\pi[\delta(\omega-\beta)+\delta(\omega+\beta)]$
Sine (ac)	$\sin\beta t$	$-j\pi[\delta(\omega-\beta)-\delta(\omega+\beta)]$

TABLE 16-2

Basic Fourier Transform Properties

Property	Time Domain	Frequency Domain
Signal	$f(t)$	$F(\omega)$
Linearity	$Af_1(t)+Bf_2(t)$	$AF_1(\omega)+BF_2(\omega)$
Differentiation	$\frac{df(t)}{dt}$	$j\omega F(\omega)$
Integration	$\int_{-\infty}^{t} f(x)dx$	$\frac{F(\omega)}{j\omega}+\pi F(0)\delta(\omega)$
Reversal	$f(-t)$	$F(-\omega)$
Symmetry	$F(t)$	$2\pi f(-\omega)$
Time shift	$f(t-T)$	$e^{-j\omega T}F(\omega)$
Frequency shift	$e^{j\beta t}f(t)$	$F(\omega-\beta)$
Scaling	$\|a\|f(at)$	$F(\omega/a)$

connection and device constraints. The connection constraints are based on KVL and KCL. A typical KVL constraint might be of the form

$$v_1(t) + v_2(t) - v_3(t) = 0$$

Because of the linearity property of the Fourier transformation this KCL equation transforms as follows:

$$V_1(\omega) + V_2(\omega) - V_3(\omega) = 0$$

This example generalizes to any number of KVL or KCL constraints. We conclude that the Fourier transformation changes the waveforms into transforms but leaves the form of the connection constraints unchanged.

The frequency-domain element constraints are found by transforming the *i-v* characteristics of the passive elements. Using the differentiation property of the Fourier transformation, we write the time-domain and frequency-domain element constraints as follows:

	TIME DOMAIN	FREQUENCY DOMAIN
Resistor:	$v(t) = Ri(t)$	$V(\omega) = RI(\omega)$
Inductor:	$v(t) = L\dfrac{di}{dt}$	$V(\omega) = j\omega LI(\omega)$
Capacitor:	$i(t) = C\dfrac{dv}{dt}$	$I(\omega) = j\omega CV(\omega)$

As might be expected, in the frequency domain the element constraints are algebraic equations similar in form to Ohm's law. The proportionality factors relating the voltage and current transforms are the ac impedance of the elements:

$$Z_R = R \qquad Z_L = j\omega L \qquad Z_C = \frac{1}{j\omega C} \tag{16-51}$$

These results have a familiar ring because we have seen them before. The central points are: (1) The Fourier transformation does not change the form of the connection constraints and (2) the transformed element constraints are similar to Ohm's law. From our previous experience we conclude that algebraic circuit analysis techniques can be used to solve for responses in Fourier-transformed circuits.

To transform a circuit, we change waveforms into Fourier transforms and replace passive elements by what amounts to their steady-state ac impedance. We then use algebraic techniques such as voltage division, equivalence, node analysis, or mesh analysis to solve for the unknown current or voltage transforms. The inverse Fourier transform is then used to obtain the response waveform in the time domain.

The discussion above closely parallels the development in Chapter 10, except for one thing—there is no mention of *initial conditions*. The reason for this omission is that the lower limit on the integral defining the direct Fourier transformation is $t = -\infty$. The Fourier transformation takes into account the entire history of a circuit beginning at $t = -\infty$ and ending at $t = +\infty$. For Laplace transformation the lower limit on the defining integral [see Eq. (9-2)] is $t = 0$, and the effect of inputs prior to $t = 0$ become the initial conditions.

In summary, the Fourier circuit analysis method inherently takes into account all inputs to the circuit including those that may have happened for $t < 0$. As a result, Fourier analysis cannot handle circuit analysis problems in which events prior to $t = 0$ are represented by initial conditions at $t = 0$. The Fourier transforms exist for noncausal waveforms such as $sgn(t)$ and the double-sided exponential, so Fourier methods produce response waveforms that are valid from $t = -\infty$ to $t = \infty$.

EXAMPLE 16-11

Use Fourier transforms to find $v_2(t)$ in the transformed circuit in Fig. 16-16 for a unit step function input $v_1(t) = u(t)$.

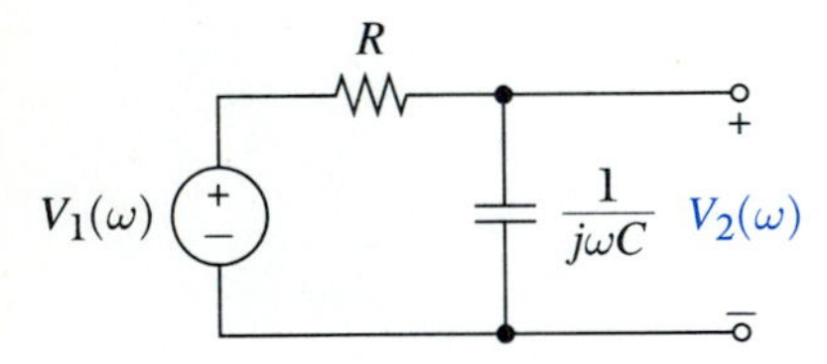

Figure 16-16

SOLUTION:

We first encountered the step response of a first-order RC circuit in Chapter 8 and revisited it again in Chapters 9. From this previous experience we know that the step response is $v_2(t) = [1 - e^{-t/RC}]u(t)$. The purpose of analyzing this circuit again here is to show that the Fourier transform method produces the same result.

The circuit in Fig. 16-16 has been transformed into the frequency domain so we proceed directly to the analysis process. We find the output transform using voltage division:

$$V_2(\omega) = \left[\frac{\dfrac{1}{j\omega C}}{R + \dfrac{1}{j\omega C}}\right] V_1(\omega) = \left[\frac{1/RC}{j\omega + 1/RC}\right] V_1(\omega)$$

For a unit step function input $V_1(\omega) = 1/j\omega + \pi\delta(\omega)$ and so the response transform is

$$V_2(\omega) = \frac{1/RC}{j\omega(j\omega + 1/RC)} + \frac{\pi\delta(\omega)/RC}{j\omega + 1/RC}$$

The first term on the right-hand side can be expanded by partial fractions. The second term on the right reduces to $\pi\delta(\omega)$ because, in general, a function $F(\omega)\delta(\omega) = F(0)\delta(\omega)$, since the impulse is zero everywhere except at $\omega = 0$. The partial fraction expansion of the response yields three terms:

$$V_2(\omega) = \frac{1}{j\omega} - \frac{1}{j\omega + 1/RC} + \pi\delta(\omega)$$

Each term in this expansion is the transform of a basic signal listed in Table 16-1. The first term is the transform of $\frac{1}{2}sgn(t)$, the second is the transform of the causal exponential $[e^{-t/RC}]u(t)$, and the third is the transform of a constant (dc) waveform $\frac{1}{2}$. Because of the uniqueness property the inverse transform $v_2(t)$ is

$$v_2(t) = \frac{1}{2}sgn(t) - \left[e^{-t/RC}\right]u(t) + \frac{1}{2}$$

This result contains two noncausal and one causal waveform. However, a dc constant can be written as $1 = u(t) + u(-t)$, so the two noncausal waveforms reduce to a causal step function:

$$\frac{1}{2}sgn(t) + \frac{1}{2} = \frac{1}{2}[u(t) - u(-t) + u(t) + u(-t)] = u(t)$$

As expected the step response waveform reduces to

$$v_2(t) = \left[1 - e^{-t/RC}\right]u(t)$$

We could have saved ourselves some work if we had observed that the first and third terms in the partial fraction expansion of $V_2(\omega)$ combine to produce the Fourier transform of a unit step function, that is, $\mathscr{F}\{u(t)\} = 1/j\omega + \pi\delta(\omega)$. The second term is the Fourier transform of a causal exponential, so we arrive at the same response more directly.

EXAMPLE 16-12

Find the output waveform of the RC circuit in Fig. 16-16 when the input is $v_1(t) = sgn(t)$.

SOLUTION:

This example uses the circuit in Example 16-11 with a signum input. From Table 16-1 the input transform is $V_1(\omega) = 2/j\omega$. Using the voltage division result from Example 16-11 we write the output transform in the form

$$V_2(\omega) = \left[\frac{1/RC}{j\omega + 1/RC}\right]\frac{2}{j\omega}$$

The output $V_2(\omega)$ can be expanded by partial fractions:

$$V_2(\omega) = \frac{2/RC}{j\omega(j\omega + 1/RC)} = \frac{2}{j\omega} - \frac{2}{j\omega + 1/RC}$$

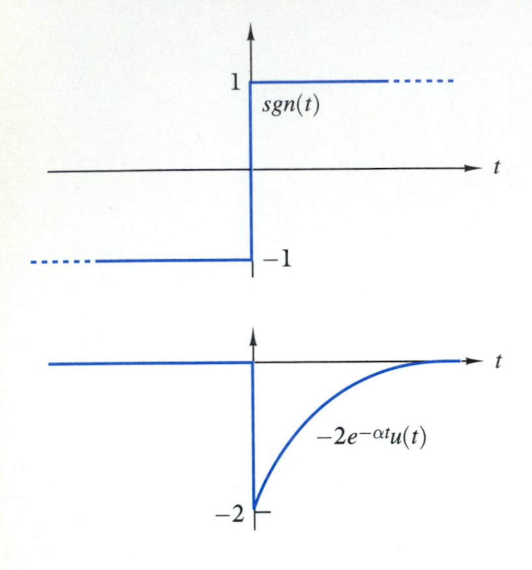

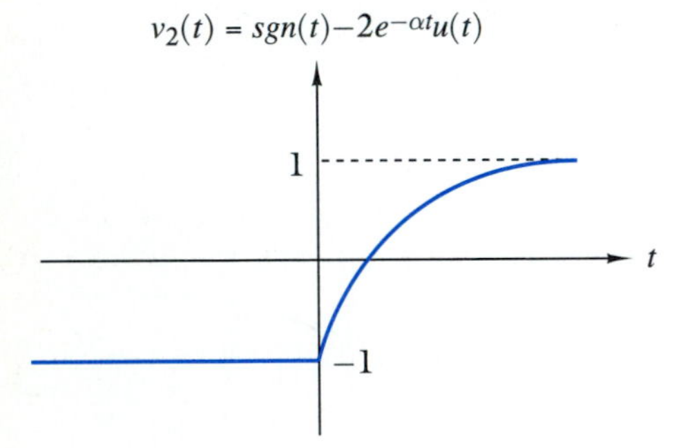

Figure 16-17

Each term in this expansion is listed in Table 16-1. The first term is the transform of $sgn(t)$, and the second is the transform of a causal exponential of amplitude $A = -2$. Performing the inverse transformation produces the response waveform:

$$v_2(t) = sgn(t) - 2\left[e^{-t/RC}\right]u(t)$$

Figure 16-17 shows how the noncausal $sgn(t)$ and the causal exponential combine to produce the response $v_2(t)$. In effect, the signum waveform applies a −1-V input at $t = -\infty$. As a result, the output was driven to a steady-state condition of $v_2(t) = -1$ V for $t < 0$. At $t = 0$ the signum input applies a 2-V input, which drives the output response from −1 V at $t = 0$ to a final condition of 1 V as $t \to \infty$. Both causal and noncausal waveforms are required to show how a circuit responds to the signum input.

EXERCISE 16-15

Find the output waveform of the RC circuit in Fig. 16-16 when the input is $v_1(t) = sgn(t) - 1$.

ANSWER:

$v_2(t) = -2u(-t) + 2[e^{-t/RC}]u(t)$

Convolution and Transfer Functions

Examples 16-11 and 16-12 show how Fourier transforms can be used to solve for circuit time-domain responses. Once we become accustomed to the idea of noncausal waveforms, the process is quite similar to Laplace transform methods. However, the Fourier transformation is not just another way to find time-domain circuit responses due to causal inputs. We already have classical methods (Chapter 8) and Laplace transforms (Chapters 9, 10, and 11) to handle such problems. We do not need another method.

Fourier transforms are interesting because they provide a way to represent idealized signals and signal processors. These ideal elements are often noncausal and cannot be represented by Laplace transforms, nor actually realized as engineering hardware. Nevertheless, ideal models are very useful because they are benchmarks for describing and evaluating the performance of signals and systems.

To illustrate ideal signal-processing models we need the Fourier transform version of the s-domain transfer function. To develop transfer function concept we use the *convolution integral* from Chapter 9:

$$y(t) = \int_0^t h(\lambda)x(t-\lambda)d\lambda \tag{16-52}$$

In the convolution integral $h(t)$ is the system impulse response and $y(t)$ is the output caused by the input $x(t)$. The limits of integration are $\lambda = 0$ to $\lambda = t$ because the derivation in Chapter 9 assumes that $h(t)$ and $x(t)$ are causal waveforms. To overcome this limitation the integration limits are extended backward to $t = -\infty$ and forward to $t = \infty$:

$$y(t) = \int_{-\infty}^{\infty} h(\lambda)x(t-\lambda)d\lambda \tag{16-53}$$

Equation (16-53) is a general form of the convolution integral that is valid for causal and noncausal waveforms.

We now take the Fourier transformation on both sides of Eq. (16-53):

$$\begin{aligned}\mathscr{F}\{y(t)\} = Y(\omega) &= \mathscr{F}\left\{\int_{\lambda=-\infty}^{\infty} h(\lambda)x(t-\lambda)d\lambda\right\} \\ &= \int_{t=-\infty}^{\infty}\left[\int_{\lambda=-\infty}^{\infty} h(\lambda)x(t-\lambda)d\lambda\right]e^{-j\omega t}dt\end{aligned} \tag{16-54}$$

Next we interchange the order of integration and factor out $h(\lambda)$, since it does not depend on t:

$$Y(\omega) = \int_{\lambda=-\infty}^{\infty} h(\lambda)\left[\int_{t=-\infty}^{\infty} x(t-\lambda)e^{-j\omega t}dt\right]d\lambda \tag{16-55}$$

Finally, we change the integration variable within the bracket in Eq. (16-55) to $z = t - \lambda$ to obtain

$$\begin{aligned}Y(\omega) &= \int_{\lambda=-\infty}^{\infty} h(\lambda)\left[\int_{z=-\infty}^{\infty} x(z)e^{-j\omega z}e^{-j\omega\lambda}dz\right]d\lambda \\ &= \int_{\lambda=-\infty}^{\infty} h(\lambda)e^{-j\omega\lambda}d\lambda \int_{z=-\infty}^{\infty} x(z)e^{-j\omega z}dz\end{aligned} \tag{16-56}$$

The two integrals in the last line of Eq. (16-56) both conform to the definition of the direct Fourier transformation in Eq. (16-22). We conclude that

$$Y(\omega) = H(\omega)X(\omega) \tag{16-57}$$

That is, the output transform $Y(\omega)$ equals the Fourier transform of the system impulse $H(\omega)$ times the input transform $X(\omega)$.

Equation (16-57) looks suspiciously like the s-domain relationship $Y(s) = T(s)X(s)$, where $T(s)$ is the s-domain transfer function. Since the two input-output relationships have the same form we call $H(\omega)$ the *Fourier-domain transfer function*. The differences between $T(s)$ and $H(\omega)$ are: (1) $T(s)$ does not exist for

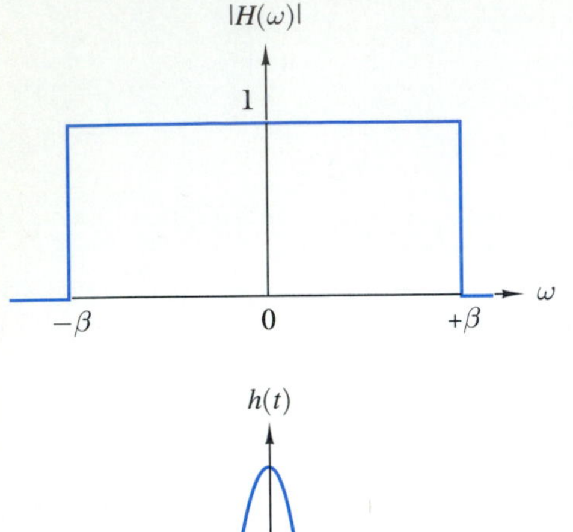

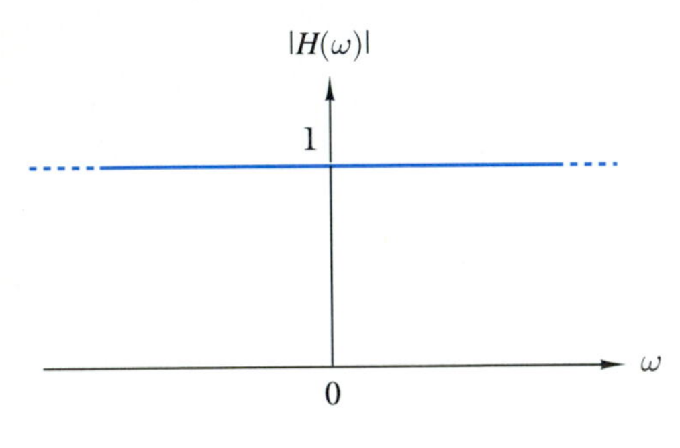

Figure 16-18 Gain and phase response of an ideal low-pass filter.

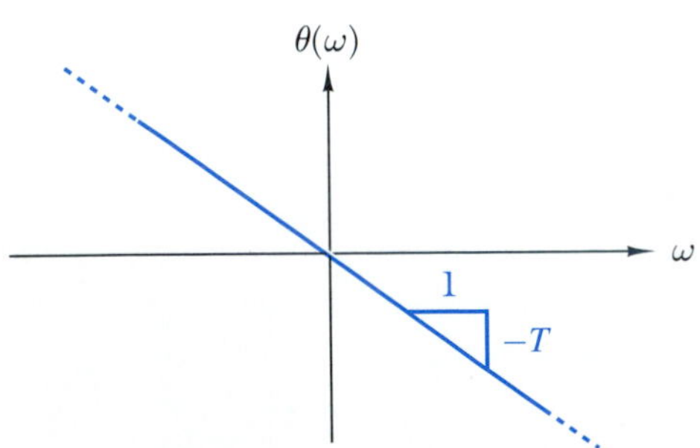

Figure 16-19 Gain and phase response of an ideal delay.

noncausal impulse response waveforms for which $H(\omega)$ exists. (2) $H(\omega)$ does not exist for unstable or unbounded impulse response waveforms for which $T(s)$ exists. Beyond these difference we can say that for a stable causal system $H(\omega) = T(s)]_{\sigma=0}$.

In other words, the transfer function $H(\omega)$ represents the frequency response of a system, including systems that cannot be treated using Laplace transforms. For example, consider the $H(\omega)$ shown in Fig. 16-18. In the filter terminology of Chapter 15 the gain $|H(\omega)|$ is an *ideal low-pass filter* with a passband gain of unity, a stopband gain of zero, and a bandwidth of β. The filter impulse response is the inverse transform of $H(\omega)$. In Example 16-5 we found that the waveform corresponding to a rectangular transform had the form $K(\sin x)/x$. The resulting noncausal $h(t)$ in Fig. 16-18 means that an ideal low-pass filter responds before the impulse is applied at $t = 0$. Such anticipatory behavior is physically impossible, so actually building an ideal filter using physical hardware is an impossibility. Nevertheless, the ideal low-pass filter model is used in conceptual design work because it is easy to understand and can be implemented in the software simulating the proposed design. Moreover, the ideal filter provides a standard for evaluating the performance of real filters.

Next, consider the *ideal delay* in which the output is a delayed replica of the input. The impulse response of such a system is $\delta(t-T)$, which means that an impulse applied at $t = 0$ appears at the output T seconds later. Using the time-shift property of the Fourier transformation, we obtain the transfer function of the ideal delay:

$$H(\omega) = \mathscr{F}\{\delta(t-T)\} = e^{-j\omega T}\mathscr{F}\{\delta(t)\} = e^{-j\omega T} \qquad (16\text{-}58)$$

The gain and phase of the transfer function are

$$|H(\omega)| = |e^{-j\omega T}| = 1$$

and

$$\theta(\omega) = \tan^{-1}[\tan(-\omega T)] = -\omega T$$

The ideal delay frequency response has a gain of unity at all frequencies and a phase that decreases linearly with frequency, as shown in Fig. 16-19. The unity gain and linear phase features of an ideal delay can only be approximated in real systems, yet the ideal model is a useful tool in conceptual design.

The Fourier transformation highlights the relationship between impulse response and frequency response. For example, a modern approach to filter design involves generating an impulse response $h(t)$ whose Fourier transform $H(\omega)$ has desired frequency response characteristics. The filtering action is actually carried out by implementing the convolution of $h(t)$ with an input $x(t)$ in a digital computer. Fourier transforms play a key role in the analysis and design of signals and systems because they are a conceptually simple and computationally efficient way to relate responses in the time and frequency domains.

EXAMPLE 16-13

The impulse response of a system is $h(t) = 1/2\alpha e^{-\alpha|t|}$.

(a) Characterize the system frequency response.

(b) Find the system transient response for $x(t) = sgn(t)$.

SOLUTION:

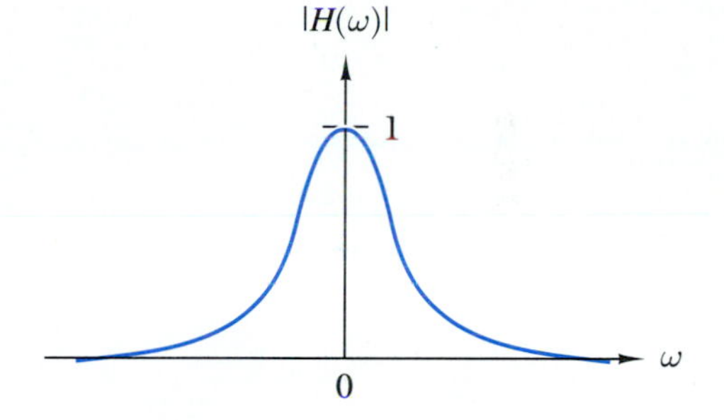

Figure 16-20

(a) The given $h(t)$ is a two-sided exponential whose transform is listed in Table 16-1 so the system transfer function is

$$H(\omega) = \frac{\alpha}{2}\mathscr{F}\{e^{-\alpha|t|}\} = \frac{\alpha^2}{\alpha^2 + \omega^2}$$

Figure 16-20 shows a plot of the magnitude of the transfer function. The system acts like a low-pass filter, since the dc gain $H(0) = 1$ and the high-frequency gain approaches zero. The high-frequency gain rolls off as $1/\omega^2$, or -40 dB/decade. The 3-dB cutoff frequency occurs when $|H(\omega)| = H(0)/\sqrt{2}$, which happens at

$$\omega_C = \alpha\sqrt{\sqrt{2} - 1} = 0.644\alpha$$

(b) From Table 16-1 the Fourier transform of the signum input is $X(\omega) = 2/j\omega$. Therefore, the output transform is

$$Y(\omega) = \frac{\alpha^2}{\alpha^2 + \omega^2}\frac{2}{j\omega} = \frac{2\alpha^2}{j\omega(\alpha + j\omega)(\alpha - j\omega)}$$

The output $Y(\omega)$ can be expanded by partial fractions:

$$Y(\omega) = \frac{k_1}{j\omega} + \frac{k_2}{\alpha + j\omega} + \frac{k_3}{\alpha - j\omega}$$

where the unknown constants in the expansion are obtained using the cover-up algorithm:

$$k_1 = j\omega Y(\omega)|_{j\omega=0} = 2$$

$$k_2 = (\alpha + j\omega)Y(\omega)]_{j\omega=-\alpha} = -1$$

$$k_3 = (\alpha - j\omega)Y(\omega)]_{j\omega=+\alpha} = +1$$

which yields the expanded output transform in the form

$$Y(\omega) = \frac{2}{j\omega} - \frac{1}{\alpha + j\omega} + \frac{1}{\alpha + j(-\omega)}$$

The first term in the expansion is the transform of $sgn(t)$, the second is a causal exponential, and the third is the reversal of a causal exponential. Performing the inverse Fourier transform yields the output waveform:

$$\begin{aligned} y(t) &= sgn(t) - u(t)e^{-\alpha t} + u(-t)e^{-\alpha(-t)} \\ &= (1 - e^{-\alpha t})u(t) - (1 - e^{\alpha t})u(-t) \end{aligned}$$

EXERCISE 16-16

The impulse response of a system is $h(t) = \delta(t) - \alpha e^{-\alpha|t|}$.
(a) Characterize its frequency response.
(b) Find its step response.

ANSWERS:
(a) Bandstop filter with $|H(0)| = |H(\infty)| = 1,\ |H(\alpha)| = 0$
(b) $y(t) = -sgn(t) + 2u(t)e^{-\alpha t} - 2u(-t)e^{-\alpha(-t)}$

16-8 PARSEVAL'S THEOREM AND FILTERING

Parseval's theorem relates the energy carried by a waveform to the spectral content of its transform. The total energy carried by any waveform is defined to be

$$W_T = \int_{-\infty}^{+\infty} p(t)dt$$

where $p(t)$ is the power the waveform delivers to a specified load. For a resistor $p(t) = v^2(t)/R = i^2(t)R$ so the energy delivered to a 1-Ω resistive load can be written in the form

$$W_{1\Omega} = \int_{-\infty}^{+\infty} f^2(t)\, dt \qquad (16\text{-}59)$$

where $f(t)$ can be either a voltage or a current waveform. Equation (16-59) is often used as the definition of the energy carried by a waveform, although an implied 1-Ω resistance is required for the result to have the dimensions of energy.

Parseval's theorem states that the total energy carried by a waveform can be calculated in either the time domain or the frequency domain:

$$W_{1\Omega} = \int_{-\infty}^{+\infty} f^2(t)dt = \frac{1}{2\pi}\int_{-\infty}^{+\infty} |F(\omega)|^2 d\omega \qquad (16\text{-}60)$$

That is, the total 1-Ω energy can be found from either the waveform $f(t)$ or its transform $F(\omega)$. In the time domain, the total energy is the integral of the square of the waveform over all time. In the frequency domain total energy equals $1/2\pi$ times the integral over all frequencies of the squared magnitude of the transform. Because the squared magnitude $|F(\omega)|^2$ is an even function of ω, the integral over all frequencies is twice the integral over the positive frequencies. Thus, the energy also can be calculated in the frequency domain by

$$W_{1\Omega} = \frac{1}{\pi}\int_{0}^{+\infty} |F(\omega)|^2 d\omega \tag{16-61}$$

Parseval's theorem assumes that the integrals in Eqs. (16-60) and (16-61) converge to finite values. Signals for which $W_{1\Omega}$ is finite are called *energy signals*. Examples of energy signals are the causal exponential, the two-sided exponential, the rectangular pulse, and the damped sine. Finite energy is a stronger requirement than the absolutely integrable requirement in Dirichlet conditions. As a result, not all signals that have Fourier transforms are energy signals. For example, some signals that have Fourier transforms but do not have finite energy are the impulse, step, $sgn(t)$, and eternal sinusoid.

The derivation of Parseval's theorem begins with energy in the time domain. Assuming the Fourier transform of $f(t)$ exists, this energy can be written in the form

$$\begin{aligned} W_{1\Omega} &= \int_{-\infty}^{+\infty} [f(t)][f(t)]dt \\ &= \int_{-\infty}^{+\infty} [f(t)]\left[\frac{1}{2\pi}\int_{-\infty}^{+\infty} F(\omega)e^{j\omega t}d\omega\right]dt \end{aligned}$$

The integration within the bracket does not involve time, so $f(t)$ can be moved inside the second integral and the $1/2\pi$ moved outside the first integral to produce

$$W_{1\Omega} = \int_{-\infty}^{+\infty} [f(t)][f(t)]dt = \frac{1}{2\pi}\int_{-\infty}^{\infty}\int_{-\infty}^{+\infty} f(t)F(\omega)e^{j\omega t}d\omega\, dt$$

We now reverse the order of integration and factor out $F(\omega)$:

$$\begin{aligned} W_{1\Omega} &= \int_{-\infty}^{+\infty} [f(t)][f(t)]dt \\ &= \frac{1}{2\pi}\int_{-\infty}^{\infty} F(\omega)\left[\int_{-\infty}^{+\infty} f(t)e^{-j(-\omega)t}dt\right]d\omega \end{aligned} \tag{16-62}$$

The function inside the bracket is $F(-\omega)$. In our study of the even and odd properties of $F(\omega)$, we found that $F(\omega)F(-\omega) = |F(\omega)|^2$. The right side of Eq. (16-62) reduces to the right side of the statement of Parseval's theorem in Eq. (16-60).

EXAMPLE 16-14

(a) Find the 1-Ω energy carried by $v(t) = V_A[e^{-\alpha t}]u(t)$ in the time domain.

(b) Repeat (a) in the frequency domain.

SOLUTION:

(a) The given waveform is causal, so the time integration in Eq. (16-59) extends from $t = 0$ to $t = \infty$.

$$W_{1\Omega} = \int_0^{+\infty} \left(V_A e^{-\alpha t}\right)^2 dt = \frac{V_A^2}{-2\alpha} e^{-\alpha t}\Big|_0^{\infty} = \frac{V_A^2}{2\alpha}$$

This result applies provided $\alpha > 0$. If $\alpha < 0$ the waveform is not an energy signal because it is unbounded.

(b) The Fourier transform of the signal is $V(\omega) = V_A/(j\omega + \alpha)$ provided $\alpha > 0$. In the frequency domain the 1-Ω energy is found to be

$$W_{1\Omega} = \frac{1}{2\pi}\int_{-\infty}^{+\infty} \frac{V_A^2}{\alpha^2 + \omega^2} d\omega = \frac{V_A^2}{2\pi\alpha} \tan^{-1} \frac{\omega}{\alpha}\Big|_{-\infty}^{+\infty}$$

$$= \frac{V_A^2}{2\pi\alpha}\left[\frac{\pi}{2} - \left(-\frac{\pi}{2}\right)\right] = \frac{V_A^2}{2\alpha}$$

which is the same as the result in (a). This result applies only for $\alpha > 0$ because $F(\omega)$ does not exist for $\alpha < 0$.

Applications of Parseval's Theorem

The function $|F(\omega)|^2$ is called the *energy spectral density* because it describes how the total energy carried by $f(t)$ is distributed among the frequencies in its spectrum. In this sense we think of the integral

$$W_{12} = \frac{1}{\pi}\int_{\omega_1}^{\omega_2} |F(\omega)|^2 d\omega$$

as the amount of energy carried by frequencies in the range from ω_1 to ω_2. The energy viewpoint offers another way to describe filtering. Since the output of a filter is $Y(\omega) = H(\omega)X(\omega)$, the energy spectral density of the output signal is

$$|Y(\omega)|^2 = |H(\omega)|^2 |X(\omega)|^2 \tag{16-63}$$

The squared magnitude of the filter transfer function $|H(\omega)|^2$ multiplies the input spectral density $|X(\omega)|^2$ to produce the output spectral density $|Y(\omega)|^2$. We can calculate the amount of energy in the output signal by combining Eqs. (16-61) and (16-63):

$$W_{1\Omega} = \frac{1}{\pi}\int_0^{+\infty} |H(\omega)|^2 |X(\omega)|^2 d\omega \tag{16-64}$$

We can view filtering as a process that rejects the input energy carried by frequencies in the stopband and transfers energy carried by frequencies in the passband. Note that we can calculate these energies directly using Fourier transforms without finding signal waveforms.

EXAMPLE 16-15

(a) Find the percentage of the total 1-Ω energy in the causal exponential that is carried by frequencies between $\omega = -\alpha$ and $\omega = \alpha$.

(b) Find the bandwidth required to pass 90% of the total 1-Ω energy in the causal exponential.

SOLUTION:

(a) In Example 16-14, the total 1-Ω energy of this signal was found to be $V_A^2/(2\alpha)$. Using Eq. (16-61) the amount of energy carried by frequencies between $-\alpha$ and α is

$$W_\alpha = \frac{1}{\pi}\int_0^{\alpha} \frac{V_A^2}{\alpha^2+\omega^2} d\omega = \frac{V_A^2}{\pi\alpha}\tan^{-1}\frac{\omega}{\alpha}\bigg|_0^{\alpha}$$
$$= \frac{V_A^2}{4\alpha}$$

The ratio of the energy in this band to the total energy is

$$\frac{W_\alpha}{W_{1\Omega}} = \frac{V_A^2/4\alpha}{V_A^2/2\alpha} = \frac{1}{2}$$

That is, 50% of the total energy of the causal exponential is carried by the frequencies between $\omega = -\alpha$ and $\omega = \alpha$.

(b) Let $\omega = \beta$ be the bandwidth required to pass 90% of the total energy. The total energy in this band is

$$W_\beta = \frac{1}{\pi}\int_0^{\beta} \frac{V_A^2}{\alpha^2+\omega^2} d\omega = \frac{V_A^2}{\pi\alpha}\tan^{-1}\frac{\omega}{\alpha}\bigg|_0^{\beta}$$
$$= \frac{V_A^2}{\pi\alpha}\tan^{-1}(\beta/\alpha) = 0.9\frac{V_A^2}{2\alpha}$$

The last line of this equation yields

$$\tan^{-1}(\beta/\alpha) = 0.9\frac{\pi}{2} \quad \text{or} \quad \beta = \alpha\tan(9\pi/20) = 6.31\alpha$$

EXAMPLE 16-16

The input to an ideal, unity-gain low-pass filter is $x(t) = 10[e^{-20t}]u(t)$. Find the percentage of the input 1-Ω energy contained in the output signal when the filter cutoff frequency is $\omega_C = 100$ rad/s.

SOLUTION:

The 1-Ω energy in the input signal is

$$W_{1\Omega,IN} = \int_0^{+\infty} 100e^{-40t}dt = -\frac{100e^{-40t}}{40}\bigg|_0^{\infty} = 2.5 \text{ J}$$

The Fourier transform of the input signal is $10/(20 + j\omega)$. The frequency response of the ideal low-pass filter is $H(\omega) = 1$ for $0 < \omega < 100$ and $H(\omega) = 0$ elsewhere. Using Eq. (16-64) we obtain the 1-Ω energy in the output signal:

$$W_{1\Omega,OUT} = \frac{1}{\pi}\int_0^{100} \frac{100}{400 + \omega^2}d\omega = \frac{100}{20\pi}\tan^{-1}\frac{\omega}{20}\bigg|_0^{100}$$
$$= \frac{100}{20\pi}\tan^{-1}5 = 2.19 \text{ J}$$

The fraction of the input energy in the output signal is 2.19/2.5, or about 87.6%.

EXAMPLE 16-17

The transfer function of a first-order low-pass filter is $H(\omega) = 100/(100 + j\omega)$. The input to the filter is $x(t) = 10[e^{-20t}]u(t)$.

(a) Find the percentage of the input energy in the output signal.

(b) Find the percentage of the output energy that lies within the filter passband.

SOLUTION:

(a) The input signal is the same as in Example 16-16, where we found $W_{1\Omega,IN} = 2.5$ J. Using Eq. (16-64) the energy in the output signal is

$$W_{1\Omega,OUT} = \frac{1}{\pi}\int_0^\infty |H(\omega)|^2|X(\omega)|^2 d\omega$$

$$= \frac{1}{\pi}\int_0^\infty \frac{100^2}{100^2+\omega^2}\frac{10^2}{20^2+\omega^2}d\omega$$

The numerical integrand can be expanded by partial fractions:

$$\frac{10^6}{(100^2+\omega^2)(20^2+\omega^2)} = \frac{k_1}{100^2+\omega^2} + \frac{k_2}{20^2+\omega^2}$$

The constants in this expansion are

$$k_2 = \left.\frac{10^6}{10^4+\omega^2}\right|_{\omega^2=-20^2} = \frac{10^6}{9600}$$

$$k_1 = \left.\frac{10^6}{20^2+\omega^2}\right|_{\omega^2=-10^4} = -k_2$$

The output energy is

$$W_{1\Omega,OUT} = \frac{10^6}{9600\pi}\left[\int_0^\infty \frac{d\omega}{20^2+\omega^2} - \int_0^\infty \frac{d\omega}{100^2+\omega^2}\right]$$

$$= \frac{10^6}{9600\pi}\left[\frac{\pi}{40} - \frac{\pi}{200}\right] = 2.08 \text{ J}$$

The fraction of the input energy in the output signal is 2.08/2.5, or about 83.2%.

(b) The cutoff frequency of the low-pass filter is 100 rad/s. The output energy in the passband is

$$W_{1\Omega,\text{OUT}} = \frac{10^6}{9600\pi}\left[\int_0^{100} \frac{d\omega}{20^2+\omega^2} - \int_0^{100} \frac{d\omega}{100^2+\omega^2}\right]$$

$$= \frac{10^6}{9600\pi}\left[\frac{\tan^{-1} 100/20}{20} - \frac{\tan^{-1} 100/100}{100}\right]$$

$$= 2.02 \text{ J}$$

The fraction of the output energy in the filter passband is 2.02/2.08, or about 97.1%.

EXERCISE 16-17

(a) Find the total 1-Ω energy carried by $v(t) = V_A e^{-\alpha|t|}$.

(b) Find the fraction of the energy found in (a) that is carried by frequencies in the range from $\omega = -\alpha$ to $\omega = \alpha$.

ANSWERS:

(a) V_A^2/α

(b) $0.5 + 1/\pi$

EXERCISE 16-18

Repeat Exercise 16-17 for $v(t) = V_A \alpha t e^{-\alpha t} u(t)$.

ANSWERS:

(a) $V_A^2/4\alpha$

(b) $0.5 + 1/\pi$

EXERCISE 16-19

The current in a 5-kΩ resistor is $i(t) = 12[e^{-200t}]u(t)$ mA. Find the total energy delivered to the resistor.

ANSWER:

1.8 mJ

EXERCISE 16-20

The input $x(t) = 5[e^{-10t}]u(t)$ drives a unity-gain high-pass filter with $\omega_C = 1$ krad/s. Find the percentage of the input energy that appears in the filter passband.

ANSWER:

0.636%

SUMMARY

- The Fourier series resolves a periodic waveform into a dc component and an ac component containing an infinite series of harmonic sinusoids. Even waveform symmetry eliminates the sine terms from the ac component. Odd waveform symmetry eliminates cosine terms and half-wave symmetry eliminates the odd harmonics.
- The exponential Fourier series describes the spectrum of a periodic waveform in terms of the amplitude and phase angle of ac components at positive and negative harmonic frequencies.
- The direct Fourier transformation converts an aperiodic waveform $f(t)$ into a transform $F(\omega)$ whose magnitude and angle describes the spectrum of the signal. The inverse transform converts a transform into a waveform.
- Sufficient conditions for the existence of $F(\omega)$ are that $f(t)$ be absolutely integrable and have a finite number of discontinuities in any finite interval. Signals not meeting these conditions may still have a Fourier transform.
- If $f(t)$ is causal and absolutely integrable, then its Fourier transform can be obtained from its Laplace transform $F(s)$ by replacing s by $j\omega$. A causal waveform $f(t)$ is absolutely integrable if all of the poles of $F(s)$ lie in the left half of the s plane.

- The basic Fourier transform properties are uniqueness, linearity, integration, differentiation, and reversal. Other useful properties are symmetry, scaling, time shifting, and frequency shifting.
- Finding circuit responses using Fourier transforms involves transforming the circuit into the frequency domain, solving for the unknown response transforms, and performing the inverse transform to obtain the response waveform.
- Fourier transforms can be used to describe signal and filter models that are useful in system design and evaluation. Parseval's theorem relates the energy carried by a waveform $f(t)$ to the energy spectral density of $|F(\omega)|^2$.

EN ROUTE OBJECTIVES AND ASSOCIATED PROBLEMS

ERO 16-1 The Fourier Series (Sect. 16-1)

(a) Given an equation or graph of a periodic waveform, find the Fourier coefficients.

(b) Construct a periodic waveform whose Fourier coefficients have prescribed characteristics.

16-1 The triangular wave in Fig. 16-1 has an amplitude of 5 V and a period of 1 ms. Write an expression for the first five nonzero terms in the Fourier series and plot the amplitude spectrum of the signal.

16-2 The sawtooth wave in Fig. 16-1 has an amplitude of 25 kV and a fundamental frequency of 13.2 kHz. Write an expression for the first five nonzero terms in the Fourier series and plot the amplitude spectrum of the signal.

16-3 A half-wave sine in Fig. 16-4 has an amplitude of 250 V and a fundamental frequency of 60 Hz. Write an expression for the first five nonzero terms in the Fourier series and plot the amplitude spectrum of the signal.

16-4 A composite waveform is formed by adding a 5-V dc waveform to a 1-kHz square wave with a peak-to-peak amplitude of 5 V. Write an expression for the first five nonzero terms in the Fourier series and plot the amplitude spectrum of the signal.

16-5 The first cycle of a periodic pulse train is described by

$$f(t) = 2u(t) - 2u(t-1) + u(t-2) - u(t-3)$$

(a) Sketch the first cycle of the waveform.

(b) Solve for the Fourier coefficients and plot the amplitude spectrum of the signal.

16-6 A sinusoidal signal $f(t) = A\ \sin(2\pi t/T_0)$ is processed by a limiting circuit that clips off the negative portion of the waveform below $-A/2$. Solve for the Fourier coefficients and plot the amplitude spectrum of the processed signal.

16-7 The first half cycle of a periodic waveform is described by $f(t) = A(1 - 2t/T_0)$ for $0 \le t \le T_0/2$.

(a) Construct a waveform for the second half cycle so that the Fourier series for the overall periodic signal has $a_n = 0$ for all n.

(b) Repeat part (a) for $b_n = 0$ for all n.

16-8 The first quarter cycle of a periodic waveform is $f(t) = A[u(t) - u(t - T_0/4)]$ for $0 \le t \le T_0/4$.

(a) Construct a waveform for the rest of the period so that the Fourier series of the overall periodic signal has $a_n = 0$ for all n.

(b) Repeat part (a) for $b_n = 0$ for all n.

(c) Repeat part (a) for quarter-wave symmetry.

16-9 By direct application of Eqs. (16-3) show that the Fourier coefficients of the periodic impulse train

$$f(t) = \sum_{n=-\infty}^{\infty} \delta(t - nT_0)$$

are $a_0 = 1/T_0$, $a_n = 2/T_0$, and $b_n = 0$.

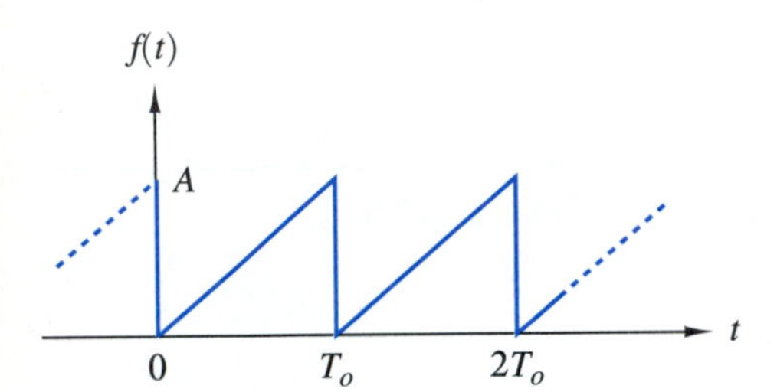

Figure P16-10

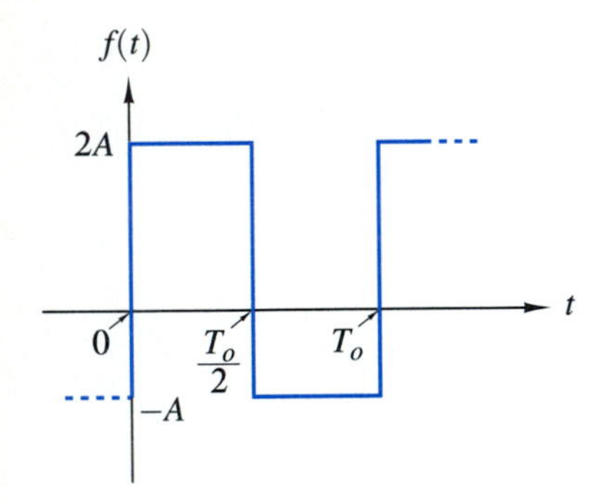

Figure P16-11

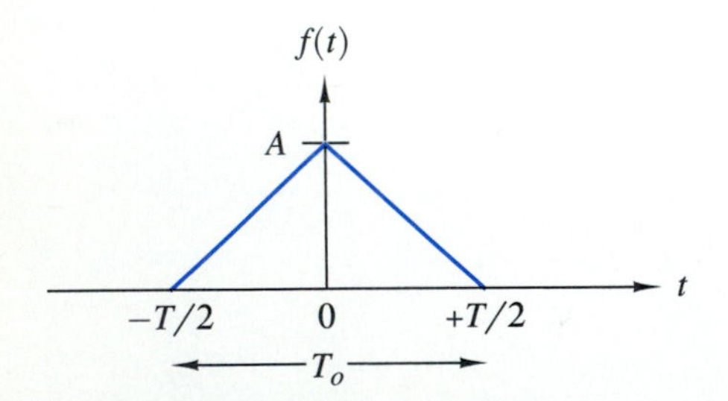

Figure P16-12

ERO 16-2 The Exponential Fourier Series (Sect. 16-2)

Given an equation or graph or the Fourier coefficients of a periodic waveform, derive expressions for the coefficients of the exponential Fourier series and plot spectrum of the signal.

16-10 Derive an expression for c_n of the exponential Fourier series representing the periodic waveform in Fig. P16-10. Sketch the amplitude and phase line spectra.

16-11 Repeat Prob. 16-10 for the periodic waveform in Fig. P16-11.

16-12 Repeat Prob. 16-10 for the periodic waveform in Fig. P16-12.

16-13 Derive an expression for c_n of the exponential Fourier series representing the full-wave rectified sine defined as $f(t) = |A\sin(2\pi t/T_0)|$. Sketch the amplitude and phase spectra.

16-14 The coefficients of a trigonometric Fourier series are

$$a_0 = 0 \qquad a_n = 0 \qquad b_n = \frac{8A}{(n\pi)^2}\sin\frac{n\pi}{2} \qquad n = 1,\ 2,\ 3,\ \ldots$$

Derive an expression for the coefficients of the exponential Fourier series and sketch the amplitude and phase spectra.

16-15 The coefficients of a trigonometric Fourier series are

$$a_0 = 0 \qquad a_n = \frac{4A}{(n\pi)^2}\sin\left(\frac{n\pi}{2}\right) \qquad b_n = -\frac{2A}{\pi n}\sin\left(\frac{n\pi}{2}\right)$$

$$n = 1,\ 2,\ 3,\ \ldots$$

Derive an expression for the coefficients of the exponential Fourier series and sketch the amplitude and phase spectra.

16-16 The signal $f(t)$ has a periodic waveform with a fundamental frequency ω_0 and exponential Fourier series coefficients c_n, $n = 0, \pm1, \pm2, \ldots$. Derive an expression for the Fourier coefficients of the signal $g(t)$ in terms of c_n when

(a) $g(t) = f(t - T)$ where T is a real positive constant

(b) $g(t) = \dfrac{df(t)}{dt}$

(c) $g(t) = \displaystyle\int_{-\infty}^{t} f(x)dx$ (assume $c_0 = 0$)

(d) $g(T) = f(-t)$

ERO 16-3 Fourier Transforms (Sects. 16-3, 16-4, 16-5, 16-6)

(a) Given a signal waveform (transform), find its transform (waveform) using the integral definitions of the Fourier transformation or basic transform properties and pairs.

(b) Derive properties of Fourier transforms using the integral definitions or basic transform properties.

16-17 Use the integral definition to find the Fourier transform of the sawtooth pulse defined by

$$f(t) = \frac{At}{T}[u(t) - u(t - T)]$$

16-18 Use the integral definition to find the Fourier transform of the cosine pulse defined by

$$f(t) = A\cos(2\pi t/T)[u(t + T/4) - u(t - T/4)]$$

16-19 Use the integral definition to find the Fourier transform of the raised cosine defined by

$$f(t) = \frac{A}{2}[1 + \cos(2\pi t/T)][u(t + T/2) - u(t - T/2)]$$

16-20 Use the integral definition to find the Fourier transform of the triangular pulse defined by

$$f(t) = A[1 - 2|t|/T][u(t + T/2) - u(t - T/2)]$$

16-21 Use the integral definition to find the inverse Fourier transform of

$$F(\omega) = \frac{\pi}{\alpha} e^{-\alpha|\omega|}$$

where α is real and positive.

16-22 Use the integral definition to find the inverse Fourier transform of

$$F(\omega) = \frac{\pi A}{\beta}[u(\omega + \beta) - u(\omega - \beta)]$$

where A and β are real and positive.

16-23 Given that $\mathscr{F}\{\delta(t)\} = 1$, use the reversal and time derivative properties of Fourier transforms to show that $\mathscr{F}\{sgn(t)\} = 2/j\omega$.

16-24 Given that $\mathscr{F}\{sgn(t)\} = 2/j\omega$, use the time derivative property of Fourier transforms to show that $\mathscr{F}\{|t|\} = -2/\omega^2$.

16-25 Given that $\mathscr{F}\{Au(t)e^{-\alpha t}\} = A/(\alpha + j\omega)$, use the time integration property to find $\mathscr{F}\{(1 - e^{-\alpha t})u(t)\}$.

16-26 Given that $\mathscr{F}\{Au(t)e^{-\alpha t}\} = A/(\alpha + j\omega)$, use the linearity and reversal properties to find $\mathscr{F}\{A\ sgn(t)e^{-\alpha|t|}\}$. Then let $\alpha \to 0$ to verify that your answer reduces to $\mathscr{F}\{A\ sgn(t)\} = 2A/j\omega$.

16-27 Given that $\mathscr{F}\{u(t)\} = 1/j\omega + \pi\delta(\omega)$:

(a) Use the frequency shifting property to show that

$$\mathscr{F}\{e^{j\beta t}u(t)\} = \frac{1}{j(\omega - \beta)} + \pi\delta(\omega - \beta)$$

(b) Use the result in part (a) and linearity to show that

$$\mathscr{F}\{[\cos\beta t]u(t)\} = \frac{j\omega}{\beta^2 - \omega^2} + \frac{\pi}{2}[\delta(\omega - \beta) + \delta(\omega + \beta)]$$

(c) Since $\mathscr{L}\{\cos\beta t\ u(t)\} = s/(s^2 + \beta^2)$, explain why replacing s by $j\omega$ in the Laplace transform does not yield the result in (b).

16-28 Given $\mathscr{L}\{e^{-\alpha t}\sin\beta t\ u(t)\} = s/[(s + \alpha)^2 + \beta^2]$, use the reversal property of Fourier transforms to find $\mathscr{F}\{e^{-\alpha|t|}\sin\beta t\}$.

16-29 Given the function $f(t) = \delta(t) - \delta(t - T)$, use the time shifting property of Fourier transforms to show that $|F(\omega)| = 0$ at $\omega = 2n\pi/T$, $n = 1, 2, 3, \ldots$.

16-30 Given the function $F(\omega) = 2\pi\delta(\omega) + \pi[\delta(\omega + \beta) - \delta(\omega - \beta)]$, use the frequency shifting property of Fourier transforms to show that $f(t) = 0$ at $t = 2n\pi/\beta$, $n = 1, 2, 3, \ldots$.

16-31 Use the integral definition of the Fourier transform to derive the frequency derivative property.

$$\text{If } \mathscr{F}\{f(t)\} = F(\omega) \text{ then } \mathscr{F}^{-1}\left\{\frac{dF(\omega)}{d\omega}\right\} = -jtf(t)$$

16-32 Use the frequency derivative property in Prob. 16-31 to show that

$$\mathscr{F}\{te^{-\alpha t}u(t)\} = \frac{1}{(j\omega + \alpha)^2}$$

16-33 Use the integral definition of the inverse Fourier transform to show that an upper bound on the peak amplitude of the waveform $f(t)$ is

$$|f(t)| \leq \frac{1}{\pi}\int_0^{\infty} |F(\omega)| d\omega$$

Evaluate the tightness of the bound using $f(t) = Ae^{-\alpha|t|}$.

16-34 Use the integral definition of the direct Fourier transform to show that an upper bound on the magnitude of the transform $F(\omega)$ is

$$|F(\omega)| \leq \int_{-\infty}^{\infty} |f(t)|\, dt$$

Evaluate the tightness of the bound using $f(t) = Ae^{-\alpha t}u(t)$.

16-35 Use the integral definition of the direct Fourier transformation to derive (a) the time shifting property and (b) the frequency shifting property.

(a) If $\mathscr{F}\{f(t)\} = F(\omega)$ then $\mathscr{F}\{f(t-T)\} = e^{-j\omega T}F(\omega)$

(b) If $\mathscr{F}^{-1}\{F(\omega)\} = f(t)$ then $\mathscr{F}^{-1}\{F(\omega - \beta)\} = e^{j\beta t}f(t)$

16-36 Use the integral definitions of the direct Fourier transformation to derive the time-domain sampling property of the impulse function.

$$\text{If } \mathscr{F}\{f(t)\} = F(\omega) \text{ then } \mathscr{F}\{f(t)\delta(t-T)\} = f(T)e^{-j\omega T}$$

16-37 Use the integral definitions of the inverse Fourier transformation to derive the frequency-domain sampling property of the impulse function.

$$\text{If } \mathscr{F}^{-1}\{F(\omega)\} = f(t) \text{ then } \mathscr{F}^{-1}\{F(\omega)\delta(\omega-\beta)\} = F(\beta)e^{-j\beta t}$$

16-38 Use partial fraction expansion and the basic transform pairs in Table 16-1 to find the inverse transforms of the following functions.

(a) $F_1(\omega) = \dfrac{1000}{(j\omega + 20)(j\omega + 50)}$

(b) $F_2(\omega) = \dfrac{j\omega}{(j\omega + 20)(j\omega + 50)}$

(c) $F_3(\omega) = \dfrac{1000}{j\omega(j\omega + 20)(j\omega + 50)}$

(d) $F_4(\omega) = \dfrac{-\omega^2}{(j\omega + 20)(j\omega + 50)}$

16-39 Use partial fraction expansion and the basic transform pairs in Table 16-1 to find the inverse transforms of the following functions.

(a) $F_1(\omega) = \dfrac{500j\omega}{(-j\omega + 50)(j\omega + 50)}$

(b) $F_2(\omega) = \dfrac{5j\omega}{(j\omega + 50)(j\omega + 50)}$

(c) $F_3(\omega) = \dfrac{5000}{j\omega(-j\omega + 50)(j\omega + 50)}$

(d) $F_4(\omega) = \dfrac{5000\delta(\omega)}{-\omega^2 + j200\omega + 2500}$

16-40 Use the transform pairs in Table 16-1 and the transform properties in Table 16-2 to find the Fourier transforms of the following waveforms.

(a) $f_1(t) = 2 + 2u(t)$.
(b) $f_2(t) = 2sgn(-t) + 6u(t)$.
(c) $f_3(t) = [2e^{-2t}u(t) + 2sgn(t)]\delta(t + 2)$
(d) $f_4(t) = 2e^{-2(t-2)}u(t - 2) + 2e^{-2(t+2)}u(t + 2)$

16-41 Use the transform pairs in Table 16-1 and the transform properties in Table 16-2 to find the waveforms corresponding to the following Fourier transforms.

(a) $F_1(\omega) = 4\pi\delta(\omega) + e^{-2\omega}$
(b) $F_2(\omega) = 4\pi\delta(\omega - 4)e^{-j2\omega}/j\omega$
(c) $F_3(\omega) = 4\pi\delta(\omega) + 4(j\omega + 1)/[j\omega(2 + j\omega)]$
(d) $F_4(\omega) = 4\pi\delta(\omega) + 4\pi\delta(\omega - 2) + 4\pi\delta(\omega + 2)$

ERO 16-4 Circuit Analysis Using Fourier Transforms (Sect. 16-7)

Given a linear circuit or its impulse response and a Fourier transformable input waveform:

(a) Use the transformed circuit to solve for transfer functions and output transforms.

(b) Use inverse Fourier transforms to find output waveforms.

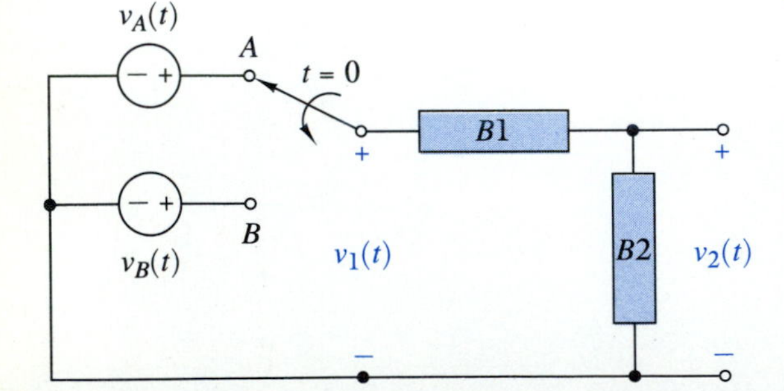

Figure P16-42

The next series of problems use the circuit in Fig. P16-42. In this circuit the switch has been in position A, since $t = -\infty$ and is moved to position B at $t = 0$ where it remains until $t = \infty$.

For each problem:

(a) Write an expression for $v_1(t)$ and find its transform $V_1(\omega)$.

(b) Fourier transform the circuit and solve for $V_2(\omega)$.

(c) Use the inverse transform to solve for $v_2(t)$.

PROB.	$v_A(t)$ IS	$v_B(t)$ IS	BRANCH 1 IS	BRANCH 2 IS
16-42	-10 V	$+10$ V	1-μF capacitor	10-kΩ resistor
16-43	-10 V	$+5$ V	10-kΩ resistor	1-μF capacitor
16-44	$10e^{100t}$ V	$10e^{-100t}$ V	1-H inductor	500-Ω resistor
16-45	$10e^{100t}$ V	$-10e^{-100t}$ V	500-Ω resistor	1-H inductor
16-46	$+10$ V	$10e^{-100t}$ V	1-H inductor	100-Ω resistor

16-47 The circuit in Fig. P16-47 is driven by $v_1(t) = 10\ sgn(t)$. Use Fourier transforms to find $v_2(t)$.

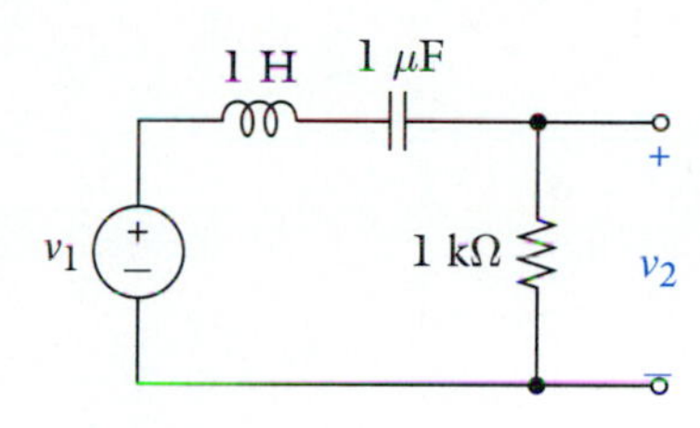

Figure P16-47

16-48 The impulse response of a linear circuit is $h(t) = sgn(t)$. Use Fourier transforms to find the output $y(t)$ for $x(t) = \delta(t) - \alpha e^{-\alpha t}u(t)$.

16-49 The impulse response of a linear circuit is

$$h(t) = -u(t)e^{-\alpha t} + u(-t)e^{-\alpha(-t)}$$

(a) Use the reversal property to find $H(\omega)$ and characterize the circuit frequency response.

(b) Use Fourier transforms to find the output $y(t)$ for $x(t) = 2u(t)$.

(c) Use Fourier transforms to show that $x(t) = sgn(t)$ produces the same output $y(t)$ as in part (b).

16-50 The impulse response of a linear circuit is

$$h(t) = \delta(t) - 2\alpha u(t)e^{-\alpha t}$$

(a) Characterize the frequency response of the circuit.

(b) Use Fourier transforms to find output $y(t)$ for $x(t) = sgn(t)$.

ERO 16-5 Parseval's Theorem and Filtering (Sect. 16-8).

Given a signal waveform or transform: Find the total 1-Ω energy carried by the signal and the percentage of the total energy in specified frequency bands.

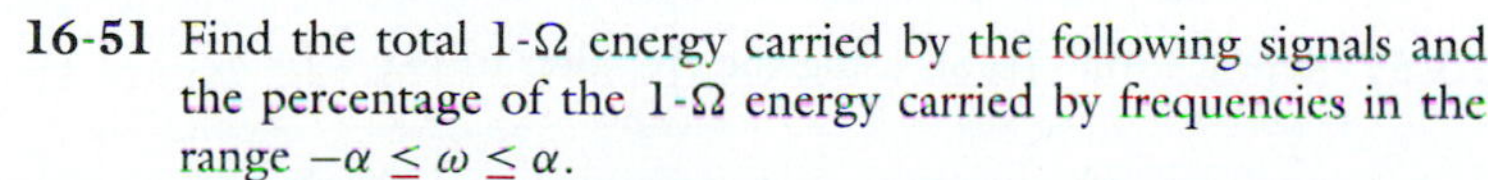

16-51 Find the total 1-Ω energy carried by the following signals and the percentage of the 1-Ω energy carried by frequencies in the range $-\alpha \leq \omega \leq \alpha$.

(a) $v_1(t) = V_A \alpha t e^{-\alpha t}u(t)$.

(b) $v_2(t) = V_A(1 - \alpha t)e^{-\alpha t}u(t)$.

16-52 Find the total 1-Ω energy carried by the following signal transforms and the percentage of the 1-Ω energy carried by frequencies in the range $-\alpha \leq \omega \leq \alpha$.

(a) $V_1(\omega) = \dfrac{2\alpha V_A}{\omega^2 + \alpha^2}$

(b) $V_2(\omega) = \dfrac{j\omega V_A}{\omega^2 + \alpha^2}$

16-53 An ideal unity-gain high-pass filter has cutoff frequencies at $\omega_C = \pm 50$ rad/s.

(a) Determine the percentage of the input 1-Ω energy that lies in the filter passband when the input signal is $x(t) = 10\ e^{-2000|t|}$.

(b) Repeat for $x(t) = 10\ sgn(t)e^{-2000|t|}$.

16-54 An ideal unity-gain low-pass filter has cutoff frequencies at $\omega_C = \pm 2$ krad/s.

(a) Determine the percentage of the input 1-Ω energy that lies in the filter passband when the input signal is $x(t) = 10e^{-50|t|}$.

(b) Repeat for $x(t) = 10e^{-500|t|}$.

16-55 An ideal unity-gain bandpass filter has cutoff frequencies at $\omega_{C1} = \pm 10$ rad/s and $\omega_{C2} = \pm 11$ rad/s and an input signal $x(t) = 10e^{-500t}u(t)$.

(a) Sketch the squared magnitude of the filter frequency response and the input signal energy spectral density on the same frequency scale.

(b) Estimate the 1-Ω energy that lies in the filter passband without formally integrating the output spectral density.

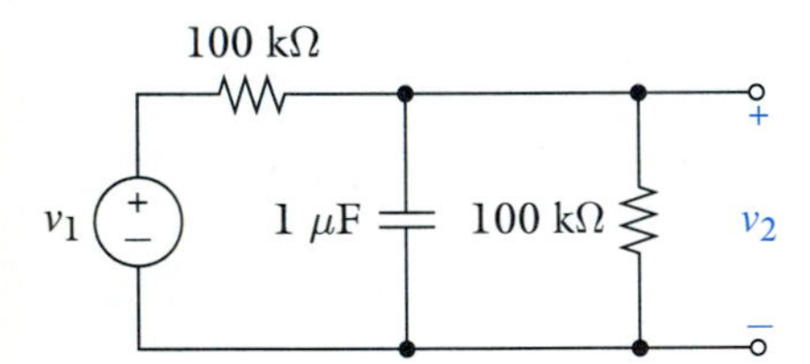

Figure P16-56

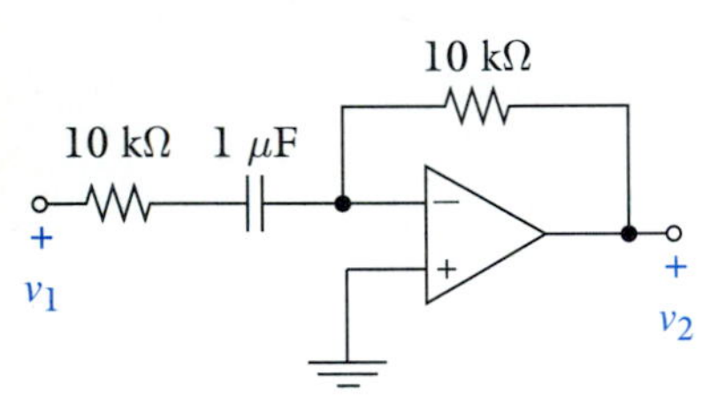

Figure P16-57

16-56 The input to the RC filter circuit in Fig. P16-56 is $v_1(t) = 20e^{-5t}u(t)$ V.

Determine:

(a) the total 1-Ω energy in the input signal,

(b) the percentage of the input 1-Ω energy in the filter output, and

(c) the percentage of the output 1-Ω energy contained in the filter passband.

16-57 Repeat Prob. 16-56 using the RC filter in Fig. P16-57.

16-58 This problem deals with the amount of energy in the output of a filter for inputs $v_1(t)$ and $v_2(t)$.

(a) Sketch the energy spectral density functions for inputs $v_1(t) = 5e^{-20t}u(t)$ and $v_1(t) = 5e^{-20|t|}$.

(b) Sketch the squared magnitude of the filter frequency response of the RC circuit in Fig. P16-56.

(c) Without formally integrating the energy density, determine which input produces the most 1-Ω energy in the output.

16-59 Repeat Prob. 16-58 using the RC filter in Fig. P16-57.

CHAPTER INTEGRATING PROBLEMS

16-60 **(A)** The RC circuit in Fig. P16-57 is driven by $v_1(t) = 10\cos 100t$. Use the Fourier transform method to find $v_2(t)$ on the interval $-\infty < t < \infty$. Check your answer using phasor circuit analysis.

16-61 **(A)** This problem uses Bode plot concepts to estimate the 1-Ω energy carried by the casual exponential $v(t) = V_A e^{-\alpha t} u(t)$ V.

(a) Find the low- and high-frequency asymptotes of the energy density function $|V(\omega)|^2$. Show that these intercepts intersect at $|\omega| = \alpha$. Sketch the asymptotic approximation of $|V(\omega)|^2$.

(b) Use Parseval's theorem and the asymptotic approximation in part (a) to estimate the 1-Ω energy carried by the signal.

(c) Show that the estimate found in (b) overstates the true value of the 1-Ω energy by a multiplicative factor of $4/\pi$.

16-62 **(A)** A filter circuit can be classified as low-pass, bandpass, high-pass, or bandstop depending on the shape of its transfer function magnitude $|T(\omega)|$. Analogously, a signal can be similarly classified depending on the shape of the magnitude of its Fourier transform $|F(\omega)|$. For the following signals:

$$f_1(t) = Ae^{-\alpha|t|} \qquad f_2(t) = A\,sgn(t)e^{-\alpha|t|}$$

$$f_3(t) = A[\delta(t)/\alpha - e^{-\alpha t}]u(t) \qquad f_4(t) = A[\delta(t)/\alpha - e^{-\alpha|t|}]$$

(a) Sketch $|F(\omega)|$ and identify the passbands and stopbands. Classify the signal as low-pass, bandpass, high-pass, or bandstop.

(b) Find the slope of the stopband roll off in dB/decade and the 3-dB cutoff frequencies in terms of α.

16-63 **(A)** The equivalent cutoff frequencies of a low-pass signal $f(t)$ are sometimes defined as

$$\omega_C = \pm\frac{1}{|F(0)|}\int_0^{\infty} |F(\omega)|d\omega$$

(a) Show that the low-pass signals $f_1(t) = Ae^{-\alpha|t|}$ and $f_2(t) = 2A\alpha te^{-\alpha t}u(t)$ have the same equivalent cutoff frequency.

(b) Compare the cutoff frequency found in (a) with the 3-dB cutoff frequencies of the two signals.

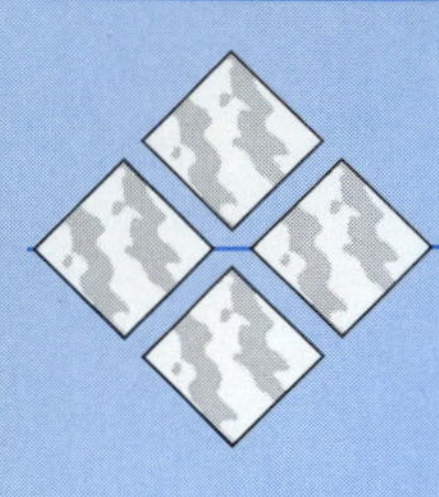

PART III

Applications

Chapters 12 to 16

COURSE INTEGRATING PROBLEMS

Course Objectives

Analysis **(A)** Given the time-domain response of a circuit, find its frequency-domain response, and vice versa.

Design **(D)** Devise a circuit or modify an existing circuit to obtain a specified time-domain or frequency-domain response.

Evaluation **(E)** Given two or more circuits with similar time-or frequency-domain responses, evaluate the circuits using specified criteria.

❖ **I-19** **(D)** A linear circuit is driven by an input $v_1(t) = \cos(2000t)$. The zero-state response due to this input is

$$v_2(t) = -e^{-1000t} + e^{-4000t} + 1.5\sin(2000t)$$

(a) **(A)** Determine the circuit transfer function.
(b) **(A)** Find the phasor output for an input $v_1(t) = \cos(4000t)$ and construct a phasor diagram showing the input and output voltages.
(c) **(A)** Construct the Bode plot of the circuit gain as a function of frequency.
(d) **(D)** Design a circuit that produces these responses.

❖❖ **I-20** **(E)** The step responses of the two second-order bandpass filters are

$$g_1(t) = e^{-5t}\sin(100t)u(t)$$

$$g_2(t) = (e^{-10t} - e^{-1000t})u(t)$$

Which filter best meets the following requirements?
(a) Bandwidth of 10 rad/s centered at 100 rad/s.
(b) Bandwidth of 1000 rad/s centered at 100 rad/s.
(c) Step response settling time less than 500 ms.
(d) Gain less than −5 dB one octave above and below the center frequency.
(e) Gain greater than −5 dB one decade above and below the center frequency.
(f) Phase shift within ±15° one octave above and below the center frequency.
(g) Pass more than 80% of the 1-Ω energy carried by an input signal $V_A e^{-100t}u(t)$

(h) Reject more than 80% of the 1-Ω energy carried by an input signal $V_A e^{-100t} u(t)$.

I-21 Figure PI-21 shows the Bode plots of three first-order filters available to reduce high-frequency noise in a low-speed digital communication channel. The filter performance requirements are:

Steady-state output for 5-V dc input:	From 2 to 7 V dc
Rise time for step-function input:	Less than 1 ms
Stopband gain at 5 kHz:	At least 20 dB below passband gain

(a) (E) Which filter best meets the requirements?

(b) (D) Design a circuit to realize the selected filter response.

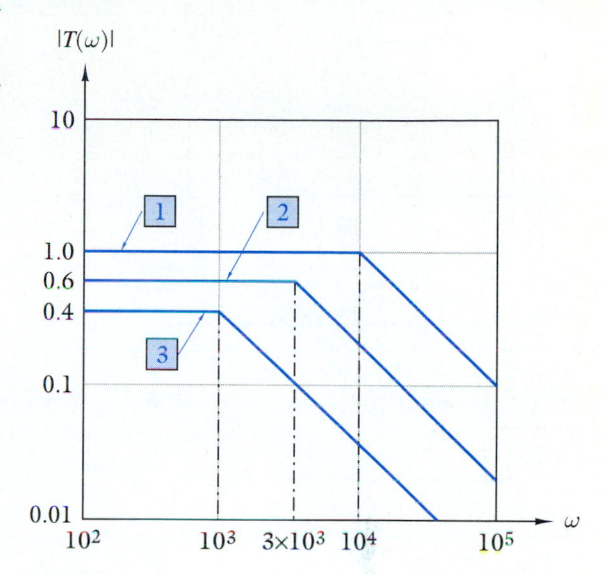

Figure PI-21

I-22 (E) The RonAl Corporation has issued Application Note No. 2 on the USPTS integrated circuit described in Prob. I-16 of Chapter 11. The note claims that interconnecting three USPTS circuits as indicated in Fig. PI-22 produces a third-order Butterworth low-pass filter with $\omega_C = 1/RC$. Verify this claim.

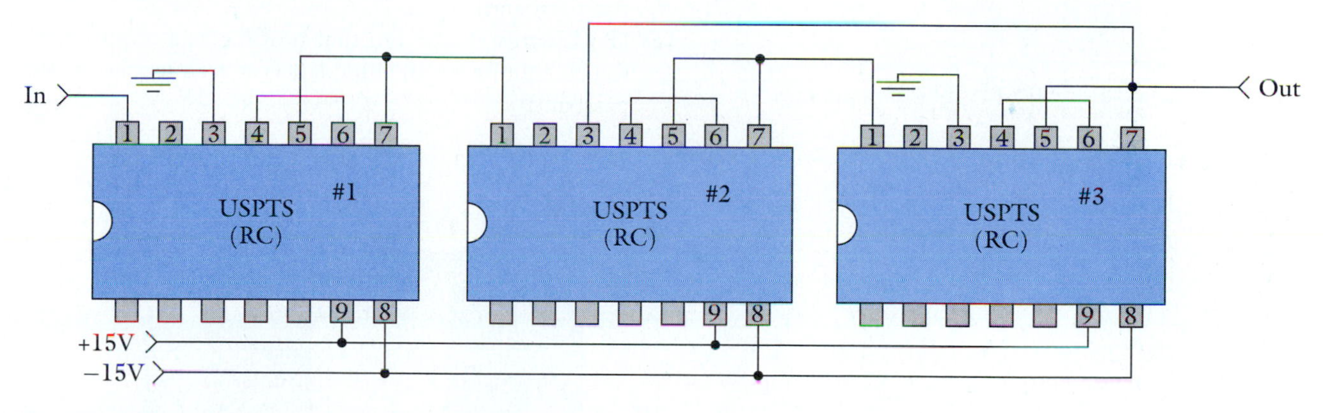

Figure PI-22

I-23 (A) Filter characteristics are sometimes specified in terms of insertion loss, defined as the ratio of average power delivered to a load without the filter inserted over the average power with the filter inserted.

(a) Circuit C1 in Fig. PI-23 shows a sinusoidal source connected directly to a resistive load. Use phasor circuit analysis to derive an expression for the average power delivered to the load in the sinusoidal steady state in terms of V_A, R_O.

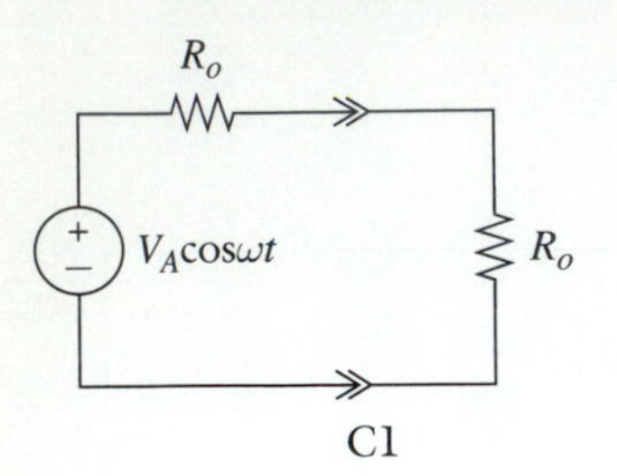

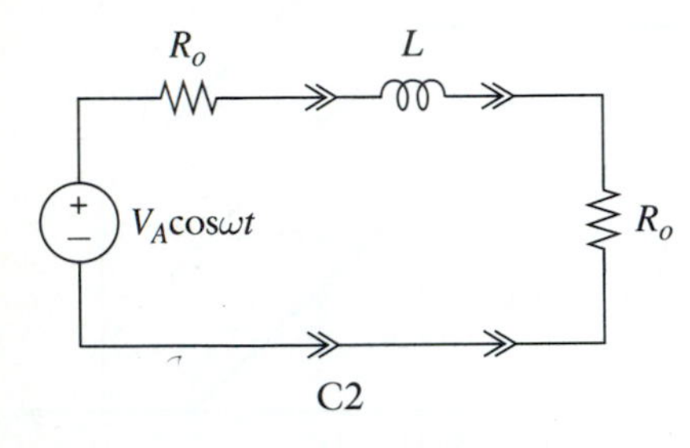

Figure PI-23

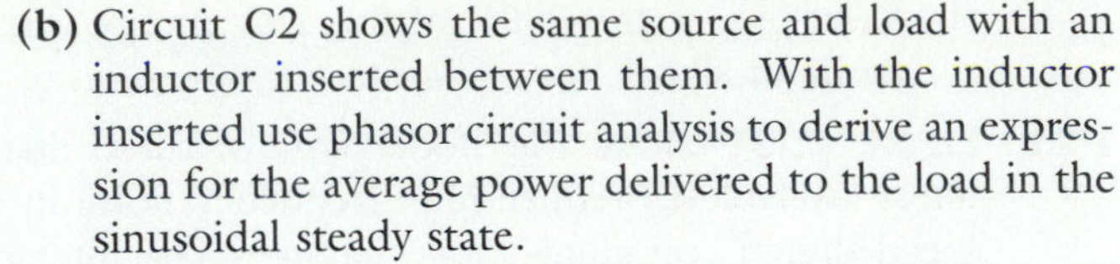

(b) Circuit C2 shows the same source and load with an inductor inserted between them. With the inductor inserted use phasor circuit analysis to derive an expression for the average power delivered to the load in the sinusoidal steady state.

(c) Insertion loss is the ratio of power found in (a) and the power found in (b) above. Derive an expression for the insertion loss of the inductor in terms of R_O, L, and ω.

(d) Show that the insertion loss is unity at zero frequency and explain this result in terms of the inductor impedance. Show that the insertion loss is infinite at infinite frequency and explain this result physically. Find the frequency at which the insertion loss is +3 dB.

(e) Show that replacing the series inductor by a parallel capacitor produces the same insertion loss as the inductor when $L = 4R_OC$.

❖❖ **I-24**

(a) **(D)** Design two different circuits whose gain responses lie entirely within the unshaded region in Fig. PI-24.

(b) **(E)** Identify the design which best meets a requirement for a 50-Ω input impedance at frequencies in the passband.

(c) **(E)** Identify the design that best meets a requirement for an infinite input impedance at frequencies in the passband.

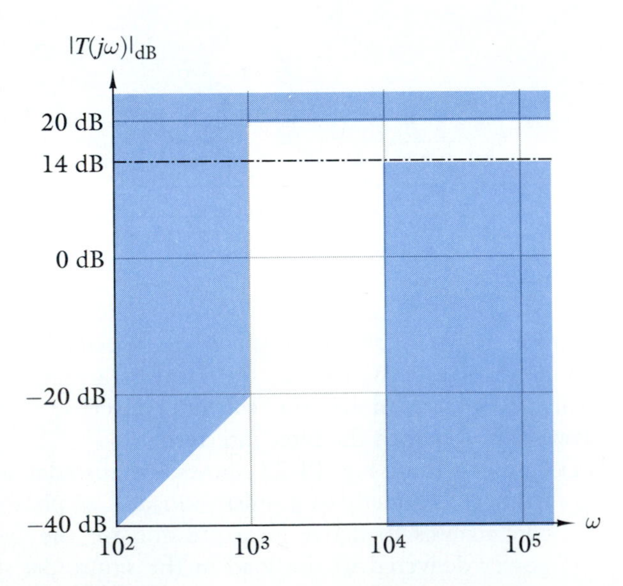

Figure PI-24

❖ **I-25** **(A,D)** During a missile flight test a telemetry system monitors a quantity $x(t)$ and records an output signal $y(t)$. The transfer function relating $x(t)$ and $y(t)$ has been calibrated and shown to have a bandpass characteristic of the form

$$\frac{Y(s)}{X(s)} = \frac{Ks}{s^2 + 2\zeta\omega_0 s + \omega_0^2}$$

where K, ζ, and ω_0 are known parameters. During post-test analysis it is necessary to estimate the quantity $x(t)$ from the recorded output $y(t)$.

(a) **(A)** For the given transfer function show that $x(t)$ can be recovered from $y(t)$ as follows

$$x(t) = \frac{1}{K}\frac{dy(t)}{dt} + \frac{2\zeta\omega_0}{K}y(t) + \frac{\omega_0^2}{K}\int_0^t y(x)dx$$

(b) **(D)** Design an RC OP AMP circuit whose input is $y(t)$ and whose output is $x(t)$.

APPENDIX A: COMPONENTS

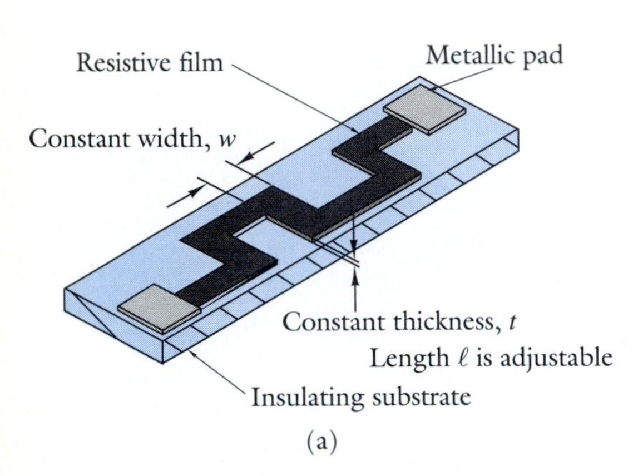

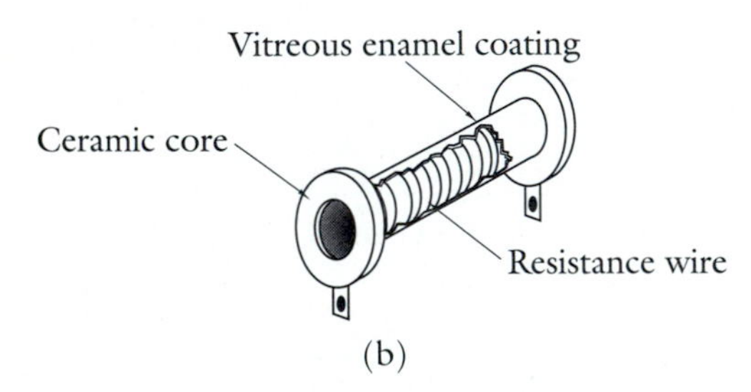

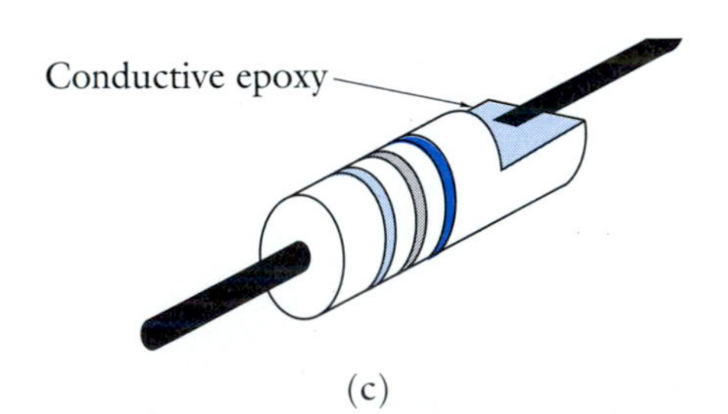

Figure A-1 Resistor types: (a) Thin or thick film. (b) Wirewound. (c) Composition.

This appendix describes some of the physical characteristics and limitations of resistors, OP AMPs, capacitors, and inductors. Circuit analysis generally involves ideal models of these devices. Some knowledge of the limitations of ideal models is necessary to appreciate the restrictions of circuit analysis. Circuit design involves choosing circuit component types and values. Some knowledge of the physical characteristics influencing these choices is necessary to appreciate the scope of the circuit design problem.

This appendix provides some background to those who may not have had hands-on experience with electronic circuits. This appendix discusses only a few key features of electronic devices. It is not intended to be a comprehensive treatment of device characteristics or circuit design considerations.[1]

Resistors

The prototype physical model for a resistor is a homogeneous conductor with a constant cross section. The *resistance* of a uniform bar of length ℓ (m) and cross-sectional area $A(m^2)$ is

$$R = \frac{\rho \ell}{A} \qquad \text{ohms}$$

where ρ is the *resistivity* of the conducting material in Ω-m. Although resistors are realized by several methods all approaches basically adjust device geometry (ℓ and A) or material properties (ρ) or both to obtain a given value of resistance.

Figure A-1 shows the three basic resistor types. A wirewound resistor consists of a length of wire wrapped around an insulating cylinder. The length of the wire or its resistance per unit length are varied to control the amount of resistance. Composition resistors are by far the most commonly used type. These devices are made from epoxy material molded into a cylindrical shape. Impregnating the epoxy with a known amount of highly

[1]For example see, Z. H. Meiksin and P. C. Thackray, *Electronic Design with Off-the-Shelf Integrated Circuits*, 2nd ed., Prentice Hall, Englewood Cliffs, NJ, 1984.

conductive material (usually graphite) changes its resistivity and controls the amount of resistance. Film resistors are made by vacuum depositing a thin layer of metal, or silk screening a thicker layer of conductive paste onto an insulating substrate. The resistivity of the conductive layer is held constant while either its thickness, width, or length is varied to control the resistance. These thin and thick film techniques can be combined with integrated circuits to produce hybrid circuits.

To reduce inventory costs the electrical industry has agreed on standard values and tolerances for commonly used components such as resistors, capacitors, and zener diodes. The standard or preferred values for 5%, 10%, and 20% tolerances are shown in Table A-1. The values listed span a range of one decade (factor of 10). Standard resistances are obtained by multiplying the values in the table by different powers of 10. For example, multiplying the preferred values for ±20% tolerance by 10^4 yields a decade range of standard resistance values of 100 kΩ, 150 kΩ, 220 kΩ, 330 kΩ, 470 kΩ, and 680 kΩ. Standard values for tolerances down to ±0.1% are also defined. The number of standard values grows rapidly as the tolerance is reduced and eventually the proliferation of values defeats the purpose of standardization.

TABLE A-1 STANDARD VALUES FOR RESISTORS AND CAPACITORS

Value	Tolerances	Value	Tolerances	Value	Tolerances
10	±5% ±10% ±20%	22	±5% ±10% ±20%	47	±5% ±10% ±20%
11	±5%	24	±5%	51	±5%
12	±5% ±10%	27	±5% ±10%	56	±5% ±10%
13	±5%	30	±5%	62	±5%
15	±5% ±10% ±20%	33	±5% ±10% ±20%	68	±5% ±10% ±20%
16	±5%	36	±5%	75	±5%
18	±5% ±10%	39	±5% ±10%	82	±5% ±10%
20	±5%	43	±5%	91	±5%

The resistances of commercially available resistors span a range of many decades. Table A-2 lists resistor types with their range, typical tolerances, power ratings, and relative cost. Not all types are available in all possible combinations of resistance, tolerance, and power ratings. A lower tolerance or higher power rating usually means higher initial component costs. The relative costs in the table are only rough guidelines. The total cost of

TABLE A-2 TYPICAL CHARACTERISTICS OF RESISTORS

Type and Range	Tolerances (±%)	Power Ratings(W)	Relative Cost
Composition 1 Ω–20 MΩ	5, 10, 20	1/8, 1/4, 1/2, 1, 2	Low
Carbon Film 1 Ω–20 MΩ	1, 2, 5	1/2–2	Medium
Metal Film 10 Ω–10 MΩ	0.01–1	1/20–1/4	Medium
Wire-Wound 0.1 Ω–200 kΩ	0.1–2	1, 2, 5, 10, 25	High
Diffused on IC 20 Ω–50 kΩ	20	—	Very low as part of IC

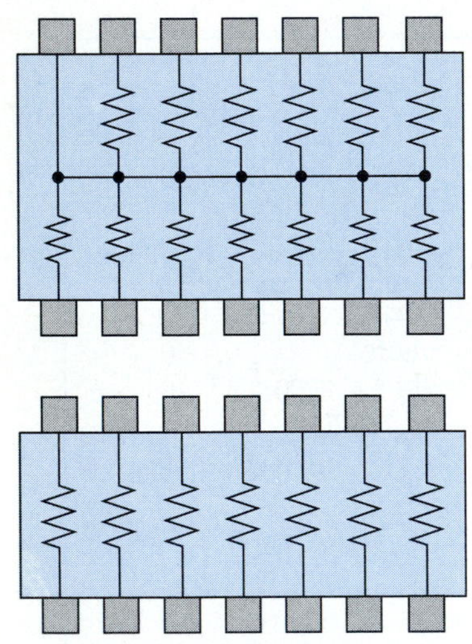

Figure A-2 Examples of resistance array packages.

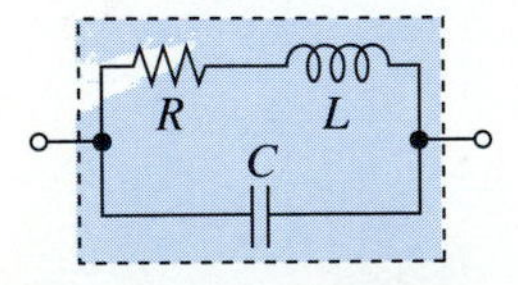

Figure A-3 Circuit model for a resistor.

including a resistor in a system depends on many other factors such as quality control testing, circuit design and manufacturing methods, and spare part requirements. These value-added costs usually greatly exceed the catalog price of the resistor.

The advent of integrated circuits has given rise to packaged resistance arrays. Figure A-2 shows some examples of arrays available in 14-pin dual-in-line packages (DIP). Resistance arrays can be fabricated using either thick or thin film technology. They generally offer tighter tolerances and closer tracking than discrete resistors of comparable price. These packages also are better suited to automated manufacturing and can be less costly than discrete resistors in large production runs.

Resistors emulate their ideal model (Ohm's law) reasonably well as long as the device is operated within nominal ratings, especially power rating. The *equivalent circuit* in Fig. A-3 may be needed to adequately model a resistor in high-frequency applications. The inductance L and capacitance C are parasitic elements that unavoidably accompany the desired resistance R. These parasitic elements cause high-frequency effects that limit the frequency range over which Ohm's law adequately describes the device. These high-frequency effects may cause puzzling malfunctions that are difficult to diagnose by studying the nominal circuit schematic. Except for low values of resistance and wire-wound resistors, the effect of the series inductance is generally negligible compared with the parallel capacitance.

TABLE A-3 CHARACTERISTICS OF POTENTIOMETERS

Type and Range	Tolerances (±%)	Power Ratings(W)	Relative Cost
Composition 50 Ω-5 MΩ	10	2	Low
Metal Film 50 Ω-10 kΩ	2.5	1/2–1	Medium
Wire-Wound 10 Ω–100 kΩ	2.5	1–1000	High

Variable Resistors

A variable resistor, or potentiometer, is a three-terminal device with two terminals connected to the ends of a resistive element, and a third terminal making movable contact with the resistive element at an intermediate point. By connecting the two fixed ends of the potentiometer to a voltage source, an adjustable fraction of the source voltage is available at the movable terminal. Basically a variable resistor functions as an adjustable voltage divider.

Figure A-4 shows the two most common types of potentiometer. The rotary devices are used in applications requiring frequent adjustments and the linear devices (trim "pots" being the most common) for infrequent adjustments. Both kinds can be fabricated using composition, film, or wire-wound techniques. Some typical characteristics of potentiometers are given in Table A-3.

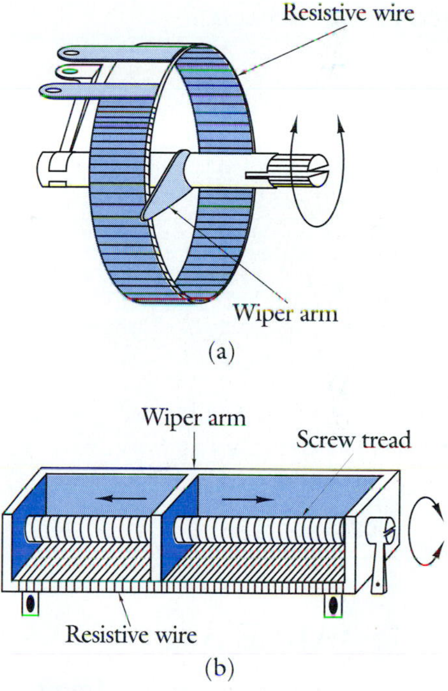

Figure A-4 Variable resistors: (a) Rotary. (b) Linear.

Operational Amplifiers

A typical integrated circuit (IC) operational amplifier (OP AMP) consists of a dozen or so transistors, roughly the same number of resistors, and perhaps one or two capacitors. All of these components are fabricated and interconnected on a tiny piece of silicon, and then packaged as a complete amplifier with only a few key terminals brought out to the external world. The finished package may contain more than one amplifier as illustrated by the 16-pin quad OP AMP package shown in Fig. A-5.

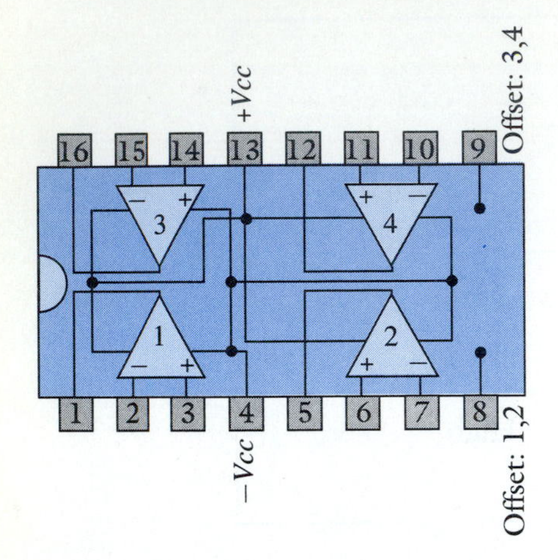

Figure A-5 Quad OP AMP integrated circuit package.

Important OP AMP parameters are: input offset, voltage gain, unity gain bandwidth, slew rate, and maximum output current. Offset refers to the fact that the amplifier output is not always zero when the input is zero. Offset is specified in terms of the input voltage (usually in mV) required to reduce the output voltage to zero. The input offset current (usually in pA) is the input current that exists when the output offset voltage has been reduced to zero. Offset is important in low-level applications where the signal output may be small compared with the no signal output, i.e., the offset. Offset can be adjusted to zero by applying dc signals to the offset control pins. However, offset varies with temperature and from unit to unit.

The amplifier voltage gain is the parameter μ discussed in Chapter 4. It is usually specified in volts per millivolt. Thus, a gain of 50 V/mV means $\mu = 50,000$. The frequency range over which an OP AMP voltage follower has a closed-loop gain of unity is called the unity gain bandwidth (usually in MHz), or sometimes the gain bandwidth product. Slew rate refers to the maximum allowable rate of change of the output voltage (usually in V/μs). If the output rate of change exceeds the slew rate the output becomes a ramp waveform. Slew rate limiting is a nonlinear effect that must be avoided in linear applications. Finally, the maximum output current (usually in mA) is given for a specified load.

Table A-4 lists some representative values of these parameters for different amplifier types. The table is not intended to be definitive, but only illustrative of the characteristics of several classes of OP AMPs. As can be seen from the table, some

TABLE A-4

TYPICAL CHARACTERISTICS OF OP AMPS

Type	Input Offset Voltage (mV)	Input Offset Current (pA)	Voltage Gain (V/mV)	Unity Gain Bandwith (MHz)	Slew Rate (V/μs)	Maximum Output Current (mA)
High power	5	200	75	1	3	500
Wideband	3	200	15	30	30	50
High slew rate	5	100	4	50	400	50
Precision input	0.05	2	500	0.4	0.06	1
General purpose	2	50	100	1	3	10

of these parameters can be optimized for specific applications, but generally only at the expense of one or more of the other characteristics. For example, a high-power OP AMP can deliver a large output current but has relatively large offset parameters. Conversely, a precision input amplifier has low offset and high gain, but at the price of bandwidth, slew rate, and output current. As the name suggests, the general-purpose amplifier strikes a balance between these extremes.

Capacitors

The prototype physical model for a capacitor is two parallel conducting plates separated by an insulating dielectric material. The capacitance of two parallel plates of area A separated a distance d is

$$C = \frac{\epsilon A}{d} \qquad \text{farads}$$

where ϵ is the permittivity of the dielectric material between the plates. Although capacitors are realized by several methods, all approaches adjust device geometry (A and d) or dielectric properties (ϵ) or both to obtain a given value of capacitance.

The disk capacitor in Fig. A-6 consists of two disk-shaped metal plates separated by a dielectric insulator. This geometry closely resembles the parallel plate model. In the final package the disk structure is epoxy coated for mechanical strength and electrical isolation. Disk capacitors are suitable for small values of capacitance, usually in the picofarad range.

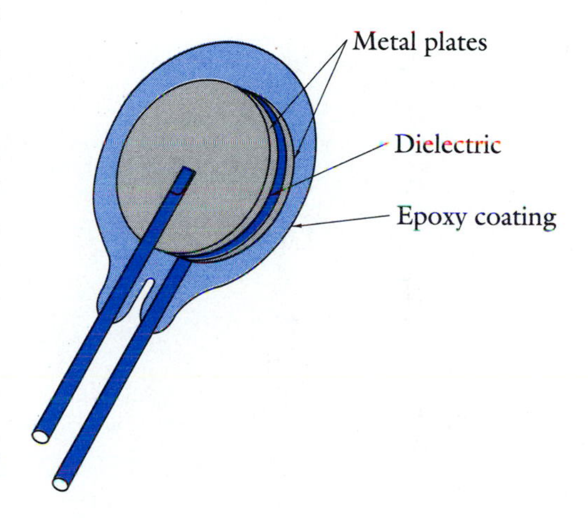

Figure A-6 Disk capacitor.

The tubular capacitor in Fig. A-7 produces capacitances from a few hundred picofarads to a microfarad or so. The device is constructed by rolling up two strips of metal foil that are separated by a dielectric strip such as mylar. The two external leads of the capacitor are connected to the foil strips and the completed roll is packaged in a cylindrical tube (hence the name) to provide mechanical strength and electrical isolation.

Larger capacitances (up to several thousand μF) are made by forming an oxide coating electrochemically on the surface of a metal plate (usually aluminum). These electrolytic devices achieve larger capacitance because the very thin oxide coating serves as the dielectric separating the capacitor plates. The oxide layer is polarized and voltages applied to these devices must agree with reference polarity marks. An example device with polarity marks is shown in Fig. A-8.

Breakdown voltage is a basic limitation of all capacitors. Above the breakdown level, the dielectric becomes conductive and supports an arc that disrupts the capacitive function of the

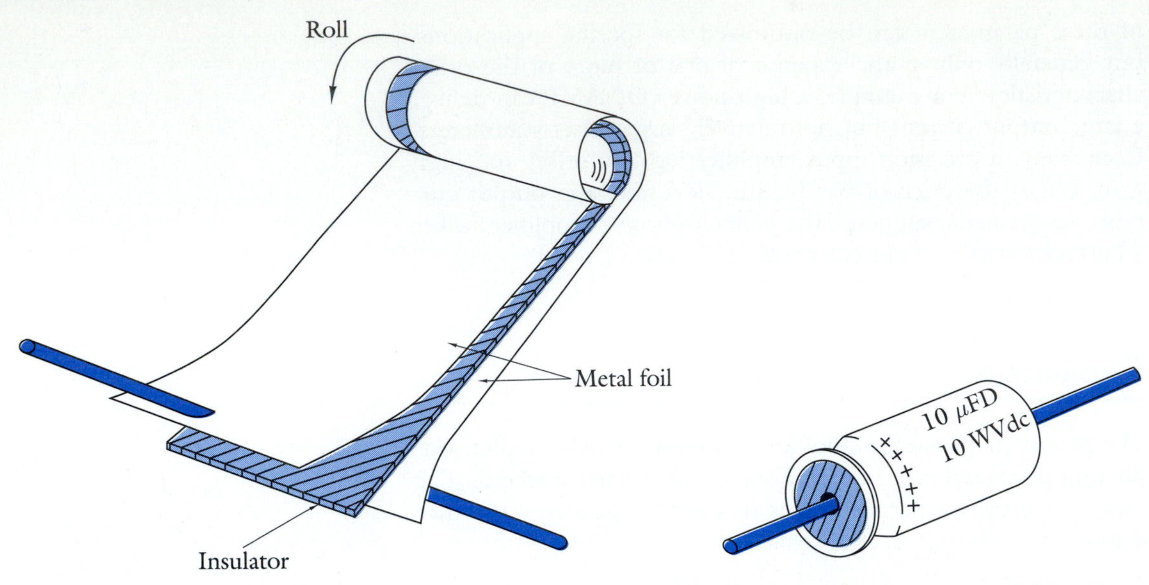

Figure A-7 Tubular capacitor.

Figure A-8 Electrolytic capacitor.

device. For this reason capacitor markings indicate their capacitance and their breakdown voltage, where the latter is usually expressed in dc working volts (DWV). *Electrolytic capacitors* also have polarity marks that must be complied with to avoid improper operation or even damage to the device.

Table A-5 lists the range of capacitances, tolerances, and breakdown voltages for capacitors using different types of dielectric materials. The values in the table give typical ranges for general-purpose capacitors. Capacitors come in standard values that are negative powers of 10 times the preferred values listed in Table A-1. For example, using a multiplicative factor 10^{-7} on values for ±20% tolerance in Table A-1 yields standard capacitances of 1.0 μF, 1.5 μF, 2.2 μF, 3.3 μF, 4.7 μF, and 6.8 μF. Not all capacitor types are available in every possible combination of capacitance, tolerance, and breakdown voltage.

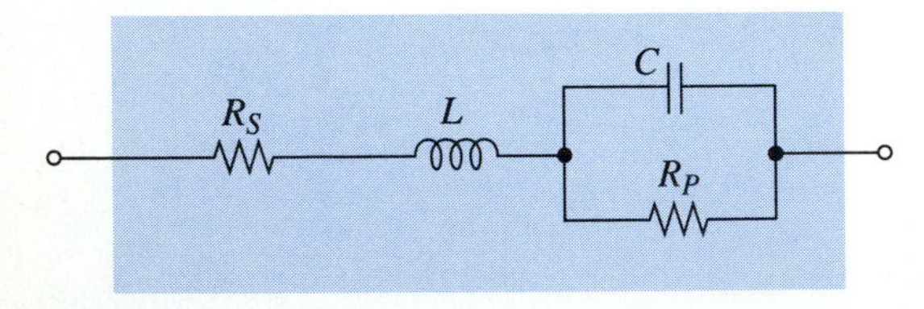

Figure A-9 Circuit model for a capacitor.

Capacitors follow their ideal element model rather closely in most applications as long as they are operated within their voltage ratings. The equivalent circuit in Fig. A-9 may be needed to adequately model a capacitive device. The values of R_P, R_S, and L vary widely between capacitor types. For example, the parallel resistance R_P ranges from as low as 10^7 Ω for some electrolytic capacitors to as high as 10^{12} Ω for high-quality disk capacitors. The parallel resistance models the insulation resistance of the device. The series resistance and inductance limit the frequency range over which the ideal element model adequately describes the device.

TABLE A-5 CHARACTERISTICS OF CAPACITORS

Dielectric	Range	Tolerances (±%)	Rated Voltage (DWV)
Glass	1–10^4 pF	5	100–1250
Mica	1–10^5 pF	1, 2, 5	50–500
Paper	10 pF–10 μF	10	50–400
Plastic	1 pF–1 μF	2, 5, 10	50–600
Ceramic	10–10^6 pF	5, 10,20	50–1600
Electrolytic			
Aluminum oxide	1 μF–2 mF	−10, +5, +30, +50	5–450
Solid tantalum	2 pF–300 μF	5, 10, 20	6–100

Inductors

The physical prototype for an inductor is a coil of wire wrapped around a central core. A long helical coil with N turns around a concentric core of cross section A and length ℓ has an inductance of

$$L = \mu \frac{N^2 A}{\ell}$$

where μ is the magnetic permeability of the core material. The value of inductance, like resistance and capacitance, depends on the device geometry and material properties.

Inductors are found in a wide range of applications that generate a vast assortment of specialized types. For example, there are more than 1500 types and sizes of radio frequency coils alone. Although inductors have many applications, the total demand does not even remotely approach the consumption of resistors and capacitors. Thus, inductors do not lend themselves to industry wide standardization to the same degree as the more frequently used devices. Inductors are not amenable to fabrication by integrated circuit technology except in very special and limited applications. Unlike resistors and capacitors, inductors are often custom designed for an application. For all of these reasons inductors tend to be relatively bulky and expensive, especially in low-frequency applications.

Inductors depart from the ideal element model to a greater degree than resistors or capacitors. The circuit model in Fig. A-10 is often needed to adequately represent an inductor across all frequencies. The series resistance represents an inherent feature of a

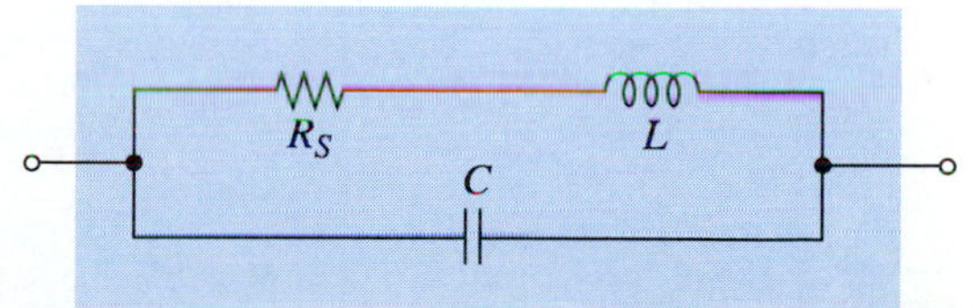

Figure A-10 Circuit model for an inductor.

coil of wire. The parallel capacitance is a lumped representation of the distributed capacitance between individual coil turns. At low frequencies the series resistance is the dominant element in the model, while at high frequencies the capacitance dominates. Thus, at best there is an intermediate frequency range in which inductance is the primary feature of an inductor.

APPENDIX B: SOLUTION OF LINEAR EQUATIONS

The purpose of this appendix is to review the methods of solving systems of linear algebraic equations. Circuit analysis often requires solving linear algebraic equations of the type

$$\begin{aligned} 5x_1 - 2x_2 - 3x_3 &= 4 \\ -5x_1 + 7x_2 - 2x_3 &= -10 \\ -3x_1 - 3x_2 + 8x_3 &= 6 \end{aligned} \tag{B-1}$$

where x_1, x_2, and x_3 are unknown voltages or currents. Often some of the unknowns may be missing from one or more of the equation. For example, the equations

$$\begin{aligned} 5x_1 - 2x_2 &= 5 \\ -4x_1 + 7x_2 &= 0 \\ -3x_2 + 8x_3 &= 0 \end{aligned}$$

involve three unknowns with one variable missing in each equation. Such equations can always be put in the standard square form by inserting the missing unknowns with a coefficient of zero.

$$\begin{aligned} 5x_1 - 2x_2 - 0x_3 &= 5 \\ -4x_1 + 7x_2 - 0x_3 &= 0 \\ 0x_1 - 3x_2 + 8x_3 &= 0 \end{aligned} \tag{B-2}$$

Equations (B-1) and (B-2) will be used to illustrate the different methods of solving linear equations.

Cramer's Rule

Cramer's rule states that the solution of a system of linear equations for any unknown x_k is found as the ratio of two determinants

$$x_k = \frac{\Delta_k}{\Delta} \tag{B-3}$$

where Δ and Δ_k are determinants derived from the given set of equations. A *determinant* is a square array of numbers or symbols called *elements*. The elements are arranged in horizontal rows and vertical columns, and are bordered by two vertical straight lines. In general, a determinant contains n^2 elements arranged in n rows and n columns. The value of the determinant is a function of the value and position of its n^2 elements.

The *system determinant* Δ in Eq. (B-3) is made up of the coefficients of the unknowns in the given system of equations. For example, the system determinant for Eqs. (B-1) is

$$\Delta = \begin{vmatrix} 5 & -2 & -3 \\ -5 & 7 & -2 \\ -3 & -3 & 8 \end{vmatrix}$$

and for Eq. (B-2) is

$$\Delta = \begin{vmatrix} 5 & -2 & 0 \\ -4 & 7 & 0 \\ 0 & -3 & 8 \end{vmatrix}$$

These two equations are examples of the general 3×3 determinant

$$\Delta = \begin{vmatrix} a_{11} & a_{12} & a_{13} \\ a_{21} & a_{22} & a_{23} \\ a_{31} & a_{32} & a_{33} \end{vmatrix} \tag{B-4}$$

where a_{ij} is the element in the ith row and jth column.

The determinant Δ_k in Eq. (B-3) is derived from the system determinant by replacing the kth column by the numbers on the right side of the system of equations. For example, Δ_1 for Eq. (B-1) is

$$\Delta_1 = \begin{vmatrix} 4 & -2 & -3 \\ -10 & 7 & -2 \\ 6 & -3 & 8 \end{vmatrix}$$

and Δ_3 for Eq. (B-2) is

$$\Delta_3 = \begin{vmatrix} 5 & -2 & 5 \\ -4 & 7 & 0 \\ 0 & -3 & 0 \end{vmatrix}$$

These examples of Δ_k's are 3×3 determinants because the system determinants from which they are derived are 3×3.

In summary, using Cramer's rule to solve linear equations boils down to evaluating the determinants formed using the coefficients of the unknowns and the right side of the system of equations.

Evaluating Determinants

The *diagonal rule* gives the value of a 2 × 2 determinant as the difference in the product of the elements on the main diagonal ($a_{11}a_{22}$) and the product of the elements on the off diagonal ($a_{21}a_{12}$). That is, for a 2 × 2 determinant

$$\Delta = \begin{vmatrix} a_{11} & a_{12} \\ a_{21} & a_{22} \end{vmatrix} = a_{11}a_{22} - a_{21}a_{12} \tag{B-5}$$

The value of 3 × 3 and higher-order determinants can be found using the *method of minors*. Every element a_{ij} has a *minor* M_{ij}, which is formed by deleting the row and column containing a_{ij}. For example, the minor M_{21} of the general 3 × 3 determinant in Eq. (B-4) is

$$M_{21} = \begin{vmatrix} a_{12} & a_{13} \\ a_{32} & a_{33} \end{vmatrix} = a_{12}a_{33} - a_{32}a_{13}$$

The *cofactor* C_{ij} of the element a_{ij} is its minor M_{ij} multiplied by $(-1)^{i+j}$:

$$C_{ij} = (-1)^{i+j} M_{ij}$$

The sign of the cofactors alternate along any row or column. The appropriate sign for cofactor C_{ij} is found by starting in position a_{11} and counting plus, minus, plus, minus ... along any combination of rows or columns leading to the position a_{ij}.

To use the method of minors we select one (and only one) row or column. The determinant is the sum of the products of the elements in the selected row or column and their cofactors. For example, selecting the first column in Eq. (B-4) we obtain Δ as follows:

$$\begin{aligned}
\Delta &= a_{11}C_{11} + a_{21}C_{21} + a_{31}C_{31} \\
&= a_{11}(-1)^2 \begin{vmatrix} a_{22} & a_{23} \\ a_{32} & a_{33} \end{vmatrix} + a_{21}(-1)^3 \begin{vmatrix} a_{12} & a_{13} \\ a_{32} & a_{33} \end{vmatrix} \\
&\quad + a_{31}(-1)^4 \begin{vmatrix} a_{12} & a_{13} \\ a_{22} & a_{23} \end{vmatrix} \\
&= a_{11}(a_{22}a_{33} - a_{32}a_{23}) - a_{21}(a_{12}a_{33} - a_{32}a_{13}) \\
&\quad + a_{31}(a_{12}a_{23} - a_{22}a_{13})
\end{aligned}$$

An identical expression for Δ is obtained using any other row or column. For determinants greater than 3 × 3 the minors themselves can be evaluated using this approach. However, systems of equations leading to determinants larger than 3 × 3 are probably better handled by computer-based methods.

EXAMPLE B-1

Solve for the three unknowns in Eqs. (B-1) using Cramer's rule.

SOLUTION:

Expanding the system determinant about the first column yields

$$\Delta = \begin{vmatrix} 5 & -2 & -3 \\ -5 & 7 & -2 \\ -3 & -3 & 8 \end{vmatrix} = 5\begin{vmatrix} 7 & -2 \\ -3 & 8 \end{vmatrix} - (-5)\begin{vmatrix} -2 & -3 \\ -3 & 8 \end{vmatrix}$$

$$+ (-3)\begin{vmatrix} -2 & -3 \\ 7 & -2 \end{vmatrix}$$

$$= 5[7 \times 8 - (-2)(-3)] - (-5)[(-2) \times 8 - (-3)(-3)]$$

$$+ (-3)[(-2)(-2) - (7)(-3)]$$

$$= 250 - 125 - 75 = 50$$

Expanding Δ_1 about the first column yields

$$\Delta_1 = \begin{vmatrix} 4 & -2 & -3 \\ -10 & 7 & -2 \\ 6 & -3 & 8 \end{vmatrix} = 4\begin{vmatrix} 7 & -2 \\ -3 & 8 \end{vmatrix} - (-10)\begin{vmatrix} -2 & -3 \\ -3 & 8 \end{vmatrix}$$

$$+ (6)\begin{vmatrix} -2 & -3 \\ 7 & -2 \end{vmatrix}$$

$$= 200 - 250 + 150 = 100$$

Expanding Δ_2 about the first column yields

$$\Delta_2 = \begin{vmatrix} 5 & 4 & -3 \\ -5 & -10 & -2 \\ -3 & 6 & 8 \end{vmatrix} = 5\begin{vmatrix} -10 & -2 \\ 6 & 8 \end{vmatrix} - (-5)\begin{vmatrix} 4 & -3 \\ 6 & 8 \end{vmatrix}$$

$$+ (-3)\begin{vmatrix} 4 & -3 \\ -10 & -2 \end{vmatrix}$$

$$= -340 + 250 + 114 = 24$$

Expanding Δ_3 about the first column yields

$$\Delta_3 = \begin{vmatrix} 5 & -2 & 4 \\ -5 & 7 & -10 \\ -3 & -3 & 6 \end{vmatrix} = 5\begin{vmatrix} 7 & -10 \\ -3 & 6 \end{vmatrix} - (-5)\begin{vmatrix} -2 & 4 \\ -3 & 6 \end{vmatrix}$$

$$+ (-3)\begin{vmatrix} -2 & 4 \\ 7 & -10 \end{vmatrix}$$

$$= 60 - 0 + 24 = 84$$

Now applying Cramer's rule we solve for the three unknowns.

$$x_1 = \frac{\Delta_1}{\Delta} = \frac{100}{50} = 2$$

$$x_2 = \frac{\Delta_2}{\Delta} = \frac{24}{50} = 0.48$$

$$x_3 = \frac{\Delta_3}{\Delta} = \frac{84}{50} = 1.68$$

EXERCISE B-1

Evaluate Δ, Δ_1, Δ_2, and Δ_3 for Eqs. (B-2)

ANSWER:

216, 280, 160, and 60

Gaussian Elimination

Gaussian elimination is the basis for most computer programs for solving linear equations. Basically this method reduces the system of equations to a triangular form by systematically eliminating one variable at a time. Given three equations in three unknowns

$$a_{11}x_1 + a_{12}x_2 + a_{13}x_3 = b_1$$

$$a_{21}x_1 + a_{22}x_2 + a_{23}x_3 = b_2$$

$$a_{31}x_1 + a_{32}x_2 + a_{33}x_3 = b_3$$

We multiply or divide these equations by constants, and add or subtract equations to reduce the equations to an equivalent form

$$x_1 + c_{12}x_2 + c_{13}x_3 = d_1$$

$$x_2 + c_{23}x_3 = d_2$$

$$x_3 = d_3$$

The last of these equations yields x_3. The value of x_3 is then back-substituted into the second equation to solve for x_2. Finally, the values of x_2 and x_3 are back-substituted into the first equation to find x_1.

Gaussian elimination produces the solution of the original equations because multiplying equations by nonzero constants or adding equations yields an equivalent set of equations that has the same solution.

EXAMPLE B-2

Solve for the three unknowns in Eq. (B-1) using Gaussian elimination.

SOLUTION:

The given equations are

$$5x_1 - 2x_2 - 3x_3 = 4$$
$$-5x_1 + 7x_2 - 2x_3 = -10$$
$$-3x_1 - 3x_2 + 8x_3 = 6$$

Dividing the first equation by 5, the second by -5, and the third by -3 yields

$$x_1 - \frac{2}{5}x_2 - \frac{3}{5}x_3 = \frac{4}{5}$$
$$x_1 - \frac{7}{5}x_2 + \frac{2}{5}x_3 = 2$$
$$x_1 + x_2 - \frac{8}{3}x_3 = -2$$

Subtracting the first equation from the second and third equations eliminates x_1 from these equations

$$x_1 - \frac{2}{5}x_2 - \frac{3}{5}x_3 = \frac{4}{5}$$
$$-x_2 + x_3 = \frac{6}{5}$$
$$\frac{7}{5}x_2 - \frac{31}{15}x_3 = -\frac{14}{5}$$

Dividing the second equation by -1 and the third equation by $5/7$ yields

$$x_1 - \frac{2}{5}x_2 - \frac{3}{5}x_3 = \frac{4}{5}$$
$$x_2 - x_3 = -\frac{6}{5}$$
$$x_2 - \frac{31}{21}x_3 = -2$$

Subtracting the second equation from the third eliminates x_2

from the last equation

$$x_1 - \frac{2}{5}x_2 - \frac{3}{5}x_3 = \frac{4}{5}$$

$$x_2 - x_3 = -\frac{6}{5}$$

$$-\frac{10}{21}x_3 = -\frac{4}{5}$$

The third equation yields $x_3 = 84/50$. Back-substituting x_3 into the second equation gives

$$x_2 - \frac{84}{50} = -\frac{6}{5}$$

which yields $x_2 = 24/50$. Back-substituting x_2 and x_3 into the first equation gives

$$x_1 - \frac{2}{5}\left(\frac{24}{50}\right) - \frac{3}{5}\left(\frac{84}{50}\right) = \frac{4}{5}$$

which yields $x_1 = 2$. These are the same answers found in Example B-1 using Cramer's rule.

EXERCISE B-2

Use Gaussian elimination to solve the following linear equations:

$$2x_1 - x_2 - x_3 = 6$$

$$-x_1 + 3x_2 = 3$$

$$-x_1 + 6x_3 = 0$$

ANSWERS:

$x_1 = 4.667$, $x_2 = 2.556$, $x_3 = 0.7778$

Matrices and Linear Equations

Circuit equations can be formulated and solved in matrix format. By definition a *matrix* is a rectangular array written as

$$\boldsymbol{A} = \begin{bmatrix} a_{11} & a_{12} & a_{13} & \dots & a_{1n} \\ a_{21} & a_{22} & a_{23} & \dots & a_{2n} \\ \dots & \dots & \dots & \dots & \dots \\ a_{m1} & a_{m2} & a_{m3} & \dots & a_{mn} \end{bmatrix} \quad \text{(B-6)}$$

The matrix $\boldsymbol{A}$ is said to be of order m by n (or $m \times n$) because it contains m rows and n columns. The matrix in Eq. (B-6) can be abbreviated as follows:

$$\boldsymbol{A} = [a_{ij}]_{mn} \quad \text{(B-7)}$$

where a_{ij} is the element in the ith row and jth column.

Some Definitions

Different types of matrices have special names. A *row matrix* has only one row ($m = 1$) and any number of columns. A *column matrix* has only one column ($n = 1$) and any number of rows. A *square matrix* has the same number of rows as columns ($m = n$). A *diagonal matrix* is a square matrix in which all elements not on the main diagonal are zero ($a_{ij} = 0$ for $i \neq j$). An *identity matrix* is a diagonal matrix for which the main diagonal elements are all unity ($a_{ii} = 1$).

For example, given

$$\boldsymbol{A} = [1 \quad -2 \quad 0 \quad 4] \quad \boldsymbol{B} = \begin{bmatrix} 3 \\ -2 \\ 6 \\ 0 \end{bmatrix} \quad \boldsymbol{C} = \begin{bmatrix} 1 & 0 & -7 \\ -3 & 12 & 0 \\ 0 & 0 & -4 \end{bmatrix}$$

$$\boldsymbol{U} = \begin{bmatrix} 1 & 0 & 0 & 0 \\ 0 & 1 & 0 & 0 \\ 0 & 0 & 1 & 0 \\ 0 & 0 & 0 & 1 \end{bmatrix}$$

we say that $\boldsymbol{A}$ is a 1×4 row matrix, $\boldsymbol{B}$ is a 4×1 column matrix, $\boldsymbol{C}$ is a 3×3 square matrix, and $\boldsymbol{U}$ is a 4×4 identity matrix.

The *determinant* of a square matrix A (denoted det A) has the same elements as the matrix itself. For example, given

$$\boldsymbol{A} = \begin{bmatrix} 4 & -6 \\ 1 & -2 \end{bmatrix} \quad \text{then} \quad \det \boldsymbol{A} = \begin{vmatrix} 4 & -6 \\ 1 & -2 \end{vmatrix} = -8 + 6 = -2$$

The *transpose* of a matrix $\boldsymbol{A}$ (denoted $\boldsymbol{A}^T$) is formed by interchanging the rows and columns. For example, given

$$\boldsymbol{A} = \begin{bmatrix} 1 & 2 & 0 & 8 \\ 4 & 7 & -1 & -3 \end{bmatrix} \quad \text{then} \quad \boldsymbol{A}^T = \begin{bmatrix} 1 & 4 \\ 2 & 7 \\ 0 & -1 \\ 8 & -3 \end{bmatrix}$$

The *adjoint* of a square matrix $\boldsymbol{A}$ (denoted $adj\ \boldsymbol{A}$) is formed by replacing each element a_{ij} by its cofactor C_{ij} and then transposing:

$$adj\ \boldsymbol{A} = [C_{ij}]^T = [C_{ji}] \qquad \text{(B-8)}$$

For example, if

$$\boldsymbol{A} = \begin{bmatrix} -3 & 2 \\ 0 & 5 \end{bmatrix} \quad \text{then} \quad C_{11} = 5 \quad C_{12} = 0 \quad C_{21} = -2 \quad C_{22} = -3$$

and therefore

$$adj\ \boldsymbol{A} = \begin{bmatrix} 5 & 0 \\ -2 & -3 \end{bmatrix}^T = \begin{bmatrix} 5 & -2 \\ 0 & -3 \end{bmatrix}$$

Matrix Algebra

The matrices $\boldsymbol{A}$ and $\boldsymbol{B}$ are equal if and only if they have the same number of rows and columns, and $a_{ij} = b_{ij}$ for all i and j. Matrix *addition* is only possible when two matrices have the same number of rows and columns. When two matrices are of the same order, their sum is obtained by adding the corresponding elements; that is,

$$\text{If} \quad \boldsymbol{C} = \boldsymbol{A} + \boldsymbol{B} \quad \text{then} \quad c_{ij} = a_{ij} + b_{ij} \tag{B-9}$$

For example, given

$$\boldsymbol{A} = \begin{bmatrix} -1 & 4 \\ -3 & -2 \end{bmatrix} \quad \text{and} \quad \boldsymbol{B} = \begin{bmatrix} 3 & 0 \\ 2 & -4 \end{bmatrix}$$

$$\text{then} \quad \boldsymbol{C} = \boldsymbol{A} + \boldsymbol{B} = \begin{bmatrix} 2 & 4 \\ -1 & -6 \end{bmatrix}$$

Multiplying a matrix $\boldsymbol{A}$ by a scalar constant k is accomplished by multiplying every element by k, that is, $k\boldsymbol{A} = [ka_{ij}]$. In particular, if $k = -1$ then $-\boldsymbol{B} = [-b_{ij}]$, and applying the matrix addition rule yields matrix *subtraction*:

$$\text{If} \quad \boldsymbol{C} = \boldsymbol{A} - \boldsymbol{B} \quad \text{then} \quad c_{ij} = a_{ij} - b_{ij} \tag{B-10}$$

Multiplication of two matrices $\boldsymbol{AB}$ is defined only if the number of columns in $\boldsymbol{A}$ equals the number of rows in $\boldsymbol{B}$. In general, if $\boldsymbol{A}$ is of order $m \times n$ and $\boldsymbol{B}$ is of order $n \times r$, then the product $\boldsymbol{C} = \boldsymbol{AB}$ is a matrix of order $m \times r$. The element c_{ij} is found by summing the products of the elements in ith row of $\boldsymbol{A}$ and the jth column of $\boldsymbol{B}$. In other words, matrix multiplication is a row by column operation:

$$c_{ij} = [a_{i1}a_{i2}\ldots a_{in}] \begin{bmatrix} b_{1j} \\ b_{2j} \\ \cdots \\ \cdots \\ \cdots \\ b_{nj} \end{bmatrix} = a_{i1}b_{1j} + a_{i2}b_{2j} + \ldots a_{in}b_{nj}$$

$$= \sum_{k=1}^{n} a_{ik}b_{kj} \tag{B-11}$$

Matrix multiplication is not commutative, so usually $\boldsymbol{AB} \neq \boldsymbol{BA}$. Two important exceptions are: (1) the product of a square matrix $\boldsymbol{A}$ and an identity matrix $\boldsymbol{U}$ for which $\boldsymbol{UA} = \boldsymbol{AU} = \boldsymbol{A}$, and (2) the product of a square matrix $\boldsymbol{A}$ and its *inverse* (denoted $\boldsymbol{A}^{-1}$) for which $\boldsymbol{A}^{-1}\boldsymbol{A} = \boldsymbol{AA}^{-1} = \boldsymbol{U}$. A closed form formula for the inverse of a square matrix is

$$\mathbf{A}^{-1} = \frac{adj\ \mathbf{A}}{\det \mathbf{A}} \tag{B-12}$$

That is, the inverse can be found by multiplying the adjoint matrix of $\mathbf{A}$ by the scalar $1/\det \mathbf{A}$. If $\det \mathbf{A} = 0$ then $\mathbf{A}$ is said to be *singular* and $\mathbf{A}^{-1}$ does not exist. Equation (B-12) is useful for deriving properties of the matrix inversion. It is not, however, a very efficient way to calculate the inverse of a matrix of order greater than 3×3.

EXERCISE B-3

Given:

$$\mathbf{A} = \begin{bmatrix} -5 & 7 \\ 7 & 11 \end{bmatrix} \quad \text{and} \quad \mathbf{B} = \begin{bmatrix} 3 & -1 \\ 6 & -2 \end{bmatrix}$$

Calculate $\mathbf{AB}$, $\mathbf{BA}$, $\mathbf{A}^{-1}$, and $\mathbf{B}^{-1}$.

ANSWERS:

$$\mathbf{AB} = \begin{bmatrix} 27 & -9 \\ 87 & -29 \end{bmatrix} \quad \mathbf{BA} = \begin{bmatrix} -22 & 10 \\ -44 & 20 \end{bmatrix}$$

$$\mathbf{A}^{-1} = \frac{1}{107}\begin{bmatrix} -11 & 7 \\ 7 & 5 \end{bmatrix} \quad \mathbf{B}^{-1} \text{ does not exist}$$

Matrix Solution of Linear Equations

The three linear equation in Eq. (B-1) are

$$5x_1 - 2x_2 - 3x_3 = 4$$
$$-5x_1 + 7x_2 - 2x_3 = -10$$
$$-3x_1 - 3x_2 + 8x_3 = 6$$

These equations are expressed in matrix form as follows:

$$\begin{bmatrix} 5 & -2 & -3 \\ -5 & 7 & -2 \\ -3 & -3 & 8 \end{bmatrix}\begin{bmatrix} x_1 \\ x_2 \\ x_3 \end{bmatrix} = \begin{bmatrix} 4 \\ -10 \\ 6 \end{bmatrix} \tag{B-13}$$

The left side of Eq. (B-13) is the product of a 3×3 square matrix and a 3×1 column matrix of unknowns. The elements in the square matrix are the coefficients of the unknown in the given equations. The matrix product on the left side in Eq. (B-13) produces a 3×1 matrix, which equals the 3×1 column matrix on the right side. The elements of the 3×1 on the right side are the constants on the right sides of the given equations.

In symbolic form we write the matrix equation in Eq. (B-13) as

$$\mathbf{AX} = \mathbf{B} \tag{B-14}$$

where

$$\boldsymbol{A} = \begin{bmatrix} 5 & -2 & -3 \\ -5 & 7 & -2 \\ -3 & -3 & 8 \end{bmatrix}, \qquad \boldsymbol{X} = \begin{bmatrix} x_1 \\ x_2 \\ x_3 \end{bmatrix} \qquad \text{and} \qquad \boldsymbol{B} = \begin{bmatrix} 4 \\ -10 \\ 6 \end{bmatrix}$$

Left multiplying Eq. (B-14) by $\boldsymbol{A}^{-1}$ yields

$$\boldsymbol{A}^{-1}\boldsymbol{A}\boldsymbol{X} = \boldsymbol{A}^{-1}\boldsymbol{B}$$

But by definition $\boldsymbol{A}^{-1}\boldsymbol{A} = \boldsymbol{U}$ and $\boldsymbol{U}\boldsymbol{X} = \boldsymbol{X}$; therefore

$$\boldsymbol{X} = \boldsymbol{A}^{-1}\boldsymbol{B} \qquad \text{(B-15)}$$

To solve linear equations by matrix methods we calculate the product $\boldsymbol{A}^{-1}\boldsymbol{B}$.

We use the matrix approach to solve for the three unknowns in Eq. (B-13). We first calculate the quantities needed to use Eq. (B-12) to obtain $\boldsymbol{A}^{-1}$. The determinant of the coefficient matrix is

$$\det \boldsymbol{A} = \begin{vmatrix} 5 & -2 & -3 \\ -5 & 7 & -2 \\ -3 & -3 & 8 \end{vmatrix} = 50$$

The cofactors of the coefficient matrix are

$$C_{11} = -\begin{vmatrix} +7 & -2 \\ -3 & 8 \end{vmatrix} = 50 \qquad C_{12} = -\begin{vmatrix} -5 & -2 \\ -3 & 8 \end{vmatrix} = 46$$

$$C_{13} = -\begin{vmatrix} -5 & +7 \\ -3 & -3 \end{vmatrix} = 36$$

and

$$C_{21} = 25 \qquad C_{22} = 31 \qquad C_{23} = 21$$

$$C_{31} = 25 \qquad C_{32} = 25 \qquad C_{33} = 25$$

Now using Eq. (B-12) we get $\boldsymbol{A}^{-1}$ as

$$\boldsymbol{A}^{-1} = \frac{adj\ \boldsymbol{A}}{\det \boldsymbol{A}} = \frac{1}{50}\begin{bmatrix} 50 & 46 & 36 \\ 25 & 31 & 21 \\ 25 & 25 & 25 \end{bmatrix}^T = \frac{1}{50}\begin{bmatrix} 50 & 25 & 25 \\ 46 & 31 & 25 \\ 36 & 21 & 25 \end{bmatrix}$$

Using Eq. (B-15) we solve for the column matrix of unknowns as

$$\begin{bmatrix} x_1 \\ x_2 \\ x_3 \end{bmatrix} = \boldsymbol{X} = \boldsymbol{A}^{-1}\boldsymbol{B} = \frac{1}{50}\begin{bmatrix} 50 & 25 & 25 \\ 46 & 31 & 25 \\ 36 & 22 & 25 \end{bmatrix}\begin{bmatrix} 4 \\ -10 \\ 6 \end{bmatrix} = \frac{1}{50}\begin{bmatrix} 100 \\ 24 \\ 84 \end{bmatrix}$$

which yields $x_1 = 2$, $x_2 = 24/50$, and $x_3 = 84/50$. These are, of course, the same results previously obtained using Cramer's rule and Gaussian elimination.

EXERCISE B-4

Find the inverse of the coefficient matrix for Eq. (B-2).

ANSWER:

$$A^{-1} = \frac{1}{216}\begin{bmatrix} 56 & 16 & 0 \\ 32 & 40 & 0 \\ 12 & 15 & 27 \end{bmatrix}$$ ■

Using MATLAB to Solve Linear Equations

MATLAB™ is an interactive software package for matrix-based numerical computations and data analysis.[1] The original MATLAB was main-frame resident software written in the 1970s to support linear algebra and matrix-theory courses. Since then personal computer versions have been developed, including a student edition, which easily fits on very modest platforms.[2]

The only objects MATLAB manipulates are matrices whose elements are numerical (possibly complex) data. The only capabilities we will discuss are the matrix operations required to implement the matrix equation $\boldsymbol{X} = \boldsymbol{A}^{-1}\boldsymbol{B}$. To implement this equation we need to know (1) how to enter a matrix into MATLAB, (2) how to find the inverse of a square matrix, and (3) how to multiply two matrices.

The student version of MATLAB operates in an interactive mode that supports keyboard entry of data and commands. The command line prompt "≫" appears on screen after MATLAB is invoked. A matrix can then be entered by typing in its elements one row at a time. The elements in a given row are separated by blanks or commas. The end of a row is indicated by a semicolon (;). The row by row entries are enclosed by a left bracket ([) at the beginning of the first row and a right bracket (]) at the end of the last row.

For example, to enter the $\boldsymbol{A}$ matrix for Eq. (B-13) we type

```
>>        A = [5 -2 -3;-5 7 -2;-3 -3 8]
```

Following a carriage return (enter) keystroke MATLAB responds with a listing of the elements of $\boldsymbol{A}$:

```
A =
        5    -2    -3
       -5     7    -2
       -3    -3     8
```

[1]MATLAB is a registered trademark of the MathWorks, Inc.

[2]System requirements and program capabilities are described in, The Student Edition of MATLAB™, Prentice Hall, Englewood Cliffs, NJ, 1992.

Similarly, we enter the $\boldsymbol{B}$ matrix for Eq. (B-13) by entering

```
>>  B = [4;-10;6]
```

Following a carriage return MATLAB responds with

```
B =

      4
    -10
      6
```

We now have the $\boldsymbol{A}$ and $\boldsymbol{B}$ matrices entered into MATLAB and can proceed to implement the equation $\boldsymbol{X} = \boldsymbol{A}^{-1}\boldsymbol{B}$. In the MATLAB language the matrix inverse is written inv($\boldsymbol{A}$) and the matrix product as $\boldsymbol{A}*\boldsymbol{B}$. Entering the command

```
>>   X = inv(A)*B
```

causes MATLAB to calculate $\boldsymbol{A}^{-1}$, multiply $\boldsymbol{A}^{-1}$ times $\boldsymbol{B}$ and list the elements of the matrix $\boldsymbol{X}$:

```
X =

     2.0000
     0.4800
     1.6800
```

These results agree with the solutions previously found. Thus, MATLAB requires only three command-line entries to solve a system of linear equations.

Two features of MATLAB need to be mentioned here. First, the MATLAB language is case sensitive so it treats $\boldsymbol{a}$ and $\boldsymbol{A}$ as different matrices. Second, MATLAB automatically assigns memory space to matrix variables so there is no need for "dimension" or "type" statements. To find out what variables are active type "whos" at the command-line prompt. MATLAB responds to this command with a current list of variables, their size, total number of elements, and whether the elements are complex. For example, entering "whos" following the data and command entries given above produces

```
Name    Size      Total   Complex
A       3 by 3    9       No
B       3 by 1    3       No
X       3 by 1    3       No
```

This listing can be checked to see if two matrices can be added or multiplied, or to verify that the dimensions of a matrix are what you expect.

APPENDIX C: COMPLEX NUMBERS

The use of complex numbers to represent signals and circuits is a fundamental tool in electrical engineering. This appendix reviews complex number representations and arithmetic operations. These procedures, though rudimentary, must be second nature to all who aspire to be electrical engineers. Exercises are provided to confirm your mastery of these basic skills.

Complex Numbers Representations

A complex number z can be written in *rectangular form* as

$$z = x + jy \tag{C-1}$$

where j represents $\sqrt{-1}$. Mathematicians customarily use i to represent $\sqrt{-1}$, but i represents current in electrical engineering, so we use the symbol j instead.

The quantity z is a two-dimensional number represented as a point in the complex plane as shown in Fig. C-1. The x component is called the *real part* and y (not jy) the *imaginary part* of z. A special notation is sometimes used to indicate these two components:

$$x = Re\{z\} \quad \text{and} \quad y = Im\{z\} \tag{C-2}$$

where $Re\{z\}$ means the real part and $Im\{z\}$ for the imaginary part of z.

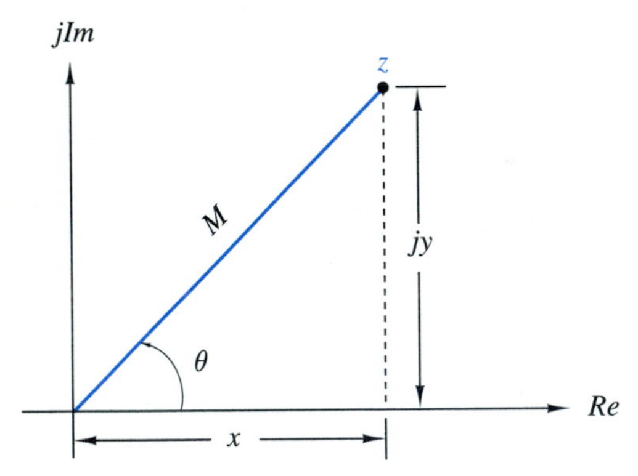

Figure C-1 Graphical representation of complex numbers.

Figure C-1 also shows the polar representation of the complex number z. In *polar form* a complex number is written

$$z = M\angle\theta \tag{C-3}$$

where M is called the *magnitude* and θ the *angle* of z. A special notation is also used to indicate these two components:

$$|z| = M \quad \text{and} \quad \angle z = \theta \tag{C-4}$$

where $|z|$ means the magnitude and $\angle z$ the angle of z.

The real and imaginary parts, and magnitude and angle of z are all shown geometrically in Fig. C-1. The relationships between the rectangular and polar forms are easily derived from the geometry in Fig. C-1:

$$\text{Rectangular-to-polar} \quad M = \sqrt{x^2 + y^2} \quad \theta = \tan^{-1}\frac{y}{x}$$
$$\text{Polar-to-rectangular} \quad x = M\cos\theta \quad y = M\sin\theta \tag{C-5}$$

Caution: The inverse tangent relation for θ involves an ambiguity,[1] which can be resolved by identifying the correct quadrant in the z plane using the signs of the two rectangular components. [See Exercise C-1.]

Another version of the polar form is obtained using Euler's relationship:

$$e^{j\theta} = \cos\theta + j\sin\theta \tag{C-6}$$

Using Euler's identity we can write the polar form as

$$z = Me^{j\theta} = M\cos\theta + jM\sin\theta \tag{C-7}$$

This polar form is equivalent to Eq. (C-3), since the right side yields the same polar-to-rectangular relationships as Eq. (C-5). Thus, a complex number can be represented in three ways:

$$z = x + jy \qquad z = M\angle\theta \qquad z = Me^{j\theta} \tag{C-8}$$

The relationships between these forms are given in Eq. (C-5).

The quantity z^* is called the conjugate of the complex number z. The asterisk indicates the *conjugate of a complex number* that is formed by reversing the sign of the imaginary component. In rectangular form the conjugate of $z = x + jy$ is written as $z^* = x - jy$. In polar form the conjugate is obtained by reversing the sign of the angle of z, $z^* = Me^{-j\theta}$. The geometric interpretation in Fig. C-2 shows that conjugation simply reflects a complex number across the real axis in the complex plane.

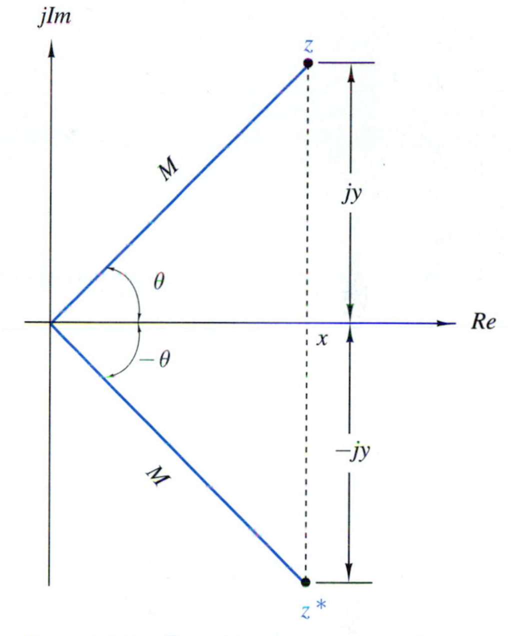

Figure C-2 Graphical representation of conjugate complex numbers.

EXERCISE C-1

Convert the following complex numbers to polar form.

(a) $1 + j\sqrt{3}$
(b) $-10 + j20$
(c) $-2000 - j8000$
(d) $60 - j80$

ANSWERS:

(a) $2e^{60^\circ}$
(b) $22.4\,e^{j117^\circ}$
(c) $8246e^{256^\circ}$
(d) $100e^{j307^\circ}$

[1] Calculators usually cannot tell the difference between $\tan^{-1}(-y/x)$ and $\tan^{-1}(y/-x)$ or $\tan^{-1}(y/x)$ and $\tan^{-1}(-y/-x)$. The student is cautioned to note the quadrant that the angle is in to determine the proper numerical value of the angle.

■ EXERCISE C-2

Convert the following complex numbers to rectangular form.

(a) $12e^{j90^\circ}$
(b) $3e^{j45^\circ}$
(c) $400\angle\pi$
(d) $8e^{-j60^\circ}$
(e) $15e^{j\frac{\pi}{6}}$

ANSWERS:

(a) $0 + j12$
(b) $2.12 + j2.12$
(c) $-400 + j0$
(d) $4 - j6.93$
(e) $13 + j7.5$ ■

■ EXERCISE C-3

Evaluate the following expressions:

(a) $Re\left(12e^{j\pi}\right)$
(b) $Im\left(100\angle 60^\circ\right)$
(c) $\angle(-2 + j6)$
(d) $Im\left[\left(4e^{j\frac{\pi}{4}}\right)^*\right]$

ANSWERS:

(a) -12
(b) 86.6
(c) 108.4°
(d) -2.83 ■

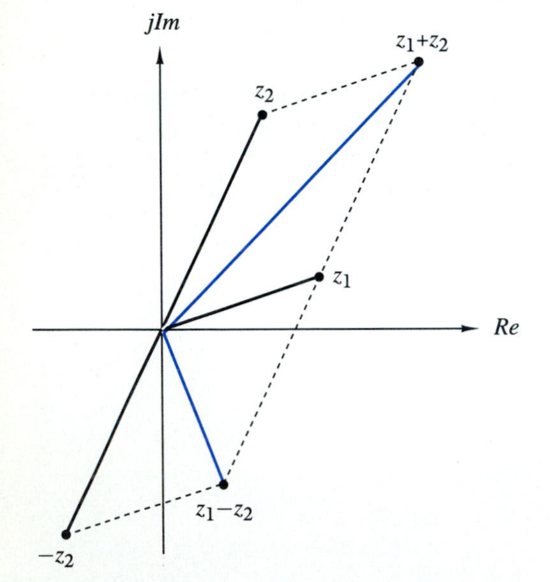

Figure C-3 Graphical representation of addition and subtraction of complex numbers.

Arithmetic Operations–Addition and Subtraction

Addition and subtraction are defined in terms of complex number in rectangular form. Given two complex numbers

$$z_1 = x_1 + jy_1 \qquad \text{and} \qquad z_2 = x_2 + jy_2 \tag{C-9}$$

Complex numbers are added by separately adding the real parts and imaginary parts to obtain the corresponding parts of their sum. The sum $z_1 + z_2$ is defined as

$$z_1 + z_2 = (x_1 + x_2) + j(y_1 + y_2) \tag{C-10}$$

Subtraction follows the same pattern except that the components are subtracted:

$$z_1 - z_2 = (x_1 - x_2) + j(y_1 - y_2) \tag{C-11}$$

Figure C-3 shows a geometric interpretation of addition and subtraction. In particular note that $z + z^* = 2x$ and $z - z^* = j2y$.

Multiplication and Division

Multiplication and division of complex numbers can be accomplished with the numbers in either rectangular or polar form. For complex numbers in rectangular form the multiplication operation yields

$$\begin{aligned} z_1 z_2 &= (x_1 + jy_1)(x_2 + jy_2) \\ &= (x_1 x_2 + j^2 y_1 y_2) + j(x_1 y_2 + x_2 y_1) \\ &= (x_1 x_2 - y_1 y_2) + j(x_1 y_2 + x_2 y_1) \end{aligned} \tag{C-12}$$

For numbers in polar form the product is

$$\begin{aligned} z_1 z_2 &= \left(M_1 e^{j\theta_1}\right)\left(M_2 e^{j\theta_2}\right) \\ &= (M_1 M_2)\, e^{j(\theta_1+\theta_2)} \end{aligned} \tag{C-13}$$

Multiplication is somewhat easier to carry out with the numbers in polar form, although both methods should be understood. In particular, the product of a complex number z and it conjugate z^* is the square of its magnitude:

$$zz^* = \left(Me^{j\theta}\right)\left(Me^{-j\theta}\right) = M^2 \tag{C-14}$$

For complex numbers in polar form the division operation yields

$$\begin{aligned} \frac{z_1}{z_2} &= \frac{M_1 e^{j\theta_1}}{M_2 e^{j\theta_2}} \\ &= \left(\frac{M_1}{M_2}\right) e^{j(\theta_1-\theta_2)} \end{aligned} \tag{C-15}$$

When the numbers are in rectangular form the numerator and denominator of the quotient are multiplied by the conjugate of the denominator.

$$\frac{z_1}{z_2}\frac{z_2^*}{z_2^*} = \frac{(x_1 + jy_1)(x_2 - jy_2)}{(x_2 + jy_2)(x_2 - jy_2)}$$

Applying the multiplication rule from Eq. (C-12) to the numerator and denominator yields

$$\frac{z_1}{z_2} = \frac{(x_1 x_2 + y_1 y_2) + j(x_2 y_1 - x_1 y_2)}{x_2^2 + y_2^2} \tag{C-16}$$

Complex division is somewhat easier to carry out with the numbers in polar form, although both methods should be understood.

EXERCISE C-4

Evaluate the following expressions using $z_1 = 3 + j4$, $z_2 = 5 - j7$, $z_3 = -2 + j3$, and $z_4 = 5\angle -30°$.

(a) $z_1 z_2$

(b) $z_3 + z_4$

(c) $z_2 z_3 / z_4$
(d) $z_1^* + z_3 z_1$
(e) $z_2 + (z_1 z_4)^*$

ANSWERS:
(a) $43 - j$
(b) $2.33 + j0.5$
(c) $-0.995 + j6.12$
(d) $-15 - j3$
(e) $28 - j16.8$ ■

■ EXERCISE C-5

Given $z = x + jy = Me^{j\theta}$, evaluate the following statements.
(a) $z + z^*$
(b) $z - z^*$
(c) z/z^*
(d) z^2
(e) $(z^*)^2$
(f) zz^*

ANSWERS:
(a) $2x$
(b) $j2y$
(c) $e^{j2\theta}$
(d) $x^2 - y^2 + j2xy$
(e) $x^2 - y^2 - j2xy$
(f) $x^2 + y^2$ ■

■ EXERCISE C-6

Given $z_1 = 1$, $z_2 = -1$, $z_3 = j$, and $z_4 = -j$, evaluate:
(a) z_1/z_3
(b) z_1/z_4
(c) $z_3 z_4$
(d) $z_3 z_3$
(e) $z_4 z_4$
(f) $z_2 z_3^*$

ANSWERS:
(a) $-j$
(b) $+j$
(c) 1
(d) -1
(e) -1
(f) j ■

■ EXERCISE C-7

Evaluate the expression $T(\omega) = j\omega/(j\omega + 10)$ at $\omega = 5, 10, 20, 50, 100$.

ANSWERS:
$0.447\angle 63.4°$, $0.707\angle 45°$, $0.894\angle 26.6°$, $0.981\angle 11.3°$, $0.995\angle 5.71°$ ■

APPENDIX D: AN ABRIDGED SPICE SYNTAX

The purpose of this appendix is to summarize the SPICE syntax introduced at various places in this book. The book only treats those features of SPICE needed to perform basic dc, ac, and transient analysis of linear circuit. The syntax covered is an abridged subset of the SPICE syntax supported by commercially available programs.[1] The SPICE subset described and used in this book is supported by the student version of PSpice.

General Information

A SPICE circuit file is a sequence of statements that accomplishes three things:

1. Defines the circuit elements and how they are interconnected
2. Identifies the type of analysis to be performed
3. Controls the number and type of calculated responses in the output file

The first statement in a circuit file is a title and the last is an .END statement. The statements in between can be in almost any order. However, we choose to always list the statements in the following order:

```
TITLE STATEMENT
ELEMENT STATEMENTS
COMMAND STATEMENTS
OUTPUT STATEMENTS
.END
```

SPICE treats the first statement in a circuit file as a title statement that is used to label output listings for documentation purposes. The last statement in a file must be an .END statement. *Note:* the period is part of the .END statement.

The following general rules apply to SPICE statements:

1. Statements cannot exceed 80 characters unless a continuation statement is used. A plus sign "+" in column 1 indicates that a statement is a continuation of the preceding statement.

[1]See for example, Paul W. Tuinenga, *SPICE: A Guide to Circuit Simulation & Analysis Using PSpice,* 2nd ed., Prentice Hall, Englewood Cliffs, NJ, 1992, Appendices B and C.

2. An asterisk "*" in column 1 indicates comment statement. Comment statements are not executable and are ignored by SPICE.
3. Uppercase letters "A" through "Z" and numbers are used to assign names to circuit elements. SPICE only uses the first eight characters in a name. The first character must be a letter that indicates the element type as follows:

   R = resistor

   V = independent voltage source

   I = independent current source

   E = voltage-controlled voltage source

   F = current-controlled current source

   G = voltage-controlled current source

   H = current-controlled voltage source

   C = capacitor

   L = inductor

   K = mutual inductance
4. Numerical values can be expressed as integers (2, 5, −12), real numbers (4.32, −1216.2), floating point numbers (3.45E-2, −6.789E+4), or as integer or real numbers followed by a suffix scale factor defined as follows:

Suffix	Scale Factor
T	10^{12}
G	10^{9}
MEG	10^{6}
K	10^{3}
M	10^{-3}
U	10^{-6}
N	10^{-9}
P	10^{-12}
F	10^{-15}

5. Circuit nodes are numbered using positive integers and zero. Although the nodes do not need to be numbered sequentially, all circuits must contain a Node zero (0), which SPICE uses as the reference (ground) node.
6. There must be at least two elements connected to every node. Every node must have a dc path to the ground node via resistors, inductors, and independent voltage sources.
7. The plus reference mark for an element voltage is assigned to the first node listed in an element statement and the passive sign convention used to assign the reference direction for the element current.

Element Statements

Element statements describe the devices in a circuit and the way they are interconnected. The following element statements are introduced in the section indicated.

Resistor (Section 5-5)

```
R<name> <1st node> <2nd node> <value>
```

The <name> field is a user-defined alphanumeric string labeling the element. The <1st node> and <2nd node> fields are integers defining the nodes to which the element is connected. The <value> field defines the numerical value of the resistance in ohms. For example,

```
RLOAD  3  7  1.2MEG
```

defines a 1.2-MΩ resistor called RLOAD connected between Nodes 3 and 7.

Independent Voltage Source (Section 5-5)

```
V<name> <1st node> <2nd node> <type> <value>
```

The <name> field is a user-defined alphanumeric string labeling the element. The <1st node> and <2nd node> fields are integers defining the nodes to which the element is connected. The <type> and <value> fields define the voltage waveform produced by the source. The following types are available in SPICE:

```
<type> = DC, <value> = dc voltage
<type> = AC, <values> = ac voltage amplitude and phase
<type> = PULSE, <value> := See Section 6-8
<type> = EXP, <value> := See Section 6-8
<type> = SIN, <value> := See Section 6-8
<type> = PWL, <value> := See Section 6-8
```

For example, the statements

```
VBAT 1 0 DC 15V
VGEN 2 0 AC 256 -30
VS32 8 7 PULSE 1V 5V 0 10N 20N 100N 500N
```

define a 15-V dc source VBAT connected between Nodes 1 and 0, an ac source VGEN with peak voltage of 256 V and a phase angle of $-30°$ connected between Nodes 2 and 0, and a pulse source VS32 connected between Nodes 8 and 7. The default source type is dc.

Independent Current Source (Section 5-5)

```
I<name> <1st node> <2nd node> <type> <value>
```

The <name> field is a user-defined alphanumeric string labeling the element. The <1st node> and <2nd node> fields are integers defining the nodes to which the element is connected. The <type> and <value> fields define the current waveform produced by the source. The waveform types available in SPICE are summarized above under the independent voltage source statement. The default source type is dc.

Voltage-Controlled Voltage Source (Section 5-5)

```
E<name> <1st node> <2nd node> <3rd node> <4th node> <value>
```

The <name> field is a user-defined alphanumeric string labeling the element. The <1st node> and <2nd node> fields are integers defining the nodes to which the voltage source is connected. The <3rd node> and <4th node> fields define the nodes across which the controlling voltage is defined. SPICE assigns plus reference marks for voltage to the 1st and 3rd nodes and assigns the reference directions of the currents using the passive sign convention. The <value> field defines the controlled source gain in V/V. For example, the statement

```
EEG  3  1  6  9  10K
```

defines a dependent voltage source EEG with a gain of 10^4 connected between Nodes 3 and 1, and controlled by the voltage between Nodes 6 and 9.

Current-Controlled Current Source (Section 5-5)

```
F<name> <1st node> <2nd node> <voltage source name> <value>
```

The <name> field is a user-defined alphanumeric string labeling the element. The <1st node> and <2nd node> fields are integers defining the nodes to which the current source is connected. <Voltage source name> defines the zero-volt independent voltage source whose current controls the dependent source. The required voltage source must be defined in a separate element statement. SPICE assigns the plus reference mark for the source voltage to the 1st node and assigns the reference direction for the source current using the passive sign convention. The <value> field defines the controlled source gain in A/A. For example, the statements

```
VICON  27  35  DC  0
IALPHA  12  0  VICON  50
```

define a dependent current source IALPHA with a gain of 50 connected between nodes 12 and 0 and controlled by the current through the voltage source VICON, which is connected between Nodes 27 and 35.

Voltage-Controlled Current Source (Section 5-5)

```
G<name> <1st node> <2nd node> <3rd node> <4th node> <value>
```

The <name> field is a user-defined alphanumeric string labeling the element. The <1st node> and <2nd node> fields are integers defining the nodes to which the current is connected. The <3rd node> and <4th node> fields define the nodes across which the controlling voltage is defined. SPICE assigns the plus reference marks for the voltages to the 1st and 3rd node and assigns the reference directions of the currents using the passive sign convention. The <value> field defines the controlled source gain in $A/V = S$. For example, the statement

```
G32  2  23  1  6  2M
```

defines a dependent current source G32 with a gain of 2×10^{-3} S connected between Nodes 2 and 23 and controlled by the voltage between Nodes 1 and 6.

Current-Controlled Voltage Source (Section 5-5)

```
H<name> <1st node> <2nd node> <voltage source name> <value>
```

The <name> field is a user-defined alphanumeric string labeling the element. The <1st node> and <2nd node> fields are integers defining the nodes to which the voltage source is connected. <Voltage source name> defines the zero-volt independent voltage source whose current controls the dependent source. The required voltage source must be defined in a separate element statement. SPICE assigns the plus reference mark for the source voltage to the 1st node. The <value> field defines the controlled source gain in $V/A = \Omega$. For example, the statements

```
VIH21  11  3  DC 0
H21  0  12  VIH21  250
```

define a dependent voltage source H21 with a gain of 250 Ω connected between Nodes 0 and 12 and controlled by the current through the voltage source VIH21 connected between Nodes 11 and 3.

Capacitor (Section 7-7)

```
C<name> <1st node> <2nd node> <value> [IC=<initial voltage>]
```

The <name> field is a user-defined alphanumeric string labeling the element. The <1st node> and <2nd node> fields are integers defining the nodes to which the element is connected. The <value> field defines the numerical value of the capacitance in farads. The [IC=<initial voltage>] field is an optional feature used to define the capacitor voltage at $t = 0$. The default assignment is IC = 0. For example, the statement

```
C17   23   6   52P   IC = 5
```

defines a 52-pF capacitor called C17 connected between Nodes 23 and 6 with an initial voltage of 5 V.

Inductor (Section 7-7)

```
L<name> <1st node> <2nd node> <value> [IC=<initial current>]
```

The <name> field is a user-defined alphanumeric string labeling the element. The <1st node> and <2nd node> fields are integers defining the nodes to which the element is connected. The <value> field defines the numerical value of the inductance in henrys. The [IC=<initial current>] field is an optimal feature defining the inductor current at $t = 0$. The default assignment is IC = 0. For example, the statement

```
LPC   2   0   33U   IC = 7M
```

defines a 33-μH inductor called LPC connected between Nodes 2 and 0 with an initial current of 7 mA.

Mutual Inductance (Section 7-7)

```
K<name> <1st Inductor> <2nd Inductor> <Value>
```

The <name> field is a user-defined alphanumeric string labeling the element. The statement defines coupling coefficient k =<value> for the mutual inductance coupling between the <1st inductor> and <2nd inductor>. The <value> field defines the numerical value of the coupling coefficient k, which be in the range $0 < k < 1$. The K statement must be accompanied by two separate inductor statements. SPICE always associates the mutual inductance reference marks (dots) with the first nodes in the two inductor statements. For example, the statements

```
L1   7   3   12.5M
L2   5   8   8M
K12   L1   L2   0.99
```

define a coupling coefficient of $k = 0.99$ between inductors L1 and L2. The mutual inductance "dots" are assigned to Nodes 7 and 5.

Command Statements

Command statements begin with a period and define the type of analysis to be performed. The following command statements are discussed in the sections indicated.

Operating Point Analysis (Section 5-5)

```
.OP
```

The .OP statement causes SPICE to calculate the dc operating point of a circuit and is the default command when no other command it given in an input file.

Transfer Function Analysis (Section 5-5)

```
.TF <Output Variable> <Input Source>
```

The .TF statement causes SPICE to calculate three numbers: (1) the circuit gain from the input source to the output variable, (2) the input resistance seen by the input source, and (3) the output resistance at the terminals defining the output variable. The gain is defined as gain = output variable/input source. The units of gain depend on the dimensions of the source and output. Numerical values of the input and output resistance are in ohms.

DC Analysis (Section 5-5)

```
.DC <source name> <start value> <stop value> <step value>
```

The .DC statement causes SPICE to perform a dc analyses as the value of the independent source identified in the <source name> field is incrementally stepped across the range from <start value> to <stop value>. The size of the increments is specified by <step value>, which must be greater than zero. For example, the statement

```
.DC  I22  0.2M  2M  0.1M
```

produces dc analyses as the output of independent current source I22 is stepped from 0.2 mA to 2 mA in steps of 0.1 mA. The incremental values assigned to I22 by the .DC statement override the value assigned in the I22 element statement.

AC Analysis (Section 12-6)

```
.AC <sweep> NFREQ FSTART FSTOP
```

The .AC statement causes SPICE to perform sinusoidal steady-state analysis as the frequency of all independent ac sources is varied (swept) from FSTART to FSTOP (in hertz). There are three types of sweeps:

1. <sweep> = LIN is a linear sweep. The frequency is varied linearly from FSTART to FSTOP generating circuit responses at NFREQ equally spaced frequencies.
2. <sweep> = OCT is an octave sweep. The frequency is varied logarithmically from FSTART to FSTOP generating circuit responses at NFREQ frequencies per octave.
3. <sweep> = DEC is an decade sweep. The frequency is varied logarithmically from FSTART to FSTOP generating circuit responses at NFREQ frequencies per decade.

The integer NFREQ controls the number of frequencies at which the calculated responses are written in the output file. For a linear sweep the output file will contain responses at NFREQ frequencies. For a decade or octave sweep the number of output frequencies will be NFREQ times the number of decades or octaves between FSTART and FSTOP plus 1. For example, the statement

```
.AC   DEC   21   1K   100K
```

produces ac analyses as the frequency of all ac sources is logarithmical stepped from 1 kHz to 100 kHz generating ac responses at 21 frequencies per decade.

Transient Analysis (Section 8-8)

```
.TRAN TSTEP TSTOP [TSTART] [TMAX] [UIC]
```

The .TRAN statement causes SPICE to calculate the circuit transient response beginning at $t = 0$ and ending at t = TSTOP. SPICE writes the responses designated in a .PRINT statement to the output file every TSTEP seconds starting at t = TSTART and ending at t = TSTOP. The parameter TMAX specifies a maximum internal step size for the computation. The UIC instruction tells SPICE to *U*se the *I*nitial *C*onditions specified in the capacitor and inductor element statement.

The TSTEP and TSTOP parameters are mandatory and the others are optional. If TSTART and TMAX are left blank, the default values are TSTART = 0 and TMAX = TSTOP/50. If the UIC instruction is not included, SPICE uses dc analysis to calculate the initial capacitor voltages and inductor currents prior to beginning the transient analysis.

Output Statements

Output statements control the types and number of calculated responses included in the output file. The output file will contain the numerical values of the gain, input resistance and output resistance when a .TF statement is included in the input file. By default the output file will contain all of the node voltages and currents through independent voltage sources when there is no output statement in an input file. The .PRINT output statement can be used to specify the calculated responses in the output file. The syntax of the .PRINT command statement is

```
.PRINT <Analysis Type> <Output Variable List>
```

The <Analysis Type> field can be DC, AC, or TRAN. The .PRINT statement is not executed unless the input file also contains an analysis command statement corresponding to the analysis type listed. The .PRINT statement restricts the calculated responses in the output file to those given in the output variable list. The output variable list identifies responses using the notation V(N), V(N1,N2), and I(Vxxxx). V(N) is the node voltage at Node N. V(N1,N2) is the voltage between Nodes N1 and N2 with the plus reference mark at N1. I(Vxxxx) is the current through the element Vxxxx. The reference direction for I(Vxxxx) is determined by the 1st node listed in the element statement defining Vxxxx.

For example, the statements

```
.TRAN 0.1U 10U
.PRINT TRAN V(1) V(7) I(VAMP)
```

directs SPICE to perform a transient analysis beginning at $t = 0$ and ending at t = TSTOP = 10 μs. The time and responses V(1), V(7), and I(VAMP) listed in the .PRINT statement are written to the output files every TSTEP = 0.1 μs.

With ac analysis the type of output data is controlled by adding one of the following suffixes to the voltage or current name:

M Amplitude of the sinusoid in V or A

DB Amplitude in decibels: $20 \times \log_{10}(\text{Amplitude})$

P Phase angle of the sinusoid in degrees

For example, the print command

```
.PRINT AC VM(4) VDB(4) IP(VIN)
```

causes SPICE to list the following responses in the output file: (1) amplitude of node voltage V(1), (2) 20 × $\log_{10}$(Amplitude) of node voltage V(4), and (3) the phase angle of the current through voltage source VIN.

APPENDIX E: TWO-PORT CIRCUITS

The purpose of this appendix is to define and illustrate the network functions used to characterize linear two-port circuits. The circuit in Fig. E-1 is in the zero state and contains only linear resistors, capacitors, inductors, mutual inductance, and controlled sources. The two pairs of external terminals are called ports, with Port 1 referred to as the input and Port 2 as the output. The variables $V_1(s)$ and $I_1(s)$ are associated with the input port and the variables $V_2(s)$ and $I_2(s)$ with the output port. Note that the reference marks for the port variables comply with the passive sign convention.

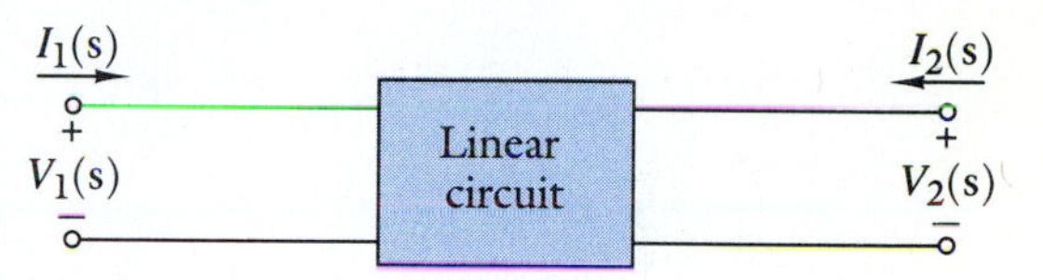

Figure E-1 Two-port circuit.

Two-port parameters are the network functions that relate the port variables. A specific set of two-port parameters is obtained by selecting two of the four port variables as independent variables and using the other two as dependent variables. The *i-v* characteristics relating dependent and independent variables involve network functions of four types: (1) driving-point functions at the input port, (2) driving-point functions at the output port, (3) forward transfer functions from the input port to the output port, and (4) reverse transfer functions from the output port to the input port. Although there are six possible sets of two-port parameters, we discuss only the impedance, admittance, hybrid, and transmission parameters.

Impedance Parameters

The impedance parameters are defined by selecting $I_1(s)$ and $I_2(s)$ as the independent variables and $V_1(s)$ and $V_2(s)$ as the dependent variables. Because the circuit in Fig. E-1 is linear, the *i-v* relationships of the two-port can be written in the form

$$\begin{aligned} V_1(s) &= z_{11}I_1(s) + z_{12}I_2(s) \\ V_2(s) &= z_{21}I_1(s) + z_{22}I_2(s) \end{aligned} \tag{E-1}$$

The network functions in these relationships are defined as

follows:

$$
\begin{aligned}
z_{11} &= \left.\frac{V_1}{I_1}\right|_{I_2=0} && \text{input driving-point impedance with Port 2 open} \\
z_{12} &= \left.\frac{V_1}{I_2}\right|_{I_1=0} && \text{reverse transfer impedance with Port 1 open} \\
z_{21} &= \left.\frac{V_2}{I_1}\right|_{I_2=0} && \text{forward transfer impedance with Port 2 open} \\
z_{22} &= \left.\frac{V_2}{I_2}\right|_{I_1=0} && \text{output driving-point impedance with Port 1 open}
\end{aligned}
\tag{E-2}
$$

Collectively these network functions are called the *impedance parameters* because they all have dimensions of ohms. The impedance parameters are found by first measuring or calculating the input driving-point impedance and the forward transfer impedance with the output port open ($I_2 = 0$), and then measuring or calculating the output driving-point impedance and reverse transfer impedance with the input port open ($I_1 = 0$).

EXAMPLE E-1

Find the impedance parameters of the circuit in Fig. E-2.

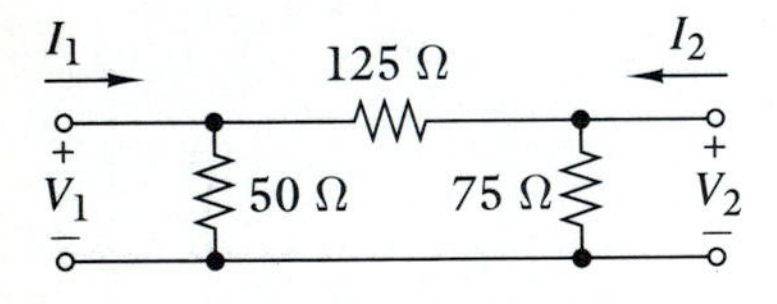

Figure E-2

SOLUTION:

The equivalent resistance looking in at Port 1 with Port 2 open is

$$z_{11} = 50 \parallel (125 + 75) = 40\ \Omega$$

The current through the 75-Ω resistor with Port 2 open is found using current division:

$$I_{75} = \frac{50}{50 + 125 + 75} I_1 = 0.2 I_1$$

By Ohm's law the output voltage is $V_2 = I_{75} \times 75$. Therefore, the forward transfer impedance is

$$z_{21} = \left.\frac{V_2}{I_1}\right|_{I_2=0} = \frac{(0.2 I_1) \times 75}{I_1} = 15\ \Omega$$

The current through the 50-Ω resistor with Port 1 open is found using current division:

$$I_{50} = \frac{75}{50 + 125 + 75} I_2 = 0.3 I_2$$

By Ohm's law the input voltage is $V_1 = I_{50} \times 50$. Therefore, the reverse transfer impedance is

$$z_{12} = \left.\frac{V_1}{I_2}\right|_{I_1=0} = \frac{(0.3I_2) \times 50}{I_2} = 15\ \Omega$$

The equivalent resistance looking in at Port 2 with Port 1 open is

$$z_{22} = 75||(125 + 50) = 52.5\ \Omega$$

Using the impedance parameters the i-v relationships of the two port in Fig. E-2 are

$$V_1 = 40I_1 + 15I_2$$
$$V_2 = 15I_1 + 52.5I_2$$

Admittance Parameters

The admittance parameters are defined by selecting $V_1(s)$ and $V_2(s)$ as the independent variables and $I_1(s)$ and $I_2(s)$ as the dependent variables. Because the circuit in Fig. E-1 is linear the i-v relationships of the two port can be written in the form

$$\begin{aligned} I_1(s) &= y_{11}V_1(s) + y_{12}V_2(s) \\ I_2(s) &= y_{21}V_1(s) + y_{22}V_2(s) \end{aligned} \tag{E-3}$$

The network functions in these relationships are defined as follows.

$$y_{11} = \left.\frac{I_1}{V_1}\right|_{V_2=0} \quad \text{input driving-point admittance with Port 2 shorted}$$

$$y_{12} = \left.\frac{I_1}{V_2}\right|_{V_1=0} \quad \text{reverse transfer admittance with Port 1 shorted}$$

$$y_{21} = \left.\frac{I_2}{V_1}\right|_{V_2=0} \quad \text{forward transfer admittance with Port 2 shorted} \tag{E-4}$$

$$y_{22} = \left.\frac{I_2}{V_2}\right|_{V_1=0} \quad \text{output driving-point admittance with Port 1 shorted}$$

Collectively these network functions are called the *admittance parameters* because they all have dimensions of siemens. The admittance parameters are found by first measuring or calculating the input driving-point admittance and the forward transfer admittance with the output port shorted ($V_2 = 0$), and then measuring or calculating the output driving-point admittance and reverse transfer admittance with the input port shorted ($V_1 = 0$).

EXAMPLE E-2

Find the admittance parameters of the circuit in Fig. E-2.

SOLUTION:

The equivalent conductance looking in at Port 1 with Port 2 shorted is

$$y_{11} = \frac{1}{50} + \frac{1}{125} = 0.028 \text{ S}$$

The output current with Port 2 shorted is $I_2 = -V_1/125$. Therefore, the forward transfer admittance is

$$y_{21} = \left.\frac{I_2}{V_1}\right|_{V_2=0} = -\frac{1}{125} = -0.008 \text{ S}$$

The input current with Port 1 shorted is $I_1 = -V_2/125$. Therefore, the reverse transfer admittance is

$$y_{12} = \left.\frac{I_1}{V_2}\right|_{V_1=0} = -\frac{1}{125} = -0.008 \text{ S}$$

The equivalent conductance looking in at Port 2 with Port 1 shorted is

$$y_{22} = \frac{1}{75} + \frac{1}{125} = -.0213 \text{ S}$$

Using admittance parameters the i-v relationships of the two port in Fig. E-2 are

$$I_1 = 0.028V_1 - 0.008V_2$$
$$I_2 = -0.008V_1 + 0.0213V_2$$

Hybrid Parameters

The hybrid parameters are defined by selecting $I_1(s)$ and $V_2(s)$ as the independent variables and $V_1(s)$ and $I_2(s)$ as the dependent variables. Because the circuit is linear the i-v relationships of the two port can be written in the form

$$\begin{aligned} V_1(s) &= h_{11}I_1(s) + h_{12}V_2(s) \\ I_2(s) &= h_{21}I_1(s) + h_{22}V_2(s) \end{aligned} \quad \text{(E-5)}$$

The network functions in these relationships are defined as follows:

$$h_{11} = \left.\frac{V_1}{I_1}\right|_{V_2=0} \quad \text{input driving-point impedance with Port 2 shorted}$$

$$h_{12} = \left.\frac{V_1}{V_2}\right|_{I_1=0} \quad \text{reverse voltage transfer function with Port 1 open}$$

$$h_{21} = \left.\frac{I_2}{I_1}\right|_{V_2=0} \quad \text{forward current transfer function with Port 2 shorted} \qquad \text{(E-6)}$$

$$h_{22} = \left.\frac{I_2}{V_2}\right|_{I_1=0} \quad \text{output driving-point admittance with Port 1 open}$$

Collectively these network functions are called the *hybrid parameters* because they involve a mixture of dimensionless and dimensioned parameters. The hybrid parameters are found by first measuring or calculating the input driving-point impedance and forward current gain with the output port shorted ($V_2 = 0$) and then measuring or calculating the output driving-point admittance and reverse voltage gain with the input port open ($I_1 = 0$).

EXAMPLE E-3

Figure E-3 shows a low-frequency small-signal transistor model. Find the hybrid parameters of this circuit when $r_{bc} \gg r_{be}$.

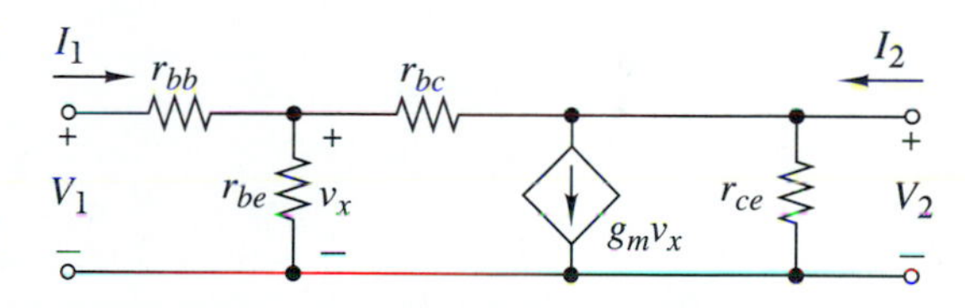

Figure E-3

SOLUTION:

The equivalent resistance looking in at Port 1 with Port 2 shorted is

$$h_{11} = r_{bb} + r_{be} \parallel r_{bc} \approx r_{bb} + r_{be}$$

The last approximation applies because $r_{bc} \gg r_{be}$. The output current with Port 2 shorted is $I_2 = g_m V_x$. The input current I_1 divides between r_{be} and r_{bc}. But since $r_{bc} \gg r_{be}$ almost all of I_1 enters r_{be}. Using Ohm's law $V_x \approx I_1 r_{be}$ and the forward current gain is

$$h_{21} = \left.\frac{I_2}{I_1}\right|_{V_2=0} \approx \frac{g_m(I_1 r_{be})}{I_1} = g_m r_{be}$$

Using voltage division with Port 1 open produces

$$V_1 = V_x = \frac{r_{be}}{r_{be} + r_{bc}} V_2$$

The reverse voltage gain is

$$h_{12} = \left.\frac{V_1}{V_2}\right|_{I_1=0} = \frac{r_{be}}{r_{be}+r_{bc}} \approx \frac{r_{be}}{r_{bc}}$$

The last approximation applies because $r_{bc} \gg r_{be}$. Writing a node equation at the output port with the input port open produces

$$I_2 = \frac{V_2}{r_{ce}} + g_m V_x + \frac{V_2}{r_{bc}+r_{be}}$$

Since $V_x = h_{12}V_2$, the output conductance with the input port open can be written as

$$h_{22} = \left.\frac{I_2}{V_2}\right|_{I_1=0} = \frac{1}{r_{ce}} + g_m h_{12} + \frac{1}{r_{be}+r_{bc}}$$

$$\approx \frac{1}{r_{ce}} + \frac{g_m r_{be}+1}{r_{bc}}$$

Transmission Parameters

The transmission parameters are obtained by selecting $V_2(s)$ and $I_2(s)$ as the independent variables and $V_1(s)$ and $I_1(s)$ as the dependent variables. Because the circuit is linear the i-v relationships of the two port can be written in the form

$$\begin{aligned} V_1(s) &= AV_2(s) + BI_2(s) \\ I_1(s) &= CV_2(s) + DI_2(s) \end{aligned} \tag{E-7}$$

The network functions in these relationships are defined as follows:

$A = \left.\dfrac{V_1}{V_2}\right|_{I_2=0}$ reciprocal of the forward voltage gain with Port 2 open

$B = \left.\dfrac{V_1}{I_2}\right|_{V_2=0}$ reciprocal of the forward transfer admittance with Port 2 shorted

$C = \left.\dfrac{I_1}{V_2}\right|_{I_2=0}$ reciprocal of the forward transfer impedance with Port 2 open

$D = \left.\dfrac{I_1}{I_2}\right|_{V_2=0}$ reciprocal of the forward current gain with Port 2 shorted

(E-8)

Collectively these network functions are called the ***transmission parameters*** because all four are reciprocals of forward transfer functions. The transmission parameters are found by first measuring or calculating the forward voltage gain and transfer impedance with the output port open ($I_2 = 0$), and then measuring or calculating the forward current gain and transfer admittance with the output port shorted ($V_2 = 0$).

EXAMPLE E-4

Figure E-4 shows the s-domain model of a transformer. Find the impedance parameters and transmission parameters of this circuit.

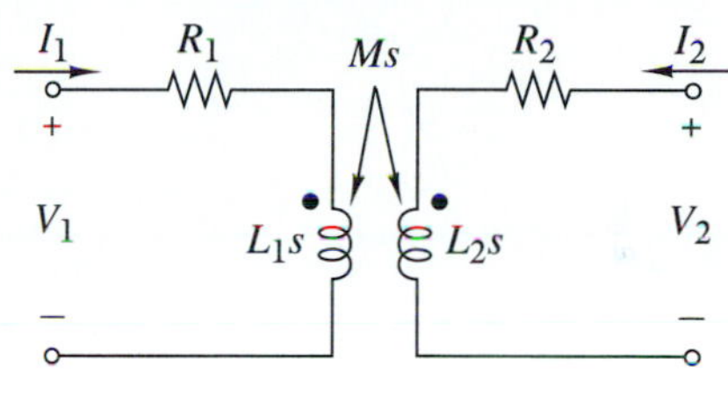

Figure E-4

SOLUTION:

The coupling in Fig. E-4 is additive so the i-v relationships of the transformer are

$$V_1 = (L_1s + R_1)I_1 + MsI_2$$

$$V_2 = MsI_1 + (L_2s + R_2)I_2$$

By inspection the impedance parameters are

$$z_{11} = L_1s + R_1$$

$$z_{12} = z_{21} = Ms$$

$$z_{22} = L_2s + R_2$$

With the output port open $I_2 = 0$ and the transformer i-v relationships yield

$$A = \left.\frac{V_1}{V_2}\right|_{I_2=0} = \frac{(L_1s + R_1)I_1}{MsI_1} = \frac{L_1s + R_1}{Ms}$$

$$C = \left.\frac{V_2}{I_1}\right|_{I_2=0} = Ms$$

With the output port shorted $V_2 = 0$ and the second i-v relationship yields

$$I_1 = -\frac{L_2s + R_2}{Ms}I_2$$

Substituting this value of I_1 into the first i-v relationship produces

$$V_1 = \left[(L_1s + R_1)\left(-\frac{L_2s + R_2}{Ms}\right) + Ms\right]I_2$$

The remaining transmission parameters are

$$B = \left.\frac{V_1}{I_2}\right|_{V_2=0} = \frac{[-(L_1 s + R_1)(L_2 s + R_2)/Ms + Ms]I_2}{I_2}$$

$$= \frac{M^2 s^2 - (L_1 s + R_1)(L_2 s + R_2)}{M^2 s^2}$$

and

$$D = \left.\frac{I_1}{I_2}\right|_{V_2=0} = -\frac{L_2 s + R_2}{Ms}$$

ANSWERS TO SELECTED PROBLEMS

Listed below are answers to selected analysis problems. In general, answers for design problems are not given because they involve alternative decisions and may have many possible correct answers. The reader can always verify a design solution by analyzing the proposed circuit to see if it performs the specified signal-processing function.

Chapter One

ERO 1-1 ELECTRICAL QUANTITIES (Sect. 1-2)

1-1 (a) 4 THz; (b) 16.6 ps; (c) 0.47 μH; (d) 150 kΩ; (e) 5600 mA

1-3 (a) 1.2 MΩ; (b) 5.55556 kHz; (c) 1 nF; (d) 6.2 μA; (e) 0.245 MV

ERO 1-2 CIRCUIT VARIABLES (Sect. 1-3)

1-7 10^{11} electrons

1-9 $t = 0$ s: 0 A; $t = 1$ s: 6 A; $t = 3$ s: 18 A

1-10 0.625 C

1-12 0.625 A; 4.32 cents

1-13 5A; 4.32 MJ; 20 hours

1-15 36 kC

1-18 100 mW absorbing

1-22 Dev 1: $p = -45$ W, delivering; Dev 2: $i = 3$ A, absorbing; Dev 3: $v = 5$ V, absorbing; Dev 4: $p = 37.5$ W, absorbing; Dev 5: $v = -10$ V, delivering; Dev 6: $v = -10$ V, absorbing

1-24 $t = 0$ s: $p = -25$ mW; $t = 5$ ms: $p = -9.20$ mW; $t = 25$ ms: $p = -0.168$ mW; $w = 125$ mJ

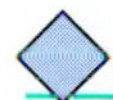

Chapter Two

ERO 2-1 ELEMENT CONSTRAINTS (Sect. 2-1)

2-1 (a) $i = 2.5$ mA; $p = 12.5$ mW; (b) $v = -1.5$ V; -4.5 mW; (c) $v = -0.6$ V; $i = 5$ mA; (d) $i = 0$ A; $p = 0$ W

2-3 (a) $i = 2$ mA; $v = 2$ V; $p = 4$ mW; (b) $i = 5$ A; $v = 0$ V; $p = 0$ W; (c) $i = 5$ A; $v = -1$ kV; $p = -5$ kW

2-6 (a) $i = +0.5$ A; $v = 50.05$ V; $p = 25.025$ W; $i = -0.5$ A; $v = -49.95$ V; $p = 24.975$ W; $i = +1$ A; $v = 100.2$ V; $p = 100.2$ W; $i = -1$ A; $v = -99.8$ V; $p = 99.8$ W; $i = +2$ A; $v = 200.8$ V; $p = 401.6$ W; $i = -2$ A; $v = -199.2$ V; $p = 398.4$ W; $i = +5$ A; $v = 505$ V; $p = 2525$ W; $i = -5$ A; $v = -495$ V; $p = 2475$ W; $i = +10$ A; $v = 1020$ V; $p = 10.2$ kW; $i = -10$ A; $v = -980$ V; $p = 9.80$ kW; (b) 0.1%

2-7 (a) 5-V voltage source; (b) 10-mA current source; (c) 3.3-Ω resistor; (d) open circuit; (e) short circuit

ERO 2-2 CONNECTION CONSTRAINTS (Sect. 2-2)

2-8 (a) C1: 1,2,4,5; 1,2,3,6; 3,6,5,4; C2: 4,1; 4,3,2; 3,2,1; C3: 1,2,4; 2,3,6; 1,3,5; C4: 1,2,5; 4,3,2; 5,3,6; (b) C1: series 1 & 2; 4 & 5; 3 & 6; parallel none; C2: series 2 & 3; parallel 1 & 4; C3: series none; parallel none; C4: series none; parallel none

2-10 (a) C1: 1,2,4,5; 1,2,6; 6,5,4; C2: 2,3; C3: 1,2; 2,3,6; 1,3,5; C4: 1,2; 4,3,2; 3,6; (b) C1: series 1 & 2; 4 & 5; parallel none; C2: series 2 & 3; parallel 2 & 3; 4 & 1; C3: series none; parallel 1 & 2; 5 & 6; C4: series none; parallel 1 & 2; 3 & 6

2-13 (a) A, B, C, & ground; (b) series: none; parallel: none

ERO 2-3 COMBINED CONSTRAINTS (Sect. 2-3)

2-16 C1: $v_X = 5$ V; C2: $v_X = 12$ V; C3: $v_X = 7.89$ V; C4: $v_X = 17.2$ V

2-20 C1: $v = -I_O R_1(R_2 + R_3)/(R_1 + R_2 + R_3)$; C2: $i = -0.707$ A; $v = -58.7$ V; C3: $i_1 = -i_2 = I_O$; $v = V_2 - V_1$

ERO 2-4 EQUIVALENT RESISTIVE CIRCUITS AND SOURCE TRANSFORMATIONS (Sect. 2-4)

2-22 $R_{AB} = 60$ Ω; $R_{BC} = 0$ Ω; $R_{AC} = 60$ Ω; $R_{CD} = 50$ Ω; $R_{BD} = 50$ Ω; $R_{AD} = 110$ Ω

2-23 $R_{AB} = 90$ Ω; $R_{AC} = 60$ Ω; $R_{AD} = 40$ Ω; $R_{BC} = 130$ Ω; $R_{BD} = 110$ Ω; $R_{CD} = 80$ Ω

2-25 C1: $R_S = 5$ Ω; $I_S = 2$ A; C2: $R_S = 15$ Ω; $V_S = 10$ V; C3: $R_S = 50$ Ω; $V_S = 10$ V

2-29 100 Ω

ERO 2-5 VOLTAGE AND CURRENT DIVISION (Sect. 2-5)

2-34 C1: $i_X = 2.5$ A; C2: $v_X = 0.666$ V; C3: $v_X = 2.5$ V; C4: $i_X = 0.6$ A

2-35 C1: $i_X = 1$ A; C2: $v_X = 5$ V

2-37 $R_{SHUNT} = 50.25$ mΩ

2-40 $x = 61.8\%$

ERO 2-6 CIRCUIT REDUCTION (Sect. 2-6)

2-42 C1: $v_X = -12.5$ V; $i_X = -1.25$ A

2-44 C3: $v_X = 5/3$ V; $i_X = 5/6$ A

2-47 C2: $v_X = 28$ V

2-49 $v_{D\text{-}B} = 3.6$ V; $v_{A\text{-}C} = 1.5$ V; $v_{C\text{-}D} = 0.9$ V

CHAPTER INTEGRATING PROBLEMS

2-56 $v_R = 3.18$ V; $v_{NLE} = 6.82$ V

2-57 $V_{C\text{-}D} = 46.89$ MV

Chapter Three

ERO 3-1 PROPORTIONALITY (Sect. 3-1)

3-1 $K = R_1/(R_1 + R_2)$

3-4 $K = (R_2 \| R_4)/(R_1 + R_2 \| R_4 + R_3)$

3-6 $K = (R_2 R_3 - R_1 R_4)/(R_1 + R_2 + R_3 + R_4)$

3-9 $i_O = 64$ mA

ERO 3-2 SUPERPOSITION (Sect. 3-2)

3-12 $V_O = V_1/4 + V_2/4$

3-14 $V_O = R(I_1/3 - I_2)$

3-16 $V_O = 0.4V_1 + 0.6V_2 + 0.2V_3$

3-17 $I_1 = 0$ A; $I_2 = 1$ A

ERO 3-3 THÉVENIN AND NORTON EQUIVALENT CIRCUITS (Sect. 3-3)

3-20 $V_T = I_S R_1 \| R_2$; $I_N = I_S$; $R_T = R_N = R_1 \| R_2$

3-22 $V_T = \dfrac{V_S (R_2 R_3 - R_1 R_4)}{R_1 R_2 + R_1 R_4 + R_2 R_3 + R_3 R_4}$;

$R_T = R_N = R_1 \| R_3 + R_2 \| R_4$; $I_N = \dfrac{V_T}{R_T}$

3-25 switch open: $V_T = I_S R_1 R_3/(R_1 + R_2 + R_3)$; $R_T = R_3 \| (R_1 + R_2)$;
switch closed: $V_T = 0$; $R_T = 0$

3-28 $V_T = (V_1 + V_2)/2$; $R_T = 2R$

3-30 $V_T = (I_1 - 3I_2)R$; $R_T = 4R$

3-33 $V_T = 21$ V; $I_N = 1.167$ A; $R_T = R_N = 18\ \Omega$

3-35 $V_S = 1.5$ V: $p_D = 50\ \mu$W; $V_S = 2.0$ V: $p_D = 1.5$ mW

ERO 3-4 MAXIMUM SIGNAL TRANSFER (Sect. 3-4)

3-38 (a) C1: $R_L = 10\ \Omega$; $P_{MAX} = 2.5$ W; C2: $R_L = 15$ kΩ; $P_{MAX} = 41.7$ mW; (b) $R_L = 0\ \Omega$;
C1: $I_{MAX} = 1$ A; C2: $I_{MAX} = -3.33$ mA; (c) $R_L = \infty\ \Omega$; C1: $V_{MAX} = 10$ V; C2: $V_{MAX} = -50$ V

3-40 (a) C5: $R_L = 13.3\ \Omega$; $P_{MAX} = 33.3$ mW; C6: $R_L = 600\ \Omega$; $P_{MAX} = 38.4$ mW;
(b) $R_L = 0\ \Omega$; C5: $I_{MAX} = 0.1$ A; C6: $I_{MAX} = 16$ mA; (c) $R_L = \infty\ \Omega$; C5: $V_{MAX} = 1.33$ V;
C6: $V_{MAX} = 9.6$ V

3-41 Amplifier 1: one speaker; Amplifier 2: two speakers in parallel; Amplifier 3: two speakers in series; Amplifier 4: four speakers in parallel; Amplifier 5: two speakers in series

ERO 3-5 INTERFACE CIRCUIT DESIGN (Sect. 3-5)

3-49 (a) $R_L = 12.5\ \Omega$; (b) $R_L = \infty\ \Omega$; (c) $R_L = 19.1\ \Omega$ or $131\ \Omega$; (d) $R_L = 50\ \Omega$; (e) $R_L = 0\ \Omega$; (f) $R_L = 50\ \Omega$; (g) $R_L = \infty\ \Omega$; (h) $R_L = \infty\ \Omega$; (i) no value $p > p_{MAX}$

3-50 (a) $R_S = 2973\ \Omega$; (b) $R_S = 8653\ \Omega$

3-52 (a) Series resistance of $150\ \Omega$ or a parallel resistance of $16.67\ \Omega$
(b) Series resistance of $100\ \Omega$ or a parallel resistance of $25\ \Omega$

CHAPTER INTEGRATING PROBLEMS

3-55 From 0.571 mA to 1.33 mA

3-57 $V_T = -8$ V; $R_T = 200\ \Omega$

3-61 $R_L = 500\ \Omega$; $p_{MAX} = 1.25$ W; $p_{SOURCE} = 85$ W

Chapter Four

ERO 4-1 LINEAR DEPENDENT SOURCES (Sects. 4-1, 4-2)

4-1 (a) C1: $-R_1 R_L \mu/[(R_S + R_1)(R_O + R_L)]$; C2: $-R_1 R_L g R_O/\{(R_S + R_1)(R_O + R_L)\}$

4-3 (a) C1: $-R_S R_O \beta/[(R_S + R_1)(R_O + R_L)]$; C2: $-R_S R_O g R_1/\{(R_S + R_1)(R_O + R_L)\}$

4-5 C1: $-R_L \beta/[R_S + R_E(\beta + 1)]$; C2: $G_S(\beta + 1)/[G_L + G_E + G_S(\beta + 1)]$

4-7 $R_{A\text{-}B} = r_O$

4-11 $\beta = 10$

4-12 $R_{A\text{-}B} = r_\pi + R_E[(\beta + 1)r_O + R_L]/(r_O + R_E + R_L)$

ERO 4-2 THE TRANSISTOR (Sect. 4-3)

4-15 $R_B = 43$ kΩ

4-16 $5R_B \geq 93R_C$

4-17 $R_B < 575\ \Omega$

ERO 4-3 OP AMP CIRCUIT ANALYSIS (Sects. 4-4, 4-5, 4-6, 4-7)

4-19 (a) C1: $i_1 = 1.67$ mA; $v_1 = 1.67$ V; $p_1 = 2.78$ mW; C2: $i_1 = 5$ mA; $v_1 = 5$ V; $p_1 = 25$ mW; C3: $i_1 = -10$ mA; $v_1 = -10$ V; $p_1 = 100$ mW; C4: $i_1 = 15$ mA; $v_1 = 15$ V; $p_1 = 225$ mW; (b) C1: 5-V source; C2, C3, C4: OP AMP power supply; (c) C1: $500\ \Omega$; C2: 4/3 kΩ; C3: 3 kΩ

4-20 C1: Gain $= -R_2/R_1$; $R_{IN} = R_1$; C2: Gain $= (R_1 + R_2)/R_1$; $R_{IN} = \infty\ \Omega$;
C3: Gain $= -R_2/(R_1 + R_4 \| R_5)$; $R_{IN} = R_5 + R_1 \| R_4$;
C4: Gain $= (R_1 + R_2)R_4/[R_1(R_4 + R_5)]$; $R_{IN} = R_4 + R_5$

4-22 $V_O = 2V_{S2} - 2V_{S1}$; C1 uses 1 OP AMP but loads sources unequally. C2 uses 2 OP AMPs but does not load sources.

4-25 $V_O = -3.3V_1 + 0.5 + V_2$

4-26 $V_O = 1.46V_S$

4-28 C1: ± 7.5 V; C2: ± 10 V; C3: ± 150 mV

4-30 $v_O = -Ri_S$

ERO 4-5 THE COMPARATOR (Sect. 4-8)

4-45

	$v_O = V_H$	$v_O = V_L$
C1:	$v_S < 0$	$v_S > 0$
C2:	$v_S < KV_A$	$v_S > KV_A$ where $K = R_2/(R_1 + R_2)$
C3:	$v_S < -2$ V	$v_S > -2$ V
C4:	$v_S < V_A$	$v_S > V_A$ for $V_A > 0$
C5:	no value	every value

CHAPTER INTEGRATING PROBLEMS

4-52 $I_C/I_B = \beta(\beta + 2)$

4-53

V_A	V_B	V_{FA}	V_{FB}
0	0	5	5
0	5	0	5
5	0	0	5
5	5	0	0

Chapter Five

ERO 5-1 GENERAL CIRCUIT ANALYSIS (Sects. 5-1 to 5-4)

5-5 $V_X = (I_S - G_1V_S)/(G_1 + G_2 + G_3)$; $I_X = -G_3V_X$

5-7 $V_X = 0.105$ V; $I_X = 0.420$ mA; $P_{R2} = 411\ \mu$W

5-9 $V_X = -5.77$ V; $I_X = 37.8$ mA; $P_{IS} = 71.2$ W

5-13 $V_X = 21.05$ V; $I_X = 73.7$ mA; $P_{TOTAL} = 7.37$ W

5-15 $V_X = 9.09$ V; $I_X = 0.364$ mA; $P_{IN} = -10.7$ mW

5-17 $V_0 = -94.3V_S$

5-18 $V_0 = 48.4V_S$

5-19 $V_0 = -\beta R_L V_S/r_\pi$; $R_{IN} = (\beta + 2)r_\pi$

5-20 $V_O/V_S = (g_m^2 - g_m g_d)/(g_m^2 + 2G_L g_d - g_m G_L)$

5-23 $V_O/V_S = -G_1G_3/(G_1G_5 + G_2G_5 + G_3G_5 + G_4G_5 + G_3G_4)$

5-24 $V_O/V_S = -G_1(G_2 + G_4 + G_5)/(G_2G_5 + G_2G_3 + G_3G_4 + G_3G_5)$

5-25 $V_O/V_S = KG_3^2/[G_3^2 + (3 - K)G_3G_4 + G_4^2]$ where $K = (G_1 + G_2)/G_2$

5-29 $(G_1 + G_3 + G_5)V_A = G_3 V_S$

5-30 Set $R_2 = 0$

ERO 5-2 COMPUTER AIDED CIRCUIT ANALYSIS (Sect. 5-5)

5-33 Example 1: no ground node $V_L = 100$ V;
Example 2: only one element connected to Node 2 $V_L = 100$ V

5-35 $V(1) = 10.00$ V; $V(3) = -20.00$ V; $I(VS) = -10$ mA; $R_{IN} = 1$ kΩ; GAIN = 2 V/V

5-37 $R_F = 3.328$ kΩ

5-39 $P_{Q1} = 2.08$ mW; $P_{Q2} = 21.3$ mW

ERO 5-3 COMPARISON OF ANALYSIS METHODS (Sect. 5-6)

5-48 (a) circuit reduction; (b) mesh-current analysis; (c) node-voltage analysis

5-49 (a) series equivalence; (b) superposition; (c) node-voltage analysis

CHAPTER INTEGRATING PROBLEMS

5-59 For $\beta \to \infty : I_{C1} = I_{C2} = (V_{CC} - V_T)/R_{C1}$; $V_{CE1} = V_T$; $V_{CE2} = V_{CC} - R_{C2}(V_{CC} - V_T)/R_{C1}$

COURSE INTEGRATING PROBLEMS

I-2 (a) $K = (G - G_X)/G_F$; (b) $R = 1.82$ kΩ; $R_F = 1.48$ kΩ; (c) C1 dissipates 620 mW at $R_X = 1$ kΩ; C2 requires two more resistors but dissipates only 126 mW at $R_X = 1$ kΩ.

I-3 (a) $R_L = 667$ Ω; (b) $R_L = \infty$ Ω; (c) $R_L = 600$ Ω; (d) Series resistance of 867 Ω;
(e) OP AMP voltage follower followed by a series resistance of 267 Ω

Chapter Six

ERO 6-1 BASIC WAVEFORMS (Sects. 6-2, 6-3, 6-4)

6-4 $v(t) = 10e^{-50t}$ V

6-9 $v(t) = 20 \cos[200\pi(t - 0.005)]$ V

6-10 $v(t) = 14.14 \cos(100\pi t - 45°) = 10 \cos(100\pi t) + 10 \sin(100\pi t)$ V

6-11 (a) 1 kHz; 1 ms; 14.1 V; 125 μs; −45°; (b) 1 kHz; 1 ms; 36.1 V; −406 μs; 146°; (c) 0.1 Hz; 10 s; 14.1 V; −1.25 s; 45°; (d) 400 Hz; 2.5 ms; 36.1 V; 0.859 ms; −124°

6-12 1 kHz; 1 ms; 22.4 V; 426 μs; −153°

6-14 (a) 2 kHz; 0.5 ms; $a = -20$ V; $b = 0$ V; (b) 2 kHz; 0.5 ms; $a = 0$ V; $b = 20$ V; (c) 2.5 mHz; 400 s; $a = 21.2$ V; $b = 21.2$ V; (d) 1 kHz; 1 ms; $a = 42.4$ V; $b = -42.4$ V

6-15 2 kHz; 500 μs; −135°; 28.3 V

ERO 6-2 COMPOSITE WAVEFORMS (Sect. 6-5)

6-22 $v(t) = u(t - 1) + u(t - 2) + u(t - 3) - 3u(t - 4)$ V

6-23 $v(t) = \{2[u(t) - u(t-1)] - [u(t-2) - u(t-3)]\} \times 10^{-3}$ V

6-24 $v(t) = u(t) + 1.5 \times 10^6 r(t) - 4.5 \times 10^6 r(t-1) + 4.5 \times 10^6 r(t-2)$ V

6-25 $v(t) = -1.5 - 3.5 \cos(1000\pi t)$ V

6-26 $v(t) = 2.5t[u(t) - u(t-2)] - 2.5(t-3)[u(t-3) - u(t-5)]$ V

6-27 $v(t) = 5r(t) - 10r(t-1) + 10r(t-3)$ V

ERO 6-3 WAVEFORM PARTIAL DESCRIPTORS (Sect. 6-6)

6-33

	V_p	V_{pp}	V_{rms}	V_{avg}	
(a)	10	10	na	na	causal
(b)	14.1	28.2	10	0	noncausal
(c)	100	199.9	na	na	causal
(d)	10	20	na	na	causal

6-34

	V_p	V_{pp}	V_{rms}	V_{avg}	
(a)	14.1	28.2	10	0	noncausal
(b)	36.1	72.2	25.5	0	noncausal
(c)	14.1	28.2	10	0	noncausal
(d)	36.1	72.2	25.5	0	noncausal

6-35

	V_p	V_{pp}	V_{rms}	V_{avg}	
(a)	20	20	na	na	causal
(b)	20	10	na	na	causal
(c)	60	40	na	na	causal
(d)	39	36.2	na	na	causal

6-37 $V_p = V_{pp} = V_A$; $V_{rms} = V_A/\sqrt{2}$; $V_{avg} = 2V_A/\pi$

6-38 $V_p = 2$ mV; $V_{pp} = 3$ mV; $V_{rms} = 1.29$ mV; $V_{avg} = 1$ mV

6-40 $V_p = 5$ V; $V_{pp} = 10$ V; $V_{rms} = 2.36$ V; $V_{avg} = 0$ V

6-41 $-5 \text{ V} < V_0 < 5 \text{ V}$

6-42 $V_{rms} = 28.28$ V; $V_{pp} = 80$ V

6-44 Sinusoid $CF = \sqrt{2}$; Square wave $CF = 1$

6-46 (a) continuous, aperiodic, causal. (b) discontinuous, aperiodic, causal. (c) discontinuous, aperiodic, causal. (d) discontinuous, aperiodic, causal.

6-48 continuous, periodic, noncausal, period $= T_0/2$

ERO 6-4 SIGNAL SPECTRA (Sect. 6-7)

6-56 $BW = 10$ kHz; $v(t) = 4.05 \cos(10^4\pi t) + 0.450 \cos(3 \times 10^4 \pi t)$ V

6-58 $v(t) = 2 + 12 \cos(200\pi t + 90°) + 6 \cos(400\pi t - 90°) + 3 \cos(1000\pi t + 45°) + \cos(2000\pi t - 30°)$ V; $BW = 1000$ Hz

6-59 $BW_{IN} = 4$ kHz; $BW_{OUT} = 0.5$ kHz; eliminates high-frequency components.

ERO 6-5 SPICE WAVEFORMS (Sect. 6-8)

6-65

```
*PROBLEM 6-65
VS 1 0 PWL( 0 2M 1 2M 1.001 0 2 0 +2.001-1M 3-1M 3.001 0 4 0)
RS 1 0 10K
.TRAN 40M 4
.PRINT TRAN V(1)
.END
```

CHAPTER INTEGRATING PROBLEMS

6-74 $V_{rms} = 62.2$ V

6-75 (b) noncausal, aperiodic; (c) $10/\alpha$; (d) $V_p = V_{pp} = V_A$.

Chapter Seven

ERO 7-1 CAPACITOR AND INDUCTOR RESPONSES (Sects. 7-1, 7-2)

7-1 $i_C(t) = 6[e^{-200t}]u(t)$ mA; $p_C(t) = 18[e^{-200t} - e^{-400t}]u(t)$ mW; $w_C(t) = 45[1 - 2e^{-200t} + e^{-400t}]u(t)$ μJ

7-3 $v_C(t) = 2000r(t) - 2000r(t - 0.005)$ V; $p_C(t) = 100t[u(t) - u(t - 0.005)]$ mW; $w_C(t) = 50t^2u(t) + [0.00125 - 50t^2]u(t - 0.005)$ J

7-5 $v_C(t) = 20 \sin 10^4t + 30$ V; $p_C(t) = 4[1 + 3 \cos 10^4t + \sin 2 \times 10^4t]$ W; $w_C(t) = 90 + 12 \sin 10^4t + 40 \sin^2 10^4t$ μJ

7-7 $v_L(t) = 4 \cos 2 \times 10^5t$ V; $p_L = 20 + 20 \sin 4 \times 10^5t$ mW; $w_L(t) = 50 - 50 \cos 8 \times 10^5t$ nJ

7-10 500 μJ; 500 μJ; 0 J

7-12 $i_C(t) = -0.1e^{-1000t}$A $t > 0$; $p_C(t) = -e^{-2000t}$W $t > 0$; delivering.

7-14 $K = 15$ V; -6667 rad/s

7-16 $C = 1$ μF

ERO 7-2 DYNAMIC OP AMP CIRCUITS (Sect. 7-3)

7-23 $v_O(t) = -500tu(t)$ V; $t = 30$ ms

7-25 $v_O(t) = -500r(t) + 500r(t - 0.002)$ V

7-27 $v_O(t) = -0.001V_A \cos 1000t$ V; 15 kV

7-28 $v_O(t) = -5000RC \cos 1000t$ V; 300 MΩ

7-29 $v_O(t) = 10^{-2}[1 - \cos 1000t]/RC$; $RC > 1.33$ ms

7-31 $v_O(t) = \dfrac{1}{RC}\displaystyle\int_0^t v_S(x)dx - v_S(t)$

7-33 $v_O(t) = \dfrac{v_S(t)}{4} - \dfrac{RC}{2}\dfrac{dv_S(t)}{dt} + \left(\dfrac{RC}{2}\right)^2 \dfrac{d^2v_S(t)}{dt^2}$

7-35 $v_O(t) = v_{O1} - v_{O2}$

$$= 50 \int_0^t v_S(x)dx - 13v_S(t) - 10^{-3}\frac{dv_S(t)}{dt}$$

ERO 7-3 EQUIVALENT CAPACITANCE AND INDUCTANCE (Sect. 7-4)

7-43 C1: 3.5 μF; C2: 1.065 mH

7-48 (a) 1.03 mF; (b) 11.9 kJ; (c) 11.9 MW

7-49 C1: $v_C = 11.25$ V; C2: $i_L = 0.75$ mA; C3: $i_L = 0$: $v_C = 600$ mV; C4: $i_L = 1$ mA: $v_C = 0$

7-51 C1: $v_O = 0$ V; C2: $v_O = -5$ V

ERO 7-4 MUTUAL INDUCTANCE (Sect. 7-5)

7-52 (b) $i_1(t) = 5[1 - \cos\ 1000t]u(t)$ A; $v_2 = 35[\sin\ 1000t]u(t)$ V

7-54 (b) $i_1(t) = [2000t + 2e^{-1000t} - 2]u(t)$ A; $v_2 = 14[1 - e^{-1000t}]u(t)$ V

7-55 (b) $v_1(t) = 0.833[\cos\ 1000t]u(t)$ V; $i_2 = 0.333[\sin\ 1000t]u(t)$ A

7-58 C1: $L_{EQ} = L_1 + L_2 + 2M$; C2: $L_{EQ} = (L_1 + L_2 - 2M)/(L_1L_2 - M^2)$

ERO 7-5 THE IDEAL TRANSFORMER (Sect. 7-6)

7-60 $n = 1/10$; $v_1(t) = 170\ \cos\ 120\pi t$ V; $i_1(t) = 33.9\ \cos\ 120\pi t$ mA; $p_1(t) = 2.88(1 + \cos\ 240\pi t)$ W

7-61 $n = 2.71$

7-62 $V_2 = 5.81$ V (rms); $V_3 = 265$ V (rms).

7-63 $n = 2.29$

7-65 $v_O = -0.2\ \cos\ 10t$ V

CHAPTER INTEGRATING PROBLEMS

7-77 (a) $d = 3.19 \times 10^{-6}$ m; (b) $C = 10$ nF; (c) yes.

7-79 (a) $-e^{-t}$ A; (b) 67.7 mJ; (c) 271 mJ.

Chapter Eight

ERO 8-1 FIRST-ORDER CIRCUITS (Sects. 8-1, 8-2, 8-3, 8-4)

8-1 (a) $v(t) = 5[e^{-10t}]u(t)$ V; (b) $i(t) = 10^{-2}[1 - 2e^{-10000t}]u(t)$ A

8-3 C1: $4RC$; C2: $4RC/3$; C3: $4L/R$; C4: $5L/3R$

8-5 (a) C1: $v_C(0) = 0$; C2: $i_L(0) = 0$; (b) C1: $v_C(\infty) = R_2V_A/(R_1 + R_2)$; C2: $i_L(\infty) = V_A/R_1$;
(c) C1: $T_C = R_1R_2C/(R_1 + R_2)$; C2: $T_C = L(R_1 + R_2)/(R_1R_2)$
(d) C1: $v_C(t) = [R_2V_A(1 - e^{-t/T_C})/(R_1 + R_2)]\ t > 0$; C2: $i_L(t) = V_A(1 - e^{-t/T_C})/R_1\ t > 0$

8-7 (a) C1: $v_C(0) = V_A$; C2: $i_L(0) = V_A/R_1$; (b) C1: $v_C(\infty) = R_2V_A/(R_1 + R_2)$; C2: $i_L(\infty) = V_A/R_1$;
(c) C1: $T_C = R_1R_2C/(R_1 + R_2)$; C2: $T_C = L(R_1 + R_2)/(R_1R_2)$
(d) C1: $v_C(t) = V_A(R_2 + R_1e^{-t/T_C})/(R_1 + R_2)\ t > 0$; C2: $i_L(t) = V_A/R_1$

8-9 C1: $v_C(t) = V_A(2e^{-t/RC} - 1)u(t)$; C2: $i_L(t) = V_A(2e^{-Rt/L} - 1)u(t)/R$

8-11 C1: $v_O(t) = \dfrac{R_2 V_A}{R_1 + R_2}\left[1 - e^{-(R_1+R_2)t/L}\right]u(t)$; C2: $v_O(t) = \dfrac{R_2 V_A}{R_1 + R_2}\left[e^{-t/(R_1+R_2)C}\right]u(t)$

8-13 $v_C(t) = 8[1 - e^{-10000t}]u(t)$ V

8-15 C1: $v_O = 3.125[1 - e^{-2500t}]u(t)$ V; C2: $i_O(t) = 0.278[e^{-1667t}]u(t)$ mA

8-17 C1: $v_O(t) = \dfrac{RV_A}{\sqrt{R^2 + (\omega L)^2}}\cos[\omega t - \tan^{-1}(\omega L/R)]$

C2: $v_O(t) = \dfrac{V_A}{\sqrt{2 + (\omega RC)^2}}\cos[\omega t - \tan^{-1}(\omega RC/2)]$

8-19 (a) Response 1: $(s + 3000)$; 1/3 ms; 5 V; Response 2: $(s + 6000)$; 1/6 ms; 5 mA; (b) Response 1: $10u(t)$ V; Response 2: 0 A

ERO 8-2 SECOND-ORDER CIRCUITS (Sects. 8-5, 8-6, 8-7)

8-25 (a) $v(t) = [5e^{-2t} - 5e^{-5t}]u(t)$ V; (b) $v(t) = [2te^{-2t}]u(t)$ V

8-27 C1: $\omega_0 = 50$ krad/s; $\zeta = 0.005$; underdamped; C2: $\omega_0 = 31.6$ Mrad/s; $\zeta = 0.158$; underdamped

8-29 (a) $v_C(0) = 10$ V; $i_L(0) = 0$ A; (b) $v_C(\infty) = 0$ V; $i_L(\infty) = 0$ A; (c) $LCd^2v_C/dt^2 + RCdv_C/dt + v_C = 0$; (d) $v_C(t) = e^{-3750t}[10 \cos 3307t + 11.34 \sin 3307t]u(t)$ V; (e) underdamped.

8-31 (a) $v_C(0) = 0$ V; $i_L(0) = 5/3$ mA; (b) $v_C(\infty) = 0$ V; $i_L(\infty) = 0$ A; (c) $LCd^2i_L/dt^2 + 4GLdi_L/dt + i_L = 0$ V; (d) $i_L(t) = e^{-1667t}[1.667 \cos 9860t + 0.2816 \sin 9860t]u(t)$ mA; (e) underdamped

8-33 $i_L(t) = [380e^{-1340t} - 27.3e^{-18660t}]u(t)$ μA; overdamped

8-35 $v_C(t) = e^{-625t}[15 \cos 4961t + 1.890 \sin 4961t]u(t)$ V; underdamped

8-37 $v_O(t) = [100 - 107.7e^{-5359t} + 7.735e^{-74641t}]u(t)$ V

8-39 $v_O(t) = 25.2[e^{-2500t} \sin 19843t]u(t)$ V

8-42 (a) $(s + 3000)(s + 6000)$; $\zeta = 1.0607$; $\omega_0 = 4243$ rad/s; $v_C(0) = 0$ V; $i_L(0) = 6$ mA; (b) $v_T(t) = 10u(t)$ V; (c) $C = 133$ nF; $L = 418$ mH; $R_T = 3.75$ kΩ.

8-45 (a) parallel; (b) $(s + 1000)^2 + (2000)^2$; $\zeta = 0.447$; $\omega_0 = 2236$ rad/s; $v_C(0) = 2$ V; $i_L(0) = 5$ mA; (c) $R_T = 1$ kΩ; $L = 400$ mH; $C = 0.5$ μF.

CHAPTER INTEGRATING PROBLEMS

8-57 (a) $(RCs)^2 + (3 - \mu)RCs + 1$; (b) $\mu < 3$; (c) $RC = 10^{-3}s$; $\mu = 2$.

Chapter Nine

ERO 9-1 LAPLACE TRANSFORMS (Sects. 9-1, 9-2, 9-3)

9-1 (a) $F(s) = A\alpha/[s(s + \alpha)]$; (b) $F(s) = A(s + 3\alpha)/(s + \alpha)^2$; (c) $F(s) = A(\beta - \alpha)/[(s + \alpha)(s + \beta)]$; (d) $F(s) = A(\alpha - \beta)s/[(s + \alpha)(s + \beta)]$

9-3 (a) $F(s) = -A[(s+\alpha)(\sin\ \phi) - \beta(\cos\ \phi)]/[(s+\alpha)^2 + \beta^2]$; (b) $F(s) = A(s+\alpha-\beta)/[(s+\alpha)^2+\beta^2]$; (c) $F(s) = A(s+\alpha)(\gamma^2-\beta^2)/\{[(s+\alpha)^2+\gamma^2][(s+\alpha)^2+\beta^2]\}$; (d) $F(s) = A(s+\alpha+\beta)^2/\{(s+\alpha)[(s+\alpha)^2+\beta^2]\}$

9-8 (a) $F(s) = A(s^2+2\beta^2)/[s(s^2+4\beta^2)]$; (b) $F(s) = A/[2(s^2+4\beta^2)]$

9-9 (a) $F(s) = A\beta/(s^2-\beta^2)$; (b) $F(s) = 2A\beta^2/(s^2+\beta^2)^2$

ERO 9-2 INVERSE TRANSFORMS (Sects. 9-4, 9-5)

9-11 (a) $f(t) = \beta^{-1}e^{-\alpha t}[\sin\ \beta t]u(t)$; (b) $f(t) = \beta^{-1}e^{-\alpha t}[\beta\ \cos\ \beta t - \alpha\ \sin\ \beta t]u(t)$; (c) $f(t) = \delta(t) + \beta^{-1}e^{-\alpha t}[(\alpha^2-\beta^2)\sin\ \beta t - 2\alpha\beta\ \cos\ \beta t]u(t)$

9-13 (a) $f(t) = 1.33(e^{-t} - e^{-4t}]u(t)$; (b) $f(t) = 4[te^{-2t}]u(t)$; (c) $f(t) = 2.31[e^{-t}\sin\sqrt{3}t]u(t)$

9-15 (a) $f(t) = [0.375 + 0.25e^{-2t} + 0.375e^{-4t}]u(t)$; (b) $f(t) = [0.333 + 0.238\ \cos\ 3t + 0.429\ \cos\ 4t]u(t)$

9-17 (a) $f(t) = [t + 0.75 - e^{-t}(0.75\ \cos\sqrt{3}t + 0.433\ \sin\sqrt{3}t)]u(t)$; (b) $f(t) = [23.4 - 13.4e^{-50t} - 120te^{-50t}]u(t)$

ERO 9-3 CIRCUIT RESPONSE USING LAPLACE TRANSFORMS (Sect. 9-6)

9-19 (a) $y(t) = 5[e^{-10t}]u(t)$; (b) $y(t) = 10[1 - 2e^{-10000t}]u(t)$

9-21 (a) $Ldi_L(t)/dt + Ri_L(t) = 0$, $i_L(0) = V_A/R$; (b) $V_A[e^{-Rt/L}]u(t)/R$; (c) all natural response

9-23 (a) $0.0004dv_C/dt + v_C = 0.5v_S(t)$, $v_C(0) = 0$ V; (b) For $v_S(t) = 10u(t)$ V: $v_O(t) = 3.125[1 - e^{-2500t}]u(t)$ V; For $v_S(t) = 10e^{-1000}u(t)$ V: $v_O(t) = 5.21[e^{-1000t} - e^{-2500t}]u(t)$ V

9-25 (a) $y(t) = 5[e^{-2t} - e^{-5t}]u(t)$; (b) $y(t) = 2[te^{-2t}]u(t)$

9-27 (a) $y(t) = 2e^{-0.4t}[\cos\ 0.2t - 2\ \sin\ 0.2t]u(t)$; (b) $y(t) = 10e^{-0.5t}[\sin\ 0.5t]u(t)$

9-29 $y(t) = [2 - 6e^{-t} + 6e^{-2t} - 2e^{-3t}]u(t)$

9-31 (a) $2RLCd^2i_L/dt^2 + Ldi_L/dt + 2Ri_L = V_Ou(t)\ i_L(0) = 0,\ i'_L(0) = 0$; (b) $i_L(t) = [1.67 - e^{-833t}(1.67\ \cos\ 4930t - 0.282\ \sin\ 4930t]u(t)$ mA

ERO 9-4 LAPLACE TRANSFORMATION PROPERTIES (Sects. 9-7, 9-8, 9-9)

9-33 (a) $F(s) = 10s^{-2}[1 - e^{-s}]^2$; (b) $F(s) = 10s^{-2}[1 - 2e^{-s} + e^{-2s}]$

9-35 (a) $f(t) = 2Au(t)t/T - 2Au(t - T/2)(t - T/2)/T - 2Au(t - T)$; (b) $F(s) = 2A/(Ts^2)[1 - e^{-Ts/2} - Tse^{-Ts}]$

9-41 (a) $IV = 0$, $FV = 2/3$; (b) $IV = 0$, $FV = 0$

9-43 (a) IV Theorem does not apply since $F(s)$ is not a proper rational function, $FV = 0$; (b) $IV = 0$, FV Theroem does not apply since $F(s)$ has a pole in the RHP.

9-46 (a) $f(t) = 0.5[t^2e^{\alpha t}]u(t)$ (b) $f(t) = (2\beta)^{-1}[\sin\ \beta t - \beta t\ \cos\ \beta t]u(t)$

Chapter Ten

ERO 10-1 EQUIVALENT IMPEDANCE (Sects. 10-1, 10-2)

10-1 $Z(s) = R(RCs + 2)/(RCs + 1)$

10-3 $Z(s) = 5000(s^2 + 10^{12})/(s^2 + 10^4 s + 10^{12})$

10-5 $Z(s) = [(RCs)^2 + 3RCs + 1]/[Cs(RCs + 2)]$

10-7 $Z(s) = 2R/(2RCs + 1)$

10-9 $Z(s) = R(RCs + 2)/[(RCs)^2 + 3RCs + 1]$

ERO 10-2 BASIC CIRCUIT ANALYSIS TECHNIQUES (Sects. 10-2, 10-3)

10-12 (b) $V_C(s) = \dfrac{V_A(s + 1/R_1C)}{s\,[s + (R_1 + R_2)/R_1R_2C)]}$;

$$v_C(t) = V_A\left[\frac{R_2 + R_1 e^{-(R_1+R_2)t/R_1R_2C}}{R_1 + R_2}\right]u(t)$$

10-14 (b) $I_L(s) = \dfrac{V_A}{R}\left[\dfrac{s - R/L}{s(s + R/L)}\right]$;

$$i_L(t) = \frac{V_A}{R}\left[2e^{-Rt/L} - 1\right]u(t)$$

10-16 (b) $I_L(s) = -\dfrac{V_AC}{LCs^2 + R_2Cs + 1}$;

$$V_C(s) = \frac{V_AC(Ls + R_2)}{LCs^2 + R_2Cs + 1};$$

(c) $i_L(t) = -60\left[e^{-1000t}\sin 1000t\right]u(t)$ mA;

$$v_C(t) = 15e^{-1000t}\left[\cos 1000t + \sin 1000t\right]u(t)\text{ V}$$

10-18 (b) $V_C(s) = \dfrac{Ls}{RLCs^2 + Ls + R}V_S(s)$;

$$v_C(t) = 20.1\left[e^{-2000t}\sin 19900t\right]u(t)\text{ V}$$

10-20 (b) $V_T = (I_AR - V_O)/s$, $Z_T = (RCs + 1)/Cs$; (c) $I_{ZS}(s) = I_ARC/(2RCs + 1)$; $I_{ZI}(s) = -CV_O/(2RCs + 1)$; $i_{ZS}(t) = 0.5I_A[e^{-t/2RC}]u(t)$; $i_{ZI}(t) = -(0.5V_O/R)[e^{-t/2RC}]u(t)$

10-22 (b) $V_T = LsI_A/(s + \alpha) - LI_O$, $Z_T = Ls + R$;
(c) $I_{ZS}(s) = sI_A/[(s + \alpha)(s + 2R/L)]$; $I_{ZI}(s) = -I_O/(s + 2R/L)$

10-24 $V(s) = \dfrac{V_A/2}{s + R/2L} + \dfrac{I_BR\beta s/2}{(s + R/2L)(s^2 + \beta^2)}$

10-26 $V_2(s) = RCsV_1(s)/[(RCs)^2 + 3RCs + 1]$

10-28 (b) $V_2(s) = \dfrac{R_1V_1(s)}{(R_1 + R_2)LCs^2 + (R_1R_2C + L)s + R_1}$; $I_2(s) = CsV_2(s)$; (c) $i_2(t) = \left[e^{-100t} - e^{-500000t}\right]u(t)$ A

10-29 (b) $I_2(s) = \dfrac{R_1CsV_1(s)}{(R_1 + R_2)LCs^2 + (R_1R_2C + L)s + R_1}$; $V_2(s) = I_1(s)/Cs$; (c) $i_2(t) = \left[e^{-100t} - e^{-500000t}\right]u(t)$ A

ERO 10-3 GENERAL CIRCUIT ANALYSIS (Sects. 10-4, 10-5, 10-6)

10-34 (a) $[G_1 + G_2 + (C_1 + C_2)s]V_A - C_1 s V_2 = G_1 V_1;\ C_2 s V_A + G_3 V_2 = 0$
(b) $\Delta(s) = (RCs)^2 + 2RCs + 2 = (RCs + 1)^2 + 1$

10-37 (b) $V_2(s) = -\dfrac{\beta R_0 R_L C s V_1(s)/(R_1 + R_\pi)}{(R_0 + R_L)Cs + 1}$

(c) $v_2(t) = -826\left[e^{-90.0t}\right]u(t)$ mV

10-41 $M = 5.66$ mH

CHAPTER INTEGRATING PROBLEMS

10-44 $i_{50\Omega} = 0.1[e^{-500t}]u(t)$ A

10-45 $i(t) = 0.1[1 - 2000te^{-2000t}]u(t)$ A

Chapter Eleven

ERO 11-1 NETWORK FUNCTIONS (Sects. 11-1, 11-2)

11-1 (a) $T_V = R_2/(Ls + R_2);\ Z_{IN} = (R_1 Cs + 1)(Ls + R_2)/[LCs^2 + (R_1 + R_2)Cs + 1]$;
(b) I_1: Forced pole at $s = 0$; natural pole at $s = -1000$; natural pole at $s = -2000$; natural zeros at $s = -1000 \pm j1000$; V_2: forced pole at $s = 0$; natural pole at $s = -1000$

11-5 (a) $Z_{IN} = R_1/(R_1 C_1 s + 1);\ T_V = -(R_2/R_1)(R_1 C_1 s + 1)/(R_2 C_2 S + 1)$;
(b) Select $R_1 = 10$ kΩ then $R_2 = 100$ kΩ; $C_1 = 50$ nF; $C_2 = 10$ nF

11-7 (b) $T_V = R_2/(R_0 R_2 C_2 s + R_0 + R_2)$;
(c) forced pole at $s = 0$; natural pole at $s = -(R_0 + R_2)/(R_0 R_2 C_2)$

11-9 (b) $T_I = R_0/(L_3 s + R_0 + R_3)$; (c) forced pole at $s = 0$; natural pole at $s = -(R_0 + R_3)/L_3$

11-11 (b) $T_V = [(R_0 + R_1)C_1 s + 1]/(R_1 C_1 s + 1)$; (c) forced pole at $s = -250$; natural pole at $s = -1/R_1 C_1$; natural zero at $s = -1/[(R_0 + R_1)C_1]$

11-13 (b) $T_I = R_2/\Delta(s),\ \Delta(s) = R_2 L_3 C_2 s^2 + (R_2 R_3 C_2 + L_3)s + R_2 + R_3$;
(c) forced zero at $s = 0$; forced poles at $s = \pm j1000$; natural poles at roots of $\Delta(s)$

11-17 (b) $T_V = R_0 C_5 s/\Delta(s),\ \Delta(s) = L_5 C_5 s^2 + R_0 C_5 s + 1$; (c) forced pole at $s = 0$; natural zero at $s = 0$; natural poles at roots of $\Delta(s)$

11-19 (b) $T_I = R_0(L_6 C_6 s^2 + 1)/\Delta(s),\ \Delta(s) = R_0 L_6 C_6 s^2 + L_6 s + R_0$; (c) forced poles at $s = \pm j1000$; forced zero at $s = 0$; natural zero at $s = \pm j/[L_6 C_6]^{\frac{1}{2}}$; natural poles at roots of $\Delta(s)$

11-23 (b) $T_V = \left[\dfrac{-R_2/R_0}{R_2 C_2 s + 1}\right]\left[\dfrac{R_2}{R_0 R_2 C s + R_0 + R_2}\right]$;

(c) 1st-stage pole at $s = -1/R_2 C_2$; 2nd-stage pole at $s = -(R_0 + R_2)/(R_0 R_2 C_s)$

11-27 (b) $T_V = [T_{V1}][T_{V2}][T_{V3}]$;

$$T_{V1} = \frac{R_0(R_2C_2s + 1)}{R_0R_2C_2s + R_0 + R_2};$$

$$T_{V2} = \frac{(R_0 + R_1)C_1s + 1}{R_1C_1s + 1};$$

$$T_{V3} = \frac{R_0C_1s}{R_1C_1s + 1}$$

ERO 11-2 PROPERTIES OF NETWORK FUNCTIONS (Sect. 11-2)

11-31 $T_V(s) = 2000/(s + 20000)$; $g(t) = 0.1(1 - e^{-20000t})u(t)$; $h(t) = 2000[e^{-20000t}]u(t)$

11-32 $T_V(s) = (s + 100)/(s + 200)$; $g(t) = 0.5(1 + e^{-200t})u(t)$; $h(t) = \delta(t) - 100[e^{-200t}]u(t)$

11-33 $T_V(s) = -2000/(s + 20000)$; $g(t) = -0.1(1 - e^{-20000t})u(t)$; $h(t) = -2000[e^{-20000t}]u(t)$

11-35 $T_V(s) = 2.5 \times 10^6/(s^2 + 1.625 \times 10^6 s + 5 \times 10^{12})$;
$g(t) = 0.5[1 - e^{-1625000t}(\cos\ 1536000t + 1.06\ \sin\ 1536000t)]u(t)$;
$h(t) = 1628000[e^{-1625000t}\sin\ 1536000t]u(t)$

11-37 (a) $T_V(s) = -s/(s + 2000)$; $h(t) = -\delta(t) + 2000[e^{-2000t}]u(t)$;
(b) $T_V(s) = 2000/(s + 2000)$; $h(t) = 2000[e^{-2000t}]u(t)$;
(c) $T_V(s) = -(0.2s + 2000)/(s + 2000)$; $h(t) = -0.2\delta(t) - 1600[e^{-2000t}]u(t)$;
(d) $T_V(s) = (s - 2000)/(s + 2000)$; $h(t) = \delta(t) - 4000[e^{-2000t}]u(t)$

11-40 (a) $T(s) = 20000/(s^2 + 2000s + 5 \times 10^6)$; $g(t) = 0.004[1 - e^{-1000t}(\cos\ 2000t + 0.5\ \sin\ 2000t]u(t)$;
(b) $T(s) = 10(s + 1000)/(s^2 + 2000s + 5 \times 10^6)$;
$g(t) = 0.002[1 - e^{-1000t}(\cos\ 2000t - 2\ \sin\ 2000t]u(t)$

11-41 $y(t) = 10[r(t) - r(t - 5) - r(t - 10) + r(t - 15)]$, where $r(t - a) = (t - a)u(t - a)$

11-44 $y(t) = 0.01[1 - 1000t - e^{-1000t}]u(t)$

ERO 11-4 TRANSFER FUNCTIONS AND STEP-RESPONSE DESCRIPTORS (Sect. 11-4)

11-62 $T_R = 110\ \mu$s; $T_S = 230\ \mu$s; $IV = 0$; $FV = 0.1$

11-63 $T_R = 110\ \mu$s; $T_S = 230\ \mu$s; $IV = 0$; $FV = -0.1$

11-64 $T_S = 1.84\ \mu$s; $PO = 3.60\%$; $IV = 0$; $FV = 0.5$

11-65 $T_R = 4.39$ ms; $T_S = 5.99$ ms; $IV = 0$; $FV = -4$

11-66 $T_S = 2.99$ ms; $PO = 4.32\%$; $IV = 0$; $FV = 1$

11-67 (a) $T(s) = 2 \times 10^{13}/[(s + 3.09 \times 10^5)^2 + (1.8 \times 10^6)^2]$

11-68 (a) $T(s) = 500/(s + 500)$

CHAPTER INTEGRATING PROBLEMS

11-80 (a) $x(t) = [2\ \cos\ t + 4\ \sin\ t]u(t)$; (b) no; (c) no

11-83 (a) $T_V = -1/RCs$, unstable; (b) $h(t) = -u(t)/RC$, $g(t) = -[t/RC]u(t)$, yes;
(c) $T_V = 1/(RCs)^2$, yes, unstable; (d) $T_V = 1/[(RCs)^2 + 1]$, unstable

COURSE INTEGRATING PROBLEMS

I-10 (a) $v_O(t) = (250/9)(e^{-100t} - e^{-1000t})u(t)$; Forced response $= (250/9)(e^{-100t})u(t)$, Natural response $= (250/9)(-e^{-1000t})u(t)$ (b) $C = 2\ \mu F$

I-12 (a) Forced $= 10u(t)$ V; Natural $= -5.77[e^{-5000t} \sin\ 8660t]u(t)$ V; (b) $v_C(0) = 10$ V; (c) $v_1(t) = [\sin\ 10^4 t]u(t)$ V

I-14 (a) $T_{VC1} = -R_2Cs/(R_1Cs + 1)$; $T_{VC2} = [(R_1 + R_2)Cs + 1]/(R_1C_1s + 1)$

Chapter Twelve

ERO 12-1 THE SINUSOIDAL STEADY STATE (Sect. 12-1)

12-1 (a) $v_2(t) = 0.477\ \cos(10000t - 26.6^\circ)$ V; (b) $v_2(t) = 0.999\ \cos(1000t - 92.9^\circ)$ V

12-3 (a) $v_2(t) = 0.477\ \cos(10000t + 153^\circ)$ V; (b) $v_2(t) = 0.999\ \cos(1000t + 87.1^\circ)$ V

12-5 (a) $v_2(t) = 4.714\ \cos(10^6 t - 45^\circ)$ V; (b) $v_2(t) = 1.67\ \cos(2 \times 10^6 t - 90^\circ)$ V

12-7 (a) $v_2(t) = \cos(100t - 90^\circ)$ V; (b) $v_2(t) = \cos(200t + 143^\circ)$ V

12-9 (a) $v_2(t) = 0.555\ \cos(1000t + 146^\circ)$ V; (b) $v_2(t) = 2\ \cos(2000t + 90^\circ)$ V

12-11 $y(t) = 3.54\ \cos(1000t + 135^\circ)$

12-13 $y(t) = 18.0\ \cos(2000t + 33.7^\circ)$

12-15 $y(t) = 1333\ \cos(2000t)$

12-17 $y(t) = 25.1\ \cos(1000t + 87.2^\circ)$

ERO 12-2 BASIC FREQUENCY RESPONSE (Sects. 12-2, 12-3, 12-4)

12-18 (a) $T(s) = 2000/(s + 20000)$; (b) $|T(0)| = 0.1$; $|T(\infty)| = 0$; $T_{MAX} = 0.1$; $\omega_C = 20$ krad/s

12-20 (a) $T(s) = 2 \times 10^{12}/(s^2 + 3 \times 10^6 s + 4 \times 10^{12})$; (b) $|T(0)| = 0.5$; $|T(\infty)| = 0$; $T_{MAX} = 0.5$; $\omega_C = 1.88$ Mrad/s

12-21 (a) $T(s) = s^2/(s^2 + 2000s + 4 \times 10^6)$; (b) $|T(0)| = 0$; $|T(\infty)| = 1$; $T_{MAX} = 1.16$; $\omega_C = 1.7$ krad/s

12-23 (a) $T(s) = -1000/(s + 1000)$; (b) $|T(0)| = 1$; $|T(\infty)| = 0$; $T_{MAX} = 1$; $\omega_C = 1$ krad/s

12-25 (a) $T(s) = 8 \times 10^5 s/[(s + 1200)^2 + (1600)^2]$; (b) $|T(0)| = 0$; $|T(\infty)| = 0$; $T_{MAX} = .167$; $\omega_{C1,2} = 1132,\ 3532$ rad/s

12-26 (a) $T(s) = 5s/(s + 2000)$

12-28 (a) $T(s) = 2\pi \times 10^7 s/(s^2 + 2\pi \times 10^6 s + 4\pi^2 \times 10^{12})$

12-30 $g(t) = 10(1 - e^{-62832t})u(t)$

12-31 $g(t) = 2.025(e^{-100t} - e^{-8000t})u(t)$

12-34 (a) $T_V(s) = \dfrac{-R_2Cs}{R_1R_2C^2s^2 + 2R_1Cs + 1}$; (b) $\omega_O = \dfrac{1}{\sqrt{R_1R_2}C}$; $\zeta = \sqrt{R_1/R_2}$

12-35 (a) $T_V(s) = \dfrac{K(RCs)^2}{(RCs)^2 + (3 - K)RCs + 1}$; (b) $\omega_O = \dfrac{1}{RC}$; $\zeta = (3 - K)/2$

ERO 12-3 BODE PLOTS (Sect. 12-5)

12-50 (a) $T(s) = (s + 200)/(s + 20)$; (b) $g(t) = (10 - 9e^{-20t})u(t)$

CHAPTER INTEGRATING PROBLEMS

12-64 (a) $T_V(s) = R_L Ms/[(L_1L_2 - M^2)s^2 + (R_L L_1 + R_1 L_2)s + R_1 R_L]$;
(b) $\omega_O = 3.164$ krad/s; $BW = 100.1$ krad/s; $\omega_{C1} = 99.9$ rad/s; $\omega_{C2} = 100.2$ krad/s

Chapter Thirteen

ERO 13-1 SINUSOIDS AND PHASORS (Sect. 13-1)

13-1 (a) $\mathbf{V}_1 = 100\angle{-90°}$ V; (b) $\mathbf{V}_2 = 206\angle{-75.96°}$ V; $v_1(t) + v_2(t) = 304\ \cos(\omega t - 80.5°)$ V

13-3 (a) $v_1(t) = 10\ \cos(100t + 30°)$ V; (b) $v_2(t) = 60\ \cos(100t + 90°)$ V; (c) $i_1(t) = -5\ \cos\ 100t$ A;
(d) $i_2(t) = 0.01\ \cos(100t - 70°)$ A

13-5 (a) $v_1(t) = 3.92\ \cos(200t + 101°)$ V; (b) $v_2(t) = 42.7\ \cos(200t - 129°)$ V;
(c) $i_1(t) = 3.16\ \cos(200t - 71.6°)$ A; (d) $i_2(t) = \cos(200t + 143°)$ A

13-8 $\mathbf{V}_1 = -6 + j2$ V

13-9 $v_2(t) = 10\ \cos(\omega t - 143°)$ V

ERO 13-2 BASIC PHASOR CIRCUIT ANALYSIS (Sects. 13-2, 13-3, 13-4)

13-11 $i(t) = 70.7\ \cos(2500t - 45°)$ mA

13-13 $i_R(t) = 140\ \cos(377t + 62.05°)$ mA; $i_C(t) = 265\ \cos(377t + 27.95°)$ mA

13-15 $Z = 0.219 + j124.4 = 124.4\angle 89.9°\ \Omega$

13-17 (a) $Z = 10\angle 135°$ kΩ; (b) $i(t) = 15\ \cos(1000t + 65°)$ mA

13-19 $v_R(t) = \dfrac{RI_m}{\sqrt{1 + (2\omega RC)^2}} \cos(\omega t - 90° - \tan^{-1} 2\omega RC)$ V

13-21 $Z = 136 - j32\ \Omega$; $v_X(t) = 107\ \cos(2000t + 26.6°)$ V

13-23 $\mathbf{V}_X = 8.575\angle 121°$ V

13-25 $i_X(t) = 994\ \cos(2000t + 6.52°) + 4.33\ \cos(200t - 4.97°)$ mA;
$v_X(t) = 227\ \cos(2000t + 6.52°) + 0.866\ \cos(200t + 85°)$ V

13-33 $\mathbf{V}_T = -45 - j55$ V: $Z_T = 25 - j25\ \Omega$; $\boldsymbol{V} = 64.2\angle{-111°}$ V; $\boldsymbol{I} = 0.287\angle{-174°}$ A

ERO 13-3 GENERAL CIRCUIT ANALYSIS (Sect. 13-5)

13-35 $\mathbf{V}_X = 2.394\angle 118.6°$ V

13-37 $\mathbf{I}_X = 2.644\angle{-38.41°}$ A

13-39 (a) $\mathbf{V}_2 = -j\omega RC\mathbf{V}_1/(1 + j\omega RC)^2$; (b) $\mathbf{V}_2(0) = 0$; $\mathbf{V}_2(\infty) = 0$; $\mathbf{V}_2(1/RC) = \mathbf{V}_1(-0.5 - j0.5)$;
bandpass filter

13-41 $\boldsymbol{I}_A = 0.142\angle{-18.4°}$ A; $\boldsymbol{I}_B = 0.3705\angle 31.8°$ A

13-42 (a) $C = 2\ \mu F$; $Z = 50 + j0\ \Omega$

13-44 $\omega_O = 100$ krad/s; $BW = 20$ krad/s; $Q = 5$; $\omega_{C2} = 110.5$ krad/s; $\omega_{C1} = 90.5$ krad/s

13-46 (a) $C = 1.02\ \mu F$; (b) $Q = 2$; $BW = 3500$ rad/s

13-48 (a) 150 V; $\mathbf{I}_C = j30$ mA; $\mathbf{I}_L = -j30$ mA; (b) upper: $75 - j75$ V; $\mathbf{I}_C = 17.7 + j17.7$ mA; $\mathbf{I}_L = -12.7 - j12.7$ mA; lower: $75 + j75$ V; $\mathbf{I}_C = -12.7 + j12.7$ mA; $\mathbf{I}_L = 17.7 - j17.7$ mA

13-50 $L = 30.75\ \mu H$; $C = 3.979$ nF

ERO 13-5 LINEAR TRANSFORMER CIRCUITS (Sect. 13-7)

13-55 (a) $v_1(t) = 11.0\ \cos(2000t + 28.1°)$ V; (b) $Z = 64.4 + j54.3\ \Omega$

13-57 (a) $Z = 7.24 + j22.0\ \Omega$; $V_L = 22.9\angle 38.4°$V; (b) $k = 0.833$

13-61 (a) $k = 0.559$; (b) $96 - j128\ \Omega$

ERO 13-6 AC CIRCUIT ANALYSIS USING SPICE (Sect. 13-8)

13-64 See 13-21 above.

13-66 See 13-33 above.

13-68 See 13-57 above.

Chapter Fourteen

ERO 14-1 COMPLEX POWER (Sects. 14-1, 14-2)

14-1 (a) $P = 264$ W; $Q = 985$ VAR; $PF = 0.259$; absorbing; (b) $P = -61.3$ W; $Q = -35.4$ VAR; $PF = -0.866$; delivering; (c) $P = 16.9$ W; $Q = 22.8$ VAR; $PF = 0.5$; absorbing; (d) $P = -1.63$ W; $Q = 1.37$ VAR; $PF = -0.766$; delivering

14-3 (a) $PF = 0.8$, lagging; (b) $PF = -0.6$, lagging; (c) $PF = 0.465$, unknown; (d) $PF = 0.6$, leading

14-5 (a) $P = 8$ kW; $Q = 6$ kVAR; $I = 4.17$ A (rms); (b) $Z = 461 + j346\ \Omega$

14-7 $PF = 0.606$; $Z = 22.2 + j29.2\ \Omega$

ERO 14-2 AC POWER ANALYSIS (Sect. 14-3)

14-9 $\mathbf{V} = 240\angle 0°$ V (rms); $\mathbf{I} = 3.83\angle -37°$ A (rms); $S = 734 + j554$ VA

14-11 $R = 19.4\ k\Omega$; $C = 1.73\ \mu F$

14-13 $I_L = 1.104$ A (rms); $Z_L = 1845 + j894\ \Omega$; $Z_W = 34 + j68\ \Omega$

14-15 (a) $\boldsymbol{I}_A = 20.6\angle -6.34°$ A; $\boldsymbol{I}_B = 4.31\angle -161°$ A; $\boldsymbol{I}_N = 16.8\angle 167°$ A; (b) $S_{upper} = 2250 + j250$ VA; $S_{lower} = 1800 + j400$ VA

ERO 14-3 MAXIMUM POWER TRANSFER (Sect. 14-4)

14-17 (a) $S = 0.288 + j0$ VA; (b) $P_{MAX} = 0.9$ W; (c) $Z_L = 1000 - j2000\ \Omega$

14-19 (a) $P_{MAX} = 41.7$ mW; (b) $R = 611\ \Omega$; $C = 136$ nF

14-21 (a) $S = 5.03 + j1.01$ VA; $P_{MAX} = 17.6$ W; (c) $Z_L = 17.5 - j83.4\ \Omega$

ERO 14-4 LOAD-FLOW ANALYSIS (Sect. 14-5)

14-23 (a) $\mathbf{I}_L = 68.9\angle{-36.9^\circ}$ A; $\mathbf{V}_S = 906\angle 1.48^\circ$ V; (b) $S = 49.0 + j38.7$ kVA; $\eta = 99.1\%$

14-25 (a) $\mathbf{I}_L = 139\angle{-16.0^\circ}$ A; $\mathbf{V}_S = 364\angle 20.2^\circ$ V; (b) $S_W = 3.86 + j19.3$ kVA; $PF = 0.807$

14-29 $S_1 = 0.48 + j0.3$ MVA; $S_2 = 0.6 + j0.344$ MVA

ERO 14-5 THREE-PHASE POWER (Sects. 14-6, 14-7)

14-33 (a) $\mathbf{V}_{\mathbf{AN}} = 277\angle 0^\circ$ V; $\mathbf{V}_{\mathbf{BN}} = 277\angle{-120^\circ}$ V; $\mathbf{V}_{\mathbf{CN}} = 277\angle{-240^\circ}$ V; $\mathbf{V}_{\mathbf{AB}} = 480\angle 30^\circ$ V; $\mathbf{V}_{\mathbf{BC}} = 480\angle{-90^\circ}$ V; $\mathbf{V}_{\mathbf{CA}} = 480\angle{-210^\circ}$ V

14-35 (a) $\mathbf{I}_{\mathbf{A}} = 100\angle 0^\circ$ A; $\mathbf{I}_{\mathbf{B}} = 100\angle{-240^\circ}$ A; $\mathbf{I}_{\mathbf{C}} = 100\angle{-120^\circ}$ A; $\mathbf{I}_1 = 57.7\angle{-30^\circ}$ A; $\mathbf{I}_2 = 57.7\angle{-270^\circ}$ A; $\mathbf{I}_3 = 57.7\angle{-150^\circ}$ A

14-37 (a) $\mathbf{I}_{\mathbf{A}} = 47.8 + j19.1$ A; $\mathbf{I}_{\mathbf{B}} = -40.4 + j31.8$ A; $\mathbf{I}_{\mathbf{C}} = -7.33 - j50.9$ A; (b) $S_L = 39.7 - j15.9$ kVA

14-39 $I_L = 102$ A; $PF = 0.813$

14-41 (a) $I_L = 60.1$ A; (b) $Z = 4.15 + j2.01\ \Omega$

14-43 (a) $I_L = 8.90$ A; (b) $Z = 374 + j280\ \Omega$

14-45 $V = 585$ V; $PF = 0.647$

ERO 14-6 AC POWER ANALYSIS USING SPICE (Sect. 14-8)

14-49 See 14-15 above.

14-50 See 14-17 above.

14-51 See 14-39 above.

14-53 See 14-45 above.

CHAPTER INTEGRATING PROBLEMS

14-57 (a) $\mathbf{I}_{\mathbf{A}} = 1.2\angle 0^\circ$ A; $\mathbf{I}_{\mathbf{B}} = 1.2\angle{-120^\circ}$ A; $\mathbf{I}_{\mathbf{C}} = 1.2\angle 30^\circ$ A; $S = 288 + j144$ VA; (b) $\mathbf{I}_{\mathbf{A}} = 0.481\angle{-18.4^\circ}$ A; $\mathbf{I}_{\mathbf{B}} = 1.80\angle{-138^\circ}$ A; $\mathbf{I}_{\mathbf{C}} = 1.61\angle 56.6^\circ$ A; $S = 346 + j259$ VA

Chapter Fifteen

ERO 15-1 CASCADE DESIGN WITH FIRST-ORDER CIRCUITS (Sect. 15-2)

15-1 (a) $T(s) = 62500/[(s + 50)(s + 250)]$

15-3 (a) $T(s) = K(s + 100)/(s + 10000)$

15-5 (a) $T(s) = 10s^2/[(s + 20)(s + 500)]$

15-7 $T(s) = 100(s + 2000)/(s + 20000)$

15-9 (a) $T(s) = 2000/(s + 20000)$

ERO 15-2 CASCADE DESIGN WITH SECOND-ORDER CIRCUITS (Sect. 15-3)

15-11 $T(s) = K/(s^2 + 400s + 10^6)$

15-13 $T(s) = Ks^2/(s^2 + 707s + 250000)$

15-15 $T(s) = Ks^2/(s^2 + 30s + 400)$

15-17 $T(s) = 100s^2/(s^2 + 500s + 250000)$

15-21 (a) $\zeta_1\zeta_2 = \sqrt{2}/4$

ERO 15-3 LOW-PASS AND HIGH-PASS FILTER DESIGN (Sects. 15-4, 15-5)

15-22 $T(s) = K/(s + 100)^3$

15-23 $T(s) = K/b_4(s/2000)$

15-24 $T(s) = K/b_2(s/50)$ or $K/c_2(s/50)$

15-25 $T(s) = K/b_3(s/25)$ or $K/c_2(s/25)$

15-26 $T(s) = K/c_2(s/200)$

15-27 $T(s) = Ks^3/b_3(s/4000)$

15-28 $T(s) = Ks^4/c_4(s/500)$

15-29 $T(s) = Ks^3/b_3(s/90)$ or $Ks^2/c_2(s/90)$

ERO 15-4 BANDPASS AND BANDSTOP FILTER DESIGN (Sect. 15-5)

15-34 $T(s) = Ks^4/[b_4(s/100)b_4(s/1000)$ or $T(s) = Ks^3/[c_3(s/100)c_3(s/1000)]$

15-35 $T(s) = Ks^4/[b_4(s/2000)b_4(s/30000)]$ or $T(s) = Ks^3/[c_3(s/2000)c_3(s/30000)]$

15-36 $T(s) = K/b_3(s/100) - Ks^3/b_3(s/3000)$ or $T(s) = K/c_2(s/100) - Ks^2/c_2(s/3000)$

15-37 $T(s) = K/b_2(s/200) - Ks^2/b_2(s/20000)$ or $T(s) = K/c_2(s/200) - Ks^2/c_2(s/20000)$

CHAPTER INTEGRATING PROBLEMS

15-54 (a) $T(s) = 1/(R^2C_1C_2s^2 + 2RC_2s + 1)$; (b) $\omega_O = 1/[R(C_1C_2)^{\frac{1}{2}}]$; $\zeta = (C_2/C_1)^{\frac{1}{2}}$

Chapter Sixteen

ERO 16-1 THE FOURIER SERIES (Sect. 16-1)

16-1 $v(t) = 4.05\cos 2000\pi t + 0.450\cos 6000\pi t + 0.162\cos 10^4\pi t + 0.0827\cos 14000\pi t + 0.0500\cos 18000\pi t + \ldots$ V

16-3 $v(t) = 79.6 + 125\sin 377t - 53\cos 754t - 10.6\cos 1508t - 4.55\cos 2262t - \ldots$ V

ERO 16-2 THE EXPONENTIAL FOURIER SERIES (Sect. 16-2)

16-11 $c_0 = A/4$; $c_n = jA/(2n\pi)$, n odd

16-13 $c_0 = A/\pi$; $c_n = 2A/[\pi(1 - n^2)]$, n even

16-15 $c_n = A[2/(n\pi)^2 - j/(n\pi)]$

ERO 16-3 FOURIER TRANSFORMS (Sects. 16-3, 16-4, 16-5, and 16-6)

16-17 $F(\omega) = (A/T)(1 - e^{-j\omega T} - j\omega T e^{-j\omega T})/(j\omega)^2$

16-19 $F(\omega) = \dfrac{AT}{2}\dfrac{\sin \omega T/2}{\omega T/2\left[1 - (\omega T/2\pi)^2\right]}$

16-21 $f(t) = 1/(\alpha^2 + t^2)$

16-25 $F(\omega) = \alpha/(-\omega^2 + j\omega\alpha)$

16-38 (a) $f_1(t) = 33.3(e^{-20t} - e^{-50t})u(t)$; (b) $f_2(t) = (-0.667e^{-20t} + 1.667e^{-50t})u(t)$; (c) $f_3(t) = (1 - 1.667e^{-20t} + 0.667e^{-50t})u(t)$; (d) $f_4(t) = \delta(t) + (13.3e^{-20t} - 83.3e^{-50t})u(t)$

16-40 (a) $F_1(\omega) = 6\pi\delta(\omega) + 2/j\omega$; (b) $F_2(\omega) = 6\pi\delta(\omega) + 10/j\omega$; (c) $F_3(\omega) = -2e^{-2j\omega}$; (d) $F_4(\omega) = [4\ \cos(2\omega)]/(2 + j\omega)$

16-41 (a) $f_1(t) = 2 + \delta(t-2)$; (b) $f_2(t) = -j0.5e^{j(t-8)}$; (c) $f_3(t) = 2 + 2\ \mathrm{sgn}(t) + 2[e^{-2t}]u(t)$; (d) $f_4(t) = 2 + 4\ \cos\ 2t$

ERO 16-4 CIRCUIT ANALYSIS USING FOURIER TRANSFORMS (Sect. 16-7)

16-42 (a) $v_1(t) = 10\ \mathrm{sgn}(t)$; $V_1(\omega) = 20/j\omega$; (c) $v_2(t) = 20[e^{-100t}]u(t)$

16-44 (a) $v_1(t) = 10\ e^{-100|t|}$; $V_1(\omega) = 10/(100 - j\omega) + 10/(100 + j\omega)$; (c) $v_2(t) = 8.33[e^{100t}]u(-t) - 4.17[e^{-500t}]u(t) + 12.5[e^{-100t}]u(t)$

16-46 (a) $v_1(t) = 10\ u(-t) + 10e^{-100t}u(t)$; $V_1(\omega) = 10/(-j\omega) + 10\pi\delta(-\omega) + 10/(100 + j\omega)$; (c) $v_2(t) = 10u(-t) + 10(1 + 100t)e^{-100t}u(t)$

16-48 $y(t) = 2e^{-\alpha t}u(t)$

16-50 (a) $H(\omega) = (j\omega - \alpha)/(j\omega + \alpha)$; allpass function; (b) $y(t) = -2\mathrm{sgn}(t) + 4e^{-\alpha t}u(t)$

ERO 16-5 PARSEVAL'S THEOREM AND FILTERING (Sect. 16-8)

16-51 (a) $V_A^2/(4\alpha)$; 81.8%; (b) $V_A^2/(4\alpha)$; 18.2%

16-53 (a) 96.8%; (b) 99.999%

16-56 (a) 40 V^2s; (b) 8 V^2s; (c) 95.9%

CHAPTER INTEGRATING PROBLEMS

16-60 $v_2(t) = 7.07\ \cos(100t - 135°)$ V

16-61 (b) $W_{1\text{-}\Omega}$(estimated) $= 2V_A^2/(\pi\alpha)$

COURSE INTEGRATING PROBLEMS

I-19 (a) $T(s) = 15 \times 10^6/[(s + 1000)(s + 4000)]$; (b) $\mathbf{V}_2 = 0.643\angle{-121°}$

I-21

	dc Gain	Rise Time	Stopband
Filter 1	yes	yes	no
Filter 2	yes	yes	yes
Filter 3	yes	no	yes

I-23 (a) $V_A^2/(8R_O)$; (c) $1 + [\omega L/(4R_O)]^2$; (d) $\omega_{3\mathrm{dB}} = 4R_O/L$

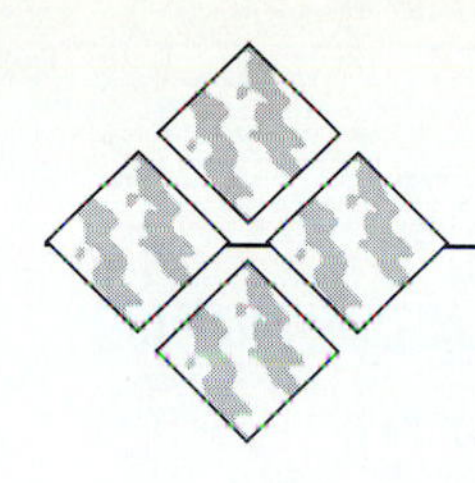

INDEX

PLACE
STAMP
HERE

Tom McElwee
Senior Marketing Manager
Engineering/Computer Science
PRENTICE HALL
Simon & Schuster Education Group
113 Sylvan Avenue, Route 9W
Englewood Cliffs, NJ 07632

Name Phone

School Address

Department

Office Hours

Course Title and Number

Current Text

Length of Course

Enrollment Fall Spring Other

Are you likely to change books? Decision Date

Are you currently using software? If yes, what kind?

Do you want to see supplements? (circle) Instructor's Manual Lab Manual

Transparency Masters All

Comments

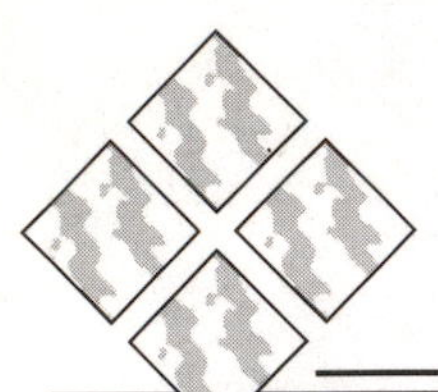

BASIC LAPLACE TRANSFORMATION PROPERTIES

PROPERTIES	TIME DOMAIN	FREQUENCY DOMAIN
Independent Variable	t	s
Signal Representation	$f(t)$	$F(s)$
Uniqueness	$\mathcal{L}^{-1}\{F(s)\}(=)[f(t)]u(t)$	$\mathcal{L}\{f(t)\} = F(s)$
Linearity	$Af_1(t) + Bf_2(t)$	$AF_1(s) + BF_2(s)$
Integration	$\int_0^t f(\tau)d\tau$	$\dfrac{F(s)}{s}$
Differentiation	$\dfrac{df(t)}{dt}$	$sF(s) - f(0-)$
	$\dfrac{d^2f(t)}{dt^2}$	$s^2F(s) - sf(0-) - f'(0-)$
	$\dfrac{d^3f(t)}{dt^3}$	$s^3F(s) - s^2f(0-) - sf'(0-) - f''(0-)$
t-Translation	$[f(t-a)]u(t-a)$	$e^{-as}F(s)$
s-Translation	$e^{-\alpha t}f(t)$	$F(s+\alpha)$
Scaling	$f(at)$	$\dfrac{1}{a}F\left(\dfrac{s}{a}\right)$
Final Value	$\lim_{t\to\infty} f(t)$	$\lim_{s\to 0} sF(s)$
Initial Value	$\lim_{t\to 0+} f(t)$	$\lim_{s\to\infty} sF(s)$